LET US KNOW WHAT YOU THINK

In order to produce a directory that will serve you best, we ask that you take the time to fill out this short questionnaire. Thank you for your continued support and please feel free to photocopy this form or attach additional sheets.

Is this the first time you have purchased the *Conservation Directory*? ❑ Yes ❑ No

If no, how long have you been purchasing the *Conservation Directory*?_______________

How would you categorize yourself? (check one)

❑ Environmental Professional ❑ Environmental Lawyer
❑ College or University Library ❑ Conservation Organization
❑ Career Center ❑ Educator
❑ Public Library ❑ Student
❑ Environmental Activist ❑ Scientist
❑ Business or Corporation ❑ Other

What resources would you like to see added to the *Conservation Directory*?

____

What would you like to see removed from the *Conservation Directory*?

____

Are you pleased with the format of the *2002 Conservation Directory*? ❑ Yes ❑ No

If no, what suggestions do you have for the future organization of the *Conservation Directory*?

____

____

Which index is most useful to you?
❑ Geographic Index ❑ Organization Name Index
❑ Staff Name Index ❑ I do not use the indices
❑ Keyword Index

If organization listings contained less information would the directory
be as useful? ❑ Yes ❑ No

Please be sure to fill in the information on this page and to send both sides if faxing.

MAIL TO: NATIONAL WILDLIFE FEDERATION
ATTN: CONSERVATION DIRECTORY
11100 WILDLIFE CENTER DRIVE
RESTON, VA 20190-5362
FAX: 703-438-6061

There are forms in the back of this book for information updates and to suggest new organizations

You may also visit the *Conservation Directory* online at www.nwf.org to update or add information.

| Prefix | First Name/MI | Last Name | Suffix |

Are you interested in any of the following?

- ❏ After school programs for youth
- ❏ Children's publications (like Ranger Rick® magazine)
- ❏ Educator Workshops
- ❏ Gardening for Wildlife
- ❏ NWF television and film programs
- ❏ Volunteerism

- ❏ After School programs for teens
- ❏ Distance learning
- ❏ Inclusive environmental education curricula
- ❏ Programs for college students
- ❏ Turning your work or school grounds into a home for wildlife

Would you like to be on our education mailing list?　❏ Yes　❏ No

Would you like to receive free Wildlife Week materials and information?　❏ Yes　❏ No

Organization: ____

Are you a:
- ❏ School Administrator
- ❏ College/University Faculty
- ❏ Green Coordinator
- ❏ Non-Formal Educator
- ❏ Librarian or Resource Specialist
- ❏ Teacher (if so, what do you teach):　　Grade _______________
  - ❏ All Disciplines
  - ❏ AP Education
  - ❏ Art/Fine Arts
  - ❏ Computer Science
  - ❏ English/Reading
  - ❏ Foreign Language
  - ❏ Home School
  - ❏ Mathematics
  - ❏ Music
  - ❏ Physical Education
  - ❏ Science
  - ❏ Social Studies
  - ❏ Special Education

Primary Address (work or school)

____
Address

____
City　　　　　State　　　　　Zip　　　　　Country

Secondary Address (home)

____
Address

____
City　　　　　State　　　　　Zip　　　　　Country

Day Phone _______________　Evening Phone _______________

Fax _______________　E-mail _______________

<u>Office Use Only</u>
Original Source: Conservation 2002 Directory Evaluation

2002 CONSERVATION DIRECTORY

47TH EDITION

The Guide to Worldwide Environmental Organizations

Bill Street, *Editor*

ISLAND PRESS

Washington • Covelo • London

NATIONAL WILDLIFE FEDERATION

11100 Wildlife Center Drive • Reston • VA 20190-5362

www.nwf.org

The mission of the National Wildlife Federation is to educate, inspire, and assist individuals and organizations of diverse cultures to conserve wildlife and other natural resources, and to protect the Earth's environment in order to achieve a peaceful, equitable, and sustainable future.

The *Conservation Directory* is published as a public service. Organization are included on the basis of their stated objectives and other information provided. Inclusion does not imply confirmation of the information nor does it imply any endorsement of the organizations listed by the National Wildlife Federation.

To purchase the *Conservation Directory*, call Island Press at (800) 828-1302. If you have any general questions about the *Conservation Directory*, call the National Wildlife Federation at (703) 438-6000. To update your listing, please visit our Web site at www.nwf.org.

Library of Congress Cataloging-in-Publication Data is available on file. British Cataloguing-in-Publication Data is also available.

LETTER FROM THE EDITOR

Dear Conservation Directory Reader:

The National Wildlife Federation is pleased to present the 47th edition of the ***Conservation Directory***. The National Wildlife Federation is the only conservation education organization currently providing a yearly update of the kind of information you need.

Since its debut in 1955, the information contained in this publication has helped a wide variety of people. Over the years we have heard from research scientists, professional resource managers, wildlife biologists, non-profit groups and their staff members, citizen activists, librarians, and students looking for internships and other job opportunities who have found what they were looking for within these pages. We sincerely hope that this will be your experience as well.

The ***2002 Conservation Directory***, by its very nature, displays constant growth and change in natural resource-related organizations, both private and governmental. When the first directory was published, it was only one-fourth the size it is today. In many ways, pages added over the years track the popular development of personal commitment to environmental stewardship in America.

This commitment is what the National Wildlife Federation is all about and we will be satisfied if our publication helps you to make a vital contact in your search to contribute to this essential stewardship ethic.

I also welcome Island Press as the new publisher of the ***Conservation Directory***. It is only fitting that this valuable resource is published by a leader in environmental publications.

As always, we welcome your comments, suggestions, additions, or corrections for next year's edition. This year, you may visit the online ***Conservation Directory*** at www.nwf.org and update your organization's information automatically.

Sincerely,

Bill Street
Director of Classroom Programs and
Electronic Education
National Wildlife Federation

National Wildlife Federation Executive Staff

Conservation Directory Editorial and Production Staff

TABLE OF CONTENTS

INTRODUCTION

This is the forty-seventh edition of the National Wildlife Federation's *Conservation Directory*. It has been published every year since 1955. The first directory listed National Wildlife Federation's state affiliate organizations and was expanded in 1960 to include other conservation organizations. The directory now lists U.S. and state government agencies, international, national, regional organizations and commissions, international government agencies non-governmental organizations from around the world. The Directory has over 3,500 entries, nearly twice the number of the 2001 edition.

USER'S GUIDE

The Conservation Directory is divided into three parts:

• Part One: Introduction

The introduction provides information about the Directory and other National Wildlife Federation programs. The Table of Contents is an excellent resource for beginning a search, listing all section headings with corresponding page numbers.

• Part Two: Descriptive Listings

Entries are arranged alphabetically according to sections:

- *US Congress, Committees, and Subcommittees*
- *US Federal and International Government Agencies*: Consists of executive branch organizations, independent government agencies, Commissions and international Government Agencies.
- *State and Provincial Government Agencies*: consists of state agencies that deal with conservation issues, organized by state.
- *Non-Governmental Organizations (Non-Profit & For-Profit)*: consists of American and international organizations that are not affiliated with any government agency whose mission is to help protect, preserve and defend the natural world. Subdivided into non-profit organizations. Listed in alphabetical order.
- *Educational Institutions*: consists of colleges and universities with conservation programs and research centers, organized by state.
- *Federally Protected Areas*: consists of National Forests, National Marine Sanctuaries, National Estuarine Research Reserves, National Parks, National Seashores, National Grasslands, Bureau of Land Management Districts and National Wildlife Refuges.

National Wildlife Federation Affiliates are listed in the front of the book. Governors are listed first for each state in address, founding date, membership, senior staff by name and title and description of the organization's primary goals and mission as provided by the organizations.

• Part Three: Indices

Organization Index

This is a quick and easy way to locate an organization. The index includes the name of every organization included in the directory in alphabetical order with the corresponding page number.

Keyword Index

This useful reference tool lists various subject areas and gives the name of those organizations whose work is related to that keyword. Index citations contain page numbers.

Staff Name Index

The Conservation Directory is very helpful if you know the name of an individual involved with an organization but do not know the specific name of the organization. This index lists all individuals cited in directory listings. Each citation contains the individual's name and page numbers where it appears.

Geographic Index

This index lists agencies and organizations by geographic regions. It is a great way to locate organizations that can be found in a certain state or province.

NWF AFFILIATES

National Wildlife Federation Affiliates are autonomous, statewide organizations that support the purposes and objectives of the National Wildlife Federation. Each affiliate is governed by its own board of directors and develops its own membership on a local level. Affiliates provide NWF with an organized grassroots network nationwide. The elected delegates from the state affiliates determine the conservation policy for NWF through a resolution process at the NWF Annual Meeting. The delegates also elect NWF's Chair, Vice Chairs, and 13 Regional Directors. Detailed descriptions can be found in the Non-Governmental Organization section.

ALABAMA WILDLIFE FEDERATION
46 Commerce Street
Montgomery, AL 36104
phone: (334) 832-9453
fax: (334) 832-9454
email: awf@mindspring.com
web: www.alawild.org

ARIZONA WILDLIFE FEDERATION
644 N. Country Club Drive, Suite E
Mesa, AZ 85201-4983
phone: (480) 644-0077
fax: (480) 644-0078
email: awf@azwildlife.org
web: www.azwildlife.org

ARKANSAS WILDLIFE FEDERATION
9700 Rodney Parham Road, Suite I-2
Little Rock, AR 72227-6212
phone: (501) 224-9200
fax: (501) 224-9214
email: arkwildlifefed@aristotle.net

PLANNING AND CONSERVATION LEAGUE
926 J Street, Suite 612
Sacramento, CA 95814
phone: (916) 444-8726
fax: (916) 448-1789
email: pclmail@pcl.org
web: www.pcl.org

COLORADO WILDLIFE FEDERATION
445 Union Blvd, #302
Lakewood, CO 80228-1243
phone: (303) 987-0400
fax: (303) 987-0200
email: cwf@coloradowildlife.org
web: www.coloradowildlife.org

CONNECTICUT FOREST AND PARK ASSOCIATION, INC.
Middlefield, 16 Meriden Road
Rockfall, CT 06481-2961
phone: (860) 346-2372
fax: (860) 347-7463
email: conn.forest.assoc@snet.net
web: www.ctwoodlands.org

DELAWARE NATURE SOCIETY
P.O. Box 700
Hockessin, DE 19707-0700
phone: (302) 239-2334
fax: (302) 239-2473
email: email@dnsashland.org
web: www.delawarenaturesociety.org

FLORIDA WILDLIFE FEDERATION
P.O. Box 6870
Tallahassee, FL 32314-6870
phone: (850) 656-7113
fax: (850) 942-4431
email: wildfed@aol.com
web: www.flawildlife.org

GEORGIA WILDLIFE FEDERATION
11600 Hazelbrand Road
Covington, GA 30014
phone: (770) 787-7887
fax: (770) 787-9229
email: gwf@gwf.org
web: www.gwf.org

CONSERVATION COUNCIL FOR HAWAII
PMB-203, 111 E. Puainako Street
Suite 585 250 Ward Ave., Suite 585
Hilo, HI 96720
phone: (808) 968-6360
fax: (808) 968-0896
email: cch@aloha.net
web: www.conservation-hawaii.org

IDAHO WILDLIFE FEDERATION
P.O. Box 6426
Boise, ID 83707-6426
phone: (208) 342-7055
fax: (208) 342-7097
email: iwfboise@micron.net

PRAIRIE RIVERS NETWORK
809 South Fifth Street
Champaign, IL 61820
phone: (217) 344-2371
fax: (217) 344-2381
email: robmoore@prairierivers.org
web: www.prairierivers.org

INDIANA WILDLIFE FEDERATION
950 North Rangeline Rd, Suite A
Carmel, IN 46032-1315
phone: (317) 571-1220
fax: (317) 571-1223
email: iwf@indy.net
web: www.indianawildlife.org

IOWA WILDLIFE FEDERATION
P.O. Box 3332
Des Moines, IA 50316-0332
phone: (319) 624-3107

KANSAS WILDLIFE FEDERATION
P.O. Box 8237
Wichita, KS 67208-0237

LEAGUE OF KENTUCKY SPORTSMEN, INC.
P.O. Box 8527
Lexington, KY 40533
phone: (859) 276-3318
email: office@kentuckysportsmen.com
web: www.kentuckysportsmen.com

LOUISIANA WILDLIFE FEDERATION, INC.
P.O. Box 65239 Audubon Station
Baton Rouge, LA 70896-5239
phone: (225) 344-6762
fax: (225) 344-6707
email: lawildfed@aol.com

NATURAL RESOURCES COUNCIL OF MAINE
3 Wade Street
Augusta, ME 04330-6351
phone: (207) 622-3101
fax: (207) 622-4343
email: nrcm@nrcm.org
web: www.maineenvironment.org

ENVIRONMENTAL LEAGUE OF MASSACHUSETTS
14 Beacon Street, Suite 714
Boston, MA 02108
phone: (617) 742-2553
fax: (617) 742-9656
email: elm@environmentalleague.org
web: www.environmentalleague.org

MICHIGAN UNITED CONSERVATION CLUBS, INC.
2101 Wood St.
Lansing, MI 48912-3728
phone: (517) 371-1041
fax: (517) 371-1505
email: mucc@mucc.org
web: www.mucc.org

MINNESOTA CONSERVATION FEDERATION
551 Snelling Avenue South, Suite B
St. Paul, MN 55116-1525
phone: (651) 690-3077
fax: (651) 690-3077
email: mncf@mtn.org
web: www.mncf.org

MISSISSIPPI WILDLIFE FEDERATION
855 S. Pear Orchard Road, Suite 500
Ridgeland, MS 39157-5138
phone: (601) 206-5703
fax: (601) 206-5705
email: mboyd@mswf.org
web: www.mswildlife.org

CONSERVATION FEDERATION OF MISSOURI
728 West Main Street
Jefferson City, MO 65101-1159
phone: (573) 634-2322
fax: (573) 634-8205
email: confedmo@socket.net
web: www.confedmo.com

MONTANA WILDLIFE FEDERATION
P.O. Box 1175
Helena, MT 59624-1175
phone: (406) 458-0227
fax: (406) 458-0373
email: mwf@mtwf.org
web: www.montanawildlife.com

NEBRASKA WILDLIFE FEDERATION, INC.
P.O. Box 81437
Lincoln, NE 68501-1437
phone: (402) 994-2001
fax: (402) 994-2021
email: NebraskaWildlife@alltel.net

NEVADA WILDLIFE FEDERATION, INC.
P.O. Box 71238
Reno, NV 89570
phone: (775) 885-7965
fax: (775) 885-0405
email: nvwf@nvwf.org
web: www.nvwf.org

NEW HAMPSHIRE WILDLIFE FEDERATION
54 Portsmouth Street
Concord, NH 03766
phone: (603) 224-5953
fax: (603) 226-7147
email: nhwf@aol.com
web: www.nhwf.org

NEW MEXICO WILDLIFE FEDERATION, INC.
3240-A Juan Tabo NE, Suite 204
Albuquerque, NM 87111
phone: (505) 299-5404

ENVIRONMENTAL ADVOCATES
353 Hamilton Street
Albany, NY 12210
phone: (518) 462-5526
fax: (518) 449-4937
email: info@eany.org
web: www. eany.org

NORTH CAROLINA WILDLIFE FEDERATION
P.O. Box 10626
Raleigh, NC 27605
phone: (919) 833-1923
fax: (919) 829-1192
email: ncwf_chuck@mindspring.com
web: www.ncwildlifefed.org

NORTH DAKOTA WILDLIFE FEDERATION
P.O. Box 7248
Bismarck, ND 58507-7248
phone: (701) 222-2557
fax: (701) 222-0334
email: ndwf@gcentral.com
web: www.ndwf.org

LEAGUE OF OHIO SPORTSMEN
3953 Indianola Avenue
Columbus, OH 43214
phone: (614) 268-9924
fax: (614) 268-9924
email: info@leaugeofohiosportsmen.org
web: www.leagueofohiosportsmen.org

OKLAHOMA WILDLIFE FEDERATION
P.O. Box 60126
Oklahoma City, OK 73146-0126
phone: (405) 524-7009
fax: (405) 521-9270
web: www.okwildlife.org

PENNSYLVANIA FEDERATION OF SPORTSMEN'S CLUBS
2426 N. Second Street
Harrisburg, PA 17110-1104
phone: (717) 232-3480
fax: (717) 231-3524
web: www.pfsc.org

ENVIRONMENT COUNCIL OF RHODE ISLAND
P.O. Box 9061
Providence, RI 02940
phone: (401) 621-8048
fax: (401) 331-5266
email: environmentcouncil@earthlink.net
web: www.environmentcouncilri.org

SOUTH CAROLINA WILDLIFE FEDERATION
2711 Middleburg Drive, Suite 104
Columbia, SC 29204
phone: (803) 256-0670
fax: (803) 256-0690
email: mail@scwf.org
web: www.scwf.org

SOUTH DAKOTA WILDLIFE FEDERATION
P.O. Box 7075
Pierre, SD 57501-7075
phone: (605) 224-7524
fax: (605) 224-7524
email: sdwf@sbtc.net
web: www.sdwf.org

TENNESSEE CONSERVATION LEAGUE
300 Orlando Avenue
Nashville, TN 37209-3257
phone: (615) 353-1133
fax: (615) 353-0083
email: tcl@conservetn.com
web: www.conservetn.com

TEXAS COMMITTEE ON NATURAL RESOURCES
1301 South IH-35, Suite 301
Austin, TX 78741
phone: (512) 441-1122
fax: (512) 328-3399
email: tconr@texas.net
web: tconr.home.texas.net

UTAH WILDLIFE FEDERATION
P.O. Box 526367
Salt Lake City, UT 84152-6367
phone: (801) 487-1946
fax: (801) 773-0412
email: uwfhall@xmission.com

VERMONT NATURAL RESOURCES COUNCIL
9 Bailey Avenue
Montpelier, VT 05602-2100
phone: (802) 223-2328
fax: (802) 223-0287
email: info@vnrc.org
web: www.vnrc.org

VIRGIN ISLANDS CONSERVATION SOCIETY, INC.
Arawak Building, Suite 3, Gallows Bay
Christiansted, VI 00820
phone: (340) 773-1989
fax: (340) 773-7545
email: sea@viaccess.net

WASHINGTON WILDLIFE FEDERATION
P.O. Box 1966
Olympia, WA 98507-1966
phone: (360) 705-1903
email: wwf@washingtonwildlife.org
web: www.washingtonwildlife.org

WEST VIRGINIA WILDLIFE FEDERATION, INC.
P.O. Box 275
Paden City, WV 26159
phone: (304) 782-3685
email: pleinbach@aol.com
web: www.wvwf.org

WISCONSIN WILDLIFE FEDERATION, INC.
2036 West 9th Street
Oshkosh, WI 54904
phone: (920) 235-9136
fax: (920) 235-6030
email: wiwf@execpc.com
web: www.execpc.com/~wiwf/

WYOMING WILDLIFE FEDERATION
P.O. Box 106
Cheyenne, WY 82003-0106
phone: (307) 637-5433
fax: (307) 637-6629
web: www.wyomingwildlife.org

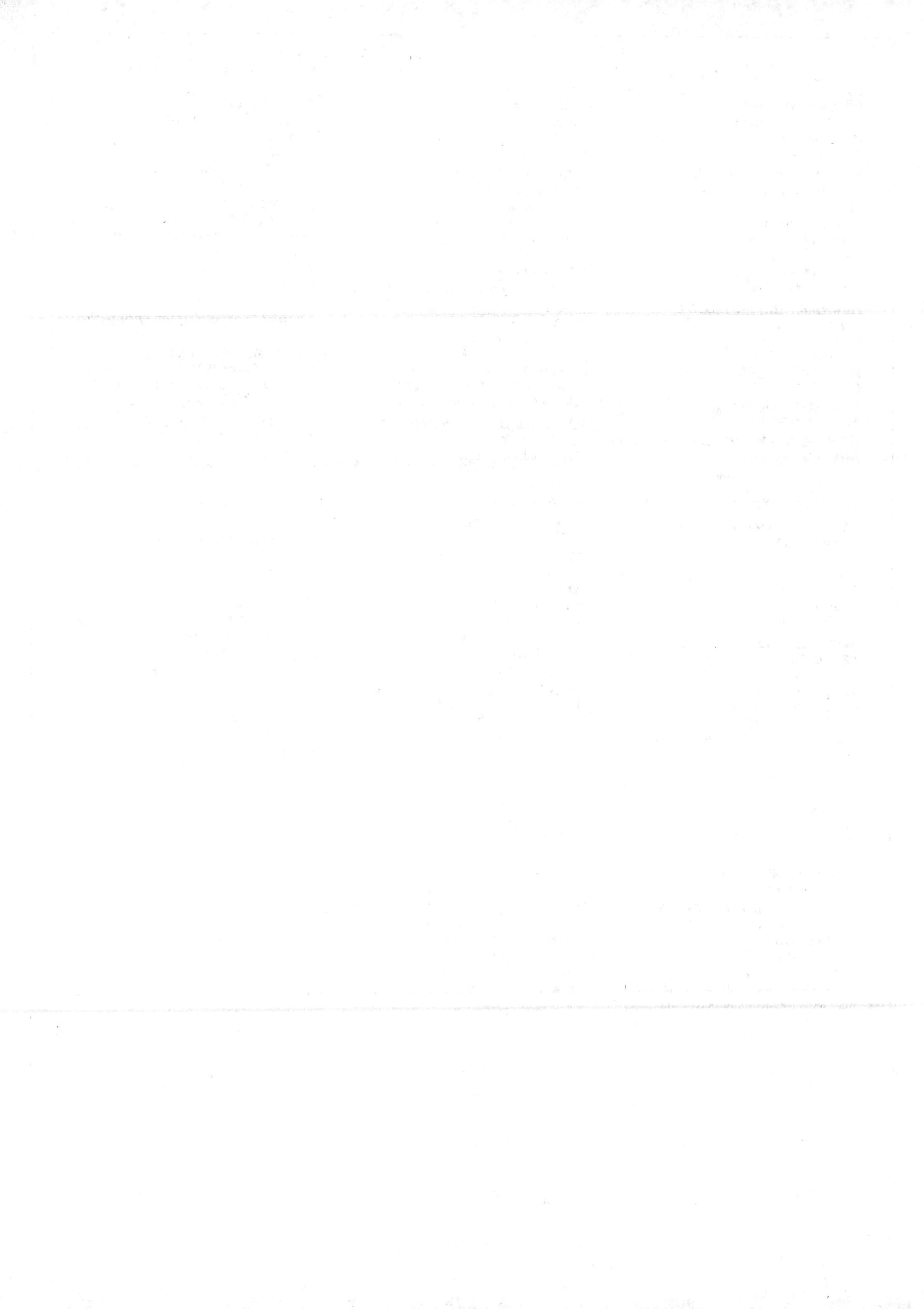

U.S. CONGRESS, COMMITTEES, AND SUBCOMMITTEES

ALABAMA

Senators:

Jeff Sessions
Phone: 202-224-4124   Fax: 202-224-3149
Website: www.sessions.senate.gov

Richard Shelby
Phone: 202-224-5744   Fax: 202-224-3416
Website: www.shelby.senate.gov

Representatives:

Sonny Callahan
Phone: 202-225-4931   Fax: 202-225-0562
Website: www.house.gov/callahan

Terry Everett
Phone: 202-225-2901   Fax: 202-225-8913
E-mail: terry.everett@mail.house.gov
Website: www.house.gov/everett

Bob Riley
Phone: 202-225-3261   Fax: 202-225-5827
E-mail: bob.riley@mail.house.gov
Website: www.house.gov/riley

Robert B. Aderholt
Phone: 202-225-4876   Fax: 202-225-5587
E-mail: robert.aderholt@mail.house.gov
Website: www.house.gov/aderholt

Robert E. "Bud" Cramer Jr.
Phone: 202-225-4801   Fax: 202-225-4392
E-mail: budmail@mail.house.gov
Website: www.house.gov/cramer

Spencer Bachus
Phone: 202-225-4291   Fax: 202-225-2082
Website: www.house.gov/bachus

Earl F. Hillard
Phone: 202-225-2665   Fax: 202-226-0772
Website: www.house.gov/hillard

ALASKA

Senators:

Ted Stevens
Phone: 202-224-3004   Fax: 202-224-2354
Website: www.stevens.senate.gov

Frank H. Murkowski
Phone: 202-224-6665   Fax: 202-224-5301
Website: www.murkowski.senate.gov

Representatives:

Don Young
Phone: 202-225-5765   Fax: 202-225-0425
E-mail: don.young@mail.house.gov
Website: www.house.gov/donyoung

ARIZONA

Senators:

John McCain
Phone: 202-224-2235   Fax: 202-228-2862
E-mail: john_mccain@nccain.senate.gov
Website: mccain.senate.gov

Jon L. Kyl
Phone: 202-224-4521   Fax: 202-224-2207
Website: kyl.senate.gov

Representatives:

Jeff Flake
Phone: 202-225-2635   Fax: 202-226-4386
E-mail: jeff.flake@mail.house.gov
Website: www.house.gov/flake

Ed Pastor
Phone: 202-225-4065   Fax: 202-225-1655
Website: www.house.gov/pastor

Bob Stump
Phone: 202-225-4576   Fax: 202-225-6328
Website: www.house.gov/stump

John B. Shadegg
Phone: 202-225-3361   Fax: 202-225-3462
E-mail: j.shadegg@mail.house.gov
Website: www.house.gov/shadegg

Jim Kolbe
Phone: 202-225-2542   Fax: 202-225-0378
Website: www.house.gov/kolbe

J.D. Hayworth
Phone: 202-225-2190   Fax: 202-225-3263
E-mail: jdhayworth@mail.house.gov
Website: www.house.gov/hayworth

ARKANSAS

Senators:

Tim Hutchinson
Phone: 202-224-2353   Fax: 202-228-3973
E-mail: senator.hutchinson@hutchinson.senate.gov
Website: hutchinson.senate.gov

Blanche L. Lincoln
Phone: 202-224-4843   Fax: 202-228-1371
E-mail: blanche_lincoln@lincoln.senate.gov
Website: lincoln.senate.gov

Representatives:

Marion Berry
Phone: 202-225-4076   Fax: 202-225-5602
Website: www.house.gov/berry

Vic Snyder
Phone: 202-225-2506   Fax: 202-225-5713
E-mail: snyder.congress@mail.house.gov
Website: www.house.gov/snyder

Asa Hutchinson
Phone: 202-225-4301   Fax: 202-225-5713
E-mail: asa.hutchinson@mail.house.gov
Website: www.house.gov/hutchinson

Mike Ross
Phone: 202-225-3772   Fax: 202-225-1314
Website: www.house.gov/ross

CALIFORNIA

Senators:

Dianne Feinstein
Phone: 202-224-3841   Fax: 202-228-3954
Website: feinstein.senate.gov

Barbara Boxer
Phone: 202-224-3553   Fax: 415-956-6701
E-mail: senator@boxer.senate.gov
Website: boxer.senate.gov

Representatives:

Mike Thompson
Phone: 202-225-3311   Fax: 202-225-4335
E-mail: m.thompson@mail.house.gov
Website: www.house.gov/mthompson

Wally Herger
Phone: 202-225-3076   Fax: 202-225-1740
Website: www.house.gov/herger

Doug Ose
Phone: 202-225-5716   Fax: 202-226-1298
E-mail: doug.ose@mail.house.gov
Website: www.house.gov/ose

John T. Doolittle
Phone: 202-225-2511   Fax: 202-225-5444
E-mail: doolittle@mail.house.gov
Website: www.house.gov/doolittle

Robert T. Matsui
Phone: 202-225-7163   Fax: 202-225-0566
Website: www.house.gov/matsui

Lynn Woolsey
Phone: 202-225-5161   Fax: 202-225-5163
E-mail: lynn.woolsey@mail.house.gov
Website: www.house.gov/woolsey

George Miller
Phone: 202-225-2095   Fax: 202-225-5609
E-mail: george.miller@mail.house.gov
Website: www.house.gov/georgemiller

Nancy Pelosi
Phone: 202-225-4965   Fax: 202-225-8259
E-mail: sf.nancy@mail.house.gov
Website: www.house.gov/pelosi

Barbara Lee
Phone: 202-225-2661   Fax: 202-225-9817
E-mail: barbara.lee@mail.house.gov
Website: www.house.gov/lee

Ellen O. Tauscher
Phone: 202-225-1880   Fax: 202-225-5914
Website: www.house.gov/tauscher

Richard Pombo
Phone: 202-225-1947   Fax: 202-225-0861
E-mail: rpombo@mail.house.gov
Website: www.house.gov/pombo

Tom Lantos
Phone: 202-225-3531   Fax: 202-226-9789
Website: www.house.gov/lantos

Fortney H. "Pete" Stark
Phone: 202-225-5065   Fax: 202-226-3805
Website: www.house.gov/stark

Anna Eshoo
Phone: 202-225-8104   Fax: 202-225-8890
E-mail: annagram@mail.house.gov
Website: www.house.gov/eshoo

Michael M. Honda
Phone: 202-225-2631   Fax: 202-225-2699
E-mail: mike.honda@mail.house.gov
Website: www.house.gov/honda

Zoe Lofgren
Phone: 202-225-3072   Fax: 202-225-3336
E-mail: zoe@lofgren.house.gov
Website: www.house.gov/lofgren

Sam Farr
Phone: 202-225-2861   Fax: 202-225-6791
Website: www.house.gov/farr

Gary Condit
Phone: 202-225-6131   Fax: 202-225-0819
Website: www.house.gov/condit

George P. Randovich
Phone: 202-225-4540   Fax: 202-225-3402
Website: www.house.gov/radanovich

Calvin Dooley
Phone: 202-225-3341   Fax: 202-225-9308
Website: www.house.gov/dooley

William M. Thomas
Phone: 202-225-2915   Fax: 202-225-8798
Website: www.house.gov/billthomas

Lois Capps
Phone: 202-225-3601   Fax: 202-225-5632
Website: www.house.gov/capps

Elton Gallegly
Phone: 202-225-5811   Fax: 202-225-1100
Website: www.house.gov/gallegly

Brad Sherman
Phone: 202-225-5911   Fax: 202-225-5879
Website: www.house.gov/sherman

Howard "Buck" McKeon
Phone: 202-225-2956   Fax: 202-226-0683
E-mail: tellbuck@mail.house.gov
Website: www.house.gov/mckeon

Howard L. Berman
Phone: 202-225-4695   Fax: 202-225-3196
E-mail: howard.berman@mail.house.gov
Website: www.house.gov/berman

Adam Schiff
Phone: 202-225-4176   Fax: 202-225-5828
E-mail: congressman.schiff@mail.house.gov
Website: www.house.gov/schiff

David Dreier
Phone: 202-225-2305   Fax: 202-225-7018
Website: www.house.gov/dreier

Henry A. Waxman
Phone: 202-225-3976   Fax: 202-225-4099
Website: www.house.gov/waxman

Xavier Becerra
Phone: 202-225-6235   Fax: 202-225-2202
Website: www.house.gov/becerra

Hilda L. Solis
Phone: 202-225-5464   Fax: 202-225-5467
Website: www.house.gov/solis

Diane Watson
Phone: 202-225-7084   Fax: 202-225-2422

Lucille Roybal-Allard
Phone: 202-225-1766   Fax: 202-226-0350
Website: www.house.gov/roybal-allard

Grace F. Napolitano
Phone: 202-225-5256   Fax: 202-225-0027
E-mail: grace@mail.house.gov
Website: www.house.gov/napolitano

Maxine Waters
Phone: 202-225-2201   Fax: 202-225-7854
Website: www.house.gov/waters

Jane F. Harman
Phone: 202-225-8220   Fax: 202-226-7290
E-mail: jane.harman@mail.house.gov
Website: www.house.gov/harman

Juanita Millender-McDonald
Phone: 202-225-7924   Fax: 202-225-7926
E-mail: millender.mcdonald@mail.house.gov
Website: www.house.gov/millender-mcdonald

Steve Horn
Phone: 202-225-6676   Fax: 202-226-1012
E-mail: steve.hon@mail.house.gov
Website: www.house.gov/horn

Edward Royce
Phone: 202-225-4111   Fax: 202-226-0335
Website: www.house.gov/royce

Jerry Lewis
Phone: 202-225-5861   Fax: 202-225-6498
Website: www.house.gov/jerrylewis

Gary G. Miller
Phone: 202-225-3201   Fax: 202-226-6962
E-mail: publicca41@mail.house.gov
Website: www.house.gov/garymiller

Joe Baca
Phone: 202-225-6161   Fax: 202-225-8671
E-mail: cong.baca@mail.house.gov
Website: www.house.gov/baca

Ken Calvert
Phone: 202-225-1986   Fax: 202-225-2004
Website: www.house.gov/calvert

Mary Bono
Phone: 202-225-5330   Fax: 202-225-2961
Website: www.house.gov/bono

Dana Rohrabacher
Phone: 202-225-2415   Fax: 202-225-0145
E-mail: dana@mail.house.gov
Website: www.house.gov/rohrabacher

Loretta L. Sanchez
Phone: 202-225-2965   Fax: 202-225-5859
E-mail: loretta@mail.house.gov
Website: www.house.gov/sanchez

Christopher Cox
Phone: 202-225-5611   Fax: 202-225-9177
E-mail: christopher.cox@mail.house.gov
Website: cox.house.gov

Darrell Issa
Phone: 202-225-3906   Fax: 202-225-3303
E-mail: congressman.issa@mail.house.gov
Website: www.house.gov/issa

Susan A. Davis
Phone: 202-225-2040   Fax: 202-225-2948
E-mail: susan.davis@mail.house.gov
Website: www.house.gov/susandavis

Bob Filner
Phone: 202-225-8045   Fax: 202-225-9073
Website: www.house.gov/filner

Randy "Duke" Cunningham
Phone: 202-225-5452   Fax: 202-225-2558
Website: www.house.gov/cunningham

Duncan Hunter
Phone: 202-225-5672   Fax: 202-225-0235
Website: www.house.gov/hunter

COLORADO

Senators:

Ben Nighthorse Campbell
Phone: 202-224-5852   Fax: 202-224-1933
Website: campbell.senate.gov

Wayne Allard
Phone: 202-224-5941   Fax: 202-224-6471
Website: allard.senate.gov

Representatives:

Diana L. DeGette
Phone: 202-225-4431   Fax: 202-225-5657
E-mail: degette@mail.house.gov
Website: www.house.gov/degette

Mark Udall
Phone: 202-225-2161   Fax: 202-226-7840
Website: www.house.gov/markudall

Scott McInnis
Phone: 202-225-4761   Fax: 202-226-0622
Website: www.house.gov/mcinnis

Bob Schaffer
Phone: 202-225-4676   Fax: 202-225-5870
E-mail: rep.schaffer@mail.house.gov
Website: www.house.gov/schaffer

Joel Hefley
Phone: 202-225-4422   Fax: 202-225-1942
Website: www.house.gov/hefley

Thomas G. Tancredo
Phone: 202-225-7882   Fax: 202-226-4623
E-mail: tom.tancredo@mail.house.gov
Website: www.house.gov/tancredo

CONNECTICUT

Senators:

Christopher J. Dodd
Phone: 202-224-2823   Fax: 202-228-1683
E-mail: senator@dodd.senate.gov
Website: dodd.senate.gov

Joseph I. Lieberman
Phone: 202-224-4041   Fax: 202-224-9750
E-mail: senator_lieberman@liberman.senate.gov
Website: lieberman.senate.gov

Representatives:

John B. Larson
Phone: 202-225-2265   Fax: 202-225-1031
Website: www.house.gov/larson

Robert R. Simmons
Phone: 202-225-2076   Fax: 202-225-4977
Website: www.house.gov/simmons

Rosa DeLauro
Phone: 202-225-3661   Fax: 202-225-4890
Website: www.house.gov/delauro

Christopher Shays
Phone: 202-225-5541   Fax: 202-225-9629
E-mail: rep.shays@mail.house.gov
Website: www.house.gov/shays

James H. Maloney
Phone: 202-225-3822   Fax: 202-225-5746
Website: www.house.gov/jimmaloney

Nancy L. Johnson
Phone: 202-225-4476   Fax: 202-225-4488
Website: www.house.gov/nancyjohnson

U.S. Congress

DELAWARE

Senators:

Joseph R. Biden Jr.
Phone: 202-224-5402   Fax: 202-224-0139
E-mail: senator@biden.senate.gov
Website: biden.senate.gov

Thomas R. Carper
Phone: 202-224-2441   Fax: 202-228-2190
Website: carper.senate.gov

Representatives:

Michael Castle
Phone: 202-225-4165   Fax: 202-225-2291
E-mail: delaware@mail.house.gov
Website: www.house.gov/castle

FLORIDA

Senators:

Bob Graham
Phone: 202-224-3041   Fax: 202-224-2237
E-mail: bob_graham@graham.senate.gov
Website: graham.senate.gov

Bill Nelson
Phone: 202-224-5274   Fax: 202-228-2183
E-mail: senator@billneslon.senate.gov
Website: billnelson.senate.gov

Representatives:

Joe Scarborough
Phone: 202-225-4136   Fax: 202-225-3414
E-mail: flo1@mail.house.gov
Website: www.house.gov/scarborough

F. Allen Boyd Jr.
Phone: 202-225-5235   Fax: 202-225-5615
Website: www.house.gov/boyd

Corrine Brown
Phone: 202-225-0123   Fax: 202-225-2256
Website: www.house.gov/corrinebrown

Ander Crenshaw
Phone: 202-225-2501   Fax: 202-225-2504
Website: www.house.gov/crenshaw

Karen Thurman
Phone: 202-225-1002   Fax: 202-226-0329
Website: www.house.gov/thurman

Cliff Stearns
Phone: 202-225-5744   Fax: 202-225-3973
E-mail: csterns@mail.house.gov
Website: www.house.gov/stearns

John Mica
Phone: 202-225-4035   Fax: 202-226-0821
E-mail: john.mica@mail.house.gov
Website: www.house.gov/mica

Ric Keller
Phone: 202-225-2176   Fax: 202-225-0999
Website: www.house.gov/keller

Michael Bilirakis
Phone: 202-225-5755   Fax: 202-225-4085
Website: www.house.gov/bilirakis

C.W. "Bill" Young
Phone: 202-225-5961   Fax: 202-225-9764
Website: www.house.gov/young

Jim Davis
Phone: 202-225-3376   Fax: 202-225-5652
Website: www.house.gov/jimdavis

Adam Putnam
Phone: 202-225-1252   Fax: 202-226-0585
E-mail: ask.adam@mail.house.gov
Website: www.house.gov/putnam

Dan Miller
Phone: 202-225-5015   Fax: 202-226-0828
Website: www.house.gov/danmiller

Porter J. Goss
Phone: 202-225-2536   Fax: 202-225-6820
E-mail: porter.goss@mail.house.gov
Website: www.house.gov/goss

Dave Weldon
Phone: 202-225-3671   Fax: 202-225-3516
Website: www.house.gov/weldon

Mark Foley
Phone: 202-225-5792   Fax: 202-225-3132
E-mail: mark.foley@mail.house.gov
Website: www.house.gob/foley

Carrie Meek
Phone: 202-225-4506   Fax: 202-226-0777
Website: www.house.gov/meek

Ileana Ros-Lehtinen
Phone: 202-225-3931   Fax: 202-225-5620
Website: www.house.gov/ros-lehtinen

Robert I. Wexler
Phone: 202-225-3001   Fax: 202-225-5974
Website: www.house.gov/wexler

Peter Deutsch
Phone: 202-225-7931   Fax: 202-225-8456
Website: www.house.gov/deutsch

Lincoln Diaz-Blart
Phone: 202-225-4211   Fax: 202-225-8576
Website: www.house.gov/diaz-balart

E. Clay Shaw Jr.
Phone: 202-225-3026   Fax: 202-225-8398
Website: www.house.gov/shaw

Alcee L. Hastings
Phone: 202-225-1313   Fax: 202-225-1171
E-mail: alcee.pubhastings@mail.house.gov
Website: www.house.gov/alceehastings

GEORGIA

Senators:

Max Cleland
Phone: 202-224-3521   Fax: 202-224-0072
Website: cleland.senate.gov

Zell B. Miller
Phone: 202-224-3643   Fax: 202-228-2090
Website: miller.senate.gov

Representatives:

Jack Kingston
Phone: 202-225-5831   Fax: 202-226-2269
E-mail: jack.kingston@mail.house.gov
Website: www.house.gov/kingston

Sanford Bishop Jr.
Phone: 202-225-3631   Fax: 202-226-3601
E-mail: bishop.email@mail.house.gov
Website: www.house.gov/bishop

Michael "Mac" Collins
  Phone: 202-225-5901   Fax: 202-225-2515
  E-mail: mac.collins@mail.house.gov
  Website: www.house.gov/maccollins

Cynthia McKinney
  Phone: 202-225-1605   Fax: 202-226-0691
  E-mail: cymck@mail.house.gov
  Website: www.house.gov/mckinney

John Lewis
  Phone: 202-225-3801   Fax: 202-225-0351
  E-mail: john.lewis@mail.house.gov
  Website: www.house.gov/johnlewis

Johnny Isakson
  Phone: 202-225-4501   Fax: 202-225-4656
  E-mail: ga06@mail.house.gov
  Website: www.house.gov/isakson

Bob Barr
  Phone: 202-225-2931   Fax: 202-225-2944
  E-mail: barr.ga@mail.house.gov
  Website: www.house.gov/barr

Saxby Chambliss
  Phone: 202-225-6531   Fax: 202-225-3013
  E-mail: saxby.chambliss@mail.house.gov
  Website: www.house.gov/chambliss

Nathan Deal
  Phone: 202-225-5211   Fax: 202-225-8272
  Website: www.house.gov/deal

Charles Norwood
  Phone: 202-225-4101   Fax: 202-226-5995
  E-mail: rep.charlie.norwood@mail.house.gov
  Website: www.house.gov/norwood

John Linder
  Phone: 202-225-4272   Fax: 202-226-4696
  E-mail: john.linder@mail.house.gov
  Website: www.house.gov/linder

HAWAII

Senators:

Daniel K. Inouye
  Phone: 202-224-3934   Fax: 202-224-6747
  E-mail: senate@inouye.senate.gov
  Website: inouye.senate.gov

Daniel K. Akaka
  Phone: 202-224-6361   Fax: 202-224-2126
  E-mail: senator@akaka.senate.gov
  Website: akaka.senate.gov

Representatives:

Neil Abercrombie
  Phone: 202-225-2726   Fax: 202-225-4580
  E-mail: neil.abercrombie@mail.house.gov
  Website: www.house.gov/abercrombie

Patsy T. Mink
  Phone: 202-225-4906   Fax: 202-225-4987
  Website: www.house.gov/mink

IDAHO

Senators:

Larry E. Craig
  Phone: 202-224-2752   Fax: 202-228-1067
  Website: craig.senate.gov

Michael D. Crapo
  Phone: 202-224-6142   Fax:
  Website: crapo.senate.gov

Representatives:

C.L. "Butch" Otter
  Phone: 202-225-6611   Fax: 202-225-3029
  E-mail: butch.otter@mail.house.gov
  Website: www.house.gov/otter

Mike Simpson
  Phone: 202-225-5531   Fax: 202-225-8216
  E-mail: mike.simpson@mail.house.gov
  Website: www.house.gov/simpson

ILLINOIS

Senators:

Richard J. Durbin
  Phone: 202-224-2152   Fax: 202-228-0400
  E-mail: dick@durbin.senate.gov
  Website: durbin.senate.gov

Peter G. Fitzgerald
  Phone: 202-224-2854   Fax: 202-228-1372
  E-mail: senator_fitzgerald@fitzgerald.senate.gov
  Website: fitzgerald.senate.gov

Representatives:

Bobby L. Rush
  Phone: 202-225-4372   Fax: 202-226-0333
  E-mail: bobby.rush@mail.house.gov
  Website: www.house.gov/rush

Jesse L. Jackson, Jr.
  Phone: 202-225-0773   Fax: 202-225-0899
  Website: www.jessejacksonjr.org

William O. Lipinski
  Phone: 202-225-5701   Fax: 202-225-1012
  Website: www.house.gov/lipinski

Luis V. Gutierrez
  Phone: 202-225-8203   Fax: 202-225-7810
  Website: www.house.gov/gutierrez

Rod R. Blagojevich
  Phone: 202-2225406   Fax: 202-225-5603
  E-mail: rod.blagojevich@mail.house.gov
  Website: www.house.gov/blagojevich

Henry J. Hyde
  Phone: 202-225-4561   Fax: 202-225-1166
  Website: www.house.gov/hyde

Danny K. Davis
  Phone: 202-225-5006   Fax: 202-225-5641
  Website: www.house.gov/davis

Philip M. Crane
  Phone: 202-225-3711   Fax: 202-225-7830
  Website: www.house.gov/crane

Janice D. Schakowsky
  Phone: 202-225-2111   Fax: 202-226-6890
  E-mail: jan.schakowsky@mail.house.gov
  Website: www.house.gov/schakowsky

Mark S. Kirk
  Phone: 202-225-4835   Fax: 202-225-3521
  E-mail: rep.kirk@mail.house.gov
  Website: www.house.gov/kirk

Jerry Weller
  Phone: 202-225-3635   Fax: 202-225-3521
  Website: www.house.gov/weller

Jerry F. Costello
Phone: 202-225-5661   Fax: 202-225-0285
Website: www.house.gov/costello

Judy Biggert
Phone: 202-225-3515   Fax: 202-225-9420
Website: www.house.gov/biggert

J. Dennis Hastert
Phone: 202-225-2976   Fax: 202-225-0697
E-mail: dhastert@mail.house.gov
Website: www.house.gov/hastert

Timothy V. Johnson
Phone: 202-225-2371   Fax: 202-225-0791
E-mail: rep.johnson@mail.house.gov
Website: www.house.gov/timjohnson

Donald A. Manzullo
Phone: 202-225-5676   Fax: 202-225-5284
Website: www.house.gov/manzullo

Lane Evans
Phone: 202-225-5905   Fax: 202-225-5396
E-mail: lane.evans@mail.house.gov
Website: www.house.gov/evans

Ray LaHood
Phone: 202-225-6201   Fax: 202-225-9249
Website: www.house.gov/lahood

David Phelps
Phone: 202-225-5201   Fax: 202-225-1541
E-mail: david.phelps@mail.house.gov
Website: www.house.gov/phelps

John M. Shimkus
Phone: 202-225-5271   Fax: 202-225-5880
Website: www.house.gov/shimkus

INDIANA

Senators:

Richard G. Lugar
Phone: 202-224-4814   Fax: 202-228-0360
E-mail: senator_lugar@lugar.senate.gov
Website: lugar.senate.gov

Evan Bayh
Phone: 202-224-5623   Fax: 202-228-1377
E-mail: senator@bayh.senate.gov
Website: bayh.senate.gov

Representatives:

Peter J. Visclosky
Phone: 202-225-2461   Fax: 202-225-2493
Website: www.house.gov/visclosky

Mike Pence
Phone: 202-[illegible]   Fax: 202-225-[illegible]
E-mail: mike.pence@mail.house.gov
Website: www.house.gov/mikepence

Tim J. Roemer
Phone: 202-225-3915   Fax: 202-225-6798
E-mail: tim.roemer@mail.house.gov
Website: www.house.gov/roemer

Mark Souder
Phone: 202-225-4436   Fax: 202-225-3479
E-mail: souder@mail.house.gov
Website: www.house.gov/souder

Steve Buyer
Phone: 202-225-5037   Fax: 202-225-2267
Website: www.house.gov/buyer

Dan Burton
Phone: 202-225-2276   Fax: 202-225-0016
Website: www.house.gov/burton

Brian D. Kerns
Phone: 202-225-5805

John N. Hostettler
Phone: 202-225-4636   Fax: 202-225-3284
E-mail: john.hostettler@mail.house.gov
Website: www.house.gov/hostettler

Baron Hill
Phone: 202-225-5315   Fax: 202-226-6866
Website: www.house.gov/baronhillf

Julia M. Carson
Phone: 202-225-4011   Fax: 202-225-5633
E-mail: rep.carson@mail.house.gov
Website: www.house.gov/carson

IOWA

Senators:

Charles E. Grassley
Phone: 202-224-3744   Fax: 202-224-6020
E-mail: chuck_grassley@grassley.senate.gov
Website: grassley.senate.gov

Tom Harkin
Phone: 202-224-3254   Fax: 202-224-9369
E-mail: tom_harkin@harkin.senate.gov
Website: harkin.senate.gov

Representatives:

Jim Leach
Phone: 202-225-6576   Fax: 202-226-1278
E-mail: talk2jim@mail.house.gov
Website: www.house.gov/leach

Jim Nussle
Phone: 202-225-2911   Fax: 202-225-9129
E-mail: nussleia@mail.house.gov
Website: www.house.gov/nussle

Leonard L. Boswell
Phone: 202-225-3806   Fax: 202-225-5608
E-mail: rep.boswell.ia03@mail.house.gov
Website: www.house.gove/boswell

Greg Ganske
Phone: 202-225-4426   Fax: 202-225-3193
E-mail: rep.ganske@mail.house.gov
Website: www.house.gov/ganske

Tom Latham
Phone: 202-225-5476   Fax: 202-225-3301
E-mail: latham.ia05@mail.house.gov
Website: www.house.gov/latham

KANSAS

Senators:

Sam Brownback
Phone: 202-224-6521   Fax: 202-228-1265
Website: brownback.senate.gov

Pat Roberts
Phone: 202-224-4774   Fax: 202-224-3514
Website: roberts.senate.gov

Representatives:

Jerry Moran
Phone: 202-225-2715   Fax: 202-225-5124
E-mail: jerry.moran@mail.house.gov
Website: www.house.gov/moranks01

Jim R. Ryun
Phone: 202-225-6601   Fax: 202-225-7986
Website: www.house.gov/ryun

Dennis Moore
Phone: 202-225-2865   Fax: 202-225-2807
Website: www.house.gov/moore

Todd Tiahrt
Phone: 202-225-6216   Fax: 202-225-1489
E-mail: tiahrt@mail.house.gov
Website: www.house.gov/tiahrt

KENTUCKY

Senators:

Mitch McConnell
Phone: 202-224-2541   Fax: 202-224-2499
E-mail: senator@mcconnell.senate.gov
Website: mcconnell.senate.gov

Jim Bunning
Phone: 202-224-4343   Fax: 202-228-1373
E-mail: jim_bunning@bunning.senate.gov
Website: bunning.senate.gov

Representatives:

Edward Whitfield
Phone: 202-225-3115   Fax: 202-225-3547
Website: www.house.gov/whitfield

Ron Lewis
Phone: 202-225-3501   Fax: 202-226-2019
E-mail: ron.lewis@mail.house.gov
Website: www.house.gov/ronlewis

Anne M. Northup
Phone: 202-225-5401   Fax: 202-225-5776
E-mail: rep.northup@mail.house.gov
Website: www.house.gov/northup

Ken R. Lucas
Phone: 202-225-3465   Fax: 202-225-0003
E-mail: 24i53.k3nludqw@mail.house.gov
Website: www.house.gov/kenlucas

Harold Rogers
Phone: 202-225-4601   Fax: 202-225-0940
Website: www.house.gov/rogers

Ernest Lee Fletcher
Phone: 202-225-4706   Fax: 202-225-2122
Website: www.house.gov/fletcher

LOUISIANA

Senators:

John B. Breaux
Phone: 202-224-4623   Fax: 202-228-2577
E-mail: senator@breaux.senate.gov
Website: breaux.senate.gov

Mary Landrieu
Phone: 202-224-5824   Fax: 202-224-9735
Website: landrieu.senate.gov

Representatives:

David Vitter
Phone: 202-225-3015   Fax: 202-225-0739
E-mail: david.vitter@mail.house.gov
Website: www.house.gov/vitter

William J. Jefferson
Phone: 202-225-6636   Fax: 202-225-1988
E-mail: jeffersonmc@mail.house.gov
Website: www.house.gov/jefferson

W.J. "Billy" Tauzin
Phone: 202-225-4031   Fax: 202-225-0563
Website: www.house.gov/tauzin

Jim McCrery
Phone: 202-225-2777   Fax: 202-225-8039
E-mail: jim.mccrery@mail.house.gov
Website: www.house.gov/mccrery

John C. Cooksey
Phone: 202-225-8490   Fax: 202-225-5639
E-mail: congressman.cooksey@mail.house.gov
Website: www.house.gov/cooksey

Richard H. Baker
Phone: 202-225-3901   Fax: 202-225-7313
Website: www.house.gov/baker

Chris John
Phone: 202-225-2031   Fax: 202-225-5724
E-mail: christopher.john@mail.house.gov
Website: www.house.gov/john

MAINE

Senators:

Olympia J. Snowe
Phone: 202-224-5344   Fax: 202-224-1946
E-mail: olympia@snowe.senate.gov
Website: snowe.senate.gov

Susan M. Collins
Phone: 202-224-2523   Fax: 202-224-2693
E-mail: senator@collins.senate.gov
Website: collins.senate.gov

Representatives:

Thomas H. Allen
Phone: 202-225-6116   Fax: 202-225-5590
E-mail: rep.tomallen@mail.house.gov
Website: www.house.gov/allen

John E. Baldacci
Phone: 202-225-6306   Fax: 202-225-2943
E-mail: baldacci@me02.house.gov
Website: www.house.gov/baldacci

MARYLAND

Senators:

Paul S. Sarbanes
Phone: 202-224-4524   Fax: 202-224-1651
Website: sarbanes.senate.gov

Barbara A. Mikulski
Phone: 202-224-4654   Fax: 202-224-8858
Website: mikulski.senate.gov

Representatives:

Wayne Gilchrest
Phone: 202-225-5311   Fax: 202-225-0254
Website: www.house.gov/gilchrest

Robert Ehrlich Jr
Phone: 202-225-3061   Fax: 202-225-3094
E-mail: ehrlich@mail.house.gov
Website: www.house.gov/ehrlich

Benjamin L. Cardin
Phone: 202-225-4061   Fax: 202-225-9219
E-mail: rep.cardin@mail.house.gov
Website: www.house.gov/cardin

Albert Wynn
Phone: 202-225-8699   Fax: 202-225-8714
Website: www.house.gov/wynn

Steny H. Hoyer
Phone: 202-225-4131   Fax: 202-225-4300
Website: www.house.gov/hoyer

Roscoe Bartlett
Phone: 202-225-2721   Fax: 202-225-2193
Website: www.house.gov/bartlett

Elijah Cummings
Phone: 202-225-4741   Fax: 202-225-3178
Website: www.house.gov/cummings

Connie A. Morella
Phone: 202-225-5341   Fax: 202-225-1389
E-mail: rep.morella@mail.house.gov
Website: www.house.gov/morella

MASSACHUSETTS

Senators:

Edward M. Kennedy
Phone: 202-224-4543   Fax: 202-224-2417
Website: kennedy.senate.gov

John F. Kerry
Phone: 202-224-2742   Fax: 202-224-8525
E-mail: john_kerry@kerry.senate.gov
Website: kerry.senate.gov

Representatives:

John W. Olver
Phone: 202-225-5335   Fax: 202-226-1224
Website: www.house.gov/olver

Richard E. Neal
Phone: 202-225-5601   Fax: 202-225-8112
Website: www.house.gov/neal

James P. McGovern
Phone: 202-225-6101   Fax: 202-225-5759
Website: www.house.gov/mcgovern

Barney Frank
Phone: 202-225-5931   Fax: 202-225-0182
Website: www.house.gov/frank

Marty Meehan
Phone: 202-225-3411   Fax: 202-226-0771
Website: www.house.gov/meehan

John F. Tierney
Phone: 202-225-8020   Fax: 202-225-5915
Website: www.house.gov/tierney

Edward J. Markey
Phone: 202-225-2836   Fax:
Website: www.house.gov/markey

Michael Capuano
Phone: 202-225-5111   Fax: 202-225-9322
Website: www.house.gov/capuano

William Delahunt
Phone: 202-225-3111   Fax: 202-225-5658
E-mail: willian.delhunt@mail.house.gov
Website: www.house.gov/delahunt

MICHIGAN

Senators:

Carl Levin
Phone: 202-224-6221   Fax: 202-224-1388
E-mail: senator2@levin.senate.gov
Website: levin.senate.gov

Debbie A. Stabenow
Phone: 202-224-4822   Fax: 202-228-0325
E-mail: senator@stabenow.senate.gov
Website: stabenow.senate.gov

Representatives:

Bart Stupak
Phone: 202-225-4735   Fax: 202-225-4744
E-mail: stupak@mail.house.gov
Website: www.house.gov/stupak

Peter Hoekstra
Phone: 202-225-4401   Fax: 202-226-0779
E-mail: tellhoek@mail.house.gov
Website: www.house.gov/hoekstra

Vernon Ehlers
Phone: 202-225-3831   Fax: 202-225-5144
E-mail: rep.ehlers@mail.house.gov
Website: www.house.gov/ehlers

Dave Camp
Phone: 202-225-3561   Fax: 202-225-9679
Website: www.house.gov/camp

James Barcia
Phone: 202-225-8171   Fax: 202-225-2168
E-mail: jim.barcia-pub@mail.house.gov
Website: www.house.gov/barcia

Fred Upton
Phone: 202-225-3761   Fax: 202-225-4986
E-mail: tellupton@mail.house.gov
Website: www.house.gov/upton

Nick Smith
Phone: 202-225-6276   Fax: 202-225-6281
E-mail: rep.smith@mail.house.gov
Website: www.house.gov/nicksmith

Michael J. Rogers
Phone: 202-225-4872   Fax: 202-225-5820
Website: www.house.gov/mikerogers

Dale E. Kildee
Phone: 202-225-3611   Fax: 202-225-6393
E-mail: dkildee@mail.house.gov
Website: www.house.gov/kildee

David E. Bonior
Phone: 202-225-2106   Fax: 202-226-1169
E-mail: david.bonior@mail.house.gov
Website: davidbonior.house.gov

Joseph Knollenberg
Phone: 202-225-5802   Fax: 202-226-2356
E-mail: rep.knollenberg@mail.house.gov
Website: www.house.gov/knollenberg

Sander M. Levin
Phone: 202-225-4961   Fax: 202-226-1033
E-mail: slevin@mail.house.gov
Website: www.house.gov/levin

Lynn Rivers
Phone: 202-225-6261   Fax: 202-225-2404
E-mail: lynn.rivers@mail.house.gov
Website: www.house.gov/rivers

John Conyers Jr.
Phone: 202-225-5126   Fax: 202-225-0072
E-mail: john.conyers@mail.house.gov
Website: www.house.gov/conyers

Carolyn C. Kilpatrick
Phone: 202-225-2261   Fax: 202-225-5730
Website: www.house.gov/kilpatrick

John D. Dingell
Phone: 202-225-4071   Fax: 202-226-0371
E-mail: public.dingell@mail.house.gov
Website: www.house.gov/dingell

MINNESOTA

Senators:

Paul David Wellstone
Phone: 202-224-5641   Fax: 202-224-8438
Website: wellstone.senate.gov

Mark Dayton
Phone: 202-224-3244   Fax: 202-228-2186
Website: dayton.senate.gov

Representatives:

Gil Gutknecht
Phone: 202-225-2472   Fax: 202-225-3246
E-mail: gil@mail.house.gov
Website: www.gil.house.gov

Mark R. Kennedy
Phone: 202-225-2331   Fax: 202-225-6475
E-mail: mark.kennedy@mail.house.gov
Website: markkennedy.house.gov

Jim Ramstad
Phone: 202-225-2871   Fax: 202-225-6351
E-mail: mn03@mail.house.gov
Website: www.house.gov/ramstad

Betty McCollum
Phone: 202-225-6631   Fax: 202-225-1968
Website: www.house.gov/mccollum

Martin Olav Sabo
Phone: 202-225-4755   Fax: 202-225-4886
E-mail: martin.sabo@mail.house.gov
Website: www.house.gov/sabo

Bill Luther
Phone: 202-225-2271   Fax: 202-225-3368
E-mail: bill.luther@mail.house.gov
Website: www.house.gov/luther

Collin Peterson
Phone: 202-225-2165   Fax: 202-225-1593
Website: www.house.gov/collinpeterson

James L. Oberstar
Phone: 202-225-6211   Fax: 202-225-0699
Website: www.house.gov/oberstar

MISSISSIPPI

Senators:

Thad Cochran
Phone: 202-224-5054   Fax: 202-224-9450
E-mail: senator@cochran.senate.gov
Website: cochran.senate.gov

Trent Lott
Phone: 202-224-6253   Fax: 202-224-2262
E-mail: senatorlott@lott.senate.gov
Website: lott.senate.gov

Representatives:

Roger F. Wicker
Phone: 202-225-4306   Fax: 202-225-3549
E-mail: roger.wicker@mail.house.gov
Website: www.house.gov/wicker

Bennie G. Thompson
Phone: 202-225-5876   Fax: 202-225-5898
E-mail: thompsonms2nd@mail.house.gov
Website: www.house.gov/thompson

Charles "Chip" Pickering Jr.
Phone: 202-225-5031   Fax: 202-225-5797
Website: www.house.gov/pickering

Ronnie Shows
Phone: 202-225-5865   Fax: 202-225-5886
E-mail: ronnie.shows@mail.house.gov
Website: www.house.gov/shows

Gene Taylor
Phone: 202-225-5772   Fax: 202-225-7074
Website: www.house.gov/genetaylor

MISSOURI

Senators:

Christopher S. "Kit" Bond
Phone: 202-224-5721   Fax: 202-224-8149
E-mail: kit_bond@bond.senate.gov
Website: bond.senate.gov

Jean Carnahan
Phone: 202-224-6154   Fax: 202-228-1518
E-mail: senator_carahan@carnahan.senate.gov
Website: carnahan.senate.gov

Representatives:

William L. Clay Jr
Phone: 202-225-2406   Fax: 202-225-1725
Website: www.house.gov/clay

Todd Akin
Phone: 202-225-2561   Fax: 202-225-2563
E-mail: rep.akin@mail.house.gov
Website: www.house.gov/akin

Richard A. Gephardt
Phone: 202-225-2671   Fax: 202-225-7452
E-mail: gephardt@mail.house.gov
Website: www.house.gov/gephardt

Ike Skelton
Phone: 202-225-2876   Fax: 202-225-2695
E-mail: ike.skelton@mail.house.gov
Website: www.house.gov/skelton

Karen McCarthy
Phone: 202-225-4535   Fax: 202-225-4403
Website: www.house.gov/karenmccarthy

Samuel B. Graves
Phone: 202-225-7041   Fax: 202-225-8221
E-mail: sam.graves@mail.house.gov
Website: www.house.gov/graves

Roy Blunt
Phone: 202-225-6536   Fax: 202-225-5604
E-mail: blunt@mail.house.gov
Website: www.house.gov/blunt

Jo Ann H. Emerson
Phone: 202-225-4404   Fax: 202-226-0326
Website: www.house.gov/emerson

Kenny C. Hulshof
Phone: 202-225-2956   Fax: 202-225-5712
Website: www.house.gov/hulshof

MONTANA

Senators:

Max Baucus
Phone: 202-224-2651   Fax: 202-228-3687
E-mail: max@baucus.senate.gov
Website: baucus.senate.gov

Conrad Burns
Phone: 202-224-2644   Fax: 202-224-8594
Website: burns.senate.gov

Representative:

Dennis Rehberg
Phone: 202-225-3211   Fax: 202-225-5687
Website: www.house.gov/rehberg

NEBRASKA

Senators:

Chuck Hagel
Phone: 202-224-4224   Fax: 202-224-5213
E-mail: chuck_hagel@hagel.senate.gov
Website: hagel.senate.gov

Ben Nelson
Phone: 202-224-6511   Fax: 202-228-0012
E-mail: senator@bennelson.senate.gov
Website: bennelson.senate.gov

Representatives:

Doug Bereuter
Phone: 202-225-4806   Fax: 202-225-5686
Website: www.house.gov/bereuter

Lee Terry
Phone: 202-225-4155   Fax: 202-226-5452
E-mail: talk2lee@mail.house.gov
Website: www.house.gov/terry

Thomas W. Osborne
Phone: 202-225-6435   Fax: 202-226-1385
Website: www.house.gov/osborne

NEVADA

Senators:

Harry Reid
Phone: 202-224-3542   Fax: 202-224-7327
E-mail: senator_reid@reid.senate.gov
Website: reid.senate.gov

John Ensign
Phone: 202-224-6244   Fax: 202-228-2193
Website: ensign.senate.gov

Representatives:

Shelley Berkley
Phone: 202-225-5965   Fax: 202-225-3119
E-mail: shelley.berkley@mail.house.gov
Website: www.house.gov/berkley

James A. Gibbons
Phone: 202-225-6155   Fax: 202-225-5679
E-mail: mail.gibbons@mail.house.gov
Website: www.house.gov/gibbons

NEW HAMPSHIRE

Senators:

Robert C. Smith
Phone: 202-224-2841   Fax: 202-224-1353
E-mail: opinion@smith.senate.gov
Website: smith.senate.gov

Judd Gregg
Phone: 202-224-3324   Fax: 202-224-4952
E-mail: mailbox@gregg.senate.gov
Website: gregg.senate.gov

Representatives:

John E. Sununu
Phone: 202-225-5456   Fax: 202-225-5822
E-mail: rep.sununu@mail.house.gov
Website: www.house.gov/sununu

Charles Bass
Phone: 202-225-5206   Fax: 202-225-2946
E-mail: cbass@mail.house.gov
Website: www.house.gov/bass

NEW JERSEY

Senators:

Robert G. Torricelli
Phone: 202-224-3224   Fax: 202-228-5803
Website: torricelli.senate.gov

Jon Corzine
Phone: 202-224-4744   Fax: 202-228-2197
Website: corzine.senate.gov

Representatives:

Robert E. Andrews
Phone: 202-225-6501   Fax: 202-225-6583
E-mail: rob.andrews@mail.house.gov
Website: www.house.gov/andrews

Frank A. LoBiondo
Phone: 202-225-6572   Fax: 202-225-3318
E-mail: lobiondo@mail.house.gov
Website: www.house.gov/lobiondo

Jim Saxton
Phone: 202-225-4765   Fax: 202-225-0778
E-mail: jim.saxton@mail.house.gov
Website: www.house.gov/saxton

Christopher H. Smith
Phone: 202-225-3765   Fax: 202-225-7768
Website: www.house.gov/chrissmith

Marge Roukema
Phone: 202-225-4465   Fax: 202-225-9048
E-mail: rep.roukema@mail.house.gov
Website: www.house.gov/roukema

Frank Pallone Jr.
Phone: 202-225-4671   Fax: 202-225-9665
E-mail: frank.pallone@mail.house.gov
Website: www.house.gov/pallone

Michael A. Ferguson
Phone: 202-225-5361   Fax: 202-225-9460
Website: www.house.gov/ferguson

William J. Pascrell, Jr.
Phone: 202-225-5751   Fax: 202-225-5782
E-mail: bill.pascrell@mail.house.gov
Website: www.house.gov/pascrell

Steven R. Rothman
Phone: 202-225-5061   Fax: 202-225-5851
E-mail: steven.rothman@mail.house.gov
Website: www.house.gov/rothman

Donald M. Payne
Phone: 202-225-3436   Fax: 202-225-4160
Website: www.house.gov/payne

Rodney Frelinghuysen
Phone: 202-225-5034   Fax: 202-225-3186
E-mail: rodney.frelinghuysen@mail.house.gov
Website: www.house.gov/frelinghuysen

Rush Holt
Phone: 202-225-5801   Fax: 202-225-6025
E-mail: rush.holt@mail.house.gov
Website: www.house.gov/rholt

Robert Menendez
Phone: 202-225-7919   Fax: 202-226-0792
E-mail: menendez@mail.house.gov
Website: www.house.gov/menendez

NEW MEXICO

Senators:

Pete V. Domenici
Phone: 202-224-6621   Fax: 202-228-0900
Website: domenici.senate.gov

Jeff Bingaman
Phone: 202-224-5521   Fax: 202-224-2852
E-mail: senator_bingaman@bingman.senate.gov
Website: bingaman.senate.gov

Representatives:

Heather A Wilson
Phone: 202-225-6316   Fax: 202-225-4975
E-mail: ask.heather@mail.house.gov
Website: www.house.gov/wilson

Joe Skeen
Phone: 202-225-2365   Fax: 202-225-9599
E-mail: joe.skeen@mail.house.gov
Website: www.house.gov/skeen

Tom Udall
Phone: 202-225-6190   Fax: 202-226-1331
E-mail: tom.udall@mail.house.gov
Website: www.house.gov/tomudall

NEW YORK

Senators:

Charles E. Schummer
Phone: 202-224-6542   Fax: 202-228-3027
E-mail: senator@schummer.senate.gov
Website: schummer.senate.gov

Hillary Rodham Clinton
Phone: 202-224-4451   Fax: 202-228-0282
Website: clinton.senate.gov

Representatives:

Felix J. Grucci, Jr.
Phone: 202-225-3826   Fax: 202-225-3143
Website: www.house.gov/grucci

Steve J. Israel
Phone: 202-225-3335   Fax: 202-225-4669
Website: www.house.gov/israel

Peter King
Phone: 202-225-7896   Fax: 202-226-2279
E-mail: peter.king@mail.house.gov
Website: www.house.gov/king

Carolyn McCarthy
Phone: 202-225-5516   Fax: 202-225-5758
Website: www.house.gov/carolynmccarthy

Gary L. Ackerman
Phone: 202-225-2601   Fax: 202-225-1589
E-mail: gary_ackerman@mail.house.gov
Website: www.house.gov/ackerman

Gregory W. Meeks
Phone: 202-225-3461   Fax: 202-226-4169
E-mail: congmeeks@mail.house.gov
Website: www.house.gov/meeks

Joseph Crowley
Phone: 202-225-3965   Fax: 202-225-1909
E-mail: write2joecrowley@mail.house.gov
Website: www.house.gov/crowley

Jerrold Nadler
Phone: 202-225-5635   Fax: 202-225-6923
E-mail: jerrold.nadler@mail.house.gov
Website: www.house.gov/nadler

Anthony David Weiner
Phone: 202-225-6616   Fax: 202-226-7253
E-mail: weiner@mail.house.gov
Website: www.house.gov/weiner

Edolphus Towns
Phone: 202-225-5936   Fax: 202-225-1018
E-mail: edolphus.towns@mail.house.gov
Website: www.house.gov/towns

Major R. Owens
Phone: 202-225-6231   Fax: 202-226-0112
Website: www.house.gov/owens

Nydia Velazquez
Phone: 202-225-2361   Fax: 202-226-0327
Website: www.house.gov/velazquez

Vito Fossella
Phone: 202-225-3371   Fax: 202-226-1272
E-mail: vito.fossella@mail.house.gov
Website: www.house.gov/fossella

Carolyn Maloney
Phone: 202-225-7944   Fax: 202-225-4709
E-mail: rep.carolyn.maloney@mail.house.gov
Website: www.house.gov/maloney

Charles B. Rangel
Phone: 202-225-4365   Fax: 202-225-0816
Website: www.house.gov/rangel

Jose E. Serrano
Phone: 202-225-4361   Fax: 202-225-6001
Website: www.house.gov/serrano

Eliot Engel
Phone: 202-225-2464 Fax: 202-225-5513
Website: www.house.gov/engel

Nita M. Lowey
Phone: 202-225-6506 Fax: 202-225-0546
E-mail: nita.lowey@mail.house.gov
Website: www.house.gov/lowey

Sue W. Kelly
Phone: 202-225-5441 Fax: 202-225-3289
E-mail: dearsue@mail.house.gov
Website: www.house.gov/suekelly

Benjamin A. Gilman
Phone: 202-225-3776 Fax: 202-225-2541
Website: www.house.gov/gilman

Michael R. McNulty
Phone: 202-225-5076 Fax: 202-2235077
E-mail: mike.mcnulty@mail.house.gov
Website: www.house.gov/mcnulty

John E. Sweeney
Phone: 202-225-5614 Fax: 202-225-6231
E-mail: john.sweeney@mail.house.gov
Website: www.house.gov/sweeney

Sherwood L. Boehlert
Phone: 202-225-3665 Fax: 202-225-1891
E-mail: rep.boehlert@mail.house.gov
Website: www.house.gov/boehlert

John McHugh
Phone: 202-225-4611 Fax: 202-226-0621
Website: www.house.gov/mchugh

James T. Walsh
Phone: 202-225-3701 Fax: 202-225-4042
E-mail: rep.james.walsh@mail.house.gov
Website: www.house.gov/walsh

Maurice Hinchey
Phone: 202-225-6335 Fax: 202-226-0774
Website: www.house.gov/hinchey

Thomas Reynolds
Phone: 202-225-5265 Fax: 202-225-5910
Website: www.house.gov/reynolds

Louise McIntosh Slaughter
Phone: 202-225-3615 Fax: 202-225-7822
E-mail: louiseny@mail.house.gov
Website: www.house.gov/slaughter

John J. LaFalce
Phone: 202-225-3231 Fax: 202-226-9911
Website: www.house.gov/lafalce

Jack Quinn
Phone: 202-225-3306 Fax: 202-226-0347
Website: www.house.gov/quinn

Amory Houghton Jr.
Phone: 202-225-3161 Fax: 202-225-5574
Website: www.house.gov/houghton

NORTH CAROLINA

Senators:

Jesse Helms
Phone: 202-224-6342 Fax: 202-228-1339
E-mail: jeese_helms@helms.senate.gov
Website: helms.senate.gov

John R. Edwards
Phone: 202-224-3154 Fax: 202-228-1374
Website: edwards.senate.gov

Representatives:

Eva M. Clayton
Phone: 202-225-3101 Fax: 202-225-3354
Website: www.house.gov/clayton

Bob Etheridge
Phone: 202-225-4531 Fax: 202-225-5662
E-mail: bob.etheridge@mail.house.gov
Website: www.house.gov/etheridge

Walter Jones Jr.
Phone: 202-225-3415 Fax: 202-225-3286
E-mail: conjones@mail.house.gov
Website: www.house.gov/jones

David E. Price
Phone: 202-225-1784 Fax: 202-225-2014
E-mail: david.price@mail.house.gov
Website: www.house.gov/price

Richard M. Burr
Phone: 202-225-2071 Fax: 202-225-2995
E-mail: richard.burrnc05@mail.house.gov
Website: www.house.gov/burr

Howard Coble
Phone: 202-225-3065 Fax: 202-225-8611
E-mail: howard.coble@mail.house.gov
Website: www.house.gov/coble

Mike McIntyre
Phone: 202-225-2731 Fax: 202-225-5773
Website: www.house.gov/mcintyre

Robin Hayes
Phone: 202-225-3715 Fax: 202-225-4036
Website: www.house.gov/hayes

Sue Myrick
Phone: 202-225-1976 Fax: 202-225-3389
E-mail: myrick@mail.house.gov
Website: www.house.gov/myrick

Cass Ballenger
Phone: 202-225-2576 Fax: 202-225-0316
E-mail: cass.ballenger@mail.house.gov
Website: www.house.gov/ballenger

Charles H. Taylor
Phone: 202-225-6401 Fax:
E-mail: repcharles.taylor@mail.house.gov
Website: www.house.gov/charlestaylor

Melvin L. Watt
Phone: 202-225-1510 Fax: 202-225-1512
E-mail: n61L.public@mail.house.gov
Website: www.house.gov/watt

NORTH DAKOTA

Senators:

Kent Conrad
Phone: 202-224-2043 Fax: 202-224-7776
E-mail: senator@conrad.senate.gov
Website: conrad.senate.gov

Byron L. Dorgan
Phone: 202-224-2551 Fax: 202-224-1193
E-mail: senator@dorgan.senate.gov
Website: dorgan.senate.gov

Representatives:

Earl Pomeroy
Phone: 202-225-2611   Fax: 202-226-0893
E-mail: rep.earl.pomeroy@mail.house.gov
Website: www.house.gov/pomeroy

OHIO

Senators:

Mike DeWine
Phone: 202-224-2315   Fax: 202-224-6519
E-mail: senator_dewine@dewine.senate.gov
Website: dewine.senate.gov

George V. Voinovich
Phone: 202-224-3353   Fax: 202-228-1382
E-mail: senator_voinovich@voinovich.senate.gov
Website: voinovich.senate.gov

Representatives:

Steve Chabot
Phone: 202-225-2216   Fax: 202-225-3012
Website: www.house.gov/chabot

Rob J. Portman
Phone: 202-225-3164   Fax: 202-225-1992
E-mail: portmail@mail.house.gov
Website: www.house.gov/portman

Tony P. Hall
Phone: 202-225-6465   Fax: 202-226-1443
Website: www.house.gov/tonyhall

Michael G. Oxley
Phone: 202-225-2676   Fax: 202-226-0577
E-mail: mike.oxley@mail.house.gov
Website: www.house.gov/oxley

Paul E. Gilmor
Phone: 202-225-6405   Fax: 202-225-1985
E-mail: paul.gilmor@mail.house.gov
Website: www.house.gov/gilmor

Ted Strickland
Phone: 202-225-5705   Fax: 202-225-5907
Website: www.house.gov/strickland

David Hobson
Phone: 202-225-4324   Fax: 202-225-1984
Website: www.house.gov/hobson

John A. Boehner
Phone: 202-225-6205   Fax: 202-225-0704
E-mail: john.boehner@mail.house.gov
Website: www.house.gov/boehner

Marcy Kaptur
Phone: 202-225-4146   Fax: 202-225-7711
E-mail: rep.kaptur@mail.house.gov
Website: www.house.gov/kaptur

Dennis J. Kucinich
Phone: 202-225-5871   Fax: 202-225-5745
Website: www.house.gov/kucinch

Stephanie Tubbs Jones
Phone: 202-225-7032   Fax: 202-225-1339
E-mail: stephanie.tubbs.jone@mail.house.gov
Website: www.house.gov/tubbsjones

Patrick J. Tiberi
Phone: 202-225-5355   Fax: 202-226-4523
Website: www.house.gov/tiberi

Sherrod Brown
Phone: 202-225-3401   Fax: 202-225-2266
E-mail: sherrod@mail.house.gov
Website: www.house.gov/sherrodbrown

Thomas C. Sawyer
Phone: 202-225-5231   Fax: 202-225-5278
Website: www.house.gov/sawyer

Deborah Pryce
Phone: 202-225-2015   Fax: 202-225-3529
E-mail: pryce.oh15@mail.house.gov
Website: www.house.gov/pryce

Ralph Regula
Phone: 202-224-3876   Fax: 202-225-3059
Website: www.house.gov/regula

James A. Traficant Jr
Phone: 202-225-5261   Fax: 202-225-3719
E-mail: telljim@mail.house.gov
Website: www.house.gov/traficant

Bob Ney
Phone: 202-225-6265   Fax: 202-225-3394
E-mail: bobney@mail.house.gov
Website: www.house.gov.ney

Steven C. laTourette
Phone: 202-225-5731   Fax: 202-225-3307
Website: www.house.gov/latourette

OKLAHOMA

Senators:

Don Nickles
Phone: 202-224-5754   Fax: 202-224-6008
E-mail: senator@nickles.senate.gov
Website: nickles.senate.gov

James M. Inhofe
Phone: 202-224-4721   Fax: 202-228-0380
E-mail: jim_inhofe@inhofe.senate.gov
Website: inhofe.senate.gov

Representatives:

Steve Largent
Phone: 202-225-2211   Fax: 202-225-9187
Website: www.house.gov/largent

Brad Carson
Phone: 202-225-2701   Fax: 202-225-3038
E-mail: brad.carson@mail.house.gov
Website: www.house.gov/bradcarson

Wes W. Watkins
Phone: 202-225-4565   Fax: 202-225-5966
E-mail: wes.watkins@mail.house.gov
Website: www.house.gov/watkins

J.C. Watts Jr.
Phone: 202-225-6165   Fax: 202-225-3512
E-mail: rep.jcwatts@mail.house.gov
Website: www.house.gov/watts

Ernest Istook Jr.
Phone: 202-225-2132   Fax: 202-226-1463
E-mail: istook@mail.house.gov
Website: www.house.gov/istook

Frank D. Lucas
Phone: 202-225-5565   Fax: 202-225-8698
E-mail: repulcas@mail.house.gov
Website: www.house.gov/lucas

OREGON

Senators:

Ron Wyden
Phone: 202-224-5244   Fax: 202-228-2717
Website: wyden.senate.gov

Gordon Smith
Phone: 202-224-3753   Fax: 202-228-3997
E-mail: oregon@gsmith.senate.gov
Website: gsmith.senate.gov

Representatives:

David Wu
Phone: 202-225-0855   Fax: 202-225-9497
Website: www.house.gov/wu

Greg Walden
Phone: 202-225-6730   Fax: 202-225-5774
E-mail: greg.walden@mail.house.gov
Website: www.house.gov/walden

Earl Blumenauer
Phone: 202-225-4811   Fax: 202-225-8941
Website: www.house.gov/blumenauer

Peter A. DeFazio
Phone: 202-225-6416   Fax: 202-225-0032
Website: www.house.gov/defazio

Darlene Hooley
Phone: 202-225-5711   Fax: 202-225-5699
Website: www.house.gov/hooley

PENNSYLVANIA

Senators:

Arlen Specter
Phone: 202-224-4254   Fax: 202-228-1229
Website: specter.senate.gov

Rick Santorum
Phone: 202-224-6324   Fax: 202-228-0604
Website: santorum.senate.gov

Representatives:

Robert A. Brady
Phone: 202-225-4731   Fax: 202-225-0088
E-mail: robert.a.brady@mail.house.gov
Website: www.house.gov/robertbrady

Chaka Fattah
Phone: 202-225-4001   Fax: 202-225-5392
Website: www.house.gov/fattah

Robert A. Borski
Phone: 202-225-8251   Fax: 202-225-4628
E-mail: robert.borski@mail.house.gov
Website: www.house.gov/borski

Melissa A. Hart
Phone: 202-225-2565   Fax: 202-226-2274
E-mail: rep.hart@mail.house.gov
Website: www.house.gov/hart

John E. Peterson
Phone: 202-225-5121   Fax: 202-225-5796
E-mail: john.peterson@mail.house.gov
Website: www.house.gov/johnpeterson

Tim Holden
Phone: 202-225-5546   Fax: 202-226-0996
Website: www.house.gov/holden

Curt Weldon
Phone: 202-225-2011   Fax: 202-225-8137
E-mail: curtpa07@mail.house.gov
Website: www.house.gov/curtweldon

Jim Greenwood
Phone: 202-225-4276   Fax: 202-225-9511
E-mail: greenwoodpa@mail.house.gov
Website: www.house.gov/greenwood

Bill Shuster
Phone: 202-225-2431   Fax: 202-225-2486
Website: www.house.gov/shuster

Don Sherwood
Phone: 202-225-3731   Fax: 202-225-9594
Website: www.house.gov/sherwood

Paul E. Kanjorski
Phone: 202-225-6511   Fax: 202-225-0764
E-mail: paul.kanjorski@mail.house.gov
Website: www.house.gov/kanjorski

John P. Murtha
Phone: 202-225-2065   Fax: 202-225-5709
E-mail: murtha@mail.house.gov
Website: www.house.gov/murtha

Joseph M. Hoeffel III
Phone: 202-225-6111   Fax: 202-226-0611
Website: www.house.gov/hoeffel

William J. Coyne
Phone: 202-225-2301   Fax: 202-225-1844
Website: www.house.gov/coyne

Pat Toomey
Phone: 202-225-6411   Fax: 202-226-0778
E-mail: rep.toomey.pa15@mail.house.gov
Website: www.house.gov/toomey

Joseph R. Pitts
Phone: 202-225-2411   Fax: 202-225-2013
E-mail: pitts.pa16@mail.house.gov
Website: www.house.gov/pitts

George W. Gekas
Phone: 202-225-4315   Fax: 202-225-8440
E-mail: askgeorge@mail.house.gov
Website: www.house.gov/gekas

Mike Doyle
Phone: 202-225-2135   Fax: 202-225-3084
E-mail: rep.doyle@mail.house.gov
Website: www.house.gov/doyle

Todd R. Platts
Phone: 202-225-5836   Fax: 202-226-1000
Website: www.house.gov/platts

Frank R. Mascara
Phone: 202-225-4665   Fax: 202-225-3377
Website: www.house.gov/mascara

Philip S. English
Phone: 202-225-5406   Fax: 202-225-3103
E-mail: phil.english@mail.house.gov
Website: www.house.gov/english

RHODE ISLAND

Senators:

Jack Reed
Phone: 202-224-4642   Fax: 202-224-4680
E-mail: jack@reed.senate.gov
Website: reed.senate.gov

Lincoln D. Chafee
Phone: 202-224-2921 Fax: 202-228-2853
E-mail: senator_chafee@chafee.senate.gov
Website: chafee.senate.gov

Representatives:

Patrick J. Kennedy
Phone: 202-225-4911 Fax: 202-225-3290
E-mail: patrick.kennedy@mail.house.gov
Website: www.house.gov/patrickkennedy

James R. Langevin
Phone: 202-225-2735 Fax: 202-225-5976
E-mail: james.langevin@mail.house.gov
Website: www.house.gov/langevin

SOUTH CAROLINA

Senators:

Strom Thurmond
Phone: 202-224-5972 Fax: 202-224-1300
E-mail: senator@thurmond.senate.gov
Website: thurmond.senate.gov

Ernest F. Hollings
Phone: 202-224-6121 Fax: 202-224-4293
Website: hollings.senate.gov

Representatives:

Henry E. Brown, Jr.
Phone: 202-225-3175 Fax: 202-225-3407
E-mail: writehenrybrown@mail.house.gov
Website: www.house.gov/henrybrown

Floyd Spence
Phone: 202-225-2452 Fax: 202-225-2455
Website: www.house/gov/spence

Lindsey Graham
Phone: 202-225-5301 Fax: 202-225-3216
Website: www.house.gov/graham

Jim DeMint
Phone: 202-225-6030 Fax: 202-226-1177
E-mail: jim.demint@mail.house.gov
Website: www.demint.house.gov

John M. Spratt, Jr.
Phone: 202-225-5501 Fax: 202-225-0464
Website: www.house.gov/spratt

James Clyburn
Phone: 202-225-3315 Fax: 202-225-2313
E-mail: jclyburn@mail.house.gov
Website: www.house.gov/clyburn

SOUTH DAKOTA

Senators:

Thomas A. Daschle
Phone: 202-224-2321 Fax: 202-224-7895
E-mail: tom_daschle@daschle.senate.gov
Website: daschle.senate.gov

Tim Johnson
Phone: 202-224-5842 Fax: 202-228-5765
E-mail: tim@johnson.senate.gov
Website: johnson.senate.gov

Representatives:

John R. Thune
Phone: 202-225-2801 Fax: 202-225-5823
E-mail: jthune@mail.house.gov
Website: www.housae.gov/thune

TENNESSEE

Senators:

Fred Thompson
Phone: 202-224-4944 Fax: 202-228-3679
E-mail: senator_thompson@thompson.senate.gov
Website: thompson.senate.gov

Bill Frist
Phone: 202-224-3344 Fax: 202-228-1264
E-mail: senator_frist@frist.senate.gov
Website: frist.senate.gov

Representatives:

William L. Jenkins
Phone: 202-225-6356 Fax: 202-225-5714
Website: www.house.gov/jenkins

John J. Duncan, Jr.
Phone: 202-225-5435 Fax: 202-225-6440
Website: www.house.gov/duncan

Zach Wamp
Phone: 202-225-3271 Fax: 202-225-3494
Website: www.house.gov/wamp

Van Hilleary
Phone: 202-225-6831 Fax: 202-225-3272
E-mail: van.hilleary@mail.house.gov
Website: www.house.gov/hilleary

Bob Clement
Phone: 202-225-4311 Fax: 202-226-1035
E-mail: bob.clement@mail.house.gov
Website: www.house.gov/clement

Bart Gordon
Phone: 202-225-4231 Fax: 202-225-6887
Website: www.house.gov/gordon

Ed Bryant
Phone: 202-225-2811 Fax: 202-225-2989
Website: www.house.gov/bryant

John S. Tanner
Phone: 202-225-4714 Fax: 202-225-1765
Website: www,house.gov/tanner

Harold E. Ford, Jr.
Phone: 202-225-3265 Fax: 202-225-5663
E-mail: rep.harold.ford.jr.@mail.house.gov
Website: www.house.gov/ford

TEXAS

Senators:

Phil Gramm
Phone: 202-224-2934 Fax: 202-228-2856
E-mail: phil_gramm@gramm.senate.gov
Website: gramm.senate.gov

Kay Bailey Hutchison
Phone: 202-224-5922 Fax: 202-224-0776
E-mail: senator@hutchison.senate.gov
Website: hutchison.senate.gov

Representatives:

Max A. Sandlin
Phone: 202-225-3035 Fax: 202-225-5866
Website: www.house.gov/sandlin

Jim Turner
Phone: 202-225-2401 Fax: 202-225-5955
E-mail: tx02wyr@mail.house.gov
Website: www.house.gov/turner

Sam Johnson
Phone: 202-225-4201 Fax: 202-225-1485
Website: www.house.gov/samjohnson

Ralph M. Hall
Phone: 202-225-6673 Fax: 202-225-3332
E-mail: rmhall@mail.house.gov
Website: www.house.gov/ralphhall

Pete Sessions
Phone: 202-225-2231 Fax: 202-225-5878
E-mail: petes@mail.house.gov
Website: wwww.house.gov/sessions

Joe Barton
Phone: 202-225-2002 Fax: 202-225-3052
Website: www.house.gov/barton

John A. Culberson
Phone: 202-225-2571 Fax: 202-225-4381
Website: www.house.gov/culberson

Kevin P. Brady
Phone: 202-225-4901 Fax: 202-225-5524
E-mail: rep.brady@mail.house.gov
Website: www.house.gov/brady

Nicholas V. Lampson
Phone: 202-225-6565 Fax: 202-225-5547
E-mail: nick.lampson@mail.house.gov
Website: www.house.gov/lampson

Lloyd Doggett
Phone: 202-225-4865 Fax: 202-225-3073
E-mail: lloyd.doggett@mail.house.gov
Website: www.house.gov/doggett

Chet Edwards
Phone: 202-225-6105 Fax: 202-225-0350
Website: www.house.gov/edwards

Kay Granger
Phone: 202-225-5071 Fax: 202-225-5683
E-mail: texas.granger@mail.house.gov
Website: www.house.gov/granger

Willliam "Mac" Thornberry
Phone: 202-225-3706 Fax: 202-225-3486
Website: www.house.gov/thornberry

Ron E. Paul
Phone: 202-225-2831 Fax: 202-226-4871
E-mail: rep.paul@mail.house.gov
Website: www.house.gov/paul

Ruben E. Hinojosa
Phone: 202-225-2531 Fax: 202-225-5688
E-mail: rep.hinojosa@mail.house.gov
Website: www.house.gov/hinojosa

Silvestre Reyes
Phone: 202-225-4831 Fax: 202-225-2016
E-mail: talk2silver@mail.house.gov
Website: www.house.gov/reyes

Charles W. Stenholm
Phone: 202-225-6605 Fax: 202-225-2234
Website: www.hosue.gov/stenholm

Sheila Jackson Lee
Phone: 202-225-3816 Fax: 202-225-3317
E-mail: tx18@mail.house.gov
Website: www.house.gov/jacksonlee

Larry Combest
Phone: 202-225-4005 Fax: 202-225-9615
Website: www.house.gov/combest

Charles A. Gonzalez
Phone: 202-225-3236 Fax: 202-225-1915
Website: www.house.gov/gonzalez

Lamar S. Smith
Phone: 202-225-4236 Fax: 202-225-8628
Website: www.house.gov/lamarsmith

Tom DeLay
Phone: 202-225-5951 Fax: 202-225-5241
Website: tomdelay.house.gov

Henry Bonilla
Phone: 202-225-4511 Fax: 202-225-2237
Website: www.house.gov/bonilla

Martin Frost
Phone: 202-225-3605 Fax: 202-225-4951
E-mail: martin.frost@mail.house.gov
Website: www.house.gov/frost

Ken Bentsen
Phone: 202-225-7508 Fax: 202-225-2947
E-mail: ken.bentsen@mail.house.gov
Website: www.house.gov/bentsen

Richard K. Armey
Phone: 202-225-7772 Fax: 202-226-8100
Website: armey.house.gov

Solomon P. Ortiz
Phone: 202-225-7742 Fax: 202-226-1134
Website: www.house.gov/ortiz

Ciro D. Rodriguez
Phone: 202-225-1640 Fax: 202-225-1641
Website: www.house.gov/rodriguez

Gene Green
Phone: 202-225-1688 Fax: 202-225-9903
Website: www.house.gov/green

Eddie Bernice Johnson
Phone: 202-225-8885 Fax: 202-226-1477
E-mail: rep.e.b.johnson@mail.house.gov
Website: www.house.gov/ebjohnson

UTAH

Senators:

Orrin G. Hatch
Phone: 202-224-5251 Fax: 202-224-6331
E-mail: senator_hatch@hatch.senate.gov
Website: hatch.senate.gov

Robert Bennett
Phone: 202-224-5444 Fax: 202-228-1168
E-mail: senator@bennett.senate.gov
Website: bennett.senate.gov

Representatives:

James V. Hansen
Phone: 202-225-0453 Fax: 202-225-5857
Website: www.house.gov/hansen

James David Matheson
Phone: 202-225-3011 Fax: 202-225-5638
E-mail: jim.matheson@mail.house.gov
Website: www,house.gov/matheson

Chris Cannon
Phone: 202-225-7751 Fax: 202-225-5629
E-mail: cannon.ut03@mail.house.gov
Website: www.house.gov/cannon

VERMONT

Senators:

Patrick J. Leahy
Phone: 202-224-4242   Fax: 202-224-3479
E-mail: senator_leahy@leahy.senate.gov
Website: leahy.senate.gov

James M. Jeffords
Phone: 202-224-5141   Fax: 202-228-0776
E-mail: vermont@jeffords.senate.gov
Website: jeffords.senate.gov

Representatives:

Bernard Sanders
Phone: 202-225-4115   Fax: 202-225-6790
E-mail: bernie@mail.house.gov
Website: bernie.house.gov

VIRGINIA

Senators:

John W. Warner
Phone: 202-224-2023   Fax: 202-224-6295
E-mail: senator@warner.senate.gov
Website: warner.senate.gov

George Allen
Phone: 202-224-4024   Fax: 202-224-5432
E-mail: senator_allen@allen.senate.gov
Website: allen.senate.gov

Representatives:

Jo Ann S. Davis
Phone: 202-225-4261   Fax: 202-225-4382
E-mail: joann.davis@mail.house.gov
Website: www.house.gov/joanndavis

Edward L. Schrock
Phone: 202-225-4215   Fax: 202-225-4218
E-mail: ed.schrock@mail.house.gov
Website: schrock.house.gov

Bobby Scott
Phone: 202-225-8351   Fax: 202-225-8354
Website: www.house.gov/scott

Randy Forbes
Phone: 202-225-6365   Fax: 202-226-1170
Website: www.house.gov/forbes

Virgil H. Goode, Jr.
Phone: 202-225-4711   Fax: 202-225-5681
E-mail: rep.goode@mail.house.gov
Website: www.house.gov/goode

Bob Goodlatte
Phone: 202-225-5431   Fax: 202-225-9681
E-mail: talk2bob@mail.house.gov
Website: www.house.gov/goodlatte

Eric I. Cantor
Phone: 202-225-2815   Fax: 202-225-0011
E-mail: eric.cantor@mail.house.gov
Website: www.house.gov/cantor

James P. Moran
Phone: 202-225-4376   Fax: 202-225-0017
Website: www.house.gov/moran

Rick Boucher
Phone: 202-225-3861   Fax: 202-225-0442
E-mail: ninthnet@mail.house.gov
Website: www.house.gov/boucher

Frank R. Wolf
Phone: 202-225-5136   Fax: 202-225-0437
Website: www.house.gov/wolf

Thomas M. Davis III
Phone: 202-225-1492   Fax: 202-225-3071
E-mail: tom.davis@mail.house.gov
Website: www.house.gov/tomdavis

WASHINGTON

Senators:

Patty Murray
Phone: 202-224-2621   Fax: 202-224-0238
E-mail: senator_murray@murray.senate.gov
Website: murray.senate.gov

Maria Cantwell
Phone: 202-224-3441   Fax: 202-228-0514
Website: cantwell.senate.gov

Representatives:

Jay Inslee
Phone: 202-225-6311   Fax: 202-226-1606
E-mail: jay.inslee@mail.house.gov
Website: www.house.gov/inslee

Richard R. Larsen
Phone: 202-225-2605   Fax: 202-225-4420
E-mail: rick.larsen@mail.house.gov
Website: www.house.gov/larsen

Brian Baird
Phone: 202-225-3536   Fax: 202-225-3478
E-mail: brian.baird@mail.house.gov
Website: www.house.gov/baird

Doc Hastings
Phone: 202-225-5816   Fax: 202-225-3251
Website: www.house.gov/hastings

George R. Nethercutt, Jr.
Phone: 202-225-2006   Fax: 202-225-3392
E-mail: george.nethercutt-pub@mail.house.gov
Website: www.house.gov/nethercutt

Norman D. Dicks
Phone: 202-225-5916   Fax: 202-226-1176
Website: www.house.gov/dicks

Jim McDermott
Phone: 202-225-3106   Fax: 202-225-6197
Website: www.house.gov/mcdermott

Jennifer Dunn
Phone: 202-225-7761   Fax: 202-225-8673
E-mail: dunnwa08@mail.house.gov
Website: www.house.gov/dunn

Adam Smith
Phone: 202-225-8901   Fax: 202-225-5893
E-mail: adam.smith@mail.house.gov
Website: www.house.gov/adamsmith

WEST VIRGINIA

Senators:

Robert C. Byrd
Phone: 202-224-3954   Fax: 202-228-0002
E-mail: senator_byrd@byrd.senate.gov
Website: byrd.senate.gov

John D. Rockefeller
Phone: 202-224-6472   Fax: 202-224-7665
E-mail: senator@rockefeller.senate.gov
Website: rockefeller.senate.gov

Representatives:

Alan B. Mollohan
Phone: 202-225-4172   Fax: 202-225-7564
Website: www.house.gov/mollohan

Shelley Moore Capito
Phone: 202-225-2711   Fax: 202-225-7856
Website: www.house.gov/capito

Nick J. Rahall II
Phone: 202-225-3452   Fax: 202-225-9061
E-mail: nrahall@mail.house.gov
Website: www.house.gov/rahall

WISCONSIN

Senators:

Herbert H. Kohl
Phone: 202-224-5653   Fax: 202-224-9787
E-mail: senator_kohl@kohl.senate.gov
Website: kohl.senate.gov

Russ Feingold
Phone: 202-224-5323   Fax: 202-224-2725
Website: feingold.senate.gov

Representatives:

Paul D. Ryan
Phone: 202-225-3031   Fax: 202-225-3393
Website: www.house.gov/ryan

Tammy Baldwin
Phone: 202-225-2906   Fax: 202-225-6942
E-mail: tammy.baldwin@mail.house.gov
Website: www.house.gov/baldwin

Ron J. Kind
Phone: 202-225-5506   Fax: 202-225-5739
E-mail: ron.kind@mail.house.gov
Website: www.house.gov/kind

Jerry Kleczka
Phone: 202-225-4572   Fax: 202-225-8135
Website: www.house.gov/kleczka

Thomas Barrett
Phone: 202-225-3571   Fax: 202-225-2185
E-mail: telltom@mail.house.gov
Website: www.house.gov/barrett

Thomas E. Petri
Phone: 202-225-2476   Fax: 202-225-2356
Website: www.house.gov/petri

David R. Obey
Phone: 202-225-3365
Website: www.house.gov/obey

Mark Green
Phone: 202-225-6666   Fax: 202-225-6720
E-mail: mark.green@mail.house.gov
Website: www.house.gov/markgreen

F. James Sensenbrenner, Jr.
Phone: 202-225-5101   Fax: 202-225-3190
E-mail: sensen09@mail.house.gov
Website: www.house.gov/sensenbrenner

WYOMING

Senators:

Craig Thomas
Phone: 202-224-6441   Fax: 202-224-1724
E-mail: craig@thomas.senate.gov
Website: thomas.senate.gov

Michael B. Enzi
Phone: 202-224-3424   Fax: 202-228-0359
E-mail: senator@enzi.senate.gov
Website: enzi.senate.gov

Representative:

Barbara Cubin
Phone: 202-225-2311   Fax: 202-225-3057
E-mail: barbara.cubin@mail.house.gov
Website: www.house.gov/cubin

DISTRICT OF COLUMBIA

Representative:

Eleanor Holmes Norton
Phone: 202-224-8050   Fax: 202-225-3002
Website: www.house.gov/norton

AMERICAN SAMOA

Representative:

Eni F.H. Faleomavaega
Phone: 202-225-8577   Fax: 202-225-8757
E-mail: faleomavaega@mail.house.gov
Website: www.house.gov/faleolavaega

GUAM

Representative:

Robert A. Underwood
Phone: 202-225-1188   Fax: 202-226-0341
E-mail: guamtodc@mail.house.gov
Website: www.house.gov/underwood

PUERTO RICO

Representative:

Anibal Acevedo-Vila
Phone: 202-225-2615   Fax: 202-225-2154
E-mail: anibal@mail.house.gov
Website: www.house.gov/acevedo-vila

VIRGIN ISLANDS

Representative:

Donna M. Christian-Christensen
Phone: 202-225-1790   Fax: 202-225-5517
E-mail: donna.christensen@mail.house.gov
Website: www.house.gov/christian-christensen

HOUSE COMMITTEES

HOUSE COMMITTEE ON AGRICULTURE
DEPARTMENT OPERATIONS, OVERSIGHT,
NUTRITION AND FORESTRY; GENERAL FARM
COMMODITIES, RESOURCE CONSERVATION AND
CREDIT; LIVESTOCK AND HORTICULTURE; RISK
MANAGEMENT, RESEARCH AND SPECIALTY CROPS
Washington, DC 20515
Phone: 202-225-2171   Fax: 202-225-0917
Website: www.agriculture.house.gov

Founded: 1820
Membership: 52
Scope: National

Description: Adulteration of seeds, insect pests, and protection of
birds and animals in forest reserves; agriculture generally; agricultural and industrial chemistry; agricultural colleges and
experiment stations; agricultural economics and research; agri-

cultural education extension services; agricultural production and marketing and stabilization of prices of agricultural products; animal industry and diseases of animals; crop insurance and soil conservation; dairy industry; entomology and plant quarantine; extension of farm credit and farm security; forestry in general, and forest reserves other than those created from the public domain; human nutrition and home economics; inspection of livestock and meat products; plant industry, soils, and agricultural engineering; rural electrification; commodities exchanges and rural development.

Contact(s):
Larry Combest, CHAIR

HOUSE COMMITTEE ON APPROPRIATIONS
AGRICULTURE, RURAL DEVELOPMENT, FOOD AND DRUG ADMINISTRATION; COMMERCE, JUSTICE, STATE, AND JUDICIARY; DISTRICT OF COLUMBIA; ENERGY AND WATER DEVELOPMENT; FOREIGN OPERATIONS, EXPORT FINANCING, AND RELATED PROGRAMS; INTERIOR; LABOR, HEALTH AND HUMANS
Washington, DC 20515
Phone: 202-225-2771
Website: www.house.gov

Founded: NA
Membership: 60
Scope: National

Description: Consists of 60 members: Appropriation of the revenue for the support of the government, rescissions of appropriations contained in appropriation acts, and transfers of unexpended balances.

Contact(s):
Bill Young, CHAIR

HOUSE COMMITTEE ON COMMERCE
TELECOMMUNICATIONS, TRADE, AND CONSUMER PROTECTION; FINANCE AND HAZARDOUS MATERIALS; HEALTH AND ENVIRONMENT; ENERGY AND POWER; OVERSIGHT AND INVESTIGATIONS
Washington, DC 20515
Phone: 202-225-2927   Fax: 202-225-1919

Founded: NA
Membership: 125
Scope: National

Description: Jurisdiction: Interstate and foreign commerce generally; national energy policy generally; measures relating to the exploration, production, storage, supply, marketing, pricing, and regulation of energy resources, including all fossil fuels, solar energy, and other unconventional or renewable energy resources; measures relating to the conservation of energy resources; measures relating to the commercial application of energy technology; measures relating to energy information generally; measures relating to: (A) the generation and marketing of power (except by federally chartered or federal regional power marketing authorities), (B) the reliability and interstate transmission of, and ratemaking for, all power, and (C) the citing of generation facilities (except the installation of interconnections between government waterpower projects); interstate energy compacts; measures relating to general management of the Department of Energy, and the management and all functions of the Federal Energy Regulatory Commission; regulation of interstate and foreign communications; securities and exchanges; consumer affairs

and consumer protection; travel and tourism; public health and quarantine; health and health facilities, except health care supported by payroll deductions; and biomedical research and development. The committee shall have the same jurisdiction with respect to regulation of nuclear facilities and of use of nuclear energy as it has with respect to regulation of non-nuclear facilities and of use of non-nuclear energy.

Contact(s):
Billy Tauzin, CHAIR
Dave Marventano, CHIEF OF STAFF
James Barnette, GENERAL COUNSEL

HOUSE COMMITTEE ON EDUCATION AND THE WORKFORCE
Washington, DC 20515
Phone: 202-225-4527   Fax: 202-225-9571
Website: edwrksorce.house.gov

Founded: NA
Membership: 47
Scope: national

Description: Jurisdiction: Measures relating to education or labor generally; child labor; Gallaudet College; Howard University; convict labor and the entry of goods made by convicts into interstate commerce; labor standards; labor statistics; mediation and arbitration of labor disputes; regulation or prevention of importation of foreign laborers under contract; food programs for children in schools; United States Employees' Compensation Commission; vocational rehabilitation; wages and hours of labor; welfare of miners; and work incentive programs.

Contact(s):
John Boehner, CHAIRMAN
202-225-6205

HOUSE COMMITTEE ON INTERNATIONAL RELATIONS
AFRICA; ASIA AND THE PACIFIC; THE WESTERN HEMISPHERE; INTERNATIONAL ECONOMIC POLICY AND TRADE; INTERNATIONAL OPERATIONS AND HUMAN RIGHTS
Washington, DC 20515
Phone: 202-225-5021   Fax: 202-225-0225
E-mail: hirc@mail.house.gov

Founded: NA
Membership: 50
Scope: National

Description: Jurisdiction: Foreign policy; international economic and environmental policy; international conferences and congresses; United Nations organizations; fishing agreements; nuclear export policy.

Contact(s):
Henry Hyde, CHAIRMAN

HOUSE COMMITTEE ON RESOURCES
Washington, DC 20515
Phone: 202-225-2761
Website: www.resourcecommittee.house.gov

Founded: NA
Membership: 52
Scope: National

Description: Consists of 52 members: Forest reserves and

national parks created from the public domain; national parks lands; forfeiture of land grants and alien ownership, including alien ownership of mineral lands; geological survey; interstate compacts relating to apportionment of waters for irrigation purposes; irrigation and reclamation, including water supply for reclamation projects, and easements on public lands for irrigation projects, and acquisition of private lands when necessary to complete irrigation projects; measures relating to the care and management of Indians, including the care and allotment of Indian lands and general and special measures relating to Indian claims; measures (including funding measures) relating generally to the U.S. territories, common-wealths, and successor governments of the Trust Territory of the Pacific Islands, except measures concerning the federal tax system and federal appropriations; military parks and bat-tlefields; national cemeteries administered by the Secretary of the Interior, and parks within the District of Columbia; mineral land laws and claims and entries thereunder; mineral resources of the public lands; mining interests generally; mining schools and experimental stations; petroleum conser-vation on the public lands and conservation of the radium supply in the U.S.; preservation of prehistoric ruins and objects of interest on the public domain; public lands generally, including entry, easements, and grazing thereon; relations of the U.S. with the Indians and the Indian tribes; regulation of the domestic nuclear energy industry, including regulation of research and development of reactors and nuclear regulatory research. Also special oversight functions with respect to all programs affecting Indians and nonmilitary nuclear energy and research and development, including the disposal of nuclear waste.

Contact(s):
James Hansen, CHAIR
Allen Freemyer, CHIEF OF STAFF

HOUSE COMMITTEE ON RULES
Washington, DC 20515
Phone: 202-225-9191   Fax: 202-225-6763
Website: www.house.gov/rules

Founded: NA
Membership: 13
Scope: National

Description: Consists of 13 members: Grants rules outlining conditions for floor debate on legislation reported by regular standing committees, which includes granting emergency waivers under the Congressional Budget Act of 1974; also has legislative authority to create committees, change the rules of the House, and provide order of business of the House.

Contact(s):
David Drier, CHAIR
Porter Gross, VICE CHAIR

HOUSE COMMITTEE ON TRANSPORTATION AND INFRASTRUCTURE
Washington, DC 20515
Phone: 202-225-4472   Fax: 202-226-1270
Website: www.house.gov/transportation

Founded: NA
Membership: 20
Scope: National
Description: Consists of 73 members.

Contact(s):
Don Young, CHAIRMAN

SENATE COMMITTEE ON AGRICULTURE, NUTRITION, AND FORESTRY
PRODUCTION AND PRICE COMPETITIVENESS; MARKETING, INSPECTION, AND PRODUCT PROMOTION; FORESTRY, CONSERVATION, AND RURAL REVITALIZATION; RESEARCH, NUTRITION, AND GENERAL LEGISLATION
Washington, DC 20510
Phone: 202-224-2035
Website: www.senate.gov/~agriculture

Founded: NA
Scope: National

Description: Concerned with agriculture and agricultural commodities; inspection of livestock, meat, and agricultural products; animal industry and diseases; pests and pesticides; agricultural extension services and experiment stations; forestry in general and forest reserves and wilderness areas other than those created from the public domain; agricultural economics and research; human nutrition; home economics; farm credit and farm security; rural development, rural electrifi-cation and watersheds; agricultural production, marketing, and stabilization of prices; crop insurance and soil conservation; school nutrition programs; food stamp programs; food from fresh waters; plant industry, soils, and agricultural engineering. Such committee shall also study and review, on a comprehen-sive basis, matters relating to food, nutrition, and hunger, both in the United States and foreign countries, and rural affairs and report thereon from time to time.

Contact(s):
Tom Harkin, CHAIRMAN
Robert Sturm, CHIEF CLERK
Mark Halverson, CHIEF OF STAFF
David Johnson, COUNSEL
Keith Luse, MINORITY STAFF DIRECTOR
Richard Lugar, RANKING REPUBLICAN MEMBER

SENATE COMMITTEE ON APPROPRIATIONS
Washington, DC 20510
Phone: 202-224-3471   Fax: 202-224-8553
Website: www.appropriations.senate.gov

Founded: NA
Scope: National

Description: Concerned with all proposed legislation, messages, petitions, memorials, and other matters relating to appropria-tion of the revenue for the support of the federal government.

Contact(s):
Robert Byrd, CHAIRMAN
Terry Sauvain, STAFF DIRECTOR

SENATE COMMITTEE ON COMMERCE SCIENCE AND TRANSPORTATION
AVIATION; COMMUNICATIONS; CONSUMER AFFAIRS, FOREIGN COMMERCE AND TOURISM; SCIENCE, TECHNOLOGY, AND SPACE; SURFACE TRANSPORTA-TION AND MERCHANT MARINE; OCEANS AND FISHERIES
Washington, DC 20510
Phone: 202-224-5115   Fax: 202-228-5769
Website: www.senate.gov/~commerce

Founded: NA

Description: Concerned with interstate commerce; transportation; regulation of interstate common carriers, including railroads, buses, trucks, vessels, pipelines, and civil aviation; merchant marine and navigation; marine and ocean navigation, safety and transportation, including navigational aspects of deepwater ports; Coast Guard; inland waterways, except construction; communications; regulation of consumer products and services, except for credit, financial services, and housing; the Panama Canal, except for maintenance, operation, administration, sanitation, and government, and interoceanic canals generally; standards and measurements; highway safety; science, engineering and technology research, and development and policy; nonmilitary aeronautical and space sciences; transportation and commerce aspects of Outer Continental Shelf lands; marine fisheries; coastal zone management; oceans, weather, and atmospheric activities; sports.

Contact(s):
John McCain, CHAIR

SENATE COMMITTEE ON ENERGY AND NATURAL RESOURCES
Washington, DC 20510
Phone: 202-224-4971   Fax: 202-224-6163

Founded: NA
Membership: 22
Scope: National

Description: Concerned with the comprehensive study and review of matters relating to energy and resources development. Jursdiction: Coal production, distribution, and utilization; energy policy; energy regulation and conservation; energy related aspects of deepwater ports; energy research and development; extraction of minerals from oceans and Outer Continental Shelf lands; hydroelectric power, irrigation, and reclamation; mining education and research; mining, mineral lands, mining claims, and mineral conservation; national parks, recreation areas, wilderness areas, wild and scenic rivers, historical sites, military parks and battlefields, and on the public domain, preservation of prehistoric ruins and objects of interest; naval petroleum reserves in Alaska; nonmilitary development of nuclear energy; oil and gas production and distribution; public lands and forests, including farming and grazing thereon, and mineral extraction therefrom; solar energy systems; and territorial possessions of the United States, including trusteeships.

Contact(s):
Frank Murkowski, CHAIR

SENATE COMMITTEE ON ENVIRONMENT AND PUBLIC WORKS
TRANSPORTATION AND INFRASTRUCTURE; SUPERFUND; WASTE CONTROL AND RISK ASSESMENT; CLEAN AIR, WETLANDS, PRIVATE PROPERTY, AND NUCLEAR SAFETY; DRINKING WATER, FISHERIES, AND WILDLIFE
Washington, DC 20510
Phone: 202-224-6176   Fax: 202-224-1273
Website: epw.senate.gov

Founded: NA
Membership: 50
Scope: National

Description: Committee on Environment and Public Works, to which shall be referred all proposed legislation, messages, petitions, memorials, and other matters relating to the following subjects: environmental policy; environmental research and development; ocean dumping; fisheries and wildlife; environmental aspects of Outer Continental Shelf lands; solid waste disposal and recycling; environmental effects of toxic substances, other than pesticides; water resources; flood control and improvements of rivers and harbors, including environmental aspects of deepwater ports; public works, bridges, and dams; water pollution; air pollution; noise pollution; nonmilitary environmental regulation and control of nuclear energy; regional economic development; construction and maintenance of highways; public buildings and improved grounds of the United States generally, including federal buildings in the District of Columbia. Such committee shall also study and review on a comprehensive basis matters relating to environmental protection and resource utilization and conservation, and report thereon from time to time.

Contact(s):
Jim Jeffords, CHAIR
Ken Connolly, DEMOCRAT STAFF DIRECTOR
J. Sliter, MINORITY STAFF DIRECTOR
Dave Conover, REPUBLICAN STAFF DIRECTOR

SENATE COMMITTEE ON FOREIGN RELATIONS
AFRICAN AFFAIRS; EAST ASIAN AND PACIFIC AFFAIRS; EUROPEAN AFFFAIRS; INTERNATIONAL ECONOMIC POLICY, EXPORT AND TRADE PROMOTION; INTERNATIONAL OPERATIONS; NEAR EASTERN AND SOUTH ASIAN AFFAIRS; WESTERN HEMISPHERE, PEACE CORPS, NARCOTICS AND TERRORISM
Washington, DC 20510-6225
Phone: 202-224-4651   Fax: 202-228-1608
Website: www.foreign.state.gov

Founded: NA
Membership: 10
Scope: National

Description: Jurisdiction: Foreign and national security policy; international treaties, conferences, and congresses; World Bank and International Monetary Fund; oceans and international environmental and scientific affairs; humanitarian assistance and hunger; and United Nations and its affiliated organizations.

Contact(s):
Bill Frist, CHAIR, SUBCOMMITTEE ON AFRICAN AFFAIRS
Chuck Hagel, CHAIR, SUBCOMMITTEE ON INTERNATIONAL ECONOMIC POLICY, EXPORT AND TRADE PROMOTION
Rod Grams, CHAIR, SUBCOMMITTEE ON INTERNATIONAL OPERATIONS
Sam Brownback, CHAIR, SUBCOMMITTEE ON NEAR EASTERN AND SOUTH ASIAN AFFAIRS
Christopher Dodd, CHAIR, SUBCOMMITTEE ON WESTERN HEMISPHERE, PEACE CORPS,NARCOTICS AND TERRORISM

SENATE COMMITTEE ON HEALTH, EDUCATION, LABOR, AND PENSIONS

AGING; CHILDREN, FAMILY, DRUGS, AND
ALCOHOLISM; EDUCATION, ARTS, AND HUMANITIES;
EMPLOYMENT AND PRODUCTIVITY; HANDICAPPED;
LABOR
Washington, DC 20510
Phone: 202-224-5375
Website: www.access.gpo.gov

Founded: NA

Contact(s):
    James Jeffords, CHAIR

U.S. FEDERAL AND INTERNATIONAL GOVERNMENT AGENCIES

A

ADVISORY COUNCIL ON HISTORIC PRESERVATION

1100 Pennsylvania Ave., NW,
#809, The Old Post Office Bldg.
Washington, DC 20004 USA
Phone: 202-606-8503   Fax: 202-606-8672
E-mail: achp@achp.gov
Website: www.achp.gov

Founded: NA
Membership: 30
Scope: National

Description: An independent federal agency, the Council is the primary policy advisor to the President and Congress on historic preservation matters and guides, and other federal agencies to ensure their actions do not result in unnecessary harm to the nation's historic properties. The Council was established by the National Historic Preservation Act of 1966, is made up of the heads of seven federal departments whose actions regularly affect historic properties; eight members, a governor and a mayor appointed by the President; and representatives of the National Trust for Historic Preservation and the National Conference of State Historic Preservation Officers. The Council is supported by a small professional staff. Offices are in Washington, DC and Denver.

Keyword(s): Protected Areas, Public Health Protection, Solid Waste Management, Toxic Substances, Nuclear-free, Water quantity, Water export and diversion

Contact(s):
John Fowler, EXECUTIVE DIRECTOR

APPALACHIAN REGIONAL COMMISSION

1666 Connecticut Ave., NW., Suite 700
Washington, DC 20009 USA
Phone: 202-884-7700   Fax: 202-884-7691
Website: www.arc.gov

Founded: 1965
Membership: 60
Scope: Regional

Description: To promote economic and human development in the 13-state Appalachian region and to provide a framework for joint federal and state efforts. Includes 406 counties in Alabama, Georgia, Kentucky, Maryland, Mississippi, New York, North Carolina, Ohio, Pennsylvania, South Carolina, Tennessee, Virginia and West Virginia.

Publication(s): Appalachia-Quarterly

Contact(s):
Tom Hunter, EXECUTIVE DIRECTOR
Phone: 202-884-7700
Jesse White, FEDERAL CO-CHAIRMAN
Phone: 202-884-7660
Michael Kiernan, PUBLIC INFORMATION
Phone: 202-884-7771
Paul Patton, STATES CO-CHAIRMAN
Bill Walker, STATES' WASHINGTON REPRESENTATIVE
Phone: 202-884-7746

ARMY CORPS OF ENGINEERS

U.S. ARMY CORPS OF ENGINEERS
U.S. ARMY ENGINEER DISTRICT, MOBILE
P.O. Box 2288
Mobile, AL 36628-0001 USA
Phone: 334-690-2511   Fax: 334-690-2525
Website: www.sam.usace.army.mil

Founded: NA
Scope: National

Contact(s):
Janet Shelby, PUBLIC AFFAIRS SPECIALIST

ATLANTIC STATES MARINE FISHERIES COMMISSION

1444 Eye St., NW, 6th Fl.
Washington, DC 20005 USA
Phone: 202-289-6400   Fax: 202-289-6051
E-mail: info@asmfc.org
Website: www.asmfc.org

Founded: 1942
Membership: 30
Scope: Regional

Description: The Commission was established by the Atlantic States Marine Fisheries Compact to promote better utilization of the fisheries, marine, and shell of the 15 Atlantic seaboard states, Maine to Florida, through the development of a joint program for the promotion and protection of such fisheries, and by the prevention of physical waste of the fisheries from any cause.

Publication(s): Focus

Contact(s):
Susan Shipman, CHAIRMAN
John Dunnigan, EXECUTIVE DIRECTOR
David Borden, PAST-CHAIRMAN
John Nelson, VICE-CHAIRMAN

C

CANADIAN WILDLIFE SERVICE

3rd Fl., Place Vincent Massey, 351 St. Joseph Blvd.
Hull, Quebec K1A 0H3 Canada
Phone: 819-997-1301   Fax: 819-953-7177

Founded: NA
Membership: 100
Scope: Regional

Publication(s): THE CANADIAN FIELD NATURALIST

Contact(s):
David Brackett, DIRECTOR GENERAL
Phone: 819-997-1301
Fax: 819-953-7177

CANADIAN WILDLIFE SERVICE

ENVIRONMENT CANADA
Ottawa, Ontario K1A 0H3 Canada
Phone: 819-997-1095   Fax: 819-997-2754
Website: http://www.cws-scf.ec.gc.ca/

Founded: NA

Description: Canada Wildlife Service is a federal agency devoted to the protection and management of migratory birds and nationally important wildlife habitat, endangered species, research on nationally important wildlife issues, control of international trade in endangered species, and international treaties.

Publication(s): Full list of publications is available on the organization's website.

Contact(s):
Ken Sato, DIRECTOR GENERAL
Phone: 819-953-8065
Fax: 819-994-2724

COLUMBIA RIVER INTER-TRIBAL FISH COMMISSION

729 NE Oregon, Suite 200
Portland, OR 97232 USA
Phone: 503-238-0667   Fax: 503-235-4228
Website: www.critfc.org

Founded: 1977
Scope: Regional

Description: The Commission was formed to return salmon to Columbia basin rivers and to protect the Indian tribes' treaty-reserved fishing rights.

Contact(s):
Don Sampson, EXECUTIVE DIRECTOR

COLUMBIA RIVER INTER-TRIBAL FISH COMMISSION

STREAMNET LIBRARY
729 NE Oregon St., Suite 190
Portland, OR 97232 USA
Phone: 503-731-1304   Fax: 503-731-1260
E-mail: fishmail@critfc.org
Website: www.fishlib.org

Founded: NA

Description: The StreamNet Library is a cooperative venture of the region's fish and wildlife agencies and tribes and serves these organizations. It is a fisheries and aquatic species library emphasizing management and restoration of the Columbia River salmon and sturgeon, providing data and data services. Open to the public.

Keyword(s): Librarians/Information Professionals, Libraries, Salmon Recovery

Contact(s):
Laurie Nock, ASSISTANT LIBRARIAN
nocl@critfc.org
Lenora Ofterdahl, HEAD LIBRARIAN
oftl@critfc.org
David Liberty, LIBRARY TECHNICIAN
libd@critfc.org

CONSERVATION COUNCIL OF WESTERN AUSTRALIA, INC.

2 Delhi St.
West Perth, West Australia 6000 Australia
Phone: 08-9420-7266   Fax: 08-9420-7273
E-mail: conswa@iinet.net.au
Website: http://members.iinet.net.au/conswa

Founded: NA

Description: To promote the cause of conservation and environmentalism throughout the state of Western Australia; and to serve as a liaison to other bodies dealing with conservation and environmental issues.

Contact(s):
Rachel Siewert, CONTACT

CORP OF ENGINEERS US ARMY

U.S. ARMY CORPS OF ENGINEERS
U.S. ARMY ENGINEER DISTRICT, VICKSBURG
4155 Clay Street
Vicksburg, MS 39183 USA
Phone: 601-631-5010   Fax: 601-631-5296
E-mail: www.usace.army.mil
Website: www.usace.army.mil

Founded: NA
Membership: 800
Scope: National

CORP. OF ENGINEERS - ARMY

U.S. ARMY CORPS OF ENGINEERS
U.S. ARMY ENGINEER DISTRICT, KANSAS CITY
601 E. 12th Street
Kansas City, MO 64106-2896 USA
Phone: 816-983-3201   Fax: 806-426-5575

Founded: NA
Membership: 900
Scope: Regional

COUNCIL ON ENVIRONMENTAL QUALITY

722 Jackson Pl., NW
Washington, DC 20503 USA
Phone: 202-456-6224   Fax: 202-456-2710
Website: www.eop.gov/ceq

Founded: 1970
Membership: 20
Scope: International

Description: CEQ serves as the source of environmental expertise and policy analysis for the President and other organizations within the Executive Office of the President, and provides for coordination between departments and agencies. It is also charged with implementing statutory or regulatory requirements and programs.

Publication(s): CEQ Annual Report

Contact(s):
Elizabeth Slotpe, ASSOCIATE DIRECTOR
V. Stevens, ASSOCIATE DIRECTOR
Dave Anderson, ASSOCIATE DIRECTOR CONGRESSIONAL AFFAIRS
Bill Leary, ASSOCIATE DIRECTOR FOR NATURAL RESOURCES
Sam Thurstrom, ASSOCIATE DIRECTOR OF COMMUNICATIONS
Jim Connaughton, CHAIRMAN
Philip Cooney, CHIEF OF STAFF
Dinah Bear, GENERAL COUNSEL
Cameron Bailey, SPECIAL ASSISTANT

D

DELAWARE RIVER BASIN COMMISSION

West Trenton, NJ 08628 USA
Phone: 609-883-9500   Fax: 609-883-9522
E-mail: drbc@drbc.state.nj.us
Website: www.drbc.net

Founded: 1961
Membership: 40
Scope: Regional

Description: The Delaware River Basin Compact created the Delaware River Basin Commission to develop and implement plans, policies, and projects relating to the water resources of the Delaware River Basin. The Commission is responsible for adopting and promoting "uniform and coordinated policies for water conservation, control, use, and management in the basin."

Publication(s): Delaware River Basic Compact, Annual Report, Water Resources Program, Administrative Manual and Water Code

Contact(s):
Carol Collier, EXECUTIVE DIRECTOR
Christopher Roberts, PUBLIC INFORMATION OFFICER

DEPARTMENT FOR ENVIRONMENT, HERITAGE AND ABORIGINAL AFFAIRS
Level 9, Chesser House, 91-97 Grenfell Street
Adelaide, 5000 Australia
Phone: 08-8204-9322   Fax: 08-8204-9321
E-mail: environmentshop@saugov.sa.gov.au
Website: www.dehaa.sa.gov.au/index.html

Founded: NA

Contact(s):
John Scanlon, CHIEF EXECUTIVE

DEPARTMENT OF CANADIAN HERITAGE
PORTFOLIO AND CORPORATE AFFAIRS
15 Eddy St. Room 12G (15-12-8)
Hull, Quebec K1A 0M5 Canada
Phone: 819-994-3046   Fax: 819-953-4796

Founded: NA
Scope: National

Contact(s):
Yazmime Laroche, ASSISTANT DEPUTY MINISTER
Phone: 819-994-3046

DEPARTMENT OF COMMERCE
NATIONAL OCEANIC AND ATMOSPHERIC ADMINISTRATION
OFFICE OF GLOBAL PROGRAM
1100 Wayne Ave.
Silver Spring, MD 20910 USA
Phone: 301-427-2089   Fax: 301-427-2222
Website: www.ogp.noaa.gov

Founded: NA
Scope: State

Description: Provides the primary focus for coordination with national and international scientific communities in the areas of global warming, Tropical Oceans and Global Atmosphere Project, and worldwide climate research.

Contact(s):
J. Hall, DIRECTOR
Phone: 301-427-2089

DEPARTMENT OF FISHERIES AND OCEANS
CANADIAN COAST GUARD
200 Kent St.
Ottawa, Ontario K1A 0E6 Canada
Phone: 613-998-1571   Fax: 613-990-2780
Website: www.ccg-gcc.gc.ca

Founded: NA
Membership: 325
Scope: International

Contact(s):
John Adams, ASSISTANT DEPUTY MINISTER, MARINE SERVICES/COMMISSIONER
Phone: 613-998-1571
Fax: 613-990-2780
Guy Bujold, DEPUTY COMMISSIONER
Phone: 613-998-1570
Fax: 613-990-2780
Anne O'toole, DIRECTOR GENERAL: INTEGRATED BUSINESS MANAGEMENT
Phone: 613-998-1440
Fax: 613-990-3480
Dave Faulkner, DIRECTOR GENERAL: INTEGRATED TECHNICAL SUPPORT
Phone: 613-998-1638
Fax: 613-993-5333
Debra Normoyle, DIRECTOR GENERAL: MARINE PROGRAMS
Phone: 613-990-5508
Fax: 613-991-4982
Charles Gadula, DIRECTOR OF FLEET
Phone: 613-993-1849
gadulac@dfo-mpo-gc.ca

DEPARTMENT OF FISHERIES AND OCEANS
CORPORATE SERVICES
200 Kent St.
Ottawa, Ontario K1A 0E6 Canada
Phone: 613-993-0868   Fax: 613-990-3604
Website: www.intra.dfo-mpo.gc.ca

Founded: NA
Membership: 300
Scope: Nationwide

Contact(s):
Donna Petrachenko, ASSISTANT DEPUTY MINISTER
Phone: 613-993-0868
Mike Hawkes, DIRECTOR GENERAL: FINANCE AND ADMINISTRATION
Phone: 613-993-9372
Eves Dupuis, DIRECTOR GENERAL: HUMAN RESOURCES
Phone: 613-990-0023
Paul Hession, DIRECTOR GENERAL: INFORMATION MANAGEMENT AND TECHNOLOGY SERVICES
Phone: 613-993-2051
Robert Bergeron, DIRECTOR GENERAL: SMALL CRAFT HARBOURS
Phone: 613-993-1937

DEPARTMENT OF FISHERIES AND OCEANS
LEGAL SERVICES
200 Kent St.
Ottawa, Ontario K1A 0E6 Canada
Phone: 613-993-0966   Fax: 613-990-9385
Website: www.intra.dfo-mpo.gc.ca

Founded: NA
Scope: Regional

Contact(s):
Liliana Longo, GENERAL COUNSEL
Phone: 613-993-0966

DEPARTMENT OF FISHERIES AND OCEANS
OCEANS
200 Kent St.
Ottawa, Ontario K1A 0E6 Canada
Phone: 613-993-0850   Fax: 613-990-2768
Website: www.intra.dfo-mpo.gc.ca

Founded: NA

Contact(s):
Matthew King, ASSISTANT DEPUTY MINISTER
Phone: 613-993-0850
Paul Cuillerier, DIRECTOR GENERAL: HABITAT
MANAGEMENT AND ENVIRONMENTAL SCIENCE
Phone: 613-991-1280
Daniel Mcdougall, DIRECTOR GENERAL: OCEANS
DIRECTORATE
Phone: 613-990-0001

DEPARTMENT OF FISHERIES AND OCEANS
POLICY
200 Kent St.
Ottawa, Ontario K1A 0E6 Canada
Phone: 613-993-1808   Fax: 613-993-6958

Founded: NA
Membership: 75
Scope: National

Contact(s):
Liseanne Forand, ASSISTANT DEPUTY MINISTER
Phone: 613-993-1808
Lori Ridgeway, DIRECTOR GENERAL: ECONOMIC AND
POLICY ANALYSIS
Phone: 613-993-1914
Richard Wex, DIRECTOR GENERAL: OFFICE OF
SUSTAINABLE AQUACULTURE
Phone: 613-993-1872
Sharon Ashley, DIRECTOR GENERAL: POLICY,
COORDINATION, AND LIAISON
Phone: 613-990-0007
Paul Thompson, DIRECTOR GENERAL: STRATEGIC
PRIORITIES AND PLANNING
Phone: 613-990-0146

DEPARTMENT OF FISHERIES AND OCEANS
SCIENCE
200 Kent St.
Ottawa, Ontario K1A 0E6 Canada
Phone: 613-990-5123   Fax: 613-990-5113
Website: http://intra.dfo-mpo.gc.ca

Founded: NA
Scope: National

Contact(s):
John Davis, ASSISTANT DEPUTY MINISTER
Phone: 613-990-5123
Tony O'Connor, DIRECTOR GENERAL: CANADIAN
HYDROGRAPHY SERVICES
Phone: 613-995-4413
Serge Labonte, DIRECTOR GENERAL: FISHERIES AND
BIODIVERSITY SCIENCE DIRECTORATE
Phone: 613-990-9082
Elizabeth Marsollier, DIRECTOR GENERAL: OCEAN AND
AQUACULTURE SCIENCE DIRECTORATE
Phone: 613-990-0271

Brian Wilson, DIRECTOR GENERAL: PROGRAM
PLANNING AND COORDINATION
Phone: 613-990-0149

DEPARTMENT OF FISHERIES AND OCEANS/CANADA DIVISION
200 Kent St. Centennial Towers, 13 Floor
Ottawa, Ontario K1A 0E6 Canada
Phone: 613-993-0999   Fax: 613-990-1866
Website: www.dfo-mpo.gc.ca

Founded: NA
Scope: Provincial

Description: Fisheries and Oceans Canada is responsible for policies and programs in support of Canada's economic, ecological, and scientific interests in oceans and inland water; and for safe, effective and environmentally sound marine services responsive to the needs of Canadian in a global economy.

Contact(s):
Wayne Wouters, DEPUTY MINISTER
Phone: 613-993-2200
Herb Dhaliwal, MINISTER
Phone: 613-992-3474

E

EGYPTIAN ENVIRONMENTAL AFFAIRS AGENCY
30, Misr Helwan St., Maadi
Cairo, Egypt
Phone: 02-525-6442   Fax: 02-525-6451

Founded: NA

Publication(s): Annual Report, "EAS, National Workplan and Law"

Contact(s):
Nirvana Khadr, 02-525-6447
Nadia Ebeid, H.E. MINISTER

ENVIRONMENT CANADA
351 St. Joseph Boulevard
Hull, Quebec K1A 0H3 Canada
E-mail: enviromfocec.ga.ca
Website: www.ec.ga.ca

Founded: NA

Description: Purpose is to formulate and take action to meet threats to environment arising through adverse impacts of human activities. Priority responsibilities include toxic chemicals, acid rain ozone depletion, urban smog, and the ongoing management of concerns such as hazardous wastes.

Contact(s):
David Anderson, MINISTER OF THE ENVIRONMENT
Phone: 819-997-1441
David.Anderson@ec.ga.ca

ENVIRONMENTAL CONSERVATION SERVICE
4905 Dufferin St.
Dawnsview, Ontario M3H 5T4 Canada
Phone: 416-739-5839   Fax: 416-739-5840

Founded: NA
Scope: International

Contact(s):
Simon Llewellyn, REGIONAL DIRECTOR

ENVIRONMENTAL CONSERVATION SERVICE
1 PLACE VINCENT MASSEY
351 St. Joseph Blvd.
Hull, Quebec K1A 0H3 Canada
Phone: 819-994-4750　　Fax: 819-997-1541
Website: wwe.infolane.ec.gc.ca

Founded: NA
Membership: 5
Scope: International

Description: In Environmental Conservation Service (ECS) our goal is to ensure that future generations of Canadians inherit a natural environment as rich as the one we enjoy today. We work with many partners—individual Canadians, environmental and community groups, Aboriginal peoples, industry, other levels of government, and international organizations. We provide information on the natural environment to Canadians.

Contact(s):
Karen Brown, ASSISTANT DEPUTY MINISTER
Phone: 819-997-2161
Fax: 819-997-1541
Ken Sato, DIRECTOR GENERAL
Phone: 819-953-8065
Fax: 819-994-2724

ENVIRONMENTAL CONSERVATION SERVICE
ATLANTIC REGION ENVIRONMENT CANADA
17 Waterfowl Lane
Sackville, NBrunswick E4L 4N1 Canada
Phone: 506-364-5044　　Fax: 506-364-5062
E-mail: nature@ec.gc.ca
Website: www.ec.gc.ca

Founded: NA
Scope: International

Contact(s):
George Finney, REGIONAL DIRECTOR

ENVIRONMENTAL CONSERVATION SERVICE
ECOSYSTEM AND ENVIRONMENTAL RESOURCES DIRECTORATE
6th Fl., Place Vincent Massey, 351 St. Joseph Blvd.
Hull, Quebec K1A 0H3 Canada
Phone: 819-997-5674　　Fax: 819-994-2541

Founded: NA
Scope: International

Contact(s):
Jennifer Moore, DIRECTOR GENERAL
Phone: 819-997-5674
Fax: 819-994-2541

ENVIRONMENTAL CONSERVATION SERVICE
PACIFIC AND YUKON REGION: ENVIRONMENT CANADA
Suite 700, 1200 West 73rd Ave.
Vancouver, British Columbia V6P 6H9 Canada
Phone: 604-664-4065　　Fax: 604-664-9195
Website: www.ec.gc.ca

Founded: NA

Contact(s):
Don Fast, REGIONAL DIRECTOR (ACTING)

ENVIRONMENTAL CONSERVATION SERVICE
PRAIRIE AND NORTHERN REGION
CANADIAN WILDLIFE SERVICE
4999-98th Ave.
Edmonton, Alberta T6B 2X3 Canada
Phone: 780-951-8853　　Fax: 780-495-2615
Website: www.ec.gc.ca

Founded: NA
Membership: 100
Scope: Regional

Contact(s):
Gerald McKeating, REGIONAL DIRECTOR

ENVIRONMENTAL CONSERVATION SERVICE
QUEBEC REGION ENVIRONMENT CANADA
Canadian Wildlife Service, 1141 Route de l'Eglise,
P.O. Box 10100, 9th Floor
Sainte-Foy, Quebec G1V 4H5 Canada
Phone: 418-648-2543　　Fax: 418-649-6475

Founded: NA

Contact(s):
Albin Tramblay, REGIONAL DIRECTOR
Phone: 418-648-7808

ENVIRONMENTAL PROTECTION AGENCY
1200 Pennsylvania Ave NW
Washington, DC 20460 USA
Phone: 202-260-2090　　Fax: 202-564-4613
Website: www.epa.gov

Founded: NA
Membership: 18,000
Scope: Statewide

Description: The Environmental Protection Agency (EPA) was established as an independent agency in the Executive Branch of the U.S. Government, pursuant to Reorganization Plan No. 3 of 1970, effective December 2, 1970. EPA endeavors to achieve systematic control and abatement of pollution, by properly administering and integrating a variety of research, monitoring, standard-setting, and enforcement activities.

Contact(s):
Christie Whitman, ADMINISTRATOR
Loretta Ucelli, ASSOCIATE ADMINISTRATOR FOR COMMUNICATIONS, EDUCATION, AND MEDIA RELATIONS
Phone: 202-260-9828
Joseph Crapa, ASSOCIATE ADMINISTRATOR FOR CONGRESSIONAL AND INTERGOVERNMENTAL RELATIONS
Phone: 202-260-5200
Jay Benforado, ASSOCIATE ADMINISTRATOR FOR REINVENTION (ACTING)
Phone: 202-260-1849
Sallyanne Harper, CHIEF FINANCIAL OFFICER (ACTING)
Phone: 202-260-1151
W. Ryan, COMPTROLLER
Phone: 202-260-9674
Linda Fisher, DEPUTY ADMINISTRATOR
Scott Fulton, GENERAL COUNSEL (ACTING)
Phone: 202-260-8064
Nikki Tinsley, INSPECTOR GENERAL (ACTING)
Phone: 202-260-3137

Donald Barnes, SCIENCE ADVISORY BOARD DIRECTOR
Phone: 202-260-4125
Jeanette Brown, SMALL AND DISADVANTAGED BUSINESS
UTILIZATION DIRECTOR
Phone: 202-260-4100

ENVIRONMENTAL PROTECTION AGENCY
901 N. 5th St.
Kansas City, KS 66101 USA
Phone: 913-551-7000
Website: www.epa.gov/region07/

Founded: NA
Membership: 700
Scope: Regional

Contact(s):
William Rice, REGIONAL ADMINISTRATOR

ENVIRONMENTAL PROTECTION AGENCY
ADMINISTRATION AND RESOURCES MANAGEMENT
1200 Pennsylvania Avenue NW
Washington, DC 20460 USA
Phone: 202-564-4700
Website: http://www.epa.gov

Founded: NA

Contact(s):
Christie Whitman, ADMINISTRATOR
Phone: 202-260-4600
William Laxton, DIRECTOR OF ADMINISTRATION AND
RESOURCES MANAGEMENT, RESEARCH TRIANGLE
PARK, NC
Phone: 919-541-2258
William Henderson, DIRECTOR OF ADMINISTRATION,
CINCINNATI, OH
Phone: 513-569-7910
Mark Day, DIRECTOR OF INFORMATION RESOURCES
MANAGEMENT (ACTING)
Phone: 202-260-4465
David O'Connor, HUMAN RESOURCES AND
ORGANIZATIONAL SERVICES DIRECTOR
Phone: 202-260-4467

ENVIRONMENTAL PROTECTION AGENCY
AIR AND RADIATION
1200 Pennsylvania Ave. NW
Washington, DC 20460 USA
Phone: 202-564-4700
Website: http://www.epa.gov/oeca

Founded: NA

Contact(s):
John Seitz, AIR QUALITY PLANNING AND STANDARDS
DIRECTOR
Phone: 919-541-5504
Robert Perciasepe, ASSISTANT ADMINISTRATOR
Phone: 202-260-7400
Paul Stoplman, ATMOSPHERIC PROGRAMS DIRECTOR
Phone: 202-564-9150
Richard Wilson, DEPUTY ADMINISTRATOR
Phone: 202-260-7400
Margo Oge, MOBILE SOURCES DIRECTOR
Phone: 202-233-7645
Lawrence Weinstock, RADIATION AND INDOOR AIR
DIRECTOR (ACTING)
Phone: 202-564-9370

ENVIRONMENTAL PROTECTION AGENCY
ENFORCEMENT AND COMPLIANCE
1200 Pennsylvania Avenue NW
Washington, DC 20460 USA
Phone: 202-564-4700
Website: http://www.epa.gov/oeca/

Founded: NA

Contact(s):
Gregory Snyder, COMPLIANCE DIRECTOR
Phone: 202-564-2461
Barry Hill, ENVIRONMENTAL JUSTICE DIRECTOR
(ACTING)
Phone: 202-564-2515
Fax: 202-501-0740
Richard Sanderson, FEDERAL ACTIVITIES DIRECTOR
Phone: 202-564-2400
Eric Schaffer, REGULATORY ENFORCEMENT DIRECTOR
Phone: 202-564-2220
Barry Breen, SITE REMEDIATION ENFORCEMENT
DIRECTOR
Phone: 202-564-5110

ENVIRONMENTAL PROTECTION AGENCY
PREVENTION, PESTICIDES, AND TOXIC
SUBSTANCES
1200 Pennsylvania Avenue NW
Ariel Rios Bldg.
Washington, DC 20460 USA
Website: www.epa.gov/opptsfrs
http://www.epa.gov/opptsfrs

Founded: NA

Contact(s):
Stephen Johnson, ASSISTANT ADMINISTRATOR
Phone: 202-260-2902
Marcia Mulkey, PESTICIDE PROGRAMS DIRECTOR
Phone: 703-305-7090
William Sanders, POLLUTION PREVENTION AND TOXICS
DIRECTOR
Phone: 202-260-3810

ENVIRONMENTAL PROTECTION AGENCY
REGION I (CT, ME, MA, NH, RI, VT)
1 Congress Street, Suite 1100
Boston, MA 02203-0001 USA
Phone: 617-918-1111
Website: http://www.epa.gov/region01/

Founded: NA

Contact(s):
John Devillars, REGIONAL ADMINISTRATOR
Phone: 617-918-1010

ENVIRONMENTAL PROTECTION AGENCY
REGION II (NJ, NY, PR, VI)
290 Broadway
New York, NY 10007-1866 USA
Phone: 212-637-3000   Fax: 212-637-5046
Website: www.epa.gov/region02

Founded: NA
Scope: Statewide

Contact(s):
William Muszynski, REGIONAL ADMINISTRATOR

ENVIRONMENTAL PROTECTION AGENCY
REGION IX (GU, AS, NV, HI, CA, AZ)
75 Hawthorne St.
San Francisco, CA 94105 USA
Phone: 415-744-1702
Website: www.epa.gov/region09/

Founded: NA
Scope: International

Contact(s):
Laura Yoshii, REGIONAL ADMINISTRATOR
yoshii.laura@epa.gov

ENVIRONMENTAL PROTECTION AGENCY
REGION V (IL, IN, MI, NM, OH, WI)
77 West Jackson Blvd.
Chicago, IL 60604-3507 USA
Phone: 312-353-2000   Fax: 312-353-1120
Website: www.epa.gov/region05/

Founded: NA
Scope: Regional

Contact(s):
David Ulrich, DEPUTY REGIONAL ADMINISTRATOR
Phone: 312-886-3000
Thomas Skinner, REGIONAL ADMINISTRATOR

ENVIRONMENTAL PROTECTION AGENCY
REGION VI (AR, LA, NM, OK, TX)
Fountain Place, 12th Fl., Suite 1200, 1445 Ross Ave.
Dallas, TX 75202-2733 USA
Phone: 214-665-6444
Website: www.epa.gov/region06/

Founded: NA
Membership: 1000
Scope: International

Contact(s):
Gregg Cooke, REGIONAL ADMINSTRATOR
Phone: 214-665-2100

ENVIRONMENTAL PROTECTION AGENCY
REGION VIII (CO, MT, ND, SD, UT, WY)
999 18th St.
Denver, CO 80202-2466 USA
Phone: 303-312-6312   Fax: 303-312-6363
Website: www.epa.gov/region08/

Founded: NA
Membership: 920
Scope: Regional

Contact(s):
Jack McGraw, REGIONAL ADMINISTRATOR (ACTING)

ENVIRONMENTAL PROTECTION AGENCY
REGION X (WA, OR, ID, AK)
1200 Sixth Ave.
Seattle, WA 98101 USA
Phone: 206-553-1200   Fax: 206-553-0149
E-mail: epa-seattle@epamail.epa.gov
Website: www.epa.gov/region10/

Founded: NA
Membership: 700
Scope: Regional

Contact(s):
Charles Findley, REGIONAL ADMINISTRATOR
Phone: 206-553-0479

ENVIRONMENTAL PROTECTION AGENCY
RESEARCH AND DEVELOPMENT
USEPA Ariel Rios Building (5101)
1200 Pennsylvania Avenue NW
Washington, DC 20460 USA
Phone: 202-564-4700
Website: www.epa.gov/ord

Founded: NA
Scope: National

Contact(s):
William Farland, ACTING DEPUTY ASSISTANT
ADMINISTRATOR

ENVIRONMENTAL PROTECTION AGENCY
SCIENCE POLICY
1200 Pennsylvania Avenue NW
Washington, DC 20460 USA
Phone: 202-564-4700
Website: www.epa.gov/science
http://www.epa.gov/science

Founded: NA

Contact(s):
Albert McGarland, ECONOMY AND ENVIRONMENT
DIRECTOR
Phone: 202-260-3354
Pamela Sterling, PROGRAM SUPPORT AND RESOURCE
MANAGEMENT DIRECTOR
Phone: 202-260-4335
Thomas Kelly, REGULATORY MANAGEMENT AND
INFORMATION DIRECTOR
Phone: 202-260-4335
Leonard Fleckenstein, SUSTAINABLE ECOSYSTEMS AND
COMMUNITIES DIRECTOR (ACTING)
Phone: 202-260-4002

ENVIRONMENTAL PROTECTION AGENCY
SOLID WASTE AND EMERGENCY RESPONSE
1200 Pennsylvania Avenue NW
Washington D.C., DC 20460 USA
Phone: 202-564-4700
Website: www.epa.gov/swerrins

Founded: NA

Contact(s):
Timothy Fields, ASSISTANT ADMINISTRATOR (ACTING)
Phone: 202-260-4610
James Makris, CHIEF PREPAREDNESS AND PREVENTION
DIRECTOR
Phone: 202-260-8600
Stephen Luftig, EMERGENCY AND REMEDIAL RESPONSE
(SUPERFUND/OIL PROGRAMS) DIRECTOR
Phone: 703-603-8960
Elizabeth Cotsworth, SOLID WASTE DIRECTOR (ACTING)
Phone: 703-308-8895
Walter Kovalick, TECHNOLOGY INNOVATION DIRECTOR
Phone: 703-603-9910

ENVIRONMENTAL PROTECTION AGENCY
WATER
USA
Website: http://epa.gov/ow

Founded: NA

Contact(s):
Jeff Besougloff, AMERICAN INDIAN ENVIRONMENTAL DIRECTORY
Phone: 202-260-7939
Tracy Mehan, ASSISTANT ADMINISTRATOR
Phone: 202-260-5700
Cynthia Dougherty, GROUND WATER AND DRINKING WATER DIRECTOR
Phone: 202-260-5543
Geoffery Grubbs, SCIENCE AND TECHNOLOGY DIRECTOR
Phone: 202-260-5400
Michael Cook, WASTEWATER MANAGEMENT DIRECTOR
Phone: 202-260-5850
Robert Wayland, WETLANDS, OCEANS, AND WATERSHEDS DIRECTOR
Phone: 202-260-7166

ENVIRONMENTAL PROTECTION SERVICE
AIR POLLUTION PREVENTION DIRECTORATE
P.O. Box 1320
Yellowknife, Northwest Territories X1A 2L9 Canada
Phone: 867-873-7654   Fax: 867-873-0221
E-mail: lisette_self@gov.nt.ca
Website: http://www.gov.nt.ca/RWED/eps/index.htm

Founded: NA

Contact(s):
D. Egar, DIRECTOR GENERAL
Place Vincent Massey, 10e et., 351 Blvd. St. Joseph, Hull, Quebec K1A 0H3
Phone: 819-997-1298
Fax: 819-953-9547

F

FISHERIES AND OCEANS CANADA
COMMUNICATIONS
200 Kent St.
Ottawa, Ontario K1A 0E6 Canada
Phone: 613-993-0999   Fax: 613-990-1866
E-mail: min@dfo-mpo.gc.ca
Website: http://www.dfo-mpo.gc.ca/index.htm

Founded: NA

Contact(s):
Paul Schubert, DIRECTOR GENERAL
Phone: 613-990-0000

FISHERIES AND OCEANS CANADA
FISHERIES AND MANAGEMENT
200 Kent St., 13th Floor Station 13228
Ottawa, Ontario K1A 0E6 Canada
Phone: 613-993-0999   Fax: 613-990-1866
E-mail: info@dfo-mpo.gc.ca
Website: http://www.dfo-mpo.gc.ca

Founded: NA

Contact(s):
Pat Chamut, ASSISTANT DEPUTY MINISTER
Phone: 613-990-9864

Paul Sprout, ASSOCIATE ASSISTANT DEPUTY MINISTER
Phone: 613-990-7203
Mike Alexander, DIRECTOR GENERAL: ABORIGINAL AFFAIRS
Phone: 613-993-8598
Dennis Brock, DIRECTOR GENERAL: CONSERVATION AND PROTECTION
Phone: 613-990-6012
Earl Wiseman, DIRECTOR GENERAL: INTERNATIONAL
Phone: 613-993-1873
David Balfour, DIRECTOR GENERAL: PROGRAM PLANNING AND COORDINATION
Phone: 613-993-2574
David Bezan, DIRECTOR GENERAL: RESOURCE MANAGEMENT
Phone: 613-990-0189

G

GENERAL SERVICES ADMINISTRATION
GSA Bldg., 1800 F St., NW
Washington, DC 20405 USA
Phone: 202-501-1231
Website: www.gsa.gov/

Founded: NA

Description: Concerned with the conveyance of surplus real property for wildlife conservation purposes to the Secretary of Interior or to a state, pursuant to Public Law 537, 80th Congress.

Contact(s):
David Barram, ADMINISTRATOR
Ronald Rice, DIRECTOR OF PROGRAM DEVELOPMENT AND OUTREACH
Phone: 202-501-0052
John Martin, DIRECTOR OF REDEPLOYMENT SERVICES DIVISION/PROPERTY DISPOSAL DIVISION
Phone: 202-501-4671

GREAT LAKES FISHERY COMMISSION
2100 Commonwealth Blvd.
Ann Arbor, MI 48105 USA
Phone: 734-662-3209   Fax: 734-741-2010
Website: www.glfc.org

Founded: NA
Membership: 15
Scope: International

Description: The 1955 Canada-U.S. Convention on Great Lakes Fisheries established the Commission to advise governments on ways to improve the fisheries, to develop and coordinate fishery research programs, to develop measures and implement programs to manage sea lamprey, and to improve and perpetuate fishery resources.

Publication(s): Economics of Great Lakes Fisheries: A 1985 Assessment, "Fish Community Objectives for Lake Superior, The State of Lake Superior in 1992", Fish-Community Objectives for Lake Huron.

Contact(s):
Chris Goddard, EXECUTIVE SECRETARY
cgoddard@glfc.org

GREAT LAKES INDIAN FISH AND WILDLIFE COMMISSION

P.O. Box 9
Odanah, WI 54861 USA
Phone: 715-682-6619   Fax: 715-682-9294
Website: www.glifwc.org

Founded: 1983

Description: Provide biological, enforcement, and legal services to our member tribes in matters related to off-reservation treaty gathering rights in Wisconsin, Michigan, and Minnesota.

Publication(s): Masinaigan, Technical reports

Contact(s):
Tom Maulson, CHAIRMAN OF THE BOARD
Gerald Deperry, DEPUTY ADMINISTRATOR
James Schlender, EXECUTIVE ADMINISTRATOR

GULF STATES MARINE FISHERIES COMMISSION

P.O. Box 726
Ocean Springs, MS 39566-0726 USA
Phone: 228-875-5912   Fax: 228-875-6604
E-mail: lsimpson@gsmfc.org
Website: www:gsmfc.org

Founded: 1949
Membership: 15
Scope: National

Description: The GSMFC is an interstate compact of the states of Alabama, Florida, Louisiana, Mississippi, and Texas. The compact was authorized by the U.S. Congress. The Commission has 15 commissioners. The purpose of the Commission is to promote better utilization of the fisheries, marine, shell, and anadromous, of the seaboard of the Gulf of Mexico by cooperative programs for the promotion and protection of such fisheries and the prevention of the physical waste of the fisheries from any cause.

Publication(s): Publications on line

Contact(s):
Ronald Lukens, ASSISTANT DIRECTOR
Virginia Vail, CHAIRMAN
Larry Simpson, EXECUTIVE DIRECTOR

H

HELSINKI COMMISSION/ BALTIC MARINE ENVIRONMENT PROTECTION COMMISSION

Katajanokanlaituri 6 B FIN- 00160
Helsinki, Finland
Phone: 358-9-6220220   Fax: 358-9-62202239
E-mail: helcom@helcom.fi

Founded: 1980

Description: To protect the marine environment of the Baltic Sea against pollution from all sources.

Publication(s): Baltic Sea Environment Proceedings (BSEP), HELCOM News

Contact(s):
Ritva Kostakow-Kampe, ADMINISTRATIVE OFFICER
Kjell Grip, ENVIRONMENT SECRETARY
Mieczyslaw Ostojski, EXECUTIVE SECRETARY
Anne Brusendorff, MARITIME SECRETARY

Ulrich Kremser, PROGRAMME IMPLEMENTATION COORDINATOR
Ain Lääne, TECHNOLOGICAL SECRETARY

I

INSTITUTO NACIONAL DE BIODIVERSIDAD (INBIO)

Apdo. 22-3100
Santo Domingo, Costa Rica
Phone: 506-244-0690   Fax: 506-244-2816
E-mail: inbioparque@inbio.ac.cr
Website: www.inbio.ac.cr

Founded: 1989

Description: INBIO is conducting a biodiversity inventory in Costa Rica's protected areas. Through this knowledge, society will be able to appreciate and value the resources contained in the wildlands, and use this information in a sustainable manner.

Publication(s): Biodiversity Prospecting, Biodiversidad de Costa Rica: Lecturas para Ecoturistas, Guia de Aves de Costa Rica, Mariposas Heliconius de Costa Rica.

Contact(s):
Natalia Zamora, BIODIVERSITY & GARDEN
nzamora@inbio.ac.cr
Sonia Rojas, BIODIVERSITY EDUCATION
srojas@inbio.ac.cr
Jesus Ugalde, BIODIVERSITY INVENTORY
jugalde@inbio.ac.cr
Karla Zanabria
COMMUNICATIONS
kzanabria@inbio.ac.cr
Alfio Piva, DEPUTY DIRECTOR
apiva@inbio.ac.cr
Rodrigo Gamez, GENERAL DIRECTOR
rgamez @inbio.ac.cr
Eric Mata, MANAGEMENT COORDINATOR
emata@inbio.ac.cr
Ana Gueuvara, PROSPECTING COORDINATOR
agueuvara@inbio.ac.cr
Vanessa Matamorros, PUBLIC RELATIONS
vmatamorros@inbio.ac.cr

INTER-AMERICAN TROPICAL TUNA COMMISSION

c/o Scripps Institution of Oceanography, 8604 La Jolla Shores Dr.
La Jolla, CA 92037-1508 USA
Phone: 858-546-7100   Fax: 858-546-7133
Website: www.ihttc.org

Founded: 1949

Description: Charged with the investigation and conservation of the tuna and dolphin resources of the eastern Pacific Ocean. Member nations: U.S., Costa Rica, El Salvador, Ecuador, France, Japan, Mexico, Nicaragua, Panama, Vanuatu and Venezuela. Established by convention between the U.S. and Costa Rica.

Publication(s): Bulletin of the Inter-American Tropical Tuna Commission, Annual Report, Special Report of the Inter-American Tropical Tuna Commission

Contact(s):
Robin Allen, DIRECTOR
Phone: 858-546-7019

William Bayliff, EDITOR
Phone: 858-546-7025

INTERNATIONAL JOINT COMMISSION

CANADIAN SECTION
234 Laurier Ave. W., 22nd Floor
Ottawa, Ontario K1P 6K6 Canada
Phone: 613-995-2984   Fax: 613-993-5583
Website: www.ijc.org

Founded: NA
Membership: 20
Scope: International

Publication(s): Focus

Contact(s):
Mary Gusella, CHAIRMAN, CANADIAN SECTION

INTERNATIONAL JOINT COMMISSION

National Headquarters, 1250 23rd St., NW, Suite 100
Washington, DC 20440 USA
Phone: 202-736-9000   Fax: 202-736-9015
Website: www.ijc.org

Founded: 1909
Membership: 12
Scope: International

Description: Established by the Boundary Waters Treaty of 1909 to prevent and resolve disputes regarding the use of the waters on the U.S.- Canadian Boundary, and to act as an independent advisor on issues referred by both countries. Regional office monitors, evaluates, and reports on compliance with the Great Lakes Water Quality Agreement of November 22, 1978. Commission functions in quasi-judicial, investigative, and coordination capacities.

Publication(s): Focus

Contact(s):
Murray Clamen, CANADIAN SECTION SECRETARY
Phone: 613-995-2984
Thomas Behlen, DIRECTOR, REGIONAL OFFICE
Phone: 519-257-6700
Fax: 519-257-6740
Frank Bevacqua, PUBLIC INFORMATION OFFICER
Phone: 202-736-9024
Fax: 202-736-9015
bevacquaf@washington.ijc.org
Jennifer Day, PUBLIC INFORMATION OFFICER
Phone: 313-226-2170
Phone: 519-257-6740
DayJ@windsor.ijc.org.

INTERNATIONAL JOINT COMMISSION

GREAT LAKES REGIONAL OFFICE
100 Ouellette Ave.
Windsor, Ontario N9A 6T3 Canada
Phone: 519-257-6700   Fax: 519-257-6740
E-mail: commission@windsor.ijc.org
Website: www.ijc.org

Founded: NA
Membership: 30
Scope: Regional

Publication(s): Biannual Reports, Focus (newsletter)

Contact(s):
Jennifer Day, DIRECTOR OF PUBLIC AFFAIRS

INTERNATIONAL PACIFIC HALIBUT COMMISSION

P.O. Box 95009
Seattle, WA 98145-2009 USA
Phone: 206-634-1838   Fax: 206-632-2983
Website: www.iphc.washington.edu

Founded: 1923

Description: Scientific investigation and management of the Pacific halibut resource. Established by a convention between Canada and the United States.

Contact(s):
Bruce Leaman, DIRECTOR

INTERNATIONAL WHALING COMMISSION

The Red House 135 Station Rd.
Impington, Cambridge+132, CB4 9NP United Kingdom
Phone: 01223-233971   Fax: 01223-232876
E-mail: iwcoffice@compuserve.com

Founded: 1946

Description: Established under the International Convention for the Regulation of Whaling in 1946 to provide for the conservation of whale stocks and the orderly development of the whaling industry. Member Nations: USA, Antigua and Barbuda, Argentina, Australia, Austria, Brazil, Chile, People's Republic of China, Costa Rica, Denmark, Dominica, Finland, France, Germany, Grenada, India, Ireland, Italy, Japan, Kenya, Republic of Korea, Mexico, Monaco, Netherlands, New Zealand, Norway, Oman, Peru, Russia Federation, Saint Kitts and Nevis, Saint Lucia, Saint Vincent and the Grenadines, Senegal, Solomon Islands, South Africa, Spain, Sweden, Switzerland, United Kingdom, and Venezuela.

Publication(s): Annual reports of the Commission (including reports and papers of the Scientific Committee), Special Issues Series on specialist cetacean subjects, International Journal of Cetacean Research and Management

Contact(s):
M. Canny, CHAIRMAN (IRELAND)
M. Harvey, EXECUTIVE OFFICER (CAMBRIDGE)
R. Gambell, SECRETARY (CAMBRIDGE)
J. Baker, U.S. COMMISSIONER
U.S. Department of Commerce, Rm. 5128, Herbert C. Hoover Bldg., 14th and Constitution Ave., NW, Washington, DC 20230
Bo Fernholm, VICE CHAIRMAN (SWEDEN)

INTERSTATE COMMISSION ON THE POTOMAC RIVER BASIN

6110 Executive Blvd., Suite 000
Rockville, MD 20852-3903 USA
Phone: 301-984-1908   Fax: 301-984-5841
E-mail: info@potomac/commission.org
Website: www.potomacriver.org

Founded: 1940
Scope: Regional

Description: Interstate compact, established by Maryland, Pennsylvania, Virginia, West Virginia, and the District of Columbia. Coordinates, tabulates, and summarizes existing data on condition of streams in Potomac Watershed; promotes uniform legislation; disseminates information; cooperates in studies; promotes coordination of program in Basin states. Areas of interest are water quality, water supply, and land resources associated with the Potomac and its tributaries.

Publication(s): Potomac Basin Reporter, In the Anacostia Watershed

Contact(s):
James Cummins, ASSOCIATE DIRECTOR OF LIVING RESOURCES
Carlton Haywood, ASSOCIATE DIRECTOR OF WATER QUALITY
Joseph Hoffman, EXECUTIVE DIRECTOR
Curtis Dalpra, PUBLIC AFFAIRS OFFICER

M

MARINE MAMMAL COMMISSION

4340 East-West Highway, Rm. 905
Bethesda, MD 20814 USA
Phone: 301-504-0087   Fax: 301-504-0099
E-mail: firstinitialandlastname@mmc.gov

Founded: 1972
Membership: 12
Scope: State

Description: Established by the Marine Mammal Protection Act of 1972, P.L. 92-522, the Marine Mammal Commission, in consultation with its Committee of Scientific Advisors on Marine Mammals, periodically reviews the status of marine mammal populations; manages a research program concerned with their conservation; and develops, reviews, and makes recommendations on federal activities and policies which affect the protection and conservation of marine mammals.

Publication(s): Annual Report, Research Reports

Contact(s):
Robert Mattlin, EXECUTIVE DIRECTOR
rmattlin@mmc.gov
Michael Gosliner, GENERAL COUNSEL
David Laist, POLICY AND PROGRAM ANALYST
Timothy Ragen, SCIENTIFIC PROGRAM DIRECTOR

MIGRATORY BIRD CONSERVATION COMMISSION

1849 C St., NW (ARL SQ. 622)
Washington, DC 20240 USA
Phone: 703-358-1716   Fax: 703-358-2223
E-mail: Jeffrey-M-Donahoe@fws.gov

Founded: 1929
Membership: 8
Scope: National

Description: Considers, passes upon, and fixes the prices for lands recommended by the Secretary of the Interior for purchase or lease by him under the Migratory Bird Conservation Act of February 18, 1929, as amended, as migratory bird refuges in the National Wildlife Refuge System.

Contact(s):
Jeffery Donahoe, SECRETARY
Phone: 703-358-1716
Fax: 703-358-2223

MINISTRY OF THE ENVIRONMENT OF THE CZECH REPUBLIC

Vrsovicka 65
100 10 Prague 10, Czech Republic
Phone: 420-2-672-2769   Fax: 420-2-6731-0370
E-mail: roudna@env.cz

Founded: NA

Contact(s):
Peter Roth, CHAIRMAN
Phone: 420-2-6712
Fax: 420-2-67311096
roth@env.cz
Milena Roudna, MINISTER
Phone: 420-2-6712
Fax: 420-2-67310307
roudna@env.cz

MINISTRY OF THE ENVIRONMENT OF THE CZECH REPUBLIC

ENVIRONMENTAL COMMISSION
Academy of Sciences of The Czech Republic, Narodni 3
11720 Praha 1, Czech Republic
Phone: 420-2-2420538   Fax: 420-2-2422094
E-mail: info@env.cz

Founded: NA

Contact(s):
Milos Kuzvart, SECRETARY
petr.kuzvart@enc.cz

MINNESOTA-WISCONSIN BOUNDARY AREA COMMISSION

619 2nd St.
Hudson, WI 54016 USA
Phone: 651-436-7131   Fax: 715-386-9571
E-mail: mwbac@mwbac.org
Website: www.mwbac.org

Founded: 1965
Membership: 4
Scope: Regional

Description: To conduct studies, develop recommendations, and coordinate planning for protection, use, and development in the public interest of lands, river valleys, and waters that form the boundary between Minnesota and Wisconsin, principally on the St. Croix and Mississippi rivers.

Publication(s): River Steward Journal

Contact(s):
Buck Malick, EXECUTIVE DIRECTOR
Robin Grawe, MISSISSIPPI VALLEY DIRECTOR
Rosetta Herricks, OFFICE MANAGER
Judith Olson, SECRETARY

N

NATIONAL INTERAGENCY FIRE CENTER

BUREAU OF LAND MANAGEMENT
NATIONAL OFFICE OF FIRE AND AVIATION
National Interagency Fire Center
3833 S. Development Ave.
Boise, ID 83705 USA
Phone: 208-387-5512   Fax: 208-387-5797
Website: www.nifc.gov

Founded: NA
Membership: 550
Scope: National

Publication(s): Burning Issues-newsletter Bi-quarterly

Contact(s):
Larry Hamilton, DIRECTOR
Phone: 208-387-5446

NATIONAL SCIENCE FOUNDATION

4201 Wilson Blvd.
Arlington, VA 22230 USA
Phone: 703-191-5111
Website: http://www.nsf.gov

Founded: 1950

Description: Responsible for the support of science and engineering research and the development of science education programs. Policy is set by the National Science Board, which is composed of 24 part-time members appointed by the President, with the consent of the Senate, and includes the Director of the Foundation.

Publication(s): Guide to Programs, Grant Proposal Guide, NSF in a Changing World: The National Science Foundation Strategic Plan, Where Discoveries Begin

Keyword(s): Ecological Education

Contact(s):
Eamon Kelly, CHAIRMAN OF NATIONAL SCIENCE BOARD
Phone: 703-306-2000
Joseph Bordogna, DEPUTY DIRECTOR
Phone: 703-306-1000
Rita Colwell, DIRECTOR
Phone: 703-306-1000
Julia Moore, DIRECTOR OF OFFICE OF LEGISLATIVE AND PUBLIC AFFAIRS
Phone: 703-306-1070
Stephanie Bianchi, HEAD LIBRARIAN
Phone: 703-306-0658

NATIONAL TRANSPORTATION SAFETY BOARD

490 L'Enfant Plaza East, SW
Washington, DC 20594 USA
Phone: 202-314-6000   Fax: 202-314-6148
Website: www.ntsb.gov

Founded: NA
Scope: National

Description: The Safety Board is an independent federal accident investigation agency. The Board's mission is to determine the "probable cause" of transportation accidents and to formulate safety recommendations to improve transportation safety.

Contact(s):
Ron Battocchi, DIRECTOR OF OFFICE OF GENERAL COUNSEL
Jamie Finch, DIRECTOR OF OFFICE OF GOVERNMENT, PUBLIC, AND FAMILY AFFAIRS

NATURAL RESOURCES CANADA, CANADIAN FOREST SERVICE

PACIFIC FORESTRY CENTRE
580 Booth St.
Ottawa, Ontario K1A 0E4 Canada
Phone: 250-363-0600   Fax: 250-363-0775
Website: www.NRCan.gc.ca/cfs

Founded: NA

Description: CFS promotes sustainable development of Canada's forests and competitiveness of the Canadian forest sector for the well-being of present and future generations of Canadians. The CFS also establishes links with other non-governmental organizations to better address issues such as international trade, market access and the sustainable management of forests world-wide.

Contact(s):
Yvan Hardy, ASSISTANT DEPUTY MINISTER OF CANADIAN FOREST SERVICE
Phone: 613-947-7400
Fax: 613-947-7395
Jean McLoskey, DEPUTY MINISTER
Phone: 613-992-3456
Fax: 613-992-3828
Normand Lafreniere, DIRECTOR GENERAL
1055 du P.E.P.S. St., P.O. Box 3800, Sainte-Foy, Quebec G1V-4C7
Phone: 418-648-3957
Fax: 418-648-7317
Paul Addison, DIRECTOR GENERAL
506 West Burnside Rd., Victoria, British Columbia V8Z 1M5
Phone: 604-363-0608
Fax: 604-363-6088
Boyd Case, DIRECTOR GENERAL
5320 122 St., Edmonton, Alberta T6H 3S5
Phone: 403-435-7202
Fax: 403-435-7396
Ed Kondo, DIRECTOR GENERAL
P.O. Box 490, 1219 Queen St. East, Sault Ste. Marie, Ontario P6A 5M7
Phone: 705-949-9461
Fax: 705-759-5714
Gerrit Van Raalte, DIRECTOR GENERAL
P.O. Box 4000, Regent St., Fredericton, New Brunswick E3B 5P7
Phone: 506-452-3508
Fax: 506-452-3140
Doug Ketcheson, DIRECTOR GENERAL FOR INDUSTRY, ECONOMICS AND PROGRAMS BRANCH (IEPB)
Phone: 613-947-9052
Fax: 613-947-9038
Jacques Carette, DIRECTOR GENERAL OF POLICY, PLANNING AND INTERNATIONAL AFFAIRS BRANCH (PPIAB)
Phone: 613-947-9100
Fax: 613-947-9038
Gordon Miller, DIRECTOR GENERAL OF SCIENCE BRANCH
Phone: 613-947-8984
Fax: 613-947-9090
Sylvie Letellier, DIRECTOR OF COMMUNICATIONS AND EXECUTIVE SERVICES
Phone: 613-947-7404
Fax: 613-947-7396
Sylvia Frehner, MANAGER OF SECTOR HUMAN RESOURCES UNIT
Phone: 613-947-7386
Fax: 613-947-7409
Anne McLellan, MINISTER
Phone: 613-996-2007
Fax: 613-996-4516

NEW ENGLAND WATER POLLUTION CONTROL COMMISSION

Boott Mills South, 100 Foot of John St.
Lowell, MA 01852-1124 USA
Phone: 978-323-7929   Fax: 978-323-7919
E-mail: general@neiwpcc.org
Website: www.neiwpcc.org

Founded: 1947
Scope: Regional

Description: The Commission provides a forum for interstate communication on high priority water-related environmental issues; provides training opportunities for state environmental staff and wastewater treatment plant operators; and provides the public with outreach and training materials on a wide range of environmental issues.

Publication(s): Water Connection (newsletter), LUSTLine (bulletin on underground storage tanks), Annual Report, NEI Environmental Information Catalog

Keyword(s): Coral Reefs

Contact(s):
Ronald Poltak, EXECUTIVE DIRECTOR

NORTH AMERICAN WETLANDS CONSERVATION COUNCIL

DIVISION OF BIRD HABITAT CONSERVATION
4401 North Fairfax Dr., Suite 110
Arlington, VA 22203 USA
Phone: 703-358-1784   Fax: 703-358-2282
E-mail: r9arw_nawwo@mail.fws.gov
Website: www.birdhabitat.fws.gov

Founded: 1989
Membership: 11
Scope: International

Description: The North American Wetlands Conservation Council encourages public-private partnerships to conserve wetland ecosystems for waterfowl, other migratory birds, fish, and wildlife. Grant projects with a 1-1 match are funded to acquire, restore, and enhance wetlands and associated habitats in Canada, the U.S., and Mexico.

Publication(s): North American Wetlands Conservation Act Progress Report for 1994-1995 and 1996-1997, North American Wetlands Conservation Act Grant Application Instructions

Keyword(s): Waterfowl, Wetlands, Aquatic Habitats, Birds, Coasts, Grants

Contact(s):
David Smith, COORDINATOR
Steve Funderburk, DEPUTY CHIEF OF DIVISION OF BIRD HABITAT CONSERVATION
Phone: 703-358-1784

NORTH PACIFIC ANADROMOUS FISH COMMISSION

889 W. Pender St.
Vancouver, British Columbia V6C 3B2 Canada
Phone: 604-775-5550   Fax: 604-775-5577
E-mail: secretariat@npafc.org
Website: www.npafc.org

Founded: 1993
Membership: 4

Scope: International

Description: Established by a Convention between Canada, Japan, Russia, and the U.S. for the conservation of the anadromous fish resources of the North Pacific Ocean.

Publication(s): North Pacific Anadromous Commission-newsletter

Contact(s):
Vladimir Fedorenko, EXECUTIVE DIRECTOR
Fran Ulmer, PRESIDENT
V. Izmailov, VICE PRESIDENT (RUSSIA)

NORTHEAST ATLANTIC FISHERIES COMMISSION

22 Berners St.
London, WIP 4DY UK
Phone: 0207-631-0016   Fax: 207-636-9225
E-mail: info@neafc.org

Founded: 1980

Description: To promote the conservation and optimum utilization of the fishery resources of the northeast Atlantic, within a framework appropriate to the regime of extended coastal state jurisdiction over fisheries, and to encourage international cooperation and consultation with respect to these resources.

Publication(s): Annual Report, Handbook of Basic Texts

Contact(s):
O. Tougaard, PRESIDENT
Sigmund Engesaeter, SECRETARY
V. Sokolov, VICE PRESIDENT
E. Lemche, VICE PRESIDENT

NORTHEASTERN FOREST FIRE PROTECTION COMMISSION

36 Roslyn Ave.
Warner, NH 03278-4021 USA
Phone: 603-456-3474   Fax: 603-456-3474
Website: www.nffpc.com

Founded: NA
Membership: 5000
Scope: International

Description: International forest fire protection mutual aid organization composed of three commissioners each from CT, ME, MA, NH, RI, VT, NY, and the Canadian Provinces of Quebec, New Brunswick, and Nova Scotia plus New England national forests (Green Mountain and White Mountain). Uniform fire organization planning and suppression technique training carried out annually by the members. The Northeastern Interstate Forest Fire Protection Compact is the governing document that established the organization.

Contact(s):
Clark Davis, EXECUTIVE DIRECTOR
36 Roslyn Ave., Warner, NH 03278-4021
Phone: 603-456-3474

NUCLEAR REGULATORY COMMISSION

2120 L Street
Washington, DC 20555 USA
Phone: 301-415-7000

Founded: 1975
Scope: National

Description: Five-member commission responsible for regulating all commercial uses of nuclear energy to protect the health and safety of the public and the environment.

Keyword(s): Energy

Contact(s):

Linda Portner, ASSOCIATE DIRECTOR
Phone: 301-415-1776
Edward Halman, ASSOCIATE DIRECTOR OF CONTRACT, SECURITY, F01 AND PUBLICATIONS
Phone: 301-415-7305
Michael Springer, ASSOCIATE DIRECTOR OF FACILITIES AND PROPERTY MANAGEMENT
Phone: 301-415-8080
Patricia Norry, ASSOCIATE DIRECTOR OF OFFICE OF ADMINISTRATION
Phone: 301-415-7443
Roy Zimmerman, ASSOCIATE DIRECTOR OF PROJECTS
Phone: 301-415-1284
Ashok Thadani, ASSOCIATE DIRECTOR, INSP. AND TECH. REVIEW
Phone: 301-415-1274
Stephen Burns, ASSOCIATE GENERAL COUNSEL FOR HEARING, ENFORCEMENT, AND ADMINISTRATION
Phone: 301-415-1740
Shirley Jackson, CHAIRMAN
Phone: 301-415-1820
Martin Steindler, CHAIRMAN OF ADVISORY COMMITTEE ON NUCLEAR WASTE
Phone: 301-415-7360
Richard Meserve, CHAIRMAN OF COMMISSION
B. Cotter, CHIEF ADMINISTRATIVE JUDGE AND CHAIRMAN OF ATOMIC SAFETY AND LICENSING BOARD PANEL
Phone: 301-415-7450
Greta Dirus, COMMISSIONER
Phone: 301-415-1820
Kenneth Rogers, COMMISSIONER
Phone: 301-415-1855
Samuel Collins, DEPUTY
Phone: 817-860-8226
William Kane, DEPUTY
Phone: 610-337-5340
Arthur Beach, DEPUTY
Phone: 708-829-9658
Luis Reyes, DEPUTY
Phone: 404-331-5610
Frederick Shon, DEPUTY CHIEF ADMINISTRATIVE JUDGE
Phone: 301-415-7468
Ronald Scroggins, DEPUTY CHIEF FINANCIAL OFFICER/CONTROLLER
Phone: 301-415-7501
Jesse Funches, DEPUTY CONTROLLER
Phone: 301-415-7322
Elizabeth Hayden, DEPUTY DIRECTOR
Phone: 301-415-8200
Malcolm Knapp, DEPUTY DIRECTOR
Phone: 301-415-8468
James McDermott, DEPUTY DIRECTOR
Phone: 301-415-7516
Themis Speis, DEPUTY DIRECTOR
Phone: 301-415-6802
Denwood Ross, DEPUTY DIRECTOR
Phone: 301-415-7473
Roger Fortuna, DEPUTY DIRECTOR
Phone: 301-415-3476

Thomas Martian, DEPUTY DIRECTOR OF DIVISION OF REACTOR PROGRAMS
Phone: 301-415-1199
Arnold Levin, DEPUTY DIRECTOR OF LICENSING SUPPORT SYSTEMS ADMINISTRATOR
Phone: 301-415-7458
Richard Bangart, DEPUTY DIRECTOR OF OFFICE OF STATE PROGRAMS
Phone: 301-415-3340
Hugh Thompson, DEPUTY EXECUTIVE DIRECTOR FOR NUCLEAR MATERIALS, SAFETY, SAFEGUARDS, AND OPERATIONS SUPPORT
Phone: 301-504-1713
James Milhoan, DEPUTY EXECUTIVE DIRECTOR FOR NUCLEAR REACTOR REGULATION, REGIONAL OPERATIONS AND RESEARCH
Phone: 301-415-1705
Martin Malsch, DEPUTY GENERAL COUNSEL
Phone: 301-415-1740
Paul Lohaus, DIRECTOR
Phone: 301-415-2326
Brian Grimes, DIRECTOR
Phone: 301-415-1193
John Cordes, DIRECTOR (ACTING) OF OFFICE OF COMMISSION APPELLATE ADJUDICATION
Phone: 301-415-1600
Frank Miraglia, DIRECTOR (ACTING) OF OFFICE OF NUCLEAR REACTOR REGULATION
Phone: 301-415-1270
Brian Sheron, DIRECTOR OF DIVISION OF ENGINEERING
Phone: 301-415-2722
Lawrence Shao, DIRECTOR OF DIVISION OF ENGINEERING TECHNOLOGY
Phone: 301-415-5678
Elizabeth Teneyck, DIRECTOR OF DIVISION OF FUEL CYCLE SAFETY AND SAFEGUARDS
Phone: 301-415-7212
Donald Cool, DIRECTOR OF DIVISION OF INDUSTRIAL AND MEDICAL NUCLEAR SAFETY
Phone: 301-415-7197
Francis Gillespie, DIRECTOR OF DIVISION OF INSPECTION AND SUPPORT PROGRAMS
Phone: 301-415-1275
Bruce Boger, DIRECTOR OF DIVISION OF REACTOR CONTROLS AND HUMAN FACTORS
Phone: 301-415-1004
Bill Morris, DIRECTOR OF DIVISION OF REGULATORY APPLICATIONS
Phone: 301-415-6207
Gary Holahan, DIRECTOR OF DIVISION OF SYSTEMS SAFETY AND ANALYSIS
Phone: 301-415-2001
M. Hodges, DIRECTOR OF DIVISION OF SYSTEMS TECHNOLOGY
Phone: 301-415-5728
John Greeves, DIRECTOR OF DIVISION OF WASTE MANAGEMENT
Phone: 301-415-7358
Lloyd Donnelly, DIRECTOR OF FINANCIAL MANAGEMENT, PROCUREMENT, AND ADMINISTRATION STAFF
Phone: 301-415-5828
Frank Congel, DIRECTOR OF INCIDENT RESPONSE DIVISION
Phone: 301-415-7476
Edward Jordan, DIRECTOR OF OFFICE FOR ANALYSIS AND EVALUATION OF OPERATIONAL DATA

Phone: 301-415-7472
Dennis Rathbun, DIRECTOR OF OFFICE OF
CONGRESSIONAL AFFAIRS
Phone: 301-415-1776
James Lieberman, DIRECTOR OF OFFICE OF
ENFORCEMENT
Phone: 301-415-2741
Gerald Cranford, DIRECTOR OF OFFICE OF INFORMATION
RESOURCES MANAGEMENT
Phone: 301-415-7585
Carlton Stoiber, DIRECTOR OF OFFICE OF
INTERNATIONAL PROGRAMS
Phone: 301-415-1780
Guy Caputo, DIRECTOR OF OFFICE OF INVESTIGATIONS
Phone: 301-415-2373
Carl Paperiello, DIRECTOR OF OFFICE OF NUCLEAR
MATERIAL SAFETY AND SAFEGUARDS
Phone: 301-415-7800
David Morrison, DIRECTOR OF OFFICE OF NUCLEAR
REGULATORY RESEARCH
Phone: 301-415-6641
Paul Bird, DIRECTOR OF OFFICE OF PERSONNEL
Phone: 301-415-7516
William Beecher, DIRECTOR OF OFFICE OF PUBLIC
AFFAIRS
Phone: 301-415-8200
Irene Little, DIRECTOR OF OFFICE OF SMALL BUSINESS
AND CIVIL RIGHTS
Phone: 301-415-7380
John Linehan, DIRECTOR OF PROGRAM MANAGEMENT,
POLICY DEVELOPMENT, AND ANALYSIS STAFF
Phone: 301-415-7780
C. Rossi, DIRECTOR OF SAFETY PROGRAMS DIVISION
Phone: 301-415-7499
William Travers, DIRECTOR OF SPENT FUEL PROJECT
OFFICE
Phone: 301-415-8500
Kenneth Raglin, DIRECTOR OF TECHNICAL TRAINING
DIVISION
Phone: 423-855-6500
John Larkins, EXECUTIVE DIRECTOR
Phone: 301-415-7360
James Taylor, EXECUTIVE DIRECTOR FOR OPERATIONS
Phone: 301-415-1700
Karen Cyr, GENERAL COUNSEL
Phone: 301-415-1743
Hubert Bell, INSPECTOR GENERAL
Phone: 301-415-5930
William Olmstead, LICENSING AND REGULATION
Phone: 301-415-1740
Hubert Miller, REGIONAL ADMINISTRATOR, REGION 1
475 Allendale Rd., King of Prussia, PA 19406-1415
Phone: 610-337-5299
Stewart Ebneter, REGIONAL ADMINISTRATOR, REGION 2
101 Marietta St., Suite 2900, Atlanta, GA 30323-0199
Phone: 404-331-5500
Arthur Beach, REGIONAL ADMINISTRATOR, REGION 3
801 Warrenville Rd., Lisle, IL 60532-4351
Phone: 708-829-9500
Leonard Callan, REGIONAL ADMINISTRATOR, REGION 4
611 Ryan Plaza Dr., Suite. 4000, Arlington, TX 76011-8064
Phone: 817-860-8225
John Hoyle, SECRETARY OF THE COMMISSION
Phone: 301-415-1969
Robert Seale, VICE CHAIRMAN
Phone: 301-415-7360

Paul Pomeroy, VICE CHAIRMAN
Phone: 301-415-7360

O

OHIO RIVER VALLEY WATER SANITATION COMMISSION

5735 Kellogg Ave.
Cincinnati, OH 45228-1112 USA
Phone: 513-231-7719   Fax: 513-231-7761
E-mail: info@orsanco.org
Website: www.orsanco.org

Founded: 1948
Scope: Statewide, Regional

Description: An interstate agency representing Illinois, Indiana, Kentucky, New York, Ohio, Pennsylvania, Virginia, and West Virginia for control of water pollution in the Ohio River Valley Compact District.

Publication(s): ORSANCO Quality Monitor, Annual Report, publications of general or technical interest such as Ohio River fish populations

Contact(s):
Rhonda Barnes-Cloth
COMMUNICATIONS COORDINATOR
Alan Vicory, EXECUTIVE DIRECTOR AND CHIEF
ENGINEER
Jeanne Ison, PUBLIC INFORMATION PROGRAMS
MANAGER

P

PACIFIC SALMON COMMISSION

1155 Robson St., Suite 600
Vancouver, British Columbia V6E 1B5 Canada
Phone: 604-684-8081   Fax: 604-666-8707
Website: www.psc.org

Founded: NA
Membership: 25
Scope: International

Description: Charged with implementation of the Pacific Salmon Treaty signed by Canada and the United States in 1985, the Commission provides regulatory advice and recommendations to the U.S. and Canada relative to their management of salmon originating in one country, but subject to interception by the other. The Commission is also charged with conserving Pacific Salmon stocks in order to achieve optimum production, and dividing salmon harvests so each nation receives benefits equivalent to salmon produced in its waters. Each nation appoints four commissioners and four alternates to serve on the Commission. The Commission is the body through which the U.S. and Canada can work to resolve complex salmon management problems.

Contact(s):
Don Kowal, EXECUTIVE SECRETARY

PACIFIC STATES MARINE FISHERIES COMMISSION

45 SE 82nd Dr., Suite 100
Gladstone, OR 97027-2522 USA
Phone: 503-650-5400   Fax: 503-650-5426
Website: www.psmfc.org

Founded: 1947

Membership: 45
Scope: Statewide

Description: The Commission serves the Pacific states of Alaska, California, Idaho, Oregon, and Washington to promote conservation, development, and management of marine and anadromous fisheries of mutual concern through a coordinated regional approach to fisheries research, monitoring, and utilization. Activities focus on multistate databases, interjurisdiction fishery management plans, marine debris, saving fisheries habitat, and marine mammal/fishery interactions.

Contact(s):
Randy Fisher, EXECUTIVE DIRECTOR

PEACE CORPS
1111 20th St., NW
Washington, DC 20526 USA
Phone: 202-692-2100
Website: www.peacecorps.gov

Founded: 1961
Membership: 540
Scope: International

Description: The Peace Corps was established by President John F. Kennedy to promote world peace and friendship. Since 1961, more than 155,000 Americans have joined the Peace Corps and have served in 134 countries around the world. The Peace Corps has three goals: to provide volunteers who contribute to the social, economic, and human development of interested countries; to promote a better understanding of Americans among the people whom Peace Corps volunteers serve; and to strengthen Americans' understanding of the other peoples and cultures—to help bring the world back home.

Publication(s): The Peace Corps Times

Contact(s):
Charles Baquet, ACTING DIRECTOR
Phone: 202-692-2100

PEACE CORPS
ECUADOR
Avenida G de Diciembre 2269
Quito, Ecuador
Phone: 593-256-1224
E-mail: fgarces@ec.peacecorps.gov

Founded: 1960

Description: Peace Corps Ecuador provides technical assistance and cultural exchange in rural communities, through conservation NGO's and government agencies. Technical assistance is provided in the fields of EE, Agroforestry, and Conservation of Protected Areas.

Publication(s): Agroforesteria en los Audes del Ecuador, Algunes alternativas para el desarrollo sosteuible eu govas cercauas al Parque nacional poolocarpus, Remedios Naturales Contra Plagas

Contact(s):
Francisco Garces, NATURAL RESOURCES PROGRAM DIRECTOR
Phone: 593-256-1224
fgarces@ec.peacecorps.gov
Marcy Kelly, PEACE CORPS ECUADOR DIRECTOR
Phone: 593-256-1224
mkelly@ec.peacecorps.gov

Tim Criste, PROGRAM AND TRAINING OFFICER
Phone: 593-256-1224
tcriste@ec.peacecorps.gov

S

ST. CROIX INTERNATIONAL WATERWAY COMMISSION
Box 610
Calais, ME 04619 USA
Phone: 506-466-7550   Fax: 506-466-7551
E-mail: staff@stcroix.org

Founded: NA
Membership: 3
Scope: International

Description: A commission of the state of Maine and province of New Brunswick to help implement a cooperative international management plan for the St. Croix River system, which forms 110 miles of the U.S. and Canada border.

Publication(s): Management Plan for the St. Croix International Waterway, Annual Report, St. Croix Heritage Brochure

Contact(s):
Don Carson, CO-CHAIRMAN
Ken Gordon, CO-CHAIRMAN
Lee Sochasky, EXECUTIVE DIRECTOR

SUSQUEHANNA RIVER BASIN COMMISSION
1721 N. Front St.
Harrisburg, PA 17102 USA
Phone: 717-238-0422   Fax: 717-238-2436
E-mail: crbc@srcb.net
Website: www.srbc.net

Founded: NA
Membership: 35
Scope: Statewide

Description: Conservation and development of water resources and water-related resources in the river basin, comprising parts of Maryland, New York, and Pennsylvania.

Publication(s): Annual Report, Susquehanna Guardian (newsletter)

Contact(s):
Paul Swartz, EXECUTIVE DIRECTOR
1721 N. Front St., Harrisburg, PA 17102
Phone: 717-238-0422

SWAZILAND ENVIRONMENT AUTHORITY (SEA)
Ministry of Natural Resources Building
Mbabane, Swaziland
Phone: 268-404-1719

Founded: 1993

Description: SEA is the responsible government agency for environmental management in Swaziland (policy, EIA, legislation, monitoring, developing standards, environmental ed/public awareness).

Publication(s): Swaziland's Environment Action Plan, Solid Waste Regulations, etc., Environment Management Act

Contact(s):
Irma Allen, CHAIR/BOARD
Phone: 268-404-1719
Fax: 268-518-6284
szallen@iafrica.sz

J. Vilakati, DIRECTOR
Phone: 268-404-1719
Stephen Zuke, SENIOR ENVIRONMENTAL OFFICER
Phone: 268-404-1719

T

TANZANIA COASTAL MANAGEMENT PARTNERSHIP

Haile Selassie St., P.O. Box 71886
Dar Es Salaam, Tanzania
Phone: +255+51-667589   Fax: - 66-8611
E-mail: gluhikula@epog.or.tz

Founded: NA

Description: Integrated coastal management policy process in Tanzania.

Keyword(s): Coasts

Contact(s):
G. Luhikula
CONTACT

TENNESSEE VALLEY AUTHORITY

400 W. Summit Hill Dr.
Knoxville, TN 37902-1499 USA
Phone: 865-632-2101
E-mail: tvainfo@tva.gov
Website: http://www.tva.gov

Founded: 1933

Description: TVA was created by an Act of Congress for the regional development of the Tennessee Valley region in Tennessee, Kentucky, Mississippi, Alabama, Virginia, Georgia, and North Carolina. In 1964, TVA opened Land Between The Lakes as a national demonstration project for outdoor recreation, environmental education, and resource management.

Publication(s): RiverPulse, Recreation on TVA Lakes, various others on different subjects

Contact(s):
Craven Crowell, CHAIRMAN OF THE BOARD
Norman Zigrossi, CHIEF ADMINISTRATIVE OFFICER
David Smith, EXECUTIVE VICE PRESIDENT AND CHIEF FINANCIAL OFFICER
Kathryn Jackson, EXECUTIVE VICE PRESIDENT OF RIVER SYSTEM OPERATIONS AND ENVIRONMENT
Mark Medford, EXECUTIVE VICE PRESIDENT, CUSTOMER SERVICE AND MARKETING
Joseph Bynum, EXECUTIVE VICE PRESIDENT, FOSSIL POWER GROUP
William Boston, EXECUTIVE VICE PRESIDENT, TRANSMISSION/POWER SUPPLY GROUP
John Scalice, EXECUTIVE VICE PRESIDENT/CHIEF OFFICER AND NUCLEAR OFFICER OF TVA NUCLEAR
Oswald Zeringue, PRESIDENT AND CHIEF OPERATING OFFICER
Edward Christenbury, SENIOR VICE PRESIDENT AND GENERAL COUNSEL
Peyton Harriston, SENIOR VICE PRESIDENT FOR STRATEGIC INITIATIVES

TENNESSEE VALLEY AUTHORITY

KNOXVILLE AND CHATTANOOGA CORPORATE LIBRARY
400 W. Summit Hill Dr., ET PC
Knoxville, TN 37902-1499 USA
Phone: 865-632-2101
Website: http://www.tra.gov

Founded: NA

TENNESSEE VALLEY AUTHORITY

MUSCLE SHOALS TECHNICAL LIBRARY
CTR 1E
Muscle Shoals, AL 35662 USA
Phone: 256-386-2601

Founded: NA

U

U S ARMY CORPS OF ENGINEERS

U.S. ARMY CORPS OF ENGINEERS
U.S. ARMY ENGINEER DISTRICT, PHILADELPHIA
Wanamaker Building, 100 Penn Square East
Philadelphia, PA 19107-3390 USA
Phone: 215-656-6501   Fax: 215-656-6899
Website: www.nap.usace.army.mil

Founded: NA
Scope: National

Publication(s): District Observer

Contact(s):
John Vickers, DEPUTY DISTRICT COMMANDER
Timothy Brown, DISTRICT COMMANDER

U S ARMY CORPS OF ENGINEERS

U.S. ARMY CORPS OF ENGINEERS
U.S. ARMY ENGINEER DISTRICT, DETROIT
P.O. Box 1027
Detroit, MI 48231-1027 USA
Phone: 313-226-6762   Fax: 313-226-6009

Founded: NA
Membership: 595
Scope: Statewide

Contact(s):
Richard Polo, COMMANDER & DISTRICT ENGINEER

U S ARMY CORPS OF ENGINEERS

U.S. ARMY ENGINEER DIVISION, GREAT LAKES AND OHIO
P.O. Box 1159
Cincinnati, OH 45201-1159 USA
Phone: 513-684-3002   Fax: 513-684-2085
E-mail: celrd-de@usace.army.mil
Website: www.lrd.usace.army.mil

Founded: NA
Membership: 100
Scope: State

Contact(s):
Steven Hawkins, DIVISION COMMANDER OF THE GREAT LAKES & OHIO RIVER

U.S. ARMY- CORPS OF ENGINEERS
U.S. ARMY CORPS OF ENGINEERS
U.S. ARMY ENGINEER DISTRICT, NEW YORK
Jacob K. Javits Federal Building, 26 Federal Plaza, Room 2109
New York, NY 10278-0090 USA
Phone: 212-264-0100   Fax: 212-264-5947
Website: www.nan.usace.army.mil

Founded: NA
Membership: 700
Scope: Statewide

U.S. ARMY CORPS OF ENGINEER HONOLULU DISTRICT
U.S. ARMY CORPS OF ENGINEERS
U.S. ARMY ENGINEER DISTRICT, HONOLULU
Building 230
Fort Shafter, HI 96858-5440 USA
Phone: 808-438-1069   Fax: 808-438-8351
Website: www.poh.usace.army.mil

Founded: NA
Membership: 350
Scope: Statewide

Contact(s):
Alex Skinner, EXECUTIVE ASSISTANT

UNITED STATES DEPARTMENT OF THE AIR FORCE
MAJOR U.S. INSTALLATIONS
DYESS AFB, TX
USA
Phone: 915-696-5049   Fax: 915-696-2899
Website: https://www.mil.dyess.af.mil

Founded: NA

Contact(s):
Jim Robertson, CULTURAL/NATURAL RESOURCES DIRECTOR

UNITED NATIONS RESEARCH INSTITUTE FOR SOCIAL DEVELOPMENT (UNRISD)
Palais des Nations, CH 1211
Geneva 10, Switzerland
Phone: 41-22-798-8400, 798-5850
Fax: 41-22-740-0791
E-mail: info@UNRISD.org
Website: www.unrisd.org

Founded: 1963

Description: UNRISD is an autonomous agency that researches the social dimensions of contemporary development problems. The Institute provides governments, development agencies, grassroots organizations, and scholars with a better understanding of how development policies and processes of economic, social, and environmental change affect different social groups. UNRISD promotes original research and strengthens research capacity in developing countries.

Publication(s): UNRISD News, The Challenge of Peace, Focus on Integrating Gender into The Politics of Development, Discussion Papers

Keyword(s): Environment, Wildlife, Forest Management, Internships, Land Use Planning, Research, Rural Development, Sustainable Development, Agriculture, Biodiversity, Conservation, Developing Countries

Contact(s):
Dharam Ghai, DIRECTOR

UNITED STATES ARMY CORPS OF ENGINEERS
U.S. ARMY CORPS OF ENGINEERS
U.S. ARMY ENGINEER DISTRICT, CHICAGO
111 N. Canal Street, Suite 600
Chicago, IL 60606-7206 USA
Phone: 312-353-6400   Fax: 312-353-2525
Website: www.lrc.usace.army.mil

Founded: NA
Membership: 215
Scope: National

Contact(s):
Mark Roncoli, DISTRICT COMMANDER

UNITED STATES ARMY CORPS OF ENGINEERS
U.S. ARMY CORPS OF ENGINEERS
U.S. ARMY ENGINEER DISTRICT, JACKSONVILLE
P.O. Box 4970
Jacksonville, FL 32232-0019 USA
Phone: 904-232-2241   Fax: 904-232-1213

Founded: NA
Membership: 700
Scope: Statewide

Contact(s):
James May, DISTRICT ENGINEER

UNITED STATES CHAMBER OF COMMERCE
1615 H St., NW
Washington, DC 20062 USA
Phone: 202-659-6000   Fax: 202-463-5839
E-mail: mbrsvcs@uschamber.com
Website: www.uschamber.com

Founded: 1912
Membership: 200,000
Scope: National

Description: Created to provide business representation on major national issues. Membership includes 3,000 chambers of commerce, 1,200 trade and professional associations, and 215,000 business firms.

Keyword(s): Trade/Business, Agriculture, Energy, Pollution Control

Contact(s):
William Kovacs, VICE PRESIDENT, ENVIRONMENT AND REGULATORY AFFAIRS

UNITED STATES DEPARTMENT OF AGRICULTURAL RESEARCH SERVICES
RESEARCH EDUCATION AND ECONOMICS
ARS MIDWEST OFFICE
1815 N. University St.
Peoria, IL 61604 USA
Phone: 309-681-6602   Fax: 309-681-6684
Website: www.mva.arf.usda.gov

Founded: NA
Membership: 8
Scope: Regional

Publication(s): Midwest Area Research Highlights 2000

Contact(s):
Adrianna Hewings, AREA DIRECTOR

UNITED STATES DEPARTMENT OF AGRICULTURE
1400 Independence Ave. SW
Washington, DC 20250 USA
Phone: 202-720-8732
Website: www.usca.gov

Founded: 1862
Scope: National

Description: Created by Congress to acquire and disperse "useful" information on subjects connected with agriculture in the most general and comprehensive sense of that word, and to procure, propagate, and distribute among the people new and valuable seeds and plants. Today, in addition to managing the national forests and grasslands, USDA manages a variety of research, regulatory, domestic and foreign marketing, food and nutrition, and many other programs.

Contact(s):
Ann Beneman

UNITED STATES DEPARTMENT OF AGRICULTURE
AGRICULTURAL RESEARCH SERVICE
1400 Independence Ave., SW, Suite 302A
Washington, DC 20250 USA
Phone: 202-720-3656   Fax: 202-720-5427
Website: www.ars.usda.gov

Founded: NA
Scope: Local, Regional

Description: Conducts research in natural resources, plant sciences, animal sciences, food sciences and human nutrition.

Contact(s):
Floyd Horn, ADMINISTRATOR
Phone: 202-720-3656
Edward Knipling, ASSOCIATE ADMINISTRATOR
Phone: 202-720-3658
Caird Rexroad, ASSOCIATE DEPUTY ADMINISTRATOR FOR ANIMAL PRODUCTION, PRODUCT VALUE, AND SAFETY (NPS)
Phone: 301-504-7050
Judith St. John, ASSOCIATE DEPUTY ADMINISTRATOR FOR CROP PRODUCTION, PRODUCT VALUE AND SAFETY (NPS)
Phone: 301-504-6252
Allen Dedrick, ASSOCIATE DEPUTY ADMINISTRATOR FOR NATURAL RESOURCES AND SUSTAINABLE AGRICULTURAL SYSTEMS (NPS)
Phone: 301-504-7987
Dwayne Buxton, DEPUTY ADMINISTRATOR (NPS)
Phone: 301-504-5084

UNITED STATES DEPARTMENT OF AGRICULTURE
ANIMAL AND PLANT HEALTH INSPECTION SERVICE
ANIMAL CARE
4700 River Road, Unit 84
Riverdale, MD 20737-0123 USA
Phone: 301-734-4980   Fax: 301-734-4978

E-mail: ace@usda.gov
Website: www.aphif.usda.gov/ac

Founded: NA
Membership: 16
Scope: Local, Regional, National

Description: Investigates and prosecutes violations of federal laws governing the movement of animals and plants between states or into and out of the United States and regulates the humane care and treatment of warm-blooded animals used for purposes of research or exhibition, for sale as pets at the wholesale level, or transported in commerce.

Publication(s): List of Registered Research Facilities, List of Licensed Exhibitors, List of Licensed Dealers

Contact(s):
Richard Watkins, ASSISTANT DEPUTY ADMINISTRATOR FOR ANIMAL CARE
Phone: 301-734-7833
W. Dehaven, DEPUTY ADMINISTRATOR FOR ANIMAL CARE
Phone: 301-734-4980

UNITED STATES DEPARTMENT OF AGRICULTURE
ANIMAL AND PLANT HEALTH INSPECTION SERVICE
ANIMAL CARE EASTERN REGIONAL OFFICE
920 Main Campus Drive, Suite 200
Raleigh, NC 27606 USA
Phone: 919-716-5532   Fax: 919-716-5696
E-mail: ace@usda.gov
Website: www.aphis.usda.gov/animalcare

Founded: NA
Membership: 50
Scope: Regional

Contact(s):
Elizabeth Goldentyer, REGIONAL DIRECTOR

UNITED STATES DEPARTMENT OF AGRICULTURE
ANIMAL AND PLANT HEALTH INSPECTION SERVICE
ANIMAL CARE REGIONAL CENTRAL OFFICE
PO BOX 915004
Ft. Worth, TX 76115-9104 USA
Phone: 817-885-6923   Fax: 817-885-6917
E-mail: ACE@USDA.GOV
Website: www.aphis.usda.gov/ac

Founded: NA
Membership: 10
Scope: National

Contact(s):
Walter Christensen, ASSISTANT REGIONAL DIRECTOR

UNITED STATES DEPARTMENT OF AGRICULTURE
ANIMAL AND PLANT HEALTH INSPECTION SERVICE
ANIMAL CARE WESTERN REGIONAL OFFICE
9580 Micron Ave., Suite J
Sacramento, CA 95827-2623 USA   Fax: 916-857-6212
Website: www.aphif.usda.gov/ac

Founded: NA

Contact(s):
Robert Gibbons, REGIONAL DIRECTOR
Phone: 916-857-6205

UNITED STATES DEPARTMENT OF AGRICULTURE

ANIMAL AND PLANT HEALTH INSPECTION SERVICE
INTERNATIONAL SERVICES ASIA AND PACIFIC
OFFICE
USDA, APHIS, IS
American Embassy, Tokyo
Unit 45004, Box 226
APO AP
Tokyo, 96337-5004 Japan
Phone: 8133-224-5457   Fax: 8133-224-5291

Founded: NA

Contact(s):
Gary Green, REGIONAL DIRECTOR

UNITED STATES DEPARTMENT OF AGRICULTURE

ANIMAL AND PLANT HEALTH INSPECTION SERVICE
INTERNATIONAL SERVICES CENTRAL AMERICA,
CARIBBEAN & PANAMA OFFICE
USDA-APHIS-IS, American Embassy, Guatemala, Unit
3319
APO, 34024-3319 USA
Phone: 011-502-331-2036

UNITED STATES DEPARTMENT OF AGRICULTURE

ANIMAL AND PLANT HEALTH INSPECTION SERVICE
INTERNATIONAL SERVICES EUROPE, AFRICA,
RUSSIA, NEAR EAST OFFICE
USDA-APHIS-IS FAS-USEU, PSC 82, Box 002
APO, 09724 USA
Phone: 011-322-508-2762

UNITED STATES DEPARTMENT OF AGRICULTURE

ANIMAL AND PLANT HEALTH INSPECTION SERVICE
INTERNATIONAL SERVICES MEXICO OFFICE
USDA-APHIS-IS, P.O. Box 3087
Laredo, TX 78044 USA
Phone: 11502-331-2036

Founded: NA

Contact(s):
Gordon Tween, REGIONAL DIRECTOR

UNITED STATES DEPARTMENT OF AGRICULTURE

ANIMAL AND PLANT HEALTH INSPECTION SERVICE
INTERNATIONAL SERVICES SCREWWORM
ERADICATION PROGRAM OFFICE
USDA-APHIS-IS, P.O. Box 3087
Laredo, TX 78044 USA
Phone: 11507-232-6683

Founded: NA

Contact(s):
John Wyss, REGIONAL DIRECTOR

UNITED STATES DEPARTMENT OF AGRICULTURE

ANIMAL AND PLANT HEALTH INSPECTION SERVICE
INTERNATIONAL SERVICES SOUTH AMERICA
OFFICE: USDA/APHSIS
American Embassy Santiago, Unit 4113
APO, NY 34033 USA
Phone: 011-562-638-1989

UNITED STATES DEPARTMENT OF AGRICULTURE

ANIMAL AND PLANT HEALTH INSPECTION SERVICE
PLANT PROTECTION AND QUARANTINE
4700 River Road
Riverdale, MD 20737 USA
Phone: 301-734-8261
Website: http://www.aphis.usda.gov/ppq/

Founded: NA

Description: Regulates the importation of plants, plant products,
and animal products from foreign countries. Regulates the
movement of such products between U.S. possessions and the
mainland and the importation and interstate movement of plant
pests. Inspects and certifies plants and plant products for
export. Administers cooperative programs with states to control
and eradicate insects, diseases, weeds, and nematodes of
economic importance. Enforces the Convention on
International Trade in Endangered Species of Flora and Fauna
(CITES) for plants.

Contact(s):
Richard Dunkle, DEPUTY ADMINISTRATOR
Phone: 202-720-5601
Richard.L.Dunkle@usda.gov

UNITED STATES DEPARTMENT OF AGRICULTURE

ANIMAL AND PLANT HEALTH INSPECTION SERVICE
REGULATORY ENFORCEMENT EASTERN REGIONAL
OFFICE
2568A Riva Rd., Suite 302
Annapolis, MD 21401-7400 USA
Phone: 410-571-9480   Fax: 410-962-0008
Website: www.aphis.usda.gov/ws/nwrc

Founded: NA

Contact(s):
John Kinsella, REGIONAL DIRECTOR
Phone: 919-716-5626
john.s.kinsella@usda.gov

UNITED STATES DEPARTMENT OF AGRICULTURE

ANIMAL AND PLANT HEALTH INSPECTION SERVICE
VETERINARY SERVICES
4700 River Road
Riverdale, MD 20737 USA
Phone: 301-734-8093
Website: www.aphis.usda.gov

Founded: NA

Description: Regulates the importation of animals, animal semen,
embryos, and animal products from foreign countries and the
interstate movement of animals. Inspects and certifies animals
for export. Administers cooperative federal-state programs to

control and eradicate animal pests and diseases. Provides laboratory support for animal health programs and diagnostic referral assistance for private and state laboratories.

Contact(s):
Gary Colgrove, ASSISTANT DIRECTOR
Phone: 301-734-4356
gary.s.colgrove@usda.gov
Joseph Annelli, CHIEF STAFF VETERINARIAN
Phone: 301-734-8073
joseph.f.annelli@usda.gov
Alfonso Torres, DEPUTY ADMINISTRATOR FOR VETERINARY SERVICES
Phone: 202-720-5193
alfonso.torres@usda.gov

UNITED STATES DEPARTMENT OF AGRICULTURE
CSREES-NATURAL RESOURCES AND ENVIRONMENT
Waterfront, Mail Stop 2210, 800 9TH St.
Washington, DC 20250-2210 USA
Phone: 202-401-4555   Fax: 202-401-1706

Founded: NA
Membership: 27
Scope: Local

Description: NRE is the operational staff of the CSREES. NRE is responsible for providing leadership and administering research and educational programs that ensure the efficient use and conservation of the Nation's natural resources and protection of the environment.

Contact(s):
Ralph Otto, DEPUTY ADMINISTRATOR
rotto@reeusda.gov
Chuck Krueger, LIAISON, STATE AG. EXPERIMENT STATION/CSREES NATIONAL ENVIRONMENT INITIATIVE (SUNEI)
Phone: 202-401-6516
ckrueger@psu.edu
Jane Dodds, NATIONAL PROGRAM LEADER, ENVIRONMENTAL HEALTH
Phone: 202-401-4044
jdodds@reeusda.gov
Catalino Blanche, NATIONAL PROGRAM LEADER, FORESTRY BIOLOGY
Phone: 202-401-4190
cblanche@reeusda.gov
Larry Biles, NATIONAL PROGRAM LEADER, FORESTRY MANAGEMENT
Phone: 202-401-4926
ibiles@reeusda.gov
Joe Wysocki, NATIONAL PROGRAM LEADER, HOUSING AND ENVIRONMENT
Phone: 202-401-4980
jwysocki@reeusda.gov
Raymond Knighton, NATIONAL PROGRAM LEADER, SOILS SCIENCE
Phone: 202-401-6417
knighton@reeusda.gov
Greg Crosby, NATIONAL PROGRAM LEADER, SUSTAINABLE DEVELOPMENT AND ENVIRONMENTAL EDUCATION
Phone: 202-401-6050
gcrosby@reeusda.gov
Michael O'Neill, NATIONAL PROGRAM LEADER, WATER QUALITY
Phone: 202-205-5952
Mary Rozum, NATIONAL PROGRAM LEADER, WATER QUALITY AND ANIMAL WASTE INITIATIVE
Phone: 202-401-4533
mrozum@reeusda.gov
Fred Swader, NATIONAL PROGRAM LEADER, WATER QUALITY AND EXTENSION INDIAN RESERVATION PROGRAM
Phone: 202-401-5853
fswader@reeusda.gov

UNITED STATES DEPARTMENT OF AGRICULTURE
ECONOMIC RESEARCH CENTER
1800 M St., NW
Washington, DC 20036 USA
Phone: 202-694-5050   Fax: 202-694-5734
Website: www.ers.usda.gov

Founded: NA
Scope: Regional

Description: Provides a program of agricultural, economic, and social research and analysis, statistical programs, technical consultation, planning assistance, and associated services. Conducts research and staff work relating to natural resources and environmental quality, including supplies, uses, and projected future requirements for land and water; effects of environmental quality improvement measures on agricultural production and agricultural resource use; achievement of environmental goals in rural areas; ownership and control of land and water resources; methods for natural resource planning; and evaluation of natural resource plans and projects.

Publication(s): Various research monographs on natural resources use, Pest control, Agricultural inputs, Conservation

Contact(s):
Susan Offutt, ADMINISTRATOR
Paul Chen, DIRECTOR OF INFORMATION SERVICES DIVISION (ACTING)
Adrie Custer, DIRECTOR OF PUBLICATION SERVICES BRANCH
Kitty Smith, DIRECTOR OF RESOURCE ECONOMICS DIVISION

UNITED STATES DEPARTMENT OF AGRICULTURE
FARM SERVICE AGENCY (FSA)
USDA, FSA, CEPD, STOP 0513, 1400 Independence Ave., SW
Washington, DC 20250-0513 USA
Phone: 202-720-6221   Fax: 202-720-4619

Founded: NA
Membership: 32
Scope: National

Description: Formerly known as (AGRICULTURAL STABILIZATION AND CONSERVATION SERVICE). Administers the following: Conservation Reserve Program, Emergency Conservation Program, various commodity loan programs, production flexibility contracts, and various farm loan and disaster assistance programs.

Contact(s):
James Little
ACTING ADMINISTRATOR

Carolyn Cooksie, DEPUTY ADMINISTRATOR FOR FARM
LOAN PROGRAMS
Larry Mitchell, DEPUTY ADMINISTRATOR FOR FARM
PROGRAMS
Grady Bilberry, DIRECTOR OF PRICE SUPPORT DIVISION
Diane Sharp, DIRECTOR OF PRODUCTION,
EMERGENCIES AND COMPLIANCE DIVISION
Tade Sullivan, DIRECTOR OF PUBLIC AFFAIRS
Robert Stephenson, DIRECTOR, CONSERVATION AND
ENVIRONMENTAL PROGRAMS

UNITED STATES DEPARTMENT OF AGRICULTURE

FOREST SERVICE
P.O. Box 96090
Washington, DC 20090-6090 USA
Phone: 202-205-8333   Fax: 202-205-1599
Website: www.fs.fed.us

Founded: NA
Scope: National

Description: Administers National Forests and National
Grasslands and is responsible for the management of their
resources. Cooperates with federal and state officials in the
enforcement of game laws on the National Forests and in the
development and maintenance of wildlife resources;
cooperates with the state and private owners in the application
of sound forest management practices, in protection of forest
lands against fire, insects, diseases, and in the distribution of
planting stock. Conducts research in the entire field of forestry
and wildland management.

Publication(s): see publications on website

Contact(s):
Paul Brouha, ASSOCIATE DEPUTY NATIONAL FOREST
SYSTEM
Phone: 202-205-1465
Barbara Weber, ASSOCIATE DEPUTY RESEARCH AND
DEVELOPMENT
Phone: 202-205-1702
Clyde Thompson, BUSINESS OPERATIONS
Phone: 202-205-1707
Dale Bosworth, CHIEF
Phone: 202-205-1661
Vincette Goerl, CHIEF FINANCIAL OFFICER
Phone: 202-205-1784
Kathy Gause, CIVIL RIGHTS DIRECTOR
Phone: 202-205-1585
Sally Collins, DEPUTY CHIEF FOR NATURAL RESOURCES
Phone: 202-205-1465
Jim Furnish, DEPUTY NATIONAL FOREST SYSTEM
Phone: 202-205-1523
Robert Lewis, DEPUTY RESEARCH AND DEVELOPMENT
Phone: 202-205-1665
Michael Rains, DIRECTOR NORTH EAST RESEARCH
STATION
Phone: 202-205-1657
George Lennon, DIRECTOR OF OFFICE
COMMUNICATIONS
Phone: 202-205-8333
Hilda Diaz-Stero, DIRECTOR OF PACIFIC SW RESEARCH
STATION REGION 5
Phone: 202-205-1491
Phil Janik, DIRECTOR OF WILDLIFE, FISH, WATER & AIR
RESEARCH
Phone: 202-205-1661

Randy Phillips, EXECUTIVE DIRECTOR OF FOREST
COUNTY PAYMENTS ADVISORY COMMITTEE
Phone: 202-205-1663
Valdis Mezainis, INTERNATIONAL PROGRAMS
Phone: 202-205-1650
Bill Wasley, LAW ENFORCEMENT AND INVESTIGATIONS
Phone: 703-605-4690

UNITED STATES DEPARTMENT OF AGRICULTURE

FOREST SERVICE
SOUTHERN RESEARCH STATION
P.O. Box 2750
Asheville, NC 28802 USA
Phone: 828-257-4832   Fax: 828-257-4263
Website: www.cs.unca.edu/nfsnc

Founded: NA
Scope: Statewide

Contact(s):
Rob McClanahan, FOREST WILDLIFE BIOLOGIST

UNITED STATES DEPARTMENT OF AGRICULTURE

RESEARCH EDUCATION AND ECONOMICS
1400 Independence Ave. SW
Washington, DC 20250-0110 USA
Phone: 202-720-5923   Fax: 202-690-2842
Website: www.usda.gov

Founded: NA
Membership: 5
Scope: National

Contact(s):
Joseph Jen, UNDER SECRETARY

UNITED STATES DEPARTMENT OF AGRICULTURE

RESEARCH EDUCATION AND ECONOMICS
ARS BELTSVILLE AREA
10300 Baltimore Rm. 223, B-003 BARC-West
Beltsville, MD 20705 USA
Phone: 301-504-6078   Fax: 301-504-5863
Website: www.ba.ars.usda.gov

Founded: NA
Scope: National

Contact(s):
Phyllis Johnson, AREA DIRECTOR
Ronald Korcak, ASSOCIATE AREA DIRECTOR

UNITED STATES DEPARTMENT OF AGRICULTURE

RESEARCH EDUCATION AND ECONOMICS
ARS MID SOUTH OFFICE
Stoneville, MS 38776 USA
Phone: 662-686-5265   Fax: 662-686-5459
E-mail: atuckeramsa-stoneville.ars.usda.gov
Website: www.ars.usda.gov

Founded: NA
Membership: 900
Scope: Regional

Contact(s):
Edgar King, AREA DIRECTOR

UNITED STATES DEPARTMENT OF AGRICULTURE

RESEARCH EDUCATION AND ECONOMICS
ARS NORTH ATLANTIC OFFICE
600 E. Mermaid Ln.
Windmoor, PA 19038 USA
Phone: 215-233-6593   Fax: 215-233-6719

Founded: NA
Membership: 600
Scope: National

Contact(s):
Wilda Martinez, AREA DIRECTOR

UNITED STATES DEPARTMENT OF AGRICULTURE

RESEARCH EDUCATION AND ECONOMICS
ARS NORTHERN PLAINS AREA OFFICE
1201 Oakridge Rd.
Fort Collins, CO 80525 USA
Phone: 970-229-5557   Fax: 970-229-5565
Website: www.npa.ars.usda.gov

Founded: NA
Membership: 32
Scope: Statewide, National

Contact(s):
Wilbert Blackburn, AREA DIRECTOR

UNITED STATES DEPARTMENT OF AGRICULTURE

RESEARCH EDUCATION AND ECONOMICS
ARS PACIFIC WEST OFFICE
800 Buchanan St.
Albany, CA 94710 USA
Phone: 510-559-6060   Fax: 510-559-5779
Website: www.pwa.ars.usda.gov

Founded: NA

Contact(s):
A. Betschart, AREA DIRECTOR

UNITED STATES DEPARTMENT OF AGRICULTURE

RESEARCH EDUCATION AND ECONOMICS
ARS SOUTH ATLANTIC OFFICE
Russell Agr. Res. Center, P.O. Box 5677, College Station Rd.
Athens, GA 30604-5677 USA
Phone: 706-546-3311   Fax: 706-546-3398
Website: www.ars-grin.gov/ars/soatlantic

Founded: NA

Contact(s):
Karl Narang, AREA DIRECTOR

UNITED STATES DEPARTMENT OF AGRICULTURE

RESEARCH EDUCATION AND ECONOMICS
ARS SOUTHERN PLAINS OFFICE: USDA
7607 Eastmark Dr., Suite 230
College Station, TX 77840 USA

Founded: NA

UNITED STATES DEPARTMENT OF AGRICULTURE

RESEARCH EDUCATION AND ECONOMICS
COOPERATIVE STATE RESEARCH, EDUCATION, AND EXTENSION SERVICE
1400 Independence Ave., SW., 305A
Washington, DC 20250 USA
Phone: 202-720-7441   Fax: 202-720-8987
Website: www.reeusda.gov

Founded: NA

Description: The Cooperative State Research, Education, and Extension Service links the research and education resources and programs of the U.S. Department of Agriculture and works with land-grant institutions in each state, territory, and the District of Columbia.

Contact(s):
Colien Hefferan, ADMINISTRATOR
Phone: 202-720-7441
colien.hefferan@usda.gov
Sally Rockey, DEPUTY ADMINISTRATOR FOR COMPETITIVE RESEARCH GRANTS AND AWARDS MANAGEMENT
Phone: 202-401-1761
Alma Hobbs, DEPUTY ADMINISTRATOR FOR FAMILIES 4-H AND NUTRITION
Phone: 202-720-2908
alma.hobbs@usda.gov
George Cooper, DEPUTY ADMINISTRATOR FOR PARTNERSHIPS
Phone: 202-720-5623
george.cooper@usda.gov
Jane Coulter, DEPUTY ADMINISTRATOR FOR SCIENCE AND EDUCATION RESOURCES DEVELOPMENT
Phone: 202-720-3377

UNITED STATES DEPARTMENT OF AGRICULTURE

UNITED STATES FOREST SERVICE
ALASKA REGION 10
P.O. Box 21628
709 W. 9th St.
Juneau, AK 99802-1628 USA
Phone: 907-586-8863   Fax: 907-586-7840
Website: www.fs.fed.us

Founded: NA
Scope: Regional

Contact(s):
Paul Forward, ACTING REGIONAL FORESTER

UNITED STATES DEPARTMENT OF AGRICULTURE

UNITED STATES FOREST SERVICE
EASTERN REGION
310 W. Wisconsin Avenue, Rm. 580
Milwaukee, WI 53203 USA
Phone: 414-297-3600

Founded: NA

UNITED STATES DEPARTMENT OF AGRICULTURE

UNITED STATES FOREST SERVICE
EASTERN REGION 9
310 W. Wisconsin Ave., Suite 580
Milwaukee, WI 53203 USA
Phone: 414-297-3600   Fax: 414-297-3778
Website: www.fs.fed.us.

Founded: NA
Scope: Statewide

Contact(s):
Robert Jacobs, REGIONAL FORESTER

UNITED STATES DEPARTMENT OF AGRICULTURE

UNITED STATES FOREST SERVICE
INTERMOUNTAIN REGION 4
Federal Office Bldg. 324, 25th St.
Ogden, UT 84401 USA
Phone: 801-625-5605   Fax: 801-625-5127
Website: www.fs.fed.us/r4

Founded: NA
Scope: Regional

Contact(s):
Jack Blackwell, REGIONAL FORESTER

UNITED STATES DEPARTMENT OF AGRICULTURE

UNITED STATES FOREST SERVICE
NORTH CENTRAL FOREST EXPERIMENT STATION
1992 Folwell Ave.
St. Paul, MN 55108 USA
Phone: 651-649-5249   Fax: 651-649-5256

Founded: NA

Contact(s):
Linda Donoghue, DIRECTOR

UNITED STATES DEPARTMENT OF AGRICULTURE

UNITED STATES FOREST SERVICE
NORTHEASTERN RESEARCH STATION
100 Matsonford Rd., 5 Radnor Corporate Center, Suite 200
Radnor, PA 19087-4585 USA
Phone: 610-975-4017

Founded: NA

Contact(s):
Bob Eav, DIRECTOR

UNITED STATES DEPARTMENT OF AGRICULTURE

UNITED STATES FOREST SERVICE
NORTHERN REGION 1
200 E. Broadway, P.O. Box 7669
Missoula, MT 59807 USA
Phone: 406-329-3316   Fax: 406-329-3411
Website: www.fs.fed.us

Founded: NA
Scope: National

Contact(s):
Brad Powell, REGIONAL FORESTER

UNITED STATES DEPARTMENT OF AGRICULTURE

UNITED STATES FOREST SERVICE
PACIFIC NORTHWEST REGION 6
333 SW 1st Ave.
Portland, OR 97208 USA
Phone: 503-808-2200
Website: www.fs.fed.us

Founded: NA
Scope: Regional

Contact(s):
Harvey Forsgren, REGIONAL FORESTER

UNITED STATES DEPARTMENT OF AGRICULTURE

UNITED STATES FOREST SERVICE
PACIFIC NORTHWEST RESEARCH STATION
333 SW 1st Ave.
Portland, OR 97208-3890 USA
Phone: 503-808-2592
Website: http://www.fs.fed.us.pnw

Founded: NA

Contact(s):
Thomas Mills, DIRECTOR

UNITED STATES DEPARTMENT OF AGRICULTURE

UNITED STATES FOREST SERVICE
PACIFIC SOUTHWEST REGION 5
Mare Island, 1323 Club Dr.
Vallejo, CA 94592 USA
Phone: 707-562-9000
Website: www.r5.fs.fed.us

Founded: NA

Contact(s):
Bradley Powell, REGIONAL FORESTER

UNITED STATES DEPARTMENT OF AGRICULTURE

UNITED STATES FOREST SERVICE
PACIFIC SOUTHWEST RESEARCH STATION
800 Buchanan St., West Annex Building
Albany, CA 94710-0011 USA
Phone: 510-559-6310   Fax: 510-559-0440
E-mail:
mailroom_pacificsouthwest_research_station@fs.fed.us
Website: www.psw.fs.fed.us

Founded: NA
Scope: National

Publication(s): see publications on website

Contact(s):
Garland Mason, (ACTING) DIRECTOR

UNITED STATES DEPARTMENT OF AGRICULTURE

UNITED STATES FOREST SERVICE
ROCKY MOUNTAIN RESEARCH STATION
2150 Centre Ave., Bldg A
Suite 376
Ft. Collins, CO 80526-2098 USA
Phone: 970-295-5926   Fax: 970-295-5927

Founded: NA
Membership: 450
Scope: National

Contact(s):
Marcia Patton-Mallory, (ACTING) DIRECTOR

UNITED STATES DEPARTMENT OF AGRICULTURE

UNITED STATES FOREST SERVICE
SOUTHERN REGION 8
1720 Peachtree Rd., NW,
Atlanta, GA 30309 USA
Phone: 404-347-4178   Fax: 404-347-4821

Founded: NA
Scope: Regional

Contact(s):
Elizabeth Estill, REGIONAL FORESTER

UNITED STATES DEPARTMENT OF AGRICULTURE

UNITED STATES FOREST SERVICE
SOUTHWESTERN REGION 3
517 Gold Ave., SW
Albuquerque, NM 87102 USA
Phone: 505-842-3300   Fax: 505-842-3800

Founded: NA

Contact(s):
Eleanor Towns, REGIONAL FORESTER

UNITED STATES DEPARTMENT OF AGRICULTURE FOREST SERVICE

UNITED STATES FOREST SERVICE
ROCKY MOUNTAIN REGION 2
P.O. Box 25127
Lakewood, CO 80225 USA
Phone: 303-275-5450

Founded: NA
Membership: 250
Scope: Regional

Contact(s):
Rick Cables, REGIONAL FORESTER

UNITED STATES DEPARTMENT OF AGRICULTURE NATURAL RESOURCES CONSERVATION SERVICES

NATURAL RESOURCES CONSERVATION SERVICE
USDA, 14th and Independence Ave., SW, P.O. Box 2890
Washington, DC 20013 USA
Phone: 202-720-3210   Fax: 202-720-1564
Website: www.nrcs.usda.gov

Founded: 1935
Membership: 175

Scope: National

Description: (formerly Soil Conservation Service) NRCS has national responsibility for helping America's farmers, ranchers, and other private landowners develop and carry out voluntary efforts to conserve and protect our natural resources. NRCS is the technical delivery arm for conservation of the United States Department of Agriculture. It provides technical assistance and conservation programs through a unique partnership with America's soil and water conservation districts and state conservation agencies.

Contact(s):
Sonja Coderre, AR PUBLIC AFFAIRS SPECIALIST
Federal Office Bldg., Rm. 5404, 700 W. Capitol Ave., Little Rock, AR 72201-3228
Phone: 501-301-3133
Fax: 501-301-3189
scoderre@ar.usda.gov
Mary McQuinn, AZ PUBLIC AFFAIRS SPECIALIST
3003 N. Central Ave., Suite 800, Phoenix, AZ 85012-2945
Phone: 602-280-8778
Fax: 602-280-8809
mmcquinn@az.nrcs.usda.gov
Anita Brown, CA PUBLIC AFFAIRS SPECIALIST
2121-C 2nd St., Suite 102, Davis, CA 95616-5475
Phone: 530-792-5644
Fax: 530-792-5791
anita.brown@ca.usda.gov
Petra Barnes, CO PUBLIC AFFAIRS SPECIALIST
655 Parfet St., Rm. E 200C, Lakewood, CO 80215-5517
Phone: 303-236-2886
Fax: 303-236-2896
pbarnes@co.nrcs.usda.gov
Carol Donzella, CT PUBLIC AFFAIRS SPECIALIST
344 Merrow Rd, Tolland, CT 06084
Phone: 203-787-0390
caroldonzella@ct.usda.gov
Paul Petrichenko, DE PUBLIC AFFAIRS SPECIALIST
1203 College Park Dr., Suite 101, Dover, DE 19904-8713
Phone: 302-678-4178
Fax: 302-678-0843
ppetrichencko@de.nrcs.usda.gov
Terry Bish, DIRECTOR, CONSERVATION COMMUNICATIONS STAFF
Rm. 6121-S, Washington, DC 20013
Dorothy Staley, FL PUBLIC AFFAIRS SPECIALIST
2614 NW, 43rd St., Gainesville, FL 32606-6611
Phone: 352-338-9565
Fax: 352-338-9574
dstaley@fl.nrcs.usda.gov
Lynn Howell, HI PUBLIC AFFAIRS SPECIALIST
300 Ala Moana Blvd., Rm. 4316, Honolulu, HI 96850-0002
Phone: 808-541-2600
Fax: 808-541-2652
lhowell@hi.nrcs.usda.gov
Lynn Betts, IA PUBLIC AFFAIRS SPECIALIST
693 Federal Bldg., 210 Walnut St., Des Moines, IA 50309-2180
Phone: 515-284-4262
Fax: 515-284-4394
lynn.betts@ia.nrcs.usda.gov
Sharon Norris, ID PUBLIC AFFAIRS SPECIALIST
3244 Elder St., Rm. 124, Boise, ID 83705-4711
Phone: 208-378-5725
Fax: 208-378-5735
snorris@id.nrcs.usda.gov

Paige Buck, IL PUBLIC AFFAIRS SPECIALIST
1902 Fox Dr., Champaign, IL 61820-7335
Phone: 217-398-5273
Fax: 217-398-5310
paige.mitchell@il.nrcs.usda.gov
Michael Mcgovern, IN PUBLIC AFFAIRS SPECIALIST
6013 Lakeside Blvd., Indianapolis, IN 46278-2933
Phone: 317-290-3222
Fax: 317-290-3225
mmcgover@in.nrcs.usda.gov
Mary Schaffer, KS PUBLIC AFFAIRS SPECIALIST
760 S. Broadway, Salina, KS 67401
Phone: 785-823-4571
Fax: 785-823-4540
mary.shaffer@ks.nrcs.usda.gov
Lois Jackson, KY PUBLIC AFFAIRS SPECIALIST
771 Corporate Dr., Suite 110, Lexington, KY 40503-5479
Phone: 606-224-7372
Fax: 606-224-7399
ljackson@kystate.ky.nrcs.usda.gov
Herb Bourque, LA PUBLIC AFFAIRS SPECIALIST
3737 Government St., Alexandria, LA 71302-3727
Phone: 318-473-7762
Fax: 318-473-7682
hbourque@laso2.la.nrcs.usda.gov
Wendi Kroll, MA PUBLIC AFFAIRS SPECIALIST
451 West St., Amherst, MA 01002-2995
Phone: 413-253-4351
Fax: 413-253-4375
wkroll@ma.nrcs.usda.gov
Carol Hollingsworth, MD PUBLIC AFFAIRS SPECIALIST
John Hanson Business Center, 339 Busch's Frontage Rd.,
Suite 30, Annapolis, MD 21401-5534
Phone: 410-757-0861
Fax: 410-757-0687
carol.hollingsworth@md.usda.gov
Elaine Tremble, ME PUBLIC AFFAIRS SPECIALIST
5 Godfrey Dr., Orono, ME 04473
Phone: 207-866-7241
Fax: 207-866-7262
etremble@me.nrcs.usda.gov
Christina Coulon, MI PUBLIC AFFAIRS SPECIALIST
1405 S. Harrison Rd., Rm. 101, East Lansing, MI 48823-5243
Phone: 517-337-6701
Fax: 517-337-6905
ccoulon@miso.mi.nrcs.usda.gov
Sylvia Rainford, MN PUBLIC AFFAIRS SPECIALIST
600 Farm Credit Services Bldg., 375 Jackson St., St. Paul,
MN 55101-1854
Phone: 612-602-7859
Fax: 612-602-7914
sr@mn.nrcs.usda.gov
Norm Klopfenstein, MO PUBLIC AFFAIRS SPECIALIST
Parkade Center, Suite 250, 601 Business Loop, 70 West,
Columbia, MO 65203-2546
Phone: 573-876-0911
Fax: 573-876-0913
normk@mo.nrcs.usda.gov
Jeanine May, MS PUBLIC AFFAIRS SPECIALIST
Federal Bldg., Suite 1321, 100 W. Capitol St., Jackson, MS
39269-1399
Phone: 601-965-4337
Fax: 601-965-4536
jbm@ms.nrcs.usda.gov

Lori Valdez, MT PUBLIC AFFAIRS SPECIALIST
Federal Bldg., Rm. 443, 10 E. Babcock St., Bozeman, MT
59715-4704
Phone: 406-587-6842
Fax: 406-587-6761
lvaladez@mt.nrcs.usda.gov
Andrew Smith, NC PUBLIC AFFAIRS SPECIALIST
4405 Bland Rd., Suite 205, Raleigh, NC 27609-6293
Phone: 919-873-2107
Fax: 919-873-2156
asmith@nc.nrcs.usda.gov
Arlene Deutscher, ND PUBLIC AFFAIRS SPECIALIST
Federal Bldg., 220 E. Rosser Ave., Rm. 278, 220 E. Rosser
Ave., Bismarck, ND 58502-1458
Phone: 701-250-4768
Fax: 701-250-4778
ajd@nd.nrcs.usda.gov
Pat McGrane, NE PUBLIC AFFAIRS SPECIALIST
Federal Bldg., Rm. 152, 100 Centennial Mall, N., Lincoln, NE
68508-3866
Phone: 402-437-5328
Fax: 402-437-5327
pat.mcgrane@ne.usda.gov
Laura Morton, NH PUBLIC AFFAIRS SPECIALIST
Federal Bldg 2 Madbury Rd, Durham, NH 03824
Phone: 732-246-1171
Irene Lieberman, NJ PUBLIC AFFAIRS SPECIALIST
1370 Hamilton St., Somerset, NJ 08873-3157
Phone: 732-246-1171
Fax: 732-246-2358
ilieberman@nj.nrcs.usda.gov
Betty Joubert, NM PUBLIC AFFAIRS SPECIALIST
6200 Jefferson St., NE, Suite 305, Albuquerque, NM 87109-
3734
Phone: 505-761-4406
Fax: 505-761-4463
betty.joubert@nm.usda.gov
Liz Warner, NV PUBLIC AFFAIRS SPECIALIST
5301 Langley Ln., Bldg. F, Suite 201, Reno, NV 89511
Phone: 775-784-5288
Fax: 702-784-5939
ewarner@nv.nrcs.usda.gov
Kathy Carpenter, NY PUBLIC AFFAIRS SPECIALIST
441 S Salilna St., Ste. 354, Syracuse, NY 13202
Phone: 315-477-6524
kathy.carpenter@ny.nrcs.usda.gov
Dwain Phillips, OK PUBLIC AFFAIRS SPECIALIST
100 USDA Agriculture Center Bldg., Suite 203, Stillwater, OK
74074-2624
Phone: 405-742-1243
Fax: 405-742-1201
dwain.phillip@ok.usda.gov
Gayle Norman, OR PUBLIC AFFAIRS SPECIALIST
101 SW Main Street, Suite 1300, Portland, OR 97204-3221
Phone: 503-414-3236
Fax: 503-414-3101
gnorman@or.nrcs.usda.gov
Stacy Mitchell, PA PUBLIC AFFAIRS SPECIALIST
One Credit Union Pl., Suite 340, Harrisburg, PA 17110-2993
Phone: 717-237-2208
Fax: 717-237-2238
smitchell@pa.nrcs.usda.gov
Marie Mundheim, PAC BAS PUBLIC AFFAIRS SPECIALIST
Phone: 671-472-7490
Fax: 671-472-7298
pacbas@ite.net

Joyce Hawkins, PROGRAM ASSISTANT
Rm. 6121-S, Washington, DC 20013
Phone: 202-720-3210
Fax: 202-720-1564
joyce.hawkins@usda.gov
Becky Fraticelli, PUERTO RICO PUBLIC AFFAIRS
SPECIALIST
IBM Bldg., 6th Flr., 654 Munoz Rivera Ave., Hato Rey, PR
00918-7013
Phone: 787-766-5206
Fax: 787-766-5987
becky@pr.nrcs.usda.gov
Jeanne Comerford, RI PUBLIC AFFAIRS SPECIALIST
60 Quaker Ln., Suite 46, Warwick, RI 02886-0111
Phone: 401-828-1300
Fax: 401-828-0433
jcomerford@ri.nrcs.usda.gov
Joyce Watkins, SD PUBLIC AFFAIRS SPECIALIST
200 4th St., SW, Federal Bldg., Huron, SD 57350-2475
Phone: 605-352-1228
Fax: 605-352-1261
joyce.watkins@sdso1.sd.nrcs.usda.gov
Larry Blick, TN PUBLIC AFFAIRS SPECIALIST
675 U.S. Courthouse, 801 Broadway St., Nashville, TN
37203-3878
Phone: 615-736-5490
Fax: 615-736-7764
lblick@tn.nrcs.usda.gov
Harold Bryant, TX PUBLIC AFFAIRS SPECIALIST
W.R. Poage Federal Bldg., 101 S. Main St., Temple, TX
76501-7682
Phone: 254-742-9811
Fax: 254-742-9819
hbryant@tx.nrcs.usda.gov
Pat Paul, VA PUBLIC AFFAIRS SPECIALIST
Culpeper Bldg., 1606 Santa Rosa Rd., Suite 209, Richmond,
VA 23229-5014
Phone: 804-287-1681
Fax: 804-287-1737
ppaul@va.nrcs.usda.gov
Anne Hillard, VT PUBLIC AFFAIRS SPECIALIST
69 Union St., Winooski, VT 05404-1999
Phone: 802-951-6796
Fax: 802-951-6327
ahillard@vt.nrcs.usda.gov
Chris Bieker, WA PUBLIC AFFAIRS SPECIALIST
Rock Pointe Tower 2, Suite 450, West 316 Boone Ave.,
Spokane, WA 99201-2348
Phone: 509-323-2912
Fax: 509-323-2909
cbieker@wa.nrcs.usda.gov
Renae Anderson, WI PUBLIC AFFAIRS SPECIALIST
6515 Watts Rd., Suite 200, Madison, WI 53719-2726
Phone: 608-276-8732
Fax: 608-276-5890
randerso@wi.nrcs.usda.gov
Peg Reese, WV PUBLIC AFFAIRS SPECIALIST
75 High St., Rm. 301, Morgantown, WV 26505
Phone: 304-291-4152
Fax: 304-291-4628
preese@wv.nrcs.usda.gov
Nancy Atkinson, WY PUBLIC AFFAIRS SPECIALIST
Federal Office Bldg., 100 East B St., Room 3124, Casper,
WY 82601
Phone: 307-261-6482
Fax: 307-261-6490
nla@wy.nrcs.usda.gov

UNITED STATES DEPARTMENT OF AGRICULTURE, NATIONAL WILDLIFE RESEARCH CENTER

ANIMAL AND PLANT HEALTH INSPECTION SERVICE
4101 LaPorte Ave., Fort Collins, CO 80521-2154 USA
Phone: 970-266-6000   Fax: 970-266-6032
Website: www.aphis.usda.gov/ws/nwrc

Founded: NA
Scope: National

Contact(s):
Richard Curnow, DIRECTOR

UNITED STATES DEPARTMENT OF COMMERCE

Herbert C. Hoover Bldg., Rm. 5610, 14th St. and
Constitution Ave., NW
Washington, DC 20230 USA
Phone: 202-219-3605   Fax: 202-482-5168
Website: www.doc.gov

Founded: NA
Scope: International

Description: The Department of Commerce promotes job
creation, economic growth, sustainable development, and
improved living standards for all Americans, by working in
partnership with business, universities, communities, and
workers.

Contact(s):
Robert Mallett, DEPUTY SECRETARY
Donald Evans, SECRETARY
Phone: 202-482-4883

UNITED STATES DEPARTMENT OF COMMERCE

ECONOMIC DEVELOPMENT ADMINISTRATION
Department of Commerce, Herbert C. Hoover Bldg.,
14th St. and Constitution Ave., NW
Washington, DC 20230 USA
Phone: 202-482-5081   Fax: 202-273-4781
E-mail: edawebmaster@eda.eoc.gov
Website: www.doc.gov/eda

Founded: NA
Scope: National

Description: Conducts programs to help stimulate private
enterprise and create permanent jobs in economically
distressed areas of the Nation. Provides public works grants
and planning and technical assistance in areas with high
unemployment or low median family income.

Contact(s):
David Sampson, ASSISTANT SECRETARY

UNITED STATES DEPARTMENT OF COMMERCE

NATIONAL ENVIRONMENTAL SATELLITE, DATA, AND
INFORMATION SERVICE
1335 East-West Highway
Silver Spring, MD 20910-3284 USA
Phone: 301-713-3578   Fax: 301-713-1249
Website: www.noah.gov

Founded: NA
Scope: State

Description: Manages satellites which observe the natural variability of the global Earth systems - the ocean, atmosphere, features of the solid earth, and the near-space system.

Contact(s):
Gregory Withee, ASSISTANT ADMINISTRATOR
Phone: 301-713-3578
Fax: 301-713-1249
greg.withee@noaa.gov

UNITED STATES DEPARTMENT OF COMMERCE
NATIONAL OCEANIC AND ATMOSPHERIC ADMINISTRATION
Department of Commerce, Herbert C. Hoover Bldg., Rm. 5128, 14th and Constitution Ave., NW
Washington, DC 20230 USA
Phone: 202-482-3436   Fax: 202-408-9674
Website: www.noaa.gov

Founded: 1970
Membership: 40
Scope: Regional

Description: NOAA was created within the Department of Commerce to promote global environmental stewardship and to describe and predict changes in the Earth's environment. NOAA conducts oceanic and atmospheric research; maintains environmental databases and disseminates environmental information products; manages living marine resources and the marine environment; and operates environmental satellites, ships, aircraft, and buoys. NOAA provides the environmental information, science, technology, and resource management expertise necessary for our nation to build a future sustained by both environmental stewardship and economic growth.

Contact(s):
Scott Gudes, DEPUTY UNDER SECRETARY
Phone: 202-482-4569

UNITED STATES DEPARTMENT OF COMMERCE
NATIONAL OCEANIC AND ATMOSPHERIC ADMINISTRATION
NATIONAL MARINE FISHERIES SERVICE
Silver Spring Metro Center 3, 1315 East-West Hwy.
Silver Spring, MD 20910 USA
Phone: 301-713-2239   Fax: 301-703-1940
Website: www.nmfs.noaa.gov

Founded: NA
Scope: International

Description: Provides management, research, and services for the protection and rational use of living marine resources for their aesthetic, economic, and recreational value. Determines the consequences of the natural environment and human activities on living marine resources and provides knowledge and services to achieve efficient and judicious domestic and international management, use, and conservation of the resources.

Publication(s): Publications on line

Contact(s):
William Hogarth, ASSISTANT ADMINISTRATOR

UNITED STATES DEPARTMENT OF COMMERCE
NATIONAL OCEANIC AND ATMOSPHERIC ADMINISTRATION
NATIONAL OCEAN SERVICE
1305 East-West Highway
Silver Spring, MD 20910 USA
Phone: 301-713-3074   Fax: 301-713-4269
Website: www.noaa.gov

Founded: NA
Scope: International

Description: Administers the National Geodetic Survey, Nautical and Aeronautical Charting, National Estuarine Research Reserves, National Marine Sanctuaries, Coastal Zone Management, Marine Assessments, and Coastal Ocean Programs.

Contact(s):
Nancy Foster, ASSISTANT ADMINISTRATOR
Phone: 301-713-3074
Dan Dewell, PUBLIC AFFAIRS OFFICER
Phone: 301-713-3070

UNITED STATES DEPARTMENT OF COMMERCE
NATIONAL OCEANIC AND ATMOSPHERIC ADMINISTRATION
NATIONAL WEATHER SERVICE
Silver Spring Metro Center 2, 1325 East-West Hwy.
Silver Spring, MD 20910 USA
Phone: 301-713-0689   Fax: 301-713-0662
Website: www.nws.noaa.gov

Founded: NA
Scope: Local, Regional

Description: Observes, describes, and predicts the natural variability of the atmosphere, and to some extent, the ocean and the earth, in order to protect life and property and enhance the national economy.

Contact(s):
Louis Boezi, DEPUTY ASSISTANT ADMINISTRATOR FOR MODERNIZATION
Phone: 301-713-0397
Susan Zevin, DEPUTY ASSISTANT ADMINISTRATOR FOR OPERATIONS
Phone: 301-713-0711
John Kelly, DIRECTOR AND ASSISTANT ADMINISTRATOR FOR WEATHER SERVICE
Ronald McPherson, DIRECTOR OF NATIONAL CENTERS FOR ENVIRONMENTAL PREDICTION
Phone: 301-713-8016
Jerry McCall, DIRECTOR OF NATIONAL DATA BUOY CENTER
Phone: 601-688-2800
Robert Burpee, DIRECTOR OF NATIONAL HURRICANE CENTER
Phone: 305-229-4470
Frederick Ostby, DIRECTOR OF SEVERE STORM FORECAST CENTER
Phone: 816-426-5922
Richard Hutcheon, NATIONAL WEATHER SERVICE ALASKA REGION
Phone: 907-271-5136
Richard Augulis, NATIONAL WEATHER SERVICE CENTRAL REGION
Phone: 816-426-5400

John Forsing, NATIONAL WEATHER SERVICE EASTERN REGION
Phone: 516-244-0100
Richard Hagemeyer, NATIONAL WEATHER SERVICE PACIFIC REGION
Phone: 808-541-1641
Thomas Potter, NATIONAL WEATHER SERVICE WESTERN REGION
Phone: 801-524-5122
Randee Exter, PUBLIC AFFAIRS OFFICER OF WEATHER
Phone: 301-713-0622
Harry Hassel, SOUTHERN REGION
Phone: 817-334-2651

UNITED STATES DEPARTMENT OF COMMERCE
OFFICE OF OCEANIC AND ATMOSPHERIC RESEARCH
Silver Spring Metro Center 3, 1315 East-West Hwy.
Silver Spring, MD 20910 USA
Phone: 301-713-2458   Fax: 301-713-0163
Website: www.oar.noah.gov

Founded: NA
Membership: 200
Scope: International

Description: Conducts environmental research in the oceans, atmosphere, and space. Administers the National Sea Grant College Program, which provides grants to academic institutions for research, education, and advisory/extension services in the marine environment.

Contact(s):
David Evans, ASSISTANT ADMINISTRATOR
David.Evans@noaa.gov
Louisa Koch, DEPUTY ASSISTANT ADMINISTRATOR
Ronald Baird, DIRECTOR OF NATIONAL SEA GRANT COLLEGE PROGRAM OF EXTENSION SERVICE (ACTING)
Phone: 301-713-2448
Barbara Moore, DIRECTOR OF NATIONAL UNDERSEA RESEARCH PROGRAM
Phone: 301-713-2427
Dane Konop, PUBLIC AFFAIRS OFFICER
Phone: 301-713-2483
Maryann Whitcomb, RESOURCE MANAGEMENT
Phone: 301-713-2454

UNITED STATES DEPARTMENT OF DEFENSE
The Pentagon, Office of the Secretary, 3400 Defense
Pentagon
Washington, DC 20301-3400 USA
Phone: 703-697-1013   Fax: 703-693-7011
Website: www.denix.osd.mil

Founded: NA
Scope: International

Description: Responsible for the security of the U.S. by establishing policies and procedures relating to national defense. The Department of Defense conducts programs to prevent pollution, enhance the environment, and conserve the natural and cultural resources on military lands.

Publication(s): Natural Resources in the Department of Defense, Cultural Resources in the Department of Defense, Legacy Resource Management Program Report to Congress, DOD Commanders' Guide to Biodiversity

Contact(s):
Sarah Hagan, KEY CONTACT

UNITED STATES DEPARTMENT OF EDUCATION
400 Maryland Ave., SW
Washington, DC 20202-0498 USA
Phone: 202-401-3000
E-mail: customerservice@inet.ed.gov
Website: www.ed.gov

Founded: NA

Contact(s):
Roderick Paige, SECRETARY
Phone: 202-401-3000

UNITED STATES DEPARTMENT OF ENERGY
Forrestal Bldg., 1000 Independence Ave., SW
Washington, DC 20585 USA
Phone: 202-586-5000   Fax: 202-586-5049
Website: http://www.energy.gov/

Founded: NA

Description: Provides the framework for a comprehensive and balanced national energy strategy through the coordination and administration of the energy functions of the federal government. The department is responsible for research, development, and demonstration of energy technology; the marketing of federal power; energy conservation programs; the nuclear weapons program; energy regulatory programs; and a central energy data collection and analysis program. Established by the Department of Energy Organization Action: 1977.

Publication(s): National Energy Strategy, National Energy Strategy: One Year Later, Report to the Congress of the United States: Limiting New Greenhouse Gas Emissions in the United States, Assessment of Costs and Benefits of Flexible and Alternative Fuel Use in the United States Transportation Sector

UNITED STATES DEPARTMENT OF ENERGY
CARBON DIOXIDE INFORMATION ANALYSIS CENTER
Oak Ridge National Laboratory, P.O. Box 2008 MS-6335
Oak Ridge, TN 37831-6335 USA
Phone: 865-574-0390   Fax: none
Website: cdiac.esd.ornl.gov

Founded: 1982
Membership: 15
Scope: International

Description: The Carbon Dioxide Information Analysis Center (CDIAC) provides data and information support for the United States Department of Energy's global change research program and makes these data and information products available to a multidisciplinary community of researchers, policymakers, and educators at no cost.

Publication(s): Annual Report, Trends Online, CDIAC Communications

Contact(s):
Robert Cushman, DIRECTOR
Phone: 423-574-4791
Sonja Jones, USER SERVICES
Phone: 423-574-3645

UNITED STATES DEPARTMENT OF ENERGY
FEDERAL ENERGY REGULATORY COMMISSION
888 First St., NE
Washington, DC 20426 USA
Phone: 202-208-1088
Website: www.SERC.GOVE

Founded: 1977
Membership: 1000+
Scope: REGIONAL

Description: The Federal Energy Regulatory Commission regulates the interstate aspects of the electric power and natural gas industries and establishes rates for transporting oil by pipeline. The Commission issues and enforces licenses for construction and operation of nonfederal hydroelectric power projects. The FERC also advises federal agencies on the merits of proposed federal multiple-purpose water development projects.

Contact(s):
James Hoecker, CHAIR
Phone: 202-208-0000
Curtis Wagner, CHIEF ADMINISTRATIVE LAW JUDGE
Phone: 202-219-2500
Thomas Herlihy, CHIEF INFORMATION OFFICER (ACTING)
Phone: 202-208-1055
Curtis Herbert, COMMISSIONER
Phone: 202-208-0601
William Massey, COMMISSIONER
Phone: 202-208-0366
Vicky Bailey, COMMISSIONER
Phone: 202-208-0388
Linda Breathitt, COMMISSIONER
Phone: 202-208-0377
Thomas Herlihy, DIRECTOR AND CHIEF FINANCIAL OFFICER OF FINANCE, ACCOUNTING, AND OPERATIONS
Phone: 202-208-0300
Richard O'Neill, DIRECTOR OF ECONOMIC POLICY
Phone: 202-208-0100
Shelton Cannon, DIRECTOR OF ELECTRIC POWER REGULATION
Phone: 202-208-1200
Carol Sampson, DIRECTOR OF HYDROPOWER LICENSING
Phone: 202-219-2700
Virginia Strasser, DIRECTOR OF OFFICE OF ADMINISTRATIVE LITIGATION
Phone: 202-219-2600
Rebecca Schaffer, DIRECTOR OF OFFICE OF EXTERNAL AFFAIRS
Phone: 202-208-0004
Kevin Madden, DIRECTOR OF PIPELINE REGULATION
Phone: 202-208-0700
Douglas Smith, GENERAL COUNSEL
Phone: 202-208-1000
David Boergers, SECRETARY
Phone: 202-208-0400

UNITED STATES DEPARTMENT OF HEALTH AND HUMAN SERVICES
200 Independence Ave., SW
Washington, DC 20201 USA
Phone: 202-690-7000   Fax: 202-690-7203
Website: www.hhs.gov

Founded: NA
Scope: National

Description: The Department of Health and Human Services is the United States government's principal agency for protecting the health of all Americans and providing essential human services, especially for those who are least able to help themselves.

Contact(s):
William Corr, CHIEF OF STAFF
Phone: 202-690-7431
Tommy Thompson, SECRETARY
Phone: 202-690-7000

UNITED STATES DEPARTMENT OF HEALTH AND HUMAN SERVICES
FOOD AND DRUG ADMINISTRATION
5600 Fishers Ln.
Rockville, MD 20857 USA
Phone: 410-433-1544
Website: www.fda.gov

Founded: NA

Description: Protects the health of American consumers by enforcing federal laws which require that foods must be safe, pure, and wholesome; human and veterinary drugs, biologies, and therapeutic devices must be safe and effective; cosmetics and radiation-emitting products must be harmless; and that all these products must be honestly and informatively labeled and packaged.

Contact(s):
R. Grant, ASSOCIATE COMMISSIONER FOR CONSUMER AFFAIRS
Phone: 410-443-5006
Stuart Nightingale, ASSOCIATE COMMISSIONER FOR HEALTH AFFAIRS
Phone: 410-433-6143
Dianne Thompson, ASSOCIATE COMMISSIONER FOR LEGISLATIVE AFFAIRS
Phone: 410-443-3793
Sharon Holston, ASSOCIATE COMMISSIONER FOR MANAGEMENT AND OPERATIONS
Phone: 410-443-3370
Paul Coppinger, ASSOCIATE COMMISSIONER FOR PLANNING AND EVALUATION
Phone: 410-433-4230
James O'Hara, ASSOCIATE COMMISSIONER FOR PUBLIC AFFAIRS
Phone: 410-443-1130
Ronald Chesemore, ASSOCIATE COMMISSIONER FOR REGULATORY AFFAIRS
Phone: 410-433-1594
Mary Porter, CHIEF COUNSEL FOR OFFICE OF GENERAL COUNSEL
Phone: 410-443-4370
Carol Scheman, DEPUTY COMMISSIONER FOR EXTERNAL AFFAIRS
Phone: 410-443-2400
Mary Veverka, DEPUTY COMMISSIONER FOR MANAGEMENT AND SYSTEMS
Phone: 410-443-1263
Michael Taylor, DEPUTY COMMISSIONER FOR POLICY
Phone: 410-443-2854
D. Burlington, DIRECTOR
Phone: 410-443-4690
Richard Teske, DIRECTOR
Phone: 410-594-1740

Bernard Schwetz, DIRECTOR
Phone: 501-543-7517
Gerald Meyer, DIRECTOR
Phone: 410-443-2894
Fred Shank, DIRECTOR
Phone: 202-205-4850
Kathryn Zoon, DIRECTOR
Phone: 410-496-3556
Mary Danello, DIRECTOR
Phone: 410-443-1565
Randolph Wykoff, DIRECTOR OF AIDS COORDINATION
STAFF
Henry Miller, DIRECTOR OF OFFICE OF BIOTECHNOLOGY
Phone: 410-443-7573
Rosamelia De la Rocha, DIRECTOR OF OFFICE OF EQUAL
EMPLOYMENT AND CIVIL RIGHTS
Phone: 410-443-5541
Joseph Levitt, DIRECTOR OF OFFICE OF EXECUTIVE
OPERATIONS
Phone: 410-443-5004
Marlene Haffner, DIRECTOR OF OFFICE OF ORPHAN
PRODUCTS DEVELOPMENT
Betsy Adams, DIRECTOR OF PRESS RELATIONS FOR
STAFF OF OFFICE OF PUBLIC AFFAIRS
Phone: 410-443-4177
Amanda Pedersen, OMBUDSMAN
Phone: 410-443-1306
Jack Martin, SPECIAL ASSISTANT TO THE
COMMISSIONER FOR PROGRAM POLICY
Phone: 410-443-6776

UNITED STATES DEPARTMENT OF HOUSING AND URBAN DEVELOPMENT

HUD Bldg., 451 7th St., SW
Washington, DC 20410 USA
Phone: 202-708-1600
Website: www.hud.gov

Founded: NA
Scope: National
Contact(s):
John Weicher, ASSISTANT SECRETARY
Phone: 202-708-3600
Michael Liu, ASSISTANT SECRETARY FOR PUBLIC &
INDIAN HOUSING
Phone: 202-708-0950
Jacquie Lawing, DEPUTY CHIEF OF STAFF FOR POLICY
AND PROGRAMS
Phone: 202-708-2236
Carol Jefferson, GENERAL DEPUTY ASSISTANT
SECRETARY FOR ADMINISTRATION
Phone: 202-708-0940
carole_jefferson@hud.gov
Joseph Smith, GENERAL DEPUTY ASSISTANT
SECRETARY FOR ADMINISTRATION
Phone: 202-708-0940
Susan Gaffney, INSPECTOR GENERAL
Phone: 202-708-0430
Mel Martinez, SECRETARY
Phone: 202-708-0417

UNITED STATES DEPARTMENT OF JUSTICE

ENVIRONMENT & NATURAL RESOURCES
10th St. and Constitution Ave., NW
Washington, DC 20530 USA
Phone: 202-514-2701　　　Fax: 202-514-0557
Website: www.doj.gov

Founded: NA
Membership: 700
Scope: International

Description: The Environment and Natural Resources Division
handles litigation involving American's pollution control laws;
central resources laws, the protection and enhancement of the
American environment and wildlife resources; the acquisition,
administration, and disposition of public land, water, and
mineral resources; and the safeguarding of Indian rights and
property.

Contact(s):
James Kilbourne, APPELLATE SECTION CHIEF
Phone: 202-514-2748
John Ashcroft, ATTORNEY GENERAL
Phone: 202-514-2001
Eileen Sobeck, DEPUTY ASSISTANT ATTORNEY GENERAL
Phone: 202-514-0943
John Cruden, DEPUTY ASSISTANT ATTORNEY GENERAL
Phone: 202-514-2718
Letitia Grishaw, ENVIRONMENTAL DEFENSE SECTION
CHIEF
Phone: 202-514-2219
Bruce Gilber, ENVIRONMENTAL ENFORCEMENT SECTION
CHIEF
Robert Bruffy, EXECUTIVE OFFICER
Phone: 202-616-3147
Jack Haugrud, GENERAL LITIGATION SECTION CHIEF
Craig Alexander, INDIAN RESOURCES CHIEF
James Clear, INDIAN RESOURCES SECTION CHIEF
Phone: 202-305-0259
Virginia Butler, LAND ACQUISITION SECTION CHIEF
Phone: 202-305-0316
Pauline Milius, POLICY OF LEGISLATION AND SPECIAL
LITIGATION SECTION CHIEF
Phone: 202-514-2586
Jean Williams, WILDLIFE AND MARINE RESOURCES
SECTION CHIEF
Phone: 202-305-0228

UNITED STATES DEPARTMENT OF LABOR

200 Constitution Ave., NW
Washington, DC 20210 USA
Phone: 202-219-5000
Website: www.dol.gov

Founded: NA

Contact(s):
Edward Montgomery, DEPUTY SECRETARY (ACTING)
Phone: 202-693-6002
J. McAteer, MINE SAFETY AND HEALTH ADMINISTRATOR
Phone: 703-235-1385
Alexis Herman, SECRETARY
Phone: 202-693-6000

UNITED STATES DEPARTMENT OF LABOR
JOB CORPS
Department of Labor, Employment and Training
Administration, Frances Perkins Bldg.
200 Constitution Ave., NW
Washington, DC 20210 USA
Phone: 202-219-8550
Website: http://www.jobcorps.org

Founded: NA

Description: Authorized by the Job Training Partnership Act, the program includes conservation centers known as Civilian Conservation Centers, located primarily in rural areas and operated for the Department of Labor by the departments of Agriculture and Interior conservation agencies. In addition to providing training and other assistance to young people, the programs of work experience, training, and remedial education are focused upon activities to conserve, develop, or manage public resources or public recreational areas or to assist in developing community projects in the public interest.

UNITED STATES DEPARTMENT OF LABOR
MINE SAFETY AND HEALTH ADMINISTRATION
Department of Labor, Ballston Tower 3
Arlington, VA 22203 USA
Phone: 703-235-1452   Fax: 703-235-4323
E-mail: asmsha@msha.gov
Website: www.msha.gov

Founded: NA
Scope: national

Description: Objectives are to administer the Federal Mine Safety and Health Act, thereby promoting safety and health in the mining industry, preventing disasters, and protecting the health and safety of the nation's miners.

Contact(s):
Earnest Teaster, ADMINISTRATOR OF METAL AND NONMETAL MINE HEALTH AND SAFETY
Phone: 703-235-1565
Dave Laurishi, ASSISTANT SECRETARY
Robert Elam, DEPUTY ASSISTANT SECRETARY
Phone: 703-235-9423
Patricia Silvey, DIRECTOR OF ADMINISTRATION AND MANAGEMENT
Phone: 703-235-1383
Jeffrey Duncan, DIRECTOR OF EDUCATIONAL POLICY AND DEVELOPMENT
Phone: 703-235-1515
Katherine Snyder, DIRECTOR OF OFFICE INFORMATION
Carol Jones, METALS & NON METALS CHIEF HEALTH DIVISION
Phone: 703-235-1910
Sylvia Milanese, OFFICE OF CONGRESSIONAL AND LEGISLATIVE AFFAIRS
Phone: 703-235-1392

UNITED STATES DEPARTMENT OF STATE
Harry S. Truman Bldg., 2201 C St., NW
Washington, DC 20520 USA
Phone: 202-647-4000   Fax: 202-736-7720
E-mail: secretary@state.gov
Website: www.state.gov

Founded: NA
Membership: 4,799
Scope: State

Contact(s):
Colin Powell, US SECRETARY OF STATE
Phone: 202-647-6575
secretary@state.gov

UNITED STATES DEPARTMENT OF STATE
BUREAU OF OCEANS AND INTERNATIONAL
ENVIRONMENTAL AND SCIENTIFIC AFFAIRS
Department of State, 2201 C St., NW
Washington, DC 20520 USA
Phone: 202-647-3004   Fax: 202-647-0217
Website: http://www.state.gov/g/oes/

Founded: NA

Description: OES has the principal responsibility for formulating and implementing U.S. policies for oceans, environmental, scientific, and technological aspects of U.S. relations with other governmental and multilateral institutions. The Bureau's activities cover a broad range of foreign policy issues relating to environment, pollution, tropical forests, biological diversity, wildlife, oceans policy, fisheries, global climate change, atmospheric ozone-depletion, space, and advanced technologies.

Contact(s):
Kenneth Brill, ACTING ASSISTANT SECRETARY
Phone: 202-647-1554
Rafe Pomerance, DEPUTY ASSISTANT SECRETARY OF ENVIRONMENT AND DEVELOPMENT (OES/E)
Phone: 202-647-2232
Mary West, DEPUTY ASSISTANT SECRETARY OF OCEANS, FISHERIES AND SPACE (OES/O)
Phone: 202-647-2396
Stephanie Kinney, EXECUTIVE ASSISTANT/EXECUTIVE DIRECTOR OF ADMINISTRATION
Phone: 202-647-3622
Mary Mclead, OFFICE OF ECOLOGY AND TERRESTRIAL CONSERVATION (OES/ETC) DIRECTOR
Phone: 202-647-2418
Nancy Foster, OFFICE OF EMERGING INFECTIOUS DISEASES (OES/EID) DIRECTOR
Phone: 202-647-2435
Michael Mtelits, OFFICE OF ENVIRONMENT POLICY (OES/ENV) DIRECTOR
Phone: 202-647-9266
Daniel Reifsnyder, OFFICE OF GLOBAL CHANGE (OES/EGC) DIRECTOR
Phone: 202-647-4069
David Balton, OFFICE OF MARINE CONSERVATION (OES/OMC) DIRECTOR
Phone: 202-647-2335
R. Scully, OFFICE OF OCEANS AFFAIRS (OES/OA) DIRECTOR
Phone: 202-647-3282
Leslie Gerson, OFFICE OF SCIENCE AND ENVIRONMENTAL INITIATIVE (OES/SCI) DIRECTOR
Phone: 202-647-3625
Ralph Braibanti, OFFICE OF SPACE AND ADVANCED TECHNOLOGY (OES/SAT) DIRECTOR
Phone: 202-647-2433
Melinda Kimble, PRINCIPAL DEPARTMENT ASSISTING SECRETARY
Mark Hambley, SPECIAL NEGOTIATOR
Roger Soles, U.S. MAN AND THE BIOSPHERE PROGRAM (MAB) DIRECTOR
Phone: 703-235-2948

UNITED STATES DEPARTMENT OF STATE

UNITED STATES MAN AND THE BIOSPHERE
PROGRAM (U.S. MAB)
U.S. MAB
USDA-Forest Service
Yates Federak Bldg. (1-NW)
P.O. Box 96090
Washington, DC 20090 USA
Phone: 202-776-8318   Fax: 202-776-8367
E-mail: usmab@state.gov
Website: http://www.mabnet.org

Founded: NA

Description: The mission of the United States Man and the
Biosphere Program (U.S. MAB) is to explore, demonstrate,
promote, and encourage harmonious relationships between
people and their environments, building on the MAB network of
Biosphere Reserves and interdisciplinary research. The long-
term goal of the U.S. MAB Program is to contribute to
achieving a sustainable society early in the 21st century. The
MAB mission and long-term goal will be implemented, in the
U.S. and internationally, through public-private partnerships
and interdisciplinary research, experimentation, education, and
information exchange on options by which societies can
achieve sustainability.

Publication(s): U.S. MAB Bulletin, research reports from U.S.
MAB, " proceedings of symposia, conferences

Contact(s):
David Hales, CHAIRMAN OF U.S. MAB NATIONAL
COMMITTEE
Roger Soles, EXECUTIVE DIRECTOR OF U.S. MAB
Phone: 202-776-8318

UNITED STATES DEPARTMENT OF THE AIR FORCE

AIR EDUCATION AND TRAINING COMMAND
266 F St., West, Bldg. 901
Randolph AFB, TX 78150-4321 USA
Phone: 210-652-3959
Website: http://www.aetc.randolph.af.mil/

Founded: NA

Contact(s):
Carl Lahsher, NATURAL RESOURCES MANAGER (HQ
AETC/CEV)

UNITED STATES DEPARTMENT OF THE AIR FORCE

AIR FORCE CENTER FOR ENVIRONMENTAL
EXCELLENCE
3207 North Rd.
Brooks AFB, TX 78235-5363 USA
Phone: 210-536-3334
Website: http://www.afcee.brooks.af.mil/

Founded: NA

Contact(s):
Mary Anderson, BOTANIST
Phone: 203-536-3808
Edward Bakunas, CHIEF, CONSULTANT DIVISION,
ENVIRONMENTAL CONSERVATION AND PLANNING
(AFCEE/ECC):
Kevin Porteck, FORESTER
Phone: 210-536-5631

Daniel Freeze, NATURAL RESOURCE SPECIALIST
Phone: 210-536-3823

UNITED STATES DEPARTMENT OF THE AIR FORCE

AIR FORCE CIVIL ENGINEERING SUPPORT AGENCY
HQ AFCESA
139 Barnes Dr.
Tyndall AFB, FL 32403-5319 USA
Phone: 850-283-6465   Fax: 850-283-6219

Founded: NA
Membership: 1
Scope: International

Contact(s):
Wayne Fordham, MANAGEMENT AGRONOMIST
(AFCESA/CEM)

UNITED STATES DEPARTMENT OF THE AIR FORCE

BIRD AIRCRAFT STRIKE HAZARD
(BASH) TEAM
9700 Avenue G, Bld. 24499
Kirtland AFB, NM 87117-5671 USA
Phone: 505-846-5674

Founded: NA

Contact(s):
Peter Windler, CHIEF, USAF BASH TEAM (HQ AFSA/SEFW)

UNITED STATES DEPARTMENT OF THE AIR FORCE

MAJOR AIR COMMANDS
(HQ USAFE)
Unit 3050, Box 10
APO, AE, 09094-5010 Germany
Phone: 011-49-6371-47-6482

Founded: NA

Contact(s):
Edwin Worth, NATURAL/CULTURAL RESOURCES
MANAGER (HQ USAFE/CEVP)

UNITED STATES DEPARTMENT OF THE AIR FORCE

MAJOR AIR COMMANDS
AIR COMBAT COMMAND
129 Andrews St., Suite 102, Major Air Commands
Langley AFB, VA 23665-2769 USA
Phone: 757-764-9338
Website: http://www.acc.af.mil/

Founded: NA

Contact(s):
Roy Barker, NATURAL RESOURCES MANAGER
(HQ ACC/CEVA)

UNITED STATES DEPARTMENT OF THE AIR FORCE

MAJOR AIR COMMANDS
AIR FORCE BASE CONVERSION AGENCY (AFBCA)
USA
Phone: 703-696-5500

Founded: NA

Contact(s):
> Jerry Cleaver, CONSERVATION MANGER (HQ AFBCA/EV)
> HQ AFBCA/EV, 1700 N. Moore St., Suite 2300, Arlington, VA 22209-2802
> Phone: 703-696-5536

UNITED STATES DEPARTMENT OF THE AIR FORCE

MAJOR AIR COMMANDS
AIR FORCE DISTRICT OF WASHINGTON
3700 Brookley Ave.
Washington, DC 20332 USA
Phone: 202-767-8600   Fax: 202-767-1160

Founded: NA

Contact(s):
> Mark Dickerson, CHIEF OF ENVIRONMENTAL PLANNING BRANCH (HQ 11WG/CEV)

UNITED STATES DEPARTMENT OF THE AIR FORCE

MAJOR AIR COMMANDS
AIR FORCE MATERIAL COMMAND
4225 Logistics Ave., Rm. A128
Wright Patterson, OH 45433-5747 USA
Phone: 937-656-1409   Fax: 937-587-5875
E-mail: mike.cornelius@wpafb.af.mil

Founded: NA
Scope: National

Contact(s):
> Michael Cornelius, NATURAL RESOURCES MANAGER (HQ AFRES/CEVP)

UNITED STATES DEPARTMENT OF THE AIR FORCE

MAJOR AIR COMMANDS
AIR FORCE RESERVES (AFRES)
155 Richard Ray Blvd.
Robins AFB, GA 31098-1635 USA

Founded: NA

UNITED STATES DEPARTMENT OF THE AIR FORCE

MAJOR AIR COMMANDS
AIR FORCE SPACE COMMAND (AFSPC)
150 Vandenberg St., Suite 1105
Peterson AFB, CO 80914-4150 USA
Phone: 719-554-9915   Fax: 719-554-3810

Founded: NA
Scope: State

Contact(s):
> Stanley Rogers, NATURAL RESOURCES MANAGER (HQ AFSPC/CEVP)
> stanley.rogers@peterson.af.mil

UNITED STATES DEPARTMENT OF THE AIR FORCE

MAJOR AIR COMMANDS
AIR FORCE SPECIAL OPERATIONS COMMAND
HQ/AFSOC/CEV
HQ AFSOC CEV, 427 Cody Ave.
Hurlburt Field, FL 32544 USA
Phone: 850-884-2260   Fax: 850-884-5982

Founded: NA
Scope: International

Contact(s):
> Michael Applegate, NATURAL RESOURCE MANAGER AND ENTOMOLOGIST
> Phone: 850-884-2562

UNITED STATES DEPARTMENT OF THE AIR FORCE

MAJOR AIR COMMANDS
AIR MOBILITY COMMAND (AMC)
HQ AMC/CEVP, 507 Symington Drive
Scott AFB, IL 62225-5022 USA
Phone: 618-229-0842   Fax: 618-229-0257

Founded: NA
Membership: 250
Scope: National

Contact(s):
> William Summers, NATURAL RESOURCES MANAGER (HQ AMC/CEVP)

UNITED STATES DEPARTMENT OF THE AIR FORCE

MAJOR AIR COMMANDS
AIR NATIONAL GUARD (ANG)
3500 Fetchet Ave.
Andrews AFB, MD 20331-5157 USA
Phone: 301-836-8798

Founded: NA

Contact(s):
> Pat Richerson, NATURAL RESOURCES MANAGER (HQ ANG/CEVP)

UNITED STATES DEPARTMENT OF THE AIR FORCE

MAJOR AIR COMMANDS
PACIFIC AIR FORCES (PACAF)
25 E St.
Hickam AFB, HI 96853-5412 USA
Phone: 808-449-0605   Fax: 808-449-4000
E-mail: pacf.csv@exchange.hickam.af.mil
Website: www.hqpacif.af.mil/ce/cevindx/cevindx.htm

Founded: NA
Membership: 19
Scope: National

Contact(s):
> Arthur Buckman, NATURAL RESOURCES MANAGER (HQ PACAF/CEVEP)

UNITED STATES DEPARTMENT OF THE AIR FORCE

MAJOR AIR COMMANDS
U.S. AIR FORCE ACADEMY
8120 Edgerton Dr., Suite 40
USAF Academy, CO 80840-2400 USA
Phone: 719-333-3336   Fax: 719-333-3337
Website: www.usafa.af.mil

Founded: NA
Scope: Local

Contact(s):
Brian Mihlbachler, NATURAL RESOURCE MANAGER (HQ USAFA/CEVP)
Jim McDermott, NATURAL RESOURCE PLANNER

UNITED STATES DEPARTMENT OF THE AIR FORCE

MAJOR U.S. INSTALLATIONS
ALASKAN REMOTE SITES (611 SUPPORT GROUP)
USA

Founded: NA

Contact(s):
Gene Augustine, NATURAL RESOURCES MANAGER
Phone: 907-552-0788

UNITED STATES DEPARTMENT OF THE AIR FORCE

MAJOR U.S. INSTALLATIONS
ALTUS AFB, OK
USA

Founded: NA

Contact(s):
Jim Bellon, NATURAL RESOURCES MANAGER
Phone: 580-481-7606

UNITED STATES DEPARTMENT OF THE AIR FORCE

MAJOR U.S. INSTALLATIONS
ANDERSON AFB, GUAM
USA

Founded: NA

Contact(s):
Heidi Hirsh, NATURAL RESOURCES MANAGER
Phone: 671-366-2549

UNITED STATES DEPARTMENT OF THE AIR FORCE

MAJOR U.S. INSTALLATIONS
ANDREWS AFB, MD
USA

Founded: NA

Contact(s):
Carol Devier-Heemey, CULTURAL/NATURAL RESOURCES MANAGER
Phone: 301-981-2579

UNITED STATES DEPARTMENT OF THE AIR FORCE

MAJOR U.S. INSTALLATIONS
ARNOLD AFB, TN
USA

Founded: NA

Contact(s):
Clark Brandon, NATURAL/CULTURAL RESOURCES MANAGER
Phone: 615-454-7115

UNITED STATES DEPARTMENT OF THE AIR FORCE

MAJOR U.S. INSTALLATIONS
AVON PARK AFB, FL
USA

Founded: NA

Contact(s):
Paul Ebersbach, CHIEF OF CONSERVATION PROGRAMS
Phone: 941-452-7119

UNITED STATES DEPARTMENT OF THE AIR FORCE

MAJOR U.S. INSTALLATIONS
BARKSDALE AFB, LA
USA

Founded: NA

Contact(s):
Bruce Holland, NATURAL RESOURCES MANAGER
Phone: 318-456-1981

UNITED STATES DEPARTMENT OF THE AIR FORCE

MAJOR U.S. INSTALLATIONS
BEALE AFB, CA
USA

Founded: NA

Contact(s):
Kristen Christopherson, NATURAL RESOUCES MANAGER
Phone: 916-634-2643

UNITED STATES DEPARTMENT OF THE AIR FORCE

MAJOR U.S. INSTALLATIONS
BOLLING AFB, WASHINGTON, DC
USA

Founded: NA

Contact(s):
Fioravante Gaetano, NATURAL RESOURCES MANAGER
Phone: 202-767-8603

UNITED STATES DEPARTMENT OF THE AIR FORCE

MAJOR U.S. INSTALLATIONS
BROOKS AFB, TX
USA

Founded: NA

Contact(s):
Hamid Kamalpour, NATURAL RESOURCES MANAGER
Phone: 210-536-6703

UNITED STATES DEPARTMENT OF THE AIR FORCE
MAJOR U.S. INSTALLATIONS
CANNON AFB, NM
USA

Founded: NA

Contact(s):
Rick Crow, CULTURAL/NATURAL RESOURCES MANAGER
Phone: 505-784-6383

UNITED STATES DEPARTMENT OF THE AIR FORCE
MAJOR U.S. INSTALLATIONS
CHARLESTON AFB, SC
USA

Founded: NA

Contact(s):
Al Urrutia, CULTURAL/NATURAL RESOURCES MANAGER
Phone: 843-963-4978

UNITED STATES DEPARTMENT OF THE AIR FORCE
MAJOR U.S. INSTALLATIONS
COLUMBUS AFB, MS
USA

Founded: NA

Contact(s):
Mark Knarr, CULTURAL/NATURAL RESOURCES MANAGER
Phone: 601-434-7315

UNITED STATES DEPARTMENT OF THE AIR FORCE
MAJOR U.S. INSTALLATIONS
DAVIS-MONTHAN AFB, AZ
USA

Founded: NA

Contact(s):
Gwen Lisa, CULTURAL/NATURAL RESOURCES MANAGER
Phone: 520-228-3215

UNITED STATES DEPARTMENT OF THE AIR FORCE
MAJOR U.S. INSTALLATIONS
DOVER AFB, DE
USA

Founded: NA

Contact(s):
Charles Mikula, CULTURAL/NATURAL RESOURCES MANAGER
Phone: 302-677-6820

UNITED STATES DEPARTMENT OF THE AIR FORCE
MAJOR U.S. INSTALLATIONS
EDWARDS AFB, CA
USA

Founded: NA

Contact(s):
Mark Hagan, NATURAL RESOURCES MANAGER
Phone: 805-277-1418

UNITED STATES DEPARTMENT OF THE AIR FORCE
MAJOR U.S. INSTALLATIONS
EGLIN AFB, FL
USA

Founded: NA

Contact(s):
Rick McWhite, NATURAL RESOURCES MANAGER
Phone: 850-822-4164

UNITED STATES DEPARTMENT OF THE AIR FORCE
MAJOR U.S. INSTALLATIONS
EIELSON AFB, AK
USA

Founded: NA

Contact(s):
Gerald Von Rueden, NATURAL RESOURCES MANAGER
Phone: 907-377-5182

UNITED STATES DEPARTMENT OF THE AIR FORCE
MAJOR U.S. INSTALLATIONS
ELLSWORTH AFB, SD
USA

Founded: NA

Contact(s):
Jim Stengler, CULTURAL/NATURAL RESOURCES MANAGER
Phone: 605-385-6677

UNITED STATES DEPARTMENT OF THE AIR FORCE
MAJOR U.S. INSTALLATIONS
ELMENDORF AFB, AK
USA

Founded: NA

Contact(s):
Alan Richmond, NATURAL RESOURCES MANAGER
Phone: 907-552-1609

UNITED STATES DEPARTMENT OF THE AIR FORCE
MAJOR U.S. INSTALLATIONS
F.E. WARREN AFB, WY
USA
Phone: 307-773-5494

Founded: NA

Contact(s):
Catherine Pazenti, NATURAL RESOURCES MANAGER
307-773-5494

UNITED STATES DEPARTMENT OF THE AIR FORCE
MAJOR U.S. INSTALLATIONS
FAIRCHILD, AFB, WA
USA

Founded: NA

Contact(s):
Gerald Johnson, NATURAL RESOURCES MANAGER
Phone: 509-247-2313

UNITED STATES DEPARTMENT OF THE AIR FORCE
MAJOR U.S. INSTALLATIONS
GOODFELLOW AFB, TX
USA

Founded: NA

Contact(s):
Lyndal Fisher, NATURAL RESOURCES MANAGER
Phone: 915-654-3451

UNITED STATES DEPARTMENT OF THE AIR FORCE
MAJOR U.S. INSTALLATIONS
GRAND FORKS AFB, ND
USA

Founded: NA

UNITED STATES DEPARTMENT OF THE AIR FORCE
MAJOR U.S. INSTALLATIONS
HANSCOM AFB, MA
USA

Founded: NA

Contact(s):
Don Morris, NATURAL RESOURCES MANAGER
Phone: 617-377-4667

UNITED STATES DEPARTMENT OF THE AIR FORCE
MAJOR U.S. INSTALLATIONS
HICKAM AFB, HI
USA

Founded: NA

Contact(s):
Gary O'Donnell, NATURAL RESOURCES MANAGER
Phone: 808-449-9695

UNITED STATES DEPARTMENT OF THE AIR FORCE
MAJOR U.S. INSTALLATIONS
HILL AFB, UT
USA
Phone: 801-777-4618

Founded: NA

Contact(s):
Marcus Blood, CULTURAL/NATURAL RESOURCES MANAGER
Phone: 801-777-4618

UNITED STATES DEPARTMENT OF THE AIR FORCE
MAJOR U.S. INSTALLATIONS
HOLLOMAN AFB, NM
USA

Founded: NA

Contact(s):
Hildy Reiser, NATURAL RESOURCES MANAGER
Phone: 505-475-3931

UNITED STATES DEPARTMENT OF THE AIR FORCE
MAJOR U.S. INSTALLATIONS
HURLBURT FIELD, FL
USA

Founded: NA

Contact(s):
Philip Pruit, NATURAL/CULTURAL RESOURCES MANAGER
Phone: 850-884-4651

UNITED STATES DEPARTMENT OF THE AIR FORCE
MAJOR U.S. INSTALLATIONS
KEESLER AFB, MS
USA
Phone: 228-377-2489

Founded: NA

Contact(s):
George Daniels, NATURAL RESOURCES MANAGER
Phone: 228-377-2489

UNITED STATES DEPARTMENT OF THE AIR FORCE
MAJOR U.S. INSTALLATIONS
KELLY AFB, TX
USA
Phone: 210-925-3100

Founded: NA

Contact(s):
Russ Rean, NATURAL RESOURCES MANAGER
Phone: 210-925-3100

UNITED STATES DEPARTMENT OF THE AIR FORCE
MAJOR U.S. INSTALLATIONS
KIRTLAND AFB, NM
USA
Phone: 505-846-6857

Founded: NA

Contact(s):
Bob Dow, NATURAL RESOURCES MANAGER
Phone: 505-846-6857

UNITED STATES DEPARTMENT OF THE AIR FORCE

MAJOR U.S. INSTALLATIONS
LACKLAND AFB, TX
USA
Phone: 210-671-4843

Founded: NA

Contact(s):
Gabriel Gonzales, NATURAL RESOURCES MANAGER
Phone: 210-671-4843

UNITED STATES DEPARTMENT OF THE AIR FORCE

MAJOR U.S. INSTALLATIONS
LANGLEY AFB, VA
USA
Phone: 757-764-1090

Founded: NA

Contact(s):
Patsy Kerr, NATURAL RESOURCES MANAGER
Phone: 757-764-1090

UNITED STATES DEPARTMENT OF THE AIR FORCE

MAJOR U.S. INSTALLATIONS
LAUGHLIN AFB, TX
USA
Phone: 830-298-5694

Founded: NA

Contact(s):
Jadee Bell, NATURAL RESOURCES MANAGER
Phone: 830-298-4298

UNITED STATES DEPARTMENT OF THE AIR FORCE

MAJOR U.S. INSTALLATIONS
LITTLE ROCK AFB, AR
USA
Phone: 501-987-3681

Founded: NA

Contact(s):
James Popham, CULTURAL/NATURAL RESOURCES
MANAGER
Phone: 501-987-3681

UNITED STATES DEPARTMENT OF THE AIR FORCE

MAJOR U.S. INSTALLATIONS
LUKE AFB (AND THE BARRY M. GOLDWATER AFR),
AZ
USA
Phone: 623-856-3823

Founded: NA

Contact(s):
Robert Barry, CHIEF OF CONSERVATION PROGRAMS
Phone: 623-856-3823

UNITED STATES DEPARTMENT OF THE AIR FORCE

MAJOR U.S. INSTALLATIONS
MACDILL AFB, FL
USA
Phone: 813-828-0459

Founded: NA

Contact(s):
Jason Kirkpatrick, NATURAL RESOURCES MANAGER
Phone: 813-828-2567

UNITED STATES DEPARTMENT OF THE AIR FORCE

MAJOR U.S. INSTALLATIONS
MALMSTROM AFB, MT
USA
Phone: 406-731-6437

Founded: NA

Contact(s):
Rudy Berzuh, CULTURAL/NATURAL RESOURCES
MANAGER
Phone: 406-731-6437

UNITED STATES DEPARTMENT OF THE AIR FORCE

MAJOR U.S. INSTALLATIONS
MAXWELL AFB, AL
USA
Phone: 334-953-3892

Founded: NA

Contact(s):
Dennis Tates, NATURAL RESOURCES MANAGER
Phone: 334-953-3892

UNITED STATES DEPARTMENT OF THE AIR FORCE

MAJOR U.S. INSTALLATIONS
MCCHORD AFB, WA
USA
Phone: 253-982-3913

Founded: NA

Contact(s):
Valerie Elliott, NATURAL RESOURCES MANAGER

UNITED STATES DEPARTMENT OF THE AIR FORCE

MAJOR U.S. INSTALLATIONS
MCCLELLAN AFB, CA
USA
Phone: 916-643-1742

Founded: NA

Contact(s):
Molly Enloe, NATURAL RESOURCES MANAGER
Phone: 919-643-1742

UNITED STATES DEPARTMENT OF THE AIR FORCE

MAJOR U.S. INSTALLATIONS
MCCONNELL AFB, KS
USA
Phone: 316-759-3884

Founded: NA

Contact(s):
John Hafker, CULTURAL/NATURAL RESOURCES
MANAGER
Phone: 316-759-3884

UNITED STATES DEPARTMENT OF THE AIR FORCE

MAJOR U.S. INSTALLATIONS
MCGUIRE AFB, NJ
USA

Founded: NA

UNITED STATES DEPARTMENT OF THE AIR FORCE

MAJOR U.S. INSTALLATIONS
MOODY AFB, GA
USA
Phone: 229-257-3070

Founded: NA

Contact(s):
Greg Lee, NATURAL RESOURCES MANAGER
Phone: 229-257-3070

UNITED STATES DEPARTMENT OF THE AIR FORCE

MAJOR U.S. INSTALLATIONS
MOUNTAIN HOME AFB, ID
USA
Phone: 208-828-6351

Founded: NA

Contact(s):
Angelia Martin, CULTURAL/NATURAL RESOURCES
MANAGER
Phone: 208-828-6351

UNITED STATES DEPARTMENT OF THE AIR FORCE

MAJOR U.S. INSTALLATIONS
NELLIS AFB, NV (AND NELLIS AIR FORCE RANGE)
USA

Founded: NA

Contact(s):
Eric Watkins, NATURAL RESOURCES MANAGER
Phone: 702-652-3173

UNITED STATES DEPARTMENT OF THE AIR FORCE

MAJOR U.S. INSTALLATIONS
NEW BOSTON AFB, NH
USA

Founded: NA

Contact(s):
Ralph Mitchell, CULTURAL/NATURAL RESOURCES
MANAGER

UNITED STATES DEPARTMENT OF THE AIR FORCE

MAJOR U.S. INSTALLATIONS
OFFUT AFB, NE
USA

Founded: NA

Contact(s):
Gene Svensen, CULTURAL/NATURAL RESOURCE
MANAGER
Phone: 402-294-7619

UNITED STATES DEPARTMENT OF THE AIR FORCE

MAJOR U.S. INSTALLATIONS
PATRICK AFB, FL
USA
Phone: 321-494-7288

Founded: NA

Contact(s):
Mike Camardese, CULTURAL/NATURAL RESOURCES
MANAGER
Phone: 321-853-0910

UNITED STATES DEPARTMENT OF THE AIR FORCE

MAJOR U.S. INSTALLATIONS
PETERSON AFB, CO
USA

Founded: NA

Contact(s):
Dana Green, CULTURAL/NATURAL RESOURCE MANAGER
Phone: 719-556-9328
Gale Ellett, NATURAL RESOURCES MANAGER
Phone: 719-554-9915

UNITED STATES DEPARTMENT OF THE AIR FORCE

MAJOR U.S. INSTALLATIONS
POPE AFB, SC
USA
Phone: 910-394-1495

Founded: NA

Contact(s):
Viola Walker, CULTURAL/NATURAL RESOURCES
MANAGER
Phone: 910-394-1495

UNITED STATES DEPARTMENT OF THE AIR FORCE

MAJOR U.S. INSTALLATIONS
RANDOLPH AFB, TX
USA
Phone: 210-652-4668

Founded: NA

Contact(s):
Catherine Vornberg, NATURAL RESOURCES MANAGER
Phone: 210-652-4668

UNITED STATES DEPARTMENT OF THE AIR FORCE
MAJOR U.S. INSTALLATIONS
SCOTT AFB, IL
USA

Founded: NA

Contact(s):
William Calvert, CULTURAL/NATURAL RESOURCES
MANAGER
Phone: 618-256-2092

UNITED STATES DEPARTMENT OF THE AIR FORCE
MAJOR U.S. INSTALLATIONS
SEYMOUR JOHNSON AFB (AND DARE COUNTY AFR),
NC
USA

Founded: NA

Contact(s):
Brian Henderson, CULTURAL/NATURAL RESOURCES
MANAGER
Phone: 919-722-5173

UNITED STATES DEPARTMENT OF THE AIR FORCE
MAJOR U.S. INSTALLATIONS
SHAW AFB, SC
USA

Founded: NA

Contact(s):
Terry Madewell, CULTURAL/NATURAL RESOURCES
MANAGER
Phone: 803-895-5193

UNITED STATES DEPARTMENT OF THE AIR FORCE
MAJOR U.S. INSTALLATIONS
SHEPPARD AFB, TX
USA
Phone: 940-676-5698

Founded: NA

Contact(s):
Tim Hunter, CULTURAL/NATURAL RESOURCES MANAGER
Phone: 940-676-5698

UNITED STATES DEPARTMENT OF THE AIR FORCE
MAJOR U.S. INSTALLATIONS
SHRIEVER AFB, CO
USA
Phone: 719-556-2075

Founded: NA

Contact(s):
Ralph Mitchell, CULTURAL/NATURAL RESOURCES
MANAGER
Phone: 719-556-2075
Steve Najarr, CULTURAL/NATURAL RESOURCES
MANAGER
Phone: 719-556-2075

UNITED STATES DEPARTMENT OF THE AIR FORCE
MAJOR U.S. INSTALLATIONS
TINKER AFB, OK
USA

Founded: NA

Contact(s):
John Krupovage, NATURAL RESOURCES MANAGER
Phone: 405-734-3093

UNITED STATES DEPARTMENT OF THE AIR FORCE
MAJOR U.S. INSTALLATIONS
TRAVIS AFB, CA
USA
Phone: 707-424-3897

Founded: NA

Contact(s):
Robert Holmes, NATURAL RESOURCES MANAGER
Phone: 707-424-3897

UNITED STATES DEPARTMENT OF THE AIR FORCE
MAJOR U.S. INSTALLATIONS
TYNDALL AFB, FL
USA
Phone: 850-283-2641

Founded: NA

Contact(s):
Bob Bates, NATURAL RESOURCES MANAGER
Phone: 850-283-2641

UNITED STATES DEPARTMENT OF THE AIR FORCE
MAJOR U.S. INSTALLATIONS
VANCE AFB, OK
USA

Founded: NA

Contact(s):
Shannon Elledge, CULTURAL/NATURAL RESOURCES
MANAGER

UNITED STATES DEPARTMENT OF THE AIR FORCE
MAJOR U.S. INSTALLATIONS
VANDENBERG AFB, CA
USA

Founded: NA

Contact(s):
Allan Naydol, NATURAL RESOURCES MANAGER

UNITED STATES DEPARTMENT OF THE AIR FORCE

MAJOR U.S. INSTALLATIONS
WHITEMAN AFB, MO
USA

Founded: NA

Contact(s):
Neil Bass, CULTURAL/NATURAL RESOURCE MANAGER
Angela Corson, NATURAL RESOURCES MANAGER

UNITED STATES DEPARTMENT OF THE AIR FORCE

MAJOR U.S. INSTALLATIONS
WRIGHT-PATTERSON AFB, OH
USA

Founded: NA

Contact(s):
Terri Lucas, NATURAL RESOURCES PLANNER
Phone: 937-257-5535

UNITED STATES DEPARTMENT OF THE AIR FORCE

OFFICE OF THE CIVIL ENGINEER
AF/ILE, 1260 Air Force Pentagon
Washington, DC 20330-1260 USA
Phone: 703-604-0632

Founded: NA

UNITED STATES DEPARTMENT OF THE AIR FORCE, ENVIRONMENTAL DIVISION

Environmental Division, HQ USAF/ILEV, 1260 Air Force
Pentagon
Washington, DC 20330-1260 USA
Phone: 703-604-0632   Fax: 703-604-3740

Founded: NA
Scope: National

Description: A comprehensive natural resources conservation
program focusing on fish and wildlife management, forestry,
outdoor recreation, and soil and water conservation has been
conducted on Air Force lands since the mid-1950's. Current
policy requires all installations with significant land and water
resources to develop integrated natural resource management
plans as part of the base comprehensive planning process.

Contact(s):
Alan Holck, NATURAL AND CULTURAL RESOURCES
PROGRAM MANAGER

UNITED STATES DEPARTMENT OF THE ARMY

Pentagon Environmental Dept
Washington, DC 20310 USA
Phone: 703-695-7824   Fax: 703-693-8149

Founded: NA
Scope: State
Contact(s):
Phil Huber, ASSISTANT FOR ENVIRONMENTAL QUALITY
Phone: 703-614-9555
Raymond Fatz, DEPUTY ASSISTANT SECRETARY OF THE
ARMY (ENVIRONMENT OF SAFETY OF AND
OCCUPATIONAL HEALTH)
Phone: 703-695-7824

UNITED STATES DEPARTMENT OF THE ARMY

ARMY TRAINING AND DOCTRINE COMMAND
Department of the Army, HQ TRADOC, ATBO-SE,
Environmental Division
Fort Monroe, VA 23651 USA

Founded: NA

Description: Manages conservation programs for 2 million acres
at 16 Army installations nationwide. It also provides for
compliance with federal, state, and local environmental
regulations.

Publication(s): Historic Preservation Sourcebook, Army Leader's
Guide to NEPA, Endangered Species Law Sourcebook

Contact(s):
Shawn Holsinger, CONSERVATION AND ANALYSIS
BRANCH
Phone: 757-727-3045
Frances Doyle, GENERAL/TECHNICAL LIBRARIES OF HQ
TRADOC (ATBO-NT) OF DIRECTOR
Ft. Monroe, VA 23651
Robert Anderson, NATURAL RESOURCES SPECIALIST
Phone: 757-727-2077
Jim White, NEPA CONSULTANT
Phone: 757-727-5896
Jack Damron, NEPA CONSULTANT
Phone: 757-727-4135
John Esson, NEPA CONSULTANT
Phone: 757-727-3335

UNITED STATES DEPARTMENT OF THE ARMY

ASSISTANT CHIEF OF STAFF FOR INSTALLATION
MANAGEMENT, OFFICE OF THE DIRECTOR OF
ENVIRONMENTAL PROGRAMS, AND CONSERVATION
TEAM
Attn: DAIM-ED-N, 600 Army Pentagon
Washington, DC 20310-0600 USA

Founded: NA

Description: Natural and cultural resources professionals are
responsible for the management of approximately 12 million
acres of land on Army military installations. Management
objectives include: Compliance with environmental laws, conservation and protection of resources, support to the military
mission uses of the land, and contributions to programs which
support the public needs. Resources managed include: Land,
forest, wildlife, soils, vegetation, and historical and archaeological sites.

Contact(s):
Vic Diersing, CONSERVATION TEAM LEADER
Phone: 703-693-0677
Chuck Wright, CULTURAL RESOURCES SPECIALIST
Phone: 703-693-0675
Bill Woodson, NATURAL RESOURCE SPECIALIST
Phone: 703-693-0680

UNITED STATES DEPARTMENT OF THE ARMY

CORP OF ENGINEERS
U.S. ARMY ENGINEER DISTRICT, ROCK ISLAND
Clock Tower Building
Rock Island, IL 61204-2004 USA
Phone: 309-794-5759   Fax: 309-794-5181
E-mail: mvr@usace.army.mil
Website: www.mvr.usace.army.mil

Founded: NA

Scope: Regional, State

Contact(s):
   Bob Romic, KEY CONTACT

UNITED STATES DEPARTMENT OF THE ARMY
ENGINEER RESEARCH & DEVELOPMENT CENTER
Champaign, IL 61826-9005 USA
Phone: 217-352-6511   Fax: 217-373-7222

Founded: 1969
Scope: National

Description: CERL conducts research on infrastructure and environmental problems facing the operations of military facilities. CERL also conducts research on innovative materials and engineering procedures; energy reduction measures and equipment; management systems; air and water pollution; environmental compliance; and natural resource management.

Publication(s): CERL Abstracts, Index to Publications, The Cutting Edge

Contact(s):
   Dana Finney, CHAMPAIGN PUBLIC AFFAIRS

UNITED STATES DEPARTMENT OF THE ARMY
HEADQUARTERS, U.S. ARMY TRAINING AND
DOCTRINE COMMAND
ATBO-SE
Fort Monroe, VA 23651 USA

Founded: NA

Contact(s):
   Roger Smith, AGRONOMIST OF FORT DIX
   Phone: 609-562-2040
   Marvin Myers, AGRONOMIST OF FORT LEONARD WOOD
   Phone: 314-596-0871
   Delarie Parmer, AGRONOMIST OF FORT RUCKER
   Phone: 205-255-9363
   William Pittman, AGRONOMIST OF HEALTH SERVICES
   COMMAND, ACADEMY OF HEALTH SCIENCES
   Phone: 512-221-4411
   Patrick Ching, AGRONOMIST OF SCHOFIELD BARRACKS
   Phone: 808-655-6383
   James Murphy, AGRONOMIST OF U.S. ARMY MILITARY
   DISTRICT OF WASHINGTON
   Phone: 202-696-3815
   Robert Jones, AGRONOMIST OF U.S. MILITARY ACADEMY,
   NATURAL RESOURCES BRANCH
   Phone: 914-938-3467
   Marie Cottrell, ARCHEOLOGIST
   Phone: 804-727-2389
   Glen Degarmo, ARCHEOLOGIST OF FORT BLISS
   Phone: 915-568-5140
   Paul Lukowski, ARCHEOLOGIST OF FORT BLISS
   Phone: 915-568-6999
   Kevin McCurdy, BIOLOGICAL TECH./GAME WARDEN OF
   FORT SILL
   Phone: 405-351-4324
   James McCracken, BIOLOGIST OF DA HEADQUARTERS
   Phone: 803-751-4622
   James Loewen, BIOLOGIST OF FORT LEE
   Phone: 804-734-5080
   Dorothy Keough, CHIEF (ACTING) OF ENVIRONMENTAL
   AND NATURAL RESOURCES DIVISION OF FORT BELVOIR
   Phone: 703-806-4007

   Bob Coleman, CHIEF ENVIRONMENT BRANCH OF FORT
   CHAFFEE
   Phone: 501-484-2516
   Al Freeland, CHIEF ENVIRONMENTAL MANAGEMENT
   DIVISION OF FORT KNOX
   Phone: 502-624-3629
   Charles Ford, CHIEF NATURAL RESOURCES MANAGER
   OF FORT BENNING
   Phone: 706-544-7319
   Jerry Sturdy, CHIEF NATURAL RESOURCES SECTION OF
   FORT CHAFFEE
   Phone: 501-484-2231
   Thomas Shafer, CHIEF OF ENVIRONMENT OF FORT
   BENJAMIN HARRISON
   Phone: 317-549-5386
   Steve Willard, CHIEF OF ENVIRONMENTAL AND NATURAL
   RESOURCES OF FORT GORDON
   Phone: 706-791-2403
   Robert Mcguire, CHIEF OF ENVIRONMENTAL RESOURCES
   MANAGEMENT OF ARMY NATIONAL GUARD BUREAU
   Phone: 703-756-5794
   Mark Dutton, CHIEF OF NATURAL RESOURCES OF DA
   HEADQUARTERS
   Phone: 803-751-4103
   Joe Deschenes, CHIEF OF NATURAL RESOURCES OF
   U.S. MILITARY ACADEMY, NATURAL RESOURCES
   BRANCH
   Phone: 914-938-2314
   Gene Stout, CHIEF OF NATURAL/ENVIRONMENTAL
   RESOURCES OF FORT SILL
   Phone: 405-351-4324
   Ron Levy, DIRECTOR OF ENVIRONMENT OF FORT
   MCCLELLAN
   Phone: 205-848-3539
   Kevin Von Finger, ECOLOGIST OF FORT BLISS
   Phone: 915-568-7031
   Donald Teig, ENTOMOLOGIST
   Phone: 804-727-2366
   Chris Dunn, ENTOMOLOGIST OF FORT BENNING
   Phone: 706-545-3224
   John Schenck, ENTOMOLOGIST OF FORT EUSTIS
   Phone: 804-878-2585
   Robert Turnbow, ENTOMOLOGIST OF FORT RUCKER
   Phone: 205-255-3710
   Stuart Hayashi, ENTOMOLOGIST OF HEADQUARTERS OF
   U.S. ARMY PACIFIC
   Phone: 808-438-2180
   Joe Tarnopol, ENTOMOLOGIST OF U.S. ARMY MILITARY
   DISTRICT OF WASHINGTON
   Phone: 202-475-1003
   Edna Barber, ENVIRONMENTAL OFFICER OF U.S. ARMY
   MILITARY DISTRICT OF WASHINGTON
   Phone: 202-696-3815
   Doug Dasher, ENVIRONMENTAL PROTECTION
   SPECIALIST OF FORT GREELY
   Phone: 907-451-2172
   Joyce Beelman, ENVIRONMENTAL PROTECTION
   SPECIALIST OF FORT GREELY
   Phone: 907-451-2141
   Brad Fristoe, ENVIRONMENTAL PROTECTION SPECIALIST
   OF FORT GREELY
   Phone: 907-451-2159
   Lawrence Hirai, ENVIRONMENTAL PROTECTION
   SPECIALIST OF HEADQUARTERS OF U.S. ARMY PACIFIC
   Phone: 808-438-8997

Mark Salley, ENVIRONMENTAL PROTECTION SPECIALIST OF SCHOFIELD BARRACKS
Phone: 808-656-2878
Bill Quirk, ENVIRONMENTAL SPECIALIST OF FORT RICHARDSON
Phone: 907-384-3021
Glen Wampler, FISH AND WILDLIFE ADMINISTRATOR OF FORT SILL
Phone: 405-442-8111
Mike Hudson, FORESTER OF FORT BELVOIR
Phone: 703-806-4007
Jack Greenlee, FORESTER OF FORT BENNING
Phone: 706-544-7319
Tony Rizzio, FORESTER OF FORT EUSTIS
Phone: 804-878-4152
Allen Braswell, FORESTER OF FORT GORDON
Phone: 706-791-2327
Dave Apsley, FORESTER OF FORT KNOX
Phone: 502-624-8147
Matt Nowak, FORESTER OF FORT LEAVENWORTH
Phone: 913-684-2749
Steve Thurman, FORESTER OF FORT LEONARD WOOD
Phone: 314-596-0871
Bill Garland, FORESTER OF FORT MCCLELLAN
Phone: 205-848-3758
Bob Shuffield, FORESTER OF FORT RUCKER
Phone: 205-255-9368
John Miller, FORESTER OF INFORMATION SYSTEMS COMMAND
Phone: 602-533-7083
Hershel Gaw, FORESTER OF MILITARY TRAFFIC MANAGEMENT COMMAND, MILITARY OCEAN TERMINAL
Phone: 919-457-8292
Mark Imlay, NATURAL RESOURCES MANAGER OF ARMY NATIONAL GUARD BUREAU
Phone: 703-756-5794
Ronald Moore, NATURAL RESOURCES MANAGER OF CAMP ATTERBURY
Phone: 812-526-1250
Don Hack, NATURAL RESOURCES MANAGER OF NAVAJO DEPOT ACTIVITY
Phone: 602-774-7161
Bob Anderson, NATURAL RESOURCES SPECIALIST
Phone: 804-727-2077
Wayne Johndrown, NATURAL RESOURCES SPECIALIST OF FORT CHAFFEE
Phone: 501-484-2231
Luther Owen, NATURAL RESOURCES SPECIALIST OF FORT MCCLELLAN
Phone: 205-848-5663
Steve Sekscienski, NATURAL RESOURCES TEAM OF U.S. ARMY ENVIRONMENTAL CENTER
Phone: 410-612-6832
Jerry Williamson, NATURAL RESOURCES TEAM OF U.S. ARMY ENVIRONMENTAL CENTER
Phone: 410-612-6833
William Herb, NATURAL RESOURCES TEAM OF U.S. ARMY ENVIRONMENTAL CENTER
Phone: 410-671-1234
Scott Belfit, NATURAL RESOURCES TEAM OF U.S. ARMY ENVIRONMENTAL CENTER
Phone: 410-612-6831
Eric Seaborn, NATURAL RESOURCES TEAM OF U.S. ARMY ENVIRONMENTAL CENTER
Phone: 410-612-6833

Bob Decker, NATURAL RESOURCES TEAM OF U.S. ARMY ENVIRONMENTAL CENTER
Phone: 410-612-6831
Pamela Klinger, NATURAL RESOURCES TEAM OF U.S. ARMY ENVIRONMENTAL CENTER
Phone: 410-612-6832
Bill Gates, WILDLIFE BIOLOGIST OF DA HEADQUARTERS
Phone: 803-751-4793
Robert King, WILDLIFE BIOLOGIST OF FORT BENNING
Phone: 706-544-7319
Clark Reames, WILDLIFE BIOLOGIST OF FORT CHAFFEE
Phone: 501-484-2231
Roger Meyers, WILDLIFE BIOLOGIST OF FORT DIX
Phone: 609-562-2040
Kenneth Boyd, WILDLIFE BIOLOGIST OF FORT GORDON
Phone: 706-791-2403
Tom Glueck, WILDLIFE BIOLOGIST OF FORT LEONARD WOOD
Phone: 314-596-0871
William Gossweiler, WILDLIFE BIOLOGIST OF FORT RICHARDSON
Phone: 907-384-3017
Sheridan Stone, WILDLIFE BIOLOGIST OF INFORMATION SYSTEMS COMMAND
Phone: 602-538-7340
Donald Sheroan, WILDLIFE BIOLOGIST TECH. OF FORT KNOX
Phone: 502-624-7373

UNITED STATES DEPARTMENT OF THE ARMY
HQ ARMY MATERIAL COMMAND
Alexandria, VA 22333-0001 USA

Founded: NA

Contact(s):
Tom Coleman, AGRONOMIST OF RED RIVER ARMY DEPOT (TEXAS)
Phone: 903-334-2385
Bob Wardwell, AGRONOMIST OF U.S. ARMY RESEARCH LABORATORY (MARYLAND)
Phone: 301-394-1060
Robert Burton, ARCHEOLOGIST OF WHITE SANDS MISSILE RANGE (NEW MEXICO)
Phone: 505-678-8731
John Martin, CHIEF OF CONSERVATION AND PRESERVATION OF DUGWAY PROVING GROUND (UTAH)
Phone: 801-831-2986
Abdul Shiek, ENTOMOLOGIST OF U.S. ARMY ABERDEEN PROVING GROUND SUPPORT ACTIVITIES (MARYLAND)
Phone: 410-278-3303
Steve Wampler, ENVIRONMENTAL PROTECTION SPECIALIST OF U.S. ARMY ABERDEEN PROVING GROUND SUPPORT ACTIVITIES (MARYLAND)
Directorate of Safety, Health and Environment, 410-671-4843
Timothy McNamara, ENVIRONMENTAL PROTECTION SPECIALIST OF U.S. ARMY ABERDEEN PROVING GROUND SUPPORT ACTIVITIES (MARYLAND)
Directorate of Safety, Health and Environment, 410-278-5622
William Burns, FORESTER OF ANNISTON ARMY DEPOT (ALABAMA)
Phone: 205-235-4217
Terry Ruth, FORESTER OF RED RIVER ARMY DEPOT (TEXAS)
Phone: 903-334-2379

Jesse Horton, FORESTER OF REDSTONE ARSENAL SUPPORT ACTIVITY (ALABAMA)
Phone: 205-876-3122
Roger Stoflet, FORESTER OF U.S. ARMY ABERDEEN PROVING GROUND SUPPORT ACTIVITIES (MARYLAND)
Phone: 410-278-4915
Bennie Murray, FORESTER, CHIEF LM OF RED RIVER ARMY DEPOT (TEXAS)
Phone: 903-334-2379
Tom Vorac, FORESTER, INSTALLATIONS AND SERVICES ACTIVITY (ILLINOIS)
Phone: 309-782-4062
Billye Haslett, LAND MANAGER OF BLUE-GRASS ARMY DEPOT
Phone: 606-625-6669
Ken Knouf, NATURAL RESOURCES MANAGER OF JEFFERSON PROVING GROUND (INDIANA)
Phone: 812-273-7436
Randy Quinn, NATURAL RESOURCES MANAGER OF LETTERKENNY ARMY DEPOT (OHIO)
Phone: 717-267-8438
Bob Speaker, NATURAL RESOURCES MANAGER OF SAVANNA ARMY DEPOT (ILLINOIS)
Phone: 815-273-8533
Richard Clewell, NATURAL RESOURCES SPECIALIST, INSTALLATIONS AND SERVICES ACTIVITY (ILLINOIS)
Phone: 309-782-8252
Mason Walker, PROJECT ENGINEER OF TOOELE ARMY DEPOT (UTAH)
Phone: 801-833-2891
James Bailey, WILDLIFE BIOLOGIST OF U.S. ARMY ABERDEEN PROVING GROUND SUPPORT ACTIVITIES (MARYLAND)
Phone: 410-278-6748
James Pottie, WILDLIFE BIOLOGIST OF U.S. ARMY ABERDEEN PROVING GROUND SUPPORT ACTIVITIES (MARYLAND)
Phone: 410-278-6772
Patrick Morrow, WILDLIFE BIOLOGIST OF WHITE SANDS MISSILE RANGE (NEW MEXICO)
Phone: 505-678-7095
Daisan Taylor, WILDLIFE BIOLOGIST OF WHITE SANDS MISSILE RANGE (NEW MEXICO)
Phone: 505-678-6140
Junior Kerns, WILDLIFE BIOLOGIST OF YUMA PROVING GROUND (ARIZONA)
Phone: 602-328-2148
Valerie Morrill, WILDLIFE BIOLOGIST OF YUMA PROVING GROUND (ARIZONA)
Phone: 602-328-2244

UNITED STATES DEPARTMENT OF THE ARMY

U.S. ARMY CORPS OF ENGINEERS
441 G St.
Washington, DC 20314-1000 USA
Website: www.usace.army.mil

Founded: NA

Description: The mission of the Corps of Engineers is to provide quality, responsive engineering and environmental services to the nation. The Corps plans, designs, builds, and operates water resources and other civil works projects. The Corps designs and manages the construction of military facilities and activities for the Army and Air Force and provides design and construction management support for other defense and federal agencies. In addition to military and civilian engineers, the Corps has a diverse workforce of biologists, geologists, hydrologists, natural resource managers and other professionals.

Contact(s):
James Wolcott, CHIEF, ENVIRONMENTAL COMPLIANCE
Phone: 202-761-0200
Darrell Lewis, CHIEF, NATURAL RESOURCES
Phone: 202-761-0247
Robert Soots, CHIEF, OFFICE OF ENVIRONMENTAL POLICY
Phone: 703-428-6491
Robert Mirelson, CHIEF, PUBLIC AFFAIRS
Phone: 202-761-0010
John Studt, CHIEF, REGULATORY
Phone: 202-761-1785
Paul Rubenstein, CULTURAL RESOURCES COORDINATOR
Phone: 202-761-1257
John Bellinger, ENDANGERED SPECIES/NEPA COORDINATOR
Phone: 202-761-0166
Lloyd Saunders, EXECUTIVE SECRETARY, ENVIRONMENTAL ADVISORY BOARD
Phone: 202-761-8731
Timothy Toplisek, FISH AND WILDLIFE COORDINATOR
Phone: 202-761-1789

UNITED STATES DEPARTMENT OF THE ARMY

U.S. ARMY CORPS OF ENGINEERS
U.S. ARMY COLD REGIONS RESEARCH AND ENGINEERING LABORATORY
72 Lyme Road
Hanover, NH 03755-1290 USA
Phone: 603-646-4200   Fax: 603-646-4178
Website: www.crrel.usace.army.mil

Founded: NA
Membership: 290
Scope: International

Publication(s): Technical Reports

Contact(s):
Barbara Sotirin, DIRECTOR
Phone: 603-646-4200

UNITED STATES DEPARTMENT OF THE ARMY

U.S. ARMY CORPS OF ENGINEERS
U.S. ARMY CONSTRUCTION ENGINEERING RESEARCH LABORATORIES
P.O. Box 9005
Champaign, IL 61826-9005 USA
Phone: 217-373-7201   Fax: 217-373-7222
Website: www.cecer.army.mil

Founded: NA
Membership: 500
Scope: International

UNITED STATES DEPARTMENT OF THE ARMY

U.S. ARMY CORPS OF ENGINEERS
U.S. ARMY CORPS OF ENGINEERS WATER RESOURCES SUPPORT CENTER
7701 Telegraph Road, Casey Building
Alexandria, VA 22315-3868 USA
Phone: 703-428-8250   Fax: 703-428-8171
E-mail: www.iwr.usce.army.mil
Website: www.iwr.usce.army.mil

Founded: NA
Membership: 157
Scope: Statewide

UNITED STATES DEPARTMENT OF THE ARMY
U.S. ARMY CORPS OF ENGINEERS
U.S. ARMY ENGINEER DISTRICT, ALASKA
Anchorage, AK 99506-0898 USA
Phone: 907-753-2520   Fax: 907-753-2526
Website: www.poa.usace.army.mil

Founded: NA
Membership: 500
Scope: National

UNITED STATES DEPARTMENT OF THE ARMY
U.S. ARMY CORPS OF ENGINEERS
U.S. ARMY ENGINEER DISTRICT, ALBUQUERQUE
4101 Jefferson Plaza NE
Albuquerque, NM 87109-3435 USA
Phone: 505-342-3432   Fax: 505-342-3199

Founded: NA
Membership: 300
Scope: National

UNITED STATES DEPARTMENT OF THE ARMY
U.S. ARMY CORPS OF ENGINEERS
U.S. ARMY ENGINEER DISTRICT, BUFFALO
1766 Niagara Street
Buffalo, NY 14207-3199 USA
Phone: 716-879-4200   Fax: 716-879-4195
Website: www.lrb.usace.army.mil

Founded: NA
Scope: State

Contact(s):
Nancy Sticht, PUBLIC AFFAIRS OFFICER
Phone: 716-879-4410
nancy.j.sticht@usace.army.mil

UNITED STATES DEPARTMENT OF THE ARMY
U.S. ARMY CORPS OF ENGINEERS
U.S. ARMY ENGINEER DISTRICT, CHARLESTON
P.O. Box 919
Charleston, SC 29401-0919 USA
Phone: 843-329-8000

Founded: NA

Contact(s):
Peter Mueller

UNITED STATES DEPARTMENT OF THE ARMY
U.S. ARMY CORPS OF ENGINEERS
U.S. ARMY ENGINEER DISTRICT, FORT WORTH
Fort Worth, TX 76102-0300 USA
Phone: 817-978-2300   Fax: 817-978-3311
Website: www.usace.army.mil

Founded: NA
Membership: 1000
Scope: Regional

UNITED STATES DEPARTMENT OF THE ARMY
U.S. ARMY CORPS OF ENGINEERS
U.S. ARMY ENGINEER DISTRICT, GALVESTON
Galveston, TX 77553-1229 USA
Phone: 409-766-3001   Fax: 409-766-3951
Website: www.usace.army.mil

Founded: NA
Membership: 360
Scope: Local

Contact(s):
Maryann Patlan, SECRETARY

UNITED STATES DEPARTMENT OF THE ARMY
U.S. ARMY CORPS OF ENGINEERS
U.S. ARMY ENGINEER DISTRICT, LITTLE ROCK
P.O. Box 867
Little Rock, AR 72203-0867 USA
Phone: 501-324-5531   Fax: 501-324-6968
Website: www.swl.usace.army.mil

Founded: NA
Membership: 800
Scope: Regional

Contact(s):
Dale Leggett, CHIEF NATURAL RESOURCES BRANCH

UNITED STATES DEPARTMENT OF THE ARMY
U.S. ARMY CORPS OF ENGINEERS
U.S. ARMY ENGINEER DISTRICT, LOUISVILLE
P.O. Box 59
Louisville, KY 40201-0059 USA
Phone: 502-315-6768   Fax: 502-582-5475
E-mail: todd.j.hornback@lrl02usace.army.mil
Website: http://www.orl.usace.army.mil/default.htm

Founded: NA

UNITED STATES DEPARTMENT OF THE ARMY
U.S. ARMY CORPS OF ENGINEERS
U.S. ARMY ENGINEER DISTRICT, MEMPHIS
167 N. Main Street, Rm. B202 Attn: Environmental
Branch
Memphis, TN 38103-1894 USA
Phone: 901-544-3221   Fax: 901-544-3628

Founded: NA
Scope: International

UNITED STATES DEPARTMENT OF THE ARMY
U.S. ARMY CORPS OF ENGINEERS
U.S. ARMY ENGINEER DISTRICT, NEW ORLEANS
New Orleans, LA 70160-0267 USA
Phone: 504-862-2204   Fax: 504-862-1259
E-mail: Cemvn-de@mvn02.usace.army.mil
Website: www.mvn.usace.army.mil

Founded: NA
Membership: 1200
Scope: Regional

UNITED STATES DEPARTMENT OF THE ARMY
U.S. ARMY CORPS OF ENGINEERS
U.S. ARMY ENGINEER DISTRICT, OMAHA
106 S. 15th St.
Omaha, NE 68102-1618 USA
Phone: 402-221-3900   Fax: 402-221-3229
E-mail: karenl.stefeno@usace.army.mil

Founded: NA
Membership: 1300
Scope: National

UNITED STATES DEPARTMENT OF THE ARMY
U.S. ARMY CORPS OF ENGINEERS
U.S. ARMY ENGINEER DISTRICT, PITTSBURGH
US Army Corps of Engineers William S. Moorhead Fed.
Bldg. 1000 Liberty Avenue
Pittsburgh, PA 15222-4186 USA
Phone: 412-395-7103   Fax: 412-644-4093

Founded: NA
Membership: 900
Scope: Regional

UNITED STATES DEPARTMENT OF THE ARMY
U.S. ARMY CORPS OF ENGINEERS
U.S. ARMY ENGINEER DISTRICT, PORTLAND
P.O. Box 2946
Portland, OR 97208-2946 USA
Phone: 503-808-4500   Fax: 503-808-4505

Founded: NA

UNITED STATES DEPARTMENT OF THE ARMY
U.S. ARMY CORPS OF ENGINEERS
U.S. ARMY ENGINEER DISTRICT, SACRAMENTO
1325 J Street
Sacramento, CA 95814-2922 USA
Phone: 916-557-7490   Fax: 916-557-7859

Founded: NA

UNITED STATES DEPARTMENT OF THE ARMY
U.S. ARMY CORPS OF ENGINEERS
U.S. ARMY ENGINEER DISTRICT, SAN FRANCISCO
211 Main Street
San Francisco, CA 94105-2195 USA
Phone: 415-744-3021   Fax: 415-977-8316

Founded: NA

Contact(s):
   Timothy O'Rourke, LTC

UNITED STATES DEPARTMENT OF THE ARMY
U.S. ARMY CORPS OF ENGINEERS
U.S. ARMY ENGINEER DISTRICT, SEATTLE
Seattle, WA 98124-3755 USA
Phone: 206-764-3690   Fax: 206-764-6544
E-mail: paoteam@usace.army.mil
Website: www.nws.usace.army.mil/index.cfm

Founded: NA
Membership: 750
Scope: Regional

Publication(s): Flagship

UNITED STATES DEPARTMENT OF THE ARMY
U.S. ARMY CORPS OF ENGINEERS
U.S. ARMY ENGINEER DISTRICT, ST. PAUL
Army Corps of Army Corps of Engineers Centre, 190 5th
Street East
St. Paul, MN 55101-1638 USA
Phone: 651-290-5300   Fax: 651-290-5478
Website: www.mvtusave.army.mil

Founded: NA
Membership: 600
Scope: Local

UNITED STATES DEPARTMENT OF THE ARMY
U.S. ARMY CORPS OF ENGINEERS
U.S. ARMY ENGINEER DISTRICT, WALLA WALLA
201 North 3rd Avenue
Walla Walla, WA 99362-1876 USA
Phone: 509-527-7700   Fax: 509-527-7804

Founded: NA

UNITED STATES DEPARTMENT OF THE ARMY
U.S. ARMY CORPS OF ENGINEERS
U.S. ARMY ENGINEER DIVISION, GREAT LAKES AND
OHIO RIVER
P.O. Box 1159
Cincinnati, OH 45201-1159 USA
Phone: 513-684-3016   Fax: 513-684-2085
E-mail: CELRD-DE@usace.army.mil

Founded: NA

UNITED STATES DEPARTMENT OF THE ARMY
U.S. ARMY CORPS OF ENGINEERS
U.S. ARMY ENGINEER DIVISION, NORTH ATLANTIC
General Lee Ave., Fort Hamilton Military Community,
Building 302
Brooklyn, NY 11252-6000 USA
Phone: 718-491-8707   Fax: 718-765-7168

Founded: NA

UNITED STATES DEPARTMENT OF THE ARMY
U.S. ARMY CORPS OF ENGINEERS
U.S. ARMY ENGINEER DIVISION, NORTHWESTERN
Regional Headquarters, P.O. Box 2870
Portland, OR 97208-2870 USA
Phone: 503-808-3700   Fax: 503-808-3706

Founded: NA

UNITED STATES DEPARTMENT OF THE ARMY
U.S. ARMY CORPS OF ENGINEERS
U.S. ARMY ENGINEER DIVISION, SOUTH ATLANTIC
Room 9M15, 60 Forsyth Street, SW
Atlanta, GA 30303-8801 USA
Phone: 404-562-5003   Fax: 404-562-5002
Website: www.sad.usace.army.mil

Founded: NA
Membership: 120
Scope: Regional

UNITED STATES DEPARTMENT OF THE ARMY

U.S. ARMY CORPS OF ENGINEERS
U.S. ARMY ENGINEER DIVISION, SOUTH PACIFIC
333 Market Street, Room 1101
San Francisco, CA 94105-2195 USA
Phone: 415-977-8001   Fax: 415-977-8316

Founded: NA

UNITED STATES DEPARTMENT OF THE ARMY

U.S. ARMY CORPS OF ENGINEERS
U.S. ARMY ENGINEER DIVISION, SOUTHWESTERN
1100 Commerce Street
Dallas, TX 75242-0216 USA
Phone: 214-767-2502   Fax: 214-767-6499

Founded: NA
Scope: National

UNITED STATES DEPARTMENT OF THE ARMY

U.S. ARMY CORPS OF ENGINEERS
U.S. ARMY ENGINEER WATERWAYS EXPERIMENT
STATION
3909 Halls Ferry Road
Vicksburg, MS 39180-6199 USA
Phone: 601-634-2664   Fax: 601-634-2388
Website: www.erdc.usace.army.mil

Founded: NA
Scope: National

Contact(s):
    Dr. James Houston, DIRECTOR

UNITED STATES DEPARTMENT OF THE ARMY

U.S. ARMY CORPS OF ENGINEERS
U.S.A.E.R.D.C. TOPOGRAPHIC ENGINEERING
CENTER
7701 Telegraph Rd.
Alexandria, VA 22315-3864 USA
Phone: 703-428-6600   Fax: 703-428-8154
Website: www.tech.army.mil

Founded: NA
Membership: 320
Scope: National

UNITED STATES DEPARTMENT OF THE ARMY

U.S. ARMY ENGINEER DISTRICT, LOS ANGELES
P.O. Box 532711
Los Angeles, CA 90053-2325 USA
Phone: 213-452-3967   Fax: 213-452-4214
Website: www.spl.usace.army.mil

Founded: NA
Membership: 650
Scope: International

UNITED STATES DEPARTMENT OF THE ARMY

U.S. ARMY FORCES COMMAND
Forester, HQ FORSCOM, Attn: AFPI-ENE
Fort McPherson, GA 30330-1062 USA
Phone: 404-464-5762   Fax: 404-669-7827
E-mail: cannons@forscom.army.mil

Founded: NA
Membership: 20
Scope: National

Contact(s):
    Stuart Cannon, FORESTER
    Albert Bivings, WILDLIFE BIOLOGIST
    bivingsb@forscom.army.mil

UNITED STATES DEPARTMENT OF THE ARMY

U.S. MILITARY ACADEMY
NATURAL RESOURCES BRANCH
DHPW
West Point, NY 10996-1592 USA
Phone: 845-938-2314   Fax: 845-938-2324

Founded: NA
Membership: 4
Scope: Local

Contact(s):
    Robert Jones, AGRONOMIST
    Phone: 914-938-6789
    Joe Deschenes, BRANCH CHIEF AND FORESTER
    James Beemer, FISH AND WILDLIFE BIOLOGIST
    Catherine Coleman, ITAM PROGRAM MANAGER
    Phone: 914-938-5453

UNITED STATES DEPARTMENT OF THE INTERIOR

U.S. Department of the Interior, 1849 C St., NW
Washington, DC 20240 USA
Phone: 202-208-6843   Fax: 202-219-0910
E-mail: waso-public-affairs@nps.gov
Website: www.nps.gov

Founded: NA
Scope: National

Description: Administers 1,378 parks, monuments, and other administrative classifications of national significance for their recreational, historical, and natural values. Manages landmarks programs for natural and historic properties; coordinates Wild and Scenic Rivers System and National Trail System; administers study and grants programs.

Contact(s):
    Rob Arnberger, ALASKA REGIONAL DIRECTOR
    Sue Masica, ASSOCIATE DIRECTOR OF BUDGET AND ADMINISTRATION
    Phone: 202-208-6953
    Kate Stevenson, ASSOCIATE DIRECTOR OF CULTURAL RESOURCES
    Phone: 202-208-7625
    Michael Soukup, ASSOCIATE DIRECTOR OF NATURAL RESOURCES
    Phone: 202-208-3884
    Dick Ring, ASSOCIATE DIRECTOR OF PARK OPERATIONS
    Terrell Emmons, ASSOCIATE DIRECTOR OF PROFESSIONAL SERVICES
    Bill Shaddox, ASSOCIATE DIRECTOR OF PROFESSIONAL SERVICES
    Phone: 202-208-3264
    David Barna, CHIEF OF OFFICE OF PUBLIC AFFAIRS
    Phone: 202-208-6843
    C. Sheaffer, COMPTROLLER
    Phone: 202-208-4566
    Denis Galvin, DEPUTY DIRECTOR
    Phone: 202-208-3818
    Fran Mainella, DIRECTOR OF THE NATIONAL PARK SERVICE
    Karen Wade, INTERMOUNTAIN REGIONAL DIRECTOR

William Schenk, MIDWEST REGIONAL DIRECTOR
1709 Jackson St., Omaha, NE 68102
Phone: 402-221-3471
Terry Carlstrom, NATIONAL CAPITAL REGIONAL
DIRECTOR
1100 Ohio Dr., SW, Washington, DC 20242
Phone: 202-619-7256
Marie Rust, NORTHEAST REGION DIRECTOR
U.S. Customs House, 5th Fl., 200 Chestnut St., Philadelphia,
PA 19106
Phone: 215-597-7013
John Reynolds, PACIFIC WEST REGIONAL DIRECTOR
600 Harrison St., Suite 600, San Francisco, CA 94107
Phone: 415-427-1300
Jerry Belson, SOUTHEAST REGIONAL DIRECTOR
100 Alabama St., SW, Atlanta Federal Center, Atlanta, GA
30303
Phone: 404-562-3100

UNITED STATES DEPARTMENT OF THE INTERIOR

Interior Bldg., 1849 C St., NW
Washington, DC 20240 USA
Phone: 202-208-3100
Website: www.doi.gov

Founded: NA

Description: The mission of the Department of the Interior is to
protect and provide access to our Nation's natural and cultural
heritage and honor our trust responsibilities to tribes.

UNITED STATES DEPARTMENT OF THE INTERIOR

BUREAU OF INDIAN AFFAIRS
1849 C St., NW
Washington, DC 20240 USA
Phone: 202-208-5116   Fax: 202-208-6334
E-mail: jamesmcdivitt@bia.gov
Website: www.doi.gov

Founded: 1824
Membership: 35
Scope: National

Description: An agency charged with carrying out the major
portion of the trust responsibility of the United States to Indian
tribes. This trust includes the protection and enhancement of
Indian lands and the conservation and development of natural
resources, including fish, wildlife, and outdoor recreation
resources.

Contact(s):
Daphne Berwald, ADMINISTRATIVE ASSISTANT
Phone: 202-208-7163
Gary Rankel, CHIEF OF BRANCH OF FISH, WILDLIFE AND
RECREATION
Phone: 202-208-4088
Sharon Blackwell, DEPUTY COMMISSIONER
Phone: 202-208-5116
Terry Virden, DIRECTOR OF OFFICE OF TRUST
RESPONSIBILITIES
Phone: 202-208-5831

UNITED STATES DEPARTMENT OF THE INTERIOR

BUREAU OF LAND MANAGEMENT
NATIONAL APPLIED RESOURCE CENTER
Denver Federal Center, Bldg. 50
Denver, CO 80225 USA
Phone: 303-236-6454   Fax: 303-236-6450
Website: www.blm.gov/nstc

Founded: NA
Membership: 97
Scope: Regional

Contact(s):
Francis Cherry, AK STATE DIRECTOR
222 W. 7th Ave., #13, Anchorage, AK 99513
Phone: 907-271-5076
Denise Meredith, AZ STATE DIRECTOR
222 North Central Avenue, Phoenix, AZ 85004
Phone: 602-417-9500
Michael Pool, CA STATE DIRECTOR
28 Cottage Way, Rm. W-1834, Sacramento, CA 95825
Phone: 916-978-4600
Fax: 916-978-4699
Ann Morgan, CO STATE DIRECTOR
2850 Youngfield St., Lakewood, CO 80215
Phone: 303-239-3700
Lee Barkow, DIRECTOR
lee_barkow@blm.gov
Gayle Gordon, EASTERN STATES DIRECTOR
7450 Boston Blvd, Springfield, VA 22153
Phone: 703-440-1700
Fax: 703-440-1599
Martha Hahn, ID STATE DIRECTOR
1387 S. Vinnell Way, Boise, ID 83709-1657
Phone: 208-373-4001
Mat Millenbach, MT STATE DIRECTOR
5001 Southgate Dr, Billings, MT 59107
Phone: 406-896-5012
Michelle Chavez, NM STATE DIRECTOR
1474 Rodeo Rd., Santa Fe, NM 87505
Phone: 505-438-7501
Fax: 505-438-7435
Robert Abbey, NV STATE DIRECTOR
1340 Financial Blvd., Reno, NV 89502-7147
Phone: 702-861-6590
Elaine Zielinski, OR STATE DIRECTOR
1515 SW 5th Ave., Portland, OR 97208
Phone: 503-952-6024
Sally Wisely, UT STATE DIRECTOR
324 S. State St. Ste. 301, Salt Lake City, UT 84145
Phone: 801-539-4010
Al Pierson, WY STATE DIRECTOR
5353 Yellowstone Rd., Cheyenne, WY 82003
Phone: 307-775-6001

UNITED STATES DEPARTMENT OF THE INTERIOR

BUREAU OF LAND MANAGEMENT
PUBLIC AFFAIRS
1849 C St., NW, LS-406
Washington, DC 20240 USA
Phone: 202-208-3801   Fax: 202-208-5242
Website: www.blm.gov

Founded: 1946
Membership: 500+

Scope: National

Description: Administers the public lands which are located primarily in the Western states and which amount to about 48 percent over 272 million acres of all federally owned lands. These lands and resources are managed under multiple-use principles, including outdoor recreation, fish and wildlife production, livestock grazing, timber, industrial development, watershed protection, and onshore mineral production.

Contact(s):
Bob Doyle, ASSISTANT DIRECTOR OF BUSINESS AND FISCAL SERVICES
Phone: 202-208-4864
Larry Finfer, ASSISTANT DIRECTOR OF COMMUNICATIONS
Phone: 202-208-6913
Warren Johnson, ASSISTANT DIRECTOR OF HUMAN RESOURCES
Phone: 202-501-6724
Gayle Gordon, ASSISTANT DIRECTOR OF INFORMATION RESOURCES MANAGEMENT
Carson Culp, ASSISTANT DIRECTOR OF MINERALS, REALTY, & RESOURCE PROTECTION
Phone: 202-208-4201
Henri Bisson, ASSISTANT DIRECTOR OF RENEWABLE RESOURCES & PLANNING
Phone: 202-208-4896
Nina Hatfield, DIRECTOR (ACTING)
Carol Macdonald, ENVIRONMENTAL (NEPA) ISSUES
Phone: 202-452-5111
W. Tipton, INFORMATION RESOURCE

UNITED STATES DEPARTMENT OF THE INTERIOR
BUREAU OF RECLAMATION
U.S. Department of the Interior 1849 C St., NW
Washington, DC 20240 USA

Founded: NA

Description: The Bureau of Reclamation was created by the Reclamation Act of 1902 to reclaim arid lands in the 17 Western states. This has been accomplished by the development of a system of works for the storage, diversion, and development of water. Reclamation's future role entails a shift in emphasis from development to total resource management and more effective use of existing facilities. Nonstructural means of meeting future water and power needs is now being emphasized.

Contact(s):
Paul Bledsoe, CHIEF OF PUBLIC AFFAIRS DIVISION
Phone: 202-208-4662
Steven Richardson, CHIEF OF STAFF
Phone: 202-208-4292
Eluid Martinez, COMMISSIONER
Phone: 202-208-4157
Stephen Magnussen, DIRECTOR OF OPERATIONS
Phone: 202-208-4082

UNITED STATES DEPARTMENT OF THE INTERIOR
BUREAU OF RECLAMATION
DENVER OFFICE
Bldg. 67, Denver Federal Center, P.O. Box 25007
Denver, CO 80225 USA

Founded: NA

Contact(s):
Wayne Deason, DEPUTY DIRECTOR OF OFFICE OF POLICY
Phone: 303-445-2781
David Montoya, DIRECTOR OF HUMAN RESOURCES
Phone: 303-445-2670
Kathy Gordon, DIRECTOR OF MANAGEMENT SERVICES
Phone: 303-445-3002
Neal Stessman, DIRECTOR OF RECLAMATION SERVICE CENTER
Phone: 303-445-2692

UNITED STATES DEPARTMENT OF THE INTERIOR
BUREAU OF RECLAMATION
LOWER COLORADO REGION
Boulder City, NV 89006-1470 USA
Phone: 702-293-8411   Fax: 702-293-8614
E-mail: rwalsh@oc.usbr.gov
Website: www.lc.usbr.gov

Founded: NA
Membership: 1000
Scope: Regional

Contact(s):
Bob Johnson, DIRECTOR
Bob Walsh, DIRECTOR OF PUBLIC AFFAIRS OFFICER

UNITED STATES DEPARTMENT OF THE INTERIOR
BUREAU OF RECLAMATION
MID PACIFIC REGION
Federal Office Bldg., 2800 Cottage Way
Sacramento, CA 95825 USA
Phone: 916-978-5000   Fax: 916-978-5599
E-mail: mp.usbr.gov

Founded: NA
Membership: 800
Scope: Regional

Contact(s):
Kirk Rodgers, DIRECTOR (ACTING)

UNITED STATES DEPARTMENT OF THE INTERIOR
BUREAU OF RECLAMATION
PACIFIC NORTHWEST REGION
1150 N. Curtis Rd., Suite 100
Boise, ID 83706-1234 USA
Phone: 208-378-5012   Fax: 208-378-5129
Website: www.pn.usbr.gov

Founded: NA
Membership: 1100
Scope: National

Contact(s):
Bill Macdonald, DIRECTOR

UNITED STATES DEPARTMENT OF THE INTERIOR
BUREAU OF RECLAMATION
UPPER COLORADO REGION
125 South State St., Rm. 6107
Salt Lake City, UT 84138 USA
Phone: 801-524-3600 Fax: 801-524-5499
Website: www.uc.usbr.gov

Founded: NA
Membership: 170
Scope: Regional

Contact(s):
Christine Karas, CHIEF OF ENVIRONMENTAL RESOURCES
GROUP
Phone: 801-524-3679
Barry Wirth, PUBLIC AFFAIRS OFFICER
Phone: 801-524-3774
Rick Gold, REGIONAL DIRECTOR

UNITED STATES DEPARTMENT OF THE INTERIOR
GREAT PLAINS REGION
P.O. Box 36900
Billings, MT 59107-6900 USA
Phone: 406-247-7600 Fax: 406-247-7393
Website: www.gp.usbr.gov

Founded: NA
Scope: Regional

Contact(s):
Maryanne Bach, DIRECTOR
Mark Andersen, DIRECTOR OF PUBLIC AFFIAIRS

UNITED STATES DEPARTMENT OF THE INTERIOR
OFFICE OF SURFACE MINING RECLAMATION AND
ENFORCEMENT
Department of Interior, Interior South Bldg.
1951 Constitution Ave., NW
Washington, DC 20240 USA
Phone: 202-208-2719
E-mail: getinfo@osmre.gov
Website: http://www.osmre.gov/

Founded: NA

Description: Established by the Surface Mining Control and
Reclamation Act of 1977 to administer the nationwide program
to protect society and the environment from adverse effects of
coal mining operations, to establish national standards for
regulating the surface environmental effects of coal mining, to
support state implementation of such regulatory programs, and
to promote reclamation of abandoned mine lands.

UNITED STATES DEPARTMENT OF THE INTERIOR
UNITED STATES FISH AND WILDLIFE SERVICE
Department of Interior, 1849 C St., Rm. 3359
Washington, DC 20240 USA
Phone: 202-208-5634 Fax: 202-208-7407
Website: www.fws.gov

Founded: NA
Scope: National

Description: Effective July 1, 1974, an act of Congress (Public
Law 93-271, April 22, 1974) renamed the Bureau of Sport
Fisheries and Wildlife, the United States Fish and Wildlife
Service, under the Assistant Secretary for Fish and Wildlife and
Parks. The Service is the lead federal agency in the conservation of the nation's migratory birds, threatened and endangered
species, certain marine mammals, and sport fishing. The
Service administers fish and wildlife restoration grant programs
to state governments, provides technical assistance to state
and foreign governments, serves as lead federal agency in
international conventions on wildlife conservation, and
operates a program of public affairs and education to enhance
the public's understanding and appreciation of America's fish
and wildlife resources.

Contact(s):
Paul Henne, ASSISTANT DIRECTOR OF ADMINISTRATION
Phone: 202-208-4888
Gary Frazer, ASSISTANT DIRECTOR OF ECOLOGICAL
SERVICES
Phone: 202-208-4646
Thomas Melius, ASSISTANT DIRECTOR OF EXTERNAL
AFFAIRS
Phone: 202-208-4500
Cathleen Short, ASSISTANT DIRECTOR OF FISHERIES
Phone: 202-208-6394
Marshall Jones, ASSISTANT DIRECTOR OF
INTERNATIONAL AFFAIRS
Phone: 202-208-6393
Daniel Ashe, ASSISTANT DIRECTOR OF REFUGES AND
WILDLIFE
Phone: 202-208-5333
Juanita Williams, CHIEF OF DIVISION OF CONTRACTING
AND GENERAL SERVICES
Phone: 703-358-1901
Nancy Gloman, CHIEF OF DIVISION OF ENDANGERED
SPECIES
Phone: 703-358-2171
Paul Camp, CHIEF OF DIVISION OF ENGINEERING
Phone: 303-275-2300
Everett Wilson, CHIEF OF DIVISION OF ENVIRONMENTAL
CONTAMINANTS
Phone: 703-358-2148
David Holland, CHIEF OF DIVISION OF FINANCE
Phone: 703-358-1742
Hannibal Bolton, CHIEF OF DIVISION OF FISH AND
WILDLIFE MANAGEMENT ASSISTANCE
Phone: 703-358-1718
Benjamin Tuggle, CHIEF OF DIVISION OF HABITAT
CONSERVATION
Phone: 703-358-2161
William Brooks, CHIEF OF DIVISION OF INFORMATION
RESOURCES MANAGEMENT
Phone: 703-358-1729
Kevin Adams, CHIEF OF DIVISION OF LAW
ENFORCEMENT
Phone: 703-358-1949
William Knapp, CHIEF OF DIVISION OF NATIONAL FISH
HATCHERIES
Phone: 703-358-1715
Kent Baum, CHIEF OF DIVISION OF PERSONNEL
MANAGEMENT
Phone: 202-208-6104
Jeffery Donahoe, CHIEF OF DIVISION OF REALTY
Phone: 703-358-1713
Jim Kurth, CHIEF OF DIVISION OF REFUGE
Phone: 703-358-1744

Arthur Ford, CHIEF OF FWS FINANCE CENTER
Denver Federal Center, P.O. Box 25207, Denver, CO
80225-0207
Jerome Butler, CHIEF OF OFFICE FOR HUMAN
RESOURCES
Phone: 202-208-3195
Alexandra Pitts, CHIEF OF OFFICE OF CONGRESSIONAL
AND LEGISLATIVE SERVICES
Phone: 202-208-5403
Robert Lange, CHIEF OF OFFICE OF FEDERAL AID
Phone: 703-358-2156
Herbert Raffaele, CHIEF OF OFFICE OF INTERNATIONAL
AFFAIRS
Phone: 703-358-1754
Kenneth Stansell, CHIEF OF OFFICE OF MANAGEMENT
AUTHORITY
Phone: 703-358-2093
Jon Andrew, CHIEF OF OFFICE OF MIGRATORY BIRD
MANAGEMENT
Phone: 703-358-1714
Megan Durham, CHIEF OF OFFICE OF PUBLIC AFFAIRS
Phone: 202-208-4131
Marshall Jones, DEPUTY DIRECTOR
Phone: 202-208-4545
Robert Lesino, PROGRAM MANAGER OF FEDERAL DUCK
STAMP PROGRAM
Phone: 202-208-4354

UNITED STATES DEPARTMENT OF THE INTERIOR

UNITED STATES FISH AND WILDLIFE SERVICE
ALASKA REGIONAL OFFICE 7
1011 E. Tudor Rd.
Anchorage, AK 99503 USA
Phone: 907-786-3542   Fax: 907-786-3306
Website: www.fws.gov

Founded: NA
Scope: National

Publication(s): see publications on website

Contact(s):
David Allen, REGIONAL DIRECTOR
Phone: 907-786-3542

UNITED STATES DEPARTMENT OF THE INTERIOR

UNITED STATES FISH AND WILDLIFE SERVICE
CALIFORNIA-NEVADA OPERATIONS
2800 Cottage Way, Rm. W-2606
Sacramento, CA 95825 USA
Phone: 916-414-6464   Fax: 916-414-6486

Founded: NA

Contact(s):
Michael Spear, MANAGER

UNITED STATES DEPARTMENT OF THE INTERIOR

UNITED STATES FISH AND WILDLIFE SERVICE
DELAWARE BAY ESTUARY PROJECT
2610 Whitehall Neck Rd.
Smyrna, DE 19977 USA
Phone: 302-653-9152   Fax: 302-653-9421

Founded: NA

Scope: National

Description: The Delaware Estuary Project was established to coordinate, complement, and support existing U.S. Fish and Wildlife Service programs, focusing on important natural resource issues in the Delaware River watershed. The office provides technical assistance to the EPA's National Estuary Program for the Delaware Bay and Delaware's Inland Bays Estuary programs (started in 1988).

UNITED STATES DEPARTMENT OF THE INTERIOR

UNITED STATES FISH AND WILDLIFE SERVICE
GREAT LAKES-BIG RIVERS REGIONAL OFFICE 3
1 Federal Dr., Federal Bldg.
Fort Snelling, MN 55111 USA
Phone: 612-713-5301   Fax: 612-713-5284

Founded: NA
Scope: Regional

Contact(s):
Brian Norris, ASSISTANT REGIONAL DIRECTOR OF
EXTERNAL AFFAIRS
Phone: 612-713-5310
Marvin Moriarty, DEPUTY REGIONAL DIRECTOR
Phone: 612-713-5201
William Hartwig, REGIONAL DIRECTOR
Phone: 612-713-5301

UNITED STATES DEPARTMENT OF THE INTERIOR

UNITED STATES FISH AND WILDLIFE SERVICE
NATIONAL CONSERVATION TRAINING CENTER
Rt. 1 Box 166
Shepherdstown, WV 25443 USA
Phone: 304-876-1600   Fax: 304-876-7227
Website: www.fws.gov

Founded: NA
Membership: 100
Scope: National

Description: The mission of the Center is to advance conservation of fish, wildlife, and their habitats through leadership in conservation education for the public, training for the conservation and resource management community, and fostering alliances among diverse interests.

Contact(s):
Todd Jones, CHIEF, DIVISION OF TRAINING
Phone: 304-876-7431
Mona Womack, DEPUTY DIRECTOR
Phone: 304-876-7263
John Lemon, DIRECTOR
Phone: 304-876-7263

UNITED STATES DEPARTMENT OF THE INTERIOR

UNITED STATES FISH AND WILDLIFE SERVICE
NATIONAL FISH AND WILDLIFE FORENSICS
LABORATORY
1490 East Main St.
Ashland, OR 97520 USA
Phone: 541-482-4191   Fax: 541-482-4989
Website: http://www.labs.fws.gov

Founded: NA

Description: The mission of the Laboratory is to provide forensic crime lab, support for wildlife law enforcement investigations at the federal, state, and international levels.

Contact(s):
Ken Goddard, DIRECTOR

UNITED STATES DEPARTMENT OF THE INTERIOR

UNITED STATES FISH AND WILDLIFE SERVICE
NORTHEAST REGIONAL OFFICE 5
300 Westgate Center Dr.
Hadley, MA 01035 USA
Phone: 413-253-8200   Fax: 413-253-8482
Website: www.fws.gov

Founded: NA
Membership: 220
Scope: Regional

Contact(s):
Mamie Parker, REGIONAL DIRECTOR
Phone: 413-253-8300

UNITED STATES DEPARTMENT OF THE INTERIOR

UNITED STATES FISH AND WILDLIFE SERVICE
PACIFIC REGIONAL OFFICE 1
Eastside Federal Complex, 911 NE 11th Ave.
Portland, OR 97232-4181 USA

Founded: NA

Contact(s):
Joan Jewett, CHIEF OFFICER OF PUBLIC AFFAIRS
Phone: 503-231-6121
Ann Badgley, REGIONAL DIRECTOR
Phone: 503-231-6828

UNITED STATES DEPARTMENT OF THE INTERIOR

UNITED STATES FISH AND WILDLIFE SERVICE
SOUTHEAST REGIONAL OFFICE 4
1875 Century Blvd.
Atlanta, GA 30345 USA
Phone: 404-679-4000   Fax: 404-679-4006
Website: www.fws.gov

Founded: NA
Membership: 250
Scope: National

Contact(s):
Sam Hamilton, REGIONAL DIRECTOR
Phone: 404-679-4000

UNITED STATES DEPARTMENT OF THE INTERIOR

UNITED STATES FISH AND WILDLIFE SERVICE
SOUTHWEST REGIONAL OFFICE 2
Albuquerque, NM 87103 USA
Phone: 505-248-6911   Fax: 505-248-6915
Website: www.southwest.fws.gov

Founded: NA
Membership: 200
Scope: Regional

Contact(s):
Tom Bauer, ASSISTANT DIRECTOR OF EXTERNAL AFFAIRS
Phone: 505-248-6911
Fax: 505-248-6915
Nancy Kaufman, REGIONAL DIRECTOR
Phone: 505-248-6282

UNITED STATES DEPARTMENT OF THE INTERIOR

UNITED STATES GEOLOGICAL SURVEY
12201 Sunrise Valley Drive National Center
Reston, VA 20192 USA
Phone: 703-648-4000
Website: www.usgs.gov

Founded: 1879
Scope: National

Description: The Geological Survey works in cooperation with more than 2,000 organizations across the country to provide reliable, impartial, scientific information to resource managers, planners, and other customers.

Contact(s):
Dennis Fenn, ASSOCIATE DIRECTOR FOR BIOLOGY
Phone: 703-648-4050
P. Leahy, ASSOCIATE DIRECTOR FOR GEOLOGY
Phone: 703-648-6600
Robert Hirsch, ASSOCIATE DIRECTOR FOR WATER
Phone: 703-648-5215
Barbara Ryan, ASSOCIATE DIRECTOR OF OPERATIONS
Phone: 703-648-7413
Thomas Casadevall, CENTRAL REGIONAL DIRECTOR
Denver Federal Center, Mail Stop 150, Denver, CO 80225
Phone: 303-202-4740
Barbara Wainman, CHIEF, OFFICE OF COMMUNICATIONS
Phone: 703-648-5750
S. Cook, CHIEF, OFFICE OF EQUAL OPPORTUNITY
Phone: 703-648-7770
Robert Hosenfeld, CHIEF, OFFICE OF PERSONNEL
Phone: 703-648-7442
Timothy West, CONGRESSIONAL LIAISON OFFICER
Phone: 703-648-4300
Charles Groat, DIRECTOR
Phone: 703-648-7411
Bonnie McGregor, EASTERN REGIONAL DIRECTOR
1700 Leetown Rd., Kearneysville, WV 25430
Phone: 304-724-4521
Trudy Harlow, PUBLIC AFFAIRS OFFICER
Phone: 703-648-4483
Amy Holley, SENIOR ADVISOR TO THE DIRECTOR
Phone: 703-648-4411
James Devine, SENIOR ADVISOR, SCIENCE APPLICATIONS
Phone: 703-648-4423
Carl Shapiro, STRATEGIC CHIEF PLANNING, ANALYSIS
John Buffington, WESTERN REGIONAL DIRECTOR
909 First Ave., 8th Flr., Seattle, WA 98104
Phone: 206-220-4600

UNITED STATES DEPARTMENT OF THE INTERIOR

UNITED STATES GEOLOGICAL SURVEY
BIOLOGICAL RESOURCES DIVISION
12201 Sunrise Valley Dr., MS-300
Reston, VA 20192 USA
Phone: 703-648-4050   Fax: 703-648-4042
Website: http://biology.usgs.gov/

Founded: NA

Description: The Biological Resources Division works with others to provide the scientific understanding and technologies needed to manage the Nation's biological resources.

Contact(s):
Dennis Fenn, CHIEF BIOLOGIST
Phone: 703-648-4050
William Gregg, CHIEF OF INTERNATIONAL AFFAIRS
Phone: 703-648-4067
Susan Haseltine, DEPUTY CHIEF BIOLOGIST FOR SCIENCE
Phone: 703-648-4060
Suzzette Kimball, REGIONAL BIOLOGIST
Eastern Regional Office, 1700 Leestown Rd., Kearneysville, WV 25430
Phone: 304-724-4500
J. Ludke, REGIONAL BIOLOGIST
Central Regional Office, Denver Federal Center, P.O. Box 25046, Bld. 20, Mailstop 300, Denver, CO 80225-0046
Phone: 303-236-2730

UNITED STATES DEPARTMENT OF THE NAVY

1000 Navy Pentagon, Department of the Navy
Washington, DC 20350-1000 USA
Website: www.navy.mil

Founded: NA

Description: The mission of the Navy is to maintain, train, and equip combat-ready Naval forces capable of winning wars, deterring aggression and maintaining freedom of the seas.

Contact(s):
Richard Danzig, SECRETARY
Jerry Hultin, UNDERSECRETARY

UNITED STATES DEPARTMENT OF THE NAVY

U.S. MARINE CORPS
Headquarters, U.S. Marine Corps, 2 Navy Annex
Washington, DC 20380-4775 USA
Phone: 703-695-8332

Founded: NA

Description: The Marine Corps, as America's premier crisis response force, trains as it fights. Accordingly, Marine Corps cultural and natural resources managers provide and maintain a variety of landscapes to support military training, while protecting and preserving the cultural and natural resources the American people cherish for their intrinsic value.

Contact(s):
Jim Omans, HEAD OF NATURAL RESOURCES SECTION
Phone: 703-695-8232

UNITED STATES DEPARTMENT OF THE NAVY

U.S. MARINE CORPS
MARINE CORPS INSTALLATIONS
USA

Founded: NA

Contact(s):
Roy Madden, MCAGCC TWENTYNINE PALMS, CA: HEAD OF NATURAL RESOURCES BRANCH
Phone: 619-830-5719
Alice Howard, MCAS BEUFORT, SC: HEAD OF ENVIRONMENTAL MANAGEMENT DEPARTMENT
Phone: 803-522-7370
Mark Brannan, MCAS CHERRY POINT, NC: HEAD OF ENVIRONMENTAL DEPARTMENT
Ron Pearce, MCAS YUMA, AZ: HEAD OF NATURAL RESOURCES BRANCH
Phone: 802-341-3318
Bob Warren, MCB CAMP LEJEUNE, NC: HEAD OF ENVIRONMENTAL MANAGEMENT DEPARTMENT
Phone: 910-451-5003
Lupe Armas, MCB CAMP PENDLETON, CA: HEAD OF ENVIRONMENTAL MANAGEMENT DEPARTMENT
Phone: 619-725-3561
Bruce Frizzell, MCCDC QUANTICO, VA: HEAD OF ENVIRONMENTAL MANAGEMENT DEPARTMENT
Phone: 703-640-4030
Jerry Palmer, MCLB ALBANY, GA: HEAD OF ENVIRONMENTAL MANAGEMENT DEPARTMENT
Phone: 912-439-6261
Jack Stormo, MCLB BARSTOW, CA: HEAD OF ENVIRONMENTAL MANAGEMENT DEPARTMENT
Phone: 619-577-6111
Johnsie Nabors, MCRD PARRIS ISLAND, SC: HEAD OF ENVIRONMENTAL MANAGEMENT DEPARTMENT
Phone: 803-525-2779

UNITED STATES DEPARTMENT OF TRANSPORTATION

Office of Public Affairs, Nassif Bldg., 400 7th St., SW
Washington, DC 20590 USA
Phone: 202-366-4000
Website: www.dot.gov

Founded: 1967
Membership: 100,000
Scope: National

Description: Composed of these main elements: The United States Coast Guard, Federal Aviation Administration, Federal Highway Administration, Federal Railroad Administration, Maritime Administration, St. Lawrence Seaway Development Corporation, National Highway Traffic Safety Administration, Federal Transit Administration, and Research and Special Programs Administration. Major objectives are to develop and improve a coordinated national transportation system consistent with other national objectives, such as environmental protection, and to stimulate technological advances in the industry, preserving the nation's free enterprise transportation network.

Publication(s): Public Roads (Federal Highway Administration), Merchant Vessels of the United States (United States Coast Guard)

Contact(s):
Norm Mineta, ASSISTANT TO THE SECRETARY AND DIRECTOR OF PUBLIC AFFAIRS
Mortimer Downey, DEPUTY SECRETARY
Phone: 202-366-2222

UNITED STATES DEPARTMENT OF TRANSPORTATION

FEDERAL AVIATION ADMINISTRATION
FOB 10A 800 Independence Ave., SW
Washington, DC 20591 USA
Phone: 202-267-3484
Website: http://www.faa.gov

Founded: NA

Description: Charged with regulating air commerce to foster aviation safety; promoting civil aviation and a national system of airports; achieving efficient use of navigable airspace; and developing and operating a common system of air traffic control and air navigation for both civilian and military aircraft.

Contact(s):
Jane Garvey, ADMINISTRATOR
Phone: 202-267-3111

UNITED STATES DEPARTMENT OF TRANSPORTATION

FEDERAL HIGHWAY ADMINISTRATION
Attn: Executive Director, Washington Headquarters, 400 7th St., SW
Washington, DC 20590 USA

Founded: NA

Description: Charged with carrying out the Department of Transportation responsibilities concerned with the highway mode of land transport, including intermodal connections, has the primary missions of ensuring the safety of the motor carrier industry and that the Nation's highway transportation system is safe, economic, and efficient with respect to the movement of people and goods, while giving full consideration to the highway's impact on the environment and social and economic conditions.

Keyword(s): Engineering, Historic Preservation, Planning Management, Transportation

Contact(s):
Kenneth Wykle, ADMINISTRATOR
Phone: 202-366-0650
Karen Skelton, CHIEF COUNSEL
Phone: 202-366-0740
Gloria Jeff, DEPUTY ADMINISTRATOR
Phone: 202-366-2240
George Moore, DIRECTOR, OFFICE OF ADMINISTRATION
Phone: 202-366-0604
Edward Morris, DIRECTOR, OFFICE OF CIVIL RIGHTS
Phone: 202-366-0693
Fred Hempel, DIRECTOR, OFFICE OF CORPORATE MANAGEMENT
Phone: 202-366-9393
Walter Sutton, DIRECTOR, OFFICE OF POLICY
Phone: 202-366-0585
Joseph Toole, DIRECTOR, OFFICE OF PROFESSIONAL DEVELOPMENT
Phone: 703-235-0519
Gail Shibley, DIRECTOR, OFFICE OF PUBLIC AFFAIRS
Phone: 202-366-0660
Dennis Judycki, DIRECTOR, OFFICE OF RESEARCH, DEVELOPMENT, AND TECHNOLOGY
Phone: 202-493-3165
Anthony Kane, EXECUTIVE DIRECTOR
Phone: 202-366-2242

Arthur Hamilton, PROGRAM MANAGER, OFFICE OF FEDERAL LANDS
Phone: 202-366-9494
Vincent Schimmoler, PROGRAM MANAGER, OFFICE OF INFRASTRUCTURE
Phone: 202-366-0371
Julie Cirillo, PROGRAM MANAGER, OFFICE OF MOTOR CARRIER AND HIGHWAY SAFETY
Phone: 202-366-2519
Christine Johnson, PROGRAM MANAGER, OFFICE OF OPERATIONS
Phone: 202-366-0408
Cynthia Burbank, PROGRAM MANAGER, OFFICE OF PLANNING AND ENVIRONMENT
Phone: 202-366-0342

UNITED STATES DEPARTMENT OF TRANSPORTATION

FEDERAL TRANSIT ADMINISTRATION
400 7th St., SW
Washington, DC 20590 USA
Phone: 202-366-4043
Website: http://www.fta.dot.gov/

Founded: NA

Description: Seeks to improve the environmental standards of American cities through grant programs which extend and modernize existing urban mass transit equipment and facilities and which study, develop, and test new equipment and concepts in urban mass transit applications and operations.

Keyword(s): Transportation

Contact(s):
Gordon Linton, ADMINISTRATOR OF OFFICE OF THE ADMINISTRATOR
Phone: 202-366-4040
Dorrie Aldrich, ASSOCIATE ADMINISTRATOR FOR ADMINISTRATION
Phone: 202-366-4007
Michael Winter, ASSOCIATE ADMINISTRATOR FOR BUDGET AND POLICY
Phone: 202-366-4050
Charlotte Adams, ASSOCIATE ADMINISTRATOR FOR PLANNING
Phone: 202-366-4033
Hiram Walker, ASSOCIATE ADMINISTRATOR FOR PROGRAM MANAGEMENT
Phone: 202-366-4020
Edward Thomas, ASSOCIATE ADMINISTRATOR FOR RESEARCH DEMONSTRATION AND INNOVATION
Phone: 202-366-4052
Patrick Reilly, CHIEF COUNSEL
Phone: 202-366-4063
Nuria Fernandez, DEPUTY ADMINISTRATOR OF OFFICE OF ADMINISTRATION
Phone: 202-366-4325
John Spencer, DEPUTY ASSOCIATE ADMINISTRATOR
Phone: 202-366-1691
A. Yen, DEPUTY ASSOCIATE ADMINISTRATOR
Phone: 202-366-4991
Timothy Wolgast, DEPUTY ASSOCIATE ADMINISTRATOR
Phone: 202-366-4007
Janet Sahaj, DEPUTY ASSOCIATE ADMINISTRATOR
Phone: 202-366-4020
Gregory McBride, DEPUTY CHIEF COUNSEL
Phone: 202-366-4063

Aurther Lopez, DIRECTOR, OFFICE OF CIVIL RIGHTS
Phone: 202-366-4018
Bruce Frame, DIRECTOR, OFFICE OF PUBLIC AFFAIRS
Phone: 202-366-4319
Mary Knapp, EXECUTIVE INFORMATION, OFFICE OF
PUBLIC AFFAIRS
Phone: 202-366-9788

UNITED STATES DEPARTMENT OF TRANSPORTATION

FEDERAL RAILROAD ADMINISTRATION
1120 Vermont Ave, NW
Washington, DC 20590 USA
Phone: 202-366-4000   Fax: 202-493-6169
Website: www.fra.dot.gov

Founded: 1967
Scope: National

Description: The FRA promulgates and enforces rail safety
regulations, administers financial assistance programs for
designated railroads, conducts research and development in
support of improved railroad safety and national rail transportation policy, as well as monitors rail passenger service
nationwide, and consolidates government support of rail transportation activities.

Contact(s):
Allan Rutter, ADMINISTRATOR OF FEDERAL RR
ADMINSTRATIONS
S. Lindsey, CHIEF COUNSEL

UNITED STATES DEPARTMENT OF TRANSPORTATION

NATIONAL HIGHWAY TRAFFIC SAFETY
ADMINISTRATION
Nassif Bldg., 400 7th St., SW
Washington, DC 20590 USA
Phone: 202-366-9550   Fax: 202-366-7402

Founded: NA
Scope: State

Contact(s):
Ricardo Martinez, ADMINISTRATOR
Phone: 202-366-1836
Herman Simms, ASSOCIATE ADMINISTRATOR FOR
ADMINISTRATION (ACTING)
Phone: 202-366-1788
William Walsh, ASSOCIATE ADMINISTRATOR FOR PLANS
AND POLICY
Phone: 202-366-2550
Raymond Owings, ASSOCIATE ADMINISTRATOR FOR
RESEARCH AND DEVELOPMENT
Phone: 202-366-1537
Kenneth Weinstein, ASSOCIATE ADMINISTRATOR FOR
SAFETY ASSURANCE
Phone: 202-366-9700
L. Shelton, ASSOCIATE ADMINISTRATOR FOR SAFETY
PERFORMANCE STANDARDS
Phone: 202-366-1810
James Nichols, ASSOCIATE ADMINISTRATOR FOR
TRAFFIC SAFETY PROGRAMS (ACTING)
Phone: 202-366-1755
Frank Seales, CHIEF COUNSEL
Phone: 202-366-9511
Philip Recht, DEPUTY DIRECTOR
Phone: 202-366-2775

Donald Bischoff, EXECUTIVE DIRECTOR
Phone: 202-366-2111

UNITED STATES DEPARTMENT OF TRANSPORTATION

SAINT LAWRENCE SEAWAY DEVELOPMENT
CORPORATION
400 7th St. SW,
Washington, DC 20590 USA
Phone: 202-366-0091   Fax: 202-366-7047
Website: www.greatlakes-seaway.com

Founded: NA
Membership: 15
Scope: Statewide

Publication(s): Seaway Compass

Contact(s):
Albert Jacquez, ADMINISTRATOR
Phone: 202-366-0091
Tim Downey, DEPUTY DIRECTOR OF OFFICE OF
CONGRESSIONAL AND PUBLIC AFFAIRS
P.O. Box 44090, Washington, DC 20026-4090
Phone: 202-366-0110

UNITED STATES DEPARTMENT OF TRANSPORTATION

UNITED STATES COAST GUARD
2100 2nd St., SW
Washington, DC 20593-0001 USA
Phone: 202-267-2229   Fax: 202-267-4696

Founded: NA
Scope: Statewide

Contact(s):
Timothy Josiah, CHIEF OF STAFF
Phone: 202-267-1642

UNITED STATES DEPARTMENT OF TREASURY

1500 Pennsylvania Ave., NW
Washington, DC 20220 USA
Phone: 202-622-2000   Fax: 202-622-1999
Website: www.ustreas.gov

Founded: NA
Scope: National

Description: The basic functions of the Department of the
Treasury include: economic and fiscal policy; government
accounting, cash, and debt management; international
economic policy; and enforcement of customs and trade laws.

UNITED STATES DEPARTMENT OF TREASURY

U.S. CUSTOMS SERVICE
1300 Pennsylvania Ave., NW,
Washington, DC 20229 USA
Phone: 202-927-1770   Fax: 202-927-1393
Website: www.customs.gov

Founded: NA
Scope: State

Description: The United States Customs Service is responsible
for the enforcement of the U.S. laws regarding the importation
and exportation of injurious and endangered species.

Publication(s): US Customs Today - Monthly Newsletter

Contact(s):
Charles Winwood, ACTING COMMISSIONER
Bonni Tischler, ASSISTANT COMMISSIONER FOR OFFICE
OF INVESTIGATIONS, RM. 6-5E
Phone: 202-927-1600
Charlie Simonsen, SPECIAL AGENT IN CHARGE
San Francisco, CA 94111
Ken Kilroy, SPECIAL AGENT IN CHARGE
Seattle, WA 98104-1048
Gary Hillberry, SPECIAL AGENT IN CHARGE
Denver, CO 80112-5131
Jeremiah Sullivan, SPECIAL AGENT IN CHARGE
Buffalo, NY 14202
Leonard Lindheim, SPECIAL AGENT IN CHARGE
San Antonio, TX 78216
Allan Doody, SPECIAL AGENT IN CHARGE
Baltimore, MD 21202
Gary Waugh, SPECIAL AGENT IN CHARGE
Detroit, MI 48226-2568
Bruce Murray, SPECIAL AGENT IN CHARGE (ACTING)
Tampa, FL 33607
Frank Figueroa, SPECIAL AGENT IN CHARGE (ACTING)
Miami, FL 33166
James Lewis, SPECIAL AGENT IN CHARGE (ACTING)
Chicago, IL 60607
Bobby Fernandez, SPECIAL AGENT IN CHARGE (ACTING)
Old San Juan, PR 00901
Joe Webber, SPECIAL AGENT IN CHARGE (ACTING)
El Paso, TX 79925
Awilda Villafane, SPECIAL AGENT IN CHARGE (ACTING)
555 E. River Rd., Tucson, AZ 85704
Ben Devane, SPECIAL AGENT IN CHARGE (ACTING)
Atlanta, GA 30354

UNITED STATES DEPARTMENT OF TREASURY
U.S. CUSTOMS SERVICE
EAST TEXAS CMC
2323 S. Shepard St., Suite 1200
Houston, TX 77019 USA
Phone: 713-387-7200   Fax: 713-387-7202

Founded: NA

Contact(s):
John Babb, DIRECTOR

UNITED STATES DEPARTMENT OF TREASURY
U.S. CUSTOMS SERVICE
GULF CMC
423 Canal St.
New Orleans, LA 70130 USA
Phone: 504-670-2404   Fax: 504-670-2286

Founded: NA
Membership: 150
Scope: Regional

Contact(s):
Leticia Moran, DIRECTOR

UNITED STATES DEPARTMENT OF TREASURY
U.S. CUSTOMS SERVICE
MID AMERICA CMC
610 S. Canal St., Suite 900
Chicago, IL 60607 USA
Phone: 312-983-9100   Fax: 312-886-4921
Website: www.uscustoms.treas.gov

Founded: NA
Membership: 22
Scope: Regional

UNITED STATES DEPARTMENT OF TREASURY
U.S. CUSTOMS SERVICE
NEW YORK CMC
6 World Trade Center, Rm. 716
New York, NY 10048 USA
Phone: 212-637-7900
Website: www.customs.treas.gov

Founded: NA
Membership: 2,000
Scope: International

Contact(s):
John Martuge, DIRECTOR

UNITED STATES DEPARTMENT OF TREASURY
U.S. CUSTOMS SERVICE
NORTH ATLANTIC CMC
10 Causeway St.
Boston, MA 02222 USA
Phone: 617-565-6210   Fax: 617-565-6277
Website: www.uscustoms.treas.gov

Founded: NA
Scope: International

Contact(s):
Philip Spayd, DIRECTOR

UNITED STATES DEPARTMENT OF TREASURY
U.S. CUSTOMS SERVICE
SOUTH PACIFIC CMC
One World Trade Center, P.O. Box 32639
Long Beach, CA 90831 USA
Phone: 562-980-3100

Founded: NA
Scope: National

Contact(s):
Audrey Adams, DIRECTOR

UNITED STATES TREASURY DEPARTMENT
U.S. CUSTOMS SERVICE
SOUTHEAST
909 SE 1st Ave.
Miami, FL 33131 USA
Phone: 305-810-5120   Fax: 305-810-5143

Founded: NA

UPPER COLORADO RIVER COMMISSION
355 S. 4th East St.
Salt Lake City, UT 84111 USA
Phone: 801-531-1150   Fax: 801-531-9705

Founded: 1949
Scope: Regional

Description: An administrative agency composed of commissioners appointed by the states of the Upper Division of the
Colorado River - Colorado, New Mexico, Utah, and Wyoming,
and by the President of the U.S.

Contact(s):
Frank Maynes, CHAIRMAN
P.O. Drawer 2717, Durango, CO 81501
Wayne Cook, EXECUTIVE DIRECTOR AND SECRETARY

U.S. ARMY CORPS OF ENGINEERS

U.S. ARMY ENGINEER DISTRICT, HUNTINGTON
502 8th Street
Huntington, WV 25701-2070 USA
Phone: 304-529-5395　　　Fax: 304-529-5591
Website: www.intra.lrh.usace.army.mil

Founded: NA
Membership: 500
Scope: Regional

Contact(s):
John Ridenburg
COLONEL

U.S. ARMY CORPS OF ENGINEERS

U.S. ARMY ENGINEER DISTRICT, NASHVILLE
P.O. Box 1070
Nashville, TN 37202-1070 USA
Phone: 615-736-5626　　　Fax: 615-736-5497
E-mail: edward.m.evans@lrn02.usace.army.mil
Website: www.lrn.usace.army.mil

Founded: NA
Membership: 125
Scope: National

Publication(s): District Digest

Contact(s):
Edward Evans, PUBLIC AFFAIRS OFFICER

U.S. ARMY CORPS OF ENGINEERS

U.S. ARMY ENGINEER DISTRICT, NEW ENGLAND
696 Virginia Rd.
Concord, MA 01742-2751 USA
Phone: 978-318-8220　　　Fax: 978-318-8821

Founded: NA
Membership: 6
Scope: International

Contact(s):
Joe Bocchino, EXECUTIVE ASSISTANT

U.S. ARMY CORPS OF ENGINEERS

U.S. ARMY ENGINEER DISTRICT, NORFOLK
Waterfield Building, 803 Front Street
Norfolk, VA 23510-1096 USA
Phone: 757-441-7601　　　Fax: 757-441-7678
E-mail: neo@usace.army.mil
Website: www.nao.usace.army.mil

Founded: NA
Membership: 450
Scope: Regional

Contact(s):
Bob Hume, CHIEF REGULARTORY BRANCH

U.S. ARMY CORPS OF ENGINEERS

U.S. ARMY ENGINEER DISTRICT, WILMINGTON
P.O. Box 1890
Wilmington, NC 28402-1890 USA
Phone: 910-251-4501　　　Fax: 910-251-4185
Website: ww.saw.usace.army.mil

Founded: NA
Membership: 200
Scope: Statewide

Publication(s): Wilmington District Newsletter

Contact(s):
Penny Schmidt, CHIEF OF PUBLIC AFFAIRS

U.S. ARMY CORPS OF ENGINEERS

U.S. ARMY ENGINEER DIVISION, PACIFIC OCEAN
Building 525
Fort Shafter, HI 96858-5440 USA
Phone: 808-438-1500　　　Fax: 808-438-8387

Founded: NA
Membership: 70
Scope: Regional

Contact(s):
Frank Oliva, KEY CONTACT

U.S. ARMY CORPS OF ENGINEERS

U.S. ARMY ENGINEER DISTRICT, ST. LOUIS
1222 Spruce Street
St. Louis, MO 63103-2833 USA
Phone: 314-331-8010　　　Fax: 314-331-8770
Website: www.mvs.usce.army.mil

Founded: NA
Membership: 300
Scope: National

Contact(s):
Linda Collins, EXECUTIVE SECRETARY

U.S. ARMY CORPS OF ENGINEERS

U.S. ARMY ENGINEER DISTRICT, TULSA
1645 South 101st East Avenue
Tulsa, OK 74128-4609 USA
Phone: 918-669-7201　　　Fax: 918-669-7207
Website: www.swt.usafe.army.mil

Founded: NA
Membership: 800
Scope: National

U.S. ARMY CORPS OF ENGINEERS

U.S. ARMY ENGINEER DIVISION, MISSISSIPPI VALLEY
Vicksburg, MS 39181-0080 USA
Phone: 601-634-5750　　　Fax: 601-634-5666
E-mail: cemvd-de@mvd02.usace.army.mil
Website: www.mvd.usace.army.mil

Founded: NA
Membership: 130
Scope: National

Contact(s):
Patti Beard, EXECUTIVE SECRETARY

U.S. ARMY CORPS OF ENGINEERS
U.S. ARMY ENGINEER DISTRICT, BALTIMORE
P.O. Box 1715
Baltimore, MD 21203-1715 USA
Phone: 410-962-4545   Fax: 410-962-7516

Founded: NA
Membership: 1200
Scope: National

US ENVIRONMENTAL PROTECTION AGENCY
REGION IV (AL, FL, GA, KY, MS, NC, SC, TN)
61 Forsyth St., S.W.
Atlanta, GA 30303 USA
Phone: 404-562-9900   Fax: 404-562-8174
Website: www.epa.gov/region04

Founded: NA
Membership: 1200
Scope: Regional

Contact(s):
    Stanley Meiburg, CONTACT

US ENVIRONMENTAL PROTECTION AGENCY
REGION III (DC, DE, MD, PA, VA, WV)
1650 Arch St.
Philadelphia, PA 19103 USA
Phone: 215-814-5000   Fax: 215-814-5103
Website: www.epa.gov/region03/

Founded: NA
Membership: 1700
Scope: Regional

Contact(s):
    Donald Welsh, REGIONAL ADMINISTRATOR

US FISH AND WILDLIFE SERVICES
UNITED STATES FISH AND WILDLIFE SERVICE
MOUNTAIN-PRAIRIE REGIONAL OFFICE 6
134 Union Blvd., P.O. Box 25486
Denver, CO 80225 USA
Phone: 303-236-7920   Fax: 303-236-8295
Website: www.fws.gov

Founded: NA
Scope: Regional

Contact(s):
    Ralph Morgenweck, REGIONAL DIRECTOR
    Phone: 303-236-7920
    Fax: 303-236-8295

USAID/TANZANIA
So Mirambo
Dar-Es-Salaam, Tanzania
Phone: 255-111-7537   Fax: 255-111-6559

Description: International Development

Contact(s):
    Z. Kristos Minja, TRAINING OFFICER
    Phone: 255-511-1753
    Fax: 255-111-6559

USDA FOREST PRODUCTS LABORATORY
UNITED STATES FOREST SERVICE
FOREST PRODUCTS LABORATORY STATION
One Gifford Pinchot Dr.
Madison, WI 53705-2398 USA
Phone: 608-231-9200   Fax: 608-231-9592
E-mail: mailroom_forest_products_laboratory@fs.fed.us
Website: www.fpl.fs.fed.us./

Founded: NA
Membership: 260
Scope: International, National

Contact(s):
    Theodore Wagner, DIRECTOR

USGS FOREST AND RANGELAND ECOSYSTEM SCIENCE CENTER
3200 SW Jefferson Way
Corvallis, OR 97331 USA
Phone: 541-750-7307
Website: http://fresc.usgs.gov

Contact(s):
    Ruth Jacobs, TECHNICAL INFORMATION SPECIALIST
    ruth_jacobs@usgs.gov

A

ADIRONDACK PARK AGENCY
State Route 86
Ray Brook, NY 12977 USA
Phone: 518-891-4050   Fax: 518-891-3938
E-mail: vxhristo@gw.dec.state.ny.us
Website: www.northnet.org/adirondackparkagency/

Founded: 1971
Membership: 64
Scope: International

Description: Created by state law and charged with developing a state Land Master Plan for the 40% of the park that is public land and a Private Land Use and Development Plan for the private lands within the six-million-acre Adirondack Park. The agency also administers the state's Wild, Scenic, and Recreational Rivers System Act for private lands within the park and the state's Freshwater Wetlands Act for both state and private lands within the park. The agency also operates two Adirondack Park Visitor Interpretive centers (nature education and tourism information centers) at Paul Smiths and Newcomb which are open year-round.

Publications: publications list available upon request., Adirondack Park, Land Use Planning for the Adirondack Park, Annual Report, The

Keyword(s): Wetlands, Rivers, Land Use Planning, Lakes, Environmental and Conservation Education

Contact(s):
Richard Lefebvre, CHAIRMAN
Sandra Bureau, DEPUTY DIRECTOR, INTERPRETIVE PROGRAMS
William Curran, DEPUTY DIRECTOR, REGULATORY PROGRAMS
John Banta, DIRECTOR OF PLANNING
Daniel Fitts, EXECUTIVE DIRECTOR
Victoria Hristovski, PUBLIC INFORMATION DIRECTOR

AGENCY OF NATURAL RESOURCES
103 S. Main St.
Waterbury, VT 05671 USA
Phone: 802-241-3600   Fax: 802-241-1102
Website: www.anr.state.vt.us

Founded: 1970
Scope: State

Description: The Agency's mission is to act as a steward of Vermont's natural resources. We work to manage Vermont's natural systems and to foster public understanding so that the integrity, vitality, and diversity of these natural systems are sustained or restored.

Contact(s):
James Bressor, DIRECTOR OF MEDIA AND PUBLIC RELATIONS
Phone: 802-241-3600
Stephen Sease, DIRECTOR OF PLANNING
Phone: 802-241-3620
Salvatore Spinosa, ENFORCEMENT
Phone: 802-241-3820
Scott Johnstone, SECRETARY
Phone: 802-241-3600

AGENCY OF NATURAL RESOURCES
DEPARTMENT OF ENVIRONMENTAL CONSERVATION
Waterbury Complex, 103 S. Main St. Bdg. 10 N
Waterbury, VT 05671 USA
Phone: 802-241-3770   Fax: 802-241-3287
Website: www.decweb.anr.state.vt.us

Founded: NA
Scope: Statewide

Contact(s):
Canute Dalmasse, COMMISSIONER
Phone: 802-241-3800
Richard Valentinetti, DIRECTOR OF AIR QUALITY
Phone: 802-241-3860
Richard Phillips, DIRECTOR OF ENVIRONMENTAL ASSISTANCE
Phone: 802-241-3470
Larry Fitch, DIRECTOR OF FACILITIES
Phone: 802-241-3742
P. Flanders, DIRECTOR OF WASTE MANAGEMENT
Phone: 802-241-3888
Marilyn Davis, DIRECTOR OF WASTEWATER
Phone: 802-241-3822
Wallace McLean, DIRECTOR OF WATER QUALITY
Phone: 802-241-3770
Jay Rutherford, DIRECTOR OF WATER SUPPLY
Phone: 802-241-3434

AGENCY OF NATURAL RESOURCES
DEPARTMENT OF FISH AND WILDLIFE
103 S. Main, 10 South
Waterbury, VT 05671-0501 USA
Phone: 802-241-3700   Fax: 802-241-3295
Website: www.vt.fishandwildlife.com

Founded: NA
Scope: National

Keyword(s): Exotic species, Aquatic nuisance species, Solid Waste Management, Wildlife, Air Quality and Pollution

Contact(s):
Linda Eldredge, BUSINESS MANAGER
Dave Mallory, CHAIR OF FISH AND WILDLIFE BOARD
Ron Regan, COMMISSIONER
Eric Palmer, DIRECTOR OF FISHERIES
Angelo Incerpi, DIRECTOR OF OPERATIONS
Lisa Wright, HUNTER EDUCATION
John Hall, INFORMATION AND EDUCATION

AGENCY OF NATURAL RESOURCES
DEPARTMENT OF FORESTS, PARKS, AND RECREATION
Commissioners Office, 103 South Main St.
Waterbury, VT 05671 USA
Phone: 802-241-3670   Fax: 802-244-1481
Website: www.vtstateparks.com

Founded: NA
Membership: 300
Scope: Statewide

Contact(s):
M. Stone, CHIEF OF FOREST RESOURCE MANAGEMENT
Phone: 802-241-3675
H. Teillon, CHIEF OF FOREST RESOURCE PROTECTION
Phone: 802-241-3676
Ed Leary, CHIEF OF OPERATIONS

Craig Whipple, CHIEF OF PARK OPERATIONS
Phone: 802-241-3663
Conrad Motyka, COMMISSIONER
David Stevens, DIRECTOR OF FORESTS
Phone: 802-241-3678
Larry Simino, DIRECTOR OF STATE PARKS
Phone: 802-241-3644
Michael Fraysier, STATELAND DIRECTOR

AGENCY OF NATURAL RESOURCES
VERMONT GEOLOGICAL SURVEY
103 S. Main St., Old Laundry Bldg.
Waterbury, VT 05671-0301 USA
Phone: 802-241-3496   Fax: 802-241-3273
E-mail: laurence.becker@anrmail.anr.state.vt.us
Website: www.anr.state.vt.us/geologyvgshmpg.htm

Founded: 1844
Membership: 4
Scope: Statewide

Description: The Vermont Geological Survey encompasses two
divisions. The State Geologist provides surveys of the geology,
mineral resources, topography and geological information
services to citizens, industry, and state and federal agencies.
The Radioactive Waste Management Program manages the
disposal of low-level radioactive waste generated in Vermont.

Publications: Price list sent upon request or visit the website.

Keyword(s): Toxic Substances, Nuclear-free, Water quantity,
Water export and diversion, Nuclear/Radiation, Protected
Areas, Geology, Communications

Contact(s):
Marjorie Gale, GEOLOGIST
marjieg@dec.anr.state.vt.us
Laurence Becker, STATE GEOLOGIST
laurencebecker@anrmail.anr.state.vt.us

AGRICULTURE & FISHERIES
P.O. Box 2223
Halifax, Nova Scotia B3J 3C4 Canada
Phone: 902-424-4560   Fax: 902-424-4671
Website: www.gov.ns.ca/nsaf/home.htm

Founded: NA
Scope: International

Description: The Department is involved in almost all aspects of
the province's fishing industry. It has significant input into some
of the policies and programs legislated and administered by
the federal government, which has jurisdiction over much of
the fishery. The department has jurisdictional responsibility for
developing and regulating aquaculture and freshwater recreational fisheries. It is also responsible for the licensing and
inspection of fish processing plants. The department provides
training, marketing, and loan assistance to the industry.
Department goals include conservation development and
enhancement of renewable fisheries resources for the benefit
of all Nova Scotians.

Publications: See publication website

Contact(s):
Janis Raymond, ACTING DIRECTOR OF MARKETING
SERVICES
Phone: 902-424-0330

Jim Sarty, CEO/FISHERIES & AQUACULTURE LOAN
BOARD
Phone: 902-424-0312
Peter Underwood, DEPUTY MINISTER
Phone: 902-424-0300
Leo Muise, DIRECTOR OF AQUACULTURE
Phone: 902-424-3664
Murray Hill, DIRECTOR OF INLAND FISHERIES
Phone: 902-485-7021
Dave Hansen, EXECUTIVE DIRECTOR
Phone: 902-424-0337
Greg Roach, EXECUTIVE DIRECTOR, FISHERIES &
AQUACULTURE SERVICES
Phone: 902-424-0348
Ernest Fage, MINISTER
Phone: 902-424-8953

ALABAMA COOPERATIVE EXTENSION SYSTEM
109 Duncan Hall
Auburn, AL 36849-5612 USA
Phone: 334-844-4444
Website: www.aces.edu

Founded: 1914
Scope: National

Description: The extension system (Alabama A&M and Auburn
Universities), through its statewide network of County
Extension Offices, conducts informal education programs
using research-based knowledge and techniques. Programs
are offered in agriculture and forestry profitability; developing,
conserving, and managing natural resources; enhancing family
and individual well being; developing human resources; and
community development.

Publications: Contact Extension Communications

Keyword(s): Agriculture, Wildlife, Forests and Forestry,
Gardening and Horticulture

Contact(s):
Martha Johnson, ASSISTANT DIRECTOR/FAMILY
PROGRAMS
ACES, 107-A Duncan Hall, Auburn University, AL 36849
Phone: 334-844-5514
mjohnson@acesag.auburn.edu
Carolyn Whatley, CO-LEADER COMMUNICATIONS
ACES, 122 Duncan Hall Annex, Auburn University, AL 36849
Phone: 334-844-5690
cwhatley@acesag.auburn.edu
James Armstrong, EXTENSION WILDLIFE SPECIALIST
Dept. of Forestry & Wildlife Science, 331 Funchess Hall,
Auburn University, AL 36849
Phone: 334-844-9233
Fax: 334-844-0234

ALABAMA COOPERATIVE FISH AND WILDLIFE
RESEARCH UNIT (USDI)
108 White Smith Hall, Auburn University
Auburn, AL 36849 USA
Phone: 334-844-4796   Fax: 334-887-4509
Website: www.ag.auburn.edu/alcfwru

Founded: 1935
Membership: 3
Scope: National

Description: The unit is sponsored by the Biological Resources Division, U.S. Geological Survey; Alabama Department of Conservation and Natural Resources, Division of Game and Fish; Auburn University; and the Wildlife Management Institute. Fish and wildlife research, graduate education, and technical assistance are the unit's primary purposes.

Keyword(s): Biodiversity, Endangered Species, Wildlife, training

Contact(s):
Elise Irwin, ASSISTANT LEADER OF FISHERIES
Michael Mitchell, ASSISTANT LEADER OF WILDLIFE
James Grand, LEADER

ALABAMA DEPARTMENT OF AGRICULTURE AND INDUSTRIES
The Richard Beard Bldg., P.O. Box 3336
Montgomery, AL 36109-0336 USA
Phone: 334-240-7100   Fax: 334-240-7190
E-mail: commtwo@agi.state.al.us
Website: www.agi.state.al.us

Founded: NA
Membership: 380
Scope: State

Description: The Alabama Department of Agriculture and Industries is responsible for enforcing the laws of Alabama relating to agriculture. It also works to provide agribusiness assistance such as marketing, loan mediation, and trade information. The department strives to ensure consumer safety and to promote all of Alabama agriculture.

Publications: Livestock Mkt News, Alabama Farmers Bulletin

Contact(s):
Charles Bishop, COMMISSIONER
commone@agi.state.al.us

ALABAMA DEPARTMENT OF CONSERVATION AND NATURAL RESOURCES
64 N. Union St.
Montgomery, AL 36130 USA
Phone: 334-242-3486   Fax: 334-242-3489
E-mail: commissioner@dcnr.state.al.us
Website: www.dcnr.state.al.us
Founded: 1905
Scope: Statewide

Description: The program goal of the Alabama Department of Conservation and Natural Resources is to protect and, where possible, enhance or restore Alabama's natural resources for this and succeeding generations.

Publications: Outdoor Alabama, Alabama's Coastal Connection

Keyword(s): Conservation, Coasts, Planning Management

Contact(s):
Richard Liles, ASSISTANT COMMISSIONER
Riley Smith, COMMISSIONER
Gil Gilder, DIVISION DIRECTOR OF COASTAL PROGRAMS
Phone: 334-242-5502
William Garner, DIVISION DIRECTOR OF MARINE POLICE
Phone: 334-353-2628
R. Vernon Minton, DIVISION DIRECTOR OF MARINE RESOURCES
PO Box 458, Gulf Shores, AL 36542
Phone: 251-968-7576
James Griggs, DIVISION DIRECTOR OF STATE LANDS
Phone: 334-242-3484

Don Cooley, DIVISION DIRECTOR OF STATE PARKS
Phone: 334-242-3334
Corky Pugh, DIVISION DIRECTOR OF WILDLIFE AND FRESHWATER FISHERIES
Phone: 334-242-3848

ALABAMA DEPARTMENT OF ENVIRONMENTAL MANAGEMENT
P.O. Box 301463
Montgomery, AL 36130-1463 USA
Phone: 334-271-7700   Fax: 334-271-7950
E-mail: Cab@adem.state.al.us
Website: http://www.adem.state.al.us/

Founded: 1982

Description: To respond in an efficient, comprehensive, and coordinated manner to environmental problems, thereby assuring a safe, healthful, and productive environment. Encompasses water quality, public water supply, underground injection control, solid waste, hazardous waste, air pollution control, well water standards, operator certification, and coastal area functions.

Publications: Environmental Update

Keyword(s): Air Quality and Pollution, Coasts, Solid Waste Management, Water Pollution Management, Wetlands

Contact(s):
Marilyn Elliott, DEPUTY DIRECTOR
James Warr, DIRECTOR
Olivia Jenkins, OFFICE OF GENERAL COUNSEL
Ron Gore, STAFF OF AIR DIVISION
Steve Jenkins, STAFF OF FIELD OPERATIONS
Gerald Hardy, STAFF OF LAND DIVISION,
Jim Moore, STAFF OF OFFICE OF EDUCATION AND OUTREACH
John Poole, STAFF OF PERMITS AND SERVICES
Clark Bruner, STAFF OF PUBLIC AFFAIRS
Charles Horn, STAFF OF WATER DIVISION

ALABAMA FORESTRY COMMISSION
Montgomery, AL 36130 USA
Phone: 334-240-9300   Fax: 334-240-9390
E-mail: forestry.state.al.us
Website: www.forestry.state.al.us

Founded: NA
Scope: Local, Regional, State

Description: The FC was created by the 1969 regular session of the Alabama Legislature and is charged by law to protect, conserve, and increase the timber and forest resources of this state. A seven-member board is the policymaking body of the Commission. The State Forester is Chief Administrative Officer. Fire prevention and suppression, educational programs and materials, and free forest management assistance are some of the services which the Commission offers to the general public.

Publications: Alabama's TREASURED Forests

Keyword(s): Environmental and Conservation Education, Forest Management, Forests and Forestry, Renewable Resources, Sustainable Development, Urban Forestry

Contact(s):
Richard Cumbie, ASSISTANT STATE FORESTER
Phone: 334-240-9367
David Long, COMMISSIONER

Gary Cole, DIRECTOR OF ADMINISTRATION DIVISION
Phone: 334-240-9333
Coleen Vansant, EDITOR
Phone: 334-240-9355
David Frederick, FIRE DIVISION DIRECTOR
Phone: 334-240-9335
Timothy Boyce, STATE FORESTER
Phone: 334-240-9304

ALABAMA SEA GRANT PROGRAM

Mississippi and Alabama Sea Grant Consortium, Caylor
Bldg., Gulf Coast Research Lab., P.O. Box 7000
Ocean Springs, MS 39566-7000 USA
Phone: 228-875-9341   Fax: 228-875-0528
Website: www.masgc.org

Founded: NA
Membership: 8
Scope: Statewide

Contact(s):
Richard Wallace, COORDINATOR AND EXTENSION
MARINE SPECIALIST
Alabama Sea Grant Extension Program: 4170 Commanders
Dr., Mobile, AL 36615
Phone: 334-438-5690
Fax: 334-438-5670

ALABAMA SOIL AND WATER CONSERVATION COMMITTEE

Executive Director
Montgomery, AL 36130 USA
Phone: 334-242-2620   Fax: 334-242-0551

Founded: NA
Membership: 6
Scope: Statewide

Contact(s):
Micky Smith, CHAIR
Rt. 1, Box 85A, Emelle, AL 35459
Phone: 205-652-7459
Beverly Riker, EXECUTIVE ASSISTANT
Stephen Cauthen, EXECUTIVE DIRECTOR
P.O. Box 304800, Montgomery, AL 36130-4800

ALASKA COOPERATIVE FISH AND WILDLIFE RESEARCH UNIT

209 Irving I Bldg., P.O. Box 757020, University of Alaska
Fairbanks
Fairbanks, AK 99775-7020 USA
Phone: 907-474-7661   Fax: 907-474-6716

Founded: 1950
Membership: 5
Scope: Statewide

Description: Sponsored jointly by the U.S. Geological Service,
Alaska Department of Fish and Game, University of Alaska
Fairbanks, U.S. Fish and Wildlife Service and Wildlife
Management Institute, the Unit conducts graduate education
and research programs on the ecology and management of
Alaskan fish and wildlife, and their habitats.

Keyword(s): Aquatic Habitats, Ecology, Wildlife, training

Contact(s):
A. McGuire, ASSISTANT LEADER OF ECOLOGY
Brad Griffith, ASSISTANT LEADER OF WILDLIFE
F. Margraf, LEADER

ALASKA DEPARTMENT OF ENVIRONMENTAL CONSERVATION

410 Willoughby Ave.
Juneau, AK 99801-1795 USA
Phone: 907-465-5000   Fax: 907-465-5770
Website: www.state.ak.us/dec

Founded: 1971
Scope: Statewide

Description: Created by the Seventh Alaska Legislature to protect
the quality of the state's natural resources and the health and
quality of life of its people. The department has broad
regulatory authority in the areas of water quality, drinking water,
air quality, solid waste disposal, oil spills, subsurface pollution,
pesticides, food safety, seafood wholesomeness, sanitation,
and radiation. The department also has programs for construction of water, sewer, and solid waste facilities in Alaskan cities
and villages.

Keyword(s): Environmental Planning, Air Quality and Pollution,
Public Health Protection, Solid Waste Management, Water
Pollution Management, Water Quality, Environmental
Protection, Environmental Health

Contact(s):
Michele Brown, COMMISSIONER
Phone: 907-465-5065
Kurt Fredriksson, DEPUTY COMMISSIONER (ACTING)
Phone: 907-465-5065
Barbara Frank, DIRECTOR OF DIVISION OF
ADMINISTRATIVE SERVICES
Phone: 907-465-5256
Tom Chapple, DIRECTOR OF DIVISION OF AIR AND
WATER QUALITY
Phone: 907-269-7686
Janice Adair, DIRECTOR OF DIVISION OF
ENVIRONMENTAL HEALTH
Phone: 907-269-7644
Dan Easton, DIRECTOR OF DIVISION OF FACILITIES
CONSTRUCTION AND OPERATIONS
Phone: 907-465-5180
Larry Dietrick
DIRECTOR OF SPILL PREVENTION AND RESPONSE
Phone: 907-465-5250
Mike Conway, DIRECTOR OF STATEWIDE PUBLIC
SERVICE
Phone: 907-465-5337
Charles Fedullo, PUBLIC INFORMATION
Phone: 907-269-3784

ALASKA DEPARTMENT OF FISH AND GAME

P.O. Box 25526
Juneau, AK 99801 USA
Phone: 907-465-4100   Fax: 907-465-2332
Website:
www.state.ak.us/local/akpages/fish.game/adfghome.htm

Founded: NA
Scope: Statewide

Description: A research and management agency whose mission
is to develop and organize its technical, human, and fiscal
assets to maintain, rehabilitate, and enhance the fish and
wildlife resources of the state, and to provide for their sustained
optimum use consistent with the social, cultural, environmental, and economic needs of the public.

Keyword(s): Wildlife, Hunting, Renewable Resources, Sport
Fishing, training

Contact(s):
John White, BOARD OF FISHERIES CHAIRMAN
Bering Sea Dental Center, P.O. Box 190, Bethel, AK 99559
Dan Coffey, BOARD OF FISHERIES VICE CHAIR
207 E. Northern Lights Blvd., Suite 200, Anchorage, AK
99503,
Lori Quakenbush, BOARD OF GAME CHAIR
P.O. Box 82391, Fairbanks, AK 99708
Greg Roczicka
BOARD OF GAME VICE CHAIRMAN
P.O. Box 513, Bethel, AK 99559
Frank Rue, COMMISSIONER
Robert Bosworth, DEPUTY COMMISSIONER
Kevin Duffy, DEPUTY COMMISSIONER
Doug Mecom, DIRECTOR OF COMMERCIAL FISHERIES
MANAGEMENT AND DEVELOPMENT DIVISION
Phone: 907-465-4210
Kevin Brooks, DIRECTOR OF DIVISION OF
ADMINISTRATION
Phone: 907-465-5999
Kelly Hepler, DIRECTOR OF DIVISION OF SPORT FISH
Phone: 907-465-4180
Mary Pete, DIRECTOR OF DIVISION OF SUBSISTENCE
Phone: 907-465-4147
Wayne Regelin, DIRECTOR OF DIVISION OF WILDLIFE
CONSERVATION
Phone: 907-465-4190
Ken Taylor, DIRECTOR OF HABITAT AND RESTORATION
DIVISION
Phone: 907-465-4105
Diana Cote, EXECUTIVE DIRECTOR OF BOARD OF GAME
AND BOARD OF FISH
Phone: 907-465-6095
Nancy Long, PUBLIC COMMUNICATIONS SECTION
Phone: 907-465-6167
Gordy Williams, SPECIAL ASSISTANT FOR LEGISLATIVE
LIAISION
Phone: 907-465-6143

ALASKA DEPARTMENT OF NATURAL RESOURCES
400 Willoughby
Juneau, AK 99801 USA
Phone: 907-465-3400   Fax: 907-586-2954
Website: www.dnr.state.ak.us

Founded: NA
Scope: Statewide

Keyword(s): Agriculture, Forests and Forestry, Land Use
Planning, Public Lands, Renewable Resources

Contact(s):
Pat Pourchot, COMMISSIONER
Rob Wells, DIRECTOR OF DIVISION OF AGRICULTURE
1800 Glenn Hwy. Suite 12, P.O. Box 949, Palmer, AK
99645-0949
Phone: 907-745-7200
Milton Wiltse, DIRECTOR OF DIVISION OF GEOLOGICAL
AND GEOPHYSICAL SURVEYS
794 University Ave. Suite 200, Fairbanks, AK 99709-3654
Phone: 907-451-5000
Bob Loeffler, DIRECTOR OF DIVISION OF MINING AND
WATER MANAGEMENT
3601 C St., Anchorage, AK 99503-5935
Phone: 907-269-8600

Mark Myers, DIRECTOR OF DIVISION OF OIL AND GAS
3601 C St. Suite 1380, Anchorage, AK 99503-5948
Phone: 907-269-8800
Jim Stratton, DIRECTOR OF DIVISION OF PARKS AND
OUTDOOR RECREATION
3601 C St. Suite 1200, Anchorage, AK 99503-5921
Phone: 907-269-8700
Jeff Jahnke, STATE FORESTER OF DIVISION OF
FORESTRY
3601 C St. Suite 1030, Anchorage, AK 99503
Phone: 907-269-8463

ALASKA DEPARTMENT OF PUBLIC SAFETY
P.O. Box 111200
Juneau, AK 99811 USA
Phone: 907-465-4322   Fax: 907-465-4362
Website: www.dps.state.ak.us

Founded: NA
Scope: Statewide

Description: Responsible for enforcing all of the Fish and Game
laws and regulations of the state.

Publications: The Quarterly

Keyword(s): Hunting, Sport Fishing, Trapping, Resource Law
Enforcement

Contact(s):
Glenn Godfrey, COMMISSIONER
James Cockrell, DEPUTY DIRECTOR
Joel Hard, DIRECTOR OF FISH AND WILDLIFE
PROTECTION
Al Cain, OPERATIONS COMMANDER

ALASKA DEPARTMENT OF PUBLIC SAFETY
DIVISION OF FISH AND WILDLIFE PROTECTION
5700 E. Tudor Rd.
Anchorage, AK 99507 USA
Phone: 907-269-5509   Fax: 907-269-5616
Website: www.dps.state.ak.us/fwp

Founded: NA
Membership: 50
Scope: Statewide

Contact(s):
Joel Hard, DIRECTOR

ALASKA HEALTH PROJECT
218 East 4th Avenue
Anchorage, AK 99501 USA
Phone: 907-276-2864   Fax: 907-279-3089

Founded: 1980

Description: To provide information and advocacy on occupational
and environmental health issues in Alaska, the Pacific Northwest,
and Canada.

Publications: Involve waste management and worker health, and
are inclusive of a 22 edition list.

Keyword(s): Communications, Engineering, Environmental and
Conservation Education, Solid Waste, Training

Contact(s):
Daniel Middaugh, EXECUTIVE DIRECTOR
R. Gryder, INSTRUCTOR

ALASKA SEA GRANT COLLEGE PROGRAM

University of Alaska, P.O. Box 755040
Fairbanks, AK 99775-5040 USA
Phone: 907-474-7086   Fax: 907-474-6285
E-mail: fygrant@uaf.edu
Website: www.uaf.edu/seagrant/

Founded: NA
Scope: Statewide

Description: A state/federal partnership administered by the National Oceanic and Atmospheric Administration and the University of Alaska that sponsors and conducts marine research, graduate education, marine industry advisory services, and formal and nonformal public education aimed at promoting the wise use and conservation of Alaska's coastal and marine resources.

Publications: Guide to Marine Mammals of Alaska, Management Strategies for Exploited Fish Populations, Biennial Program Report, SeaWeek Curriculum Series, posters, videos, free catalog

Keyword(s): Coasts, Environmental and Conservation Education, Wildlife, Marine Mammals, Whale, Dolphin, Seal, Oceanography

Contact(s):
Kurt Byers, COMMUNICATIONS MANAGER
Alaska Sea Grant College Program, University of Alaska, P.O. Box 755040, Fairbanks, AK 99775-5040
Phone: 907-474-6702
fnkmb1@uaf.edu
Ronald Dearborn, DIRECTOR
fnrkd@uaf.edu
Donald Kramer, DIRECTOR OF MARINE ADVISORY PROGRAM
University of Alaska, Carlton Trust Bldg., Suite 110, 2221 E. Northern Lights Blvd., Anchorage, AK 99508-4140
Phone: 907-274-9691
Fax: 907-277-5242
afdek@uaa.alaska.edu
Sue Keller, PUBLICATIONS MANAGER
P.O. Box 755040, Fairbanks, AK 99775-5040
Phone: 907-474-6703
fnsk@uaf.edu
Sherri Pristash, PUBLICATIONS MANAGER
Phone: 888-789-0090
fypubs@uaf.edu

ALBERTA DEPARTMENT OF ENVIRONMENTAL PROTECTION

Main Fl., Petroleum Plaza, North Tower, 9945-108 St.
Edmonton, Alberta T5K 2G0 Canada
Phone: 780-427-2391   Fax: 780-422-6259
E-mail: infocent@gov.ab.ca
Website: http://www.gov.ab.ca/env/

Founded: 1992

Description: The Department of Environmental Protection is responsible for protecting, enhancing, and ensuring the wise use of Alberta's environment. The services within the department work cooperatively to meet the needs of Albertans by protecting wildlife, forests, parks, and other natural resources through enforcement of provincial legislation and ensuring the sustainable management of all these resources.

Publications: State of Environment Report, Annual Regulation Guides to Sportfishing, Hunting, and Trapping, Timber Supply Report, State of Alberta's Wildlife Report

Contact(s):
Lorne Taylor, MINISTER

ALBERTA DEPARTMENT OF ENVIRONMENTAL PROTECTION

COMMUNICATIONS DIVISION
9th Floor, Petroleum Plaza, S. Tower, 9945-108 St.
Edmonton, Alberta T5K 2C6 Canada

Founded: NA

Contact(s):
Bob Scott, DIRECTOR
Phone: 403-427-8636

ALBERTA DEPARTMENT OF ENVIRONMENTAL PROTECTION

CORPORATE MANAGEMENT SERVICE
9th Floor, Petroleum Plaza, S. Tower, 9945-108 St.
Edmonton, Alberta T5K 2C6 Canada

Founded: NA

ALBERTA DEPARTMENT OF ENVIRONMENTAL PROTECTION

ENVIRONMENTAL SERVICE
Alberta, Canada
Phone: 780-427-6247   Fax: 780-427-6247
E-mail: doug.tupper@gov.abc.ca
Website: http:ww.gov.ab.ca/env/

Founded: NA

Contact(s):
Doug Tupper, ASSISTANT DEPUTY MINISTER
Phone: 403-427-6247

ALBERTA DEPARTMENT OF ENVIRONMENTAL PROTECTION

LAND AND FOREST SERVICE
Alberta, Canada
Phone: 780-427-3542   Fax: 780-422-6068
E-mail: cliff.henderson@gov.ab.ca
Website: www.gov.ab.ca/env/

Founded: NA

Contact(s):
Cliff Henderson, ASSISTANT DEPUTY MINISTER

ALBERTA DEPARTMENT OF ENVIRONMENTAL PROTECTION

NATURAL RESOURCES SERVICE
Alberta, Canada
Phone:
Website: http://www.gov.ab.ca/env/

Founded: NA

Contact(s):
Morley Barret, DEPUTY MINISTER
Phone: 780-427-6749
Fax: 780-427-8884
morley.barrett@gov.ab.ca

ALBERTA TRAPPERS ASSOCIATION

Commmunication Division, 9th floor
Petroleum Plaza, S. Tower, 9945-108 St.
Edmonton, Alberta T5K 2C6 Canada
Fax: 780-427-6247
E-mail: bob.scott@gov.ab.cainfo@albertatrappers.com
Website: www.albertatrappers.com

Founded: 1974
Membership: 1300
Scope: State

Description: Cooperates with all trappers associations and government agencies for a sensible conservation program.

Publications: Alberta Trapper, The

Contact(s):
Gus Deisting, 1ST VICE PRESIDENT
Ed Graham, 2ND VICE PRESIDENT
Bob Scott, DIRECTOR
Phone: 780-427-6247
Reed Gauthier, PRESIDENT
Phone: 780-826-5026

ARIZONA STATE LAND DEPARTMENT

1616 W. Adams St.
Phoenix, AZ 85007 USA
Phone: 602-542-4621   Fax: 602-542-2590
Website: www.land.state.az.us

Founded: 1915
Membership: 175
Scope: Regional

Description: The purpose of the Arizona State Land Department is to manage 9.4 million acres of Trust lands, through leasing and sale, in order to generate revenue for 14 state institutions. Resource protection and preservation is an integral part of trust land management.

Keyword(s): Land Use Planning, Exotic species, Aquatic nuisance species, Renewable Resources, Sustainable Development, Public Lands

Contact(s):
Lynn Larson, DIRECTOR OF ADMINISTRATION AND RESOURCE ANALYSIS DIVISION
Phone: 602-542-4621
Kirk Rowdabaugh, DIRECTOR OF FIRE MANAGEMENT DIVISION
Phone: 602-255-4059
Fax: 602-255-1781
T. Hart, DIRECTOR OF FORESTRY DIVISION
Phone: 602-542-4627
Fax: 602-542-2590
Bill Dowdle, DIRECTOR OF NATURAL RESOURCES DIVISION
Phone: 602-542-4625
Fax: 602-542-3507
Richard Oxford, DIRECTOR OF OPERATIONS DIVISION
Phone: 602-542-4602
Fax: 602-542-5223
Ron Ruzifka, DIRECTOR OF PLANNING AND LAND DISPOSITION DIVISION
Phone: 602-542-1704
Jody Latimer, NRCD ADMINISTRATOR
Phone: 602-542-2699
Fax: 602-542-3507
Michael Anable, STATE LAND COMMISSIONER

ARIZONA COOPERATIVE FISH AND WILDLIFE RESEARCH UNIT (USDI)

Rm. 104, Biological Sciences East, University of Arizona
Tucson, AZ 85721 USA
Phone: 520-621-1959   Fax: 520-621-8801

Founded: NA
Membership: 3
Scope: Statewide

Description: The Unit is a cooperative effort by the U.S. Department of Interior, the Arizona Game and Fish Department, the University of Arizona, and the Wildlife Management Institute. The Unit conducts research on fish and wildlife questions for client agencies.

Contact(s):
Scott Bonar, LEADER

ARIZONA COOPERATIVE STATE EXTENSION SERVICES

University of Arizona, P.O. Box 210036
Tucson, AZ 85721-0036 USA
Phone: 520-621-7205   Fax: 520-621-1314
Website: ag.arizona.edu/extension/

Founded: NA
Membership: 5
Scope: Statewide
Publications: see publication website

Contact(s):
Kevin Fitzsimmons, AQUACULTURE SPECIALIST
Soil, Water, and Environmental Science, P.O. Box 210038, University of Arizona, Tucson, AZ 85721
Phone: 520-626-3324
kevfitz@ag.arizona.edu
Bill Peterson, ASSISTANT DIRECTOR, 4-H
Phone: 520-621-3623
bpeters@ag.arizona.edu
James Christenson, ASSOCIATE DEAN AND DIRECTOR OF COOPERATIVE EXTENSION
Phone: 520-621-7209
jimc@ag.arizona.edu
Deborah Young, ASSOCIATE DIRECTOR, PROGRAMS
Phone: 520-621-5308
djyoung@ag.arizona.edu
Larry Sullivan, NATURAL RESOURCES SPECIALIST, WILDLIFE
School of Renewable Natural Resources: College of Agriculture, P.O. Box 210043, University of Arizona, Tucson, AZ 85721
Phone: 520-621-7998
Fax: 520-621-8801
sullivan@ag.arizona.edu
George Ruyle, RANGE MANAGEMENT AND FOREST RESOURCES PROGRAM CHAIR
School of Renewable Natural Resources: College of Agriculture, P.O. Box 210043, University of Arizona, Tucson, AZ 85721
Phone: 520-621-1384
gruyle@ag.arizona.edu
Richard Hawkins, WATERSHED MANAGEMENT SPECIALIST
School of Renewable Natural Resources: College of Agriculture, P.O. Box 210043, University of Arizona, Tucson, AZ 85721
Phone: 520-621-7273
rhawkins@ag.arizona.edu

ARIZONA DEPARTMENT OF AGRICULTURE

1688 W. Adams
Phoenix, AZ 85007 USA
Phone: 602-542-4373   Fax: 602-542-5420
E-mail: http:agriculture.state.az.us
Website: www.agriculture.state.az.us

Founded: NA
Scope: International

Description: The ADA regulates and supports Arizona agriculture in a manner that encourages farming, ranching, and agribusiness while protecting consumers and natural resources. The ADA provides a number of services to its regulated industry, as well as the general public.

Contact(s):
Sheldon Jones, DIRECTOR
Phone: 602-542-0998

ARIZONA DEPARTMENT OF AGRICULTURE

ANIMAL SERVICES DIVISION
1688 W. Adams
Phoenix, AZ 85007 USA
Phone: 602-542-6309   Fax: 602-542-3244
Website: http://www.agriculture.state.az.us/ASD/asd.htm

Founded: NA
Scope: Statewide

Description: The Animal Services Division is responsible for the protection of livestock from theft and disease and for the regulation of the state's aquaculture, dairy, egg, and slaughtering and meat-processing industries.

Contact(s):
Al Davis, ASSOCIATE DIRECTOR
Sheldon Jones, DIRECTOR

ARIZONA DEPARTMENT OF AGRICULTURE

ENVIRONMENTAL SERVICES DIVISION
1688 W. Adams
Phoenix, AZ 85007 USA
Phone: 602-542-2579   Fax: 602-542-0466
Website: www.agriculture.state.az.us

Founded: NA
Scope: Statewide

Description: The Environmental Services Division is responsible for regulating the agricultural industry to ensure the safe use of pesticides, and to ensure the quality of feed, fertilizer, and pesticide formulations.

Contact(s):
Jack Peterson, ASSOCIATE DIRECTOR

ARIZONA DEPARTMENT OF AGRICULTURE

PLANT SERVICES DIVISION
1688 W. Adams St.
Phoenix, AZ 85007 USA
Phone: 602-542-0998   Fax: 602-542-0999
Website: www.agriculture.state.ac.us

Founded: NA
Membership: 100
Scope: State

Description: The Plant Services Division is responsible for enforcement of state plant regulatory statutes, state agricultural industry plants, and plant health service programs.

Contact(s):
John Caravetta, ASSOCIATE DIRECTOR
Phone: 602-542-0994

ARIZONA DEPARTMENT OF ENVIRONMENTAL QUALITY

3033 N. Central Ave.
Phoenix, AZ 85012 USA
Phone: 602-207-2300   Fax: 602-207-2218
Website: www.adeq.state.az.us

Founded: 1987
Membership: 800
Scope: Regional

Description: The Arizona Department of Environmental Quality shall preserve, protect, and enhance the environment and the public health, and shall be a leader in the development of public policy to maintain and improve the quality of Arizona's air, land, and water resources.

Keyword(s): Air Quality and Pollution, Environmental Protection, Coral Reefs, Solid Waste Management, Toxic Substances, Nuclear-free, Water quantity, Water export and diversion, Pollution Prevention, Water Quality, Environmental and Conservation Education

Contact(s):
Jacquelin Schafer, DIRECTOR
Phone: 602-207-2203
Robert Rocha, DIRECTOR OF ADMINISTRATION SERVICES DIVISION
Phone: 602-207-4867
Nancy Wrona, DIRECTOR OF AIR QUALITY DIVISION
Phone: 602-207-2308
Jean Calhoun, DIRECTOR OF WASTE PROGRAMS DIVISION
Phone: 602-207-2381
Karen Smith, DIRECTOR OF WATER QUALITY DIVISION
Phone: 602-207-2306
Charles Matthewson, SOUTHERN REGIONAL OFFICE
400 W. Congress Suite 433, Tucson, AZ
Phone: 520-628-6733

ARIZONA GAME AND FISH DEPARTMENT

2221 W. Greenway Rd.
Phoenix, AZ 85023-4312 USA
Phone: 602-942-3000   Fax: 602-789-3924
Website: www.azgfd.com

Founded: NA
Membership: 300
Scope: Statewide

Description: The mission of the Arizona Game and Fish Department is to conserve, enhance, and restore Arizona's diverse wildlife resources and habitats through aggressive protection and management programs, and to provide wildlife resources and safe watercraft and off-highway vehicle recreation for the enjoyment, appreciation, and use by present and future generations.

Publications: Arizona Wildlife Views

Keyword(s): Environmental and Conservation Education, Sport Fishing, training, Nongame Wildlife

Contact(s):
Mike Senn, ASSISTANT DIRECTOR OF FIELD
OPERATIONS DIVISION
Jim Burton, ASSISTANT DIRECTOR OF INFORMATION AND
EDUCATION DIVISION
Richard Rico, ASSISTANT DIRECTOR OF SPECIAL
SERVICES DIVISION
Bruce Taubert, ASSISTANT DIRECTOR OF WILDLIFE
MANAGEMENT DIVISION
W. Gilstrap, COMMISSIONER
Steve Ferrell, DEPUTY DIRECTOR
Duane Shroufe, DIRECTOR
Alan Silverberg, FUNDS AND CONTRACTS
Dick Mays, HERITAGE PROGRAM COORDINATOR
Diana Shaffer, PERSONNEL MANAGER
Bob Miles, PUBLICATIONS EDITOR
Heidi Vasiloff, REGIONAL COORDINATOR

ARIZONA GEOLOGICAL SURVEY

416 W. Congress St.
Tucson, AZ 85701 USA
Phone: 520-770-3500   Fax: 520-770-3505
E-mail: www.azgs.state.az.us
Website: www.azgs.state.az.us

Founded: 1881
Membership: 20
Scope: Statewide

Description: Develops, maintains, and disseminates information related to the geologic framework, geological hazards and limitations, and mineral and energy resources. Provides staff support for the Arizona Oil and Gas Conservation Commission, which regulates the drilling and production of oil, gas, geothermal, carbon dioxide and helium resources.

Publications: Down to Earth, Arizona Geology, circulars, special papers, bulletins, maps, open-file reports, miscellaneous maps, contributed maps and reports, digital information.

Keyword(s): Land Use Planning, Environmental and Conservation Education, Engineering, Research, Energy, Mineral Resources, Public Information, Exotic species, Aquatic nuisance species, Geology

Contact(s):
Larry Fellows, DIRECTOR AND STATE GEOLOGIST
Phone: 520-770-3500
fellows_larry@pop.state.az.us

ARIZONA STATE PARKS BOARD

1300 W. Washington Ave.
Phoenix, AZ 85007 USA
Phone: 602-542-4174   Fax: 602-542-4188
Website: www.pr.state.az.us

Founded: 1957
Scope: Statewide

Description: The purposes and objectives of the Arizona State Parks Board are to select, acquire, preserve, establish, and maintain areas of natural features, scenic beauty, historical and scientific interest, zoos, and botanical gardens, for the education, pleasure, recreation, and health of the people.

Publications: Arizona Rivers and Streams Guide, Arizona State Trails Guides, Arizona Wildlife Viewing Guides, Access Arizona (disabled/seniors)

Keyword(s): Botanical Gardens, Outdoor Recreation, Natural Areas, Nongame Wildlife, Cultural Preservation

Contact(s):
Walter Armer, CHAIRPERSON
Kenneth Travous, EXECUTIVE DIRECTOR
Ellen Bilbrey, PUBLIC INFORMATION OFFICER
Phone: 602-542-1996

ARKANSAS COOPERATIVE RESEARCH UNIT

Department of Interior, U.S. Geological Survey, Biological Sciences SCEN 617, University of Arkansas
Fayetteville, AR 72701 USA
Phone: 501-575-6709   Fax: 501-575-3330

Founded: 1988
Membership: 3
Scope: Statewide

Description: Primary purpose is field research, graduate research training in fisheries and wildlife resources, technical assistance, and extension activities. Areas of research include habitat selection, life history and demographics, ecology, animal behavior, and fisheries and wildlife biology.

Keyword(s): Ecology, training, Wildlife, Research

Contact(s):
David Krementz, UNIT LEADER

ARKANSAS DEPARTMENT OF PARKS AND TOURISM

One Capitol Mall
Little Rock, AR 72201 USA
Phone: 501-682-7777   Fax: 501-682-1364
E-mail: info@arkansas.com
Website: www.arkansas.com

Founded: NA
Membership: 650
Scope: Statewide

Description: Develop, maintain and operate 50 state parks and four museums; advertise and promote all the state's recreation and travel potentials; provide information to attract retirees and others wanting to relocate; and assist communities in establishing local litter prevention and recycling projects.

Publications: The Arkansas Calendar of Events—brochure, Arkansas State Parks Guide—brochure, The Arkansas Adventure Guide—brochure

Keyword(s): Cultural Preservation, Outdoor Recreation, Protected Areas, Historic Preservation, Environmental and Conservation Education

Contact(s):
R. Cargile, DIRECTOR OF ADMINISTRATION
Phone: 501-682-2039
rl.cargile@mail.state.ar.us
Nancy Clark, DIRECTOR OF GREAT RIVER ROAD
DIVISION
Phone: 501-682-1120
nancy.clark@mail.state.ar.us
Patricia Murphy, DIRECTOR OF HISTORICAL RESOURCES
AND MUSEUM SERVICES SECTION
Phone: 501-682-3603
patriciam.murphy@mail.state.ar.us
John Ferguson, DIRECTOR OF HISTORY COMMISSION
Phone: 501-682-6900
johnl.ferguson@mail.state.ar.us
Greg Butts, DIRECTOR OF PARKS DIVISION
Phone: 501-682-7743
greg.butts@mail.state.ar.us

Joe Rice, DIRECTOR OF TOURISM DIVISION
Phone: 501-682-1088
joedavid.rice@mail.state.ar.us
Richard Davies, EXECUTIVE DIRECTOR
Phone: 501-682-2535
richardw.davies@mail.state.ar.us
Robert Phelps, KEEP ARKANSAS BEAUTIFUL
Phone: 501-682-3507
robert.phelps@mail.state.ar.us
Bryan Kellar, OUTDOOR RECREATION GRANTS
Phone: 501-682-1301
bryan.kellar@mail.stste.ar.us

ARKANSAS GAME AND FISH COMMISSION

2 Natural Resources Dr.
Little Rock, AR 72205 USA
Phone: 501-223-6300   Fax: 501-223-6444
Website: www.agfc.com

Founded: 1915
Membership: 8
Scope: Statewide

Description: The mission of the Arkansas Game and Fish Commission is to wisely manage the fish and wildlife resources of Arkansas while providing maximum enjoyment for the people.

Publications: Arkansas Outdoors Newsletter, Explore Arkansas, Arkansas Wildlife

Contact(s):
Scott Henderson, ASSISTANT DIRECTOR
Phone: 501-223-6309
Fax: 501-223-6448
shenderson@agfc.state.ar.us
Marion McCollum, CHAIRMAN
POB 8509, Pinebluff, AK 71611
Phone: 807-535-9000
Fax: 870-535-8544
Loren Hitchcock, CHIEF OF ENFORCEMENT
Phone: 501-223-6384
Bill Robinson, CHIEF OF COMPUTER SERVICES
Phone: 501-223-6368
Marcus Kilburn, CHIEF OF EDUCATIONAL SERVICES
Phone: 501-223-6402
Ray Sebren, CHIEF OF FISCAL SERVICES
Phone: 501-223-6341
Mike Gibson, CHIEF OF FISHERIES MANAGEMENT
Phone: 501-223-6371
Donny Harris, CHIEF OF WILDLIFE MANAGEMENT
Phone: 501-223-6359
Len Pitcock, DIRECTOR OF COMMUNICATIONS
Phone: 501-223-6473
Keith Sutton, EDITOR
Phone: 501-223-6406
Jim Spencer, EDITOR
Phone: 501-223-6336
Mary Smith, EDUCATIONAL DIRECTOR
Phone: 501-223-6317
Jim Goodhart, LEGAL COUNSEL
Phone: 501-223-6327
Stephen Wilson, PUBLIC AFFAIRS COORDINATOR
Phone: 501-223-6408

ARKANSAS NATURAL HERITAGE COMMISSION

1500 Tower Bldg., 323 Center St.
Little Rock, AR 72201 USA
Phone: 501-324-9619   Fax: 501-324-9618
E-mail: info@arkansasheritage.org
Website: www.naturalheritage.org

Founded: 1973
Membership: 13
Scope: Statewide

Description: Responsible for system of natural areas; acquires and holds both lands and interests in land; maintains a registry of natural areas in other ownerships; maintains a Heritage Inventory System and environmental review and information sharing program; and defends natural areas from adverse influences.

Keyword(s): Biodiversity, Natural Areas, Protected Areas, Land Purchase, Sustainable Ecosystems, Rivers, Endangered Species

Contact(s):
Edwin Lawson, CHAIRMAN
Tom Foti, CHIEF OF RESEARCH
Phone: 501-324-9761
tom@arkansasheritage.or
Michael Warriner, COORDINATOR OF ENVIRONMENTAL APPLICATIONS
Phone: 501-324-9613
bills@arkansasheritage.org
Doug Fletcher, COORDINATOR OF STEWARDSHIP
Phone: 501-324-9612
douglas@arkansasheritage.org
Karen Smith, EXECUTIVE DIRECTOR
Phone: 501-324-9614
karen@arkansasheritage.org

ARKANSAS STATE EXTENSION SERVICES

FOUR H CENTER
1 4-H Way
Little Rock, AR 72223 USA
Phone: 501-821-4444   Fax: 501-821-2545
Website: www.arkansas4hcenter.org

Founded: NA
Membership: 36
Scope: Statewide

Description: An off-campus education organization with faculty and offices in each county with the basic mission to disseminate and encourage the application of research-generated knowledge and leadership techniques to individuals, families, and communities. The county faculty is backed by subject matter specialists and their research counterparts.

Publications: Brochures

Keyword(s): Agriculture, Forests and Forestry, Gardening and Horticulture, Renewable Resources, training

Contact(s):
J. Pitman, 4-H ASSISTANT SPECIALIST-OUTDOOR EDUCATION
Leslie Gall, ASSISTANT EXTENSION SPECIALIST, ENVIRONMENTAL EDUCATION
#1, Four-H Way, Little Rock, AR 72211
Phone: 501-821-4444
lgall@uaex.edu

Lucy Moreland, NATURAL RESOURCES INSTRUCTOR
#1, Four-H Way, Little Rock, AR 72211
Phone: 501-821-4444
lmoreland@uaex.edu
Kevin Jones, PROGRAM SPECIALIST FOR OUTDOOR
RECREATION

B

BOARD OF MINERALS ENVIRONMENT

523 E. Capitol Avenue
Pierre, SD 57501 USA
Phone: 605-773-3151   Fax: 605-773-6035
Website: www.state.sd.us\denr

Founded: 1981
Scope: Statewide

Description: The Board of Minerals and Environment
promulgates rules and issues permits in the areas of air quality,
solid waste, hazardous waste, mineral exploration and mining,
and oil and gas exploration and production.

Contact(s):
Steven Pirner, SECRETARY OF THE DEPARTMENT
Phone: 605-773-5559

BOTANICAL SOCIETY OF WESTERN PENNSYLVANIA

5837 Nicholson St.
Pittsburgh, PA 15217-2309 USA
Phone: 412-521-9425
E-mail: speedy@kiski.net
Website: http://home.kiski.net/~speedy/b1.html

Founded: NA
Scope: Local

Description: Botanical Society of Western Pennsylvania brings
together those who are interested in botany and encourages
the study of botany and knowledge of plants.

Publications: Wildflowers, Wildflowers of Pennsylvania

Keyword(s): Flowers, Plants, and Trees, Conservation,
Gardening and Horticulture, Environmental and Conservation
Education

Contact(s):
Mary Haywood, PRESIDENT
Phone: 412-578-6175
Loree Speedy, SECRETARY
Walter Gardill, TREASURER
Phone: 412-364-5308

BUREAU OF ECONOMIC GEOLOGY

University of Texas at Austin, University Station, Box X
Austin, TX 78713-7508 USA
Phone: 512-471-1534   Fax: 512-471-0140
E-mail: beg@utexas.edu
Website: www.beg.utexas.edu

Founded: 1909
Membership: 150
Scope: International, Statewide

Description: Functions as a state geological survey. Program
includes basic research; application of geology to resources,
conservation, and engineering problems; and publication of
varied reports and maps. Maintains an extensive environmental mapping program.

Publications: Geological Circulars, Special Publications,
Environmental Geologic Atlases, Geological Atlas of Texas,
Geological Quadrangle Maps, guidebooks, handbooks, annual
reports, mineral resource circula, Publications, University of
Texas Report of Investigations

Keyword(s): Geology, Exotic species, Aquatic nuisance species,
Energy

Contact(s):
D. Ratcliff, ASSOCIATE DIRECTOR FOR ADMINISTRATION
Scott Tinker, DIRECTOR
Phone: 512-471-1534

C

CALIFORNIA COOPERATIVE FISHERY RESEARCH UNIT (USGS)

CO-OP FISH UNIT
Humboldt State University
Arcata, CA 95521 USA
Phone: 707-826-3268   Fax: 707-826-3269
E-mail: cuca@humboldt.edu
Website: www.humboldt.edu/~cuca

Founded: NA
Membership: 3
Scope: National
Publications: see publication website

Contact(s):
Margaret Wilzbach, ASSISTANT LEADER
Phone: 707-826-5645
Walter Duffy, LEADER
Phone: 707-826-5644
Kenneth Cummins, SENIOR ADVISORY SCIENTIST

CALIFORNIA DEPARTMENT OF EDUCATION

OFFICE OF ENVIRONMENTAL EDUCATION
721 Capitol Mall, P.O. Box 944272
Sacramento, CA 94244-2720 USA
Website: http://www.cde.ca.gov

Founded: 1970

Description: The California Department of Education provides
technical assistance and curriculum leadership in environmental education for counties and schools in California. The
department offers an annual competitive grant program and
provides curriculum and other publications related to environmental education in California.

Publications: Environmental Education Compendia, A Child's
Place in the Environment, Greatest Hits of Environmental
Education, Endangered Species Resource Guide

Keyword(s): Air Quality and Pollution, Endangered Species,
Environmental and Conservation Education, Scholarships and
Grants, Education

Contact(s):
Bill Andrews, ENVIRONMENTAL EDUCATION CONSULTANT
Phone: 916-657-5374
Fax: 916-657-3682
bandrews@cde.ca.gov

CALIFORNIA DEPARTMENT OF FOOD AND AGRICULTURE

1220 N St.
Sacramento, CA 95814 USA
Phone: 916-654-0462 Fax: 916-657-4240
E-mail: officeofpublicaffairs@bdfa.ca.gov
Website: www.cdfa.ca.gov

Founded: NA
Membership: 2500
Scope: National

Description: To assure public health, safety, and welfare; protects agriculture by administering, directing and enforcing the state's agricultural laws and regulations.

Publications: See publications on website

Contact(s):
Francine Kammeyer, CHIEF COUNSEL
Phone: 916-654-1393
Dan Webb, DEPUTY SECRETARY
Phone: 916-654-0321
Richard Breitmeyer, DIRECTOR OF ANIMAL INDUSTRY
Phone: 916-654-0881
Elizabeth Houser, DIRECTOR OF FAIRS AND EXPOSITIONS
Phone: 916-263-2952
John Donahue, DIRECTOR OF INSPECTION SERVICES
Phone: 916-654-0792
Kelly Krug, DIRECTOR OF MARKETING SERVICES
Phone: 916-654-1240
Mike Cleary, DIRECTOR OF MEASUREMENT STANDARDS
Phone: 916-229-3000
Don Henry, DIRECTOR OF PLANT INDUSTRY
Phone: 916-654-0317
Steve Lyle, DIRECTOR OF PUBLIC AFFAIRS
Phone: 916-654-0462
Janice Strong, LEGISLATIVE DIRECTOR
Phone: 916-654-0326
Les Lombardo, PLANNING INFORMATION TECHNOLOGY DIRECTOR
Phone: 916-653-7643
William Lyons, SECRETARY
Phone: 916-654-0433

CALIFORNIA ENERGY COMMISSION (AKA)

ENVIRONMENTAL DEPT.
1516 9th St.
Sacramento, CA 95814 USA
Phone: 916-654-4287 Fax: 916-654-4420
E-mail: www.energy.ca.gov
Website: www.energy.ca.gov

Founded: 1975
Membership: 400
Scope: Statewide

Description: To ensure continuation of a reliable and affordable supply of energy for California at a level consistent with the state's needs.

Contact(s):
William Keese, CHAIRMAN
Phone: 916-654-5000
Robert Pernell, COMMISSIONER
Phone: 530-654-5036

Scott Matthews, DEPUTY DIRECTOR OF ENERGY EFFICIENCY DIVISION
Phone: 916-654-5013
Fax: 916-654-4304
Robert Therkelsen, DEPUTY DIRECTOR OF ENERGY FACILITIES SITING AND ENVIRONMENTAL PROTECTION DIVISION
Phone: 916-654-3924
Fax: 916-654-3882
Nancy Deller, DEPUTY DIRECTOR OF ENERGY TECHNOLOGY DIVISION
Phone: 916-654-4628
Fax: 916-654-4676
Steve Larson, EXECUTIVE DIRECTOR
Phone: 916-654-4996
Art Rosenfeld, VICE CHAIR
Phone: 916-654-4930

CALIFORNIA ENVIRONMENTAL PROTECTION AGENCY

P. O. Box 2815
Sacramento, CA 95812 USA
Phone: 916-445-3846 Fax: 916-445-6401
Website: www.calepa.ca.gov

Founded: 1991
Scope: Statewide

Description: The Secretary for Environmental Protection, a member of the Governor's Cabinet, serves as the Governor's principal advisor on environmental protection issues and oversees the activities of the Air Resources Board, Water Resources Control Board, and Integrated Waste Management Board, the Department of Toxic Substances Control, the Office of Environmental Health Hazard Assessment, and the Department of Pesticide Regulation.

Contact(s):
Deborah Barnes, DEPUTY SECRETARY OF LAW ENFORCEMENT AND COUNSEL
Phone: 916-327-2064
James Spagnole, DIRECTOR OF COMMUNICATIONS
Phone: 916-324-9670
Patty Zwarts, DIRECTOR OF LEGISLATIVE (ACTING)
Phone: 916-322-7315
Winston Hickox, SECRETARY
Phone: 916-445-3846
C. Haddix, UNDERSECRETARY

CALIFORNIA ENVIRONMENTAL PROTECTION AGENCY

CALIFORNIA AIR RESOURCES BOARD
P.O. Box 2815
Sacramento, CA 95812 USA
Phone: 916-322-2990 Fax: 916-445-5025
Website: www.arb.ca.gov

Founded: 1968
Scope: National

Description: The California Air Resources Board is responsible for the adoption and enforcement of the state's ambient air quality standards, rules, and regulations for the control of vehicular air pollution and toxic air contaminants throughout the state. Oversees the efforts of 34 air pollution control districts that regulate emissions from industrial facilities. Conducts studies of the causes of air pollution and evaluates its effort upon human, plant, and animal life. The Board is comprised of 11 members.

Contact(s):
Larry Morris, ADMINISTRATIVE SERVICES DIVISION CHIEF
OF THE CALIFORNIA AIR RESOURCES BOARD
Phone: 916-322-8198
Lynn Terry, ASSISTANT EXECUTIVE OFFICER OF THE
CALIFORNIA AIR RESOURCES BOARD
Phone: 916-322-2739
Alan Lloyd, CHAIRMAN OF CALIFORNIA AIR RESOURCES
BOARD
Tom Cackette, CHIEF DEPUTY EXECUTIVE OFFICER OF
THE CALIFORNIA AIR RESOURCES BOARD
Phone: 916-322-2892
James Morgester, COMPLIANCE DIVISION CHIEF OF THE
CALIFORNIA AIR RESOURCES BOARD
Phone: 916-322-6022
Michael Scheible, DEPUTY EXECUTIVE OFFICER OF THE
CALIFORNIA AIR RESOURCES BOARD
Phone: 916-322-2890
Michael Kenny, EXECUTIVE OFFICER OF THE
CALIFORNIA AIR RESOURCES BOARD
Phone: 916-445-4383
Ron Oglesby, LEGISLATIVE OFFICE CHIEF OF THE
CALIFORNIA AIR RESOURCES BOARD
Phone: 916-322-2896
Bob Cross, MOBILE SOURCE DIVISION CHIEF OF THE
CALIFORNIA AIR RESOURCES BOARD
Phone: 818-575-6820
William Loscutoff, MONITORING AND LABORATORY
DIVISION CHIEF OF THE CALIFORNIA AIR RESOURCES
BOARD
Phone: 916-445-3742
Kathleen Walsh, OFFICE OF LEGAL AFFAIRS GENERAL
COUNSEL OF THE CALIFORNIA AIR RESOURCES BOARD
Phone: 916-322-2884
Jim Schoning, OFFICE OF OMBUDSMAN CHIEF OF THE
CALIFORNIA AIR RESOURCES BOARD
Phone: 916-323-6791
Jerry Martin, PUBLIC INFORMATION OFFICER OF THE
CALIFORNIA AIR RESOURCES BOARD
Phone: 916-322-2990
John Holmes, RESEARCH DIVISION CHIEF OF THE
CALIFORNIA AIR RESOURCES BOARD
Phone: 916-445-0753
Peter Venturini, STATIONARY SOURCE DIVISION CHIEF
OF THE CALIFORNIA AIR RESOURCES BOARD
Phone: 916-445-0650
Terry McGuire, TECHNICAL SUPPORT DIVISION CHIEF OF
THE CALIFORNIA AIR RESOURCES BOARD
Phone: 916-322-5350

CALIFORNIA ENVIRONMENTAL PROTECTION AGENCY

DEPARTMENT OF PESTICIDE REGULATION
Sacramento, CA 95812-4015 USA
Phone: 916-445-4300   Fax: 916-324-1452
Website: www.cdpr.ca.gov

Founded: NA
Scope: Statewide

Description: The Department of Pesticide Regulation regulates
all aspects of pesticides sales and use, recognizing the need
to control pests, while protecting public health and the
environment and fostering reduced-risk pest management
strategies.

Keyword(s): Air Quality and Pollution, Pesticides, Solid Waste
Management, Toxic Substances, Nuclear-free, Water quantity,
Water export and diversion, Water Pollution Management

Contact(s):
Paul Helliker, DIRECTOR
Phone: 916-445-4000

CALIFORNIA ENVIRONMENTAL PROTECTION AGENCY

DEPARTMENT OF TOXIC SUBSTANCES CONTROL
Sacramento, CA 95812-0806 USA
Phone: 916-323-9723   Fax: 916-324-1788
Website: www.etsc.ca.gov

Founded: NA
Membership: 1000
Scope: Statewide

Description: Responsible for overseeing the cleanup of
hazardous waste sites; monitoring and regulatory
management of hazardous waste transportation, treatment,
storage and disposal; and promotion of hazardous waste
reduction in California.

Contact(s):
Bob Borzeueri, CHIEF DEPUTY DIRECTOR
Phone: 916-322-0449
bborzeuri@etsc.ca.gov
Edwin Lowry, DIRECTOR
Phone: 916-322-0504
elowry@etsc.ca.gov

CALIFORNIA ENVIRONMENTAL PROTECTION AGENCY

INTEGRATED WASTE MANAGEMENT BOARD, IWMB
1001 I Street, P.O.B. 4025
Sacramento, CA 95812 USA
Phone: 916-341-6000
Website: http://www.calepa.ca.gov

Founded: 1990

Description: The CIWMB is comprised of six members; four
appointed by the governor and two by the legislature. CIWMB's
goal is to protect the public's health and safety and the
environment through waste prevention, waste diversion, and
safe waste processing and disposal. Also, minimizing waste
generation and disposal in California while facilitating the
development of industries that use recyclable materials, will be
realized by establishing sustainable markets for recyclable
materials, reducing reliance on land disposal, and effectively
educating the public.

Contact(s):
Linda Moulton-Patterson, CHAIRMAN
Phone: 916-341-6024
lmoulton@ciwmb.ca.gov

CALIFORNIA ENVIRONMENTAL PROTECTION AGENCY

OFFICE OF ENVIRONMENTAL HEALTH HAZARD
ASSESSMENT
P.O. Box 4010
Sacramento, CA 95812-4010 USA
Phone: 916-324-7572   Fax: 916-327-1097
Website: www.oehha.ca.gov

Founded: NA
Membership: 100

Scope: Statewide

Description: The Office of Environmental Health Hazard Assessment is charged with assessing human health risks posed by chemicals in the environment. The office is also the lead agency for implementation of the Safe Drinking Water and Toxic Enforcement Act of 1986 (Proposition 65).

Contact(s):
Joan Denton, DIRECTOR

CALIFORNIA ENVIRONMENTAL PROTECTION AGENCY
WATER RESOURCES CONTROL BOARD
1001 I St.
Sacramento, CA 95814- USA
Phone: 916-341-5254   Fax: 916-341-5252
Website: www.swrcb.ca.gov

Founded: NA
Scope: Statewide

Description: To protect water quality and allocate water rights. These objectives are achieved through two action programs: water quality and water rights.

Contact(s):
Peter Silva, BOARD MEMBER
Richard Katz, BOARD MEMBER
Arthur Baggett, CHAIRMAN (ACTING)
Thomas Howard, DEPUTY DIRECTOR
Edward Anton, EXECUTIVE DIRECTOR (ACTING)
Roger Briggs, REGIONAL EXECUTIVE OFFICER OF THE CENTRAL COAST REGION
Phone: 805-549-3147
Gary Carlton, REGIONAL EXECUTIVE OFFICER OF THE CENTRAL VALLEY REGION
Phone: 916-255-3000
Phil Gruenberg, REGIONAL EXECUTIVE OFFICER OF THE COLORADO RIVER BASIN REGION
Phone: 619-346-7491
Harold Singer, REGIONAL EXECUTIVE OFFICER OF THE LAHONTAN REGION
Phone: 916-542-5400
Dennis Dickerson, REGIONAL EXECUTIVE OFFICER OF THE LOS ANGELES REGION
Phone: 213-266-7500
Lee Michlin, REGIONAL EXECUTIVE OFFICER OF THE NORTH COAST REGION
Phone: 707-576-2220
John Robertus, REGIONAL EXECUTIVE OFFICER OF THE SAN DIEGO REGION
Phone: 619-467-2952
Loretta Barsamian, REGIONAL EXECUTIVE OFFICER OF THE SAN FRANCISCO BAY REGION
Phone: 510-286-1255
Gerald Thibeault, REGIONAL EXECUTIVE OFFICER OF THE SANTA ANA REGION
Phone: 909-782-4130

CALIFORNIA SEA GRANT COLLEGE SYSTEM
California Sea Grant College, University of California,
9500 Gilman Dr.
La Jolla, CA 92093-0232 USA
Phone: 858-534-4440   Fax: 858-534-2231
E-mail: caseagrant@ucsd.edu
Website: www-csgc.ucsd.edu/

Founded: NA

Description: A multiuniversity program of marine research, extension services, and education that contributes to the growing body of knowledge about coastal and oceanic resources and helps solve contemporary problems in the marine sphere.

Publications: Sea Grant in Brief

Keyword(s): Coasts, Wildlife, Exotic species, Aquatic nuisance species, Wetlands, Biotechnology

Contact(s):
Christopher Dewees, COORDINATOR OF SEA GRANT EXTENSION PROGRAM
Cooperative Extension: University of California, Davis, CA 95616
Phone: 530-752-1497
Linda Duguay, DIRECTOR OF USC SEA GRANT PROGRAM
University of Southern California: University Park, Los Angeles, CA 90089-0373
Phone: 213-740-1961
Fax: 213-740-5936
seagrant@mizar.usc.edu
Clinton Winant, INTERIM DIRECTOR
California Sea Grant College System, University of California, 9500 Gilman Drive, La Jolla, CA 92093-0232
Phone: 619-534-4440
Fax: 619-534-2231
Judy Lemus, LEADER OF USC SEA GRANT MARINE ADVISORY PROGRAM AND ASSOCIATE DIRECTOR OF USC SEA GRANT PROGRAM
USC Sea Grant Program Marine Advisory Service: University of Southern California: University Park, Los Angeles, CA 90089-0373
Phone: 213-740-1965
Fax: 213-740-5936

CALIFORNIA STATE EXTENSION SERVICES
Cooperative Extension and Agricultural Experiment Station, University of California, 1111 Franklin St., 6th Fl.
Oakland, CA 94607-5200 USA
Phone: 510-987-0036
Website: danr.ucop.edu/regional.htm

Founded: NA

Contact(s):
Lanny Lund, ASSISTANT VICE PRESIDENT, PROGRAMS AND ASSISTANT DIRECTOR AES & STATE LEADER CE
Henry Vaux, ASSOCIATE VICE PRESIDENT AND ASSOCIATE DIRECTOR AES & EC
Phone: 415-987-0026
Neal Alfen, DEAN
College of Agricultural and Environmental Sciences, Associate Director of AES Programs, University of California, Davis, CA 95616
Phone: 916-752-1605
Bennie Osburn, DEAN
Veterinary Medicine, University of California, Davis, CA 95616
Phone: 916-752-1360
Michael Clegg, DEAN
College of Natural and Agricultural Sciences; Associate Director: AES & CE: University of California, Riverside, CA 92521
Phone: 714-787-3101
Gordon Rausser, DEAN
College of Natural Resources: Associate Director: AES & CE: University of California, Berkeley, CA 94720
Phone: 154-2171

Susan Laughlin, REGIONAL DIRECTOR, CENTRAL COAST AND SOUTH
125 Highlander Hall, University of California, Riverside, CA 92521
Phone: 714-787-3321
Linda Manton, REGIONAL DIRECTOR, CENTRAL VALLEY
Kearney Agricultural Center, 9240 S. Riverbend Ave., Parlier, CA 93648
Phone: 209-891-2566
Nicelma King, REGIONAL DIRECTOR, NORTH COAST AND MOUNTAIN
University of California, Davis, CA 95616
Phone: 916-754-8509
W. Gomes, VICE PRESIDENT
Agricultural and Natural Resources, Director AES & CE
Phone: 415-987-0060
E. Fitzhugh, WILDLIFE SPECIALIST
Wildlife and Conservation Biology, University of California, 1 Shields Ave., Davis, CA 95616-8751
Phone: 530-752-1496
Fax: 530-752-4154

CALIFORNIA STATE LANDS COMMISSION

100 Howe Ave., Suite 100 S
Sacramento, CA 95825-8202 USA
Phone: 916-574-1800　　　Fax: 916-574-1810
Website: www.slc.ca.gov

Founded: NA
Scope: State

Description: Jurisdiction over and management responsibility for state-owned sovereign and legislatively granted lands. Handles related land leases, exchanges, and associated transactions. Conducts oil, gas, geothermal, and leasing of other mineral onstate-owned lands. Related activities include boundary and ownership determination, granted lands administration, and maintaining land information systems.

Keyword(s): Public Lands

Contact(s):
Jack Rump, CHIEF COUNSEL
Phone: 916-574-1850
Dwight Sanders, CHIEF, ENVIRONMENTAL PLANNING AND MANAGEMENT DIVISION
Phone: 916-574-1890
Robert Lynch, CHIEF, LAND MANAGEMENT DIVISION
Phone: 916-574-1940
Gary Gregory, CHIEF, MARINE FACILITIES DIVISION
330 Golden Shore, Suite 210, Long Beach, CA 90802
Paul Mount, CHIEF, MINERAL RESOURCES DIVISION
200 Oceangate, 12th Fl., Long Beach, CA 90802
Phone: 562-590-5205
Kathleen Connell, COMMISSIONER
Paul Thayer, EXECUTIVE OFFICER
Phone: 916-574-1800
William Morrison, LEGISLATIVE LIAISON
Phone: 916-574-1800
Cruz Bustamante, LIEUTENANT GOVERNOR & COMMISSIONER
B. Gage, STATE DIRECTOR OF FINANCE

CLEMSON UNIVERSITY EXTENSION SERVICE

Clemson University, Rm. 103 Barre Hall
Clemson, SC 29634-0110 USA
Phone: 864-656-3382　　　Fax: 864-656-5819
Website: www.clemson.edu/extension/

Founded: NA
Scope: Statewide

Contact(s):
Daniel Smith, DIRECTOR OF EXTENSION SERVICE
103 Barre Hall, Clemson University, Clemson, SC 29634-0310
Phone: 864-656-3382
dbsmith@clemson.edu
Allen Dunn, DIRECTOR OF SCHOOL OF NATURAL RESOURCES
130 Lehotsky Hall, Clemson University, Clemson, SC 29634
Phone: 864-656-3215
adunn@clemson.edu
P. Horton, EXTENSION ENTOMOLOGIST
103 Barre Hall, Clemson, SC 29634
Phone: 864-656-3382
mhorton@clemson.edu
John Sweeney, EXTENSION FISH SPECIALIST
Department Head of Aquaculture, Fisheries, and Wildlife, Lehotsky Hall, Clemson University, Clemson, SC 29634-0362
Phone: 864-656-3117
jswny@clemson.edu
Larry Nelson, EXTENSION FORESTER
272-E Lehotsky Hall, Clemson University, Clemson, SC 29634-1003
Phone: 864-656-4866
lnelson@clemson.edu
Greg Yarrow, EXTENSION WILDLIFE SPECIALIST
Phone: 864-656-7370
gyarrow@clemson.edu

COLLEGE OF TROPICAL AGRICULTURE AND HUMAN RESOURCES

University of Hawaii, 3050 Maile Way
Honolulu, HI 96822 USA
Phone: 808-956-8131　　　Fax: 808-956-9105
E-mail: research@ctahr.hawaii.edu

Founded: 1901
Membership: 200
Scope: Statewide

Description: Plan and implement research and extension in agriculture, natural resources, and human resources relevant to Hawaii and the tropics, with emphasis on the Pacific and Asia.

Publications: Various research and extension publications

Keyword(s): Conservation Tillage, Biotechnology, Research, Engineering, Environment, Gardening and Horticulture, Pesticides, Public Health Protection, Renewable Resources, Flowers, Plants, and Trees, Health and Nutrition, Insects and Butterflies, Landscape Architecture, Agriculture, Chemical Pollution Control

Contact(s):
Catherine Chanhalbrandt, ASSOCIATE DEAN AND ASSOCIATE DIRECTOR FOR RESEARCH

COLORADO COOPERATIVE FISH AND WILDLIFE RESEARCH UNIT (USDI)

201 Wagar Bldg., Dept. of Fishery and Wildlife Biology, Colorado State University
Ft. Collins, CO 80523-1484 USA
Phone: 970-491-5396　　　Fax: 970-491-1413

Founded: 1947

Membership: 5
Scope: International

Description: Offers expertise and training facilities in fish and wildlife population ecology, aquatic habitat analysis, conservation biology, sampling and analysis theory, and biostatistics.

Keyword(s): Whirling Disease, Endangered Species, Wildlife, Nongame Wildlife, Fish Wildlife Management, Birds

Contact(s):
Kenneth Burnham, ASSISTANT LEADER
Eric Bergersen, ASSISTANT LEADER
David Anderson, LEADER

COLORADO DEPARTMENT OF AGRICULTURE

700 Kipling St., Suite 4000
Lakewood, CO 80215 USA
Phone: 303-239-4100   Fax: 303-239-4176
Website: www.ag.state.co.us

Founded: 1949
Scope: State

Description: Strives to meet the increasingly complex needs of agriculture through work on marketing problems, technological changes in pest and insect control, and rapidly changing patterns in crop and livestock operations.

Keyword(s): Environmental and Conservation Education, Pesticides, Public Lands, Toxicology, Agriculture

Contact(s):
Gary Shoun, BRAND COMMISSIONER OF BOARD OF STOCK INSPECTION DIVISION
Don Ament, COMMISSIONER
Robert McLavey, DEPUTY COMMISSIONER
Jerry Bohlender, DIRECTOR OF ANIMAL INDUSTRY DIVISION
Ronald Turner, DIRECTOR OF DIVISION OF INSPECTION AND CONSUMER SERVICES
Jim Rubingh, DIRECTOR OF MARKETS DEVELOPMENT DIVISION
John Gerhardt, DIRECTOR OF PLANT INDUSTRY DIVISION
David Carlson, RESOURCE ANALYST

COLORADO DEPARTMENT OF EDUCATION

STATE OFFICE
201 E. Colfax Ave.
Denver, CO 80203 USA
Phone: 303-866-6600   Fax: 303-830-0793
Website: www.cde.state.co.us

Founded: NA
Scope: Statewide

Description: Conservation Education Services, jointly with the Colorado Division of Wildlife.

Keyword(s): Environmental Ethics, Pollution Prevention, Urban Environment, training, Schoolyard Habitats, Environmental and Conservation Education

Contact(s):
Don Hollums, ENVIRONMENTAL EDUCATION CONSULTANT
Phone: 303-866-6787
hollums_d@cde.state.co.us

COLORADO DEPARTMENT OF NATURAL RESOURCES

1313 Sherman St.
Denver, CO 80203 USA
Phone: 303-866-3311   Fax: 303-866-2115
Website: www.dnr.state.co.us

Founded: 1968
Membership: 1800
Scope: Statewide

Description: Responsible for mineral and energy, land, water, wildlife, and park resources management for the state. Also responsible for major environmental conservation and management programs.

Publications: see publication website

Contact(s):
Bill Daley, DEPUTY DIRECTOR
Ronald Cattany, DEPUTY DIRECTOR
Greg Walcher, EXECUTIVE DIRECTOR
Cindy Horiuchi, HUMAN RESOURCES DIRECTOR

COLORADO DEPARTMENT OF NATURAL RESOURCES

COLORADO GEOLOGIC SURVEY
1313 Sherman St., Rm. 715
Denver, CO 80203 USA
Phone: 303-866-2611   Fax: 303-866-2461
E-mail: cgspubs@state.co.us
Website: www.geosurvey.state.co.us

Founded: NA
Membership: 25
Scope: State
Publications: Rock Talk - quarterly newsletters

Contact(s):
Vicki Cowart, STATE GEOLOGIST

COLORADO DEPARTMENT OF NATURAL RESOURCES

DIVISION OF MINERALS AND GEOLOGY
1313 Sherman St., Rm. 215 Geology
Denver, CO 80203 USA
Phone: 303-866-3567   Fax: 303-832-8106
E-mail: dmg_pio@state.co.us
Website: www.mining.state.co.us

Founded: NA
Membership: 70
Scope: State

Contact(s):
Michael Long, DIRECTOR
michael.long@state.co.us

COLORADO DEPARTMENT OF NATURAL RESOURCES

DIVISION OF PARKS AND OUTDOOR RECREATION
1313 Sherman St., Rm. 618
Denver, CO 80203 USA
Phone: 303-866-3437   Fax: 303-866-3206
Website: www.parks.state.co.us

Founded: NA
Scope: State

Contact(s):
Tom Kenyon, ACTING DIRECTOR

COLORADO DEPARTMENT OF NATURAL RESOURCES
DIVISION OF WATER RESOURCES
1313 Sherman St.
Denver, CO 80203 USA
Phone: 303-866-3581　　　Fax: 303-866-3589
Website: www.water@state.co.us

Founded: NA
Membership: 80
Scope: Statewide
Publications: Streamline—newsletter

Contact(s):
Wil Burt, DEPUTY STATE ENGINEER
Harold Simpson, STATE ENGINEER

COLORADO DEPARTMENT OF NATURAL RESOURCES
DIVISION OF WILDLIFE
6060 Broadway
Denver, CO 80216 USA
Phone: 303-297-1192　　　Fax: 303-294-0894
E-mail: AskDOW@state.co.us
Website: www.wildlife.state.co.us

Founded: NA
Scope: Statewide

Contact(s):
Russell George, DIRECTOR

COLORADO DEPARTMENT OF NATURAL RESOURCES
OIL AND GAS CONSERVATION COMMISSION
1120 Lincoln St., Suite 801
Denver, CO 80203 USA
Phone: 303-894- 210　　　Fax: 303-894-2109
Website: http://www.dnr.state.co.us

Founded: NA

Contact(s):
Richard Griebling, DIRECTOR

COLORADO DEPARTMENT OF NATURAL RESOURCES
STATE BOARD OF LAND COMMISSIONERS
1313 Sherman St., Rm. 621
Denver, CO 80203 USA
Phone: 303-866-3454　　　Fax: 303-866-3152
Website: www.trustlands.state.co.us

Founded: NA
Membership: 35
Scope: Statewide
Publications: available on web

Contact(s):
John Brejcha, ACTING DIRECTOR
Phone: 303-866-3454

COLORADO DEPARTMENT OF PUBLIC HEALTH AND ENVIRONMENT
4300 Cherry Creek Dr., S.
Denver, CO 80246-1530 USA
Phone: 303-692-2035　　　Fax: 303-691-7702
E-mail: cdphe.information@state.co.us
Website: http://www.cdphe.state.co.us/

Founded: NA
Scope: State

Description: The Colorado Department of Public Health and Environment has the responsibility for improving and protecting the health and environment for Colorado's citizens by: assuring a healthy working and living environment, protecting people against exposure to diseases, establishing preventive health services, and providing a quality environment through air, waste, water, radiation, and other environmental protection activities.

Contact(s):
Jane Norton, EXECUTIVE DIRECTOR

COLORADO STATE FOREST SERVICE
203 Forestry Building
Ft. Collins, CO 80523-5060 USA
Phone: 970-491-6303　　　Fax: 970-491-7736
Website: www.colostate.edu/Depts/CSFS

Founded: 1885
Membership: 110
Scope: Statewide

Description: The mission of the State Forest Service is to achieve stewardship of Colorado's environment through forestry outreach and service.

Keyword(s): Forests and Forestry, Urban Forestry, Land Use Planning, Sustainable Ecosystems, Environmental and Conservation Education

Contact(s):
Phil Hoefer, COMMUNITY FORESTRY
Bob Sturtevant, CONSERVATION EDUCATION
Phil Schwolert, FOREST MANAGEMENT
James Hubbard, STATE FORESTER
Rich Homann, WILDFIRE PROTECTION

COLORADO STATE SOIL CONSERVATION BOARD
COLORADO DEPARTMENT OF AGRICULTURE
1313 Sherman St., Rm. 219
Denver, CO 80203-2243 USA
Phone: 303-866-3351　　　Fax: 303-832-8106
Website: www.ag.state.co.us

Founded: NA
Membership: 3
Scope: Statewide

Contact(s):
Robert Zebroski, DIRECTOR
robert.zebroski@ag.state.co.us

COLORADO STATE UNIVERSITY COOPERATIVE EXTENSION
1 Administration Bldg., Colorado State University
Ft. Collins, CO 80523 USA
Phone: 970-491-6281　　　Fax: 970-491-6208
Website: www.colostate.edu/Depts/CoopExt/

Founded: 1914
Membership: 370
Scope: State

Description: A branch of Colorado State University. Conducts statewide noncredit educational programs off campus.

Publications: Publications listed on website; www.cerc1@ur.colostate.edu

Keyword(s): Hunting, Biotechnology, Environmental and Conservation Education, Gardening and Horticulture, Health and Nutrition, Precision Farming, Sustainable Development, Sustainable Ecosystems, Urban Environment, Outdoor Recreation, Pesticides, Renewable Resources, Rural Development, Solid Waste Management, Agriculture

Contact(s):
Mary Gray, ASSOCIATE DIRECTOR OF PROGRAMS
gray@coop.ext.colostate.edu
Milan Rewerts, DIRECTOR OF COOPERATIVE EXTENSION
mrewerts@coop.ext.colostate.edu
Shelley Stanley, EXTENSION AGENT-NATURAL RESOURCES
15200 W. Sixth Ave., Golden, CO 80401
Phone: 303-271-6620
William Andelt, EXTENSION WILDLIFE SPECIALIST: ANIMAL DAMAGE CONTROL
Dept. of Fishery and Wildlife Biology: 109 Wagar: Colorado State University, Ft. Collins, CO 80523
Phone: 970-491-7093
Delwin Benson, EXTENSION WILDLIFE SPECIALIST: WILDLIFE MANAGEMENT
Dept. of Fishery and Wildlife Biology: 109 Wagar: Colorado State University, Ft. Collins, CO 80523
Phone: 970-491-6411
Fax: 970-491-5091

COLORADO WATER CONSERVATION BOARD

WATER CONSERVATION BOARD
1313 Sherman St.,
Denver, CO 80203 USA
Phone: 303-866-3441   Fax: 303-866-4474
Website: www.dnr@state.co.us

Founded: NA
Scope: Statewide

Keyword(s): Outdoor Recreation, Public Lands, Exotic species, Aquatic nuisance species, Environmental and Conservation Education

Contact(s):
Rod Kuharich, DIRECTOR

COLUMBIA RIVER GORGE COMMISSION

P.O. Box 730
White Salmon, WA 98672 USA
Phone: 509-493-3323   Fax: 509-493-2229
E-mail: crgc@gorge.net
Website: www.gorgecommission.org

Founded: NA
Membership: 20
Scope: Regional

Description: Established by the states of Oregon and Washington to implement the Columbia River Gorge National Scenic Area Act by developing a regional management plan, in cooperation with the U.S. Forest Service. The commission is composed of three members from Oregon, three from Washington, and one from each of the six local Gorge counties. A Secretary of Agriculture appointee is a thirteenth nonvoting member. Purpose of the National Scenic Area Act is to protect and enhance scenic, natural, cultural, and recreation resources, while encouraging economic development within 13 established urban areas.

Keyword(s): training, Outdoor Recreation, Land Use Planning, Natural Areas, Environmental Planning

Contact(s):
Anne Squier, CHAIRMAN
Maratha Bennett, EXECUTIVE DIRECTOR
bennett@gorgecommission.org
Wayne Wooster, VICE CHAIRMAN
Phone: 509-493-3724
wooster@gorge.net

CONNECTICUT COUNCIL ON ENVIRONMENTAL QUALITY

79 Elm Street
Hartford, CT 06106 USA
Phone: 860-424-4000   Fax: 860-424-4070
Website: www.ceq.state.ct.us

Founded: 1971
Membership: 9
Scope: Statewide

Description: Prepares annual reports to the Governor on the status of Connecticut's environment; receives and investigates citizen complaints pertaining to the environment; and reviews environmental assessments of construction activities of state agencies. The council is composed of nine appointed members who serve without compensation.

Publications: Environmental Quality in Connecticut (Annual Report)

Keyword(s): Environmental Planning, Environmental Protection, Land Use Planning, Environment

Contact(s):
Donal O'Brien, CHAIRMAN
Karl Wagener, EXECUTIVE DIRECTOR

CONNECTICUT DEPARTMENT OF AGRICULTURE

765 Asylum Ave.
Hartford, CT 06105 USA
Phone: 860-713-2503   Fax: 860-713-2516
E-mail: ctdeptag@po.state.ct.us
Website: www.state.ct.us/doag

Founded: NA
Scope: National

Publications: Connecticut Weekly

Contact(s):
Shirley Ferris, COMMISSIONER
Phone: 860-713-2514
Bruce Gresczyk, DEPUTY COMMISSIONER
Phone: 860-713-2526
Frank Intino, DEPUTY DIRECTOR: MARKETING AND TECHNOLOGY
Phone: 860-713-2503
Gabriel Moquin, DEPUTY DIRECTOR: REGULATION AND INSPECTION
Phone: 860-713-2508

Dawn Cassada, DIRECTOR: ADMINISTRATION
Phone: 860-713-2502
John Volk, DIRECTOR: AQUACULTURE DIVISION
P.O. Box 97, Milford, CT 06460
Phone: 203-874-2855
Joseph Dippel, DIRECTOR: FARMLAND PRESERVATION
Phone: 860-713-2511
Robert Pellegrino, DIRECTOR: MARKETING AND
TECHNOLOGY
Phone: 860-713-2503
Emilie Andrews, DIRECTOR: PERSONNEL
Phone: 860-713-2501
Bruce Sherman, DIRECTOR: REGULATION AND
INSPECTION
Phone: 860-713-2504
David Carey, EXECUTIVE DIRECTOR: CONNECTICUT
MARKETING AUTHORITY
101 Reserve Rd., Hartford, CT 06114
Phone: 860-566-3699

CONNECTICUT DEPARTMENT OF ENVIRONMENTAL PROTECTION

79 Elm St.
Hartford, CT 06106-5127 USA
Phone: 860-424-3000   Fax: 860-424-4078
Website: www.dep.state.ct.us

Founded: NA
Membership: 10
Scope: Statewide

Description: Created by the Connecticut General Assembly to conserve, protect, and improve the state's environment and to manage the basic resources of air, water, and land for the benefit of present and future generations.

Publications: Connecticut Wildlife

Contact(s):
Jane Stahl, ASSISTANT COMMISSIONER, AIR, WATER
AND WASTE
Phone: 860-424-3009
Carmine Dibattista, CHIEF, BUREAU OF AIR MANAGEMENT
Phone: 860-424-3026
Edward Parker, CHIEF, BUREAU OF NATURAL
RESOURCES
Phone: 860-424-3010
Richard Clifford, CHIEF, BUREAU OF OUTDOOR
RECREATION
Phone: 860-424-3200
Richard Barlow, CHIEF, BUREAU OF WASTE
MANAGEMENT
Phone: 860-424-3021
Robert Smith, CHIEF, BUREAU OF WATER MANAGEMENT
Phone: 860-424-3704
Arthur Rocque, COMMISSIONER
Phone: 860-424-3001
David Leff, DEPUTY COMMISSIONER ENVIRONMENTAL
CONSERVATIONS
Phone: 860-424-3005
Michele Sullivan, DIRECTOR, COMMUNICATIONS,
EDUCATION AND PUBLICATIONS
Phone: 860-424-4100
Ernest Beckwith, DIRECTOR, FISHERIES DIVISION
Phone: 860-424-3474
Donald Smith, DIRECTOR, FORESTRY DIVISION
Phone: 860-424-3630
Charles Reed, DIRECTOR, LAND ACQUISITION AND
MANAGEMENT
Phone: 860-424-3016
George Barone, DIRECTOR, LAW ENFORCEMENT
DIVISION
Phone: 860-424-3012
Pamela Adams, DIRECTOR, PARKS DIVISION
Phone: 860-424-3200
Dale May, DIRECTOR, WILDLIFE DIVISION
Phone: 860-424-3011

CONNECTICUT SEA GRANT COLLEGE PROGRAM UNIVERSITY OF CONNECTICUT

1080 Shennecossett Rd.
Groton, CT 06340-6048 USA
Phone: 860-405-9110   Fax: 860-405-9109
Website: www.SEAGRANT.UCONN.EDU

Founded: NA
Membership: 9
Scope: National

Contact(s):
Edward Monahan, DIRECTOR
sgoadm01@uconnvm.uconn.edu

CONSERVATION AND SURVEY DIVISION (NEBRASKA)

University of Nebraska-Lincoln, 113 Nebraska Hall, 901
N. 17th St.
Lincoln, NE 68588 USA
Phone: 402-472-3471   Fax: 402-472-4608
Website: www.csd.unl.edu

Founded: NA
Scope: Statewide

Description: CSD, the state geological, water, soil and land cover survey, has state-mandated responsibilities to inventory and investigate geologically related natural resources of the state; to record the results of these investigations; to assist non-profit, private and governemental agencies working to conserve the state's natural resources; to study the geologic history and geography of the state to aid sustainable economic development; and to publish maps, reports and electronic information about these activities.

Publications: see publication website

Contact(s):
Mark Kuzila, DIRECTOR
Phone: 402-472-7537

COOPERATIVE EXTENSION SERVICE

UNIVERSITY OF ALASKA FAIRBANKS
COLLEGE OF RURAL ALASKA
CES Bldg., University of Alaska
Fairbanks, AK 99775-6180 USA
Phone: 907-474-7246   Fax: 907-474-6971
E-mail: fyace@uaf.edu
Website: www.uaf.edu/coop-ext

Founded: NA
Membership: 70
Scope: Statewide

Contact(s):
Anthony Nakazawa, DIRECTOR
anatn@uaa.alaska.edu

Peter Stortz, FISH AND NR SPECIALIST
CES/Palmer Research Center, 533 E. Fireweed Ln., Palmer, AK 99645
Phone: 907-746-9459
Fax: 907-746-2677
ffpjs@ufa.edu
Michele Hebert, LAND RESOURCES AGENT
Phone: 907-474-1530
ffmah@uaf.edu
Robert Gorman, LAND RESOURCES PROGRAM CHAIR
Phone: 907-786-6323
ffrfg@uaf.edu

COUNCIL ON RESOURCES AND DEVELOPMENT

c/o Office of State Planning, 2 1/2 Beacon St.
Concord, NH 03301 USA
Phone: 603-271-2155   Fax: 603-271-1728
Website: www.state.nh.us/osp/planning

Founded: 1963
Membership: 40
Scope: Statewide

Description: The ten members on the council represent the state's development and resource agencies. The council conducts studies and presents recommendations concerning problems in the fields of environmental protection, natural resources, and growth management; consults with, negotiates with, and obtains information from other state and federal agencies; offers guidance and recommendations to the Governor and Council or the General Court; recommends disposition or lease of state-owned surplus real property; and resolves differences or conflicts concerning development, resource management, and the implementation of the state policy.

Contact(s):
Jeffrey Taylor, CHAIRMAN

D

DELAWARE COOPERATIVE EXTENSION SERVICES

Delaware Cooperative Extension, Townsend Hall, University of Delaware
Newark, DE 19717-1303 USA
Phone: 302-831-2504   Fax: 302-831-6758
E-mail: pbarber@udel.edu
Website: bluehen.ags.udel.edu/deces/

Founded: NA
Scope: Statewide

Keyword(s): Environmental Protection, Conservation Tillage, Engineering, Environmental and Conservation Education, Gardening and Horticulture, Pesticides, Precision Farming, Public Health Protection, Flowers, Plants, and Trees, Forest Management, Health and Nutrition, Insects and Butterflies, Land Use Planning, Agriculture, Energy Conservation

Contact(s):
John Ewart, AQUACULTURE SPECIALIST
University of Delaware Aquac. Research Center, 700 Pilottown Rd., Lewes, DE 19958
Phone: 302-645-4060
Fax: 302-645-4007

Patricia Barber, ASSOCIATE DEAN FOR EXTENSION AND OUTREACH
pbarber@udel.edu
Robin Morgan, DEAN, COLLEGE OF AGRICULTURAL SCIENCES AND DIRECTOR, AGRICULTURAL EXPERIMENT STATION

DELAWARE DEPARTMENT OF NATURAL RESOURCES AND ENVIRONMENTAL CONTROL

89 Kings Highway
Dover, DE 19901 USA
Phone: 302-739-4403   Fax: 302-739-6242
Website: www.dnrec@state.de.us

Founded: 1970
Membership: 800
Scope: Statewide

Description: The mission of the Delaware Department of Natural Resources and Environmental Control is to protect and manage the state's natural resources, protect public health and safety, provide quality outdoor recreation and to serve and educate the citizens of Delaware to promote the wise use, conservation, and enhancement of Delaware's environment.

Publications: Outdoor Delaware, DNREC News

Keyword(s): Air Quality and Pollution, Environmental and Conservation Education, Soil Conservation, Beaches, Water Quality, Waste Management

Contact(s):
David Small, DEPUTY SECRETARY
Melinda Carl, EDITOR, DNREC NEWS
Phone: 302-739-4506
Kathleen Jamison, EDITOR, OUTDOOR DELAWARE
Phone: 302-739-4506
Nicholas Dipasquale, SECRETARY
ndipasquale@state.de.us

DELAWARE DEPARTMENT OF NATURAL RESOURCES AND ENVIRONMENTAL CONTROL

DIVISION OF AIR AND WASTE MANAGEMENT
89 Kings Hwy., P.O. Box 1401
Dover, DE 19901 USA
Phone: 302-739-4403   Fax: 302-739-6242

Founded: NA
Scope: Statewide

Contact(s):
William Hill, ADMINISTRATOR ENFORCEMENT
Phone: 302-739-5072
Denise Ferguson-Southard, DIRECTOR
Phone: 302-739-4764
Nancy Marker, MANAGER, HAZARDOUS WASTE
Phone: 302-739-3689
Christina Wirtz, MANAGER, SITE INVESTIGATION AND REMEDIATION
Phone: 302-395-2600
Jamie Rutherford, MANAGER, SOLID WASTE
Phone: 302-739-3820
Kathleen Stiller, MANAGER, UNDERGROUND STORAGE TANKS
Phone: 302-323-4588

DELAWARE DEPARTMENT OF NATURAL RESOURCES AND ENVIRONMENTAL CONTROL
DIVISION OF FISH AND WILDLIFE
89 Kings Hwy.
Dover, DE 19901 USA
Phone: 302-739-5295   Fax: 302-739-6157
Website: www.dnrec.state.de.us-fw-index.htm

Founded: NA
Membership: 126
Scope: Local
Publications: The Observer

Contact(s):
Dr. William Meredith, ADMINISTRATOR OF MOSQUITO CONTROL
Phone: 302-739-3493
James Graybeal, ADMINISTRATOR, ENFORCEMENT
Phone: 302-739-3440
Charles Lesser, ADMINISTRATOR, FISHERIES
Phone: 302-739-3441
H. Alexander, ADMINISTRATOR, WILDLIFE
Phone: 302-739-5297
Andrew Manus, DIRECTOR
Phone: 302-739-5295
Lynn Herman, FEDERAL AID COORDINATOR AND SENIOR PLANNER
Phone: 302-739-5296
Phil Carpenter, MANAGER, ACQUISITIONS
Phone: 302-739-3441
Lacy Nichols, MANAGER, CONSTRUCTION
Phone: 302-739-3441

DELAWARE DEPARTMENT OF NATURAL RESOURCES AND ENVIRONMENTAL CONTROL
DIVISION OF PARKS AND RECREATION
89 Kings Hwy.
Dover, DE 19901 USA
Phone: 302-739-4401   Fax: 302-739-3817

Founded: NA
Scope: Statewide

Contact(s):
Charles Salkin, DIRECTOR
Phone: 302-739-4401
James Oneill, MANAGER, CULTURAL & RECREATION SERVICES
Phone: 302-739-4413
Clyde Shipman, MANAGER, PARK OPERATIONS
Phone: 302-739-4406
Mark Chura, MANAGER, PLANNING, PRESERVATION AND DEVELOPMENT
Phone: 302-739-5285

DELAWARE DEPARTMENT OF NATURAL RESOURCES AND ENVIRONMENTAL CONTROL
DIVISION OF SOIL AND WATER CONSERVATION
89 Kings Highway
Dover, DE 19901 USA
Phone: 302-739-4411   Fax: 302-739-6724

Founded: NA
Scope: Statewide

Keyword(s): Environmental and Conservation Education, Soil Conservation, Water Pollution Management, Air Quality and Pollution

Contact(s):
Sarah Cooksey, ADMINISTRATOR: DELAWARE COASTAL MANAGEMENT PROGRAM
Phone: 302-739-3451
Robert Henry, ADMINISTRATOR: SHORELINE AND WATERWAY MANAGEMENT
John Hughes, DIRECTOR
jhughes@dnrec.state.de.us

DELAWARE DEPARTMENT OF NATURAL RESOURCES AND ENVIRONMENTAL CONTROL
DIVISION OF WATER RESOURCES
89 Kings Hwy.
Dover, DE 19901 USA
Phone: 302-739-4403
Website: www.dnrc.state.de.us

Founded: NA
Scope: Statewide

Contact(s):
Sergio Huerta, ADMINISTRATOR, ENVIRONMENTAL SERVICES
Kevin Donnelly, DIRECTOR
Phone: 302-739-4860
Rodney Wyatt, MANAGER, GROUND WATER DISCHARGES
Phone: 302-739-4761
Peder Hansen, MANAGER, SURFACE WATER DISCHARGES
Phone: 302-739-5731
Stewart Lovell, MANAGER, WATER SUPPLY
Phone: 302-739-4793
John Schneider, MANAGER, WATERSHED ASSESSMENT
Phone: 302-739-4590
William Moyer, MANAGER, WETLANDS AND SUBAQUEOUS LANDS
Phone: 302-739-4691

DELAWARE FOREST SERVICE
2320 S. DuPont Highway
Dover, DE 19901-5515 USA
Phone: 302-739-4811   Fax: 302-697-6245
E-mail: AUSTIN@DDA.STATE.DE.US
Website: www.state.de.us/deptagri/

Founded: NA
Membership: 24
Scope: Statewide

Description: The DDA works to provide mandated services which protect the health and welfare of Delaware consumers and to advertise those services; to promote the sound utilization of resources, especially agricultural lands; and to advance the economic viability of the food, fiber, and agricultural industries of Delaware.

Keyword(s): Forests and Forestry, Land Use Planning, Pesticides, Urban Forestry, Agriculture

Contact(s):
Teresa Crenshaw, AGRICULTURE COMPLIANCE LABORATORY
teresa@dda.state.de.us
Anne Fitzgerald, COMMUNITY RELATIONS OFFICER
anne.dda.state.de.us
Susan Stuchlik-Edwards, DEPUTY SECRETARY
susan@dda.state.de.us

Bruce Walton, EXECUTIVE ASSISTANT
brucew@dda.state.de.us
Michael Scuse, SECRETARY

DELAWARE GEOLOGICAL SURVEY
DGS Bldg., University of Delaware
Newark, DE 19716 USA
Phone: 302-831-2833   Fax: 302-831-3579
E-mail: delgeosurvey@udel.edu
Website: www.udel.edu/dgs

Founded: 1951
Membership: 15
Scope: State

Description: The survey was formed to study the geology, water, and other earth resources of Delaware; also to prepare reports, maps, and otherwise disseminate its findings, and to provide assistance in its area to other agencies and individuals.

Keyword(s): Research, Geology, Exotic species, Aquatic nuisance species

Contact(s):
John Talley, ASSOCIATE DIRECTOR
waterman@udel.edu
Dorothy Windish, LIBRARIAN
Robert Jordan, STATE GEOLOGIST AND DIRECTOR

DELAWARE SEA GRANT PROGRAM
University of Delaware
Newark, DE 19716-3501 USA
Phone: 302-831-8083   Fax: 302-831-4389
E-mail: marine.com@udel.edu
Website: www.ocean.udel.edu/seagrant/

Founded: NA

Contact(s):
Carolyn Thoroughgood, DIRECTOR
C.Thoroughgood@mvs.udel.edu
David McCarren, EXECUTIVE DIRECTOR
Phone: 302-831-8255

DELAWARE SOLID WASTE AUTHORITY
1128 S. Bradford St., P.O. Box 455
Dover, DE 19903 USA
Phone: 302-739-5361   Fax: 302-739-4287
E-mail: dra@dswa.com
Website: www.dswa.com

Founded: 1975
Membership: 120
Scope: Local

Description: To define, develop, and implement cost-effective plans and programs for solid waste management which best serve Delaware and protect our public health and environment.

Publications: Great Waste Mystery Curriculum, The, Marketing Research Findings and Executive Summary, Trash Tracks (DSWA Newsletter), Statewide Solid Waste Management Plan and Executive Summary

Keyword(s): Environmental Planning, Public Health Protection, Research, Solid Waste Management, Environmental and Conservation Education

Contact(s):
N. Vasuki, CHIEF EXECUTIVE OFFICER (P.E., DEE)
ncv@dswa.com

Thomas Houska, CHIEF OF ADMINISTRATIVE/ SERVICES OFFICER (P.E.)
teh@dswa.com
Pasquale Canzano, CHIEF OPERATING OFFICER (P.E., DEE)
psc@dswa.com

DEPARTMENT OF CONSERVATION MAINE FOREST SERVICE
22 State House Station
Augusta, ME 04333 USA
Phone: 207-287-2791   Fax: 207-287-8422
Website: www.state.me.us/doc/mfs

Founded: NA
Membership: 200
Scope: Statewide

Contact(s):
Tom Doak, DIRECTOR
Don Mansius, FOREST POLICY AND MANAGEMENT
Tom Parent, FOREST PROTECTION
Peter Beringer, RESOURCE ADMINISTRATOR
Dave Struble, STATE ENTOMOLOGIST

DEPARTMENT OF ENVIRONMENT AND CONSERVATION (TENNESSEE)
401 Church St., 21st Fl.
Nashville, TN 37243 USA
Phone: 615-532-0109   Fax: 615-532-0120
E-mail: www.@dec.net
Website: www.tdec.org

Founded: NA
Membership: 3000
Scope: State

Description: To plan, promote, protect, and conserve this state's natural, cultural, recreational, and historical resources, and to enforce environmental laws and regulations which protect the state's land and water.

Contact(s):
Tom Callery, ASST. COMMISSIONER FOR CONSERVATION
Phone: 615-532-4511
Fax: 615-532-0231
John Leonard, ASST. COMMISSIONER FOR ENVIRONMENT
Phone: 615-532-0225
Fax: 615-532-0120
Mark Williams, ASST. COMMISSIONER FOR STATE PARKS
Phone: 615-532-0022
Fax: 615-532-0732
Milton Hamilton, COMMISSIONER
Phone: 615-532-0109
Fax: 615-532-0120
Barry Stephens, DIRECTOR OF AIR POLLUTION CONTROL
Phone: 615-532-0554
Fax: 615-532-0614
Ron Zurawski, DIRECTOR OF GEOLOGY
Phone: 615-532-1500
Fax: 615-532-1517
Kent Taylor, DIRECTOR OF GROUNDWATER PROTECTION
Phone: 615-532-0762
Fax: 615-532-0778
Herbert Harper, DIRECTOR OF HISTORICAL COMMISSION
Phone: 615-532-1550
Fax: 615-532-1549

Toye Heape, DIRECTOR OF INDIAN AFFAIRS
Phone: 615-532-0745
Fax: 615-532-0732
Tim Eagle, DIRECTOR OF LAND RECLAMATION
Phone: 865-594-5609
Reggie Reeves, DIRECTOR OF NATURAL HERITAGE
DIVISION
Phone: 615-532-0431
Fax: 615-532-0231
Eddie Nanney, DIRECTOR OF RADIOLOGICAL HEALTH
Phone: 615-532-0364
Joyce Hoyle, DIRECTOR OF RECREATION RESOURCES
Phone: 615-742-0748615-532-0778
Charles Brewton, DIRECTOR OF RESORT OPERATIONS
Phone: 615-532-0263
Fax: 615-532-0740
Mike Apple, DIRECTOR OF SOLID WASTE MANAGEMENT
Phone: 615-532-0780
Fax: 615-532-0886
Jim Haynes, DIRECTOR OF SUPERFUND
Phone: 615-532-0900
Fax: 615-532-0938
Wayne Gregory, DIRECTOR OF UNDERGROUND
STORAGE TANKS
Phone: 615-532-0945
Fax: 615-532-0938
Paul Davis, DIRECTOR OF WATER POLLUTION CONTROL
Phone: 615-532-0625
Fax: 615-532-0046
David Draughon, DIRECTOR OF WATER SUPPLY
Phone: 615-532-0191
Fax: 615-532-0503
Dodd Galbreagh, ENVIRONMENTAL POLICY OFFICE
Phone: 615-532-8545
Fax: 615-532-0120
Joe Sanders, GENERAL COUNSEL
Phone: 615-532-0131
Fax: 615-532-0145
Kim Olson, PUBLIC INFORMATION OFFICER
Phone: 615-532-0288
Fax: 615-532-0740
Nick Fielder, STATE ARCHAEOLOGIST OF ARCHAEOLOGY
DIVISION
Phone: 615-741-1588
Fax: 615-741-7329

DEPARTMENT OF ENVIRONMENT AND WILDLIFE (QUEBEC)

Edifice Marie-Guyart, 675, Blvd. Rene-Levesque East
Quebec City, G1R 5V7 Canada
Phone: 418-521-3830   Fax: 418-646-5974
E-mail: info @menv.gouv.qc.ca
Website: http://www.menv.gouv.qc.ca

Founded: NA
Contact(s):
Diane Gian, DEPUTY MINISTER
Phone: 418-521-3860
Andre Taillon, DIRECTOR GENERAL OF WILDLIFE
PROTECTION
Phone: 819-623-1981
Luc Berthiaume, DIRECTOR OF INTERNAL AFFAIRS
Phone: 418-521-3828
Andre Martel, DIRECTOR OF WILDLIFE PROTECTION
Phone: 418-622-0313

Herve Bolduc, GENERAL SECRETARY
Phone: 418-521-3850
Paul Begin, MINISTER
Phone: 418-521-3911
Claudette Blais, VICE-PRESIDENT OF PARKS FOR
QUEBEC
Phone: 418-521-3850
George Arsenault, VICE-PRESIDENT OF SOCIETY OF
FAUNE AND PARKS OF QUEBEC
Phone: 418-521-3851

DEPARTMENT OF ENVIRONMENTAL MANAGEMENT (RHODE ISLAND)

235 Promenade St.
Providence, RI 02908 USA
Phone: 401-222-2774   Fax: 401-222-6174
Website: www.state.ri.us/dem

Founded: NA
Membership: 565
Scope: Statewide

Description: The Department of Environmental Management's
top priorities include the preservation and protection of the
environmental quality of Rhode Island. Air pollution, water
pollution, and waste disposal problems are handled by the
DEM. The DEM develops, administers, and enforces programs
designed to preserve and manage Rhode Island's forests,
parks, farms, wildlife, fisheries, and coastline. DEM is also
responsible for providing, on the average, 750 full-time jobs for
the people of Rhode Island.

Contact(s):
Alicia Good, ASSISTANT DIRECTOR OF WATER
RESOURCES
235 Promenade St., Providence, RI 02908
Phone: 401-222-3961
Malcolm Grant, ASSOCIATE DIRECTOR FOR NATURAL
RESOURCE MANAGEMENT (DEM)
235 Promenade St., Providence, RI 02908
Phone: 401-222-6605
Frederick Vincent, ASSOCIATE DIRECTOR FOR PLANNING
AND ADMINISTRATION
235 Promenade St., Providence, RI 02908
Phone: 401-222-2776
Terrence Gray, ASST. DIRECTOR FOR AIR, WASTE &
COMPLIANCE
235 Promenade St., Providence, RI 02908
Phone: 401-222-6677
Kathleen Lanphear, CHIEF HEARING OFFICER OF
ADMINISTRATIVE ADJUDICATION
235 Promenade, Providence, RI 02908
Phone: 401-222-1357
Kenneth Ayers, CHIEF OF AGRICULTURE
83 Park St., Providence, RI 02903
Phone: 401-222-2781
Stephen Majkut, CHIEF OF AIR RESOURCES
235 Promenade St., Providence, RI 02908
Phone: 401-222-2808
Donald McGovern, CHIEF OF COASTAL RESOURCES
83 Park St., Providence, RI 02903
Phone: 401-222-3429
Kurt Schatz, CHIEF OF CRIMINAL INVESTIGATION OFFICE
235 Promenade St., Providence, RI 02908
Phone: 401-222-6768
Steven Hall, CHIEF OF ENFORCEMENT
83 Park St., Providence, RI 02903
Phone: 401-222-2284

John Stolgitis, CHIEF OF FISH AND WILDLIFE
Stedman Government Center, Wakefield, RI 02879
Phone: 401-222-3075
Thomas Dupree, CHIEF OF FOREST ENVIRONMENT
R.F.D. #2 Box 851, North Scituate, RI 02859
Phone: 401-222-1414
Glenn Miller, CHIEF OF MANAGEMENT SERVICES
235 Promenade St., Providence, RI 02908
Phone: 401-222-6825
Melanie Marcaccio, CHIEF OF OFFICE OF HUMAN
RESOURCES
235 Promenade St., Providence, RI 02908
Phone: 401-222-2774
Larry Mouradjian, CHIEF OF PARKS AND RECREATION
2321 Hartford Ave., Johnston, RI 02919
Phone: 401-222-2632
Russell Chateauneuf, CHIEF OF PERMITTING
235 Promenade St., Providence, RI 02908
Phone: 401-222-2306
Robert Sutton, CHIEF OF PLANNING AND DEVELOPMENT
235 Promenade St., Providence, RI 02908
Phone: 401-222-2776
Janet Keller, CHIEF OF STRATEGIC PLANNING AND
POLICY
235 Promenade St., Providence, RI 02908
Phone: 401-277-3434
Ronald Gagnon, CHIEF OF TECHNICAL AND CUSTOMER
ASSISTANCE
291 Promenade St., Providence, RI 02908
Phone: 401-277-2797
Leo Hellested, CHIEF OF WASTE MANAGEMENT
235 Promenade St., Providence, RI 02908
Phone: 401-277-2797
Susan Bundy, CHIEF OF WATERSHED AND STANDARDS
235 Promenade St., Providence, RI 02908
Dean Albro, COMPLIANCE AND INSPECTION
235 Promenade St., Providence, RI 02908
Phone: 401-277-6820
Jan Reitsma, DIRECTOR
235 Promenade St., Providence, RI 02908
Phone: 401-222-2771

DEPARTMENT OF ENVIRONMENTAL QUALITY (ARKANSAS)

8001 National Dr., P.O. Box 8913
Little Rock, AR 72219-8913 USA
Phone: 501-682-0744   Fax: 501-682-0798
E-mail: help-cuspsvs@adeq.state.ar.us
Website: www.adeq.state.ar.us

Founded: 1949
Membership: 100
Scope: Statewide

Description: To prevent, abate, and control all types of pollution
and maintain the state's natural environment.

Publications: Arkansas Waste Line

Keyword(s): Air Quality and Pollution, Coral Reefs, Environmental
Protection, Solid Waste, Ecology

Contact(s):
Ed Morris, ADMINISTRATOR OF MANAGEMENT SERVICES
Mary Leath, CHIEF DEPUTY DIRECTOR
Phone: 501-682-0959
Keith Michaels, CHIEF OF AIR DIVISION
Robert Gage, CHIEF OF COMPUTER SERVICES DIVISION

James Gilson, CHIEF OF CUSTOMER SERVICE DIVISION
Sandy Formica, CHIEF OF ENVIRONMENTAL
PRESERVATION DIVISION
Leigh Ann Chrouch, CHIEF OF FISCAL DIVISION
Mike Bates, CHIEF OF HAZARDOUS WASTE DIVISION
Al Eckert, CHIEF OF LEGAL DIVISION
Jim Shell, CHIEF OF REGULATED STORAGE TANK
DIVISION
Dennis Burks, CHIEF OF SOLID WASTE DIVISION
Richard Cassat, CHIEF OF TECHNICAL SERVICES
DIVISION
Chuck Bennett, CHIEF OF WATER DIVISION
Becky Keogh, DEPUTY DIRECTOR
Richard Weiss, DIRECTOR

DEPARTMENT OF FISH AND WILDLIFE (WASHINGTON)

600 Capitol Way, N.
Olympia, WA 98501-1091 USA
Phone: 360-902-2200   Fax: 360-902-2947
E-mail: Webmaster@dfw.wa.gov
Website: www.wa.gov/wdfw

Founded: 1933
Scope: Statewide

Description: The mission of the Department is "Sound
Stewardship of Fish and Wildlife". The Department is
responsible for the preservation, protection, and perpetuation
of wildlife, fish, shellfish, and fish and wildlife habitat. It also
must maximize fishing, hunting, and recreational opportunities
compatible with healthy and diverse fish and wildlife
populations. The Department is also charged with maintaining
the economic well-being and stability of the fishing industry,
promote orderly fisheries, and enhance recreational and
commercial fishing.

Keyword(s): training, Sport Fishing, Endangered Species,
Hunting, Aquatic Habitats

Contact(s):
Jim Lux, ASSISANT DIRECTOR OF BUSINESS SERVICES
PROGRAM
Phone: 360-902-2200
Dave Brittell, ASSISTANT DIRECTOR OF ADMINISTRATIVE
SERVICES
Phone: 360-902-2206
Bruce Bjork, ASSISTANT DIRECTOR OF ENFORCEMENT
PROGRAM
Phone: 360-902-2936
Lew Atkins, ASSISTANT DIRECTOR OF FISH PROGRAM
Phone: 360-902-2700
Greg Hueckel, ASSISTANT DIRECTOR OF HABITAT
PROGRAM
Phone: 360-902-2534
Dave Brittell, ASSISTANT DIRECTOR OF WILDLIFE
MANAGEMENT PROGRAM
Phone: 360-902-2515
Larry Peck, DEPUTY DIRECTOR
Phone: 360-902-2650
Jeff Koenings, Ph.D., DIRECTOR
Phone: 360-902-2225
Penny Cusick, PERSONNEL MANAGER
Phone: 360-902-2276
Steve Keller, REGIONAL DIECTOR, REGION 6
48 Devonshire Rd., Montesano, WA 98563-6918
Phone: 360-249-4628

John Anderson, REGIONAL DIRECTOR, REGION 1
8702 N.Division St., Spokane, WA 99218-1199
Phone: 509-754-4624
Dennis Beich, REGIONAL DIRECTOR, REGION 2
1550 Alder St., Ephrata, WA 98823-9699
Phone: 509-754-4624
Jeff Tayer, REGIONAL DIRECTOR, REGION 3
1701 S. 24th Ave., Yakima, WA 98823-5720
Phone: 509-575-2740
Bob Everitt, REGIONAL DIRECTOR, REGION 4
16018 Mill Creek Blvd., Mill Creek, WA 98012-1296
Phone: 425-775-1311
Lee Tussenbrook, REGIONAL DIRECTOR, REGION 5
2108 Grand Blvd., Vancouver, WA 98661-4624
Phone: 360-696-6211
Tim Smith, SPECIAL ASSISTANT OF LEGISLATIVE AND
EXTERNAL AFFAIRS
Phone: 360-902-2223
Tim Waters, SPECIAL ASSISTANT OF PUBLIC AFFAIRS
Phone: 360-902-2250
Sara Laborde, SPECIAL ASSISTANT OF QUALITY
INITIATIVES
Phone: 360-902-2224
Phil Anderson, STAFF DIRECTOR OF
INTERGOVERNMENTAL POLICY GROUP
Phone: 360-902-2720

DEPARTMENT OF GEOLOGY AND MINERAL INDUSTRIES

800 NE Oregon St., Suite 965, #28
Portland, OR 97232-2162 USA
Phone: 503-731-4100   Fax: 503-731-4066
Website: www.oregongeology.com

Founded: NA
Membership: 20
Scope: Statewide
Publications: Oregon Geology - quarterly

Contact(s):
 Klaus Nevendorf, LIBRARIAN
 800 NE Oregon St., Suite 965, #28, Portland, OR
 97232-2162
 John Beaulieu, STATE GEOLOGIST
 john.beaulieu@state.or.us

DEPARTMENT OF INTERIOR, U.S.G.S/B.R.D, SOUTH CAROLINA COOPERATIVE FISH AND WILDLIFE RESEARCH UNIT

G27 Lehotsky Hall, Clemson University
Clemson, SC 29634 USA
Phone: 864-656-0168   Fax: 864-656-1034
E-mail: sccoop_l@clemson.edu
Website: www.clemson.edu

Founded: 1988
Scope: National

Description: The Unit conducts ecological research of importance
 to its cooperators, i.e., the Department of Interior, Clemson
 University, and the state of South Carolina. Its mission also
 involves training of graduate students in fish and wildlife
 biology and related fields.

Keyword(s): training, Wetlands, Wildlife, Nongame Wildlife,
 Endangered Species, Research, Sport Fishing, Hunting,
 Landscape Ecology, Birds

Contact(s):
 J. Isely, ASSISTANT LEADER OF FISHERIES
 Craig Allen, ASSISTANT LEADER OF WILDLIFE

DEPARTMENT OF LAND AND NATURAL RESOURCES

DIVISION OF BOATING AND OCEAN RECREATION
(DOBOR)
333 Queen Street
Honolulu, HI 96813 USA
Phone: 808-587-1963   Fax: 808-587-1977
Website: www.hawaii.gov/dlnr/ddor

Founded: NA
Scope: Statewide

Contact(s):
 Mason Young, ADMINISTRATOR (ACTING)

DEPARTMENT OF LAND AND NATURAL RESOURCES

DIVISION OF WATER RESOURCE MANAGEMENT,
P.O. Box 621
Honolulu, HI 96809 USA
Phone: 808-587-0214   Fax: 808-587-0219
Website: www.state.hi.us/dlnr/cwrm

Founded: NA
Scope: Statewide

Description: protect and enhance the water resources of the state
 of Hawaii through wise and responsible management.

Keyword(s): Exotic species, Aquatic nuisance species, Wildlife,
 Environmental Protection, Land Preservation

Contact(s):
 Linnel Nishioka, DEPUTY

DEPARTMENT OF LAND AND NATURAL RESOURCES (HAWAII)

P.O Box 621
Honolulu, HI 96809 USA
Phone: 808-587-0400   Fax: 808-587-0390
E-mail: dlnr@pixie.com

Founded: NA
Scope: Statewide

Contact(s):
 Gilbert Coloma-Agaran, CHAIRMAN, COMMISSION ON
 WATER RESOURCES MANAGEMENT
 Phone: 808-587-0401
 Linnel Nishioka, DEPUTY DIRECTOR COMMISSION ON
 WATER RESOUCE MANAGEMENT
 Phone: 808-587-0214
 Janet Kawelo, DEPUTY TO CHAIRPERSON
 Phone: 808-587-0403

DEPARTMENT OF LAND AND NATURAL RESOURCES (HAWAII)

601 Kamokila Blvd.
Kapolei, HI 96707 USA
Phone: 808-692-8015   Fax: 808-692-8020
Website: www.state.hi.us-dlnr-hpd-hpgreeting.htm

Founded: NA
Membership: 18
Scope: Local

Contact(s):
Don Hibbard, ADMINISTRATOR
Phone: 808-587-0045

DEPARTMENT OF LAND AND NATURAL RESOURCES (HAWAII)
DIVISION OF AQUATIC RESOURCES
AQUATIC RESOURCES
1151 Punchbowl St.
Honolulu, HI 96813 USA
Phone: 808-587-0100   Fax: 808-587-0115
Website: www.state.hi.us/dlnr/dnr

Founded: NA
Scope: Statewide

Contact(s):
William Devick, ADMINISTRATOR (ACTING)
Phone: 808-587-0110
Michael Fugimoto, PROGRAM MANAGER: COMMERCIAL
FISHERIES AQUACULTURE BRANCH
Phone: 808-587-0085

DEPARTMENT OF LAND AND NATURAL RESOURCES (HAWAII)
DIVISION OF CONSERVATION AND RESOURCES
ENFORCEMENT
1151 Punchbowl St., Rm. 311
Honolulu, HI 96813 USA
Phone: 808-587-0077   Fax: 808-587-0080
Website: www.state.hi.us/dlnr.html

Founded: NA
Scope: Statewide

Publications: Hunter Education

Contact(s):
Gary Moniz, ADMINISTRATOR (ACTING)
Wendell Kam, MANAGER: HUNTER EDUCATION
PROGRAM
Phone: 808-587-0200

DEPARTMENT OF LAND AND NATURAL RESOURCES (HAWAII)
DIVISION OF FORESTRY AND WILDLIFE
1151 Punchbowl St.
Honolulu, HI 96813 USA
Phone: 808-587-0166   Fax: 808-587-0160

Founded: NA
Membership: 23
Scope: Statewide

Contact(s):
Michael Buck, ADMINISTRATOR
Carl Masaki, MANAGER, FORESTRY PROGRAM

DEPARTMENT OF LAND AND NATURAL RESOURCES (HAWAII)
DIVISION OF STATE PARKS
P.O. Box 621
Honolulu, HI 96809 USA
Phone: 808-587-0300

Founded: NA

Contact(s):
Ralston Nagata, ADMINISTRATOR: STATE PARKS

DEPARTMENT OF LAND AND NATURAL RESOURCES (HAWAII)
LAND DIVISION
P. O. Box 621
Honolulu, HI 96809 USA
Phone: 808-587-0446   Fax: 808-587-0455
Website: www.state.hi.us

Founded: NA
Membership: 40
Scope: Statewide

Contact(s):
Harry Yada, ADMINISTRATOR

DEPARTMENT OF LANDS (IDAHO)
P.O. Box 83720
Boise, ID 83720-0050 USA
Phone: 208-334-0200   Fax: 208-334-2339
E-mail: boise@idl.state.id.us
Website: www.state.id.us.lands

Founded: NA
Membership: 300
Scope: Statewide

Description: The State Board of Land Commissioners is a constitutional board charged with administering the trust under which endowment lands are held. These lands were granted to the state at the time of statehood for the financial support of nine beneficiaries, the largest being the common schools.

Publications: Public Involvement Brochure, Sentinel Newsletter Quarterly

Contact(s):
Alan Lance, ATTORNEY GENERAL
Phone: 208-334-2400
Winston Wiggins, SECRETARY OF BOARD & DIRECTOR
OF THE IDAHO DEPT. OF LANDS
Pete Cenarrusa, SECRETARY OF STATE
Phone: 208-334-2300
Dirk Kempthorne, STATE BOARD OF LAND
COMMISSIONER PRESIDENT
Phone: 208-334-2100
J. Williams, STATE CONTROLLER
Phone: 208-334-3100
Marilyn Howard, SUPERINTENDENT OF PUBLIC
INSTRUCTION
Phone: 208-332-6800

DEPARTMENT OF PARKS AND RECREATION
DEPARTMENT OF PARKS AND RECREATION
1416 9th St.
Sacramento, CA 94206-0001 USA
Phone: 916-653-8380   Fax: 916-657-3903
Website: www.cal-parks.ca.gov

Founded: NA
Scope: Local, State

Description: Responsible for the acquisition, preservation, development, interpretation, and operation of the state park system; also responsible for the administration of grants for recreation to local government and for development of the California Outdoor Recreation Resources Plan.

Keyword(s): Outdoor Recreation

Contact(s):
Bill Berry, CENTRAL FIELD DIVISION CHIEF
Phone: 916-653-2021
Denzil Verardo, CHIEF DEPUTY DIRECTOR OF
ADMINISTRATION
Phone: 916-653-0528
Dick Troy, CHIEF DEPUTY DIRECTOR OF PARK
STEWARDSHIP
Phone: 916-653-8288
Mark Schrader, CHIEF OF ENVIRONMENTAL DESIGN
DIVISION
Phone: 916-653-7475
mschrader@parks.ca.gove
Ron Brean, CHIEF OF NORTHERN DIVISION
Phone: 916-657-4042
Richard Rayburn, CHIEF OF RESOURCE MANAGEMENT
DIVISION
Phone: 916-653-6745
Steven Treanor, CHIEF OF SOUTHERN DIVISION
Phone: 916-657-4042
strea@parks.ca.gov
Dan Abeyta, DEPUTY DIRECTOR OF HISTORIC
PRESERVATION OFFICE
Phone: 916-653-6624
Ruth Coleman, DEPUTY DIRECTOR OF LEGISLATION
Phone: 916-653-6887
rcole@parks.ca.gov
John McMahon, DEPUTY DIRECTOR OF MARKETING AND
REVENUE GENERATION
Phone: 916-653-5841
jmcma@parks.ca.gov
Dave Widell, DEPUTY DIRECTOR OF OFF HIGHWAY
MOTOR VEHICLE RECREATION
Phone: 916-324-5801
Rusty Areias, DIRECTOR
Phone: 916-653-8380
Ray Watson, HUMAN RESOURCES
Phone: 916-653-9990
Tim Lafranchi, LEGAL OFFICE
Phone: 916-653-6884

DEPARTMENT OF PARKS, RECREATION AND TOURISM
EDGAR A. BROWN BLDG.
1205 Pendleton St.
Columbia, SC 29201 USA
Phone: 803-734-1700   Fax: 803-734-0138
E-mail: fulfillment@scprt.com
Website: www.discoversouthcarolina.com

Founded: NA
Scope: Statewide
Contact(s):
Marion Edmonds, AGENCY SPOKESPERSON
Ronald Carter, DEPUTY DIRECTOR
Amy Duffy, DEPUTY DIRECTOR
John Durst, DIRECTOR
Toni Nance, DIRECTOR OF BUSINESS DEVELOPMENT
OFFICE
Charles Harrison, DIRECTOR OF DIVISION OF PARKS AND
RECREATION
Isabel Hill, DIRECTOR OF DIVISION OF TOURISM
DEVELOPMENT
Yvette Sistare, DIRECTOR OF FINANCE OFFICE
Curt Cottle, DIRECTOR OF HERITAGE TOURISM
DEVELOPMENT OFFICE

David Elwart, DIRECTOR OF INFORMATION
TECHNOLOGY
Roger Deaton, DIRECTOR OF INTERNAL OPERATIONS
Terri Cowling, DIRECTOR OF MARKETING OFFICE
Robert Liming, DIRECTOR OF NEW MARKET
DEVELOPMENT
Beth McClure, DIRECTOR OF OFFICE OF RECREATION,
PLANNING, AND ENGINEERING
Beverly Shelley, DIRECTOR OF SALES OFFICE
Van Stickles, DIRECTOR OF STATE PARK SERVICE
R. McGowan, DIRECTOR OF TOURISM

DEPARTMENT OF PLANNING AND NATURAL RESOURCES
Suite 231, Nisky Center
St. Thomas, VI 00803 USA
Phone: 809-774-3320

Founded: 1970

Description: Responsible for: Fish and wildlife; trees, vegetation and water resources; air and water pollution control; flood control; sewers and sewage disposal; culture and the arts; libraries and museums; minerals and other natural resources; historical preservation; submerged lands; earth change permits; and oil spill prevention and control.

Publications: Zone Management Notes (CZM Notes), Blue Book, Proceedings—Fisheries in Crisis Conference (Division of Fish and Wildlife), Wildlife Plant booklet, Natural History Atlas to the Cays of the Virgin Islands, Species Technical Bulletin (Bureau, Annual Report)

Keyword(s): Mammals, Environmental and Conservation Education, Birds, Endangered Species, Aquatic Habitats

Contact(s):
David Nellis, CHIEF OF WILDLIFE BUREAU
Dean Plaskett, COMMISSIONER
Janice Hodge, DIRECTOR OF COASTAL ZONE
MANAGEMENT
Lucia Roberts, DIRECTOR OF ENVIRONMENTAL
ENFORCEMENT DIVISION
Hollis Griffin, DIRECTOR OF ENVIRONMENTAL
PROTECTION DIVISION
Barbara Kojis, DIRECTOR OF FISH AND WILDLIFE
DIVISION
Phone: 809-775-6762

DEPARTMENT OF PLANNING AND NATURAL RESOURCES
DIVISION OF FISH AND WILDLIFE
6291 Estate Nazareth, 101
St. Thomas, VI 00802 USA
Phone: 340-775-6762   Fax: 340-775-3972

Founded: NA
Membership: 16
Scope: Regional
Publications: Wildlife Viewing Guide

Contact(s):
Judy Pierce, CHIEF OF WILDLIFE
sula@vitelcom.net
Barbara Kojis, DIRECTOR
bkojis@hotmail.com

DEPARTMENT OF PUBLIC WORKS

2000 14th St., NW
Washington, DC 20009 USA
Phone: 202-727-1000

Founded: NA

Keyword(s): Solid Waste, Urban Environment, Coral Reefs, Pedestrian Environment, Transportation

Contact(s):
Leslie Hotaling, DIRECTOR

DEPARTMENT OF RENEWABLE RESOURCES

Box 2703
Whitehorse, Yukon Territory Y1A 2C6 Canada
Phone: 867-667-5652   Fax: 867-393-6213
Website: http://www.renres.gov.yk.ca

Founded: NA

Contact(s):
Jim Connell, ACTING ASSISTANT DEPUTY MINISTER
Phone: 402-667-8955
Don Toews, ACTING DIRECTOR OF FISH AND WILDLIFE
Phone: 403-667-5715
Karyn Armour, ACTING DIRECTOR OF POLICY AND PLANNING
Phone: 403-667-5634
Bill Oppen, DEPUTY MINISTER
Phone: 403-667-5460
Dave Beckman, DIRECTOR OF AGRICULTURE
Phone: 403-667-5838
Joe Ballantyne, DIRECTOR OF ENVIRONMENTAL PROTECTION AND ASSESSMENT
Phone: 403-667-8177
Stan Marinoske, DIRECTOR OF FINANCE AND ADMINISTRATION
Phone: 403-667-5197
Jim McIntyre, DIRECTOR OF PARKS AND OUTDOOR RECREATION
Phone: 403-667-5261

DEPARTMENT OF RESOURCES AND ECONOMIC DEVELOPMENT

172 Pembroke Rd.
Concord, NH 03302-1856 USA
Phone: 603-271-2411   Fax: 603-271-2629
E-mail: gbald@dred.state.nh.us
Website: www.dred.state.nh.us

Founded: NA
Scope: Statewide

Contact(s):
Paul Gray, CHIEF OF BUREAU OF OFF-HIGHWAY RECREATIONAL VEHICLES
Phone: 603-271-3254
George Bald, COMMISSIONER
Phone: 603-271-2411
Stuart Arnett, DIRECTOR OF DIVISION OF ECONOMIC DEVELOPMENT
Phone: 603-271-2341
Philip Bryce, DIRECTOR OF DIVISION OF FORESTS AND LANDS
Phone: 603-271-2214
Richard McLeod, DIRECTOR OF DIVISION OF PARKS
Phone: 603-271-3556

Lauri Klefos, DIRECTOR OF DIVISION OF TRAVEL AND TOURISM DEVELOPEMENT
J. Cullen, URBAN FORESTER OF URBAN FORESTRY CENTER
Phone: 603-431-6774

DEPARTMENT OF THE ENVIRONMENT

2500 Broening Highway
Baltimore, MD 21224 USA
Phone: 410-631-3000   Fax: 410-631-3966
Website: www.mde.state.md.us

Founded: 1987
Scope: State

Description: The Department of the Environment is charged with protection of the state's land, air, and water resources, to ensure the long-term protection of public health and quality of life.

Publications: Annual Air Quality Data Report, Biennial Water Report, List of Potential Hazardous Waste Sites, Regulatory Calendar

Keyword(s): Coral Reefs, Solid Waste Management, Air Quality and Pollution, Environmental Protection, Pollution Prevention

Contact(s):
Denise Ferguson-Southard, ASSISTANT SECRETARY
Merrylin Zaw-Mon, DEPUTY DIRECTOR
Allan Jensen, DIRECTOR OF ADMINISTRATIVE AND EMPLOYEE SERVICES
Phone: 410-631-3116
Ann Marie Debiase
DIRECTOR OF AIR AND RADIATION MANAGEMENT ADMINISTRATION
Richard Collins, DIRECTOR OF WASTE MANAGEMENT ADMINISTRATION
Phone: 410-631-3304
Robert Summers, DIRECTOR OF WATER MANAGEMENT ADMINISTRATION
Etta Lyles, LIBRARIAN
Phone: 410-631-3818
Jane Nishida, SECRETARY
Phone: 410-631-3084

DEPARTMENT OF TOURISM CULTURE & RECREATION

Commerce Court
Cornerbrook, Newfoundland A2H 6J8 Canada
Phone: 709-729-2817   Fax: 709-637-2004

Founded: NA
Membership: 15
Scope: Statewide

Contact(s):
Keith Healey, ASSOCIATE DEPUTY MINISTER
Clyde Granter, DEPUTY MINISTER
Kevin Aylward, MINISTER
Phone: 709-729-4715

DEPARTMENT OF TRANSPORTATION (OREGON)

Mgr. Of Envir., 1158 Chemeketa St., NE
Salem, OR 97301-2528 USA
Phone: 503-986-3477   Fax: 503-986-3524

Founded: NA
Membership: 50
Scope: Statewide
Keyword(s): Transportation, Research, Environment, Environmental Planning, Engineering
Contact(s):
Lori Sundstrom, MANAGER OF ENVIRONMENTAL SERVICES SECTION
Oregon Department of Transportation, 1158 Chemeketa St., NE, Salem, OR 97310
Phone: 503-986-3477

DEPARTMENT OF TRANSPORTATION (RHODE ISLAND)
Two Capitol Hill
Providence, RI 02903 USA
Phone: 401-222-1362

Founded: NA

Description: To provide a safe, efficient, effective, and environmentally responsible intermodal transportation system that supports economic development and improves our quality of life.

Keyword(s): Transportation, Engineering
Contact(s):
William Ankner, DIRECTOR

DEPARTMENT OF WILDLIFE AND FISHERY SCIENCES
Box 2140B, SD State University
Brookings, SD 57007 USA
Phone: 605-688-6121   Fax: 605-688-4515
E-mail: wildlifefish@abs.sdstate.edu
Website: www.wff.sdstate.edu

Founded: NA
Scope: State

Description: Conducts fish and wildlife research and provides training for fishery and wildlife biologists. Cooperating Agents: South Dakota Department of Game, Fish and Parks, South Dakota State University, U.S. Geological Survey, U.S.D.I., Wildlife Management Institute.

Keyword(s): Wetlands, Rivers, Wildlife, Grasslands, Prairies, Endangered Species
Contact(s):
Steven Chipps, ASSISTANT LEADER FOR FISHERIES
Kenneth Higgins, ASSISTANT LEADER FOR WILDLIFE
Charles Berry, LEADER

DEPARTMENT OF WILDLIFE CONSERVATION
1801 N. Lincoln, P.O. Box 53465
Oklahoma City, OK 73152 USA
Phone: 405-521-3851   Fax: 405-521-6535
E-mail: pmoore@odwc.state.ok.us
Website: www.wildlifedepartment.com

Founded: 1909
Membership: 350
Scope: Regional
Publications: Outdoor Oklahoma
Keyword(s): Sport Fishing, Hunting, Endangered Species, Environmental and Conservation Education, Aquatic Habitats

Contact(s):
Richard Hatcher, ASSISTANT DIRECTOR
Phone: 405-522-6279
Melinda Sturgess-Striech, CHIEF OF ADMINISTATION
Phone: 405-521-4640
Kim Erickson, CHIEF OF FISHERIES
Phone: 405-521-3721
David Warren, CHIEF OF INFORMATION-EDUCATION
Phone: 405-521-3855
John Streich, CHIEF OF LAW ENFORCEMENT
Phone: 405-521-3719
Alan Peoples, CHIEF WILDLIFE DIVISION
Phone: 405-521-2739
Ed Abel, COMMISSION SECRETARY
Harlan Stonecipher, COMMISSIONER CHAIRMAN
Greg Duffy, DIRECTOR
Phone: 405-521-4660
Nels Rodefeld, EDITOR
Phone: 405-521-4635
Kyle Eastham, HUMAN RESOURCES ADMINISTRATOR
Phone: 405-521-4640
Ron Suttles, NATURAL RESOURCES COORDINATOR
Phone: 405-521-4616
Vyrl Keeter, VICE CHAIRMAN

DISTRICT OF COLUMBIA DEPARTMENT OF HEALTH
ENVIRONMENTAL HEALTH ADMINISTRATION, WATERSHED PROTECTION DIVISION
51 N Street NE 5th Floor
Washington, DC 20002 USA
Phone: 202-535-2240
Website:
http://www.dchealth.com/eha/watersheds/welcome.htm

Founded: NA

E

ENERGY, MINERALS, AND NATURAL RESOURCES DEPARTMENT
Pinon Bldg., 1220 St. Francis Dr.
Santa Fe, NM 87505 USA
Phone: 505-476-3200   Fax: 505-476-3220
Website: http://www.emnrd.state.nm.us/default.htm

Founded: NA
Scope: Statewide

Description: As the steward for New Mexico's natural resources, the department seeks to preserve the unique natural beauty of New Mexico and to facilitate the beneficial development and use of its resources in an environmentally responsible manner.

Contact(s):
Jennifer Salisbury, CABINET SECRETARY
2040 S. Pacheco, Santa Fe, NM 87505
Phone: 505-827-5950

ENERGY, MINERALS, AND NATURAL RESOURCES DEPARTMENT
ADMINISTRATIVE SERVICES DIVISION
1220 South St. Francis Drive
Santa Fe, NM 87505 USA
Phone: 505-476-3200   Fax: 505-476-3220
Website: http://www.emnrd.state.nm.us/default.htm

Founded: NA
Scope: Statewide

Description: Provides clerical, recordkeeping, and administrative support to the department in the areas of personnel, budget, procurement and contracting, and administration of federal and state grants.

Keyword(s): Renewable Resources, Outdoor Recreation, Environmental and Conservation Education, Protected Areas, Energy

Contact(s):
Dale Lucero, DIRECTOR

ENERGY, MINERALS, AND NATURAL RESOURCES DEPARTMENT
ENERGY CONSERVATION AND MANAGEMENT DIVISION
Villagra Bldg., 408 Galisteo St.
Santa Fe, NM 87505 USA
Phone: 505-827-5900

Founded: NA

Description: Administers state and federally funded energy conservation and alternative energy technology programs to state agencies, political subdivisions, regional organizations, nonprofit community service agencies, and New Mexico energy consumers, by providing engineering and technical assistance, and informational, financial, and programmatic support.

Contact(s):
Diane Caron, DIRECTOR

ENERGY, MINERALS, AND NATURAL RESOURCES DEPARTMENT
FORESTRY DIVISION
1220 St. Francis Dr., Rm 112
Santa Fe, NM 87505 USA
Phone: 505-476-3325   Fax: 505-476-3330
Website: http://www.emnrd.state.nm.us/forestry/

Founded: NA
Scope: Statewide

Description: Provides management and protection of New Mexico's renewable forest, rangeland, soil, and water resources through professional forest, pest, fire, and land management; provides law enforcement and administration, public education in conservation; and supports to enhance the environment and quality of resources to protect jobs and maintain social and economic benefits.

Publications: see publication website

Contact(s):
Toby Martinez, STATE FORESTER

ENERGY, MINERALS, AND NATURAL RESOURCES DEPARTMENT
MINING AND MINERALS DIVISION
1220 S. St. Francis Drive
Santa Fe, NM 87505 USA
Phone: 505-476-3405
Website: http://www.emnrd.state.nm.us/mining/

Founded: NA
Scope: Statewide

Description: Provides for the study, development, and optimum production of the mineral and energy resources within the state; the reduction of hazards associated with these processes consistent with the conservation of these resources; the protection of public health, safety, and the environment, and the economic well-being of the citizens.

Contact(s):
Douglas Bland, DIRECTOR

ENERGY, MINERALS, AND NATURAL RESOURCES DEPARTMENT
OIL CONSERVATION DIVISION
1240 S. Pacheco
Santa Fe, NM 87505 USA
Phone: 505-827-7133   Fax: 505-827-8177
Website: www,emnrd.state.nm.us

Founded: NA
Scope: Statewide

Description: Regulates and sets standards for operations related to the drilling and production of crude oil, natural gas, and geothermal resources and promotes the development and conservation of these resources while ensuring the prevention of waste and protection. Cares for the prevention of loss and contamination of freshwater supplies.

Contact(s):
Lori Wrotenberry, DIRECTOR

ENERGY, MINERALS, AND NATURAL RESOURCES DEPARTMENT
STATE PARKS AND RECREATION DIVISION
1220 St. Francis, P.O. Box 1147
Santa Fe, NM 87505 USA
Phone: 505-827-7173   Fax: 505-827-1376
E-mail: nmparks@state.nm.us
Website: http://www.emnrd.state.nm.us/nmparks/

Founded: NA
Scope: Statewide

Description: Provides and cares for the recreational resources, facilities, and opportunities, and promotes user safety on recreational land and water to benefit and enrich the lives of New Mexico residents and visitors alike.

Contact(s):
Tom Trujillo, DIRECTOR

ENVIRONMENTAL BOARD
WATER RESOURCE BOARD
National Life Records Center Building, Drawer 20
Montpelier, VT 05620 USA
Phone: 802-828-3309
E-mail: bbartlett@envboard.state.vt.us
Website: http://www.state.vt.us/envboard/

Founded: NA
Scope: State

Contact(s):
Marcy Harding, CHAIRMAN OF THE BOARD
Phone: 802-828-5440

ENVIRONMENTAL CENTER

UNIVERSITY OF HAWAII
KRAUSS ANNEX 19 WATER RESOURCES RESEARCH CNTR.
2500 Dole St.
Honolulu, HI 96822 USA
Phone: 808-956-7361　　　Fax: 808-956-3980
E-mail: envctr@hawaii.edu
Website: www.2.hawaii.edu/~.envctr/

Founded: 1970
Membership: 4
Scope: Statewide

Description: To stimulate, expand, and coordinate education, research, and service efforts of the university related to ecological relationships, natural resources, and environmental quality, with special relation to human needs and social institutions, particularly with regard to the state. For information on Liberal Studies BA Degree in Environmental Studies, check the University of Hawaii website: www.hawaii.edu/catalog/special-pgms-files/inter-progs.html

Keyword(s): Research, Environmental and Conservation Education, Environmental Law, Water Quality

Contact(s):
Jacquelin Miller, ASSOCIATE ENVIRONMENTAL COORDINATOR
jackiem@hawaii.edu
James Moncur, DIRECTOR
John Harrison, ENVIRONMENTAL COORDINATOR
jth@hawaii.edu

ENVIRONMENTAL PROTECTION BUREAU

DEPARTMENT OF LAW
120 Broadway
New York City, NY 10271 USA
Phone: 518-474-8096　　　Fax: 212-416-6007
Website: www.oag.state.ny.us

Founded: NA
Scope: Statewide

Description: Institutes legal actions on behalf of the people of the state in cases involving air and water pollution, protection of wildlife, waste site remediation, and protection of scenic and natural resources. Has responsibility for enforcement of laws protecting endangered species of wildlife, as well as public nuisance actions to restrain pollution and other environmental damage.

Contact(s):
Peter Lehner, BUREAU CHIEF
Peter Skinner, ENVIRONMENTAL SCIENTIST
Phone: 518-474-2432

ENVIRONMENTAL PROTECTION MASSACHUSETTS

DEPARTMENT OF ENVIRONMENTAL PROTECTION
One Winter St. Dept of Environmental Affairs
Boston, MA 02108 USA
Phone: 617-292-5500　　　Fax: 617-556-1049
Website: www.state.ma.us/dep

Founded: NA
Membership: 1200
Scope: State
Publications: Publications on line

Contact(s):
James Coleman, ASSISTANT COMMISSIONER: WASTE PREVENTION
Phone: 617-292-5570
Lauren Liss, COMMISSIONER
Barbara Kwetz, DIRECTOR OF PLANNING AND EVALUATION
Phone: 617-292-5593

ENVIRONMENTAL QUALITY DEPARTMENT

122 W. 25th St., Herschler Bldg.
Cheyenne, WY 82002 USA
Phone: 307-777-7937　　　Fax: 307-777-7682
E-mail: deqwyo@state.wy.us
Website: www.deq.state.wy.us

Founded: 1973
Membership: 300
Scope: Statewide

Description: Established to plan the development, use, reclamation, preservation, and enhancement of the air, land, and water resources of the state.

Publications: Outreach

Contact(s):
Evon Green, ADMINISTRATOR OF ABANDONED MINE LAND
Phone: 307-777-6145
Dan Olson, ADMINISTRATOR OF AIR QUALITY
Phone: 307-777-7391
Richard Chancellor, ADMINISTRATOR OF LAND QUALITY
Phone: 307-777-7756
James Uzzell, ADMINISTRATOR OF MANAGEMENT SERVICES
Phone: 307-777-7937
Gary Beach, ADMINISTRATOR OF WATER QUALITY
Phone: 307-777-7781
Dennis Hemmer, DIRECTOR
Phone: 307-777-7938
David Finley, MANAGER OF SOLID WASTE PROGRAM
Phone: 307-777-7752

ENVIRONMENTAL REVIEW APPEALS COMMISSION

236 E. Town St., Rm. 300
Columbus, OH 43215 USA
Phone: 614-466-8950

Founded: NA
Membership: 6
Scope: Regional

Description: The Environmental Review Appeals Commission is an administrative commission designed to review the actions of the Ohio EPA, State Fire Marshal, and the various county boards of health charged with environmental jurisdiction in order to determine that the agencies' actions have been reasonable and lawful.

Contact(s):
Maria Armstrong, MEMBER

EXECUTIVE OFFICE OF ENVIRONMENTAL AFFAIRS (MASSACHUSETTS)

Leverett Saltonstall Bldg., 100 Cambridge St., Rm. 2000
Boston, MA 02202 USA
Phone: 617-626-1000   Fax: 617-626-1181
E-mail: env.internet@state.ma.us
Website: http://www.state.ma.us/envir/

Founded: NA

Description: The cabinet-level environmental agency in the state and includes within the secretariat all state environmental agencies.

Contact(s):
Tom Skinner, DIRECTOR: COASTAL ZONE MANAGEMENT
Phone: 617-727-9530
Joel Lerner, DIRECTOR: CONSERVATION SERVICES
Phone: 617-727-1552
Jay Wickersham, DIRECTOR: IMPACT REVIEW UNIT (MEPA)
Phone: 617-727-5830
Bob Durand, SECRETARY
Steve Bernard, UNDER SECRETARY: ADMINISTRATION & FINANCE

EXECUTIVE OFFICE OF ENVIRONMENTAL AFFAIRS (MASSACHUSETTS)

251 Causeway St., 9th Floor
Boston, MA 02114 USA
Phone: 617-626-1000   Fax: 617-626-1181
Website: http://state.ma.us/envir

Founded: NA

Contact(s):
Bob Durand, SECRETARY

EXECUTIVE OFFICE OF ENVIRONMENTAL AFFAIRS (MASSACHUSETTS)

BUREAU OF PESTICIDES
251 Causeway St., Suite 900
Boston, MA 02114 USA
Phone: 617-626-1000   Fax: 617-262-1181
E-mail: www.state.ma.us/envir
Website: www.state.ma.us/envir

Founded: NA
Scope: Statewide

Publications: State of Our Environment

Contact(s):
Brad Mitchell, CHIEF
Phone: 617 727 7710

EXECUTIVE OFFICE OF ENVIRONMENTAL AFFAIRS (MASSACHUSETTS)

DEPARTMENT OF ENVIRONMENTAL MANAGEMENT
251 Causeway St. Suite 700
Boston, MA 02114 USA
Phone: 617-973-8700

Founded: NA

Contact(s):
Peter Webber, COMMISSIONER
Richard Thibedeau, DEPUTY COMMISSIONER (ACTING): RESOURCE CONSERVATION
Phone: 617-727-3267

Ralph Silva, DIRECTOR OF ENGINEERING
Todd Fredericks, DIRECTOR: DIVISION OF FORESTS AND PARKS
Phone: 617-626-1000

EXECUTIVE OFFICE OF ENVIRONMENTAL AFFAIRS (MASSACHUSETTS)

DEPARTMENT OF FOOD AND AGRICULTURE
100 Cambridge St., Rm. 2103
Boston, MA 02202 USA
Phone: 617-727-3000
Website: http://www.state.ma.us/envir/

Founded: NA

EXECUTIVE OFFICE OF ENVIRONMENTAL AFFAIRS (MASSACHUSETTS)

DIVISION OF CONSERVATION SERVICES
251 Causeway St., 9th Floor
Boston, MA 02114 USA
Website: http://state.ma.us/envir/conservation

Founded: NA

EXECUTIVE OFFICE OF ENVIRONMENTAL AFFAIRS (MASSACHUSETTS)

ENVIRONMENTAL TRUST
33 Union Street, 4th Floor
Boston, MA 02108 USA
Phone: 617-727-0249   Fax: 617-367-1616

Founded: NA

EXECUTIVE OFFICE OF ENVIRONMENTAL AFFAIRS (MASSACHUSETTS)

GEOGRAPHIC INFORMATION SYSTEM
251 Causeway St., 9th Floor
Boston, MA 02114 USA
Website: http://state.ma.us/mgis

Founded: NA

Contact(s):
Christian Jacqz, DIRECTOR
Phone: 617-626-1056

EXECUTIVE OFFICE OF ENVIRONMENTAL AFFAIRS (MASSACHUSETTS)

MASSACHUSETTS COASTAL ZONE MANAGEMENT
251 Causeway St. Suite 900
Boston, MA 02114 USA
Phone: 617-626-1200   Fax: 617-626-1240
Website: http://state.ma.us/czm

Founded: NA

Contact(s):
Tom Skinner, DIRECTOR
Phone: 617 6-26 -1201

EXECUTIVE OFFICE OF ENVIRONMENTAL AFFAIRS (MASSACHUSETTS)

MASSACHUSETTS ENVIRONMENTAL POLICY ACT
251 Causeway St.-Suite 900
Boston, MA 02114 USA
Phone: 617-626-1020   Fax: 617-626-1181
Website: http://state.ma.us/envir/mepa

Founded: NA

Contact(s):
Jay Wickersham, DIRECTOR
Phone: 617-626-1022

EXECUTIVE OFFICE OF ENVIRONMENTAL AFFAIRS (MASSACHUSETTS)

OFFICE OF TECHNICAL ASSISTANCE FOR TOXIC USE REDUCTION
251 Causeway St., 9th Floor
Boston, MA 02114 USA
Phone: 617-626-1060   Fax: 617-626-1095
Website: http://state.ma.us/ota

Founded: NA

Contact(s):
Paul Richard, DIRECTOR
Phone: 617 6-26 -1042

EXECUTIVE OFFICE OF ENVIRONMENTAL AFFAIRS (MASSACHUSETTS)

WETLANDS AND WATERWAYS PROGRAM
1 Winter St.
Boston, MA 02108 USA
Phone: 617-292-5695   Fax: 617-292-5696
Website: www.state.ma.us/dep

Founded: NA
Scope: Statewide

Contact(s):
Michael Stroman, ACTING DIRECTOR OF WETLAND
Glenn Haas, DIRECTOR

F

FISH AND GAME COMMISSION

FISH AND GAME COMMISSION
1416 9th St., Rm. 1320, P.O. Box 944209
Sacramento, CA 94244 USA
Phone: 916-653-4899   Fax: 916-653-5040
E-mail: fgc@dfg.ca.gov
Website: www.dfg.ca.gov/fg_comm/

Founded: 1870
Membership: 15
Scope: Regional

Description: Adopts fish, game, and plant regulations as authorized by the Fish and Game Code and sets policies for the Department of Fish and Game.

Contact(s):
Robert Treanor, EXECUTIVE DIRECTOR
rtreanor@dfg.ca.gov
Mike Chrisman, PRESIDENT
Sam Schuchat, VICE PRESIDENT

FLORIDA COOPERATIVE EXTENSION SERVICE

1038 McCarty Hall, P.O. Box 110210, University of Florida
Gainesville, FL 32611-0210 USA
Phone: 352-392-1761   Fax: 352-392-3583
Website: www.ifas.ufl.edu

Founded: NA
Scope: Regional

Publications: see publications on website

Keyword(s): Lakes, Wildlife, Environmental and Conservation Education, Forests and Forestry

Contact(s):
Nat Frazer, ASSISTANT EXTENSION SCIENTIST, WILDLIFE
Pinellas County Extension Office, 12175 125th St. North, Largo, FL 33774-3695
Phone: 813-582-2100
Fax: 813-582-2149
whk@gnv.ifas.ufl.edu
Christine Waddill, DEAN OF EXTENSION
1038 McCarty Hall, P.O. Box 110210, University of Florida, Gainesville, FL 32611-0210
Pierce Jones, DIRECTOR OF ENERGY EXTENSION SERVICE (ACTING)
Box 110570, 102 Rogers Hall, University of Florida, Gainesville, FL 32611-0570
Phone: 352-392-8074
Fax: 352-392-4092
ez@agen.ufl.edu
Joseph Schaefer, DIRECTOR OF NATURAL RESOURCES
Univ. of Florida, Wildlife Ecology and Conservation, P.O. Box 110430, Gainesville, FL 32611-0430
Phone: 352-846-0568
Fax: 352-392-6984

FLORIDA COOPERATIVE FISH AND WILDLIFE RESEARCH UNIT (USDI)

P.O. Box 110450, 117 Newins-Ziegler Hall,
University of Florida
Gainesville, FL 32611-0450 USA
Phone: 904-392-1861   Fax: 352-846-0841
E-mail: rayc@zoo.ufl.edu
Website: http://www.wec.ufl.edu/coop/

Founded: 1979

Description: Established by cooperative agreement among the National Biological Survey, Florida Game and Fresh Water Fish Commission, and the University of Florida. Primary purpose is research, graduate education, and extension activities integrating fish and wildlife ecology and management in Florida's unique ecosystems, particularly wetlands.

Keyword(s): Biodiversity, Reptiles and Amphibians, Wetlands, Aquatic Habitats, Endangered Species

Contact(s):
H. Percival, ASSISTANT UNIT LEADER: WILDLIFE
Wiley Kitchens, UNIT LEADER

FLORIDA DEPARTMENT OF AGRICULTURE & CONSUMER SERVICES OFFICE OF AGRICULTURAL WATER POLICY

1203 Governor Square Blvd.
Tallahassee, FL 32301 USA
Phone: 850-488-6249   Fax: 850-921-2153

Founded: NA
Membership: 2500
Scope: Statewide

Description: Provides administrative, legislative, and promotional assistance to 63 Soil and Water Conservation Districts in Florida.

Contact(s):
Chuck Aller, DIRECTOR
John Folks, ENVIRONMENTAL ADMINISTRATOR
Phone: 850-488-6249

FLORIDA DEPARTMENT OF AGRICULTURE AND CONSUMER SERVICES
The Capitol, PL 10
Tallahassee, FL 32399-0800 USA
Phone: 850-488-3022   Fax: 850-488-7585
Website: www.doacs.state.fl.us

Founded: NA
Membership: 30
Scope: Statewide

Contact(s):
Charles Bronson, COMMISSIONER
Phone: 850-488-3022

FLORIDA DEPARTMENT OF AGRICULTURE AND CONSUMER SERVICES
DIVISION OF FORESTRY
3125 Conner Blvd.
Tallahassee, FL 32399-1650 USA
Phone: 850-488-4274   Fax: 850-488-0863
Website: www.fl-dof.com

Founded: 1927
Membership: 1200
Scope: Regional

Description: To protect and manage Florida's forest resources through a stewardship ethic to assure these resources will be available for future generations. Current number of employees: 1,100.

Contact(s):
Mike Long, ASSISTANT DIRECTOR
Phone: 850-414-9967
James Karels, BUREAU OF FOREST PROTECTION
Phone: 850-488-6106
C. Maynard, CHIEF OF FOREST MANAGEMENT
Phone: 850-488-6611
Rich Ashley, CHIEF OF PLANNING AND SUPPORT SERVICES
Phone: 850-414-0843
Raymond Geiger, CHIEF, FIELD OPERATIONS
Phone: 850-414-9969
L. Peterson, DIRECTOR
Phone: 850-922-0135

FLORIDA DEPARTMENT OF AGRICULTURE AND CONSUMER SERVICES
SOIL AND WATER CONSERVATION COUNCIL
3125 Conner Blvd., Suite C, Mail Stop C28
Tallahassee, Fl 32399 USA
Phone: 850-488-5321   Fax: 850-921-2153

Founded: NA
Scope: Statewide

Contact(s):
Richard Machek, CHAIR
17 NW 16th St., Delray Beach, FL 33444
Clegg Hooks, SWC ADMINISTRATOR
3125 Conner Blvd., Suite C, Conner Bldg, Tallahassee, FL 32399
Phone: 850-488-6249
Fax: 850-921-2153

FLORIDA DEPARTMENT OF ENVIRONMENTAL PROTECTION
3900 Commonwealth Blvd.
Tallahassee, FL 32399-3000 USA
Phone: 850-921-1222   Fax: 850-487-3267
Website: http://DEP.STATE.FL.US

Founded: NA

Contact(s):
Denver Stutler, CHIEF OF STAFF
Phone: 850-488-7131
Teri Donalson, GENERAL COUNSEL
Phone: 850-488-9314
Pinky Hall, INSPECTOR GENERAL
Phone: 850-488-2287
David Struhs, SECRETARY
Phone: 850-488-1154

FLORIDA DEPARTMENT OF ENVIRONMENTAL PROTECTION
AIR RESOURCES MANAGEMENT DIVISION
2600 Blair Stone Rd. MS 5500
Tallahassee, FL 32399-2400 USA
Phone: 850-488-0114   Fax: 850-922-6979

Founded: NA

FLORIDA DEPARTMENT OF ENVIRONMENTAL PROTECTION
BEACHES AND COASTAL SYSTEMS
3900 Commonwealth Blvd.
Tallahassee, FL 32399-3000 USA
Phone: 850-487-4475
Website: http://dep.statel.fl.us/beach

Founded: NA

Contact(s):
Tom Watters, COASTAL DATA ACQUISITION
Phone: 850-487-4475
Alfred Devereaux, DIRECTOR
Phone: 850-488-3181
Paden Woodruff, ENVIRONMENTAL BEACH PROGRAM ADMINISTRATOR
Phone: 850-487-4475
Mark Leadon, RESEARCH ANALYSIS & POLICY PE
Phone: 850-487-4475

FLORIDA DEPARTMENT OF ENVIRONMENTAL PROTECTION
COASTAL & AQUATIC MANAGED AREAS
3900 Commonwealth Blvd. Mail St. 235
Tallahassee, FL 32399 USA
Phone: 850-488-3456   Fax: 850-488-3896

Founded: NA

Contact(s):
Anna Hartman, DIRECTOR
Phone: 850 4-88 -3456
Dennis Raley, PROGRAM ADMINISTRATOR
Phone: 850 4-88 -3456

FLORIDA DEPARTMENT OF ENVIRONMENTAL PROTECTION
LAW ENFORCEMENT DIVISION
3900 Commonwealth Blvd. MS 600
Tallahassee, FL 32399 USA
Phone: 850-488-5757
Website: http://dep.state.fl.us/law

Founded: NA

Contact(s):
Thomas Tramell III, DIVISION DIRECTOR
Phone: 850 4-88 -5757

FLORIDA DEPARTMENT OF ENVIRONMENTAL PROTECTION
RECREATION AND PARKS DIVISION
3900 Commonwealth Blvd.
Tallahassee, FL 32399 USA
Website:http://myflorida.com/communities/learn/stateparks

Founded: NA

Contact(s):
Tom Brown, STAFF OF AQUATIC PLANT MANAGEMENT
Phone: 904-488-5631
Walter Schmidt, STAFF OF GEOLOGY
Phone: 904-488-4191
Joe Bakker, STAFF OF MINE RECLAMATION
Phone: 904-488-8217

FLORIDA DEPARTMENT OF ENVIRONMENTAL PROTECTION
RESOURCE ASSESSMENT & MANAGEMENT
3900 Commonwealth Blvd
Tallahassee, FL 32399 USA
Phone: 850-922-6407
Website: http://dep.state.fl.us/resource

Founded: NA

Contact(s):
Ed Conklin, DIRECTOR
Phone: 850-922-6407

FLORIDA DEPARTMENT OF ENVIRONMENTAL PROTECTION
STATE LANDS DIVISION
3900 Commonwealth Blvd.
Tallahassee, FL 32399 USA
Website: http://dep.state.A.us/stland

Founded: NA

Contact(s):
Rob Lovern, ASSISTANT DIRECTOR
Eva Armstrong, DIRECTOR

FLORIDA DEPARTMENT OF ENVIRONMENTAL PROTECTION
WASTE MANAGEMENT DIVISION
3900 Commonwealth Blvd.
Tallahassee, FL 32399 USA
Phone: 850-488-0300   Fax: 850-414-0414
Website: http://dep.state.fl.us/dwn

Founded: NA

Contact(s):
John Ruddell, STAFF
Phone: 904-487-3299
Bill Hinkley, STAFF OF SOLID AND HAZARDOUS WASTE
Phone: 904-488-0300
Doug Jones, STAFF OF WASTE CLEANUP
Phone: 904-488-0190

FLORIDA DEPARTMENT OF ENVIRONMENTAL PROTECTION
WATER RESOURCE MANAGEMENT
3900 Commonwealth Blvd.
Tallahassee, FL 32399 USA
Phone: 850-487-1855

Founded: NA

Contact(s):
Jerry Brooks, DEPUTY DIRECTOR
Phone: 850 4-87 -1855
Mimi Drew, DIRECTOR
Phone: 850 4-87 -1855

FLORIDA FISH AND WILDLIFE CONSERVATION COMMISSION
620 S. Meridian St.
Tallahassee, FL 32399-1600 USA
Phone: 850-488-3641   Fax: 850-414-8212
Website: www.state.fl.us/fwc

Founded: NA
Membership: 100
Scope: Statewide
Publications: Florida Wildlife
Keyword(s): Hunting, Birds, Environment, Wildlife, Sport Fishing, Lakes, Outdoor Recreation, training, Aquatic Habitats, Endangered Species

Contact(s):
Victor Heller, ASSISTANT EXECUTIVE DIRECTOR
Phone: 850-488-3084
Susan Weaver, BUREAU OF LICENSING AND PERMITTING
Sandra Porter, DIRECTOR: DIVISION OF ADMINISTRATIVE SERVICES
Phone: 850-488-6551
Edwin Moyer, DIRECTOR: DIVISION OF FRESHWATER FISHERIES
Phone: 850-488-0331
Robert Edwards, DIRECTOR: DIVISION OF LAW ENFORCEMENT
Phone: 850-488-6251
Frank Montalbano, DIRECTOR: DIVISION OF WILDLIFE
Phone: 850-488-3831
Brad Hartman, DIRECTOR: OFFICE OF ENVIRONMENTAL SERVICES
Phone: 850-488-6661
Dick Sublette, EDITOR
Phone: 850-488-5564

Allan Egbert, EXECUTIVE DIRECTOR
Phone: 850-487-3796
James Antista, GENERAL COUNSEL
Phone: 850-487-1764

FLORIDA SEA GRANT COLLEGE

Florida Sea Grant College Program, P.O. Box 110400,
University of Florida
Gainesville, FL 32611-0400 USA
Phone: 352-392-5870   Fax: 352-392-5113
E-mail: www.flseagrant.org
Website: www.flseagrant.org

Founded: NA
Membership: 12
Scope: Statewide

Description: A statewide university-based program of coastal and ocean research, education, and public service to enhance productivity, conservation, and long-term use and management of marine systems and resources.

Publications: listing of various publications available, Fathom Magazine

Keyword(s): Environmental and Conservation Education, Coasts, Wildlife

Contact(s):
Michael Spranger, ASSISTANT DEAN AND COORDINATOR: SEA GRANT EXTENSION PROGRAM
P.O. Box 110405, University of Florida, Gainesville, FL 32611-0405
Phone: 352-392-1837
msspranger@mail.ifas.ufl.
William Seaman, ASSOCIATE DIRECTOR: FLORIDA SEA GRANT COLLEGE PROGRAM
seaman @mail.ikfas.ufl.ed
James Cato, DIRECTOR: FLORIDA SEA GRANT COLLEGE PROGRAM
jcato@mail.ifas.ufl.edu

FLORIDA STATE DEPARTMENT OF HEALTH

State Health Office, 2020 Capital Circle SE, BIN # AOO
Tallahassee, FL 32399-1701 USA
Phone: 904-487-2945   Fax: 850-410-1375

Founded: NA

Keyword(s): Toxicology, Public Health Protection, Toxic Substances, Nuclear-free, Water quantity, Water export and diversion

Contact(s):
Richard Hunter, DEPUTY STATE HEALTH OFFICER FOR PREVENTION AND CONTROL PROGRAMS
Phone: 850-487-2945
Sharon Heber, DIVISION DIRECTOR OF ENVIRONMENTAL HEALTH
Phone: 850-488-6811
Roger Inman, ENVIRONMENTAL HAZARDS
Phone: 850-488-3385
Bart Bibler, ENVIRONMENTAL PROGRAMS
Phone: 850-488-4070
Eric Grimm, ENVIRONMENTAL PROGRAMS
Phone: 850-487-0004
Russell Mardon, EPIDEMIOLOGY PROGRAMS, CHIEF (ACTING)
Phone: 850-488-3370

FORESTRY COMMISSION (ARKANSAS)

3821 W. Roosevelt Rd.
Little Rock, AR 72204-6395 USA
Phone: 501-296-1940   Fax: 501-296-1949
Website: www.forestry.state.ar.us

Founded: NA
Membership: 24
Scope: Regionally

Description: To prevent and suppress forest fires; control forest insects and diseases; grow and distribute forest planting stock; and collect and disseminate information concerning growth, utilization, and renewal of forests.

Keyword(s): Forests and Forestry

Contact(s):
Alan Murray, BAUCUM NURSERY
Phone: 501-907-2485
Tom Curtner, COMMISSION CHAIRMAN
Larry Nance, DEPUTY STATE FORESTER
Phone: 501-296-1943
Don McBride, FIRE CONTROL
Phone: 501-296-1940
Faustine McDowell, FISCAL DEPARTMENT
Phone: 501-296-1931
George Rheinhardt, FOREST MANAGEMENT
Phone: 501-296-1940
James Grant, INFORMATION AND EDUCATION
John Shannon, STATE FORESTER

FORESTRY COMMISSION (SOUTH CAROLINA)

Box 21707
Columbia, SC 29221-1707 USA
Phone: 803-896-8800   Fax: 803-798-8097
E-mail: scsc@forestry.state.sc.us
Website: www.state.sc.us\forest

Founded: 1927
Scope: State

Description: Provides basic forest fire protection on all state and private forest lands in South Carolina; assists landowners in proper management and utilization of forest lands; promotes forest fire prevention and other forestry practices through an information and education program; and operates forest tree nursery, seed orchards, and state forests.

Keyword(s): Wetlands, Urban Forestry, Forests and Forestry, Rural Development, Environmental and Conservation Education

Contact(s):
Cecil Campbell, COASTAL REGIONAL FORESTER
Phone: 843 638 3708
Ed Muckenfuss, COMMISSION CHAIRMAN
P.O. Box 1950, Summerville, SC 29484
Phone: 843-871-5000
Ken Hill, COMMISSION VICE CHAIRMAN
308 Fuller St., Manning, SC 29102
Phone: 803-435-8133
Bill Boykin, DEPUTY STATE FORESTER
Phone: 803-896-8832
Joe Richbourg, DIRECTOR OF ADMINISTRATION DIVISION
Phone: 803-896-8858
Tim Adams, DIRECTOR OF FIELD OPERATIONS SUPPORT
Phone: 803-896-8802

Judy Weston, EXECUTIVE ASSISTANT TO STATE
FORESTER
Phone: 803-896-8875
Steve Scott, PEE DEE REGIONAL FORESTER
Phone: 843-662-5571
Charles Ramsey, PIEDMONT REGIONAL FORESTER
Phone: 803-276-0205
Robert Showalter, STATE FORESTER
C. Carson, TECHNICAL ASSISTANT TO THE STATE
FORESTER
Phone: 803-896-8822

FORESTRY, FIRE & STATE LANDS
DIVISION OF FORESTRY, FIRE AND STATE LANDS
1594 W. North Temple, Suite 3520, P.O. Box 145703
Salt Lake City, UT 84114-5703 USA
Phone: 801-538-5555   Fax: 801-533-4111
Website: www.nr.utah.gov/slf/slfhome.htm

Founded: NA
Scope: Statewide

Description: Legislation enacted July 1994 dissolved the Division
and Board of State Lands and Forestry and created the
Division of Sovereign Lands and Forestry, and the Sovereign
Lands and Forestry Advisory Council. Division and council
names were legislatively changed to Division of Forestry, Fire
and State Lands and Forestry, Fire and State Lands Advisory
Council in July 1995.

Contact(s):
Arthur Dufault, STATE FORESTER AND DIRECTOR
Karl Kappe, STRATEGIC PLANNER

G

GAME AND PARKS COMMISSION
GAME AND PARKS COMMISSION
1212 Bob Gibson Blvd.
Omaha, NE 68108 USA
Phone: 402-595-2144
Website: www.ngpc.state.ne.us

Founded: NA
Scope: Local

GAME AND PARKS COMMISSION-NEBRASKA
AK-SAR-BEN AQUARIUM
21502 W Hwy. 31
Gretna, NE 68028 USA
Phone: 402-332-3901   Fax: 402-332-5853
Website: www.ngpc.state.ne.us

Founded: NA
Membership: 7
Scope: Statewide

Contact(s):
Darrell Feit, DIRECTOR
dfeit@ngpc.state.nd.us

GEORGIA COOPERATIVE FISH AND WILDLIFE RESEARCH UNIT (USDI)
Warnell School of Forest Resources, University of
Georgia
Athens, GA 30602-2152 USA
Phone: 706-542-5260   Fax: 706-542-8356
Website: www.uga.edu/~gacoop

Founded: 1984
Membership: 20
Scope: Statewide

Description: The Unit is supported by the Biological Resources
Division, USGS; Georgia Department of Natural Resources;
and the Wildlife Management Institute; and the University of
Georgia. Fisheries and wildlife research, graduate education
and training, technical assistance, and extension are the main
missions of the Unit.

Keyword(s): training, Protected Areas, , Aquatic Habitats, Wildlife

Contact(s):
Glynis Habeck, ADMINISTRATIVE SECRETARY
Michael Conroy, ASSISTANT LEADER: WILDLIFE
James Peterson, ASSISTANT UNIT LEADER (FISHERIES)
Cecil Jennings, UNIT LEADER

GEORGIA DEPARTMENT OF AGRICULTURE
19 Martin Luther King Dr., Capitol Sq.
Atlanta, GA 30334 USA
Phone: 404-656-3600   Fax: 404-656-9380
Website: www.agr.state.ga.us

Founded: 1874
Membership: 900
Scope: Statewide

Description: The department serves farmers and consumers in
the state by verifying and enforcing the accuracy and quality of
both products and services in many areas including food
products, seed, fertilizers, pesticides, fuel, weights and
measures, and bedding, and by overseeing the health and
well-being of Georgia's livestock, poultry, and commercial pet
industry.

Publications: Georgia Agricultural Facts, Georgia Poultry Facts,
Farmers and Consumers Market Bulletin

Keyword(s): Consumer Protection, Pesticides, Food Safety,
Agriculture, Consumer Services

Contact(s):
Cameron Smoak, ASSISTANT COMMISSIONER FOR
CONSUMER PROTECTION FIELD FORCES
Phone: 404-656-3627
Brenda James-Griffin, ASSISTANT COMMISSIONER FOR
PUBLIC AFFAIRS
Phone: 404-656-3689
Earl Harris, ASSISTANT COMMISSIONER OF
ADMINISTRATION
Phone: 404-656-3608
Lee Meyers, ASSISTANT COMMISSIONER OF ANIMAL
INDUSTRY
Phone: 404-656-3671
Phil Kea, ASSISTANT COMMISSIONER OF FINANCE
Phone: 404-656-3608
Bobby Harris, ASSISTANT COMMISSIONER OF
MARKETING
Phone: 404-656-3368
Tommy Irvin, COMMISSIONER OF AGRICULTURE

GEORGIA DEPARTMENT OF AGRICULTURE
CONSUMERS SERVICES LIBRARY
Agriculture Bldg., Capitol Square
Atlanta, GA 30334 USA
Phone: 404-656-3645   Fax: 404-651-7957
Website: www.agr.state.ga.us

Founded: NA
Scope: Statewide

Publications: Farmers and Consumers Market Bulletin

Contact(s):
Brenda Griffin, ASSISTANT COMMISSIONER
bgriffin@agr.state.ga.us
Tommy Irvin, COMMISSIONER
bgriffin@agr.state.ga.us
Arty Schronce, DIRECTOR OF PUBLIC AFFAIRS
aschronce@agr.state.ga.us
Rosa Bundrage, INFORMATION SPECIALIST SENIOR
rbattle@agr.state.ga.us

GEORGIA DEPARTMENT OF EDUCATION

1766 Twin Towers, East
Atlanta, GA 30334-5040 USA
Phone: 404-656-0913   Fax: 404-651-8582
E-mail: bmoore@doe.k12.ga.us
Website: www.doe.k12.ga.us

Founded: NA
Membership: 1500
Scope: Statewide
Publications: Georgia Science Teacher, Observation—newsletter

Contact(s):
Bob Moore, SCIENCE PROGRAM SPECIALIST

GEORGIA DEPARTMENT OF NATURAL RESOURCES

2070 U.S. Highway 278, SE
Social Circle, GA 30025 USA
Phone: 770-918-6400   Fax: 706-557-3030
Website: www.georgiawildlife.com

Founded: NA
Membership: 35
Scope: Statewide
Publications: see publications on website

Contact(s):
Chuck Coomer, CHIEF: FISHERIES MANAGEMENT
Phone: 770-918-6406
Todd Holbrook, CHIEF: GAME MANAGEMENT
Phone: 770-918-6404
Ron Bailey, CHIEF: LAW ENFORCEMENT
Phone: 770-918-6408
Mike Harris, CHIEF: NONGAME WILDLIFE/NATURAL HERITAGE
Phone: 770-761-3035
David Waller, DIRECTOR

GEORGIA DEPARTMENT OF NATURAL RESOURCES

205 Butler St., SE, East Tower
Atlanta, GA 30334 USA
Phone: 404-656-3500   Fax: 404-656-0770
Website: www.dnr.state.ga.us

Founded: NA
Scope: Statewide

Contact(s):
Lonice Barrett, COMMISSIONER

GEORGIA DEPARTMENT OF NATURAL RESOURCES

COASTAL RESOURCES DIVISION
One Conservation Way
Brunswick, GA 31520 USA
Phone: 912-264-7218   Fax: 912-262-3143
E-mail: www.grant..org/dnr/crd
Website: www.dnr.state.ga.us

Founded: NA
Membership: 50
Scope: Statewide

Contact(s):
Duane Harris, DIRECTOR
Phone: 912-264-7218

GEORGIA DEPARTMENT OF NATURAL RESOURCES

ENVIRONMENTAL PROTECTION DIVISION
205 Butler Street SE, Suite 1152 - East Tower
Atlanta, GA 30334 USA
Phone: 404-657-5947   Fax: 404-657-5778
Website: http://ganet.org/dnr/environ

Founded: NA

Contact(s):
Harold Reheis, DIRECTOR
Phone: 404-656-4713

GEORGIA DEPARTMENT OF NATURAL RESOURCES

ENVIRONMENTAL PROTECTION DIVISION
USA
Phone: 912-264-7218

Founded: NA

Contact(s):
David Word, ASSISTANT DIRECTOR
Ron Methier, CHIEF: AIR PROTECTION BRANCH
Jennifer Kaduck, CHIEF: HAZARDOUS WASTE BRANCH
Mark Smith, CHIEF: LAND PROTECTION BRANCH
Jim Setser, CHIEF: PROGRAM COORDINATION BRANCH
Alan Hallum, CHIEF: WATER PROTECTION BRANCH
Nolton Johnson, CHIEF: WATER RESOURCES BRANCH
Harold Reheis, DIRECTOR
Phone: 404-656-4713

GEORGIA DEPARTMENT OF NATURAL RESOURCES

HISTORIC PRESERVATION DIVISION
156 Trinity Ave., SW, Suite 101
Atlanta, GA 30303 USA
Phone: 404-656-2840   Fax: 404-651-8739
Website: www.gashpo.org

Founded: NA
Membership: 35
Scope: Statewide
Publications: Publications on line

Contact(s):
Ray Luce, DIRECTOR
Phone: 404-656-2840

GEORGIA DEPARTMENT OF NATURAL RESOURCES

PARKS, RECREATION AND HISTORIC SITES DIVISION
205 Butler, SE, Suite 1352
Atlanta, GA 30334 USA
Phone: 404-656-2770   Fax: 404-651-5871

Founded: NA
Scope: Statewide

Contact(s):
David Freedman, CHIEF, MAINTENANCE AND
ENGINEERING SECTION
Wayne Escoe, CHIEF, PARKS OPERATION SECTION
Burt Weerts, DIRECTOR
Phone: 404-656-2770

GEORGIA DEPARTMENT OF NATURAL RESOURCES

POLLUTION PREVENTION ASSISTANCE DIVISION
7 Martin Luther King Jr. Dr., SW, Suite 450
Atlanta, GA 30334-9004 USA
Phone: 404-651-5120   Fax: 404-651-5130
Website: www.p2ad.org

Founded: NA
Scope: Statewide

Contact(s):
Robert Kerr, DIRECTOR
Phone: 404-651-5120

GEORGIA FORESTRY COMMISSION

P.O. Box 819
Macon, GA 31202-0819 USA
Phone: 912-751-3500   Fax: 478-751-3465
Website: www.gfc.state.ga.us

Founded: 1925
Scope: Statewide

Description: To foster, improve, and encourage reforestation; to engage in research and other projects for better forestry practices; to inform the public of the values and benefits of forestry; and to detect, prevent, and combat forest fires.

Publications: Wood Using Industries, Georgia Forestry

Contact(s):
Jim Gillis, BOARD OF COMMISSIONER CHAIRMAN
Garland Nelson, CHIEF OF FOREST ADMINISTRATION
Phone: 478-751-3464
Lynn Hooven, CHIEF OF FOREST MANAGEMENT
Phone: 478-751-3458
Alan Dozier, CHIEF OF FOREST PROTECTION
Phone: 478-751-3488
Sharon Dolliver, CHIEF OF INFORMATION AND
EDUCATION
Phone: 478-751-3530
Johnny Branan, CHIEF OF REFORESTATION
Phone: 478-751-3530
William Lazenby, DEPUTY DIRECTOR
Phone: 478-751-3480
J. Allen, DIRECTOR
Phone: 478-751-3480
Lynn Walton, EDITOR
Phone: 478-751-3530

Robert Farris, FIELD SUPERVISOR
Randall Perry, PERSONNEL OFFICER
Phone: 404-298-4949

GEORGIA SEA GRANT COLLEGE PROGRAM

The University of Georgia, Marine Sciences Bldg., Rm. 220
Athens, GA 30602-3636 USA
Phone: 706-542-5954   Fax: 706-542-3652
Website: www.alpha.marsci.uga.edu/gaseagrant.html

Founded: 1971
Scope: Statewide

Description: A part of the National Sea Grant College Program, the Georgia program fosters the sustainable development and environmental stewardship of the nation's marine resources. It is a competitive grant program funding, applied marine research, education, and advisory service projects at universities in Georgia.

Keyword(s): Renewable Resources, Environmental and Conservation Education, Sustainable Development, Coasts, Oceanography

Contact(s):
Mac Rawson, DIRECTOR, SEA GRANT COLLEGE
PROGRAM
mrawson@arches.uga.edu
David Bryant, LEADER OF MARINE ADVISORY SERVICE
COMMUNICATOR
William Hayes, LEADER, EDUCATION PROGRAM
Keith Gates, LEADER, MARINE ADVISORY SERVICE
William Potter, LIBRARIAN
Phone: 706-542-0621

GEORGIA STATE EXTENSION SERVICE

College of Agricultural and Environmental Sciences, 101
Conner Hall, The University of Georgia
Athens, GA 30602-7501 USA
Phone: 706-542-3924   Fax: 706-542-0803
E-mail: caesdean@arches.uga.edu
Website: www.uga.edu/caes/

Founded: NA
Membership: 3
Scope: International

Contact(s):
George Lewis, AQUACULTURE AND FISHERIES
SPECIALIST
Phone: 706-542-9038
Gale Buchanan, DEAN AND DIRECTOR
Phone: 706-542-3924
caesdean@arches.uga.edu
Tony Tyson, INTERIM ASSOCIATE DEAN FOR EXTENSION
111 Conner Hall, Athens, GA 30602
Phone: 706-542-3824
Fax: 706-542-8815
caesext@arches.uga.edu
Jeffery Jackson, WILDLIFE SPECIALIST
University of Georgia, Warnell School of Forest Resources,
Athens, GA 30602-2152
Phone: 706-542-9054
Fax: 706-542-3342

GOVERNOR OF ALABAMA

State Capitol, 600 Dexter Ave.
Montgomery, AL 36130 USA
Phone: 334-242-7100   Fax: 334-353-0004
E-mail: governor@governor.state.al.us
Website: www.governor.state.al.us

Founded: NA
Membership: 70
Scope: Statewide

Contact(s):
    Don Siegelman, GOVERNOR

GOVERNOR OF ALASKA

P.O. Box 110001
Juneau, AK 99811-0001 USA
Phone: 907-465-3500   Fax: 907-465-3532
E-mail: office_of_the-governor@gov.state.ak.us
Website: www.state.ak.us

Founded: NA
Scope: Statewide

Publications: The Talking Points - newsletter weekly

Contact(s):
    Tony Knowles, GOVERNOR

GOVERNOR OF ARIZONA

State House, 1700 W. Washington
Phoenix, AZ 85007 USA
Phone: 602-542-4331   Fax: 602-542-7601
E-mail: azgov@az.gov
Website: www.governor.state.az.us

Founded: NA
Scope: Statewide
Publications: see publication website

Contact(s):
    Jane Hull, GOVERNOR
    azgov@az.gov

GOVERNOR OF ARKANSAS

250 State Capitol
Little Rock, AR 72201 USA
Phone: 501-682-2345   Fax: 501-682-1382
E-mail: mike.huckabee@state.ar.us
Website: www.state.ar.us

Founded: NA
Scope: Statewide

Contact(s):
    Mike Huckabee, GOVERNOR
    mike.huckabee@state.ar.us

GOVERNOR OF CALIFORNIA

State Capitol Building
Sacramento, CA 95814 USA
Phone: 916-445-2841   Fax: 916-445-4633
Website: www.governor.ca.gov

Founded: NA
Scope: Statewide

Contact(s):
    Gray Davis, GOVERNOR

GOVERNOR OF COLORADO

136 State Capitol
Denver, CO 80203 USA
Phone: 303-866-2471   Fax: 303-866-2003
E-mail: governorowens@state.co.us
Website: www.state.co.us

Founded: NA
Scope: Statewide

Contact(s):
    Roy Palmer, CHIEF OF STAFF
    Sean Tonner, DEPUTY CHIEF OF STAFF
    Bill Owens, GOVERNOR

GOVERNOR OF CONNECTICUT

210 Capitol Ave.
Hartford, CT 06106 USA
Phone: 860-566-4840   Fax: 860-524-7396
E-mail: governor.roland@po.state.ct.us
Website: www.po.state.ct.us

Founded: NA
Scope: Statewide

Contact(s):
    John Rowland, GOVERNOR

GOVERNOR OF DELAWARE

Tatnall Bldg., 600 William Penn St.
Dover, DE 19901 USA
Phone: 302-739-4101   Fax: 302-739-2775
Website: www.state.ge.us

Founded: NA
Scope: Statewide

Contact(s):
    Ruth Minner, GOVERNOR

GOVERNOR OF FLORIDA

State Capitol
Tallahassee, FL 32399-0001 USA
Phone: 850-488-2272   Fax: 850-487-0801
E-mail: jeb.bush@myflorida.com
Website: www.myflorida.com

Founded: NA
Scope: Statewide

Contact(s):
    Jeb Bush, GOVERNOR

GOVERNOR OF GEORGIA

205 Butler St., Suite 1252
Atlanta, GA 30334 USA
Phone: 404-656-3500   Fax: 404-656-0770
Website: www.dnr.state.ga.us

Founded: NA
Scope: Statewide

Contact(s):
    David Waller, DIRECTOR WILD RESOURCE DIVISION
    2070 US Highway 278, Social Circle, GA 30025
    Phone: 770-918-6401
    Fax: 706-557-3030
    Roy Barnes, GOVERNOR

GOVERNOR OF HAWAII
235 S. Beretania St., State Capitol
Honolulu, HI 96813 USA
Phone: 808-586-0034   Fax: 808-586-0006
Website: www.gov.state.hi.us

Founded: NA
Scope: Statewide

Contact(s):
Benjamin Cayetano, GOVERNOR

GOVERNOR OF IDAHO
700 West Jefferson
Boise, ID 83720-0034 USA
Phone: 208-334-2100   Fax: 208-334-3454
E-mail: governor@gov.state.id.us
Website: www.state.id.us

Founded: NA
Scope: Statewide

Contact(s):
Dirk Kempthorne, GOVERNOR

GOVERNOR OF ILLINOIS
State Capitol, Rm. 207
Springfield, IL 62706 USA
Phone: 217-782-6830   Fax: 217-524-4049
E-mail: governor@state.il.us
Website: www.state.il.us/

Founded: NA
Scope: Regional

Contact(s):
George Ryan, GOVERNOR

GOVERNOR OF INDIANA
Rm. 206, Statehouse
Indianapolis, IN 46204 USA
Phone: 317-232-4567   Fax: 317-232-3443
E-mail: fobannon@state.in.us
Website: www.state.in.us/gov/

Founded: NA
Scope: Statewide

Contact(s):
Frank Obannon, GOVERNOR

GOVERNOR OF IOWA
State Capitol
Des Moines, IA 50319 USA
Phone: 515-281-5211   Fax: 515-281-6611
Website: www.state.ia.us/governor

Founded: NA
Scope: Statewide

Contact(s):
Tom Vilsack, GOVERNOR

GOVERNOR OF KANSAS
State Capitol, 2nd Fl.
Topeka, KS 66612-1590 USA
Phone: 785-296-3232

Founded: NA

Contact(s):
Bill Graves, GOVERNOR

GOVERNOR OF KENTUCKY
State Capitol, 700 Capitol St.
Frankfort, KY 40601 USA
Phone: 502-564-2611   Fax: 502-564-2517
Website: www.state.ky.us

Founded: NA
Scope: Statewide

Contact(s):
Paul Patton, GOVERNOR

GOVERNOR OF LOUISIANA
State Capitol
Baton Rouge, LA 70804 USA
Phone: 225-342-7015   Fax: 225-342-7099
Website: www.gov.state.la.us

Founded: NA
Scope: Local

Contact(s):
M. Foster, GOVERNOR

GOVERNOR OF MAINE
State House, Station 1
Augusta, ME 04333 USA
Phone: 207-287-3531   Fax: 207-287-1034
E-mail: governor@state.me.us
Website: www.state.me.us

Founded: NA
Scope: Statewide

Contact(s):
Angus King, GOVERNOR

GOVERNOR OF MARYLAND
State House, 100 State Cir.
Annapolis, MD 21401 USA
Phone: 410-974-3901   Fax: 410-974-3275
Website: www.govenor.gov.state.md.us

Founded: NA
Scope: Statewide

Contact(s):
Parris Glendening, GOVERNOR

GOVERNOR OF MASSACHUSETTS
State House, Rm. 360
Boston, MA 02133 USA
Phone: 617-727-3600   Fax: 617-727-9725
E-mail: goffice@state.ma.us
Website: www.state.ma.us

Founded: NA
Scope: Statewide

Contact(s):
Jane Swift, ACTING GOVERNOR
jane.swift@gov.state.ma.us

GOVERNOR OF MICHIGAN
State Capitol, P.O. Box 30013
Lansing, MI 48909 USA
Phone: 517-373-3400   Fax: 517-335-6863
Website: www.state.mi.us/mi.gov

Founded: NA
Membership: 75
Scope: State

Contact(s):
John Engler, GOVERNOR

GOVERNOR OF MINNESOTA
130 State Capitol, 75 Constitution Ave.
St.Paul, MN 55155 USA
Phone: 651-296-3391   Fax: 681-296-2089

Founded: NA
Scope: Statewide

Contact(s):
Molly Hoffman, CHIEF OF STAFF
Jesse Ventura, GOVERNOR

GOVERNOR OF MISSISSIPPI
P.O. Box 139
Jackson, MS 39205 USA
Phone: 877-405-0733   Fax: 601-359-3741
Website: http://www.governor.state.ms.us

Founded: NA

Contact(s):
Ronnie Musgrove, GOVERNOR

GOVERNOR OF MISSOURI
State Capitol
Jefferson City, MO 65102 USA
Phone: 573-751-3222   Fax: 573-751-1495
Website: www.gov.state.mo.us

Founded: NA
Scope: Statewide

Contact(s):
Bob Holden, GOVERNOR

GOVERNOR OF MONTANA
P.O. Box 0801
Helena, MT 59620 USA
Phone: 406-444-3111   Fax: 406-444-5529
E-mail: governor@state.mt.us
Website: www.state.mt.us

Founded: NA
Scope: Statewide

Contact(s):
Judy Martz, GOVERNOR

GOVERNOR OF NEBRASKA
Lincoln, NE 68509 USA
Phone: 402-471-2244   Fax: 402-471-6031
E-mail: mjohanns@notes.state.ne.us
Website: www.gov.nol.org

Founded: NA
Scope: Statewide

Contact(s):
Mike Johanns, GOVERNOR

GOVERNOR OF NEVADA
State Capitol
Carson City, NV 89701 USA
Phone: 775-684-5670   Fax: 775-684-5683
Website: http://gov.state.nv.us/

Founded: NA

Contact(s):
Kenny Guinn, GOVERNOR

GOVERNOR OF NEW HAMPSHIRE
State House, Rm. 208
Concord, NH 03301 USA
Phone: 603-271-2121   Fax: 603-271-6998
Website: www.state.nh.us/governor

Founded: NA
Scope: Statewide

Contact(s):
Jeanne Shaheen, GOVERNOR

GOVERNOR OF NEW JERSEY
State House, 125 W. State St., Office of the Governor,
CN-001
Trenton, NJ 08625 USA
Phone: 609-292-6000   Fax: 609-292-3454
Website: www.state.nj.us

Founded: NA
Scope: Statewide

Contact(s):
Donald Difrancesco, (ACTING) GOVERNOR

GOVERNOR OF NEW MEXICO
State Capitol, Suite 400
Santa Fe, NM 87503 USA
Phone: 505-827-3000   Fax: 505-827-3026
Website: www.gov.state.nm.us

Founded: NA
Scope: Statewide

Contact(s):
Gary Johnson, GOVERNOR

GOVERNOR OF NEW YORK
Executive Chamber State Capitol
Albany, NY 12224 USA
Phone: 518-474-8390
Website: www.state.ny.us

Founded: NA
Scope: Statewide

Contact(s):
George Pataki, GOVERNOR

GOVERNOR OF NORTH CAROLINA
20301 Mail Service Center
Raleigh, NC 27699-0301 USA
Phone: 919-733-4240   Fax: 919-733-2120
E-mail: gov@ncmail.net
Website: www.ncgov.com

Founded: NA
Scope: State

Contact(s):
Michael Easley, GOVERNOR

GOVERNOR OF NORTH DAKOTA

State Capitol, 600 E. Blvd. Ave.
Bismarck, ND 58505-0001 USA
Phone: 701-328-2200   Fax: 701-328-2205
E-mail: governor@state.nd.us
Website: www.discovernd.com

Founded: NA
Scope: Statewide

Contact(s):
John Hoeven, GOVERNOR
Phone: 701-328-2200

GOVERNOR OF OHIO

Governors Office 77 S. High St., 30th Fl.
Columbus, OH 43215-6117 USA
Phone: 614-466-3555   Fax: 614-466-9354
E-mail: Governor.Taft@das.state.oh.us
Website: www.state.oh.us/gov

Founded: NA
Membership: 70
Scope: Statewide

Contact(s):
Bob Taft, GOVERNOR
Phone: 614-466-3555
Fax: 614-466-9354
Governor.Taft@das.state.oh.us

GOVERNOR OF OKLAHOMA

State Capitol Bldg., Suite 212
Oklahoma City, OK 73105 USA
Phone: 405-521-2342   Fax: 405-521-3353
E-mail: governor@gov.state.ok.us
Website: www.governor.state.ok.us

Founded: NA
Scope: Statewide

Contact(s):
Frank Keating, GOVERNOR

GOVERNOR OF OREGON

DEPT. OF NATURAL RESOURCES
900 Court St., NE, Room 254
Salem, OR 97301 USA
Phone: 503-378-3111   Fax: 503-378-3225
Website: www.governor.state.or.us/email.htm

Founded: NA
Scope: Statewide

Contact(s):
John Kitzhaber, GOVERNOR

GOVERNOR OF PENNSYLVANIA

Rm. 225, Main Capitol Bldg.
Harrisburg, PA 17120 USA
Phone: 717-787-2500   Fax: 717-772-8284
E-mail: governor@state.pa.us
Website: www.state.pa.us

Founded: NA
Scope: State

Contact(s):
Tom Ridge, GOVERNOR

GOVERNOR OF PUERTO RICO

La Fortaleza,
San Juan, PR 00901 USA
Phone: 787-723-8297   Fax: 787-721-7328
E-mail: lafortaleza@gov.com
Website: http://www.fortaleza.gobierno.pr

Founded: NA
Scope: State

Contact(s):
Pedro Rossello, GOVERNOR

GOVERNOR OF RHODE ISLAND

State House
Providence, RI 02903 USA
Phone: 401-222-2080   Fax: 401-272-0860
E-mail: rigov@giv.state.ri.us
Website: www.gov.state.ri.us

Founded: NA
Scope: Statewide

Contact(s):
Lincoln Almond, GOVERNOR

GOVERNOR OF SOUTH CAROLINA

Columbia, SC 29211 USA
Phone: 803-734-9400   Fax: 803-734-9413
E-mail: governor@govoepp.state.sc.us
Website: www.myscgov.com

Founded: NA
Scope: Statewide

Keyword(s): Wetlands, Rivers, Environmental Law, Forests and
Forestry, Agriculture

Contact(s):
Jim Hodges, GOVERNOR

GOVERNOR OF SOUTH DAKOTA

500 E. Capitol
Pierre, SD 57501 USA
Phone: 605-773-3212   Fax: 605-773-4711
E-mail: sdgov@state.sd.us

Founded: NA
Scope: Statewide

Contact(s):
William Janklow, GOVERNOR

GOVERNOR OF TENNESSEE

State Capitol, 1st Fl.
Nashville, TN 37243 USA
Phone: 615-741-2001   Fax: 615-532-9712
E-mail: dsundquist@sunmail.state.tn.us
Website: www.state.tn.us

Founded: NA
Scope: Statewide

Contact(s):
Don Sundquist, GOVERNOR

GOVERNOR OF TEXAS

POLICY OFFICE
P.O. Box 12428
Austin, TX 78711 USA
Phone: 512-463-2000   Fax: 512-463-1849
Website: www.governor.state.tx.us

Founded: NA
Scope: Regional

Contact(s):
Rick Perry, GOVERNOR
Phone: 512-463-2000

GOVERNOR OF THE VIRGIN ISLANDS

DEPT. OF PLANNING & NATURAL RESOURCES
Government House, Charlotte Amalie
St.Thomas, VI 00802 USA
Phone: 340-774-0001   Fax: 340-774-1361
Website: http://www.gov.vi/html/gov.html

Founded: NA
Scope: Statewide

Contact(s):
Dean Plaskett, COMMISSIONER
Charles Turnbull, GOVERNOR

GOVERNOR OF UTAH

210 State Capitol
Salt Lake City, UT 84114 USA
Phone: 801-538-1000   Fax: 801-538-1528
Website: www.governor@state.ut.us

Founded: NA
Scope: Statewide

Contact(s):
Mike Leavitt, GOVERNOR

GOVERNOR OF VERMONT

Pavilion Office Bldg.
Montpelier, VT 05609 USA
Phone: 802-828-3333   Fax: 802-828-3339

Founded: NA
Scope: Statewide

Contact(s):
Howard Dean, GOVERNOR

GOVERNOR OF VIRGINIA

DEPT. OF NATURAL RESOURCES
State Capitol
Richmond, VA 23219 USA
Phone: 804 786 2211   Fax: 004-071-0333
Website: http://snr.vipnet.org/

Founded: NA
Scope: Statewide

Contact(s):
Ronald Hamm, DEPUTY SECRETARY OF NATURAL
RESOURCES
Phone: 804-786-0044
Mark Warner, GOVERNOR
John Woodley, SECRETARY OF NATURAL RESOURCES
Phone: 804-786-0044

GOVERNOR OF WASHINGTON

Legislative Bldg., P.O. Box 40002
Olympia, WA 98504-0002 USA
Phone: 360-902-4111   Fax: 360-753-4110
E-mail: governor.locke@governor.was.gov
Website: www.access.wa.gov

Founded: NA
Scope: Statewide

Contact(s):
Gary Locke, GOVERNOR

GOVERNOR OF WISCONSIN

State Capitol
Madison, WI 53707 USA
Phone: 608-266-1212   Fax: 608-267-8983
E-mail: wisgov@mail.state.wi.us
Website: www.wisgov.state.wi.us

Founded: NA
Scope: Statewide

Contact(s):
Scott McCallum, GOVERNOR

GOVERNOR OF WYOMING

State Capitol Bldg., Rm. 124
Cheyenne, WY 82002 USA
Phone: 307-777-7434   Fax: 307-632-3909
Website: www.state.wy.us

Founded: NA
Scope: Statewide

Contact(s):
Jim Geringer, GOVERNOR

GOVERNOR'S OFFICE OF ENERGY, MANAGEMENT, AND CONSERVATION (COLORADO)

225 E. 16th Ave., Ste. 650
Denver, CO 80203 USA
Phone: 303-894-2383   Fax: 303-894-2388
E-mail: oemc@state.co.us
Website: http://www.state.co.us/oemc/

Founded: 1977

Description: OEC's mission includes leading the citizens of
Colorado by promoting the efficient use of energy and
resources. OEC develops, implements, and monitors energy
conservation programs and offers services for individuals,
community organizations, institutions, businesses, and
government. Those services are designed to reduce energy
consumption and increase awareness of the environmental,
economic, and personal benefit to efficient energy use.

Publications: Recycle Colorado Bulletin

Keyword(s): Environmental and Conservation Education,
Renewable Resources, Solar Energy, Solid Waste
Management, Composting

Contact(s):
Rick Grice, DIRECTOR

GOVERNORS OFFICE OF PLANNING AND RESEARCH (CALIFORNIA)

STATE CLEARINGHOUSE DEPT.
1400 10th St.
Sacramento, CA 95812-3044 USA
Phone: 916-322-2318　　　Fax: 916-323-3018
Website: www.my.ca.gov

Founded: 1970
Scope: Statewide

Description: Primary areas of concentration are the development of environmental and related land use goals and policies; growth management; evaluation of state plans and programs; and preparation of statewide environmental goals and policies statements.

Keyword(s): Environmental Planning, Geography, Land Use Planning, Planning Management, Population Growth

Contact(s):
Steve Nissen, ACTING DIRECTOR
Terry Roberts, STATE CLEARINGHOUSE SENIOR PLANNER
Phone: 916-445-0613

GRANT TECH CONSULTING AND CONSERVATION SERVICES

9564 Cheyenne Rd.
Meriden, KS 66512 USA
Phone: 785-876-0106　　　Fax: 785-876-0106
E-mail: kellyhiesberger@hotmail.com
Website: www.granttechconsulting.com

Description: At Grant Tech Consulting and Conservation Services we offer a broad range of services to assist you in the development of your conservation project. Whether you are a private land owner, school, community organization, or non-profit organization, we have what you need to create, preserve, or enhance your natural resources. We offer funding proposal development, trail design and construction services, design for outdoor education and recreation sites, and information on the most up to date resources available.

GUADALUPE-BLANCO RIVER AUTHORITY

933 East Ct.
Seguin, TX 78155 USA
Phone: 830-379-5822　　　Fax: 830-379-9718
Website: www.gbra.org

Founded: 1935
Membership: 130
Scope: Regional

Description: Responsibility to develop, conserve, and protect the water resources within a ten county statutory district and to aid in the prevention of soil erosion and flooding. Actively engaged in water supply, irrigation, hydroelectric power generation, water and wastewater treatment, and outdoor recreation operations.

Publications: Water Resources Report—Quarterly

Contact(s):
Thomas Hill, CHIEF ENGINEER
Fred Blumberg, DEPUTY GENERAL MANAGER
Alvin Schuerg, DIRECTOR OF ACCOUNTING AND FINANCE
David Welsch, DIRECTOR OF PROJECT DEVELOPMENT
Todd Volteler, DIRECTOR OF WATER POLICY

Debbie Magin, DIRECTOR OF WATER QUALITY
W. West, GENERAL MANAGER
Rose Degargo, HUMAN RESOURCES
Judy Gardner, MANAGER OF COMMUNICATIONS AND EDUCATION
Bryan Serold, MANAGER OF LOWER BASIN
John Smith, MANAGER OF UPPER BASIN

GULF COAST RESEARCH LABORATORY

703 East Beach Drive
Ocean Springs, MS 39566-7000 USA
Phone: 228-872-4200　　　Fax: 228-872-4204
Website: http://www.cms.usm.edu/gindex.htm

Founded: 1947

Description: Conducts research in marine biology, fisheries, geology, chemistry, and oceanography, and conducts an academic program in the marine sciences.

Publications: Gulf Research Reports-Scientific Journal, Marine Briefs Newsletter

Contact(s):
William Walker, ASSISTANT DIRECTOR: RESEARCH
Robert Vanaller, EDITOR
Robert Vanaller, INTERIM DIRECTOR

H

HAWAII COOPERATIVE FISHERY RESEARCH UNIT (USDI)

2538 The Mall, University of Hawaii
Honolulu, HI 96822 USA
Phone: 808-956-8350　　　Fax: 808-956-4238

Founded: NA
Scope: State, National, Internet

Description: Activities include research, graduate program teaching, and public service regarding inshore marine and inland waters with emphasis on native fishes and invertebrates.

Publications: Scientic Reports (irregular)

Keyword(s): Oceanography, Wildlife, Sustainable Ecosystems, Aquatic Habitats, Islands

Contact(s):
Dr. Charles Birkeland, ASSISTANT LEADER
James Parrish, LEADER

HAWAII DEPARTMENT OF AGRICULTURE

P.O. Box 22159
Honolulu, HI 96823-2159 USA
Phone: 808-973-9560
E-mail: holoa-info@exec.state.hi.us
Website: http://hawaiiag.org/hdoa

Founded: NA

Description: Promotes the best use of Hawaii's agricultural resources. Concerned with the protection of agricultural lands and water and diversification of the state's agricultural economy. Functions include agricultural planning, agricultural credit, product promotion and market development, plant and animal quarantine, plant and animal disease and pest control, milk control, livestock and market reporting service, commodities grading, pesticide use enforcement, and enforcement of weights and measures standards.

Contact(s):
James Nakatani, CHAIRPERSON
Phone: 808-973-9551
Elaine Abe, CHIEF: ADMINISTRATIVE SERVICES
Phone: 808-973-9606
Letitia Uyehara, DEPUTY TO THE CHAIRPERSON
Phone: 808-973-9553
Doreen Shishido, HEAD: AGRICULTURAL LOAN DIVISION
Phone: 808-973-9460
Paul Matsuo, HEAD: AGRICULTURAL RESOURCE
MANAGEMENT DIVISION
Phone: 808-973-9475
Samuel Camp, HEAD: AGRICULTURE DEVELOPMENT
DIVISION
Phone: 808-973-9566
Samuel Camp, HEAD: QUALITY ASSURANCE DIVISION
Phone: 808-586-0870

HAWAII DEPARTMENT OF HEALTH
OFFICE OF ENVIRONMENTAL QUALITY CONTROL
O235 S. Beretania St.
Honolulu, HI 96813 USA
Phone: 808-586-4185   Fax: 808-586-4186
E-mail: OEQC@HEALTH.STATE.HI.US
Website: www.state.hi.us/health/oeqc

Founded: NA
Membership: 5
Scope: Local

Description: OEQC advises the Governor on environmental
quality control matters; implements Hawaii's EIS law; reviews
all documents required by Hawaii's EIS process; and informs
the public of proposed actions through The Environmental
Notice (OEQC Bulletin). The director of OEQC is also
responsible for environmental education projects, and
proposing and encouraging legislation supporting the preservation of environmental resources.

Publications: OEQC Bulletin, Annual Report—Environmental
Indicators and Report Card, A Guidebook for the Hawaii State
Environmental Review Process

Keyword(s): Environmental and Conservation Education,
Environmental Law

Contact(s):
Genevieve Salmonson, DIRECTOR

HAWAII INSTITUTE OF MARINE BIOLOGY
University of Hawaii
Kaneohe, HI 96744-1346 USA
Phone: 808-236-7401   Fax: 808-236-7443
Website: www.hawaii.edu/HIMB

Founded: NA
Scope: Statewide

Description: Concerned with research in tropical marine biology
and oceanography with emphasis on coral reef biology,
aquaculture, fish endocrinology, and behavior of reef
organisms. Provides research facilities for investigations in
tropical marine biology. Offers annual summer program in
selected topics for graduate students.

Contact(s):
Joann Leong, DIRECTOR

HAWAII SEA GRANT COLLEGE PROGRAM
UNIVERSITY OF HAWAII
2525 Correa Rd., HIG 238
Honolulu, HI 96822 USA
Phone: 808-956-7031   Fax: 808-956-3014
E-mail: seagrant@soest.hawaii.edu
Website: www.soest.hawaii.edu/seagrant/

Founded: NA
Membership: 27
Scope: International

Description: The University of Hawaii Sea Grant College
Program supports research projects in marine-related areas.
Its extension arm has agents and specialists located in
Honolulu, Maui, the Big Island of Hawaii, Pohnpei (Federated
States of Micronesia) and Saipan Commonwealth of the
Northern Mariana Islands. Agents help marine users benefit
from the Sea Grant-supported research, especially in the areas
of commercial and recreational fishing, aquaculture, marine
recreation and tourism development, marine education and
conservation.

Contact(s):
Clyde Tamaru, AQUACULTURE EXTENSION SPECIALIST
Phone: 808-956-2869
Fax: 808-956-2858
Raymond Tabata, COASTAL RECREATION & TOURISM
EXTENSION AGENT
Phone: 808-956-2866
Fax: 808-956-2858
Christine Woolaway, COASTAL RECREATION & TOURISM
EXTENSION AGENT
Phone: 808-956-2872
Fax: 808-956-2858
Peter Rappa, COASTAL RESOURCE MANAGEMENT
EXTENSION AGENT
Phone: 808-956-2868
Fax: 808-956-2858
Priscilla Billig, DIRECTOR OF COMMUNICATIONS
PROGRAM
Phone: 808-956-2414
Fax: 808-956-2880
Dr. Richard Brock, DIRECTOR OF SEA GRANT EXTENSION
SERVICE
Gordon Grau, DIRECTOR: SEA GRANT COLLEGE
PROGRAM
sg-dir@soest.hawaii.edu
Elizabeth Kumabe, EDUCATION SPECIALIST
Phone: 808-956-2860
Fax: 808-956-2858
Richard Brock, FISHERIES EXTENSION AGENT
Phone: 808-956-2859
Fax: 808-956-2858
Alan Kam, FISHERIES EXTENSION AGENT
Phone: 808-956-2865
Fax: 808-956-2858
Jeff Kuwabara, HANAUMA BAY EDUCATIONAL PROGRAM
VOLUNTEER COORDINATOR
Phone: 808-396-1319
Fax: 808-956-2858

I

IDAHO COOPERATIVE EXTENSION

P.O. Box 442338
Moscow, ID 83844-2338 USA
Phone: 208-885-6639   Fax: 208-885-6654
E-mail: extdir@uidaho.edu
Website: www.uidaho.edu/extension/

Founded: NA
Membership: 220
Scope: Statewide

Contact(s):
Lou Riesenberg, AGRICULTURE AND EXTENSION
EDUCATION
1134 W 6th Street, Moscow, ID 83844
Phone: 208-885-6358
lriesenb@uidaho.edu
Paul McCawley, ASSOCIATE EXTENSION DIRECTOR
(INTERIM)
Phone: 208-885-5883
anauman@uidaho.edu
Leroy Luft, EXTENSION DIRECTOR
extdir@uidaho.edu
Ronald Mahoney, EXTENSION FORESTER
Univ. of Idaho, College of Natural Resources, Moscow, ID
83844-1140
Phone: 208-885-6356
Phone: 208-885-6226

IDAHO COOPERATIVE FISH AND WILDLIFE RESEARCH UNIT (USDI)

COLLEGE OF NATURAL RESOURCES
Po Box 44-1141
Moscow, ID 83844-1141 USA
Phone: 208-885-2750   Fax: 208-885-9080
Website: www.its.uidaho.edu/coop

Founded: 1963
Membership: 200
Scope: National, International

Description: An interagency organization which conducts
research, graduate level training, and extension in the fields of
fish, wildlife, and conservation biology.

Keyword(s): Wildlife, Birds, training, Biodiversity, Endangered
Species

Contact(s):
James Congleton, ASSISTANT LEADER
R. Wright, ASSISTANT LEADER
J. Scott, LEADER

IDAHO DEPARTMENT FISH & GAME

600 S. Walnut, Box 25
Boise, ID 83707 USA
Phone: 208-334-3700   Fax: 208-334-2114
Website: www.state.id.us/fishgame

Founded: 1938
Scope: Statewide

Description: To preserve, protect, perpetuate, and manage all
wildlife within the state of Idaho; to make and declare such
rules and regulations, and to employ personnel necessary to
administer and enforce the harvest of wildlife.

Publications: Idaho Fish & Game News

Keyword(s): Sport Fishing, Wildlife, Environmental and
Conservation Education, Hunting

Contact(s):
Al Vanvooren, ASSISTANT DIRECTOR
Phone: 208-334-5159
Stephen Barton, BUREAU CHIEF OF ADMINISTRATION
Phone: 208-334-3782
Bob Royce, CHIEF OF DP MANAGEMENT
Phone: 208-334-3700
Al Nicholson, CHIEF OF ENFORCEMENT
Phone: 208-334-3736
Phil Jeppson, CHIEF OF ENGINEERING
Phone: 208-334-3730
Virgil Moore, CHIEF OF FISHERIES
Phone: 208-334-3791
Tracey Trent, CHIEF OF NATURAL RESOURCES POLICY
Phone: 208-334-2595
Steve Huffaker, CHIEF OF WILDLIFE
Phone: 208-334-2920
Jack Trueblood, EDITOR
Phone: 208-334-3746
John Gahl, INTERIM CHIEF OF INFORMATION AND
EDUCATION
Phone: 208-334-2633

IDAHO DEPARTMENT OF PARKS AND RECREATION

Boise, ID 83720-0065 USA
Phone: 208-334-4199   Fax: 208-334-3741
Website: www.idahoparks.org

Founded: 1965
Membership: 200
Scope: Regional

Description: To formulate and put into execution a long-range
program for the acquisition, planning, protection, operation,
maintenance, development, and wise use of parks; and to
provide state leadership in recreation.

Keyword(s): Outdoor Recreation, Environmental and
Conservation Education, Rivers, Cultural Preservation,
Protected Areas

Contact(s):
Glenn Shewmaker, CHAIRMAN OF THE BOARD
Phone: 208-736-3608
William Dokken, DEPUTY DIRECTOR
Rick Collignon, DIRECTOR

IDAHO DEPARTMENT OF WATER RESOURCES

Statehouse
Boise, ID 83720-0098 USA
Phone: 208-327-7900   Fax: 208-327-7866
Website: www.idwr.state.id.us

Founded: NA
Membership: 200
Scope: Statewide

Description: Administration of State Water Plan and Energy Plan;
allocation and planning of water resources and energy
programs and projects; permit and license procedures for
water rights, dams, and mine tailing impoundment structures,
well construction, injection wells, and stream channel
alterations.

Publications: Rules, State Water Plan, and Newsletter, Water
and Energy Information Bulletins

Contact(s):
Robert Hoppie, ADMINISTRATOR OF ENERGY DIVISION
Phone: 208-327-7910
Hal Anderson, ADMINISTRATOR OF PLANNING AND
TECHNICAL SERVICES DIVISION
Phone: 208-327-7910
Norman Young, ADMINISTRATOR OF WATER
MANAGEMENT DIVISION
Phone: 208-327-7910
Joe Jordan, CHAIRMAN OF THE BOARD
Phone: 208-253-1103
Karl Dreher, DIRECTOR
Phone: 208-327-7910

IDAHO FISH AND WILDLIFE FOUNDATION

P.O. Box 2254
Boise, ID 83701 USA
Phone: 208-334-2648   Fax: 208-334-2148
Website: www.idfishnhunt.com\~ifwf

Founded: 1990
Membership: 100
Scope: Statewide

Description: To facilitate the organization and funding of natural resource projects: fish, wildlife, habitat, and education. Work with Idaho Department of Fish and Game and other entities to build public and private partnerships for wildlife projects.

Publications: Steelhead Fishing Economic Values brochure, Salmon Fishing Economic Survey (1998), Steelhead Fishing Economic Survey (1996)

Keyword(s): Environmental and Conservation Education, training, Nature Centers, Wildlife, Natural Areas

Contact(s):
Gayle Valentine, EXECUTIVE DIRECTOR

IDAHO GEOLOGICAL SURVEY

Morrill Hall, Third Floor, University of Idaho
Moscow, ID 83844-3014 USA
Phone: 208-885-7991   Fax: 208-885-5826
E-mail: igf@uidaho.edu
Website: www.idahogeology.org

Founded: 1919
Membership: 15
Scope: Statewide

Description: The Survey is the lead state agency for the collection, interpretation, and dissemination of all geologic and mineral data for Idaho. Conducts field investigations and laboratory studies; assists in preparation of geologic maps, derivative land-use planning, and geologic hazards maps; provides expertise to Individuals and governmental and private groups in planning land use.

Contact(s):
Kurt Othberg, ASSOCIATE DIRECTOR
Roy Breckenridge, ASSOCIATE DIRECTOR
Phone: 208-885-7991
Earl Bennett, DIRECTOR
Phone: 208-885-7991

IDAHO STATE DEPARTMENT OF AGRICULTURE

P.O. Box 790
Boise, ID 83701 USA
Phone: 208-332-8500
Website: http://www.agri.state.id.us

Founded: NA

Keyword(s): Endangered Species, Biotechnology, Land Preservation, Land Use Planning, Public Health Protection, Soil Conservation, Water Quality, Watersheds, Pesticides, Solid Waste Management, Toxic Substances, Nuclear-free, Water quantity, Water export and diversion, Coral Reefs, Agriculture, Chemical Pollution Control

Contact(s):
Patrick Takasugi, DIRECTOR
Lane Jolliffe, DIVISION OF AGRICULTURAL INSPECTIONS
ADMINISTRATOR
Phone: 208-332-8666
Mike Everett, DIVISION OF AGRICULTURAL RESOURCES,
MARKETING, AND DEVELOPMENT
Phone: 208-332-8531
Bob Hillman, DIVISION OF ANIMAL INDUSTRIES
ADMINISTRATOR
Phone: 208-332-8541
Roger Vega, DIVISION OF PLANT INDUSTRIES AND LABS
ADMINISTRATOR
Phone: 208-332-8627

IDAHO STATE SOIL CONSERVATION COMMISSION

P.O. Box 790
Boise, ID 83701-0790 USA
Phone: 208-332-8650   Fax: 208-334-2386
Website: www.scc.state.id.us

Founded: 1939
Membership: 5
Scope: Statewide

Description: Coordinates programs and activities of Soil Conservation Districts in Idaho. Concerned with overall leadership and administration of districts in development, wise use, and conservation of soil and water and other closely related resources. Participates in the National Cooperative Soil Survey Program through employment of soil scientists and has been designated the state water quality management agency for private and state agricultural lands. Administers low-interest loan program for conservation improvements on private and public lands.

Keyword(s): Soil Conservation, Environmental and Conservation Education, Exotic species, Aquatic nuisance species, Agriculture, Renewable Resources

Contact(s):
Jerry Nicolescu, ADMINISTRATOR
Phone: 208-332-8649
Fax: 208-334-2386
Tom Johnston, CHAIRMAN
22410 Ten Davis Rd., Parma, ID 83660
Phone: 208-722-6224
Fax: 208-722-6090
tjohn@micron.net

ILLINOIS DEPARTMENT OF AGRICULTURE

State Fairgrounds, P.O. Box 19281
Springfield, IL 62794-9281 USA
Phone: 217-782-2172
Website: http://www.agr.state.il.us/

Founded: 1917

Description: The Illinois Department of Agriculture protects and promotes the state's agricultural and natural resources. The agency provides services that benefit consumers, farmers, and agribusinesses.

Publications: Illinois Agricultural Organizations Directory, Illinois Grain and Livestock Market News, Illinois Food Products, Illinois Agricultural Guide

Keyword(s): Renewable Resources, Land Use Planning, Soil Conservation, Agriculture, Pesticides

Contact(s):
Chet Boruff, DEPUTY DIRECTOR FOR NATURAL RESOURCE AND AGRI-INDUSTRY REGULATION
Becky Doyle, DIRECTOR
Dave Bender, EXECUTIVE OFFICE: ASSISTANT DIRECTOR
Jim Reynolds, SUPERINTENDENT FOR FAIRS AND PROMOTIONS

ILLINOIS DEPARTMENT OF AGRICULTURE
BUREAU OF LAND AND WATER RESOURCES
Springfield, IL 62794-9281 USA
Phone: 217-782-6297   Fax: 217-524-4882
Website: www.agr.state.il.us

Founded: NA
Scope: Statewide

Contact(s):
Steve Frank, BUREAU CHIEF
Phone: 217-557-0993

ILLINOIS DEPARTMENT OF NATURAL RESOURCES
524 S. 2nd St., Rm. 400 LTP
Springfield, IL 62701-1787 USA
Phone: 217-782-6302   Fax: 217-785-9236
Website: www.dnr.state.il.us

Founded: NA
Scope: Statewide

Description: The mission of the Illinois Department of Natural Resources is to promote an understanding and appreciation of the state's natural resources and work with the people of Illinois to protect and manage those resources to ensure a high quality of life for present and future generations.

Publications: Illinois Fishing Information Book, State Park Magazine, Outdoor Illinois Magazine, Digest of Hunting and Trapping Regulations

Keyword(s): Endangered Species, Wildlife Rehabilitation, Environmental and Conservation Education, Biodiversity, Conservation

Contact(s):
John Bandy, CHIEF FISCAL OFFICER
Phone: 217-785-8552
Robert Lawley, CHIEF LEGAL COUNSEL
Phone: 217-782-1809
Tom Wakolbinger, CHIEF OF LAW ENFORCEMENT OFFICE
Phone: 217-782-6431
John Schmitt, CONSERVATION FOUNDATION EXECUTIVE DIRECTOR
100 W. Randolph, Chicago, IL 60601
Phone: 312-814-7237
Dick Mottershaw, DEPUTY DIRECTOR
Jim Garner, DEPUTY DIRECTOR
Brent Manning, DIRECTOR
Larry Closson, DIRECTOR OF LAW ENFORCEMENT OFFICE

Phone: 217-782-6431
Diane Hendren, DIRECTOR OF LEGISLATION AND CONSTITUENCY SERVICES
Phone: 217-785-0073
Kim Underwood, DIRECTOR OF OFFICE MINES AND MINERALS
Phone: 217-782-0031
Kevin Sronce, DIRECTOR OF OFFICE OF ADMINISTRATION
Phone: 217-782-0179
Bruce Clark, DIRECTOR OF OFFICE OF CAPITAL DEVELOPMENT
Phone: 217-782-1807
Jerry Beverlin, DIRECTOR OF OFFICE OF LAND MANAGEMENT AND EDUCATION
Phone: 217-782-6752
Carol Knowles, DIRECTOR OF PUBLIC AFFAIRS OFFICE
Phone: 217-785-0970
Jim Fulgenzi, DIRECTOR OF PUBLIC SERVICES OFFICE
Phone: 217-782-7454
Tom Flattery, DIRECTOR OF REALTY AND ENVIRONMENTAL PLANNING OFFICE
Phone: 217-782-7940
Kirby Cottrell, DIRECTOR OF RESOURCE CONSERVATION OFFICE
Phone: 217-785-8547
Don Vonnahme, DIRECTOR OF WATER RESOURCES OFFICE
Phone: 217-782-2152
Brad Hammond, DIVISION MANAGER OF INTERNAL AUDIT OFFICE
Phone: 217-785-0853
Theresa Cummings, EQUAL EMPLOYMENT OPPORTUNITY OFFICER
Phone: 217-785-0067

ILLINOIS DEPARTMENT OF TRANSPORTATION
2300 S. Dirksen Pkwy.
Springfield, IL 62764 USA
Phone: 217-782-5597
Website: www.dot.state.il.us

Founded: NA
Scope: Statewide

Keyword(s): Transportation

Contact(s):
Mike Hines, CHIEF: BUREAU OF DESIGN AND ENVIRONMENT
Phone: 217-782-7526
Fax: 217-524-0989
James Slifer, DIRECTOR: DIVISION OF HIGHWAYS
Phone: 217-782-2151
Fax: 217-524-2972
Kirk Brown, SECRETARY
Phone: 217-782-6828

ILLINOIS ENVIRONMENTAL PROTECTION AGENCY
1021 North Grand Ave., E.
Springfield, IL 62794-9276 USA
Phone: 217-782-3397   Fax: 217-782-9039
Website: www.epa.state.us./

Founded: 1970
Membership: 1400
Scope: Statewide

Description: Responsible for implementing the environmental program for the state of Illinois. Administers a variety of programs to protect the air, land, and water.

Publications: Environmental Progress, Digester/Over the Spillway

Keyword(s): Toxic Substances, Nuclear-free, Water quantity, Water export and diversion, Environmental Law, Coral Reefs, Air Quality and Pollution, Protected Areas

Contact(s):
Marcia Willhite, BUREAU CHIEF
Dave Kolaz, CHIEF: BUREAU OF AIR
Phone: 217-785-4140
William Child, CHIEF: BUREAU OF LAND
Phone: 217-785-9407
James Park, CHIEF: BUREAU OF WATER
Phone: 217-782-1654
William Seith, DEPUTY DIRECTOR
Renee Cipriano, DIRECTOR
Joan Muraro, EDITOR
Phone: 217-785-7209
Nancy Simpson, HEAD LIBRARIAN
1021 N. Grand Ave. E., P.O. Box 19276, Springfield, IL 62794-9276
Phone: 217-782-9691
Dennis McMurray, MANAGER: PUBLIC INFORMATION

ILLINOIS NATURE PRESERVES COMMISSION (INPC)

524 S. Second St., Lincoln Tower Plaza
Springfield, IL 62701-1787 USA
Phone: 217-785-8686   Fax: 217-785-6040
Website: www.dnr.state.il.us/inpc/index.htm

Founded: 1963
Membership: 25
Scope: Statewide

Description: The mission of the Illinois Nature Preserves Commission (INPC) is to assist private and public landowners in protecting high quality natural areas and habitats of endangered and threatened species in perpetuity, through voluntary dedication of such lands into the Illinois Nature Preserves System. The commission promotes the preservation of these significant lands, and once dedicated as nature preserves, oversees their stewardship, management, and protection.

Publications: Directory of Illinois Nature Preserves Volume 1 & 2

Keyword(s): Land Preservation, Protected Areas, National Parks

Contact(s):
Joyce O'Keefe, CHAIRPERSON
Don McFall, DEPUTY DIRECTOR FOR PROTECTION
Randy Heidorn, DEPUTY DIRECTOR FOR STEWARDSHIP
Carolyn Grosboll, DIRECTOR
Carolyn Grosboll, LEGAL COUNSEL
Jonathon Schwegman, SECRETARY
Jonathan Ellis, VICE CHAIRPERSON

ILLINOIS-INDIANA SEA GRANT COLLEGE PROGRAM

Purdue University, Department of Forestry and Natural Resources, 1200 Forest Products Bldg.
West Lafayette, IN 47907-1200 USA
Phone: 765-494-3573   Fax: 765-496-6026
Website: www.iisgcp.org

Founded: NA
Membership: 17
Scope: Regional

Contact(s):
Brian Miller, ASSOCIATE DIRECTOR
bmiller@fnr.purdue.edu
Robin Goettel, COMMUNICATIONS COORDINATOR
University of Illinois, Urbana, IL 61801
Phone: 217-333-9448
Fax: 217-333-2614
goettel@uiuc.edu
Richard Warner, INTERIM DIRECTOR
Phone: 217-333-5199

INDIANA DEPARTMENT OF ENVIRONMENTAL MANAGEMENT

Indianapolis, IN 46206 USA
Phone: 317-232-8560   Fax: 317-233-6647
Website: www.in.gov/idem

Founded: NA
Membership: 1000
Scope: Statewide

Description: The Indiana Department of Environmental Management is dedicated to conserving, protecting, enhancing, restoring, and managing Indiana's environment. We strive to fairly but vigorously enforce laws and standards; promulgate regulations consistent with the law and public policy; and promote conservation, pollution prevention, and a healthy and sustainable ecosystem. We are committed to making Indiana a cleaner, healthier place to live.

Contact(s):
Jim Mahern, ASSISTANT COMMISSIONER FOR OFFICE OF POLLUTION PREVENTION
Phone: 317-233-6658
Cynthia Collier, ASSISTANT COMMISSIONER FOR PUBLIC POLICY AND PLANNING
Phone: 317-233-5965
Janet McCabe, ASSISTANT COMMISSIONER OF AIR MANAGEMENT
Phone: 317-233-6861
Bill Divine, ASSISTANT COMMISSIONER OFFICE OF LEGAL COUNSEL
Phone: 317-233-5546
Mary Tuohy, ASSISTANT COMMISSIONER, OFFICE OF LAND QUALITY
Phone: 317-234-0337
Dana Reed-Wise, CHIEF OF STAFF
Phone: 317-233-2773
Lori Kaplan, COMMISSIONER
Phone: 317-232-8611
Timothy Method, DEPUTY COMMISSIONER FOR ENVIRONMENTAL AND REGULATORY AFFAIRS
Phone: 317-233-3706
Felicia Robinson, DEPUTY COMMISSIONER OF LEGAL AFFAIRS
Phone: 317-233-3706
frobinso@dem.state.in.us

Bruce Palin, DEPUTY COMMISSIONER, OFFICE OF LAND QUALITY
Phone: 317-233-6591
Kristin Whittington, DIRECTOR OF AGRICULTURAL RELATIONS
Phone: 317-232-8587
Ericka Seydel, DIRECTOR OF BUSINESS & LEGISLATIVE RELATIONS
Phone: 317-232-8598
Karen Terrell, DIRECTOR OF COMMUNITY RELATIONS
Phone: 317-233-6648
kterrell@dem.state.in.us
Phillip Schermerhorn, DIRECTOR OF MEDIA & COMMUNICATION SERVICES
Phone: 317-232-8560
Terry Coleman, DIRECTOR OF NORTHERN REGIONAL OFFICE
Phone: 219-245-4870
Adriane Blaesing, DIRECTOR OF NORTHWEST REGIONAL OFFICE
Phone: 219-881-6712
ablaesin@dem.state.in.us
Paula Smith, DIRECTOR OF PLANNING & ASSESSMENT
Phone: 317-233-1210
Judy Thomann, DIRECTOR OF SOUTHWEST REGIONAL OFFICE
Phone: 812-436-2570

INDIANA DEPARTMENT OF NATURAL RESOURCES

402 W. Washington St., Rm. W255B
Indianapolis, IN 46204-2748 USA
Phone: 317-232-4200
Website: www.state.in.us/dnr

Founded: NA
Scope: Statewide

Description: The DNR administers more than 100 properties throughout Indiana, comprising more than 400,000 acres. The DNR provides recreational opportunities for millions of Hoosiers and out-of-state visitors annually at its state parks, forests, reservoirs, and fish and wildlife areas. The DNR also has wide-ranging responsibilities for various programs such as maintaining the Indiana State Museum and more than a dozen historic sites throughout the state; ensuring that coal mining and reclamation of those mines take place in a manner that is in the best interests of the citizens of Indiana, and making sure that proper use is made of, and adequate protection is given to, the state's natural resources such as water, soil, forests, wildlife and historic resources.

Publications: Outdoor Indiana

Contact(s):
Jerry Miller, CHAIRMAN: LANDS AND CULTURAL RESOURCES ADVISORY COUNCIL
Phone: 317-232-4020
Michael Kiley, CHAIRMAN: NATURAL RESOURCES COMMISSION
Phone: 317-232-4020
Joseph Siener, CHAIRMAN: WATER AND RESOURCE REGULATION ADVISORY COUNCIL
Phone: 317-232-4020
Lori Kaplan, CHIEF COUNSEL
Phone: 317-232-4020
Steven Lucas, CHIEF HEARINGS OFFICER
Phone: 317-232-0156

John Costello, DEPUTY DIRECTOR: BUREAU OF LANDS AND CULTURAL RESOURCES
Phone: 317-232-4020
Paul Ehret, DEPUTY DIRECTOR: BUREAU OF MINE RECLAMATION
Phone: 317-232-4020
David Herbst, DEPUTY DIRECTOR: BUREAU OF WATER AND RESOURCE REGULATION
Phone: 317-232-4020
David Vice, DEPUTY DIRECTOR: LAW ENFORCEMENT AND ADMINISTRATION
Phone: 317-232-4020
Patrick Ralston, DIRECTOR
Phone: 317-232-4020
Thomas Barton, DIRECTOR: DIVISION OF ACCOUNTING
Phone: 317-232-4041
Gary Doxtater, DIRECTOR: DIVISION OF FISH AND WILDLIFE
Phone: 317-232-4080
Daniel Fogerty, DIRECTOR: DIVISION OF HISTORIC PRESERVATION AND ARCHAEOLOGY
Phone: 317-232-1646
James Liverett, DIRECTOR: DIVISION OF INTERNAL AUDIT
Phone: 317-232-8092
John Davis, DIRECTOR: DIVISION OF LAND ACQUISITION
Phone: 317-232-4050
Charles Walker, DIRECTOR: DIVISION OF LAW ENFORCEMENT
Phone: 317-232-4010
John Bacone, DIRECTOR: DIVISION OF NATURE PRESERVES
Phone: 317-232-4052
Jim Slutz, DIRECTOR: DIVISION OF OIL AND GAS
Phone: 317-232-4055
Emily Kress, DIRECTOR: DIVISION OF OUTDOOR RECREATION (ADMINISTERS LAND AND WATER CONSERVATION FUND)
Phone: 317-232-4070
Stephen Sellers, DIRECTOR: DIVISION OF PUBLIC INFORMATION AND EDUCATION
Phone: 317-232-4200
Mike Sponsler, DIRECTOR: DIVISION OF RECLAMATION
Phone: 812-665-2207
J. Taylor, DIRECTOR: DIVISION OF RESERVOIR MANAGEMENT
Phone: 317-232-4060
Philip Wagner, DIRECTOR: DIVISION OF SAFETY AND TRAINING
Phone: 317-232-4145
Richard Gantz, DIRECTOR: DIVISION OF STATE MUSEUM AND HISTORICAL SITES
Phone: 317-232-1637
Gerald Pagac, DIRECTOR: DIVISION OF STATE PARKS
Phone: 317-232-4124
John Simpson, DIRECTOR: DIVISION OF WATER
Phone: 317-232-4160
Mike Quigley, DIRECTOR: MANAGEMENT INFORMATION SYSTEMS
Phone: 317-232-4007
Stephen Sellers, EDITOR
Phone: 317-232-4200
Tom Hohman, HEAD CHIEF ENGINEER: DIVISION OF ENGINEERING
Phone: 317-232-4150

Robert Waltz, STATE ENTOMOLOGIST DIRECTOR: DIVISION OF ENTOMOLOGY AND PLANT PATHOLOGY
Phone: 317-232-4120
Burnell Fischer, STATE FORESTER: HEAD: DIVISION OF FORESTRY
Phone: 317-232-4105

INDIANA DEPARTMENT OF NATURAL RESOURCES

402 W. Washington St., Rm. W265
Indianapolis, IN 46204-2782 USA
Phone: 317-233-3870   Fax: 317-233-3882
Website: www.state.in.us/dnr

Founded: NA
Membership: 85
Scope: Regional

Description: The Division's mission is to facilitate the protection, wise use, and enhancement of Indiana's soil and water resources by: coordinating implementation of the state's T-by-2000 soil conservation/water quality protection program and providing assistance to local soil and water conservation districts.

Publications: Topsoil, Lake and River Enhancement Program, Urban Conservation Program, Erosion Control for the Home Builder, Indiana Handbook for Erosion Control in Developing Areas*

Keyword(s): Water Quality, Soil Conservation, Lakes, Agriculture, Rivers

Contact(s):
Peter Hippensteel, CHAIRMAN OF THE BOARD
Harry Nikides, DIRECTOR
David Avery, VICE CHAIRMAN

INDIANA GEOLOGICAL SURVEY

Institute of Indiana University, 611 N. Walnut Grove
Bloomington, IN 47405 USA
Phone: 812-855-7636   Fax: 812-855-2862
E-mail: igsinfo@indiana.edu
Website: www.indiana.edu/~igs/

Founded: 1869
Membership: 70
Scope: Statewide

Description: Conducts basic and applied research in geology and disseminates geologic information as published reports and maps; consults with industry, academia, and the public on the geologic makeup, mineral and energy resources, and geologic hazards of the state.

Publications: Geologic Publications

Keyword(s): Environmental Planning, Geology, Mineral Resources

Contact(s):
John Hill, ASSISTANT DIRECTOR
Phone: 812-855-6067
John Stenmetz, DIRECTOR AND STATE GEOLOGIST
Phone: 812-855-5067

INDIANA STATE DEPARTMENT OF HEALTH

Two North Meridian St.
Indianapolis, IN 46204 USA
Phone: 317-233-1325
E-mail: rfeldman@isdh.state.in.us
Website: www.state.in.us/isdh

Founded: NA
Scope: Statewide

Keyword(s): Toxic Substances, Nuclear-free, Water quantity, Water export and diversion, Toxicology, Nuclear/Radiation, Health and Nutrition, Public Health Protection

Contact(s):
Richard Feldman, STATE HEALTH COMMISSIONER

INDUSTRIAL SITING DIVISION/DEPARTMENT OF ENVIRONMENTAL QUALITY

State of Wyoming, 3rd Fl, E Herschler Bldg.
Cheyenne, WY 82002 USA
Phone: 307-777-4369
E-mail: VFORSE@missc.state.wy.us
Website: http://dequ.state.wy.us/

Founded: 1975

Description: Administers the Wyoming Industrial Development Information and Siting Act, which deals with the social, economic, and environmental impacts of large-scale industrial development. Responsibilities consist of investigating, reviewing, processing, and serving notice of permit applications.

Contact(s):
Gary Beach, ADMINISTRATOR
Phone: 307-777-7369

INSTITUTE FOR ECOLOGICAL STUDIES UNIVERSITY OF NORTH DAKOTA

P.O. Box 7110,
Grand Forks, ND 58202 USA
Phone: 701-777-2851   Fax: 701-777-2623

Founded: 1965
Membership: 15
Scope: Local

Description: A nonprofit university research center devoted to ecology, policy analysis, and environmental biology. A interdisciplinary staff composed of university faculty, biologists, and associates conducts basic and applied research centering in the upper Midwest, and provides technical services for government and corporate agencies, and the public.

Publications: Research Reports, Contributions

Keyword(s): training, Wetlands, Endangered Species, Environmental and Conservation Education, Biodiversity

Contact(s):
Richard Crawford, DIRECTOR

INTERAGENCY COMMITTEE FOR OUTDOOR RECREATION (IAC)

1111 Washington St., SE, P.O. Box 40917
Olympia, WA 98504-0917 USA
Phone: 360-902-3000   Fax: 360-902-3026
E-mail: info@iac.wa.gov
Website: www.wa.gov/iappleconnie

Founded: 1965
Scope: Statewide

Description: IAC administers grants and technical assistance programs for public recreation, open space, and conservation projects in Washington state. The agency assists local, state, federal, and nonprofit organizations in planning, acquiring, and developing recreation resources. IAC also writes the state's outdoor recreation and open space plan, as well as plans on trails and nonhighway off-road vehicle recreation.

Publications: Refer to Website for publication listings:www.wa.gov/iappleconnie

Keyword(s): Public Lands, Outdoor Recreation, Land Purchase, Land Use Planning, Protected Areas

Contact(s):
Debra Wilhelmi, DEPUTY DIRECTOR OF MANAGEMENT SERVICES
Phone: 360-902-3005
Laura Johnson, DIRECTOR
Phone: 360-902-3000
Gregory Lovelady, MANAGER OF APPLIED PLANNING
Phone: 360-902-3008
Jim Fox, SPECIAL ASSISTANT TO THE DIRECTOR
Phone: 360-902-3021

IOWA ASSOCIATION OF COUNTY CONSERVATION BOARDS

405 SW 3rd, Suite 1
Ankeny, IA 50021 USA
Phone: 515 9-63 -9582   Fax: 515-963-9582
E-mail: iaccb@ecity.net
Website: http://george.ecity.net/iacob

Founded: NA

Description: Promotes the objectives of Iowa's County Conservation Boards, board member education, information exchange, legislation, and public awareness.

Publications: IACCB Newsletter, IACCB Legislative Update, Outdoor Adventure Guide (Area Directory), Board Member Handbook, Iowa Board Member

Contact(s):
Don Brazelton, EXECUTIVE SECRETARY
Dan Heissel, PRESIDENT
Ann Adkins, SECRETARY
Steve Lekwa, VICE PRESIDENT

IOWA COOPERATIVE FISH AND WILDLIFE RESEARCH UNIT

Animal EcologyScience Hall II Iowa State University
Ames, IA 50011-3221 USA
Phone: 515-294-3056   Fax: 515-294-5468
E-mail: coopunit@iastate.edu
Website: www.icfwru.iastate.edu

Founded: 1932
Membership: 20
Scope: Statewide
Publications: Annual Report

Keyword(s): training, Sustainable Ecosystems, Wildlife, Gap Analysis, Biodiversity, Research

Contact(s):
Rolf Koford, ASSISTANT LEADER OF WILDLIFE
David Otis, UNIT LEADER

IOWA DEPARTMENT OF AGRICULTURE AND LAND STEWARDSHIP

BUREAU OF FIELD SERVICES
E. 9th and Grand Ave., Wallace Bldg
Des Moines, IA 50319-0034 USA
Phone: 515-281-5321
Website: http://state.ia.us/agriculture

Founded: NA

Contact(s):
James Gillespie, 515-281-5258

IOWA DEPARTMENT OF AGRICULTURE AND LAND STEWARDSHIP

BUREAU OF FINANCIAL INCENTIVE PROGRAM
E. 9th and Grand Ave., Wallace Bldg
Des Moines, IA 50319-0034 USA
Phone: 515-281-5851   Fax: 515-281-6170

Founded: NA
Scope: Statewide

Contact(s):
William McGill, BUREAU CHIEF
Phone: 515-281-5851

IOWA DEPARTMENT OF AGRICULTURE AND LAND STEWARDSHIP

BUREAU OF MINES AND MINERALS
E. 9th and Grand Ave., Wallace Bldg.
Des Moines, IA 50319-0034 USA
Phone: 515-281-6142   Fax: 515-281-6170
Website: www.state.ia.us/agriculture

Founded: NA
Membership: 30
Scope: Statewide

Contact(s):
Kenneth Tow, BUREAU CHIEF
Phone: 515-281-6142
Dean Lemke, BUREAU CHIEF
Phone: 515-281-6146
Jim Gillespie, BUREAU CHIEF, MINES & MINERALS BUREAU
Phone: 515-281-5258
Bill McGill, BUREAU CHIEF, WATER RESOUCE BUREAU

IOWA DEPARTMENT OF AGRICULTURE AND LAND STEWARDSHIP

BUREAU OF WATER RESOURCES
E. 9th and Grand Ave., Wallace Bldg
Des Moines, IA 50319-0050 USA
Phone: 515-281-5851   Fax: 515-281-6170
Website: www.state.ia.us/agriculture

Founded: NA
Membership: 5
Scope: Statewide

Contact(s):
Dean Lemke, CHIEF WATER RESOURCE BUREAU
Phone: 515-281-3963

IOWA DEPARTMENT OF AGRICULTURE AND LAND STEWARDSHIP

DIVISION OF SOIL CONSERVATION
Wallace State Office Bldg.
Des Moines, IA 50319 USA
Phone: 515-281-5851   Fax: 515-281-6170
Website: www.state.ia.us-agriculture

Founded: NA
Membership: 25
Scope: Regional

Description: Administers soil and water conservation district laws. Allocates state appropriations to 100 soil and water conservation districts for personnel, commissioners' expense, and financial incentives for erosion control and water quality measures. Reviews watershed and RC & D project applications. Oversees erosion control law. Involved with water resources and nonpoint-source pollution control planning. Division licenses mine operations and administers federal surface mining and Abandoned Mine Land Program regulations.

Contact(s):
Kenneth Tow, ASSISTANT DIRECTOR
Phone: 515-281-6142
James Gulliford, DIRECTOR
Phone: 515-281-5851
Fax: 515-281-6170
jgull@sela.osmre.gov
Russell Brandes, VICE CHAIRPERSON
37333 Mahogany Rd., Hancock, IA 51536-4012

IOWA DEPARTMENT OF NATURAL RESOURCES

E. 9th and Grand Ave., Wallace Bldg.
Des Moines, IA 50319-0034 USA
Phone: 515-281-5145   Fax: 515-281-8895
Website:
http://www.state.ia.us/government/dnr.organize/forest/forest.htm

Founded: 1986
Scope: Statewide

Description: Established with the merging of the following state agencies: Iowa Conservation Commission, Department of Water, Air and Waste Management; Iowa Geological Survey; and the resources/conservation functions of the Energy Policy Council. The seven-member Natural Resources Commission is a policy and rule-setting authority over the Fish and Wildlife Division, Parks, Recreation, and Preserves Division, and the Forestry Division; the nine-member Environmental Protection Commission is the policy and rule-setting authority over the Environmental Protection Division and the Waste Management Division.

Keyword(s): Conservation, Chemical Pollution Control, Aquatic Habitats, Endangered Species, Environment, Wildlife, Lakes, Land Purchase, Natural Areas, Energy, Geology, Hunting, Air Quality and Pollution, Birds

Contact(s):
Ross Harrison, CHIEF OF INFORMATION-EDUCATION BUREAU
Larry Wilson, DEPUTY DIRECTOR
Lyle Asell, DIRECTOR
Julie Sparks, EDITOR OF IOWA CONSERVATIONIST
William Ehm, ENVIRONMENTAL PROTECTION COMMISSION CHAIR

Joan Schneider, NATURAL RESOURCE COMMISSION CHAIR

IOWA DEPARTMENT OF NATURAL RESOURCES

ADMINISTRATIVE SERVICES DIVISION
502 E. 9th St, Wallace Star Office Building
Des Moines, IA 50319-0034 USA
Phone: 515-281-5918
E-mail: webmaster@dnr.state.ia.us
Website: http://state.ia.us/dnr

Founded: NA

Contact(s):
Linda Hanson, ADMINISTRATOR
Sally Jagnandan, CHIEF OF ADMINISTRATIVE SUPPORT BUREAU
Mark Slatterly, CHIEF OF BUDGET AND FINANCE BUREAU
Basil Nimry, CHIEF OF CONSTRUCTION SERVICES BUREAU
Larry Bartleman, CHIEF OF LAND ACQUISITION AND MANAGEMENT BUREAU (ACTING)
Judy Pawell, CHIEF OF LICENSING BUREAU AND DATA PROCESSING BUREAU

IOWA DEPARTMENT OF NATURAL RESOURCES

COOPERATIVE NORTH AMERICAN SHOTGUNNING EDUCATION PROGRAM
Wallace State Office Bldg.
Des Moines, IA 50319 USA
Phone: 515-281-5918   Fax: 503-884-2974
Website: www.state.ia.us/dnr

Founded: 1982
Scope: Regional

Description: The Cooperative North American Shotgunning Program is a research, information, and education program designed to assist wildlife professionals, hunters, and sportsmen in making a successful transition from lead shot to nontoxic shot, as well as educating sportsmen on improving shooting skills and harvest efficiency, thereby reducing wounding losses.

Publications: Periodic Ballistics Reports, CONSEP Newsletter

Keyword(s): Waterfowl, Hunting

Contact(s):
Tom Roster, CONSULTANT
1190 Lynnewood Blvd., Klamath Falls, OR 97601
Phone: 503-884-2974
Lloyd Alexander, CONTACT FOR ATLANTIC FLYWAY
89 Kings Highway, P.O. Box 1401, Dover, DE 302-739-5287
Bob McLean, CONTACT FOR CANADIAN WILDLIFE SERVICE
17th Fl., Place Vincent Massey, Ottawa, Ontario K1A 0H3
Phone: 819-997-2957
George Vandel, CONTACT FOR CENTRAL FLYWAY
445 E. Capitol, Pierre, SD 605-773-3381
John Smith, CONTACT FOR MISSISSIPPI FLYWAY
P.O. Box 180, Jefferson City, MO 65102-0180
Phone: 573-751-4115
Don Childress, CONTACT FOR PACIFIC FLYWAY
1420 E. 6th, Box 20071, Helena, MT 59601
Phone: 406-444-2612
Jeffrey Zonk, DIRECTOR
Phone: 515-281-5918
jeff.zonk@dnr.state.ia.us

Richard Bishop, WILDLIFE BUREAU CHIEF
Wallace State Bldg., Des Moines, IA 50319
Phone: 515-281-6156
richard.bishop

IOWA DEPARTMENT OF NATURAL RESOURCES
ENERGY AND GEOLOGICAL RESOURCES DIVISION
Wallace State Office Building
East 9 & Grand Avenue
Des Moines, IA 50319 USA
Phone: 515-281-4308

Founded: NA

Contact(s):
Larry Bean, ADMINISTRATOR
Sharon Tahtinen, CHIEF OF ENERGY BUREAU
Don Koch, CHIEF OF GEOLOGICAL SURVEY BUREAU

IOWA DEPARTMENT OF NATURAL RESOURCES
ENVIRONMENTAL PROTECTION DIVISION
Wallace State Office Building
East 9 & Grand Avenue
Des Moines, IA 50319 USA
Phone: 515-281-8694

Founded: NA

Contact(s):
Michael Valde, ADMINISTRATOR
Pete Hamlin, CHIEF OF AIR QUALITY BUREAU
Mike Murphy, CHIEF OF COMPLIANCE & ENFORCEMENT
BUREAU
Jack Riessen, CHIEF OF WATER QUALITY BUREAU

IOWA DEPARTMENT OF NATURAL RESOURCES
FISH AND WILDLIFE DIVISION
Wallace State Office Building
East 9 & Grand Avenue
Des Moines, IA 50319 USA
Phone: 515-281-4687   Fax: 515-281-6794

Founded: NA

Contact(s):
Allen Farris, ADMINISTRATOR
Marion Conover, CHIEF OF FISHERIES BUREAU
Lowell Joslin, CHIEF OF LAW ENFORCEMENT BUREAU
Richard Bishop, CHIEF OF WILDLIFE BUREAU

IOWA DEPARTMENT OF NATURAL RESOURCES
FORESTS AND PRAIRIES DIVISION
Wallace State Office Building
East 9 & Grand Avenue
Des Moines, IA 50319 USA
Phone: 515-281-8657
E-mail: mike.brandrup@dnr.state.ia.us

Founded: NA

Contact(s):
Mike Brandrup, ADMINISTRATOR
John Walkowiak, CHIEF OF FORESTRY SERVICES
BUREAU
Jim Bulman, CHIEF OF STATE FORESTS MANAGEMENT
BUREAU

IOWA DEPARTMENT OF NATURAL RESOURCES
PARKS
Wallace State Office Building
East 9 & Grand Avenue
Des Moines, IA 50319 USA
Phone: 515-281-8563
E-mail: janet.ott@dnr.state.ia.us

Founded: NA

Contact(s):
Mike Carrier, ADMINISTRATOR
Stephen Pennington, CHIEF OF FIELD OPERATIONS
BUREAU
Arnie Sohn, CHIEF OF PROGRAM ADMINISTRATION
BUREAU

IOWA DEPARTMENT OF NATURAL RESOURCES
WASTE MANAGEMENT DIVISION
Wallace State Office Building
East 9 & Grand Avenue
Des Moines, IA 50319 USA
Phone: 515-281-4367   Fax: 515-281-8895
E-mail: teresa.barrie@dnr.state.ia.us

Founded: NA

Contact(s):
Roya Stanley, ADMINISTRATOR
Brent Laning, EXECUTIVE OFFICER

IOWA STATE EXTENSION SERVICES
1099 Court Avenue PO Box 146
Marengo, IA 52301 USA
Phone: 319-642-5504   Fax: 319-642-5505
E-mail: xiowa@iastate.edu
Website: www.extension.iastate.edu

Founded: NA
Scope: Statewide

Description: ISU Extension is a client-centered organization that
provides research-based, unbiased information and education
to help people make better decisions in their personal,
community, and professional lives.

Contact(s):
Paul Wray, EXTENSION FORESTER
251 Bessey Hall, Iowa State University, Ames, IA 50011
Phone: 515-294-1168
James Pease, EXTENSION WILDLIFE CONSERVATIONIST
103 Science II, Iowa State University, Ames, IA 50011
Phone: 515-294-7429
Fax: 515-294-7874

K

KANSAS BIOLOGICAL SURVEY
2021 Constant Ave. Foley Hall
Lawrence, KS 66047-2729 USA
Phone: 785-864-7725   Fax: 785-864-5093
Website: www.kbs.ukans.edu

Founded: 1959
Membership: 60
Scope: Statewide

Description: A research and development branch of the
University of Kansas whose purpose is to survey and inventory
the native plants and animals of Kansas, report on its findings,

and develop and administer lands for the study and preservation of native animal and plant resources.

Publications: Publications on website

Keyword(s): Sustainable Ecosystems, Prairies, Biodiversity, Aquatic Habitats, Endangered Species

Contact(s):
Paul Liechti, ASSISTANT DIRECTOR
Frank Denoyelles, ASSOCIATE DIRECTOR
Edward Martinko, DIRECTOR AND STATE BIOLOGIST

KANSAS COOPERATIVE FISH AND WILDLIFE RESEARCH UNIT

205 Leasure Hall, Kansas State University
Manhattan, KS 66506-3501 USA
Phone: 785-532-6070   Fax: 785-532-7159
Website: http://www.ksu.edu/kscfwru/

Founded: 1991
Scope: National

Contact(s):
Christopher Guy, ASSISTANT LEADER OF FISHERIES
Jack Cully, ASSISTANT LEADER OF WILDLIFE
Philip Gipson, LEADER

KANSAS DEPARTMENT OF AGRICULTURE

109 SW 9th St., 2nd Fl.
Topeka, KS 66612-1280 USA
Phone: 785-296-3717   Fax: 785-296-1176
Website: http://www.ink.org/public/kda/dwr

Founded: NA
Membership: 70
Scope: Statewide

Keyword(s): Wetlands, Exotic species, Aquatic nuisance species, Environmental and Conservation Education, Agriculture, Renewable Resources

Contact(s):
David Pope, CHIEF ENGINEER
Phone: 785-296-3717
Steve Stankiewicz, OPERATIONS MANAGER FOR WATER RESOURCES DIVISION
Jamie Adams, SECRETARY OF AGRICULTURE
Phone: 785-296-3902
Fax: 785-296-8389
Tom Huntzinger, WATER APPROPRIATIONS PROGRAM MANAGER

KANSAS DEPARTMENT OF HEALTH AND ENVIRONMENT

1000 SW Jackson Blvd
Topeka, KS 66612-1367 USA
Phone: 785-296-1535   Fax: 785-296-8464
Website: www.kdhe.state.ks.us

Founded: NA
Membership: 700
Scope: Statewide

Description: The Kansas Department of Health and Environment is responsible for administering a diverse collection of programs that enhance public health and state wildlife protection efforts. The path of the department is defined by strengthening programs and developing initatives on pollutant releases, spill cleanup, air and water quality, water resources, pollution prevention, waste management, and general health and environmental protection.

Keyword(s): Exotic species, Aquatic nuisance species, Solid Waste, Health and Nutrition, Environment, Water Quality, Pollution Prevention, Air Quality and Pollution, Public Health Protection

Contact(s):
Theresa Hodges, DIRECTOR: BUREAU OF ENVIRONMENTAL FIELD SERVICES
Phone: 785-296-6603
Michael Moser, DIRECTOR: BUREAU OF ENVIRONMENTAL HEALTH SERVICES
Phone: 785-296-1343
Gary Blackburn, DIRECTOR: BUREAU OF ENVIRONMENTAL REMEDIATION
Phone: 785-296-1660
Bill Bider, DIRECTOR: BUREAU OF WASTE MANAGEMENT
Karl Mueldener, DIRECTOR: BUREAU OF WATER
Phone: 785-296-5500
Ron Hammerschmidt, DIRECTOR: DIVISION OF ENVIRONMENT
Phone: 785-296-1535
Sharon Watson, DIRECTOR: OFFICE OF PUBLIC INFORMATION
Phone: 785-296-5795
Clyde Graber, SECRETARY
Phone: 785-296-0461

KANSAS DEPARTMENT OF WILDLIFE AND PARKS

900 SW Jackson St., Suite 502
Topeka, KS 66612 USA
Phone: 785-296-2281   Fax: 785-296-6953
Website: www.kdwp.state.ks.us

Founded: NA
Membership: 17
Scope: State

Description: Charged with the conservation of state wildlife and fishery resources, provision of environmental services and habitat protection, and park development and management. Administers state boating law, hunter education programs, Land and Water Conservation Funds, and other related functions.

Publications: Kansas Wildlife & Parks Magazine

Keyword(s): Birds, Wildlife Rehabilitation, Aquatic Habitats, Communications, Conservation Tillage, Ecology, Wildlife, Hunting, Lakes, Conservation, Endangered Species, Environment, Agriculture, Biodiversity

Contact(s):
Richard Koerth, ASSISTANT SECRETARY FOR ADMINISTRATION
Cheryl Swayne, BOATING EDUCATION (TOPEKA OFFICE)
John Dykes, COMMISSIONER MEMBERS CHAIRMAN
Phone: 913-831-3058
Terry Denker, FEDERAL AID COORDINATOR
Steven Williams, SECRETARY

KANSAS DEPARTMENT OF WILDLIFE AND PARKS

1001 W. McArtor
Dodge City, KS 67801 USA
Phone: 316-227-8609   Fax: 620-227-8600
Website: www.kdwp.state.ks.us

Founded: NA

Membership: 8
Scope: Statewide

KANSAS DEPARTMENT OF WILDLIFE AND PARKS

3300 SW 29th St.
Topeka, KS 66614 USA
Phone: 785-273-6740   Fax: 785-273-6757
Website: www.wp.state.ks.us

Founded: NA
Scope: State

KANSAS DEPARTMENT OF WILDLIFE AND PARKS

OPERATIONS OFFICE
512 SE 25th Ave.
Pratt, KS 67124-8174 USA
Phone: 316-672-5911   Fax: 316-672-6020
E-mail: seedback@wp.state.ks.us
Website: www.kdwp.state.ks.us

Founded: NA
Membership: 395
Scope: Statewide
Publications: Kansas Wildlife and Parks

Keyword(s): Sport Fishing, Outdoor Recreation, Wildlife, Public
Lands

Contact(s):
Mike Theurer, ADMINISTRATIVE SERVICES DIVISION
Keith Sexson, ASSISTANT SECRETARY FOR OPERATIONS
Wayne Doyle, COORDINATOR OF HUNTER EDUCATION
AND FUR HARVESTER EDUCATION SECTIONS
Roland Stein, COORDINATOR OF WILDLIFE EDUCATION
SERVICE
Kevin Jones, DIRECTOR OF LAW ENFORCEMENT
Mike Miller, EDITOR
Joe Kramer, FISHERIES AND WILDLIFE DIVISION
Bob Mathews, INFORMATION AND EDUCATION
Jerold Hover, PARKS DIVISION
Bob Mathews, PUBLIC INFORMATION

KANSAS DEPARTMENT OF WILDLIFE AND PARKS

REGION 1
1426 Hwy. 183 Alt.
Hays, KS 67601 USA
Phone: 785-628-8614   Fax: 785-623-2945
Website: www.kdwp.state.ks.us

Founded: NA
Scope: Statewide

Publications: Kansas Wildlife and Parks

Contact(s):
Mary Jane Pfannenstiel, SECRETARY

KANSAS DEPARTMENT OF WILDLIFE AND PARKS

REGION 4
6232 E. 29th St., N Region 4
Wichita, KS 67220 USA
Phone: 316-683-8069   Fax: 316-683-4664
Website: www.kdwp.state.ks.us

Founded: NA

Membership: 20
Scope: State

KANSAS DEPARTMENT OF WILDLIFE AND PARKS

REGION 5
1500 W. 7th, P.O. Box 777
Chanute, KS 66720-0777 USA
Phone: 316-431-0380   Fax: 620-431-0381
Website: www.kdwp.state.ks.us

Founded: NA
Membership: 7
Scope: Statewide

Contact(s):
Victoria Payne, REGIONAL SECRETARY

KANSAS FOREST SERVICE

2610 Claflin Rd.
Manhattan, KS 66502-2798 USA
Phone: 785-532-3300   Fax: 785-532-3305
E-mail: kfs@lists.osnet.ksu.edu
Website: www.kansasforests.org

Founded: NA
Membership: 21
Scope: Statewide

Description: Provides technical forestry assistance to
landowners, wood industries, and communities; conducts a
tree distribution program, and a rural fire protection program.

Keyword(s): Exotic species, Aquatic nuisance species, Urban
Forestry, Forests and Forestry, Forest Management, Water
Quality, Environmental and Conservation Education,
Renewable Resources

Contact(s):
William Loucks, CONSERVATION FORESTER
Casey McCoy, FIRE MANAGER
Robert Atchison, RURAL FORESTRY COORDINATOR
Raymond Aslin, STATE FORESTER

KANSAS GEOLOGICAL SURVEY

1930 Constant Ave., Campus West, Kansas University
Lawrence, KS 66047 USA
Phone: 785-864-3965   Fax: 785-864-5317
Website: www.kgs.ku.edu

Founded: 1889
Membership: 150
Scope: Statewide

Description: Purpose is to research and develop information
about minerals, water resources, and geologic hazards of
Kansas, and to publish reports on those subjects.

Publications: journals, maps, technical series, educational
series, public information circulars, Bulletin

Keyword(s): Exotic species, Aquatic nuisance species, Coral
Reefs, Environmental and Conservation Education, Energy,
Geology

Contact(s):
Don Whittemore, CHIEF: GEOHYDROLOGY
Pieter Berendsen, CHIEF: GEOLOGIC INVESTIGATIONS
John Davis, CHIEF: MATHEMATICAL GEOLOGY
Timothy Carr, CHIEF: PETROLEUM RESEARCH
William Harrison, DEPUTY DIRECTOR

M. Allison, DIRECTOR AND STATE GEOLOGIST
lallison@kgs.ku.edu

KANSAS STATE CONSERVATION COMMISSION

109 SW Ninth St.
Topeka, KS 66612-1215 USA
Phone: 785-296-3600　　Fax: 785-296-6172
E-mail: tstreeter@scc.state.ks.us
Website: www.ink.org/public/kscc

Founded: 1937
Membership: 15
Scope: Statewide

Description: The SCC administrative responsibility is to provide leadership, direction, and support to the conservation districts, watershed districts, and other special purpose districts for the protection and enhancement of Kansas' natural resources. It administers a total of ten programs: seven are financial assistance programs funded by appropriations from the Special Revenue Fund of the State Water Plan.

Keyword(s): Wetlands, Exotic species, Aquatic nuisance species, Soil Conservation, Water Quality, Watersheds, Environmental and Conservation Education, Coral Reefs

Contact(s):
Tracy Streeter, EXECUTIVE DIRECTOR
Phone: 913-296-3600
Fax: 913-296-6172

KANSAS STATE EXTENSION SERVICES

Wildlife Damage Control, Department of Animal Sciences and Industry, 131 Call Hall, Kansas State University
Manhattan, KS 66506-1600 USA
Phone: 785-532-5734　　Fax: 785-532-5681
Website: www.oznet.ksu.edu

Founded: 1914
Scope: Local, State, Regional.

Keyword(s): Youth Organizations, Trapping, Predators, Wetlands, Hunting, training

Contact(s):
Charles Lee, EXTENSION SPECIALIST
clee@oznet.ksu.edu

KANSAS WATER OFFICE

901 S. Kansas Ave.
Topeka, KS 66612-1249 USA
Phone: 785-296-3185　　Fax: 785-296-0878
Website: http://www.kwo.org/

Founded: NA

Description: State water planning, policy, and coordination agency. Prepares state plan of water resources management; conservation; fish and wildlife, and recreation and development; reviews water laws, and recommends new or amendatory legislation. Administers the state water monitoring program.

Keyword(s): Exotic species, Aquatic nuisance species, Sustainable Development, Planning Management, Lakes, Rivers

Contact(s):
Clark Duffy, ASSISTANT DIRECTOR
Al Ledoux, DIRECTOR

KENTUCKY DEPARTMENT OF AGRICULTURE

7th Fl., 500 Mero St.
Frankfort, KY 40601 USA
Phone: 502-564-4696　　Fax: 502-564-2133
Website: www.kyagr.com

Founded: 1876
Scope: Statewide

Description: The service, regulatory, and promotional agency for Kentucky's agriculture industry.

Publications: see publications on website, Kentucky Agricultural Statistics (Yearly), Kentucky Agricultural News

Keyword(s): Wetlands, Soil Conservation, Environmental and Conservation Education, Agriculture, Pesticides

Contact(s):
Billy Smith, COMMISSIONER
Doug Thomas, DIRECTOR: DIVISION OF COMMUNICATIONS
Phone: 502-564-4696
Ted Sloan, EDITOR
Phone: 502-564-4696

KENTUCKY DEPARTMENT OF FISH AND WILDLIFE RESOURCES

#1 Game Farm Rd.
Frankfort, KY 40601 USA
Phone: 800-858-1549　　Fax: 502-564-6508
E-mail: info.center@mail.state.ky.us
Website: http://kdfwr.state.ky.us/

Founded: 1944

Description: We are stewards of Kentucky's fish and wildlife resources and their habitats. We manage for the perpetuation of these resources and their use by present and future generations. Through partnerships, we will enhance wildlife diversity and promote sustainable use, including hunting, fishing, boating, and other nature-related recreation.

Publications: Kentucky Afield Magazine, Kentucky Fish, Hunting and Fishing Regulation Guides, Kentucky Wildlife Viewing Guide

Keyword(s): training, Environmental and Conservation Education, Biodiversity, Wildlife

Contact(s):
C. Bennett, COMMISSIONER
Don Walker, COORDINATOR OF PITTMAN-ROBERTSON SECTION
James Axon, COORDINATOR OF SPORT FISH RESTORATION SECTION
Thomas Young, DEPUTY COMMISSIONER
Robert Bates, DIRECTOR OF ADMINISTRATIVE SERVICES DIVISION
Peter Pfeiffer, DIRECTOR OF DIVISION OF FISHERIES
Lee Carolan, DIRECTOR OF DIVISION OF INFORMATION AND EDUCATION
David Loveless, DIRECTOR OF DIVISION OF LAW ENFORCEMENT
Roy Grimes, DIRECTOR OF DIVISION OF WILDLIFE
Charles Bush, DIRECTOR OF ENGINEERING DIVISION
Lynn Garrison, DIRECTOR OF PUBLIC AFFAIRS DIVISION
Mike Boatwright, DISTRICT 1 COMMISSIONER
2601 N. 10th St., Paducah, KY 42001
Tom Baker, DISTRICT 2 COMMISSIONER
661 A U.S. 31 W. By-Pass, Bowling Green, KY 42101

Allen Gailor, DISTRICT 3 COMMISSIONER
730 W. Market, Louisville, KY 40202
Charles Bale, DISTRICT 4 COMMISSIONER
855 Parkers Grove Rd., Hodgenville, KY 42748
James Rich, DISTRICT 5 COMMISSIONER
5975 Taylor Mill Rd., Covington, KY 41015
Frank Brown, DISTRICT 6 COMMISSIONER
124 Lancaster Ave., Richmond, KY 40475
Phone: 606-624-0820
Doug Hensley, DISTRICT 7 COMMISSIONER
P.O. Box 480, Hazard, KY 41701
Phone: 606-436-5180
Robert Webb, DISTRICT 8 COMMISSIONER
45 Webb Circle, Grayson, KY 41143
David Godby, DISTRICT 9 COMMISSIONER
P.O. Box 1277, Somerset, KY 42502
Phone: 606-677-0115
Dennis Watson, REGIONAL BOATING SUPERVISOR
Route 1 Sand Knob, Falls of Rough, KY 40119
K. Henderson, REGIONAL BOATING SUPERVISOR
P.O. Box 131, Clarkson, KY 42726
Reed Sanders, REGIONAL BOATING SUPERVISOR
185 Gwinn Island Circle, Danville, KY 40422
Steve Owens, REGIONAL BOATING SUPERVISOR
338 Candlelite Drive, Almo, KY 42020
Gerald Alexander, REGIONAL LAW ENFORCEMENT
SUPERVISOR
Phone: 6575 Beech Grove Rd., Farmington, KY 42040
John Akers, SUPERINTENDENT OF STATE GAME FARM

KENTUCKY DEPARTMENT OF PARKS

10th Fl., Capital Plaza Tower
Frankfort, KY 40601 USA
Phone: 502-564-2172　　　Fax: 502-564-9015
Website: www.kystateparks.com

Founded: NA
Membership: 2500
Scope: Regional

Publications: Kentucky Hiking Trails, Kentucky State Parks
Booklet

Keyword(s): Outdoor Recreation, Land Preservation, Public
Lands

Contact(s):
Kenny Rapier, COMMISSIONER
Bob Bender, DEPUTY COMMISSIONER
Jim Goodman, DIRECTOR OF RESORT PARKS
Danny Reed, DIRECTOR: RANGERS
Carey Tichenor, STATE NATURALIST

KENTUCKY GEOLOGICAL SURVEY

228 Mining and Mineral Resources Bldg., University of
Kentucky
Lexington, KY 40506-0107 USA
Phone: 859-257-5500　　　Fax: 859-257-1147
Website: www.uky.edu/kgs

Founded: 1854
Membership: 70
Scope: Statewide

Description: Investigates the geology and mineral and water
resources of Kentucky and makes this information available to
the public. It is a research and service organization.

Publications: see publication website

Contact(s):
John Kiefer, ASSISTANT STATE GEOLOGIST FOR
ADMINISTRATION
kiefer@kgs.mm.uky.edu
James Cobb, DIRECTOR AND STATE GEOLOGIST
cobb@kgs.mm.uky.edu
Steven Cordiviola, HEAD: COMPUTER AND LABORATORY
SERVICES SECTION
cordiviola@kgs.mm.uky.edu
James Drahovzal, HEAD: ENERGY & MINERAL SECTION
drahovzal@kgs.mm.uky.edu
James Dinger, HEAD: WATER RESOURCES SECTION
dinger@kgs.mm.uky.edu

KENTUCKY SOIL AND WATER CONSERVATION COMMISSION

663 Teton Trail
Frankfort, KY 40601 USA
Phone: 502-564-3080　　　Fax: 502-564-9195
E-mail: coleman@mail.nr.state.ky.us

Founded: NA

Contact(s):
David Gerrein, CHAIR
Phone: 606-623-3960
Stephen Coleman, DIRECTOR OF DIVISION OF
CONSERVATION
Phone: 502-564-3080
Fax: 502-564-9195

KENTUCKY STATE COOPERATIVE EXTENSION SERVICES

307 WP Garrigus Building
Lexington, KY 40546-0091 USA
Phone: 859-257-4302　　　Fax: 859-323-1991
Website: www.ca.uky.edu/coopext/

Founded: NA
Membership: 500
Scope: Statewide
Publications: Extension Today

Contact(s):
Rick Maurer, ASSISTANT DIRECTOR, RURAL AND
ECONOMIC DEVELOPMENT
500 Garrigus Bldg., University of Kentucky, Lexington, KY
40546-0215
Phone: 606-257-7585
rmaurer@ca.uky.edu
Curtis Absher, ASSISTANT EXTENSION DIRECTOR OF
AGRICULTURE
309 Garrigus Bldg., University of Kentucky, Lexington, KY
40546
Phone: 606-257-1846
cabsher@ca.uky.edu
Paul Warner, INTERIM ASSOCIATE DIRECTOR OF
EXTENSION SERVICE
wwalla@ca.uky.edu
Thomas Barnes, WILDLIFE SPECIALIST
Univ. of Kentucky, Dept. of Forestry, Lexington, KY
40546-0073
Phone: 606-257-8633
Fax: 606-323-1031

KENTUCKY STATE NATURE PRESERVES COMMISSION

801 Schenkel Ln.
Frankfort, KY 40601 USA
Phone: 502-573-2886   Fax: 502-573-2355
E-mail: nrepc.ksnpcemail@mail.state.ky.us
Website: www.kynaturepreserves.org

Founded: 1976
Membership: 23
Scope: Statewide

Description: KSNP's mission is to protect Kentucky's natural heritage by (1) identifying, acquiring, and managing natural areas that represent the best known occurrences of rare native species, natural communitites, and significant natural features in a statewide nature preserve system; (2) working with others to protect biological diveristy; and (3) educating Kentuckians as to the value and purpose of nature preserves and biodiversity conservation.

Publications: Naturally Kentucky - Quarterly Newsletter

Contact(s):
Clara Wheatley, CHAIR OF COMMISSION
Phone: 502-358-8643
Donald Dott, DIRECTOR
Phone: 502-573-2886
Fax: 502-573-2355
don.dott@mail.state.ky.us
Ken Jackson, SECRETARY OF COMMISSION
Phone: 606-734-4436

L

LEE COUNTY PARKS AND RECREATION SERVICES

Regional Park Program Office, 7330 Gladiolus Dr.
Fort Myers, FL 33908 USA
Phone: 941-432-2004   Fax: 941-432-2032
Website: http://www.lee-county.com/parksandrec/

Founded: 1990

Description: To promote and develop environmental awareness in Southwest Florida by conducting educational programs which teach ecological concepts and outdoor skills, and by coordinating informational events which alert citizens and community leaders of environmental concerns.

Publications: Explorers Companion Brochure, Elements Newsletter

Keyword(s): Protected Areas, Wetlands, training, Biodiversity, EcoAction, Public Lands, National Parks, Environmental Protection, Birds, Conservation, Endangered Species, Energy Conservation, Land Preservation, Environmental and Conservation Education, Outdoor Recreation

Contact(s):
John Kiseda, ENVIRONMENTAL EDUCATOR
kisedajb@leegov.com
Mary Rude, INTERPRETIVE NATURALIST
rudeme@leegov.com
Nancy Macphee, OUTDOOR RECREATION SPECIALIST
macpheen@leegov.com

LOUISIANA COOPERATIVE FISH AND WILDLIFE RESEARCH UNIT (USDI)

U.S. Geological Survey, School of Forestry, Wildlife and Fisheries, FWF Building, Rm. 124, Louisiana State University
Baton Rouge, LA 70803-6202 USA
Phone: 225-388-4179   Fax: 225-388-4144

Founded: NA
Membership: 3
Scope: National

Contact(s):
Alan Afton, ASSISTANT LEADER
Phone: 225-388-4212
aafton@lsu.edu
Megan Lapeyre, ASSITANT LEADER FISHERIES
Phone: 225-578-4180
Charles Bryan, LEADER
Phone: 225-388-4184
Fax: 225-388-4144
cbryan@lsu.edu

LOUISIANA DEPARTMENT OF AGRICULTURE AND FORESTRY

P.O. Box 631
Baton Rouge, LA 70821-0631 USA
Phone: 225-922-1234   Fax: 225-922-1253
E-mail: info@ldaf.state.la.us
Website: http://www.ldaf.state.la.us/

Founded: NA

Contact(s):
Skip Rhorer, ASSISTANT COMMISSIONER: OFFICE OF MANAGEMENT AND FINANCE
Bob Odom, COMMISSIONER
Bud Courson, DEPUTY COMMISSIONER
Phone: 504-922-1238
Fax: 504-922-1253

LOUISIANA DEPARTMENT OF AGRICULTURE AND FORESTRY

OFFICE OF FORESTRY
Office of Forestry
Baton Rouge, LA 70821-1628 USA
Phone: 225-925-4500   Fax: 225-922-1356
Website: www.ldaf.state.la.us

Founded: 1944
Membership: 17
Scope: Statewide

Description: Charged with: Detection and suppression of wildfire on forest lands; providing technical management assistance to forest landowners; and dissemination of materials and information for education of the public. Produces approximately 50 million seedlings annually (pine and hardwood) for Louisiana landowners, operates a 400-acre seed orchard that produces improved slash and loblolly pine seed that are genetically improved. Actively engaged in promoting urban forestry activities.

Publications: Publications on website

Keyword(s): Renewable Resources, Outdoor Recreation, Forest Management, Sustainable Ecosystems, Urban Forestry, Environmental and Conservation Education

Contact(s):
Cyril Lejeune, ASSOCIATE STATE FORESTER
Phone: 225-952-8002
cyril_l@ldaf.state.la.us
Burton Weaver, CHAIRMAN
Donald Feduccia, CHIEF: FOREST MANAGEMENT
Phone: 225-925-8010
Louis Heaton, CHIEF: FOREST PROTECTION
Phone: 225-952-8013
louis_h@ldaf.state.la.us
Charles Matherne, CHIEF:REFORESTATION
Phone: 225-925-4515
Paul Frey, STATE FORESTER
Phone: 225-952-8002
paul_f@ldaf.state.la.us

LOUISIANA DEPARTMENT OF AGRICULTURE AND FORESTRY

OFFICE OF SOIL AND WATER CONSERVATION, STATE
SOIL AND WATER CONSERVATION COMMITTEE
P.O. Box 3554
Baton Rouge, LA 70821-3554 USA
Phone: 225-922-1269   Fax: 225-922-2577

Founded: 1938

Description: To assist soil and water conservation districts in
carrying out their conservation programs, to coordinate
activities among districts, and to secure the cooperation and
assistance of state and federal agencies in the work of such
districts.

Contact(s):
Pedro Angelle, CHAIRMAN
4879 Main Hwy., St. Martinville, LA 70582
Phone: 318-332-2910
Fax: 318-332-6563
Bradley Spicer, EXECUTIVE DIRECTOR
Phone: 504-922-1269
Fax: 504-922-2577
A. Allee, SECRETARY AND TREASURER
Thad Spurlock, VICE CHAIRMAN

LOUISIANA DEPARTMENT OF NATURAL RESOURCES

P.O. Box 94396
Baton Rouge, LA 70804 USA
Phone: 225-342-4500   Fax: 225-342-4471
Website: www.sunrise.com

Founded: NA
Membership: 250
Scope: Regional

Contact(s):
Katherine Vaughan, ASSISTANT SECRETARY
Phone: 225-342-1375
Fax: 225-342-5861

LOUISIANA DEPARTMENT OF NATURAL RESOURCES

625 N. 4th St.
Baton Rouge, LA 70802 USA
Phone: 225-342-8955   Fax: 224-342-3442
Website: www.dnr.state.la.us

Founded: NA
Scope: Statewide

Contact(s):
Katherine Vaughan, DEPUTY SECRETARY
Phone: 225-342-1375
Jack Caldwell, SECRETARY
Phone: 225-342-4503
Fax: 225-342-5861
Robert Harper, UNDERSECRETARY

LOUISIANA DEPARTMENT OF NATURAL RESOURCES

OFFICE OF CONSERVATION
P.O. Box 94275
Baton Rouge, LA 70804-9275 USA
Phone: 225-342-5540   Fax: 225-342-3705
Website: http://www.dnr.state.la.us/cons/conserv.ssi

Founded: 1901

Contact(s):
Philip Asprodites, COMMISSIONER OF CONSERVATION

LOUISIANA DEPARTMENT OF NATURAL RESOURCES

OFFICE OF MINERAL RESOURCES
625 North Fourth Street, P.O. Box 2827
Baton Rouge, LA 70804 USA
Phone: 225-342-4615
Website: http://www.dnr.state.la.us

Founded: NA

Contact(s):
Gus Rodemacher, ASSISTANT SECRETARY

LOUISIANA DEPARTMENT OF WILDLIFE AND FISHERIES

P.O. Box 98000
Baton Rouge, LA 70898-9000 USA
Phone: 225-765-2800
Website: www.wlf.state.la.us

Founded: 1872
Scope: Statewide

Description: Established as a part of state government to protect,
conserve, and replenish the natural resources of the state,
including wild game and nongame quadrupeds or animals,
oysters, fish, and other aquatic life.

Publications: Louisiana Conservationist

Keyword(s): training, Fish Wildlife Management

Contact(s):
Winton Vidrine, ADMINISTRATOR: COLONEL: LAW
ENFORCEMENT DIVISION
Phone: 225-765-2989
Brandt Savioe, ADMINISTRATOR: FUR & REFUGE
DIVISION
Phone: 225-765-2811
Bennie Fontenot, ADMINISTRATOR: INLAND FISHERIES
DIVISION
Phone: 225-765-2330
Karen Foote, ADMINISTRATOR: MARINE FISHERIES
DIVISION
Phone: 225-765-2384
Tommy Prickett, ADMINISTRATOR: WILDLIFE DIVISION
Phone: 225-765-2346

John Roussel, ASSISTANT SECRETARY: OFFICE OF FISHERIES
Phone: 225-765-2801
Phil Bowman, ASSISTANT SECRETARY: OFFICE OF WILDLIFE
Phone: 225-765-2806
Bill Busbice, CHAIRMAN
James Jenkins, SECRETARY
Phone: 225-765-2623
James Patton, UNDERSECRETARY: OFFICE OF MANAGEMENT AND FINANCE
Phone: 225-765-2860

LOUISIANA DEPT OF WILDLIFE & FISHERIES
WILDLIFE
P. O. Box 98000
Baton Rouge, LA 70898 USA
Phone: 225-765-2800
Website: www.wlf.state.la.us

Founded: NA
Membership: 250
Scope: Statewide

Description: The State Office of Conservation is an oil and gas regulatory agency.

Keyword(s): Transportation, Pollution Prevention, Engineering, Energy, Geology

Contact(s):
James Jenkins, SECRETARY

LOUISIANA GEOLOGICAL SURVEY
P.O. Box G, University Station
Baton Rouge, LA 70893 USA
Phone: 225-578-5320   Fax: 225-578-3662
E-mail: hammer@isu.edu
Website: www.lgs.lsu.edu

Founded: 1934

Description: The Survey is charged with conducting geologic investigations and preparing technical reports that assist in finding and developing new reserves of natural resources in the state and in protecting the state's environment.

Keyword(s): Wetlands, Exotic species, Aquatic nuisance species, Energy, Coasts, Geology

Contact(s):
Chacko John, DIRECTOR

LOUISIANA SEA GRANT COLLEGE PROGRAM
Louisiana State University
Baton Rouge, LA 70803 USA
Phone: 225-388-6710   Fax: 225-578-6331
Website: www.laseagrant.org/

Founded: 1968
Scope: National

Description: The Louisiana Sea Grant College Program is a research, education, and public service organization supported by federal, state, and private sector funds. The Program provides the knowledge, trained personnel, and public awareness needed to wisely and effectively develop and manage coastal and marine areas and resources in a manner that will assure sustainable economic and societal benefits.

Publications: Coast and Sea: Marine and Coastal Research in Louisianas Universities

Keyword(s): Wetlands, Sustainable Development, Environmental and Conservation Education, Coasts, Renewable Resources

Contact(s):
Michael Liffmann, ASSISTANT DIRECTOR
Ronald Becker, ASSOCIATE DIRECTOR
Sea Grant College Program, Louisiana State University, Baton Rouge, LA 70803
Phone: 225-388-6345
Elizabeth Coleman, COMMUNICATIONS COORDINATOR
Louisiana Sea Grant College Program, Louisiana State University, Baton Rouge, LA 70803
Phone: 225-388-6448
Jack Vanlopik, EXECUTIVE DIRECTOR
Louisiana State University, Baton Rouge, LA 70803
Phone: 225-388-6710
Fax: 225-388-6331
jvl@lsu.edu

LSU AGCENTER - LOUISIANA COOPERATIVE EXTENSION SERVICE
P.O. Box 25100
Baton Rouge, LA 70894-5100 USA
Phone: 225-578-6083   Fax: 504-578-2478
Website: www.lsu.edu.center.edu

Founded: NA
Scope: State

Publications: see publications on website

Contact(s):
Todd Shupe, ASSISTANT SPECIALIST FOR FORESTRY
Phone: 225-388-4087
Fax: 225-388-2478
tshupe@agctr.lsu.edu
Donald Reed, ASSISTANT SPECIALIST FOR FORESTRY AND WILDLIFE
Phone: 225-388-4087
Fax: 225-388-2478
dreed@agctr.lsu.edu
Rex Caffey, ASSISTANT SPECIALIST FOR WETLAND AND COASTAL RESOURCES
Phone: 225-388-2266
Fax: 225-388-2478
rcaffey@agctr.lsu.edu
Charles Lutz, ASSOCIATE SPECIALIST OF AQUACULTURE
Phone: 225-388-2152
Fax: 225-388-2478
glutz@agctr.lsu.edu
Paul Coriel, DIRECTOR OF EXTENSION SERVICE
jlngent@agctr.lsu.edu
Kenneth Roberts, PROJECT LEADER OF AQUACULTURE: FISHERIES: WETLAND & COASTAL MANAGEMENT AND SEA GRANT AND SPECIALIST FOR MARINE RESOURCE ECONOMICS
Phone: 225-388-2145
Fax: 225-388-2478
kroberts@agctr.lsu.edu
Michael Moody, SPECIALIST FOR SEAFOOD TECHNOLOGY
Phone: 225-388-2152
Fax: 225-388-2478
mmoody@agctr.lsu.edu

Michael Dunn, SPECIALIST: FORESTRY
Phone: 225-578-4087
Fax: 225-388-2478
mdunn@agctr.lsu.edu

M

MAINE ATLANTIC SALMON COMMISSION

650 State St.
Bangor, ME 04401-5654 USA
Phone: 207-941-4449   Fax: 207-941-4443
Website: www.state.me.us/asa

Founded: 1948
Membership: 8
Scope: Statewide

Description: (formerly Maine Atlantic Salmon Authority) The Atlantic Salmon Commission was established for the purposes of undertaking research, planning, management, restoration, and propagation of the Atlantic sea run salmon in the state. The Commission has authority to adopt and amend regulations to promote the conservation and propagation of Atlantic salmon in all Maine waters.

Keyword(s): Atlantic Salmon, Sport Fishing, Wildlife, Endangered Species, Wildlife Rehabilitation, Rivers

Contact(s):
Lee Perry, COMMISSIONER OF INLAND FISHERIES AND WILDLIFE
George Lapointe, COMMISSIONER OF MARINE RESOURCES
Frederick Kircheis, EXECUTIVE DIRECTOR
Paul Frinsko, MEMBER AT LARGE
Henry Nichols, POLICY DEVELOPMENT SPECIALIST

MAINE COOPERATIVE FISH AND WILDLIFE RESEARCH UNIT (USDI)

USGS Biological Resources Division, 5755 Nutting Hall, University of Maine
Orono, ME 04469-5755 USA
Phone: 207-581-2870   Fax: 207-581-2858

Founded: 1935
Membership: 3
Scope: Regional

Description: Provide graduate training and research experience in wildlife and fish ecology and management. Supported cooperatively by the University of Maine in Orono, ME, Maine Department of Inland Fisheries and Wildlife, U.S. Geological Survey, and the Wildlife Management Institute.

Contact(s):
John Moring, ASSISTANT LEADER: FISHERIES
310 Murray Hall, University of Maine, Orono, ME 04469
Phone: 207-581-2582
Cynthia Loftin, ASSISTANT LEADER: WILDLIFE
230 Nutting Hall, University of Maine, Orono, ME 04469
Phone: 207-581-2843
Fax: 207-581-2858
William Krohn, LEADER
258 Nutting Hall, University of Maine, Orono, ME 04469
Phone: 207-581-2870
Fax: 207-581-2858

MAINE DEPARTMENT OF AGRICULTURE, FOOD, AND RURAL RESOURCES

DEPT. OF AGRICULTURE & ENVIRONMENTAL RESOURCES
Office of Agricultural Natural
28 State House Station
Augusta, ME 04333-0028 USA
Phone: 207-287-1132   Fax: 207-287-7548
Website: www.state.me.us/agriculture

Founded: NA
Membership: 100
Scope: Statewide

Contact(s):
Robert Spear, COMMISSIONER
Phone: 207-287-3419
Peter Mosher, DIRECTOR
Phone: 207-287-1132
Fax: 207-287-7548
peter.mosher@state.me.us

MAINE DEPARTMENT OF CONSERVATION

22 State House Station
Augusta, ME 04333-0022 USA
Phone: 207-287-2211   Fax: 208-287-2216
Website: www.doc.state.me.us

Founded: 1973
Scope: State

Description: To preserve, protect, and enhance the land resources of the State of Maine; to encourage the wise use of the scenic, mineral, and forest resources; to ensure that coordinated planning for the future allocation of lands for recreational, forest production, mining, and other public and private uses is effectively accomplished; and to provide for the effective management of public lands.

Contact(s):
Gale Ross, ADMINISTRATIVE ASSISTANT
Ronald Lovaglio, COMMISSIONER
Dawn Gallagher, DEPUTY COMMISSIONER
Will Harris, DIRECTOR OF GENERAL SERVICES
Susan Benson, DIRECTOR OF PUBLIC INFORMATION

MAINE DEPARTMENT OF CONSERVATION

BUREAU OF GEOLOGY & NATURAL AREAS
22 State House Station
Augusta, ME 04333 USA
Phone: 207-287-2801   Fax: 207-287-2353
E-mail: mgs@state.me.us
Website: www.state.me.us/doc/nrimc.mgs/mgs.htm

Founded: NA
Membership: 15
Scope: State
Publications: see publications on website

Contact(s):
Robert Tucker, DIRECTOR, EARTH RESOURCES INFORMATION
Molly Docherty, DIRECTOR, MAINE NATURAL AREAS PROGRAM
Phone: 207-287-8045
Tom Weddle, HYDROGEOLOGIST & DIVISON DIRECTOR
Robert Marvinney, STATE GEOLOGIST AND DIRECTOR
Phone: 207-287-2804
Fax: 207-287-2353
robert.g.marvinney@state,me.us

MAINE DEPARTMENT OF CONSERVATION
LAND USE REGULATION COMMISSION
State House, Station #22
Augusta, ME 04333 USA
Phone: 207-287-2631   Fax: 207-287-7439
Website: www.state.me.us/doc/lurc/lurchome.htm

Founded: NA
Membership: 30
Scope: Statewide
Publications: available on web

Contact(s):
John Williams, DIRECTOR
Peggy Dwyer, RESOURCE ADMINISTRATOR
Phone: 207-287-4924

MAINE DEPARTMENT OF CONSERVATION (BUREAU OF PARKS AND LANDS)
22 State House Station
Augusta, ME 04333 USA
Phone: 207-287-3821   Fax: 207-287-3823
Website: www.state.me.us/doc

Founded: NA
Scope: Statewide

Publications: Outdoors & Maine - brochure

Contact(s):
Marilyn Tourtelotte, ALLAGASH WILDERNESS WATERWAY
Phone: 207-941-4014
marilyn.tourtelotte@state.me.us
Richard Skinner, BOATING FACILITIES
Phone: 207-287-4953
richard.skinner@state.me.us
Herb Hartman, DEPUTY DIRECTOR
Phone: 207-287-4961
herb.hartman@state.me.us
Tom Morrison, DIRECTOR
Phone: 207-287-3821
tom.morrison@state.me.us
Scott Ramsay, OFF-ROAD VEHICLE PROGRAM
Phone: 207-287-4956
scott.ramsay@state.me.us
Ralph Knoll, PLANNING & LAND USE ACQUISITION
Phone: 207-287-4911
ralph.knoll@state.me.us
Marlene Bowman, RESOURCE ADMINISTRATOR
Phone: 207-287-4912
marlene.bowman@state.me.us

MAINE DEPARTMENT OF ENVIRONMENTAL PROTECTION
State House Station 17
Augusta, ME 04333 USA
Phone: 207-287-7688   Fax: 207-287-7826
Website: www.state.me.us/dep

Founded: 1972
Membership: 475
Scope: Statewide

Description: DEP is charged with the protection and improvement of Maine's natural environment and acting in the best interests of the citizens' health and quality of life.

Publications: The Quality of Maine Waters—A Condensed Version on the 1996 Maine Water Quality Assessment, Watershed: An Action Guide to Improving Maine Waters, Planning Guides for Municipalities (series), Issue profiles and, A Citizen's Guide to Lake Watershed Surveys

Keyword(s): Water Quality, Solid Waste Management, Environmental Protection, Air Quality and Pollution, Pollution Prevention

Contact(s):
Martha Kirkpatrick, COMMISSIONER
Brooke Barnes, DEPUTY COMMISSIONER
James Brooks, DIRECTOR: BUREAU OF AIR QUALITY
David Van Wie, DIRECTOR: BUREAU OF LAND AND WATER QUALITY
David Lenette, DIRECTOR: BUREAU OF REMEDIATION AND WASTE MANAGEMENT

MAINE DEPARTMENT OF INLAND FISHERIES AND WILDLIFE
284 State St
Augusta, ME 04333-0041 USA
Phone: 207-287-8000   Fax: 207-287-6395
E-mail: ifw-webmaster@state.me.us
Website: www.mefishwildlife.com

Founded: 1880
Membership: 40
Scope: Statewide
Publications: Maine Fish and Wildlife

Keyword(s): training, Enforcement, Sport Fishing, Wildlife, Recreational Boating, Education, Snowmobiling, Endangered Species, Hunting

Contact(s):
Ron Taylor, CHIEF, ENGINEERING & REALTY DIVISION
G. Stadler, CHIEF: WILDLIFE RESEARCH AND MANAGEMENT DIVISION
Phone: 207-287-5252
Timothy Peabody, COLONEL: BUREAU OF WARDEN SERVICE
Phone: 207-287-2766
Lee Perry, COMMISSIONER
Phone: 207-287-5202
Frederick Hurley, DEPUTY COMMISSIONER
Phone: 207-287-5202
Peter Bourque, DIRECTOR OF FISHERIES AND HATCHERIES DIVISION
Phone: 207-287-5261
Richard Record, DIRECTOR: BUREAU OF ADMINISTRATIVE SERVICE
Phone: 207-287-5210
Donald Kleiner, DIRECTOR: BUREAU OF INFORMATION AND EDUCATION
Phone: 207-287-5244
Kenneth Elowe, DIRECTOR: BUREAU OF RESOURCE MANAGEMENT
Phone: 207-287-5252
Vesta Billing, DIRECTOR: LICENSING AND REGISTRATION DIVISION
Phone: 207-287-5225
Andrea Erskine, RULES AND REGULATIONS OFFICER
Phone: 207-287-5201

MAINE DEPARTMENT OF MARINE RESOURCES

21 State House Station
Augusta, ME 04333-0021 USA
Phone: 207-624-6550   Fax: 207-624-6024
Website: www.state.me.us/dmr

Founded: 1867
Membership: 30
Scope: Statewide

Description: Responsible for research, development, promotion, planning, and enforcement of laws relating to conservation of Maine's marine resources. The department was established to conserve and develop marine and estuarine resources of the state of Maine by conducting and sponsoring scientific research, promoting and developing the Maine commercial fishing industry, and by advising agencies of government concerned with development or activity in coastal waters.

Keyword(s): Public Health Protection, Wildlife, Ecology, Aquatic Habitats, Environmental and Conservation Education

Contact(s):

Joseph Fessenden, CHIEF OF BUREAU OF MARINE PATROL
Phone: 207-624-6024
George Lapointe, COMMISSIONER
E. Estabrook, DEPUTY COMMISSIONER
Dr. Linda Mercer, DIRECTOR OF BUREAU OF RESOURCE MANAGEMENT
West Boothbay Harbor, ME 04575
Phone: 207-633-9500
Fax: 207-633-9579
linda.mercer@state.me.us
Gilbert Bilodeau, DIRECTOR OF DIVISION OF ADMINISTRATIVE SERVICES
gilbert.m.bilodeau@state.me.us

MAINE SEA GRANT PROGRAM

UNIVERSITY OF MAINE
5715 Coburn Hall #14
Orono, ME 04469-5715 USA
Phone: 207-581-1435   Fax: 207-581-1426
E-mail: umseagrant@main.edu
Website: www.seagrant.umaine.edu

Founded: NA
Membership: 15
Scope: National

Keyword(s): Oceanography, Wildlife, Coasts, Aquatic Habitats, Environmental and Conservation Education

Contact(s):

Paul Anderson, ASSOCIATE DIRECTOR & MARINE EXTENSION LEADER
Phone: 207-581-1422
panderson@maine.edu
Ian Davison, DIRECTOR
davison@maine.maine.edu
Kristen Whiting grant, EXTENSION AGENT
Wells Reserve, Wells, ME 04090
Phone: 207-646-1555
Fax: 207-646-2930
kristen.whiting-grant@maine.edu
Dana Morse, EXTENSION ASSOCIATE, DARLING MARINE CENTER
Clarks Cove, Walpole, ME 04573
Phone: 207-563-3146
Fax: 207-563-3119

dana.l.morse@umit.maine.edu
Chris Bartlett, FINFISH AQUACULTURE SPECIALIST
Marine Technology Center, Washington County Technical College, 16 Deep Cove Road, Eastport, ME 04631-0618
Phone: 207-853-2518
Fax: 207-853-0940
chris.bartlett@umit.maine.edu
Natalie Springuel, SCIENCE WRITER

MARINE LABORATORY (FLORIDA)

Florida State University, Rt. 1, Box 219A
Sopchoppy, FL 32358 USA
Phone: 904-697-4095   Fax: 904-697-4098
Website: http://www.fsu.edu/~fsuml/

Founded: NA

Description: Includes studies on the biology, chemistry, and geology of coastal communities, physical oceanography of near-shore waters, aquatic and terrestrial ecosystems, and aquaculture.

Keyword(s): Wildlife, Wildlife Rehabilitation, Oceanography, Aquatic Habitats, Coasts

Contact(s):

Nancy Marcus, DIRECTOR

MARYLAND DEPARTMENT OF AGRICULTURE

50 Harry S. Truman Pkwy.
Annapolis, MD 21401 USA
Phone: 410-841-5700   Fax: 410-841-5914
Website: www.mda.state.md.us

Founded: 1972
Membership: 600
Scope: Statewide

Description: Created as a cabinet-level state agency, the Department is charged with assisting soil conservation districts to protect state waters from agricultural nonpoint source pollution, overseeing numerous inspection, testing, grading, and marketing programs, mosquito control and gypsy moth control, and forest pest management under various laws. The department also has responsibility for regulatory functions, such as pesticide applicators, weights and measures, nursery inspection, seed and turf regulation, certification; and agricultural chemical and product registration.

Publications: MDA News, Agricultural in Maryland Brochure, MDA Annual Report

Contact(s):

Robert Halman, ASSISTANT SECRETARY: MARKETING, ANIMAL INDUSTRIES AND CONSUMER SERVICES
Charles Puffinberger, ASSISTANT SECRETARY: OFFICE OF PLANT INDUSTRIES AND PEST MANAGEMENT
Phone: 410-841-5870
Royden Powell, ASSISTANT SECRETARY: OFFICE OF RESOURCE CONSERVATION
Phone: 410-841-5865
Craig Nielsen, COUNSEL
Phone: 410-841-5883
Bradley Powers, DEPUTY SECRETARY
Don Vandrey, DIRECTOR OF COMMUNICATIONS
Hagner Mister, SECRETARY
Roger Olson, STATE VETERINARIAN
Phone: 410-841-5810

MARYLAND DEPARTMENT OF AGRICULTURE

50 Harry S. Truman Pkwy.
Annapolis, MD 21401 USA
Phone: 410-841-5882   Fax: 410-841-5914
Website: www.mda.state.md.us

Founded: 1961
Membership: 250
Scope: Statewide

Description: Established by law as the Agricultural Advisory Board to the Governor, the Maryland Agricultural Commission was renamed in 1968 and placed within the Department of Agriculture in 1973. The commission formulates and makes proposals for the advancement of Maryland agriculture by serving as an advisory body to the Secretary of Agriculture. Composed of 24 members, one member is the principal administrative officer for agricultural affairs at the University of Maryland; one appointee represents consumer interests and serves a three-year term; the remaining 22 members are appointed for three-year terms by the Governor from nominations submitted by commodity and agricultural organizations.

Contact(s):
Henry Passi, CHAIRMAN
Gilbert Bowling, EXECUTIVE DIRECTOR
Phone: 410-841-5882

MARYLAND DEPARTMENT OF AGRICULTURE

STATE SOIL CONSERVATION COMMITTEE
50 Harry S. Truman Pkwy.
Annapolis, MD 21401 USA
Phone: 410-841-5863   Fax: 410-841-5736
Website: www.mda.sate.md.us

Founded: 1937
Membership: 350
Scope: Statewide

Description: Established to organize soil conservation districts and to establish policy, resolve problems to give guidance and assistance to districts. The SSCC membership includes representatives from the Maryland Departments of Natural Resources, Agriculture, and Environment, Maryland Agricultural Commission, University of Maryland, Maryland Association of Soil Conservation Districts, and five soil conservation district supervisors. The committee is a unit of the Maryland Dept. of Agriculture.

Publications: SSCC Reporter Newsletter

Keyword(s): Pesticides, Environmental and Conservation Education, Agriculture, Wildlife

Contact(s):
Royden Powell, ASSISTANT SECRETARY
Phone: 410-841-5865
Fax: 410-841-5914
Dave Thomas, CHAIRMAN
203 Middleton Road, Oakland, MD 21550
Phone: 301-334-3952
Louise Lawrence, EXECUTIVE SECRETARY
Phone: 410-841-5863
Fax: 410-841-5914
lawrenl@mda.state.md.us

MARYLAND DEPARTMENT OF NATURAL RESOURCES

Tawes State Office Bldg.
Annapolis, MD 21401 USA
Phone: 410-260-8101   Fax: 410-260-8111
E-mail: www.dnr.state.nd.us
Website: www.dnr.state.md.us

Founded: NA
Membership: 1300
Scope: Statewide

Contact(s):
Stanley Arthur, DEPUTY SECRETARY
Phone: 410-260-8102
sarthur.dnr.state.md.us
Chuck Porcari, DIRECTOR: PUBLIC COMMUNICATIONS OFFICE
Phone: 410-260-8001
cporcari.dnr.state.md.us
Nita Settina, LEGISLATIVE OFFICER: INTERGOVERNMENTAL AND COMMUNITY RELATIONS
Phone: 410-260-8110
nsettina.dnr.state.md.us
Joseph Gill, PRINCIPAL COUNSEL
Phone: 410-260-8350
jgill.dnr.state.md.us
J. Decker, SECRETARY

MARYLAND DEPARTMENT OF NATURAL RESOURCES

CHESAPEAKE BAY AND WATERSHED PROGRAMS
Tawes State Office Bldg.
Annapolis, MD 21401 USA
Phone: 410-260-8116
Website: www.dnr.state.md.us

Founded: NA
Membership: 200
Scope: Statewide

Contact(s):
Verna Harrison, ASSISTANT SECRETARY
Phone: 410-260-8116
David Burke, DIRECTOR OF CHESAPEAKE AND COASTAL WATERSHED SERVICE
Phone: 410-260-8705
Mark Bundy, DIRECTOR OF EDUCATION, BAY POLICY, AND GROWTH MANAGEMENT
Phone: 410-260-8720
Paul Massicot, DIRECTOR OF RESOURCE ASSESSMENT SERVICE
Phone: 410-260-8680

MARYLAND DEPARTMENT OF NATURAL RESOURCES

MANAGEMENT SERVICES
580 Taylor
Annapolis, MD 21401 USA
Phone: 410-260-8113   Fax: 410-260-8111
Website: www.dnr@state.md.us

Founded: NA
Membership: 500
Scope: Regional

Keyword(s): Watersheds, Forests and Forestry, Wildlife, Public Lands

Contact(s):
Sumita Chaudhuri, ASSISTANT SECRETARY
David Minges, DIRECTOR OF CHESAPEAKE BAY TRUST
Phone: 410-974-2941
Steven Powell, DIRECTOR OF FINANCE AND
ADMINISTRATIVE SERVICES
Kathryn Marr, DIRECTOR OF HUMAN RESOURCES
SERVICES
Phone: 410-260-8081
John Bernstein, DIRECTOR OF MARYLAND
ENVIRONMENTAL TRUST
Phone: 410-514-7900

MARYLAND DEPARTMENT OF NATURAL RESOURCES
RESOURCE MANAGEMENT SERVICES
580 Taylor Ave.
Annapolis, MD 21401 USA
Phone: 410-260-8100   Fax: 410-260-8111
Website: www.dnr@state.md.us

Founded: NA
Scope: Regional

Contact(s):
J. Charles Fox, SECRETARY
Phone: 410-260-8113
Ray Dintaman, DIRECTOR OF ENVIRONMENTAL REVIEW
Phone: 410-260-8331
Eric Schwaab, DIRECTOR OF FISHERIES SVC.
Phone: 410-260-8281
James Mallow, DIRECTOR OF FOREST SERVICE
Phone: 410-260-8501
Bruce Gilmore, DIRECTOR OF LICENSING REGISTRATION
SERVICES
Phone: 410-260-8233
Rich Dolesh, DIRECTOR OF WILDLIFE AND NATURAL
HERITAGE
Phone: 410-260-8582
Ren Serey, EXECUTIVE DIRECTOR OF CHESAPEAKE BAY
CRITICAL AREAS COMMISSION
Phone: 410-974-2426

MARYLAND DEPARTMENT OF NATURAL RESOURCES
WILDLIFE & HERITAGE DIVISION
580 Taylor Ave
Annapolis, MD 21401 USA
Phone: 410-260-8540   Fax: 410-260-8595
Website: www.dnr.state.md.us

Founded: NA
Scope: Statewide

Contact(s):
Robert Gaudette, DIRECTOR OF ENGINEERING AND
CONSTRUCTION
Phone: 410-260-8897
Michael Nelson, DIRECTOR OF LAND AND WATER
CONSERVATION
Phone: 410-260-8446
H. Dehart, DIRECTOR OF PROGRAM OPEN SPACE
Phone: 410-260-8425
Gene Piotrowski, DIRECTOR OF RESOURCE PLANNING
Phone: 410-260-8405
John Rhoads, SUPERINTENDENT OF NATURAL
RESOURCES POLICE

Phone: 410-260-8881
Rick Barton, SUPERINTENDENT OF STATE FOREST AND
PARK SERVICE
Phone: 410-260-8186

MARYLAND SEA GRANT COLLEGE
UNIVERSITY OF MARYLAND
Sea Grant College, 112 Skinner Hall
College Park, MD 20742-7640 USA
Phone: 301-405-6371   Fax: 301-314-9581
E-mail: mdsg@mdsg.umd.edu
Website: www.mdsg.umd.edu/

Founded: 1977
Membership: 12
Scope: National

Description: Maryland Sea Grant supports marine research,
education, and outreach activities, especially in connection
with the Chesapeake Bay. It currently supports research at four
of the region's marine laboratories, and on the campuses of the
University System of Maryland, the Johns Hopkins University,
and other institutions of higher learning.

Publications: Aquafarmer - quarterly newsletters, Watershed,
Maryland Sea Grant Books and Videos, Maryland Marine
Notes

Keyword(s): Oceanography, Wildlife, Biotechnology, Aquatic
Habitats, Environmental and Conservation Education

Contact(s):
Jack Greer, ASSISTANT DIRECTOR
Phone: 301-405-6377
Jonathan Kramer, DIRECTOR
Phone: 301-405-6371
kramer@mdsg.umd.edu

MARYLAND-NATIONAL CAPITAL PARK AND PLANNING COMMISSION
6611 Kenilworth Ave.
Riverdale, MD 20737 USA
Phone: 301-454-1740   Fax: 301-454-1750
Website: www.mncppc.org

Founded: 1927
Membership: 1000
Scope: National

Description: Established by the General Assembly of the state of
Maryland to provide for the orderly development of
Montgomery and Prince George's counties; to provide a
system of parks to serve the residents of this bi-county region;
and to provide recreation programs and services in Prince
George's county.

Contact(s):
Mary Wells-Harley, ACTING DIRECTOR OF PRINCE
GEORGE'S COUNTY PARKS AND RECREATION
Phone: 301-699-2582
Charles Loehr, DIRECTOR OF MONTGOMERY COUNTY
DEPARTMENT OF PARKS AND PLANNING
Phone: 301-495-4500
Fern Piret, DIRECTOR: PRINCE GEORGE'S COUNTY
PLANNING
Phone: 301-952-3595
Trudye Johnson, EXECUTIVE DIRECTOR
Phone: 301-454-1740

Richard Romine, GENERAL COUNSEL, LEGAL
DEPARTMENT
Phone: 301-454-1670
Donald Cochran, MONTGOMERY COUNTY DIRECTOR:
PARKS
Phone: 301-495-2500
Patricia Barney, SECRETARY OF TREASURER
William Hussmann, VICE CHAIRMAN
8787 Georgia Ave., Silver Spring, MD 20910
Phone: 301-495-4605
Elizabeth Hewlett, VICE CHAIRMAN
14741 Governor Oden Bowie Dr., Upper Marlboro, MD 20772
Phone: 301-952-3560

MASSACHUSETTS COOPERATIVE FISH AND WILDLIFE RESEARCH UNIT (USDI)

Box 34220, Holdsworth Natural Resources Ctr., University
of Massachusetts
Amherst, MA 01003-4220 USA
Phone: 413-545-0398   Fax: 413-545-4358

Founded: 1948
Scope: National

Description: Provides graduate training and research experience
in fisheries and wildlife research management, ecology,
habitat, population dynamics, and management. Supported
cooperatively by the University of Massachusetts,
Massachusetts Division of Fisheries and Wildlife,
Massachusetts Division of Marine Fisheries, the U.S.
Department of Interior, U.S.G.S.-BRD, and the Wildlife
Management Institute.

Contact(s):
Martha Mather, ASSISTANT LEADER: FISHERIES
Paul Sievert, ASSISTANT LEADER: WILDLIFE
Steve Destefano, LEADER

MASSACHUSETTS HIGHWAY DEPARTMENT

10 Park Plaza
Boston, MA 02116 USA
Phone: 617-973-7800   Fax: 617-973-8040
Website: www.state.ma.us/mhd

Founded: NA
Scope: Statewide

Description: The mission of the Massachusetts Highway
Department is to provide a safe, efficient, quality highway
system in a cost-effective and environmentally sensitive
manner that continuously meets the diverse needs of its users.

Keyword(s): Wetlands, Cultural Preservation, Environment,
Water Quality, Noise, Hazardous Materials & Waste, Solid
Waste, Transportation, Air Quality

Contact(s):
Thomas Broderick, CHIEF ENGINEER
Phone: 617-973-7830
thomas.broderick@state.ma.us
Matthew Amorello, COMMISSIONER
Phone: 617-973-7800
matthew.amorello@state.ma.us
David Anderson, DEPUTY CHIEF ENGINEER:
CONSTRUCTION
Phone: 817-973-7491
david.anderson@state.ma.us

Gregory Prendergast, DEPUTY CHIEF ENGINEER:
ENVIRONMENTAL DIVISION
Phone: 617-973-7484
gregory.prendergast@state.ma.us
John Blundo, DEPUTY CHIEF ENGINEER: HIGHWAY
ENGINEERING
Phone: 617-973-7521
john.blundo@state.ma.us
Gordon Broz, DEPUTY CHIEF ENGINEER: OPERATIONS
Phone: 617-973-7741
gordon.broz@state.ma.us
Kevin Walsh, PROJECT DEVELOPMENT
Phone: 617-973-7529
kevin.walsh@state.ma.us
James Elliott, SUPERVISOR: CULTURAL RESOURCES
UNIT
Phone: 617-973-7494
james.elliott@state.ma.us
Steven Miller, SUPERVISOR: PERMITTING AND
REGULATORY COMPLIANCE
Phone: 617-973-7582
steven.miller@state.ma.us
Henry Barbaro, SUPERVISOR: WETLANDS AND WATER
RESOURCES
Phone: 617-973-7419
henry .barbaro@state.ma.us

METROPOLITAN DISTRICT COMMISSION

METROPOLITAN DISTRICT COMMISSION
20 Somerset St.
Boston, MA 02108 USA
Phone: 617-727-5114   Fax: 617-727-0891
Website: www.state.ma.us

Founded: 1919
Membership: 900
Scope: Statewide

Description: Operates and maintains 19 swimming pools, 17 salt
water beaches, 3 fresh water beaches, 23 skating rinks, and
various other recreational facilities; also maintains a network of
parkways and main traffic roadways and a police force for
protection of its property and people using its facilities.

Contact(s):
David Balfour, COMMISSIONER
Gary Doak, DIRECTOR: DIVISION OF RECREATION
Phone: 617-727-9547
Brian Broderick, DIRECTOR: RESERVATIONS AND
HISTORIC SITES UNIT
Phone: 617-727-2744

MICHIGAN DEPARTMENT OF AGRICULTURE

P.O. Box 30017
Lansing, MI 48909 USA
Phone: 517-373-1052   Fax: 517-373-9146
Website: www.mda.state.mi.us

Founded: NA
Scope: Regional

Keyword(s): Soil Conservation, Pesticides, Biodiversity,
Agriculture, Flowers, Plants, and Trees

Contact(s):
Dan Wyant, DIRECTOR
Ken Rauscher, DIRECTOR OF PESTICIDE & PLANT PEST
MANAGEMENT DIVISION
Phone: 517-373-1087

MICHIGAN DEPARTMENT OF ENVIRONMENTAL QUALITY

106 West Allegan St., Hollister Building, 6th Fl., P.O. Box 30473
Lansing, MI 48909-7973 USA
Phone: 517-373-7917　　　Fax: 517-241-7401
Website: www.deq.state.mi.us

Founded: 1995
Scope: State

Description: Our mission is to drive improvements in environmental quality for the protection of public health and natural resources to benefit current and future generations. This will be accomplished through effective administration of agency programs, and providing for the use of innovative strategies, while helping to foster a strong and sustainable economy.

Publications: see publications on website

Keyword(s): Environmental Protection, Environmental and Conservation Education, Chemical Pollution Control, Coasts, Geology, Mining, Pollution Prevention, Toxicology, Toxic Substances, Nuclear-free, Water quantity, Water export and diversion, Coral Reefs, Exotic species, Aquatic nuisance species, Lakes, Drinking Water Protection, Solid Waste Management, Air Quality and Pollution

Contact(s):
Keith Harrison, (ACTING) DIRECTOR OF OFFICE OF THE GREAT LAKES
Phone: 517-335-4056
Fax: 517-335-4053
Dennis Fedewa, CHIEF OF FINANCIAL AND BUSINESS SERVICES DIVISION
Phone: 517-241-7427
Fax: 517-241-7428
Gary Hughes, DEPUTY DIRECTOR FOR OPERATIONS
Phone: 517-241-7394
Fax: 517-241-7401
Arthur Nash, DEPUTY DIRECTOR FOR PROGRAMS AND REGULATIONS
Phone: 517-241-7392
Fax: 517-241-7401
Russell Harding, DIRECTOR
Phone: 517-373-7917

MICHIGAN DEPARTMENT OF NATURAL RESOURCES

Box 30444
Lansing, MI 48909 USA
Phone: 517-373-1263　　　Fax: 517-373-6705
E-mail: dnr-wld-webpages@state.mi.us
Website: www.michigandnr.com

Founded: 1921
Scope: Statewide

Description: State agency for administration, including enforcement of laws and regulations, regarding the state's natural resources; and for enhancing recreational opportunities and quality. Derived from the Department of Conservation.

Contact(s):
Rob Abent, CHIEF OF FINANCE AND OPERATIONS SERVICE BUREAU
Phone: 517-373-1750
Kelly Smith, CHIEF OF FISHERIES
Phone: 517-373-1280

Garald Harris, CHIEF OF HUMAN RESOURCES
Phone: 517-373-1207
Rick Asher, CHIEF OF LAW ENFORCEMENT
Phone: 517-373-1230
Thomas Benson, CHIEF OF OFFICE OF INTERNAL AUDIT
Phone: 517-373-0755
Rodney Stokes, CHIEF OF PARKS AND RECREATION
Phone: 517-373-9900
Lowen Schuett, CHIEF OF PROPERTY MANAGEMENT DIVISION
Phone: 517-241-2438
Guy Gordon, CHIEF OF STAFF
Phone: 517-373-2329
Rebecca Humphries, CHIEF OF WILDLIFE
Phone: 517-373-1263
Kelli Sobel, DEPUTY FOR ADMINISTRATIVE SERVICES
Phone: 517-373-2425
George Burgoyne, DEPUTY FOR RESOURCE MANAGEMENT
Phone: 517-373-0046
James Ekdahl, DEPUTY UPPER PENINSULA FIELD HEADQUARTERS
Phone: 906-228-6561
K. Cool, DIRECTOR
Phone: 517-373-2329
Teresa Gloden, EXECUTIVE SECRETARY TO THE NATURAL RESOURCES COMMISSION
Phone: 517-373-2352
Mindy Koch, FOREST, MINERAL AND FIRE MANAGEMENT
Phone: 517-373-1275
Cordree McConnell, LEGAL SERVICES
Phone: 517-373-3503
Carol Bambery, LEGISLATIVE LIAISON
Phone: 517-373-0023

MICHIGAN DEPARTMENT OF NATURAL RESOURCES

Lansing, MI 48909 USA
Phone: 517-373-2329　　　Fax: 517-335-4242
Website: www.dnr.state.mi.us

Founded: 1996
Scope: State

Description: The department is responsible for health policy and management of the state's publicly-funded health service systems. It was created by Governor Engler in order to provide a more holistic approach to health care in Michigan. It is the largest department in Michigan government.

Keyword(s): Public Health Protection

Contact(s):
K. Cool, DIRECTOR

MICHIGAN SEA GRANT COLLEGE PROGRAM

University of Michigan, 4108 IST. 2200 Bonisteel Blvd.
Ann Arbor, MI 48109-2099 USA
Phone: 734-763-1437　　　Fax: 734-647-0768
Website: www.miceegrant.org

Founded: 1969
Membership: 30
Scope: National

Description: To promote the understanding and wise use of the Great Lakes through research, education and extension.

Contact(s):
William Taylor, ASSOCIATE DIRECTOR
College of Ag and Natural Resources, Michigan State University, 104 Agriculture Hall, East Lansing, MI 48824
Phone: 517-355-0233
Elizabeth Laporte, COMMUNICATION COORDINATOR
George Carignan, DIRECTOR
Phone: 734-763-1437
Joyce Daniels, EDITOR
Phone: 734-647-0766
John Schwartz, PROGRAM LEADER: SEA GRANT EXTENSION
Michigan Sea Grant College Program, Michigan State University, 334 Natural Resources Bldg., East Lansing, MI 48824
Phone: 517-355-9637
Minuet Henderson, PUBLICATIONS ASSISTANT
Phone: 734-764-1118

MICHIGAN STATE UNIVERSITY EXTENSION

Bulletin Office, 10-B Agric. Hall
East Lansing, MI 48824 USA
Phone: 517-355-0240
E-mail: msue@msue.msu.edu
Website: www.msue.msu.edu/msue/

Founded: NA
Membership: 1
Scope: Statewide

Description: Helps people improve their lives through an educational process that applies knowledge to critical issues, needs, and opportunities. Publications, instructional videos, and microcomputer software are listed in a catalogue which is available by writing to the Bulletin Office.

Contact(s):
Margaret Bethel, ACTING DIRECTOR

MINSTRY OF WATER, LAND & AIR PROTECTION

P.O. Box 9360
Victoria, British Columbia V8W 9M2 Canada
Phone: 250-387-9422   Fax: 250-356-6464
E-mail: www.wlapmail@gems5.gov.bc.ca
Website: www.giv.bc.ca/wlap

Founded: NA
Membership: 26
Scope: Province

Description: The Ministry of Environment's mission is to provide leadership in building environmental principles into day-to-day decisions of governments, corporations, and private individuals; to monitor and report on the state of the environment, and to ensure that defensible environmental standards are set and complied with; and to manage natural habitats, fish, wildlife, and water resources for ecological diversity and the economic and recreational opportunities they provide.

Contact(s):
Jim Walker, ASSISTANT DEPUTY MINISTER FOR WILDLIFE, HABITAT, AND ENFORCEMENT
Phone: 250-356-0139
Dana Hayden, ASSISTANT DEPUTY MINISTER, CORPORATE SERVICES
Denis O'Gorman, ASSISTANT DEPUTY MINISTER, PARKS DIVISION
Phone: 250-387-9997
Dick Roberts, ASSISTANT DEPUTY MINISTER, REGIONS DIVISION
Margaret Eckenfelder, ASSISTANT DEPUTY MINISTER, ENVIRONMENT & LAND HEADQUARTERS
Derek Thompson, DEPUTY MINISTER
Phone: 250-387-5429
Rod Davis, DIRECTOR OF RESOURCE STEWARDSHIP BRANCH
Phone: 250-356-7725
Jim Mattison, DIRECTOR OF RESOURCES INVENTORY BRANCH
Phone: 250-387-1112
Doug Dryden, DIRECTOR OF WILDLIFE
Phone: 250-387-9731
Joyce Murray, MINISTER

MINISTRY OF AGRICULTURE, FOOD AND FISHERIES

BC FISHERIES
780 Blanshard St., 3rd. Fl.
Victoria, British Columbia V8V 1X4 Canada
Phone: 250-387-3190   Fax: 250-387-3291
Website: www.gov.bc.ca/fish

Founded: NA
Membership: 40
Scope: Statewide

Contact(s):
Jamie Alley, DIRECTOR OF FISHERIES MANAGEMENT
Phone: 250-387-9711
Joyce Murray, MINISTER OF WATER, LAND, AIR PROTECTION

MINISTRY OF NATURAL RESOURCES

300 Water St., 2nd Fl., North Tower
Peterborough, ON K9J 8M5 Canada
Phone: 705-755-2363   Fax: 705-755-1640
E-mail: nric@mnr.gov.on.ca
Website: www.mnr.gov.on.ca

Founded: NA
Scope: Statewide

Description: Provides the ministry with leadership in the development and application of scientific knowledge, information management, and information technology. The division also plays a lead role in the provision of land-related information.

Contact(s):
Des McKee, ADM OF SCIENCE OF INFORMATION DIVISION
Phone: 705-755-1401
Collin Turnpenny, ASSOCIATE DIRECTOR OF ZIMBABWE NATURAL RESOURCE MANAGEMENT
Phone: 416-314-1550
Jim Hamilton, DIRECTOR
Phone: 705-755-2139

MINISTRY OF NATURAL RESOURCES

ALGONQUIN FORESTRY AUTHORITY
222 Main St. W
Huntsville, Ontario P1H 1Y1 Canada
Phone: 705-789-9647   Fax: 705-789-3353
E-mail: huntsville.office@algonquinforestry.on.ca
Website: www.algonquinforestry.on.ca

Founded: NA
Scope: Provincial

Contact(s):
Carl Corbett, GENERAL MANAGER

MINISTRY OF NATURAL RESOURCES
CORPORATE SERVICES DIVISION
POB 7000 300 Water Street
Peterborough, Ontario K9J 8M5 Canada
Phone: 705-755-2505   Fax: 705-755-2508

Founded: NA
Scope: Regional

Description: This division facilitates the delivery of ministry
programs by providing leadership, strategic advice, and
responsive results-oriented services to ministry clients. These
services include business planning, audit and evaluation,
financial, administrative, legal, and human resources. The
division also develops corporate and administrative policies
and gives advice on standards, guidelines, planning, and
management. It is the primary liaison with the central agencies
of government for corporate policy and the functions
associated with the Chief Administrative Officer.

Contact(s):
George Hutchinson, COMMUNICATIONS SERVICES
BRANCH
Phone: 416-314-2119
John Kenrick, DIRECTOR OF FINANCE AND BUSINESS
BRANCH
Phone: 705-755-2505
Dave Lynch, DIRECTOR OF HUMAN RESOURCES
BRANCH
Phone: 705-755-3131
Ann-Marie Gutierrez, DIRECTOR OF LEGAL SERVICES
BRANCH
Phone: 416-314-2025

MINISTRY OF NATURAL RESOURCES
FIELD SERVICES DIVISION
435 S James St., Ste. 221
Ontario, P7E 6S8 Canada

Founded: NA

Description: Delivering resource management programs for
Ontario's fisheries, wildlife, forests and provincial lands is the
responsibility of this division. It is also responsible for the
Aviation, Flood, and Fire Management Branch and the
Provincial Enforcement Section. The division's structure is
highly decentralized with three regional offices, 25 district
offices, and 17 area offices located across the province.

Contact(s):
Jack McFadden, DIRECTOR OF AVIATION, FLOOD, AND
FIRE MANAGEMENT
Phone: 705-945-5937
Gregg Sons, DIRECTOR OF ENFORCEMENT BRANCH
(ACTING)
Phone: 705-755-1750
Charlie Lauer, DIRECTOR OF NORTHWEST REGION
Phone: 807-475-1264

MINISTRY OF NATURAL RESOURCES
NATURAL RESOURCE MANAGEMENT DIVISION
99 Wellesley St. W
Toronto, Ontario M7A 1W3 Canada
Phone: 416-314-2624   Fax: 416-314-1994
E-mail: mnr.nric@mnr.gov.on.ca
Website: www.mnr.gov.on.ca/MNR/

Founded: NA
Scope: Regional

Description: The division is responsible for ensuring that natural
resource programs are responsive to the needs of Ontarians
and consistent with the ministry's vision of sustainable
development and its mission of ecological sustainability. Its
mandate covers lands, waters, forests, fish, wildlife, and parks,
and includes fish hatcheries, tree nurseries, and the
management of the Great Lakes.

Publications: Hunt Ontario, Fish Ontario, Hunting Regulations
Summary, Fishing Regulations Summary, Annual Parks Guide

Contact(s):
Peter Wallace, ASSISTANT DEPUTY MINISTER
Toronto, 416-314-6131
Cameron Mack, DIRECTOR OF FISH AND WILDLIFE
BRANCH
Phone: 705-755-1909
David De-Lay, DIRECTOR OF LANDS WATER BRANCH
Phone: 705-755-1629
Adair Ireland-Smith, MANAGING DIRECTOR OF ONTARIO
PARKS
Phone: 705-755-1702

MINISTRY OF NATURAL RESOURCES
NORTHEAST REGION
Ontario Government Complex, Highway 101 East,
P.O. Bag 3020
South Porcupine, Ontario P0N 1H0 Canada
Phone: 705-235-1154   Fax: 705-235-1226

Founded: NA
Scope: International

Publications: North Science & Technology Newsletter

Contact(s):
Mary Ellen Stoll, MANAGER OF SCIENCE & TECHNOLOGY
Rob Galloway, REGIONAL DIRECTOR

MINISTRY OF NATURAL RESOURCES
NORTHWEST REGION
435 James St., S
Thunder Bay, Ontario P7E 6S8 Canada
Phone: 807-475-1261   Fax: 807-473-3023
Website: www.mnr.gov.on.ca

Founded: NA
Membership: 40
Scope: Regional

Contact(s):
Charlie Lauer, REGIONAL DIRECTOR

MINISTRY OF NATURAL RESOURCES
SOUTH CENTRAL REGION
P.O. Box 7000
4th Floor, South Tower,
300 Water Street
Huntsville, Ontario K9J 8M5 Canada
Phone: 705-755-2500

Founded: NA

Contact(s):
Allan Stewart, REGIONAL DIRECTOR

MINISTRY OF NATURAL RESOURCES FISH & WILDLIFE BRANCH

FISH AND WILDLIFE BRANCH
300 Water St.
Peterborough, Ontario K9J 8M5 Canada
Phone: 705-755-1909   Fax: 705-755-1900

Founded: NA
Membership: 75
Scope: Province

Contact(s):
Cameron Mack, DIRECTOR
Phone: 705-755-1909
cameron.mack@mnr.gov.on.ca
Dave Maraldo, MANAGER OF FISHERIES
dave.maraldo@mnr.gov.on.ca
Deborah Stetson, MANAGER OF WILDLIFE
Phone: 705-755-1925
deb.stetson@mnr.gov.on.ca

MINISTRY OF COMMUNITY ABORIGINAL & WOMEN SERVICES

P.O. Box 9805, Stn. Prov. Govt.
Victoria, British Columbia V8W 9W1 Canada
Phone: 250-356-6305   Fax: 250-387-3798
Website: www.gov.bc.ca/mcaws

Founded: NA
Membership: 5
Scope: Province

Contact(s):
George Abbott, MINISTER
Phone: 250-356-3089

MINNESOTA BOARD OF WATER AND SOIL RESOURCES

One W. Water St., Suite 200
St. Paul, MN 55107 USA
Phone: 651-296-3767   Fax: 651-297-5615
Website: www.bwsr.state.mn.us

Founded: 1987
Membership: 70
Scope: Statewide

Description: Formed under M.S. chapter 103B to develop the capabilities of local governments in resource management. Works most often with soil and water conservation districts, watershed districts, watershed management organizations, and counties. Provides these local governments with financial and technical assistance. Administers programs focusing on erosion control and water quality.

Publications: Directory of local governments, various brochures, reports, and fact sheets, Water BillBoard, The

Keyword(s): Wetlands, Exotic species, Aquatic nuisance species, Land Use Planning, Conservation Tillage, Erosion Control, Watersheds, Training, Lakes, Soil Conservation

Contact(s):
Lee Coe, CHAIRMAN
Ronald Harnack, DIRECTOR
Phone: 651-297-5615
ron.harnack@bwsr.state.mn.us

MINNESOTA COOPERATIVE FISH AND WILDLIFE RESEARCH UNIT

U.S. Geological Survey, Biological Resources Division, University of Minnesota, Department of Fisheries and Wildlife, 200 Hodson Hall, 1980 Folwell Ave.
St. Paul, MN 55108 USA
Phone: 612-624-3421   Fax: 612-625-5299
Website: www.fw.umn.edu/co-op/co-op.html

Founded: 1987
Scope: Regional

Description: The research mission of the Minnesota Cooperative Fish and Wildlife Research Unit (MNCFWRU) is to address the biological, social, and economic aspects of both game and nongame wildlife and fisheries management in the context of conservation of biological diversity, and integrity and sustainability of ecosystems.

Keyword(s): training, Toxicology, Wildlife, Aquatic Habitats, Nongame Wildlife

Contact(s):
Bruce Vondracek, ASSISTANT LEADER: FISHERIES
David Fulton, ASSISTANT LEADER: WILDLIFE
Loralee Kerr, LIBRARIAN
375 Hodson Hall, 1980 Folwell Ave., St. Paul, MN 55108
Phone: 612-624-9288
David Andersen, UNIT LEADER

MINNESOTA DEPARTMENT OF AGRICULTURE

90 W. Plato Blvd.
St. Paul, MN 55107 USA
Phone: 651-297-2200   Fax: 651-297-7868
Website: www.mda.state.mn.us

Founded: 1919
Scope: State

Description: Enforces laws to protect the public health, promote family farming and marketing of Minnesota farm products, conserve soil and water, and prevent fraud and deception in the manufacture and distribution of foods, animal feeds, fertilizers, pesticides, seeds, and other items.

Keyword(s): Coral Reefs, Sustainable Development, Biodiversity, Agriculture, Pesticides

Contact(s):
Tom Masso, ASSISTANT COMMISSIONER
Perry Aasness, ASSISTANT COMMISSIONER
Gene Hugoson, COMMISSIONER
Sharon Clark, DEPUTY COMMISSIONER
Becky Leschner, DIRECTOR OF ACCOUNTING DIVISION
Phone: 651-215-5770
Shirley Bohm, DIRECTOR OF DAIRY AND FOOD INSPECTION
Phone: 651-296-1590
Dale Heimermann, DIRECTOR OF GRAIN AND PRODUCE INSPECTION
Phone: 612-341-7190
Mike Hunst, DIRECTOR: AGRICULTURAL STATISTICS
Phone: 651-296-3896

James Gryniewski, DIRECTOR: AGRICULTURE
CERTIFICATION
Phone: 651-297-2230
Curtis Pietz, DIRECTOR: AGRICULTURE FINANCE
Phone: 651-297-3557
Gerald Heil, DIRECTOR: AGRICULTURE MARKETING AND
DEVELOPMENT
Phone: 651-296-1486
Greg Buzicky, DIRECTOR: AGRONOMY AND PLANT
PROTECTION
Phone: 651-297-7121
Larry Palmer, DIRECTOR: INFORMATION SERVICES
Phone: 651-296-4659
William Krueger, DIRECTOR: LABORATORY SERVICES
Phone: 651-296-3273

MINNESOTA DEPARTMENT OF NATURAL RESOURCES

500 Lafayette Rd.
St. Paul, MN 55155-4040 USA
Phone: 651-296-6157
E-mail: info@dnr.state.mn.us
Website: http://www.dnr.state.mn.us/

Founded: 1931

Description: The Department of Conservation was renamed the
Department of Natural Resources (DNR) in 1971. The DNR's
goal is to achieve optimum natural resources planning,
protection, and development responsive to public need,
consistent with resource potentials, and for the social and
economic well-being of both present and future generations.

Publications: Minnesota Conservation Volunteer, The

Contact(s):
John Ernster, ADMINISTRATOR: BUREAU OF
ENGINEERING
Phone: 651-296-2119
Mary O'Neill, ADMINISTRATOR: BUREAU OF HUMAN
RESOURCES
Phone: 651-296-6493
Margaret Winkel-Ledin, ADMINISTRATOR: BUREAU OF
LICENSES
Phone: 651-296-4507
Henry May, ADMINISTRATOR: BUREAU OF MANAGEMENT
INFORMATION SERVICES
Phone: 651-297-3906
Peggy Adelmann, ADMINISTRATOR: MANAGEMENT AND
BUDGET SERVICES
Phone: 651-296-8340
Norm Kordell, ADMINISTRATOR: BUREAU OF FIELD
SERVICES
Phone: 651-297-3758
Brad Moore, ASSISTANT COMMISSIONER: OPERATIONS
Phone: 651-296-5229
Lee Pfannmuller, CHIEF: ECOLOGICAL SERVICES
SECTION
Phone: 651-296-2835
Ron Payer, CHIEF: FISHERIES SECTION
Phone: 651-296-3325
Tim Bremicker, CHIEF: WILDLIFE SECTION
Phone: 651-296-3344
Allen Garber, COMMISSIONER
Phone: 651-296-2549
Steve Morse, DEPUTY COMMISSIONER
Phone: 651-296-2540

Wayne Edgerton, DIRECTOR: AGRICULTURAL POLICY
Phone: 651-297-8341
Bill Bernhjelm, DIRECTOR: DIVISION OF ENFORCEMENT
Phone: 651-296-4828
Roger Holmes, DIRECTOR: DIVISION OF FISH AND
WILDLIFE
Phone: 651-297-1308
Gerald Rose, DIRECTOR: DIVISION OF FORESTRY
Phone: 651-296-4491
William Brice, DIRECTOR: DIVISION OF LANDS AND
MINERALS
Phone: 651-296-4807
William Morrissey, DIRECTOR: DIVISION OF PARKS AND
RECREATION
Phone: 651-296-9223
Kent Lokkesmoe, DIRECTOR: DIVISION OF WATERS
Phone: 651-296-4800
Michelle Beeman, DIRECTOR: REGULATORY AND
LEGISLATIVE SERVICES
Phone: 651-296-0915
Dennis Asmussen, DIRECTOR: TRAILS AND WATERWAYS
UNIT
Phone: 651-297-1151
Colleen Mlecoch, LIBRARY DIRECTOR
Phone: 651-296-1305
Paul Swenson, REGIONAL ADMINISTRATOR
Phone: 218-755-3955
John Guenther, REGIONAL ADMINISTRATOR
Phone: 218-327-4455
Cheryl Heide, REGIONAL ADMINISTRATOR
Phone: 507-359-6000
Larry Nelson, REGIONAL ADMINISTRATOR
Phone: 507-285-7418
Kathleen Wallace, REGIONAL ADMINISTRATOR
Phone: 612-772-7900

MINNESOTA ENVIRONMENTAL QUALITY BOARD

3rd Fl. Centennial Bldg., 658 Cedar St.
St. Paul, MN 55155 USA
Phone: 651-296-6157

Founded: 1973

Description: The EQB is Minnesota's principal forum for
discussing environmental issues. The EQB provides an
opportunity for the public to have direct input into the
development of the state's environmental policy. The EQB is
an independent decisionmaking body and is staffed by the
Minnesota Office of Strategic and Long Range Planning.

Publications: EQB Monitor

Keyword(s): Exotic species, Aquatic nuisance species,
Sustainable Ecosystems, Environmental Planning, Energy,
Sustainable Development

Contact(s):
Rod Sando, COMMISSIONER
3rd Fl. Centennial Bldg., 658 Cedar St., St.Paul, MN 55155
Phone: 612-297-1257
Michael Sullivan, EXECUTIVE DIRECTOR
3rd Fl. Centennial Bldg., 658 Cedar St., St.Paul, MN 55155
Phone: 612-296-9027

MINNESOTA GEOLOGICAL SURVEY

University of Minnesota, 2642 University Ave.
St. Paul, MN 55114 USA
Phone: 612-627-4780   Fax: 612-627-4778
E-mail: mgs@umn.edu
Website: www.geo.umn.edu/mgs

Founded: 1872
Membership: 4
Scope: State

Description: Established as a Geological and Natural History Survey, reconstituted in 1911 as the Minnesota Geological Survey to investigate the geology of the state; describe, classify and map the geological formations and mineral and water resources; and investigate all aspects of the geology affecting the environment.

Publications: List available on request.

Keyword(s): Exotic species, Aquatic nuisance species, Geology

Contact(s):
David Southwick, DIRECTOR

MINNESOTA POLLUTION CONTROL AGENCY

520 Lafayette Rd.
St. Paul, MN 55155 USA
Phone: 612-296-6300   Fax: 651-297-8687
Website: www.pca.state.mn.us

Founded: 1967
Membership: 700
Scope: State

Description: Administers the state statutes covering water pollution, air pollution, and solid and hazardous waste control.

Contact(s):
Gordon Wegwart, ASSISTANT COMMISSIONER
Phone: 612-296-7319
Karen Studders, CHAIRMAN OF THE BOARD AND COMMISSIONER
Phone: 612-296-7301
Lisa Thorvig, DEPUTY COMMISSIONER
Phone: 612-296-7331
James Warner, DIRECTOR: DIVISION OF GROUNDWATER AND SOLID WASTE
Phone: 612-296-7777

MINNESOTA POLLUTION CONTROL AGENCY

18 Wood Lake Dr., SE
Rochester, MN 55904 USA
Phone: 507-285-7343   Fax: 507-280-5513
Website: www.pca.state.mn.us

Founded: NA
Membership: 30
Scope: Regional

Contact(s):
Larry Landherr, DIRECTOR

MINNESOTA POLLUTION CONTROL AGENCY

1420 East College
Marshall, MN 56258 USA
Phone: 507-537-7146   Fax: 507-537-6001
E-mail: www.pca.state.mn.us
Website: www.pca.state.mn.us

Founded: NA
Membership: 11
Scope: Regional

Contact(s):
Mark Jacobs, SUPERVISOR

MINNESOTA POLLUTION CONTROL AGENCY

1800 College Road South
Baxter, MN 56425 USA
Phone: 218-828-2492   Fax: 218-828-2594
Website: www.pca.state.mn.us

Founded: NA
Scope: Statewide

Publications: see publications on website

Contact(s):
Reed Larson, REGIONAL MANAGER

MINNESOTA POLLUTION CONTROL AGENCY

DETROIT LAKES, MN
Lake Avenue Plaza, 714 Lake Ave. Suite 220
Detroit Lakes, MN 56501 USA
Phone: 218-847-1519   Fax: 218-846-0719
Website: www.pca.state.mn.us

Founded: NA
Membership: 35
Scope: State

Contact(s):
Jeff Lewis, DIRECTOR
Phone: 218-846-0730
jeff.lewis@pca.state.mn.us

MINNESOTA POLLUTION CONTROL AGENCY

DULUTH, MN
525 S. Lake Ave., Suite 400
Duluth, MN 55802 USA
Phone: 218-723-4660   Fax: 218-723-4727
Website: www.pca.state.mn.us

Founded: NA
Membership: 40
Scope: Regional, Local, State

Contact(s):
Suzanne Hanson, DIRECTOR

MINNESOTA SEA GRANT COLLEGE PROGRAM

Univ. of MN, 208 Washburn Hall, 2305 E. 5th St.
Duluth, MN 55812-1445 USA
Phone: 218-726-8100   Fax: 218-726-6556
E-mail: seagr@d.umn.edu
Website: www.seagrant.umn.edu

Founded: NA
Membership: 16
Scope: Statewide

Description: A statewide program that supports research, outreach, and educational programs related to Lake Superior and Minnesota's inland waters. Research areas include: water quality, fisheries, biotechnology, aquaculture, exotic species, and coastal tourism.

Publications: see publication website, Seiche, The Newsletter

Contact(s):
Doug Jensen, COORDINATOR, EXOTIC SPECIES
INFORMATION CENTER
Phone: 218-726-8712
djensen1@d.umn.edu
Carl Richards, DIRECTOR
crichard@d.umn.edu
Sharon Moen, EDITOR
Phone: 218-726-6195
smoen@d.umn.edu

MINNESOTA STATE EXTENSION SERVICES

University of Minnesota, 240 Coffey Hall, 1420 Eckles
Ave.
St. Paul, MN 55108-6070 USA
Phone: 612-625-1915   Fax: 612-625-6227
E-mail: info@mes.umn.edu
Website: www.extension.umn.edu

Founded: NA
Membership: 100
Scope: State

Contact(s):
Steven Daley Laursen, ASSOCIATE DEAN AND
COLLEGIATE PROGRAM LEADER
Phone: 612-624-9298
Charles Casey, DEAN AND DIRECTOR EXTENSION
SERVICE
Phone: 612-624-2703
casey002@umn.edu
Melvin Baughman, FOREST RESOURCES SPECIALIST
Phone: 612-624-0734
Patrick Huelman, HOUSING SPECIALIST
Phone: 612-624-1286
Steven Taff, PUBLIC POLICY SPECIALIST
Phone: 612-625-3103
Jeffrey Gunderson, SEA GRANT EXTENSION AND
FISHERIES EDUCATOR
Cynthia Hagley, WATER QUALITY EDUCATOR SPECIALIST
James Cooper, WILDLIFE SPECIALIST
104 Hodson, University of Minnesota, St. Paul, MN 55108
Phone: 612-624-1223
Fax: 612-625-5299
jac@umn.edu
Stephan Carlson, YOUTH SPECIALIST
Phone: 612-626-1259

MISSISSIPPI STATE UNIVERSITY EXTENSION SERVICES

WILDLIFE FISHERIES
Box 9690
Mississippi State, MS 39762 USA
Phone: 662-325-3133   Fax: 662-325-8750
E-mail: wildlife@ext.msstate.edu
Website: msucares.com

Founded: NA
Membership: 5
Scope: National

Contact(s):
Joe McGilberry, DIRECTOR INTERIM
Martin Brunson, EXTENSION LEADER PROFESSOR
WILDLIFE & FISHERIES
martyb@ext.mstate.edu

MISSISSIPPI-ALABAMA SEA GRANT CONSORTIUM

Mississippi-Alabama Sea Grant Consortium, Caylor Bldg.,
Gulf Coast Research Laboratory, P.O. Box 7000
Ocean Springs, MS 39566-7000 USA
Phone: 228-875-9341   Fax: 228-875-0528
Website: www.masgc.org

Founded: NA
Scope: Statewide

Contact(s):
Ladon Swann, ASSOCIATE DIRECTOR OFFICE OF
ALABAMA PROG
P.O. Box 369-370 D, Dauphin Island, Al 36528
Phone: 334-861-7544
Fax: 334-861-4646
swanndl@aubern.edu
Barry Costa-Pierce, DIRECTOR
b.costapierce@usm.edu

MISSISSIPPI COOPERATIVE FISH AND WILDLIFE RESEARCH UNIT

MISSISSIPPI STATE UNIVERSITY
Mailstop 9691
Mississippi State, MS 39762 USA
Phone: 662-325-2643   Fax: 662-325-8276

Founded: 1978
Membership: 3
Scope: Statewide

Description: The Unit is sponsored by the U.S.G.S. Biological
Resources Division; Mississippi Department of Wildlife,
Fisheries, and Parks; Mississippi State University; and the
Wildlife Management Institute. Fisheries and wildlife research,
graduate education, technical assistance, and extension are
the Unit's main missions.

Keyword(s): Sport Fishing, Rivers, Wildlife, Endangered Species,
Nongame Wildlife

Contact(s):
L. Miranda, ASSISTANT LEADER: FISHERIES
Francisco Vilella, ASSISTANT LEADER: WILDLIFE
Harold Schramm, LEADER
hschramm@cfr.msstate.edu

MISSISSIPPI DEPARTMENT OF AGRICULTURE AND COMMERCE

P.O. Box 1609
Jackson, MS 39215-1609 USA
Phone: 601-354-7050   Fax: 601-354-6290
E-mail: www.mdac.state.ms.us
Website: www.mdac.state.me.us

Founded: 1906
Membership: 200+
Scope: Statewide

Description: The department was created to foster and promote
the business of agriculture. Duties include: Regulatory,
consumer protection, marketing, and a wide range of service
activities.

Publications: Mississippi Market Bulletin

Contact(s):
Lester Spell, COMMISSIONER
Chris Sparkman, DEPUTY COMMISSIONER
Phone: 601-359-1138

Rodney Sanders, DIRECTOR OF ADMINISTRATION AND FINANCE
Phone: 601-359-1132
Jim Watson, DIRECTOR OF BOARD OF ANIMAL HEALTH
Phone: 601-359-1170
John Tillson, DIRECTOR OF CONSUMER PROTECTION
Phone: 601-359-1148
Billy Carter, DIRECTOR OF FARMERS MARKET
Phone: 601-354-6818
Donnis Roberson, DIRECTOR OF FRUIT & VEGETABLE INSPECTIONS
Phone: 601-354-6573
Keith Pouncy, DIRECTOR OF GRAIN INSPECTION
Umesh Sanjanwala, DIRECTOR OF INFORMATION SYSTEMS
Phone: 601-359-1151
Roger Barlow, DIRECTOR OF MARKET DEVELOPMENT
Phone: 601-359-1158
Billy Carter, DIRECTOR OF MARKET NEWS
Phone: 601-354-6818
Jim Meadows, DIRECTOR OF MEAT INSPECTION
Tommy Gregory, DIRECTOR OF NATIONAL AGRICULTURAL STATISTICS
Phone: 601-965-4575
Robert Louys, DIRECTOR OF PETROLEUM
Phone: 601-359-1101
Julia McLemore, DIRECTOR OF REGULATORY SERVICES
Phone: 601-359-1144
Russell Robbins, DIRECTOR OF WEIGHTS AND MEASURES
Phone: 601-359-1117
Claude Nash, EDITOR
Phone: 601-359-1123
Stella Cessna, PERSONNEL OFFICER
Phone: 601-359-1152

MISSISSIPPI DEPARTMENT OF ENVIRONMENTAL QUALITY

OFFICE OF LAND AND WATER RESOURCES
Southport Mall, P.O. Box 10631
Jackson, MS 39289 USA
Phone: 601-961-5200   Fax: 601-354-6938
Website: www.deq.state.ms.us

Founded: 1956
Scope: Statewide

Description: Administers Water Use Permitting Act of 1985, licensing of water well drillers, and the 1978 Dam Safety Act; inventories water resources; coordinates water and land resources planning; and conducts reviews of proposed water resources development.

Contact(s):
Patricia Phillips, CHIEF: DIVISION OF HYDROLOGIC INVESTIGATION AND REPORTING
Phone: 601-961-5213
Charles Branch, HEAD

MISSISSIPPI DEPARTMENT OF ENVIRONMENTAL QUALITY

OFFICE OF POLLUTION CONTROL
P.O. Box 10385
Jackson, MS 39289-0385 USA
Phone: 601-961-5171   Fax: 601-354-6612
Website: www.deq.state.ms.us

Founded: NA

Scope: State
Contact(s):
P. Harkins, HEAD
Phone: 601-961-5002

MISSISSIPPI DEPARTMENT OF WILDLIFE, FISHERIES, AND PARKS

1505 Eastover Drive
Jackson, MS 39211 USA
Phone: 601-432-2400
Website: http://www.mdwfp.com

Founded: NA

Description: The purpose of the MDWFP is to manage, conserve, develop, and protect Mississippi's outdoors, state parks, wildlife and marine resources, and their habitats; and to provide continuing recreational, economic, educational, ecological, aesthetic, social and scientific benefits for present and future generations.

Publications: Mississippi Soundings, Mississippi Outdoors
Contact(s):
Jimmy Laird, BOATING ENFORCEMENT
Phone: 601-364-2182
Ron Garavelli, CHIEF: FISHERIES
Phone: 601-364-2202
Bill Thomason, CHIEF: GAME
Phone: 601-364-2212
Randall Miller, CHIEF: LAW ENFORCEMENT
Phone: 601-364-2232
Tommy Shropshire, COORDINATOR: PLANNING AND POLICY
Phone: 601-364-2107
Bob Tyler, DEPUTY ADMINISTRATOR
Phone: 601-364-2004
Ellen Morgan
DIRECTOR OF MARKETING
Phone: 601-364-2152
Robert Cook, DIRECTOR: ADMINISTRATIVE SERVICES
Phone: 601-364-2006
Libby Hartfield, DIRECTOR: MUSEUM OF NATURAL SCIENCE
Phone: 601-354-7303
Jim Walker, DIRECTOR: PUBLIC INFORMATION
Phone: 601-364-2124
Al Tuck, DIRECTOR: SUPPORT SERVICES DIVISION
Phone: 601-364-2046
David Watts, EDITOR
Phone: 601-364-2129
Bill Quisenberry, EXECUTIVE ASSISTANT
Phone: 601-364-2005
Sam Polles, EXECUTIVE DIRECTOR
Phone: 601-364-2000
Mary Stevens, HEAD LIBRARIAN
Museum of Natural Science, 111 N. Jefferson, Jackson, MS 39202-2897
Phone: 601-354-7303
Steve Adcock, HUNTER EDUCATION
Phone: 601-364-2192
Mitiz Stubbs, OUTDOOR RECREATION GRANTS
Phone: 601-364-2156

MISSISSIPPI FORESTRY COMMISSION

301 N. Lamar St., Suite 300
Jackson, MS 39201 USA
Phone: 601-359-1386   Fax: 601-359-1349
Website: www.mfc.state.ms.us

Founded: 1926
Scope: Statewide

Description: Basic duties are forest protection against wildfire, insects, and disease; operation of tree-seedlings nurseries for reforestation; provision of forest resource management assistance to private landowners; and creation of interest in forestry.

Publications: various forest management brochures, Forestry Forum Magazine

Keyword(s): Urban Forestry, Renewable Resources, Forests and Forestry, Environmental and Conservation Education, Public Lands

Contact(s):
James Mordica, DEPUTY STATE FORESTER SERVICES
Everard Baker, DEPUTY STATE FORESTER: MANAGEMENT CHIEF
William Lambert, DEPUTY STATE FORESTER: PROTECTION CHIEF
Lezlin Proctor, DIRECTOR OF FINANCE & ADMINISTRATION
Harold Anderson, EDITOR AND EDUCATION DIRECTOR
Kent Grizzard, INFORMATION DIRECTOR
James Sledge, STATE FORESTER

MISSISSIPPI SOIL AND WATER CONSERVATION COMMISSION

Attn: Public Relations Director, P.O. Box 23005
Jackson, MS 39225 USA
Phone: 601-354-7645   Fax: 601-354-6628
E-mail: gmartin@mswcc.state.ms.us
Website: www.mswcc.state.ms.us

Founded: 1938
Scope: Statewide

Description: Originally established as the state agency for the control of soil erosion. Current statutory responsibilities include assistance to local soil and water conservation districts in the areas of water and soil quality projects, qualifications and elections of Commissioners, and administration of programs. Other responsibilities include reviewing and commenting on surface mining reclamation efforts. Also serves as the state resource agency for agricultural nonpoint source pollution issues and projects by assisting individual landowners, operators and other organized groups through demonstrations and educational programs.

Publications: Conservation Comments

Keyword(s): Water Quality, Exotic species, Aquatic nuisance species, Soil Conservation, Nonpoint Source Pollution, Erosion Control, Environmental and Conservation Education, Coral Reefs

Contact(s):
Ross McGehee, CHAIRMAN
176 McGehee Road, Natchez, MS 39120
Gale Martin, EXECUTIVE DIRECTOR
P.O. Box 23005, Jackson, MS 39225-3005
Phone: 601-354-7645
Fax: 601-354-6628

Emma Connolly, PUBLIC RELATIONS DIRECTOR
P.O. Box 23005, Jackson, MS 39225-3005

MISSISSIPPI STATE DEPARTMENT OF HEALTH

P.O. Box 1700
Jackson, MS 39215-1700 USA
Phone: 601-576-7400   Fax: 601-576-7364
Website: www.msdh.state.ms.us

Founded: NA
Scope: Statewide

Contact(s):
Mary Currier, DIRECTOR: DIVISION OF EPIDEMIOLOGY / STATE EPIDEMIOLOGIST
Phone: 601-576-7725
Rick Harrington, DIRECTOR: ENVIRONMENTAL HEALTH
Phone: 601-576-7680
F. Thompson, STATE HEALTH OFFICER
Phone: 601-576-7633

MISSOURI COOPERATIVE FISH AND WILDLIFE RESEARCH UNIT (USDI)

302 Anheuser-Busch Natural Resources Building, Fisheries and Wildlife, University of Missouri
Columbia, MO 65211-7240 USA
Phone: 573-882-3634   Fax: 573-884-5070
E-mail: CoopUnit@missouri.edu

Founded: 1985
Membership: 3
Scope: National, International

Description: Established by cooperative agreement among the Biological Resources Division of the U.S. Geological Survey, Missouri Department of Conservation, University of Missouri, and Wildlife Management Institute. Primary purpose is research and graduate student education in wildlife conservation, aquatic ecology, and fisheries management areas.

Keyword(s): training, Wetlands, Rivers, Wildlife, Waterfowl

Contact(s):
Sandy Clark, ADMINISTRATIVE OFFICER
Phone: 573-882-3634
David Galat, ASSISTANT LEADER: FISHERIES
Phone: 573-882-9426
Ronald Drobney, ASSISTANT LEADER: WILDLIFE
Phone: 573-882-9420
Charles Rabeni, LEADER
Phone: 573-882-3524

MISSOURI DEPARTMENT OF AGRICULTURE

P.O. Box 630, 1616 Missouri Blvd.
Jefferson City, MO 65102 USA
Phone: 573-751-4211   Fax: 573-751-1784
Website: www.mda.state.mo.us

Founded: NA
Membership: 450
Scope: Statewide

Contact(s):
Peter Hosherr, DEPUTY DIRECTOR
Phone: 573-751-3376
Lowell Mohler, DIRECTOR
lowell_mohler@mail.mda.state.mo.us
Sally Oxenhandler, PUBLIC INFORMATION OFFICER
Phone: 573-751-4645

MISSOURI DEPARTMENT OF CONSERVATION
P.O. Box 180
Jefferson City, MO 65102-0180 USA
Phone: 573-751-4115   Fax: 573-751-4467
Website: www.conservation.state.mo.us/

Founded: 1937
Scope: Statewide

Description: The department is responsible for the control, management, restoration, conservation and regulation of the bird, fish, game, forestry, and all wildlife resources of the state. These responsibilities are met through a wide variety of programs encompassing fish, wildlife, and forest management, regulations and enforcement, conservation education and interpretation, endangered species, and policy development.

Publications: Missouri Conservationist

Contact(s):
Gerald Ross, ASSISTANT TO DIRECTOR
John Smith, DEPUTY DIRECTOR
Jerry Conley, DIRECTOR
Jane Smith, GENERAL COUNSEL
Robbie Briscoe, INTERNAL AUDITOR

MISSOURI DEPARTMENT OF CONSERVATION
ADMINISTRATIVE SERVICES DIVISION
USA

Founded: NA

Contact(s):
David Erickson, ADMINISTRATOR

MISSOURI DEPARTMENT OF CONSERVATION
DESIGN AND DEVELOPMENT DIVISION
USA

Founded: NA

Contact(s):
William Lueckenhoff, ADMINISTRATOR

MISSOURI DEPARTMENT OF CONSERVATION
FISHERIES DIVISION
USA

Founded: NA

Contact(s):
Norman Stucky, ADMINISTRATOR

MISSOURI DEPARTMENT OF CONSERVATION
FORESTRY DIVISION
USA

Founded: NA

Contact(s):
Marvin Brown, ADMINISTRATOR

MISSOURI DEPARTMENT OF CONSERVATION
HUMAN RESOURCES SECTION
USA

Founded: NA

Contact(s):
Deborah Goff, CHIEF

MISSOURI DEPARTMENT OF CONSERVATION
NATURAL HISTORY SECTION
USA

Founded: NA

Contact(s):
Richard Thom, CHIEF

MISSOURI DEPARTMENT OF CONSERVATION
OUTREACH AND EDUCATION DIVISION
USA

Founded: NA

Contact(s):
Kathryn Love, ADMINISTRATOR

MISSOURI DEPARTMENT OF CONSERVATION
PROTECTION DIVISION
USA

Founded: NA

Contact(s):
Ronald Glover, ADMINISTRATOR

MISSOURI DEPARTMENT OF CONSERVATION
WILDLIFE DIVISION
USA

Founded: NA

Contact(s):
Oliver Torgerson, ADMINISTRATOR

MISSOURI DEPARTMENT OF NATURAL RESOURCES
P.O. Box 176
Jefferson City, MO 65102 USA
Phone: 573-751-4422   Fax: 573-526-3878
Website: www.dnr.state.mo.us

Founded: NA
Scope: State

Description: The Missouri Department of Natural Resources is the state resource management agency responsible for addressing environmental and natural resource-related issues. Areas of responsibility include: protecting Missouri's air, land and water resources, enforcing related laws where applicable; managing and maintaining the state's 80 state parks and state historic sites while protecting and promoting Missouri's cultural heritage and recreational opportunities; assisting citizens and government in the area of energy-efficiency management and developing the state's mineral resources in an environmentally conscious and safe manner.

Publications: Missouri Resources

Keyword(s): Public Lands, Outdoor Recreation, Protected Areas, Energy, Geology

Contact(s):
Connie Patterson, COMMUNICATIONS DIRECTOR
Phone: 573-751-1010
nrpattc@mail.dnr.state.mo.us
Gary Heimericks, DIRECTOR OF DIVISION OF ADMINISTRATIVE SUPPORT
Phone: 573-751-7961

John Young, DIRECTOR OF DIVISION OF AIR AND LAND
RESOURCES
Phone: 573-751-0763
younj@mail.dnr.state.mo.us
Douglas Eiken, DIRECTOR OF DIVISION OF STATE PARKS
Phone: 573-751-9392
Thomas Welch, DIRECTOR OF ENVIRONMENTAL
IMPROVEMENT AND ENERGY RESOURCES AUTHORITY
P.O. Box 744, Jefferson City, MO 65102-0176
Phone: 573-751-4919
Mimi Garstang, DIRECTOR OF GEOLOGICAL SURVEY AND
RESOURCE ASSESSMENT
Phone: 573-368-2101
garsm@mail.dnr.state.mo.us
Scott Totten, DIVISION OF WATER QUALITY SOIL
CONSERVATION
Phone: 573-751-5998
totts@mail.dnr.state.mo.us
Sara Parker, OUTREACH AND ASSISTANCE
Phone: 573-522-8796
parks@mail.dnr.state.mo.us

MISSOURI STATE EXTENSION SERVICES

University of Missouri, 309 University Hall
Columbia, MO 65211 USA
Phone: 573-882-7754   Fax: 573-884-4204
Website: http://outreach.missouri.edu

Founded: NA
Membership: 10
Scope: Statewide

Contact(s):
Ronald Turner, DIRECTOR: EXTENSION SERVICE
John Slusher, EXTENSION FORESTER
203 Natural Resources Bldg., Columbia, MO 65211
Phone: 573-882-4444
Fax: 573-882-1977

MIT SEA GRANT COLLEGE PROGRAM

E38-330, 292 Main St.
Cambridge, MA 02139-9910 USA
Phone: 617-253-7131   Fax: 617-258-5730
Website: www.mit.edu/seagrant/

Founded: NA
Scope: National

Contact(s):
Chrys Chryssostomidis, DIRECTOR
chrys@deslab.mit.edu

MONTANA BUREAU OF MINES AND GEOLOGY

GOELOGY SURVEY
Montana Tech of the University of Montana
Butte, MT 59701-8997 USA
Phone: 406-496-4167   Fax: 406-496-4451
E-mail: pubsales@mtech.edu
Website: mbmgsun.mtech.edu

Founded: 1919
Membership: 60
Scope: Statewide

Description: Established by law to aid the development and wise
use of the state's mineral, energy, and groundwater resources
by geologic and hydrogeologic studies of their occurrence and
potential. Publishes formal reports and maps on Montana
geology and groundwater.

Keyword(s): Geology, Energy, Exotic species, Aquatic nuisance
species

Contact(s):
Edmond Deal, DIRECTOR AND STATE GEOLOGIST
Phone: 406-496-4180
edeal@mbmgsun.mtech.edu

MONTANA COOPERATIVE FISHERY RESEARCH UNIT (USDI)

Dept. Ecology, Montana State University
Bozeman, MT 59717-3460 USA
Phone: 406-994-4549   Fax: 406-994-7479
E-mail: dtopp@montana.edu
Website: www.montana/ecology/facility

Founded: NA
Scope: National

Keyword(s): Rivers, Research, Endangered Species, Aquatic
Habitats, Wildlife

Contact(s):
Dee Topp, KEY CONTACT
Phone: 406-994-4549

MONTANA COOPERATIVE WILDLIFE RESEARCH UNIT (USGS/BRD)

University of Montana
Missoula, MT 59812 USA
Phone: 406-243-5372   Fax: 406-243-6064
E-mail: mtcwru@selway.umt.edu
Website: www.pica.wru.umt.edu/mtcwru

Founded: 1950
Membership: 6
Scope: International

Description: Conducts basic and applied research, trains
graduate students in wildlife biology and management, and disseminates information. Research specialties include breeding
productivity, nest predation, and habitat use by birds (particularly nongame and waterfowl species) in relation to land use
practices, predator populations, and natural variation in the
environment.

Keyword(s): Grasslands, Forests and Forestry, Nongame
Wildlife, Wetlands, Birds, Waterfowl

Contact(s):
Thomas Martin, ASSISTANT LEADER
tmartin@selway.umt.edu
I. Ball, LEADER
ball1@selway.umt.edu

MONTANA DEPARTMENT OF AGRICULTURE

P.O. Box 200201
Helena, MT 59620-0201 USA
Phone: 406-444-3144   Fax: 406-444-5409
E-mail: agr@state.mt.us
Website: www.agr.state.mt.us

Founded: NA
Membership: 60
Scope: Statewide

Keyword(s): Pest Management, Ground Water Protection,
Endangered Species, Agriculture, Pesticides

Contact(s):
Gregory Ames, ADMINISTRATOR, AGRICULTURAL
SCIENCES DIVISION

Will Kissinger, ADMINISTRATOR: AGRICULTURAL
DEVELOPMENT DIVISION
Phone: 406-444-2402
W. Peck, DIRECTOR

MONTANA DEPARTMENT OF FISH, WILDLIFE, AND PARKS

P.O. Box 200701
Helena, MT 59620-0701 USA
Phone: 406-444-3186   Fax: 406-444-4952
E-mail: fwpgen@state.mt.us
Website: www.swp.state.mt.us

Founded: NA
Membership: 560
Scope: State
Publications: Montana Outdoors

Keyword(s): training, Sport Fishing, Hunting, Environmental and
  Conservation Education, Outdoor Recreation

Contact(s):
  Dave Mott, ADMINISTRATOR: ADMINISTRATION AND
  FINANCE
  Phone: 406-444-4786
  Ron Aasheim, ADMINISTRATOR: CONSERVATION
  EDUCATION
  Phone: 406-444-4038
  Beata Galda, ADMINISTRATOR: ENFORCEMENT
  Phone: 406-444-5657
  Doug Monger, ADMINISTRATOR: PARKS
  Phone: 406-444-3750
  Don Childress, ADMINISTRATOR: WILDLIFE
  Phone: 406-444-2612
  Larry Peterman, ADMINISTRATOR:FISHERIES
  Phone: 406-444-2449
  Christian Smith, CHIEF OF STAFF
  Phone: 406-444-3186
  Jeff Hagener, DIRECTOR
  Phone: 406-444-3186
  Dave Books, EDITOR
  Phone: 406-444-2474
  Paul Sihler, FIELD SERVICE ADMINISTRATOR
  Spence Hegstad, FOUNDATION LIAISON

MONTANA DEPARTMENT OF NATURAL RESOURCES AND CONSERVATION

1625 11th Ave., P.O. Box 201601
Helena, MT 59620-1601 USA
Phone: 406-444-2074   Fax: 406-444-2684
Website: www.dnrc.state.mt.us

Founded: 1971
Membership: 260
Scope: Statewide

Description: Administers state-owned water projects; plans,
  regulates, and coordinates the development and use of state
  school trust, land, and forest resources; wildland fire
  protection; service forestry; water-right adjudication; floodplain
  management; supervision, assistance, and coordination for
  local conservation and grazing districts; and regulation of oil
  and gas production.

Keyword(s): Soil Conservation, Forestry, Oil and Gas, Agriculture,
  Exotic species, Aquatic nuisance species

Contact(s):
  Tom Schultz, ADMINISTRATION TRUST LAND

MANAGEMENT DIVISION
Ann Bauchman, ADMINISTRATOR FOR CENTRAL
SERVICES DIVISION
Phone: 406-444-6734
Ray Beck, ADMINISTRATOR: CONSERVATION AND
RESOURCE DEVELOPMENT DIVISION
1520 E. 6th Ave., Helena, MT 59620
Phone: 406-444-6667
Fax: 406-444-6721
rbeck@mt.gov
Don Artley, ADMINISTRATOR: FORESTRY DIVISION
2705 Spurgin Rd., Missoula, MT 59801
Phone: 406-542-4300
Terri Perrigo, ADMINISTRATOR: OIL AND GAS
CONSERVATION DIVISION
Susan Cottingham, ADMINISTRATOR: RESERVED WATER
RIGHTS COMPACT COMMISSION
Phone: 406-444-6841
Jack Stults, ADMINISTRATOR: WATER RESOURCES
DIVISION
1520 E. 6th Ave., Helena, MT 59620
Phone: 406-444-6605
Donald MacIntyre, CHIEF LEGAL COUNSEL
Phone: 406-444-6713
Bud Clinch, DIRECTOR
Phone: 406-444-2074
Mike Mikota, PERSONNEL AND EEO OFFICER
Carole Massman, SUPERVISOR AND EDITOR FOR
INFORMATION SERVICES
Phone: 406-444-6737

MONTANA ENVIRONMENTAL QUALITY COUNCIL

State Capitol
Helena, MT 59620-1704 USA
Phone: 406-444-3742   Fax: 406-444-3971
Website: www.led.state.mt.us

Founded: NA
Membership: 22
Scope: Statewide

Keyword(s): Exotic species, Aquatic nuisance species, Water
  Quality, Environmental Law, Air Quality and Pollution, Land
  Use Planning

Contact(s):
  Bea McCarthy, CHAIR
  Todd Everts, LEGISLATIVE ENVIRONMENTAL ANALYST
  Doug Mood, VICE-CHAIR

MONTANA NATURAL HERITAGE PROGRAM

1515 E 6th Ave.
Helena, MT 59620-1800 USA
Phone: 406-444-3009   Fax: 406-444-0581
E-mail: mtnhp@nris.state.mt.us
Website: http://nhp.nris.state.mt.us/

Founded: 1985
Scope: Local

Description: A centralized repository and clearinghouse of
  information on Montana's biodiversity, emphasizing features
  and species that are rare, threatened, endangered, or in need
  of further research.

Publications: Montana Plant Species of Special Concern,
  Montana Animal Species of Special Concern, Montana Bird
  Distribution

Keyword(s): Zoology

Contact(s):
Susan Crispin, DIRECTOR

MONTANA STATE EXTENSION SERVICES

P.O. Box 172230, Montana State University
Bozeman, MT 59717-2230 USA
Phone: 406-994-1750   Fax: 406-994-1756
E-mail: iciad@montana.edu
Website: www.extn.msu.montana.edu/

Founded: NA
Membership: 180
Scope: Statewide
Publications: MontGuides - Newsletter

Contact(s):
David Bryant, EXECUTIVE DIRECTOR
dbryant@montana.edu
Jim Knight, EXTENSION WILDLIFE SPECIALIST
Dept. of Animal and Range Science, Montana State
University, Bozeman, MT 59717
Phone: 406-994-5579
Fax: 406-944-5589
Jim Johannes
STATEWIDE DIRECTOR OF PROGRAMMING

N

NATIVE AMERICAN HERITAGE COMMISSION

915 Capitol Mall, Rm. 364
Sacramento, CA 95814 USA
Phone: 916-653-4082   Fax: 916-657-5390
E-mail: nahc@pacbell.net
Website: www.nahc.ca.gov

Founded: NA
Membership: 4
Scope: Statewide

Contact(s):
Larry Myers, EXECUTIVE SECRETARY

NATURAL RESOURCES AND ENVIRONMENTAL PROTECTION CABINET

14 Reilly Rd.
Frankfort, KY 40601 USA
Phone: 502-564-2150   Fax: 502-564-4245
Website: www.nr.state.ky.us

Founded: NA
Membership: 1200
Scope: Statewide

Contact(s):
Robert Logan, COMMISSIONER
Ralph Collins, DEPUTY COMMISSIONER
John Hornback, DIRECTOR: DIVISION FOR AIR QUALITY
Phone: 502-573-3382
Fax: 502-573-3787
William Davis, DIRECTOR: DIVISION OF ENVIRONMENTAL
SERVICES
Phone: 502-564-6120
Fax: 502-564-8930
Robert Daniell, DIRECTOR: DIVISION OF WASTE
MANAGEMENT
Phone: 502-564-6716
Fax: 502-564-4049
Jack Wilson, DIRECTOR: DIVISION OF WATER

Phone: 502-564-3410
Fax: 502-564-4245

NATURAL RESOURCES AND ENVIRONMENTAL PROTECTION CABINET

Capital Plaza Tower
Frankfort, KY 40601 USA
Phone: 502-564-3350   Fax: 502-564-3354
Website: www.kyenvironment.org

Founded: NA
Scope: Statewide

Contact(s):
Hank List, DEPUTY SECRETARY
Barbara Foster, GENERAL COUNSEL: OFFICE OF LEGAL
SERVICES
Phone: 502-564-5576
Fax: 502-564-6131
James Bickford, SECRETARY

NATURAL RESOURCES AND ENVIRONMENTAL PROTECTION CABINET

2 Hudson Hollow
Frankfort, KY 40601 USA
Phone: 502-564-6940   Fax: 502-564-6764
E-mail: blaine.fennell@mail.state.ky.us

Founded: NA
Membership: 100
Scope: Statewide

Contact(s):
Carl Campbell, COMMISSIONER
Allen Luttrell, DEPUTY COMMISSIONER
Stephen Hohmann, DIRECTOR: DIVISION OF ABANDONED
LANDS
Phone: 502-564-2141
Fax: 502-564-6544
stevehohmann@mail.state.ky.us
Mark Thompson, DIRECTOR: DIVISION OF FIELD
SERVICES
Phone: 502-564-2340
Fax: 502-564-5848
markw.thompson@mail.state.ky.us
Larry Adams, DIRECTOR: DIVISION OF PERMITS
Phone: 502-564-2320
Fax: 502-564-6764
larry.adams@mail.state.ky.us

NATURAL RESOURCES AND ENVIRONMENTAL PROTECTION CABINET

DEPARTMENT FOR NATURAL RESOURCES
663 Teton Trail
Frankfort, KY 40601 USA
Phone: 502-564-2184   Fax: 502-564-9195
E-mail: steve.coleman@mail.state.ky.us
Website: www.nr.state.ky.us/nrepc/dnr/dnrhome2.htm

Founded: NA
Membership: 25
Scope: Regional
Publications: Publications on website

Contact(s):
Hugh Archer, COMMISSIONER
Steve Coleman, DIRECTOR: DIVISION OF CONSERVATION
Phone: 502-564-3080

Fax: 502-564-9195
John Davies, DIRECTOR: DIVISION OF ENERGY
Phone: 502-564-7192
Fax: 502-564-7484
Leah Macswords, DIRECTOR: DIVISION OF FORESTRY
Phone: 502-564-4496
Fax: 502-564-6553

NATURAL RESOURCES AND ENVIRONMENTAL PROTECTION CABINET

ENVIRONMENTAL QUALITY COMMISSION
14 Reilly Rd.
Frankfort, KY 40601 USA
Phone: 502-564-2150   Fax: 502-567-4245
Website: www.kyeqc.net

Founded: NA
Scope: Statewide

Publications: State of Kentucky's Environment - biyearly book, Kentucky's Environment—bimonthly newsletter

Contact(s):
Aloma Dew, CHAIR
Leslie Cole, EXECUTIVE DIRECTOR

NATURAL RESOURCES AND ENVIRONMENTAL PROTECTION CABINET

KENTUCKY STATE NATURE PRESERVES COMMISSION
801 Schenkel Ln.
Frankfort, KY 40601 USA
Phone: 502-573-2886   Fax: 502-573-2355
E-mail: nrepc.ksnpcmail@mail.state.ky.us
Website: www.kynaturepreserves.org

Founded: NA
Scope: Statewide

Publications: Naturally Kentucky (newsletter quarterly)

Contact(s):
Clara Wheatley, CHAIRMAN
Don Dott, DIRECTOR
Phone: 502-573-2355

NATURAL RESOURCES CANADA

ONTARIO
Toronto, Ontario M7A 1W3 Canada
Phone: 705-755-2000   Fax: 416-314-2102
Website: http://www.NRCan-RNCan.gc.ca/inter/index.html

Founded: NA
Scope: Statewide

Description: The ministry's business plan establishes the following as MNR's core businesses: natural resource management; Crown land management; public safety and enforcement; parks and protected areas; and geographic information. In pursuing these core businesses, the ministry contributes to the environmental, social, and economic well-being of Ontario through the biological features of provincial interest, and protects human life, the resource base, and physical property from the threats of forest fires, floods, and erosion.

Contact(s):
Ann-Marie Guttier, ASSISTANT DEPUTY MINISTER FOR CORPORATE SERVICES
Gail Beggs, ASSISTANT DEPUTY MINISTER FOR FIELD SERVICES
Patricia Malcolmson, ASSISTANT DEPUTY MINISTER FOR SCIENCE AND INFORMATION RESOURCES
Linda Kemerman, COMMISSIONER OF MINING AND LANDS
700 Bay St., 24th Fl.,
John Burke, DEPUTY MINISTER
John McHugh, DIRECTOR OF COMMUNICATIONS SERVICES
Phone: 416-314-2119
Jeff Krantzberg, DIRECTOR OF COMMUNICATIONS SERVICES BRANCH
Phone: 416-314-2119
John Snobelen, MINISTER
Ted Chudleigh, PARLIAMENTARY ASSISTANT
Phone: 416-314-2193

NEBRASKA DEPARTMENT OF AGRICULTURE

301 Centennial Mall S., P.O. Box 94947
Lincoln, NE 68509 USA
Phone: 402-471-2341   Fax: 402-471-6876
Website: www.agr.state.ne.us

Founded: NA
Membership: 95
Scope: Statewide

Contact(s):
Greg Ibach, ASSISTANT DIRECTOR
Phone: 402-471-6876
Merlyn Carlson, DIRECTOR

NEBRASKA DEPARTMENT OF ENVIRONMENTAL QUALITY

1200 N St.
Lincoln, NE 68509-8922 USA
Phone: 402-471-2186   Fax: 402-471-2909
Website: www.deq.state.ne.us

Founded: 1971
Scope: Statewide

Description: Created by the Nebraska Environmental Protection Act. Administers and enforces rules and regulations, and monitors the quality of the environment in Nebraska.

Publications: Environmental Update

Contact(s):
Tom Lamberson, DEPUTY DIRECTOR (ADMINISTRATION) AND HEARING OFFICER, STATE ENVIRONMENTAL QUALITY COUNCIL
Jay Ringenberg, DEPUTY DIRECTOR (PROGRAMS)
Michael Linder, DIRECTOR
Brian McManus, PUBLIC INFORMATION OFFICER AND PUBLICATIONS EDITOR

NEBRASKA DEPARTMENT OF NATURAL RESOURCES

301 Centennial Mallox
Lincoln, NE 68509-4676 USA
Phone: 402-471-2363   Fax: 402-471-2900
E-mail: dnr@dnr.state.ne.us
Website: www.dnr.state.ne.us

Founded: 1937
Membership: 70
Scope: Statewide

Description: The state agency responsible for comprehensive water resources planning, flood plain management, administration of state financial assistance for water resources, flood control, and soil and water conservation. It also has advisory and administrative responsibility for Natural Resources Districts throughout the state.

Publications: Nebraska Resources - quarterly newsletter

Keyword(s): Exotic species, Aquatic nuisance species, Soil Conservation, Planning Management, Rivers, Environmental and Conservation Education

Contact(s):
Richard Jiskra, CHAIRPERSON
2342 County Rd. 1600, Swanton, NE 68445
Phone: 402-448-5305

NEBRASKA DEPARTMENT OF NATURAL RESOURCES
State House Station
Lincoln, NE 68509 USA
Phone: 402-471-2363   Fax: 402-471-2900
Website: www.dnr.state.ne.us

Founded: NA
Membership: 60
Scope: Statewide

Description: Administers and enforces the state water laws and all matters pertaining to water rights; measuring and recording the flow of various streams and canals; approving plans and specifications for dam construction; inspection of dams; and registration of wells.

Publications: Biannual Report, Hydrographic Report, Nebraska Resources, newsletter

Keyword(s): Exotic species, Aquatic nuisance species, Lakes, Rivers, Engineering

Contact(s):
Ann Bleed, DEPUTY DIRECTOR
Roger Patterson, DIRECTOR
James Cook, LEGAL COUNSEL
Susan France, PERMITS AND ADJUDICATIONS

NEBRASKA GAME AND PARKS COMMISSION
2200 N. 33rd St., P.O. Box 30370
Lincoln, NE 68503-0370 USA
Phone: 402-471-0641   Fax: 402-471-5528
E-mail: ngpc@state.ne.us
Website: www.ngpc.state.ne.us

Founded: NA
Membership: 150
Scope: Statewide

Description: The commission has sole charge of state parks, game and fish, and all things pertaining thereto; boating; and administration of the Land and Water Conservation Fund. Complete information on Game and Parks Commission facilities is available on the WWW at: www.ngpc.state.ne.us.

Publications: Nebraskaland Magazine

Contact(s):
Jim Sheffield, ADMINISTRATOR: ENGINEERING
Phone: 402-471-5557
jsheff@ngpc.state.ne.us
Mark Brohman, ADMINISTRATOR: ADMINISTRATION
Phone: 402-471-5539
mbrohman@ngpc.state.ne.us
Patrick Cole, ADMINISTRATOR: BUDGET AND FISCAL
Phone: 402-471-5523
pcole@ngpc.state.ne.us
James Carney, ADMINISTRATOR: CENTRAL REGIONAL PARKS MANAGER
Phone: 402-471-5547
jcarney@ngpc.state.ne.us
Don Gabelhouse, ADMINISTRATOR: FISHERIES
Phone: 402-471-5515
gabel@ngpc.state.ne.us
Paul Horton, ADMINISTRATOR: INFORMATION AND EDUCATION
Phone: 402-471-5481
phorton@ngpc.state.ne.us
Ted Blume, ADMINISTRATOR: LAW ENFORCEMENT
Phone: 402-471-4010
tblume@ngpc.state.ne.us
Earl Johnson, ADMINISTRATOR: OPERATIONS AND CONSTRUCTION
Phone: 402-471-5525
James Fuller, ADMINISTRATOR: PARKS
Phone: 402-471-5550
jfuller@ngpc.state.ne.us
Duane Westerholt, ADMINISTRATOR: PLANNING AND DEVELOPMENT
Phone: 402-471-5511
dwester@ngpc.state.ne.us
Bruce Sackett, ADMINISTRATOR: REALTY
Phone: 402-471-5536
bsackett@ngpc.state.ne.us
James Douglas, ADMINISTRATOR: WILDLIFE
Phone: 402-471-5411
jdouglas@ngpc.state.ne.us
Roger Kuhn, ASSISTANT DIRECTOR
Phone: 402-471-5512
rkuhn@ngpc.state.ne.us
Kirk Nelson, ASSISTANT DIRECTOR
Phone: 402-471-5539
knelson@ngpc.state.ne.us
Noelyn Isom, ASSISTANT DIRECTOR
Phone: 402-471-5539
nisom@ngpc.state.ne.us
Rex Amack, DIRECTOR
Phone: 402-471-5539
Jim Swenson, EASTERN REGIONAL PARKS MANAGER
Phone: 402-471-5499
jswenson@ngpc.state.ne..us
Tom White, EDITOR
Phone: 402-471-5471
twhite@ngpc.state.ne.us
Barbara Voeltz, LIBRARIAN
2200 N 33rd St., P.O. Box 30370, Lincoln, NE 68503
Phone: 402-471-5587
Fax: 402-471-5528
bvoeltz@ngpc.state.ne.us
Steve Kemper, WESTERN REGIONAL PARKS MANAGER
Phone: 308-665-2900

NEBRASKA STATE EXTENSION SERVICES
211 Agricultural Hall, University of Nebraska
Lincoln, NE 68583-0703 USA
Phone: 402-472-2966   Fax: 402-472-5557
Website: www.unl.edu/ianr/coopext/coopext.htm

Founded: NA

Membership: 300
Scope: International

Keyword(s): training, Sustainable Ecosystems, Environmental and Conservation Education, Renewable Resources, Agriculture, Rural Development

Contact(s):
Elbert Dickey, DEAN AND DIRECTOR OF COOPERATIVE EXTENSION
edickey1@unl.edu
Susan Fritz, DEPT. HEAD OF AGRICULTURAL LEADERSHIP, EDUCATION AND COMMUNICATION
300 Agricultural Hall, University of Nebraska, Lincoln, Nebraska 68583-0709
Phone: 402-472-9559
Fax: 402-472-5863
John Allen, DIRECTOR, CENTER FOR RURAL COMMUNITY REVITALIZATION AND DEVELOPMENT
58C H.C. Filley Hall, University of Nebraska, Lincoln, NE 68583-0947
Phone: 402-472-8012
Fax: 402-472-3460
jallen1@unl.edu
Scott Hygnstrom, VERTEBRATE PEST SPECIALIST
202 Natural Resources Hall, University of Nebraska, Lincoln, NE 68583-0819
Phone: 402-472-6822
Fax: 402-472-2946

NEVADA BUREAU OF MINES AND GEOLOGY

Mail Stop 178, University of Nevada, Reno
Reno, NV 89557-0088 USA
Phone: 775-784-6691   Fax: 775-784-1709
E-mail: nbmginfo@unr.edu
Website: www.nbmg.unr.edu

Founded: NA
Membership: 34
Scope: Statewide

Description: Conducts research on Nevada geology and mineral resources. Collects and disseminates information (including published maps and reports) on Nevada geology, mineral resources, base maps, and airphotos.

Publications: Living with Earthquakes in Nevada, Major Mines of Nevada, Nevada Mineral Industry

Keyword(s): Exotic species, Aquatic nuisance species, Public Lands, Geology, Land Use Planning, Geography

Contact(s):
Jonathan Price, DIRECTOR AND STATE GEOLOGIST
jprice@unr.edu
David Davis, GEOLOGIC INFORMATION SPECIALIST

NEVADA COOPERATIVE EXTENSION

Universtiy of Nevada, 2345 Red Rock St,
Las Vegas, NV 89146 USA
Phone: 702-251-7531   Fax: 702-251-7536
Website: www.unce.unr.edu

Founded: NA
Scope: Regional

Keyword(s): Sustainable Ecosystems, Exotic species, Aquatic nuisance species, Renewable Resources

Contact(s):
John Burton, ASSISTANT DIRECTOR

Phone: 702-784-7070
Jason Davidson, CENTRAL AREA AGRONOMY AND RANGE SPECIALIST
Fallon, NV 702-428-0212
Karen Hinton, DIRECTOR
Sherman Swanson, STATE RANGE SPECIALIST AND RIPARIAN SCIENTIST
1000 Valley Rd., Reno, NV 89512
Phone: 702-784-4057
Fax: 702-784-4583
Mark Walker, STATE WATER SPECIALIST
Phone: 702-784-1938
Ed Smith, WESTERN AREA NATURAL RESOURCES SPECIALIST
Phone: 702-782-9960
John Cobourn, WESTERN AREA WATER SPECIALIST
Phone: 702-832-4150

NEVADA DEPARTMENT OF AGRICULTURE

350 Capitol Hill Ave.
Reno, NV 89502-2923 USA
Phone: 775-688-1180   Fax: 775-688-1178
Website: www.agri.state.nv.us

Founded: NA
Membership: 53
Scope: Statewide
Publications: Test Alert

Contact(s):
David Thain, ADMINISTRATOR AND STATE VETERINARIAN, DIVISION OF ANIMAL INDUSTRY
Edward Hoganson, ADMINISTRATOR DIVISION OF MEASURMENT STANDARDS
Phone: 775-688-1166
Fax: 775-688-2533
Robert Gronowski, ADMINISTRATOR, DIVISION OF PLANT INDUSTRY
Paul Iverson, DIRECTOR

NEVADA DEPARTMENT OF CONSERVATION AND NATURAL RESOURCES

123 W. Nye Ln.
Carson City, NV 89706-0818 USA
Phone: 775-687-4360   Fax: 775-687-6122
Website: www.state.nv.us/cnr

Founded: NA
Membership: 1100
Scope: Statewide

Contact(s):
Pamela Wilcox, ADMINISTRATOR AND STATE LAND REGISTRAR AND ADMINISTRATOR OF CONSERVATION DISTRICTS DIVISION
Phone: 775-687-4363
Allen Biaggi, ADMINISTRATOR: DIVISION OF ENVIRONMENTAL PROTECTION
Phone: 775-687-4670
Wayne Perock, ADMINISTRATOR: DIVISION OF STATE PARKS
Phone: 775-687-4384
Terry Crawforth, ADMINISTRATOR: DIVISION OF WILDLIFE
Phone: 775-688-1500
Freeman Johnson, ASSISTANT DIRECTOR
R. Michael Turnipseed P.E., DIRECTOR
Colleen Murphy, EXECUTIVE SECRETARY
Glenn Clemmer, PROGRAM MANAGER: NEVADA NATURAL

HERITAGE
Phone: 775-687-4245
Hugh Ricci, STATE ENGINEER: DIVISION OF WATER
RESOURCES
Phone: 775-687-4278
Roy Trenoweth, STATE FORESTER: DIVISION OF
FORESTRY
Phone: 775-684-2500

NEVADA DIVISION OF WILDLIFE

1100 Valley Rd.
Reno, NV 89512 USA
Phone: 775-688-1500   Fax: 775-688-1595
Website: www.nevadadivisionofwildlife.org

Founded: NA
Scope: Statewide

Description: A regulatory and policymaking body, administering laws, regulations, and policies. Mission is the protection, propagation, restoring, introduction, transplanting, and management of wildlife throughout the state.

Contact(s):
Steve Bremer, ADMINISTRATIVE
Terry Crawforth, ADMINISTRATOR
Boyd Spratling, BOARD OF WILDLIFE COMMISSIONER
VICE CHAIRMAN
David Rice, CHIEF, CONSERVATION EDUCATION
Thomas Atkinson, CHIEF, ENFORCEMENT
Gene Weller, CHIEF, FISHERIES
Gregg Tanner, CHIEF, GAME

NEVADA NATURAL HERITAGE PROGRAM

E. College Parkway,
Carson City, NV 89706-7921 USA
Phone: 775-687-4245   Fax: 775-687-1288
Website: www.state.nv.us/nvnhp/

Founded: 1986
Membership: 6
Scope: Statewide

Description: The program represents an ongoing effort to collect and standardize data on Nevada's sensitive biodiversity and share this information with developers, researchers, and decision-makers for environmentally wise planning.

Publications: Nevada Rare Plant Atlas, Nevada's Sensitive Species List, Scorecard-Highest Priority Conservation Sites, Endangered, Threatened, and Sensitive Vascular Plants of Nevada

Keyword(s): Biodiversity, Conservation, Endangered Species, Sensitive Species

Contact(s):
Glenn Clemmer, PROGRAM MANAGER

NEW HAMPSHIRE DEPARTMENT OF AGRICULTURE, MARKETS, AND FOOD

P.O. Box 2042
Concord, NH 03302-2042 USA
Phone: 603-271-3551   Fax: 603-271-1109
E-mail: marketbulletin@agr.state.nh.us

Founded: 1913
Scope: Statewide

Description: The department is responsible for a broad range of activities, including protecting the environment, food safety, market integrity, animal and plant health, and the economic security of the New Hampshire agricultural industry.

Publications: Weekly Market Bulletin

Keyword(s): Sustainable Development, Rural Development, Environmental Justice, Land Protection, Agriculture

Contact(s):
Stephen Taylor, COMMISSIONER
Phone: 603-271-3551
Fax: 603-271-1109

NEW HAMPSHIRE DEPARTMENT OF AGRICULTURE, MARKETS, AND FOOD

STATE CONSERVATION COMMITTEE
P.O. Box 2042
Concord, NH 03302-2042 USA
Phone: 603-271-3551

Founded: 1945
Scope: Statewide

Description: The SCC consists of twelve members. Six members represent state agencies, five are appointed, and one represents the NH Association of Conservation Commissions. Duties are to offer assistance to supervisors of the ten conservation districts, keep supervisors of each district informed of other district activities, and coordinate the conservation of New Hampshire activities.

Keyword(s): Wetlands, Exotic species, Aquatic nuisance species, Environmental and Conservation Education, Soil Conservation, Agriculture

Contact(s):
Samuel Doyle, CHAIR
P.O. Box 4, North Sutton, NH 03260
Phone: 603-927-4163
Fax: 603-224-8260
Joanna Pellerin, COORDINATOR
118 North Rd., Brentwood, NH 03833-6614
Phone: 603-679-2790
Fax: 603-679-2860

NEW HAMPSHIRE DEPARTMENT OF ENVIRONMENTAL SERVICES

6 Hazen Dr.
Concord, NH 03302-0095 USA
Phone: 603-271-3503   Fax: 603-271-2867
E-mail: pip@des.state.nh.us
Website: www.des.state.nh.us

Founded: 1987
Scope: International

Description: The DES is a result of a legislatively-mandated state environmental agency. The DES consists of three divisions: Water Division; Waste Management Division; and Air Resources Division.

Keyword(s): Wetlands, Exotic species, Aquatic nuisance species, Solid Waste, Coral Reefs, Air Quality and Pollution

Contact(s):
Robert Varney, ADMINISTRATOR, PUBLIC INFORMATION AND PERMITTING UNIT
G. Bisbee, ASSISTANT COMMISSIONER
Kenneth Colburn, DIRECTOR OF AIR RESOURCES DIVISION
Phone: 603-271-1370
Fax: 603-271-1381

Philip O'Brien, DIRECTOR OF WASTE MANAGEMENT
DIVISION
Phone: 603-271-2900
Fax: 603-271-2456
Harry Stewart, DIRECTOR OF WATER DIVISION
Phone: 603-271-3503
Fax: 603-271-2982

NEW HAMPSHIRE FISH AND GAME DEPARTMENT

2 Hazen Dr.
Concord, NH 03301 USA
Phone: 603-271-3422　　　Fax: 603-271-1438
E-mail: info@wildlife.state.nh.us
Website: www.wildlife.state.nh.us

Founded: NA
Scope: State

Publications: New Hampshire Wildlife Journal, see publication
website

Contact(s):
Daniel Lynch, ASSISTANT DIRECTOR
dlynch@wildlife.state.nh.us
Richard Cunningham, BUSINESS ADMINISTRATOR
Phone: 603-271-2741
rcunningham@wildlife.state.nh.us
Charles Miner, CHIEF OF ACCESS AND ENGINEERING
cminer@wildlife.state.nh.us
Stephen Perry, CHIEF: INLAND FISHERIES
Ronald Alie, CHIEF: LAW ENFORCEMENT DIVISION
Phone: 603-271-3127
ralie@wildlife.state.nh.us
John Nelson, CHIEF: MARINE FISHERIES DIVISION
Durham, NH 603-868-1095
Judy Stokes, CHIEF: PUBLIC AFFAIRS DIVISION
Phone: 603-271-3211
jstokes@wildlife.state.nh.us
Steven Weber, CHIEF: WILDLIFE DIVISION
Phone: 603-271-2461
sweber@wildlife.state.nh.us
Richard Moquin, COMMISSION CHAIRMAN
212 Coolidge Ave., Manchester, NH 03102
Ellis Hatch, COMMISSION VICE CHAIRMAN
31 Harding Street, Rochester, NY 03867
Jim Jones, COMMISSIONER SECRETARY
501 Beanhill Rd., Norfield, NH 03276
Wayne Vetter, EXECUTIVE DIRECTOR
wvetter@wildlife.state.nh.us

NEW HAMPSHIRE NATURAL HERITAGE INVENTORY

P.O. Box 1856
Concord, NH 03302-1856 USA
Phone: 603-271-3623　　　Fax: 603-271-2629
Website: www.dred.state.nh.us

Founded: 1987
Scope: Regional

Description: New Hampshire Natural Heritage Inventory is
responsible for finding, tracking, and providing information
about the state's rare species and exemplary ecosystems.

Publications: List of New Hampshire's Rare Plant Species, List of
New Hampshire's Rare Animal Species, Checklist of New
Hampshire's Vascular Plants

Keyword(s): Biodiversity, Wildlife Rehabilitation, Endangered
Species, Ecology, Flowers, Plants, and Trees, Terrestrial
Habitats

Contact(s):
Lionel Chute, COORDINATOR

NEW HAMPSHIRE SEA GRANT PROGRAM

Kingman Farm, University of New Hampshire
Durham, NH 03824-3512 USA
Phone: 603-749-1565　　　Fax: 603-743-3997
Website: www.seagrant.unh.edu

Founded: NA
Scope: National

Keyword(s): Wetlands, Coral Reefs, Wildlife, Oceanography,
Aquatic Habitats, Marine Mammals, Whale, Dolphin, Seal, Sea
Grass, Sport Fishing, Sustainable Development, Sustainable
Ecosystems, Coasts, Renewable Resources, Research,
Environmental and Conservation Education

Contact(s):
Brian Doyle, ASSOCIATE DIRECTOR AND PROGRAM
LEADER
brian.doyle@unh.edu
Steve Adams, COORDINATOR, COMMUNICATIONS
steve.adams@unh.edu
Ann Bucklin, DIRECTOR
Phone: 603-862-0122
acb@cisunix.unh.edu

NEW JERSEY DEPARTMENT OF AGRICULTURE

P.O. Box 330
Trenton, NJ 08625 USA
Phone: 609-292-5530　　　Fax: 609-292-3978
Website: www.state.nj.us

Founded: NA
Membership: 210
Scope: Local
Keyword(s): Agriculture

Contact(s):
Samuel Garrison, ASSISTANT SECRETARY
Phone: 609-292-5530
Carol Shipp, CHIEF OF STAFF
Phone: 609-633-7794
John Gallagher, DIRECTOR OF DIVISION OF
ADMINISTRATION
Phone: 609-292-6931
Ernest Zirkle, DIRECTOR OF DIVISION OF ANIMAL HEALTH
Phone: 609-292-3965
P. Mullen, DIRECTOR OF DIVISION OF DAIRY AND
COMMODITY REGULATION, ACTING
Phone: 609-292-5575
A. Murray, DIRECTOR OF DIVISION OF MARKETS, ACTING
Phone: 609-292-5536
Robert Balaam, DIRECTOR OF DIVISION OF PLANT
INDUSTRY
Phone: 609-292-5441
George Horzepa, DIRECTOR OF DIVISION OF RURAL
RESOURCES
Phone: 609-292-5532
Gregory Romano, EXECUTIVE DIRECTOR OF STATE
AGRICULTURE DEVELOPMENT COMMITTEE
Phone: 609-984-2504

NEW JERSEY DEPARTMENT OF AGRICULTURE STATE SOIL & CONSERVATION COMMITTEE

P.O. Box 330
Trenton, NJ 08625 USA
Phone: 609-292-5540   Fax: 609-633-7229
Website: www.state.nj.us

Founded: 1937
Membership: 10
Scope: Statewide

Description: A unit of state government administered by the state Dept. of Agriculture. Responsible for conservation of soil resources and control and prevention of soil erosion and nonpoint source pollution, prevention of damage by floodwater or sediment, and conservation of water for agricultural purposes. Provides direction, leadership, standards, rules, funding, and administrative assistance; coordinates local district conservation programs; and is interagency with 12 members.

Publications: ON FARM STRATEGIES TO PROTECT WATER QUALITY, Standards for Soil Erosion and Sediment Control in New Jersey

Keyword(s): Exotic species, Aquatic nuisance species, Coral Reefs, Renewable Resources, Soil Conservation, Water Quality, Watersheds, Urban Environment, Agriculture

Contact(s):
Arthur Brown, CHAIRMAN
Phone: 609-292-3976
Fax: 609-292-3978
Samuel Race, EXECUTIVE SECRETARY
Phone: 609-292-5540
Fax: 609-633-7229
agurace@ag.state.nj.us

NEW JERSEY DEPARTMENT OF ENVIRONMENTAL PROTECTION

401 E State St.,
Trenton, NJ 08625-0402 USA
Phone: 609-292-2885   Fax: 609-292-7695
E-mail: askdep@dep.state.nj.us
Website: www.state.nj.us/dep

Founded: NA
Membership: 400
Scope: Statewide

Description: To assist the residents of New Jersey in preserving, sustaining, protecting and enhancing the environment to ensure the integration of high environmental quality, public health and economic vitality.

Publications: New Jersey Outdoors

Contact(s):
Marlen Dooley, ASSISTANT COMMISSIONER: ENFORCEMENT
Phone: 609-984-3285
Ray Cantor, ASSISTANT COMMISSIONER: LAND USE MANAGEMENT
401 E. State St., P.O. Box 439, 609-292-2178
Cari Wild, ASSISTANT COMMISSIONER: NATURAL AND HISTORIC RESOURCES
Phone: 609-292-3541
Sue Boyle, ASSISTANT COMMISSIONER: SITE REMEDIATION
Phone: 609-292-1250
Fax: 609-777-1914
Gary Sondermeyer, CHIEF OF STAFF
Phone: 609-292-2795
Fax: 609-292-7695
Robert Shinn, COMMISSIONER
Leslie McGeorge, DIRECTOR, DIV. OF SCIENCE & RESEARCH/ASSISTANT COMMISSIONER OF POLICY & PLANNING
Phone: 609-984-6070
Peter Page, DIRECTOR: COMMUNICATIONS
Phone: 609-777-1344
Fax: 609-292-1410
Denise Mikics, EDITOR
Phone: 609-777-4182

NEW JERSEY DEPARTMENT OF ENVIRONMENTAL PROTECTION

DIVISION OF FISH AND WILDLIFE
P.O. Box 400
Trenton, NJ 08625-0400 USA
Phone: 609-292-2965   Fax: 609-984-1414
Website: www.Njfishandwildlife.com

Founded: NA
Membership: 300
Scope: Statewide

Contact(s):
Martin McHugh, ASSISTANT DIRECTOR
Phone: 609-292-0891
David Chanda, ASSISTANT DIRECTOR
Phone: 609-292-0891
Lawrence Herrighty, CHIEF RURAL WILDLIFE MANAGEMENT
Phone: 609-292-6685
James Sciascia, CHIEF WILDLIFE INFORMATION EDUCATION
Phone: 609-292-9450
Tom McCloy, CHIEF: BUREAU OF MARINE FISHERIES
Phone: 609-984-5546
Robert Soldwedel, CHIEF: FRESHWATER FISHERIES
Phone: 609-292-8642
Tony Petrongolo, CHIEF: LANDS MANAGEMENT
Phone: 609-292-1599
Rob Winkel, CHIEF: LAW ENFORCEMENT
Phone: 609-292-9430
Jim Joseph, CHIEF: SHELL FISHERIES
Phone: 609-984-5546
Larry Niles, CHIEF: ENDANGERED AND NONGAME SPECIES PROGRAM
Phone: 609-292-9101
Robert McDowell, DIRECTOR
Phone: 609-292-9410

NEW JERSEY DEPARTMENT OF ENVIRONMENTAL PROTECTION

DIVISION OF PARKS AND FORESTRY
P.O. Box 404
Trenton, NJ 08625-0404 USA
Phone: 609-292-2733   Fax: 609-984-0503

Founded: NA
Scope: Regional

Contact(s):
Frank Gallagher, ADMINISTRATOR, OFFICE OF INTERPRETIVE & EDUCATIONAL SERVICES
Phone: 609-292-8190

Dorothy Guzzo, ADMINISTRATOR: OFFICE OF HISTORIC PRESERVATION
Phone: 609-984-0176
Richard Barker, ASSISTANT DIRECTOR: STATE PARK SERVICE
Phone: 609-292-2772
James Barresi, ASSISTANT DIRECTOR: STATE PARK SERVICE
Phone: 609-292-2530
Maris Gabliks, CHIEF: BUREAU OF FOREST FIRE MANAGEMENT AND STATE FIREWARDEN
Phone: 609-292-2977
Edward Lempicki, CHIEF: BUREAU OF FOREST MANAGEMENT: STATE FORESTER
Phone: 609-292-2531
Carl Nordstrom, DEPUTY DIRECTOR
Phone: 609-292-5990
Gregory Marshall, DIRECTOR

NEW JERSEY DEPARTMENT OF ENVIRONMENTAL PROTECTION

DIVISION OF PUBLICLY FUNDED SITE REMEDIATION
P.O. Box 402
Trenton, NJ 08625-0402 USA
Phone: 609-984-3081   Fax: 609-777-0756
Website: www.state.nj.us/dep/index.html

Founded: NA
Scope: Statewide

Description: To assist the residents of New Jersey in preserving, sustaining, protecting and enhancing the environment to ensure the integration of high environmental quality, public health and economic vitality.

Contact(s):
Anthony Farro, DIRECTOR

NEW JERSEY DEPARTMENT OF ENVIRONMENTAL PROTECTION

DIVISION OF SOLID AND HAZARDOUS WASTE
P.O. Box 414
Trenton, NJ 08625-0414 USA
Phone: 609-984-6880   Fax: 609-984-6874
E-mail: dshwed@vep.state.nj.us
Website: www.statenj.us/dep

Founded: NA
Scope: Statewide

Contact(s):
John Castler, DIRECTOR

NEW JERSEY DEPARTMENT OF ENVIRONMENTAL PROTECTION

GEOLOGICAL SURVEY
P.O. Box 427
Trenton, NJ 08625-0427 USA
Phone: 609-292-1185   Fax: 609-633-1004
Website: www.state.nj.us/dep/njgs/

Founded: 1835
Membership: 43
Scope: Regional

Description: Formed to study, evaluate, and prepare maps and reports on New Jersey's resources. In addition to a geologic map and information on the mineral industry and water resources, the survey provides geologic and ground water reports, geologic and topographic maps, ground water monitoring, and other resource information.

Keyword(s): Exotic species, Aquatic nuisance species, Geology

Contact(s):
Richard Dalton, CHIEF OF BUREAU OF GEOLOGY AND TOPOGRAPHY
Phone: 609-292-2576
Thomas Seckler, EDITOR
Phone: 609-292-2576
Carl Muessig, STATE GEOLOGIST
Phone: 609-292-1185

NEW JERSEY DEPARTMENT OF ENVIRONMENTAL PROTECTION

GREEN ACRES AND RECREATION PROGRAM
Trenton, NJ 08625-0412 USA
Phone: 609-984-0500   Fax: 609-984-0608
Website: www.state.nj.us\dep\greenacres

Founded: NA
Membership: 60
Scope: Statewide

Contact(s):
Thomas Wells, ADMINISTRATOR
Martin McHugh, CHIEF OF OFFICE OF NATURAL RESOURCE DAMAGES
Phone: 609-984-5475
Robert Stokes, CHIEF OF OUTDOOR RECREATION PLANNING
Phone: 609-984-0495
Dennis Davidson, DEPUTY ADMINISTRATOR
Phone: 609-984-0555

NEW JERSEY PINELANDS COMMISSION

P.O Box 7
New Lisbon, NJ 08064 USA
Phone: 609-894-7300   Fax: 609-894-7330
E-mail: info@njpines.state.nt.us
Website: www.state.nj.us/pinelands/

Founded: 1979
Membership: 50
Scope: Statewide

Description: State planning and regulatory agency with jurisdiction over land use and development in the million-acre Pinelands national reserve; 53 municipalities in the state Pinelands area have and revise local master plans and zoning ordinances to incorporate standards of regional conservation plan

Publications: a list of reports and studies is available upon request., Pinelander, The Newsletter

Keyword(s): Exotic species, Aquatic nuisance species, Sustainable Ecosystems, Land Use Planning, Planning Management, Protected Areas

Contact(s):
William Harrison, ASSISTANT DIRECTOR OF DEVELOPMENT REVIEW AND ENFORCEMENT
John Stokes, ASSISTANT DIRECTOR OF PLANNING AND MANAGEMENT
Jerrold Jacobs, CHAIRMAN
Elizabeth Carpenter, EDUCATIONAL COORDINATOR
Annette Barbaccia, EXECUTIVE DIRECTOR

NEW JERSEY SEA GRANT COLLEGE PROGRAM

New Jersey Marine Sciences Consortium, Bldg. 22
Fort Hancock, NJ 07732 USA
Phone: 732-872-1300　　　Fax: 732-291-4483
Website: www.njmsc.org/seagrant.htm

Founded: NA

Description: The New Jersey Marine Sciences Consortium is an alliance of 29 institutions from New Jersey, New York, and Pennsylvania formed for the purposes of conducting sponsored research in marine and coastal sciences, technology development through group action; and assembling material resources which lie beyond the capabilities of the individual member institutions. The consortium manages the Sea Grant College Program and the Sea Grant Extension Program.

Contact(s):
Eleanor Bochenek, ASSOCIATE SEA GRANT DIRECTOR/SEA GRANT EXTENSION DIRECTOR
Eleanor@njmsc.org
Michael Weinstein, DIRECTOR
mikew@njmsc.org

NEW MEXICO BUREAU OF GEOLOGY AND MINERAL RESOURCES

801 Leroy Pl.
Socorro, NM 87801 USA
Phone: 505-835-5420　　　Fax: 505-835-6333
E-mail: pubsosc@gis.nmt.edu
Website: www.geoinfo.nmt.edu

Founded: 1927
Membership: 75
Scope: Local

Description: Charged with investigating and reporting on all types of mineral resources and the geology of the state, including environmental geology, water resources, and geological hazards; responsible for conducting applied research on all aspects of geology and mineral resources.

Publications: Open File Report, Topo Maps, Circulars, Memoirs, Ground Water Reports, Geologic Maps, New Mexico Geology, Lite Geology, Scenic Trips to the Geologic Past, Databases on CD-ROM and home page, Bulletins

Keyword(s): Exotic species, Aquatic nuisance species, education, Energy, Geology, Chemistry

Contact(s):
Peter Scholle, DIRECTOR AND STATE GEOLOGIST
Phone: 505-835-5302
pscholle@gis.nmt.edu
Jane Love, EDITOR
jane@gisnmt.edu
David Love, ENVIRONMENTAL GEOLOGIST
Phone: 505-835-5146
dave@gis.nmt.edu
Bruce Allen, ENVIRONMENTAL GEOLOGIST
Phone: 505-255-0317
allenb@gis.nmt.edu
Susan Welch, MANAGER OF GEOLOGICAL EXTENSION SERVICE
Phone: 505-835-5112
susie@nmt.edu

NEW MEXICO BUREAU OF GEOLOGY AND MINERAL RESOURCES

GEOLOGICAL INFORMATION CENTER LIBRARY
801 Leroy Place
Socorro, NM 87801 USA
Phone: 505-835-5145　　　Fax: 505-835-6333
E-mail: bureau@gis.nmt.edu
Website: www.geoinfo.nmt.edu/

Founded: NA
Membership: 65
Scope: Statewide
Publications: New Mexico Geology—quarterly newsletter
Keyword(s): Librarians/Information Professionals

Contact(s):
Peter Scholle, CONTACT

NEW MEXICO COOPERATIVE FISH AND WILDLIFE RESEARCH UNIT

P.O. Box 30003, MSC 4901, New Mexico State University
Las Cruces, NM 88003-0003 USA
Phone: 505-646-6053　　　Fax: 505-646-1281
E-mail: coopunit@nmsu.edu
Website: www.leopold.nmsu/fwscoop

Founded: 1988
Membership: 20
Scope: National

Description: Supported cooperatively by the U.S.G.S. Biological Resources Division, New Mexico State University, New Mexico Department of Game and Fish, and the Wildlife Management Institute, the New Mexico Fish and Wildlife Research Unit's primary purpose is research on management and conservation of fish and wildlife species and graduate research training in fisheries and wildlife resources.

Keyword(s): training, Toxic Substances, Nuclear-free, Water quantity, Water export and diversion, Arid Lands, Endangered Species, Conservation Plannning, Aquatic Habitats

Contact(s):
Colleen Caldwell, ASSISTANT LEADER FOR FISHERIES
ccaldwel@nmsu.edu
Lewis Bender, ASSISTANT UNIT LEADER
Bruce Thompson, UNIT LEADER
bthompso@nmsu.edu

NEW MEXICO DEPARTMENT OF AGRICULTURE

MSC 3189, P.O. Box 30005
Las Cruces, NM 88003-8005 USA
Phone: 505-646-3007
Website: nmdaweb.nmsu.edu

Founded: 1955
Membership: 120
Scope: Statewide

Description: Organized to protect state agriculture from importation of plant diseases and insects and help control those that gain entrance; to ensure products offered for sale meet quality standards as advertised and labeled; maintain inspection of agricultural products for interstate shipping; laboratory analyses of animal diseases and deaths on fee basis; promote state agricultural commodities; provide market news; and conduct consumer and producer service activities designated by law.

Publications: New Mexico Agricultural Statistics, Biennial Report

Contact(s):
  Jeff Witte, ASSISTANT DIRECTOR
  Phone: 505-646-3007
  Frank Dubois, DIRECTOR AND SECRETARY
  Phone: 505-646-3007
  Larry Dominguez, DIRECTOR OF AGRICULTURAL AND
  ENVIRONMENTAL SERVICES DIVISION
  Phone: 505-646-3208
  Ronald White, DIRECTOR OF AGRICULTURAL PROGRAMS
  AND RESOURCES DIVISION
  Phone: 505-646-2642
  Edward Avalos, DIRECTOR OF MARKETING AND
  DEVELOPMENT DIVISION
  Phone: 505-646-4929
  Gary West, DIRECTOR OF STANDARDS AND CONSUMER
  SERVICES DIVISION
  Phone: 505-646-1616
  Richard Larock, DIRECTOR OF VETERINARY DIAGNOSTIC
  SERVICES
  Phone: 505-841-2576
  Rick Janecka, STATE CHEMIST OF LABORATORY
  Phone: 505-646-3318
  Richard Kochevar, STATE SEED ANALYST OF
  LABORATORY
  Phone: 505-646-3407

NEW MEXICO DEPARTMENT OF GAME AND FISH

P.O. Box 25112
Santa Fe, NM 87504 USA
Phone: 505-827-7911   Fax: 505-476-8124
E-mail: iispa@state.nm.us
Website: www.gmfsh.state.nm.us

Founded: NA
Membership: 200
Scope: Statewide

Description: The State Game Commission and the Game and
Fish Department are administratively attached to the Energy,
Minerals, and Natural Resources Department. The responsibil-
ity of the State Game Commission is to develop policy for the
Game and Fish Department.

Publications: Publications on line

Contact(s):
  Roberta Salazar-Henry, ASSISTANT DIRECTOR OF
  ADMINISTRATIVE SERVICES AND INFORMATION
  SYSTEMS SERVICES
  Phone: 505-827-6333
  rhenry@state.nm.us
  Scott Brown, ASSISTANT DIRECTOR OF RESOURCE
  DIVISIONS
  Phone: 505-827-6333
  sbrown@state.nm.us
  Jennifer Salisbury, CABINET SECRETARY OF ENERGY,
  MINERALS, AND NATURAL RESOURCES
  Phone: 505-827-5950
  jsalisbury@state.nm.us
  Steven Emery, CHAIRMAN OF STATE GAME COMMISSION
  Phone: 505-856-0963
  semery@state.nm.us
  Lydia Duran, CHIEF OF ADMINISTRATIVE SERVICES
  (ACTING)
  Phone: 505-827-7920
  lduran@state.nm.us
  Tod Stevenson, CHIEF OF CONSERVATION SERVICES
  Phone: 505-827-7882
  tstevenson@state.nm.us
  Jack Kelly, CHIEF OF FISH MANAGEMENT
  Phone: 505-827-7905
  jkelly@state.nm.us
  Dan Brook, CHIEF OF LAW ENFORCEMENT
  Phone: 505-827-7934
  dbrook@state.nm.us
  Don MacCarter, CHIEF OF PUBLIC AFFAIRS
  Barry Hale, CHIEF OF WILDLIFE DIVISION
  Phone: 505-827-7885
  bhale@state.nm.us
  Larry Bell, DIRECTOR
  P.O. Box 25112, SantaFe, NM 87504
  Phone: 505-827-6333
  lbell@state.nm.us

NEW MEXICO DEPARTMENT OF GAME AND FISH

ALBUQUERQUE NM OFFICE
3481 Midway Pl., NE
Albuquerque, NM 87109 USA
Phone: 505-841-8881   Fax: 505-841-8885
Website: www.gmfsh.state.nm.us

Founded: NA
Membership: 34
Scope: Statewide

Contact(s):
  Luke Shelby, CHIEF OF OPERATIONS NW AREA
  Chris Chadwick, PUBLIC AFFAIRS SPECIALIST

NEW MEXICO DEPARTMENT OF GAME AND FISH

LAS CRUCES NM OFFICE
566 N. Telshor Blvd.
Las Cruces, NM 88011 USA
Phone: 505-522-9796
Website: www.gmfsh.state.nm.us

Founded: NA
Membership: 30
Scope: Regional

Contact(s):
  Steve Henry, CHIEF

NEW MEXICO DEPARTMENT OF GAME AND FISH

RATON NM OFFICE
P.O. Box 1145, 215 York Canyon Rd.
Raton, NM 07740 USA
Phone: 505-445-2311   Fax: 505-445-5651
E-mail: gmfsh@state.nm.us
Website: www.gmfsh.state.nm.us

Founded: NA
Scope: Statewide

Contact(s):
  Joanna Lackey, CHIEF
  jlackey@state.nm.us

NEW MEXICO DEPARTMENT OF GAME AND FISH

ROSWELL NM OFFICE
1912 West 2nd St.
Roswell, NM 88201 USA
Phone: 505-624-6135   Fax: 505-624-6136
Website: www.gmfsh.state.nm.us

Founded: NA
Scope: Statewide

Contact(s):
Roy Hayes, CHIEF

NEW MEXICO ENVIRONMENT DEPARTMENT

1190 Saint Francis Dr., P.O. Box 26110
Santa Fe, NM 87502 USA
Phone: 505-827-2855   Fax: 505-827-2836
Website: www.nmenv.state.nm.us

Founded: NA
Membership: 650
Scope: Statewide

Description: To preserve, protect, and perpetuate New Mexico's environment for present and future generations.

Contact(s):
John Parker, CHIEF OF DEPARTMENT OF ENERGY OVERSIGHT BUREAU
Phone: 505-827-4252
john_parker@nmenv.state.nm.us
Butch Tongate, CHIEF OF SOLID WASTE BUREAU
Phone: 505-827-2775
butch_tongate@nmenv.state.nm.us
Sandra Eli, CHIEF OF AIR QUALITY BUREAU
Phone: 505-827-1494
sandra_eli@nmenv.state.nm.us
Marcy Leavitt, CHIEF OF GROUND WATER PROTECTION AND REMEDIATION BUREAU
Phone: 505-827-2919
marcy_leavitt@nmenv.state.nm.us
James Bearzi, CHIEF OF HAZARDOUS & RADIOACTIVE MATERIALS BUREAU
Phone: 505-827-1557
james_bearzi@nmenv.state.nm.us
Sam Rogers, CHIEF OF OCCUPATIONAL HEALTH AND SAFETY BUREAU
Phone: 505-827-2877
sam_rogers@nmenv.state.nm.us
Cliff Hawley, CHIEF OF PROGRAM SUPPORT BUREAU
Phone: 505-827-2844
cliff_hawley@nmenv.state.nm.us
Jerry Scheppner, CHIEF OF UNDERGROUND STORAGE TANK BUREAU
Phone: 505-827-0188
jerry_sheppner@nmenv.state.nm.us
Robert Horwitz, DIRECTOR OF ADMINISTRATIVE SERVICES DIVISION
Phone: 505-827-2773
robert_horwitz@nmenv.state.nm.us
Arie Gruebel, DIRECTOR OF ENVIRONMENTAL PROTECTION DIVISION
Mike Koranda, DIRECTOR OF FIELD OPERATIONS DIVISION
Phone: 505-827-1080
mike_koranda@nmenv.state.nm.us
Greg Lewis, DIRECTOR OF WATER AND WASTE MANAGEMENT DIVISION
Phone: 505-827-2886
greg_lewis@nmenv.state.nm.us
Courte Vohres, MANAGER OF DISTRICT II
Phone: 505-827-1840
courte_vohres@nmenv.state.nm.us
Ken Smith, MANAGER OF DISTRICT III
Phone: 505-524-6300
ken_smith@nmenv.state.nm.us
Darwin Pattengale, MANAGER OF DISTRICT IV
Phone: 505-624-6046
darwin_pattengale@nmenv.state.nm.us
Pete Maggiore, SECRETARY
Phone: 505-827-2855
pete_maggiore@nmenv.state.nm.us

NEW MEXICO SOIL AND WATER CONSERVATION COMMISSION

MSC APR PO BOX 30005
Las Cruces, NM 88003-8005 USA
Phone: 505-646-2642   Fax: 505-646-1540
E-mail: acoleman@nmda-bubba.nmsu.edu

Founded: NA
Membership: 7
Scope: Statewide

Contact(s):
Ron White, DIVISION DIRECTOR
P.O. BOX 30005 MSC APR, Las Cruces, NM 88003-8005
Phone: 505-524-6210
Fax: 505-524-6211

NEW MEXICO STATE UNIVERSITY

COOPERATIVE EXTENSION SERVICES
COLLEGE OF AG AND HOME ECONOMICS
Box 30003, Campus Box 3AG
Las Cruces, NM 88003 USA
Phone: 505-646-3748   Fax: 505-646-5975
E-mail: agdean@nmsu.edu
Website: www.cahe.nmsu.edu/ces/

Founded: NA
Scope: Statewide

Contact(s):
Jerry Schickedanz, DEAN AND CHIEF ADMINISTRATIVE OFFICER
Billy Dictson, ASSOCIATE DEAN AND DIRECTOR CES
NM State University, Box 3AE, Las Cruces, NM 88003
Phone: 505-646-3015
Ron Parker, EXTENSION DEPARTMENT HEAD OF ANIMAL RESOURCES
Box 3AE, NM State University, Las Cruces, NM 88003
Phone: 505-646-1709
Ron Byford, EXTENSION DEPARTMENT HEAD OF PLANT SCIENCES
Box 3AE, NM State University, Las Cruces, NM 88003
Phone: 505-646-2458
Chris Allison, EXTENSION RANGE MANAGEMENT SPECIALIST
Box 3AE, NM State University, Las Cruces, NM 88003
Phone: 505-646-1944
Jon Boren, EXTENSION WILDLIFE SPECIALIST
Box 3AE, NM State University, Las Cruces, NM 88003
Phone: 505-646-1164
Fax: 505-646-1281

NEW YORK COOPERATIVE FISH AND WILDLIFE RESEARCH UNIT

Department of Natural Resources, Fernow Hall, Cornell University
Ithaca, NY 14853 USA
Phone: 607-255-2839   Fax: 607-255-1895
E-mail: dnrcru-mailbox@cornell.edu
Website: www.dnr.cornell.edu/f@wres/nycf@wru.htm

Founded: 1961 (Wildlife Unit), 1963 (Fisheries Unit), 1984
Scope: National

Description: Primary purpose is field and laboratory research on management and conservation of a variety of fish and wildlife species, and graduate research training in fisheries and wildlife resources. Supported cooperatively by U.S. Geological Survey, Cornell University, New York State Department of Environmental Conservation, and the Wildlife Management Institute.

Publications: see publication website

Contact(s):
Mark Bain, ASSISTANT LEADER OF FISHERIES
Phone: 607-255-2840
mbb1@cornell.edu
Richard Malecki, ASSISTANT LEADER OF WILDLIFE
Phone: 607-255-2836
ram26@cornell.edu
Milo Richmond, LEADER
Phone: 607-255-2151
mer6@cornell.edu

NEW YORK DEPARTMENT OF AGRICULTURE AND MARKETS

1 Winners Cir.
Albany, NY 12235 USA
Phone: 518-457-3880   Fax: 518-457-3087
E-mail: info@agmkt.state.ny.us
Website: www.agmkt.state.ny.us

Founded: 1884
Membership: 400
Scope: Statewide

Description: Promotes and regulates production, manufacturing, marketing, storing, and distribution of food. Supervises quality of plant materials, health of animals, and regulates dogs. Also, represents agricultural interests before NY Public Service Commission on siting of transmission lines and power plants.

Publications: available on website

Keyword(s): Soil Conservation, Pesticides, Biotechnology, Flowers, Plants, and Trees, Agriculture

Contact(s):
Nathan Rudgers, COMMISSIONER
Kim Blot, DIRECTOR OF DIVISION OF AGRICULTURAL PROTECTION AND SUPPORT SERVICES
Robert Mungari, DIRECTOR OF DIVISION OF PLANT INDUSTRY
Jessica Chittenden, PUBLIC INFORMATION OFFICER

NEW YORK DEPARTMENT OF AGRICULTURE AND MARKETS

STATE SOIL AND WATER CONSERVATION COMMITTEE
1 Winners Circle
Albany, NY 12235 USA
Phone: 518-457-3738   Fax: 518-457-3412
Website: www.agmkt.state.ny.us

Founded: NA
Membership: 9
Scope: Statewide
Publications: newsletter - Down to Earth (once a month)

Contact(s):
Philip Griffen, CHAIR
28 Spook Hollow Rd., Stillwater, NY 12170
Phone: 518-664-5038
John Wildeman, DIRECTOR
Phone: 518-457-3738
Fax: 518-457-1204

NEW YORK DEPARTMENT OF ENVIRONMENTAL CONSERVATION

625 Broadway 12th Fl.
Albany, NY 12233 USA
Phone: 518-474-2121   Fax: 518-402-9392
Website: www.dec.state.ny.us

Founded: 1970
Scope: Statewide

Description: The mission of the New York State Department of Environmental Conservation is to conserve, improve, and protect its natural resources and environment, and control water, land and air pollution, in order to enhance the health, safety and welfare of the people of the state and their overall economic and social well-being.

Publications: Environmental Notice Bulletin

Contact(s):
John Kelly, ADIRONDACKS
Phone: 518-623-3671
Susan Taluto, ASSISTANT COMMISSIONER FOR ADMINISTRATIVE SERVICES
Phone: 518-457-6533
John McKeon, ASSISTANT COMMISSIONER FOR OFFICE OF BOND ACT
Phone: 518-402-9401
James Tuftey, ASSISTANT COMMISSIONER FOR THE OFFICE OF PUBLIC PROTECTION
James Ferreira, ASSISTANT COMMISSIONER OF OFFICE OF HEARINGS AND MEDIATION SERVICES
Erin Crotty, COMMISSIONER
Phone: 518-402-8540
Frank Bifera, DEPUTY COMMISSIONER AND COUNSEL
Phone: 518-457-4415
Peter Duncan, DEPUTY COMMISSIONER FOR NATURAL RESOURCES
Phone: 518-457-0975
Carl Johnson, DUPUTY COMMISSIONER AIR AND WASTE MANAGEMENT
Thomas Kelly, ENVIRONMENTAL FACILITIES CORPORATION
Phone: 518-402-6951
Fran Dunwell, HUDSON RIVER
Phone: 914-256-3017

Gordon Colvin, MARINE RESOURCES
Phone: 631-444-0430
Francis Sheehan, NATURAL RESOURCES PLANNING
Phone: 518-457-4208
Fran Verdoliva, SALMON RIVER
Phone: 315-298-7605
Linda Frick, SPECIAL ASSISTANT TO THE COMMISSIONER
Phone: 518-457-0904
Tom Kunkel, SPECIAL PROJECTS
Phone: 718-482-4949
Jim Austin, SPECIAL PROJECTS
Phone: 518-485-8437

NEW YORK DEPARTMENT OF ENVIRONMENTAL CONSERVATION

DIVISION OF AIR RESOURCES
50 Wolf Rd.
Albany, NY 12233 USA
Phone: 518-457-7230

Founded: NA
Scope: Statewide

Contact(s):
Robert Warland, DIRECTOR
Phone: 518-457-7230

NEW YORK DEPARTMENT OF ENVIRONMENTAL CONSERVATION

DIVISION OF ENVIRONMENTAL PERMITS REGION 4
1150 N. Westcott Rd.
Schenectady, NY 12306 USA
Phone: 518-357-2069   Fax: 518-357-2460

Founded: NA
Scope: Statewide

NEW YORK DEPARTMENT OF ENVIRONMENTAL CONSERVATION

DIVISION OF ENVIRONMENTAL REMEDIATION
625 Broadway
Albany, NY 12233 USA
Phone: 800-342-9296
Website: www.dec.state.ny.us

Founded: NA
Scope: Statewide

Contact(s):
Michael Otoole, DIRECTOR
Phone: 518-457-5861

NEW YORK DEPARTMENT OF ENVIRONMENTAL CONSERVATION

DIVISION OF FISH, WILDLIFE AND MARINE RESOURCES
1150 Westcott Road
Schenectady, NY 12306 USA
Phone: 508-357-2234
Website: http://www.dec.state.ny.us/website/dfwmr/

Founded: NA
Scope: Statewide

Contact(s):
Gerry Barnhart, DIRECTOR
Phone: 518-457-5690

NEW YORK DEPARTMENT OF ENVIRONMENTAL CONSERVATION

DIVISION OF FOREST PROTECTION & FIRE MANAGEMENT
50 Wolf Rd
Albany, NY 12233-2560 USA
Phone: 518-402-8839   Fax: 518-485-8458
Website:
www.dec.state.ny.us/website/protection/rangers/index.html

Founded: NA
Scope: Statewide

Contact(s):
Thomas Rinaldi, DIRECTOR (ACTING)
Phone: 518-457-5740

NEW YORK DEPARTMENT OF ENVIRONMENTAL CONSERVATION

DIVISION OF INFORMATION SERVICES
625 Broadway
Albany, NY 12233 USA
Phone: 518-457-6367
Website: www.dec.state.ny.us

Founded: NA
Scope: Statewide

NEW YORK DEPARTMENT OF ENVIRONMENTAL CONSERVATION

DIVISION OF LAW ENFORCEMENT
50 Wolf Rd.
Albany, NY 12233-2500 USA
Phone: 518-402-8829   Fax: 518-485-8449
Website: http://www.dec.state.ny.us/website/dle/index.htm

Founded: NA
Scope: Statewide

Contact(s):
Lawrence Johnson, ACTING DIRECTOR
Dave Egelston, COLONEL
Phone: 518-457-5681

NEW YORK DEPARTMENT OF ENVIRONMENTAL CONSERVATION

DIVISION OF MINERAL RESOURCES
50 Wolf Rd.
Albany, NY 12233-6500 USA
Phone: 518-457-9337   Fax: 518-457-9298
Website: www.dec.state.ny.us/website/dmn

Founded: NA
Scope: Statewide

Contact(s):
Gregory Sovas, DIRECTOR
Phone: 518-457-9337

NEW YORK DEPARTMENT OF ENVIRONMENTAL CONSERVATION

DIVISION OF OPERATIONS
50 Wolf Rd.
Albany, NY 12233-5250 USA
Phone: 518-457-6310
Website: www.dec.state.ny.us

Founded: NA
Scope: Statewide

Contact(s):
Michael Turley, DIRECTOR
Phone: 518-457-6310

NEW YORK DEPARTMENT OF ENVIRONMENTAL CONSERVATION
DIVISION OF PUBLIC AFFAIRS AND EDUCATION
50 Wolf Rd., Rm 548
Albany, NY 12233 USA
Phone: 518-457-1141
Website: www.dec.state.ny.us

Founded: NA
Scope: Statewide

Contact(s):
Laurel Remus, DIRECTOR
Phone: 518-457-0840

NEW YORK DEPARTMENT OF ENVIRONMENTAL CONSERVATION
DIVISION OF SOLID & HAZARDOUS MATERIALS
625 Broadway
Albany, NY 12232 USA
Phone: 518-402-8540
Website: http://dec.state.ny.us

Founded: NA

Contact(s):
Erin Crotty, COMMISSIONER
Phone: 518-402-8540
Stephen Hammond, DIRECTOR, DIVISION OF SOLID & HAZARDOUS MATERIALS
Phone: 518-402-8651

NEW YORK DEPARTMENT OF ENVIRONMENTAL CONSERVATION
DIVISION OF WATER
50 Wolf Rd.
Albany, NY 12233 USA
Phone: 518-402-8233   Fax: 518-402-8230

Founded: NA
Scope: Statewide

Contact(s):
N. Kaul, DIRECTOR
Phone: 518-457-6674

NEW YORK DEPARTMENT OF ENVIRONMENTAL CONSERVATION
PRESS OFFICE
625 Broadway
Albany, NY 12232 USA
Phone: 518-402-8000   Fax: 518-402-2209
Website: http://dec.state.ny.us

Founded: NA

Contact(s):
Jennifer Post, PUBLIC INFORMATION OFFICER

NEW YORK DEPARTMENT OF ENVIRONMENTAL CONSERVATION
REGIONAL DIRECTORS
USA

Founded: NA

Contact(s):
Raymond Cowen, REGION 1
Bldg. 40, State University of New York, Stony Brook, NY 11794
Phone: 516-444-0354
Mary Kris, REGION 2
Hunters Point Plaza, Long Island City, NY 11101
Phone: 718-482-4900
Marc Moran, REGION 3
21 S. Putt Corners Rd., New Paltz, NY 12561
Phone: 845-256-3000
Steve Schassler, REGION 4
11 North Westcott Road, Schenectady, NY 12306
Phone: 518-357-2234
Stuart Buchanan, REGION 5
Route 86, P.O. Box 296, Ray Brook, NY 12977
Phone: 518-897-1200
Sandy Lebarron, REGION 6
317 Washington Street, Watertown, NY 13204
Phone: 315-785-2239
Kenneth Lynch, REGION 7
615 Erie Blvd., W, Syracuse, NY 13204
Phone: 315-426-7400
John Hicks, REGION 8
6274 E. Avon-Lima Road, Avon, NY 14414
Phone: 716-226-2466
Gerald Mikol, REGION 9
270 Michigan Avenue, Buffalo, NY 14203
Phone: 716-851-7000

NEW YORK DEPARTMENT OF HEALTH
Tower Bldg., Empire State Plaza
Albany, NY 12237 USA
Phone: 1 800 4-58 -1158
Website: http://health.state.ny.us

Founded: NA

Contact(s):
Ronald Trammontano, DIRECTOR OF CENTER FOR ENVIRONMENTAL HEALTH (P.E.)
Phone: 518-458-6400
1 800-458-1158

NEW YORK GEOLOGICAL SURVEY AND STATE MUSEUM
Cultural Education Center
Albany, NY 12230 USA
Phone: 518-474-5816   Fax: 518-486-2034
Website: www.nysm.nysed.gov

Founded: 1836
Scope: State

Description: The Geological Survey and State Museum serves as a clearinghouse for information concerning bedrock and surficial geology within the state. The survey conducts regular mapping projects and investigations in basic, environmental, and applied geology and publishes maps and reports of investigations.

Publications: publications list of the New York State Geological

Survey, New York State Geogram

Keyword(s): Exotic species, Aquatic nuisance species, Museum, Land Use Planning, Energy, Geology, Coasts

Contact(s):
Robert Fickies, ENGINEERING AND ENVIRONMENTAL GEOLOGY, GEOLOGIC INFORMATION-OPEN FILE
Phone: 518-474-5810
Richard Nyahay, OIL AND GAS OFFICE DIRECTOR
Phone: 518-486-2161
Robert Fakundiny, STATE GEOLOGIST AND CHIEF SCIENTIST
Phone: 518-474-5816
rnyahay@mail.nysev.gov

NEW YORK SEA GRANT
121 Discovery Hall, SUNY at Stony Brook
Stony Brook, NY 11794-5001 USA
Phone: 631-632-6905　　　Fax: 631-632-6917
E-mail: NYSeaGrant@notes.cc.sunysb.edu
Website: www.seagrant.sunysb.edu/

Founded: 1971

Description: A cooperative program of the State University of New York and Cornell University fostering the wise use and development of coastal resources through research grants, extension advisory services, education, training, and informational materials.

Keyword(s): Exotic species, Aquatic nuisance species, Scholarships and Grants, Environmental and Conservation Education, Research, Coasts

Contact(s):
Paul Focazio, ASSISTANT COMMUNICATOR
Phone: 516-632-6910
Cornelia Schlenk, ASSISTANT DIRECTOR
Dale Baker, ASSOCIATE DIRECTOR AND PROGRAM LEADER
New York Sea Grant, 348 Roberts Hall, Cornell University, Ithaca, NY 14853-4203
Phone: 607-255-2832
Barbara Branca, COMMUNICATOR
Phone: 516-632-6956
Jack Mattice, DIRECTOR
jmattice@ccmail.sunysb.edu
Stefanie Massucci, FISCAL OFFICER OF NEW YORK SEA GRANT
David White, GREAT LAKES PROGRAM COORDINATOR
New York Sea Grant, 101 Rich Hall, SUNY College at Oswego, Oswego, NY 13126-3599
Phone: 315-341-3042
Robert Kent, MARINE PROGRAM COORDINATOR
New York Sea Grant, Cornell University Lab, 3059 Sound Ave., Riverhead, NY 11901-1098
Phone: 516-727-3910

NEW YORK STATE COOPERATIVE EXTENSION
New York State College of Agriculture and Life Sciences, and Human Ecology, 365 Roberts Hall, Cornell University
Ithaca, NY 14853-4203 USA
Phone: 607-255-2237　　　Fax: 607-255-2473
E-mail: cce@cornell.edu
Website: www.cce.cornell.edu/

Founded: NA
Membership: 10
Scope: Statewide

Keyword(s): Exotic species, Aquatic nuisance species, Solid Waste Management, Environmental and Conservation Education, Renewable Resources, Wildlife, Planning Management, Sustainable Ecosystems, Forest Management, Wetlands, Agriculture

Contact(s):
R. Smith, AGRICULTURE
D. Ewert, DIRECTOR OF COOPERATIVE EXTENSION
Marianne Krasny, ENVIRONMENTAL/CONSERVATION YOUTH EDUCATION
Associate Professor, Dept. of Natural Resources, 16 Fernow Hall, Cornell University, Ithaca, NY 14853-3001
Phone: 607-255-2827
Peter Smallidge, FORESTRY RESOURCE MANAGEMENT
Sr. Extension Associate, Dept. of Natural Resources, 116 Fernow Hall, Cornell University, Ithaca, NY 14853-3001
Phone: 607-255-4696
Gary Goff, FORESTRY/WILDLIFE
Extension Associate/Director Master Forest Owners/COVERTS Volunteer Program, Dept. of Natural Resources, 104 Fernow Hall, Cornell University, Ithaca, NY 14853-3001
Phone: 607-255-2824
Tommy Brown, HUMAN DIMENSIONS RESEARCH UNIT
Sr. Res. Assoc., Dept. of Natural Resources, 122B Fernow Hall, Cornell University, Ithaca, NY 14853-3001
Phone: 607-255-7695
David Gross, PROTECTED AREA PLANNING AND MANAGEMENT
Sr. Extension Associate & Environmental Program Leader, Department of Natural Resources, 112 Fernow Hall, Cornell University, Ithaca, NY 14853-3001
Phone: 607-255-2825
Steve Brown, SPORTFISHING AND AQUATIC RESOURCES EDUCATION (SAREP)
Director, Extension Associate, Dept. of Natural Resources, 120 Fernow Hall, Cornell University, Ithaca, NY 14853-3001
Phone: 607-255-9370
Rebecca Schneider, WETLANDS
Assistant Professor, Dept. of Natural Resources, 122C Fernow Hall, Cornell University, Ithaca, NY 14853-3001
Phone: 607-255-2110
Paul Curtis, WILDLIFE MANAGEMENT
Sr. Extension Associate, Dept. of Natural Resources, 114 Fernow Hall, Cornell University, Ithaca, NY 14853-3001
Phone: 607-255-2835
Fax: 607-255-2815

NEW YORK STATE DEPARTMENT OF ENVIRONMENTAL CONSERVATION
DIVISION OF ENVIRONMENTAL ENFORCEMENT
50 Wolf Rd.
Albany, NY 12233 USA
Phone: 518-402-8829　　　Fax: 518-485-8478

Founded: NA
Scope: Statewide

Contact(s):
Charles Sullivan, DIRECTOR
Phone: 518-457-4348

NEW YORK STATE DEPARTMENT OF ENVIRONMENTAL CONSERVATION

DIVISION OF LANDS AND FORESTS
625 Broadway
Albany, NY 12233 USA
Phone: 518-402-9425
Website: www.dec.state.ny.us

Founded: NA
Scope: Statewide

Contact(s):
    Frank Dunstan, DIRECTOR
    Phone: 518-457-2475

NEW YORK STATE DEPARTMENT OF ENVIRONMENTAL CONSERVATION

DIVISION OF LEGAL AFFAIRS
50 Wolf Rm. 638
Albany, NY 12233 USA
Phone: 518-457-3551   Fax: 518-457-3978

Founded: NA
Scope: Statewide

Contact(s):
    Alison Smith, DIRECTOR
    Phone: 518-457-3551

NEW YORK STATE DEPARTMENT OF ENVIRONMENTAL CONSERVATION

DIVISION OF MANAGEMENT AND BUDGET SERVICES
1150 Westcott Road
Schenectady, NY 12306 USA
Website: www.dec.state.ny.us

Founded: NA
Scope: Statewide

Contact(s):
    Richard Randless, DIRECTOR
    Phone: 518-457-1141

NEW YORK STATE DEPARTMENT OF ENVIRONMENTAL CONSERVATION

REGION 9
182 E. Union,
Allegany, NY 14706 USA
Phone: 716-372-0645   Fax: 716-372-2113
Website: www.gw.dec.state.ny.us

Founded: NA
Scope: Regional

Contact(s):
    Emery Green, FWMA BOARD CHAIRMAN
    Steve Mooradian, REGIONAL FISH MANAGER
    Russ Biss, REGIONAL WILDLIFE MANAGER

NEW YORK STATE FISH AND WILDLIFE MANAGEMENT BOARD

625 Broadway
Albany, NY 12233 USA
Phone: 518-402-8924   Fax: 518-402-9027
E-mail: fwinfo@gw.dec:state.ny.us
Website: www.dec.state.ny.us

Founded: 1957
Scope: Statewide

Description: Membership composed of sportsmen, landowners, and local government representatives. State and regional boards advise the Department of Environmental Conservation in programs designed to improve resource management by landowners and increase public access to private lands.

Keyword(s): training, Exotic species, Aquatic nuisance species, Outdoor Recreation, Public Lands, Hunting

Contact(s):
    Emory Green, CHAIRMAN
    519 Rte. 247, Rushville, NY 14544
    Phone: 716-554-3362
    Clark Pell, SECRETARY
    50 Wolf Rd., Albany, NY 12233
    Phone: 518-457-5420
    Lewis Nagy, VICE CHAIRMAN
    RTE 1, Box 271-A1, Glenfield, NY 13343
    Phone: 315-376-3389

NEW YORK STATE FISH AND WILDLIFE MANAGEMENT BOARD

REGION 3
2 Ridgeway
Goshen, NY 10924 USA
Phone: 914-294-9360

Founded: NA

Contact(s):
    Rudy Vallet, BOARD CHAIRMAN

NEW YORK STATE FISH AND WILDLIFE MANAGEMENT BOARD

REGION 4
1150 Westcott Rd.
Schenectady, NY 12306 USA
Phone: 518-357-2234   Fax: 607-547-8814

Founded: NA
Scope: Statewide

Contact(s):
    Dean Winsor, BOARD CHAIRMAN, REGION 4

NEW YORK STATE FISH AND WILDLIFE MANAGEMENT BOARD

REGION 5
P.O. Box 123
Paradox, NY 12858 USA
Phone: 518-585-7250   Fax: 518-585-9799

Founded: NA
Scope: Regional, State

Contact(s):
    Don Sage, BOARD CHAIRMAN

NEW YORK STATE FISH AND WILDLIFE MANAGEMENT BOARD

REGION 6
Harrisville, NY 13648 USA
Phone: 315-543-2781   Fax: 315-543-2781

Founded: NA
Membership: 30
Scope: Statewide

Contact(s):
    Kelley Dickinson, BOARD CHAIRMAN

NEW YORK STATE FISH AND WILDLIFE MANAGEMENT BOARD

REGION 7
2365 Olanco Rd.(Region 7) Citizens Advisory Board
Marietta, NY 13110 USA
Phone: 315-636-8891
E-mail: adkwldexp@aol.com

Founded: NA
Membership: 15
Scope: Regional

Contact(s):
Craig Tryon, BOARD CHAIRMAN

NEW YORK STATE FISH AND WILDLIFE MANAGEMENT BOARD

REGION 8
6274 E. Avon-Lima Rd.
Avon, NY 14414 USA
Phone: 315-539-2820   Fax: 716-226-3905

Founded: NA
Membership: 65
Scope: Regional

Contact(s):
John Andrews, BOARD CHAIRMAN
Ron Schroder, SECRETARY
6274 E. Avon-Lima Rd., Avon, NY 14414

NEW YORK STATE OFFICE OF PARKS, RECREATION AND HISTORIC PRESERVATION

Empire State Plaza Agency Bldg 1
Albany, NY 12238 USA
Phone: 518-474-0456   Fax: 518-486-2924
Website: www.nysparks.state.ny.us

Founded: NA
Membership: 8600
Scope: Regional

Description: Administers and operates 151 parks, park preserves, and recreational facilities, three arboretums, and 35 historic sites throughout the state; administers 15 heritage areas in partnership with local communities. Acquires and protects public lands and open space; coordinates athletic programs; develops environmental interpretive programs; maintains a field services bureau which oversees historic resources and National Historic Register entries; administers boating and snowmobiling laws.

Publications: Historic Sites and Their Programs, Exploring New York's Past, New York State Boater's Guide, New York State Boat Launching Sites, New York State Operated Parks

Keyword(s): Outdoor Recreation, Historic Preservation, Environmental and Conservation Education, Protected Areas, Cultural Preservation

Contact(s):
Bernadette Castro, COMMISSIONER
Phone: 518-474-0443
Nancy Palumbo, DEPUTY COMMISSIONER FOR ADMINISTRATION
Phone: 518-474-0430
Winthrop Aldrich, DEPUTY COMMISSIONER FOR HISTORIC PRESERVATION
Phone: 518-473-5385
Albert Caccese, DEPUTY COMMISSIONER FOR LAND MANAGEMENT
Phone: 518-474-0402
Julia Stokes, DEPUTY COMMISSIONER FOR OPERATIONS, SARATOGA/TACONIC/PALISADES
Phone: 518-584-2000
Anthony Ellis, DIRECTOR OF LAW ENFORCEMENT
Phone: 518-474-0456
Dominic Jacangelo, DIRECTOR OF MARINE, COASTAL, AND LEGISLATIVE PROGRAM DEVELOPMENT
Phone: 518-474-7336
Wendy Gibson, DIRECTOR OF PUBLIC AFFAIRS
Phone: 518-486-1868
Margaret Reilly, REGIONAL DIRECTOR LONG ISLAND REGION
Phone: 631-669-1000

NEWFOUNDLAND DEPARTMENT OF FOREST RESOURCES AND AGRIFOODS

ECOSYSTEM HEALTH DIVISION
P.O. Box 8700
St. John's, Newfoundland A1B 4J6 Canada
E-mail: info@gov.nf.ca
Website: http://gov.nf.ca/forest

Founded: NA

Contact(s):
D. Fong, DIRECTOR
Phone: 709-729-1804
J. Brazil, SENIOR BIOLOGIST (ENDANGERED SPECIES)
Phone: 709-729-3773
C. Butler, SENIOR BIOLOGIST (ENVIRONMENTAL/LAND USE)
Phone: 709-729-2543

NEWFOUNDLAND DEPARTMENT OF FOREST RESOURCES AND AGRIFOODS

INLAND FISH AND WILDLIFE DIVISION
Bldg. 810, Pleasantville, P.O. Box 8700
St. John's, Newfoundland A1B 4J6 Canada

Founded: NA

Description: Objective is to maintain diverse and abundant wildlife populations and wildlife habitat; provide for the safe and sustainable use of wildlife, both consumptive and nonconsumptive; and help create a social environment conducive to effective wildlife conservation.

Publications: Trappers Guide, Newfoundland and Labrador Hunter Education Manual (student and instructor editions), Trapper's Update, Endangered Species Poster and brochure series, Newfoundland and Labrador Hunting and Trapping Guide

Contact(s):
K. Curnew, CHIEF OF INLAND FISH
Phone: 709-729-2540
S. Mahoney, CHIEF OF RESEARCH AND INVENTORY
Phone: 709-729-3593
M. Cahill, CHIEF OF WILDLIFE MANAGEMENT PLANNING
Phone: 709-729-2548
J. Hancock, DIRECTOR
Phone: 709-729-2817
J. Blake, MANAGER OF CONSERVATION SERVICES
Phone: 709-729-3509
R. Jarvis, MANAGER OF SALMONIER NATURE PARK AND ENVIRONMENTAL EDUCATION
Phone: 709-729-6974

M. Vanzyll de Jong, SENIOR BIOLOGIST (INLAND FISH)
Phone: 709-729-4306
M. McGrath, SENIOR BIOLOGIST (SMALL GAME/FUR)
Phone: 709-729-0748
L. Croke, SUPERVISOR OF ADMINISTRATION
Phone: 709-729-2636
R. Gulliver, SUPERVISOR OF LICENCING
Phone: 709-729-2630

NEWFOUNDLAND DEPARTMENT OF FOREST RESOURCES AND AGRIFOODS

REGIONAL OFFICES
P.O. Box 2222
Gander, NF A1V 2N9 Canada
Phone: 709-256-1450   Fax: 709-256-1459

Founded: NA
Membership: 14
Scope: Statewide

Contact(s):
David Fong, EASTERN DIRECTOR (GANDER)
Phone: 709-256-1451
K. Colbert, LABRADOR DIRECTOR (GOOSE BAY)
Phone: 709-896-3405
D. Leboubon, REGIONAL COMPLIANCE MANAGER
Phone: 709-896-2541
R. Trask, REGIONAL COMPLIANCE MANAGER
Phone: 709-256-1461
A. Masters, WESTERN DIRECTOR (CORNER BROOK)
Phone: 709-637-2370

NIAGARA ESCARPMENT COMMISSION

232 Guelph St. 3rd floor
Georgetown, Ontario L7G4B1 Canada
Phone: 905-877-5191   Fax: 905-873-7452
E-mail: nec@escarpment.org
Website: www.escarpment.org

Founded: 1973
Scope: Statewide

Description: Maintains the Niagara Escarpment and land in its vicinity substantially as a continuous natural environment, and ensures that only such development occurs as is compatible with that natural environment. The commission was established under the Niagara Escarpment Planning and Development Act. In 1990, the Niagra Escarpment was designated a World Biosphere Reserve.

Publications: Annual Reports, Ontario Niagara Escarpment, Explorer Brochures

Contact(s):
Don Scott, CHAIR
Mark Frawley, DIRECTOR
Richard Murzin, MANAGER OF COMMISSION
Shannon Cassidy, PUBLIC AFFAIRS OFFICER
webmaster@escarpment.org

NORTH CAROLINA DEPARTMENT OF ENVIRONMENT AND NATURAL RESOURCES

1601 Mail Services Center
Raleigh, NC 27699-1601 USA
Phone: 919-733-4984   Fax: 919-715-3060
Website: www.enr.state.nc.us

Founded: NA
Scope: Statewide

Contact(s):
Robin Smith, ASSISTANT SECRETARY FOR ENVIRONMENT
Phone: 919-715-4141
Donna Moffitt, DIRECTOR OF COASTAL MANAGEMENT
Phone: 919-733-2293
Preston Pate, DIRECTOR OF DIVISION OF MARINE FISHERIES
Phone: 919-726-7021
Mel Frye, DIRECTOR OF DIVISION OF RADIATION PROTECTION
Phone: 919-571-4141
Bill Meyer, DIRECTOR OF DIVISION OF SOLID WASTE MANAGEMENT
Phone: 919-733-4996
John Morris, DIRECTOR OF DIVISION OF WATER RESOURCES
Phone: 919-733-4064
Tommy Stevens, DIRECTOR OF ENVIRONMENTAL WATER QUALITY
Phone: 919-733-7015
Stanford Adams, DIRECTOR OF FOREST RESOURCES
Phone: 919-733-2162
Betsy Bennett, DIRECTOR OF MUSEUM OF NATURAL SCIENCES
Phone: 919-733-7450
Rhett White, DIRECTOR OF NORTH CAROLINA AQUARIUMS
Phone: 919-733-2290
Anne Taylor, DIRECTOR OF OFFICE OF ENVIRONMENTAL EDUCATION
Phone: 919-733-0711
Gary Hunt, DIRECTOR OF OFFICE OF POLLUTION PREVENTION AND ENVIRONMENTAL ASSISTANCE
Phone: 919-715-4100
Don Reuter, DIRECTOR OF PUBLIC AFFAIRS
Phone: 919-715-4112
David Vogel, DIRECTOR OF SOIL AND WATER CONSERVATION
Phone: 919-733-2302
Phil McKnelly, DIRECTOR OF STATE PARKS AND RECREATION
Phone: 919-733-4181
David Jones, DIRECTOR OF ZOOLOGICAL PARK
Phone: 910-879-7102
Charles Fullwood, EXECUTIVE DIRECTOR OF WILDLIFE RESOURCES COMMISSION
Phone: 919-733-3391
Charles Gardner, LAND RESOURCES STAFF
Phone: 919-733-3833
Fiona Clem, LIBRARIAN
Bill Holman, SECRETARY
Phone: 919-715-4101

NORTH CAROLINA COOPERATIVE EXTENSION SERVICE

North Carolina State University,
Raleigh, NC 27695 USA
Phone: 919-515-2811   Fax: 919-515-3135
Website: www.ces.ncsu.edu

Founded: NA
Scope: Statewide

Contact(s):

Thomas Losordo, AQUACULTURE SPECIALIST
Box 7646, North Carolina State University, Raleigh,
NC 29695
Phone: 919-515-7587
Fax: 919-515-5110
tlosordo@unity.ncsu.edu
Harry Daniels, AQUACULTURE SPECIALIST
Vernon James Research and Extension Center, 207
Research Station Rd., Plymouth, NC 27962
Phone: 252-793-4428
Fax: 252-793-5142
harry_daniels@ncsu.edu
Roger Crickenberger, ASSISTANT DIRECTOR AND STATE
PROGRAM LEADER
NCSU, Box 7602, Raleigh, NC 27695-7602
Phone: 919-515-3252
Fax: 919-515-5950
roger_crickenberger@ncsu.edu
Jon Ort, DIRECTOR OF EXTENSION SERVICE
jon_ort@ncsu.edu
Ronald Hodson, DIRECTOR OF SEA GRANT PROGRAM
Sea Grant, Box 8605, North Carolina State University,
Raleigh, NC 27695
Phone: 919-515-2455
Fax: 919-515-7095
ronald_hodson@ncsu.edu
James Rice, EXTENSION FISHERIES SPECIALIST
Box 7617, North Carolina State University, Raleigh, NC 27695
Phone: 919-515-4592
Fax: 919-515-5327
jim_rice@ncsu.edu
Jeffrey Hinshaw, EXTENSION TROUT SPECIALIST
Research and Extension Center, Box 9628, 2016 Fanning
Bridge Rd., Fletcher, NC 28732-9216
Phone: 828-684-3562
Fax: 828-684-8715
jeff_hinshaw@ncsu.edu
Chris Moorman, EXTENSION WILDLIFE SPECIALIST
North Carolina State University, Box 8003, Raleigh, NC
27695
Phone: 919-515-5578
Fax: 919-515-6883
chris_moorman@ncsu.edu
Peter Bromley, WILDLIFE SPECIALIST
Zoology Dept., Box 7646, North Carolina State University,
Raleigh, NC 27695
Phone: 919-515-7587
Fax: 919-515-5110
pete_bromley@ncsu.edu

NORTH CAROLINA COOPERATIVE FISH AND WILDLIFE RESEARCH UNIT (USDI)

201 David Clark Lab
Raleigh, NC 27695 USA
Phone: 919-515-2631   Fax: 919-515-4454
Website: www.4.scsu.edu:8010/nccoopunit

Founded: NA
Membership: 5
Scope: Statewide

Contact(s):

Wendy Moore, ADMINISTRATIVE ASSISTANT
Tom Kwak, UNIT LEADER

NORTH CAROLINA DEPARTMENT OF AGRICULTURE & CONSUMER SERVICES

P.O. Box 27647
Raleigh, NC 27611 USA
Phone: 919-733-7125   Fax: 919-733-1141
Website: www.agr.state.nc.us

Founded: NA
Membership: 1400
Scope: State

Keyword(s): Soil Conservation, Health and Nutrition, Pesticides,
Agriculture

Contact(s):

Richard Reich, AGRONOMIC SERVICES STAFF
Phone: 919-733-2655
Tom Ellis, AQUACULTURE & NATURAL RESOURCES
STAFF
Phone: 919-733-7125
Meg Phipps, COMMISSIONER
Phone: 919-733-7125
Mike Blanton, DIRECTOR
Phone: 919-733-4216
Bruce Williams, FOOD AND DRUG PROTECTION STAFF
Phone: 919-733-7366
David McLeod, LEGAL STAFF
Phone: 919-733-7125
William McClelland, PESTICIDE DISPOSAL STAFF
Phone: 919-733-7366
Jim Burnette, PESTICIDE SECTION STAFF
Phone: 919-733-3556
Cecil Frost, PLANT CONSERVATION PROGRAM STAFF
Phone: 919-733-3610
Bill Dickerson, PLANT INDUSTRY DIVISION STAFF
Phone: 919-733-3933
Carl Tart, RESEARCH STATIONS STAFF
Phone: 919-733-3236
Carl Falco, STRUCTURAL PEST STAFF
Phone: 919-733-6100
David Marshall, VETERINARY SERVICES STAFF
Phone: 919-733-7601

NORTH CAROLINA DIVISION OF SOIL & WATER

STATE SOIL AND WATER CONSERVATION
COMMISSION
1614 Mail Service Center
Raleigh, NC 27699-1614 USA
Phone: 919-733-2302   Fax: 919-715-3559
Website: www.dern.com

Founded: 1937
Membership: 50
Scope: State

Description: A unit of state government administered by the
Division of Soil and Water Conservation in the Department of
Environment and Natural Resources. To organize soil and
water conservation districts; grant funds for operations,
technical assistance, and the NC Agriculture Cost-Share
Program for Nonpoint Source Pollution Control—a water
quality program; provide for control of soil erosion and
improvement of water quality; accept PL566 Small Watershed
applications. Support staff provided through the Division of Soil
and Water Conservation Commission.

Contact(s):
James Ferguson, CHAIRMAN
11571 Betsy Gap Rd., Clyde, NC 28721
Phone: 704-627-6458
David Vogel, DIRECTOR
Phone: 919-715-6097
Fax: 919-715-3559
david.vogel@ncmail.net

NORTH CAROLINA SEA GRANT PROGRAM
NCSU Box 8605, 100B 1911 Bldg.
Raleigh, NC 27695-8605 USA
Phone: 919-515-2454   Fax: 919-515-7095
Website: www.ncsu.edu/seagrant

Founded: NA
Scope: Statewide

Publications: Coast Watch, Water Wise, Marine Extension News

Keyword(s): Coral Reefs, Oceanography, Ecotourism, Coasts, Wildlife, Aquatic Species, Harmful Algal Blooms, Estuaries, Coastal Construction and Erosion, Water Quality, Aquaculture, Seafood Technology, Aquatic Habitats

Contact(s):
Katie Mosher, ASSISTANT DIRECTOR FOR COMMUNICATIONS
Steve Rebach, ASSOCIATE DIRECTOR
Ronald Hodson, DIRECTOR
ronald_hodson@ncsu.edu
Jack Thigpen, EXTENSION DIRECTOR

NORTH CAROLINA WILDLIFE RESOURCES COMMISSION
1701 Mail Service Center
Raleigh, NC 27699-1701 USA
Phone: 919-733-3391   Fax: 919-733-7083
Website: www.ncwildlife.org

Founded: 1947
Membership: 525
Scope: Statewide

Description: The commission has the function, purpose, and duty to manage, restore, develop, cultivate, conserve, protect, and regulate the wildlife resources of the state, and to administer the laws relating to boating, hunting, fishing, and other wildlife resources, including nongame.

Publications: Wildlife in North Carolina

Contact(s):
Richard Hamilton, ASSISTANT DIRECTOR
Cecilia Edgar, CHIEF OF DIVISION OF ADMINISTRATIVE SERVICES
Phone: 919-733-4566
Ginger Williams, CHIEF OF DIVISION OF CONSERVATION EDUCATION
Roger Lequire, CHIEF OF DIVISION OF ENFORCEMENT
Phone: 919-733-7191
Fred Harris, CHIEF OF DIVISION OF INLAND FISHERIES
Phone: 919-733-3633
David Cobb, CHIEF OF DIVISION OF WILDLIFE MANAGEMENT
Phone: 919-733-7291
John Pechmann, COMMISSION CHAIRMAN
Phone: 910-483-0107
Wes Seegars, COMMISSION VICE CHAIRMAN
Phone: 919-735-8211

Gordon Myers, DIVISION OF ENGINEERING SERVICES
Phone: 919-715-3155
Charles Fullwood, EXECUTIVE DIRECTOR
Carol Batker, HEAD OF PERSONNEL SECTION
Phone: 919-733-2241
Rodney Fousshee, PUBLICATION COORDINATOR AND EDITOR

NORTH DAKOTA FOREST SERVICE
307 First St. E.
Bottineau, ND 58318-1100 USA
Phone: 701-228-5422   Fax: 701-228-5448
E-mail: forest@state.nd.us
Website: www.state.nd.us/forest

Founded: 1891
Membership: 29
Scope: Statewide

Description: Mission Statement: Caring for, protecting, and improving forest resources for future generations.

Publications: Prairie Forester, The

Keyword(s): Urban Forestry, Public Lands, Flowers, Plants, and Trees, Forests and Forestry, Environmental and Conservation Education

Contact(s):
W. Jackson Bird, COMMUNITY FORESTRY COORDINATOR
1511 E. Interstate Ave., Bismarck, ND 58501
Phone: 701-328-9945
Fax: 701-250-4454
Michael Santucci, FIRE MANAGEMENT COORDINATOR
1511 E. Interstate Ave, Bismarck, ND 58501
Phone: 701-328-9946
Glenda Fauske, INFORMATION AND EDUCATION COORDINATOR
Phone: 701-228-5446
Thomas Berg, STAFF FORESTER
Phone: 701-228-5483
Thomas Karch, STATE FOREST COORDINATOR
Larry Kotchman, STATE FORESTER
Phone: 701-228-5422
Thomas Claeys, SUSTAINABLE FORESTRY COORDINATOR
Phone: 701-228-5486
Fax: 701-228-5448
Roy Laframboise, TOWNER NURSERY MANAGER
878 Nursery Rd., Towner, ND 58788
Phone: 701-537-5636
Fax: 701-537-5680

NORTH DAKOTA DEPARTMENT OF AGRICULTURE
600 E. Blvd. Ave., Department 602
Bismarck, ND 58505-0020 USA
Phone: 701-328-2231   Fax: 701-328-4567
Website: www.agdepartment.com

Founded: NA
Membership: 55
Scope: Statewide

Description: The North Dakota Department of Agriculture is the regulating and licensing agency for the agricultural industry in North Dakota.

Contact(s):
Roger Johnson, COMMISSIONER
Phone: 701-328-2231
Fax: 701-328-4567
rojohnso@state.nd.us

NORTH DAKOTA DEPARTMENT OF HEALTH

600 E. Blvd. Ave.
Bismarck, ND 58506-5520 USA
Phone: 701-328-2372　　Fax: 701-328-4727
Website: www.health.state.nd.us

Founded: NA
Membership: 308
Scope: Statewide

Description: State pollution control programs.

Keyword(s): Water Quality, Solid Waste Management, Environmental Justice, Public Health Protection, Air Quality and Pollution

Contact(s):
Francis Schwindt, CHIEF ENVIRONMENTAL HEALTH SECTION
P.O. Box 5520, Bismarck, ND 58506-5520
Phone: 701-328-5150
Fax: 701-328-5200
Jack Long, DIRECTOR OF DIVISION OF MUNICIPAL FACILITIES
P.O. Box 5520, Bismarck, ND 58506-5520
Phone: 701-328-5211
Fax: 701-328-5200
L. David Glatt, DIRECTOR OF DIVISION OF WASTE MANAGEMENT
P.O. Box 5520, Bismarck, ND 58506-5520
Phone: 701-328-5166
Fax: 701-328-5200
Dennis Fewless, DIRECTOR OF DIVISION OF WATER QUALITY
P.O. Box 5520, Bismarck, ND 58506-5520
Phone: 701-328-5210
Fax: 701-328-5200
Terry O'Clair, DIRECTOR OF DIVISION OF AIR QUALITY

NORTH DAKOTA GAME AND FISH DEPARTMENT

100 N. Bismarck Expressway
Bismarck, ND 58501 USA
Phone: 701-328-6300　　Fax: 701-328-6352
E-mail: discovernd.com/gnf/
Website: www.discovernd.com/gnf/

Founded: NA
Membership: 130
Scope: Regional
Publications: North Dakota Outdoors

Contact(s):
Paul Schadewald, CHIEF OF ADMINISTRATIVE SERVICES
Mike McKenna, CHIEF OF CONSERVATION & COMMUNICATON
Ray Goetz, CHIEF OF ENFORCEMENT
Terry Steinwand, CHIEF OF FISHERIES
Randy Kreil, CHIEF OF WILDLIFE
Roger Rostvet, DEPUTY DIRECTOR
srrostvet@state.nd.us
Dean Hildebrand, DIRECTOR
Harold Umber, EDITOR

NORTH DAKOTA GEOLOGICAL SURVEY

600 E. Blvd.
Bismarck, ND 58505-0840 USA
Phone: 701-328-8000　　Fax: 701-328-8010
Website: www.state.nd.us/ndgs

Founded: 1895
Membership: 23
Scope: Regional

Description: Responsible for collecting and disseminating geologic information.

Publications: NDGS Newsletter

Keyword(s): Coral Reefs, Land Use Planning, Protected Areas, Geology, Energy

Contact(s):
John Bluemle, STATE GEOLOGIST

NORTH DAKOTA PARKS AND RECREATION DEPARTMENT

1835 Bismarck Expressway
Bismarck, ND 58504 USA
Phone: 701-328-5357　　Fax: 701-328-5363
E-mail: parkrec@state.nd.us
Website: www.ndparks.com

Founded: 1993
Membership: 25
Scope: Statewide

Description: Plan and coordinate government programs encouraging the full development and preservation of existing and future parks, outdoor recreation areas, nature preserves, rare plant and animal species, and unique natural communities.

Publications: Discover Newspaper

Keyword(s): National Parks, Land Preservation, Protected Areas, Historic Preservation, Biodiversity

Contact(s):
Kathy Duttenhefner, COORDINATOR OF NATURE PRESERVE/NATURAL HERITAGE PROGRAMS
Jesse Hanson, COORDINATOR OF PLANNING AND NATURAL RESOURCES
Douglass Prchal, DIRECTOR

NORTH DAKOTA STATE SOIL CONSERVATION COMMITTEE

4023 North State St., Suite 30
Bismarck, ND 58503-0620 USA
Phone: 701-328-5128　　Fax: 701-328-5123
Website: www.ag.ndsu.nodak.edu/ndsscc/sscc

Founded: 1937
Scope: Statewide

Description: To organize soil conservation districts and provide for control and prevention of soil erosion; represent the state in soil conservation matters; accept P.L. 566 Small Watershed applications and, assign planning priority; and administer the Surface Mining Reports Law; and soil conservation technician grants program.

Keyword(s): Wetlands, Soil Conservation, Environmental and Conservation Education, Pesticides, Agriculture

Contact(s):
Curtiss Klein, CHAIRPERSON
Phone: 701-984-2669
cklein@daktel.com
Scott Hochhalter, SOIL CONSERVATION COORDINATOR
shochhal@ndsuext.nodak.edu

NORTH DAKOTA STATE UNIVERSITY EXTENSION SERVICE

North Dakota State University,
Fargo, ND 58105-5437 USA
Phone: 701-231-7173   Fax: 701-231-8378
Website: www.ndsuext.nodak.edu

Founded: NA
Scope: Statewide

Contact(s):
Sharon Anderson, DIRECTOR, EXTENSION SERVICE
ext-dir@ndsuext.nodak.edu
Cole Gustafson, DIRECTOR, NORTH DAKOTA
AGRICULTURAL EXPERIMENT STATION
NDSU, Box 5655, Fargo, ND 58105-5655
Phone: 701-231-7655
Fax: 701-231-8520
exp-dir@ndsuext.nodak.edu
Kevin Sedivec, NATURAL RESOURCE INFORMATION
NDSU, P.O. Box 5053, Fargo, ND 58105

NORTH DAKOTA WATER COMMISSION

900 E. Blvd.
Bismarck, ND 58505-0850 USA
Phone: 701-328-2750   Fax: 701-328-3696
Website: www.water.swc.state.nd.us

Founded: NA
Membership: 85
Scope: Statewide

Contact(s):
Dale Frink, STATE ENGINEEER

NORTHERN VIRGINIA REGIONAL PARK AUTHORITY

5400 Ox Rd.
Fairfax Station, VA 22039 USA
Phone: 703-352-5900   Fax: 703-273-0905
E-mail: info@nvrpa.org
Website: www.nvrpa.org

Founded: 1959
Scope: Regional

Description: To preserve open and wooded areas and provide
outdoor recreation to meet the needs of a growing population.

Publications: Calendar of Events, Policy Plan, Washington and
Old Dominion Railroad Regional Park Trail Guide, Discover
Your Regional Parks

Keyword(s): Outdoor Recreation, Flowers, Plants, and Trees,
Environmental and Conservation Education, Protected Areas,
Botanical Gardens

Contact(s):
Walter Mess, CHAIRMAN
Gary Fenton, EXECUTIVE DIRECTOR
Phone: 703-359-4605
gfenton@nvrpa.org

NOVA SCOTIA DEPARTMENT OF NATURAL RESOURCES

P.O. Box 698
Halifax, Nova Scotia B3J 2T9 Canada
Phone: 902-424-4103   Fax: 902-424-7735
Website: www.gov.nf.ca

Founded: NA
Scope: Statewide

Contact(s):
Nancy Leek, DIRECTOR OF FORESTRY
P.O. Box 68, Truro, Nova Scotia B2N 5B8
Phone: 902-893-5749
Fax: 902-893-5662
Harold Carroll, DIRECTOR OF PARKS AND RECREATION
Barry Sabean, DIRECTOR OF WILDLIFE
136 Exhibition St., Kentville, Nova Scotia B4N 4E5
Phone: 902-679-6139
Fax: 902-679-6176
Ed Macaulay, EXECUTIVE DIRECTOR
Phone: 902-424-4103
Bert Vissers, MANAGER OF WILDLIFE PARKS
P.O. Box 299, Shubenacadie, Nova Scotia B0N 2H0
Phone: 902-758-2040
Fax: 902-758-7011

NOVA SCOTIA DEPARTMENT OF NATURAL RESOURCES

P.O. Box 698
Halifax, Nova Scotia B3J 2T9 Canada
Phone: 902-424-5935   Fax: 902-424-7735
Website: www.gov.ns.ca/natr

Founded: NA
Scope: State

Description: Charged with the administration of the Wildlife Act,
Endangered Species Act and various other statutes. Inherent
in the legislation and incumbent upon the Department are
responsibilities pertaining to the productivity of the forests
generally, the supply of forest products, the conservation of
wildlife, the development of mineral resources, the administration of Crown Lands, and the enhancement of recreational
areas.

Publications: Nova Scotia Trappers Newsletter, Nature's
Resources

Contact(s):
Daniel Graham, DEPUTY MINISTER
Phone: 902-424-4121
Barry Sabean, DIRECTOR OF WILDLIFE
Ernie Fage, MINISTER
Phone: 902-424-4037

NOVA SCOTIA DEPARTMENT OF NATURAL RESOURCES

CORPORATE SERVICE UNIT
P.O. Box 698
Halifax, NS B3J 2T9 Canada
Phone: 902-424-5935   Fax: 902-424-7735
Website: www.gov.ns.ca

Founded: NA
Membership: 100
Scope: Statewide

Publications: magazine—Natures Resources (comes out 3 or 4 times a yr)

Contact(s):
Frank Dunn, DIRECTOR OF FINANCE
Phone: 902-424-3288
Susan Zinck, INFORMATION OFFICER
Phone: 902-424-2354

NOVA SCOTIA DEPARTMENT OF NATURAL RESOURCES
LAND SERVICES BRANCH
Halifax, Nova Scotia B3J 2T9 Canada
Phone: 902-424-4267   Fax: 902-424-7735
E-mail: rcpenfou@gov.ns.ca
Website: www.gov.ns.ca/natr

Founded: NA
Scope: Statewide

Contact(s):
Jo-Ane Himmelman, DIRECTOR LAND ADMINISTRATION
Phone: 902-424-4267
himmelgj.gov.ns.ca
Rosalind Penfound, EXECUTIVE DIRECTOR
Phone: 902-424-4267
rcpenfou.gov.ns.ca

NOVA SCOTIA DEPARTMENT OF NATURAL RESOURCES
REGIONAL SERVICES BRANCH
P.O. Box 698
Halifax, Nova Scotia B3J 2T9 Canada
Phone: 902-424-3949   Fax: 902-424-7735
Website: www.gov.ns.ca

Founded: NA
Membership: 1000
Scope: Regional

Contact(s):
John Mombourquette, DIRECTOR
Phone: 902-424-5254
Dan Eidt, DIRECTOR OF CROWN LANDS MANAGEMENT
Phone: 902-424-7594
Bill Smith, DIRECTOR OF EXTENSION SERVICES
Phone: 902-424-4445
Arden Whidden, DIRECTOR OF PRIVATE LANDS MANAGEMENT
Phone: 902-424-5703
Brian Gilbert, EXECUTIVE DIRECTOR
Phone: 902-424-3949

O

OFFICE OF ENERGY EFFICIENCY AND ENVIRONMENT
New York State Dept. of Public Service, 3 Empire State Plaza
Albany, NY 12223 USA
Phone: 518-473-7248

Founded: 1970

Description: The Office of Energy Efficiency and Environment provides staff support in developing and administering policies that assure appropriate consideration of energy efficiency and environmental protection in utility regulation, management, and restructuring. The office also plays a major role in the development of systems and procedures necessary to introduce retail competition in the state.

Keyword(s): Renewable Resources, Land Use Planning, Energy, Environmental Planning, Acid Rain

Contact(s):
Paul Powers, DIRECTOR

OFFICE OF STATE PARKS, DEPARTMENT OF CULTURE, RECREATION, AND TOURISM
P.O. Box 44426
Baton Rouge, LA 70804 USA
Phone: 225-342-8111   Fax: 225-342-8107
E-mail: parks@crt.state.la.us
Website: www.crt.state.la.us

Founded: 1934
Scope: State

Description: Created to plan, design, construct, operate, and maintain the state's parks, natural areas, recreational facilities, and commemorative sites. Office has 17 parks or recreational areas, 15 commemorative sites, and one preservation area open to the public. The Office is assisted by the Parks and Recreation Commission, an advisory board appointed by the Governor.

Keyword(s): Public Lands, Outdoor Recreation, Historic Preservation, Cultural Preservation, National Parks

Contact(s):
Dwight Landreneau, ASSISTANT SECRETARY
dlandreneau @crt.state.la.us

OHIO DEPARTMENT OF AGRICULTURE
8995 E. Main Street
Reynoldsburg, OH 43068 USA
Phone: 614-466-2732   Fax: 614-728-6226
Website: www.state.oh.us/agr.

Founded: NA
Membership: 450
Scope: Regional

Contact(s):
Mark Anthony, COMMUNICATION DIRECTOR
Phone: 614-752-4505
Fred Dailey, DIRECTOR

OHIO DEPARTMENT OF DEVELOPMENT
OFFICE OF ENERGY EFFICIENCY
77 S. High St.
Columbus, OH 43215 USA
Phone: 614-466-6797   Fax: 614-466-1864
Website: www.odod.state.oh.us

Founded: 1979
Membership: 30
Scope: Regional

Description: The Office of Energy Efficiency, Ohio's state energy office, develops policies and programs that use energy efficiency and renewable energy to enhance economic benefits and better Ohio's environment.

Publications: Ohio's Home Weatherization Assistance Program: An Independent Evaluation, Ohio's Home Weatherization Assistance Program (brochure), 1999 Energy Efficiency Bookmark Contest Winners

Keyword(s): Air Quality, Consumer Services, Energy, Energy

Conservation, Energy Efficiency, Green Building, Healthy Home, Renewable Resources, Solar Energy, Sustainable Energy, Transportation

Contact(s):
Dawn Smith, ASSISTANT CHIEF
Phone: 614-466-1835
dsmith@odod.state.oh.us
Sara Ward, CHIEF
Phone: 614-466-8396
sward@odod.state.oh.us
Bill Manz, COMMERCIAL/INDUSTRIAL PROGRAMS
Phone: 614-466-7429
wmanz@odod.state.oh.us
Stjepan Vlahovich, EDUCATION PROGRAMS
Phone: 614-466-0545
svlahovich@odod.state.oh.us
Tim Lenahan, RESIDENTIAL PROGRAMS
Phone: 614-466-8434
tlenahan@odod.state.oh.us

OHIO DEPARTMENT OF NATURAL RESOURCES
1930 Belcher Dr. Fountain Square Bldg. D
Columbus, OH 43224 USA
Phone: 614-265-6565   Fax: 614-261-9601
Website: www.dnr.state.oh.us

Founded: 1949
Scope: State

Description: The mission of the Ohio Department of Natural Resources is to provide for the preservation, conservation, utilization, and enjoyment of our natural resources through the wise management, careful planning, efficient and effective delivery of services, and the collection and dissemination of data and information needed for environmental protection and natural resource management decisions. We recognize that these actions impact the social, recreational, and economic well-being of our citizens and that education is the key to the realization of an environmental ethic and the preservation of our natural and cultural hertiage.

Publications: Publications on line

Contact(s):
J. Moody, ASSISTANT DIRECTOR
14-265-6877
Sally Prouty, CHIEF OF DIVISION OF CIVILIAN CONSERVATION
Phone: 614-265-6423
Ronald Abraham, CHIEF OF DIVISION OF FORESTRY
Phone: 614-265-6694
Thomas Berg, CHIEF OF DIVISION OF GEOLOGICAL SURVEY
Phone: 614-265-6876
Stuart Lewis, CHIEF OF DIVISION OF NATURAL AREAS AND PRESERVES (ACTING)
Phone: 614-265-6453
Tom Tugned, CHIEF OF DIVISION OF OIL AND GAS
Phone: 614-265-6917
Dan West, CHIEF OF DIVISION OF PARKS AND RECREATION
Phone: 614-265-6561
Wayne Warren, CHIEF OF DIVISION OF REAL ESTATE AND LAND MANAGEMENT
Phone: 614-265-6395
Lawrence Vance, CHIEF OF DIVISION OF SOIL AND WATER CONSERVATION
Phone: 614-265-6610

Fax: 614-262-2064
larry.vance@dnr.ohio.gov.us
Jim Morris, CHIEF OF DIVISION OF WATER
Phone: 614-265-6717
Jeff Hoedt, CHIEF OF DIVISION OF WATERCRAFT
Phone: 614-265-6480
Peter King, CHIEF OF OFFICE OF HUMAN RESOURCES
Phone: 614-265-6859
Nancy Manacke, CHIEF OF OFFICE OF MARKETING SERVICES
Phone: 614-265-6787
Michael Canfield, CHIEF OF OFFICE OF RECYCLING AND LITTER PREVENTION
Phone: 614-265-6333
Ron Kolbash, DEPUTY DIRECTOR
Phone: 614-265-6875
Scott Zody, DEPUTY DIRECTOR
Phone: 614-265-6845
Lori Soupe, DEPUTY DIRECTOR
Phone: 614-265-6845
Samuel Speck, DIRECTOR
Phone: 614-265-6875

OHIO ENVIRONMENTAL PROTECTION AGENCY
122 South Front Street
Columbus, OH 43215 USA
Phone: 614-644-3020   Fax: 614-644-3184
Website: www.epa.state.oh.us

Founded: 1972
Membership: 1300
Scope: State

Description: The Mission of the Ohio Environmental Protection Agency is to protect the environment and public health by ensuring compliance with environmental laws and demonstrating leadership in envirnmental stewardship. The agency has authority to implement laws and regulations regarding air and water quality standards; solid, hazardous and infectious waste disposal standards; water quality planning, supervision of sewage treatment and public drinking water supplies; and cleanup of unregulated hazardous waste sites.

Keyword(s): Wetlands, Coral Reefs, Environmental and Conservation Education, Solid Waste, Air Quality and Pollution

Contact(s):
Joseph Koncelik, ASSISTANT DIRECTOR
Phone: 614-644-2782
Bruce Coleman, CHIEF OF CENTRAL DISTRICT OFFICE
Phone: 614-782-3778
John Albrecht, CHIEF OF DATA AND SYSTEMS
Phone: 614-644-2990
Robert Hodanbosi, CHIEF OF DIVISION OF AIR POLLUTION CONTROL
Phone: 614-644-2270
Michael Baker, CHIEF OF DIVISION OF DRINKING AND GROUND WATERS
Phone: 614-644-2752
Cindy Hafner, CHIEF OF DIVISION OF EMERGENCY AND REMEDIAL RESPONSE
Phone: 614-644-2924
Greg Smith, CHIEF OF DIVISION OF ENVIRONMENTAL AND FINANCIAL ASSISTANCE
Phone: 614-644-2798
Linda Friedman, CHIEF OF DIVISION OF ENVIRONMENTAL SERVICES
Phone: 614-644-4247

Michael Savage, CHIEF OF DIVISION OF HAZARDOUS
WASTE MANAGEMENT
Phone: 614-644-2917
Daniel Harris, CHIEF OF DIVISION OF SOLID AND
INFECTIOUS WASTE
Phone: 614-644-2621
Lisa Morris, CHIEF OF DIVISION OF SURFACE WATER
Phone: 614-644-2001
Jacqueline Hymes, CHIEF OF EQUAL EMPLOYMENT
OPPORTUNITY
Phone: 614-644-3553
Edmund Tormey, CHIEF OF LEGAL AFFAIRS
Phone: 614-644-1037
Bill Skowronski, CHIEF OF NORTHEAST DISTRICT OFFICE
Phone: 330-425-9171
Edwin Hammett, CHIEF OF NORTHWEST DISTRICT
OFFICE
Phone: 419-352-8461
William Kirk, CHIEF OF OFFICE OF EMPLOYEE SERVICES
Phone: 614-644-2100
Carolyn Watkins, CHIEF OF OFFICE OF ENVIRONMENTAL
EDUCATION
Phone: 614-644-2873
Chris Geyer, CHIEF OF OFFICE OF FISCAL
ADMINISTRATION
Phone: 614-644-2339
Mike Kelley, CHIEF OF OFFICE OF POLLUTION
PREVENTION
Phone: 614-644-3467
Bonnie Crocket, CHIEF OF OPERATIONS AND FACILITIES
Phone: 614-644-2089
Carol Hester, CHIEF OF PUBLIC INTEREST CENTER
Phone: 614-644-2166
Stuart Bruny, CHIEF OF SOUTHEAST DISTRICT OFFICE
Phone: 740-385-8501
Tom Winston, CHIEF OF SOUTHWEST DISTRICT OFFICE
Phone: 937-285-6357
Al Franks, CHIEF OF STRATEGIC MANAGEMENT
Phone: 614-644-2782
Laura Powell, DEPUTY DIRECTOR OF POLICY
Phone: 614-644-2782
Patricia Madigan, DEPUTY DIRECTOR OF POLICY
Phone: 614-644-2782
Christopher Jones, DIRECTOR
Phone: 614-644-2782

OHIO SEA GRANT COLLEGE PROGRAM

1314 Kinnear Rd.
Columbus, OH 43212-1194 USA
Phone: 614-292-8949   Fax: 614-292-4364
Website: www.sg.ohio-state.edu/

Founded: NA
Membership: 100
Scope: State

Description: The Ohio Sea Grant College Program is dedicated
to the goal of promoting the understanding and management,
development, utilization, and conservation of ocean, coastal,
and Great Lakes resources, specifically Lake Erie, through
research, education, outreach, and communications. The
program is administrated by The Ohio State University. Stone
Laboratory is Ohio's biological field station located on Gibraltar
Island at Put-in-Bay, Ohio.

Publications: Twine Line

Keyword(s): Coral Reefs, Toxicology, Environmental and

Conservation Education, Wildlife, Lakes, Research, Aquatic
Species, Aquatic Habitats

Contact(s):
Karen Ricker, ASSISTANT DIRECTOR AND
COMMUNICATIONS COORDINATOR
ricker.15@osu.edu
Jeffrey Reutter, DIRECTOR
reutter.1@osu.edu

OHIO STATE UNIVERSITY EXTENSION

2120 Fyffe Rd.
Columbus, OH 43210 USA
Phone: 614-292-6181   Fax: 614-688-3807
Website: www.ag.ohio-state.edu/

Founded: NA
Membership: 20
Scope: Statewide

Description: To help people improve their lives through an
educational process using scientific knowledge focused on
identified issues and needs.

Contact(s):
Robert Roth, ASSOCIATE DIRECTOR OF
ENVIRONMENTAL EDUCATION
School of Natural Resources, 207 Kottman Hall, 2021 Coffey
Rd., Columbus, OH 43210
Phone: 614-292-2265
Fax: 614-292-7432
roth.3@osu.edu
Keith Smith, DIRECTOR OF EXTENSIONS
smith.150@osu.edu
Gary Mullins, DIRECTOR OF SCHOOL OF NATURAL
RESOURCES
Ohio State University, 210 Kottman Hall, 2021 Coffey Rd.,
Columbus, OH 43210
Phone: 614-292-2265
Fax: 614-292-7432
mullins.2@osu.edu
Erik Norland, EXTENSION SPECIALIST OF NATURAL
RESOURCES
School of Natural Resources, 2021 Coffey Rd., Columbus,
OH 43210
Phone: 614-292-6544
Fax: 614-292-7432
norland.1@osu.edu

OKLAHOMA DEPARTMENT OF AGRICULTURE

P.O. Box 528804
Oklahoma City, OK 73152-8804 USA
Phone: 405-521-3864   Fax: 405-521-4912
Website: www.state.ok.us/~okag/

Founded: NA
Membership: 500
Scope: Statewide

Description: The Oklahoma Department of Agriculture is
principally a service agency, but it is also a promotional and
cooperative agency for segments of agriculture and forestry.
Major divisions of the department are forestry, animal industry,
legal, plant industry, marketing, water quality, agriculture
laboratory, and the federal-state cooperative programs of
wildlife services and agricultural statistics.

Keyword(s): Urban Environment, Pesticides, Environmental Law,
Forests and Forestry, Agriculture

Contact(s):
David Ligon, ADMINISTRATION STAFF
Phone: 405-521-3864
Lynn Davis, ASSISTANT DIRECTOR
Charles Freeman, DEPUTY COMMISSIONER
Phone: 405-521-3864
Mike Talkington, DIRECTOR OF AGRICULTURAL LABORATORY
Barry Bloyd, DIRECTOR OF AGRICULTURAL STATISTICS
Phone: 405-525-9226
Burke Healey, DIRECTOR OF ANIMAL INDUSTRY SERVICES
Phone: 405-521-3864
Roger Davis, DIRECTOR OF FORESTRY SERVICES
Phone: 405-521-3864
Janet Stewart, DIRECTOR OF LEGAL SERVICES
Rick Maloney, DIRECTOR OF MARKET DEVELOPMENT
414 SERVICES
John Steuber, DIRECTOR OF WILDLIFE SERVICES
Phone: 405-521-4039
Jack Carson, INFORMATION OFFICER
Phone: 405-521-3864
Dennis Howard, SECRETARY OF AGRICULTURE

OKLAHOMA BIOLOGICAL SURVEY

111 E. Chesapeake St., University of Oklahoma
Norman, OK 73019 USA
Phone: 405-325-4034   Fax: 405-325-7702
Website: www.biosurvey.ou.edu

Founded: 1927
Membership: 12
Scope: State

Description: State office and organized research unit of university. Acquires information on biological resources and natural areas, conducts research on natural biota, jointly maintains Bebb Herbarium, has responsibility for Oklahoma Natural Heritage Inventory, and provides training for students. Jointly operates Oklahoma Fishery Research Laboratory with Oklahoma Department of Wildlife Conservation.

Keyword(s): Zoology, Nongame Wildlife, Endangered Species, Flowers, Plants, and Trees, Wildlife Rehabilitation, Ecology, training, Conservation, Botany, Geographic Information Systems, Biodiversity

Contact(s):
Bruce Hoagland, COORDINATOR OF OKLAHOMA NATURAL HERITAGE INVENTORY
Phone: 405-325-1985
bhoagland@ou.edu
Caryn Vaughn, DIRECTOR
Phone: 405-325-4034
cvaughn@ou.edu
Steve Sherrod, EXECUTIVE DIRECTOR OF SUTTON AVIAN RESEARCH CENTER
Phone: 918-336-7778
gmsarc@aol.com
Wayne Elisens, INTERIM CURATOR

OKLAHOMA CONSERVATION COMMISSION

2800 N. Lincoln Blvd., Suite 160
Oklahoma City, OK 73105 USA
Phone: 405-521-2384   Fax: 405-521-6686
Website: www.okcc.state.ok.us

Founded: 1938
Membership: 25

Scope: State

Description: To assist and supervise conservation districts in carrying out conservation practices of all renewable natural resources.

Publications: Geographic Information Systems Newsletter, Conservation Conversation Newsletter

Keyword(s): Upstream Flood Prevention, Nonpoint Source Pollution, Soil Conservation, Abandoned Mine Land Reclamation, Environmental and Conservation Education

Contact(s):
Ben Pollard, ASSISTANT DIRECTOR
benp@cc.state.ok.us
Jim Leach, ASSISTANT DIRECTOR OF WATER QUALITY
jiml@ok.cc.state.ok.us
Sandra Drummond, CONTACT
Mike Kastl, DIRECTOR OF ABANDONED MINE LAND RECLAMATION PROGRAM
mikek@ok.cc.state.ok.us
Dan Sebert, DISTRICT OPERATIONS DIRECTOR
dans@cc.state.ok.us
Mike Thralls, EXECUTIVE DIRECTOR
miket@ok.cc.state.ok.us
Mark Harrison, INFORMATION OFFICER
markh@ok.cc.state.ok.us

OKLAHOMA COOPERATIVE FISH AND WILDLIFE RESEARCH UNIT (USDI)

404 Life Sciences West Bldg., Oklahoma State University
Stillwater, OK 74078-3051 USA
Phone: 405-744-6342   Fax: 405-744-5006

Founded: NA
Membership: 6
Scope: National, Statewide

Keyword(s): training, Research, Wildlife, Nongame Wildlife, Endangered Species

Contact(s):
William Fisher, ASSISTANT LEADER (ECOLOGY)
Dana Winkleman, ASSISTANT LEADER (FISHERIES)
David Leslie, LEADER

OKLAHOMA DEPARTMENT OF ENVIRONMENTAL QUALITY

1000 NE 10th St.
Oklahoma City, OK 73117-1212 USA
Phone: 405-271-8056
Website: http://www.deq.state.ok.us

Founded: NA

Description: The Department of Environmental Quality is dedicated to providing quality service to the people of Oklahoma through comprehensive environmental protection and management programs. Those programs are designed to assist the people of the state in sustaining a clean sound environment and in preserving and enhancing our natural surroundings.

Publications: Certified Operator News Letter (Waterworks and Wastewater), Superfund Program Sites Status Report, Air Quality Annual Report, Clear View

Keyword(s): Water Quality, Solid Waste Management, Environmental Protection, Pollution Prevention, Air Quality and Pollution

Contact(s):
Steven Thompson, DEPUTY EXECUTIVE DIRECTOR
Phone: 405-271-8056
Larry Byrum, DIRECTOR OF AIR QUALITY
Phone: 405-271-5220
Larry McKee, DIRECTOR OF COMPLAINTS AND LOCAL
SERVICES
Phone: 405-271-7363
Judy Duncan, DIRECTOR OF CUSTOMER SERVICES
Phone: 405-271-1400
Ellen Bussert, DIRECTOR OF PUBLIC INFORMATION AND
EDUCATION
Phone: 405-271-8056
Lawrence Gales, DIRECTOR OF SUPPORT SERVICES
Phone: 405-271-8062
H.a. Caves, DIRECTOR OF WASTE MANAGEMENT
Phone: 405-271-5338
Jon Craig, DIRECTOR OF WATER QUALITY
Phone: 405-271-5205
Mark Coleman, EXECUTIVE DIRECTOR
Phone: 405-271-8056
Bob Kellogg, GENERAL COUNSEL
Phone: 405-271-8056

OKLAHOMA GEOLOGICAL SURVEY

University of Oklahoma, Sarkeys Energy Center, 100 E.
Boyd, Rm. N-131
Norman, OK 73019-0628 USA
Phone: 405-325-3031   Fax: 405-325-7069
Website: www.ou.edu/special/ogs-pttc/

Founded: 1908
Membership: 40
Scope: Statewide

Description: To investigate and disseminate information on the
geology of the state, with special reference to mineral
resources and environmental issues. Investigations include:
geologic mapping, evaluation of metallic and nonmetallic
mineral deposits, and studies of earthquakes, groundwater,
and fossil fuels, plus basic research and environmental
studies.

Publications: bulletins, circulars, guidebooks, geologic map
series, educational publications series, hydrologic atlases,
special publications, etc., Oklahoma Geology Notes

Keyword(s): Exotic species, Aquatic nuisance species, Land Use
Planning, Environment, Geology, Mapping, Energy

Contact(s):
Joyce Stiehler, CHIEF PUBLICATIONS CLERK
Phone: 405-360-2886
ogssales@ou.edu
Charles Mankin, DIRECTOR
Claren Kidd, LIBRARIAN
100 E. Boyd, Rm. 220, Norman, OK 73019-0628
Connie Smith, PROMOTION AND INFORMATION
SPECIALIST

OKLAHOMA STATE EXTENSION SERVICES

DIVISION OF AGRICULTURE
Oklahoma State University, Rm. 139, Agricultural Hall
Stillwater, OK 74078 USA
Phone: 405-744-5398   Fax: 405-744-5339
Website: http://www.dasnr.okstate.edu

Founded: 1946
Scope: International

Description: Extension Forestry and Wildlife are a unit of the
Oklahoma Cooperative Extension Service and the Department
of Forestry, Division of Agricultural Sciences and Natural
Resources, Oklahoma State University. The Department of
Forestry provides accredited education in forest resources
management, conducts forestry research through the
Oklahoma Agricultural Experiment Station, and brings forest
resources education to the citizens of Oklahoma through its
extension efforts.

Publications: Oklahoma Forest Industry bulletin, Natural
Resources Speakers Bureau, Oklahoma Renewable
Resources newsletter

Contact(s):
Craig McKinley, DEPARTMENT HEAD OF FORESTRY
Oklahoma State University, Rm. 011, Agricultural Hall,
Stillwater, OK 74078
Phone: 405-744-5438
Kenneth Hitch, ASSISTANT EXTENSION FORESTER AND
WILDLIFE SPECIALIST
Oklahoma State University, Rm. 008C, Agricultural Hall,
Stillwater, OK 74078
Phone: 405-744-5442
D. Coston, ASSOCIATE DIRECTOR FOR OKLAHOMA
AGRICULTURAL EXPERIMENT STATION
Sam Curl, DIRECTOR OF COOPERATIVE EXTENSION
SERVICE
Ronald Masters, EXTENSION WILDLIFE SPECIALIST
Oklahoma State University, Rm. 240, Agricultural Hall,
Stillwater, OK 74078
Phone: 405-744-8065

OKLAHOMA TOURISM AND RECREATION DEPARTMENT

15 N. Robinson
Oklahoma City, OK 73102 USA
Phone: 405-521-2409   Fax: 405-521-3992
Website: www.touroklahoma.com

Founded: 1972
Membership: 32
Scope: State

Description: To encourage residents and travelers to "Native
America" as a vacation destination; and to develop human and
natural resources for the purpose of promoting tourism,
recreation, wildlife preservation, and environmental conservation. An annual industry conference is held each fall.

Publications: Oklahoma Vacation Guide, Oklahoma Today

Keyword(s): Public Lands, Nongame Wildlife, Outdoor
Recreation, National Parks

Contact(s):
Jayne Jayroe, CABINET SECRETARY AND EXECUTIVE
DIRECTOR
Phone: 405-521-2413
Doug Enevoldsen, CHIEF OF OPERATIONS
Phone: 405-521-4678
Betty Koehn, DIRECTOR OF ADMINISTRATION
Phone: 405-520-4031
John Ressmeyer, DIRECTOR OF STATE PARKS
Phone: 405-521-4291
Amos Moses, DIVISION OF HUMAN RESOURCES
Phone: 405-522-4523
Kristina Marek, DIVISION OF PLANNING AND
DEVELOPMENT
Phone: 405-521-2973

Kathleen Marks, DIVISION OF TRAVEL AND TOURISM
Phone: 405-521-3981
Joan Henderson, PUBLISHER
Phone: 405-521-2496
Jeff Erwin, RESORT DIVISION

OKLAHOMA WATER RESOURCES BOARD

3800 N. Classen Blvd.
Oklahoma City, OK 73118 USA
Phone: 405-530-8800   Fax: 405-530-8900
Website: www.state.ok.us/~owrb

Founded: 1957
Membership: 90
Scope: State

Description: Promulgates water quality standards for state; lead agency in Clean Lakes Program; investigates pollution complaints; assesses water quality, quantity of groundwater, and stream water; issues permits for water use; administers dam safety, floodplain management programs, and plans for adequate supplies of good quality water for all beneficial uses; updates plans; administers financial assistance programs for water and wastewater systems.

Publications: Oklahoma Water News, The Well Drillers Log Newsletter

Keyword(s): Exotic species, Aquatic nuisance species, Water Quality, Lakes, Rivers, Environmental and Conservation Education

Contact(s):
Michael Melton, ASSISTANT TO THE DIRECTOR
Duane Smith, EXECUTIVE DIRECTOR
dasmith@
Jim Schuelein, PUBLICATIONS

OREGON COOPERATIVE FISH AND WILDLIFE RESEARCH UNIT (USDI)

Department of Fisheries and Wildlife, Oregon State University
Corvallis, OR 97331-3803 USA
Phone: 503-737-4531

Founded: NA

Description: Research focus on physiological, ecological, and genetic factors affecting production and performance of freshwater fishes. The staff consists of two permanent and two-three other Ph.D., level scientists, as well as graduate students and technicians.

Keyword(s): standards, Wildlife, Biodiversity, Environmental and Conservation Education, Genetis, Physiology, Aquatic Habitats

Contact(s):
Hiram Li, ASSISTANT LEADER
Carl Schreck, LEADER

OREGON COOPERATIVE FISH AND WILDLIFE RESEARCH UNIT (USDI)

104 Oregon State University
Corvallis, OR 97331-3803 USA
Phone: 541-737-1938   Fax: 541-737-3590
E-mail: or_sfwru@orst.edu
Website: www.orst.edu/dept/fish_wild

Founded: NA
Membership: 4

Scope: International

Contact(s):
Hiram Li, ASSISTANT LEADER
Carl Schreck, LEADER
Daniel Roby, WILDLIFE ASSISTANT LEADER
Robert Anthony, WILDLIFE LEADER

OREGON DEPARTMENT OF AGRICULTURE

NATURAL RESOURCES DIVISION
635 Capitol St., NE
Salem, OR 97310-0110 USA

Founded: 1939 (Department of Agriculture), 1989 (Natural Resources Division)

Description: Supervises the organization and operation of soil and water conservation districts, approves or disapproves all projects, practices, personnel, budgets, contracts, and regulations of Oregon's 45 districts. State administrative agency for nonpoint source water quality programs dealing with agricultural lands. Also responsible for managing the state's field burning weather monitoring program, the native plant species conservation program, and the weather modification program.

Publications: Oregon Natural Resources Conservation News

Contact(s):
John Mellott, ADMINISTRATOR
Phone: 503-378-3810
jmellott@oda.state.or.us
Jack Sainsbury, ADVISORY MEMBER
ASCS, 1220 SW 3rd. Ave., 15th Fl., Portland, OR 97204-2880
Phone: 503-326-2741
Andrew Hashimoto, ADVISORY MEMBER
Department of Agriculture Engineering, Oregon State University, Corvallis, OR 97331-2213
Phone: 503-737-2041
Ray Ledgerwood, ADVISORY MEMBER
NACD, NE 1615 Eastgate Blvd., Suite B, Pullman, WA
Phone: 509-334-1823
Bob Graham, ADVISORY MEMBER
SCS, 1220 SW 3rd Ave., 16th Fl., Portland, OR 97204-2881
Phone: 503-326-2751
Charles Craig, ASSISTANT ADMINISTRATOR
Phone: 503-378-3810
Thomas Straughan, SOIL AND WATER CONSERVATION COMMISSION CHAIRMAN
1421 SW 45th Dr., Pendleton, OR 97801
Phone: 503-278-0218
tstraugh@orednet.org
Joe Brumbach, SOIL AND WATER CONSERVATION COMMISSIONER VICE CHAIRMAN
4260 Buckhorn Rd., Roseburg, OR 97470
Phone: 503-673-3998

OREGON DEPARTMENT OF ENVIRONMENTAL QUALITY (DEQ)

811 S.W. 6th Ave.
Portland, OR 97204 USA
Phone: 503-229-5696   Fax: 503-229-6124
Website: www.deq.state.or.us

Founded: 1969
Scope: State

Description: Our mission is to be an active leader in restoring, maintaining, and enhancing the quality of Oregon's air, water, and land.

Publications: Beyond Waste, Tankline, Recycling Newsletter

Keyword(s): Environmental Cleanup, Water Pollution Management, Environmental and Conservation Education, Solid Waste Management, Air Quality and Pollution

Contact(s):
Langdon Marsh, DIRECTOR
Phone: 503-229-5300
Stephanie Hallock, EASTERN REGION ADMINISTRATOR
Phone: 541-338-6146
Sally Puent, INTERN
Rick Gates, LABORATORY ADMINISTRATOR
Phone: 503-229-5983
Neil Mullane, NORTHWEST REGION ADMINISTRATOR
Phone: 503-229-5372
Michael Llewelyn, WATER QUALITY ADMINISTRATOR
Phone: 503-229-5324
Steve Greenwood, WESTERN REGION ADMINISTRATOR
Phone: 541-686-7838

OREGON DEPARTMENT OF FISH & WILDLIFE (ODFW)

2501 SW 1st Ave.
Portland, OR 97201 USA
Phone: 503-872-5310   Fax: 503-872-5302
E-mail: odfwinfo@state.or.us
Website: www.dfw.state.or.us

Founded: 1975
Scope: Statewide

Description: Responsibilities include management of fish and wildlife resources and regulation of commercial and recreational harvest.

Publications: Oregon Wildlife

Contact(s):
Ron Anglin, DEPUTY DIRECTOR
Lindsay Ball, DIRECTOR
Brian Alula, DIRECTOR OF INFORMATION SERVICES
Phone: 503-872-5267
Pat Wray, EDITOR
7118 NE Vandenberg Ave., Corvallis, OR 97330
Phone: 541-757-4186
Fax: 541-757-4252
Dave McAllister, HABITAT DIVISION DIRECTOR
Phone: 503-872-5255
Chip Dale, HIGH DESERT REGION DIRECTOR
61374 Parrell Rd., Bend, OR 97702
Phone: 541-388-6363
Fax: 541-388-6281
Carol Brown, HUMAN RESOURCES DIVISION DIRECTOR
Phone: 503-872-5262
Susan Adams Gunn, INFORMATION AND EDUCATION
Wayne Rawlins, MANAGER OF BUSINESS SERVICES/REALTY
Phone: 503-872-5310
Craig Ely, NE REGION DIRECTOR
107 20th St., LaGrande, OR 97850
Phone: 541-963-2138
Fax: 541-963-6670
Chris Wheaton, NW REGION DIRECTOR
17330 SE Evelyn St., Clackamas, OR 97015
Phone: 503-657-2000
Fax: 503-657-2050

Steve Denney, SW REGIONAL DIRECTOR (ACTING)

OREGON DEPARTMENT OF FORESTRY

2600 State St.
Salem, OR 97310 USA
Phone: 503-945-7200   Fax: 503-945-7212
Website: www.odf.state.or.us

Founded: 1911
Membership: 1000
Scope: Statewide

Description: The department is responsible for fire protection of 15.8 million acres of private and public forests; directs insect and disease management on 11 million acres of state and private forests; manages 789,000 acres of state-owned forests; provides forestry assistance to private forest landowners; enforces other Oregon forest laws; provides forestry information to schools, organizations, and individuals; and advises Governor and State Legislature on forestry matters.

Publications: CommuniTree News (Quarterly), The Oregon Forests Report (Annual), Forest Log, The

Keyword(s): Urban Forestry, Wildlife, Forests and Forestry, Endangered Species

Contact(s):
Charlie Stone, ASSISTANT STATE FORESTER OF FOREST PROTECTION
Phone: 503-945-7205
Steve Thomas, ASSISTANT STATE FORESTERS
Ted Lorensen, ASSISTANT STATE FORESTERS
Gayle Birch, BOARD OF FORESTRY
Dave Gilbert, CHAIRMAN
Roy Woo, DEPUTY STATE FORRESTER
801 Gales Creek Rd., Forest Grove, OR 97117-1199
Phone: 503-357-2191
Fax: 503-357-4548
Rick Gibson, DIRECTOR OF FIRE PREVENTION
Phone: 503-945-7440
Lanny Quackenbush, DIRECTOR OF FIRE PROTECTION
Phone: 503-945-7435
Wallace Rutledge, DIRECTOR OF FORESTRY ASSISTANCE
Phone: 503-945-7392
Cary Greenwood, DIRECTOR OF PUBLIC AFFAIRS
Phone: 503-945-7420
James Brown, SECRETARY
Phone: 503-945-7211

OREGON FISH AND WILDLIFE DIVISION/DEPARTMENT OF STATE POLICE

400 Public Service Bldg.
Salem, OR 97310 USA
Phone: 503-378-3720   Fax: 503-363-5475
E-mail: first.last@state.or.us
Website: www.osp.state.or.us

Founded: NA
Scope: State

Description: The Fish and Wildlife Division is charged with the enforcement of fish and game, commercial fish, shellfish, environmental protection laws, and all endangered species laws, rules, and regulations. Also provides general law enforcement services in rural areas. Provides law enforcement services on contract with the Oregon Department of Fish and Wildlife, the Department of Environmental Quality, and the Department of

Forestry. Enforcement priorities for fish and wildlife resources cooperatively identified with ODFW.

Publications: see publication website

Keyword(s): Law Enforcement, Hunting, Environmental Law, Wildlife, training, Endangered Species

Contact(s):
J. Hunsaker, AIRCRAFT SUPERVISOR
Phone: 503-378-3720
Lindsay Ball, C & D DIRECTOR (CAPTAIN)
Phone: 503-378-3720
K. Allison, DISTRICT I SUPERVISOR OF PORTLAND, OR
Phone: 503-731-3027
S. Lane, DISTRICT II SUPERVISOR OF SALEM, OR
Phone: 503-378-2110
S. Ross, DISTRICT III SUPERVISOR OF MEDFORD, OR
Phone: 503-776-6114
R. Scorby, DISTRICT IV SUPERVISOR OF BAKER CITY, OR
Phone: 503-523-5848
W. Markee, SPECIAL INVESTIGATIONS UNIT SUPERVISOR
Phone: 503-378-3387
D. Cleary, STAFF OF COMMERCIAL FISHERIES
Phone: 503-378-3720
C. Kok, STAFF OF WILDLIFE
Phone: 503-378-3720

OREGON PARKS AND RECREATION DEPARTMENT

1115 Commercial St., NE, Suite 1
Salem, OR 97301-1002 USA
Phone: 503-378-6305   Fax: 503-378-6447
Website: www.orgegonparks.org

Founded: 1921
Membership: 100
Scope: Statewide

Description: To provide and protect outstanding natural, scenic, cultural, historic, and recreational sites for the enjoyment and education of present and future generations.

Keyword(s): Rivers, Outdoor Recreation, Historic Preservation, Natural Areas, Ocean Conservation, Environmental and Conservation Education

Contact(s):
Michael Carrier, DIRECTOR
Phone: 503-378-5019

OREGON SEA GRANT PROGRAM

500 322 Kerr Administration Bldg., Oregon State University
Corvallis, OR 97331-2131 USA
Phone: 541-737-2714   Fax: 541-737-7958
E-mail: seagrant.admin@orst.edu
Website: seagrant.orst.edu/

Founded: NA
Scope: Statewide

Description: Oregon Sea Grant takes an integrated approach to addressing the problems and opportunities of Oregon's marine resources through three related primary activities—research, education, and extension services. Oregon Sea Grant responds to the needs of ocean users.

Publications: Catalogue available upon request from Sea Grant Communications.

Keyword(s): Oceanography, Wildlife, Coasts, Communications, Biotechnology

Contact(s):
Joseph Cone, ASSISTANT DIRECTOR FOR COMMUNICATIONS
402 Kerr Admin. Bldg. OSU, Corvallis, OR 97331-2134
Phone: 541-737-2716
Fax: 541-737-7958
joe.cone@orst.edu
Jan Auyong, ASSISTANT DIRECTOR FOR PROGRAMS
jan.auyong@orst.edu
Jay Rasmussen, EXTENSION PROGRAM LEADER
Hatfield Marine Science Center, 2030 S. Marine Science Dr., Newport, OR 97365
Phone: 541-867-0370
Fax: 541-867-0369
jay.rasmussen@hmsc.orst.edu
Robert Malouf, PROGRAM DIRECTOR
maloufr@ccmail.orst.edu

OREGON STATE EXTENSION SERVICES

Oregon State University, 101 Ballard Extension Hall
Corvallis, OR 97331-3606 USA
Phone: 541-737-2713   Fax: 541-737-4423
Website: www.osu.orst.edu/extension/

Founded: NA
Scope: Statewide

Contact(s):
Peter Bloome, ASSOCIATE DIRECTOR
peter.bloome@orst.edu
Lyla Houglum, DEAN AND DIRECTOR
William Braunworth, EXTENSION PROGRAM COORDINATOR, AGRICULTURE
Stag Hall, Room 138, Oregon State University, Corvallis, OR 97331
Phone: 541-737-4251
Fax: 541-737-3178
Bill.Braunworth@orst.edu
W. Edge, NATURAL RESOURCES INFORMATION
104 Nash Hall, Oregon State University, Corvallis, OR 97331-3803
Phone: 541-737-1953
Fax: 541-737-3590
A. Reed, PROGRAM LEADER, FORESTRY
119 Peavy Hall, Oregon State University, Corvallis, OR 97331
Phone: 541-737-3700
Fax: 541-737-3008
Deborah Maddy, REGIONAL DIRECTOR
Phone: 541-737-2711
deborah.maddy@orst.edu
Michael Stoltz, REGIONAL DIRECTOR
Phone: 541-737-2711
michael.stoltz@orst.edu

OREGON WATER RESOURCES DEPARTMENT

158 12th St., NE
Salem, OR 97310 USA
Phone: 503-378-8455   Fax: 503-378-2496
Website: www.wrd.state.or.us

Founded: NA
Membership: 250
Scope: Statewide

Description: The Water Resources Department is the steward of the state's water resources. The agency enforces state water laws and policies; promotes actions that restore and protect streamflows and watersheds in order to ensure the long-term sustainability of Oregon's ecosystems, economy, and quality of life; addresses water supply needs; and increases the understanding of the resource and the demands on it.

Contact(s):
Bruce Moyer, ADMINISTRATOR OF ADMINISTRATIVE SERVICES DIVISION
Phone: 503-378-8455
Tom Paul, ADMINISTRATOR OF FIELD SERVICES DIVISION
Phone: 503-378-8455
Barry Norris, ADMINISTRATOR OF TECHNICAL SERVICES DIVISION
Phone: 503-378-8455
Dick Bailey, ADMINISTRATOR OF WATER RIGHTS AND ADJUDICATIONS
Phone: 503-378-8455
Paul Cleary, DIRECTOR
Phone: 503-378-2982

OREGON WATER RESOURCES DEPARTMENT
WATER RESOURCES COMMISSION
P.O. Box 1588
Medford, OR 97501-0244 USA
Phone: 541-857-8222
Website: http://oda.state.or.us

Founded: 1985
Scope: Statewide

Description: The Water Resources Commission was created to oversee policy and all rulemaking activity of the department.

Keyword(s): Exotic species, Aquatic nuisance species, Sustainable Ecosystems, Rivers, Sustainable Development, Planning Management

Contact(s):
Nancy Leonard, CHAIR, COMMISSIONER, WEST CENTRAL REGION
Phone: 541-563-2187
Susie Smith, COMMISSIONER
Jim Nakano, COMMISSIONER, EASTERN REGION
Phone: 541-889-6823
Tyler Hansell, COMMISSIONER, EASTSIDE AT LARGE
Phone: 541-567-8939
Ron Nelson, COMMISSIONER, NORTH CENTRAL REGION
Phone: 541-548-6047
John Fregonese, COMMISSIONER, NORTHWEST REGION
Dan Thorndike, COMMISSIONER, SOUTHWEST REGION
Phone: 541-857-8222

P

PENNSYLVANIA COOPERATIVE FISH AND WILDLIFE RESEARCH UNIT
Merkle Bldg., Pennsylvania State University
University Park, PA 16802 USA
Phone: 814-865-4511   Fax: 814-863-4710
E-mail: klc2@psu.edu
Website: www.pacfwru.cas.psu.edu

Founded: 1938
Membership: 3
Scope: National

Description: Established as a cooperative activity among Pennsylvania State University, Pennsylvania Fish and Boat Commission, Wildlife Management Institute, Pennsylvania Game Commission, and the Department of the Interior. Areas of research are: The effects of natural and manmade forces on aquatic and terrestrial ecosystems, animal-habitat interactions, acid precipitation effects, fish and wildlife management, and health profiles of game animals. Graduate training is also provided.

Publications: Annual Report available on request.

Contact(s):
Duane Diefenbach, ASSISTANT LEADER FOR WILDLIFE
Erin Snyder, ASSISTANT UNIT LEADER FOR FISHERIES
ems19@psu.edu
Robert Carline, LEADER

PENNSYLVANIA DEPARTMENT OF AGRICULTURE
CRICKET FIELD PLAZA
1307 7th St.
Altoona, PA 16601-4701 USA
Phone: 814-946-7315   Fax: 814-946-7354

Founded: NA
Scope: Regional

Contact(s):
Kenneth Mowry, DIRECTOR

PENNSYLVANIA DEPARTMENT OF AGRICULTURE
REGION I
13410 Dunham Rd. Region 1
Meadville, PA 16335 USA
Phone: 814-332-6890   Fax: 814-333-1431
Website: www.tda.state.pa.us

Founded: NA
Membership: 31
Scope: State

Contact(s):
George Gregg, DIRECTOR

PENNSYLVANIA DEPARTMENT OF AGRICULTURE
REGION II
542 County Farm Rd., Suite 102
Montoursville, PA 17754-9685 USA
Phone: 570-433-2640   Fax: 570-433-4770
Website: www.pda.state.pa.us

Founded: NA
Scope: Statewide

Contact(s):
Dean Ely, DIRECTOR

PENNSYLVANIA DEPARTMENT OF AGRICULTURE
REGION III
Rt. 92 South, P.O. Box C
Tunkhannock, PA 18657 USA
Phone: 570-836-2181   Fax: 570-836-6266

Founded: NA
Membership: 40
Scope: Statewide

Contact(s):
Russell Gunton, DIRECTOR

PENNSYLVANIA DEPARTMENT OF AGRICULTURE

REGION IV
5349 William Flynn Hwy.
Gibsonia, PA 15044 USA
Phone: 724-443-1585   Fax: 724-443-8150

Founded: NA
Membership: 55
Scope: Local

Contact(s):
R. Nehrig, DIRECTOR

PENNSYLVANIA DEPARTMENT OF AGRICULTURE

REGION VI
2301 North Cameron Street
Harrisburg, PA 17110-9408 USA
Phone: 717-787-4737
Website: http://www.pda.state.pa.us/

Founded: NA

Contact(s):
Sam Hayes, SECRETARY

PENNSYLVANIA DEPARTMENT OF AGRICULTURE

REGION VII
P.O. Box 300
Creamery, PA 19430 USA
Phone: 610-489-1003   Fax: 610-489-6119
E-mail: PSTARR@STATE.PA.US
Website: WWW.PDA.STATE.PA.US

Founded: NA
Membership: 45
Scope: Regional

Contact(s):
Frank Stearns, DIRECTOR

PENNSYLVANIA DEPARTMENT OF AGRICULTURE

STATE CONSERVATION COMMISSION
2301 N. Cameron St.
Harrisburg, PA 17110-9408 USA
Phone: 717-787-8821   Fax: 717-705-3778
E-mail: KBROWN@STATE.PA.US
Website: www.pascc.org

Founded: 1945
Membership: 7
Scope: Statewide

Description: To establish policy for Pennsylvania's 66 local conservation districts. Programs administered by the commission include: a $2,850,000 annual grant program, which provides funds to conservation districts for the employment of managerial and technical staff; a $5.2 million annual Chesapeake Bay Program, which provides technical and financial assistance to farmers to install soil conservation and nutrient management practices, and a $4,000,000 annual grant program; which provides funds to local municipalities for the maintenance of dirt and gravel roads.

Publications: Highlights Newsletter

Keyword(s): Environmental and Conservation Education, Soil Conservation, Agriculture

Contact(s):
Karl Brown, EXECUTIVE SECRETARY
Phone: 717-705-3778
kbrown@agric.state.us

PENNSYLVANIA DEPARTMENT OF CONSERVATION AND NATURAL RESOURCES

RACHEL CARSON STATE OFFICE BLDG.
Harrisburg, PA 17105-8551 USA
Phone: 717-787-2869   Fax: 717-705-2832
Website: www.dcnr.state.pa.us

Founded: 1995
Scope: State

Description: To maintain and preserve state parks; to manage state forest lands to assure their long-term health, sustainability and economic use; to provide information on Pennsylvania's ecological and geologic resources; and to administer grant and technical assistance programs that will benefit river conservation, trails and greenways, local recreation, regional heritage conservation and environmental education programs across Pennsylvania.

Publications: PA State Park Recreation Guide, Penn's Woods, Become a Conservation Volunteer, Discover DCNR, Resource

Keyword(s): Environmental and Conservation Education, Outdoor Recreation, State Parks, Conservation, Land Preservation, Natural Areas, Forest Management, State Forests, Geology, Community Conservation, Biodiversity

Contact(s):
William Shakely, CHIEF COUNSEL
Phone: 717-772-4171
Karen Deklinski, DEPUTY SECRETARY FOR ADMINISTRATION
Phone: 717-772-9100
Richard Sprenkle, DEPUTY SECRETARY FOR CONSERVATION AND ENGINEERING SERVICES
Phone: 717-787-9306
Geralyn Umstead, DIRECTOR OF COMMUNITY RELATIONS
Phone: 717-772-9087
Jay Parrish, DIRECTOR OF DEPT OF TOPOGRAPHIC & GEOLOGICAL SURVEY
Joan Dlippinger, DIRECTOR OF ENVIRONMENTAL EDUCATION
Phone: 717-705-2008
Damon Anderson, DIRECTOR OF INFORMATION TECHNOLOGY
Phone: 717-783-9732
Frederick Carlson, DIRECTOR OF POLICY
Dana Datres, DIRECTOR, BUREAU OF ADMINISTRATIVE SERVICES
Phone: 717-787-2362
Eugene Comoss, DIRECTOR, BUREAU OF FACILITY DESIGN AND CONSTRUCTION
Phone: 717-787-7398
James Grace, DIRECTOR, BUREAU OF FORESTRY
Phone: 717-787-2703
Dennis Farley, DIRECTOR, BUREAU OF PERSONNEL
Phone: 717-787-5496

Larry Williamson, DIRECTOR, BUREAU OF RECREATION
AND CONSERVATION
Phone: 717-783-2658
Roger Fickes, DIRECTOR, BUREAU OF STATE PARKS
Phone: 717-787-6640
John Plonski, EXECUTIVE DEPUTY SECRETARY FOR
PARKS AND FORESTRY
Phone: 717-772-9104
Kurt Leitholf, EXECUTIVE DIRECTOR, CONSERVATION
AND NATURAL RESOURCES ADVISORY COUNCIL TO
DCNR
8th Fl., Rachel Carson State Office Bldg., P.O. Box 8773,
Harrisburg, PA 17105-8773
Phone: 717-705-0031
Joseph Graci, LEGISLATIVE LIAISON
Phone: 717-772-9101
Gretchen Leslie, PRESS SECRETARY
Phone: 717-772-9101
John Oliver, SECRETARY
Phone: 717-787-2869
Sally Just, SENIOR ADVISOR TO SECRETARY
Phone: 717-787-2869

PENNSYLVANIA DEPARTMENT OF ENVIRONMENTAL PROTECTION

P.O. Box 2063
Harrisburg, PA 17105-2063 USA
Phone: 717-783-7404　　　Fax: 717-783-8926
Website: www.dep.state.pa.us

Founded: 1995
Scope: State

Description: The Department of Environmental Protection's
mission is to protect Pennsylvania's air, land, and water from
pollution and to provide for the health and safety of its citizens
through a cleaner environment. We will work as partners with
individuals, organizations, governments, and businesses to
prevent pollution and restore our natural resources.

Keyword(s): Public Health Protection, Protected Areas
Contact(s):
Christopher Allen, ACTING DIRECTOR OF LOCAL
GOVERNMENT RELATIONS
Phone: 717-787-9580
Michael Bedrin, CHIEF COUNSEL
Phone: 717-787-4449
Kimberly Nelson, CHIEF INFORMATION OFFICER
Phone: 717-772-0801
Denise Chamberlain, DEPUTY SECRETARY FOR AIR,
RECYCLING AND RADIATION PROTECTION
Phone: 717-772-2724
Donald Welsh, DEPUTY SECRETARY FOR
FEDERAL/STATE RELATIONS
Phone: 717-783-1566
Terry Fabian, DEPUTY SECRETARY FOR FIELD
OPERATIONS
Phone: 717-787-5028
Kenwood Giffhorn, DEPUTY SECRETARY FOR
MANAGEMENT AND TECHNICAL SERVICES
Phone: 717-787-7116
Robert Dolence, DEPUTY SECRETARY FOR MINERAL
RESOURCES MANAGEMENT
Phone: 717-783-5338
Robert Barkanic, DEPUTY SECRETARY FOR POLLUTION
PREVENTION AND COMPLIANCE ASSISTANCE
Phone: 717-783-0540

Lawrence Tropea, DEPUTY SECRETARY FOR WATER
MANAGEMENT
Phone: 717-787-4686
Helen Olena, DIRECTOR OF ENVIRONMENTAL
EDUCATION
Phone: 717-772-1828
Roderick Fletcher, DIRECTOR, BUREAU OF ABANDONED
MINE RECLAMATION
Phone: 717-783-2267
James Salvaggio, DIRECTOR, BUREAU OF AIR QUALITY
Phone: 717-787-9702
Richard Stickler, DIRECTOR, BUREAU OF DEEP MINE
SAFETY
Phone: 724-439-7469
Ronald Flory, DIRECTOR, BUREAU OF FISCAL
MANAGEMENT
Phone: 717-787-1319
Gary Niland, DIRECTOR, BUREAU OF INVESTIGATIONS
Phone: 717-787-0453
James Snyder, DIRECTOR, BUREAU OF LAND
RECYCLING AND WASTE MANAGEMENT
Phone: 717-783-2388
Jay Roberts, DIRECTOR, BUREAU OF MINING AND
RECLAMATION
Phone: 717-787-5103
Larry Brown, DIRECTOR, BUREAU OF NETWORK
OPERATIONS
Phone: 717-772-5909
James Erb, DIRECTOR, BUREAU OF OIL AND GAS
MANAGEMENT
Phone: 717-772-2199
Richard Mather, DIRECTOR, BUREAU OF REGULATORY
COUNSEL
Phone: 717-787-7060
Stuart Gansell, DIRECTOR, BUREAU OF WATERSHED
CONSERVATION
Phone: 717-787-5267
Michael Conway, DIRECTOR, BUREAU OF WATERWAYS
ENGINEERING
Phone: 717-787-3411
Jeffrey Jarrett, DIRECTOR, DISTRICT MINING
OPERATIONS
Phone: 724-942-7204
Barbara Sexton, DIRECTOR, THE POLICY OFFICE
Phone: 717-783-8727
David Hess, EXECUTIVE DEPUTY SECRETARY
Phone: 717-772-1856
Gregory Mahon, LEGISLATIVE LIAISON
Phone: 717-783-8303
Irene Brooks, OFFICE FOR RIVER BASIN COOPERATION
EXECUTIVE DIRECTOR
Phone: 717-772-4785
James Seif, SECRETARY
Phone: 717-787-2814

PENNSYLVANIA FISH AND BOAT COMMISSION

P.O. Box 67000
Harrisburg, PA 17106-7000 USA
Phone: 717-705-7800　　　Fax: 717-705-7802
Website: www.fish.state.pa.us

Founded: 1866

Description: To conduct and support public education and
information efforts related to aquatic resource protection,
improvement, and management programs, and enhance public
understanding of the wise and safe use of our fishing and
boating resources.

Publications: Pennsylvania Angler and Boater

Keyword(s): Sport Fishing, Outdoor Recreation, Environmental and Conservation Education, Wildlife, Aquatic Habitats

Contact(s):
Ted Walke, ART DIRECTOR
Phone: 717-564-6846
Dennis Guise, DEPUTY EXECUTIVE DIRECTOR/CHIEF COUNSEL
Phone: 717-657-4525
Wasyl Polischuk, DIRECTOR OF BUREAU OF ADMINISTRATION SERVICES
Phone: 717-657-4522
John Simmons, DIRECTOR OF BUREAU OF BOATING AND EDUCATION
Phone: 717-657-4538
James Young, DIRECTOR OF BUREAU OF ENGINEERING AND DEVELOPMENT
Phone: 814-359-5152
Delano Graff, DIRECTOR OF BUREAU OF FISHERIES
Phone: 814-359-5169
Thomas Kamerzel, DIRECTOR OF BUREAU OF LAW ENFORCEMENT
Phone: 717-657-4542
Arthur Michaels, EDITOR
Phone: 717-657-4520
Peter Colangelo, EXECUTIVE DIRECTOR
Phone: 717-657-4515
Joseph Greene, LEGISLATIVE LIAISON
Phone: 717-657-4517
Thomas Ford, PLANNING COORDINATOR
Phone: 717-657-4394
Ted Keir, PRESIDENT
Leon Reed, VICE PRESIDENT

PENNSYLVANIA FISH AND BOAT COMMISSION

REGION 1, NORTHWEST
11528 State Highway 98
Meadville, PA 16335 USA
Phone: 814-337-0444　　Fax: 814-337-0579
Website: www.fish.state.pa.us

Founded: NA
Membership: 17
Scope: State

Contact(s):
Frank Parise, ASSISTANT REGIONAL MANAGER
Robert Nestor, ASSISTANT REGIONAL MANAGER
Gary Deiger, LAW ENFORCEMENT SUPERVISOR

PENNSYLVANIA FISH AND BOAT COMMISSION

REGION 2 SOUTHWEST
236 Lake Rd. Region 2 SW
Somerset, PA 15501-1644 USA
Phone: 814-445-8974　　Fax: 814-445-3497
E-mail: pfbcsw1@twd.net
Website: www.fish.state.pa.us

Founded: NA
Membership: 30
Scope: Regional

Contact(s):
Dennis Tubbs, AQUATIC RESOURCE PROGRAM SPECIALIST
Phone: 814-443-9841
Rick Lorson, AREA FISHERIES BIOLOGIST

Emil Svetahor, LAW ENFORCEMENT SUPERVISOR
Phone: 814-445-3554

PENNSYLVANIA FISH AND BOAT COMMISSION

REGION 3 NORTHEAST
P.O. Box 88
Sweet Valley, PA 18656-0008 USA
Phone: 570-477-5717　　Fax: 570-477-3221
E-mail: pfbcne1@ptd.net
Website: http://www.pda.state.pa.us

Founded: NA

Contact(s):
Kerry Messerle, LAW ENFORCEMENT SUPERVISOR

PENNSYLVANIA FISH AND BOAT COMMISSION

REGION IV SOUTHEAST
Box 9 South IV S.E.
Elm, PA 17521-0008 USA
Phone: 717-626-0228　　Fax: 717-626-0486
E-mail: pfbcse1@redrose.net
Website: www.fish.state.pa.us

Founded: NA
Membership: 15
Scope: State

Contact(s):
Jeffrey Bridi, LAW ENFORCEMENT SUPERVISOR

PENNSYLVANIA FISH AND BOAT COMMISSION

REGION V NORTHCENTRAL
1601 Elmerton Avenue
Harrisburg, PA 17110 USA
Phone: 717-705-7801
Website: http://www.fish.state.pa.us

Founded: NA

Contact(s):
Paul Swanson, LAW ENFORCEMENT SUPERVISOR

PENNSYLVANIA FISH AND BOAT COMMISSION

REGION VI SOUTHCENTRAL
1704 Pine Rd.
Newville, PA 17241-9544 USA
Phone: 717-486-7087　　Fax: 717-486-8227
E-mail: psbcsc1@epix.net
Website: www.fish.state.pa.us

Founded: NA
Scope: Statewide

Publications: Pennsylvania Angler & Boater Magazine

Contact(s):
George Geisler, REGIONAL MANAGER SOUTHCENTRAL
1704 Pine Rd, Newville, PA 17241

PENNSYLVANIA FISH AND BOAT COMMISSION: REGION 1

REGION 1, NORTHWEST, 11528 STATE HIGHWAY 98
MEADVILLE, PA 16335 USA
Phone: 814-337-0444　　Fax: 814-337-0579
Website: www.fish.state.pa.us
Membership: 6
Scope: State

Contact(s):
Gary Deiger, REGIONAL MANAGER
Phone: 814-337-0444
gdeiger@state.pa.us

PENNSYLVANIA FISH AND BOAT COMMISSION: REGION 2 SW
236 Lake Rd.
Somerset, PA 15501-1644 USA
Phone: 814-445-8974   Fax: 814-445-3497
Website: www.fish.state.pa.us
Scope: Regional

Contact(s):
Emil Svetahor, CONTACT

PENNSYLVANIA FISH AND BOAT COMMISSION: REGION 3
REGION 3 NORTHWEST
Sweet Valley, PA 18656-0008 USA
Phone: 570-477-5717   Fax: 570-477-3221
Website: www.fish.state.pa.us
Membership: 14
Scope: Statewide

PENNSYLVANIA FISH AND BOAT COMMISSION: REGION 4
REGION 4, SOUTHEAST, BOX 9
Elm, PA 17521-0008 USA
Phone: 717-626-0228   Fax: 717-626-0486
Website: www.fish.state.pa.us
Membership: 16
Scope: Regional

PENNSYLVANIA FISH AND BOAT COMMISSION: REGION 6
SOUTHCENTRAL REGION 1704 Pine Rd.
Newville, PA 17241-9544 USA
Phone: 717-486-7087   Fax: 717-486-8227
Website: www.fish.state.pa.us
Membership: 17
Scope: Regional

Contact(s):
George Geisler, REGIONAL MANAGER

PENNSYLVANIA FOREST STEWARDSHIP PROGRAM
DCNR, Bureau of Forestry, P.O. Box 8552
Harrisburg, PA 17105-8552 USA
Phone: 717-787-2106

Founded: 1990
Scope: Regional

Description: To educate Pennsylvania forest landowners and citizens about the importance of sound forest management and the need to conserve our forest resources for future generations through wise use today. Works in conjunction with the Stewardship Incentive Program, which provides cost-share assistance to landowner's forest management practices.

Publications: Sources of Information and Guidance for Forest Stewards, Teaching Youth About Forest Stewartship, Forest Stewardship Bulletin Series (forest management information)

Keyword(s): training, Exotic species, Aquatic nuisance species, Forests and Forestry, Sustainable Ecosystems, Biodiversity

Contact(s):
Gene Odato, CHIEF OF STEWARDSHIP & EDUCATION PROGRAM
James Grace, DIRECTOR

PENNSYLVANIA GAME COMMISSION
2001 Elmerton Ave.
Harrisburg, PA 17110-9797 USA
Phone: 717-787-4250   Fax: 717-772-0502
Website: www.pgc.state.pa.us

Founded: 1895
Membership: 100
Scope: Statewide

Description: The Department of Agriculture was established as an administrative agency of the Executive Department of the Commonwealth. The Secretary of Agriculture is charged with "encouraging and promoting agriculture and related industries throughout the Commonwealth." The Department's mission is accomplished through three major programs: Consumer protection, property protection, and agribusiness development. Three deputies help determine policy, program development, and overall administration. The department provides a full range of services to farmers and consumers from Harrisburg through seven regional offices located around the state.

Keyword(s): Public Health Protection, Pesticides, Flowers, Plants, and Trees, Land Preservation, Agriculture

Contact(s):
Vernon Ross, EXECUTIVE DIRECTOR

PENNSYLVANIA GAME COMMISSION
2001 Elmerton Ave.
Harrisburg, PA 17110-9797 USA
Phone: 717-787-4250   Fax: 717-772-0502
Website: www.pgc.state.pa.us

Founded: 1895

Publications: Pennsylvania Game News

Contact(s):
F. Whitman, ASSITANT DIRECTOR PUBLIC INFORMATION DIVISION
Phone: 717-787-7015
Robert D'Angelo, ASSOCIATE EDITOR
Phone: 717-787-3745
Joseph Osman, CHIEF OF AUDIO-VISUAL SERVICES DIVISION
Phone: 717-787-1434
Keith Snyder, CHIEF OF HUNTER AND TRAPPER EDUCATION DIVISION
Phone: 717-787-7015
Robert Mitchell, CHIEF OF PUBLICATIONS DIVISION
Phone: 717-787-3745
Theresa Alberici, COORDINATOR OF PROJECT WILD
Phone: 717-783-4872
Michael Schmit, DEPUTY EXECUTIVE DIRECTOR
Phone: 717-787-3633
Howard Harshaw, DEPUTY EXECUTIVE DIRECTOR
Thomas Wylie, DIRECTOR OF BUREAU OF ADMINISTRATION
Phone: 717-787-5670
Robert Strailey, DIRECTOR OF BUREAU OF AUTOMATED TECHNOLOGY SERVICES
Phone: 717-787-4076

J. Graybill, DIRECTOR OF BUREAU OF INFORMATION AND EDUCATION (ACTING)
Phone: 717-787-6286
Gregory Grabowicz, DIRECTOR OF BUREAU OF LAND MANAGEMENT
Phone: 717-787-6818
Calvin Dubrock, DIRECTOR OF BUREAU OF WILDLIFE MANAGEMENT
Phone: 717-787-5529
Robert Mitchell, EDITOR, PENNSYLVANIA GAME NEWS
Phone: 717-787-3745
Vernon Ross, EXECUTIVE DIRECTOR
Phone: 717-787-3633
William Schultz, LEGISLATIVE LIAISON
Phone: 717-783-1076
Barry Hambley, NORTH CENTRAL REGIONAL DIRECTOR
Phone: 570-398-4744
Barry Warner, NORTHEAST REGIONAL DIRECTOR
Phone: 570-675-1143
Nichols Spock, PRESIDENT
Jerry Feaser, PRESS SECRETARY
Willis Sneath, SOUTHCENTRAL REGIONAL DIRECTOR
Phone: 814-643-1831
Barry Moore, SOUTHEAST REGIONAL DIRECTOR
Phone: 610-926-3136
Harry Richards, SOUTHWEST REGIONAL DIRECTOR
Phone: 724-238-9523

PENNSYLVANIA STATE EXTENSION SERVICES

217 Agricultural Administration Bldg., Pennsylvania State University
University Park, PA 16802-2600 USA
Phone: 814-863-3438　　Fax: 814-863-3438
Website: www.cas.psu.edu/docs/COEXT/COOPEXT.HTML

Founded: NA
Scope: Statewide

Contact(s):
Diane Brown, ASSOCIATE DIRECTOR OF EXTENSION
217 Agricultural Administration Bldg., Pennsylvania State University, University Park, PA 16802-2600
Phone: 814-863-3438
Theodore Alter, DIRECTOR OF EXTENSION

PINE BLUFF COOPERATIVE FISHERY RESEARCH PROJECT

USGS-BRG-University of Arkansas, Ag. Exp. Station, 1200 N. University, P.O. Box 4005
Pine Bluff, AR 71611-2799 USA
Phone: 501-543-8165

Founded: NA

Description: The Pine Bluff Cooperative Fishery Research Project is a cooperative educational effort between the U. S. Geological Survey-Biological Resources Division, Cooperative Research Units Division, and the University of Arkansas - Pine Bluff. The project provides undergraduate training in fisheries science and biology and conducts research on environmental problems, fisheries, and related topics.

Keyword(s): Environmental Law, Environmental Protection, Environmental and Conservation Education, Wildlife, Research, Rivers, Endangered Species

Contact(s):
Steve Lochmann, PROJECT LEADER

PUERTO RICO DEPARTMENT OF AGRICULTURE

Box 10163
Santurce, PR 00908-1163 USA
Phone: 809-721-2120　　Fax: 809-722-0291
Website: HTTP://WWW.NASS.USDA.GOV/PR/DE-AG-PR.HTM

Founded: NA

Contact(s):
Brenda Marrero, EXECUTIVE SECRETARY

PUERTO RICO SEA GRANT PROGRAM

Marine Education Program, A. MONTEROMAYOR, Coordinator, Marine Education Center, Humacao
University College, HUC Station
Humacao, PR 00791-4300 USA
Phone: 787-850-9360　　Fax: 787-850-0710

Founded: 1985
Scope: International

Description: Promote marine education activities among pre-college teachers and students. Faciliate interdisciplinary teaching and learning experience using the marine environment as a resource.

Publications: Seagrant in the Caribbean, The boletin Marino

Keyword(s): Oceanography, Environmental and Conservation Education, Biodiversity, Coasts, Endangered Species, Wildlife, Mangrove Habitats, Marine Mammals, Whale, Dolphin, Seal, Natural Areas, Oceanography, EcoAction, Geology, Islands, Aquatic Habitats

Contact(s):
Manuel Valdez-Pizzini, DIRECTOR OF UPR SEA GRANT COLLEGE PROGRAM
Phone: 787-832-3585
Fax: 787-265-2880
ma_valdes@rumac.upr.clu

PUERTO RICO STATE EXTENSION SERVICES

Puerto Rico Agricultural Extension Service, P.O. Box 9031
Mayaguez, PR 00681-9031 USA
Phone: 787-833-2665　　Fax: 787-265-4130
E-mail: m_comas@seam.upr.clu.edu
Website: www.sea.upr.clu.edu/

Founded: 1934
Scope: State, Regional, National

Description: The Cooperative Extension Service helps people improve their lives through an educational process that uses scientific knowledge focused on issues and needs. It is a dynamic, ever changing organization pledged to meet the country's needs for research, knowlege, and educational programs that will enable people to make practical decisions that can improve their lives.

Publications: Related to Agriculture, Home Economics, Leadership, and Community Development

Keyword(s): Youth Organizations, Soil Conservation, Environmental and Conservation Education, Health and Nutrition, Agriculture

Contact(s):
Mirna Comas, ASSOCIATE DEAN AND SUBDIRECTOR
University of Puerto Rico, Mayaguez, PR 00681

Ernesto Riquelme, DEAN AND DIRECTOR
University of Puerto Rico, Mayaguez, PR 00681
Phone: 787-265-3850
Rafael Olmeda, SPECIALIST AGRICULTURAL PROGRAMS
Teresa Nieves, SPECIALIST OF I/C HOME ECONOMICS
AND NUTRITION

PURDUE UNIVERSITY EXTENSION SERVICES

1140 Agriculture Administration Bldg., Purdue University
West Lafayette, IN 47907-1140 USA
Phone: 888-398-4636   Fax: 765-494-5876
E-mail: extension@aes.purdue.edu
Website: www.ces.purdue.edu/

Founded: NA

Contact(s):
David Petritz, DIRECTOR, COOPERATIVE EXTENSION
SERVICE
david.petritz@ces.purdue.edu
Linda Chezem, PROGRAM LEADER, 4-H/YOUTH
Phone: 765-494-8422
lchezem@.four-h.purdue.edu
Janet Ayres, PROGRAM LEADER, LEADERSHIP AND
COMMUNITY DEVELOPMENT
Phone: 765-494-4215
ayres@agecon.purdue.edu
Brian Miller, WILDIFE SPECIALIST AND SEA GRANT
COORDINATOR
Purdue Univ., 1159 Forestry Building, West Lafayette, IN
47907-1159
Phone: 765-494-3586
Fax: 765-496-2422

R

RESOURCES AGENCY, THE

1416 9th St., Rm. 1311
Sacramento, CA 95814 USA
Phone: 916-653-5656   Fax: 916-653-8102
Website: www.ceres.ca.tov

Founded: NA
Scope: State

Description: Responsible for ensuring an adequate and properly
balanced management of government functions related to
California's natural environment.

Contact(s):
Don Wallace, ASSISTANT SECRETARY OF
ADMINISTRATION AND FINANCE
Jennifer Galehause, DEPUTY SECRETARY FOR
LEGISLATIVE AFFAIRS
Stanley Young, DEPUTY SECRETARY FOR OPERATIONS
Margaret Kim, GENERAL COUNSEL
Mary Nicholas, SECRETARY OF RESOURCES
Michael Sweeney, UNDER SECRETARY OF RESOURCES

RESOURCES AGENCY, THE

CALIFORNIA COASTAL COMMISSION
45 Fremont St.,
San Francisco, CA 94105-2219 USA
Phone: 415-904-5200   Fax: 415-904-5400
Website: www.coastal.ca.gov

Founded: NA
Scope: Statewide

Description: A coastal management agency which carries out
mandated policies on coastal conservation and development
through regulation and planning programs. These policies deal
with public access to the coast, coastal recreation, the
California marine environment, coastal land resources, and
coastal development of various types, including power plant
and other energy installation.

Contact(s):
Ralph Faust, CHIEF COUNSEL
Susan Hansch, CHIEF DEPUTY DIRECTOR
Lane Yee, CHIEF OF ADMINISTRATIVE SERVICES
DIVISION
Jaime Kooser, DEPUTY DIRECTOR FOR ENERGY, OCEAN
RESOURCES AND WATER QUALITY
Steve Scholl, DEPUTY DIRECTOR FOR NORTH COAST
DISTRICT
Peter Douglas, EXECUTIVE DIRECTOR
Chris Parry, PUBLIC EDUCATION AND ACTIVITIES
COORDINATOR

RESOURCES AGENCY, THE

CALIFORNIA COASTAL CONSERVANCY
1330 Broadway, Suite 1100
Oakland, CA 94612 USA
Phone: 510-286-1015   Fax: 510-286-0470
E-mail: dwayman@scc.ca.gov
Website: www.coastalconservancy.ca.gov

Founded: NA
Membership: 50
Scope: Statewide

Description: A state agency using planning, land-use conflict
resolution, acquisition, and development techniques in the
restoration, enhancement, and preservation of coastal
resources. Program areas include agricultural preservation, lot
consolidation, urban waterfront restoration, coastal resource
enhancement, the reservation of significant resource sites,
provision of public access, and assistance to nonprofit
organizations.

Publications: Coast and Ocean

Contact(s):
Sam Schuchat, EXECUTIVE OFFICER

RESOURCES AGENCY, THE

CALIFORNIA CONSERVATION CORPS
1719 24th St.
Sacramento, CA 95814 USA
Phone: 916-341-3100

Founded: 1976

Description: The CCC was created with a dual mission: the
employment and development of the state's youth, and the
protection and enhancement of California's natural resources.
Some 42 million hours of public service conservation work and
emergency assistance have been provided by the Corps in its
twenty years of existence.

Contact(s):
John Coleman, DEPUTY DIRECTOR OF EXTERNAL
AFFAIRS
Karen Meyerles, DEPUTY DIRECTOR OF PROGRAM
SUPPORT
Al Aramburu, DIRECTOR
Paul Carrillo, REGIONAL DEPUTY DIRECTOR
Tom Powers, REGIONAL DEPUTY DIRECTOR

Walt Hughes, REGIONAL DEPUTY DIRECTOR
Stew Ogburn, REGIONAL DEPUTY DIRECTOR
Craig Miller, SPECIAL ASSISTANT TO THE DIRECTOR

RESOURCES AGENCY, THE
CALIFORNIA WATER COMMISSION
1416 9th St., Rm. 1148
Sacramento, CA 95814 USA
Phone: 916-653-5656   Fax: 916-653-9745
Website: http://ceres.ca.gov/cra/

Founded: 1913
Scope: Statewide

Description: Serves as a policy advisory body to the Director of Water Resources on matters within the Department's jurisdiction and coordinates state and local views on federal appropriations for water projects in California. The commission also conducts public hearings and investigations statewide for the department and provides an open forum for interested citizens to voice their opinion on water development issues.

Keyword(s): Flood Control, Wildlife, Watersheds, Wetlands, Rivers, Exotic species, Aquatic nuisance species

Contact(s):
Doug Priest, EXECUTIVE OFFICER
priestd@water.ca.gov

RESOURCES AGENCY, THE
COLORADO RIVER BOARD OF CALIFORNIA
770 Fairmont Ave., Suite 100
Glendale, CA 91203-1035 USA
Phone: 818-543-4676   Fax: 818-543-4685
E-mail: crb@crb.ca.gov
Website: www.crb.ca.gov

Founded: 1937
Membership: 6
Scope: Statewide

Description: The board was established to represent California, its agencies, and citizens in matters concerning the water and power resources provided by the Colorado River and its tributaries. Working with federal and state agencies, Congress, courts, and other Colorado River Basin states, the board analyzes engineering, legal, and economic matters concerning the use of Colorado River resources within the United States.

Publications: Callifornia Journal, Western Water

Contact(s):
Gerald Zimmerman, EXECUTIVE DIRECTOR

RESOURCES AGENCY, THE
DEPARTMENT OF BOATING AND WATERWAYS
2000 Evergreen, Suite 100
Sacramento, CA 95815-3888 USA
Phone: 888-326-2822   Fax: 916-263-0648
Website: http://www.dbw.ca.gov/

Founded: NA

Description: Makes loans to public agencies and small businesses for small craft harbor development and grants to public agencies for boat-launching facilities, floating restrooms, and vessel waste disposal equipment; licenses yacht and ship brokers and for-hire vessel operators; conducts programs of boating safety, education, and regulation; and grants funds to local entities for boating law enforcement activities. Participates with the Corps of Engineers and local agencies in the construction of beach erosion control projects, assists local jurisdictions in obtaining the greatest benefits available from federal beach erosion programs, and conducts an aquatic pest control program in the Sacramento-San Joaquin Delta.

Keyword(s): Grants, Lakes, Oceanography, Outdoor Recreation, Coasts

Contact(s):
Michael Cleary, 916-263-8122
Debra Deverter, CHIEF OF ADMINISTRATIVE SERVICES DIVISION
Phone: 916-263-0842
Dolores Farrell, CHIEF OF BOATING OPERATIONS DIVISION
Phone: 916-262-8181
Carlton Moore, INTERIM DIRECTOR

RESOURCES AGENCY, THE
DEPARTMENT OF CONSERVATION
801 K St., MS 24-01
Sacramento, CA 95814 USA
Phone: 916-322-1080   Fax: 916-445-0732
Website: www.consrv.ca.gov

Founded: NA
Membership: 450
Scope: Statewide

Description: The mission of the department is to protect health and safety, ensure environmental quality, and support the states's long-term economic viability in the use of California's land and mineral resources.

Contact(s):
Pat Meehan, DEPUTY DIRECTOR
Darryl Young, DIRECTOR

RESOURCES AGENCY, THE
DEPARTMENT OF FISH AND GAME
1416 9th St. Fl 112
Sacramento, CA 95814 USA
Phone: 916-653-7664   Fax: 916-653-1856
Website: www.dfg.ca.gov

Founded: NA
Membership: 2400
Scope: State

Description: Responsible for the protection and management of fish and wildlife and threatened native plants in California. Enforces the laws pertaining to fish and game and threatened native plants enacted by the legislature and the regulations of the Fish and Game Commission.

Publications: Outdoor California Magaznine, Track Mangazine

Contact(s):
Sandra Morey, CHIEF HABITAT CONSERVATION PLANNING
Greg Laret, CHIEF OF CONSERVATION & EDUCATION & ENFORCEMENT BRANCH
Perry Herrgesell, CHIEF, CENTRAL VALLEY BAY-DELTA
Phone: 209-948-7800
Gene Fleming, CHIEF, FISHERIES PROGRAMS
Phone: 916-653-4280
Michael Harris, DEPUTY DIRECTOR OF ADMINISTRATION
Phone: 916-653-4633
Ron Rempel, DEPUTY DIRECTOR OF HABITAT CONSERVATION DIVISION
Phone: 916-653-1070

Sonke Mastrup, DEPUTY DIRECTOR OF WILDLIFE & INLAND FISHERIES DIVISION
Robert Hight, DIRECTOR
Phone: 916-653-7667
Michael Valentine, GENERAL COUNSEL (ACTING)
Phone: 916-654-3821
L. Boydstun, INTERGOVERNMENTAL AFFAIRS REPRESENTATIVE
Phone: 916-653-3136
Julie Oltmann, LEGISLATIVE REPRESENTATIVE
Phone: 916-653-5581
Larry Week, NATIVE ANADROMOUS FISH & WATERSHED BRANCH
Phone: 916-327-8847
Jim Steele, NATIVE ANADROMOUS FISH & WATERSHED BRANCH, TECHNICAL ASST TEAM- PROG MGR
Phone: 916-653-2459

RESOURCES AGENCY, THE
DEPARTMENT OF FORESTRY & FIRE PROTECTION
1416 9th St., P.O. Box 944246
Sacramento, CA 94244-2460 USA
Phone: 916-653-5121　　　Fax: 916-654-7661
Website: www.fire.ca.gov

Founded: NA
Scope: Statewide

Description: The department protects the people of California from fires, responds to emergencies, and protects and enhances forest, range, and watershed values providing social, economic, and environmental benefits to rural and urban citizens.

Contact(s):
Woody Allshouse, CHIEF DEPUTY DIRECTOR
Phone: 916-653-4175
woodyunderscoreallshouse@fire.ca.gov
Ross Johnson, DEPUTY DIRECTOR, RESOURCE MANAGEMENT
Phone: 916-653-4298
rossunderscorejohnson@fire.ca.gov
Andrea Tuttle, DIRECTOR
Phone: 916-653-7772
andreaunderscoretuttle@fire.ca.gov

RESOURCES AGENCY, THE
DEPARTMENT OF WATER RESOURCES
Sacramento, CA 94236-0001 USA
Phone: 916-653-5791　　　Fax: 916-653-5028
Website: wwwdwr.water.ca.gov

Founded: NA
Scope: Regional

Description: To manage the water resources of California in cooperation with other agencies, to benefit the state's people, and to protect, restore, and enhance the natural and human environments.

Contact(s):
L. Chipponeri, ASSISTANT DIRECTOR OF LEGISLATION
Phone: 916-653-0488
Susan Weber, CHIEF COUNSEL
Phone: 916-653-6186
Steve Macaulay, CHIEF DEPUTY DIRECTOR
Phone: 916-653-6055
Stephen Kashiwada, DEPUTY DIRECTOR
Phone: 916-653-7092

Thomas Hannigan, DIRECTOR
Phone: 916-653-7007
Karl Winkler, DISTRICT CHIEF OF CENTRAL
Phone: 916-227-7550
Louis Beck, DISTRICT CHIEF OF SAN JOAQUIN
Phone: 209-445-5222
Charles White, DISTRICT CHIEF OF SOUTHERN
Phone: 818-543-4600
Les Harder, DIVISION OF ENGINEERING
Phone: 916-653-3927
Chester Winn, DIVISION OF FISCAL SERVICES
Phone: 916-653-4413
George Qualley, DIVISION OF FLOOD MANAGEMENT
Phone: 916-653-7572
Frank Conti, DIVISION OF LAND AND RIGHT OF WAY
Phone: 916-653-7891
Linda Gage, DIVISION OF MANAGEMENT SERVICES
Phone: 916-653-6743
William Bennett, DIVISION OF PLANNING AND LOCAL ASSISTANCE
1020 9th St., Sacramento, CA 95814
Phone: 916-327-1646
Steve Verigin, DIVISION OF SAFETY OF DAMS
Phone: 916-445-7606
Randall Brown, ENVIRONMENTAL SERVICES OFFICE
3252 S St., Sacramento, CA 95816
Phone: 916-227-7531
Pete Weisser, OFFICE OF WATER EDUCATION
196-653-7431
Kathy Kelly, STATE WATER PROJECT PLANNING OFFICE
Phone: 916-653-1099

RESOURCES AGENCY, THE
SAN FRANCISCO BAY CONSERVATION AND DEVELOPMENT COMMISSION
50 California St., Ste 2600
San Francisco, CA 94111 USA
Phone: 415-352-3600　　　Fax: 415-352-3606
E-mail: info @bcdc.ca.gov
Website: http://www.bcdc.ca.gov/

Founded: 1965

Description: To implement a planning and regulatory program designed to conserve and use beneficially the environmental, economic, social, and aesthetic values of San Francisco Bay through carefully considered and democratically determined policies. Composed of 27 commissioners, representing the public and state, federal, and local governmental agencies.

Contact(s):
Barbara Kaufman, CHAIRMAN
Phone: 415-352-3663
barbarakaufman01@aol.com
Will Travis, EXECUTIVE DIRECTOR
Phone: 415-352-3653
travis@bcdc.ca.gov

RESOURCES AGENCY, THE
STATE RECLAMATION BOARD
1416 9th St., Rm. 1601
Sacramento, CA 95814 USA
Phone: 916-653-5434　　　Fax: 916-653-5805
E-mail: pamb@water.ca.gov
Website: www.water.ca.gov/RecBd

Founded: 1911
Membership: 7

Scope: Statewide

Description: Agency provides flood protection along the Sacramento and San Joaquin Rivers and their tributaries by planning, constructing, operating, and maintaining flood control projects in cooperation with local, state, and federal agencies, and by implementing nonstructural flood control measures.

Contact(s):
Peter Rabbon, GENERAL MANAGER
Barbara Levake, PRESIDENT
Brenda JAHNS-SHWICK, SECRETARY
Frances Mizuno, VICE PRESIDENT

RESOURCES AGENCY DEPARTMENT OF FISH & GAME

WILDLIFE CONSERVATION BOARD
1807 13th St., Suite 103
Sacramento, CA 95814-7117 USA
Phone: 916-445-8448   Fax: 916-323-0280
Website: www.dfg.ca.gov

Founded: 1947
Membership: 32
Scope: Statewide

Description: In concert with the Department of Fish and Game, the board authorizes the acquisition, restoration, and enhancement of land and water for wildlife conservation and related recreational purposes. The board also administers the Inland Wetlands Conservation Program and the California Riparian Habitat Conservation Program to protect, restore, and enhance wetland and riparian habitats.

Contact(s):
Georgia Lipphardt, ASSISTANT EXECUTIVE DIRECTOR OF DEVELOPMENT PROGRAM
James Sarro, ASSISTANT EXECUTIVE DIRECTOR OF LAND ACQUISITION
Al Wright, EXECUTIVE DIRECTOR

RHODE ISLAND COOPERATIVE EXTENSION SERVICE

Woodward Hall, University of Rhode Island
Kingston, RI 02881 USA
Phone: 401-874-2900   Fax: 401-874-2259
E-mail: ceec@etal.uri.edu
Website: www.uri.edu/ce/ceec

Founded: NA
Scope: Local, State, National

Publications: Publication on website

Keyword(s): Wetlands, Exotic species, Aquatic nuisance species, Pesticides, Water Pollution Management, Agriculture

Contact(s):
Joseph Dealteris, AQUACULTURE AND FISHERIES LEADER
Phone: 401-874-5333
Fax: 401-789-8930
jdealteris@uri.edu
Patrick Logan, COMMUNITY ECONOMIC DEVELOPMENT LEADER
Phone: 401-874-2970
Fax: 401-874-4017
mayfly@uri.edu
Jeffrey Seemann, DIRECTOR
Art Gold, NATURAL RESOURCES LEADER
Phone: 401-874-2903
Fax: 401-874-4561
agold@uri.edu

RHODE ISLAND SEA GRANT COLLEGE PROGRAM

129 Coastal Institute Uri Mera Campus
Narragansett, RI 02882-1197 USA
Phone: 401-874-6800   Fax: 401-789-8340
Website: www.seagrant.gso.uri.edu/riseagrant/

Founded: NA
Scope: National

Contact(s):
Ames Colt, INTERIM DIRECTOR

RUTGERS COOPERATIVE EXTENSION

Rutgers Cooperative Extension, 88 Lipman Dr.
New Brunswick, NJ 08901 USA
Phone: 732-932-5000   Fax: 732-932-6633
Website: www.rce.rutgers.edu

Founded: NA
Membership: 30
Scope: Statewide
Keyword(s): Youth Organizations, Solid Waste Management, Gardening and Horticulture, Pesticides, Agriculture

Contact(s):
Zane Helsel, DIRECTOR
helsel@aesop.rutgers.edu
David Drake, EXTENSION SPECIALIST IN WILDLIFE
Phone: 732-932-1509
Fax: 732-932-3222
Mark Vodak, SPECIALIST IN FOREST RESOURCES
Rutgers Cooperative Extension, 80 Nichol Ave, New Brunswick, NJ 08901-2828
Phone: 732-932-8993

S

SASKATCHEWAN ENVIRONMENT AND RESOURCE MANAGEMENT

CORPORATE SERVICES
3211 Albert St.
Regina, Saskatchewan S4S 5W6 Canada
Website: http://serm.gov.sk.ca

Founded: NA

Publications: Annual Reports, State of the Environment Report

Contact(s):
Ann Milton, CORPORATE DEVELOPMENT
Phone: 306-787-2336
Dave Tulloch, CORPORATE DEVELOPMENT
Phone: 306-787-1095
Al Parenteau, CORPORATE DEVELOPMENT
Phone: 306-787-8449
Mike Dumelie, DIRECTOR OF INFORMATION MANGEMENT
Phone: 306-787-3194
Donna Kellsey, DIRECTOR OF SERVICE BUREAU
Phone: 306-787-6121
Lynn Tulloch, EXECUTIVE DIRECTOR OF CORPORATE SERVICES
Phone: 306-787-1176

SASKATCHEWAN ENVIRONMENT AND RESOURCE MANAGEMENT

ENFORCEMENT AND COMPLIANCE BRANCH
Box 3003
Prince Albert, SK S6V 6G1 Canada
Phone: 306-953-2991　　Fax: 306-953-2999
Website: www.serm.gov.sk.ca

Founded: NA
Membership: 35
Scope: International

Contact(s):
Dave Harvey, REGIONAL DIRECTOR
Box 3003, Prince Albert, Saskatchewan S6V 6G1
Phone: 306-953-2993

SASKATCHEWAN ENVIRONMENT AND RESOURCE MANAGEMENT

FIRE MANAGEMENT AND FOREST PROTECTION BRANCH
P.O. Box 3003
Prince Albert, SK S6V 6G1 Canada
Phone: 306-425-7625　　Fax: 306-953-3447
E-mail: utcoop@cc.usu.edu
Website: http://www.nr.usu.edu/UTCFWRU/index.html

Founded: NA
Membership: 30
Scope: International

Contact(s):
Murdoch Carrierre, REGIONAL DIRECTOR
Box 3003, Prince Albert, Saskatchewan S6V 6G1
Phone: 306-953-2206

SASKATCHEWAN ENVIRONMENT AND RESOURCE MANAGEMENT

PROGRAMS
3211 Albert Street
Regina, SK S4S 5W6 Canada
Phone: 306-787-7812
E-mail: inquiry@serm.gov.sk.ca
Website: www.serm.gov.sk.ca

Founded: NA
Scope: International

Contact(s):
Bob Ruggles, ASSISTANT DEPUTY MINISTER
Phone: 306-787-5419
Shelly Vandermey, DIRECTOR OF ECONOMIC
DEVELOPMENT BRANCH
Phone: 306-787-5482
Joe Muldoon, DIRECTOR OF ENVIRONMENTAL
PROTECTION BRANCH
Phone: 306-787-6178
Dennis Sherratt, DIRECTOR OF FISH AND WILDLIFE
BRANCH
Phone: 306-787-2309
Al Willcocks, DIRECTOR OF FOREST ECOSYSTEMS
BRANCH
Box 3003, Prince Albert, Saskatchewan S6V 6G1
Phone: 306-953-2486
Don Macaulay, DIRECTOR OF PARKS AND SPECIAL
PLACES BRANCH
Phone: 306-787-2846

Bob Carles, DIRECTOR OF SASKATCHEWAN WETLAND
CONSERVATION CORPORATION
101-2022 Cornwall St., Regina, Saskatchewan S4P 2K5
Phone: 306-787-0779
Doug Mazur, DIRECTOR OF SUSTAINABLE LAND
MANAGEMENT BRANCH
Phone: 306-787-7024
Ed Dean, DIRECTOR OF WATER MANAGEMENT
Phone: 306-787-7812

SOUTH CAROLINA DEPARTMENT OF AGRICULTURE

Wade Hampton Office Bldg., P.O. Box 11280
Columbia, SC 29211 USA
Phone: 803-734-2210　　Fax: 803-734-2192
Website: www.scda.state.sc.us

Founded: 1904
Scope: Statewide

Description: Administers more than 30 state laws relating to agriculture and the consumer. Represents the farmer in national, regional, and state policy matters and is involved in local and international programs of commodity promotion. Enforces regulatory programs affecting the consumer on a statewide basis.

Publications: South Carolina Market Bulletin, The

Keyword(s): Wetlands, Chemistry, Gardening and Horticulture, Agriculture

Contact(s):
Daniel Breazeale, ADMINISTRATIVE MANAGER
Sidney Whalen, ASSISTANT EDITOR
D. Tindal, COMMISSIONER
Larry Boyleston, DIRECTOR OF AGRIBUSINESS
DEVELOPMENT
Carol Fulmer, DIRECTOR OF CONSUMER SERVICES
William Brooks, DIRECTOR OF LABORATORY SERVICES
Wayne Mack, DIRECTOR OF MARKETING
Becky Walton, DIRECTOR OF PUBLIC INFORMATION
Kay Rike, EXECUTIVE ASSISTANT TO THE
COMMISSIONER
David Tompkins, FARMERS MARKETS ADMINISTRATOR

SOUTH CAROLINA DEPARTMENT OF HEALTH AND ENVIRONMENTAL CONTROL

J. Marion Sims Bldg., 2600 Bull St.
Columbia, SC 29201 USA
Phone: 803-896-8940　　Fax: 803-896-8940
Website: http://scdhec.net/eqc/

Founded: NA

Publications: A General Guide to Environmental Permitting in South Carolina

Keyword(s): Exotic species, Aquatic nuisance species, Toxic Substances, Nuclear-free, Water quantity, Water export and diversion, Public Health Protection, Solid Waste Management, Air Quality and Pollution

Contact(s):
James Joy, BUREAU OF AIR QUALITY
Phone: 803-734-4750
Douglas Bryant, COMMISSIONER
Phone: 803-734-4880
R. Shaw, DEPUTY COMMISSIONER OF ENVIRONMENTAL
QUALITY CONTROL OFFICE
Phone: 803-734-5360

SOUTH CAROLINA DEPARTMENT OF HEALTH AND ENVIRONMENTAL CONTROL

OFFICE OF OCEAN AND COASTAL RESOURCE
MANAGEMENT (OCRM)
Suite 400, 1362 McMillan Avenue
Charleston, SC 29405 USA
Phone: 803-744-5838   Fax: 803-744-5847
Website: http://www.scdhec.net/

Founded: NA

Description: OCRM is a division of South Carolina's Department
of Health and Environmental Control. OCRM has the dual
responsibility of protecting the coastal environment while
promoting responsible development within the eight coastal
counties.

Publications: Legislature Update, Carolina Currents

Keyword(s): Wetlands, Protected Areas, Environmental Law,
Environmental Planning, Coasts

Contact(s):
Christopher Brooks, BUREAU CHIEF
Steve Moore, DIRECTOR OF PERMITTING
Steve Snyder, DIRECTOR OF PLANNING

SOUTH CAROLINA DEPARTMENT OF NATURAL RESOURCES

Rembert C. Dennis Bldg.
Columbia, SC 29202 USA
Phone: 803-734-3888   Fax: 803-734-4300
Website: www.dnr.state.sc.us

Founded: 1994
Membership: 300
Scope: Statewide

Description: The Department was created by Act 181 of 1993 to
provide for the conservation, management, utilization, and
protection of the state's natural resources. It also administers
the state's Heritage Trust Program for significant natural areas
and historical sites. Five state agencies combined July 1, 1994,
to form the SC Dept. of Natural Resources: the former Wildlife
and Marine Resources Dept.; SC Geological Survey; Migratory
Waterfowl Committee; and the non-regulatory portions of the
Water Resources Commission and the Land Resources
Conservation Commission.

Publications: Resource, The, South Carolina Geology, South
Carolina Weekly Climate Summary, South Carolina Wildlife

Contact(s):
Cary Chamblee, ASSOCIATE DIRECTOR
Phone: 803-734-9102
Joab Lesesne, BOARD CHAIRMAN
Phone: 864-597-4010
Carole Collins, DEPUTY DIRECTOR CONSERVATION
EDUCATION & COMMUNICATION
Phone: 803-734-3957
carolec@scdnr.state.sc.us
Alfred Vang, DEPUTY DIRECTOR OF LAND, WATER AND
CONSERVATION DIVISION
1201 Main St., Suite 1100, Columbia, SC 29201
Phone: 803-737-0800
John Miglarese, DEPUTY DIRECTOR OF MARINE
RESOURCES DIVISION
P.O. Box 12559, Charleston, SC 29422-2559
Phone: 843-762-5000

Alvin Wright, DEPUTY DIRECTOR OF NATURAL
RESOURCES LAW ENFORCEMENT DIVISION
Phone: 803-734-4021
William McTeer, DEPUTY DIRECTOR OF WILDLIFE AND
FRESHWATER FISHERIES DIVISION
Phone: 803-734-3889
Paul Sandifer, DIRECTOR
Phone: 803-734-4007
James Timmerman, DIRECTOR EMERITUS
Phone: 803-798-2858

SOUTH CAROLINA ENERGY OFFICE

1201 Main St., Suite 820
Columbia, SC 29201 USA
Phone: 803-737-8030   Fax: 803-737-9846
Website: www.state.sc.us/energy/

Founded: NA
Scope: Statewide

Description: The SC Energy Office is responsible for the
statewide promotion of energy conservation and cost effective
use of new energy sources.

Publications: Passive Solar Home Designs for South Carolina,
Landscaping for Energy Efficiency, $aving Money in Your
Manufactured Home Through Energy Efficiency—A Guide for
South Carolinians, How to Reduce Your Energy Costs—A
Guide for Business, Industry, Government, and Institutions,
Energy Savers, The Energy Factbook, Energy Connection
Newsletter, The

Keyword(s): Transportation, Solid Waste, Renewable Resources,
Solar Energy, Energy

Contact(s):
Mitch Perkins, DIRECTOR
Renee Daggerhart, PUBLIC INFORMATION COORDINATOR
rdaggerhart@drd.state.sc.us

SOUTH CAROLINA SEA GRANT CONSORTIUM

287 Meeting St.
Charleston, SC 29401 USA
Phone: 843-727-2078   Fax: 843-727-2080
Website: www.scseagrant.org

Founded: NA
Membership: 13
Scope: National

Description: A Universtiy-based state agency that supports
research, education and outreach to conserve coastal and
marine reserves and provide economic oppurtunities for the
cities of South Carolina and the region.

Publications: Inside SeaGrant, extension materials, marine
education publications and slide presentations, aquaculture
handbooks, coastal hazard information., Coastal Heritage

Keyword(s): Wetlands, Sustainable Ecosystems, Oceanography,
Sustainable Development, Wildlife

Contact(s):
Elaine Knight, ASSISTANT DIRECTOR
knightel@musc.edu
Linda Blackwell, DIRECTOR OF COMMUNICATIONS
blackwlj@musc.edu
M. Devoe, EXECUTIVE DIRECTOR
devoemr@musc.edu
Bob Bacon, EXTENSION PROGRAM LEADER
baconrh@musc.edu

SOUTH DAKOTA COOPERATIVE EXTENSION SERVICE

South Dakota State University, AgH, 0
Brookings, SD 57007 USA
Phone: 605-688-4792   Fax: 605-688-6347
Website: www.abs.sdstate.edu/CES/index2.htm

Founded: NA
Membership: 180
Scope: Regional
Publications: see publication web site

Keyword(s): Exotic species, Aquatic nuisance species, Renewable Resources, Environmental and Conservation Education, Pesticides, Agriculture

Contact(s):
Larry Tidemann, DIRECTOR OF COOPERATIVE EXTENSION SERVICE
Barry Bunn, RANGE LIVESTOCK PRODUCTION SPECIALIST
Brookings, SD 57007
Phone: 605-688-5455
Patricia Johnson, RANGE MANAGEMENT SPECIALIST
West River Ag Center, South Dakota State University, 1905 Plaza Blvd., Rapid City, SD 57702-9302

SOUTH DAKOTA DEPARTMENT OF AGRICULTURE

523 E. Capitol, Foss Bldg.
Pierre, SD 57501-3182 USA
Phone: 605-773-5425   Fax: 605-773-5926
E-mail: agmail@state.sd.us
Website: www.state.sd.us/doa/doa.html

Founded: NA
Scope: State

Publications: see publication website

Contact(s):
Larry Gabriel, SECRETARY
Raymond Sowers, STATE FORESTER
Phone: 605-773-3623
raymond.sowers@state.sd.us

SOUTH DAKOTA DEPARTMENT OF AGRICULTURE

DIVISION OF RESOURCE CONSERVATION AND FORESTRY
523 E. Capitol Ave.
Pierre, SD 57501-3182 USA
Phone: 605-773-3623   Fax: 605-773-4003
Website: www.state.sd.us/doa/forestry/index.htm

Founded: NA
Scope: Statewide

Contact(s):
Raymon Sowers, DIRECTOR
Phone: 605-773-3623

SOUTH DAKOTA DEPARTMENT OF AGRICULTURE

STATE CONSERVATION COMMISSION
523 E. Capitol Ave.
Pierre, SD 57501-3182 USA
Phone: 605-773-3623   Fax: 605-773-4003
Website: www.state.sd.us/state/doa/doa

Founded: NA
Scope: Statewide

Contact(s):
Robert Gab, CHAIRMAN

SOUTH DAKOTA DEPARTMENT OF ENVIRONMENT AND NATURAL RESOURCES

523 East Capitol Ave.
Joe Foss Office Building
Pierre, SD 57501 USA
Phone: 605-773-3151   Fax: 605-773-6035
E-mail: DENRINTERNET@state.sd.us
Website: www.state.sd.us/denr

Description: To provide environmental and natural resources assessment, financial assistance,nand regulation in a customer service manner that protects the public health, conserves natural resources, preserves the environment, and promotes economic development.

Contact(s):
Tim Tolletsrud, DIRECTOR OF DIVISION OF ENVIRONMENTAL SERVICES
Tim.Tolletsrud@state.sd.us
David Templeton, DIRECTOR OF DIVISION OF FINANCIAL AND TECHNICAL ASSISTANCE
Phone: 605-773-4216
Fax: 605-773-4068
dave.templeton@state.sd.us
Steven Pirner, SECRETARY
Phone: 605-773-3153
Fax: 605-773-6035

SOUTH DAKOTA DEPT. OF GAME FISH & PARKS

GAME, FISH & PARKS
523 E. Capitol, Joe Foss Office Bldg.
Pierre, SD 57501 USA
Phone: 605-773-3387   Fax: 605-773-6245
E-mail: WILDINFO@GFP.STATE.SD.US
Website: www.state.sd.us/

Founded: NA
Membership: 70
Scope: Statewide

Description: To provide environmental and natural resources assessment, financial assistance, and regulation in a customer service orientated manner which provides protection of public health, conservation of natural resources, preservation of the environment, and promotes economic development.

Publications: Conservation Digest

Keyword(s): Lakes, Solid Waste, Chemical Pollution Control, Environmental Protection, Grants, Water Quality, Watersheds, Solid Waste Management, Environment, Mining, Toxic Substances, Nuclear-free, Water quantity, Water export and diversion, Air Quality and Pollution

Contact(s):
Chuck Schlueter, INFORMATION OFFICER
Phone: 605-773-3485

SOUTH DAKOTA GAME, FISH, AND PARKS DEPARTMENT

523 East Capitol
Pierre, SD 57501-3182 USA
Phone: 605-773-3387

Founded: NA
Scope: State

Publications: South Dakota Conservation Digest

Keyword(s): training, Terrestrial Habitats, Wildlife, Outdoor Recreation, Aquatic Habitats

Contact(s):
William Shattuck, BOATING AND HUNTING SAFETY
Phone: 605-773-4506
Chuck Schlueter, COMMUNICATIONS MANAGER
Phone: 605-773-3485
Ken Anderson, DIRECTOR OF ADMINISTRATION DIVISION
Phone: 605-773-3396
Rollie Noem, DIRECTOR OF CUSTER STATE PARK DIVISION
Phone: 605-255-4515
Doug Hofer, DIRECTOR OF PARKS AND RECREATION DIVISION
Phone: 605-773-3391
Doug Hansen, DIRECTOR OF WILDLIFE DIVISION
Phone: 605-773-3381
Bruce Coonrod, EDITOR
Phone: 605-773-3485
Wayne Winter, FEDERAL AID MANAGER
Phone: 605-773-6228
Emmett Keyser, OPERATIONS ASSISTANT DIRECTOR OF WILDLIFE DIVISION
Phone: 605-773-4607
John Cooper, SECRETARY
Phone: 605-773-3387
Dave McCrea, SPECIALIST OF ENFORCEMENT
Phone: 605-773-4243
John Kirk, SPECIALIST OF ENVIRONMENTAL REVIEW
Phone: 605-773-4501
Ron Fowler, SPECIALIST OF GAME
Phone: 605-773-4193
Dave McGuigan, SPECIALIST OF HABITAT
Phone: 605-773-4194
Dennis Unkenholz, STAFF SPECIALIST OF FISHERIES
Phone: 605-773-4508
George Vandel, TECHNICAL SERVICES ASSISTANT DIRECTOR OF WILDLIFE DIVISION
Phone: 605-773-4192
Robert Schuurmans, TURN IN POACHERS (TIPS) TRAINING COORDINATOR
Phone: 605-773-5906

SOUTH FLORIDA WATER MANAGEMENT DISTRICT

3301 Gun Club Rd., P.O. Box 24680
West Palm Beach, FL 33416-4680 USA
Phone: 561-686-8800   Fax: 561-682-6200
Website: www.sfwmd.gov

Founded: NA
Membership: 9
Scope: Statewide

Description: Responsible for local cooperation in the Federal-State Central and Southern Florida flood Control Project. Goals include: Flood control, water supply, water quality, and environmental protection for sixteen counties in south Florida. Additional benefits are preservation of natural conditions in the Everglades, land purchases under Save Our Rivers program and enhancement of wetlands, fish, wildlife, waterfowl and public recreation.

Keyword(s): Flood Control, Everglades

Contact(s):
Anita Sewell, DIRECTOR OF COMMUNICATIONS
Phone: 561-682-6171
Henry Dean, EXECUTIVE DIRECTOR
hdean@sfw.gov
Joseph Schweigart, EXECUTIVE DIRECTOR EVERGLADES RESTORATION
cberger@sfwm.gov
Naomi Duerr, EXECUTIVE DIRECTOR FOR WATER RESOURCE MANAGEMENT
Alvin Jackson, EXECUTIVE DIRECTOR OF CORPORATE RESOURCES
Phone: 561-682-2805

SOUTHWEST FLORIDA WATER MANAGEMENT DISTRICT (SWFWMD)

2379 Broad St., U.S. 41 South
Brooksville, FL 34604-6899 USA
Phone: 352-796-7211   Fax: 352-754-6883
Website: www.watermatters.org

Founded: 1961
Membership: 700
Scope: Statewide

Description: A governmental agency dedicated to resource protection conservation programs, which are supported through regulatory and nonregulatory initiatives and cooperative funding projects.

Publications: Plant Guide and associated technical bulletins, list of vendors and manufacturers of water conservation devices and services, various residential and commercial water conservation education resources., Fifty Ways to do Your Part

Keyword(s): Exotic species, Aquatic nuisance species, Environmental Planning, Wetlands, Water Quality, Water Conservation, Natural Systems, Environmental and Conservation Education, Rivers

Contact(s):
Kathy Foley, CONSERVATION PROJECT MANAGER
Kathy Foley, WATER RESOURCE ANALYST STAFF AND SECRETARY OF THE FLORIDA WATER WISE COUNCIL

STATE ENGINEER OFFICE/INTERSTATE STREAM COMMISSION

Bataan Memorial Bldg., P.O. Box 25102
Santa Fe, NM 87504 USA
Phone: 505-827-6160   Fax: 505-827-6188
Website: www.seo.state.nm.us

Founded: NA
Membership: 20
Scope: State, Regional

Description: Administration, development, protection, and conservation of the water resources of the state of New Mexico.

Publications: Water Line

Keyword(s): Exotic species, Aquatic nuisance species

Contact(s):
Elaine Pacheco, CHIEF OF TECHNICAL DIVISION, ACTING
Paul Saavedra, CHIEF OF WATER RIGHTS DIVISION
Norman Gaume, DIRECTOR OF INTERSTATE STREAM
COMMISSION, INTERSTATE STREAM ENGINEER
Phone: 505-827-6160
Thomas Turney, STATE ENGINEER AND SECRETARY
Phone: 505-827-6160
Hoyt Pattison, VICE CHAIRMAN

STATE FORESTRY DIVISION (WYOMING)
1100 W. 22nd St.
Cheyenne, WY 82002 USA
Phone: 307-777-7586   Fax: 3077-770-5986
E-mail: Forestry@state.wy.us

Founded: 1952
Membership: 45
Scope: Nationwide

Description: Has direction of all forestry matters within the jurisdiction of the state of Wyoming; manages of state-owned forest
land; coordinates fire protection on twenty-nine million acres of
state and private rural lands; assists landowners and
communities in proper management of woody vegetation and
forested lands; and provides forestry information to schools,
organizations, and individuals.

Publications: Wyoming's Forest Wealth, Wyoming State Forest
Resource Program (Executive Summary), Wyoming State
Forest Resource Program

Keyword(s): Insects and Butterflies, Urban Forestry, Forests and
Forestry, Renewable Resources, Diseases, Flowers, Plants,
and Trees

Contact(s):
Ray Weidenhaft, ASSISTANT STATE FORESTER OF FIRE
MANAGEMENT
Howard Pickerd, ASSISTANT STATE FORESTER OF
FOREST MANAGEMENT
Daniel Perko, DEPUTY STATE FORESTER
Thomas Ostermann, STATE FORESTER

STATE MARINE BOARD (OREGON)
P.O. Box 14145
Salem, OR 97309-5065 USA
Phone: 503-378-8587   Fax: 503-378-4597
Website: www.boatoregon.com

Founded: NA
Scope: Statewide

Keyword(s): Recreational Boating, Water Quality, Outdoor
Recreation

Contact(s):
Paul Donheffner, DIRECTOR

STATE OF IDAHO DEPARTMENT OF ENVIRONMENTAL QUALITY
1410 N. Hilton St.
Boise, ID 83706-1255 USA
Phone: 208-373-0502   Fax: 208-373-0417
Website: www.2. state.id.us/deq

Founded: NA
Scope: Statewide

Description: Administers and directs programs designed to
protect and enhance the environment and public health.

Emphasis is placed on monitoring, technical assistance, and
environmental education at the community level. The agency is
additionally responsible for all permitting and permit review
functions.

Publications: Drinking Water Report, Groundwater Report,
Strategic Plan, State of the Environment Report, Performance
Partnership Agreement, Hazardous Waste Report

Keyword(s): Water Quality, Environmental Planning, Air Quality
and Pollution, Environmental Protection

Contact(s):
Katherine Kelly, AIR QUALITY ADMINISTRATOR
Phone: 208-373-0502
J. Sandoval, CHIEF OF STAFF
Phone: 208-373-0240
C. Allred, DIRECTOR OF ENVIRONMENTAL QUALITY
Phone: 208-373-0240
David Mabe, STATE WATER QUALITY PROGRAM
ADMINISTRATOR
Phone: 208-373-0502
Orville Green, WASTE MANAGERMENT & REMEDIATION
MEDIATOR
Phone: 208-373-0440

STATE PARKS & CULTURAL RESOURCES
DIVISION OF STATE PARKS AND HISTORIC SITES
1st Fl., Herschler Bldg. Park & Historic Sites
Cheyenne, WY 82002 USA
Phone: 307-777-6323   Fax: 307-777-6472
E-mail: sphs@state.wy.us
Website: www.wyo-park.com

Founded: 1967
Membership: 20
Scope: Regional

Description: Responsible for administering the state parks, state
recreation areas, historic sites, petroglyph site, archaeological
site, markers and monuments, snowmobile program, and state
trails program.

Keyword(s): Public Lands, Outdoor Recreation, Land
Preservation, National Parks, Historic Preservation

Contact(s):
Bill Gentle, DIRECTOR OF DIVISION OF STATE PARKS
AND HISTORIC SITES

STATE PARKS AND RECREATION COMMISSION (WASHINGTON)
7150 Cleanwater Ln., P.O. Box 42650
Olympia, WA 98504-2650 USA
Phone: 360-902-8500

Founded: 1912
Scope: Statewide

Description: To acquire, develop, improve, and maintain state
parks and recreation areas. Involvement includes but is not
limited to state parks, seashore conservation, water and
boating safety, snowmobile safety, and natural and historic
heritage interpretation.

Keyword(s): Public Lands, Outdoor Recreation, National Parks,
Open Space, Cultural Preservation

Contact(s):
Ann Hersley, ADMINISTRATOR OF PUBLIC AFFAIRS
Phone: 360-902-8562

Rita Cooper, ASSISTANT DIRECTOR OF ADMINISTRATIVE SERVICES
Phone: 360-902-8525
Larry Fairleigh, ASSISTANT DIRECTOR OF RESOURCES DEVELOPMENT
Phone: 360-902-8642
Tom Boyer, CHIEF ENGINEER
Phone: 360-902-8616
Jim French, CHIEF OF BOATING PROGRAMS
Phone: 360-902-8515
Marsh Taylor, CHIEF OF BUDGET SERVICES
Phone: 360-902-8532
Judy Johnson, CHIEF OF EMPLOYEE SERVICES
Phone: 360-902-8568
Bill Jolly, CHIEF OF ENVIRONMENTAL COORDINATION
Phone: 360-902-8636
Sandy Rees, CHIEF OF FISCAL SERVICES
Phone: 360-902-8575
Art Brown, CHIEF OF INFORMATION PROCESSING
Phone: 360-902-8585
Dan Ingman, CHIEF OF NATURAL RESOURCE MANAGEMENT
Phone: 360-902-8592
Paul George, CHIEF OF PARKS MAINTENANCE
Phone: 360-902-8540
James Horan, CHIEF OF PROGRAMS MANAGEMENT
Phone: 360-902-8580
William Jolly, CHIEF OF RESEARCH AND LONG RANGE PLANNING
Phone: 360-902-8641
Bill Koss, CHIEF OF SITE PLANNING
Phone: 360-902-8629
Bill Gansberg, CHIEF OF VISITOR PROTECTION AND LAW ENFORCEMENT
Phone: 360-902-8598
Pam McConkey, CHIEF OF VISITOR SERVICES
Phone: 360-902-8595
Wayne McLaughlin, CONTRACTS SPECIALIST
Phone: 360-902-8599
Frank Boteler, DEPUTY DIRECTOR
Phone: 360-902-8502
Cleve Pinnix, DIRECTOR
Phone: 360-902-8501
Rex Derr, LEGISLATIVE LIAISON
Phone: 306-902-8504

STATE PARKS AND RECREATION COMMISSION (WASHINGTON)

EASTERN REGION
2201 N. Duncan Dr.
Wenatchee, WA 98801-1007 USA
Phone: 509-662-0420   Fax: 509-663-9754
Website: www.parks.wa.gov

Founded: NA
Membership: 150
Scope: Statewide
Publications: call information number at 1-800-233-0321

Contact(s):
Jim Harris, REGIONAL MANAGER
2201 N. Duncan Dr., Wenatchee, WA 98801
Phone: 509-662-0420

STATE PARKS AND RECREATION COMMISSION (WASHINGTON)

PUGET SOUND REGION
P.O.. Box 42650
USA
Fax: 800-233-0321
Website: http://parks.gov

Founded: NA

Contact(s):
Don Simmons, STAFF
1602 29th St., SE, Auburn, WA 98002
Phone: 206-931-3907

STATE PARKS AND RECREATION COMMISSION (WASHINGTON)

SOUTHWEST REGION
11838 Tilley Rd. SW
Olympia, WA 98512 USA
Phone: 360-753-7143   Fax: 360-586-4272
Website: www.parks.wa.gov/

Founded: NA
Scope: Statewide

Contact(s):
Paul Malmberg, STAFF
11838 Tilley Rd., S., Olympia, WA 98512-9167
Phone: 360-753-7143

STATE PARKS AND RECREATION COMMISSION NORTHWEST REGION

NORTHWEST REGION
Burlington, WA 98233 USA
Phone: 360-755-9231   Fax: 360-428-1094
Website: www.parks.wa.gov

Founded: NA
Membership: 10
Scope: Statewide

Contact(s):
Terry Doran, REGION MNGR.
P.O. Box 487, Burlington, WA 98801-1007
Phone: 360-755-9231

STATE PLANT BOARD (ARKANSAS)

1 Natural Resources Dr., P.O. Box 72203
Little Rock, AR 72205 USA
Phone: 501-225-1598   Fax: 501-225-3590
Website: www.naturallyarkansas.org

Founded: NA
Membership: 150
Scope: Statewide
Publications: Plant Board News
Keyword(s): Native Plants

Contact(s):
Darryl Little, ASSISTANT DIRECTOR
Don Alexander, DIRECTOR

STATE SOIL AND WATER CONSERVATION COMMISSION (GEORGIA)

P.O. Box 8024
Athens, GA 30603 USA
Phone: 706-542-3065   Fax: 706-542-4242
Website: www.gaswcc.org

Founded: 1937
Membership: 30
Scope: Local, Regional

Description: Established under the Soil Conservation Districts Act to work with and assist the 40 Soil and Water Conservation Districts and their 370 District Supervisors throughout Georgia.

Publications: Conservation Contact, Conservation Commission

Keyword(s): Exotic species, Aquatic nuisance species, Soil Conservation, Wetlands, Agriculture, Urban Environment

Contact(s):
Garland Thompson, CHAIRMAN
David Bennett, DEPUTY DIRECTOR
F. Liles, EXECUTIVE DIRECTOR
Phone: 706-542-3065
Fax: 706-542-4242

STATE WATER RESOURCES BOARD (RHODE ISLAND)
100 North Main Street, 5th Floor
Providence, RI 02903 USA
Phone: 401-222-2217   Fax: 401-222-4707

Founded: 1967

Description: The Water Resources Board is the key agency in water-supply planning, financing, regulation, and development. The Board also plans for the future water needs of cities and towns.

Publications: RI Fish and Wildlife, RI Industrial Water, RI Legal and Legislative Aspects of Water Supply, RI Public Water Supply

Keyword(s): Water Quality, Exotic species, Aquatic nuisance species, Public Health Protection, Renewable Resources, Land Use Planning

Contact(s):
Daniel Schatz, CHAIRMAN
M. Sams, GENERAL MANAGER AND SECRETARY AND TREASURER
Maurice Trudeau, VICE CHAIRMAN

T

TENNESSEE AGRICULTURAL EXTENSION SERVICE
Knoxville, TN 37901-1071 USA
Phone: 865-974-7114   Fax: 865-974-1068
E-mail: clnorman@utk.edu
Website: www.utextension.utk.edu

Founded: NA
Membership: 850
Scope: Statewide

Contact(s):
Charles Norman, DEAN OF EXTENSION SERVICE
Phone: 865-974-7114
Fax: 865-974-1068
clnorman@utk.edu
Thomas Hill, GENERAL FISH AND WILDLIFE SPECIALIST
Phone: 865-974-7164
Fax: 865-974-4714
tkhill@utk.edu
Craig Harper, GENERAL WILDLIFE SPECIALIST
Phone: 865-974-7346
Fax: 865-974-4714

charper@utk.edu
George Hopper, HEAD OF FORESTRY, WILDLIFE & FISHERIES
Phone: 865-974-7126
Fax: 865-974-0957
gmhopper@utk.edu
George Hopper, HEAD, FORESTRY, WILDLIFE & FISHERIES
Fax: 865-974-7126
Phone: 865-974-0957

TENNESSEE COOPERATIVE FISHERY RESEARCH UNIT (USDI)
Tennessee Technological University, Box 5114
Cookeville, TN 38505 USA
Phone: 931-372-3094   Fax: 931-382-6257

Founded: NA

Keyword(s): Wildlife

Contact(s):
James Layzer, LEADER

TENNESSEE DEPARTMENT OF AGRICULTURE
Ellington Agricultural Center
Nashville, TN 37204 USA
Phone: 615-360-0103   Fax: 615-837-5333
E-mail: dwheeler@mail.state.tn.us
Website: www.state.tn.us/agriculture

Founded: NA
Scope: Statewide

Contact(s):
Dan Wheeler, COMMISSIONER

TENNESSEE DEPARTMENT OF AGRICULTURE
STATE SOIL CONSERVATION COMMITTEE
Ellington Agriculture Center, P.O. Box 40627
Nashville, TN 37204 USA
Phone: 615-360-0108

Founded: NA

Keyword(s): Water Quality, Pollution Prevention, Environmental and Conservation Education, Soil Conservation, Agriculture

Contact(s):
Barry Lake, CHAIR
P.O. Box 107, Hickory Valley, TN 38042
Phone: 901-764-2909
Jim Nance, EXECUTIVE SECRETARY
Phone: 615-360-0108

TEXAS AGRICULTURAL EXTENSION SERVICE
Texas A&M University
College Station, TX 77843-7101 USA
Phone: 979-845-7800   Fax: 979-845-9542
E-mail: agextension@tmu.edu
Website: agextension.tamu.edu/

Founded: NA

Keyword(s): Sustainable Development, Renewable Resources, Environmental and Conservation Education, Forests and Forestry, Agriculture

Contact(s):
C. Hanselka, ASSOCIATE DEPARTMENT HEAD AND EXTENSION PROGRAM LEADER

Rangeland Ecology and Management, Rt. 2 Box 589, Corpus Christi, TX 78406-9704
Phone: 361-265-9203
Fax: 361-265-9434
c-hanselka@tamu.edu
Donny Steinbach, ASSOCIATE DEPARTMENT HEAD AND EXTENSION PROGRAM LEADER
Department of Wildlife and Fisheries Sciences, 111 Nagle Hall, Texas A&M University, College Station, TX 77843
Phone: 409-845-7471
Fax: 409-845-7103
d-steinbach@tamu.edu
Alan Dreesen, ASSOCIATE DEPARTMENT HEAD AND EXTENSION PROGRAM LEADER
Department of Forest Science, 4390 FM 1488, Conroe, TX 77384-3905
Phone: 409-273-2120
Fax: 409-273-5233
a-dreesen@tamu.edu
B. Harris, ASSOCIATE DIRECTOR FOR AGRICULTURE SCIENCES
Texas A&M University, College Station, TX 77843-7101
Phone: 409-862-3932
Fax: 409-845-9542
b-harris4@tamu.edu
Edward Hiler, DIRECTOR OF EXTENSION SERVICE

TEXAS COOPERATIVE FISH AND WILDLIFE RESEARCH UNIT

Texas Tech. University P.O. Box 42120
Lubbock, TX 79409-2120 USA
Phone: 806-742-2851   Fax: 806-742-2946
E-mail: txcoop@hobbes.tcru.ttu.edu
Website: www.tcru.ttu.edu/tcru/

Founded: 1988
Membership: 5
Scope: National

Description: To conduct research, train graduate students, and provide technical assistance in the maintenance and management of fish and wildlife biodiversity, biological informatics, wetland ecology, molecular (genetic) biology, aquatic and wildlife ecology, general and reproductive physiology, and fish culture using the technical expertise of three federal staff members and collaborators.

Keyword(s): Wildlife, Environment, Biodiversity, Biotechnology, Biological informatics, Agriculture

Contact(s):
Clint Boal, ASSISTANT LEADER
Reynaldo Patino, ASSISTANT LEADER OF FISHERIES
Nick Parker, LEADER

TEXAS DEPARTMENT OF AGRICULTURE

Austin, TX 78711 USA
Phone: 512-463-7476   Fax: 512-463-1104
Website: www.agr.state.tx.us

Founded: 1904
Scope: Statewide

Description: Our mission is to make Texas the nation's leader in agriculture while providing efficient and extraordinary service.

Contact(s):
Raette Hearne, ASSISTANT COMMISSIONER, ADMINISTRATIVE SERVICES
Phone: 512-463-7582

Allen Spelce, ASSISTANT COMMISSIONER, COMMUNICATIONS
Lee Deviney, ASSISTANT COMMISSIONER, FINANCE & AGRIBUSINESS
Phone: 512-475-1762
Delane Caeser, ASSISTANT COMMISSIONER, MARKETING AND PROMOTION
Phone: 512-463-7843
Donnie Dippel, ASSISTANT COMMISSIONER, PESTICIDE DIVISION
Phone: 512-475-1618
David Kostroun, ASSISTANT COMMISSIONER, REGULATORY DIVISION
Phone: 512-463-8225
Susan Combs, COMMISSIONER
Martin Hubert, DEPUTY COMMISSIONER
Brian Murray, SPECIAL ASSISTANT OF PRODUCER RELATIONS
Phone: 512-463-7553
brian.murray@agr.state.tx.us

TEXAS DEPARTMENT OF HEALTH

1100 W. 49th St.
Austin, TX 78756 USA
Phone: 512-458-7111
Website: www.tdh.state.tx.us

Founded: 1879
Scope: Statewide

Description: The Department of Health was created to protect and promote the health of the people of Texas.

Keyword(s): Toxic Substances, Nuclear-free, Water quantity, Water export and diversion, Nuclear/Radiation, Toxicology, Health and Nutrition

Contact(s):
Joseph Fuller, ASSOCIATE COMMISSIONER OF ENVIRONMENTAL AND CONSUMER HEALTH
Phone: 512-458-7541
Debra Stabeno, DEPUTY COMMISSIONER OF PUBLIC HEALTH SCIENCES AND QUALITY
Phone: 512-458-7437
Kirk Wiles, DIRECTOR OF SEAFOOD SAFETY DIVISION
Phone: 512-719-0215

TEXAS FOREST SERVICE

301 Tarrow, Suite 364
College Station, TX 77840-7896 USA
Phone: 979-458-6600   Fax: 979-458-6610
E-mail: joverhouse@tfs.tamu.edu
Website: http://texasforestservice.tamu.edu

Founded: 1915
Membership: 10
Scope: Statewide

Description: To encourage and aid private landowners to practice multiple-use forestry; to protect private forest land against wildfire, insects, and diseases; and to inform the public of the contribution that forests make.

Keyword(s): Urban Forestry, Renewable Resources, Forests and Forestry, Public Lands, Forest Management, Water Quality, Flowers, Plants, and Trees

Contact(s):
Tom Boggus, ASSOCIATE DIRECTOR FOR ADMINISTRATION

Edwin Barron, ASSOCIATE DIRECTOR OF FOREST
RESOURCE DEVELOPMENT
Phone: 409-458-6650
Bobby Young, ASSOCIATE DIRECTOR OF FOREST
RESOURCES PROTECTION
Phone: 409-639-8100
James Hull, DIRECTOR
jim-hull@pamu.edu
I. Weldon, HEAD OF FOREST PRODUCTS DEPARTMENT
Phone: 409-639-8180
Ernest Smith, REGIONAL FORESTER OF NORTHERN
REGION
P.O. Box 3527, Longview, TX 75606-3527
Phone: 903-234-2829
William Oates, REGIONAL FORESTER OF SOUTHERN
REGION
1825 Sycamore, Huntsville, TX 77340
Phone: 409-435-0852
Robert Fewin, REGIONAL FORESTER OF WEST TEXAS
Rt. 3 Box 216, Lubbock, TX 79401
Phone: 806-746-5801

TEXAS GENERAL LAND OFFICE

Stephen F. Austin State Office Bldg., 1700 N. Congress
Ave.
Austin, TX 78701-1495 USA
Phone: 512-463-5001
Website: www.glo.state.tx.us

Founded: NA
Scope: Statewide

Description: Serves as the custodian of approximately 20.5
million acres of state-owned land including 4.25 million acres of
submerged coastal land. Responsibilities include: protecting
state land from unlawful use; managing special projects which
protects the state's natural resources; and providing the public
with information pertaining to the state's land resources.

Publications: Public Informations Office

Keyword(s): Public Lands, Coasts

Contact(s):
David Dewhurst, COMMISSIONER
Phone: 512-463-5256
david.dewhurst@glo.state.tx.us
Ashley Wadick, DEPUTY COMMISSIONER, RESOURCE
MANAGEMENT
Phone: 512-305-9121
ashley.wadick@glo.state.tx.us

TEXAS PARKS AND WILDLIFE DEPARTMENT

4200 Smith School Rd.
Austin, TX 78744 USA
Phone: 512-389-4800   Fax: 512-389-4814
E-mail: webcomments@tpwd.state.tx.us
Website: www.tpwd.state.tx.us

Founded: NA
Scope: State

Description: The agency manages and conserves natural and
cultural resources of Texas for the use and enjoyment of future
generations.

Publications: Texas Parks and Wildlife Magazine

Contact(s):
Hal Osborne, CHIEF FINANCIAL OFFICER
Phone: 512-389-4862

hal.osborne@tpwd.state.tx.us
Gene McCarty, CHIEF OF STAFF
Phone: 512-389-4418
gene.mccarty@tpwd.state.tx.us
Robert Cook, CHIEF OPERATING OFFICER
Phone: 512-389-4976
robert.cook@tpwd.state.tx.us
Lee Bass, COMMISSION CHAIRMAN
Phone: 817-390-8408
lee.bass@tpwd.state.tx.us
Carol Dinkins, COMMISSION VICE CHAIRMAN
Phone: 713-615-5311
carol.dinkins@tpwd.state.tx.us
Lydia Saldana, DIRECTOR OF COMMUNICATIONS
Phone: 512-389-4994
Annette Dominguez, DIRECTOR OF HUMAN RESOURCES
Phone: 512-389-4809
Dan Patton, DIRECTOR OF INFRASTRUCTURE
Phone: 512-389-4995
Phil Durocher, DIRECTOR OF INLAND FISHERIES
Phone: 512-389-8110
Jim Robertson, DIRECTOR OF LAW ENFORCEMENT
Phone: 512-389-4845
jim.robertson@tpwd.state.tx.us
Larry McKinney, DIRECTOR OF RESOURCE PROTECTION
Phone: 512-389-4864
Walter Dabney, DIRECTOR OF STATE PARKS
Phone: 512-389-4866
walter.dabney@tpwd.state.tx.us
Gary Graham, DIRECTOR OF WILDLIFE
Phone: 512-389-4971
gary.graham@tpwd.state.tx.us
Susan Ebert, EDITOR, TEXAS PARKS AND WILDLIFE
MAGAZINE
Phone: 512-912-7000
Andrew Sansom, EXECUTIVE DIRECTOR
Phone: 512-389-4802
andrew.sansom@tpwd.state.tx.us
Tom Harvey, MEDIA AND NEWS COORDINATOR
Phone: 512-389-4453

TEXAS SEA GRANT PROGRAM

1716 Briarcrest, Suite 702
Bryan, TX 77802 USA
Phone: 409-845-3854
Website: texas-sea-grant.tamu.edu/

Founded: NA
Scope: State, National

Publications: Marine Education, Texas Shores

Keyword(s): Communications, Coasts, Aquatic Habitats, Wildlife
Rehabilitation, Conservation, Ecology, Endangered Species,
Environmental Law, Environmental and Conservation
Education, Environmental Planning, Protected Areas,
Environment, Environmental Ethics, Agriculture

Contact(s):
Amy Broussard, ASSOCIATE DIRECTOR, MARINE
INFORMATION SERVICE
2700 Earl Rudder Frwy. S, Suite 1800, College Station,
TX 77845
Phone: 979-862-3767
Fax: 979-845-7525
abrouss@unix.tamu.edu

Robert Stickney, DIRECTOR
2700 Earl Rudder Frwy. S Suite 1800, College Station,
TX 77845
Phone: 979-845-3854
Fax: 979-845-7525
stickne@unix.tamu.edu
Jim Hiney, EDITOR, TEXAS SHORES MAGAZINE
2700 Earl Rudder Frwy. S, Suite 1800, College Station,
TX 77845
Phone: 409-862-3773
Fax: 409-862-3786
bohiney@unix.tamu.edu
Russell Miget, MARINE ADVISORY SERVICE
Texas A&M University, Natural Resources Center, 6300
Ocean Dr., Suite 2800, Corpus Christi, TX 78412
Phone: 361-825-3460
Fax: 361-825-3465
rmiget@falcon.tamucc.edu

TEXAS STATE SOIL AND WATER CONSERVATION BOARD

P.O. Box 658
Temple, TX 76503-0658 USA
Phone: 254-773-2250   Fax: 254-773-3311
Website: www.tsswcb.state.tx.us

Founded: NA
Scope: Regional

Description: The Texas State Soil and Water Conservation Board
is a state agency established to administer and carry out
Texas' soil and water conservation law. The Board is charged
with the responsibility of administering and coordinating Texas'
soil and water conservation program with the state's 216 local
soil and water conservation districts. The Board is also the
agency responsible for planning, implementing, and managing
programs and practices for abating agricultural and silvicultur-
al nonpoint source pollution within Texas.

Keyword(s): Exotic species, Aquatic nuisance species, Soil
Conservation, Environmental and Conservation Education,
Renewable Resources, Agriculture

Contact(s):
Robert Buckley, EXECUTIVE DIRECTOR
311 N. 5th St., Temple, TX 76501-3107
Phone: 254-773-3311

TEXAS WATER DEVELOPMENT BOARD

1700 N. Congress
Austin, TX 78701 USA
Phone: 512-463-7847   Fax: 512-475-2053
Website: www.twdb.state.tx.us

Founded: 1957
Membership: 6
Scope: Statewide

Description: Texas Water Development Board provides loans to
local governments for water supply projects; water quality
projects, including wastewater treatment, municipal solid waste
management, and nonpoint source pollution control; agricultural water conservation projects; and flood control projects.
Provides water related research and planning and agricultural
water conservation funding.

Publications: Rainwater Harvesting - brochure, Texas Water
Facts, ground water reports since 1957, bay and estuary
reports since 1967, Water For Texas - Today and Tomorrow

Keyword(s): Exotic species, Aquatic nuisance species, Rivers,
Environmental and Conservation Education, Planning
Management, Aquatic Habitats

Contact(s):
Ignacio Madera, BORDER PROJECT MANAGEMENT
DIVISION
Phone: 512-463-7509
ignacio.madera@twdb.state.tx.us
William Madden, CHAIRMAN
J. Ward, DEPUTY EXECUTIVE ADMINISTRATOR FOR
OFFICE OF PROJECT FINANCE AND CONSTRUCTION
ASSISTANCE
Phone: 512-463-0991
kevin.ward@twdb.state.tx.us
Tommy Knowles, DEPUTY EXECUTIVE ADMINISTRATOR
FOR PLANNING
Phone: 512-463-8043
tommy.knowles@twdb.state.tx.us
Hugh Bender, DIRECTOR OF TEXAS NATURAL
RESOURCES INFORMATION SYSTEM
Phone: 512-463-8051
hugh.bender@twdb.state.tx.us
Bill Mullican, DIRECTOR OF WATER RESOURCES
INFORMATION
Phone: 512-936-0813
bill.mullican@twdb.state.tx.us
Craig Pedersen, EXECUTIVE ADMINISTRATOR
Phone: 512-463-7850
craig.pedersen@twdb.state.tx.us
Suzanne Schwartz, GENERAL COUNSEL
Phone: 512-463-7981
suzanne.schwartz@twdb.state.tx.us
George Green, NORTHERN PROJECT MANAGEMENT
DIVISION
Phone: 512-463-7853
george.green@twdb.state.tx.us
Leonard Olson, SPECIAL ASSISTANT FOR
INTER-GOVERNMENTAL AND EXTERNAL CUSTOMER
RELATIONS
Phone: 512-463-7931
leonard.olson@twdb.state.tx..us
Wales Madden, VICE CHAIRMAN

TUG HILL COMMISSION

317 Washington St.
Watertown, NY 13601 USA
Phone: 315-785-2380   Fax: 315-785-2574
E-mail: tughill@tughill.org
Website: www.tughill.org

Founded: 1972
Membership: 20
Scope: Regional

Description: The Tug Hill Commission is a nonregulatory state
agency charged with helping local governments, organizations,
and citizens shape the future of this rural, 2,100 square mile
region, especially its environment and economy.

Publications: Tug Hill Program, The, Cooperative Rural Planning,
Issue Paper series, Headwaters

Keyword(s): Rural Development, Planning Management, Historic
Preservation, Outdoor Recreation, Environmental Planning

Contact(s):
Robert Quinn, EXECUTIVE DIRECTOR

TULANE INSTITUTE FOR ENVIRONMENTAL LAW AND POLICY

6329 Freret St.
New Orleans, LA 70118 USA
Phone: 504-862-8827

U

UNIVERSITY OF MARYLAND COOPERATIVE EXTENSION

1296 Symons Hall
College Park, MD 20742 USA
Phone: 301-405-2072　　　Fax: 301-405-2963
E-mail: tf.43@umail.umd.edu
Website: www.agnr.umc.edu

Founded: NA
Membership: 250
Scope: National
Publications: located on website

Contact(s):

James Wade, ASSOCIATE DEAN & ASSOCIATE DIRECTOR MARYLAND COOPERATIVE EXTENSION
University of Maryland, Cooperative Extension, 1200 Symons Hall, College Park, MD 20742-5565
Phone: 301-405-2907
Fax: 301-405-2963
Doug Lipton, COORDINATOR: SEA GRANT EXTENSION PROGRAM
University of Maryland, Cooperative Extension, 2218B Symons Hall, College Park, MD 20742
Phone: 301-405-1280
Thomas Fretz, DEAN AND DIRECTOR OF AGRICULTURAL EXPERIMENT STATION AND MARYLAND COOPERATIVE EXTENSION
Phone: 301-405-2072
Fax: 301-314-9146
tf43@umail.umd.edu
James Hanson, PROGRAM LEADER AND ASSISTANT DIRECTOR OF AGRICULTURE AND NATURAL RESOURCES PROGRAM
University of Maryland, Cooperative Extension, 1200 Symons Hall, College Park, MD 20742-5565
Phone: 301-405-7992
Fax: 301-405-2963
Bob Tjaden, REGIONAL NATURAL RESOURCE SPECIALIST
University of Maryland, Cooperative Extension, Wye Research and Education Center, P.O. Box 169, Queenstown, MD 21658
Phone: 410-827-8056
Fax: 410-827-9039
Jonathan Kays, REGIONAL NATURAL RESOURCE SPECIALIST
University of Maryland, Cooperative Extension, Western Maryland Research and Education Center, 18330 Keedysville Road, Keedysville, MD 21756
Phone: 301-432-2735
Fax: 301-432-4089

UNIVERSITY OF CONNECTICUT COOPERATIVE EXTENSION

COLLEGE OF AGRICULTURE & NATURAL RESOURCES
R Unit 4066 1376 Storrs Rd., Univesity of Connecticut
West Hartford, CT 06117 USA
Phone: 860-570-9010　　　Fax: 860-570-9008
Website: www.canr.uconn.edu

Founded: NA
Scope: Statewide

Description: Natural resource components includes forest management, forest stewardship, urban forestry, water resources, and wildlife managment.

Contact(s):

Stephen Broderick, EXTENSION EDUCATOR: FOREST MANAGMENT
139 Wolf Den Rd., Brooklyn, CT 06234
Phone: 860-774-9600
Glenn Warner, EXTENSION SPECIALIST: WATER RESOURCES
Natural Resources Managment and Engineering: Box U-87: University of Connecticut, Storrs, CT 06269-4087
Phone: 860-486-2840
John Barclay, EXTENSION SPECIALIST: WILDLIFE AND DIRECTOR, WILDLIFE CONSERVATION RESEARCH CENTER
Natural Resources Management and Engineering: Box U-87: University of Connecticut, Storrs, CT 06269-4087
Phone: 860-486-0143
Fax: 860-486-5875
Norman Bender, PROGRAM LEADER: MARINE ADVISORY PROGRAM
University of CT-MAS: 1084 Shennecossett Rd., Groton, CT 06340-6097
Phone: 860-445-8664
Xiusheng Yang, STATE CLIMATOLOGIST
Natural Resources Management and Engineering: Box U-87: University of Connecticut, Storrs, CT 06269-4087
Phone: 860-486-2840

UNIVERSITY OF HAWAII COOPERATIVE EXTENSION PROGRAM

COLLEGE OF TROPICAL AGRICULTURE AND HUMAN RESOURCES
College of Tropical Agriculture and Human Resources, Gilmore Hall 202, 3050 Maile Way, Univ. of Hawaii at Manoa
Honolulu, HI 96822 USA
Phone: 808-956-8234　　　Fax: 808-956-9105
E-mail: dean@cteahr.hawaii.edu
Website: www2.ctahr.hawaii.edu/extout/extout.asp

Founded: NA
Scope: International

Contact(s):

Ronald Mau, ASSOCIATE DEAN AND ASSOCIATE DIRECTOR FOR COOPERATIVE EXTENSION (INTERIM)
Andrew Hashimoto, DEAN AND DIRECTOR OF COOPERATIVE EXTENSION (INTERIM)
Richard Brock, RESEARCHER/FISHERIES SPECIALIST
Univ. of Hawaii Sea Grant Program, 1000 Pope Rd./MSB 204, Honolulu, HI 96822
Phone: 808-956-2859
Fax: 808-956-2858

UNIVERSITY OF MAINE COOPERATIVE EXTENSION

FORESTRY & WILDLIFE OFFICE
5755 Nutting Hall
Orono, ME 04469-5755 USA
Phone: 207-581-2902   Fax: 207-581-3466
Website: www.umext.maine.edu

Founded: NA
Membership: 3
Scope: State
Contact(s):
Les Hyde, FORESTRY EDUCATOR
Phone: 207-581-3466
lhyde@umext.maine.edu
James Philp, FORESTRY SPECIALIST
261 Nutting Hall, University of Maine, Orono, ME 04469-5755
Phone: 207-581-2885
John Rebar, PROGRAM ADMINISTRATION

UNIVERSITY OF MASSACHUSETTS EXTENSION

Stockbridge Hall, Box 30099, University of Massachusetts
Amherst, MA 01003 USA
Phone: 413-545-6555   Fax: 413-545-4800
E-mail: umextadm@umext.umass.edu
Website: www.umass.edu/umext/

Founded: NA

Keyword(s): Urban Forestry, Sustainable Ecosystems, Forests and Forestry, Wildlife, Renewable Resources
Contact(s):
Scott Jackson, CONSERVATION SPECIALIST
University of Massachusetts, Department of Forestry and Wildlife Management, Holdsworth Natural Resources Center, Amherst, MA 01003
Phone: 413-545-2665
John Gerber, DIRECTOR
jgerber@umext.umass.edu
Anna Hicks, NATURAL RESOURCES AND ENVIRONMENTAL CONSERVATION PROGRAM
Holdsworth Hall, University of Massachusetts, Amherst, MA 01003
Phone: 413-545-4743
Fax: 413-545-4358
ahicks@umext.umass.edu

UNIVERSITY OF NEW HAMPSHIRE COOPERATIVE EXTENSION

59 College Rd., Taylor Hall
Durham, NH 03824-3587 USA
Phone: 603-862-1520   Fax: 603-862-1585
Website: www.ceinfo.unh.edu

Founded: 1925 (Forestry Program)
Scope: Statewide

Description: The natural resource components include Forest Stewardship, Community Forestry, Rural Economic Well-Being, Agriculture, and Natural Resource Conservation Education.

Keyword(s): training, Urban Forestry, Forests and Forestry, Open Spaces, Land Use Planning, Lakes, Exotic species, Aquatic nuisance species, Environmental and Conservation Education

Contact(s):
John Pike, DEAN AND DIRECTOR OF UNH COOPERATIVE EXTENSION
UNH, 59 College Rd., Taylor Hall, Durham, NH 03824-3587
Phone: 603-862-1520
Frank Mitchell, EXTENSION SPECIALIST: WATER RESOURCES
UNH Cooperative Extension, 55 College Rd., Pettee Hall, Durham, NH 03824-3599
Phone: 603-862-1067
Jeffrey Schloss, EXTENSION SPECIALIST: WATER RESOURCES LAKES LAY MONITORING PROGRAM
UNH, 55 College Rd., Pettee Hall, Durham, NH 03824-3599
Phone: 603-862-3848
Karen Bennett, FOREST STEWARDSHIP COORDINATOR
UNH Cooperative Extension, 55 College Rd., Pettee Hall, Durham, NH 03824-3599
Phone: 603-862-2512
Robert Edmonds, PROGRAM LEADER: FORESTRY/WILDLIFE
Phone: 603-862-2619
Ellen Snyder, WILDLIFE SPECIALIST
UNH Cooperative Extension, 55 College Rd., Pettee Hall, Durham, NH 03824-3599
Phone: 603-862-3594
Fax: 603-862-2157

UNIVERSITY OF VERMONT EXTENSION

601 Main St.
Burlington, VT 05401-3439 USA
Phone: 802-656-2990   Fax: 802-656-8642
Website: www.ctr.uvm.edu/ext

Founded: 1914
Membership: 150
Scope: Statewide

Description: UVM Extension is a system of nonformal education, bringing research information in a practical form to Vermonters. Extension with the specific expertise of our state university meets the needs of agriculture, communities, families, and youth. Programs are specifically focused on natural resource conservation, sustainable agriculture and rural development, health care in rural areas, resource distribution in communities, and the contemporary stresses on the American family.

Keyword(s): Youth Organizations, Sustainable Development, Health and Nutrition, Research, Acid Rain

Contact(s):
Lawrence Forcier, DIRECTOR
Vern Grubinger, DIRECTOR OF SUSTAINABLE AGRICULTURE CENTER
Phone: 802-257-7907
Thom McEvoy, EXTENSION FORESTER
School of Natural Resources, 345 Aiken Center, Burlington, VT 05405
Phone: 802-656-2913
Fax: 802-656-8683
Lois Frey, FAMILY AND COMMUNITY RESOURCE AND ECONOMIC DEVELOPMENT
Phone: 802-223-2389
Doug Lantagne, NATURAL RESOURCES AND ENVIRONMENTAL MANAGEMENT
Phone: 802-656-2990
Dale Steen, NUTRITION, FOOD SAFETY, AND HEALTH
Phone: 802-748-8177
Robert Tyzbir, NUTRITION, FOOD SAFETY, AND HEALTH

Phone: 802-656-3374
John Winder, PROGRAM LEADER OF AGRICULTURE
Phone: 802-257-7967
Alan Gotlieb, PROGRAM LEADER OF AGRICULTURE
Phone: 802-656-0474

UNIVERSITY OF VERMONT EXTENSION

PUBLICATIONS OFFICE
Communications Technology Resources Agr. Eng. Bldg.
63 Carrigg Dr.
Burlington, VT 05405-0004 USA
Phone: 802-656-3024   Fax: 802-656-5878
Website: cpr.uvm.edu/ext

Founded: NA
Membership: 100
Scope: Local Regional

UNIVERSITY OF WISCONSIN SEA GRANT INSTITUTE

UW SEA GRANT INSTITUTE
1975 Willow Dr.
Madison, WI 53706-1177 USA
Phone: 608-262-0905   Fax: 608-262-0591
E-mail: administrator@seagrant.wisc.edu
Website: www.seagrant.wisc.edu/

Founded: 1968
Scope: National

Description: The University of Wisconsin Sea Grant Institute is a statewide program of basic and applied research, education, and technology transfer dedicated to the wise stewardship and sustainable use of Great Lakes and ocean resources.

Keyword(s): Coral Reefs, Lakes, Biotechnology, Wildlife, Aquatic Habitats

Contact(s):
Mary Reeb, ASSISTANT DIRECTOR FOR ADMINISTRATION AND INFORMATION TECHNOLOGY
Phone: 608-263-3296
mlreeb@seagrant.wisc.edu
Stephen Wittman, ASSISTANT DIRECTOR OF COMMUNICATIONS
Phone: 608-263-5371
swittman@seagrant.wisc.edu
Anders Andren, DIRECTOR
awandren@seagrant.wisc.edu

UTAH COOPERATIVE FISH AND WILDLIFE RESEARCH UNIT (USDI-USGS-BRD-CRU)

College of Natural Resources, Utah State University
Logan, UT 84322-5210 USA
Phone: 435-797-2509   Fax: 435-797-4025
E-mail: utcoop@cc.usu.edu
Website: http//ella.nr.usu.edu/utcop/index.html

Founded: 1935 (Wildlife), 1962 (Fisheries), 1985 (combined)
Membership: 3
Scope: National

Description: The unit conducts research and training in all aspects of fishery and wildlife biology and management.

Keyword(s): Sport Fishing, Renewable Resources, Birds, Nongame Wildlife, Aquatic Habitats, Endangered Species, Wildlife, Terrestrial Habitats, training, Landscape Analysis, Ecology, Mammals, Predators, Biodiversity

Contact(s):
Thomas Edwards,, ASSISTANT LEADER OF WILDLIFE
Phone: 435-797-2529
Esther Biesinger, FINANCIAL ASSISTANT
Phone: 435-797-2467
John Bissonette, LEADER
Phone: 435-797-2511
johnbissonette@cnor.usu.edu

UTAH DEPARTMENT OF AGRICULTURE

PO BOX 146500
Salt Lake City, UT 84114-6500 USA
Phone: 801-538-7100   Fax: 801-538-7126
Website: www.ag.state.ut.us

Founded: NA
Membership: 200
Scope: Statewide

Contact(s):
Miles Ferry, COMMISSIONER
Van Burgess, DEPUTY COMMISSIONER
Renee Matsuura, DIRECTOR OF ADMINISTRATIVE SERVICES
Randy Parker, DIRECTOR OF MARKETING
G. Wilson, DIRECTOR OF PLANT INDUSTRY
El Shaffer, INFORMATION OFFICER
James Christensen, STAFF OF AGRICULTURAL DEVELOPMENT AND CONSERVATION
Kyle Stephens, STAFF OF FOOD AND DAIRY
Bob Smoot, STAFF OF WEIGHTS AND MEASURES
Ahmad Salari, STATE CHEMIST
Michael Marshall, STATE VETERINARIAN

UTAH DEPARTMENT OF HEALTH

P.O. Box 14100
Salt Lake City, UT 84114-1011 USA
Phone: 801-538-6003   Fax: 801-538-6306
Website: www.dhrn.state.ut.us

Founded: NA
Scope: Statewide

Publications: Baby Your Baby, Health Data Programs

Keyword(s): Toxic Substances, Nuclear-free, Water quantity, Water export and diversion, Toxicology, Health and Nutrition, Public Health Protection, Biotechnology

Contact(s):
Jana Kettering, DEPARTMENT OF HEALTH PUBLIC INFORMATION OFFICER
Phone: 801-538-6339
Rod Betit, EXECUTIVE DIRECTOR

UTAH GEOLOGICAL SURVEY

1594 W. North Temple, Suite 3110, P.O. Box 146100
Salt Lake City, UT 84114-6100 USA
Phone: 801-537-3300   Fax: 801-537-3400
Website: www.ugs.state.ut.us

Founded: NA
Scope: Statewide

Publications: Catalogs available upon request

Contact(s):
　Richard Allis, DIRECTOR

UTAH STATE DEPARTMENT OF NATURAL RESOURCES

1594 W. North Temple, Suite 3710, P.O. Box 145610
Salt Lake City, UT 84114-5610 USA
Phone: 801-538-7200　　　Fax: 801-538-7315
Website: www.nr.state.ut.us

Founded: NA
Membership: 1200
Scope: Statewide

Contact(s):
　Darin Bird, ASSISTANT DIRECTOR
　Sherm Hoskins, DEPUTY DIRECTOR
　Hugh Thompson, DEPUTY DIRECTOR
　Kathleen Clarke, EXECUTIVE DIRECTOR

UTAH STATE DEPARTMENT OF NATURAL RESOURCES

DIVISION OF WATER RESOURCES
1594 W. North Temple, Suite 310
Salt Lake City, UT 84114-6201 USA
Phone: 801-538-7230　　　Fax: 801-538-7279
Website: NR.STATE.UT.US

Founded: NA
Membership: 50
Scope: Statewide

Contact(s):
　D. Anderson, DIRECTOR

UTAH STATE DEPARTMENT OF NATURAL RESOURCES

DIVISION OF WILDLIFE
1594 W. North Temple
Salt Lake City, UT 84114-6300 USA
Phone: 801-538-4700　　　Fax: 801-538-4745
E-mail: wcomment.nrdwr@state.ut.us
Website: www.nr.utah.gov/dwr

Founded: NA
Scope: Statewide

Publications: Wildlife News—weekly newsletter, Wildlife Review—magazine

Contact(s):
　John Kimball, DIRECTOR

UTAH STATE DEPARTMENT OF NATURAL RESOURCES

DIVISION OF WILDLIFE RESOURCES
1594 W. North Temple, Suite 2110, P.O. Box 146301
Salt Lake City, UT 84114-6301 USA
Phone: 801-538-4700　　　Fax: 801-538-4745
Website: www.nr.state.ut.us

Founded: NA
Scope: Statewide

Publications: Wildlife Review (Quarterly), Weekly Wildlife News (Weekly)

Contact(s):
　John Kimball, DIRECTOR OF DIVISION OF WILDLIFE RESOURCES

UTAH STATE DEPARTMENT OF NATURAL RESOURCES

DIVISION OF WILDLIFE RESOURCES
1594 W. North Temple, Suite 2110, P.O. Box 146301
Salt Lake City, UT 84114-6301 USA
Phone: 801-538-4700　　　Fax: 801-538-4745
E-mail: wcomment.nrdwr@state.ut.us
Website: http://www.wildlife.utah.gov

Founded: NA
Scope: Statewide

Contact(s):
　Kevin Conway, ASSISTANT DIRECTOR
　Rick Danvir, BOARD MEMBER
　Raymond Heaton, BOARD MEMBER
　Brenda Freeman, BOARD MEMBER
　J. Allan, BOARD MEMBER
　Connie Brooks, BOARD MEMBER
　B. Dastrup, BOARD MEMBER
　Max Morgan, CHAIR PERSON
　John Kimball, DIRECTOR
　Jordan Pedersen, SUPERVISOR OF CENTRAL REGION
　115 N. Main St., Springfield, UT 84663
　Phone: 435-489-5678
　Fax: 435-489-7000
　Walt Donaldson, SUPERVISOR OF NORTHEAST REGION
　152 E. 100 N., Vernal, UT 84078
　Phone: 435-789-3103
　Fax: 435-789-8343
　Robert Hasenyager, SUPERVISOR OF NORTHERN REGION
　515 E. 5300 S., Ogden, UT 84405
　Phone: 435-476-2740
　Fax: 435-479-4010
　Miles Moretti, SUPERVISOR OF SOUTHEASTERN REGION
　475 W. Price River Dr., Suite C, Price, UT 84501
　Phone: 435-636-0260
　Fax: 435-637-7361
　Jim Guymon, SUPERVISOR OF SOUTHERN REGION
　P.O. Box 606, Cedar City, UT 84721
　Phone: 435-865-6100
　Fax: 435-586-2457

UTAH STATE DEPARTMENT OF NATURAL RESOURCES

OFFICE OF ENERGY AND RESOURCE PLANNING
1594 W. North Temple, Suite 3610, P.O. Box 146480
Salt Lake City, UT 84114-6480 USA
Phone: 801-538-5428　　　Fax: 801-521-0657
E-mail: nroerp@state.ut.us

Founded: NA
Scope: Statewide

Contact(s):
　Jeffrey Burks, DIRECTOR

UTAH STATE EXTENSION SERVICES

College of Natural Resources, 4900 Old Main Hall, Utah State University
Logan, UT 84322-4900 USA
Website: www.ext.usu.edu/

Founded: NA

Keyword(s): Wetlands, Exotic species, Aquatic nuisance species, Public Lands, Renewable Resources, Forests and Forestry

Contact(s):
Robert Schmidt, ANIMAL DAMAGE MANAGEMENT EXTENSION SPECIALIST
Phone: 801-797-2459
Charles Gay, ASSISTANT DEAN FOR EXTENSION AND ADMINISTRATION
Phone: 801-797-2445
F. Busty, DEAN
Phone: 801-797-2445
Terry Messmer, EXTENSION FISH AND WILDLIFE SPECIALIST
Phone: 801-797-2459
Mike Kuhns, EXTENSION FORESTER
Phone: 801-797-4056
Stephen Burr, EXTENSION OUTDOOR RECREATION AND TOURISM SPECIALIST
Phone: 801-797-2530
Roger Banner, EXTENSION RANGE SPECIALIST
Phone: 801-797-2472
G. Rasmussen, EXTENSION RANGE SPECIALIST
Phone: 801-797-2469
Clifford Craig, UTAH GEOGRAPHIC ALLIANCE
Phone: 801-797-1790
Robert Gilliland, VICE PRESIDENT FOR EXTENSION AND CONTINUING EDUCATION
Nancy Mesner, WATER QUALITY EXTENSION SPECIALIST
Phone: 801-797-2465

UTAH STATE PARKS & RECREATION

DIVISION OF PARKS AND RECREATION
1594 W. North Temple, Suite 116, P.O. Box 146001
Salt Lake City, UT 84114-6001 USA
Phone: 801-538-7220   Fax: 801-538-7378
E-mail: parkcomments@states.ut.us
Website: http://parks.state.ut.us

Founded: NA
Membership: 175
Scope: Statewide
Publications: Utah State Field Guide

Contact(s):
Ted Woolley, BOATING COORDINATOR
Jay Christianson, CHIEF OF LAW ENFORCEMENT
Phone: 801-538-7326
Mary Tullius, DEPUTY DIRECTOR
Dave Morrow, DEPUTY DIRECTOR
Courtland Nelson, DIRECTOR
Terry Green, PARK PLANNING MANAGER
Phone: 801-538-7346
Dennis Weaver, REGIONAL MANAGER OF NORTHEAST
Phone: 435-649-9109
Jim Harland, REGIONAL MANAGER OF NORTHWEST
Phone: 801-649-9109
Tim Smith, REGIONAL MANAGER OF SOUTHEAST
Phone: 435-259-3750
Gordon Topham, REGIONAL MANAGER OF SOUTHWEST
Phone: 435-586-4497

UTAH STATE SOIL CONSERVATION COMMISSION

350 N. Redwood Rd.
Salt Lake City, UT 84116 USA
Phone: 801-538-7120   Fax: 801-538-4949
E-mail: sadginton@state.ut.us
Website: www.ag.state.ut.us

Founded: 1938
Scope: Statewide

Description: Assists Utah's 39 soil conservation districts (SCD) in encouraging land operators to implement measures and practices; to prevent soil deterioration; restore depleted soil; prevent flood damage; improve irrigation water efficiency; and to encourage nonpoint water pollution control programs. The commission has 12 members; 5 ex-officio, and 7 governor-appointed SCD members with their alternates.

Keyword(s): Sustainable Development, Soil Conservation, Environmental and Conservation Education, Renewable Resources, Agriculture

Contact(s):
Miles Ferry, CHAIRMAN
K. Jacobson, EXECUTIVE

V

VERMONT DEPARTMENT OF AGRICULTURE, FOOD, AND MARKETS

116 State St., Drawer 20
Montpelier, VT 05620-2901 USA
Phone: 802-828-2500   Fax: 802-828-2361
Website: www.state.vt.us\agric

Founded: 1908
Membership: 85
Scope: Statewide
Publications: list available on request., Agriview

Contact(s):
Rudolph Polli, BUSINESS MANAGER OF ADMINISTRATIVE SERVICES
Phone: 802-828-3567
Leon Graves, COMMISSIONER
Phone: 802-828-2430
Louise Calderwood, DEPUTY COMMISSIONER OF ADMINISTRATION
Philip Benedict, DIRECTOR OF PLANT INDUSTRY OF LABORATORIES OF AND CONSUMER ASSURANCE
Phone: 802-828-2431
Todd Johnson, STATE VETERINARIAN
Phone: 802-828-2421

VERMONT DEPARTMENT OF AGRICULTURE, FOOD, AND MARKETS

NATURAL RESOURCES CONSERVATION COUNCIL
116 State St.
Montpelier, VT 05620-2901 USA
Phone: 802-828-2416   Fax: 802-828-2361
E-mail: jwa@agr.state.vt.us

Founded: NA
Membership: 8
Scope: Local

Description: The Conservation Council is the administrative body for the 14 conservation districts in Vermont. The goal of conservation districts is to ensure the wise use, protection and enhancement of Vermont soil, water, and related natural resources; to foster public awareness and appreciation of the need for conservation; and to advance the concept that we are all stewards of the living earth.

Keyword(s): Coral Reefs, Soil Conservation, Environmental and Conservation Education, Protected Areas, Agriculture

VERMONT DEPARTMENT OF AGRICULTURE, FOOD, AND MARKETS

STATE CONSERVATION COMMISSION
48 Bushey Dr.
Shelburne, VT 05482 USA
Phone: 802-985-2048

Founded: NA
Scope: Statewide

Contact(s):
Thomas Bushey, CHAIR
Phone: 802-985-2048
Fax: 802-951-6327
Jon Anderson, EXECUTIVE SECRETARY
Phone: 802-828-3529
Fax: 802-828-2361
jwa@agr.state.vt.us

VERMONT DEPARTMENT OF HEALTH

108 Cherry St.
Burlington, VT 05402 USA
Phone: 802-863-7280   Fax: 802-863-7425
Website: www.state.vt.us/health

Founded: NA
Membership: 500
Scope: Statewide

Keyword(s): Toxic Substances, Nuclear-free, Water quantity, Water export and diversion, Toxicology, Health and Nutrition, Pesticides, Air Quality and Pollution

Contact(s):
Jan Carney, COMMISSIONER
Phone: 802-863-7280
Larry Crist, DIRECTOR OF HEALTH PROTECTION
Phone: 802-863-7223

VIRGIN ISLANDS COOPERATIVE EXTENSION SERVICE

University of Virgin Islands, R.R. 2, Box 10,000, Kingshill
St. Croix, VI 00850 USA
Phone: 340-692-4080   Fax: 340-692-4085
Website: www.rps.uvi.edu/cds

Founded: NA
Scope: National

Contact(s):
Jozef Keularts, COORDINATOR, INTEGRATED PEST MANAGEMENT OF PESTICIDE IMPACT ASSESSEMENT PROGRAM LIAISON
James Rakocy, DIRECTOR OF AGRICULTURAL EXPERIMENT STATION
Univeristy of VI, RR2, Box 10,000, Kingshill, VI 00850
Phone: 340-692-4031
Fax: 340-692-4035
Kwame Garcia, DIRECTOR OF CES
Clinton George, PROGRAM LEADER OF AGRICULTURE AND NATURAL RESOURCES

VIRGIN ISLANDS SOIL AND WATER CONSERVATION DIVISION

USA

Founded: NA

Contact(s):
Henry Schuster, COMMISSIONER
Phone: 809-778-0997

VIRGINIA COOPERATIVE EXTENSION

VIRGINIA POLYTECNIC INSTITUTE & STATE UNIVERSITY
Blacksburg, VA 24061-0402 USA
Phone: 540-231-5299   Fax: 540-231-4370
Website: www.ext.vt.edu/

Founded: NA
Membership: 700
Scope: National

Contact(s):
J. Barrett, DIRECTOR OF COOPERATIVE EXTENSION
davebarr@vt.edu
Brian Nerrie, EXTENSION AQUACULTURE SPECIALIST
Virginia State University, P.O. Box 9081, Petersburg, VA 23806
Phone: 804-524-5903
bnerrie@vsu.edu
Louis Helfrich, EXTENSION FISHERIES SPECIALIST
Department of Fisheries and Wildlife Sciences, Virginia Polytechnic Institute and State University, Blacksburg, VA 24061-0321
Phone: 540-231-5059
lhelfric@vt.edu
James Parkhurst, EXTENSION WILDLIFE SPECIALIST
Department of Fisheries & Wildlife Sciences, Virginia Polytechnic Institute and State University, Blacksburg, VA 24061-0321
Phone: 540-231-9283
Fax: 540-231-7265
jparkhur@vt.edu
Gerald Cross, EXTENSION WILDLIFE SPECIALIST
Department of Fisheries and Wildlife Sciences, Virginia Polytechnic Institute and State University, Blacksburg, VA 24061-0321
Phone: 540-231-8844
gecross@vt.edu
James Johnson, PROJECT LEADER OF FORESTRY AND WILDLIFE EXTENSION
College of Natural Resources, Virginia Polytechnic Institute and State University, Blacksburg, VA 24061-0324
Phone: 540-231-7679
jej@vt.edu
George Flick, SEA GRANT EXTENSION SEAFOOD TECHNOLOGIST
Dept. of Food Science and Technology, Virginia Polytechnic Institute and State University, Blacksburg, VA 24061-0418
Phone: 540-231-6065
flickg@vt.edu

VIRGINIA COOPERATIVE FISH AND WILDLIFE RESEARCH UNIT (USDI)

100 Cheatham Hall, Virginia Polytechnic Institute and State University
Blacksburg, VA 24061 USA
Phone: 540-231-5573   Fax: 540-231-7580
E-mail: vsutherl@vt.edu
Website: www.cnr.vt.edu/fisheries/

Founded: 1935
Scope: Statewide

Description: Founded for training graduate students in fisheries and wildlife; with teaching and extension in fisheries and wildlife biology. Cooperatively supported by the Biological Resources Division of U.S.G.S., Department of Game and Inland Fisheries, and Virginia Polytechnic Institute and State University.

Publications: journal articles, research publications., Annual reports

VIRGINIA DEPARTMENT OF AGRICULTURE AND CONSUMER SERVICES

Richmond, VA 23218 USA
Phone: 804-786-3501   Fax: 804-371-2945
Website: www.vdacs.state.va.us

Founded: 1877
Membership: 350
Scope: Statewide

Description: To promote the economic growth and development of Virginia agriculture, encourage environmental stewardship, and provide consumer protection. Thirteen-member board appointed by Governor.

Publications: Bulletin

Keyword(s): Endangered Species, Pesticides, Agriculture

Contact(s):
J. Courter, DEPARTMENT COMMISSIONER
Phone: 804-786-3501
Elaine Lidholm, DIRECTOR OF COMMUNICATION
Phone: 804-786-7686
Roy Seward, DIRECTOR OF POLICY PLANNING AND RESEARCH
Phone: 804-786-3535
Elaine Lidholm, EDITOR
Marvin Lawson, MANAGER OF PESTICIDES SERVICES
Phone: 804-371-6558

VIRGINIA DEPARTMENT OF CONSERVATION AND RECREATION

203 Governor St., Suite 302
Richmond, VA 23219 USA
Phone: 804-786-5046   Fax: 804-786-6141
E-mail: dcr@state.va.us
Website: www.dcr.state.va.us

Founded: NA
Membership: 500
Scope: Local

Publications: Virginia Outdoors Plan, Chesapeake Bay Susquehanna River & Tidal Tributaries Public Access Guide

Contact(s):
David Brickley, DIRECTOR

VIRGINIA DEPARTMENT OF CONSERVATION AND RECREATION

203 Governor St., Suite 302
Richmond, VA 23219 USA
Phone: 804-786-1712   Fax: 804-786-6141
E-mail: pco@dcr.state.va.us
Website: www.dcr.state.va.us

Founded: NA
Scope: State

Description: The Department's mission is to conserve, protect, enhance, and advocate wise use of Virginia's natural, recreational, and scenic resources in order to maintain and improve the quality of life for present and future generations. The Department is responsible for administrative support of various state collegial bodies including: The Board of Conservation and Recreation, the Virginia Cave Board, the Virginia Soil and Water Conservation Board, the Breaks Interstate Park Commission, the Conservation and Development of Public Beaches Board, Chippokes Plantation Farm Foundation, Virginia State Parks Foundation, Virginia Land Conservation Foundation and 17 state Scenic River Boards and Committees.

Contact(s):
Linda Cox, ADMINISTRATIVE STAFF SPECIALIST
Phone: 804-786-2123
Fax: 804-786-6141
ljcox@dcr.state va.us
Leon App, CHIEF DEPUTY (ACTING)
Phone: 804-786-4570
Fax: 804-786-6141
leonapp@dcr.state.va.us
David Dowling, CONSERVATION & DEVELOPMENT PROGRAMS SUPERVISOR
Phone: 804-786-2291
Fax: 804-786-2291
ddowling@dcr.state.va.us
David Brickley, DIRECTOR
Phone: 804-786-2123
dgbrickley@dcr.state.va.us

VIRGINIA DEPARTMENT OF CONSERVATION AND RECREATION

BOARD OF CONSERVATION AND RECREATION
203 Governor St., Suite 302
Richmond, VA 23219 USA

Founded: NA

Contact(s):
W. Wingo, CHAIRMAN
203 Governor St., Suite 302, Richmond, VA 23219

VIRGINIA DEPARTMENT OF CONSERVATION AND RECREATION

BREAKS INTERSTATE PARK COMMISSION
203 Governor St., Suite 302
Richmond, VA 23219 USA

Founded: NA

Contact(s):
Joseph Elton, ADVISOR
Jack Sykes, CHAIRMAN
101 Summit Drive, Pikesville, KY 41501
Phone: 606-432-1447

VIRGINIA DEPARTMENT OF CONSERVATION AND RECREATION

CHIPPOKES PLANTATION FARM FOUNDATION
203 Governor St., Suite 302
Richmond, VA 23209 USA

Founded: NA

Contact(s):
Katherine Wright, ADVISOR
Frederick Quayle, CHAIRMAN
Member, Senate of Virginia, 3808 Poplar Hill Road, Chesapeake, VA 23321

VIRGINIA DEPARTMENT OF CONSERVATION AND RECREATION

CONSERVATION AND DEVELOPMENT OF PUBLIC BEACHES BOARD
203 Governor St., Suite 302
Richmond, VA 23209 USA

Founded: NA

Contact(s):
Carlton Hill, ADVISOR
Donald Campen, CHAIRMAN
7603 Hillside Avenue, Richmond, VA 23229

VIRGINIA DEPARTMENT OF CONSERVATION AND RECREATION

DIVISION OF ADMINISTRATION
203 Governor St., Suite 302
Richmond, VA 23219 USA
Phone: 804-786-6124　　Fax: 804-786-6141
Website: www.dcr.state.va.us

Founded: NA
Membership: 700
Scope: Statewide

Contact(s):
William Price, DIRECTOR OF ADMINSTRATION
203 Governor St., Suite 204, Richmond, VA 23219
Phone: 804-786-0001
Donald Bryne, DIRECTOR OF ADP
Timothy Bishton, DIRECTOR OF FINANCE
Karen Carey, DIRECTOR OF HUMAN RESOURCES
Gary Waugh, PUBLIC RELATIONS MANAGER
Phone: 804-786-5045

VIRGINIA DEPARTMENT OF CONSERVATION AND RECREATION

DIVISION OF DAM SAFETY
203 Governor St., Suite 302
Richmond, VA 23219 USA
Phone: 804-786-1369　　Fax: 804-786-0536
E-mail: dam@dcr.state.va.us
Website: www.dcr.va.us

Founded: NA
Membership: 5
Scope: Statewide

VIRGINIA DEPARTMENT OF CONSERVATION AND RECREATION

DIVISION OF NATURAL HERITAGE
217 Governor St., 3rd Fl.
Richmond, VA 23219 USA
Phone: 804-786-7951　　Fax: 804-371-2674
Website: www.dcr.state.va.us\dnh

Founded: NA
Membership: 50
Scope: Statewide

Contact(s):
Thomas Smith, DIRECTOR
217 Governor St., 3rd Fl., Richmond, VA 23219
Phone: 804-786-7951

VIRGINIA DEPARTMENT OF CONSERVATION AND RECREATION

DIVISION OF SOIL AND WATER CONSERVATION
203 Governor St., Suite 213
Richmond, VA 23219 USA
Phone: 804-786-1712
E-mail: pco@dcr.state.va.us
Website: www.dcr.state.va.us

Founded: NA
Scope: Statewide

Contact(s):
David Brickley, DIRECTOR
dordswc@erols.com

VIRGINIA DEPARTMENT OF CONSERVATION AND RECREATION

DIVISION OF STATE PARKS
203 Governor St., Suite 306
Richmond, VA 23219 USA
Phone: 804-692-0403　　Fax: 804-786-9294
Website: www.dcr.state.va.us

Founded: NA
Scope: Statewide

Publications: Virginia State Parks

Contact(s):
Joseph Elton, DIRECTOR
203 Governor St., Suite 306, Richmond, VA 23219
Phone: 804-786-4377

VIRGINIA DEPARTMENT OF CONSERVATION AND RECREATION

VIRGINIA CAVE BOARD
USA

Founded: NA

Contact(s):
Lawrence Smith, ADVISOR
Bill Keith, CHAIRMAN
Rt. 1 Box 17, Cleveland, VA 24225

VIRGINIA DEPARTMENT OF ENVIRONMENTAL QUALITY

629 E. Main St.
Richmond, VA 23219 USA
Phone: 804-698-4000　　Fax: 804-698-4500
E-mail: vanaturally@deq.state.va.us
Website: www.deq.state.va.us

Founded: 1993
Scope: State

Description: The Department of Environmental Quality strives to provide efficient, cost-effective services that promote a proper balance between environmental improvement and economic vitality.

Keyword(s): Exotic species, Aquatic nuisance species, Environment, Solid Waste Management, Air Quality and Pollution

Contact(s):
Dennis Treacy, DIRECTOR
Phone: 804-698-4020
dhtreacy@deq.state.va.us

Ann Regan, ENVIRONMENTAL EDUCATION COORDINATOR

VIRGINIA DEPARTMENT OF FORESTRY

900 Natural Resources Dr., Suite 800
Charlottesville, VA 22903 USA
Phone: 804-977-6555   Fax: 804-296-2369
Website: www.dof.state.va.us

Founded: 1914
Scope: Statewide

Description: The mission of the Department of Forestry is to protect and develop healthy, sustainable forest resources for Virginians. The Department assists private landowners with the management and protection of forest resources. We also provide at-cost seedlings for reforestation of the state's forestlands, and management of public state forests and other state public forest lands.

Contact(s):
Bettina Ring, DEPUTY STATE FORESTER
Faye Difazio, FISCAL DIRECTOR
Lou Southard, FOREST PROTECTION
Ellie Whinnery, HUMAN RESOURCES DIRECTOR
James Garner, STATE FORESTER
James Starr, TEAM LEADER FOR FOREST MANAGEMENT
Ronald Jenkins, TEAM LEADER FOR GENERAL SERVICES
Edwina Blalock, TEAM LEADER FOR INFORMATION TECHNOLOGY
Timothy Tigner, TEAM LEADER FOR RESOURCE INFORMATION

VIRGINIA DEPARTMENT OF GAME AND INLAND FISHERIES

1320 Belman Road
Fredericksburg, VA 22401 USA
Phone: 540-899-4169   Fax: 540-899-4381
E-mail: dgifweb@dgif.state.va.us
Website: www.dgif.state.va.us

Founded: NA
Scope: Statewide

VIRGINIA DEPARTMENT OF GAME AND INLAND FISHERIES

REGION 1
5806 Mooretown Rd.
Williamsburg, VA 23188 USA
Phone: 757-253-7072   Fax: 757-253-4182
Website: www.vgif.state.va.us

Founded: NA
Scope: Regional

Contact(s):
Glen Askins, WILDLIFE BIOLOGIST, REGIONAL MANAGER

VIRGINIA DEPARTMENT OF GAME AND INLAND FISHERIES

REGION II (LYNCHBURG)
1121 Thomas Jefferson Road
Forest, VA 24551-9223 USA
Phone: 804-525-7522   Fax: 804-525-7720
E-mail: www.dgif.state.va.us
Website: www.dgif.state.va.us

Founded: NA

Scope: Statewide

Description: To provide for the management, conservation, restoration, and enhancement of the Commonwealth's fish and wildlife resources. The department also provides boat registration and titling services and boating law administration and enforcement; as well as providing public informational and educational services related to wildlife resources and recreational boating

Publications: Virginia Wildlife

Contact(s):
Larry Hart, BOAT REGISTRATION SECTION, TITLE
Phone: 804-367-1295
lhart@dgif.state.va.us
Charles Sledd, BOATING LAW ADMINISTRATOR
Phone: 804-367-6481
csledd@dgif.state.va.us
James Adams, CAPITAL PROGRAMS DIRECTOR
William Woodfin, DIRECTOR
Phone: 804-367-9231
Fax: 804-367-0405
bwoodfin@dgif.state.va.us
Raymond Davis, DIRECTOR OF ADMINISTRATION
Phone: 804-367-2387
rdavis@dgif.state.va.us
Gary Martel, DIRECTOR OF FISHERIES DIVISION
Phone: 804-367-0509
gmartel@dgif.state.va.us
Larry Harizanoff, DIRECTOR OF HUMAN RESOURCES
Phone: 804-367-8195
lharizanoff@dgif.state.va.us
Virgil Kopf, DIRECTOR OF INFORMATION MANAGEMENT SYSTEMS
Phone: 804-367-0787
vkopf@dgif.state.va.us
Jeffrey Uerz, DIRECTOR OF LAW ENFORCEMENT DIVISION
Phone: 804-367-0776
juerz@dgif.state.va.us
Charles Sledd, DIRECTOR OF PROGRAM DEVELOPMENT
Phone: 804-367-6481
csledd@dgif.state.va.us
David Whitehurst, DIRECTOR OF WILDLIFE DIVERSITY DIVISION
Phone: 804-367-4335
dwhitehurst@dgif.state.va.us
Robert Duncan, DIRECTOR OF WILDLIFE DIVISION
Phone: 804-367-9588
rduncan@dgif.state.va.us
Lee Walker, EDITOR
Phone: 804-367-0486
Fax: 804-367-0488
lwalker@dgif.state.va.us
Fred Leckie, FEDERAL AID COORDINATOR OF FISHERIES
Phone: 804-367-8629
fleckie@dgif.state.va.us
Rick Busch, FEDERAL AID COORDINATOR OF WILDLIFE
Phone: 804-367-1215
rbusch@dgif.state.va.us
Julia Smith, MEDIA RELATIONS COORDINATOR
Phone: 804-367-0991
jsmith@dgif.state.va.us
Terry Bradberry, OUTDOOR EDUCATION

VIRGINIA DEPARTMENT OF GAME AND INLAND FISHERIES

REGION II (LYNCHBURG)
1121 Thomas Jefferson Rd.
Forest, VA 24551-9223 USA
Phone: 804-525-7522   Fax: 804-525-7720
Website: www.dgif.state.va.us

Founded: NA
Scope: Statewide

VIRGINIA DEPARTMENT OF GAME AND INLAND FISHERIES

REGION III
1796 Highway Sixteen
Marion, VA 24354 USA
Phone: 540-783-4860   Fax: 540-783-6115
E-mail: vjessee@dgif.state.va.us
Website: www.dgif.state.va.us

Founded: 1917
Scope: Statewide

Description: Coordination of wildlife management, fisheries management, and wildlife law enforcement in region.

VIRGINIA DEPARTMENT OF GAME AND INLAND FISHERIES

REGION IV (STAUNTON)
Verona, VA 24482 USA
Phone: 540-248-9360   Fax: 540-248-9399
Website: www.dgif.state.va.us

Founded: NA
Membership: 25
Scope: Regional

VIRGINIA DEPARTMENT OF HEALTH

Commissioners Office, Suite 214, Main St. Station
Richmond, VA 23219 USA
Phone: 804-786-3561   Fax: 804-786-4616
Website: www.vdh.state.va.us

Founded: 1872
Scope: Statewide

Description: The Department carries out protective and preventive public health services for all citizens of the Commonwealth and provides public health care services to the indigent.

Publications: Virginia's Health

Contact(s):
Anne Peterson, COMMISSIONER (M.D., M.P.H.)
Helen Tarantino, DEPUTY COMMISSIONER OF ADMINISTRATION

VIRGINIA DEPARTMENT OF MINES, MINERALS AND ENERGY

Ninth St. Office Bldg., 8th Fl., 202 N. Ninth St.
Richmond, VA 23219 USA
Phone: 540-523-8119   Fax: 804-692-3237
Website: www.mme@state.va.us

Founded: 1985
Membership: 240
Scope: Regional

Description: The department is committed to enhancing the development and conservation of energy and mineral resources in a safe and environmentally sound manner in order to support a more productive economy in Virginia.

Contact(s):
O. Dishner, DIRECTOR

VIRGINIA DEPARTMENT OF MINES, MINERALS AND ENERGY

DIVISION OF ENERGY
USA

Founded: NA

Description: The Division of Energy promotes the efficient use and conservation of energy and the use of alternative energy sources.

Contact(s):
Stephen Walz, DIRECTOR
Ninth St. Office Bldg., 8th Fl., 202 N. Ninth St., Richmond, VA 23219
Phone: 540-692-3211

VIRGINIA DEPARTMENT OF MINES, MINERALS AND ENERGY

DIVISION OF GAS AND OIL
Abingdon, VA 24210 USA
Phone: 540-676-5423   Fax: 540-676-5459
Website: www.mme.state.va.us

Founded: NA
Membership: 9
Scope: Statewide

Description: The Division of Gas and Oil regulates the operation and reclamation of gas and oil extractions.

Contact(s):
Bob Wilson, DIRECTOR

VIRGINIA DEPARTMENT OF MINES, MINERALS AND ENERGY

DIVISION OF MINED LAND RECLAMATION
Drawer 900
Big Stone Gap, VA 24219 USA
Phone: 540-523-8100   Fax: 540-523-8148
E-mail: dmmeinfo@mme.state.va.us
Website: www.mme.state.va.us

Founded: NA
Scope: Statewide

Description: The Division of Mined Land Reclamation regulates the operation of coal surface-mining activities, enforces the reclamation laws and regulations, and administers financial resources for reclaiming abandoned coal mining sites.

VIRGINIA DEPARTMENT OF MINES, MINERALS AND ENERGY

DIVISION OF MINERAL MINING
P.O. Box 900
Big Stone Gap, VA 24219 USA
Phone: 80-495-1631
E-mail: dmmeinfo@mme.state.va.us
Website: http://www.mme.state.va.us

Founded: NA

Scope: Statewide

Description: The Division of Mineral Mining regulates the operation of noncoal mining activities for environmental protection and worker safety.

Contact(s):
Conrad Spangler, DIRECTOR
P.O. Box 3727, Charlottesville, VA 22903
Phone: 804-951-6310

VIRGINIA DEPARTMENT OF MINES, MINERALS AND ENERGY
DIVISION OF MINERAL RESOURCES
P.O. Box 3667
Charlottesville, VA 22903 USA
Phone: 804-951-6340   Fax: 804-951-6365
Website: www.@geology.state.va.us

Founded: NA
Scope: Statewide

Description: The Division of Mineral Resources provides information on Virginia's geology, mineral resources, and physical and cultural features

Publications: see publications on website

VIRGINIA DEPARTMENT OF MINES, MINERALS AND ENERGY
DIVISION OF MINES
Big Stone Gap, VA 24219 USA
Phone: 540-523-8224   Fax: 540-523-8239
Website: www.mne.state.va.us

Founded: NA
Scope: Statewide

Description: The Division of Mines enforces the coal mining laws of the Commonwealth to promote the safety and health of coal miners.

Contact(s):
Frank Linkous, CHIEF
P.O. Drawer 900, Big Stone Gap, VA 24219
Phone: 540-523-8100

VIRGINIA MARINE RESOURCES COMMISSION
2600 Washington Ave.
Newport News, VA 23607 USA
Phone: 757-247-2200   Fax: 757-247-2020
Website: www.state.va.us/vmrc

Founded: 1875
Membership: 141
Scope: Statewide

Description: This state agency holds regulatory jurisdiction over all commercial and sports fishing, marine fish, marine shellfish, and marine organisms in the tidal waters of Virginia. Holds permit jurisdiction on all projects involving use of state-owned submerged lands and authority over use or development in vegetated and nonvegetated tidal wetlands and coastal primary sand dunes.

Publications: Virginia Landings Bulletin

Contact(s):
Robert Craft, CHIEF OF ADMINISTRATION AND FINANCE
Jim Wesson, CHIEF OF CONSERVATION AND REPLENISHMENT
Jack Travelstead, CHIEF OF FISHERIES MANAGEMENT
Steven Bowman, CHIEF OF LAW ENFORCEMENT
Erik Barth, CHIEF OF MANAGEMENT INFORMATION SYSTEMS
Robert Grabb, CHIEF, HABITAT MANAGEMENT
William Pruitt, COMMISSIONER

VIRGINIA MUSEUM OF NATURAL HISTORY
1001 Douglas Ave.
Martinsville, VA 24112 USA
Phone: 540-666-8600   Fax: 276-632-6487
Website: www.vmnh.org

Founded: 1988
Membership: 35
Scope: Statewide

Description: Preserves, studies, and interprets Virginia's natural and cultural heritage. Statewide system of museum facilities, research sites, and educational programs. The museum has more than eleven million specimens in collections.

Publications: Virginia Naturally, Books, Scientific Publication Series, Children's Activity Books, Virginia Explorer, The

Keyword(s): Research, Geology, Endangered Species, Environmental and Conservation Education, Museum, Biodiversity

Contact(s):
Stephen Pike, EXECUTIVE DIRECTOR
spike@ngocomm.net
Judy Winston, RESEARCH DIRECTOR
Phone: 540-666-8609
jwinston@vmnh.org

VIRGINIA OUTDOORS FOUNDATION
203 Governor St., Suite 317
Richmond, VA 23219 USA
Phone: 804-225-2147   Fax: 804-371-4810
E-mail: ldtrew@virginiaoutdoorsfoundation.org
Website: www.virginiaoutdoorsfoundation.org

Founded: NA
Scope: Statewide

Description: To preserve Virginia's natural scenic, historic, scientific, open space, and recreational areas by means of private philanthropy. The Foundation accepts gifts of cash, stock, real property, or open spaces easements to achieve its purpose.

Keyword(s): Open Space, Land Preservation, Conservation

Contact(s):
Tamara Vance, EXECUTIVE DIRECTOR
Phone: 540-951-2822
Fax: 540-951-2695

VIRGINIA SEA GRANT PROGRAM
Virginia Graduate Marine Science Consortium, 170 Rugby Rd., Madison House, University of Virginia
Charlottesville, VA 22903 USA
Phone: 804-924-5965   Fax: 804-982-3694
Website: www.virginia.edu/virginia-sea-grant/

Founded: NA
Scope: statewide

Keyword(s): Wetlands, Renewable Resources, Coasts, Environmental and Conservation Education, Aquatic Habitats

Contact(s):
William Rickards, DIRECTOR
rickards@virginia.edu
William Dupaul, STAFF OF MARINE ADVISORY PROGRAM
Virginia Institute of Marine Science, Gloucester Point, VA
23062
Phone: 804-684-7163

VIRGINIA SOIL AND CONSERVATION BOARD

7293 Hanover Green Dr., Suite B-101
Mechanicsville, VA 23111 USA
Phone: 804-559-0324   Fax: 804-559-0325
E-mail: vaswcd@erols.com
Website: www.vaswcd@erols.com

Founded: NA
Scope: Regional

Contact(s):
Jack Frye, ADVISOR
Charles Horn, CHAIRMAN
203 Governor St., Suite 206, Richmond, VA 23219
Phone: 804-786-2064
Stephanie Martin, EXCUTIVE DIRECTOR
Mechanicville, VA 23111
Phone: 804-786-3914
Fax: 804-786-1798

W

WASHINGTON COOPERATIVE FISH AND WILDLIFE RESEARCH UNIT (USDI)

SCHOOL OF AQUATIC & FISHERY SCIENCES
Box 355020 University of Washington
Seattle, WA 98195 5020 USA
Phone: 206-543-6475   Fax: 206-616-9012
E-mail: washcoop@u.washington.edu
Website: www.fish.washington.edu/wacfwru

Founded: 1988
Scope: National

Description: The goals of the WCFWRU are: (1) conduct
research in support of the Department of the Interior and
Washington State; (2) train graduate students in fisheries and
wildlife science through research support and by teaching; and
(3) disseminate research results to the scientific community,
management agencies, and the general public.

Keyword(s): training, Toxicology, Biodiversity, Wildlife, Aquatic
Habitats

Contact(s):
Glenn Vanblaricom, ASSISTANT LEADER OF WILDLIFE
David Beauchamp, ASSISTANT UNIT LEADER OF FISHERY
Christian Grue, LEADER

WASHINGTON DEPARTMENT OF AGRICULTURE

Olympia, WA 98504-2560 USA
Phone: 360-902-1800
Website: www.wa.gov/agr

Founded: NA
Scope: Statewide

Keyword(s): Pesticides, Agriculture

Contact(s):
Bob Gore, ASSISTANT DIRECTOR OF COMMODITY
INSPECTION
Phone: 360-902-1827

John Daley, ASSISTANT DIRECTOR OF FOOD SAFETY
AND ANIMAL HEALTH DIVISION
Mary Toohey, ASSISTANT DIRECTOR OF LABORATORY
SERVICES
Phone: 360-902-1907
Bob Arrington, ASSISTANT DIRECTOR OF PESTICIDE
MANAGEMENT DIVISION
Phone: 360-902-2011
Bill Brookreson, DEPUTY DIRECTOR
Phone: 360-902-1810
Jim Jesernig, DIRECTOR
Phone: 360-902-1801
Linda Waring, INFORMATION OFFICER
Phone: 360-902-1815
Robert Mead, STATE VETERINARIAN
Phone: 360-902-1881

WASHINGTON DEPARTMENT OF ECOLOGY

P.O. Box 47600
Olympia, WA 98504-7600 USA
Phone: 360-407-6000
Website: www.ecy.wa.gov

Founded: 1970

Description: Charged with programs of air quality control, water
pollution control, solid waste management, management of
water resources, hazardous waste management, reduction,
and cleanup, shoreline management, coastal zone
management, and State Environmental Policy Act (SEPA).

Contact(s):
Carol Fleskes, ADMINISTRATIVE SERVICES MANAGER
Phone: 360-407-7012
Bill Alkire, ASSISTANT DIRECTOR OF LEGISLATIVE &
INTERGOVERNMENTAL RELATIONS
Phone: 360-407-7003
David Mears, ATTORNEY GENERAL OF OFFICE OF
ATTORNEY GENERAL
Phone: 360-459-6158
Nancy Stevenson, CHIEF FINANCIAL OFFICER
Phone: 360-407-7005
Dan Silver, DEPUTY DIRECTOR
Phone: 360-407-7011
Fax: 360-407-6989
Tom Fitzsimmons, DIRECTOR
Phone: 360-407-7001
Fax: 360-407-6989
tfit@461.us.gov
Phyllis Shafer, LIBRARIAN
Phone: 206-407-6150
Sandy Howard, PUBLIC INFORMATION OFFICER

WASHINGTON DEPARTMENT OF ECOLOGY

CENTRAL REGIONAL OFFICE
15 West Yakima Ave., Suite 200
Yakima, WA 98902 USA
Phone: 509-575-2490   Fax: 509-575-2809
Website: www.ecy.wa.gov

Founded: NA
Scope: Statewide

Contact(s):
Polly Zehm, STAFF

WASHINGTON DEPARTMENT OF ECOLOGY

EASTERN REGIONAL OFFICE
4601 North Monroe
Spokane, WA 99205 USA
Phone: 509-456-2926　　　Fax: 509-456-6175
Website: www.ecy.wa.gov

Founded: NA
Scope: Statewide

Contact(s):
Tony Grover, STAFF
Phone: 509-456-6149
agro461@ecy.wa.gov

WASHINGTON DEPARTMENT OF ECOLOGY

NORTHWEST REGIONAL OFFICE
3190 160th Ave., SE
Bellevue, WA 98008 USA
Phone: 425-649-7000　　　Fax: 425-649-7098
Website: www.ecy.wa.gov

Founded: NA
Scope: Regional

Contact(s):
Ray Hellwig, REGIONAL DIRECTOR
Phone: 425-649-7010

WASHINGTON DEPARTMENT OF ECOLOGY

SOUTHWEST REGIONAL OFFICE
P.O. Box 47775
Olympia, WA 98504-7775 USA
Phone: 360-407-6300　　　Fax: 360-407-6305
Website: www.ecy.wa.gov

Founded: NA
Membership: 180
Scope: Regional

Contact(s):
Sue Mauermann, REGIONAL DIRECTOR
Phone: 360-407-6307
smau461@ecy.wa.gov

WASHINGTON DEPARTMENT OF NATURAL RESOURCES

P.O. Box 47001
Olympia, WA 98504-7001 USA
Phone: 360-902-1000　　　Fax: 360-902-1775
Website: www.wa.gov/enr

Founded: NA
Membership: 2000
Scope: Regional
Publications: DNR News

Contact(s):
Fran McNair, AQUATIC STEWARD
Terry Kirkpatrick, BUSINESS SYSTEM SUPPORT
Phone: 360-902-1600
Doug Sutherland, COMMISSIONER
Jennifer Belcher, COMMISSIONER OF PUBLIC LANDS
Phone: 360-902-1004
Fax: 360-902-1775
Sue Zemek, COMMUNICATIONS
Phone: 360-902-1023
Julie Boyer, DEPARTMENT SUPERVISOR
Phone: 360-902-1034

Kaleen Cottingham, DEPUTY COMMISSIONER
Phone: 360-902-1003
Bonnie Bunning, EXCUTIVE DIRECTOR
Michelle Benton, EXECUTIVE ASSISTANT
Phone: 360-902-1004
Pat McElroy, EXECUTIVE DIRECTOR
Connie Manson, GEOLOGY LIBRARY MANAGER
Phone: 360-902-1472
Bruce McKay, LAND STEWARD
Bill Boyum, MANAGER OF AGRICULTURAL RESOURCES DIVISION (ACTING)
Phone: 360-902-1130
Chuck Turley, MANAGER OF AQUATIC RESOURCES DIVISION
Phone: 360-902-1100
Meg Grimaldi, MANAGER OF EMPLOYEE SERVICES DIVISION
Phone: 360-902-1150
Tony Ifie, MANAGER OF ENGINEERING DIVISION
Phone: 360-920-1200
Loren Stern, MANAGER OF FINANCIAL MANAGEMENT DIVISION
Phone: 360-902-1250
Catherine Elliott, MANAGER OF FOREST PRACTICES DIVISION
Phone: 360-902-1400
Ray Lasmanis, MANAGER OF GEOLOGY AND EARTH RESOURCES DIVISION
Phone: 360-902-1450
Al Bloomberg, MANAGER OF INFORMATION TECHNOLOGY DIVISION
Phone: 360-902-1500
Randy Acker, MANAGER OF RESOURCE PROTECTION DIVISION
Phone: 360-902-1300

WASHINGTON DEPARTMENT OF NATURAL RESOURCES

CENTRAL REGION
1405 Rush Rd.
Chelias, WA 98532 USA
Phone: 360-748-2383　　　Fax: 360-748-2387
Website: www.wadnr.gov

Founded: NA
Membership: 150
Scope: Regional

Contact(s):
Vicky Christenson, CENTRAL REGION MANAGER

WASHINGTON DEPARTMENT OF NATURAL RESOURCES

NORTHEAST REGION
P.O. Box 47001
Olympia, WA 98504 USA
Phone: 360-902-1000　　　Fax: 360-902-1775
E-mail: information@wadnr.gov

Founded: NA
Scope: Statewide

WASHINGTON DEPARTMENT OF NATURAL RESOURCES

NORTHWEST REGION
919 N Township St.
Sedro Woolley, WA 98284 USA
Phone: 360-856-3500   Fax: 360-856-2150
Website: www.wadnr.gov\

Founded: NA
Membership: 150
Scope: Local

Contact(s):
Bill Wallace, REGIONAL MANAGER
Phone: 360-856-3500

WASHINGTON DEPARTMENT OF NATURAL RESOURCES

OLYMPIC REGION
411 Tillicum Ln.
Forks, WA 98331 USA
Phone: 360-374-6131   Fax: 360-374-5446

Founded: NA
Scope: Statewide

Contact(s):
Tom Robinson, REGIONAL MANAGER
Phone: 360-374-6131

WASHINGTON DEPARTMENT OF NATURAL RESOURCES

SOUTH PUGET SOUND REGION
P.O. Box 68
Enumclaw, WA 98022 USA
Phone: 360-825-1631   Fax: 360-825-1672

Founded: NA
Scope: Statewide

Contact(s):
Bonnie Bunning, STAFF
Phone: 360-825-1631

WASHINGTON DEPARTMENT OF NATURAL RESOURCES

SOUTHEAST REGION
713 Bowers Road
Ellensburg, WA 98926 USA
Phone: 509-925-8510

Founded: NA
Scope: Statewide

WASHINGTON DEPARTMENT OF NATURAL RESOURCES

SOUTHWEST REGION
P.O. Box 280
Castle Rock, WA 98611-0280 USA
Phone: 360-577-2025   Fax: 360-274-4196

Founded: NA
Scope: Statewide

Contact(s):
Rick Cooper, STAFF
Phone: 360-577-2025

WASHINGTON NATURAL HERITAGE PROGRAM

P.O. Box 47014
Olympia, WA 98504-7014 USA
Phone: 360-902-1600   Fax: 360-902-1789
Website: www.wa.gov/dnr/

Founded: 1978
Scope: Statewide

Description: Identify and evaluate native ecosystems and species, set conservation priorities, provide information to protect these irreplaceable resources for the benefit of current and future generations.

Publications: State of Washington Natural Heritage Plan, Endangered, Threatened, and Sensitive Vascular Plants of Washington with working list of Rare Non-vascular Species

Keyword(s): Botany, Zoology, Biodiversity

Contact(s):
John Gamon, PROGRAM MANAGER AND BOTANIST
Phone: 360-902-1661

WASHINGTON SEA GRANT PROGRAM

3716 Brooklyn Ave., NE
Seattle, WA 98105-6716 USA
Phone: 206-543-6600   Fax: 206-685-0380
E-mail: seagrant@u.washington.edu
Website: www.wsg.washington.edu/

Founded: NA
Scope: Statewide

Description: Since 1968, Washington Sea Grant Program has supported research, advisory, and communication activities for the benefit of marine resources, users, and communities. It is part of a national network of universities meeting the changing environmental and economic needs of people in our coastal and Great Lakes regions.

Publications: Ocean Ecology of North Pacific Salmonids, Guide to Manila Clam Culture in Washington, Shape and Form of Puget Sound, The, El Nino North: Nino Effects in the Eastern Subarctic Pacific Ocean

Keyword(s): Wetlands, Sustainable Development, Wildlife, Oceanography, Biotechnology

Contact(s):
Andrea Copping, ASSITANT DIRECTOR
Phone: 206-685-8209
Nancy Blanton, COMMUNICATIONS MANAGER
Phone: 206-685-9215
Louie Echols, DIRECTOR
echols@u.washington.edu
Susan Cook, PUBLICATIONS COORDINATOR/WEB MASTER
Phone: 206-685-2606
Fax: 206-685-0380

WASHINGTON STATE CONSERVATION COMMISSION

P.O. Box 47721
Olympia, WA 98504-7721 USA
Phone: 206-407-6200   Fax: 206-407-6215

Founded: 1939

Description: Assists, guides, and coordinates the programs of 48 conservation districts and encourages the cooperation and collaboration of the federal, state, regional, interstate, and local public agencies which assist them; keeps the public informed of renewable natural resource conservation activities.

Keyword(s): Exotic species, Aquatic nuisance species, Soil Conservation, Environmental and Conservation Education, Renewable Resources, Agriculture

Contact(s):
Chuck Bagley, CENTRAL WA FIELD REPRESENTATIVE
Phone: 509-664-3154
Fax: 509-665-3366
Ronald Juris, CHAIR
P.O. Box 157, Bickleton, WA 98322-0157
William Broughton, EASTERN WA FIELD REPRESENTATIVE
Phone: 509-397-4740
Fax: 509-397-4921
Steven Meyer, EXECUTIVE DIRECTOR
Phone: 206-407-6201
Phone: 360-407-6215
smey461@ecy.wa.gov
Robert Bottman, GRANTS OFFICER
Phone: 360-407-6204
Stu Trefry, PUGET SOUND FIELD REPRESENTATIVE
Phone: 360-407-6211

WASHINGTON STATE EXTENSION SERVICES

Director, WA State U., P.O. Box 646230
Pullman, WA 99164 USA
Phone: 509-335-2933   Fax: 509-335-2926
Website: www.ext.wsu.edu/

Founded: NA
Scope: Statewide

Description: Washington State University Cooperative Extension helps people develop leadership skills and use research-based knowledge to improve their economic status and quality of life.

Keyword(s): Sustainable Ecosystems, Renewable Resources, Environmental and Conservation Education, Forests and Forestry, Agriculture

Contact(s):
Michael Tate, ASSOCIATE DEAN AND ASSOCIATE DIRECTOR OF EXTENSION SERVICES
David Baumgartner, EXTENSION FORESTER
Department of Natural Resource Sciences, P.O. Box 646410, Washington State University, Pullman, WA 99164-6410
Phone: 509-335-2964
Donald Hanley, EXTENSION FORESTER
College of Forest Resources, University of Washington, Box 352100, Seattle, WA 98195-2100
Phone: 206-685-4960
John Munn, EXTENSION NATURALIST
Washington State University CES, 600 128th St., SE, Everett, WA 98208
Phone: 425-388-2400
Fax: 425-338-3994
Sheila Gray, EXTENSION URBAN AND COMMUNITY HORTICULTURE (INTERIM)
Washington State University, 207 4th Ave. N., Kelso, WA 98262-4124
Phone: 360-577-3014
Edward Adams, PROGRAM LEADER
Washington State University Cooperative Extension, 668 N. Riverpoint Blvd., Box B, Spokane, WA 99202-1662
Phone: 509-358-7960
Fax: 509-358-7900
adamse@wsu.edu

WASHINGTON STATE OFFICE OF ENVIRONMENTAL EDUCATION

Office of Superintendent of Public Instruction, 2800 NE 200th St.
Seattle, WA 98155-1418 USA
Phone: 206-365-3893   Fax: 206-367-4540
E-mail: wsoee@earthlink.net
Website: www.k12.wa.us\envedu

Founded: NA
Membership: 2
Scope: Regional

Description: To provide curriculum resources and training for teachers in environmental education, and to evaluate these programs pursuant to improving content and effectiveness. The office is responsible for E.E. program coordination and cooperation as it applies to K-12 public school programs and to state mandate requiring E.E. integrated into the K-12 curriculum.

Publications: Environmental Education Guidelines For Washington Schools, Energy Food & You, Puget Sound Habitats Teachers Guide and Charts, Closing the Achievement Gap: Using the Environment as an Integrating Context for Learning, Clean Water, Streams and Fish: A Holistic View of Watersheds

Keyword(s): training, Water Pollution Management, Solid Waste Management, Sustainable Ecosystems, Environmental and Conservation Education

Contact(s):
Michele Halfhill, ADMINISTRATIVE ASSISTANT
Tony Angell, STATE SUPERVISOR OF ENVIRONMENTAL EDUCATION

WATER RESOURCES RESEARCH CENTER

University of Hawaii, Holmes Hall 283, 2540 Dole St.
Honolulu, HI 96822 USA
Phone: 808-956-7847   Fax: 808-956-5044
E-mail: moraz@hawaii.edu
Website: www.hawaii.edu/moraz/wrcc.html

Founded: 1964
Scope: Statewide

Description: WRRC's mission is to coordinate and conduct research to identify, characterize and quantify water and environmental concerns of the state, the nation and other Pacific Islands and formulate methods for resolving these concerns. WRRC produces reports, national and international journal articles, books, newsletters and project bulletins and organizes seminars, workshops and conferences.

Publications: Technical Memorandum Report, Annual Report, Cooperative Report, Publications List, Project Reports and Special Publications, Technical Report

Keyword(s): Water Pollution Management, Oceanography, Exotic species, Aquatic nuisance species, Aquatic Habitats, Public Health Protection

Contact(s):
Philip Moravcik, COMMUNICATIONS COORDINATOR
James Moncur, DIRECTOR

WEST VIRGINIA COOPERATIVE FISH AND WILDLIFE RESEARCH UNIT

DIVISION OF FORESTRY
Division of Forestry, West Virginia University
Morgantown, WV 26506-6125 USA
Phone: 304-293-3794   Fax: 304-293-4826
E-mail: wvcoop@wvu.edu
Website: www.forestry.caf.wvu.edu

Founded: NA
Scope: State, National

Description: A cooperative research and graduate education organization sponsored by the Biological Resources Division of USGS, West Virginia Division of Natural Resources, West Virginia University, and Wildlife Management Institute. The role of the unit is to conduct natural resources research of state, regional, or national scope, and to train graduate-level researchers in natural resources.

Keyword(s): training, Rivers, Toxicology, Wildlife, Nongame Wildlife, Aquatic Habitats

Contact(s):
Stuart Welsh, ASSISTANT LEADER: FISHERIES
Petra Wood, ASSISTANT LEADER: WILDLIFE
Joseph McNeel, DIRECTOR DIVISION OF FORESTRY
jmcneel@wvu.edu
Patricia Mazik, UNIT LEADER: FISHERIES

WEST VIRGINIA DEPARTMENT OF AGRICULTURE

1900 Kanawha Blvd. E
Charleston, WV 25305 USA
Phone: 304-558-2201   Fax: 304-558-2203
Website: www.state.wv.us/agriculture

Founded: NA
Scope: State

Keyword(s): Exotic species, Aquatic nuisance species, Soil Conservation, Pesticides, Public Health Protection, Agriculture

Contact(s):
Gus Douglass, COMMISSIONER
Richard Hannah, DEPUTY COMMISSIONER
Janet Fisher, DEPUTY COMMISSIONER
Charles Coffman, DIRECTOR OF PLANT INDUSTRIES DIVISION

WEST VIRGINIA DEPARTMENT OF ENVIRONMENTAL PROTECTION

Division of Environmental Protection, #10, McJunkin Rd.
Nitro, WV 25143-2546 USA
Phone: 304-759-0515   Fax: 304-759-0562
Website: www.dp.state.wv.us

Founded: 1991
Membership: 800
Scope: Statewide

Description: The Division of Environmental Protection is charged with the protection of West Virginia's environment through the regulation and administration of the state's abandoned mine lands, air quality, mining & reclamation, oil & gas, waste management, and water resources programs.

Contact(s):
Andy Gallagher, CHIEF COMMUNICATIONS OFFICER
Phone: 304-558-4253
agallagher@mail.dep.state.wv.us
Randy Huffman, CHIEF OF ADMINISTRATION
Phone: 304-558-5529
rhuffman@mail.dep.state.wv.us
Perry McDaniel, CHIEF OF LEGAL SERVICES
Phone: 304-558-9160
pmcdaniel@mail.dep.state.wv.us
Matthew Crum, CHIEF OF MINING & RECLAMATION
Phone: 304-759-0510
mcrum@mail.dep.state.wv.us
Ken Ellison, CHIEF OF WASTE MANAGEMENT
Phone: 304-558-2508
kellison@mail.dep.state.wv.us
Alan Turner, CHIEF OF WATER RESOURCES
Phone: 304-558-2107
aturner@mail.dep.state.wv.us
Michael Callaghan, DIRECTOR
mcallaghan@mail.dep.state.wv.us

WEST VIRGINIA DIVISION OF NATURAL RESOURCES

1900 Kanawha Blvd., East Building 3
Charleston, WV 25305 USA
Phone: 304-558-2754   Fax: 304-558-3147
Website: www.dnr.state.wv.us

Founded: 1933
Scope: State

Description: The Division's objective is to provide a comprehensive program for the exploration, conservation, development, protection, enjoyment, and use of the natural resources of the state of West Virginia. The commission was the forerunner of the Department of Natural Resources, created by the legislature in 1961 and modified to the Division of Natural Resources in 1993.

Publications: Wonderful West Virginia

Contact(s):
Walt Kordek, ASSISTANT CHIEF IN CHARGE OF BIOMETRICS AND PLANNING
Phone: 304-637-0245
Michael Shingleton, ASSISTANT CHIEF IN CHARGE OF COLDWATER FISHERIES
Phone: 304-637-0245
Paul Johansen, ASSISTANT CHIEF IN CHARGE OF GAME MANAGEMENT
Phone: 304-558-2771
pjohans@dnr.state.wz.us
Donald Phares, ASSISTANT CHIEF IN CHARGE OF SPECIAL PROJECTS
Phone: 304-637-0245
Bert Pierce, ASSISTANT CHIEF IN CHARGE OF WARMWATER FISHERIES
Phone: 304-558-2771
bpierce@bnr.state.wz.us
Emily Fleming, CHIEF OF ENVIRONMENTAL RESOURCES
James Fields, CHIEF OF LAW ENFORCEMENT
Phone: 304-558-2784
John R. Pope, CHIEF OF PARKS
Curtis Taylor, CHIEF OF WILDLIFE
Bernard Dowler, CHIEF OF WILDLIFE RESOURCES
Phone: 304-558-2771
Fax: 304-558-3147
bdowler@dnr.state.wz.us
Emily Fleming, CONSERVATION EDUCATION/LITTER CONTROL
Phone: 304-558-3370

W. Daniel, DEPUTY CHIEF, LAW ENFORCEMENT
Phone: 304-558-2784
Ken Caplinger, DEPUTY CHIEF, PARKS AND RECREATION
Phone: 304-558-2764
Gordon Robertson, DEPUTY CHIEF, WILDLIFE
RESOURCES
Phone: 304-558-2771
grobert@dnr.state.wz.us
Edward Hamrick, DIRECTOR
David Milne, DISTRICT 11 COMMISSIONER
Charles Capito, DISTRICT I COMMISSIONER
Suite #3 2619 Pennsylvania Ave., Weirton, WV 36062
Phone: 304-723-3355
Jeffrey Bowers, DISTRICT I COMMISSIONER
HC 70 Box 40 A, Sugar Grove, WV 26815
Phone: 304-358-3333
Carl Gainer, DISTRICT II COMMISSIONER
P.O. Box 670, Richwood, WV 26261
Phone: 304-846-6247
Charles Hooten, DISTRICT II COMMISSIONER
1570 Summit Drive, Charleston, WV 25302
Phone: 304-346-0521
Thomas Homan, DISTRICT II COMMISSIONER
1410 Bedford Rd., Charleston, WV 25314
Phone: 304-346-3330
Arnout Hyde, EDITOR
Phone: 304-558-9152
Harry Price, EXECUTIVE SECRETARY
Phone: 304-558-3315
hprice@dnr.state.wz.us
Thomas Dotson, NATURAL RESOURCES COMMISSION
Twila Matheney, NATURAL RESOURCE COMMISSION
Hoy Murphy, PUBLIC INFORMATION OFFICER
Phone: 304-558-3380
James Jones, REAL ESTATE MANAGEMENT
Phone: 304-558-3225
Fax: 304-558-3680

WEST VIRGINIA GEOLOGICAL AND ECONOMIC SURVEY

Box 879
Morgantown, WV 26507-0879 USA
Phone: 304-594-2331   Fax: 304-594-2575
E-mail: info@geosrv.wvnet.edu
Website: www.wvgs.wvnet.edu

Founded: 1897
Membership: 55
Scope: State, International

Description: Charged with the responsibility of examining all geological formations and physical features of the state with particular emphasis on their economic importance, utilization, and conservation and preparing reports and maps of the geology and natural resources of West Virginia.

Publications: Maps, Reports of investigations, Coal-geology bulletins, Environmental geology bulletins, Mineral resources series, River basin bulletins, basic data reports, county geologic reports, educational series, state park bulletins, field trip guide

Contact(s):
Carl Smith, ASSOCIATE STATE GEOLOGIST AND DEPUTY DIRECTOR
Phone: 304-594-2331
John May, DEPUTY DIRECTOR OF FINANCE AND ADMINISTRATION

Larry Woodfork, DIRECTOR AND STATE GEOLOGIST
Phone: 304-594-2331
woodfork@geoserv.wvnet.edu
Chuck Gover, EDITOR
Phone: 304-594-2331
Gloria Rowan, PROGRAM MANAGER FOR
ADMINISTRATION
Nick Fedorko, PROGRAM MANAGER FOR COAL
Mary Behling, PROGRAM MANAGER FOR GEOLOGIC
DATA
Katherine Avary, PROGRAM MANAGER FOR OIL AND GAS,
Charles Gover, PROGRAM MANAGER FOR PUBLICATIONS
AND GRAPHICS SECTION
Steven McClelland, PROGRAM MANAGER FOR SERVICE

WEST VIRGINIA SOIL CONSERVATION AGENCY

WEST VIRGINIA SOIL CONSERVATION AGENCY
1900 Kanawha Blvd. East
Charleston, WV 25305-0193 USA
Phone: 304-558-2204   Fax: 304-340-4839
Website: www.wvsca.org

Founded: NA
Membership: 65
Scope: State

Contact(s):
Lance Tabor, EXECUTIVE DIRECTOR
Phone: 304-558-2204
Fax: 304-558-1635
taborl@wvlc.wvnet.edu

WEST VIRGINIA UNIVERSITY EXTENSION SERVICE

PO BOX 6031 KNAPP HALL
West Virginia University, 817 Knapp Hall, P. O. Box 6031
Morgantown, WV 26506 USA
Phone: 304-293-5691   Fax: 304-293-7163
Website: www.wvu.edu/~exten/

Founded: NA
Membership: 500
Scope: Statewide

Keyword(s): Exotic species, Aquatic nuisance species, Renewable Resources, Forests and Forestry, Pesticides, Agriculture

Contact(s):
Edmond Collins, ASSOCIATE DIRECTOR, CENTER FOR
AGRICULTURE AND NATURAL RESOURCES
DEVELOPMENT
West Virginia University, 2078 Agricural Sciences Bldg.,
P.O. Box 6108, Morgantown, WV 26506-6108
Phone: 304-293-6131
ecollin2@wvu.edu
Lawrence Cote, ASSOCIATE PROVOST FOR EXTENSION
AND PUBLIC SERVICE
West Virginia University, 817 Knapp Hall, P.O. Box 6031,
Morgantown, WV 26506-6125
Phone: 304-293-5691
lcote@wvu.edu
Richard Zimmerman, DIRECTOR, CENTER FOR
AGRICULTURAL AND NATURAL RESOURCES
DEVELOPMENT
West Virginia University, 2080 Agricultural Science Bldg., P.O.
Box 6108, Morgantown, WV 26506-6108
Phone: 304-293-6131
rzimmerm@wvu.edu

Jeffrey Skousen, EXTENSION SPECIALIST, LAND RECLAMATION
West Virginia University, 1106 Agricultural Science Bldg., Morgantown, WV 26506-6108
Phone: 304-293-6131
jskousen@wvu.edu
Thomas Basden, EXTENSION SPECIALIST, NUTRIENT MANAGEMENT
West Virginia University, 1058 Agricultural Sciences Bldg., P.O. Box 6108, Morgantown, WV 26506-6108
Phone: 304-293-6131
tbasden2@wvu.edu
D. Bhumbla, EXTENSION SPECIALIST, SOIL AND WATER RESOURCES
West Virginia University, 1072 Agricultural Sciences Bldg., P.O. Box 6108, Morgantown, WV 26506-6108
Phone: 304-293-6131
dbhumbla@wvu.edu
William Grafton, EXTENSION SPECIALIST, WILDLIFE
West Virginia University, 311-B Percival Hall, P.O Box 6125, Morgantown, WV 26506-6125
Phone: 304-293-4797
Fax: 304-293-7553
wgrafton@wvu.edu

WILDLIFE RESOURCES AGENCY

P.O. Box 40747, Ellington Agricultural Center
Nashville, TN 37204 USA
Phone: 615-781-6500　　Fax: 615-781-6654
Website: www.state.tn.us/twra

Founded: 1949
Scope: Regional

Description: Created to have full and exclusive jurisdiction of the duties and functions relating to wildlife and boating and to the management, protection, propagation, and conservation of wildlife, including hunting and fishing.

Publications: Tennessee Wildlife

Contact(s):
Ron Fox, ASSISTANT DIRECTOR OF FIELD OPERATIONS
Phone: 615-781-6557
Allen Gebhardt, ASSISTANT DIRECTOR OF STAFF OPERATIONS
Phone: 615-781-6555
Ken Tarkington, CHIEF OF ADMINISTRATIVE SERVICES DIVISION
Phone: 615-781-6512
Ed Carter, CHIEF OF BOATING DIVISION
Phone: 615-781-6682
Les Haun, CHIEF OF ENGINEERING DIVISION
Phone: 615-781-6646
David McKinney, CHIEF OF ENVIRONMENTAL DIVISION
Phone: 615-781-6643
Bill Reeves, CHIEF OF FISH MANAGEMENT DIVISION
Phone: 615-781-6575
Dave Woodward, CHIEF OF INFORMATION & EDUCATION
Phone: 615-781-6502
Sonny Richardson, CHIEF OF LAW ENFORCEMENT DIVISION
Phone: 615-781-6580
srichard.state.tn.us
Loy Fulford, CHIEF OF MANAGEMENT SYSTEMS
Phone: 615-781-6528
Jim Dillard, CHIEF OF PERSONNEL
Phone: 615-781-6594

Barry Summer, CHIEF OF PLANNING AND FEDERAL AID
Phone: 615-781-6599
bsummer@mail.state.tn.us
John Gregory, CHIEF OF REAL ESTATE AND FORESTRY DIVISION
Phone: 615-781-6560
Larry Marcum, CHIEF OF WILDLIFE MANAGEMENT DIVISION
Phone: 615-781-6610
Beverly Johnson, COMMISSION CHAIRMAN
Phone: 931-363-7621
Fax: 931-363-9679
G. L. Teague, COMMISSION VICE CHAIRMAN
Phone: 731-847-0848
Fax: 731-847-0748
Dave Woodward, EDITOR
Phone: 615-781-6502
Phil Neil, EDUCATION SUPERVISOR
Phone: 615-781-6538
pneil@mail.state.tn.us
Gary Myers, EXECUTIVE DIRECTOR
Phone: 615-781-6552
L. Garland, GENERAL COUNSEL
Phone: 615-781-6606
Richard Kirk, MANAGER OF NONGAME/ENDANGERED SPECIES
Phone: 615-781-6619
Clarence Coffey, REGIONAL MANAGER OF CUMBERLAND PLATEAU (REGION III)
Phone: 931-484-9571
Bob Ripley, REGIONAL MANAGER OF EAST TENNESSEE (REGION IV)
Phone: 615-587-7037
Steve Patrick, REGIONAL MANAGER OF MIDDLE TENNESSEE (REGION II)
Phone: 615-781-6622
Gary Cook, REGIONAL MANAGER OF WEST TENNESSEE (REGION I)
Phone: 901-423-5725

WISCONSIN CONSERVATION CORPS

30 W. Mifflin, Suite 406
Madison, WI 53703-2558 USA
Phone: 608-266-7730　　Fax: 608-267-2733
Website: www.dwd.state.wi.us/wcc/

Founded: 1983
Membership: 300
Scope: Statewide

Description: The WCC provides work experience and personal development opportunities to young adults ages 18-25, and valuable conservation and other services to Wisconsin communities. Approximately 550 corps members annually work at four dozen rotating project sites throughout the state. Government agencies and nonprofit organizations are eligible to apply for WCC assistance.

Publications: Bi-Annual Report, On Corps! Newsletter

Keyword(s): Exotic species, Aquatic nuisance species, Outdoor Recreation, Wildlife, Forests and Forestry, Environmental and Conservation Education

Contact(s):
Laura Degolier, EXECUTIVE DIRECTOR
degolla@dwd.state.wi.us

WISCONSIN COOPERATIVE FISHERY RESEARCH UNIT USGS

College of Natural Resources, University of Wisconsin
Stevens Point, WI 54481 USA
Phone: 715-346-2178   Fax: 715-346-3624
E-mail: coopfish@uwsp.edu
Website: www.uwsp.edu/cnr/wicfru

Founded: NA
Membership: 2
Scope: International

Description: Interagency organization on the federal, state, and university levels. It carries out research, training, and extension in biology and management of freshwater fishery resources.

Keyword(s): Sport Fishing, Rivers, Wildlife, Lakes, Aquatic Habitats

Contact(s):
Michael Bozek, LEADER

WISCONSIN COOPERATIVE WILDLIFE RESEARCH UNIT (USDI)

USGS, Department of Wildlife Ecology, 204 Laboratories, University of Wisconsin
Madison, WI 53706-1598 USA
Phone: 608-263-4519   Fax: 608-263-4519
E-mail: dlziebar@facstaff.wisc.edu
Website: www.wisc.edu/wildlife

Founded: NA
Membership: 2
Scope: International

Contact(s):
Christine Ribic, LEADER
caribic@facstaff.wisc.edu

WISCONSIN DEPARTMENT OF AGRICULTURE TRADE AND CONSUMER PROTECTION

LAND AND WATER RESOURCES BUREAU
2811 Agriculture Dr., P.O. Box 8911
Madison, WI 53708-8911 USA
Phone: 608-224-4620   Fax: 608-224-4615
Website: www.datcp.state.wi.us

Founded: NA
Membership: 30
Scope: State

Description: Responsible for administering state soil and water conservation and farmland preservation programs.

Keyword(s): Rural Development, Land Use Planning, Protected Areas, Land Preservation, Soil Conservation, Water Quality, Watersheds, Exotic species, Aquatic nuisance species, Wetlands, Environment

Contact(s):
David Jelinski, BUREAU DIRECTOR
Phone: 608-224-4621
Fax: 608-224-4615
James Harsdorf, SECRETARY

WISCONSIN DEPARTMENT OF NATURAL RESOURCES

101 South Webster St.
Madison, WI 53702 USA
Phone: 608-266-2121
E-mail: csweb.dnr.state.wi.us
Website: www.dnr.state.wi.us

Founded: NA
Scope: Statewide

Description: Responsibilities include: Fisheries, wildlife, forest, parks management, endangered resources protection, forest fire control, air and water pollution control, solid and hazardous waste management, mining regulation, enforcement of conservation and environmental laws, flood plain and shoreland zoning, water management and regulation, lake rehabilitation, and long-range planning in the broad fields of outdoor recreation and natural resources.

Publications: WI Natural Resources Magazine

Keyword(s): Exotic species, Aquatic nuisance species, Solid Waste, Forests and Forestry, Outdoor Recreation, Endangered Resources, Land Management, Air Quality and Pollution

Contact(s):
Jay Hochmuth, ADMINISTRATOR OF AIR AND WASTE DIVISION
Phone: 608-267-9521
David Meier, ADMINISTRATOR OF ENFORCEMENT AND SCIENCE DIVISION
Phone: 608-266-0015
Steven Miller, ADMINISTRATOR OF LAND DIVISION
Phone: 608-266-5782
Susan Sylvester, ADMINISTRATOR OF WATER DIVISION
Phone: 608-266-1099
Allen Shea, BUREAU OF WATERSHED MANAGEMENT
Phone: 608-267-2759
Mark McDermid, CONTACT
Mary Kopecky, DEPUTY ADMINISTRATOR OF AIR AND WASTE DIVISION
Phone: 608-261-8448
Bruce Baker, DEPUTY ADMINISTRATOR OF BUREAU OF WATER
Phone: 608-266-1902
Darrell Bazzell, DEPUTY SECRETARY
Phone: 608-266-2252
Lloyd Eagan, DIRECTOR OF BUREAU OF AIR MANAGEMENT
Phone: 608-266-0603
Laurel Steffes, DIRECTOR OF BUREAU OF COMMUNICATION AND EDUCATION
Phone: 608-266-8109
Kathryn Curtner, DIRECTOR OF BUREAU OF COMMUNITY FINANCIAL ASSISTANCE
Phone: 608-266-0860
Jill Jonas, DIRECTOR OF BUREAU OF DRINKING WATER AND GROUND WATER
Phone: 608-267-7651
Paul Delong, DIRECTOR OF BUREAU OF ENDANGERED RESOURCES
Phone: 608-264-9224
Robert Roden, DIRECTOR OF BUREAU OF FACILITIES AND LANDS
Phone: 608-266-2197
Gene Francisco, DIRECTOR OF BUREAU OF FORESTRY
Phone: 608-266-0842

James Addis, DIRECTOR OF BUREAU OF INTERGRATED SCIENCE SERVICES
Phone: 608-266-0837
Thomas Harelson, DIRECTOR OF BUREAU OF LAW ENFORCEMENT
Phone: 608-266-1115
Michael Staggs, DIRECTOR OF BUREAU OF MANAGEMENT AND HABITAT PROTECTION
Phone: 608-267-0796
Mark Giesfeldt, DIRECTOR OF BUREAU OF REMEDIATION AND REDEVELOPMENT
Phone: 608-267-7562
Suzanne Bangert, DIRECTOR OF BUREAU OF WASTE MANAGEMENT
Phone: 608-266-0014
Thomas Hauge, DIRECTOR OF BUREAU OF WILDLIFE MANAGEMENT
Phone: 608-266-2193
Susan Black, DIRECTOR OF PARKS AND RECREATION
Phone: 608-266-2185
Trygve Solberg, NATURAL RESOURCES BOARD CHAIRMAN
Phone: 715-356-7711
Ron Kazmierczak, NORTHEAST REGIONAL DIRECTOR
P.O. Box 10448, Green Bay, WI 54307
Phone: 920-492-5815
William Smith, NORTHERN REGIONAL DIRECTOR
810 W. Maple Street, Spooner, WI 54801
Phone: 715-635-4010
George Meyer, SECRETARY
Phone: 608-266-2121
Ruthe Badger, SOUTH CENTRAL REGIONAL DIRECTOR
3911 Fish Hatchery Road, Madison, WI 53711
Phone: 608-275-3260
Gloria McCutcheon, SOUTHEAST REGIONAL DIRECTOR
P.O. Box 12436, Milwaukee, WI 53212
Phone: 414-263-8510
Scott Humrickhouse, WEST CENTRAL REGIONAL DIRECTOR
P.O. Box 4001, Eau Claire, WI 54702
Phone: 715-839-3711

WISCONSIN DEPARTMENT OF PUBLIC INSTRUCTION

125 S. Webster St., P.O. Box 7841
Madison, WI 53707-7841 USA
Phone: 800-441-4563   Fax: 608-267-9110
E-mail: http://www.dpi.state.wi.us/

Founded: NA

Description: A state government agency that promotes environmental education in public schools and supervises teacher preparation programs. Conducts workshops, and provides consultant services to elementary and secondary schools, colleges and universities. Produces publications to aid in program development.

Publications: "Wisconsin's Model Academic Standards for Environmental Education", A Guide to Curriculum Planning in Environmental Education

Keyword(s): Environmental and Conservation Education

Contact(s):
Sue Grady
ENVIRONMENTAL EDUCATION CONSULTANT
Phone: 608-266-2364

WISCONSIN ENVIRONMENTAL EDUCATION BOARD (WEEB)

110 College of Natural Resources UW Stevens Point
Stevens Point, WI 54481 USA
Phone: 715-346-3805   Fax: 715-346-3025
E-mail: weeb@uwsp.edu
Website: http://www.uwsp.edu/cnr/weeb

Founded: 1990

Description: Grants board providing $200,000 annually to environmental education (EE) initiative projects within the state of Wisconsin, with a maximum grant of $20,000 per project. The board priorities are further development of previously-funded WEEB projects.

Publications: Grant Application Guidelines, Annual Report

Keyword(s): Scholarships and Grants, Environmental and Conservation Education

Contact(s):
Jack Finser, CHAIRPERSON
Ron Rueckert, PROGRAM ASSISTANT

WISCONSIN GEOLOGICAL AND NATURAL HISTORY SURVEY

University of Wisconsin Extension, 3817 Mineral Point Rd.
Madison, WI 53705 USA
Phone: 608-262-1705   Fax: 608-262-8086
Website: www.uwex.edu/wgnhs

Founded: 1897
Scope: Statewide

Description: Created by the legislature, with the responsibility to survey the state's geology, mineral, water, soil, plant, animal, and climate resources, and to coordinate topographic mapping.

Keyword(s): Exotic species, Aquatic nuisance species, education, Soil Conservation, Geology

Contact(s):
Ronald Hennings, ASSISTANT DIRECTOR
Phone: 608-263-7395
James Robertson, STATE GEOLOGIST AND DIRECTOR
Phone: 608-263-7384

WISCONSIN STATE EXTENSION SERVICES

COMMUNITY NATURAL RESOURCE AND ECONOMIC DEVELOPEMENT
University of Wisconsin Extension, 432 N. Lake St.
Madison, WI 53706 USA
Phone: 608-262-1748   Fax: 608-262-9166
Website: www.uwex.edu/ces/

Founded: NA
Scope: Statewide

Contact(s):
Carl Oconnor, DEAN AND DIRECTOR OF COOPERATIVE EXTENSION
Mark Rickenbach, FOREST ECOLOGY & MANAGEMENT SPECIALIST
Phone: 608-262-0134
mrickenbach@cals.wisc.edu

Patrick Walsh, STATEWIDE PROGRAM LEADER
Community, Natural Resource and Economic Development,
University of Wisconsin-Extension, Rm. 625, 432 N. Lake St.,
Madison, WI 53706
Phone: 608-262-1748
patrick.walsh@ces.uwex.edu
Scott Craven, WILDLIFE SPECIALIST
8233 Russel Labs, Madison, WI 53706
Phone: 608-263-6325
Fax: 608-262-6099
srcraven@facstaff.wisc.edu

WOODS HOLE OCEANOGRAPHIC INSITITUTION (WHOI) SEA GRANT PROGRAM

Woods Hole Oceanographic Institution, 193 Oyster Pond
Rd., MS #2
Woods Hole, MA 02543-1525 USA
Phone: 508-289-2398   Fax: 508-457-2172
E-mail: seagrant@whoi.edu
Website: www.whoi.edu/seagrant/

Founded: 1973
Membership: 6
Scope: National

Description: The WHOI Sea Grant Program supports research,
education, and advisory projects to promote the wise use and
understanding of ocean and coastal resources for the public
benefit. It is part of the National Sea Grant College Program of
the National Oceanic and Atmospheric Administration, a
network of 29 individual programs located in each of the
coastal and Great Lakes states to foster cooperation among
government, academia, and industry.

Keyword(s): Coral Reefs, Oceanography, Coasts, Aquaculture,
Biodiversity, Wildlife

Contact(s):
Tracey Crago, COMMUNICATOR
Phone: 508-289-2665
tcrago@whoi.edu
Judith McDowell, DIRECTOR
Phone: 508-289-2557
jmcdowell@whoi.edu
Dale Leavitt, FISHERIES AQUACULTURE SPECIALIST
Phone: 508-289-2997
dleavitt@whoi.edu
Sheri Derosa, PROGRAM ASSISTANT
Phone: 508-289-2398
sderosa@whoi.edu

WYOMING COOPERATIVE EXTENSION SERVICES

University Station, Box 3354
Laramie, WY 82071 USA
Phone: 307-766-5124   Fax: 307-766-3998
Website: www.uwyo.edu/ces/ceshome.htm

Founded: NA
Membership: 50
Scope: Statewide

Publications: Scientific Abstracts, Science Monographs,
Research Journals, Bulletins, 4-H Publications, regional publications, Journal Articles

Contact(s):
Steve Aagard, ASSOCIATE DIRECTOR
aagard@uwyo.edu

Ruth Wilson, ASSOCIATE DIRECTOR
drruth@uwyo.edu
Glen Whipple, DIRECTOR AND ASSOCIATE DEAN
glen@uwyo.edu
Barb Farmer, MANAGER OF COMMUNICATIONS &
TECHNOLGIES
Phone: 307-766-3702
barbaraa@uwyo.edu

WYOMING COOPERATIVE FISH AND WILDLIFE RESEARCH UNIT (USDI)

University of Wyoming, Box 3166, Biological Sciences
Bldg., Rm. 419
Laramie, WY 82071 USA
Phone: 307-766-5415   Fax: 307-766-5400

Founded: 1980
Membership: 40
Scope: National

Description: Conducts research under auspices of the USGS
Biological Resources Division and Wyoming Game and Fish
Department in the northern Rocky Mountain region.

Contact(s):
Wayne Hubert, ASSISTANT LEADER OF FISHERIES
drhubert@uwyo.edu
Fred Lindzey, ASSISTANT LEADER OF WILDLIFE
flindzey@uwyo.edu
Stanley Anderson, LEADER
anderson@uwyo.edu

WYOMING DEPARTMENT OF AGRICULTURE

2219 Carey Ave.
Cheyenne, WY 82002 USA
Phone: 307-777-7321   Fax: 307-777-6593
Website: www.wyagric.state.wy.us

Founded: NA
Scope: State

Contact(s):
Jim Schwartz, DEPUTY DIRECTOR
Phone: 307-777-6591
Ron Micheli, DIRECTOR
Phone: 307-777-6569
Fax: 307-777-6593
rmiche@missc.state.wy.us
Grant Stumbough, NATURAL RESOURCE AND POLICY
MANAGER
Phone: 307-777-6579

WYOMING GAME AND FISH DEPARTMENT

5400 Bishop Blvd.
Cheyenne, WY 82006 USA
Phone: 307-777-4600   Fax: 307-777-4699
E-mail: gf@state.wy.us
Website: http://gf.state.wy.us/

Founded: 1939
Membership: 475
Scope: State

Description: To provide an adequate and flexible system for the
control, propagation, management, protection, and regulation
of Wyoming wildlife for the public interest.

Keyword(s): training, Nongame Wildlife, Wildlife, Hunting,
Environmental and Conservation Education

Contact(s):
Lynda Cook, ASSISTANT ATTORNEY GENERAL
Phone: 307-777-4687
CASPER REGIONAL OFFICE, 3030 Energy Ln., Suite 100,
Casper, WY 82604
Phone: 307-473-3400
Larry Gabriele, CHIEF OF FISCAL DIVISION
lgabri@missc.state.wy.us
Mike Stone, CHIEF OF FISH DIVISION
Phone: 307-777-4559
Jay Lawson, CHIEF OF WILDLIFE DIVISION
Phone: 307-777-4579
CODY REGIONAL OFFICE
2820 State Highway 120, Cody, WY 82414
Phone: 307-527-7125
Bill Wichers, DEPUTY DIRECTOR, EXTERNAL PROGRAMS
Greg Arthur, DEPUTY DIRECTOR, INTERNAL PROGRAMS
John Baughman, DIRECTOR
GREEN RIVER REGIONAL OFFICE
351 Astle, Green River, WY 82935
Phone: 307-857-3223
JACKSON REGIONAL OFFICE
P.O. Box 67, 360 N. Cache, Jackson, WY 83001
Phone: 307-733-2321
LANDER REGIONAL OFFICE
260 Buena Vista, Lander, WY 82520
Phone: 307-332-2688
LARAMIE LAB
Room 323, Biological Sciences Bldg., P.O. Box 3312
University Station, Laramie, WY 82071
Phone: 307-766-6313
LARAMIE REGIONAL OFFICE
528 S. Adams, Laramie, WY 82070
Phone: 307-745-4046
PINEDALE REGIONAL OFFICE
P.O. Box 850, 117 S. Sublette Avenue, Pinedale, WY 82941
Phone: 307-367-4353
SHERIDAN REGIONAL OFFICE
P.O. Box 6249, 700 Valley View, Sheridan, WY 82801
Phone: 307-672-7418
Larry Kruckenberg, SPECIAL ASSISTANT FOR POLICY
Phone: 307-777-4539
SYBILLE WILDLIFE RESEARCH AND CONSERVATION
UNIT
2362 Highway 34, Wheatland, WY 82201
Phone: 307-322-2571

WYOMING STATE BOARD OF LAND COMMISSIONERS

Herschler Building 3 W
Cheyenne, WY 82002 USA
Phone: 307-777-7331   Fax: 307-777-5400
E-mail: slfmail@state.wy.us
Website: www.state.wy.us

Founded: NA
Scope: State

Keyword(s): Renewable Resources, Land Preservation, Public
Lands, Environment

Contact(s):
Jim Geringer, CHAIRMAN
Ron Arnold, SECRETARY

WYOMING STATE GEOLOGICAL SURVEY

Box 3008
Laramie, WY 82071 USA
Phone: 307-766-2286   Fax: 307-766-2605
E-mail: wsgs@wsgs.uwyo.edu; sales@wsgs.uwyo.edu
Website: www.wsgsweb.uwyo.edu

Founded: 1933
Membership: 20
Scope: Statewide

Description: Activities include surface and subsurface geologic
mapping; mineral, rock, and fossil investigations; natural
resource and natural hazards investigations; and assistance in
resources development.

Publications: bulletins, reports of investigations, public
information circulars, quarterly newsletter (Wyoming Geo-
notes), list of publications sent on request., Memoirs

Keyword(s): Mineral Resources, Geology, Land Use Planning,
Energy

Contact(s):
Robert Lyman, COAL GEOLOGIST
Richard Jones, EDITOR
James Case, GEOLOGIC HAZARDS GEOLOGIST
Alan Verploeg, GEOLOGIC MAPPING GEOLOGIST
Ray Harris, INDUSTRIAL MINERALS GEOLOGIST
Rodney Debruin, PETROLEUM GEOLOGIST
W. Hausel, SENIOR ECONOMIC GEOLOGIST OF METALS
AND PRECIOUS STONES
Lance Cook, STATE GEOLOGIST
gglass@wsgs.uwyo.edu

NON-GOVERNMENTAL NON-PROFIT ORGANIZATIONS

20/20 VISION

1828 Jefferson Pl., NW
Washington, DC 20036 USA
Phone: 202-833-2020   Fax: 202-833-5307
E-mail: vision@2020vision.org
Website: www.2020vision.org

Founded: 1986
Membership: 10,000
Scope: National

Description: 20/20 Vision is a nonprofit grassroots organization dedicated to protecting the environment and promoting peace through lobbying and citizen education and activism. 20/20 Vision empowers citizens to speak out for a clean environment and a world free of nuclear weapons. Each month we pick one issue that is critically important to the future of the planet, where your voice will make the most difference. Thousands of letters, written in just 20 minutes a month can move political mountains.

Publication(s): Viewpoint, Action Alert Postcards, Tools for Activists Fact Sheets

Keyword(s): EcoAction, Environmental and Humanitarian Education

Contact(s):
Erik Olsen, CAMPAIGNS COORDINATOR
erik@2020vision.org
James Wyerman, EXECUTIVE DIRECTOR
jwyerman@2020vision.org

A

ACADEMY FOR EDUCATIONAL DEVELOPMENT

1825 Connecticut Ave., Suite 800
Washington, DC 20009-8721 USA
Phone: 719-556-8318   Fax: 202-884-8997
E-mail: greencom@aed.org
Website: www.usaid.gov/environment/greencom

Founded: 1961
Scope: International

Description: A domestic and international development organization with a multi-million dollar environmental education, communication and environmental health population agenda working in over 25 countries.

Publication(s): Publication on website

Keyword(s): Agriculture, Air Quality, Asia Water Environment, Biodiversity, Communications, Conservation, Drinking Water Protection, Ecotourism, Environmental and Conservation Education, Natural Resource Conservation, Forests and Forestry, Public Lands, Rivers, Water Pollution Management, Wilderness, Recycling, Women in the Environment

Contact(s):
Brian Day, DIRECTOR
Phone: 202-884-8897
bday@aed.org
William Smith, EXECUTIVE VICE PRESIDENT
Phone: 202-884-8750
Fax: 202-884-8752
bsmith@aed.org
Richard Bassi, LATIN AMERICA DIRECTOR
Phone: 202-884-8898
rbassi@aed.org
Gregory Niblett, SENIOR VICE PRESDENT
niblett@aed.org

ACRES LAND TRUST

2000 N. Wells St.
Fort Wayne, IN 46808-2474 USA
Phone: 219-422-1004   Fax: 219-422-1004
E-mail: acreslt@fwi.com
Website: www.acres-land-trust.org

Founded: 1960
Scope: Regional

Description: A nonprofit organization dedicated to the acquisition and permanent preservation of natural areas in northeastern Indiana. Conducts a guided field-trip program for children and adults. Organizes canoe trips, concerts and festivals for the membership and the public. Administers 40 nature preserves totaling more than 2,600 acres.

Publication(s): Acres Brochure, Field Guide, Acres Quarterly

Keyword(s): Environmental and Conservation Education, Biodiversity, Land Preservation, Wetlands, Endangered Species

Contact(s):
Carolyn McNagny, EXECUTIVE DIRECTOR
James Haddock, PRESIDENT
Richard Walker, TREASURER
Sam Schwartz, VICE PRESIDENT
Robert Weber, VICE PRESIDENT
Theodore Heemstra, VICE PRESIDENT

ACTION FOR NATURE, INC.

2269 Chestnut St., Suite 263
San Francisco, CA 94123 USA
Phone: 415-421-2640   Fax: 415-922-5717
E-mail: action@dnai.com
Website: www.actionfornature.org

Founded: NA
Scope: National

Description: Action for Nature was organized to foster respect and affection for nature through personal environmental initiatives. AFN is a clearing-house and catalyst for personal action projects and publicizes young peoples' successful environmental initiatives through a newsletter and publication of a book of some of their stories.

Publication(s): Acting for Nature (book)

Keyword(s): Youth Organizations, Environmental Communication

Contact(s):
Sidney Hoh, BIOLIGIST, MARINE CONSERVATIONIST, ANIMAL PROTECTOR
David Brower, CHAIRMAN, EARTH ISLAND INSTITUTE
Huey Johnson, FOUNDER, PRESIDENT, RENEWAL RESOURCES INSTITUTE
Maria Luisa Cohen, MURALIST, PRESIDENT, ASSISI NATURE COUNCIL
Evelyn De Ghetaldi, PRESIDENT
Ruth Gottstein, PUBLISHER, VOLCANO PRESS
Albert Baez, SCIENCE EDUCATOR, PRESIDENT, VIVAMOS MEJOR
Mary Griffin-Jones, SECRETARY
Bethany Baugh, SYSTEMS MANAGER
David Yamakawa, TREASURER
Shimon Schwarzschild, WRITER/EDITOR/CONSULTANT

ADIRONDACK COUNCIL, THE
P.O. Box D-2
Elizabethtown, NY 12932 USA
Phone: 518-873-2240   Fax: 518-873-6675
E-mail: adkcouncil@aol.com
Website: www.adirondackcouncil.org

Founded: 1975
Membership: 18000
Scope: Local Regional National

Description: A nonprofit environmental organization working for protection and preservation of the six million acre Adirondack Park. Programs include monitoring and influencing state programs in the park, helping to promote understanding of the park and the need to protect its very special character, and supporting and advancing positive programs to enhance the park and benefit its people.

Publication(s): Acid Rain A Continuing National Tragedy, Adirondack Wildguide: A Natural History of The Adirondack Park, State of the Park, Adirondack Council Newsletter

Keyword(s): Acid Rain, Land Use Planning, Open Space, Sustainable Development, training

Contact(s):
David Skovron, CHAIRMAN
Bernard Melewski, EXECUTIVE DIRECTOR
Barbara Glaser, SECRETARY
Edward Fowler, TREASURER
Patricia Winterer, VICE CHAIRMAN
David Bronston, VICE CHAIRMAN

ADIRONDACK MOUNTAIN CLUB, INC., THE
814 Goggins Rd.
Lake George, NY 12845-4117 USA
Phone: 518-668-4447   Fax: 518-668-3746
E-mail: adkinfo@adk.org
Website: www.adk.org

Founded: 1922
Membership: 35000
Scope: State

Description: The Adirondack Mountain Club is dedicated to the protection and responsible recreational use of the New York State Forest Preserve, parks and other wild lands and waters. The Club is a member-directed organization committed to public service and stewardship. ADK employs a balanced approach to outdoor recreation, advocacy, environmental education and natural resource conservation. ADK has 26 chapters in NY and NJ.

Publication(s): Guides and maps to Adirondack and Catskill Trails, Adirondack Canoe Waters, North Flow, South and West Flow, Western and Central New York State, Adirondack cultural and literary history information, Adirondack Magazine

Keyword(s): Conservation, Bicycle, Trail, Environmental and Conservation Education, Protected Areas, Outdoor Recreation, Environmental Protection, Land Preservation, Forest Management, Natural Areas, education, Nongame Wildlife, Wilderness, Wildlands Management, Public Lands

Contact(s):
Neil Woodworth, DEPUTY EXECUTIVE DIRECTOR FOR PUBLIC AND LEGAL AFFAIRS
Phone: 518-449-3870
nwoodworth@nycap,rr,com
Neal Burdick, EDITOR
35 Woods Dr., Canton, NY 13617

Jo Benton, EXECUTIVE DIRECTOR
Phone: 518-668-4447
Timothy Tierney, NORTH COUNTRY DIRECTOR OF FIELD PROGRAMS
Greg Macdonald, NORTH COUNTY DIRECTOR OF FACILITIES
Terry Sexton, PRESIDENT

ADIRONDACK NATURE CONSERVANCY/ADIRONDACK LAND TRUST, INC.
P.O. Box 65
Keene Valley, NY 12943 USA
Phone: 518-576-2082
E-mail: dfeeley@tnc.org
Website: www.nature.org http://www.nature.org

Founded: 1984
Scope: Statewide

Description: The Adirondack Nature Conservancy and Adirondack Land Trust are separate land conservation organizations that have acted in partnership since 1988, coordinating programs and staff. The Adirondack Nature Conservancy protects the plants, animals, and natural communities that represent the diversity of life in the Adirondacks by protecting the lands and waters they need to survive.

Publication(s): Developing a Land Conservation Strategy: A Handbook for Land Trusts (1987)

Keyword(s): Agriculture, Forests and Forestry, Land Preservation, Open Space, Sustainable Development

Contact(s):
Edward McNeil, CHAIRMAN
108 Burlingame Rd., Syracuse, NY 13202-1604
Todd Dunham, DIRECTOR OF LAND PROTECTION
Phone: 518-576-2082
Fax: 518-576-4203
tdunham@tnc.org
Michael Garr, EXECUTIVE DIRECTOR
Phone: 518-576-2082
Fax: 518-576-4203
mgarr@tnc.org
Francisca Irwin, SECRETARY
Rt. 1 Box 80, Essex, NY 12936
Meredith Prime, TREASURER
Heather Hill, Lake Placid, NY 12946
Timothy Barnett, VICE PRESIDENT
Phone: 518-576-2082
Fax: 518-576-4203
tbarnett@tnc.org

ADKINS ARBORETUM
P.O. Box 100
Ridgely, MD 21660 USA
Phone: 410-634-2847   Fax: 410-634-2878
E-mail: adkinsar@intercom.net
Website: www.adkinsarboretum.org

Founded: 1979
Membership: 400
Scope: Regional

Description: Adkins Arboretum is dedicated to the appreciation, understanding and stewardship of the indigenous, nontidal plant communities of the Central Delmarva Peninsula. It strives to maintain a divers and dynamic living collection that is authentic, engaging and a model for land management. As a significant cultural, education, scientific and recreational resource, the Arboretum fosters civic pride, encourages public dialogue and contributes to the economic vitality of the region.

Publication(s): Native Seed

Keyword(s): Native Plants, Biodiversity, Conservation, Ecology, Endangered Species, Environmental and Conservation Education, Flowers, Plants, and Trees, Landscape Architecture, Wetlands, training, Botanical Gardens

Contact(s):
Ellie Altman, EXECUTIVE DIRECTOR
Phone: 410-634-2847
Fax: 410-634-2878
ealtman@intercom.net
Lorie Staber, PRESIDENT OF THE BOARD
Louise Barton, SECRETARY
Graham Donaldson, TREASURER
Kathy Carnean, VICE PRESIDENT

ADOPT-A-STREAM FOUNDATION, THE

600-128th St., SE
Everett, WA 98208-6353 USA
Phone: 425-316-8592   Fax: 425-338-1423
E-mail: aasf@streamkeeper.org
Website: www.streamkeeper.org

Founded: 1985
Membership: 250
Scope: National

Description: Adopt-A-Stream Foundation's mission is to empower people to become stewards of watersheds, wetlands and streams. The Foundation's long term goal is to ensure that all streams are adopted by watershed residents. The current focus is in the Pacific Northwest. The Foundation conducts "Streamkeeper" workshops that train volunteers and students of all ages how to conduct watershed inventories, monitor small streams and other educational programming.

Publication(s): Adopting a Stream: A Northwest Handbook, Adopting a Wetland: A Northwest Guide, A Streamkeeper's Field Guide: Watershed Inventory and Stream Monitoring Methods, The Streamkeeper (newsletter), Video: The Streamkeeper, featuring Bill Nye "The Science Guy."

Keyword(s): Aquatic Habitats, Environmental Planning, Environmental and Conservation Education, Protected Areas, Environmental Protection, Lakes, Land Use Planning, Wildlife, National Parks, Nongame Wildlife, Raptors, Reptiles and Amphibians, Urban Environment, Open Space, Rivers

Contact(s):
Darryl Williams, BOARD PRESIDENT
Tom Murdoch, EXECUTIVE DIRECTOR
Dan Crouse, SECRETARY
Grant Woodfield, TREASURER

AFRICAN WILDLIFE FOUNDATION

1400 16th St., NW
Washington, DC 20036 USA
Phone: 202-939-3333   Fax: 202-939-3332
E-mail: africanwildlife@awf.org
Website: www.awf.org

Founded: NA
Membership: 1000
Scope: International

Description: The African Wildlife Foundation recognizes that the wildlife and wild lands of Africa have no equal. We work with people-our supporters worldwide and our partners in Africa-to craft and deliver creative solutions for the long-term well-being of Africa's remarkable species, habitats, and the people who depend upon them.

Publication(s): African Wildlife News

Keyword(s): Conservation, Endangered Species, Sustainable Development, training

Contact(s):
Stuart Saunders, CHAIRMAN OF THE BOARD
R. Wright, PRESIDENT
Patrick Bergin, PRESIDENT OF AFRICA OPERATIONS
Henry McIntosh, TREASURER
William Richards, VICE CHAIR OF THE BOARD

AIR AND WASTE MANAGEMENT ASSOCIATION

One Gateway Center, 3rd Fl.
Pittsburgh, PA 15222 USA
Phone: 412-232-3444   Fax: 412-232-3450
E-mail: info@awma.org
Website: www.awma.org

Founded: 1907
Membership: 10000
Scope: National

Description: The Air and Waste Management Association is a nonprofit, technical, and environmental association that provides a neutral forum for discussing all sides of an environmental issue. The Association encourages environmental technology development, facilitates technology transfer and improves environmental management and education. As a leading forum for discussing diverse views on environmental issues, it challenges leaders, professionals and citizens worldwide to use dialogue for improving the quality of decisions affecting our environment.

Publication(s): Journal of the Air & Waste Management Association, other publications include proceedings of specialty conferences and symposia, prints and videotapes, EM, A&WMA News, quarterly newsletter

Keyword(s): Air Quality and Pollution, Engineering, Environmental Law, Greenhouse Effect/Global Warming, Solid Waste Management

Contact(s):
Robert Hall, 1ST VICE PRESIDENT
Douglas Bisset, TREASURER

ALABAMA ASSOCIATION OF SOIL AND WATER CONSERVATION DISTRICTS

Attn: Executive Director
P.O. Box 30480
Montgomery, AL 36130-4800 USA
Phone: 334-223-7257

Founded: NA
Scope: Statewide

Keyword(s): Conservation Districts

Contact(s):
Jake Harper, 1ST VICE PRESIDENT
Rt. 1 Box 468, Camden, AL 36726
Phone: 334-682-4463

Terry Poague, 2ND VICE PRESIDENT
3716 Clause Fleahop Rd., Tallahassee, AL 36078
Phone: 334-567-6183
Charles Holmes, BOARD MEMBER
Rt 1 Box 212, Marion, AL 36756
Phone: 334-683-6869
Fax: 334-583-6869
George Robertson, PRESIDENT
2181 County Rd. 22, Waverly, AL 36879
Phone: 334-887-6070
Fax: 334-826-8219
Charles Rittenour, SECRETARY/TREASURER
1144 Meriwether Rd., Pike Road, AL 36064
Phone: 334-284-5320

ALABAMA B.A.S.S. CHAPTER FEDERATION

Attn: President, P.O. Box 190
Notasulga, AL 36866 USA
Phone: 334-257-1177   Fax: 334-257-4665
Website: www.albassfed.org

Founded: NA
Membership: 3000
Scope: Statewide

Description: An organization of Bassmaster chapters, affiliated with the Bass Anglers Sportsman Society, organized to fight pollution, assist state and national conservation agencies in their efforts, and teach the young people of our country good conservation practices. Dedicated to the realistic conservation of our water resources.

Contact(s):
Jim Howard, CONSERVATION DIRECTOR
501 Five Mile Rd., Eufaula, AL 36027
Phone: 334-616-6956
Fax: 334-616-7194
bassnbuddy@aol.com
Al Redding, PRESIDENT
Phone: 334-257-1177
Fax: 334-257-4665

ALABAMA ENVIRONMENTAL COUNCIL

2717 7th Ave., S., Suite 207
Birmingham, AL 35233 USA
Phone: 205-322-3126   Fax: 205-324-3784
E-mail: stateoffice@aeconline.ws
Website: www.aeconline.ws

Founded: 1967
Membership: 5
Scope: Regional

Description: Dedicated to the preservation of Alabama's environment on all fronts: air, water, land and wildlife.

Publication(s): State News

Keyword(s): Endangered Species, Environmental and Conservation Education, Forests and Forestry, Recycling, Solid Waste, Coral Reefs

Contact(s):
Frank Juliano, DIRECTOR OF RECYCLING
Rachel Reinhart, EXECUTIVE DIRECTOR
2717 7th Ave S Suite 207, Birmingham, Alabama 35233
Phone: 205-322-3126
Fax: 205-324-3784
director@aeconline.ws
Thomas Carruthers, PRESIDENT

ALABAMA NATURAL HERITAGE PROGRAM

The Nature Conservancy, 1500 E. Fairview Ave.
Montgomery, AL 36106 USA
Phone: 334-834-4519   Fax: 334-834-5439
E-mail: alnhp@zebra.net
Website: www.heritage.tnc.org/nhp/us/al/

Founded: 1989
Membership: 7
Scope: International

Description: The mission of the Alabama Natural Heritage program is to provide the best available scientific information on the biological diversity of Alabama, guide conservation action and promote sound stewardship practices within the state and throughout the Southeast.

Publication(s): Natural Heritage News, Inventory List of Rare Threatened and Endangered Plants, Animals, and Natural Communities of Alabama

Keyword(s): Biodiversity, National Parks, Endangered Species, Conservation, Natural Areas, Land Use Planning, education, Native Plants

Contact(s):
James Godwin, ACTING DIRECTOR
Kathy Freeland, EXECUTIVE DIRECTOR, ALABAMA CHAPTER OF THE NATURE CONSERVANCY
Phone: 205-251-1155
Fax: 205-251-4444

ALABAMA WATERFOWL ASSOCIATION, INC. (AWA)

P.O. Box 67
Guntersville, AL 35768 USA
Phone: 202-682-9400
E-mail: awa@alabamawaterfowl.org
Website: alabamawaterfowl.org

Founded: 1987
Scope: Statewide

Description: To protect, enhance and create wetlands habitat for all wildlife species and other human values; and to enhance waterfowl population and protect our hunting heritage in Alabama.

Publication(s): Wetlands and Waterfowl News

Keyword(s): Historic Preservation, Hunting, Waterfowl, Wetlands, training

Contact(s):
Jerry Davis, CHIEF EXECUTIVE OFFICER
1346 County Rd. 11, Scottsboro, AL 35768
Phone: 205-260-2600
Gary Benefield, EXECUTIVE DIRECTOR
P.O. Box 67, Guntersville, AL 35976
Phone: 205-593-7712
Roger Crouch, EXECUTIVE TREASURER
P.O. Box 67, Guntersville, AL 35976

ALABAMA WILDFLOWER SOCIETY, THE

606 India Rd.
Opelika, AL 36801 USA
Phone: 334-745-2494   Fax: 334-749-5200
Website: www.auburn.edu/~deancar

Founded: 1971
Membership: 500

Scope: Statewide

Description: The society promotes knowledge, appreciation, use of native plants, preserves and propagates rare native plants, preserves areas of significant native flora and provides scholarships.

Publication(s): Newsletter of the Alabama Wildflower Society, Wildflower Brochure

Keyword(s): Conservation Biology, Conservation, Endangered Species, Native Plants, Scholarships, Wildflowers

Contact(s):
Caroline Dean, BOARD MEMBER
George Wood, EDITOR
Phone: 205-339-2541
Virginia Lusk, PAST PRESIDENT
Phone: 205-988-0299
ginny1@bellsouth.net
Shirley Fifield, PRESIDENT
Phone: 334-277-2070
rgfifi@aol.com

ALABAMA WILDLIFE FEDERATION
46 Commerce St.
Montgomery, AL 36104 USA
Phone: 334-832-9453   Fax: 334-832-9454
E-mail: awf@mindspring.com
Website: www.alawild.org

Founded: 1935
Scope: State

Description: A representative statewide organization affiliated with the National Wildlife Federation, dedicated to the protection and enhancement of wildlife and its habitat through public education and government interaction.

Publication(s): Alabama Wildlife

Contact(s):
April Lupardus, CONSERVATION PROGRAMS SPECIALIST
Jeff McCollum, EDUCATION PROGRAMS CONTACT
Tim Gothard, EXECUTIVE DIRECTOR, ALTERNATE REPRESENTATIVE & EDITOR
Robert Thornton, PRESIDENT
Rebecca Prichett, REPRESENTATIVE
Clinton Berry, TREASURER

ALASKA ASSOCIATION OF SOIL AND WATER CONSERVATION DISTRICTS
949 E. 36th Ave. Ste 400
Anchorage, AK 99508 USA
Phone: 907-271-2424   Fax: 907-271-3951
E-mail: akdistrict@customcpu.com

Founded: NA
Scope: Statewide

Keyword(s): Conservation Districts

Contact(s):
Shirley Schollenberg, 1ST VICE PRESIDENT
HC 67 Box 250, Anchor Point, AK 99556
Phone: 907-567-3467
Meribeth Crick, 2ND VICE PRESIDENT
P.O. Box 56505, North Pole, AK 56505
Phone: 907-488-2215

Mike Carlson, ALTERNATE BOARD MEMBER
P.O. Box 953, Delta Junction, AK 99737
Phone: 907-895-4819
Omar Stratman, PRESIDENT, BOARD MEMBER
P.O. Box 2376, Kodiak, AK 99615
Phone: 907-486-5578
Fax: 907-486-5578
Doug Witte, PROJECT COORDINATOR
351 W. Parks Hwy 101, Wasilla, AK 99645
Phone: 907-373-7923
Fax: 907-373-7192
Meg Burgett, SECRETARY-TREASURER
P.O. Box 874554, Wasilla, AK 99687
Phone: 907-373-0885

ALASKA CENTER FOR THE ENVIRONMENT
519 W. 8th, Suite 201
Anchorage, AK 99501 USA
Phone: 907-274-3621   Fax: 907-274-8733
E-mail: akcenter@alaska.net
Website: http.www.akcenter.org

Founded: 1971
Scope: Statewide

Description: Nonprofit organization which functions as an advocacy and citizen organizing facility for Alaskan environmental activities. With a professional staff of twelve and a corps of volunteers, the center conducts policy analyses and encourages grassroots activism to conserve and protect Alaska's natural resources, particularly its wildlands.

Keyword(s): Natural Resource Conservation, Forests and Forestry, Public Lands, Rivers, Water Pollution Management, Wilderness, Wildlands

Contact(s):
Michell Wilson, CHUGACH /KENAI/PWS PROGRAM COORDINATOR
Phone: 907-274-3665
michel@akcenter.org
Cliff Eames, DIRECTOR OF ISSUES AND STATE LANDS DIRECTOR
Tom Burek, DIRECTOR OF TRAILSIDE DISCOVERY CAMP
Phone: 907-274-5437
Joe Lebeau, DIRECTOR OF VALLEY ACE
642 S. Alaska St., Suite 201, Palmer, AK 99645
Phone: 907-745-8223
Fax: 907-745-8223
joe@akcenter.org
Randy Virgin, EXECUTIVE DIRECTOR
Dwayne Lee, FINANCIAL DIRECTOR AND OFFICER MANAGER
dwayne@akcenter.org
Beth Porterfield, MEMBERSHIP DIRECTOR AND VOLUNTEER COORDINATOR
Phone: 907-274-3650
bet@akcenter.org

ALASKA CONSERVATION ALLIANCE
750 West 2nd Ave., Suite 109
Anchorage, AK 99501 USA
Phone: 907-258-6171   Fax: 907-258-6177
E-mail: unite@akvoice.org
Website: www.akvoice.org

Founded: 1997
Scope: National

Description: An alliance dedicated to strengthening environmental organizations and empowering individuals to protect Alaska's environment through public education, training, advocacy, communication and strategy development, all with respect for communities and human dignity.

Keyword(s): Air Quality and Pollution, Aquatic Habitats, Ancient Forests, Chemical Pollution Control, Wildlife Protection, Cultural Preservation, Environment, EcoAction, Environmental Justice, Environmental and Conservation Education, Environmental Protection, Wetlands, Sustainable Development, Wilderness

Contact(s):
Mary Core, EXECUTIVE DIRECTOR
Phone: 907-258-6174
mary@akvoice.org
Paula Phillips, FIELD ORGANIZER
paula@akvoice.org

ALASKA CONSERVATION FOUNDATION

441 W. 5th Ave., Suite 402
Anchorage, AK 99501-2340 USA
Phone: 907-276-1917   Fax: 907-274-4145
E-mail: acfinfo@akcf.org
Website: www.akcf.org

Founded: NA
Scope: Statewide

Description: A community foundation providing grants for environmental conservation in Alaska. It is not a membership organization. It lists its donors as "Circle of Friends."

Publication(s): Dispatch, Annual Report, Grant Guidelines, Alaska Conservation Directory

Keyword(s): Wildlife, Forests and Forestry, Sustainable Ecosystems, Exotic species, Aquatic nuisance species, training

Contact(s):
David Rockefeller, ADVISOR
Rick Caulfield, CHAIR
Deborah Williams, EXECUTIVE DIRECTOR
Jimmy Carter, HONORARY CHAIR
David Wigglesworth, PROGRAM OFFICER
Peg Tileston, SECRETARY AND TREASURER
Eric Meyers, VICE CHAIR OF ALASKA TRUSTEES

ALASKA CONSERVATION VOTERS

750 West 2nd Ave., Suite 109
Anchorage, AK 99501 USA
Phone: 907-258-6171   Fax: 907-258-6177
E-mail: unite@akvoice.org
Website: www.akvoice.org

Founded: 1997
Membership: 44
Scope: Statewide

Description: An organization dedicated to protecting Alaska's environment through public education and advocacy in the Alaska state legislature, Congress and other forums.

Publication(s): newsletter (end of year report) - United Voices, The Score Card

Keyword(s): Air Quality and Pollution, Aquatic Habitats, Ancient Forests, Chemical Pollution Control, Wildlife Protection, Cultural Preservation, Environment, EcoAction, Environmental Justice, Environmental and Conservation Education, Environmental Protection, Wetlands, Sustainable Development, Wilderness

Contact(s):
Leah Rachocki, DEVELOPMENT DIRECTOR
Phone: 907-258-6191
leah@akvoice.org
Jason Geck, GIS ANALYST
Phone: 907-258-6148
jason@akvoice.org
Kevin Harun, INTERIM DIRECTOR
Phone: 907-258-6174
kevin@akvoice.org
Sue Schrader, LOBBYIST
Phone: 907-463-3366
sue@akvoice.org
Martha Levensaler, PRESIDENT
levensaler@nwf.org
Marlo Shedlock, PROGRAM COORDINATOR
Phone: 906-258-6181
marlo@akvoice.org

ALASKA NATURAL HISTORY ASSOCIATION

750 West Second Ave., Suite 100
Anchorage, AK 99501-2167 USA
Phone: 907-274-8440   Fax: 907-274-8343
E-mail: alaskanha@npa.gov
Website: www.alaskanha.org

Founded: 1959
Membership: 4000
Scope: Statewide

Description: The Alaska Natural History Association is dedicated to enhancing the understanding and conservation of Alaska's natural and cultural resources by providing educational materials and services.

Publication(s): membership newsletter goes out twice a yr- Northern Migratiions, other publications can be found on website, catalog (once a yr) Alaska Books

Keyword(s): Arctic, Cultural Preservation, Natural Resource Conservation, Forests and Forestry, Public Lands, Rivers, Water Pollution Management, Wilderness, Environmental and Conservation Education, Environmental Education Curriculum, education

Contact(s):
Charles Money, EXECUTIVE DIRECTOR
charles_money@nps.gov

ALASKA NATURAL RESOURCE AND OUTDOOR EDUCATION ASSOCIATION

P.O. Box 110536
Anchorage, AK 99511-0536 USA
Phone: 907-456-0558   Fax: 253-498-2399
E-mail: anroe@.org
Website: www.anroe.org

Founded: 1983
Membership: 350
Scope: Statewide

Description: ANROE is a statewide network of K-12 school teachers, state and federal agency staff, university faculty and staff, students and other concerned citizens united to promote the development, delivery and implementation of educational efforts that help people of all ages learn about and appreciate Alaska's natural resources.

Publication(s): ANROE Guide to Natural Resource Education Materials (catalog), Targeting Excellence, Flyways, Pathways, and Waterways (newsletter), Discovering Alaskas Salmon Childrens ctivity Guide

Keyword(s): Environmental and Conservation Education, education, Training

Contact(s):
Janet Warburton, PRESIDENT

ALASKA RAINFOREST CAMPAIGN

201 Lincoln St., Suite 1
Sitka, AK 99835 USA
Phone: 907-747-8292   Fax: 907-747-8873
E-mail: info@akrain.org
Website: www.akrain.org

Founded: 1992
Scope: Statewide/National

Description: Alaska Rainforest Campaign is a coalition of Alaska-based and national environmental organizations working to protect the coastal old-growth rainforests of Alaska, especially the Tongass and Chugach National Forests.

Keyword(s): Rainforests

Contact(s):
Brian McNitt, CAMPAIGN MANAGER
Corrie Bosman, NATIONAL FIELD DIRECTOR

ALASKA WILDLIFE ALLIANCE, THE

P.O. Box 202022
Anchorage, AK 99520 USA
Phone: 907-277-0897   Fax: 907-277-7423
E-mail: awa@alaska.net
Website: www.akwildlife.org

Founded: 1978
Scope: Statewide

Description: The Alliance is a nonprofit organization whose mission is the protection of Alaska's natural wildlife and habitat diversity for its intrinsic value as well as for the benefit of present and future generations.

Publication(s): The Spirit

Contact(s):
Karen Deatherage, ASSOCIATE DIRECTOR
Paul Joslin, EXECUTIVE DIRECTOR

A. E. HOWELL WILDLIFE CONSERVATION CENTER, INC.

HC 61, Box 6
N. Amity, ME 04471-9601 USA
Phone: 207-532-6880   Fax: 207-532-0910
E-mail: eagleman@mfx.net
Website: www.members.nbci.com/spruceacres/index.html

Founded: 1981
Membership: 300
Scope: International

Description: The A.E.H.W.C.C., Inc. and Spruce Acres Refuge have combined to provide a 65+ acre Refuge Rehabilitation center for people from all the world to enjoy. The Conservation Center is a nonprofit organization established for the purpose of preserving our natural resources and providing educational programs to all people to encourage proper wildlife and natural resource management.

Publication(s): Membership and Features, Coyotes in Maine, Trees - Walk The Nature Trails, Planet Earth, American Wetlands, If You Care Please Leave Them There

Keyword(s): Environmental and Conservation Education, Mammals, Raptors, Wetlands, accreditation

Contact(s):
Maxim Langstaff, CHAIRMAN OF THE BOARD
John Bushey, DIRECTOR
Arthur Howell, FOUNDER/PRESIDENT/WILDLIFE REHABILITATOR
A. Howell, GENERAL MANAGER
Amity, ME 04471
Dorothy Howell, SECRETARY AND TREASURER
Michael Michaud, VICE CHAIR
Penny Kern, VICE PRESIDENT

ALBERTA WILDERNESS ASSOCIATION

Box 6398, Station D
Calgary, Alberta T2P 2E1 Canada
Phone: 403-283-2025   Fax: 403-270-2743
E-mail: a.w.a@home.com
Website: albertawilderness.ca

Founded: 1968
Membership: 1800
Scope: Regional

Description: A province-wide, non-profit, charitable organization with a mission to be an advocate for wild Alberta through awareness and action and functioning on the values of eco-centredness, integrity, respectfulness, participation, tenacity and passion. The AWA promotes sound ideas and policies for wilderness conservation, fosters appreciation and enjoyment of wilderness, and works with government, industry, organizations and individuals to encourage careful management of Alberta's natural lands and waters.

Publication(s): Eastern Slopes Wildlands: Our Living Heritage, Landscapes of Southern Alberta, Other Pub Listings Available Upon Request, Wild Lands Advocate

Keyword(s): Conservation, Endangered Species, Protected Areas, National Parks, Public Lands, Wildlands, Wilderness, training

Contact(s):
Shirley Bray, LIBRARIAN
Cliff Wallis, PAST PRESIDENT
Cliss Wallis, PRESIDENT
Stephen Legault, VICE PRESIDENT

ALDO LEOPOLD FOUNDATION, INC.

E12919 Levee Rd.
Baraboo, WI 53913-0077 USA
Phone: 608-355-0279   Fax: 608-356-7309
E-mail: mail@aldoleopold.org
Website: www.aldoleopold.org

Founded: 1982
Membership: 300
Scope: National

Description: The mission is to promote harmony between people and land. Activities include restoration, research, education, maintaining photographic archives and acting as a clearinghouse for information about Aldo Leopold. The nonprofit organization was founded by the children of Aldo Leopold.

Publication(s): Leopold Outlook, The

Keyword(s): Ecology, Environmental and Conservation Education, Environmental Ethics, Land Conservation, Land Management, Native Plants, Prairies, Restoration, Training

Contact(s):
Teresa Searock, ADMINISTRATIVE ASSISTANT
teresa@aldoleopold.org
Estella Leopold, CHAIRMAN
Steve Swenson, ECOLOGIST
steve@aldoleopold.org
Wellington Huffaker, EXECUTIVE DIRECTOR
buddy@aldoleopold.org
Rob Nelson, OUTREACH COORDINATOR
rob@aldoleopold.org

ALLIANCE FOR THE CHESAPEAKE BAY
P.O. Box 1981
Richmond, VA 23218 USA
Phone: 804-775-0951   Fax: 804-775-0954
E-mail: mail@acb-online.org
Website: www.alliancechesbay.org

Founded: 1971
Membership: 790
Scope: State

Description: To build, maintain and serve the partnership among the general public, the private sector and the government that is essential for establishing and sustaining policy, programs and the political will to preserve and restore the resources of the Chesapeake Bay.

Publication(s): Watershed Watch, Bay Journal

Keyword(s): Environment, Environmental and Conservation Education, Environmental Protection, Wildlife, Pollution Prevention, Population Growth, Sustainable Development, Rivers, Toxic Substances, Nuclear-free, Water quantity, Water export and diversion, Coral Reefs, Exotic species, Aquatic nuisance species, Watersheds, Water Quality, Wetlands

Contact(s):
David Bancroft, EXECUTIVE DIRECTOR
6600 York Rd. Suite 100, Baltimore, MD 21212
Phone: 410-377-6270
Terry Harwood, PRESIDENT
Phone: 610-774-5043
Brigid Kenney, SECRETARY
Fax: 301-380-3106
Michael Marino, TREASURER
Phone: 410-347-6201
David Cottingham, VICE PRESIDENT
Phone: 202-466-4633
Charles Conklin, VICE PRESIDENT, MARYLAND
Phone: 410-234-5528
Marshall Kaiser, VICE PRESIDENT, PENNSYLVANIA
Phone: 717-763-4985
Susan Hansen, VICE PRESIDENT, VIRGINIA
Phone: 757-397-3481

ALLIANCE FOR THE CHESAPEAKE BAY
BALTIMORE OFFICE
6600 York Rd., Suite 100
Baltimore, MD 21212 USA
Phone: 410-377-6270   Fax: 410-377-7144
E-mail: mail@acb-online.org
Website: www.alliancechesbay.org

Founded: NA

Membership: 1500
Scope: Regional
Publication(s): BAY JOURNAL
Keyword(s): Environmental and Conservation Education, Rivers, Water Quality, Coral Reefs, Wetlands

Contact(s):
Gail Krus, OFFICE MANAGER

ALLIANCE FOR THE CHESAPEAKE BAY
CRIS OFFICE
P.O. Box 1981
Richmond, VA 23218 USA
Phone: 804-775-0951   Fax: 804-775-0954

Founded: NA
Scope: Statewide

ALLIANCE FOR THE CHESAPEAKE BAY
HARRISBURG OFFICE
600 N. Second St., #300B
Harrisburg, PA 17101 USA
Phone: 717-236-8825   Fax: 717-236-9019
E-mail: acbpa@acb-online.org
Website: www.acbpa.acb-online.org

Founded: NA
Scope: State

Publication(s): publications on website

Contact(s):
David Bancroft, EXECUTIVE DIRECTOR
dbancroft@acb-online.org

AMANAKAA AMAZON NETWORK
60 E. 13th St., 5th Fl.
New York, NY 10003 USA
Phone: 212-253-9502   Fax: 212-253-9507
E-mail: amanakaa@amanakaa.org
Website: http://www.amanakaa.org

Founded: 1990
Scope: National

Description: The Amanaka's Amazon Network is a nonprofit environmental and social justice organization. Amanaka'a serves as a liaison between the peoples of the Amazon and their allies in the U.S. We work to educate the American public about the Amazon Rainforest and its peoples and support grassroots organizations in the Amazon.

Publication(s): Letters from the Amazon, series of booklets by people of the Amazon, Amanaka'a Update

Keyword(s): Environmental Protection, Sustainable Development, Rainforests, Indigenous People, Cultural Preservation, Human Rights

Contact(s):
Christine Halvorson, EXECUTIVE DIRECTOR (ACTING)
Zeze Weiss, PRESIDENT
Christine Halvorson, SECRETARY
John Friede, VICE PRESIDENT

AMERICA THE BEAUTIFUL FUND
1730 K. St., NW, Suite 1002
Washington, DC 20006 USA
Phone: 202-638-1649   Fax: 202-204-0028
Website: www.freeseeds.org

Founded: 1965
Membership: 2000
Scope: National

Description: America the Beautiful Fund gives recognition, technical support, small seed grants, gifts of free seeds and national recognition awards to volunteers and community groups to initiate new local action projects improving the quality of the environment, including design, land preservation, local food production, arts, historical and cultural preservation and horticultural therapy.

Publication(s): Old Glory, Better Times, The Green Earth Guide

Keyword(s): Agriculture, Biodiversity, Botanical Gardens, Conservation, Culture, Cultural Preservation, Conservation Tillage, Ecology, Environmental and Conservation Education, Gardening and Horticulture, Grants, Historic Preservation, Flowers, Plants, and Trees, Health and Nutrition

Contact(s):
Thomas Farrell, CHAIRMAN OF THE BOARD
First Chicago, 153 W 51 St., New York, NY 10019
Nanine Bilski, PRESIDENT
1730 K St. Suite 1002, Washington, DC 20006
Daniel Schneider, SECRETARY
31 Mill Hills Rd, Woodstock, NY 12498
Phone: 914-679-9868
Susan Anderson, TREASURER
235 Mason Dr., Manhasset, NY 11030
Kay Lautman, VICE PRESIDENT
1730 Rhode Island Ave., NW, Suite 700, Washington, DC 20036
Jean Douglas, VICE PRESIDENT
4733 Woodway Ln., NW, Washington, DC 20016

AMERICAN ALLIANCE FOR HEALTH PHYSICAL EDUCATION AND RECREATION AND DANCE
1900 Association Dr.
Reston, VA 22091-1502 USA
Phone: 703-476-3400   Fax: 703-476-9527
Website: www.aahperd.org

Founded: NA
Membership: 3000
Scope: Statewide

Description: A voluntary professional organization for educators in the fields of physical education, sports and athletics, dance, health and safety, recreation and outdoor and environmental education. Its purpose is the improvement of education through such professional services as consultation, periodicals and special publications, conferences and workshops, leadership development, determination of standards and research.

Publication(s): Journal of Physical Education, Recreation and Dance, Research Quarterly for Exercise and Sports, Health Education, AAHPERD Update, Strategies

Contact(s):
Glenn Roswal, PRESIDENT ELECT OF BOARD OF GOVERNORS
Lucinda Adams, PRESIDENT OF BOARD OF GOVERNORS
Westfield State College, Westfield, MA 01086

AMERICAN AQUATICS, INC
TENNESSEE CHAPTER
103A Valley Court
Oak Ridge, TN 37830 USA
Phone: 865-483-0600   Fax: 865-483-0674
Website: www.american-aquatics.com

Founded: 1977
Membership: 10
Scope: Statewide

Contact(s):
Fred Heitman, PRESIDENT
Phone: 931-372-3194
sbc5959@tntech.edu

AMERICAN ASSOCIATION FOR LEISURE AND RECREATION-AALR
1900 Association Dr.
Reston, VA 20191 USA
Phone: 703-476-3472

Founded: 1939
Scope: National

Description: The AALR was formed to promote school, community and national leisure and recreation programs; to communicate to society the importance of intelligent use of leisure time; and to provide for those working or interested in recreation and leisure the opportunity to network and to join together for mutual strength and benefit.

Publication(s): Leisure Today, AAL Reporter

Keyword(s): Outdoor Recreation, Family Recreation, Leisure

Contact(s):
Randy Swedburg, PAST PRESIDENT
Concordia University, Montreal, QE H4B 1R6
Phone: 514-848-3331
Fax: 514-848-4200
Dale Adkins, PRESIDENT
Western Illinois University, Dept. of RPTA, Macomb, IL 61455
Phone: 309-298-1967
Fax: 309-298-2967
Vicki Highstreet, PRESIDENT ELECT
University of Nebraska, Lincoln, NE 68588
Phone: 402-472-4771
Fax: 402-472-8080
Gale Widow, REPRESENTATIVE TO THE BOARD OF GOVERNORS
University of South Dakota, Dakotadome 221A, Vermillion, SD 57069
Phone: 605-677-5336
Fax: 605-677-5338

AMERICAN ASSOCIATION FOR THE ADVANCEMENT OF SCIENCE
1200 New York Ave., NW
Washington, DC 20005 USA
Phone: 202-326-6400
E-mail: N\webmaster@aaas.org
Website: www.aaas.org

Founded: 1848
Scope: National

Description: Objectives are to further the work of scientists, to facilitate cooperation among them, to foster scientific freedom and responsibility, to improve the effectiveness of science in the promotion of human welfare and to increase public understanding and appreciation of the importance and promise of the methods of science in human progress.

Publication(s): Science, Science Books and Films, Science's Next Wave

Keyword(s): Wildlife Rehabilitation, Biodiversity, Communications, Greenhouse Effect/Global Warming, Population Growth

Contact(s):
Stephen Gould, CHAIRMAN OF THE BOARD
Donald Kennedy, EDITOR IN CHIEF
Richard Nicholson, EXECUTIVE OFFICER
Phone: 202-326-6639
Mary Good, PRESIDENT
William Golden, TREASURER
Phone: 212-425-0333

AMERICAN ASSOCIATION OF BOTANICAL GARDENS AND ARBORETA, INC.

351 Longwood Rd.
Kennett Square, PA 19348 USA
Phone: 610-925-2500   Fax: 610-925-2700
E-mail: aabga@aabga.org
Website: www.aabga.org

Founded: 1940
Membership: 2000
Scope: National

Description: AABGA is a nonprofit, membership organization serving North American botanical gardens, arboreta and their professional staffs.

Publication(s): Public Garden, The, AABGA Newsletter

Keyword(s): Botanical Gardens, Flowers, Plants, and Trees, Gardening and Horticulture, Internships

Contact(s):
Carla Pastore, EXECUTIVE DIRECTOR
Gerald Donnelly, PAST PRESIDENT
Executive Director of the Moton Arboretum, 4100 Illinois Rte. 53, Lisle, IL 60532-1293
Mary Matheson, PRESIDENT
Mary Casey, PUBLICATIONS MANAGER
mjcasey887@aol.com
Kathleen Socolofsky, SECRETARY, Director of UC Davis Arboretum
One Shields Ave., Davis, CA 95616
Richard Piacentini, TREASURER
Executive Director of Phipps Conservatory and Botanical Gardens, One Schenley Park, Pittsburgh, PA 15213-3830

AMERICAN ASSOCIATION OF FIELD BOTANISTS

P.O. Box 23542
Chattanoga, TN 37422
USA
4201 Gann Store Rd.
Chattanooga, TN 37422 USA

Founded: 1983
Scope: National

Description: The American Association of Field Botanists is an organization of amateur and professional botanists dedicated to the protection of native plants and the preservation of their natural habitats, the exchange of information necessary to understand and maintain biodiversity and the education of the public regarding threatened and endangered flora.

Publication(s): American Association of Field Botanists Newsletter

Keyword(s): Biodiversity, Conservation, Ecology, Endangered Species, Flowers, Plants, and Trees, Native Plants

Contact(s):
Charles Wilson, PRESIDENT
4201 Gann Store Rd., Hixson, TN 37343
Phone: 423-875-9625
Joyce Merritt, SECRETARY
327 Guild Dr., Chattanooga, TN 37421
Kurt Emmanuele, TREASURER
Baylor School, Box 1337, Chattanooga, TN 37401

AMERICAN ASSOCIATION OF ZOO KEEPERS, INC.

ADMINISTRATIVE OFFICES
3609 SW 29th St.
Topeka, KS 66614 USA
Phone: 785-273-9149   Fax: 785-273-1980
E-mail: akfeditor@kscablre.com
Website: www.aazk.org

Founded: 1967
Membership: 3600
Scope: National

Description: An international nonprofit organization of animal keepers and other persons interested in quality animal care and in promoting animal keeping as a profession. Chapters are active at zoos throughout North America. Promotes continuing education for keepers, national and international conservation projects, keeper-initiated zoo research, and educational publications.

Publication(s): Animal Keepers' Forum, Enrichment Notebook, second edition, Handbook of Zoonotic Diseases, Crisis Management Resource Notebook

Keyword(s): Endangered Species, Zoology, education, Conservation

Contact(s):
Barbara Manspeaker, ADMINISTRATIVE SECRETARY OF ADMINISTRATIVE OFFICES
Bob Hayes, BOARD OF DIRECTORS
bulletbobhayes@hotmail.com
Bruce Elkins, BOARD OF DIRECTORS
belkins@indyzoo.com
Jacque Blessington, BOARD OF DIRECTORS
Kansas City Zoological Gardens, 6700 Zoo Dr., Kansas City, MO 816-513-5700
Jan Reed-Smith, BOARD OF DIRECTORS
John Ball Zoo, 1300 W. Fulton St., Grand Rapids, MI 49504
Phone: 616-336-4301
Denise Wagner, BOARD OF DIRECTORS
dwagner@sandiegoz..org
Linda King, BOARD OF DIRECTORS
lmking83@aol.com
Susan Chan, EDITOR
Ed Hansen, EXECUTIVE DIRECTOR OF THE BOARD
3601 SW 29th, Suite 133, Topeka, KS 66614
Phone: 785-273-9149
Kevin Shelton, PRESIDENT
kshelton@flaquarium.org

AMERICAN BASS ASSOCIATION OF EASTERN PENNSYLVANIA/ NEW JERSEY, THE

Attn: President, 7 Logan Dr
Summerville, NJ 08876 USA
Phone: 908-526-7721   Fax: 908-685-0970
Website: www.emterco.com

Founded: NA
Scope: Statewide

Contact(s):
Paul Rinaldo, PRESIDENT

AMERICAN BASS ASSOCIATION, INC.

P.O Box 896
Gate City, VA 24251-0896 USA
Phone: 540-386-2109
E-mail: aba@mounet.com

Founded: 1985
Scope: National

Description: A nonprofit, tax-exempt national association dedicated to protecting and enhancing America's fishery resources; to promoting bass fishing as a major sport; and to teaching young people the fun of fishing and instilling in them an appreciation of the life-giving waters of America.

Keyword(s): Fish, Fish Wildlife Management, Water Quality, Sport Fishing

Contact(s):
Audrey Barnett, DIRECTOR
1201 Bohmen, Pueblo, CO 81006
Wayne Hood, DIRECTOR
2909 N. Bayshore Dr., LaCrosse, WI 54603
Paul Noechel, DIRECTOR OF EASTERN REGION
Rt. 2 Box 270, Lost Creek, WV 26385
John Cowan, DIRECTOR OF NORTHEAST REGION
235 Ridgeview, Weare, NH 03281
Ed Metzger, DIRECTOR-AT-LARGE
710 Edgewater Dr., Inverness, FL 34450
Bob Barker, PRESIDENT
John Keegan, SECRETARY
104 S. Cove Rd., Williamsburg, VA 23188
Phone: 757-564-0825

AMERICAN BIRD CONSERVANCY

P.O. Box 249
The Plains, VA 20198 USA
Phone: 540-253-5780   Fax: 540-253-5782
E-mail: abc@abcbirds.org
Website: www.abcbirds.org

Founded: 1994
Membership: 2600
Scope: National

Description: American Bird Conservancy (ABC), is a U.S. based, nonprofit, membership organization dedicated to the conservation of wild birds and their habitats throughout the Americas. ABC supports the Partners in Flight initiative through the magazine Bird Conservation and programs such as the Important Bird Areas Program.

Publication(s): Annual Report, Bird Conservation Directory 2001, All the Birds of North America (field guide), Communication Towers: A Deadly Hazard to Birds, Recovering Paradise: Making Pasture Land Productive for People & Biodiversity, Conservation of Land Birds of The US, see publications on website, Bird Conservation, Bird Calls

Keyword(s): Birds, Conservation, Endangered Species, Environmental Planning

Contact(s):
Howard Brokaw, CHAIRMAN
Linda Winter, DIRECTOR OF CATS INDOORS CAMPAIGN
Robert Chipley, DIRECTOR OF IMPORTANT BIRD AREAS
Kelley Tucker, DIRECTOR OF PESTICIDES AND BIRDS CAMPAIGN
George Fenwick, PRESIDENT
Cynthia Lenhart, VICE CHAIR
David Pashley, VICE PRESIDENT OF CONSERVATION
Merrie Morrison, VICE PRESIDENT OF MEMBERSHIP DEVELOPMENT
Gerald Winegrad, VICE PRESIDENT OF POLICY
Mike Parr, VICE PRESIDENT OF PROGRAM DEVELOPMENT

AMERICAN BIRDING ASSOCIATION

P.O. Box 6599
Colorado Springs, CO 80934 USA
Phone: 719-578-9703   Fax: 719-578-1480
E-mail: member@aba.org
Website: www.americanbirding.org

Founded: 1969
Membership: 22,000
Scope: National

Description: The American Birding Association provides leadership to field birders by increasing their knowledge, skills and enjoyment of birding. The ABA supports the interests of birders of all ages and experience and actively encourages the conservation of birds and their habitats.

Publication(s): Birding, A Birds Eye View, Winging It, North American Birds, ABA Checklist, ABA/Lane Series of Birdfinding Guides (15 titles), Volunteer Directory, Membership Directory

Keyword(s): Biodiversity, Birds, Environmental and Conservation Education, Outdoor Recreation, training

Contact(s):
Richard Payne, CHAIRMAN OF THE BOARD
rhp@shsu.edu
Matt Pelikan, EDITOR OF NEWSLETTER
winging@aba.org
Paul Baicich, EDITOR OF THE MAGAZINE
baicich@aba.org
Paul Green, EXECUTIVE DIRECTOR
Ann Stone, SECRETARY
104673.1143@compuserve.com
Dennis Lacoss, TREASURER
dlacoss@mindspring.com

AMERICAN CAMPING ASSOCIATION, INC.

5000 State Rd. 67N
Martinsville, IN 46151 USA
Phone: 765-342-8456   Fax: 765-342-2065
E-mail: aca@acacamps.org
Website: www.acacamps.org

Founded: 1910
Scope: National

Description: The American Camping Association is a national community of camp professionals dedicated to enriching the lives of children and adults through camp experience. ACA recognizes the camp experience as a significant contributor to positive child and youth development. It is the only organization that accredits all types of camps based on 300 standards for health, safety and program quality.

Publication(s): Camping Magazine, Guide to Accredited Camps, Various camp-related publications

Keyword(s): Environmental and Conservation Education, Health and Nutrition, Internships, Outdoor Recreation, Youth Organizations, Camp

Contact(s):
Terri Nicodemus, DIRECTOR OF PUBLIC RELATIONS
Peg Smith, EXECUTIVE DIRECTOR
Rodger Popkin, PRESIDENT

AMERICAN CANAL SOCIETY, INC.
840 Rinks Ln.
Savannah, TN 38372-6774 USA
Website: http://www.canals.com/ACS/acs.html

Founded: 1972
Scope: National

Description: A nonprofit organization dedicated to historic canal research, preservation and canal parks.

Publication(s): American Canals (Quarterly Bulletin), American Canal Guides 1, 2, 3, 4, 5, Best From American Canals 1, 2, 3, 4 5, 6, 7, 8

Keyword(s): Cultural Preservation

Contact(s):
David Ross, EDITOR
Terry Woods, PRESIDENT
6939 Eastham Circle, Canton, OH 44708
Charles Derr, SECRETARY AND TREASURER
117 Main St., Freemansburg, PA 18017
Phone: 610-691-0956

AMERICAN CAVE CONSERVATION ASSOCIATION
119 E. Main St., P.O. Box 409
Horse Cave, KY 42749 USA
Website: www.cavern.org/acca.htm

Founded: 1977
Scope: National

Description: The ACCA is a national organization formed to conserve caves and karstlands and other resources associated with them. Primary objectives are to provide information, technical assistance, and public education and management training programs; and operation of the American Cave and Karst Center, a national environmental education center and museum.

Publication(s): American Caves Magazine

Keyword(s): Endangered Species, Environmental and Conservation Education, Geology, Land Use Planning, Exotic species, Aquatic nuisance species, Museum, Cave

Contact(s):
David Foster, EXECUTIVE DIRECTOR
Phone: 502-786-1466
Roy Powers, PRESIDENT
Phone: 540-546-5386
Tom Aley, VICE PRESIDENT
Phone: 417-785-4289

AMERICAN CETACEAN SOCIETY
San Pedro, CA 90733-0391 USA
Phone: 310-548-6279   Fax: 310-548-6950
E-mail: acs@pobox.com
Website: www.acsonline.org

Founded: 1967
Scope: National

Description: A nonprofit organization that works in the areas of conservation, education and research to protect marine mammals, especially whales, dolphins and porpoises and the oceans they live in.

Publication(s): Whalewatcher: Journal of the American Cetacean Society

Keyword(s): Aquatic Habitats, Endangered Species, Environmental and Conservation Education, Marine Mammals, Whale, Dolphin, Seal, training, Dolphins, Whales

AMERICAN CHESTNUT FOUNDATION, THE
P.O. Box 4044
Bennington, VT 05201-4044 USA
Phone: 802-447-0110   Fax: 802-442-6855
E-mail: chestnut@acf.org
Website: www.acf.org

Founded: 1983
Membership: 5000
Scope: National

Description: Funded by private contributions, the purpose of The American Chestnut Foundation is to promote the preservation and restoration of the American chestnut, an important wildlife and timber tree killed by a blight early in the Twentieth Century; to operate two research breeding farms in Meadowview, VA; to provide grants for cutting edge research; and to identify surviving trees and establish satellite research plantings.

Publication(s): Bark, The, Journal of The American Chestnut Foundation

Keyword(s): Biodiversity, Endangered Species, Flowers, Plants, and Trees, training, Trees

Contact(s):
Gerrie Rousseau, MEMBERSHIP DIRECTOR
Central Lake, MI 49622
Tammy Carpenter, DEVELOPMENT DIRECTOR
469 Main St., P.O. Box 4044, Bennington, VT 05201-4044
Kelly Grundman, MEMBERSHIP DIRECTOR
469 Main St., P.O. Box 4044, Bennington, VT 05201-4044
Dennis Fulbright, SECRETARY
Dept. of Botany and Plant Pathology, 166 Plant Biology, Michigan State University, E.Lansing, MI 48824
William Macdonald, TREASURER
College of Agriculture and Forestry, West Virginia University, P.O. Box 6057, Morgantown, WV 26506
Forrest Macgregor, VICE PRESIDENT OF DEVELOPMENT
TACF Asheville Office, 46 Haywood St., Suite 213, Ashevill, NC 28801
J. Craddock, VICE PRESIDENT OF SCIENCE
U of TN - Chattanooga, 615 McCallie Ave, Chattanooga, TN 37403-2598

AMERICAN CONSERVATION ASSOCIATION, INC.

1200 New York Avè., NW, Suite 400
Washington, DC 20005 USA
Phone: 202-289-2431   Fax: 202-289-1396

Founded: 1958
Scope: National

Description: A nonmembership nonprofit, educational and scientific organization formed to advance knowledge and understanding of conservation and to preserve and develop natural resources for public use.

Keyword(s): Air Quality and Pollution, Coasts, Environmental Law, Outdoor Recreation, Urban Environment

Contact(s):
Charles Clusen, EXECUTIVE DIRECTOR
Laurance Rockefeller, FOUNDER, HONORARY TRUSTEE
Laurance Rockefeller, PRESIDENT
R. Greathead, SECRETARY
Carmen Reyes, TREASURER

AMERICAN CONSERVATION ASSOCIATION, INC.

NEW YORK OFFICE
30 Rockefeller Plaza, Rm. 5402
New York, NY 10112 USA
Phone: 212-649-5822

Founded: NA
Scope: National

Contact(s):
Carmen Meyers, MAIN CONTACT

AMERICAN COUNCIL FOR AN ENERGY-EFFICIENT ECONOMY

1001 Connecticut Ave., NW, 801
Washington, DC 20036-5525 USA
Phone: 202-429-8873   Fax: 202-429-2248
E-mail: info@aceee.org
Website: aceee.org

Founded: 1980
Scope: National

Description: Advancing energy efficiency as a means of promoting both economic prosperity and environmental protection. ACEEE conducts technical and policy assessments; advises governments and utilities; publishes books, conference proceedings and reports; organizes conferences and workshops; and informs consumers.

Publication(s): Consumer Guide to Home Energy Savings: Guide to Energy-Efficient Office Equipment, Transportation and Energy: Strategies for a Sustainable Transportation System, Using Consensus Building to Improve Utility Reulation, Energy Innovations: A Prosperous Path

Keyword(s): Transportation, Energy Efficiency, Utility Restructuring, Energy, Pollution Prevention

Contact(s):
Steve Nadel, DIRECTOR
Phone: 202-429-8873
Fax: 202-429-2248
Carl Blumstein, PRESIDENT
Phone: 202-429-8873
Fax: 202-429-2248

AMERICAN EAGLE FOUNDATION

P.O. Box 333
Pigeon Forge, TN 37868 USA
Phone: 865-429-0157   Fax: 865-429-4743
E-mail: eaglemail@eagles.org

Founded: 1985
Scope: National

Description: Dedicated to saving, restoring, and protecting America's endangered national symbol, the Bald Eagle, and preserving America's wildlife, waterways, forests, natural resources, ecosystems, and environment.

Publication(s): American Eagle News

Contact(s):
Al Cecere, PRESIDENT AND CEO
P.O. Box 120206, Nashville, TN 37212
Steven Compton, SECRETARY
Brentwood, TN
Juliette Ingram, TREASURER
Nashville, TN
Bobby Halliburton, VICE PRESIDENT
Nashville, TN

AMERICAN FARMLAND TRUST

1200 18th St., NW
Washington, DC 20036 USA
Phone: 202-331-7300   Fax: 202-659-8339
E-mail: info@farmland.org
Website: www.farmland.org

Founded: 1980
Scope: National

Description: AFT is a nonprofit conservation organization working to stop the loss of productive farmland and to promote farming practices that lead to a healthy environment. Its programs include public education, technical assistance, policy development and direct farmland-protection projects.

Publication(s): Your Land Is Your Legacy (1997), Farming on the Edge II (1997), Living on the Edge:The Costs and Risks of Scatter Development (1998), Sharing the Responsibility: What Agricultural Landowners Think About Propert, Saving American Farmland: What Works (1997)

Keyword(s): Agriculture, Environmental and Conservation Education, Land Purchase, Land Use Planning, Wetlands, Alternative Agriculture, Air Quality

Contact(s):
Robert Wagner, ASSISTANT VICE PRESIDENT OF FIELD PROGRAMS
William Reilly, CHAIRMAN OF THE BOARD
Sharon Phenneger, CONTROLLER
Julia Freedgood, DIRECTOR FOR FARMLAND ADVISORY SERVICES
jfreedgood@farmland.org
Ann Sorensen, DIRECTOR OF CENTER FOR AGRICULTURE IN THE ENVIRONMENT
Bryan Petrucci, DIRECTOR OF FARMS
Dennis Bidwell, DIRECTOR OF LAND PROTECTION
dbidwell@farmland.org
Bernadine Prince, DIRECTOR OF PUBLIC EDUCATION
bprince@farmland.org
Ralph Grossi, PRESIDENT
rgrossi@farmland.org
Edward Thompson, SENIOR VICE PRESIDENT FOR PUBLIC POLICY

Edward Harte, VICE CHAIRMAN OF THE BOARD
Jimmy Daukas, VICE PRESIDENT FOR MARKETING
jdaukas@farmland.org
Tim Warman, VICE PRESIDENT FOR PROGRAMS
twarman@farmland.org

AMERICAN FEDERATION OF MINERALOGICAL SOCIETIES

Central Office, P.O. Box 891
Oklahoma City, OK 73189-1208 USA
Phone: 405-682-2938
E-mail: central_office@amfed.org
Website: http://www.amfed.org/

Founded: 1945
Membership: 56,000
Scope: National

Description: To promote popular interest and education in the various earth sciences, in particular, the subjects of geology, mineralogy, paleontology, lapidary and other related subjects, and to sponsor and provide means of coordinating the work and efforts of all persons and groups interested therein; to sponsor and encourage the formation and international development of societies and regional federations, and by and through such means to strive toward greater international goodwill and fellowship.

Publication(s): American Federation Newsletter, AFMS Uniform Rules booklets, AFMS Safety Manual

Keyword(s): Protected Areas, Public Lands, Wilderness, Mineral Resources

Contact(s):
Mel Albright, EDITOR
Rt. 3 Box 8500, Bartlesville, OK 74003
Lewis Elrod, PRESIDENT
2699 Lascassas Pike, Murfreesboro, TN 37230
Dan McLennan, SECRETARY
P.O. Box 26523, Oklahoma City, OK 73126-0523
Toby Cozens, TREASURER
4401 SW Hill St., Seattle, WA 98116-1924

AMERICAN FISHERIES SOCIETY

5410 Grosvener Ln., Suite 110
Bethesda, WA 20814 USA
Phone: 301-897-8616   Fax: 301-897-8616
E-mail: fisheriesmain@fisheries.org
Website: www.flsherles.org

Founded: NA
Membership: 10
Scope: National

Contact(s):
Ken Bael, PRESIDENT

AMERICAN FISHERIES SOCIETY

Headquarters, 5410 Grosvenor Ln., Suite 110
Bethesda, MD 20814 USA
Phone: 301-897-8616   Fax: 301-897-8096
E-mail: main@fisheries.org
Website: www.fisheries.org

Founded: 1870
Scope: National

Description: A professional society to promote the conservation, development, and wise utilization of fisheries, both recreational and commercial.

Publication(s): Transactions of the American Fisheries Society, North American Journal of Fisheries Management

Keyword(s): Aquatic Habitats, Water Pollution Management, Professional Organization, Wildlife, Renewable Resources, Coral Reefs, Wetlands

Contact(s):
Ira Adelman, 2ND VICE-PRESIDENT
Betsy Fritz, DIRECTOR OF ADMINISTRATION AND FINANCE
Ghassan Rassam, EXECUTIVE DIRECTOR
grassam@fisheries.org
Carl Burger, PRESIDENT
carl_v_burger@r1.fws.gov
Fred Harris, PRESIDENT-ELECT
harrisfa@mail.wildlife.state.nc.us

AMERICAN FISHERIES SOCIETY

5410 Grosvenor Lane, Suite 110
Bethesda, MD 98368 USA
Phone: 301-897-8616   Fax: 301-897-8096
E-mail: main@fisheries.org
Website: http://www.fisheries.org

Founded: 1975
Scope: National

Description: Abundant Life Seed Foundation is a nonprofit, tax-exempt organization that propagates and preserves seeds of Northwest native plants and heritage (non-hybrid) vegetables, herbs, and flowers. The Foundation conducts the distribution of seeds (and related books) via a mail-order catalog. Also operates the World Seed Fund, donating seed internationally to those in need, both in the United States and internationally.

Publication(s): Seed and Book Catalog, Seed Midden

Keyword(s): Agriculture, Endangered Species, Flowers, Plants, and Trees, Native Plants, Gardening and Horticulture

Contact(s):
Kirsten Fzykitka, OUTREACH COORDINATOR
Elsa Goltf, PRESIDENT

AMERICAN FISHERIES SOCIETY

5800 A River Pond Rd.
Tuttle Creek Park
Manhattan, KS 66502 USA
Phone: 785-539-7941   Fax: 785-539-3183
Website: www.kdwp.state.ks.us

Founded: 1975
Membership: 8
Scope: Statewide

Contact(s):
Chuck Bever, PRESIDENT

AMERICAN FISHERIES SOCIETY

AGRICULTURE ECONOMICS SECTION
University of Florida
Gainesville, FL 32611 USA
Phone: 352-392-4991   Fax: 352-392-3646
Website: www.fred.ifas.ufl.edu

Founded: NA
Membership: 42
Scope: Regional

Contact(s):
Chris Andrew, ASSISTANT CHAIRMAN
Burl Long, ASSISTANT CHAIRMAN
John Gordon, CHAIRMAN
Charles Adams, PRESIDENT
adams@fred.ifas.ufl.edu

AMERICAN FISHERIES SOCIETY
BIOENGINEERING SECTION
1646 Jeannette Pl.
Bainbridge Island, WA 98110 USA
Phone: 508-829-6000   Fax: 206-842-8195
E-mail: dailydesign@bainbridge.net

Founded: NA
Scope: National

Contact(s):
Wayne Daley, PAST-PRESIDENT
wjd1163@aol.com
Ned Taft, PRESIDENT
Phone: 508-829-6000
ntaft@aldenlab.com

AMERICAN FISHERIES SOCIETY
CANADIAN AQUATIC RESOURCES SECTION
2204 Main Mall, Univ. BC
Vancouver, British Columbia V6T 1Z4 Canada
Phone: 604-222-6753   Fax: 604-660-1849
Website: www.fisheries.org/cars/index/htm

Founded: NA
Scope: National

Contact(s):
Bruce Ward, PRESIDENT
bruce.ward@gems8.gov.bc.ca
Martin Castongue, PRESIDENT-ELECT

AMERICAN FISHERIES SOCIETY
COMPUTER USER SECTION
Dept. of Natural Sciences, MS Valley State Univ.
Itta Bena, MS 38941 USA
Phone: 601-254-3383   Fax: 601-254-3668

Founded: NA
Scope: National

Contact(s):
Michael Porter, PRESIDENT
mdporter@cypress.mcsr.olemiss.edu

AMERICAN FISHERIES SOCIETY
EARLY LIFE HISTORY
NOAA - National Marine Fisheries Service, The Beaufort
Laboratory, 101 Pivers Island Rd.
Beaufort, NC 28526 USA
Phone: 206-526-4108   Fax: 252-728-8747

Founded: NA
Scope: National

Publication(s): Stages Newsletter

Contact(s):
Art Kendal, PRESIDENT
art.kendal@noaa.org

AMERICAN FISHERIES SOCIETY
EQUAL OPPORTUNITIES SECTION
402 W. Washigton St. Rm W273
Indianapolis, IN 46205 USA
Phone: 371-233-5468   Fax: 317-233-3882
Website: fisheries.org

Founded: NA
Scope: National

Contact(s):
Gwen White, PRESIDENT
gwhite@dnr.state.in.us

AMERICAN FISHERIES SOCIETY
FISH CULTURE SECTION
3425 Settlers Rd
LaPorte, CO 80535 USA
Phone: 304-724-4457

Founded: NA
Membership: 450
Scope: National
Publication(s): Quarterly Newsletter, Quarterly Journal- North
American Journal of Aquaculture

Contact(s):
Steve Flickinger, PRESIDENT
Phone: 970-484-4167
flick@worldnet.att.net

AMERICAN FISHERIES SOCIETY
FISH HEALTH SECTION
CA State Univ., Dept. of Biological Sciences
Hayward, CA 94542 USA
Phone: 541-737-1856   Fax: 541-737-0496
Website: http://www.fisheries.org/fhs/

Founded: NA
Scope: National

Publication(s): Publications on website

Contact(s):
Jerri Bartholomew, PRESIDENT
bartholj@bcc.arst.edu

AMERICAN FISHERIES SOCIETY
FISHERIES ADMINISTRATORS SECTION
5410 Grosvenor Lane
Bethesda, Maryland 20814 USA
Phone: 301-897-8616   Fax: 301-897-8096
E-mail: main@fisheries.org
Website: www.fisheries.org

Founded: NA
Scope: National

Contact(s):
Michael Staggs, PRESIDENT
Phone: 608-267-0796
Fax: 608-266-2244
staggm@dnr.state.wi.us

AMERICAN FISHERIES SOCIETY
FISHERIES HISTORY SECTION
2901 Channel Dr.
Stevens Point, WI 54481 USA
Phone: 715-344-0152   Fax: 715-346-3624
Website: www.fishieries.org

Founded: NA
Membership: 110
Scope: International
Publication(s): Fisheries History Section Newsletter

Contact(s):
Daniel Coble, PRESIDENT

AMERICAN FISHERIES SOCIETY
FISHERIES MANAGEMENT SECTION
OK Fish Res Lab, 500 E. Constellation
Norman, OK 73072 USA
Phone: 405-325-7288   Fax: 405-325-7631

Founded: NA
Scope: National

Contact(s):
Jeff Boxrucker, PRESIDENT
jboxrucker@aol.com

AMERICAN FISHERIES SOCIETY
GENETICS SECTION
4302 Underwood St.
University Park, MD 20782 USA
Phone: 301-864-2553
Website: fisheries.org

Founded: NA
Scope: National

Contact(s):
John Epifanio, PRESIDENT
jepifan@atlas.vcu.edu

AMERICAN FISHERIES SOCIETY
INDIANA CHAPTER
Attn: President, Scott Shuler Aquatic Control, P.O. Box 100
Seymour, IN 47274 USA
Phone: 812-497-2410
Website: www.bsu.edu\csh\bio\inafs\

Founded: 1970
Scope: Statewide

Contact(s):
Scott Shuler, PRESIDENT

AMERICAN FISHERIES SOCIETY
INTRODUCED FISH SECTION
1415 Green Road
Ann Arbor, MI 48105 USA
E-mail: jacqueline_savino@usgs.gov

Founded: NA
Scope: National

Contact(s):
John Cassani, PRESIDENT
jcassani@peganet.com

AMERICAN FISHERIES SOCIETY
LOUISIANA CHAPTER
Attn: President, LA Dept. of Wildlife and Fisheries, P.O. Box 98000
Baton Rouge, LA 70898 USA
Phone: 225-765-2375

Founded: 1979
Membership: 60
Scope: Local
Publication(s): Abstracts of Annual Technical Session

Contact(s):
Terry Tiersch, PRESIDENT

AMERICAN FISHERIES SOCIETY
MARINE FISHERIES SECTION
Hatfield Marine Science Center, Oregon State University, 2030 Marine Sciences Dr.
Newport, OR 97365 USA
Phone: 541-867-0135   Fax: 541-867-0138
Website: www.fisheries.org

Founded: NA
Membership: 375
Scope: National
Publication(s): www.fisheries.org/publications.html

Contact(s):
Anne Richards, PRESIDENT
Phone: 508-495-2393

AMERICAN FISHERIES SOCIETY
NATIVE PEOPLE FISHERIES SECTION
USFWS, 4401 N. Fairfax Dr., Suite 840
Arlington, VA 22203 USA
Phone: 703-358-1718   Fax: 703-358-2044

Founded: NA
Scope: National

Contact(s):
Hannibal Bolton, PRESIDENT
hannibal_bolton@fws.gov

AMERICAN FISHERIES SOCIETY
NORTHEASTERN DIVISION
NYSDEC - Bureau of Fisheries 50 Wolf Road
50 Wolf Road
Albany, NY 12233-5827 USA
Phone: 518-457-5420   Fax: 518-485-5827
E-mail: dxstang@gw.bec.state.ny.us
Website: www.fisheries.org

Founded: NA
Scope: International, Regional

Contact(s):
Douglas Stang, PRESIDENT
1156 Meadowdale Rd., Altamont, NY 518-457-9435
Phone: 518-485-5827
dxstang@gw.bec.state.ny.us

AMERICAN FISHERIES SOCIETY
PHYSIOLOGY SECTION
Towson State University
Dept. of Biological Sciences
Towson, MD 21204-7097 USA
Phone: 410-830-3945   Fax: 604-822-4400
E-mail: nelson@saber.towson.edu
Website: http://www.fisheries.org/phs/

Founded: NA
Scope: National

Contact(s):
George Iwama, PRESIDENT
giwama@unixg.ubc.ca

AMERICAN FISHERIES SOCIETY
SOUTHERN DIVISION, WESTERN DIVISION, NORTH-
EASTERN DIVISION & NORTHCENTRAL
5410 Grosvenor Lane
Bethesda, MD 20814 USA
Phone: 301-897-8616   Fax: 301-897-8096
Website: www.fisheries.org

Founded: NA
Scope: National

Contact(s):
Jeff Boxrucker, PRESIDENT
Bethesda, MD 20814
Phone: 405-325-7288
Phone: 301-897-8096
jboxrucker@aol.com

AMERICAN FISHERIES SOCIETY
WATER QUALITY SECTION
5083 Veranda Terrace
Davis, CA 695616 USA
Phone: 541-754-4516   Fax: 916-278-3071

Founded: NA
Membership: 3000
Scope: National
Publication(s): Water Quality Matters

Contact(s):
Robert Hughes, PAST PRESIDENT
Larry Brown, PRESIDENT
John Meldrum, SECRETARY- TREASURER

AMERICAN FISHERIES SOCIETY
WESTERN DIVISION
Alaska Biological Science Center 1011 East Tubar Road
Anchorage, AK 99503 USA
Phone: 907-786-3842   Fax: 907-786-3636
Website: http://www.fisheries.org/wd/

Founded: NA
Scope: National

Contact(s):
Eric Knudsen, PRESIDENT
Phone: 907-786-3842
Fax: 907-786-3636
Eric_Knudsen@usga.gov

AMERICAN FISHERIES SOCIETY,
ALABAMA CHAPTER
Alabama Chapter, 3355 Audubon Rd.
Montgomery, AL 36106-2404 USA
Phone: 334-353-7998

Founded: 1991
Scope: Statewide

Contact(s):
Gregory Lein, PRESIDENT
Phone: 334-844-9318
glein@acesag.auburn.edu

AMERICAN FISHERIES SOCIETY,
ALASKA CHAPTER
ASGS/Alaska Science Center
1011 East Tudor Road
Anchorage, AK 99703 USA
Website: http://www.fisheries.org/afs-ak/

Founded: 1973
Scope: Statewide

Description: The American Fisheries Society is dedicated to the
preservation and conservation of aquatic resources, and facilitating the information exchange between students, professionals, and the public regarding our knowledge of aquatic
resources.

Contact(s):
Carol Ann Woody, PRESIDENT
Phone: 907-786-3314
Fax: 907-786-3636
carol_woody@usgs.gov

AMERICAN FISHERIES SOCIETY,
ARIZONA-NEW MEXICO CHAPTER
2221 W.Green WY Rd
Phoenix, AZ 85023 USA
Phone: 602-789-3258
Website: www.fisheries.org

Founded: 1967
Scope: International

Publication(s): AFS journals, online fisheries magazine

Contact(s):
Larry Riley, PRESIDENT
Phone: 602-789-3258

AMERICAN FISHERIES SOCIETY,
ARKANSAS CHAPTER
Aquaculture & Fisheries Center UAPB
P.O.B. 4912
Pine Bluff, AR 71611 USA

Founded: 1986
Scope: Statewide

Description: The American Fisheries Society is dedicated to the
preservation and conservation of aquatic resources, and facilitating the information exchange between students, professionals,
and the public regarding our knowledge of aquatic resources.

Contact(s):
Steve Lochmann, PRESIDENT
Phone: 870-543-8165
slochmann@auex.edu

AMERICAN FISHERIES SOCIETY,
AUBURN UNIVERSITY CHAPTER
Attn: President, 203 Swingle Hall
Auburn University, AL 36840 USA
Phone: 334-844-4058

Founded: 1974
Scope: Statewide

Contact(s):
Joe Slaughter, PRESIDENT

AMERICAN FISHERIES SOCIETY,
BONNEVILLE CHAPTER
Attn: President, P.O. Box 305
Dutch John, UT 84023 USA
Phone: 801-789-3103

Founded: 1963
Scope: Statewide

Publication(s): bulletins, quarterly newsletter

Contact(s):
Roger Schneidervin, PRESIDENT
Phone: 801-885-3249
nrdwr.rschneid@state.ut.us

AMERICAN FISHERIES SOCIETY,
CALIFORNIA-NEVADA CHAPTER
Trust for Public Land
116 New Montgomery Street, Suite 300
San Francisco, CA 94105 USA
Website: http://www.afs-calneva.org

Founded: 1963
Scope: Regional

Description: The American Fisheries Society is dedicated to the preservation and conservation of aquatic resources, and facilitating the information exchange between students, professionals, and the public regarding our knowledge of aquatic resources.

Contact(s):
Elise Holland, PRESIDENT
Phone: 415-495-5660
Fax: 415-495-0541
elise.holland@tpl.org

AMERICAN FISHERIES SOCIETY,
COLLEGE OF ENVIRONMENTAL SCIENCE AND FORESTRY CHAPTER
Syracuse, NY USA
Website: http://www.esf.edu/org/afs/

Founded: 1975
Scope: Statewide

Description: The American Fisheries Society is dedicated to the preservation and conservation of aquatic resources, and facilitating the information exchange between students, professionals, and the public regarding our knowledge of aquatic resources.

Contact(s):
Stephen Coghlan, PRESIDENT
Phone: 315-472-0488
clapton18@hotmail.com

AMERICAN FISHERIES SOCIETY,
COLORADO-WYOMING CHAPTER
Wyoming Game and Fish Department
P.O.B. 67
Jackson, WY 83001 USA
Phone:
Website: http://www.fisheries.org/co-wy/

Founded: 1966
Scope: Statewide

Description: The American Fisheries Society is dedicated to the preservation and conservation of aquatic resources, and facilitating the information exchange between students, professionals, and the public regarding our knowledge of aquatic resources.

Contact(s):
Rob Gipson, PRESIDENT
Phone: 307-733-2321
Fax: 307-733-2276
rgipso@state.wy.us

AMERICAN FISHERIES SOCIETY,
DAKOTA CHAPTER
1200 Mossourri Ave. P.O. Box 5520
Bismarck, ND 58506 USA
Phone: 701-328-5210   Fax: 701-328-5200
Website: www.health.state.nd.us/ndhd/default.asp

Founded: 1964
Membership: 1
Scope: Statewide
Publication(s): ARI News Bulletin

Contact(s):
Scott Elstad, ENVIRONMENTAL SCIENTIST
Wade King, PRESIDENT

AMERICAN FISHERIES SOCIETY,
FLORIDA CHAPTER
Website: http://nerps.nerdc.ufl.edu/~fafs/

Founded: 1981
Scope: Statewide

Description: The American Fisheries Society is dedicated to the preservation and conservation of aquatic resources, and facilitating the information exchange between students, professionals, and the public regarding our knowledge of aquatic resources.

Contact(s):
Peter Hood, PRESIDENT
peter.hood@gulfcouncil.org

AMERICAN FISHERIES SOCIETY,
GREATER PORTLAND, OR CHAPTER
Science Division, Mt. Hood Community College
26000 SE Stark Street
Gresham, OR 97030 USA

Founded: 1962
Scope: Statewide

Publication(s): Urban Naturalist, The, Familiar Birds of the Northwest, Protecting a Vanishing Ecosystem - The Ancient Forests of the Pacific Northwest, Audubon Warbler

Contact(s):
Todd Hanna, PRESIDENT
Phone: 503-491-7163
Fax: 503-491-7481
hannat@mhcc.cc.or.us

AMERICAN FISHERIES SOCIETY,
HAWAII CHAPTER
P.O.B. 22085
Honolulu, HI 96823 USA
Website:
http://home.hawaii.rr.com/ikehara/afshi/index.html

Founded: 1982
Scope: Statewide

Description: The American Fisheries Society is dedicated to the preservation and conservation of aquatic resources, and facilitating the information exchange between students, professionals, and the public regarding our knowledge of aquatic resources.

Contact(s):
Walter Ikehara, PRESIDENT
Phone: 808-587-0096
Fax: 808-587-0115
walter_n_ikehara@exec.state.hi.us

AMERICAN FISHERIES SOCIETY,
HUMBOLDT CHAPTER
Arcata, CA 95521 USA

Founded: 1973
Scope: Local Region

Description: The American Fisheries Society is dedicated to the preservation and conservation of aquatic resources, and facilitating the information exchange between students, professionals, and the public regarding our knowledge of aquatic resources.

Contact(s):
Mary Knapp, PRESIDENT
mary_m_knapp@fws.gov

AMERICAN FISHERIES SOCIETY,
IDAHO CHAPTER
Eagle, ID 83616 USA
Website: http://www.fisheries.org/idaho/

Founded: 1963
Scope: Statewide

Description: The American Fisheries Society is dedicated to the preservation and conservation of aquatic resources, and facilitating the information exchange between students, professionals, and the public regarding our knowledge of aquatic resources.

Contact(s):
Ted Koch, PRESIDENT
ted_koch@mail.fws.gov

AMERICAN FISHERIES SOCIETY,
ILLINOIS CHAPTER
Illinois Chapter of the American Fisheries Society
In care of Illinois Dept. of Natural Resource Division of Fisheries 524 S. 2nd St.
Springfield, IL 62701-1787 USA
Phone: 309-582-5611   Fax: 309-582-5613

Founded: 1963
Scope: Regional

Contact(s):
R. Sallee, PRESIDENT
Phone: 309-582-5611

AMERICAN FISHERIES SOCIETY,
IOWA CHAPTER
110 Lake Darling Rd.
Brighton, IA 52540 USA

Founded: 1969
Scope: Statewide

Contact(s):
Don Kline, PRESIDENT

AMERICAN FISHERIES SOCIETY,
KENTUCKY CHAPTER
Kentucky Department of Fish & Wildlife Resources
1 Game Farm Road
Frankfort, KY 40601 USA
Website: http://www.kfwis.state.ky.us/AFS/kyafs.htm

Founded: 1990
Scope: Statewide

Description: The American Fisheries Society is dedicated to the preservation and conservation of aquatic resources, and facilitating the information exchange between students, professionals, and the public regarding our knowledge of aquatic resources.

Contact(s):
Kerry Prather, PRESIDENT
Phone: 502-564-5448
Fax: 502-564-4519
kprather@mail.state.ky.us

AMERICAN FISHERIES SOCIETY,
MICHIGAN CHAPTER
Attn: President, 484 Cherry Creek Rd.
Marquette, MI 49855 USA
Phone: 906-249-1611   Fax: 906-249-3190
Website: www.fw.msu.edu/orgf/mi_afs/afs.hdm

Founded: 1973
Scope: Statewide

Contact(s):
Edgar Baker, PRESIDENT
Phone: 906-249-1611

AMERICAN FISHERIES SOCIETY,
MID-ATLANTIC CHAPTER
P.O.B. 330
Little Creek, DE 19961 USA
Website: http://64.224.98.53/mid-atl/

Founded: 1983
Scope: Statewide

Description: The American Fisheries Society is dedicated to the preservation and conservation of aquatic resources, and facilitating the information exchange between students, professionals, and the public regarding our knowledge of aquatic resources.

Contact(s):
John Wright, NC VICE-PRESIDENT
1953-A Quail Ridge Rd.
Greenville, NC 27858, Jwright@skantech.net
Van Atkins, PRESIDENT
2040 Church Creek Dr.
Charleston, SC 29414, President@carolinabirdclub.org
Donna Bailey, SC VICE-PRESIDENT
176 Ravens Place
Winnsboro, SC 29180, dsbailey@conterra.com
Sue Pulsipher, SECRETARY
2441 Ramey Dr.
Linden, NC 28356—9771
puls@infi.net
Patricia Earnhardt-Tyndall, TREASURER
400 Kilmarnock Ct.
Wake Forest, NC 27587
ptearn@aol.com

AMERICAN FISHERIES SOCIETY,
MID-CANADA CHAPTER
Attn: President, 19 Acadia Bay, Canada
Winnipeg, Manitoba R3T 3J1
Phone: 204-945-7794
E-mail:
Website:

Founded: 1986
Scope: National

Contact(s):
Arthur Derksen, PRESIDENT
Phone: 204-945-7791
Fax: 204-948-2308
aderksen@nr.gov.mb.ca

AMERICAN FISHERIES SOCIETY,
MINNESOTA CHAPTER
Attn: President, 5504 Hay Creek Rd.
Fort Ripley, MN 56449 USA
Phone: 218-828-2246   Fax: 218-828-6022
Website: www.fw.umn.edu/mnafs

Founded: 1967
Membership: 250
Scope: Local
Publication(s): Ryva (quarterly newsletter)

Contact(s):
Paul Radomski, PRESIDENT
Phone: 218-828-2246
Fax: 218-828-6022
paul.radomski@dnr.state.mn.us

AMERICAN FISHERIES SOCIETY,
MISSISSIPPI CHAPTER
University, MS USA
Website: http://www.cfr.msstate.edu/msafs/msafs.htm

Founded: 1975
Scope: Statewide

Description: The American Fisheries Society is dedicated to the preservation and conservation of aquatic resources, and facilitating the information exchange between students, professionals, and the public regarding our knowledge of aquatic resources.

Contact(s):
Jim Franks, PRESIDENT
jim.franks@usm.edu

AMERICAN FISHERIES SOCIETY,
MISSOURI CHAPTER
P.O.B. 10267
Columbia, MO 65205 USA
E-mail: moafs@tranquility.net
Website: http://www.moafs.org/

Founded: 1963
Scope: Statewide

Description: The American Fisheries Society is dedicated to the preservation and conservation of aquatic resources, and facilitating the information exchange between students, professionals, and the public regarding our knowledge of aquatic resources.

Contact(s):
Mike Roell, PRESIDENT

AMERICAN FISHERIES SOCIETY,
MONTANA CHAPTER
Attn: President, MT State University, Dept. of Biology
Bozeman, MT 59717 USA
Website: www.fisheries.org/afsmontana

Founded: 1998
Scope: Statewide

Description: The American Fisheries Society is dedicated to the preservation and conservation of aquatic resources, and facilitating the information exchange between students, professionals, and the public regarding our knowledge of aquatic resources.

Contact(s):
Michael Enk, PRESIDENT
Phone: 406-791-7729
Fax: 406-761-1972
menk@fs.fed.us

AMERICAN FISHERIES SOCIETY,
NEBRASKA CHAPTER
Nebraska Game & Parks
2200 North 33rd Street
Lincoln, NE 68503 USA
Website: http://www.4w.com/nebraskaAFS/

Founded: 1969
Membership: 60
Scope: Statewide

Description: The American Fisheries Society is dedicated to the preservation and conservation of aquatic resources, and facilitating the information exchange between students, professionals, and the public regarding our knowledge of aquatic resources.

Contact(s):
Gerald Mestl, PRESIDENT
Phone: 402-471-5447
Fax: 402-471-5528
gmestl@ngpc.state.ne.us

AMERICAN FISHERIES SOCIETY,
NEW MEXICO STATE UNIVERSITY STUDENT
CHAPTER
Las Cruces, NM 88002 USA

Founded: 1972
Scope: Statewide

Description: The American Fisheries Society is dedicated to the preservation and conservation of aquatic resources, and facilitating the information exchange between students, professionals, and the public regarding our knowledge of aquatic resources.

Contact(s):
Richard Saurez, PRESIDENT
rsaurez@nmsu.edu

AMERICAN FISHERIES SOCIETY,
NEW YORK CHAPTER
Owego, NY 13827 USA

Founded: 1968
Scope: Statewide

Description: The American Fisheries Society is dedicated to the preservation and conservation of aquatic resources, and facilitating the information exchange between students, professionals, and the public regarding our knowledge of aquatic resources.

Contact(s):
John Farrell, PRESIDENT
jmfarrel@mailbox.syr.edu

AMERICAN FISHERIES SOCIETY,
NORTH CAROLINA WILDLIFE RESOURCE
COMMISSION
1721 Mail Center Service
Raleigh, NC 27699 USA
Phone: 919-362-3557
Website: www4.ncsu.edu/unity/users/j/jhncsu/public/afs

Founded: 1990
Membership: 100
Scope: Statewide
Publication(s): Annual Meeting, Quarterly Newsletter

Contact(s):
John Crutchfield, PAST PRESIDENT
Robert Curry, PRESIDENT
Phone: 919-715-7643
Shari Bryant, PRESIDENT ELECT

AMERICAN FISHERIES SOCIETY,
NORTH PACIFIC INTERNATIONAL CHAPTER
Attn: President, King Cnty. Water & Land Resources,
Div. 201, S. Jackson St., Suite 600
Seattle, WA 98104 USA
Phone: 206-296-1946

Founded: 1978
Scope: National
Contact(s):
Kurt Fresh, PRESIDENT
Phone: 360-902-2756
Fax: 360-902-2980
freshklf@dfw.wa.gov

AMERICAN FISHERIES SOCIETY,
NORTHWESTERN ONTARIO CHAPTER
Thunder Bay, Ontario, Canada

Founded: 1979
Scope: National

Description: The American Fisheries Society is dedicated to the preservation and conservation of aquatic resources, and facilitating the information exchange between students, professionals, and the public regarding our knowledge of aquatic resources.

Contact(s):
Rob Mackereth, PRESIDENT
rob.mackereth@mnr.gov.on.ca

AMERICAN FISHERIES SOCIETY,
OHIO CHAPTER
1840 Belcher Dr., G-3
Columbus, OH 43224 USA
Phone: 614-265-6349   Fax: 614-262-1143
Website: www.community.cleveland.com/cc/ocafs

Founded: 1974
Scope: Statewide

Publication(s): A Guide to Ohio Streams

Contact(s):
Debra Walters, PRESIDENT

AMERICAN FISHERIES SOCIETY,
OREGON CHAPTER
Attn: President, 2910 NW Miller Ln.
Albany, OR 97321 USA
Phone: 54-175-3044

Founded: 1964
Scope: Statewide

Contact(s):
Timothy Hardin, PRESIDENT
Phone: 541-926-2262
Fax: 541-926-1230
hardint@peak.org

AMERICAN FISHERIES SOCIETY,
POTOMAC CHAPTER
Attn: President, Dept. of Commerce NOAA, Rm. 6117, 14
& Constitution Ave.
Washington, DC 20230 USA
Phone: 202-482-3260   Fax: 202-501-3024
Website: www.poyamac-afc.org

Founded: 1976
Membership: 110
Scope: Regional

Contact(s):
Lee Beneka, PRESIDENT

AMERICAN FISHERIES SOCIETY,
SOUTH NEW ENGLAND CHAPTER
Attn: President, c/o NMFS NE Fish Center, 166 Water St.
Woods Hole, MA 02543-1097 USA
Phone: 508-495-2380   Fax: 508-495-2393
Website: http://www.nefsc.nmfs.gov/snecafs/index.html

Founded: 1967
Membership: 250

Scope: Regional

Contact(s):
Russ Brown, PRESIDENT

AMERICAN FISHERIES SOCIETY,
SOUTHERN ONTARIO CHAPTER
Beak International, 14 Abacus Rd
Brampton, Ontario L6T 5B7, Canada
Phone: 905-794-2325   Fax: 905-794-2338
E-mail: bhendley@beak.com
Website: www.beak.com

Founded: 1988
Scope: National

Contact(s):
Cynthia Mitton-Walker, PRESIDENT
Phone: 416-235-5230
Fax: 416-235-4940
mitton@mto.gov.on.ca

AMERICAN FISHERIES SOCIETY,
TIDEWATER CHAPTER
Attn: President, c/o VA Institute of Marine Science
Gloucester Point, VA 23062 USA

Founded: 1986
Scope: Statewide

Contact(s):
John Olney, PRESIDENT
Phone: 804-642-7334
Fax: 804-642-7097
olney@vims.edu

AMERICAN FISHERIES SOCIETY,
UNIVERSITY OF WYOMING STUDENT CHAPTER
Attn: President, P.O. Box 3166
Laramie, WY 82071-3166 USA
Phone: 307-766-2426
Website: www.fisheries.org/co-wy/uw%zochapter.html

Founded: NA
Scope: Statewide

Contact(s):
Seth Wite, PRESIDENT
oclark@uwyo.edu
Leisa Tooker, SECRETARY
leisatooker@hotmail.com
Jason Burckhardt, TREASURER
jburck@uwyo.edu
Mark Smith, VICE PRESIDENT
masmith@uwyo.edu

AMERICAN FISHERIES SOCIETY,
VIRGINIA CHAPTER
VIRGINIA DEPT. GAME & INLAND FISHERIES
4010 W. Broad St.
Richmond, VA 23230 USA
Phone: 804-367-8351
Website: fwie.fw.vt.edu/va-afs/index.htm

Founded: 1990
Membership: 130
Scope: Statewide

Contact(s):
Charles Gowan, PAST PRESIDENT
Becky Wajda, PRESIDENT

AMERICAN FISHERIES SOCIETY,
VIRGINIA TECH CHAPTER
Attn: President, 101 Cheatham Hall
Blacksburg, VA 24061 USA
Phone: 540-231-3329   Fax: 540-231-7580
Website: http://filebox.vt.edu/org/vtafs

Founded: 1972
Scope: Regional

Publication(s): Lab Notes

Contact(s):
Louis Helfrich, FACULTY ADVISOR
lhelfric@vt.edu
Anne Holloway, PRESIDENT
John Harris, SECRETARY
Ginnie Lintecum, TREASURER
Jamie Roberts, VICE PRESIDENT

AMERICAN FISHERIES SOCIETY,
WEST VIRGINIA
Elkins, WV 26241 USA
Phone: 304-637-0215   Fax: 304-637-0250

Founded: 1989
Scope: Statewide

Contact(s):
Michael Shingleton, PRESIDENT
Phone: 304-637-0245
Fax: 304-637-0250
mshingleton@dnr.state.wv.us

AMERICAN FISHERIES SOCIETY,
ATLANTIC INTERNATIONAL CHAPTER
ATLANTIC INTERNATIONAL CHAPTER
689 Farmington Rd
Strong, ME 04983 USA
Phone: 207-778-3322   Fax: 207-778-3323
Website: www.afs.org

Founded: 1975
Membership: 50
Scope: International
Publication(s): Atlantic International Chapter Newsletter

Contact(s):
Forrest Bonney, PRESIDENT
forrest.bonney@state.me.us

AMERICAN FISHERIES SOCIETY,
GEORGIA CHAPTER
GEORGIA CHAPTER
Attn: President, GA DNR Albany Fisheries,
2024 Newton Rd.
Albany, GA 31701-3576 USA
Phone: 229-430-4256   Fax: 229-430-5110

Founded: 1985
Membership: 23
Scope: Statewide

Contact(s):
Matthew Thomas, SENIOR BIOLOGIST
Phone: 912-430-4256
Fax: 912-430-5110
matt_thomas@mail.dnr.state.ga.us

AMERICAN FISHERIES SOCIETY, PENNSYLVANIA CHAPTER

PENNSYLVANIA CHAPTER
1259 Edward St.
State College, PA 16801 USA
Phone: 814-353-2226

Founded: 1969
Scope: Statewide

Contact(s):
Doug Nieman, PRESIDENT
Phone: 610-948-4700
dnieman@normandeau.com

AMERICAN FISHERIES SOCIETY, SOUTH CAROLINA CHAPTER

SOUTH CAROLINA CHAPTER
South Carolina American Fisheries Society, P O Box 1040
Abbeville, SC 29620 USA
Phone: 864-223-2008   Fax: 864-223-0649
Website: http://www.inlet.geol.sc.edu/scafs

Founded: 1982
Membership: 110
Scope: Statewide

Contact(s):
Wade Bales, PRESIDENT
Phone: 864-223-2008
Fax: 864-223-0649
Chris Thomason, PRESIDENT ELECT
Phone: 803-259-5474

AMERICAN FISHERIES SOCIETY, TEXAS A&M CHAPTER

TEXAS A&M CHAPTER
210 Nagle Hall- 2258 TMU
College Station, TX 77843-2258 USA
Phone: 979-845-5777   Fax: 979-845-3786
Website: www.wfscnet.tamu.edu

Founded: 1969
Membership: 50
Scope: Statewide
Publication(s): Former Student Newsletter

AMERICAN FISHERIES SOCIETY, WISCONSIN CHAPTER

WISCONSIN CHAPTER
University of Wisconsin, Eau Claire Biology Dept.
Eau Claire, WI USA
Phone: 715-836-3260
E-mail: lonzard@uwec.edu

Founded: 1972
Scope: Statewide

Publication(s): Wisconsin Chapter American Fisheries Society Newsletter (Newsletter)

Contact(s):
Dr. David Lonzarich, PRESIDENT

AMERICAN FOREST FOUNDATION

1111 19th St., NW, Suite 780
Washington, DC 20036 USA
Phone: 202-463-2462   Fax: 202-463-2461
Website: www.asfoundation.org

Founded: 1981
Membership: 11
Scope: National

Description: American Forest Foundation conducts charitable education and research programs. AFF supports American Tree Farm System — 71,000 private landowners managing 95 million acres of forests — and Project Learning Tree (PLT), award-winning pre K-12 environmental education curriculum and training program, active in U.S. and abroad. Nongrantmaking.

Publication(s): Tree Farmer Magazine, PLT Branch

Keyword(s): Communications, Environmental and Conservation Education, Forests and Forestry, Renewable Resources

Contact(s):
Laurence Wiseman, PRESIDENT
Robert Simpson, VICE PRESIDENT OF AMERICAN TREE FARM SYSTEM
Kathy Mcglauflin, VICE PRESIDENT OF PROJECT LEARNING TREE

AMERICAN FORESTS

P. O. Box 2000
Washington, DC 20013 USA
Phone: 202-955-4500   Fax: 202-955-4588
E-mail: member@amfor.org
Website: www.americanforests.org

Founded: 1875
Membership: 20000
Scope: National

Description: (formerly American Forestry Association) Building on its rich history as the oldest national citizens' conservation organization in the U.S. and conservation movement pioneer, American Forests has several programs to address today's environmental challenges: Global ReLeaf 2000, the Urban Forest Center and the Forest Policy Center.

Publication(s): American Forests (quarterly magazine)

Keyword(s): Environmental and Conservation Education, Forests and Forestry, Greenhouse Effect/Global Warming, Outdoor Recreation, Public Lands, Trees, Urban Forestry

Contact(s):
Doug Cowan, BOARD CHAIR
Steve Westcott, DIRECTOR OF COMMUNICATIONS
Christina Cromley, DIRECTOR OF FOREST POLICY
ccromley@amfor.org
Michelle Robbins, EDITOR
Phone: 202-887-1075
mrobbins@amfor.org
Deborah Gangloff, EXECUTIVE DIRECTOR
Jeff Meyer, FAMOUS & HISTORIC TREE NURSERY
Phone: 904-765-0727
Fax: 904-768-4630
jmeyer@historictrees.org
Bob Skiera, FIELD REPRESENTATIVE
Zane Smith, FIELD REPRESENTATIVE
37899 Shenandoah Loop Rd., Springfield, OR 97478

Jane Westenberger, FIELD REPRESENTATIVE
7437 Saratoga Lane, Santa Fe, NM 87505
Richard Crouse, SENIOR VICE PRESIDENT FOR
DEVELOPMENT
rcrouse@amfor.org
Stacey Mandell, SOUTHEAST REGION COORDINATOR
Phone: 305-372-6555
Doug Hall, TREASURER
Gerald Gray, VICE PRESIDENT OF FOREST POLICY
CENTER
ggray@amfor.org
Gary Moll, VICE PRESIDENT OF URBAN FOREST CENTER
gmoll@amfor.org

AMERICAN GEOGRAPHICAL SOCIETY

120 Wall St.
New York, NY 10005-3904 USA
Phone: 212-422-5456   Fax: 212-422-5480
E-mail: amgeosoc@earthlink.net
Website: www.amergeog.org

Founded: 1851
Membership: 5
Scope: National

Description: The AGS has sponsored research projects field
work, and educational travel, held symposia and lectures, and
published scientific and popular books, periodicals, and maps.
Its publications bring accurate, up-to-date information on man
and the land to more than 8,000 fellows and subscribers in
over 100 countries.

Publication(s): Ubique, Around the World Program, Geographical
Review, Focus

Keyword(s): Environmental and Conservation Education, Land
Use Planning, Urban Environment, Greenhouse Effect/Global
Warming, Sustainable Development

Contact(s):
Richard Nolte, CHAIR EMERITUS
Paul Starrs, EDITOR
Geographical Review Dept. Geography, University of Nevada-
Reno, Reno, NV 89557
Hilary Hopper, EDITOR, FOCUS AND AROUND THE
WORLD
Dept. of Geography, University of Kentucky, Lexington, KY
40506
Peter Lewis, EDITOR, UBIQUE
William Doyle, PRESIDENT
John Wilford, SECRETARY
John Mccabe, TREASURER

AMERICAN GEOLOGICAL INSTITUTE

4220 King St.
Alexandria, VA 22302-1502 USA
Phone: 703-379-2480   Fax: 703-379-7563
E-mail: agi@agiweb.org
Website: www.agiweb.org

Founded: 1948
Scope: National

Description: AGI provides information services for earth scientists
to be an advocate for the interests of the earth-science
community; plays a major role in strengthening earth-science
education; and increases public awareness of the role that
earth sciences play in mankind's use of resources and
interaction with the environment.

Publication(s): Glossary of Geology, Directory of Geoscience
Departments, Geotimes, Bibliography and Index of Geology

Keyword(s): Geology

Contact(s):
Julie Jackson, DIRECTOR OF OUTREACH AND
COMMUNICATIONS
Marcus Milling, EXECUTIVE DIRECTOR

AMERICAN GROUND WATER TRUST

P.O. Box 1796, 16 Centre St.
Concord, NH 03301 USA
Phone: 603-228-5444   Fax: 603-228-6557
E-mail: trustInfo@agwt.org
Website: www.agwt.org

Founded: 1987
Membership: 7
Scope: National

Description: The American Ground Water Trust is an independent
nonprofit, membership organization which promotes public
awareness of the environmental and economic importance of
ground water through public education programs. The Trust
promotes opportunity, cooperation and action among
individuals, groups and organizations throughout America.

Publication(s): Water Well Basics (video), Ground Water
information pamphlets, Ground Water and Wetlands in the
United States

Keyword(s): Environmental and Conservation Education,
Geology, Scholarships and Grants, Coral Reefs, Exotic
species, Aquatic nuisance species, Ground Water Protection

Contact(s):
Mike Lally, CHAIRMAN
Andrew Stone, EXECUTIVE DIRECTOR
Scott Slater, SECRETARY
Richard Schramm, TREASURER

AMERICAN HIKING SOCIETY

1422 Fenwick Ln.
Silver Spring, MD 20910 USA
Phone: 301-565-6704   Fax: 301-565-6714
E-mail: info@americanhiking.org
Website: www.americanhiking.org

Founded: 1976
Membership: 11000
Scope: National

Description: American Hiking Society (AHS) is a recreation-
based conservation organization dedicated to establishing,
protecting and maintaining foot trails in America. AHS is
comprised of over 120 member trail clubs and 10,000
individual members, represents half a million outdoorspeople
and serves as the voice of the American hiker. AHS effectively
lobbies to encourage funding for trails and promotes volunteerism in trail building and maintenance.

Publication(s): Volunteer Vacation, Helping Out in the Outdoors,
American Hiker, Pathways Across America

Keyword(s): Environmental and Conservation Education, Forests
and Forestry, Trail, Land Use Planning, Outdoor Recreation,
Public Lands

Contact(s):
Michael Hechter, MEMBERSHIP COORDINATOR
Mary Margaret Sloan, PRESIDENT
Susan Crosby, VICE PRESIDENT AND DEVELOPMENT

AMERICAN HORSE PROTECTION ASSOCIATION

1000 29th St., NW
Washington, DC 20007 USA
Phone: 202-965-0500   Fax: 202-965-9621
Website: www.americanhorseprotection.org

Founded: 1966
Scope: National

Description: A national nonprofit, tax-exempt organization dedicated entirely to the welfare of horses, both wild and domestic. Works for the enforcement of all humane legislation for both wild and domestic horses.

Publication(s): Newsletter, Special Bulletins

Keyword(s): Environmental Law, Land Use Planning, Mammals, Public Lands, Animal Welfare

Contact(s):
Robin Lohnes, EXECUTIVE DIRECTOR
Nancy Hargrave, PRESIDENT AND CHAIRMAN OF THE BOARD OF DIRECTORS
Nancy Murray, SECRETARY AND TREASURER
Michele Rydell, VICE PRESIDENT

AMERICAN HUMANE ASSOCIATION

63 Inverness Dr., E
Englewood, CO 80112 USA
Phone: 303-792-9900   Fax: 303-792-5333
E-mail: animal@americanhumane.org
Website: www.americanhumane.org

Founded: 1877
Membership: 2000
Scope: National

Description: AHA provides training and resources to 6,500 animal care and control agencies in the US and Canada; ensures the humane treatment of animals in movies and TV productions; serves as a national coordinator of emergency animal relief during natural disasters and works on legislation to protect animals.

Publication(s): Protecting Animals

Keyword(s): Animal Welfare

Contact(s):
Jack Sparks, DIRECTOR OF COMMUNICATIONS
Connie Howard, DIRECTOR OF SHELTER PROGRAMS
Tim O'Brien, PRESIDENT
Lynn Anderson, VETERINARIAN
animal@americanhumane.org

AMERICAN INSTITUTE OF BIOLOGICAL SCIENCES

1444 I St., NW
Washington, DC 20005 USA
Phone: 202-628-1500   Fax: 202-628-1509
E-mail: admin@aibs.org
Website: www.aibs.org

Founded: 1947
Scope: National

Description: A national organization for biology and biologists, combining an individual membership organization with the federation principle. Operates educational, advisory, liaison, informational, publication and editorial programs to serve biologists, promote unity and effectiveness of effort and apply knowledge of biology to human welfare.

Publication(s): BioScience

Keyword(s): Wildlife Rehabilitation, Grants, Sustainable Ecosystems, Zoology

Contact(s):
Herman Marshall, EDITOR
hmarshall@aibs.org
Richard O'Grady, EXECUTIVE DIRECTOR
rogrady@aibs.org
Judith Weis, PRESIDENT
Dept. of Biological Sciences Rutgers Univ.,
Neward, NJ 07102
Phone: 973-353-5387
Fax: 973-353-5518
jweis@andromeda.rutgers.edu
Jane Brockmann, SECRETARY AND TREASURER
Professor of Zoology, University of Fl,
Gainesville, FL 32611-8525
Phone: 352-392-1297
Fax: 352-392-3704
hjb@zoo.ufl.edu

AMERICAN INSTITUTE OF FISHERY RESEARCH BIOLOGISTS

c/o National Marine Fisheries Science Center, SW
Fisheries Science Center
P.O.B. 271
La Jolla, CA 92038 USA
Website: http://www.iattc.org/aifrb/default.htm

Founded: 1957
Scope: National

Description: The Institute was founded to advance the science of fishery biology and to promote conservation and proper use of fishery resources. It serves that goal primarily by being concerned with the professional development and performance of its members and recognition of their competence and achievement.

Publication(s): Briefs

Keyword(s): Professional Organization, Wildlife Rehabilitation, Aquatic Habitats, Wildlife, Renewable Resources

Contact(s):
Gene Huntsman, EDITOR
205 Blades Rd., Havelock, NC 28523
Phone: 704-274-7773
Gary Sakagawa, PRESIDENT
gary.sakagawa@noaa.gov
Barbara Warkentine, SECRETARY
SUNY-Maritime College, Science Dept., 6 Pennyfield Ave.,
Ft. Schuyler, Bronx, NY 10465-4198
Phone: 206-543-1101
Alan Shimada, TREASURER
7909 Sleaford Place, Bethesda, MD 20814

AMERICAN LAND CONSERVANCY

1388 Sutter St.
San Francisco, CA 94109 USA
Phone: 415-749-3010   Fax: 415-749-3011
E-mail: mail@alcnet.org
Website: www.alcnet.org

Founded: 1990
Membership: 10
Scope: National

Description: To preserve land for this and future generations; in particular, to preserve its scientific, historic, educational, ecological, geological, recreational, agricultural and scenic features, and its native plant and animal life or biotic community.

Publication(s): Fifty Wildflowers of Bear Valley, American Land Conservancy Newsletter, Statement of Opportunity Brochure

Keyword(s): Coasts, Deserts, Forests and Forestry, Land Purchase, Public Lands

Contact(s):
Harriet Burgess, PRESIDENT
mail@alcnet.org

AMERICAN LANDS
726 7th St., SE
Washington, DC 20003 USA
Phone: 202-547-9400   Fax: 202-547-9213
E-mail: wafcdc@americanlands.org
Website: www.americanlands.org

Founded: 1991
Scope: National

Description: (formerly Western Ancient Forest Campaign) The mission of American Lands is the protection and recovery of North American native forest, grassland, and aquatic ecosystems; the preservation of biological diversity; the restoration of watershed integrity; and the promotion of environmental justice in connection with these goals. This mission is accomplished by strengthening grassroots conservation networks; providing advocacy services and other assistance to local conservation groups; and helping to improve communications and coordination among these groups and other societal institutions.

Publication(s): Report from Washington

Keyword(s): Aquatic Habitats, Forests and Forestry, Public Lands

Contact(s):
Steve Holmer, CAMPAIGN COORDINATOR
Phone: 202-547-9105
Jim Jontz, EXECUTIVE DIRECTOR
Phone: 202-547-9095
Randi Spivak, PRESIDENT
CA 310-458-8869
Christopher Peters, TREASURER
P.O. Box 4569, Arcata, CA 95518
Phone: 707-825-7640

AMERICAN LEAGUE OF ANGLERS AND BOATERS
1225 New York Ave., NW
Washington, DC 20005 USA
Phone: 202-682-9530   Fax: 202-682-9529
Website: www.funoutdoors.com

Founded: 1985
Membership: 20
Scope: National

Description: ALAB was formed to be a vigilant patron of the Sport Fishing and Boating Enhancement Act (PL 98-369 and the Aquatic Resources Trust Fund created by the Act. Composed of more than 30 organizations, ALAB is dedicated to this pioneering user-pays legislation which provides some $330 million annually in funding for U.S. Coast Guard recreational boating programs and in matching grants to the states for sportfish research and enhancement, as well as wetlands conservation, boating safety and boating access improvements. Membership in ALAB is open to nonprofit organizations, businesses, corporations, and individuals seeking improvement in the scope and health of the nation's aquatic resources and expansions of opportunities for responsible utilization by the fishing and boating public.

Keyword(s): Wildlife, Outdoor Recreation, Sport Fishing, Exotic species, Aquatic nuisance species

Contact(s):
Veronica Floyd, CO-CHAIR
Phone: 703-960-2223
Derrick Crandall, CO-CHAIR
Phone: 301-897-8616
George Stewart, TREASURER
Phone: 302-678-9143

AMERICAN LITTORAL SOCIETY
Headquarters, Sandy Hook
Highlands, NJ 07732 USA
Phone: 732-291-0055
E-mail: als@netlabs.net
Website: alsnyc.org

Founded: 1961
Scope: National

Description: A national organization of professionals and amateurs interested in the study and conservation of coastal habitat, barrier beaches, wetlands, estuaries, and near-shore waters, and their fish, shellfish, bird, and mammal resources. Publishes scientific and popular material. Conducts field trips, dive and study expeditions, and a fish tag-and-release program. Special activities for scuba divers.

Publication(s): Underwater Naturalist, Coastal Reporter

Keyword(s): Aquatic Habitats, Coasts, Wildlife, Wetlands

Contact(s):
D. Bennett, EXECUTIVE DIRECTOR
Michael Huber, PRESIDENT
Angela Cristini, SECRETARY
Sheldon Abrams, TREASURER
Frank Steimle, VICE PRESIDENT

AMERICAN LITTORAL SOCIETY
DELAWARE RIVERKEEPER NETWORK
P.O. Box 326
Washington Crossing, PA 18977 USA
Phone: 215-369-1188   Fax: 215-369-1181
E-mail: drn@delawareriverkeeper.org
Website: www.delawareriverkeeper.org

Founded: NA
Scope: Regional

Publication(s): Stormwater Runoff Community Asset A Guide to Preventing Storm Water Runoff, Stream Restoration In PA, Ten Case Studies

Contact(s):
Maya Van rossum, DELAWARE RIVERKEEPER/EXECUTIVE DIRECTOR

AMERICAN LITTORAL SOCIETY

NORTHEAST CHAPTER
28 West 9th Rd.
Broad Channel, NY 11693 USA
Phone: 718-634-6467   Fax: 718-318-9345
E-mail: alsbeach@aol.com
Website: www.alsnyc.org

Founded: NA
Membership: 2800
Scope: National
Publication(s): Literally Speaking (newsletter)

Contact(s):
Barbara Toborg, EDITOR
Phone: 718-474-1127
tobytoborg@aol.com
Barbara Cohen, NEW YORK STATE BEACH CLEANUP
COORDINATOR
Phone: 718-471-2166
alsbeach@aol.com

AMERICAN LIVESTOCK BREEDS CONSERVANCY

15 Hillsboro
Pittsboro, NC 27312 USA
Phone: 919-542-5704   Fax: 919-545-0022
E-mail: albc@albc-usa.org
Website: www.albc-usa.org

Founded: 1977
Membership: 4000
Scope: National

Description: ALBC is a nonprofit membership organization
working to protect genetic diversity in domestic animals
through the conservation of nearly 100 rare breeds of livestock
and poultry in America. ALBC does research on breed status
and characteristics, operates a gene bank to preserve genetic
materials for the future and provides technical support on conservation breeding and animal use in sustainable, diversified
agriculture.

Publication(s): ALBC News, Taking Stock Of Water Fowl, A
Conservation Breeding Handbook, A Rare Breeds Album of
American Livestock, Birds Of A Feather: Saving Rare Turkeys
From Extinction, Taking Stock: The North American Livestock
Census

Keyword(s): Agriculture, Biodiversity, Endangered Species,
Historic Preservation, Sustainable Development

Contact(s):
Donald Bixby, EXECUTIVE DIRECTOR
Cindy Harman, MARKETING COORDINATOR
Marjorie Bender, PROGRAM COORDINATOR
Phillip Sponenberg, TECHNICAL COORDINATOR

AMERICAN LUNG ASSOCIATION

1740 Broadway
New York, NY 10019-4374 USA
Phone: 212-315-8700   Fax: 212-265-5642
E-mail: info@lungusa.org
Website: www.lungusa.org

Founded: 1904
Membership: 84
Scope: National

Description: Formerly known as the National Tuberculosis and
Respiratory Disease Association. The American Lung
Association is a voluntary agency concerned with the conquest
of lung disease and the promotion of lung health, which
includes preventing and controlling air pollution. National Air
Conservation Commission and local and state air conservation
committees work with citizenry and other groups for effective
air pollution control. Informational material available from
national, state and local lung associations.

Publication(s): American Journal of Respiratory and Critical Care
Medicine, American Journal of Respiratory Cell and Molecular
Biology

Keyword(s): Air Quality and Pollution

Contact(s):
John Kirkwood, CHIEF OPERATING OFFICER

AMERICAN MUSEUM OF NATURAL HISTORY

COMMUNICATIONS
Central Park West at 79th St.
New York, NY 10024 USA
Phone: 212-769-5100   Fax: 212-769-5199
Website: www..amnh.org

Founded: 1869
Scope: National

Description: Conducts research in anthropology, astronomy,
entomology, herpetology, ichthyology, invertebrates,
mammalogy, earth and planetary sciences, ornithology and
vertebrate and invertebrate paleontology using museum
collections and field studies. Publishes scientific and popular
material. Instructs the public, especially its over three million
yearly visitors, in natural sciences, including living and extinct
animals, ecological relationships, evolution of earth and life,
development of human cultures, and astronomy.

Publication(s): Natural History, American Museum Novitates,
Anthropological Papers of the American Museum of Natural
History, Micropaleontology Press, Curator, Bulletin of the
American Museum of Natural History

Keyword(s): Biodiversity, Endangered Species, Geology,
education, Zoology, Museum

Contact(s):
Ellen Futter, PRESIDENT

AMERICAN NATURE STUDY SOCIETY

c/o PEEC, R.D. Box 1010
Dingmans Ferry, PA 18328 USA
Phone: 607-749-3655
Website: http://hometown.aol.com/anssonlne/

Founded: 1908
Scope: National

Description: Promotes environmental education and avocation
by conducting meetings, workshops and field excursions,
producing and distributing publications, and contributing to
publications of other agencies; cooperates with organizations
with allied interests, and, through membership in Alliance for
Environmental Education, encourages members to contribute
consultant services; assists in training nature lay leaders.

Publication(s): ANSS Newsletter, Nature Study, A Journal of
Environmental Education and Interpretation

Keyword(s): Environmental and Conservation Education,
education, Outdoor Recreation, Urban Environment, Youth
Organizations, Nature Study

Contact(s):
Florence Mauro, EDITOR
PEEC, R.D. 2 Box 1010, Dingmans Ferry, PA 18328
Phone: 717-828-2319
Janet Hawkes, EDITOR
1420 Tanghannock Blvd., Ithaca, NY 14850
Phone: 607-273-6260
Steve Melcher, PRESIDENT
103 Kreag Rd., Fairport, NY 14450-363
Phone: 716-425-1059
Flo Mauro, RECORDING SECRETARY
PEEC, R.D. 2 Box 1010, Dingmans Ferry, PA 18328
Phone: 717-828-2319
Betty McKnight, SECRETARY
R.D. 3, Trumansburg, NY 14886
Paul Spector, TREASURER
Holden Arboretum, 9500 Sperry Rd., mentor, OH 44094
Phone: 216-256-1110

AMERICAN OCEANS CAMPAIGN

600 Pennsylvania Ave., SE, Suite 210
Washington, DC 20003 USA
Phone: 202-544-3526   Fax: 202-544-5625
E-mail: info@americanoceans.org
Website: www.americanoceans.org

Founded: 1987
Scope: National

Description: The well-being and sustainability of the Earth is dependent upon healthy oceans. The mission of American Oceans Campaign is to safeguard the vitality of the oceans and our coastal waters. AOC is committed to scientific information in advocating for sound public policy. We are equally committed to developing partnerships with all entities interested in protecting the environment. AOC seeks to ensure healthy sources of food and coastal recreation as well as to protect the ocean's grandeur for future generations.

Publication(s): Splash, Fish Briefs, Chemical Contaminant Release Into the Santa Monica Bay: A Pilot Study, Drainage to the Oceans: The Effects of Stormwater Pollution on Coastal Waters, Esturaries on the Edge: The Vital Link Between Land and Sea

Keyword(s): Aquatic Habitats, Wildlife, Public Health Protection, Coral Reefs, Water Quality, Marine Protected Areas, Pollution Prevention, Estuaries, Coasts, Beaches, Ocean Conservation

Contact(s):
Warner Chabot, BOARD CHAIR
Barbara Polo, EXECUTIVE DIRECTOR
bjpolo@americanoceans.org
Ted Danson, FOUNDING PRESIDENT
Barbara Kuhn, TREASURER
Annett Wolf, VICE PRESIDENT

AMERICAN OCEANS CAMPAIGN

WASHINGTON, DC OFFICE
600 Pennsylvania Ave., SE,
Washington, DC 20003 USA
Phone: 202-544-3526   Fax: 202-544-5625
E-mail: info@americanoceans.org
Website: www.americanoceans.org

Founded: NA
Membership: 14
Scope: National
Publication(s): Splash-Newsletter

Contact(s):
Barbara Polo, EXECUTIVE DIRECTOR
Ted Morton, POLICY DIRECTOR

AMERICAN ORNITHOLOGISTS UNION

National Museum of Natural History, MRC-116,
Smithsonian Institution
Washington, DC 20560-0116 USA
Phone: 202-357-2051   Fax: 202-633-8084
E-mail: aou@nmnh.si.edu

Founded: 1883
Scope: National

Description: Aims to advance ornithological science through its publications, annual meetings, committees and membership.

Publication(s): Ornithological Newsletter, Ornithological Monographs

Keyword(s): Birds

Contact(s):
Steven Beissinger, CHAIRMAN OF THE CONSERVATION COMMITTEE
Director of Ecosystem Science, 151 Hilgard Hall Suite 3110, University of California, Berkeley, CA 94720-3110
Phone: 313-763-5945
Kimberly Smith, EDITOR
Department of Biological Sciences, Fayetteville, AR 72701
Phone: 501-575-3251
Fax: 501-575-4010
auk@comp.uark.edu
David Wiedenfeld, MONOGRAPHS EDITOR
Sutton Avian Research Center, P.O. Box 2007, Bartlesville, OK 74005
Cheryl Trine, NEWSLETTER EDITOR
3889 E. Valley View, Berrien Springs, MI 49103
Phone: 508-224-6521
ctrine@andrews.edu
John Fitzpatrick, PRESIDENT
Laboratory of Ornithology at Cornell University, Ithaca, NY 14850
Phone: 607-254-2410
Fax: 607-254-2415
jwf7@cornell.edu
M. Lein, SECRETARY
Dept. of Biology, University of Calgary, 2500 University Dr., NW, Calgary, Alberta T2N 1N4
Jeff Brawn, TREASURER
Illinois Natural History Survey, Champaign, IL 61820
Phone: 217-244-5937
Fax: 217-333-4949
i-brawn@uiuc.edu
Mary McDonald, VICE PRESIDENT
Lewis Science Center 129, University of Central Arkansas, Conway, AR 72035
Phone: 501-450-5924

AMERICAN PIE (PUBLIC INFORMATION ON THE ENVIRONMENT)

124 High St., P.O. Box 340
South Glastonbury, CT 06073 USA
Phone: 800-320-2743   Fax: 860-633-5090
E-mail: info@americanpie.org
Website: www.AmericanPIE.org

Founded: 1993

Membership: 405
Scope: National

Description: American PIE is a 501(c)3 nonprofit group serving the nation with a 1-800 Environmental Information Line. The organization offers action programs and uniquely accessible assistance to people who have environmental questions and concerns in a wide variety of subject areas ranging from drinking water safety to wetlands preservation. Trained staff answer the information line Monday-Friday, 8:30 - 5:00 Eastern time.

Publication(s): American PIE

Keyword(s): Communications, Environmental and Conservation Education, Pesticides, EcoAction, Environmental Justice

Contact(s):
Lawrence Bacon, DIRECTOR
36 Carriage Dr., Farmington, CT 06032
Phone: 860-674-8442
Toni Easterson, PRESIDENT AND SECRETARY
Brad Easterson, VICE PRESIDENT AND TREASURER
Phone: 860-633-9786

AMERICAN PLANNING ASSOCIATION

1776 Massachusetts Ave., NW, Ste. 400
Washington, DC 20036 USA
Phone: 202-872-0611   Fax: 202-872-0643
Website: www.planning.org

Founded: 1909
Membership: 20
Scope: National

Description: Provides informational services, education and research in city and regional planning. Includes the American Institute of Certified Planners which sets professional and ethical standards and participates in the accreditation of planning degree programs. Forty-six chapters include all of the states. Sixteen divisions address planning specialties and provide placement services and studies.

Publication(s): Planning Magazine, Land Use Law and Zoning Digest, Environment and Development Newsletter, Journal of The American Planning Association

Keyword(s): Environmental Law, Environmental Planning, Land Use Planning, Planning Management, Urban Environment

Contact(s):
Sylvia Lewis, EDITOR
Frank So, EXECUTIVE DIRECTOR
Phone: 202-872-0611
Sam Casella, IMMEDIATE PAST PRESIDENT
Eric Kelley, PRESIDENT
James Shelby, SECRETARY AND TREASURER

AMERICAN RECREATION COALITION

1225 New York Ave., NW, Suite 450
Washington, DC 20005 USA
Phone: 202-682-9530   Fax: 202-682-9529
E-mail: arc@funoutdoors.com
Website: www.funoutdoors.com

Founded: 1979
Scope: National

Description: ARC is a national nonprofit, tax-exempt federation of more than 125 recreation-related trade associations, corporations and enthusiasts' organizations that provides a unified voice for American recreation interests to ensure their full participation in government policy-making on such issues as energy and public lands and waters management. ARC also initiates and supports partnerships between public and private recreation providers and conducts meetings, seminars and activities to improve public awareness of recreation opportunities.

Contact(s):
David Humphreys, CHAIRMAN
RVIA, 1896 Preston White Dr., Reston, VA 22090
Phone: 703-620-6003
Derrick Crandall, PRESIDENT
Catherine Ahern, VICE PRESIDENT OF MEMBER SERVICES

AMERICAN RESOURCES GROUP

374 Maple Ave. E., Suite 310
Vienna, VA 22180 USA
Phone: 703-255-2700   Fax: 703-281-9200
Website: www.nationalforestry.net

Founded: 1981
Membership: 3
Scope: National

Description: A conservation service organization engaged in education, monitoring, research, and related activities to promote the wise use of America's forest resources. Provides forestry, environmental inventory, conservation support services and land acquisition assistance to conservation organizations, public agencies and landowners. Programs include: Land Conservation Fund of America (land acquisition), National Forestry Network (referrals), National Historic Lookout Register, American Woodlands (demonstration forests and conservation easements), and National Forestry Association Forest Practice Certification.

Publication(s): National Woodlands Magazine, Woodland Report, Conservation News Digest

Keyword(s): Forests and Forestry, Land Purchase, Public Lands, Renewable Resources

Contact(s):
David Edson, GREEN TAG FORESTRY CERTIFICATION
Phone: 202-827-4456
Bob Spear, NORTHEAST REPRESENTATIVE OF NATIONAL HISTORIC LOOKOUT REGISTER
Phone: 973-209-7897
Ray Kresek, NORTHWEST REPRESENTATIVE OF NATIONAL HISTORIC LOOKOUT REGISTER
Phone: 509-466-9171
Keith Argow, PRESIDENT/EDITOR
Loren Larson, VICE PRESIDENT OF FORESTRY

AMERICAN RIVERS

150 Nickerson St., Suite 311 Northwest Regional
Seattle, WA 98109 USA
Phone: 206-213-0330   Fax: 206-213-0334
E-mail: arnw@amrivers.org
Website: www.americanrivers.org

Founded: NA
Scope: Regional

Description: (formerly American Rivers Conservation Council)

Contact(s):
Rob Masonis, DIRECTOR OF NORTHWEST
CONSERVATION PROGRAM
rmasonis@amrivers.org
Richard Penny, OFFICE DIRECTOR
Katherine Ransel, SENIOR COUNSEL
kransel@amrivers.org

AMERICAN RIVERS

1025 Vermont Ave. NW 720
Washington, DC 20005 USA
Phone: 202-347-7550   Fax: 202-347-9240
E-mail: americanrivers@americanrivers.org
Website: www.americanrivers.org

Founded: NA
Membership: 30000
Scope: Natioanl

Description: (formerly American Rivers Conservation Council)

Publication(s): Available on website

Contact(s):
Rebecca Wodder, PRESIDENT
Phone: 202-347-7550

AMERICAN RIVERS

1025 Vermont Ave., NW
Washington, DC 20005 USA
Phone: 202-347-7550   Fax: 202-347-9240
E-mail: amrivers@amrivers.org
Website: www.americanrivers.org

Founded: 1973
Membership: 50
Scope: National

Description: (formerly American Rivers Conservation Council)
America's leading river conservation organization. Preserves
and restores America's river systems and fosters a river
stewardship ethic. River conservation goals focused on
protecting wild rivers, restoring hometown rivers, and repairing
big rivers. Conservation programs in wild/nationally significant
rivers, hydropower reform, urban rivers, and floodplains.
Three-part strategy: develop and demonstrate community-
based solutions to protect and restore rivers; communicate
river values and build a diverse nationwide constituency for
river conservation; and advocate reform of national policies
and practices to foster river health and restore river values.

Publication(s): American Rivers Newsletter, Mississippi Monitor,
Missouri Monitor, In Harm's Way: The Costs of Floodplain
Development, Voyage of Recovery: Restoring the Rivers of
Lewis and Clark, America's Most Endangered Rivers (Annual
Report)

Keyword(s): Environmental and Conservation Education, Rivers,
Exotic species, Aquatic nuisance species, Watersheds,
Wetlands, Conservation, Riparian Restoration, Dams,
Hydropower Relicensing, Salmon Recovery

Contact(s):
Whitney Hatch, CHAIR, BOARD OF DIRECTORS
Amy Souers, COMMUNICATIONS ASSOCIATE
Betsy Otto, DIRECTOR OF RIVER RESTORATION
FINANCE
Bea Keller, MANAGER, MEMBERSHIP SERVICES
Jeff Stein, MISSISSIPPI RIVER REGIONAL
REPRESENTATIVE

Chad Smith, MISSOURI RIVER REGIONAL
REPRESENTATIVE
Andrew Fahlund, POLICY DIRECTOR OF HYDROPOWER
PROGRAMS
Rebecca Wodder, PRESIDENT
Margaret Bowman, SENIOR DIRECTOR OF DAM
PROGRAMS
Ann Mills, VICE-PRESIDENT FOR CONSERVATION
amills@amrivers.org
Pat Appel, VICE-PRESIDENT FOR RESOURCE
DEVELOPMENT
pappel@amrivers.org

AMERICAN RIVERS

MONTANA FIELD OFFICE
215 Woodland Estates
Great Falls, MT 59404 USA
Phone: 406-454-2076   Fax: 406-454-2530
E-mail: malbers@amrivers.org
Website: www.americanrivers.org

Founded: NA
Membership: 40,000
Scope: Regional

Description: (formerly American Rivers Conservation Council)

Publication(s): available on website

Contact(s):
Mark Albers, OFFICE DIRECTOR

AMERICAN RIVERS

NEBRASKA FIELD OFFICE
650 J St.
Lincoln, NE 68508 USA
Phone: 402-477-7910   Fax: 402-477-2565

Founded: NA
Scope: Regional

Description: (formerly American Rivers Conservation Council)

Contact(s):
Chad Smith, DIRECTOR

AMERICAN SOCIETY FOR ENVIRONMENTAL HISTORY

701 Vickers Ave.
Durham, NC 27701 USA
Phone: 919-682-9319   Fax: 919-682-2349
Website: www.h-net.msu.edu`aseh

Founded: 1976
Scope: National

Description: A nonprofit international society that seeks understanding of human ecology through the perspectives of history
and the humanities.

Publication(s): Environmental History, Newsletter

Keyword(s): Environmental Law, Environmental and
Conservation Education, education, Public Lands, Exotic
species, Aquatic nuisance species

Contact(s):
Ed Russell, BOOK REVIEW EDITOR
Hal Rothman, EDITOR
Department of History, University of Nevada-Las Vegas, Las
Vegas, NV 89154
Phone: 702-739-3349

Jeffrey Stine, PAST PRESIDENT
National Museum of American History, Smithsonian Institute,
Washington, DC 20560
Phone: 202-357-2058
Carolyn Merchant, PRESIDENT
Lisa Mighetto, SECRETARY
Historical Research Associates, 119 Pine St., Suite 207,
Seattle, WA 98101

AMERICAN SOCIETY OF ICHTHYOLOGISTS AND HERPETOLOGISTS

Attn: Secretary, Maureen Donnelly, Dept of Biological
Sciences
College of Arts & Science
Florida International University
North Miami, FL 33181 USA
Phone: 305-919-5651   Fax: 305-919-5964
E-mail: donnelly@fiu.edu
Website: www.lgg.245.200.110

Founded: 1913
Scope: International

Description: To advance the scientific study of fishes, amphibians
and reptiles.

Publication(s): Copeia, ASIH Special Publications

Keyword(s): Fish, Aquatic Habitats, Wildlife Rehabilitation,
Endangered Species, Reptiles and Amphibians, Zoology

Contact(s):
Margaret Stewart, HISTORIAN
Harry Greene, PRESIDENT
Brooks Burr, PRESIDENT-ELECT
Maureen Donnelly, SECRETARY
Larry Page, TREASURER

AMERICAN SOCIETY OF INTERNATIONAL LAW/WILDLIFE INTEREST GROUP

1210 Floribunda Ave. Ste 7
Burlingame, CA 94010 USA
Phone: 650-703-3280   Fax: 801-838-4710
E-mail: asilwildlife@pacbell.net
Website: www.eelink.net/~asilwildlife

Founded: 1984
Membership: 163
Scope: International

Description: The ASIL and WIG works to improve the effectiveness of international wildlife treaty regimes and national
legislation that implements such regimes.

Publication(s): Journal of International Wildlife Law and Policy

Keyword(s): Environmental Law, International Environmental Law,
Environmental Legislation, Wildlife Protection, International
Wildlife, International Trade and Environment

Contact(s):
William Burns, CO-CHAIRMAN
jwlt@pacbell.net

AMERICAN SOCIETY OF LANDSCAPE ARCHITECTS

636 Eye St., NW
Washington, DC 20001-3736 USA
Phone: 202-898-2444   Fax: 202-898-1185
Website: http://asla.org

Founded: 1899
Membership: 12,500
Scope: National

Description: ASLA is a professional organization representing the
landscape architecture profession in the United States.
Landscape architicture, comprehensive by definition, is the art
and science of analysis, planning, design, management,
preservation and rehabilitation of the land. ASLA works on
issues such as land use planning, sustainable communities,
transportation, public open space and water conservation. The
mission of ASLA is the advancement of the art and science of
landscape architecture by leading and informing the public, by
serving members and by leading the profession in achieving
quality in the natural and built environments.

Publication(s): Landscape Architecture Magazine, Landscape
Architecture News Digest (LAND)

Keyword(s): Land Use Planning, Landscape Architecture, Open
Space, Public Lands, Urban Environment

Contact(s):
Jim Tolliver, DIRECTOR OF PUBLIC AFFAIRS
Anne Powell, EDITOR
Peter Kirsch, EXECUTIVE VICE PRESIDENT

AMERICAN SOCIETY OF LIMNOLOGY AND OCEANOGRAPHY

5400 Bosque Blvd., Suite 680
Waco, TX 76710-4446 USA
Phone: 254-399-9635   Fax: 254-776-3767
E-mail: business@aslo.org
Website: www.aslo.org/

Founded: 1936
Membership: 4000
Scope: International

Description: To promote the advancement of the various aquatic
science diciplines through scientific and technical symposia,
colloquia and meetings; promotion of scientific research;
discussion, publication and education; and conducting special
programs in response to community interest.

Publication(s): Limnology and Oceanography, Bulletin

Keyword(s): Ocean Conservation, Lakes, Rivers, Streams,
Wetlands

Contact(s):
Everett Fee, EDITOR-IN-CHIEF
343 Lady MacDonald Crescent, Canmore,
Alberta T1W 1H5
Phone: 403-609-2456
Fax: 403-609-2400
efee@telusplanet.net
Jonathan Phinney, EXECUTIVE DIRECTOR
1444 I St NW Ste. 200, Washington, DC 20005
jpinney@aslo.org
William Lewis, PRESIDENT
Denise Breitburg, SECRETARY

AMERICAN SOCIETY OF MAMMALOGISTS

USA
Phone: 805-892-2504
E-mail: asm@aibs.org
Website: http://www.mammalsociety.org

Founded: 1919
Scope: National

Description: Encourages research and learning in all phases of mammalogy and by holding annual meetings for presentation and discussion of the results of research dealing with mammals, through issuing periodicals and other publications and by giving advice on matters pertaining to mammals, particularly conservation issues.

Publication(s): Journal of Mammalogy, Special Publications of American Society of Mammalogists, Mammalian Species

Keyword(s): Endangered Species, Mammals, Marine Mammals, Whale, Dolphin, Seal, Nongame Wildlife

Contact(s):
Sarah George, 2ND VICE PRESIDENT
Utah Museum of Natural History, University of Utah, Salt Lake City, UT 84112
Phone: 801-581-4889
Gordon Kirkland, CHAIRMAN OF COMMITTEE ON CONSERVATION OF LAND MAMMALS
The Vertebrate Museum, Shippensburg University, Shippensburg, PA 17257
Winston Smith, CHAIRMAN OF COMMITTEE ON LEGISLATION AND REGULATIONS
Southern Forest Experimental Station, S. Hardwoods Laboratory, P.O. Box 227, Stoneville, MS 38776
John Hayning, CHAIRMAN OF COMMITTEE ON MARINE MAMMALS
Natural History Museum of Los Angles CA, 900 Exposition Blvd., Los Angles, CA 90007
Phone: 213-746-2999
Troy Best, MANAGING EDITOR
Department of Zoology, 331 Funchess Hall, Auburn University, AL 36849
Phone: 205-844-9260
Tom Kunz, PRESIDENT
Dept. of Biology Boston University, Boston, Ma 02215
H. Smith, SECRETARY AND TREASURER
Department of Zoology, Brigham Young University, Provo, UT 84602
Phone: 801-378-2492

AMERICAN SPORT FISHING ASSOCIATION

225 Reinekers Ln
Alexandria, VA 22314 USA
Phone: 703 518 6338   Fax: 700-519-1072
E-mail: info@asafishing.org
Website: www.asafishing.org

Founded: NA
Membership: 20
Scope: National

Description: FishAmerica Foundation is the conservation arm of the American Sportfishing Association. The Foundation is a nonprofit organization dedicated to enhancing the water quality and fish populations of North America.

Publication(s): Forbes Darby, Grant Guidelines

Keyword(s): Environmental and Conservation Education, Outdoor Recreation, Sport Fishing, Coral Reefs, Exotic species, Aquatic nuisance species

Contact(s):
Tom Marshall, MANAGING DIRECTOR

AMERICAN SPORTFISHING ASSOCIATION

1033 North Fairfax St., Suite 200
Alexandria, VA 22314 USA
Phone: 703-519-9691   Fax: 703-519-1872
E-mail: info@asafishing.org
Website: www.asafishing.org

Founded: 1994
Scope: National

Description: ASA is a nonprofit industry association working to ensure healthy and sustainable fisheies resources and increase sportfishing participation through education, conservation, promotion and marketing.

Publication(s): American Sportfishing

Contact(s):
Burt Steinberg, CHAIRMAN
Mike Hayden, PRESIDENT AND CEO
Norville Prosser, VICE PRESIDENT
Michael Nussman, VICE PRESIDENT

AMERICAN WATER RESOURCES ASSOCIATION

4 West Federal St.
Middleburg, VA 20118-1626 USA
Phone: 540-687-8390   Fax: 540-687-8395
E-mail: info@awra.org
Website: www.awra.org

Founded: 1964
Membership: 7
Scope: International

Description: A nonprofit scientific organization which advances water resources research, planning, development and management; establishes a common meeting ground for engineers and physical, biological, and social scientists concerned with water resources; disseminates information in the field of water resources policy, science and technology through the publication of a scientific journal newsletter and symposium proceedings. Two specialty conferences/symposia and one Annual Conference on Water Resources are held each year.

Publication(s): Journal of the American Water Resources Association, Water Resources IMPACT, Symposium Proceedings

Keyword(s): Environmental and Conservation Education, Renewable Resources, Rivers, Exotic species, Aquatic nuisance species, Wetlands

Contact(s):
Christopher Lant, EDITOR OF THE JOURNAL
N. Spangenberg, EDITOR OF WATER RESOURCES IMPACT
Kenneth Reid, EXECUTIVE VICE PRESIDENT
John Grounds, PRESIDENT
Kenneth Lanfear, PRESIDENT ELECT
D. Adams, SECRETARY AND TREASURER

AMERICAN WATER WORKS ASSOCIATION (AWWA)

6666 W. Quincy Ave.
Denver, CO 80235 USA
Phone: 303-794-7711   Fax: 303-795-1440
Website: www.awwa.org

Founded: 1881
Scope: international

Description: The AWWA advances the science, technology, consumer awareness management, government policies and water use efficiencies related to public drinking water.

Publication(s): Mainstream, Water Week, Opflow, WaterWiser - The Water Efficiency Clearinghouse (an on-line Internet Resource (http://www.waterwiser.org), AWWA Journal

Keyword(s): Planning Management, Pollution Prevention, Public Health Protection, Exotic species, Aquatic nuisance species, Water Quality

Contact(s):
Robert Renner, DEPUTY EXECUTIVE DIRECTOR
Tom Curtis, DEPUTY EXECUTIVE DIRECTOR OF GOVERNMENT AFFAIRS DIVISION
Washington, DC 20005
Phone: 202-628-8803
Jack Hoffbuhr, EXECUTIVE DIRECTOR

AMERICAN WHITEWATER

1430 Fenwick Ln.
Silver Spring, MD 20910 USA
Phone: 301-589-9453   Fax: 301-589-6121
E-mail: nick@amwhitewater.org
Website: www.americanwhitewater.org

Founded: 1957
Membership: 8500
Scope: National

Description: American Whitewater's mission is to conserve and restore America's whitewater resources and enhance opportunities to enjoy them safely. This is achieved by means of conservation, river access, education, safety and event programs.

Publication(s): American Whitewater Journal, Inventory of Whitewater Rivers, Safty Code of American Whitewater - Pamphlet

Keyword(s): Outdoor Recreation, Rivers

Contact(s):
Nick Lipkowski, EXECUTIVE ASSISTANT
nick@amwhitewater.org
Risa Shimoba, EXECUTIVE DIRECTOR
RichB@amwhitewater.org
Jay P. K. Kenney, PRESIDENT
Phone: 303-534-5722
Fax: 303-534-5721
jaypkk@aol.com
Richard Penny, VICE-PRESIDENT
Seattle, WA 206-213-0330
Phone: 206-213-0334
rpenny@amrivers.org

AMERICAN WILDLANDS

40 East Main #2
Bozeman, MT 59715 USA
Phone: 406-516-8175   Fax: 406-516-8242
E-mail: amwild@wildlands.org
Website: http://www.wildlands.org/

Founded: 1977
Scope: Regional

Description: A nonprofit conservation organization dedicated to ecologically sustainable use and protection of America's wildland resources in the Rocky Mountains West, including wilderness, wetlands, rangelands, free-flowing rivers, wildlife and fisheries and forests.

Publication(s): On The Wild Side, Policy Reports, Forest Activist Green Papers

Keyword(s): Biodiversity, Forests and Forestry, Public Lands, Wildlands, training

Contact(s):
Jeff Larmer, EXECUTIVE DIRECTOR
Clifton Merritt, EXECUTIVE EDITOR
Sally Ranney, PRESIDENT
Clifton Morritt, SECRETARY AND TREASURER
William Cunningham, VICE CHAIRMAN

AMERICAN WILDLIFE RESEARCH FOUNDATION, INC.

50 West High St.
Balton Spa, NY 12020 USA
E-mail: wms4@cornell.edu

Founded: 1911
Membership: 51
Scope: International

Description: AWRF uses the interest income of its funds to support research of wildlife and its habitats. Its mission is to enhance fish and wildlife resources and their habitats through research, education and conservation, ensuring that present and future generations can continue to use and enjoy them.

Publication(s): Newsletter

Keyword(s): Grants, Research

Contact(s):
Stuart Free, PRESIDENT
Phone: 518-861-5357
Fax: 518-452-6392
William Schwerd, SECRETARY
Phone: 518-885-8995
Fax: 518-885-9078

AMERICAN ZOO AND AQUARIUM ASSOCIATION (AZA)

8403 Colesville Rd, Suite 710
Silver Spring, MD 20910 USA
Phone: 301-562-0777   Fax: 301-562-0888
Website: www.aza.org

Founded: 1924
Membership: 196
Scope: National

Description: Dedicated to the improvement of modern, professionally-managed zoological parks and aquariums through conservation, public education, scientific research and membership services. Administers scientifically-managed captive breeding and field conservation programs for 134

threatened and endangered species through its Species Survival Plan Program.

Publication(s): COMMUNIQUE, Annual Report on Conservation and Science, Annual and Regional Conference Proceedings, AZ A Membership Directory

Keyword(s): Aquariums, Endangered Species, Zoological Parks

Contact(s):
Kristin Vehrs, DEPUTY DIRECTOR AND DIRECTOR OF GOVERNMENT AFFAIRS
Michael Hutchins, DIRECTOR OF CONSERVATION AND SCIENCE
Bruce Carr, DIRECTOR OF CONSERVATION EDUCATION
Linda Martin-MCormic, DIRECTOR OF DEVELOPMENT AND MARKETING
Laura Benson, DIRECTOR OF FINANCE AND ADMINISTRATION
Jane Ballentine, DIRECTOR OF PUBLIC AFFAIRS
Sydney Butler, EXECUTIVE DIRECTOR
Ted Beattie, PRESIDENT

ANACOSTIA WATERSHED SOCIETY THE GEORGE WASHINGTON HOUSE

4302 Baltimore Ave.
Bladensburg, MD 20710 USA
Phone: 301-699-6204   Fax: 301-699-3317
Website: www.anacostiaws.org

Founded: 1989
Membership: 1100
Scope: Regional

Description: The Anacostia Watershed Society provides opportunities for volunteers to take part in local environmental restoration projects; and provides advocacy for environmental equity issues in the Anacostia-Washington region.

Publication(s): Voice of the River

Keyword(s): Rivers, Watersheds, Pollution Prevention, Coral Reefs, Environmental Protection

Contact(s):
James Connolly, EXECUTIVE DIRECTOR
jim@anacostiaws.org
Robert Boone, PRESIDENT
robert@anacostiaws.org
John Perhonis, SECRETARY
jperhoni@nsf.gov
David Tibbetts, TREASURER

ANCIENT FOREST INTERNATIONAL

P.O. Box 1850
Redway, CA 95560 USA
Phone: 707-923-3015   Fax: 707-923-4486
E-mail: afi@ancientforest.org
Website: www.ancientforest.org

Founded: 1989
Scope: International

Description: An alliance of conservationists dedicated to helping preserve, study and increase awareness of the Earth's few still-intact forest ecosystems, while providing habitat continuity through the creation of corridors. Old-growth forests of southern Chile, highland Mexico, Ecuador and the north Pacific coast are current projects. Work is also underway to document the distribution of ancient rainforests worldwide and to promote their preservation.

Publication(s): News of Old Growth, Chile's Native Forest: An Overview

Keyword(s): Biodiversity, Endangered Species, Forests and Forestry, Land Purchase, Ancient Forests, Rainforests

Contact(s):
Rick Klein, PRESIDENT
Suzelle Hunt, SECRETARY
Tim Metz, TREASURER

ANGLERS FOR CLEAN WATER

P.O. Box 17900
Montgomery, AL 36141-0900 USA
Phone: 334-272-9530   Fax: 334-270-8549
Website: http://www.ag.auburn.edu/grassroots/acw/

Founded: 1970
Scope: National

Description: A nonprofit organization dedicated to educating the American public on the conditions of pollution nationwide and to the danger of the failure to halt the pollution of the streams, rivers and lakes of the United States and to promote, educate and inform the American public of the need for conservation of our water and fisheries resources.

Publication(s): Living Waters

Keyword(s): Aquatic Habitats, Communications, Environmental and Conservation Education, Wildlife, Exotic species, Aquatic nuisance species

Contact(s):
Ann Lewis, COMMUNICATIONS
Bruce Shupp, CONSERVATION DIRECTOR
Matt Vincent, EDITOR
Karl Dabbs, FINANCE
Helen Sevier, PRESIDENT

ANIMAL PROTECTION INSTITUTE

1122 S Street
Sacramento, CA 95814 USA
Phone: 916-731-5521   Fax: 916-447-3070
E-mail: info@api4animals.org
Website: www.api4animals.org

Founded: 1968
Membership: 8500
Scope: National

Description: The Animal Protection Institute is a national animal advocacy nonprofit organization dedicated to protecting animals against abuse through enforcement and legislative actions, investigations, advocacy campaigns, crisis intervention, public awareness and education. Specific areas of concern are wildlife protection and habitat conservation, companion animals, marine mammals, domestic and farm animals, animals used in research and humane eucation.

Publication(s): Animal Issues

Keyword(s): Endangered Species, Hunting, Mammals, Public Lands, training, Nongame Wildlife, Predators, Marine Mammals, Whale, Dolphin, Seal, Trapping, Animal Welfare

Contact(s):
Gary Pike, CHAIRMAN OF THE BOARD
Barbara Lawrie, CREATIVE SERVICES
Gil Lamont, EDITOR
Alan Berger, EXECUTIVE DIRECTOR

ANIMAL WELFARE INSTITUTE

P.O. Box 3650
Washington, DC 20007 USA
Phone: 202-337-2332   Fax: 202-338-9478
E-mail: awi@awionline.org
Website: www.awionline.org

Founded: 1951
Membership: 25000
Scope: National

Description: Active in improvement of conditions for laboratory animals and reducing the numbers used in research, protection of endangered species, Save the Whales campaign, ending use of steel jaw traps, stopping imports of wild birds for the pet trade and humane education. Albert Schweitzer award is presented for outstanding contributions to animal welfare.

Publication(s): Animal Welfare Institute Quarterly, Endangered Species Handbook, Alternative Traps, Animals and Their Legal Rights

Keyword(s): Endangered Species, Mammals, Marine Mammals, Whale, Dolphin, Seal, Trapping, Animal Welfare

Contact(s):
Cathy Liss, EXECUTIVE DIRECTOR
Phone: 202-337-2332
Lynne Hutchison, EXECUTIVE SECRETARY
Diane Halverson, FARM ANIMAL CONSULTANT
Ben White, INTERNATIONAL COORDINATOR
Viktor Reinhardt, LABORATORY ANIMAL CONSULTANT
Nell Naughton, MAIL ORDER SECRETARY
Christine Stevens, PRESIDENT
Phone: 202-337-2332
Ava Armandarez, PUBLICATIONS COORDINATOR
Phone: 202-337-2332
Adam Roberts, RESEARCH ASSOCIATE
Phone: 202-337-2332
Fred Hutchison, TREASURER
Cynthia Wilson, VICE PRESIDENT

ANTARCTICA PROJECT

P.O. Box 76920
Washington, DC 20013 USA
Phone: 202-234-2480   Fax: 202-387-4823
E-mail: antarctica@igc.org
Website: www.asoc.org

Founded: 1982
Scope: International

Description: Works to preserve Antarctica by monitoring all activities to ensure minimal environmental impact and consulting with key users of Antarctica, including scientists, tourists, governments. Conducts legal and policy research and analysis; produces educational materials; focuses international scientific community on globally-significant research. Secretariat to Antarctic and Southern Ocean Coalition (ASOC), composed of 240 conservation groups in 50 nations.

Publication(s): ECO Newspaper (ASOC), Publications and educational resources list on request, Antarctica Project (Quarterly Newsletter)

Keyword(s): Biodiversity, Conservation, Wilderness

Contact(s):
Mark Stevens, COORDINATOR FISHERIES CAMPAIGN
Phone: 20-223-8052
mark.antarctica@igc.org
Scott Altmann, COORDINATOR OF PROTOCOL IMPLEMENTATION CAMPAIGN
scott.antarctica@igc.org
Jim Barnes, COUNSEL
Beth Clark, DIRECTOR

APPALACHIAN MOUNTAIN CLUB

5 Joy St.
Boston, MA 02108 USA
Phone: 617-523-0636   Fax: 617-523-6617
E-mail: information@amcinfo.org
Website: www.outdoors.org

Founded: 1876
Membership: 94000
Scope: Regional

Description: The AMC pursues a far-reaching conservation agenda while encouraging responsible recreation, based on the philosophy that successful, long-term conservation depends on firsthand experience and enjoyment of the natural environment. Areas of focus: Northern Forest, Sterling Forest, White Mountain N.F., NY and NJ Highlands, Berkshire and Taconics Region, Delaware Water Gap National Recreation Area, and Acadia National Park. Expertise: Conservation policy, advocacy; land, trail, river and greenway stewardship; environmental research, education, guidebook and outdoor leadership publishing.

Publication(s): Appalachia Journal, AMC guidebooks and maps, AMC Outdoors

Keyword(s): Air Quality and Pollution, Environmental and Conservation Education, Outdoor Recreation, Trail, Public Lands, Rivers

Contact(s):
Walter Graff, DEPUTY DIRECTOR
Phone: 603-466-2721
Eric Antebi, DIRECTOR OF CONSERVATION POLICY AND ADVOCACY
Phone: 617-523-0655
Peg Brady, DIRECTOR OF CONSERVATION PROGRAMS
Phone: 617-523-0655
Andrew Falender, EXECUTIVE DIRECTOR
Laurie Burt, PRESIDENT
Kenneth Kimball, RESEARCH DIRECTOR
Phone: 603-466-2721

APPALACHIAN TRAIL CONFERENCE

P.O. Box 807
Harpers Ferry, WV 25425-0807 USA
Phone: 304-535-6331   Fax: 304-535-2667
E-mail: general@appalachiantrail.org
Website: www.appalachiantrail.org

Founded: 1925
Membership: 33000
Scope: Regional

Description: Coordinates preservation and management of the Appalachian Trail, a 2,160 mile footpath and protective corridor generally following the crest of the Appalachian Mountains from Maine to Georgia. Prepares and distributes trail guidebooks and other user information.

Publication(s): Appalachian Trailway News, Trail Lands, Inside ATC, Register, The

Keyword(s): Trail, Endangered Species, Environmental and Conservation Education, Historic Preservation, Land Purchase, Outdoor Recreation

Contact(s):
Brian King, DIRECTOR OF PUBLIC AFFAIRS
bking@atconf.org
Brian Fitzgerald, CHAIR
55 Ward Hills Rd, S Duxbury, VT 05660
Robert Rubin, EDITOR
David Startzell, EXECUTIVE DIRECTOR
Terthema Martin, SECRETARY
Kennard Honick, TREASURER
1800 Second St., Suite 810, Sarasota, FL 34236
Phone: 941-366-3944
Thyra Sperry, VICE CHAIR
740 Oak Hill Dr., Boiling Springs, PA 17007-9624
Phone: 717-258-5261
Maryanne Skwen, VICE CHAIR
553 Forest Superior Ave, Decatur, GA 30033
Phone: 540-427-4536
Carl Demrow, VICE CHAIR
202 Mason Rd, West Topsham, VT 05086

ARCHAEOLOGICAL CONSERVANCY
5301 Central Ave., NE, Suite 402
Albuquerque, NM 87108 USA
Phone: 505-266-1540   Fax: 505-266-0311
E-mail: archons@nm.net
Website: www.americanarchaeology.com

Founded: 1979
Scope: National

Description: National nonprofit membership organization dedicated to the permanent preservation of the most significant archaeological sites in the United States, usually through acquisition. Cooperates with government, universities, museums, and private conservation organizations to acquire lands for permanent archaeological preserves.

Publication(s): American Archaeology

Keyword(s): Cultural Preservation, Land Purchase, Historic Preservation

Contact(s):
Earl Gadbery, CHAIRMAN OF THE BOARD
Rob Crisell, EASTERN REGIONAL DIRECTOR
1307 S. Glebe Rd., Arlington, VA 22204
Phone: 703-979-4410
Paul Gardner, MIDWEST REGIONAL OFFICE DIRECTOR
295 Acton Rd, Columbus, OH 43214
Phone: 614-267-1100
Mark Michel, PRESIDENT
Alan Gruber, SOUTHEASTERN REGIONAL OFFICE DIRECTOR
5997 Cedar Crest Rd., Acworth, GA 30101
Phone: 770-975-4344
James Walker, SOUTHWEST REGIONAL OFFICE DIRECTOR
5301 Central Ave. NE, Suite 1218, Albuquerque, NM 87108
Phone: 505-266-1540
Lynn Dunbar, WESTERN REGIONAL OFFICE DIRECTOR
1217 23rd St., Sacramento, CA 95816-4917
Phone: 916-448-1892

ARCHBOLD BIOLOGICAL STATION
P.O. Box 2057
Lake Placid, FL 33862-2057 USA
Phone: 863-465-2571   Fax: 863-699-1927
E-mail: archbold@archbold-station.org
Website: www.archbold-station.org

Founded: 1941
Membership: 60
Scope: Statewide

Description: The Station is an independent, nonprofit facility devoted to long-term ecological research and conservation. Primary focus is on organisms, including many endangered species, and environments of the unique Lake Wales Ridge and adjacent Florida.

Publication(s): Biennial Report

Keyword(s): Ecology, Endangered Species, Environmental and Conservation Education, Protected Areas, Research

Contact(s):
Patrick Bolen, ASSISTANT DIRECTOR FOR AGRO-ECOLOGY
Nancy Deyrup, EDUCATION COORDINATOR
Hilary Swain, EXECUTIVE DIRECTOR
Tina Fleischer, INTERNSHIP COORDINATOR
Fred Lohrer, LIBRARIAN

ARCHERY MANUFACTURERS AND MERCHANTS ORGANIZATION (AMO)
304 Brown St. East
Comfrey, MN 56019 USA
Phone: 703-242-8310   Fax: 507-877-2149
Website: www.amo-archery.org

Founded: 1953
Membership: 1000+
Scope: International
Keyword(s): Hunting, Outdoor Recreation, training, Youth Organizations

Contact(s):
Kelly Kelly, DIRECTOR OF OPERATIONS AND MEMBERSHIP SERVICES
kelly@amoarchery.com
Jay McAninch, PRESIDENT AND CEO

ARCTIC INSTITUTE OF NORTH AMERICA
University Library Tower, 2500 University Dr., NW
Calgary, Alberta T2N 1N4 USA
Phone: 403-220-7515   Fax: 403-282-4609
Website: www.ucalgary.ca/aine

Founded: 1945
Membership: 1600
Scope: International

Description: A nonprofit research organization dedicated to acquisition, interpretation, and dissemination of knowledge of the polar regions. Sponsors research by its thirty research associates.

Publication(s): Arctic Journal

Keyword(s): Arctic

Contact(s):
Sonja Hogg, BUSINESS MANAGER
Phone: 403-220-7517
hogg@ucalgary.ca

James Raffan, CHAIR OF THE CANADIAN BOARD OF DIRECTORS
Phone: 613-387-2866
raffanj@educ.queensun.ca
Carl Benson, CHAIRMAN OF THE AINA USA BOARD OF GOVERNORS
Karen McCullough, EDITOR
Phone: 403-220-4049
kmccullo@ucalgary.ca
Karla Jessen-Williamson, EXECUTIVE DIRECTOR
Ross Goodwin, MANAGER OF ASTIS DATABASE

ARIZONA ASSOCIATION OF CONSERVATION DISTRICTS

Attn: Executive Director, 3003 N. Central Ave., Suite 800
Phoenix, AZ 85012 USA
Phone: 602-280-8803   Fax: 602-280-8779
E-mail: aacd@az.nrcs.usda.gov
Website: http://www.aacdonline.com

Founded: NA
Scope: Statewide

Keyword(s): Conservation Districts

Contact(s):
Frank Martinez, 1ST VICE PRESIDENT
Box 1152, Parker, AZ 85344
Phone: 520-669-8459
Marcareo Herrera, EXECUTIVE DIRECTOR
Phone: 602-280-8803
Fax: 602-280-8779
Sharon Reid, PRESIDENT AND BOARD MEMBER
Rt. 1 Box 49-C, St. David, AZ 85630
Phone: 520-586-3347
Johnny Lavin, SECRETARY/TREASURER
HC 1 Box 760, Benson, AZ 83602
Phone: 520-212-3211
Fax: 520-384-2735
Robert Ahkeah, VICE PRESIDENT
P.O. Box 550, Shiprock, NM 87420
Phone: 505-368-5430

ARIZONA B.A.S.S. CHAPTER FEDERATION

P.O. Box 577
Kearny, AZ 85237 USA
Phone: 520-363-5912

Founded: NA
Membership: 630
Scope: Statewide

Description: An organization of Bassmaster chapters, affiliated with the Bass Anglers Sportsman Society, organized to fight pollution, assist state and national conservation agencies in their efforts, and teach the young people of our country good conservation practices. Dedicated to the realistic conservation of our water resources.

Contact(s):
Dave Cohen, CONSERVATION DIRECTOR
839 S. Westwood 266, Mesa, AZ 85210
Phone: 602-962-9009
Mike Johnson, PRESIDENT

ARIZONA WILDLIFE FEDERATION

644 N. Country Club Dr. - Suite E
Mesa, AZ 85201-4983 USA
Phone: 480-644-0077   Fax: 480-644-0078
E-mail: awf@azwildlife.org
Website: www.azwildlife.org

Founded: 1923
Membership: 3500
Scope: Statewide

Description: A representative statewide organization, affiliated with the National Wildlife Federation, dedicated to the protection and enhancement of wildlife and its habitat through public education and government interaction.

Publication(s): Arizona Wildlife News

Contact(s):
Don Farmer, ALTERNATE REPRESENTATIVE
Steve Galliziolli, EDITOR
Ken Haefner, EXECUTIVE DIRECTOR
Jerry Thorson, PRESIDENT & ACTING TREASURER
Jack Simon, REPRESENTATIVE
Mike Perkinson, VICE-PRESIDENT
Randy Lamb, VICE-PRESIDENT

ARKANSAS ASSOCIATION OF CONSERVATION DISTRICTS

Attn: Exec. Vice President, 101 E. Capitol Ste 350
Little Rock, AR 72201 USA

Founded: NA
Scope: Statewide

Keyword(s): Conservation Districts

Contact(s):
Paul Mayfield, 1ST VICE PRESIDENT
783 Rio Vista Rd., Bald Knob, AR 72010
Phone: 501-724-5932
mayfield@IPA.Net
Debbie Moreland, EXECUTIVE VICE PRESIDENT
20311 Lake Vista, Roland, AR 72135
Phone: 501-868-5294
Bill Rainwater, PRESIDENT
P.O. Box 2245, Jonesboro, AR 72401
Phone: 870-935-1624
Roy Mahler, SECRETARY/TREASURER
Rt. 2 Box 130, Elkins, AR 72727
Phone: 501-643-3385

ARKANSAS B.A.S.S. CHAPTER FEDERATION

500 Coles Chapel Circle
Branch, AR 72928 USA
Phone: 501-635-5951
Website: www.arkansasbass.com

Founded: NA
Membership: 800
Scope: National

Description: An organization of Bassmaster chapters, affiliated with the Bass Anglers Sportsman Society, organized to fight pollution, assist state and national conservation agencies in their efforts, and teach the young people of our country good conservation practices. Dedicated to the realistic conservation of our water resources.

Publication(s): Available on website, Arkansas Bass Newsletter

Contact(s):
Bobby Davenport, CONSERVATION DIRECTOR
Phone: 870-673-1799
Gene Carson, PRESIDENT

ARKANSAS ENVIRONMENTAL EDUCATION ASSOCIATION

P.O. Box 488
Hackett, AR 72937 USA
Phone: 501-638-7151   Fax: 501-638-7151
E-mail: arkenved@aol.com
Website: www.aeea.org

Founded: 1995
Scope: Statewide

Description: The Association promotes environmental education and supports the work of environmental eucators in Arkansas.

Publication(s): Natural State, The, Membership Directory, EE Resource Directory

Keyword(s): Environmental and Conservation Education, Education

Contact(s):
Robert Mcafee, EXECUTIVE DIRECTOR
Constance Gwinn, PRESIDENT
Phone: 501-756-5583
cgwinn@crg.org
Suzanne Hirrel, PRESIDENT (PAST)
Phone: 501-671-2288
Fax: 501-671-2110
shirrel@uaex.edu

ARKANSAS WILDLIFE FEDERATION

9700 Rodney Parham Road,
Suite I-2
Little Rock, AR 72227-6212 USA
Phone: 501-224-9200   Fax: 501-224-9214
E-mail: arkwildlifefed@aristotle.net

Founded: 1936
Membership: 3000
Scope: Statewide

Description: A representative statewide organization, affiliated with the National Wildlife Federation, dedicated to the protection and enhancement of wildlife and its habitat through public education and government interaction.

Publication(s): Arkansas Fish and Wildlife

Contact(s):
Steve Duzan, ALTERNATE REPRESENTATIVE
Bob Apple, EDITOR
Terry Horton, EXECUTIVE DIRECTOR & EDUCATION PROGRAMS CONTACT
Ducote Haynes, PRESIDENT
Jim Wood, REPRESENTATIVE & TREASURER

ARLINGTON OUTDOOR EDUCATION ASSOCIATION, INC.

PHOEBE HALL KNIPLING OUTDOOR LABORATORY
P.O. Box 5646
Arlington, VA 22205 USA
Phone: 540-347-2258   Fax: 540-349-3336

Founded: 1967
Scope: Local, Regional

Description: AOEA's Outdoor Lab annually provides approximately 9,000 northern Virginia school children, in grades kindergarten through twelve, with enriching environmental and educational opportunities in a natural setting. In addition to daily classes during the school year, the lab conducts camps during the summer and astronomical observatory sessions throughout the year.

Keyword(s): Aquatic Habitats, Environmental and Conservation Education, Environmental Protection, Wildlife, Land Preservation

Contact(s):
Neil Heinekamp, LAB DIRECTOR
Terry Rusnack, PRESIDENT
Anita Scott, SECRETARY
Maureen McManus, TREASURER
Lori Lowe, VICE PRESIDENT

ASSOCIATION FOR BIODIVERSITY INFORMATION

1101 Wilson Blvd.
Arlington, VA 22209 USA
Phone: 703-908-1800   Fax: 703-908-1917
Website: www.abi.org
Scope: Regional

Description: The Association for Biodiversity Information (ABI) is a non-profit organization dedicated to providing knowledge to protect natural heritage programs and conservation data centers, ABI is a leading source for scientific information on rare and endandered species and threatened ecosystems. ABI and its member programs operate in the United States, Canada, Latin America, and the Caribbean providing essential information for conservation action.

Contact(s):
Larry Sugarbaker, CHIEF INFORMATION OFFICER
Mark Schaefer, PRESIDENT AND CEO
Mary Klein, VICE PRESIDENT FOR NATURAL HERITAGE NETWORK OPERATIONS
Joy Gaddy, VICE PRESIDENT FOR OPERATIONS
Bruce Stein, VICE PRESIDENT FOR PROGRAMS
Dennis Grossman, VICE PRESIDENT FOR SCIENCE

ASSOCIATION FOR CONSERVATION INFORMATION, INC.

Attn: President, New Hampshire Fish and Game Department, 2 Hazen Dr.
Concord, NH 33301 USA
Website: www.aci-net.org

Founded: 1900
Scope: National

Description: Facilitates free exchange of ideas, materials, techniques, experiences, and procedures bearing on conservation information and education and establishes media furthering such exchange; promotes public understanding of basic conservation principles; informs states, territories, and provinces that do not have conservation education programs of their desirability and assists them in setting up conservation education, information and public relations programs.

Publication(s): Balance Wheel, The

Keyword(s): Environmental and Conservation Education

Contact(s):
Judy Stokes, PRESIDENT
New Hampshire Fish & Game Department, 2 Hazen Drive,
Concord, NH 03301
Phone: 603-271-3211

ASSOCIATION FOR NATURAL RESOURCES ENFORCEMENT TRAINING

Missouri Department of Conservation Box 180
Jefferson City, MO 65102 USA
Phone: 573-751-4115
E-mail: yamnil@mail.conservation.state.mo.us
Website: www.dirdid.com\anret

Founded: NA
Scope: National

Description: The goal of the association is to promote and
enhance professional standards of training in fish and wildlife
enforcement. The objectives are: to promote officer safety and
a safer working environment; exchange training information;
promote law enforcement research and development; to
encourage cost-effective training programs; to act as a
repository for catalogue agency training personnel and
materials; and to host annual workshop to facilitate the
exchange of training information. Open to Canadian and
United States agencies.

Keyword(s): Resource Law Enforcement, Law Enforcement

Contact(s):
Scottey Roxburgh, SECRETARY
Department of Fisheries and Oceans, Vancouver, British
Columbia V6B 5G3
Phone: 604-666-0123
roxburgh@dfo/moo.ga.ca
Fred Campbell, TREASURER
Natural Resources Conservation Authority, Kingston 10,
Jamaica. 1
Phone: 876-754-7567
fcampbell@nrca.org
Dave Windsor, VICE PRESIDENT
Indiana Department of Natural Resources,
Indianapolis, IN 46204
Phone: 317-232-4014
dwindsor@dnr.state.in.us

ASSOCIATION FOR THE PROTECTION OF THE ADIRONDACKS, THE

Schenectady, NY 12301 USA
Phone: 518-377-1452   Fax: 518-377-1452
Website: www.global.2000.net/protectadks/

Founded: 1901
Membership: 2000
Scope: International

Description: To protect the natural character of the state forest
preserve lands in the Adirondacks and Catskills as water-
holding and regulating forests which serve as a home for
wildlife and as wilderness recreation areas, and to protect and
enhance the natural resources of the Adirondack Park.

Publication(s): The Association News Quarterly Newsletter, The
Forest Preserve Magazine

Keyword(s): Biodiversity, Land Use Planning, Lakes, Public
Lands, Sustainable Development

Contact(s):
David Gibson, EXECUTIVE DIRECTOR
Abbey Verner, PRESIDENT

ASSOCIATION OF AMERICAN GEOGRAPHERS

1710 16th St., NW
Washington, DC 20009-3198 USA
Phone: 202-234-1450   Fax: 202-234-2744
E-mail: gaia@aag.org
Website: www.aag.org

Founded: 1904
Membership: 6000
Scope: International

Description: To further professional investigations in geography
and encourage the application of geographic findings in
education, government, and business.

Publication(s): AAG Newsletter, The Annals, Professional
Geographer, The

Keyword(s): Agriculture, Geography, Land Use Planning, Urban
Environment, Exotic species, Aquatic nuisance species

Contact(s):
Heather Baker, EDITOR OF NEWSLETTER
Ronald Abler, EXECUTIVE DIRECTOR
Amy Jo Woodruff, MANAGING EDITOR
Susan Cutter, PAST PRESIDENT
Jan Monk, PRESIDENT
Jennifier Wolch, SECRETARY
Robert Kent, TREASURER
Duane Neelis, VICE PRESIDENT

ASSOCIATION OF AVIAN VETERINARIANS

Central Office, P.O. Box 811720
Boca Raton, FL 33481 USA
Phone: 561-393-8901   Fax: 561-393-8902
E-mail: aavctrlofc@aol.com
Website: www.aav.org/aav

Founded: 1980
Membership: 2500
Scope: International

Description: The Association of Avian Veterinarians is a nonprofit
international organization dedicated to advancing and
promoting avian medicine and stewardship.

Publication(s): The Proceedings the Assoc. of Avian
Veterinarians, Journal of Avian Medicine and Surgery

Keyword(s): Birds, Endangered Species, Environmental and
Conservation Education, accreditation

Contact(s):
Robert Groskin, CONSERVATION COMMITTEE

ASSOCIATION OF CONSULTING FORESTERS OF AMERICA

732 North Washington St.,
Alexandria, VA 22314-1921 USA
Phone: 703-548-0990   Fax: 703-548-6395
E-mail: director@acf-foresters.com
Website: www.acf-foresters.com

Founded: 1948
Membership: 625
Scope: National

Description: The Association of Consulting Foresters of America, Inc. represents interests of private consulting foresters. Administers a continuing education program, enforces a code of ethics, and promotes use of private consulting foresters.

Publication(s): Membership Specialization Directory, Consultant

Keyword(s): Environmental and Conservation Education, Forests and Forestry, Renewable Resources, training, Professional Organization

Contact(s):
Lynn Wilson, EXECUTIVE DIRECTOR
Glenn Dabney, GULF DIRECTOR
William Steigerwaldt, NORTHERN DIRECTOR
William Humphries Jr., PAST PRESIDENT
David Parker, PRESIDENT
J. Tobey Wright, PRESIDENT-ELECT
Michael Lewis, SOUTHERN DIRECTOR
Richard Quarters, WESTERN DIRECTOR

ASSOCIATION OF FIELD ORNITHOLOGISTS
Attn: President, Inst. For Field Ornithology, Univ. of ME at Machias, 9 O'Brien Ave.
Machias, ME 04654 USA
Website: http://www.afonet.org/

Founded: 1922
Scope: National

Description: To promote the study of birds in their natural habitats throughout the new world and dissemination of the information obtained from this study.

Publication(s): Journal of Field Ornithology

Keyword(s): Wildlife Rehabilitation, Birds, Conservation Biology, Research

Contact(s):
C. Chandler, EDITOR
Dept. of Bio., GA Southern Univ., Statesboro, GA 30460-8042
Phone: 912-681-5657
chandler@gasou.edu
Jerome Jackson, PRESIDENT
Whitaker Center, College of Arts & Sciences, Florida Gulf Coast University, 10501 FGCU Blvd South,
Fort Myers, FL 33965
jjackson@fgcu.edu
Russ McClain, SECRETARY
Department of Biology, University of Memphis,
Memphis, TN 38152
Phone: 901-678-2581
wrmcclain@msuvxi.memphis.edu
George Mock, TREASURER
P.O. Box 395, Mattapoisett, MA 02739
Phone: 508-758-4408
gmock@nyclubricants.com

ASSOCIATION OF GREAT LAKES OUTDOOR WRITERS
Benld, IL 62009 USA
Phone: 217-839-2490   Fax: 217-839-2490
Website: www.greatlakeswriters.org

Founded: 1957
Membership: 300
Scope: Regional

Description: A nonprofit professional association of outdoor communicators dedicated to perpetuate the great outdoors through the judicious use of the written and spoken word.

Publication(s): AGLOW Horizons

Keyword(s): Environmental Communication

Contact(s):
Mike Seeling, CHAIRMAN OF THE BOARD
P.O. Box 604, Glenwood, IN 60425
Phone: 765-629-2493
Bob Schmidt, EDITOR
5016 Argyle, Chicago, IL 60630
Phone: 773-283-7871
Curt Hicken, EXECUTIVE DIRECTOR
David Mull, PRESIDENT
pondermull@aol.com
Clayton Diskerud, SECRETARY
Dan Donarski, TREASURER

ASSOCIATION OF MIDWEST FISH AND GAME LAW ENFORCEMENT OFFICERS
43 E Emerald Forest Lane
Sequim, WA 98382 USA
Phone: 360-582-1370   Fax: 360-582-1370

Founded: 1944
Scope: International

Description: To promote law enforcement cooperation among members, develop efficient cooperative law enforcement practices, establish a medium for disseminating information relating to illegal game law practices, devise legislative or regulatory changes for improving and standardizing law enforcement, and encourage the highest possible standards and practices of law enforcement among member organizations in the United States and Canada.

Keyword(s): Environmental Law, Wildlife, Outdoor Recreation, Renewable Resources

ASSOCIATION OF NEW JERSEY ENVIRONMENTAL COMMISSIONS
P.O. Box 157
Mendham, NJ 07945 USA
Phone: 973-539-7547   Fax: 973-539-7713
E-mail: anjec@aol.com
Website: www.anjec.org

Founded: 1969
Membership: 2400
Scope: State

Description: Private, nonprofit environmental organization serving the state's municipal environmental commissions, environmental organizations, and individual members by providing training programs, publications, research, reference, and liaison services.

Publication(s): Environmental Manual for Municipal Officials, Freshwater Wetlands Protection in New Jersey: A Manual for Local Officials, Keeping Our Garden State Green: A Local Government Guide for Greenway and Open Space Planning, Environmental Commission Handbook, ANJEC Report

Keyword(s): Environmental Protection, Protected Areas, Open Space, Pollution Prevention, Rural Development, Training, Sustainable Development, Exotic species, Aquatic nuisance species, Water Quality, Watersheds

Contact(s):
Sally Dudley, EXECUTIVE DIRECTOR
Gary Szelc, PRESIDENT
Michelle Gaynor, RESOURCE CENTER DIRECTOR

ASSOCIATION OF PARTNERS FOR PUBLIC LANDS

8375 Jumpers Hole Rd. Suite 104
Millersville, MD 21108 USA
Phone: 410-647-9001   Fax: 410-647-9003
E-mail: appl@appl.org
Website: www.appl.org

Founded: 1977
Membership: 80
Scope: National

Description: CNPCA is the official umbrella organization for nonprofit interpretive associations that operate bookstores and sales areas in national parks and in other federal, state, and municipal vistor centers. The Associations are the single largest contributors of donated funds for the support of education, vistor services, and research activities in our nation's parks ($17 million per year).

Publication(s): Cooperating Association Directory, Newswire

Keyword(s): Natural Areas, Outdoor Recreation, Public Lands, Training

Contact(s):
Donna Asbury, EXECUTIVE DIRECTOR
8375 Jumpers Hole Rd., Suite 104, Millersville, MD 21108
Phone: 410-647-9001

ASSOCIATION OF STATE AND TERRITORIAL HEALTH OFFICIALS

1275 K St., NW, Suite 800
Washington, DC 20005 USA
Phone: 202-371-9090   Fax: 202-371-9797
Website: www.astho.org

Founded: 1941
Scope: National

Description: ASTHO represents the directors of public health in each of the 50 states, the District of Columbia, and the U.S. Territories. Its purpose is to formulate and influence through collective action the establishment of sound national public health policy. ASTHO also assists and serves state health agencies in the development and implementation of state programs and policies in advancing the public health and prevention of disease.

Publication(s): Environmental Health News, Astho Report, Tobacco-Free Press

Keyword(s): Communications, Environment, Health and Nutrition, Public Health Protection

Contact(s):
George Hardy, EXECUTIVE VICE PRESIDENT

ATLANTIC CENTER FOR THE ENVIRONMENT

55 S. Main St.
Ipswich, MA 01938-2396 USA
Phone: 978-356-0038   Fax: 978-356-7322
E-mail: atlantic@qlf.org
Website: www.qlf.org

Founded: NA
Scope: International

Description: A regional community-based conservation organization promoting public involvement in resource management through year-round education, policy, and research programs in Atlantic Canada, Eastern Quebec, and New England (the Atlantic Region). As a technical assistance resource for local private and public agencies, the Atlantic Center provides resource assessments, conservation planning and strategy, policy analysis, and information services. It conducts many of its programs through an intern work force, which it recruits from colleges and universities across North America. The Atlantic Center also facilitates the exchange of ideas between its region and others. It has an active exchange program with organizations in Latin America and the Caribbean, the Middle East, and Europe. The Atlantic Center is a division of the Quebec-Labrador Foundation.

Publication(s): Compass

Keyword(s): Environmental and Conservation Education, Internships, Scholarships and Grants, Sustainable Development

Contact(s):
Linda Mitton, ADMINISTRATIVE ASSISTANT
Lawrence Morris, PRESIDENT
Kathleen Blanchard, PRESIDENT OF QLF CANADA
Jessica Brown, VICE PRESIDENT FOR INTERNATIONAL PROGRAMS
Thomas Horn, VICE PRESIDENT OF OPERATIONS

ATLANTIC CENTER FOR THE ENVIRONMENT

NEW ENGLAND OFFICE
P.O. Box 217
Montpelier, VT 05602 USA
Phone: 802-229-0707   Fax: 802-223-3593
Website: www.qlf.org

Founded: NA
Membership: 15
Scope: Regional
Publication(s): Compass

Contact(s):
Thomas Horn, VICE PRESIDENT

ATLANTIC CENTER FOR THE ENVIRONMENT

QLF CANADA OFFICE
1253 McGill College Ave., Suite 680
Montreal, Quebec H3B 2Y5 Canada
Phone: 514-395-6020   Fax: 514-395-4505
E-mail: montreal@qlf.org
Website: www.qlf.org

Founded: NA
Membership: 30
Scope: Internationally
Publication(s): Compass

ATLANTIC SALMON FEDERATION

International Headquarters, P.O. Box 5200
St. Andrews, New Brunswick E5B 3S8 Canada
Phone: 506-529-4581   Fax: 506-529-4438
E-mail: asf@nbnet.nb.ca
Website: www.asf.ca

Founded: 1982
Scope: International

Description: The largest international nonprofit organization dedicated to the preservation and wise management of the Atlantic salmon and its habitat. It was established upon consolidation of two leading salmon organizations, The Atlantic Salmon Association and The International Atlantic Salmon Foundation. ASF programs are directed toward research, conservation, education, and international cooperation. The Federation supports a network of regional groupings of local salmon conservation and other organizations which address a variety of salmon issues. ASF is totally dependent on contributions from individuals, foundations, and corporations in Canada, the United States, and overseas. Membership inquiries are welcomed.

Publication(s): Atlantic Salmon Journal, The

Keyword(s): Salmon Recovery, Atlantic Salmon, Endangered Species, training, Research

Contact(s):
Donald O'Brien, CHAIRMAN OF NEW YORK OFFICE
Milbank, Tweed, Hadley, and McCloy, One Chase Manhattan Plaza, 54th Floor, New York, NY 10005-1413
Phone: 212-530-5818
Bill Mallory, CONTROLLER
Phone: 506-529-1075
wmallory@nbnet.nb.ca
Charles Cusson, DIRECTOR OF QUEBEC PROGRAMS
Atlantic Salmon Federation 1253 Ave, McGill College, Bureau 680,, Montreal, Canada H3B 2Y5
Phone: 514-871-9660
fsa-asf-quebec@globetrotter.net
Andrew Goode, DIRECTOR, U.S. PROGRAMS
Phone: 207-725-2833
goodeasf@blazenetme.net
Jim Gourlay, EDITOR
P.O. Box 5200, St. Andrew., New Brunswick E5B 3S8
Phone: 506-529-4581
jgourlay@saltscapes.com
Sue Scott, EXECUTIVE DIRECTOR OF COMMUNICATIONS
Phone: 506-529-1027
policy@nbnet.nb.ca
Bill Taylor, PRESIDENT
Robert Beatty, VICE PRESIDENT OF DEVELOPMENT
Phone: 506-529-1031
rbeatty@nbnet.nb.ca
Frederick Whoriskey, VICE PRESIDENT-RESEARCH AND ENVIRONMENT
Phone: 506-529-1039
asfres@nvnet.nb.ca

ATLANTIC STATES LEGAL FOUNDATION

658 W. Onondaga St.
Syracuse, NY 13204-3757 USA
Phone: 315-475-1170   Fax: 315-475-6719
E-mail: atlantic.states@aslf.org
Website: www.aslf.org

Founded: 1982

Scope: Internationally

Description: Atlantic States Legal Foundation, Inc., enforces environmental laws, engages in public education, conducts research and promotes environmental justice for the economically disadvantaged and people of color.

Publication(s): Quarterly newsletter, Onondaga Lake Review, Superfund Review

Keyword(s): Chemical Pollution Control, Developing Countries, Environmental Justice, Environmental Law, Environmental and Conservation Education, Mining, People of Color in the Environment, Lakes, Pollution Prevention, Public Health Protection, Solid Waste Management, Sustainable Development, Urban Environment, Renewable Resources, Sustainable Ecosystems

Contact(s):
Samuel Sage, SENIOR SCIENTIST
Sean Lynch, STAFF ATTORNEY

AUDUBON MISSOURI

1001 E. Walnut Ste. 200
Columbia, MO 65201 USA
Phone: 573-442-2139   Fax: 573-443-4378
Website: www.audubon.org/chapter/mo/

Founded: 1999
Membership: 9,000
Scope: Statewide

Description: A statewide council composed of delegates from 14 National Audubon chapters and the Audubon Society of Missouri. Formed to coordinate efforts on various conservation and environmental issues in Missouri. Advised and assisted by the National Audubon Society.

Keyword(s): Birds, Biodiversity, Environment, Endangered Species, Water Quality

Contact(s):
Karen Uhlenhuth, CHAIRPERSON
3714 E. Roanoke Dr., Kansas City, MO 64111
Phone: 816-561-1371

AUDUBON SOCIETY OF OMAHA

11809 Old Maple Road
Omaha, NE 68164 USA
Phone: 402-493-0373   Fax: 402-493-0373

Founded: 1985
Membership: 2000
Scope: Local

Description: A statewide council of representatives of the eight National Audubon Society chapters in Nebraska. The Council's purpose is to coordinate efforts of the chapters on statewide environmental issues and advocate protection, preservation, and wise use of our soil, water, plants, and wildlife.

Contact(s):
Ione Werthen, CONTACT

AUDUBON ALASKA SOCIETY

308 G St., Suite 217
Anchorage, AK 99501 USA
Phone: 907-276-7034   Fax: 907-276-5069
Website: www.home.gci.net/~akaudubon

Founded: NA
Membership: 2400

Scope: Statewide

Description: The Alaska Audubon Society applies sound science and common sense to protect birds, other wildlife and their habitats in Alaska. The staff works in cooperation with five local chapters to create a culture of conservation and an environmental ethic that supports a healthy, sustainable economy and a quality of life in harmony with Alaska's natural environment.

Publication(s): Alaska Audubon News

Keyword(s): Predators, Biodiversity, Wildlife Rehabilitation, Birds, Conservation, Endangered Species, Forest Management, National Parks, Environmental and Conservation Education, Public Lands, Sustainable Ecosystems, training, Wetlands, Wildlands

Contact(s):
Stanley Senner, EXECUTIVE DIRECTOR
Catherine Dennerlein, OFFICE MANAGER/EDUCATION COORDINATOR
John Schoen, SENIOR SCIENTIST

AUDUBON COUNCIL OF CONNECTICUT

c/o Audubon Center in Greenwich, 613 Riversville Rd.
Greenwich, CT 06831 USA
Phone: 203-629-1248

Founded: 1967
Scope: Statewide

Description: The Audubon Council of Connecticut is a coalition of 16 chapters and affiliates of the National Audubon Society in Connecticut, representing close to 10,000 residents. The Council recognize humankind's dependence on the natural environment and appreciates the beauty and wondrous diversity of the natural world. The mission of the Council is, therefore, to protect and restore biodiversity in our state and on our planet.

Keyword(s): Birds, Natural Areas, Environment

AUDUBON COUNCIL OF ILLINOIS

434 N Charlotte
Palatine, IL 60067 USA
Phone: 847-797-7820

Founded: 1973
Membership: 16000
Scope: Statewide

Description: Composed of representatives of 13 National Audubon Society chapters in Illinois, the Council's purpose is to coordinate efforts of the chapters on statewide environmental issues.

Keyword(s): Biodiversity, Birds, Protected Areas, Wetlands, training

Contact(s):
Brian Herner, PRESIDENT
434 N Charlotte, Palatine, IL 60067
Phone: 630-891-4879
brian_herner@premierinc.com
Bonnie John, SECRETARY
824 S Dunton, Arlington Heights, IL 60005
Phone: 847-259-5168
hayspella@aol.com
Mary Blackmore, TREASURER
9024 W. Grove Rd., Forreston, IL 61030
Phone: 815-938-3204

Marianne Hahn, VICE PRESIDENT
18429 Gottschalk Ave., Homewood, IL 60430
Phone: 708-799-0249

AUDUBON INTERNATIONAL

Headquarters, 46 Rarick Rd.
Selkirk, NY 12158 USA
Phone: 518-767-9051   Fax: 518-767-9076
E-mail: acss@audubonintl.org
Website: www.audubonintl.org

Founded: 1897
Membership: 2500
Scope: International

Description: Audubon International is a nonprofit environmental organization that specializes in sustainable natural resource management. The mission of Audubon International is to improve the quality of life and the environment through research, education, and conservation assistance.

Publication(s): Guide to Environmental Stewardship for your Business, Managing Wildlife Habitant on Golf Courses, Stewardship News, Landscape Restoration Handbook, Principles for Sustainable Resource Management, A Guide to Environmental Stewardship on the Golf Course, Golf Course Design, Stewardship News

Keyword(s): Birds, Environmental and Conservation Education, Environmental Planning, Sustainable Development, Land Use Planning, Nongame Wildlife, Renewable Resources, Research, Water Quality, training, Biodiversity

Contact(s):
Paula Realbuto, BUSINESS MANAGER
Nancy Richardson, DIRECTOR OF AUDUBON SIGNATURE PROGRAM
Phone: 502-869-9419
Jean Mackay, DIRECTOR OF ENVIRONMENTAL EDUCATION
Miles Smart, DIRECTOR OF ENVIRONMENTAL PLANNING
Phone: 919-380-9640
Lawrence Woolbright, DIRECTOR OF RESEARCH
Phone: 518-783-2440
Mary Jack, EXECUTIVE ASSISTANT
Eric Dodson, MIS
Ronald Dodson, PRESIDENT AND CEO

AUDUBON NATURALIST SOCIETY OF THE CENTRAL ATLANTIC STATES

8940 Jones Mill Rd.
Chevy Chase, MD 20815 USA
Phone: 301-652-9188   Fax: 301-951-7179
E-mail: hq@audubonnaturalist.org
Website: www.audubonnaturalist.org

Founded: 1897
Scope: Regional

Description: One of the original independent Audubon societies active in environmental education, conservation issues, sanctuaries, and natural science studies in the greater Washington metropolitan area for 100 years. The ANS is headquartered at Woodend, a 40-acre Nature Preserve in suburban Maryland.

Publication(s): Environmental Education and Conservation Brochure, Quarterly, Audubon Naturalist News

Keyword(s): Birds, Environmental and Conservation Education, education, Sustainable Development, Water Quality, National Parks

Contact(s):
Neal Fitzpatrick, DIRECTOR OF CONSERVATION
Leslie Cronin, EDITOR
Mike Nelson, EXECUTIVE DIRECTOR
Muriel Robinson, MANAGER OF ACCOUNTING
Tara Fuad, VOLUNTEER COORDINATOR
Regina Sakaria, VOLUNTEER COORDINATOR

AUDUBON OF FLORIDA

444 Brickell Ave., Suite 850
Miami, FL 33131 USA
Phone: 305-371-6399   Fax: 305-371-6398
E-mail: info@audubonofflorida.org
Website: audubonofflorida.org

Founded: 1900
Membership: 35,000
Scope: Statewide

Description: A statewide organization formed to promote public
interest, understanding, and protection of Florida wildlife, and
of the environment and habitats that support it.

Publication(s): Florida Naturalist, The

Keyword(s): Birds, Conservation, Land Use Planning, training,
Endangered Species, Environmental and Conservation
Education, Forest Management, Sea Grass, Rivers, Public
Lands, Lakes, Raptors, Population Growth

Contact(s):
Stuart Strahl, PRESIDENT
Phone: 305 3-71 -6399

AUDUBON OF FLORIDA

EVERGLADES CAMPAIGN OFFICE
444 Brickell Ave., Suite 850
Miami, FL 33131 USA
Phone: 305-371-6399   Fax: 305-371-6398
Website: www.audubonofflorida.org

Founded: NA
Membership: 90
Scope: Regional
Publication(s): Florida Naturalist, Audubon Advocate,
Everglades Report
Keyword(s): Everglades

Contact(s):
Stuart Strahl, PRESIDENT/CEO

AUDUBON OF KANSAS

P.O. Box 156
Manhattan, KS 66505-0156 USA
Phone: 785-537-4385
Website: www.audubonofkansas.org

Founded: 1974
Membership: 5500
Scope: Statewide

Description: (formerly Kansas Audubon Council) A statewide
nonprofit organization working in partnership with eleven local
Audubon chapters, a Board of Trustees and other members.
Established in 1974, leadership was expanded in 1999 to
establish Audubon of Kansas as a broad-based alliance to
promote appreciation and stewardship of the natural
ecosystems of Kansas, with special emphasis on conservation
of prairies, grassland birds and other wildlife.

Publication(s): Prairie Wings

Keyword(s): Birds, Grasslands, Prairies, Environmental and
Conservation Education, Conservation, training

Contact(s):
Dick Seaton, CHAIRMAN OF BOARD
Ron Klataske, EXECUTIVE DIRECTOR
Patricia Marlett, SECRETARY
4406 W. 11th, Wichita, KS 67212
Carol Cumberland, TREASURER
1106 Gretchen, Wichita, KS 67206
Robert McElroy, VICE CHAIRMAN

AUDUBON PENNSYLVANIA

100 Wildwood Way
Harrisburg, PA 17110 USA
Phone: 717-213-6880   Fax: 717-213-6883
E-mail: painquiries@audubon.org
Website: www.audubon.org/chapter/pa

Founded: 1987
Membership: 27,000
Scope: Statewide

Description: The Pennsylvania Audubon Society promotes and
encourages the conservation and protection of our natural
resources through public education, communication with public
officials, and sponsorship of programs to help children and
adults become aware of their relationship to the environment.

Publication(s): Population&Habitat Newsletter, Pennsylvania
Songbirds - a K-12 Teacher's Guide, Important Bird Areas of
Pennsylvania Report, Audubon Protecting Animals Through
Habitat, Project Mayfly, Wetlands Action Guide, Quarterly
Newsletter

Keyword(s): Birds, Wetlands, Environmental and Conservation
Education, training, Outdoor Recreation, education

Contact(s):
Cindy Dunn, EXECUTIVE DIRECTOR
Carmen Santasania, PRESIDENT
1410 Charles St., State College, PA 16801
Phone: 814-359-5760
Marian Crossman, SECRETARY
6 Tussey Circle, Pittsburgh, PA 15237
Phone: 412-366-3339
Leigh Altadonna, TREASURER
161 Greenwood Ave., Wyncote, PA 15834
Phone: 215-886-0656

AUDUBON SOCIETY

850 Richards Street, 505
Honolulu, HI 06010 4709 UUA
Phone: 212-979-3000   Fax: 808-537-5294
E-mail: hiaudsoc@pixi.com
Website: www.audubon.org

Founded: 1939
Membership: 1,800
Scope: Statewide

Description: For better understanding, appreciation, and conservation of Hawaii's native wildlife resources, especially its unique
and endangered bird species and their associated ecosystems.

Publication(s): Hawaii's Birds, Voice of Hawaii's Birds (cassette
tapes), checklists, field card checklist, Map-Treasures of Oahu,
Elepaio (Journal)

Keyword(s): Birds, Endangered Species, Environmental and Conservation Education, Protected Areas, Aquatic Habitats, Biodiversity, Conservation, Islands, Natural Areas, Nongame Wildlife, Public Lands, training

Contact(s):
Wendy Johnson, PRESIDENT
Sharon Reilly, RECORDING SECRETARY

AUDUBON SOCIETY OF MISSOURI
Attn: President, 1001 SW 19th
Blue Springs, MO 64015 USA
Phone: 573-445-9115
E-mail: grosbeak@discoverynet.com
Website: www.mobirds.org

Founded: 1901
Scope: Statewide

Description: A nonprofit statewide society affiliated with National Audubon Society. Dedicated to the preservation and protection of birds and all wildlife forms and habitat; to educate citizenry toward appreciation of the natural world; and to work for wise conservation practices related to people and wildlife.

Publication(s): Annotated Checklist of the Birds of Missouri, Guide to the Birding Areas of Missouri, A, Bluebird, The

Keyword(s): Birds, Environmental and Conservation Education, education, training

Contact(s):
Jerry & Edge Wade, HOTLINE COORDINATOR
1221 Bradshaw Ave., Columbia, MO 65203-0807
Phone: 573-445-6697
Susan Hazelwood, PAST-PRESIDENT
3005 Chapel Hill Rd., Columbus, MO 65203
Mike Beck, PRESIDENT
Phone: 816-229-6811
Susan Dornfeld, SECRETARY
700 S. Weller, Springfield, MO 65208
Phone: 417-831-9702
Jean Graebner, TREASURER
1800 S. Roby Farm Rd., Rocheport, MO 65279
Phone: 314-698-2855

AUDUBON SOCIETY OF NEW HAMPSHIRE
3 Silk Farm Rd.
Concord, NH 03301-8200 USA
Phone: 603-224-9909   Fax: 603-226-0902
Website: www.nhaudobon.org

Founded: 1914
Membership: 7500
Scope: Statewide

Description: Independent statewide nonprofit organization dedicated to the preservation, understanding, and appreciation of New Hampshire's wildlife and other natural resources.

Publication(s): Newsletter, Bi-Monthly

Keyword(s): training

Contact(s):
Tupper Kinder, CHAIR OF THE BOARD OF TRUSTEES
Sheehan Phinney, Bas and Green, 1000 Elm St., Manchester, NH 03105-3701
Kent Taylor, DIRECTOR FOR MEMBERSHIP
Scott Fitzpatrick, DIRECTOR OF EDUCATION
Julian Zelazny, DIRECTOR OF ENVIRONMENTAL AFFAIRS

Harry Vogel, DIRECTOR OF LOON PRESERVATION COMMITTEE
Richard Moore, PRESIDENT
Larry Sunderland, SECRETARY
RFD 1 Box 179, Hillsboro, NH 03244
Anita Maclean, TREASURER
Sylvia Bates, VICE CHAIRPERSON
Rt. 1, Box 313, Ashland, NH 03217
Richard Cook, VICE PRESIDENT FOR CONSERVATION

AUDUBON SOCIETY OF PORTLAND
5151 NW Cornell Rd.
Portland, OR 97210 USA
Phone: 503-292-6855   Fax: 503-292-1021
E-mail: general@audubonportland.org
Website: www.audubonportland.org

Founded: 1902
Membership: 9500
Scope: Local

Description: The Audubon Society of Portland promotes the enjoyment, understanding and protection of native birds, other wildlife and their habitats, focusing on the local community and the Pacific Northwest.

Publication(s): Audubon Warbler

Keyword(s): Environmental and Conservation Education, Environmental Protection, Birds, Ancient Forests, Conservation, accreditation, Outdoor Recreation, education, Urban Environment, Wetlands, Endangered Species, Sustainable Development

Contact(s):
Jim Rapp, BOARD PRESIDENT
Sybil Ackerman, DIRECTOR OF CONSERVATION
Steve Robertson, EDUCATION DIRECTOR
Dave Eshbaugh, EXECUTIVE DIRECTOR
Bob Wilson, NATURE STORE DIRECTOR
Mitch Luckett, SANCTUARY DIRECTOR
Scott Lukens, SECRETARY

AUDUBON SOCIETY OF RHODE ISLAND
12 Sanderson Rd.
Smithfield, RI 02917-2600 USA
Phone: 401-949-5454   Fax: 401-949-5788
Website: www.asri.org

Founded: 1897
Membership: 3500
Scope: Statewide

Description: To focus attention on critical natural resource problems, provide leadership when conservation action is necessary, carry out a broad program of public conservation education, and preserve examples of unique natural areas and native wildlife habitat.

Publication(s): Checklist of Rhode Island Birds, Fields Notes of Rhode Island Birds, Audubon Society of Rhode Island Report

Keyword(s): Air Quality and Pollution, Birds, Coasts, Conservation, Endangered Species, Environment, Protected Areas, Environmental and Conservation Education, Wildlife, Flowers, Plants, and Trees, Land Preservation, Open Space, Insects and Butterflies

Contact(s):
A. Kohlenberg, 1ST VICE PRESIDENT
Dickson Boenning, 2ND VICE PRESIDENT

Lawrence Taft, DIRECTOR OF PROPERTIES AND ACQUISTIONS
Ken Weber, EDITOR
Lee Schisler, EXECUTIVE DIRECTOR
Doris Thorpe, MEMBERSHIP SECRETARY
Joseph Dimase, SECRETARY
Frank Sciuto, TREASURER

AUDUBON SOCIETY OF VERMONT

255 Sherman Hollow Rd.
Huntington, VT 05462 USA
Phone: 802-434-3068   Fax: 802-434-4686
E-mail: audubonmtn@aol.com

Founded: 1962
Scope: Statewide

Description: Green Mountain Audubon Society is a chapter of the National Audubon Society. The Society operates nature center and wildlife sanctuary areas in Huntington, Shelburne, and Popasquash Island in Lake Champlain. The Nature Center offers environmental education programs for visitors and local schools year-around.

Keyword(s): Nature Centers, Environmental and Conservation Education

Contact(s):
Shirley Johnson, BOARD PRESIDENT

AUDUBON SOCIETY OF WESTERN PENNSYLVANIA

Beechwood Farms Nature Reserve, 614 Dorseyville Rd.
Pittsburgh, PA 15238-1618 USA
Phone: 412-963-6100   Fax: 412-963-6761
E-mail: aswp@aswp.org
Website: www.aswp.org

Founded: 1916
Membership: 3000
Scope: Local

Description: The mission of the Audubon Society of Western Pennsylvania is to inspire and educate people of southwestern Pennsylvania to be respectful and responsible stewards of the natural world.

Publication(s): Seasoning, Teacher Guide, Bulletin

Keyword(s): Environmental and Conservation Education, Birds, education, Flowers, Plants, and Trees, Insects and Butterflies, Mammals, training, Reptiles and Amphibians, Raptors, Waterfowl

Contact(s):
Fred Peterson, PRESIDENT
Danforth Fales, VICE PRESIDENT

B

B.A.S.S. DIVISION OF ESPN PRODUCTIONS INC

5845 Carmichael Rd.
Montgomery, AL 36117 USA
Phone: 334-272-9530   Fax: 334-270-8549
E-mail: conservation@bassmaster.com
Website: www.bassmaster.com

Founded: 1968
Scope: National

Description: Organized to fight pollution, assist state and national conservation agencies in their efforts, and teach the young people of our country good conservation practices. Dedicated to the realistic conservation of our water resources.

Publication(s): Bassmaster Magazine, B.A.S.S. Times, Fishing Tackle Retailer, Television Show: The Bassmasters, Guns & Gear

Keyword(s): Coral Reefs, Conservation, Exotic species, Aquatic nuisance species, Sport Fishing

Contact(s):
Helen Sevier, EXECUTIVE VICE PRESIDENT
Bruce Shupp, NATIONAL CONSERVATION DIRECTOR
bruce.shupp@bassmaster.com
Al Smith, NATIONAL FEDERATION DIRECTOR
al.smith@bassmaster.com

BAMA BACKPADDLERS ASSOCIATION

307 Madison Pl.
Trussville, AL 35173 USA
E-mail: backpaddlers@aol.com
Website: http://members.aol.com/backpaddlers/

Founded: 1978
Scope: Statewide

Description: Dedicated to promoting recreation, conservation, education and safety on Alabama's waterways.

Publication(s): As the Eddy Turns

Keyword(s): Conservation, Dams, Development, Protecting Special Places, Local Resource Conservation, Exotic species, Aquatic nuisance species, Water Quality, Coral Reefs, Rivers, Outdoor Recreation

Contact(s):
Jennifer Taylor, CONSERVATION
Phone: 205-951-0320
jentay@uab.edu
Betty Harrison, NEWSLETTER
kayakbba@aol.com
Renee Clark, PRESIDENT
Nancy Cate, TREASURER
Bob Shepard, TRIP COORDINATOR

BARRIER ISLAND TRUST, INC.

P.O. Box 37310
Tallahassee, FL 32315 USA
Phone: 850-933-2761   Fax: 203-629-2453
Website: www.bit.org

Founded: 1989
Scope: Statewide

Description: To preserve the natural resources of Florida's barrier islands, initially focusing on Dog Island and Apalachicola Bay, hold and manage barrier island property to preserve it in its natural state, promote research on barrier island ecology and translate research into educational programs and effective policies for protection of barrier islands.

Keyword(s): Islands, Land Preservation, Coasts, Land Protection

Contact(s):
Guy Smith, BOARD OF TRUSTEE CHAIR
352 North St., Greenwich, CT 06830
Phone: 203-629-1264
Leroy Collins, BOARD OF TRUSTEE PRESIDENT
16 Davis Blvd. Suite 12, Tampa, FL 33606
Phone: 813-259-9484

Mitchell Smith, BOARD OF TRUSTEE TREASURER
P.O. Box 1912, Albany, GA 31702
Dianne Mellon, BOARD OF TRUSTEE VICE PRESIDENT
1515 Country Club, Tallahassee, FL 32301
Phone: 850-877-3942

BAT CONSERVATION INTERNATIONAL

P.O. Box 162603
Austin, TX 78716 USA
Phone: 512-327-9721   Fax: 512-327-9724
E-mail: batinfo@batcon.org
Website: www.batcon.org

Founded: 1982
Membership: 14000
Scope: International

Description: A nonprofit organization with members in 60 countries. BCI's purpose is to document and publicize the values and conservation needs of bats, to promote bat conservation projects, and to assist with management initiatives worldwide.

Publication(s): BATS(quarterly), Catalog available on website

Keyword(s): Endangered Species, Environmental and Conservation Education, Biodiversity, training, Mammals, Nongame Wildlife, Ecology, Research, Scholarships, Wildlife Rehabilitation, Environment, Bats

Contact(s):
Merlin Tuttle, FOUNDER AND EXECUTIVE DIRECTOR
Bob Benson, PUBLIC INFORMATION MANAGER
John Phillips, SECRETARY
Mark Ritter, TREASURER

BERKSHIRE-LITCHFIELD ENVIRONMENTAL COUNCIL, INC.

P.O. Box 552
Lakeville, CT 06039 USA
Phone: 203-435-2004

Founded: 1970
Scope: Statewide

Description: Primarily concerned with energy, invasive transportation, and land use issues in the southern Berkshires and Litchfield Hills. Offers public programs and environmental education for all ages.

Publication(s): BLEC News

Keyword(s): Agriculture, Biodiversity, Environmental and Conservation Education, Protected Areas, Wilderness

Contact(s):
William Morrill, COUNSEL
datibbetts@annapolis.net
Judy Thomas, EXECUTIVE DIRECTOR
Starling Childs, PRESIDENT
Phone: 203-542-5569
Ellery Sinclair, SECRETARY
Peter Dolan, TREASURER
Nic Osborn, VICE PRESIDENT

BEYOND PESTICIDES/NATIONAL COALITION AGAINST THE MISUSE OF PESTICIDES

701 E St., SE, Suite 200
Washington, DC 20003 USA
Phone: 202-543-5450   Fax: 202-543-4791
E-mail: info@beyondpesticides.org
Website: www.beyondpesticides.org

Founded: 1981
Membership: 6
Scope: National

Description: Nonprofit membership organization committed to assisting individuals, organizations, and communities with useful information on pesticides and their alternatives. NCAMP's information clearinghouse provides material on a wide range of both agricultural and urban issues concerning protection of children, workers' safety, food safety, lawn care safety, groundwater problems, and alternatives to pesticides, as well as legislation.

Publication(s): Expelling Pesticides from schools - Book, West Nile Virus Organizing Manual, Beyond Pesticides/ NCAMPs Technical Report (Monthly), Poison Poles: Their Toxic Trial and the Safer Alternatives, Safety at Home: A Guide to the Hazards of Lawn and Garden Pesticides and Safer Ways to Manage Pests, Unnecessary Risk, Pesticides and You Newsletter (Quarterly)

Keyword(s): Agriculture, Environment, Insects and Butterflies, Pesticides, Toxicology

Contact(s):
Jay Feldman, EDITOR
Jay Feldman, EXECUTIVE DIRECTOR
Gregg Small, PRESIDENT BOARD OF DIRECTORS

BIG BEND NATURAL HISTORY ASSOCIATION

P.O. Box 196
Big Bend National Park, TX 79834 USA
Phone: 915-477-2236
E-mail: bbnha@nps.gov
Website: www.bigbendbookstore.org

Founded: 1956
Membership: 500
Scope: National

Description: A private nonprofit organization whose main objectives are to facilitate popular interpretation of the scenic, scientific and historical values of Big Bend, and to encourage research related to those values. To accomplish these goals, the association is authorized by the National Park Service to publish, print, or otherwise provide books, maps, and illustrative material on the Big Bend region and to sponsor a Big Bend seminar program.

Publication(s): Big Bend Paisano

Keyword(s): Environmental and Conservation Education, Flowers, Plants, and Trees, Gardening and Horticulture, Natural Areas, National Parks

Contact(s):
Rob Dunagan, CHAIRMAN
Phone: 915-336-5274
Thomas Vandenberg, EDITOR
Mike Boren, EXECUTIVE DIRECTOR

BILLFISH FOUNDATION, THE

2161 E Commercial Blvd. 2nd Fl.
Ft. Lauderdale, FL 33308 USA
Phone: 954-938-0150   Fax: 954-938-5311
E-mail: tbf@billfish.org
Website: www.billfish.org

Founded: NA
Membership: 80000
Scope: International

Description: The Billfish Foundation is a nonprofit organization dedicated to the conservation of billfish worldwide through scientific research, education, and advocacy. Through scientific, economic and conservation decisions provided through research, TBF strives for sound and constructive measures to recover overfished stocks.

Publication(s): Tag & Brag - Quarterly Newsletter, TBF News, Spearfish, Billfish

Keyword(s): Research, Environmental and Conservation Education, Migration, Wildlife, Sport Fishing, Billfish

Contact(s):
Winthrop Rockefeller, CHAIRMAN OF TRUSTEES
Ellen Peel, PRESIDENT
Paxson Offield, TREASURER OF TRUSTEES
Hal Prewitt, VICE CHAIRMAN
Ralph Vicente, VICE CHAIRMAN OF TRUSTEES

BIODIVERSITY LEGAL FOUNDATION

P.O. Box 278
Louisville, CO 80027 USA
Phone: 303-926-7606   Fax: 303-440-0434
E-mail: blfrog@aol.com

Founded: 1991
Membership: 2200
Scope: National, International

Description: The Biodiversity Legal Foundation is a nonprofit, science-based tax-exempt organization dedicated to the preservation of all native plants, animals, and naturally functioning ecosystems. Through educational, administrative, and legal actions, we endeavor to encourage improved public attitudes for all living things.

Publication(s): Administrative and Legal Update, Guidelines for the Preparation of Species Status Reviews and Species Conservation Assessments, Guide to Ecosystem Management -a citizens guide to biocentric ecological sustain ability, How You Can Help Rare and Endangered Species within the Framework of Existing Conservation Law

Keyword(s): Biodiversity, Endangered Species, Environmental Law, Wildlife, Sustainable Ecosystems, Predators, Prairies, Public Lands, Nongame Wildlife, training

Contact(s):
Jasper Carlton, DIRECTOR
Edward Mudd, PRESIDENT
Joyce Hudson, SECRETARY AND TREASURER
Roger Candee, VICE PRESIDENT

BIO-INTEGRAL RESOURCE CENTER

P.O. Box 7414
Berkeley, CA 94707 USA
Phone: 510-524-2567   Fax: 510-524-1758
E-mail: birc@igc.org
Website: www.birc.org

Founded: 1979
Membership: 2000
Scope: International

Description: A nonprofit educational organization dedicated to providing information on least-toxic pest control.

Publication(s): IPM Practitioner, The, Least-toxic Pest Management for Fleas, Termites, Cockroaches, Raccoons, Ticks, and more, Common Sense Pest Control Quarterly

Keyword(s): Agriculture, Gardening and Horticulture, Insects and Butterflies, Pesticides, Urban Environment

Contact(s):
Jennifer Bates, BUSINESS MANAGER
William Quarles, EXECUTIVE DIRECTOR, MANAGING EDITOR OF PUBLICATIONS

BIOMASS USERS NETWORK

383 Franklin St.
Bloomfield, NJ 07003 USA
Phone: 973-680-9100

Founded: 1985
Scope: International

Description: To advance rural economic development in Third World countries in an environmentally sound manner, through the innovative production and efficient use of biomass resources.

Publication(s): Network News

Keyword(s): Energy, Flowers, Plants, and Trees, Forests and Forestry, Soil Conservation, Sustainable Development

Contact(s):
David Mazambani, CHAIRMAN

BIRDLIFE INTERNATIONAL

Canada Nature Federation,1 Nicholas St., Ste. 606
Ottawa, Ontario KIN 7B7 Canada
Phone: 613-562-3447   Fax: 613-562-3371
E-mail: cnf@cnf.ca
Website: www.cnf.ca

Founded: NA
Scope: National

Description: Protection of birds and their habitats in Canada, in their winter quarters in North and South America, and off Canada's coasts are among major concerns.

Publication(s): Nature Canada, Nature Matters-Newsletter

Keyword(s): Birds, Conservation, Endangered Species, Protected Areas, training, National Parks

Contact(s):
Michael Bradstreet, CONTACT
Bird Studies Canada, Box 160, Port Rowan,
Ontario N0E 1M0
Phone: 519-586-3531
Fax: 519-586-3532
Caroline Schultz, CONTACT

BLUE GOOSE ALLIANCE

10 Circle S Rd.
Edgewood, NM 87015 USA
Phone: 505-861-6895

Contact(s):
Evelyn Redfearn
sec4bga@aol.com

BLUEBIRDS ACROSS VERMONT PROJECT

255 Sherman Hollow Rd.
Huntington, VT 05462 USA
Phone: 802-434-3068

Founded: 1987
Scope: Statewide

Description: A project of the Vermont Audubon Council and Green Mountain Audubon, Bluebirds Across Vermont (BAV) was formed to help restore native eastern bluebird populations. BAV promotes the proper placement of correctly-built nestboxes by informed citizens who monitor them throughout the nesting season and send the data to BAV for yearly compilation.

Contact(s):
Mark Labarr, CONTACT

BOONE AND CROCKETT CLUB

250 Station Dr.
Missoula, MT 59801 USA
Phone: 406-542-1888   Fax: 406-542-0784
E-mail: bcclub@boone-crockett.org
Website: www.boone-crockett.org

Founded: 1887
Membership: 190
Scope: National

Description: A 501 (c) (3) organization. Established by Theodore Roosevelt and other concerned sportsmen to promote hunting ethics, foster the concept of Fair Chase, and help establish wildlife conservation practices which led to the recovery of big game animals in North America. The Club documents the records of North American big game and exhibits its National Collection of Heads and Horns in Cody, WY.

Publication(s): Fair Chase, Records of North American Elk and Mule Deer, An American Crusade for Wildlife, Records of North American Whitetail Deer, Return of Royalty, Records of North American Big Game

Keyword(s): Environmental and Conservation Education, Hunting, Public Lands, Scholarships and Grants, training

Contact(s):
George Vettas, COMMUNICATIONS COMMITTEE
Phone: 509-335-4531
Thomas Price, CONSERVATION COMMITTEE
Phone: 541-276-4246
Jack Reneau, DIRECTOR OF NORTH AMERICAN BIG GAME RECORDS
Phone: 406-542-1888
Thomas Price, FIRST VICE PRESIDENT
Frederick King, MUSEUM COMMITTEE
Phone: 406-994-2654
Earl Morgenorth, PRESIDENT
C. Byers, RECORDS OF NORTH AMERICAN BIG GAME COMMITTEE
Phone: 208-885-7341
Robert Hanson, SECRETARY

Joseph Ostervich, TREASURER
Phone: 608-486-2341
Tommy Caruthers, VICE PRESIDENT RECORDS ETHICS

BOONE AND CROCKETT FOUNDATION

250 Station Dr.
Missoula, MT 59801 USA
Phone: 406-542-1888   Fax: 406-542-0784
E-mail: bcclub@boone-crockett.org
Website: www.boone-crockett.org

Founded: NA
Scope: National

Description: The BCF owns and operates the 6,000 acre Theodore Roosevelt Memorial Ranch near Dupuyer, MT, as a working cattle ranch for research, education and demonstration. BCF supports natural resource conservation research, education, and demonstration primarily through the Boone and Crockett wildlife conservation program in conjunction with the University of Montana.

Publication(s): Fair Chase Magazine

Keyword(s): Hunting, Renewable Resources, Sustainable Development, Sustainable Ecosystems, training

Contact(s):
Lisa Flowers, CONSERVATION EDUCATION PROGRAM MANAGER
Phone: 406-466-2078
Daniel Pedrotti, PRESIDENT
Phone: 512-884-2443
Jack Thomas, PROFESSOR OF WILDLIFE CONSERVATION
Phone: 406-243-5566
Gilbert Adams, SECRETARY
Phone: 409-835-3000
Joseph Ostervich, TREASURER
Phone: 608-486-2341
John Rappold, TRMR MANAGER
Phone: 406-472-3380

BORDER ECOLOGY PROJECT (BEP)

Drawer CP
Bisbee, AZ 85603 USA
Phone: 520-432-7456   Fax: 520-432-7473
E-mail: bep@primenet.com
Website: www.borderep.org

Founded: 1983
Scope: National

Description: BEP advocates for solutions to environmental problems along the U.S. and Mexico border. Areas of focus include Right-to-Know, environmental pollution, international trade, mining, hazardous materials trucking, and bi-national environmental health issues including lupus.

Publication(s): Environmental and Health Conditions in the Interior of Mexico: Options for Transnational Safeguards, Refer to website for further publications, Environmental Protection within the Mexican Mining Sector and the Impact of the World Bank Loan 3359

Keyword(s): Air Quality and Pollution, Rivers, Solid Waste Management, Toxic Substances, Nuclear-free, Water quantity, Water export and diversion, Water Pollution Management

Contact(s):
A. Hotaling, COORDINATOR
Dick Kamp, DIRECTOR

BOTANICAL CLUB OF WISCONSIN

c/o Wisconsin Academy of Science, Arts, and Letters,
1922 University Ave.
Madison, WI 53705 USA
Phone: 608-262-5489   Fax: 608-265-2993
Website: www.wisc.edu/botany/herbariun/dcwindex.html

Founded: 1969
Membership: 200
Scope: Regional

Description: Botanical Club of Wisconsin promotes preservation
of Wisconsin's native plants and educates the public as to the
value of plants. The Club also fosters research on plant biology
and provides a means for fellowship and information
exchange.

Publication(s): Wisconsin Flora, The Bulletin of the Botanical
Club of Wisconsin

Keyword(s): Flowers, Plants, and Trees, Environmental and
Conservation Education, National Parks

Contact(s):
Emmet Judziewicz, PRESIDENT
Phone: 920-842-4620
judzie@dnr.state.wi.us
Edward Glover, TREASURER
Madison
Phone: 608-437-4578
glover@oncology.wisc.edu
James Bennett, VICE PRESIDENT

BOUNTY INFORMATION SERVICE (WILDLIFE)

WILDLIFE
4849 E. St. Charles Rd.
Columbia, MO 65201 USA
Phone: 573-474-6967

Founded: 1966
Membership: 50
Scope: National, International

Description: Promotes the removal of bounties in North America
by publishing Bounty News and studies of the bounty system
and by coordinating activities and legal aspects.

Publication(s): A History Wildlife Bounties, Bounty News, A Guide
to the Removal of Bounties

Keyword(s): Environmental and Conservation Education,
Mammals, Trapping, training

Contact(s):
H. Laun, DIRECTOR AND EDITOR

BOY SCOUTS OF AMERICA

National Office, P.O. Box 152079,
1325 West Walnut Hill Ln.
Irving, TX 75015-2079 USA
Phone: 972-580-2000   Fax: 972-580-2000
Website: www.bsa.scouting.org

Founded: 1910
Scope: National

Description: Boy Scouts of America (BSA) was chartered by
Congress in 1916 to provide an educational program for boys
and young adults that builds character and develops responsibility, citizenship, and personal fitness. Community groups with
goals compatible with BSA receive national charters to use the
Scouting program as part of their own youth work.

Keyword(s): Environmental and Conservation Education,
Outdoor Recreation, Youth Organizations

Contact(s):
Francis Olmstead, ASSISTANT TREASURER
Roy Williams, CHIEF SCOUT EXECUTIVE
David Bates, CONSERVATION DIRECTOR
Milton Ward, PRESIDENT
Erik Nystrom, REGIONAL EXECUTIVE
Raymond Blackwell, REGIONAL EXECUTIVE
P.O. Box 3085, Naperville, IL 60566-7085
Phone: 630-983-6730
Kenneth Connelly, REGIONAL EXECUTIVE OF
NORTHEAST REGION
P.O. Box 268, Jamesburg, NJ 08831-0268
Phone: 609-655-9600
John Cushman, TREASURER

BRANDYWINE CONSERVANCY INC.

P.O. Box 141
Chadds Ford, PA 19317 USA
Phone: 610-388-2700   Fax: 610-388-1575
E-mail: enc@brandywine.org
Website: www.brandywineconservancy.org

Founded: 1967
Scope: Regional

Description: A nonprofit organization providing model land use
and environmental regulations for Pennsylvania municipalities.
Brandywine Conservancy provides land, water resources and
historic site conservation and management assistance to
landowners and conservation organizations, primarily in southeastern Pennsylvania and northern Delaware.

Publication(s): Environmental Currents, Environmental
Management Handbook, Catalyst

Keyword(s): Land Conservation, Pollution Prevention, Exotic
species, Aquatic nuisance species, Wetlands, Environmental
Protection, Land Use Planning

Contact(s):
John Snook, ASSOCIATE DIRECTOR OF DESIGN
(ENVIRONMENTAL MANAGEMENT CENTER)
Wesley Horner, ASSOCIATE DIRECTOR OF MUNICIPAL
ASSISTANCE (ENVIRONMENTAL MANAGEMENT CENTER)
David Shields, ASSOCIATE OF DIRECTOR OF LAND
STEWARDSHIP (ENVIRONMENTAL MANAGEMENT
CENTER)
George Weymouth, CHAIRMAN
Kathryn Saterson, DIRECTOR OF ENVIRONMENTAL
MANAGEMENT CENTER
James Duff, EXECUTIVE DIRECTOR
Halsey Spruance, PUBLIC RELATIONS

BRITISH COLUMBIA FIELD ORNITHOLOGISTS

P.O. Box 8059
Victoria, British Columbia V8W 3R7 Canada
Phone:
Website: http://birding.bc.ca/bcfo/

Founded: 1991
Scope: Statewide

Description: To promote the study and enjoyment of birds in
British Columbia; to disseminate knowledge and appreciation
of birds by means of publications; to foster cooperation
between amateur and professional ornithologists; and to
promote conservation of birds and their habitats.

Publication(s): BC Birding (newsletter), British Columbia Birds (journal)

Keyword(s): Birds

Contact(s):
Marilyn Buhler, EDITOR
1132 Loenholm Rd,. Victoria, British Columbia V8Z 2Z6
Phone: 250-744-2521
Andy Buhler, EDITOR
1132 Loenholm Rd,. Victoria, British Columbia V8Z 2Z6
Phone: 250-744-2521
Martin McNicholl, EDITOR
4735 Canada Way, Burnaby, British Columbia V5G 1L3
Phone: 250-294-9333
Tony Greenfield, PRESIDENT
P.O. Box 319, Sechelt, British Columbia V0N 3A0
Phone: 250-885-5539
Jim Fliczuck, TREASURER
3614-1507 Queensbury Ave,. Victoria,
British Columbia V8P 5M5
Phone: 250-656-8066
Bryan Gates, VICE PRESIDENT
3085 Uplands Rd,. Victoria, British Columbia V8R 6B3
Phone: 250-598-7789

BRITISH COLUMBIA WATERFOWL SOCIETY, THE

5191 Robertson Rd.
Delta, British Columbia V4K 3N2 Canada
Phone: 604-946-6980   Fax: 604-946-6980

Founded: NA
Membership: 2200
Scope: Statewide

Description: The organization was set in 1963 on federal land leased for 30 years to be opened to the public as a bird viewing area at the mouth of the Fraser River, which supports one of the largest wintering populations of waterfowl in Canada. The organization attempts to promote awareness of all parts of the environment.

Publication(s): BirdCheck List, Marsh Notes

Contact(s):
John Ireland, MANAGER
Jack Bates, PRESIDENT
James Morrison, TREASURER

BROOKS BIRD CLUB INC., THE

P.O. Box 4077
Wheeling, WV 26003 USA
Website: http://www.brooksbirdclub.org

Founded: 1932
Scope: National

Description: A nonprofit organization formed to encourage the study and conservation of birds and other phases of natural history. Members in thirty-eight states, Canada, and eight foreign countries. Named in honor of A.B. Brooks, naturalist.

Publication(s): Redstart, The, Mail Bag, The

Keyword(s): Flowers, Plants, and Trees, Birds, Environmental and Conservation Education, National Parks, training

Contact(s):
Carl Slater, ADMINISTRATOR
57290 Mehlmen Rd,. Bellaire, OH 43906

James Bullard, IMMEDIATE PAST PRESIDENT
P.O. Box 137, Ashton, MD 20861
William Murray, MAIL-BAG EDITOR
P.O. Box 944, New Cumberland, WV 26047
Carolyn Conrad, MEMBERSHIP CHAIRMAN
423 Warwood Ave, Wheeling, WV 26003
Cindy Ellis, PRESIDENT
103A Oakwood Estates, Scott Depot, WV 25560
Fred McCullough, PRESIDENT-ELECT
A. Buckelew, REDSTART-EDITOR
Box J, Bethany, WV 26032
Virginia Cronenberger, SECRETARY
Rt. 1 Box 37, Petroleum, WV 26161
Gerald Devaul, TREASURER
17 Mozart Rd, Wheeling, WV 26003

BROTHERHOOD OF THE JUNGLE COCK, INC., THE

P.O. Box 576
Glen Burnie, MD 21061 USA
Phone: 410-761-7727   Fax: 410-553-0575

Founded: NA
Membership: 7000
Scope: National

Description: Seeks to teach youth the true meaning of conservation. Primary interest is the preservation of American game fishes, placing great emphasis on adult responsibility of personal instruction along those lines.

Keyword(s): Endangered Species, Environmental and Conservation Education, Wildlife, Coral Reefs, Youth Organizations

Contact(s):
Bosley Wright, EXECUTIVE VICE PRESIDENT
William Simms, PRESIDENT
Edward Little, SECRETARY
6623 Kenwood Av., Baltimore, MD 21237
Phone: 401-682-4631
M. Day, TREASURER
706 Orchard Way, Silver Spring, MD 20904

C

CADDO LAKE INSTITUTE, INC.

P.O. Box 2710
Aspen, CO 81612 USA
Phone: 970-925-2710   Fax: 970-923-4245
E-mail: dks@sopris.net
Website: www.caddolakeinstitute.org

Founded: 1993
Scope: Local

Description: A non-profit organization whose purpose is environmental awareness. The program director is based near Caddo Lake, Texas. The director will coordinate college programs. Students are paid a stipend to collect samples and return to the student's laboratory for analysis; will also give seminars at secondary schools, all to promote environmental awareness.

Publication(s): see publication web site

Keyword(s): Wetlands, Watersheds, training, Ecology, Internships, Youth Organizations

Contact(s):
Sara Kneipp, EDUCATION DIRECTOR
sjkneipp@aol.com

Dwight Shellman, PRESIDENT
dks@sopris.net

CALCASIEU PARISH ANIMAL CONTROL & PROTECTION DEPARTMENT

5500A Swift Plant Rd.
Lake Charles, LA 70615 USA
Phone: 337-439-8879   Fax: 337-437-3343
Website: http://cpac.cppj.net

Founded: NA
Scope: Statewide

Description: Regional branch of Elsa Wild Animal Appeal; concerned with wildlife matters, educational programs, liaison with other wildlife and governmental groups for the betterment of natural environment and wildlife protection; establishes local volunteer corps to implement programs in conjunction with the Calcasieu Parish Animal Control and Protection Department; and participates in Wildlife Rehabilitation Programs with Heck Haven and Westlake Bird Sanctuary.

Keyword(s): Birds, Endangered Species, Mammals, Urban Environment, accreditation

Contact(s):
David Marcantel, OPERATIONS SUPERVISOR

CALIFORNIA ACADEMY OF SCIENCES

Golden Gate Park
San Francisco, CA 94118 USA
Phone: 415-221-5100   Fax: 415-750-7346
Website: www.calacademy.org

Founded: 1853
Scope: Statewide

Description: The Academy of Sciences' goal is the exploration and interpretation of natural history. Maintains research collections and operates a museum-aquarium-planetarium complex to which one and one-half million visitors come each year.

Publication(s): California Wild, Proceedings, Occasional Papers, Academy Newsletter

Keyword(s): education, Reptiles and Amphibians, National Parks, Research, Environment, Museum, Aquariums

Contact(s):
John Pearse, PRESIDENT
W. Bingham, BOARD OF TRUSTEES CHAIRMAN
Patrick Kociolek, DIRECTOR
Sandra Linder, SECRETARY
Lewis Coleman, VICE CHAIRMAN
John Larson, VICE CHAIRMAN

CALIFORNIA ACADEMY OF SCIENCES

CALIFORNIA ACADEMY OF SCIENCES LIBRARY
Golden Gate Park
San Francisco, CA 94118 USA
Phone: 415-750-7102   Fax: 415-750-7145
E-mail: viodiv@calacademy.org
Website: www.calacademy.org

Founded: NA
Membership: 250
Scope: International

Description: Non-circulating, closed-stack collection open to the public. Reference requests accepted by mail, phone, fax or e-mail. Interlibrary loan requests accepted. Library holdings included in OCLC, University of CA MELVYL on-line catalog and CA Union List of Periodicals.

Keyword(s): Librarians/Information Professionals, Libraries
Contact(s):
Dianne Sands, COORDINATOR

CALIFORNIA ASSOCIATION OF RESOURCE CONSERVATION DISTRICTS

102710th St. Ste C
Sacramento, CA 95814 USA
Phone: 916-447-7237   Fax: 916-447-2532
E-mail: staff@carcd.org
Website: www.carcd.org

Founded: NA
Membership: 103
Scope: Statewide
Publication(s): CCP News, Conservation Express, CARCD Newsline
Keyword(s): Conservation Districts

Contact(s):
John Schramel, PRESIDENT
681 Main St., Greenville, CA 95947
Phone: 530-284-7954
Fax: 530-284-6211
Tom Wehri, EXECUTIVE DIRECTOR
801 K St. Suite 1318, Sacramento, CA 95814
Phone: 916-447-7237
Fax: 916-447-2532
carcd@ns.net
Donna Thomas, PAST PRESIDENT
Phone: 760-377-4525
Robert Beegle, SECRETARY-TREASURER
3911 Yellowstone Ln, El Dorado, CA 95762
Phone: 916-852-6691
Fax: 916-852-6693
Nadine Scott, VICE PRESIDENT
550 Hoover Street, Oceanside, 92054
Phone: 760-757-6685

CALIFORNIA B.A.S.S. CHAPTER FEDERATION

Attn: President, 21517 Appaloosa Ct.
Canyon Lake, CA 92587 USA
Phone: 562-498-9113
Website: www.californiabass.org

Founded: NA
Scope: Statewide

Description: An organization of Bassmaster chapters, affiliated with the Bass Anglers Sportsman Society, organized to fight pollution, assist state and national conservation agencies in their efforts, and teach the young people of our country good conservation practice. Dedicated to the realistic conservation of our water resources.

Contact(s):
Gary Bradford, PRESIDENT
Phone: 562-498-9113

CALIFORNIA NATIVE PLANT SOCIETY, THE

1722 J St., Suite 17
Sacramento, CA 95814 USA
Phone: 916-447-2677   Fax: 916-447-2727
E-mail: cnps@cnps.org
Website: www.cnps.org

Founded: 1965
Membership: 9500
Scope: Statewide

Description: A statewide nonprofit organization of amateurs and professionals with a common interest in California's native plants. The society, working through its local chapters, seeks to increase understanding of California's native flora and to preserve the rich resource for future generations. Membership is open to all.

Publication(s): Bulletin, Conservation & Management of Rare and Endangered Plants, Terrestrial Vegetation of California, California's Changing Landscape, Inventory of Rare and Endangered Vascular Plants of California, Flora of San Bruno Mountain, Plant Communities, Fremontia

Keyword(s): Biodiversity, Endangered Species, Environmental and Conservation Education, Protected Areas, Wetlands, Native Plants

Contact(s):
Sue Britting, PRESIDENT
Phone: 530-333-2679
Fax: 530-333-9178
britting@innercite.com
Jim Bishop, VICE PRESIDENT FOR ADMINISTRATION
Phone: 530-538-6761
cjbishop@cnc.net
Lorrae Fuentes, VICE PRESIDENT FOR EDUCATION
Phone: 909-625-8767
Fax: 909-626-7670
lorrae.fuentes@cgu.edu

CALIFORNIA TRAPPERS ASSOCIATION

Attn: Executive Secretary, 99 Poinsettia Gardens Dr.
Ventura, CA 93004 USA
Phone: 805-647-8903   Fax: 805-647-9970

Founded: 1969; Incorporated: 1973
Membership: 2400
Scope: Statewide

Description: Dedicated to the encouragement of conservation, enhancement, and scientific management of all our natural resources, especially furbearing mammals. Promotes state and federal wildlife projects through volunteer skilled labor and financial contributions. Gives $500 to $1,000 grants each year to college students studying furbearing mammals.

Publication(s): Legislative Alerts, Fur Facts

Keyword(s): Wildlife Rehabilitation, Endangered Species, Environmental Ethics, Environmental and Conservation Education, Environmental Planning, Protected Areas, Predators, Environmental Protection, Renewable Resources, Scholarships and Grants, Trapping, Wilderness, training, Wetlands, Wildlands

Contact(s):
Donald Stehsel, EXECUTIVE SECRETARY
Kathy Lynch, LOBBYIST
Phone: 916-537-7169

Keith Carly, PRESIDENT
P.O. Box 73, Elk Creek, CA 95939
Phone: 916-968-5038
Tom Laustalot, TREASURER
18907 Indian Creek Rd, Fort Jones, CA 96032
Phone: 916-468-2228
John Clark, VICE PRESIDENT
907 Holmes Flat Rd, Red Crest, CA 95569
Phone: 707-722-4259

CALIFORNIA TROUT, INC.

870 Market St., Suite 1185
San Francisco, CA 94102 USA
Phone: 415-392-8887   Fax: 415-392-8895
E-mail: info@caltrout.org
Website: www.caltrout.org

Founded: 1970
Membership: 6000
Scope: Statewide

Description: Statewide organization of anglers dedicated to protection and restoration of wild trout, native steelhead, and their waters in California, and to the creation of high-quality angling adventures for the public to enjoy. Motto: "Keeper of the Streams."

Publication(s): Streamkeepers Log

Keyword(s): Wildlife, Forests and Forestry, Hydropower Relicensing, Outdoor Recreation, Sport Fishing, Trout, Water Conservation

Contact(s):
Mark Bergstrom, EXECUTIVE DIRECTOR
bergstrom@caltrout.org
Katrina Kuznick, OFFICE MANAGER
Jeff Eshbaugh, STREAMKEEPER COORDINATOR

CALIFORNIA WATERFOWL ASSOCIATION

4630 Northgate Blvd., Suite 150
Sacramento, CA 95834 USA
Phone: 916-648-1406   Fax: 916-648-1665
Website: www.calwaterfowl.org

Founded: 1945
Scope: Statewide

Description: A statewide nonprofit, public benefit corporation, whose principal objectives are the conservation, protection, and enhancement of California's waterfowl resources and the waterfowling opportunities which they provide. The association directly represents the interests of over 13,000 sportsmen and conservationists throughout the state and indirectly represents the interests of other Californians who are concerned with and benefit from these unique resources.

Publication(s): Sprig Tales Newsletter, California Waterfowl Magazine

Keyword(s): Environmental and Conservation Education, Hunting, Exotic species, Aquatic nuisance species, Waterfowl, Wetlands, Habitat Conservation

Contact(s):
George Kammerer, CHAIRMAN OF THE BOARD
Becky Easter, DIRECTOR OF COMMUNICATIONS/ EDUCATION
Bill Gaines, DIRECTOR OF GOVERNMENT AFFAIRS
Greg Yarris, DIRECTOR OF WATERFOWL PROGRAMS
Dave Patterson, DIRECTOR OF WETLAND PROGRAMS
Robert McLandress, PRESIDENT

CALIFORNIA WILD HERITAGE CAMPAIGN
915 20th St.
Sacramento, CA 95814 USA
Phone: 916-442-3155   Fax: 916-442-3396
E-mail: info@californiawild.org
Website: www.californiawild.org

Founded: 1997
Scope: State

Description: The Sierra Nevada Forest Protection Campaign works to protect old growth forests, wildlands and wild rivers in the Sierra Nevada mountain range.

Publication(s): Headwaters

Keyword(s): Ancient Forests, Biodiversity, Conservation, Ecosystems, Endangered Species, Environmental and Conservation Education, Forests and Forestry, Habitat Conservation, National Parks, Preservation and Protection, Riparian Restoration, Sustainability, Wilderness, Wildlands Management, training

Contact(s):
Jean Munoz, COMMUNICATIONS DIRECTOR
Bob Schneider, DIRECTOR
Califorina Wild Heritage Campaign, Sacremento,
Craig Thomas, FOREST DEFENSE COORDINATOR
cthomas@innercite.com
Tracy Sheehan, OUTREACH DIRECTOR

CALIFORNIA WILDLIFE DEFENDERS
P.O. Box 2025
Hollywood, CA 90078 USA
Phone: 213-663-1856
Website: scced.org

Founded: NA
Scope: Statewide

Description: A nonprofit association working to eradicate the prejudice towards predator animals, especially coyotes. Responsible for the discontinuation of the removal and destruction of wildlife policies in the city of Los Angeles, halting the use of leghold traps there and author of an ordinance enacted in several California cities banning the feeding of coyotes in order to limit exacerbation of urban coyote problems.

Publication(s): How to Coexist with Urban Wildlife (brochure), Wildlife Alerts

Keyword(s): National Parks, Nongame Wildlife, Predators, Trapping, training, accreditation

Contact(s):
Alberta Burke, ASSOCIATE
Suzanne Ulman, ASSOCIATE
Jessica Gates, ASSOCIATE
Lila Brooks, DIRECTOR

CALIFORNIA WILDLIFE FEDERATION
P.O. Box 1527
Sacramento, CA 95812-1527 USA
Phone: 916-441-7563   Fax: 916-441-6490

Founded: 1952
Membership: 3000
Scope: Statewide

Description: A nonprofit statewide organization of councils, clubs, and individual members dedicated to promote the conservavation,enhancement, scientific management, and wise use of all our natural resources.

Publication(s): California Wildlife

Keyword(s): Biodiversity, Endangered Species, Wildlife, Hunting, training

Contact(s):
Cheri Fuller, EDITOR
Phone: 916-441-7563
cheri_fuller@hotmail.com
Randy Walker, PRESIDENT
4908 Sunset Dr,, Fresno, CA 93704
Phone: 559-225-9003
C. Starr, TREASURER
2105 Westhaven Ave., Bakersfield, CA 93304
Phone: 661-835-8337
Tim Leblanc, VICE PRESIDENT
P.O. Box 1343, Lake Arrowhead, CA 92352
Phone: 909-336-1048

CALIFORNIA, FOREST LANDOWNERS OF
980 9th St., Suite 1600
Sacramento, CA 95814 USA
Phone: 916-972-0273   Fax: 916-979-7892
Website: www.forestlandowners.org

Founded: 1974
Scope: Regional

Description: A statewide organization affiliated with the National Woodland Owners Association that provides educational programs, information services, and legislative representation to families who own forest land for long-term investment, recreational, and conservation reasons.

Publication(s): Forest Landowner

Keyword(s): Forests and Forestry

Contact(s):
Len Lindstrand, 1ST VICE PRESIDENT
Allen Edwards, 2ND VICE PRESIDENT
Daniel Weldon, EXECUTIVE DIRECTOR AND EDITOR
Forest Tilley, PRESIDENT
Phone: 707-964-0690
Ron Adams, SECRETARY

CALIFORNIANS FOR POPULATION STABILIZATION (CAPS)
3440 Wilshire Blvd., Suite 542
Los Angeles, CA 90010 USA
Phone: 213-387-6454   Fax: 213-387-6093
E-mail: caps@cap-s.org
Website: www.cap-s.org

Founded: 1986
Scope: Statewide

Description: CAPS is a nonprofit membership organization dedicated to stabilizing population in California to protect and preserve the state's environment, ecology, and resources. CAPS believes overpopulation is the ultimate environmental threat. Activities include: public education, media campaigns, public policy research and advocacy, and grassroots organizing.

Publication(s): CAPS Newsletters, action alerts, brochures and fact sheets, CAPS Data Reports

Keyword(s): Protected Areas, Land Preservation, Land Use Planning, Open Space, Population Growth

Contact(s):
Jo Wideman, DIRECTOR OF OPERATIONS
Diana Hull, PRESIDENT OF THE BOARD

CAMP FIRE CLUB OF AMERICA, THE

230 Campfire Rd.
Chappaqua, NY 10514 USA
Phone: 914-941-0199   Fax: 914-923-0977
E-mail: campfireclub@aol.com
Website: unknown

Founded: 1897
Scope: National

Description: Works to preserve forests and woodland; to protect and conserve the wildlife of our country; and to sponsor and support all reasonable measures to the end that present and future generations may continue to enjoy advantages and benefits of life outdoors.

Publication(s): Backlog, The

Keyword(s): Forests and Forestry, Hunting, Sport Fishing, training

Contact(s):
Leonard Vallender, CHAIRMAN OF COMMITTEE ON CONSERVATION OF FORESTS AND WILDLIFE
Lewis Jordan, PRESIDENT
David Petzal, PUBLICATIONS CHAIRMAN
Thomas Quirk, SECRETARY OF COMMITTEE ON CONSERVATION OF FORESTS AND WILDLIFE

CAMP FIRE CONSERVATION FUND

230 Camp Fire Rd.
Chappaqua, NY 10514 USA
Phone: 914-941-9681

Founded: 1977
Scope: National

Description: A tax-exempt membership organization, dedicated to the preservation of wildlife and its habitat to coordinate the efforts of sportsmen's and conservation organizations; to inform the general public and governmental agencies with regard to intelligent use of our natural resources; and to support and promote conservation research.

Keyword(s): Endangered Species, Environmental and Conservation Education, Scholarships and Grants, training

Contact(s):
George Lamb, PRESIDENT
Henry Ayres, SECRETARY
Mottell Peek, TREASURER

CAMP FIRE USA

4601 Madison Ave.
Kansas City, MO 64112-1278 USA
Phone: 816-756-1950   Fax: 816-756-0258
E-mail: info@campfire.usa.org
Website: www.campfireusa.org

Founded: 1910
Membership: 650000
Scope: National

Description: Open to boys and girls from birth to 21 years of age, without regard to race, creed, ethnic origin, sex, or income level. Provides a program of informal education that focuses on developing skills in interpersonal relationships, decision-making, leadership, creativity, citizenship, community service, and individual growth.

Keyword(s): Air Quality and Pollution, Endangered Species, Environmental and Conservation Education, Health and Nutrition, Outdoor Recreation

Contact(s):
Stewart Smith, NATIONAL EXECUTIVE DIRECTOR AND CEO
Rosie Mauk, NATIONAL PRESIDENT

CANADIAN ARCTIC RESOURCE COMMITTEE, INC.

7 Hinton Ave. N., Suite 200
Ottawa, Ontario K1Y 4P1 Canada
Phone: 613-759-4284   Fax: 613-722-3318
E-mail: info@carc.org
Website: www.carc.org

Founded: 1971
Membership: 4500
Scope: International

Description: To ensure that important social, environmental, and economic ramifications of northern development are studied and analyzed before major decisions relating to northern Canada are made; to exchange information and viewpoints among the public, government, and industry; to develop better perspectives on options available; and to inform the public.

Publication(s): see publications on website, Northern Perspectives Member's Update, list of books on request.

Keyword(s): Arctic, Urban and Rural Development

Contact(s):
John Crump, EXECUTIVE DIRECTOR
ahunter@carc.org
Melissa Douglas, INFORMATION OFFICER

CANADIAN COOPERATIVE WILDLIFE HEALTH CENTRE

Dept. of Veterinary Pathology, WCVM, Univ. of Saskatchewan, 52 Campus Dr.
Saskatoon, Saskatchewan S7N 5B4
Phone: 306-966-5099   Fax: 306-966-7439
E-mail: ccwhc@sask.usask.ca

Founded: 1992
Scope: National

Description: The Canadian Cooperative Wildlife Health Centre is a national organization that provides diagnosis of disease, investigation of disease outbreaks, information, education, and consultation to wildlife managers, veterinarians, and members of the public on matters pertaining to the health of free-living wild animals in Canada.

Publication(s): Wildlife Health Centre Newsletter, Wildlife Disease Investigation Manual, Directory of Wildlife Health Expertise, Bulletin du Centre de la Sante de la Faune

Keyword(s): standards, Research, Birds, Mammals, Reptiles and Amphibians

Contact(s):
G. Wobeser, CO-DIRECTOR
Phone: 306-966-7310

F. Leighton, CO-DIRECTOR
Phone: 306-966-7281
Pierre-Yves Daoust, CONTACT FOR ATLANTIC REGION
Phone: 902-566-0667
I. Barker, CONTACT FOR ONTARIO REGION
Phone: 519-823-8800
Daniel Martineau, CONTACT FOR QUEBEC REGION
Phone: 514-773-8521
Trent Bollinger, CONTACT FOR WEST AND NORTH REGION
Phone: 306-966-5099

CANADIAN ENVIRONMENTAL LAW ASSOCIATION

517 College St.
Suite 401
Toronto, Ontario M6G 4A2 Canada
Phone: 416-960-2284   Fax: 416-960-9392
E-mail: cela@web.ca
Website: www.cela.ca

Founded: 1970
Scope: National

Description: Nonprofit, independent, public-interest legal group formed to use current environmental laws to protect the environment, and to promote better environmental legislation throughout Canada.

Publication(s): Newsletter, Intervenor, The

Keyword(s): Environmental Law

Contact(s):
Sarah Miller, COORDINATOR
mcclenat@lao.on.ca
Ramani Nadarajah, COUNSEL
millers@lao.on.ca
Theresa Mclenaghan, COUNSEL
cela@web.ca
Richard Lindgren, COUNSEL
r.lindgren@sympahco.ca
Michelle Swenarchuk, DIRECTOR OF INTERNATIONAL PROGRAMS
muldoonp@loa.on.ca
Paul Muldoon, EXECUTIVE DIRECTOR
mschanel@lao.on.ca
Lisa McShane, LIBRARIAN
r.lindgren@sympatico.ca
Kathy Cooper, RESEARCHER
cela@web.ca

CANADIAN FEDERATION OF HUMANE SOCIETIES

30 Concourse Gate, Suite 102
Nepean, Ontario K2E 7V7 Canada
Phone: 613-224-8072   Fax: 613-723-0252
E-mail: info@cfhs.ca
Website: www.cfhs.ca

Founded: 1957
Membership: 150
Scope: National

Description: CFHS is a national body comprised of animal welfare organizations and individuals whose purpose is to promote compassion and humane treatment for all animals.

Publication(s): Publications available on website, The Humane Educator, Animal Welfare in Focus

Keyword(s): Agriculture, Endangered Species, Wildlife, Marine Mammals, Whale, Dolphin, Seal, training, Trapping, Animal Welfare

Contact(s):
Frances Rodenburg, EDITOR
Robert Van tongerloo, EXECUTIVE DIRECTOR
J. Ripley, PRESIDENT

CANADIAN FORESTRY ASSOCIATION

185 Somerset St., W., Suite 203
Ottawa, Ontario K2P 0J2 Canada
Phone: 613-232-1815   Fax: 613-232-4210
E-mail: cfa@canadianforestry.com
Website: www.canadianforestry.com

Founded: 1900
Scope: National

Description: The Canadian Forestry Association is a federation of self-governing Provincial Forestry Associations across Canada. It is nongovernmental and nonindustrial. Its purpose is to develop public understanding and cooperation in the wise use, conservation, and sustainable development of Canada's forests and related resources of land, water, and wildlife.

Publication(s): National Forest Education Resources Catalogue, Proceedings: Canadian Urban Forests Conference, National Forest Week Teaching Guide, Forest Forum, Proceedings of National Forest Congress 1992

Keyword(s): Forests and Forestry

Contact(s):
Susan Gesner, IMMEDIATE PAST PRESIDENT
Barry Waito, PRESIDENT

CANADIAN INSTITUTE FOR ENVIRONMENTAL LAW AND POLICY (CIELAP)

517 College St., Suite 400
Toronto, Ontario M6G 4A2 Canada
Phone: 416-923-3529   Fax: 416-923-5949
E-mail: cielap@cielap.org
Website: www.cielap.org

Founded: 1970
Scope: National

Description: CIELAP is an independent, not-for-profit research and education institute providing environmental law and policy analysis. CIELAP provides leadership in the development of environmental law and policy which promotes the public interest and the principles of sustainability, including the protection of the health and well-being of present and future generations, and of the natural environment.

Publication(s): Ontario's Environment and the "Common Sense Revolution", Hazardous Waste Management in Ontario: A Report and Recommendation, Environment on Trial: A Guide to Ontario Environmental Law and Policy, A Carbon Dioxide Strategy for Ontario: A Discussion paper

Keyword(s): Environmental Law

Contact(s):
Anne Mitchell, EXECUTIVE DIRECTOR
David Powell, PRESIDENT
Murray Klippenstein, SECRETARY AND TREASURER

CANADIAN INSTITUTE OF FORESTRY/
INSTITUT FORESTIER DU CANADA

151 Slater St., Suite 606
Ottawa, Ontario K1P 5H3 Canada
Phone: 613-234-2242   Fax: 613-234-6181
E-mail: cif@cif-ifc.org
Website: www.cif-ifc.org

Founded: 1908
Membership: 2800
Scope: National

Description: Our mission is to advance the stewardship of Canada's forest resources through leadership, professional competence and public awareness. Our membership includes foresters, forest technicians, academics, scientists and others with a professional interest in Forestry. CIF/IFC represents the largest professional voice for forestry in Canada.

Publication(s): The Forestry Chronicle

Keyword(s): Forests and Forestry, Professional Organization

Contact(s):
Roxanne Comeau, EXECUTIVE DIRECTOR
151 Slater St., Suite 606, Ottawa, Ontario K1P 5H3
Phone: 613-234-2242
Fax: 613-234-6181
cif@cif-ifc.org

CANADIAN NATIONAL SPORTSMENS SHOWS

703 Evans Ave., Suite 202
Toronto, Ontario M9C 5E9 Canada
Phone: 416-695-0311   Fax: 416-695-0381
E-mail: info@sportshows.ca
Website: www.sportshows.ca

Founded: 1948
Membership: 24
Scope: National

Description: A national corporation presenting outdoor shows and events. Products relate to fishing, hiking, camping, boating, skiing, and the consumer shows are produced from Vancouver to Quebec City. All net proceeds are distributed to projects which encourage Canadians to appreciate, enjoy, and protect Canada's outdoor heritage.

Keyword(s): Outdoor Recreation

Contact(s):
B. Meadows, EXECUTIVE ASSISTANT
Walter Oster, PRESIDENT

CANADIAN NATURE FEDERATION

1 Nicholas St., Suite 606
Ottawa, Ontario K1N 7B7 Canada
Phone: 613-562-3447   Fax: 613-562-3371
Website: www.cnf.ca

Founded: 1971
Membership: 25
Scope: National

Description: Canada's national naturalists' organization promotes protection of nature, its diversity and the processes that sustain it. The Federation was formed from the Canadian Audubon Society, the CNF represents over 150 affiliated conservation groups and 40,000 individual supporters across the country.

Publication(s): Nature Matters, Nature Canada Magazine

Contact(s):
Caroline Schultz, DIRECTOR OF CONSERVATION & AFFILIATE DEVELOPMENT
Barbara Stevenson, EDITOR
Julie Gelfand, EXECUTIVE DIRECTOR
Phone: 613-562-3447
Jackie Krindle, PRESIDENT

CANADIAN PARKS AND WILDERNESS SOCIETY

880 Wellington St., Suite 506
Ottawa, Ontario K1R 6K7 Canada
Phone: 613-569-7226   Fax: 613-569-7098
E-mail: info@cpaws.org
Website: www.cpaws.org

Founded: 1963
Membership: 12
Scope: National

Description: A national, nonprofit advocacy organization dedicated to the protection of wilderness areas and the preservation and proper stewardship of Canada's national and provincial parks.

Publication(s): Wilderness Activist, The

Keyword(s): National Parks, Wilderness

Contact(s):
Stephen Hazell, EXECUTIVE DIRECTOR
shazell@cpaws.org
Clayton Forrest, MANAGER OF MEMBERSHIP SERVICES
David Thomson, PRESIDENT

CANADIAN SOCIETY OF ENVIRONMENTAL BIOLOGISTS

P.O. Box 962, Station F
Toronto, Ontario M4Y 2N9 Canada
E-mail: cseb@freenet.edmonton.ab.ca
Website: http://freenet.edmonton.ab.ca

Founded: 1959
Scope: National

Description: A Canada-wide society of environmental biologists whose primary goals are: the conservation of the natural resources of Canada; the prudent management of these resources so as to minimize adverse environmental effects; the interchange of ideas among environmental biologists; and maintaining high professional standards in education, research, and management related to natural resources and the environment.

Publication(s): Canadian Society of Environmental Biologists Newsletter

Keyword(s): Professional Organization, Wildlife Rehabilitation

Contact(s):
Patrick Stewart, PRESIDENT
Phone: 902-798-4022
Fax: 902-798-4022
enviroco@ns.sympatico.ca

CANADIAN WILDLIFE FEDERATION

350 Michael Cowpland Dr.
Kanata, Ontario K2m 2W1 Canada
Phone: 613-721-2286   Fax: 613-271-9591
E-mail: info@cwf-fcf.org
Website: ww.cwf-fcf.org

Founded: 1961
Membership: 250000
Scope: National

Description: To foster understanding of natural processes so that people may live in harmony with the land and its resources for the long-term benefit and enrichment of society; to maintain a substantial program of information and education based on ecological principles; and to conduct or sponsor research and scientific investigation.

Publication(s): Wild Magazine, Your Big Backyard, Biosphere, You Can Do It, Wildlife Update, Canadian Wildlife

Contact(s):
Bob Barton, 1ST VICE PRESIDENT
Nicholas Laurin, 2ND VICE PRESIDENT
Pat Doyle, 3RD VICE PRESIDENT
Colin Maxwell, EXECUTIVE VICE PRESIDENT
Nestor Romaniuk, PAST PRESIDENT
Derek Stanley, PRESIDENT

CANVASBACK SOCIETY

P.O. Box 101
Gates Mills, OH 44040 USA

Founded: 1975
Scope: National

Description: A nonprofit, tax-exempt organization established to conserve, restore, and promote the increase of the canvasback species of duck on the North American continent.

Keyword(s): Renewable Resources, Water Pollution Management, Exotic species, Aquatic nuisance species, Waterfowl, Wetlands

Contact(s):
Keith Russell, CHAIRMAN OF THE BOARD
Oakley Andrews, PRESIDENT AND TREASURER
Phone: 216-621-0200

CARIBBEAN CONSERVATION CORPORATION

4424 NW 13St. Suite 8A1
Gainesville, FL 32609 USA
Phone: 352-373-6441   Fax: 352-375-2449
E-mail: ccc@cccturtle.org
Website: www.cccturtle.org

Founded: NA
Membership: 6
Scope: International

Description: A nonprofit international membership organization founded in 1959 to support research and conservation of marine turtles in the Caribbean and throughout the world. In addition to conservation activities, it operates a research station at Tortuguero, Costa Rica—the site of the largest green turtle nesting colony in the Caribbean Sea—and maintains a semi-natural impoundment for sea turtles on Great Inagua Island in the Bahamas.

Publication(s): Sea Turtle Migration Tracking and Coastal Habitat Education Program, Velador

Contact(s):
L. Clay, CHAIRMAN OF THE BOARD OF DIRECTORS
David Godfrey, EXECUTIVE DIRECTOR
Peggy Cavanaugh, PRESIDENT
Roger Stone, SECRETARY

CAROLINA BIRD CLUB, INC.

11 W Jones St.
Raleigh, NC 27601-1029 USA
Phone: 919-733-7450   Fax: 919-715-6439
Website: www.carolinabirdclub.org

Founded: 1937
Scope: Statewide

Description: A nonprofit, educational ornithological organization to promote bird study and conservation. Affiliated local chapters.

Publication(s): CBC Newsletter, Chat, The

Keyword(s): Birds

Contact(s):
Judy Walker, EDITOR
7639 Farm Gate Dr, Charlotte, NC 28215
jwalker@email.uncc.edu
Bob Wood, EDITOR
2421 Owl Circle, West Columbia, NC 29169
Tullie Johnson, HEADQUARTERS SECRETARY
Raleigh, NC 27626
Phone: 919-733-7450
Len Pardue, PRESIDENT
16th Circle, Asheville, NC 28801

CARRYING CAPACITY NETWORK

2000 P St., NW, Suite 310
Washington, DC 20036-5915 USA
Phone: 202-296-4548   Fax: 202-296-4609
E-mail: ccn@us.net
Website: www.carryingcapacity.org

Founded: 1989
Scope: National

Description: CCN is a nonprofit network which mobilizes many diverse individuals and groups to meet the critical challenges facing our nation with solid information and analysis, effective advocacy tools, and targeted solutions. CCN's action-oriented initiatives focus on achieving national revitalization, population stabilization, immigration limitation, resource conservation, and economic sustainability.

Publication(s): Network Bulletin, FOCUS

Keyword(s): Population Growth, Sustainable Development, Environment, Renewable Resources, Agriculture, Wilderness

Contact(s):
David Durham, PRESIDENT OF THE BOARD
Virginia Abernethy, VICE PRESIDENT

CASCADIA RESEARCH

218 1/2 W. 4th Ave.
Olympia, WA 98501 USA
Website: cascadiaresearch.org

Founded: 1979
Scope: National

Description: A nonprofit, tax-exempt organization established to conduct scientific research and education related to marine mammals and birds. Primary funding for research projects comes from federal and state agencies and environmental groups.

Keyword(s): Birds, Endangered Species, Marine Mammals, Whale, Dolphin, Seal, Nongame Wildlife, Coral Reefs

Contact(s):
    Gretchen Steiger, PRESIDENT
    John Calambokidis, SECRETARY AND TREASURER
    James Cubbage, VICE PRESIDENT

CATSKILL CENTER FOR CONSERVATION AND DEVELOPMENT, INC., THE

P O Box 504 Route 28
Arkville, NY 12406-0504 USA
Phone: 845-586-2611   Fax: 845-586-3044
E-mail: cccd@catskillcenter.org
Website: www.catskillcenter.org

Founded: 1969
Membership: 2500
Scope: Regional

Description: The Catskill Center is a non-for-profit membership organization concerned with increasing public awareness of and involvement with issues affecting human communities and the natrual environment of the Catskill Mountain Region. Its activities emphasize environmental education and regional planning advocacy, as well as development and support of programs relating to historic preservation, sustainable economic development and regional arts and culture in the Catskill.

Publication(s): Successful Catskill Communities, Summary Guide to the Terms of the Watershed Agreement, Catskill Center News

Keyword(s): Natural Resource Conservation, Forests and Forestry, Public Lands, Rivers, Water Pollution Management, Wilderness, Land Protection, Rural Development, Sustainable Development, Environmental and Conservation Education, Alternative Agriculture, Culture

Contact(s):
    Geddy Sveikauskas, PRESIDENT
    H. Shostal, SECRETARY
    Helen Chase, TREASURER
    Philip Weinberg, VICE PRESIDENT

CATSKILL FOREST ASSOCIATION

P.O. Box 336
Arkville, NY 12406 USA
Phone: 845-586-3054   Fax: 845-586-4071
E-mail: cfa@catskill.net
Website: www.catskillforest.org

Founded: 1982
Scope: Local

Description: Advocates of quality forest management practices to improve the health of the forest and prevent threats to the forest ecosystem, the Catskill Forest Association is an independent nonprofit regional organization that supports forest conservation efforts in the Catskill Mountains through the promotion of forest stewardship by landowners, foresters, timber harvesters, and the general public.

Publication(s): CFA News

Keyword(s): Biodiversity, Environmental and Conservation Education, Forests and Forestry, Sustainable Development, Sustainable Ecosystems

Contact(s):
    Richard Sloman, EXECUTIVE DIRECTOR
    Dave Elmore, PRESIDENT
    Art Rotman, SECRETARY AND TREASURER
    Robert Bishop, VICE PRESIDENT

CAVE RESEARCH FOUNDATION

P.O. Box 126
Louisville, KY 40201-0126 USA
Phone: 502-637-2030
Website: www.cave-research.org

Founded: 1957
Membership: 1000
Scope: International

Description: The Foundation is a nonprofit organization that supports and promotes research, interpretation, and conservation activities in caves and karst areas. Permanent field operations are maintained within Mammoth Cave National Park, Carlsbad Caverns National Park, Sequoia and Kings Canyon National Parks, and Lava Beds National Monument. Approximately 800 joint-venturers participate in program.

Publication(s): Cave Books, Annual Report, CRF Newsletter

Keyword(s): Protected Areas, Geology, Historic Preservation, Public Lands, Scholarships and Grants

Contact(s):
    Paul Nelson, EDITOR
    2644 S Quarry Lane D, Walnut, CA 91789-4067
    Pat Kambesis, PRESIDENT
    P.O. Box 343, Wenona, IL 61377-0343
    Phone: 815-863-5184
    kembesis@jun
    Peter Bosted, SECRETARY
    2301 Sharon Rd, Menlo Park, CA 94025-680
    Phone: 650-926-2319
    bosted@slac.spanford.edu
    Paul Cannaley, TREASURER
    Phone: 317-862-5618

CENTER FOR A SUSTAINABLE COAST

221B Mallory St.
Saint Simons Island, GA 31522 USA
Phone: 912-638-3612   Fax: 912-638-3615
E-mail: susdev@gate.net
Website: www.sustainablecoast.com

Founded: 1997
Scope: Regional

Description: To promote sustainable use, protection, enhancement, and understanding of coastal Georgia's natural, economic, historic, and cultural resources through education, advocacy, technical assistance, and research.

Publication(s): Works in Progress, Surface Water Withdrawal and Coastal Economic Issues, Fisheries and Water Resource Permit Issues in the Lower Altamaha and Other Coastal Georgia Rivers

Keyword(s): Water Quality

Contact(s):
    David Kyler, EXECUTIVE DIRECTOR

CENTER FOR BIOLOGICAL DIVERSITY

P.O. Box 710
Tuscon, AZ 85702-0710 USA
Phone: 520-623-5252   Fax: 520-623-9797
E-mail: center@biologicaldiversity.org
Website: www.biologicaldiversity.org

Founded: 1989
Membership: 5000
Scope: National

Description: The Southwest Center for Biological Diversity uses a combination of scientific research, public education, and strategic litigation to defend the forests, rivers and deserts of western North America.

Publication(s): E-mail: Newsletter - Bio Diversity Activist

Keyword(s): Biodiversity, Endangered Species, Public Lands, Watersheds, Deserts, Mining, Forest Management

Contact(s):
Robin Silver, CONSERVATION CHAIR
Phone: 602-246-4170
Fax: 602-249-2576
rsilver@biologicaldiversxity.org
Kieran Suckling, PRESIDENT AND EXECUTIVE DIRECTOR
ksuckling@biologicaldiversity.org

CENTER FOR CHESAPEAKE COMMUNITIES
209 West St., Suite 201
Annapolis, MD 21401 USA
Phone: 410-267-8595   Fax: 410-267-8597
E-mail: shall@chesapeakecommunities.org
Website: www.chesapeakecommunities.org

Founded: 1997
Scope: Regional

Description: A nonprofit, independent organization dedicated to assisting local governments in the Chesapeake Bay watershed in their environmental restoration and protection initiatives.

Keyword(s): Environment, Environmental and Conservation Education, Environmental Planning, Protected Areas, Environmental Protection, Land Management, Land Use Planning, Nonpoint Source Pollution, Sustainability

Contact(s):
Gary Allen, EXECUTIVE DIRECTOR
gallen@chesapeakecommunities.org

CENTER FOR ENVIRONMENT
1250 24th Street NW
Washington, DC 20037-1124 USA
Phone: 202-331-0664

Founded: 1985
Scope: National

Description: A nonprofit public interest organization dedicated to protecting the environment, enhancing the human ecology, and working to ensure the efficient use of natural resources. (CE)2's secondary mission is to provide opportunities for blacks and other minorities to participate in the environmental movement. Major areas of concern are: Air quality and pollution, water resources and pollution, energy, renewable resources, toxic substances, Africa and Third World environment, land use, internships, and urban environment.

Publication(s): A complete list of (CE)2's publications is available upon request, African American Environmentalist

Contact(s):
Charles Stephenson, CHAIRMAN
Norris McDonald, PRESIDENT
Phone: 202-879-3183
Jannie Pittman, TREASURER

CENTER FOR ENVIRONMENTAL EDUCATION
c/o Antioch New England, 40 Avon St.
Keene, NH 03431 USA
Phone: 603-355-3251   Fax: 603-357-0718
E-mail: cee@antiochne.edu
Website: www.schoolsgogreen.org

Founded: 1989
Membership: 5000
Scope: National

Description: A nonprofit environmental education resource center housing one of the nation's most comprehensive collections of environmental education materials. The library has over 10,000 materials—books, videos, curricula and resources that can be accessed in person, by phone, fax, or through the website.

Publication(s): Blueprint for a Green School, Natures Course, Grapevine Newsletter

Keyword(s): Environmental and Conservation Education, Internships, Youth Organizations

Contact(s):
David Sobel, CO-EXECUTIVE DIRECTOR
Cindy Thomashow, CO-EXECUTIVE DIRECTOR
Jayni Chase, FOUNDER

CENTER FOR ENVIRONMENTAL INFORMATION
55 St. Paul St.
Rochester, NY 14604-1314 USA
Phone: 716-262-2870   Fax: 716-262-4156
E-mail: ceiroch@aol.com
Website: www.rochesterenvironment.org

Founded: 1974
Membership: 300
Scope: National

Description: Provides on-call reference and referral and current awareness and educational services to scientists, educators, government agency staff, policymakers, business and industry managers and interested citizens. Sponsors conferences and seminars.

Publication(s): Proceedings of Annual Conferences

Keyword(s): Acid Rain, Communications, Energy, Environmental and Conservation Education, Greenhouse Effect/Global Warming

CENTER FOR ENVIRONMENTAL PHILOSOPHY
University of North Texas
Denton, TX 76203-0980 USA
Phone: 940-565-2727   Fax: 940-565-4439
E-mail: ee@unt.edu.
Website: www.cep.unt.edu

Founded: 1980
Membership: 2
Scope: International

Description: A nonprofit, tax-deductible organization. The Center promotes research and instruction in environmental ethics and its application in environmental policy and decision-making. The Center works with governmental and environmental organizations on conferences, workshops, and other educational projects.

Publication(s): Environmental Ethics: An Interdisciplinary Journal Dedicated to the Philosophical Aspects of Environmental Problems

Keyword(s): Environmental Ethics, Wilderness

Contact(s):

Jan Dickson, EXECUTIVE DIRECTOR
Center for Environmental Philosophy, Univ. of North TX, P.O.
Box 310980, Denton, TX 76203-0980
Phone: 940-565-2727
Fax: 940-565-4439
jdickson@unt.edu
Eugene Hargrove, PRESIDENT, EDITOR AND PUBLISHER
Center for Environmental Philosophy, Univ. of North TX, P.O.
Box 310980, Denton, TX 76203-0980
Phone: 940-565-2727
Fax: 940-565-4439
hargrove@unt.edu
Max Oelschlaeger, SECRETARY AND TREASURER
Dept. Humanities, Arts & Religion, Northern Arizona, P.O. Box
6031, Flagstaff, AZ 86011-6031
Phone: 520-523-0389
max.oelschlaeger@nau.edu
J. Callicott, VICE PRESIDENT
Dept. of Philosophy, Univ. of North TX, P.O. Box 310920,
Denton, TX 76203-0920
Phone: 940-565-2255
Fax: 940-565-4448
callicott@unt.edu

CENTER FOR ENVIRONMENTAL STUDY

Grand Rapids Community College, 143 Bostwick NE
Grand Rapids, MI 49503 USA
Phone: 616-234-3935   Fax: 616-234-3936
E-mail: ces1@iserv.net
Website: www.cesmi.org

Founded: 1969
Scope: National

Description: The Center for Environmental Study, a 501C (3)
organization, has served its community as an independent,
science-based environmental education and research
authority. It provides curricula and conducts professional
development workshops on a variety of subjects ranging from
water and air quality to tropical forest and Great Lakes issues.

Publication(s): Field Guide to Ecosystems and Habitats of the
Great Lakes Region (in prep.), The Great Lakes - An
Interactive CD-ROM Game for grade 3—adult, Mahogany: A
Research Bibliography of Swietenia

Keyword(s): Air Quality and Pollution, Communications, Ecology,
Environmental and Conservation Education, training, Youth
Organizations, Watersheds

Contact(s):

John Martin, BOARD CHAIR
Rick Sullivan, EXECUTIVE DIRECTOR
Peter Wege, FOUNDER

CENTER FOR HEALTH, ENVIRONMENT, AND JUSTICE

P.O. Box 6806
Falls Church, VA 22040-6806 USA
Phone: 703-237-2249   Fax: 703-237-8389
E-mail: chej@chej.org
Website: www.chej.org

Founded: 1981
Scope: National

Description: CHEJ provides assistance to more than 8,000
grassroots citizens groups fighting for environmental justice.
CHEJ is a nonprofit organization dedicated to helping people
nationwide by providing one-on-one assistance and
information on how to organize politically, scientifically, and
legally to prevent or clean up environmental catastrophies.
Founded by Lois Marie Gibbs.

Publication(s): Catalog of how-to guide books, Environmental
Health Monthly, Everyone's Backyard

Keyword(s): Internships, Scholarships and Grants, Toxic
Substances, Nuclear-free, Water quantity, Water export and
diversion, People of Color in the Environment, Women in the
Environment

Contact(s):

Lois Gibbs, EXECUTIVE DIRECTOR

CENTER FOR INTERNATIONAL ENVIRONMENTAL LAW (CIEL)

1367 Connecticut Ave., NW, Suite 300
Washington, DC 20036-1860 USA
Phone: 202-785-8700   Fax: 202-785-8701
E-mail: info@ciel.org
Website: www.ciel.org

Founded: 1989
Membership: 30
Scope: International

Description: The CIEL is a public interest environmental law
organization founded to focus the energy and experience of
the public interest environmental law movement on reforming
international environmental law and institutions, and on forging
stronger and more meaningful connection between the top
down diplomatic approach of international law, and the bottom
up participatory approach that has been the hallmark of the
public interest environmental law movement.

Publication(s): Biodiversity in the Seas: Implementing the
Convention on Biological Diversity in Marine and Coastal
Habitats, A Citizen's Guide to the World Bank Inspection Panel,
Trade and the Environment: Law, Economics, and P,
International Environmental Law and Policy, publications also
on website

Keyword(s): Biodiversity, Environmental Law, Environmental and
Conservation Education, Greenhouse Effect/Global Warming,
International Environmental Law, International Trade and
Environment

Contact(s):

Jeffrey Wanha, DIRECTOR FOR FINANCE &
ADMINISTRATION
David Hunter, EXECUTIVE DIRECTOR
Durwood Zaelke, PRESIDENT

CENTER FOR PLANT CONSERVATION

P.O. Box 299
St. Louis, MO 63166 USA
Phone: 314-577-9450   Fax: 314-577-9465
E-mail: cpc@mobot.org
Website: www.mobot.org/cpc

Founded: NA
Membership: 32
Scope: National

Description: A national network of 28 botanical gardens and
arboreta dedicated to the conservation and study of rare and
endangered U.S. plants. The Center establishes conservation
collections of endangered species in regional gardens and

seed banks as a resource for conservation and research efforts: The National Collection of Endangered Plants.

Publication(s): Americas Vanishing Flora, Guidelines for the Manangent of Orthodox Seeds, Plants in Peril, Plant Conservation (newsletter)

Keyword(s): Biodiversity, Botanical Gardens, Endangered Species, Environmental and Conservation Education, Flowers, Plants, and Trees, Protected Areas, Environment

Contact(s):
Kathryn Kennedy, EXECUTIVE DIRECTOR

CENTER FOR RESOURCE ECONOMICS/ ISLAND PRESS

1718 Connecticut Ave., NW, Suite 300
Washington, DC 20009 USA
Phone: 202-232-7933   Fax: 202-234-1328
E-mail: info@islandpress.org
Website: www.islandpress.org

Founded: 1978
Scope: National

Description: The Center for Resource Economics is a nonprofit organization that develops, publishes, markets and disseminates books and other information products essential for solving local and global environmental problems and planning for the future.

Publication(s): Island Press Book Distribution Center: Box 7, Covelo, CA 95428

Keyword(s): Biodiversity, Energy, Sustainable Development, Sustainable Ecosystems, Wetlands

Contact(s):
Jonathan Cobb, EDITOR, SHEARWATER BOOKS
Phone: 914-631-7088
jcobb@islandpress.org
Samuel Dorrance, MARKETING DIRECTOR
Phone: 202-232-7933
sdorrance@islandpress.org
Charles Savitt, PRESIDENT
Phone: 202-232-7933
csavitt@islandpress.org
Ken Hartzell, VP OF FINANCES AND CFO
Phone: 202-232-7933
khartzell@islandpress.org
Dan Sayre, VP/PUBLISHER
Phone: 202-232-7933
dsayre@islandpress.org

CENTER FOR RESOURCEFUL BUILDING TECHNOLOGY NCAT

P.O. Box 100
Missoula, MT 59806 USA
Phone: 406-549-7678   Fax: 406-549-4100
E-mail: crbt@ncat.org
Website: www.crbt.org

Founded: 1990
Membership: 3
Scope: National

Description: The Center for Resourceful Building Technology (CRBT) is a non-profit corporation dedicated to promoting environmentally responsible practices in construction. Its mission is to serve as both catalyst and facilitator in encouraging building technologies which realize a sustainable and efficient use of resources.

Publication(s): Recraft 90 Handbook, Affordable Resource Efficiency: Reducing Construction and Demolition Waste, Building Our Children's Future, Guide to Resource Efficient Building Elements

Keyword(s): Environmental and Conservation Education, Protected Areas, Forests and Forestry, Renewable Resources, Sustainable Development

Contact(s):
Tim Mellgren, CHAIR
151 Fairway Dr., Missoula, MT 59801
Phone: 406-728-6711
Steve Loken, FOUNDER
2605 Lincoln Hills Dr., Missoula, MT 59802
Phone: 406-728-1412
Tracy Mumma, RESEARCH DIRECTOR
Dale McCormick, TECHNICAL DIRECTOR

CENTER FOR SCIENCE IN THE PUBLIC INTEREST

1875 Connecticut Ave., NW, Suite 300
Washington, DC 20009 USA
Phone: 202-332-9110   Fax: 202-265-4954
E-mail: cspi@cspinet.org
Website: www.cspinet.org

Founded: 1971
Scope: National

Description: National consumer advocacy organization that focuses on health, nutrition, and alcohol issues. CSPI informs the public of its findings through a variety of publications, press releases, speeches, media appearances, and initiates legal actions. The Center has an intern program throughout the year.

Publication(s): reports, posters, books, and video, Nutrition Action Healthletter

Keyword(s): Agriculture, Health and Nutrition, Internships, Pesticides, Toxicology

Contact(s):
Michael Jacobson, EXECUTIVE DIRECTOR

CENTER FOR THE STUDY OF TROPICAL BIRDS, INC., ADMINISTRATIVE OFFICE

ADMINISTRATIVE OFFICE
218 Conway
San Antonio, TX 78209-1716 USA
Phone: 210-828-5306   Fax: 210-828-9732
E-mail: office@cstb.inc.org
Website: www.cstbinc.org

Founded: 1987
Scope: International

Description: A nonprofit organization devoted to the conservation of tropical birdlife through cooperative programs of scientific research, conservation and education. Current projects include raising palms to save cracids in El Cielo Biosphere Reserve in Tamaulipas, Mexico; Development of avi-tourism in Zapata, Texas and release of endangered Red Siskins on the Bocas Islands of Trinidad (www.lfpt.rwthaachen.de/~ckr/carduelan/ red-trini.html), in addition to field studies on Singing Quail, Bearded Wood partridges and other birds along the Rio Grande in Texas and NE Mexico.

Publication(s): publications available on website

Keyword(s): Birds, Endangered Species, Nongame Wildlife, Raptors, Research, training, Grants

Contact(s):
Jack Eitniear, DIRECTOR
Alvaro Tapia, MEXICO PROGRAM COORDINATOR
Michael Gartside, TREASURER

CENTER FOR WATERSHED PROTECTION

8391 Main St.
Ellicott City, MD 21043 USA
Phone: 410-461-8323　　　Fax: 410-461-8324
E-mail: center@cwp.org
Website: www.cwp.org

Founded: 1992
Membership: 15
Scope: National

Description: CWP is dedicated to new cooperative ways of protecting and restoring watersheds.

Publication(s): Watershed Protection Techniques, Site Planning for Urban Stream Protection, Environmental Indicators to Assess Stormwater Control Programs and Practices, www.cwp.org

Keyword(s): Environmental Planning, Land Use Planning, Urban Environment, Water Pollution Management, Watersheds

Contact(s):
Hye Kwon, ADMINISTRATIVE DIRECTOR
hyk@cwp.org
Thomas Schueler, EXECUTIVE DIRECTOR
Dan Oleary, PRINCIPAL ENGINEER
djo@cwp.org

CENTER FOR WILDLIFE LAW

Institute of Public Law at University of New Mexico
School of Law, 1117 Stanford NE
Albuquerque, NM 87131 USA
Phone: 505-277-5006　　　Fax: 505-277-5483
Website: www.ipl.unm.edu/cwl

Founded: 1990
Scope: National

Description: Through projects, publications, conferences, and training programs, the Center provides wildlife law and policy analysis and other educational information to legal and nonlegal communities. The Center also conducts a unique law-related wildlife education program. Staff have expertise in wildlife and environmental law and policy, biology, education, geographic information systems, publishing.

Publication(s): Federal Wildlife Laws Handbook with Related Laws, State Wildlife Laws Handbook, Wild Friends: Kids Bringing People Together on Wildife Issues, The Status of Poaching in the U.S., Wild News, Wildlife Law News Quarterly

Contact(s):
Ruth Musgrave, DIRECTOR

CENTRAL OHIO ANGLERS AND HUNTERS CLUB

P.O. Box 28224
Columbus, OH 43228 USA
Phone: 614-879-7757

Founded: NA
Scope: Local

Description: Promotion of conservation and conservation education in all their phases, with a particular reference to land, air, and water; to promote good fellowship and good citizenship; to inculcate regard for the rights of others and respect for the obedience to law; to support a safe and effective conservation program for and by the state and nation.

Keyword(s): Hunting, Sport Fishing, Conservation

Contact(s):
Doug Eakins, PRESIDENT
767 Larri Ct., W. Jefferson, OH 43162
Phone: 614-879-7757
Kevin Burke, SECRETARY
Eric Obrein, TREASURER

CENTRO DE INFORMACION, INVESTIGACION Y EDUCACION SOCIAL (CIIES)

RR-9, Buzon 1722
San Juan, PR 00926-9736 USA
Phone: 787-292-0620　　　Fax: 787-760-0496
E-mail: sctinc@coqui.net

Founded: 1989
Scope: State

Description: CIIES was founded as a part of Servicios Cientificos y Tecnicos, a nonprofit organization dealing with natural resources, environmental health, and safety issues. It provides services in the form of seminars, workshops, and a resource center to students, teachers, journalists, communities, and workers.

Keyword(s): Biodiversity, Environmental and Conservation Education, Pollution Prevention, Habitat Conservation, Environmental Justice

Contact(s):
Neftali Martinez, DIRECTOR
Jose Sepulveda, SECRETARY
Maria Vilches, TREASURER

CETACEAN SOCIETY INTERNATIONAL

P.O. Box 953
Georgetown, CT 06829 USA
Phone: 203-431-1606　　　Fax: 203-431-1606
E-mail: rossiter@csiwhalesalive.org
Website: http://csiwhalesalive.org/

Founded: 1974
Scope: International

Description: CSI is dedicated to the preservation and protection of all cetaceans (whales, dolphins, and porpoises) and the marine environment on a global basis. CSI is an all-volunteer, nonprofit conservation education and research organization with representatives in over 21 countries.

Publication(s): Meet the Great Ones (book, English and Spanish), Several education packages., Whales Alive (newsletter)

Keyword(s): Endangered Species, Environmental Protection, Marine Mammals, Whale, Dolphin, Seal, Research

Contact(s):
Robbins Barstow, DIRECTOR EMERITUS
190 Stillwold Dr., Wethersfield, CT 06109
Phone: 860-563-2565
robbinsb@aol.com
William Rossiter, PRESIDENT
21 Laurel Hill Rd., Ridgefield, CT 06877
Phone: 203-544-8902
william_rossiter@compuserve.com

Martha Fitzgerald, SECRETARY
120 Retreat Ave. C-3, Hartford, CT 06106
Phone: 860-246-3143
Robert Victor, TREASURER
57 Crossroads Ln., Glastonbury, CT 06033
rfvictor@juno.com
Barbara Kilpatrick, VICE PRESIDENT
15 Wood Pond Rd., West Hartford, CT 06107
Phone: 860-561-0187

CHARLES A. AND ANNE MORROW LINDBERGH FOUNDATION, THE

2150 Third Ave. N., Suite 310
Anoka, MN 55303 USA
Phone: 763-576-1596　　Fax: 763-576-1664
E-mail: info@lindberghfoundation.org
Website: www.lindberghfoundation.org

Founded: 1977
Membership: 4
Scope: International

Description: The Charles A. and Anne Morrow Lindbergh Foundation is a nonprofit organization, advancing Charles and Anne Morrow Lindbergh's vision of a balance between technological progress and environmental preservation by offering Lindbergh Grants to individuals for research and educational projects which will further this balance, presenting the Lindbergh Award for extraordinary contributions to the balance, and sponsoring other projects and programs. Associate membership information is available by writing the Foundation office.

Publication(s): Charles A. and Anne Morrow Lindbergh Foundation Newsletter, The

Keyword(s): Research Grants, Protected Areas, Sustainable Development, Exotic species, Aquatic nuisance species, training, Agriculture

Contact(s):
Clare Hallward, CHAIRPERSON OF GRANTS SELECTION COMMITTEE
Kelley Welf, EDITOR
Marlene White, EXECUTIVE DIRECTOR
Reeve Lindbergh, PRESIDENT
James Lloyd, SECRETARY
Gene Bratsch, TREASURER
Clare Hallward, VICE PRESIDENT
Kristina Lindbergh, VICE PRESIDENT AND CHAIRMAN OF AWARD SELECTION AND LECTURE COMMITTEE

CHELONIA INSTITUTE

3330 Washington Blvd.
Arlington, VA 22201 USA
Phone: 703-516-2600　　Fax: 703-522-1427
Website: www.truland.com

Founded: 1977
Membership: 1000
Scope: National

Description: A private operating foundation with ecological concerns focused primarily on the conservation of marine turtles. The Institute undertakes a broad range of programs including technical publications, land acquisition, and so on, and works cooperatively with other organizations.

Contact(s):
Mary Truland, ASSISTANT DIRECTOR
Robert Truland, PRESIDENT

CHESAPEAKE BAY FOUNDATION, INC.

Philips Merrill Environmental Center, 6 Herndon Ave
Annapolis, MD 21403 USA
Phone: 410-268-8833　　Fax: 410-280-3513
E-mail: chesapeake@cbf.org
Website: www.cbf.org

Founded: 1967
Membership: 90000
Scope: Regional

Description: The Chesapeake Bay Foundation's Maryland office conducts activities of the foundation specific to the state of Maryland and operates field offices in southern Maryland and Delaware.

Publication(s): Bay Savers Bulletin, Save the Bay

Keyword(s): Aquatic Habitats, Environmental and Conservation Education, Land Use Planning, Renewable Resources, Environmental Justice

Contact(s):
Terry Cummings
Phone: 410-268-8833
chesepeake@cbf.org
Theresa Pierno, EXECUTIVE DIRECTOR
tpierno@savethebay.cbf.org

CHESAPEAKE BAY FOUNDATION, INC.

6 Herndon Ave
Annapolis, MD 21403 USA
Phone: 410-268-8816　　Fax: 410-268-6687
E-mail: chesapeake@cbf.org
Website: www.savethebay.cbf.org

Founded: 1966
Scope: National

Description: A nonprofit membership organization established to promote the environmental protection and restoration of Chesapeake Bay and its full watershed. CBF operates programs in environmental education and environmental protection and restoration.

Publication(s): Megalops, Grassroots Bulletin, Save the Bay

Keyword(s): Environmental Protection, Watersheds, Environmental and Conservation Education, Restoration

Contact(s):
Wayne Mills, CHAIRMAN
William Baker, PRESIDENT
Donald Baugh, VICE PRESIDENT FOR EDUCATION
Michael Shultz, VICE PRESIDENT FOR PUBLIC AFFAIRS
Michael Hirshfield, VICE PRESIDENT FOR RESOURCE PROTECTION AND RESTORATION

CHESAPEAKE BAY FOUNDATION, INC.

PENNSYLVANIA OFFICE
The Old Waterworks Bldg., 614 N. Front St., Suite G
Harrisburg, PA 17101 USA
Phone: 717-234-5550　　Fax: 717-234-9632
E-mail: chesapeake@savethe bay.cbf.org
Website: www.savethebay.cbf.org

Founded: 1966
Membership: 90000
Scope: Statewide

Description: The Foundation conducts activities and programs specific to the Commonwealth of Pennsylvania.

Publication(s): Bay Beginnings - newsletter

Keyword(s): Air Quality and Pollution, Energy, Solid Waste, Transportation

Contact(s):
Jolene Chinchilli, PENNSYLVANIA EXECUTIVE DIRECTOR

CHESAPEAKE BAY FOUNDATION, INC., VIRGINIA OFFICE
VIRGINIA OFFICE
1108 East Main St., Suite 1600
Richmond, VA 23219 USA
Phone: 804-780-1392   Fax: 804-648-4011
Website: www.savethebay.cbf.org

Founded: 1967
Scope: State

Description: The Chesapeake Bay Foundation conducts activities of the foundation in the Commonwealth of Virginia and operates field offices in Norfolk and Tappahannock, Virginia.

Keyword(s): Birds, Environmental and Conservation Education, Environmental Protection, National Parks, Water Quality

Contact(s):
Roy Hoagland, ASSISTANT DIRECTOR
Joseph Maroon, EXECUTIVE DIRECTOR

CHESAPEAKE FARMS
7321 Remington Dr.
Chestertown, MD 21620 USA
Phone: 410-778-0141   Fax: 410-778-8405

Founded: 1956
Scope: Local

Description: Operated by Dupont Agricultural Products to demonstrate, research, and promote sustainable farming and wildlife management practices. Provides a forum for exploring agricultural issues and interactions between environmental and economic sustainability. Agricultural project conducted by coalition of Dupont, universities, government and private organizations. Wildlife research conducted through graduate fellows.

Contact(s):
Mark Conner, MANAGER

CHESAPEAKE WILDLIFE HERITAGE (CWH)
P.O. Box 1745
Easton, MD 21601 USA
Phone: 410-822-5100   Fax: 410-822-4016
E-mail: info@cheswildlife.org
Website: www.cheswildlife.org

Founded: 1980
Membership: 900
Scope: Regional

Description: A private, nonprofit conservation group working with private and public landowners to restore and protect wildlife habitat in the Chesapeake Bay watershed. CWH constructs and manages wetlands, warm season grass and wildflower meadows, nesting structures, marshes, and woodlands. CWH advises on and carries out sustainable farming techniques in order to benefit the Chesapeake Bay and its wildlife. CWH also conducts ecological research on plants and migratory birds.

Publication(s): Habitat Works

Keyword(s): training

Contact(s):
Larry Albright, BOARD OF DIRECTORS PRESIDENT
John Gerber, HABITAT ECOLOGIST
Michael Haggie, HABITAT ECOLOGIST
Susanna Engvall, SECRETARY OF PUBLIC RELATIONS

CHICAGO HERPETOLOGICAL SOCIETY
2060 N. Clark St.
Chicago, IL 60614 USA
Phone: 219-464-8514
E-mail: clech2@siu.edu
Website: www.chicagoherp.org

Founded: 1966
Membership: 1,957
Scope: Statewide

Description: The Chicago Herpetological Society is a group of reptile and amphibian enthusiasts. Its goals are education, conservation, and the advancement of herpetology.

Publication(s): Bulletin of the Chicago Herpetological Society

Keyword(s): Wildlife Rehabilitation, Environmental and Conservation Education, Reptiles and Amphibians, Scholarships and Grants, Zoology

Contact(s):
Jack Schoenfelder, PRESIDENT
c/o Ivy Tech College, 2401 Valley Dr., Valparaiso, IN 46383
Phone: 219-929-1525
Greg Brim, TREASURER
Phone: 603-834-4446
gregbrim@mediaone.net
Lori King, VICE PRESIDENT
Phone: 773-447-3645
loriguanid@aol.com

CHIHUAHUAN DESERT RESEARCH INSTITUTE
P.O. Box 905
Ft. Davis, TX 79734 USA
Phone: 915-364-2499   Fax: 915-364-2509
E-mail: maager@cdri.org
Website: http://www.cdri.org

Founded: 1974
Scope: National

Description: Nonprofit organization formed to promote human understanding and appreciation of the Chihuahuan Desert through scientific research and public education. Current studies include life history related studies, systematic zoology, systematic botany, desert ecology, anthropology, archeology, geology, and theoretical ecology.

Publication(s): Chihuahuan Desert Discovery, The, Chihuahaun Newsbriefs, CDRI Contributions

Keyword(s): Wildlife Rehabilitation, Deserts, Endangered Species, Geology, education

Contact(s):
Cathryn Hoyt, EXECUTIVE DIRECTOR
Rob Dunagan, PRESIDENT
Thomas Brunner, SECRETARY
Larry Bryant, TREASURER
James Scudday, VICE PRESIDENT

CHINA REGION LAKES ALLIANCE

RR 1, Box 970
South China, ME 04358 USA
Phone: 207-445-5021   Fax: 207-445-3208
E-mail: chiname@pivot.net

Founded: 1994
Membership: 7
Scope: Local

Description: To protect and improve water quality in 3 culturally eutrophic Maine lakes, (China Lake, Threemile Pond and Webber Pond) and to benefit our local economy through integrated watershed management.

Publication(s): Vegetative Buffer Strips, Starting a Local Youth Conservation Corps, Walk for a Rainy Day: What You Can Do For Your Camp Road

Keyword(s): Environmental and Conservation Education, Pollution Prevention, Lakes, Environmental Planning, Water Quality, Exotic species, Aquatic nuisance species, Watersheds, Coral Reefs

Contact(s):
Rebecca Manthey, EXECUTIVE DIRECTOR
Daniel Dubord, PRESIDENT
Phone: 207-872-2743
Fax: 207-872-2962

CHLORINE-FREE PAPER CONSORTIUM

1411 Ellis Ave.
Ashland, WI 54806 USA
Phone: 715-682-1847   Fax: 715-682-1308
E-mail: mail@clfree.org
Website: www.clfree.org

Founded: NA
Scope: National

Description: The CPC aims to reduce the use of chlorinated substances in the paper-making process by informing people about the effects of chlorine by-products and facilitating communication between buyers and sellers of chlorine-free paper products. A project of Northland College and the National Wildlife Federation.

Publication(s): Brochure

Keyword(s): Toxic Substances, Nuclear-free, Water quantity, Water export and diversion, Toxic Reduction, Water Pollution Management, Water Quality, Hazardous Materials & Waste, Health and Nutrition, Chemical Pollution Control, Pollution Prevention, International Trade and Environment, Research

Contact(s):
Jeffery Huxmann, EXECUTIVE COORDINATOR
huxmann@clfree.org

CHRISTINA CONSERVANCY, INC.

P.O. Box 1680
Wilmington, DE 19899-1680 USA
Phone: 302-984-3801   Fax: 302-652-5379

Founded: NA
Scope: Local

Description: The purpose of the Christina Conservancy, Inc. is to preserve, protect, and urge the wise use of the Christina River.

Keyword(s): Rivers

Contact(s):
Edward Cooch, PRESIDENT

CINCINNATI NATURE CENTER

4949 Tealtown Rd.
Milford, OH 45150-9752 USA
Phone: 513-831-1711   Fax: 513-831-8052
E-mail: cnc@cincynature.org
Website: www.cincynature.org

Founded: 1965
Membership: 5500
Scope: Local

Description: A private, non-profit organization consiting of a nature preserve and two working farms which provide environmental, natural history and agricultural education for the Greater Cincinnati area.

Publication(s): Newsleaf (newsletter)

Keyword(s): Nature Centers, Internships, education, Environmental and Conservation Education, Agriculture, Ecology, Interpretation, Professional Development

Contact(s):
Ken Friel, CHAIRMAN OF THE BOARD
tcrink@strauss-troy.com
Connie Brockman, EDUCATION DIRECTOR
Phone: 513-965-4891
William Hopple, PRESIDENT/EXECUTIVE DIRECTOR
Phone: 513-965-4246
bhopple@cincynature.org

CIRCUMPOLAR CONSERVATION UNION

1615 M3 NW 3200
Washington, DC 20036 USA
Phone: 202-429-7440   Fax: 202-429-7444
E-mail: circumpolar@igc.org
Website: www.circumpolar.org

Founded: 1993
Scope: International

Description: Circumpolar Conservation Union is a public interest initiative dedicated to protecting the ecological and cultutral integrity of the Arctic for present and future generations. CCU works nationally and internationally through policy advocacy, public education, and by building links among diverse constituencies, to achieve comprehensive legal protection for the Arctic.

Publication(s): Persistent Organic Pollutants POPS in Alaska: What does Science tell Us

Keyword(s): Arctic, Environmental Law, Indigenous People, Sustainable Development

Contact(s):
Evelyn Hurwich, EXECUTIVE DIRECTOR

CITIZENS ADVISORY COUNCIL TO PENNSYLVANIA DEPARTMENT OF ENVIRONMENTAL PROTECTION

Attn: Executive Director, Rachel Carson State Office Bldg., 5th Fl., P.O. Box 8459
Harrisburg, PA 17105-8459 USA
Phone: 717-787-4527   Fax: 717-787-2878
Website: www.cac.dep.state.pa.us

Founded: NA
Membership: 18
Scope: Regional
Publication(s): Annual Report, Regional Report, Advisory-Monthly

Keyword(s): Environmental Protection, Public Health Protection, Public Participation

Contact(s):
Stephanie Mioff, ADMINISTRATIVE ASSISTANT
mioff.stephanie@dep.state.pa.us
Daniel Snowden, ENVIRONMENTAL PLANNER
Susan Wilson, EXECUTIVE DIRECTOR
wilson.susan@dep.state.pa.us

CITIZENS FOR A SCENIC FLORIDA, INC.
CITIZENS FOR A SCENIC FLORIDA
4401 Emerson St., Suite 10
Jacksonville, FL 32207 USA
Phone: 904-396-0037   Fax: 904-398-4647
E-mail: scenicfl@bellsouth.net
Website: www.scenicflorida.org

Founded: 1998
Scope: Regional

Description: Citizens for a Scenic Florida: Preserving Florida's Scenic Heritage.

Publication(s): Florida View Points

Keyword(s): Natural Areas

Contact(s):
Lane Welch, DIRECTOR

CITIZENS NATURAL RESOURCES ASSOCIATION OF WISCONSIN, INC.
Attn: President, 3805 Paunack St.
Madison, WI 53711 USA
Phone: 608-231-9721   Fax: 608-218-1647
E-mail: ekolink@aol.com

Founded: 1951
Scope: Statewide

Description: To protect Wisconsin's natural resources through education, legislation, and the courts. The CNRA initiated and sponsored the action which resulted in the banning of DDT in Wisconsin and two years later in the United States. Recently, the CNRA has been concentrating on protecting and restoring native vegetation along Wisconsin's roads.

Publication(s): Wisconsin Roadsides, CNRA Report, The

Contact(s):
Louise Coumbe, MEMBERSHIP CHAIR
1028 Elmwood Ave., Oshkosh, WI 54901
Kira Henschel, PRESIDENT
Zaiga Maassen, SECRETARY
913 Honey Creek, Oshkosh, WI 54904
Charles Sturm, TREASURER
J-1233 Mayfair Rd. Suite 125, Milwaukee, WI 53226
Jan Calpone, VICE PRESIDENT/ EDITOR
2 Meminiee Dr, Oskosh, WI 54901
Phone: 92-023-1006

CLEAN OCEAN ACTION
MAIN OFFICE
P.O. Box 505
Sandy Hook, NJ 07732 USA
Phone: 732-872-0111   Fax: 732-872-8041
E-mail: sandyhook@cleanoceanaction.org
Website: www.cleanoceanaction.org

Founded: 1984
Scope: Regional

Description: A broad-based coalition of 175 conservation, fishing, diving, boating, real estate, student, and civic groups; over 300 businesses; and thousands of citizens concerned with the degraded waters off the New York and New Jersey coasts. COA uses education, research, and citizen action to pressure public officials to enact and enforce protective laws for our marine resources. Programs include: storm drain stenciling; regulatory reviews; contaminated sediments; and non-point source pollution.

Publication(s): Citizen Guide for Dredged Material Management, Wasting Our Waters Away, Annual Statewide Beach Sweep Report, Ocean Advocate, The Ocean is a flush away - NJ citizen guide to wastewater mgmt

Keyword(s): Coasts, Coral Reefs, Contaminated Sediments, Nonpoint Source Pollution

Contact(s):
Cindy Zipf, EXECUTIVE DIRECTOR
Dery Bennett, PRESIDENT
Pat Schneider, SECRETARY
Ben Forest, TREASURER
William Decamp, VICE PRESIDENT

CLEAN OCEAN ACTION
MID-COAST OFFICE
P.O. Box 1303
Tuckerton, NJ 08087 USA
Phone: 609-294-8040   Fax: 609-294-8044
E-mail: mmaxcoa@aol.com
Website: www.cleanoceanaction.org

Founded: NA
Scope: Local, Regional

Contact(s):
Martha Doyle, OFFICE ASSISTANT

CLEAN OCEAN ACTION
SOUTH JERSEY OFFICE
P.O. Box 1098
Wildwood, NJ 08260 USA
Phone: 609-729-7262   Fax: 609-729-3383
E-mail: AATotah@aol.com

Founded: NA
Scope: Local Region

CLEAN WATER ACTION
4455 Connecticut Ave., NW, Suite A300
Washington, DC 20008-2328 USA
Phone: 202-895-0420   Fax: 202-895-0438
E-mail: cwa@cleanwater.org
Website: www.cleanwateraction.org

Founded: NA
Scope: National

Description: The national citizen's organization working full-time for clean safe water at an affordable cost, control of toxic chemicals, and protection of our natural resources.

Contact(s):
Jim Pierce, DEVELOPMENT ASSOCIATE

CLEAN WATER FUND

4455 Connecticut Ave., NW
Washington, DC 20008 USA
Phone: 202-895-0420   Fax: 202-895-0438
E-mail: cleanwater@essential.org
Website: www.cleanwaterfund.org

Founded: NA
Scope: National

Description: Clean Water Fund is a 501 (c) (3) research, training and educational organization that advances environmental and consumer protection with a special focus on water pollution, toxic hazards, solid waste management, and natural resources.

Publication(s): SOLID WASTE: Expanding Rhode Island's Market with RI War on Waste, 1993, TOXICS: If Its Broke, Fit it, 1993, TOXICS: Toxic Metals in Batteries, 1992, WATER: Riches for Clean Up, Pennies for Prevention, 1993

Contact(s):
David Zwick, BOARD OF DIRECTORS, EXECUTIVE VICE PRESIDENT
dzwick@cleanwater.org
Peter Van Lockwood, BOARD OF DIRECTORS, PRESIDENT
Kathleen Aterno, BOARD OF DIRECTORS, TREASURER
katerno@cleanwater.org
Jim Pierce, DEVELOPMENT ASSOCIATE
jpierce@cleanwater.org

CLEAN WATER NETWORK, THE

1200 New York Avenue, NW, Suite 400
Washington, DC 20005 USA
Phone: 202-289-2395   Fax: 202-289-1060
E-mail: cleanwaternt@igc.org
Website: www.cwn.org

Founded: NA
Membership: 1000
Scope: National

Description: The Clean Water Network is a national alliance of over 1,000 organizations representing environmentalists, commercial fishers, anglers, surfers, family farmers, environmental justice advocates, faith communities, civic associations, boaters, labor unions, and recreational enthusiasts working together for cleaner waters.

Publication(s): Spills and Kills: Manure pollution in Americas livestock feedlogs, Spilling swills: A survey of factory form water pollution in 1999, Americas animal factories: House dates fail to prevent pollution from livestock waste, Wetlands for Clean Water: How wetlands protect rivers, lakes, and coastal waters from pollution, Prescription for Clean Water: How to meet the goals of the Clean Water Act

Contact(s):
Eddie Scher, CONTACT
escher@nrdc.org
Ami Grace, GRASS ROOTS DIRECTOR
Phone: 202-289-2421
agrace@nrdc.org
Merritt Frey, POLICY ANALYST/IDAHO
Phone: 208-345-7776
Lindsey Crist, PROGRAM ASSISTANT
Phone: 202-289-2422
lcrist@nrdc.org
Linda Young, SOUTHEAST FIELD COORDINATOR/FLORIDA
Phone: 850-222-9188

CLEVELAND MUSEUM OF NATURAL HISTORY, THE

1 Wade Oval Dr., University Cir.
Cleveland, OH 44106 USA
Phone: 216-231-4600   Fax: 216-231-5919
E-mail: botany@cmnh.org
Website: www.cmnh.org

Founded: 1920
Membership: 110
Scope: Local

Description: To instill an understanding of and appreciation for nature and inspire responsibility for conservation and stewardship of natural diversity. The Museum program areas include exhibits, publications, education, collections, research, and natural areas. The Museum owns a system of 22 sanctuaries.

Publication(s): Kirtlandia, Explorer

Keyword(s): Museum, Environmental and Conservation Education, Research

Contact(s):
James Bissell, COORDINATOR OF NATURAL AREAS

CLIMATE INSTITUTE

333 1/2 Pennsylvania Ave., SE
Washington, DC 20003 USA
Phone: 202-547-0104   Fax: 202-547-0111
E-mail: info@climate.org
Website: www.climate.org

Founded: 1986
Scope: International

Description: Designed to serve as a catalyst for international response and cooperation to address the threats posed by climate change and depletion of the stratospheric ozone layer. The Climate Institute operates as a bridge between scientists and policymakers with the intent of expediting policy responses to the challenges posed by human-induced climate change.

Publication(s): Coping with Climate Change, Forests in a Changing Climate, Climate Change in Asia, Environmental Exodus, Climate Alert

Keyword(s): Coasts, Energy, Sustainable Development, Urban Forestry

Contact(s):
John Topping, PRESIDENT
Phone: 202-547-0104
jtopping@climate.org

COALITION FOR CLEAN AIR

10780 Santa Monica Blvd., 210
Los Angeles, CA 90025 USA
Phone: 310-441-1544   Fax: 310-446-4362
Website: www.coalitionforcleanair.org

Founded: 1970
Membership: 1600
Scope: Statewide

Description: A nonprofit, tax-exempt organization dedicated to restoring clean, healthful air to Southern California residents through a combination of efforts including outreach and education, litigation, research, and policy advocacy.

Publication(s): Publications available on website, Clearing the Air

Keyword(s): Environmental Protection, Pollution Prevention, Air Quality and Pollution

Contact(s):
Tim Carmichael, EXECUTIVE DIRECTOR
Todd Campbell, POLICY DIRECTOR
David Allgood, PRESIDENT
Abby Arnold, TREASURER
Wendy James, VICE PRESIDENT

COALITION FOR EDUCATION IN THE OUTDOORS

E331 Park Center
S.U.N.Y. at Cortland Box 2000
Cortland, NY 13045 USA
Phone: 607-753-4968   Fax: 607-753-5982
E-mail: yaplec@snycorva.cortland.edu
Website: http://www.cortland.edu/ceo/

Founded: 1986
Scope: National

Description: The Coalition is composed of more than 100 businesses, institutions, organizations, associations, centers, agencies, and individuals affiliated in support of communicating and networking concerning education in, for and about the outdoors. The Coalition's magazine is a critically acclaimed education resource. The Coalition also conducts a biennial Outdoor Education Research Symposium.

Publication(s): Outdoor Education Research Symposium Proceedings, Taproot

Keyword(s): Environmental and Conservation Education, Environmental Ethics, National Parks, Outdoor Recreation, Wildlands, Outdoor Education

Contact(s):
Charles Yaple, EXECUTIVE COORDINATOR

COALITION FOR NATURAL STREAM VALLEYS, INC.

430 Orchard Rd.
Newark, DE 19711-5137 USA
Phone: 302-366-8059

Founded: NA
Scope: Regional

Description: The purpose of the Coalition for Natural Stream Valleys, Inc. is to promote the wise use of, and the preservation of natural stream valleys.

Contact(s):
Roland Roth, CHAIRMAN
Phone: 302-831-1300
Dorthy Miller, CORRESPONDING SECRETARY

COAST ALLIANCE

600 Pennsylvania Ave., SE
Washington, DC 20003 USA
Phone: 202-546-9554   Fax: 202-546-9609
E-mail: coast@coastalliance.org
Website: www.coastalliance.org

Founded: 1979
Membership: 600
Scope: National

Description: The Coast Alliance is a nonprofit public interest group dedicated to raising public awareness about our priceless coastal resources. Composed of concerned activists across the United States, the Coast Alliance provides information on activities affecting the nation's four coasts: the Atlantic, Pacific, Gulf of Mexico, and Great Lakes.

Publication(s): Storm on The Horizon: The National Flood Insurance Program and America's Coasts, Pointless Pollution, Muddy Waters, Mission Possible, Using Common Sense to Protect The Coasts: The Need to Expand The Coastal Barrier Resources System, Getting to The, And Two If By Sea: Fighting The Attack on America's Coasts

Keyword(s): Aquatic Habitats, Coasts, Environmental and Conservation Education, Land Use Planning, Coral Reefs, Runoff, Development, Contaminated Sediments

Contact(s):
Dery Bennett, CHAIRPERSON OF THE BOARD
American Littoral Society Bldg 18 Hartshorne Drive, Sandy Hook Highlands, NJ 07732
Phone: 732-872-8041
Jacqueline Savitz, EXECUTIVE DIRECTOR
Todd Miller, TREASURER AND SECRETARY
North Carolina Coastal Federation 3609 HWY 24, Newport, NC 28570
Phone: 252-393-8185
David Miller, VICE CHAIRPERSON
National Audubon 200 Trillium Lane, Albany, NY 12203
Phone: 518-869-9731

COASTAL CONSERVATION ASSOCIATION

4801 Woodway, Suite 220 West
Houston, TX 77056 USA
Phone: 713-626-4234   Fax: 713-626-5852
E-mail: ntl@joincca.org
Website: www.joincca.org

Founded: 1977
Membership: 75,000
Scope: National

Description: A national nonprofit corporation organized exclusively for the purpose of promoting and advancing the conservation, and protection of the marine, animal, and plant life both onshore and offshore along the coastal areas of the United States for the benefit and enjoyment of the general public.

Publication(s): Tide

Keyword(s): Aquatic Habitats, Coasts, Wildlife, Exotic species, Aquatic nuisance species, Sport Fishing, Wetlands, Water Quality

Contact(s):
Walter Fondren, CHAIRMAN OF THE BOARD
Doug Pike, EDITOR
David Dexter, EXECUTIVE DIRECTOR OF ALABAMA REGIONAL OFFICE
144 Florence Place, Mobile, AL 36607
Phone: 334-478-3474
Fax: 334-476-5214
Ted Forsgren, EXECUTIVE DIRECTOR OF FLORIDA REGIONAL OFFICE
905 East Park Ave., Tallahassee, FL 32301-2646
Phone: 850-224-3474
Fax: 850-224-5199
Jeff Angers, EXECUTIVE DIRECTOR OF LOUISIANA REGIONAL OFFICE
P.O. Box 373, Baton Rouge, LA 70821
Phone: 225-952-9200
Fax: 225-952-9204

Pat Keliher, EXECUTIVE DIRECTOR OF MAINE REGIONAL
OFFICE
40 Lafayette St., Yarmouth, ME 04096
Phone: 207-846-1015
Fax: 207-846-1168
Austin Ragsdale, EXECUTIVE DIRECTOR OF NORTH
CAROLINA REGIONAL OFFICE
3701 National Drive Suite 217, Raleigh, NC 27612
Phone: 919-781-3474
Fax: 919-781-3475
Scott Whitaker, EXECUTIVE DIRECTOR OF SOUTH
CAROLINA REGIONAL OFFICE
P.O. Box 290640, Columbia, SC 29229
Phone: 803-865-4164
Fax: 803-865-5104
Kevin Daniels, EXECUTIVE DIRECTOR OF TEXAS
REGIONAL OFFICE
4801 Woodway, Suite 220W, Houston, TX 77056
Phone: 713-626-4222
Fax: 713-961-3801
Richard Welton, EXECUTIVE DIRECTOR OF VIRGINIA
REGIONAL OFFICE
2100 Marina Shores Dr., Suite 108, Virginia Beach, VA 23451
Phone: 757-481-1226
Fax: 757-481-6910
David Cummins, PRESIDENT
Will Ohmstede, VICE CHAIRMAN
Alex Jernigan, VICE CHAIRMAN
Gus Schram, VICE PRESIDENT

COASTAL CONSERVATION ASSOCIATION GEORGIA

515 Denmark St
Suite 300
Statesboro, GA 30458 USA
Phone: 800-266-0693   Fax: 912-764-6497
E-mail: ccaga.org
Website: www.ccaga.org

Founded: 1987
Membership: 1200+
Scope: Statewide

Description: The CCAG promotes conservation through
education—promoting, protecting and enhancing the availability of marine, animal, plant life and other coastal resources for
the benefit and enjoyment of the general public.

Publication(s): Tidelines (newsletter), Tide Magazine

Keyword(s): Coasts, Conservation, Conservation Plannning,
Environmental and Conservation Education, Fish Wildlife
Management, Wildlife, Marine Conservation, Preservation and
Protection, Water Quality

Contact(s):
Martin Nesmith, CHAIRMAN
Phone: 912-739-1744
Fax: 912-739-4889
William Phillips, VICE-CHAIRMAN
Phone: 912-764-6567
Fax: 912-764-6568
ringo@bulloch.com

COASTAL GEORGIA LAND TRUST INC.

428 Bull St., Suite 210
Savannah, GA 31401 USA
Phone: 912-231-0507   Fax: 912-231-1143
E-mail: cglt@bellsouth.net
Website: www.cglt.org

Founded: 1993
Membership: 200
Scope: Regional

Description: The mission of the Coastal Georgia Land Trust, Inc.,
a nonprofit organization, is to promote the responsible
stewardship and preservation of land in coastal Georgia.

Publication(s): Coastal Georgia Land Trust, Brochure, Coastal
Georgia Land Trust Newsletter, biannual newsletter

Keyword(s): Conservation Easements, Habitat Conservation,
Historic Preservation, Land Conservation, Land Preservation,
Land Protection, Native Plants, Natural Areas, education,
National Parks, Open Space, Stewardship, Sustainable
Development, Wetlands, training

Contact(s):
Mary Elfner, EXECUTIVE DIRECTOR
Rhett Mouchet, PRESIDENT OF THE BOARD OF
DIRECTORS
Alan Bailey, VICE PRESIDENT, BOARD OF DIRECTORS
Phone: 912-925-3159
Fax: 912-927-9766
acbailey@worldnet.atf.net

COASTAL SOCIETY, THE

P.O. Box 25408
Alexandria, VA 22313-5408 USA
Phone: 703-768-1599   Fax: 703-768-1598
E-mail: coastal@aol.com
Website: www.coastalsociety.org

Founded: 1975
Membership: 300
Scope: International

Description: The Coastal Society is an organization of private
sector, academic and governmental professionals and
students dedicated to actively addressing emerging coastal
issues, fostering dialog, forging partnerships and promoting
communication and education.

Publication(s): conference proceedings, Coastal Society, The
Bulletin

Keyword(s): Coasts, Environmental and Conservation Education,
Sustainable Development, Coral Reefs, Wetlands

Contact(s):
Judy Tucker, EXECUTIVE DIRECTOR
Megan Bailiff, PAST PRESIDENT
Nicholas School for the Environment, Duke University, 135
Duke Marine Lab Rd., Beaufort, NC 28516-9720
Walter Clark, PRESIDENT
University of Washington, Box 355060,
Seattle, WA 98105-5060
Phone: 206-685-1108
John Duff, PRESIDENT-ELECT
University of Maine, Portland, ME 04102-2898
Robert Boyles, SECRETARY
University of Rhode Island, Kingston, RI 301-713-3155
William Hall, TREASURER
3635 Fremont Ave. North, 307, Seattle, WA 98103
Phone: 406-442-4002

COEREBA SOCIETY

7336 16th Ave. SW
Seattle, WA 98106 USA
Phone: 206-768-8827
E-mail: info@coereba.org
Website: www.coereba.org

Scope: International

Description: Seattle-based environmental nonprofit education about nature and its conservation in Puerto Rico region and the greater Caribbean, especially through mass media channels and environmental interpretation.

Contact(s):
Jose Placer, EXECUTIVE DIRECTOR
jplacer@coereba.org

COLORADO ASSOCIATION OF SOIL CONSERVATION DISTRICTS

Attn: President, 1858 M Rd.
Fruita, CO 81521 USA
Phone: 970-858-7165   Fax: 970-858-7165

Founded: NA
Scope: National

Keyword(s): Conservation Districts

Contact(s):
John Freziers, BOARD MEMBER/EXEC. DIRECTOR
1858 M Rd., Fruita, CO 81521
Phone: 970-858-7165
Robert Cordova, PRESIDENT, ALTERNATE BOARD MEMBER
18105 Enoch Rd., Colorado Springs, CO 80930
Phone: 719-683-2126
Lee Campbell, SECRETARY AND TREASURER
1603 Eastlawn Ave., Durango, CO 81301
Phone: 970-247-1496
Fax: 970-385-7910
Jim Rossi, VICE PRESIDENT
P.O. Box 247, Oak Creek, CO 80467
Phone: 970-638-4459

COLORADO B.A.S.S. CHAPTER FEDERATION

Attn: President, 4485 Enchanted Circle N.
Colorado Springs, CO 80917 USA
Phone: 719-597-2304
Website: www.coloradobassfederation.org

Founded: NA
Membership: 300
Scope: Statewide

Description: An organization of Bassmaster chapters, affiliated with the Bass Anglers Sportsman Society, organized to fight pollution, assist state and national conservation agencies in their efforts, and teach the young people of our country good conservation practices. Dedicated to the realistic conservation of our water resources.

Contact(s):
Bernie Stein, CONSERVATION DIRECTOR
1218 N 3rd St., Johnstown, CO 80534
Phone: 970-587-9163
John Bentz, PRESIDENT
Phone: 719-597-2304

COLORADO ENVIRONMENTAL COALITION

1536 Wynkoop
Denver, CO 80202 USA
Phone: 303-534-7066   Fax: 303-534-7063
E-mail: infocec@cecenviro.org
Website: www.ourcolorado.org

Founded: 1965
Scope: Statewide

Description: The Colorado Environmental Coalition is the grass roots action arm of Colorado's environmental movement. The Coalition coordinates the conservation community and mobilizes citizen constituencies behind environmental campaigns to preserve wilderness, wildlife, and a sustainable way of life.

Publication(s): Colorado Environmental Handbook-State of the State, Conservationist's Wilderness Proposal for BLM Lands, Colorado Environmental Report

Keyword(s): Public Lands, Wilderness, training, Sustainable Development, Environmental Health

Contact(s):
Jeff Widen, ASSOCIATE
Phone: 970-385-8509
widen@cecenviro.org
Carter Johnson, CIRCUIT RIDER
trey@cecenviro.org
Elise Jones, EXECUTIVE DIRECTOR
sjtix@cecenviro.org
Monica Piergrossi, FRONT RANGE FIELD DIRECTOR
monica@cecenviro.org
John Powers, PRESIDENT
Pete Kolbenschlag, WEST SLOPE FIELD ORGANIZER
1000 N 9th St., 29, Grand Junction, CO 81501
Phone: 970-243-0002
pete@cecenviro.org

COLORADO FORESTRY ASSOCIATION

P.O. Box 270132
Ft. Collins, CO 80527 USA
Phone: 719-566-1648

Founded: 1982
Membership: 175
Scope: Statewide

Description: A statewide organization affiliated with the National Woodland Owners Association, concerned with forest ecology and advocating a forest-perpetuating balance between preservation and harvest of Colorado forests.

Publication(s): Colorado Forestry

Keyword(s): Forests and Forestry
Contact(s):
John Oram, EDITOR
Phone: 303-477-0552
Vincent Calderon, PRESIDENT
38 Dartmouth Ave., Pueblo, CO 81005
Phone: 719-566-1648
Ken Ashley, SECRETARY
5227 South County Rd. 7, Fort Collins, CO 80528
Phone: 970-223-3255
Edwin Olmsted, TREASURER
1065 West 112th Ave. B, Northglenn, CO 80234
Phone: 303-452-8643
Ken Ashley, VICE PRESIDENT
Ft. Collins, CO

COLORADO NATURAL HERITAGE PROGRAM

254 General Services Bldg., Colorado State University
Ft. Collins, CO 80523 USA
Phone: 970-491-1309   Fax: 970-491-3349
E-mail: heritage@lamar.colostate.edu
Website: www.cnhp.colostate.edu

Founded: 1979
Scope: Statewide

Description: The mission of the Colorado Natural Heritage Program is to preserve the natural diversity of life by contributing the scientific foundation that leads to lasting conservation of Colorado's biological wealth.

Publication(s): Colorado Rare Plant Guide, Colorado Conservation Status Handbook, Rare and Imperiled Animals, Plants, and Plant Communities of Colorado

Keyword(s): Biodiversity, Land Use Planning, Land Management, Ecology

Contact(s):
Boyce Drummond, DIRECTOR
Phone: 970-491-1309
Fax: 970-491-3349
heritage@lamar.colostate.edu

COLORADO TRAPPERS ASSOCIATION

0250 County Rd 127
Glenwood Springs, CO 81601 USA
Phone: 970-945-7193   Fax: 970-945-0449

Founded: 1975
Membership: 400
Scope: Regional

Description: Associate of Fur Takers of America and National Trappers Association. Dedicated to the wise conservation and management of furbearing animals, the education of fur harvesters and public about furbearer management and the preservation of America's rich heritage in the harvest of wild furs.

Publication(s): Fur Marketing and Trappers Supply Handbook, Managing Rocky Mountain Furbearers

Keyword(s): Agriculture, Hunting, Predators, Trapping

Contact(s):
Maj. Boddicker, DIRECTOR OF PUBLICATIONS
Maj. Boddicker, EDITOR
Eddie Montoya, METRO DIRECTOR
Al Davidson, PRESIDENT
P.O. Box 625, Saguache, CO 81149
Phone: 719-655-2777
Kandy Herrman, SECRETARY
0250 County Rd 127, 19, Glenwood Springs, CO 81601
Phone: 970-945-7193
Fax: 970-945-0449
Darla Jackson, TREASURER
Phone: 719-643-5263
Marvin Miller, VICE PRESIDENT
29156 Summit Ranch Dr., Golden, CO 80401
Phone: 303-526-9207

COLORADO WATER CONGRESS

1580 Logan St., Suite 400
Denver, CO 80203 USA
Phone: 303-837-0812   Fax: 303-837-1607
E-mail: macravey@cowatercongress.org
Website: www.cowatercongress.org

Founded: 1958
Scope: Statewide

Description: To institute and advance programs for the conservation, development, protection, and efficient utilization of the water resources of Colorado.

Publication(s): Colorado Water Almanac & Directory, Water Intelligence Report, Water Legal News, Water Legislative Report, Water Research News, Water Special Report, Water Quality News, Colorado Laws Enacted of Interest to Water Users, Colorado Water Rights

Contact(s):
Richard Macravey, EXECUTIVE DIRECTOR
Rod Kuharich, PRESIDENT
Daniel Birch, VICE PRESIDENT

 ## COLORADO WILDLIFE FEDERATION

445 Union Blvd., Suite 302
P.O. Box 280967
Lakewood, CO 80228-1243 USA
Phone: 303-987-0400   Fax: 303-987-0200
E-mail: cfw@coloradowildlife.org
Website: coloradowildlife.org

Founded: 1953
Membership: 5000
Scope: Statewide

Description: A representative statewide organization, affiliated with the National Wildlife Federation, dedicated to the protection and enhancement of wildlife and its habitat through public education and government interaction.

Publication(s): Colorado Wildlife

Contact(s):
Dennis Buechler, CHAIR AND ALTERNATE REPRESENTATIVE
Barbara Young, EDITOR
Mike Brogan, EDUCATION PROGRAMS CONTACT
Jody Flemming, EXECUTIVE DIRECTOR
Colleen Gadd, REPRESENTATIVE
Jim Goddard, TREASURER

COLORADO WILDLIFE HERITAGE FOUNDATION

6060 Broadway
Denver, CO 80216 USA
Phone: 303-291-7212   Fax: 303-291-7106

Founded: 1989
Scope: Statewide

Description: The Colorado Wildlife Heritage Foundation has been endorsed by four Colorado governors. The foundation's objectives are threefold: (1) environmental education, (2) habitat acquistion and management, and (3) wildlife research. Where appropriate, the foundation pursues projects with the support and expertise of the Colorado Division of Wildlife.

Publication(s): The Colorado Wildlife Viewing Guide

Keyword(s): Endangered Species, Environmental and Conservation Education, Land Purchase, Sustainable Ecosystems, training

Contact(s):
Karin Ballard, ASSOCIATE EXECUTIVE DIRECTOR
James Cowperthwaite, CHAIRMAN
378 S. Pontiac Way, Denver, CO 80224
Phone: 303-355-3957
Charles Warren, DIRECTOR
333 Logan St., Denver, CO 80203
Phone: 303-778-7797
Fax: 303-698-5091
Edward Alexander, EXECUTIVE DIRECTOR
Colorado Wildlife Heritage Foundation, 6060 Broadway,
Denver, CO 80216
Phone: 303-291-7416
Terry Combs, PRESIDENT
American Cargo Handling, P.O. Box 17594, Denver, CO
80217
Phone: 303-398-2416
Fax: 303-322-6142
Linda Hamlin, SECRETARY
100 Dexter St., Denver, CO 80220
Phone: 303-388-8176
Buck Hutchison, TREASURER
Hutchison Western P.O. Box 1158, Adams City, CO 80022
Phone: 303-287-2826
Fax: 303-289-3286

COLUMBIA BASIN FISH AND WILDLIFE AUTHORITY

2501 SW 1st Ave. Suite 200
Portland, OR 97201 USA
Phone: 503-229-0191　　　Fax: 503-229-0443
Website: www.cbfwf.org

Founded: 1982
Scope: Regional

Description: A regional association of all the fish and wildlife
agencies (two federal, five state) and Indian tribes (13 in the
Columbia River Basin (Idaho, Montana, Oregon, and
Washington). Established to coordinate planning and implementation of the fish and wildlife provisions of the Pacific
Northwest Electric Power Planning and Conservation Act and
for oversight of fish and wildlife resource management under
the Fish and Wildlife Coordination Act and other authorities.
Current charter: 1987.

Keyword(s): Wildlife, accreditation, training, Fish Wildlife
Management

Contact(s):
Brian Allee, EXECUTIVE DIRECTOR
Phone: 503-229-0191
Fax: 503-229-0443
brian@cbfwf.org

COLUMBIA ENVIRONMENTAL RESEARCH CENTER

USGS-BRD-ECRC, 4200 New Haven Rd.
Columbia, MO 65201-9634 USA
Phone: 573-875-5399　　　Fax: 573-876-1896

Founded: NA
Membership: 125
Scope: National

Contact(s):
Pam Haverland, PAST-PRESIDENT
pamela_haverland@usgs.gov

COMMITTEE FOR NATIONAL ARBOR DAY

Attn: National Chairman, 63 Fitzrandolph Rd.
West Orange, NJ 07052 USA

Founded: 1936
Scope: National

Description: To establish a unified national observance date on
the last Friday in April.

Keyword(s): Trees

Contact(s):
Edward Scanlon, HONORARY NATIONAL CHAIRMAN
P.O. Box 38247, Olmsted Falls, OH 44138
Harry Banker, NATIONAL CHAIRMAN
63 Fitzrandolph Rd., West Orange, NJ 07052

COMMITTEE ON AGRICULTURAL SUSTAINABILITY FOR DEVELOPING COUNTRIES

10 G Street, NE, Suite 800
Washington, DC 20002 USA
Phone: 202-729-7600　　　Fax: 202-729-7610
Website: www.wri.org

Founded: 1987
Scope: National

Description: The Committee on Agricultural Sustainability aims to
increase support for farmers in developing countries working to
improve their living conditions and feed the world's expanding
population without destroying the soil and water supplies on
which agricultural productivity ultimately depends.

Keyword(s): Agriculture, Developing Countries, Soil
Conservation, Sustainable Development

Contact(s):
Jonathan Lash, PRESIDENT

COMMUNITIES FOR A BETTER ENVIRONMENT

1611 Telegraph Ave.
Suite 450
Oakland, CA 94612 USA
Phone: 510-302-0430　　　Fax: 510-302-0437
Website: cbecal.org

Founded: 1971
Membership: 15,000
Scope: Statewide

Description: The CBE is a nonprofit, multiracial environmental
health organization working to prevent public exposure to toxic
chemical pollutants. CBE has over 19 years experience in the
California environmental arena. CBE uses science-based
research, legal tactics, and organizing strategies to prevent air
and water pollution, to eliminate toxic hazards, and to improve
the health of the people of California.

Publication(s): Oil Rag, Environmental Review

Keyword(s): Air Quality and Pollution, Environmental Justice,
Environmental Law, Internships, Coral Reefs

Contact(s):
Stephanie Pincetl, BOARD PRESIDENT
Richard Drury, EXECUTIVE DIRECTOR
Everett Delano, SECRETARY

COMMUNITY CONSERVATION CONSULTANTS/HOWLERS FOREVER, INC.

50542 One Quiet Lane
Gays Mills, WI 54631 USA
Phone: 608-735-4717   Fax: 608-735-4765
E-mail: ccc@mwt.net
Website: www.communityconservation.org

Founded: 1989
Membership: 1
Scope: International

Description: Specializing in catalyzing of community-based conservation initiatives and designing for their sustainability. Active in Wisconsin, Belize, and India. Coordination of volunteers for projects.

Publication(s): check web site for pub listings

Keyword(s): Conservation, Local Resource Conservation

Contact(s):
Rob Horwich, DIRECTOR

COMMUNITY ENVIRONMENTAL COUNCIL

930 Miramonte Dr.
Santa Barbara, CA 93109 USA
Phone: 805-963-0583   Fax: 805-962-9080
E-mail: cecadmin@rain.org
Website: www.communityenvironmentalcouncil.org

Founded: 1970
Membership: 1000
Scope: Local

Description: CEC's primary goal is to serve as a connecting institution linking government agencies, business and industry, universities and regulatory bodies, environmental organizations, and the community. Using Santa Barbara as its urban laboratory, CEC conducts research and develops local programs in recycling, hazardous waste, sustainable agriculture, and environmental education.

Publication(s): Manufacturing with Recyclables, A Question of Responsibility: Recycling Market Development, Gildea Review

Keyword(s): Gardening and Horticulture, Land Use Planning, Solid Waste Management, Sustainable Development, Toxic Substances, Nuclear-free, Water quantity, Water export and diversion

Contact(s):
Laurence Laurent, EXECUTIVE DIRECTOR
Kim Kimbell, PRESIDENT
Sarita Vasquez, VICE PRESIDENT

COMMUNITY RIGHTS COUNSEL

1726 M St., NW, Suite 703
Washington, DC 20036-4524 USA
Phone: 202-296-6889   Fax: 202-296-6895
E-mail: crc@communityrights.org
Website: www.communityrights.org

Founded: 1997
Membership: 4
Scope: National

Description: CRC is a public interest law firm defending laws that make our communities healthier, more livable, and socially just.

Publication(s): Nothing for Free: How Private Judicial Seminars are Undermining Environmental Protection and Breaking the Public's Trust, Hostile Environment: How Activist Federal Judges Threaten our Air, Water, Land, Takings Litigation Handbook: The Takings Project: Using Federal Courts to Attack Community and Environmental Protections. Douglas T. Kendall, Charles P. Lord, April 1998

Keyword(s): Endangered Species, Environmental Law, Environmental Planning, Land Preservation, Land Use Planning, Planning Management, Open Space, Sustainable Development, Urban Environment, Wetlands

CONCERN, INC.

1794 Columbia Rd., NW
Washington, DC 20009 USA
Phone: 202-328-8160   Fax: 202-387-3378
E-mail: concern@igc.org
Website: www.sustainable.communitiesnetwork

Founded: 1970
Membership: 3
Scope: National

Description: A nonprofit, tax-exempt organization which provides environmental information to individuals and groups. Concern's publications give an overview of the issue and include guidelines to encourage and aid citizen participation in the community and in policy decisions at the local, state, and federal levels of government. Also are developing a community sustainability program which includes a database on sustainability.

Publication(s): Sustainability In Action, Building Sustainable communities, Community Action Guides on Pesticides, Drinking Water, Farmland, Waste, Household Waste, and Global Warming

Keyword(s): Environmental and Conservation Education, Pesticides, Solid Waste Management, Sustainable Development, Exotic species, Aquatic nuisance species

Contact(s):
Burks Lapham, CHAIR
Susan Boyd, EXECUTIVE DIRECTOR

CONFEDERATED SALISH AND KOOTENAI TRIBES

P.O. Box 278
Pablo, MT 59855 USA
Phone: 406-675-2700   Fax: 406-675-2739
E-mail: info@cskt.org
Website: www.cskt.org

Founded: NA
Membership: 7000
Scope: National

Description: The 1.25 million acre Flathead Indian Reservation was created in 1855 by the Treaty of the Hellgate as a homeland for the Salish, Kootenai, and Pend d' Oreille Tribes. The constitutional government of the Confederated Salish and Kootenai Tribes was formed in 1934 and approved by the Secretary of the Interior in 1935 to establish a more responsible organization, promote our general welfare, conserve and develop our land and resources, and secure to ourselves and our posterity the power to exercise certain rights of self-government. The Tribal Natural Resources Department includes the divisions of Water, Environmental Protection, Lands, and Fisheries, Wildlife, Recreation, and Conservation. The Tribal Forestry Department is responsible for forest management activities on the Reservation.

Publication(s): Char-Koosta News

Keyword(s): Cultural Preservation, Protected Areas, Land Use Planning, Renewable Resources

Contact(s):
Sandra Morigeau, EXECUTIVE SECRETARY
Gary Orr, FORESTRY DEPARTMENT HEAD
Rhonda Swaney, NATURAL RESOURCES DEPARTMENT HEAD
rhondas@cskt.org
Fred Matt, TRIBAL CHAIRMAN

CONNECTICUT ASSOCIATION OF CONSERVATION DISTRICTS, INC.
106 Sherman Lee Dr
Middletown, CT 06457 USA
Phone: 860-647-6379

Founded: NA
Scope: Statewide

Keyword(s): Conservation Districts

Contact(s):
Ann Hadley, PRESIDENT
Tony Inch, SECRETARY/TREASURER
134 Heather Lane, Wilton, CT 06897
Phone: 203-762-9994
John Breakell, VICE PRESIDENT
Phone: 860-491-2243

CONNECTICUT AUDUBON SOCIETY, INC.
118 Oak St.
Hartford, CT 06106 USA
Phone: 860-527-8737   Fax: 860-549-3094
E-mail: codonnell@ctaudubon.org
Website: www.ctaudubon.org

Founded: 1898
Scope: Statewide

Description: Dedicated to environmental education to conserve natural resources. Sanctuary acquisition and management, legislative action, and wildlife and natural areas research.

Publication(s): The Connecticut Audubon News - quarterly newsletter

Keyword(s): Environmental and Conservation Education, Environmental Protection, Open Space, Sustainable Development, training

Contact(s):
David Engelman, CHAIRMAN OF THE BOARD
Andrew Griswold, DIRECTOR ECO TRAVEL
67 Main St., Essex, CT 06426
Phone: 860-767-0660
Fax: 860-767-9988
Christopher Nevins, DIRECTOR OF BIRDCRAFT MUSEUM
314 Unquowa Rd., Fairfield, CT 06430
Phone: 203-259-0416
Fax: 203-259-1344
Alison Olivieri, DIRECTOR OF DEVELOPMENT
Phone: 203-254-3315
Lisa Santacroce, DIRECTOR OF ENVIRONMENTAL AFFAIRS
Phone: 860-527-6750
James Sirch, DIRECTOR OF FAIRFIELD CENTER
2325 Burr St., Fairfield, CT 06430
Phone: 203-259-6305

Judy Harper, DIRECTOR OF GLASTONBURY CENTER
1361 Main St., Glastonbury, CT 06033
Phone: 860-633-8402
Fax: 860-659-9467
Ann Guion, DIRECTOR OF POMFRET CENTER
189 Pomfret St., Pomfret Center, CT 06259
Phone: 860-928-4948
Cathy O'Donnell, EDITOR
Phone: 203-254-1092
Debbie Dubitsky, EDUCATOR ROLLING NATURE CENTER
118 Oak St., Hartford, CT 06106
Phone: 860-246-6285
Anne Harper, PRESIDENT
2325 Burr St., Fairfield, CT 06430
Phone: 203-254-4467
Fax: 203-254-7673
Eileen Fielding, STATEWIDE DIRECTOR OF EDUCATION
118 Oak St., Hartford, CT 06106
Phone: 860-524-8012
Fax: 860-549-3094
Todd Russo, TEACHER/NATURALIST
1361 Main St., Glastonbury, CT 06033
Phone: 860-633-8402
Fax: 860-659-9467
Mary Williams, TEACHER/NATURALIST
Chris Krumperman, TEACHER/NATURALIST
1361 Main St., Glastonbury, CT 06033
Jeff Weiler, TEACHER/NATURALIST
189 Pomfret St., Pomfret Center, CT 06259
Phone: 860-928-4948
Kasha Breau, TEACHER/NATURALIST
1361 Main St., Glastonbury, CT 06033
Phone: 860-633-8402
Richard Julian, TEACHER/NATURALIST
1 Milford Point Rd., Milford, CT 06460
Phone: 203-878-7440
Peter Kunkel, VICE PRESIDENT
Judith Richardson, VICE PRESIDENT
Duffy Schade, VICE PRESIDENT
Jack Tierney, VICE PRESIDENT
W. Morehouse, VICE PRESIDENT OF LEGAL

CONNECTICUT B.A.S.S. CHAPTER FEDERATION
Attn: President, 119 Straitsville Rd.
Prospect, CT 06712 USA
Phone: 203-758-0069
Website: www.geocities.com/yosemite/rapids/8723

Founded: NA
Scope: Statewide

Description: An organization of Bassmaster chapters, affiliated with the Bass Anglers Sportsman Society, organized to fight pollution, assist state and national conservation agencies in their efforts, and teach the young people of our country good conservation practices. Dedicated to the realistic conservation of our water resources.

Contact(s):
Lee Johnson, CONSERVATION DIRECTOR
155 Candlewood Lake Rd. North, New Milford, CT 06776
Phone: 860 3-50 -1368
Jon Puhalski, ENVIRONMENTAL DIRECTOR
53 Overlook Road, Winstead, CT 06098
Phone: 860 3-79 -9387
Tom Reynolds, PRESIDENT
309 Hamburg Road, Lyme, CT 06371
Phone: 863 4-34 -7677

Jim Marenzana, SECRETARY
40 South Street Unit 17c, Bristol, CT 06010
J.Haren@snet.net
Ron Murack, TOURNAMENT DIRECTOR
P.O. Box 416, Granby, CT 06035
Phone: 860 6-53 -6397
Joe Rackiewicz, TREASURER
21 Birch Place, Milford, CT 06460
Phone: 203 8-78 -8909
Ken Bell, VICE PRESIDENT
21 Cloud Street, Enfield, CT 06082
Phone: 860 7-49 -2044
gretafudge@earthlink.net

CONNECTICUT BOTANICAL SOCIETY

CBS
New Haven, CT 06532 USA
Phone: 860-633-7557
Website: www.ct-botanical-society.org

Founded: 1903
Membership: 300
Scope: Local

Description: The Society increases knowledge of the state's flora accumulate and maintains specimens and records for a permanent botanical record. The Society also recommends botanically significant areas for protection and supports scholarly botanical research.

Publication(s): Newsletter, Yearbook, The Vascular Flora of Southeastern Connecticut

Keyword(s): Conservation, Endangered Species, Environmental Protection, Flowers, Plants, and Trees, National Parks, Wetlands

Contact(s):
Casper Ultee, PRESIDENT
Phone: 860-633-7557
casperu@aol.com
Karen Sexton, SECRETARY
Phone: 860-228-4647
Paul Stetson, TREASURER
Carol Lemmon, VICE PRESIDENT
Phone: 203-488-7813

CONNECTICUT FOREST AND PARK ASSOCIATION

Middlefield, 16 Meriden Rd.
Rockfall, CT 06481-2961 USA
Phone: 860-346-2372   Fax: 860-347-7463
E-mail: conn.forest.assoc@snet.net
Website: www.ctwoodlands.org

Founded: 1895
Membership: 2000
Scope: Statewide

Description: A representative statewide organization, affiliated with the National Wildlife Federation and the National Woodland Owners Association, dedicated to the protection and enhancement of wildlife and its habitat through public education and government interaction.

Contact(s):
Adam Moore, EXECUTIVE DIRECTOR
Richard Whitehouse, PRESIDENT
Ruth Cutler, REPRESENTATIVE
Ron Manzi, TREASURER

CONNECTICUT FUND FOR THE ENVIRONMENT

1032 Chapel St., 3rd Fl.
New Haven, CT 06510 USA
Phone: 203-787-0646   Fax: 203-787-0246
Website: www.cfenv.org

Founded: NA
Scope: Statewide

Description: CFE is a nonprofit group dedicated to protecting Connecticut's natural resources through legal action, education and scientific investigation.

Publication(s): Fact Sheets, Annual Reports, Newsletter

Keyword(s): Air Quality and Pollution, Open Space, Wetlands, Energy, Toxic Substances, Nuclear-free, Water quantity, Water export and diversion, Coral Reefs

Contact(s):
Donald Strait, EXECUTIVE DIRECTOR
Nancy Faesy, SECRETARY
Thomas Holloway, TREASURER
Michael Kashgarian, VICE-PRESIDENT

CONNECTICUT PUBLIC INTEREST RESEARCH GROUP (CONN PIRG)

198 Park Rd. 2 floor
W. Hartford, CT 06119 USA
Phone: 860-233-7554   Fax: 860-233-7574
Website: http://www.connpirg.org

Founded: 1972
Membership: 31,000
Scope: Statewide

Description: Works for concrete solutions to improve and protect our environment. Engaged in public education, study, and legislative action in many areas of the environment, including water and air pollution and solid waste.

Publication(s): ConnPIRG Reports

Keyword(s): Air Quality and Pollution, Endangered Species, Solid Waste Management, Toxic Substances, Nuclear-free, Water quantity, Water export and diversion, Water Pollution Management

CONNECTICUT RIVER WATERSHED COUNCIL INC.

15 Bank Row
Greenfield, MA 01301 USA
Phone: 413-529-9500   Fax: 413-772-2090
E-mail: crwc@crocker.com
Website: www.ctriver.org

Founded: 1952
Membership: 2500
Scope: Regional

Description: A member-supported nonprofit organization, CRWC is a regional voice for improvement and protection of the Connecticut River and water resources throughout the 11,260 square-mile, four-state river basin of Vermont, New Hampshire, Massachusetts, and Connecticut. CRWC participates in relevant environmental and resource allocation issues through its land conservancy, water quality improvement, and watershed stewardship programs. Land conservancy revolving loan fund. Conservation education and research grant fund.

Publication(s): Complete Boating Guide to the Connecticut River, Currents and Eddies

Keyword(s): Watersheds, Land Use Planning, Rivers, Coral Reefs, Water Quality, Environmental Protection, Wetlands

Contact(s):
Neil Sheridan, CHAIRMAN
Tom Miner, EXECUTIVE DIRECTOR
Nancy Rogers, SECRETARY
Erling Heistad, VICE CHAIR
Andy Wizner, VICE CHAIR

CONNECTICUT WATERFOWL ASSOCIATION, INC.
P.O. Box 74
Bozrah, CT 06334-0074 USA
Phone: 860-535-8482   Fax: 203-838-8423
E-mail: ducks@uconect.net
Website: www.uconect.net/~ducks

Founded: 1967
Scope: Statewide

Description: To preserve, reclaim, and enhance wetland and wildlife habitat in the state of Connecticut in a manner that promotes the wise use of our natural resources and the progress of our society.

Publication(s): Connecticut Waterfowl and Wetlands

Keyword(s): Environmental and Conservation Education, Waterfowl, Wetlands, training

Contact(s):
N. Patricia Hochman, PRESIDENT
Patti Holmes, SECRETARY
Paul Capotosto, TREASURER
Paul Rothbart, VICE PRESIDENT

CONSERVANCY OF SOUTHWEST FLORIDA, THE
1450 Merrihue Dr.
Naples, FL 34102-3449 USA
Phone: 941-262-0304   Fax: 941-262-0672
E-mail: info@conservancy.org
Website: www.conservancy.org

Founded: 1964
Membership: 6000
Scope: Local, Regional

Description: Leading the challenge to protect and sustain Southwest Florida's natural environment through environmental policy, science and education. The Conservancy manages two nature centers, offers learning adventures, rehabilitates injured wildlife, monitors sea turtles and acquires land.

Publication(s): Eye on the Issues, Yearbook, Learning Adventures, Update

Keyword(s): Biodiversity, Conservation Biology, Endangered Species, Environmental and Conservation Education, Nature Centers, Internships, Land Preservation, Museum, Sea Turtles, accreditation

Contact(s):
Tracy Zanpaglione, DIRECTOR OF COMMUNICATIONS MARKETING
tracyz@conservancy.org
Steve Bortone, DIRECTOR, ENVIRONMENTAL SCIENCE
steveb@conservancy.org
Kathy Prosser, PRESIDENT AND CEO
kaythyp@conservancey.org

E. Louise Taylor, SCHOOL PROGRAMS MANAGER
Phone: 941-403-4239
Fax: 941-263-3019
Michael Simonik, VP, ENVIRONMENTAL POLICY
michaels@conservancy.org

CONSERVATION COUNCIL FOR HAWAII
PMB-203, 111 E. Puainako St., Suite. 585
Hilo, HI 96720 USA
Phone: 808-968-6360   Fax: 808-968-0896
E-mail: cch@aloha.net
Website: www.conservation-hawaii.org

Founded: NA
Scope: Statewide

Description: A representative statewide organization, affiliated with the National Wildlife Federation, dedicated to the protection and enhancement of wildlife and its habitat through public education and government interaction.

Publication(s): The Hawai'I Conserver

Contact(s):
Janet Dellaria, ALTERNATE REPRESENTATIVE
Karen Blue, EXECUTIVE DIRECTOR AND EDITOR
Steven Montgomery, PRESIDENT
Kate Schuerch, REPRESENTATIVE & TREASURER

CONSERVATION COUNCIL OF NORTH CAROLINA
Raleigh, NC 27605 USA
Phone: 919-839-0006   Fax: 919-839-0767
E-mail: ccnc@bellsouth.net
Website: www.serve.com/ccnc

Founded: 1968
Membership: 700
Scope: Statewide

Description: To initiate, participate in, and coordinate local, regional, state, and national action in environmental and energy matters and in conservation and environmental education.

Publication(s): Legislative Scorecard, Carolina Conservationist Newsletter

Keyword(s): Energy, Environmental and Conservation Education, Air Quality and Pollution, Coasts, Nuclear/Radiation, Rivers, Solid Waste

Contact(s):
Maureen Sutton, EDITOR
345 N. Page, Southern Pines, NC 28387
Carrie Oren, EXECUTIVE DIRECTOR
Phone: 919-839-0006
Dan Besse, POLITICAL DIRECTOR
Nina Szlosberg, PRESIDENT
Laura Lauffer, VICE PRESIDENT

CONSERVATION FEDERATION OF MARYLAND/ F.A.R.M.

P.O. Box 455
Poolesville, MD 20837 USA
Phone: 301-916-3510   Fax: 301-349-5941
E-mail: f.a.r.m@erols.com
Website: www.darnet.com

Founded: NA
Membership: 3000
Scope: Regional

Description: The Conservation Federation of Maryland is devoted to the wise use, conservation, aesthetic appreciation, and restoration of wildlife and other natural resources. The Conservation Federation of Maryland was recently merged with For A Rural Maryland to help safeguard the dwindling supply of farmland and open space in the state of Maryland.

Publication(s): Keep It Country (Newsletter), This Place We Call Home (15 minute Documentary about preservation of farmland and open space)

Keyword(s): Agriculture, Conservation, Environmental Justice, Land Preservation, Sustainable Development

Contact(s):
Caroline Taylor, EXECUTIVE DIRECTOR
15711 Hughes Rd., Poolesville, MD 20837
Phone: 301-972-7866
Dolores Milmoe, PRESIDENT
18801 River Rd., Poolesville, MD 20837
Cathy Hall, SECRETARY
17826 Walling Rd., Poolesville, MD 20837
Rudy Gole, TREASURER
17105 Oxley Farm Rd, Poolesville, MD 20837

CONSERVATION FEDERATION OF MISSOURI

728 W. Main St.
Jefferson City, MO 65101-1159 USA
Phone: 573-634-2322   Fax: 573-634-8205
E-mail: confedmo@socket.net
Website: www.confedmo.com

Founded: 1935
Scope: Statewide

Description: A representative statewide organization, affiliated with the National Wildlife Federation, dedicated to the protection and enhancement of wildlife and its habitat through public education and government interaction.

Publication(s): Missouri Wildlife

Contact(s):
Charles Davidson, EDITOR
cdfed@socket.net
Jennifer Mills, EDUCATION PROGRAMS CONTACT
Denny Ballard, EXECUTIVE DIRECTOR
Ike Lovan, PRESIDENT AND ALTERNATE REPRESENTATIVE
Abe Phillips, REPRESENTATIVE
Arnold Meysenburg, SECRETARY
Randy Washburn, TREASURER

CONSERVATION FORCE

3900 N. Causeway Blvd., Suite 1045
Metairie, LA 70002 USA
Phone: 504-837-1233   Fax: 504-837-1145
Website: www.conservationforce.org

Founded: 1997
Scope: International

Description: The force was formed to unify sportsmen's organizations, improve the profile of hunters and further the role and valve of hunting in wildlife conservation as a force.

Publication(s): Conservation Force Supplement to the Hunting Report

Keyword(s): Bears, Biodiversity, Conservation, Conservation Plannning, Endangered Species, Environmental Law, Fish Wildlife Management, Hunting, Indigenous People, International Wildlife, Legal Advocacy, Sustainability, training

Contact(s):
Bertrand Des Clers, INTERNATIONAL VICE PRESIDENT, Africa
John Jackson, PRESIDENT
James Teer, VICE PRESIDENT
TX 409-458-1359
Phone: 409-845-3786
jteer@tamu.edu

CONSERVATION FUND, THE

1800 North Kent St., Suite 1120
Arlington, VA 22209 USA
Phone: 703-525-6300   Fax: 703-525-4610
E-mail: postmaster@conservationfund.org
Website: www.conservationfund.org

Founded: 1985
Scope: National

Description: The Conservation Fund seeks sustainable conservation solutions for the 21st century, emphasizing the integration of economic and environmental goals. Through land conservation services, demonstration projects, education and community-based activities, the Fund develops innovative measures to conserve land and water. The Fund forges partnerships to protect America's irreplaceable outdoor heritage and a tangible legacy for future generations. Programs: Land Conservation Services, Sustainable Programs, American Land Conservation Program, American Greenways, Freshwater Institute, Conservation Leadership Network, Civil War Battlefield Campaign. Awards: Alexander Calder Conservation Award, Cartledge Award for Excellence in Environmental Education, American Greenways DuPont, American Land Conservation, CF Industries National Watershed, Hastings National Park Leadership awards.

Publication(s): Common Ground

Keyword(s): Environmental and Conservation Education, Land Purchase, Land Use Planning, Exotic species, Aquatic nuisance species, Training

Contact(s):
Patrick Noonan, CHAIRMAN
David Phillips, CHIEF FINANCIAL OFFICER
Mark Benedict, CONSERVATION LEADERSHIP NETWORK LIAISON
Edward McMahon, DIRECTOR OF AMERICAN GREENWAYS PROGRAM

Frances Kennedy, DIRECTOR OF CIVIL WAR BATTLEFIELD PROGRAM
Margaret Kohring, DIRECTOR OF MIDWEST OFFICE
Phone: 312-913-9459
Kiku Hanes, DIRECTOR OF MONTANA OFFICE
Phone: 406-388-9733
Nancy Bell, DIRECTOR OF VERMONT OFFICE
Phone: 802-492-3368
Mike McQueen, EDITOR
John Turner, PRESIDENT
Pamela Gray, SECRETARY
Richard Erdmann, SENIOR VICE PRESIDENT AND GENERAL COUNSEL
David Sutherland, SENIOR VICE PRESIDENT, REAL ESTATE
Lawrence Selzer, SENIOR VICE PRESIDENT, SUSTAINABLE PROGRAMS
Hadlai Hull, TREASURER
Elizabeth Madison, VICE PRESIDENT, DEVELOPMENT
Elizabeth Dowdle, VICE PRESIDENT, FLORIDA OFFICE
Phone: 561-624-4925
Rex Boner, VICE PRESIDENT, GEORGIA OFFICE
Phone: 770-414-0211
Sydney Macy, VICE PRESIDENT, WESTERN REGIONAL OFFICE
Phone: 303-444-4369

CONSERVATION INTERNATIONAL

1919 M St., NW, Suite 600
Washington, DC 20036 USA
Phone: 202-912-1000   Fax: 202-887-5188
Website: www.conservation.org

Founded: 1987
Scope: International

Description: Conservation International (CI) is a private nonprofit organization dedicated to the preservation of tropical and temperate ecosystems. CI works in partnership with indigenous peoples and with organizations to sustain biological diversity and the ecological processes that support life on earth. CI has programs in Bolivia, Botswana, Brazil, Colombia, Costa Rica, Ecuador, Ghana, Guatemala, Guyana, Indonesia, Madagascar, Mexico, Papua, New Guinea, Panama, Peru, the Philippines, the Solomon Islands, and Surinam.

Keyword(s): Biodiversity, Endangered Species, Environmental and Conservation Education, Forests and Forestry

Contact(s):
Peter Seligmann, CEO AND CHAIRMAN OF THE BOARD
Phone: 202-973-2275
p.seligmann.org
Russell Mittermeier, PRESIDENT
Phone: 202-973-2212
Fax: 202-887-0192

CONSERVATION LAW FOUNDATION (CLF)

NEW ENGLAND REGION
VERMONT OFFICE
15 E. State St., Suite 4
Montpelier, VT 05602 USA
Phone: 802-223-5992   Fax: 802-223-0060
Website: www.clf.org

Founded: NA
Scope: Regional

Publication(s): Vermont (Newsletter), Conservation Matters

CONSERVATION LAW FOUNDATION, INC. (CLF)

62 Summer St.
Boston, MA 02110 USA
Phone: 617-350-0990   Fax: 617-350-4030
Website: www.clf.org

Founded: 1966
Membership: 40
Scope: Regional

Description: CLF is a nonprofit, member-supported environmental law organization dedicated to improving resource management, environmental protection, and public health in New England. Work includes: Energy and water conservation, environmental health, transportation planning, water resources protection, land preservation, and marine resources protection.

Publication(s): Power to Spare I&II (energy conservation opportunities in New England), A Silent and Costly Epidemic (costs of childhood lead poisoning), Troubled Waters (report on the environmental health of CA), Take Back Your Streets (community transportation planning)

Keyword(s): Air Quality and Pollution, Energy, Environmental Law, Transportation, Coral Reefs

Contact(s):
Charles Cabot, CHAIRMAN OF THE BOARD
Douglas Foy, PRESIDENT
Eugene Clapp, TREASURER
Paula Gold, VICE CHAIRMAN OF THE BOARD
John Teal, VICE CHAIRMAN OF THE BOARD

CONSERVATION LAW FOUNDATION, INC. (CLF)

Main Office, 120 Tillson Ave.
Rockland, ME 04841 USA
Phone: 207-594-8107   Fax: 207-596-7706
Website: www.clf.org

Founded: NA
Membership: 5
Scope: National
Publication(s): City Routes City Rights, Effects of Fishing Gear on the Sea Floor of New England, The Wild Sea, Conservation Matters

Contact(s):
Peter Shelley, CENTER DIRECTOR

CONSERVATION MANAGEMENT INSTITUTE

203 W. Roanoke St.
Blacksburg, VA 24061 USA
Phone: 540-231-7348
E-mail: fwiexchg@vt.edu
Website: www.fwie.fw.vt.edu/

Founded: 1984
Membership: 50
Scope: National

Description: The FWIE is a clearinghouse and technical assistance center to state and federal fish and wildlife agencies in the area of fish and wildlife databases and computer applications. The FWIE is available to help agencies with biological and administrative information management for application to environmental assessment, planning, extension, research, and education. The FWIE is a unit within the Department of Fisheries and Wildlife Sciences, Virginia Tech.

Keyword(s): Aquatic Habitats, Biodiversity, Birds, Ecology, Endangered Species, Insects and Butterflies, Mammals,

Wildlife, Nongame Wildlife, Reptiles and Amphibians, Terrestrial Habitats, training, Sport Fishing

Contact(s):
Jefferson Waldon, ASST. DIRECTOR

CONSERVATION TECHNOLOGY INFORMATION CENTER

1220 Potter Dr.
West Lafayette, IN 47906-1383 USA
Phone: 765-494-9555   Fax: 765-494-5969
E-mail: ctic@ctic.purdue.edu
Website: www.ctic.purdue.edu

Founded: 1982
Membership: 8
Scope: National

Description: Conservation Technology Information Center (CTIC) is a nonprofit information and data transfer center. The national Center promotes environmentally and economically beneficial agricultural decision-making by: producing and circulating information, data, and contacts, coordinating national initiatives, and sponsoring interactive meetings and conferences. The Center is supported by members and participating governmental agencies.

Publication(s): Conservation Tillage: A Checklist for U.S. Farmers, Watershed Management, CTIC Partners Newsletter

Keyword(s): Conservation Tillage, Precision Farming, Water Quality, Watersheds

Contact(s):
Bruno Alesii, CHAIR, BOARD OF DIRECTORS
106 Pebble Creek, Boerne, TX 78006
John Hassell, EXECUTIVE DIRECTOR
hassell@ctic.purdue.edu
Dan Towery, NATURAL RESOURCES SPECIALIST
towery@ctic.purdue.edu
Ed Frye, PROJECT MANAGER

CONSERVATION TREATY SUPPORT FUND

3705 Cardiff Rd.
Chevy Chase, MD 20815 USA
Phone: 301-654-3150   Fax: 301-652-6390
E-mail: ctsf@conservationtreaty.org
Website: www.conservationtreaty.org

Founded: 1986
Scope: International

Description: CTSF provides direct support to major inter-governmental treaties, including CITES (the endangered species treaty), the wetlands and the migratory species treaty, through fund raising and education.

Publication(s): CITES Endangered Species Book, CITES Video (also Wetlands Video), Caribbean Buyer Beware Poster "Wild Treasures of the Caribbean", "Treasures of Wetlands" Poster, Bateman prints and posters

Keyword(s): Wetlands, training, Biodiversity, Habitat Conservation

Contact(s):
George Furness, PRESIDENT
Phone: 301-654-3150
Fax: 301-652-6390
ctsf@conservationtreaty.org

Faith Campbell, SECRETARY
Phone: 202-547-9120
Fax: 202-547-9213
Lawrence Mason, TREASURER
Phone: 703-241-8896
Fax: 703-241-8896
lnmason@compuserve.com
Frederick Morris, VICE PRESIDENT
Phone: 703-683-8512
Fax: 703-683-4622

CONSERVATION TRUST OF PUERTO RICO

P.O. Box 9023554, 155 Tetuan St.
Old San Juan, PR 00902-3554 PR
Phone: 787-722-5834
E-mail: fideiccomiso@fideicomiso.org
Website: www.fideicomiso.org

Founded: 1970
Scope: Statewide

Description: A private nonprofit institution created by the Governor of Puerto Rico and the U.S. Secretary of the Interior to preserve and enhance Puerto Rico's natural beauty and resources, primarily through land acquisition. Owns or manages over 14,000 acres representative of the island's major endangered habitats. It educates the public about environmental issues; manages a vast reforestation program; and finances conservation in Caribbean countries via debt-for-nature swaps. Puerto Rico's major conservation organization, the Trust acts as land acquisition agent for the Commonwealth Department of Natural Resources.

Keyword(s): Land Protection

Contact(s):
Thomas Lovejoy, CHAIRMAN
Francisco Blanco, EXECUTIVE DIRECTOR

COOK INLET KEEPER

P.O. Box 3269
Homer, AK 99603 USA
Phone: 907-235-4068   Fax: 907-235-4069
E-mail: keeper@inletkeeper.org
Website: www.inletkeeper.org

Founded: 1995
Scope: Regional

Description: The mission of Cook Inlet Keeper is to protect the Cook Inlet Watershed and the life it sustains. Keeper relies on environmental monitoring, research, education, and advocacy to give citizens the tools they need to protect water quality.

Publication(s): State of the Inlet Report, Cook Inlet Watershed Directory, Cook Inlet GIS Atlas on CD-ROM

Keyword(s): Water Quality, Toxic Substances, Nuclear-free, Water quantity, Water export and diversion, Environmental Law, Watersheds

Contact(s):
Bob Shavelson, EXECUTIVE DIRECTOR

COOPER ORNITHOLOGICAL SOCIETY

ORNITHOLOGICAL SOCIETIES OF NORTH AMERICA
P.O.B. 1897
Lawrence, KS 66044 USA
Fax: 208-378-5347
Website: http://www.cooper.org

Founded: 1893
Membership: 2,340
Scope: National

Description: Observation and cooperative study of birds; the spread of interest in bird study; the conservation of birds and wildlife in general; the publication of ornithological knowledge.

Publication(s): Studies in Avian Biology, Condor, The

Contact(s):
John Rotenberry, EDITOR, STUDIES IN AVIAN BIOLOGY
Department of Biology, University of California,
Riverside, CA 92521
Phone: 909-787-3953
rote@citrus.ucr.edu
David Dobkin, EDITOR, THE CONDOR
Terry Rich, PRESIDENT
Branch of Bird Conservation, Div. Of Migratory Birds
US FWS, 1387 S. Vinnell Way, Boise, ID 83709
Glenn Walfberg, PRESIDENT
Bonnie Bowen, PRESIDENT-ELECT/TREASURER
Department of Animal Ecology
124 Science Hall II
Iowa State Univ., Ames, IO 50011
Phone: 515-294-6391
Eileen Kirsch, SECRETARY
BRD/USGS, Upper Mississippi Science Center, P.O. Box 818,
LaCrosse, WI 54602
Phone: 608-783-6451
eileen_kirsch@usgs.gov

COOSA RIVER BASIN INITIATIVE

408 Broad St.
Rome, GA 30161 USA
Phone: 706-232-2724   Fax: 706-235-9066
E-mail: crbi@roman.net
Website: www.roman.net/~crbi

Founded: 1992
Membership: 400
Scope: Regional

Description: CRBI works to inform and empower citizens so they may become involved with the process of creating a cleaner, healthier, econonically viable Coosa River Basin.

Publication(s): Main Stream, The

Keyword(s): Environmental Cleanup, Environmental Legislation, Environmental Protection, Flood Control, Habitat Conservation, Lakes, Preservation and Protection, Restoration, Stewardship, Water Conservation, Coral Reefs, Water Quality, Exotic species, Aquatic nuisance species, Watersheds, Wetlands

Contact(s):
Joe Cook, ADVOCACY CHAIR
Mitch Lawson, COORDINATOR
Phone: 706-232-2724
Bill Davin, EDUCATION CHAIR
Jerry Jennings, PRESIDENT
Phone: 706-290-2665
Fax: 706-238-5827
jjennings@barry.edu
Monica Cook, PUBLICATION CHAIR
Phone: 706-235-1170
jmc@artfamily.com
Benjamin Harris, VICE PRESIDENT
Phone: 706-295-0858
benhar@bellsouth.net

CORAL REEF ALLIANCE, THE (CORAL)

2014 Shattuck Ave
Berkeley, CA 94704 USA
Phone: 510-848-0110   Fax: 510-848-3720
E-mail: info@coral.org
Website: http://www.coralreefalliance.org/

Founded: 1994
Scope: International

Description: The Coral Reef Alliance is a nonprofit organization that works with divers, government conservation organizations, and others to promote coral reef conservation around the the world. CORAL focuses primarily on helping local communities to establish their own marine protected area. CORAL also sponsors a number of educational programs and publications.

Publication(s): Coral Reefs - The Vanishing Rainbow, Coral News

Keyword(s): Coasts, Conservation

Contact(s):
Stephen Colwell, EXECUTIVE DIRECTOR
Janine Kraus, MANAGING DIRECTOR

CORNELL LAB OF ORNITHOLOGY

159 Sapsucker Woods Rd.
Ithaca, NY 14850 USA
Phone: 607-254-2473   Fax: 607-254-2415
E-mail: cornellbirds@cornell.edu
Website: www.birds.cornell.edu

Founded: 1917
Membership: 200
Scope: International

Description: A membership institute dedicated to the study, appreciation, and conservation of birds world-wide. The Lab maintains programs in academic research, public education, and citizen involvement. The Lab promotes science to foster understanding about nature and the importance of the earth's biological diversity.

Publication(s): Birdscope, Bird Notes, Living Bird

Keyword(s): Biodiversity, Birds, Environmental and Conservation Education, education, Nongame Wildlife

Contact(s):
Scott Sutcliffe, ASSOCIATE DIRECTOR
Phone: 607-254-2424
Brian Mingle, COMMUNICATIONS/PUBLICATIONS OUTREACH ASSISTANT
Gregory Budney, CURATOR OF LIBRARY OF NATURAL SOUNDS
Phone: 607-254-2406
Kevin McGowan, CURATORIAL ASSOCIATE FOR SYSTEMATICS AND COLLECTIONS
Phone: 607-257-8135
Christopher Clark, DIRECTOR OF BIOACOUSTICS RESEARCH PROGRAM
Phone: 607-254-2405
Andre Dhondt, DIRECTOR OF BIRD POPULATION STUDIES
Phone: 607-254-2445
Rick Bonney, DIRECTOR OF EDUCATION PROGRAM
Phone: 607-254-2440
birdeducation@cornell.edu
Tim Gallagher, EDITOR
Phone: 607-254-2443

John Fitzpatrick, LOUIS AGASSIZ FUERTES DIRECTOR
Phone: 607-254-2410

COUNCIL FOR ENVIRONMENTAL EDUCATION

Executive Director, 5555 Morningside Dr., Suite 212
c/o Josetta Hawthorne
Houston, TX 77005 USA
Phone: 713-520-1936   Fax: 713-520-8008
E-mail: info@c-e-e.org
Website: www.c-e-e.org

Founded: 1970
Scope: National

Description: The Council for Environmental Education is a nonprofit education organization creating a partnership and network between education and natural resource professionals. CEE cosponsors balanced, non-biased environmental education programs such as Project Learning Tree, Project WILD, Project WILD Aquatic and Project WET. In an effort to encourage more environmental education outreach to urban youth, CEE has launched WET in the City, a community-based water education initiative.

Publication(s): Project WILD K-12 Curriculum & Activity Guide, Project WILD Aquatic K-12 Curriculum & Activity Guide, WET in The City Curriculum & Activity Guide

Keyword(s): Environmental and Conservation Education, Urban Environment

Contact(s):
Josetta Hawthorne, EXECUTIVE DIRECTOR AND DIRECTOR, WET IN THE CITY
Bill Andrews, PRESIDENT
Phone: 916-657-5374
Kathy Mcglauflin, PROJECT LEARNING TREE DIRECTOR
1111 19TH St., NW, Suite 780, Washington, DC 20036
Phone: 202-436-2468

COUNCIL FOR PLANNING AND CONSERVATION

Box 228
Beverly Hills, CA 90213 USA
Phone: 310-276-2685
E-mail: esharris@earthlink.net
Website: www.beverlyhillscitizen.org

Founded: NA
Scope: Statewide

Description: Serves as a clearinghouse for information and gives inexperienced groups ready access to advice and assistance. Provides a center through which opportunities for southern California's environmental protection and enhancement may be communicated. Concerns include: air and water quality, water supply, energy options, waste management, land use, transportation, coastal conservation, urban planning and housing.

Keyword(s): Air Quality, Planning Management, Energy, Transportation, Urban and Rural Development, Coastal Construction and Erosion

Contact(s):
Ellen Harris, PRESIDENT AND EXECUTIVE DIRECTOR
Sam Weisz, TREASURER AND SECRETARY
Betty Harris, VICE PRESIDENT

COUSTEAU SOCIETY, INC., THE

870 Greenbrier Cir.
Chesapeake, VA 23320 USA
Phone: 757-523-9335   Fax: 727-523-2747
E-mail: cousteau@infi.net
Website: www.cousteausociety.org

Founded: 1973
Membership: 200,000
Scope: International

Description: A nonprofit, membership-supported environmental education organization dedicated to the protection and improvement of the quality of life for present and future generations. Believing that an informed and alerted public can best make the choices that will sustain the water planet, it produces television films, research, books and other publications, exploring relationships between humans and ecosystems.

Publication(s): Dolphin Log, Calypso Log

Keyword(s): Environmental and Conservation Education, Aquatic Species, Marine Conservation, Sustainable Ecosystems, Exotic species, Aquatic nuisance species

Contact(s):
Francine Cousteau, PRESIDENT
Robert Steele, VICE PRESIDENT FINANCE

CRAIGHEAD ENVIRONMENTAL RESEARCH INSTITUTE

201 S. Wallace Avenue
Bozman, MT 59715 USA
Phone: 406-585-8705   Fax: 406-585-8220
E-mail: ceri@avicom.net
Website: www.grizzlybear.org

Founded: 1955
Membership: 20
Scope: Regional, International

Description: A nonprofit professional organization of scientists, dedicated to exploring the cause-and-effect relationships of man and his environment. Activity includes research, education, and conservation, with emphasis on ecological studies and interdisciplinary approach. Originally the Outdoor Recreation Institute. Staff Members: 6.

Keyword(s): Raptors, Rivers, training

Contact(s):
Charles Craighead, MEDIA DIRECTOR
Frank Craighead, PRESIDENT, PROGRAM DIRECTOR
April Craighead, SECRETARY

CRAIGHEAD WILDLIFE-WILDLANDS INSTITUTE

5200 Upper Miller Creek Rd.
Missoula, MT 59803 USA
Phone: 406-251-3867

Founded: 1977
Scope: National

Description: A nonprofit, multidisciplinary research center in the Northern Rockies devoted to field-based ecological discovery and scientific activism. The Institute's mission is to generate new ecological information and concepts, widely communicate these insights, and influence public policy and individual behavior in directions that preserve regional biodiversity.

Publication(s): An Integrated Satellite Technique to Evaluate

Grizzly Bear Habitat Use (1997), Mapping Arctic Vegetation in Northwest Alaska Using Landsat MSS Imagery (1988), The Grizzly Bears of Yellowstone: Their Ecology in the Yellowstone Ecosystem 1959-1992 (1995)

Keyword(s): Biodiversity, Ecology, Endangered Species, Predators, Research, Sustainable Ecosystems

Contact(s):
John Craighead, CHAIRMAN OF THE BOARD
5125 Orchard Ln., Missoula, MT 59803
Phone: 406-251-3944

CRESTON VALLEY WILDLIFE MANAGEMENT AUTHORITY
Creston, BC V0B 1G0 Canada
Phone: 250-428-3260   Fax: 250-428-3276
E-mail: info@cwildlife.bc.ca
Website: www.cwildlife.bc.ca

Founded: 1968
Scope: Statewide

Description: A unique joint provincial-private agency established in 1968 to conserve, develop, and manage remaining waterfowl habitat in a mountain valley. Broader purpose is to demonstrate cooperative wetland management in action. Operates a 17,000-acre wetland and upland complex, primarily as a waterfowl management area, with public recreation facilities (campground, trails, and visitor centre).

Publication(s): Creston Valley Wildlife Management Area Annual

Contact(s):
Brian Stushnoff, AREA MANAGER
Phone: 250-428-3260
Steve Bullock, CHAIRMAN OF THE MANAGEMENT AUTHORITY
Phone: 250-428-2214

CROSBY ARBORETUM, THE
MISSISSIPPI STATE UNIVERSITY
370 Ridge Rd.
Picayune, MS 39466 USA
Phone: 601-799-2311   Fax: 601-799-2372
E-mail: crosbyar@datastar.net
Website: www.msstate.edu/dept/crec/camain.html

Founded: 1980
Membership: 700
Scope: Statewide

Description: The main activity of the Arboretum is to preserve, protect and display plants native to the pearl river drainage basin. Additionally, we provide environmental and horticultural research opportunities and offer educational, scientific, and recreational programs.

Publication(s): Native Trees for Urban Landscapes

Keyword(s): Biodiversity, Conservation, Environmental and Conservation Education, Ecology

Contact(s):
Stewart Gammill, PRESIDENT
Jennifer McKay, SECRETARY
Bob Brzuszek, SENIOR CURATOR
Phone: 601-799-2311
Fax: 601-799-2372
crosbyar@datastar.net
Norman Stevens, TREASURER
Richard Clark, VICE PRESIDENT

DAWES ARBORETUM, THE
7770 Jacksontown Rd., SE
Newark, OH 43056-9380 USA
Phone: 740-323-2355   Fax: 740-323-4058
Website: www.dawesarb.org

Founded: 1929
Membership: 40
Scope: Statewide

Description: A not-for-profit organization that promotes the planting of forest and ornamental trees, and promotes increased love and knowledge of trees, shrubs, and related subjects through over 200 annually-offered classes and programs. The 1,149-acre grounds are open daily from dawn to dusk, free of charge.

Publication(s): Dawes Arboretum Newsletter, The

Keyword(s): Botanical Gardens, Flowers, Plants, and Trees

Contact(s):
Luke Messinger, DIRECTOR
lemessinger@ee.net
Michael Ecker, HORTICULTURIST
meecker@ee.net
Timothy Mason, NATURAL RESOURCE SPECIALIST
tamason@ee.net
Lori Totman, NATURALIST AND EDUCATOR
latotman@ee.net
Laura Kaparoff, PUBLIC RELATIONS EDITOR

DEEP-PORTAGE CONSERVATION RESERVE
2197 Nature Center Dr., NW
Hackensack, MN 56452-2431 USA
Phone: 218-682-2325   Fax: 218-682-3121
E-mail: portage@uslink.net
Website: www.deep-portage.org

Founded: 1975
Membership: 25
Scope: Statewide

Description: Deep-Portage is a 6,100-acre demonstration working forest with a primary purpose of environmental education. The campus includes dormitories, classrooms, laboratory, theater, interpretive center, natural history museum, and thirty-seven miles of recreational trails. It is owned by Cass County and operated by the Deep-Portage Conservation Foundation, a nonprofit corporation.

Publication(s): Deep-Portage Log, Camp Brochures

Keyword(s): Environmental and Conservation Education, Forests and Forestry, Internships, Exotic species, Aquatic nuisance species, training

Contact(s):
Dale Yerger, EXECUTIVE DIRECTOR
Bruce Steiner, PRESIDENT

DEFENDERS OF WILDLIFE
1101 14th St., NW
Suite 1400
Washington, DC 20005 USA
Phone: 202-682-9400   Fax: 202-682-1331
E-mail: information@defenders.org
Website: www.defenders.org

Founded: 1947

Membership: 400,000
Scope: National

Description: Since 1947, Defenders of Wildlife has been one of the nation's most effective advocates for wildlife, endangered species, and habitat. Defenders works to protect and restore native species, habitats, ecosystems, and overall biological diversity. Defenders is a nonprofit, tax-exempt organization, supported by 400,000 members.

Publication(s): Defenders

Keyword(s): Biodiversity, Endangered Species, Environment, Environmental and Conservation Education, Protected Areas, National Parks, Predators, Environmental Protection, Wilderness, training, Wolves

Contact(s):
Alan Pilkington, CHAIRMAN OF THE BOARD
Nina Fascione, DIRECTOR OF CARNIVORE CONSERVATION
Martha Schumacher, DIRECTOR OF DEVELOPMENT
Robert Jones, DIRECTOR OF FINANCE AND ADMINISTRATION
Laura Hood, DIRECTOR OF HABITAT CONSERVATION DIVISON
Sajjad Ahrabi, DIRECTOR OF INFORMATION SYSTEMS
Michael Senatore, DIRECTOR OF LEGAL DEPARTMENT
Mary Beth Beetham, DIRECTOR OF LEGISLATIVE AFFAIRS
Kate Mathews, DIRECTOR OF MEMBERSHIP
Maria Cecil, DIRECTOR OF PUBLICATIONS AND EXECUTIVE EDITOR
Sara Vickerman, DIRECTOR OF WEST COAST
1637 Laurel St., Lake Oswego, OR 97034
Phone: 503-697-3222
Hank Fischer, NORTHERN ROCKIES DIRECTOR
1534 Mansfield Ave., Missoula, MT 59801
Phone: 406-549-0761
Rodger Schlickeisen, PRESIDENT
Ann Boren, SECRETARY
Charles Orasin, SENIOR VICE PRESIDENT FOR OPERATIONS
Mark Shaffer, SENIOR VICE PRESIDENT FOR PROGRAM
Alan Steinberg, TREASURER
Winsome McIntosh, VICE CHAIRMAN
Philip Rabin, VICE PRESIDENT FOR COMMUNICATIONS
Robert Dewey, VICE PRESIDENT FOR GOVERNMENT RELATIONS
William Snape, VICE PRESIDENT FOR LAW AND LITIGATION
James Deane, VICE PRESIDENT FOR PUBLICATIONS AND EDITOR, DEFENDERS
Robert Ferris, VICE PRESIDENT FOR SPECIES CONSERVATION

DEFENDERS OF WILDLIFE
1101 14th St., NW, Suite 1400
Washington, DC 20005 USA
Phone: 202-682-9400   Fax: 202-682-1331
E-mail: info@defenders.org
Website: www.defenders.org

Founded: 1974
Scope: International

Description: (United States Office) Nongovernmental organization sponsored by major Canadian and American conservation and environmental groups. Established to facilitate interchange of information and cooperative action on questions of concern in the two nations.

Keyword(s): Air Quality and Pollution, Endangered Species, Public Lands, training

Contact(s):
Julie Gelfand, COORDINATING COMMITTEE CO-CHAIRMAN
Canadian Nature Federation, Suite 606, 1 Nicholas St., Ottawa, Ontario K1N 7B7
Phone: 613-562-3447
Fax: 613-562-3371
James Deane, COORDINATING COMMITTEE CO-CHAIRMAN
Defenders of Wildlife, 1101 14th St. NW, Suite 1400, Washington, DC 20005
Phone: 202-682-9400
Fax: 202-682-1334

DELAWARE ASSOCIATION OF CONSERVATION DISTRICTS
President, P.O. Box 242
Dover, DE 19903-0242 USA
Phone: 302-739-4411   Fax: 302-739-6724

Founded: 1953
Membership: 30
Scope: Regional

Description: DACD is a voluntary nonprofit alliance that provides a forum for discussion and coordination among the Delaware Conservation Districts as they work to ensure the wise use and treatment of renewable resources.

Keyword(s): Agriculture, Land Use Planning, Soil Conservation, Urban Environment, Water Pollution Management, Conservation Districts

Contact(s):
Dariel Rakestraw, PAST PRESIDENT, BOARD MEMBER
2138 Graves Rd., Hockessin, DE 19707
Phone: 302-239-2969
Terry Pepper, PRESIDENT
104 Captain Davis Dr., Campden-Wyoming, DE 19934
Phone: 302-697-6176
Fax: 303-736-2040
kentcol@aol.com
Martha Pileggi, STAFF ASSISTANT
P.O. Box 242, Dover, DE 19903-0242
Phone: 302-739-4411
Fax: 302-739-6724
mpilegg@state.de.us
Ron Breeding, VICE PRESIDENT
Rt. 1, Box 345-B, Seaford, DE 19973
Phone: 302-629-3964
Fax: 302-739-6724

DELAWARE AUDUBON SOCIETY
P.O. Box 1713
Wilmington, DE 19899 USA
Phone: 302-428-3959
E-mail: mail@delawareaudubon.org
Website: www.delawareaudubon.org

Founded: NA
Scope: Statewide

Description: The Delaware Audubon Society promotes an appreciation and understanding of nature to preserve and protect our natural environment and to affirm the necessity for clean air and water and the stewardship of our natural resources.

Keyword(s): Water and Air Quality, Preservation and Protection

Contact(s):
Leslie Savage, PRESIDENT
Matthew Delpizzo, VICE-PRESIDENT

DELAWARE B.A.S.S. CHAPTER FEDERATION

Attn: President, 3700 South State St.
Camden, DE 19934 USA
Phone: 302-698-9257   Fax: 302-720-1230
Website: www.ezy.net/~delbass/

Founded: NA
Membership: 350
Scope: Statewide

Description: An organization of Bassmaster chapters, affiliated with the Bass Anglers Sportsman Society, organized to fight pollution, assist state and national conservation agencies in their efforts, and teach the young people of our country good conservation practices. Dedicated to the realistic conservation of our water resources.

Publication(s): Bassing on Delmarva

Contact(s):
Roger Richardson, CONSERVATION DIRECTOR
632 Fencepost Ln, Viola, DE 19979
Phone: 302-284-8383
Jim Fields, PRESIDENT
Phone: 302-698-9257

DELAWARE DEPARTMENT OF AGRICULTURE - FOREST SERVICE

2320 S. DuPont Hwy.
Dover, DE 19901 USA
Phone: 302-739-4811   Fax: 302-697-6287
E-mail: kay@dda.state.de.us
Website: www.state.de.us/deptagri/

Founded: 1982
Scope: Statewide

Description: A statewide organization affiliated with the National Woodland Owners Association, dedicated to promote good forest practices and multiple use of private forest lands in Delaware.

Publication(s): DFA Newsletter

Keyword(s): Forests and Forestry

Contact(s):
W. Jones, PRESIDENT AND EDITOR
Phone: 410-742-3163
Jim Bennett, VICE PRESIDENT

DELAWARE GREENWAYS, INC.

P.O. Box 2095
Wilmington, DE 19899 USA
Phone: 302-655-7275   Fax: 302-655-7274
E-mail: greenwalks@aol.com
Website: www.delawaregreenways.org

Founded: 1989
Membership: 400
Scope: Statewide

Description: Preserve, enhance, and connect the ecological, scenic, historical, cultural, and recreational resources in Delaware.

Publication(s): see publications on website

Keyword(s): Cultural Preservation, Protected Areas, Historic Preservation, Sustainable Development, Transportation

Contact(s):
Tim Plemmons, ASSISTANT EXECUTIVE DIRECTOR
Gail Van Gilder, EXECUTIVE DIRECTOR

DELAWARE MUSEUM OF NATURAL HISTORY

P.O. Box 3937
Wilmington, DE 19807 USA
Phone: 302-658-9111   Fax: 302-658-2610
Website: www.delmnh.org

Founded: 1957
Membership: 1000
Scope: Regional

Description: The Delaware Museum of Natural History exists to excite and inform people about the natural world. The Museum's core purpose is to help develop a caring society which respects and values our planet. The major focus of the Museum is continued leadership in research and collections in malacology and ornithology and the ecology of the Delmarva Peninsula.

Publication(s): Musenews, Nemouria

Keyword(s): Birds, Research, education, Museum

Contact(s):
Gene Hess, COLLECTION MANAGER ORNITHOLOGY
Jean Woods, CURATOR OF BIRDS
Stephen Reynolds, EDITOR AND PUBLIC RELATIONS
Geoff Halfpenny, EXECUTIVE DIRECTOR

DELAWARE NATURE SOCIETY

P.O. Box 700
Hockessin, DE 19707-0700 USA
Phone: 302-239-2334   Fax: 302-239-2473
E-mail: email@dnsashland.org
Website: www.delawarenaturesociety.org

Founded: 1964
Membership: 8000
Scope: Regional

Description: A representative statewide organization, affiliated with the National Wildlife Federation, dedicated to the protection and enhancement of wildlife and its habitat through public education and government interaction.

Publication(s): Delaware Nature Society Voice, Delaware Nature Society News

Keyword(s): Air Quality and Pollution, Endangered Species, Environmental and Conservation Education, Natural Areas, Wetlands

Contact(s):
Bernard Dempsey, ALTERNATE REPRESENTATIVE
Janice Taylor, EDITOR
Helen Fischel, EDUCATION PROGRAMS CONTACT
Mike Riska, EXECUTIVE DIRECTOR
Peter Flint, PRESIDENT
Richard Fleming, REPRESENTATIVE
George Fisher, TREASURER

DELAWARE WILD LANDS, INC.

315 Main St.
Odessa, DE 19730-0505 USA
Phone: 302-378-2736   Fax: 302-378-3629
E-mail: dwl@delanet.com

Founded: 1961
Scope: Local

Description: A nonprofit charitable land conservancy actively engaged in acquiring areas on the Delmarva Peninsula for their natural resource values and for educational purposes; presently owns and manages approximately 20,000 acres. Produced two films, "The Endangered Shore" and "Swamp," available on loan or for purchase.

Keyword(s): Wildlands, Coasts, Wetlands, Land Purchase

Contact(s):
Susan Crawford, ADMINISTRATIVE ASSISTANT
Holger Harvey, EXECUTIVE DIRECTOR

DELMARVA ORNITHOLOGICAL SOCIETY

P.O. Box 4247
Greenville, DE 19807 USA
Fax: 302-478-8300

Founded: 1963
Scope: Statewide

Description: The purpose of this society shall be the promotion of the study of birds, the advancement and diffusion of ornithological knowledge, and the conservation of birds and their environment.

Publication(s): DOS Flyer, Delmarva Ornithologist

Contact(s):
Irene Goverts
bbcdel@ezd.com
Jim White, PRESIDENT
3507 Barley Mill Rd., Hockessin, DE 19707
Phone: 302-239-7065
Mike Smith, VICE PRESIDENT
msmith10@student.vill.edu

DELTA WATERFOWL FOUNDATION

R.R. 1 Box 1
Portage la Prairie, Manitoba R1N 3A1
Phone: 204-239-1900   Fax: 203-239-5950
E-mail: canada@deltawaterfowl.org
Website: www.deltawaterfowl.org

Founded: NA
Scope: International

Description: Delta Waterfowl's primary mission is to support graduate student training and research on all aspects of waterfowl and wetlands ecology and management. Since 1938, Delta students have produced over 200 graduate theses and 600 scientific publications. In addition to graduate research, Delta is currently involved in several demonstration projects incuding Adopt-A-Pothole, a habitat easement program; Hen Houses, predator-resistant nesting structures, Voluntary Restraint, a hunter ethics program; and several policy initiatives.

Publication(s): Publications on website

Keyword(s): Communications, Environmental and Conservation Education, Waterfowl, Wetlands

Contact(s):
Daniel Hughes, CHAIR
Jonathan Scarth, PRESIDENT
George Nolte, SECRETARY
Thomas Hutchens, TREASURER
Donald Douglas, VICE CHAIR
Lloyd Jones, VICE PRESIDENT

DELTA WILDLIFE INC.

P.O. Box 276
Stoneville, MS 38776 USA
Phone: 662-686-3370   Fax: 662-686-3382

Founded: 1990
Membership: 3
Scope: Regional

Description: Delta Wildlife is committed to wildlife habitat enhancement, habitat restoration and conservation education in northwest Mississippi.

Publication(s): Delta Wildlife Magazine

Keyword(s): Agriculture, Communications, Environmental and Conservation Education, Wetlands, training

Contact(s):
Bill Kennedy, CHAIRMAN
Phone: 662-265-5828
Trey Cooke, EXECUTIVE DIRECTOR
Phone: 662-686-3370
teycoo@yahoo.com

DESERT FISHES COUNCIL

P.O. Box 337
Bishop, CA 93515 USA
Phone: 760-872-8751   Fax: 760-872-8750
E-mail: phildesfish@telis.org
Website: www.desertfishes.org

Founded: 1969
Scope: International

Description: A nationwide and international representation of state, federal, and university scientists and resource specialists and private conservation groups to provide for the exchange and transmittal of information on the status, protection, and management of the endemic fauna and flora of North American desert ecosystems.

Publication(s): Proceedings of the Desert Fishes Council

Keyword(s): Aquatic Habitats, Conservation, Deserts, Endangered Species, Natural Areas, Exotic species, Aquatic nuisance species, Zoology, Sustainable Development, Wildlife, Native Fish

Contact(s):
Dean Hendrickson, EDITOR OF THE PROCEEDINGS
Austin, TX 78758
Phone: 512-471-9774
Fax: 512-471-9775
Edwin Pister, EXECUTIVE SECRETARY
Desert Fishes Council, P.O. Box 337, Bishop, CA 93515
Phone: 760-872-8751
phildesfish@telis.org
David Propst, PRESIDENT
NM Game & Fish Dept. P. O. Box 25112, Santa Fe, NM 87504
Phone: 505-827-9906

DESERT TORTOISE COUNCIL

P.O. Box 3141
Wrightwood, CA 92397 USA
Phone: 619-431-8449
E-mail: info@deserttortoise.org
Website: deserttortoise.org

Founded: 1975
Scope: National

Description: Formed to assure the continued survival of viable populations of the desert tortoise, Gopherus agassizi, which is endemic to Arizona, California, Nevada, and Utah.

Keyword(s): Wildlife Rehabilitation, Conservation, Deserts, Endangered Species, Environment, Health and Nutrition, Land Preservation, Environmental Protection, Land Use Planning, Mining, Public Lands, Reptiles and Amphibians, National Parks, training

Contact(s):
Tim Duck, CO-CHAIRMAN
Tracy Goodlett, CO-CHAIRMAN
Ed Larue, RECORDING SECRETARY
Ed Larue, SECRETARY
Mike Coffeen, TREASURER

DESERT TORTOISE PRESERVE COMMITTEE, INC.

4067 Mission Inn Ave.
Riverside, CA 92501 USA
Phone: 909-683-3872   Fax: 909-683-6949
E-mail: dtpc@pacbell.net
Website: www.tortise-tracks.org

Founded: 1974
Scope: Regional

Description: A nonprofit organization formed to promote the welfare of the desert tortoise in the southwestern United States and to manage and establish preserves in the Western Mojave Desert.

Publication(s): Tortoise T-R-A-C-K-S, Newsletter

Keyword(s): Endangered Species, Land Purchase, Nongame Wildlife, training

Contact(s):
Michael Connor, EXECUTIVE DIRECTOR
Bob Brooks, PRESIDENT
13711 East Gaylin St., Whittier, CA 90601

DISTRICT OF COLUMBIA SOIL AND WATER CONSERVATION - DISTRICT

Attn: Chair, 800 9th St. SW 3rd Fl.
Washington, DC 20024 USA
Fax: 202-442-8989
Website: http://www.obc.dc.gov

Founded: NA
Scope: Statewide

Keyword(s): Conservation Districts

Contact(s):
Theodore Gordon, CHAIR
Phone: 202-442-8989

DIVISION DE PATRIMONIO NATURAL AREA DE DEPARTMENTO DE RECURSOS

P.O. Box 9066600 Puerta de Tierra Station
San Juan, Puerto Rico 00906-6600
Phone: 787-724-4255   Fax: 787-723-4255
Website: www.drnapr.com

Contact(s):
Aida Martinez, DIVISION DIRECTOR

DRAGONFLY SOCIETY OF THE AMERICAS, THE

2091 Partridge Ln.
Binghamton, NY 13903 USA
Phone: 607-722-4939
E-mail: tdonnel@binghamton.edu

Founded: 1989
Scope: International

Description: The organization is concerned with all factors relevant to the world species assemblage of odonata (Insecta: Dragonflies). We study their systematics, biology, and taxonomy. The organization is also concerned with maintaining and improving the environmental conditions for Odonata through better water quality management, wetlands conservation, and aquatic habitat preservation.

Publication(s): Bulletin of American Odonatology, ARGIA

Keyword(s): Aquatic Habitats, Protected Areas, Insects and Butterflies, Rivers, Wetlands

Contact(s):
T. W. Donnelly, EDITOR
2091 Partridge Lane, Binghamton, NY 13903
Michael May, PRESIDENT
Dpt. Of Entomology, Cook College, Rutgers University, New Brunswick, NJ 08903
S. W. Dunkle, SECRETARY
Biology Department, Collin County Community College, Plano, TX 75074
J. J. Daigle, TREASURER
Tallahassee, FL 32311

DUCKS UNLIMITED

CANADA
#200, 10720 - 178 St.
Edmonton, Alberta T5S 1J3 Canada
Phone: 780-489-2002   Fax: 780-489-1856
E-mail: du_edmonton@ducks.ca
Website: www.ducks.ca

Founded: NA
Scope: Province

Publication(s): Conservator

Contact(s):
Brett Calverley, ALBERTA NAWMP COORDINATOR
Gary Stewart, CONSEREVATION PROGRAMS BIOLOGIST
Gordon Edwards, DIRECTOR OF REGIONAL OPERATIONS, PRAIRIE REGIONS
Jim Wohl, ENGINEER

DUCKS UNLIMITED

QUEBEC, CANADA
Suite 260, 710 Bouvier St.
Quebec, CA G2J 1C2 Canada
Phone: 418-623-1650   Fax: 418-623-0420
E-mail: du_quebec@ducks.ca
Website: www.ducks.ca

Founded: 1000
Membership: 140,000+
Scope: International

Description: Ducks Unlimited Canada is an international, private, non-profit organization dedicated to the conservation of wetlands and associated habitats for the perpetuation of North America's waterfowl, which in turn provide healthy environments for wildlife and people.

Publication(s): Conservationniste, Conservator

Contact(s):
Paul St. George, DIRECTOR OF FUND RAISING
Patrick Plante, DIRECTOR OF REGIONAL OPERATIONS
Bernard Filion, MANAGER OF FIELD OPERATIONS

DUCKS UNLIMITED

SASKATCHEWAN OPERATION, CANADA
P.O. Box 4465, 1030 Winnipeg St.
Regina, Saskatchewan S4P 3W7 Canada
Phone: 306-569-0424   Fax: 306-565-3699
Website: www.ducks.ca

Founded: 1938
Scope: International

Description: A private, nonprofit, conservation organization dedicated to preserving waterfowl by creating and restoring breeding habitat in Canada. This organization is funded by sportsmen of United States and Canada.

Publication(s): Newsletter

Contact(s):
L. Moats, AGRICULTURAL PROGRAM SPECIALIST
D. Chekay, DIRECTOR OF PUBLIC POLICY
Tim Thiele, MANAGER OF FIELD OPERATIONS

DUCKS UNLIMITED CANADA

1 Mallard Bay at Hwy. 220, P.O. Box 11660
Stonewall, Manitoba R0C 2Z0 Canada
Phone: 204-467-3000   Fax: 204-467-9028
E-mail: webfoot@ducks.ca
Website: www.ducks.ca

Founded: 1938
Membership: 120
Scope: National

Description: Ducks Unlimited Canada's mission is to conserve wetlands and associated habitats for the benefit of North America's waterfowl, which in turn provide healthy environments for wildlife and people.

Keyword(s): Wetlands, Waterfowl

Contact(s):
L. Warren, CHIEF FINANCIAL OFFICER
Robert Kindrachuk, COMMUNICATIONS MANAGER
Brian Gray, DIRECTOR OF CONSERVATION PROGRAMS
Phone: 204-467-3349
Fax: 204-467-9028
Richard Walker, DIRECTOR OF CORPORATE
DEVELOPMENT AND MAJOR GIFTS
100-279 Midpark Way S.E., Calgary, AB T2X IM2
Phone: 403-201-5577
Fax: 403-201-5580
Rod Fowler, EXECUTIVE VICE PRESIDENT
Gary Goodwin, HUMAN RESOURCES
MANAGER/CORPORATE COUNSEL
Rick Wishart, MANAGER OF EDUCATION PROGRAM

DUCKS UNLIMITED CANADA

566 Welham Rd.
Barrie, Ontario L4N 8Z7 Canada
Phone: 705-721-4444   Fax: 705-721-4999
E-mail: du_barrie@ducks.ca
Website: www.ducks.ca

Founded: NA
Membership: 22000
Scope: Regional

Contact(s):
Ron Maher, MANAGER EASTERN ONTARIO FIELD OFFICE
Bob Clay, MANAGER WESTERN ONTARIO FIELD OFFICE

DUCKS UNLIMITED CANADA

NOVA SCOTIA, CANADA
P.O. Box 430, 64 Highway 6
Amherst, Nova Scotia B4H 3Z5 Canada
Phone: 902-667-8726   Fax: 902-667-0916
Website: www.ducks.ca

Founded: NA
Scope: Statewide

Contact(s):
Brian McCullough, ATLANTIC ENGINEER
b_mccullough@ducks.ca
Mark Gloutney, MANAGER OF CONSERVATION
PROGRAMS-ATLANTIC CANADA
m_gloutney@ducks.ca

DUCKS UNLIMITED, INC.

One Waterfowl Way
Memphis, TN 38120 USA
Phone: 901-758-3825   Fax: 901-758-3850
E-mail: nhq@ducks.org

Founded: 1937
Membership: 701000
Scope: National

Description: The mission of Ducks Unlimited is to fulfill the annual life cycle needs of North American waterfowl by protecting, enhancing, restoring, and managing important wetlands and associated uplands. Only those activities which contribute directly toward that end shall be undertaken by Ducks Unlimited, Inc.

Publication(s): Puddler Magazine, Ducks Unlimited Magazine

Contact(s):
Julius Wall, CHAIRMAN OF THE BOARD
116 W. Jefferson, P.O. Box 226, Clinton, MO 64735
Phone: 660-885-2221
Bruce Batt, CHIEF BIOLOGIST/DIRECTOR, IWWR
Randy Graves, CHIEF FINANCIAL OFFICER
Robert Mims, CONTROLLER
Eric Keszler, DIRECTOR OF COMMUNICATIONS
Steve Adair, DIRECTOR OF CONSERVATION PROGRAMS
Wayne Dierks, DIRECTOR OF HUMAN RESOURCES AND
STAFF DEVELOPMENT
Montserrat Carbonell, DIRECTOR OF LATIN AMERICAN
PROGRAM
Bill Willsey, EXECUTIVE SECRETARY
D. Young, EXECUTIVE VICE-PRESIDENT
James Flood, GENERAL COUNSEL
W. Wentz, GROUP MANAGER CONSERVATION
PROGRAMS

James Ware, GROUP MANAGER
FUNDRAISING/MEMBERSHIP & MARKETING
Jim Boyd, GROUP MANAGER MANAGEMENT
INFORMATION SYSTEMS
Tom Fulgham, MAGAZINE EDITOR-IN-CHIEF
Gary Goodpaster, NATIONAL DIRECTOR OF EVENTS &
MEMBERSHIP SUPPORT
Dan Thiel, NATIONAL DIRECTOR OF MAJOR GIFTS & GIFT
PLANNING
Linda Schoenrock, NATIONAL DIRECTOR OF MARKETING
& COMMUNICATIONS
L. Mayeux, PRESIDENT
P.O. Box 1529, Rue De Medecine, Marksville, LA 71351
Phone: 318-253-9643
Stephen Reynolds, SECRETARY
899 Madison Avenue, Memphis, TN 38146
Phone: 901-227-5117
W. Lewis, TREASURER
P.O. Box 1344, Natchez, MS 39120
Phone: 601-446-6621

DUCKS UNLIMITED, INC.

WETLANDS AMERICA TRUST, INC. OFFICE
One Waterfowl Way
Memphis, TN 38120 USA
Phone: 901-758-3825   Fax: 901-758-3850
E-mail: ducks.org
Website: www.ducks.org

Founded: NA
Scope: Regional

Description: A nonprofit trust organized to operate exclusively for
charitable, educational, scientific and conservation purposes.
The Trust seeks to protect the natural balance of our
continent's wetland ecosystems, ensuring the future viability of
waterfowl and other wetland wildlife.

Contact(s):
Bill Willsey, ASSISTANT SECRETARY
D. Young, CHIEF OPERATING OFFICER
L. Mayeux, PRESIDENT

E

EAGLE NATURE FOUNDATION, LTD.

300 East Hickory
Apple River, IL 61001 USA
Phone: 815-594-2306   Fax: 815-594-2305
E-mail: eaglenature.tni@juno.com
Website: www.eaglenature.org

Founded: 1995
Membership: 4
Scope: International

Description: ENF is a nonprofit international organization, which
develops and implements habitat preservation strategies,
conducts a wide variety of nature education and awareness
programs, and engages in and supports bald eagle research.

Publication(s): Bald Eagle News, Nature News, Bald Eagle Bus
Tours

Keyword(s): Preservation, Environmental and Conservation
Education, Endangered Species, Birds

Contact(s):
James Ronnerud, DIRECTOR
Phone: 608-776-2755

Joseph Lukascyk, DIRECTOR
Phone: 708-430-0779
Yvonne Johnson, DIRECTOR
Phone: 815-895-4487
Art Gorov, DIRECTOR
Phone: 773-262-4662
David Smith, DIRECTOR
Phone: 715-344-5084
Terrence Ingram, PRESIDENT AND EXECUTIVE DIRECTOR
Juanita. Ray, SECRETARY
Phone: 708-447-1899
Susan Ertmer, TREASURER
Phone: 815-845-2253
Eugene Small, VICE PRESIDENT
Phone: 773-434-8328

EARTH DAY NEW YORK

201 E. 42nd St.,
New York, NY 10017 USA
Phone: 212-922-0048   Fax: 212-922-1936
E-mail: earthdayny@aol.com
Website: www.earthdayny.net

Founded: 1989
Scope: Regional

Description: Earth Day New York is a low-overhead, broadly
educational nonprofir 501c(3) organization that promotes environmental awareness and solutions through a three-pronged
program: 1) involving schools, teachers, and students through
the Earth Day Education Program; 2) educating public and
private policymakers through conferences; and 3) involving the
general public in annual Earth Day events.

Publication(s): Building the Sustainable Economy Conference I
Proceedings, Earth Day Education Program 1991-2001,
Lessons Learned High Performance Buildings, Lessons
Learned Four Times Square

Keyword(s): Environmental and Conservation Education, Green
Building, Internships

Contact(s):
Douglas Durst, CHAIRMAN
Phone: 212-789-1155
Fax: 212-789-1199
Pamela Lippe, EXECUTIVE DIRECTOR AND VICE
PRESIDENT
Phone: 212-922-0048
Fax: 212-922-1936
Fred Kent, PRESIDENT
Phone: 212-620-5660
Fax: 212-620-3821
Jim Tripp, SECRETARY
Phone: 212-505-2100
Fax: 212-505-2375
Timon Malloy, TREASURER
Phone: 203-535-5326
Fax: 203-353-5329

EARTH FORCE

1908 Mount Vernon Ave.
Alexandria, VA 22301 USA
Phone: 703-299-9400   Fax: 703-299-9485
E-mail: earthforce@earthforce.org
Website: www.earthforce.org

Founded: 1993
Membership: 15

Scope: National

Description: Earth Force is a national nonprofit environmental organization. Earth Force is dedicated to young people changing their communities and caring for our environment now, while developing life-long habits of active citizenship and environmental stewardship.

Publication(s): Free Campaign Materials for Kids and Educators

Keyword(s): Environmental and Conservation Education, Youth Organizations

Contact(s):
Christine Bates, BOARD OF DIRECTORS
Thomas Martin, PRESIDENT
Tom Martin, PRESIDENT AND DIRECTOR
Phone: 703-519-6867
Fax: 703-299-9485
F. Hagele, SECRETARY
9th Flr., 1515 Market St., Philadelphia, PA 19102
Phone: 212-851-8640
Donna Power, VICE PRESIDENT FOR LOCAL PROGRAMS
Vince Meldrum, VICE PRESIDENT FOR NATIONAL PROGRAMS

EARTH FOUNDATION
5401 Mitchelldale, Suite B4
Houston, TX 77092 USA
Phone: 713-686-9453
E-mail: sales@earthfound.com
Website: www.earthfound.com

Founded: 1990
Membership: 9
Scope: National

Description: The purpose of Earth Foundation is to empower educators and students to work towards a sustainable economy, just society, and healthy environment. Our focus is on education, fundraising for conservation, and cooperative programs with conservation groups and indigenous organizations working in the race to save the planet.

Publication(s): Rainforest Rescue Campaign Teacher Update

Keyword(s): Biodiversity, Environmental and Conservation Education, Protected Areas, Land Purchase, Endangered Species, Rainforests

Contact(s):
Cynthia Everage, PRESIDENT AND DIRECTOR
5151 Mitchelldale B11, Houston, TX 77092

EARTH ISLAND INSTITUTE
300 Broadway, Suite 28
San Francisco, CA 94133 USA
Phone: 415-788-3666   Fax: 415-788-7324
E-mail: earthisland@earthisland.org
Website: www.earthisland.org

Founded: 1982
Scope: National

Description: Through education and activism, Earth Island Institute counteracts threats to the biological and cultural diversity that sustains and enriches the global environment. The Institute develops and supports projects that promote the conservation, preservation, and restoration of the Earth. The Institute was founded by David Brower, veteran environmental leader.

Publication(s): Paper Locator, Ocean Alert, Earth Island Journal

Keyword(s): Endangered Species, Marine Mammals, Whale, Dolphin, Seal, Sustainable Development, Urban Environment

Contact(s):
John Knox, EXECUTIVE DIRECTOR
David Phillips, EXECUTIVE DIRECTOR
Bob Wilkinson, PRESIDENT
Maria Moyer-Angus, SECRETARY
Tim Rands, TREASURER

EARTH JUSTICE
ENVIRONMENTAL LAW CLINIC, UNIVERSITY OF DENVER LAW SCHOOL
University of Denver, Forbes House, 1714 Poplar St.
Denver, CO 80220 USA
Phone: 303-871-6996   Fax: 303-871-6991
E-mail: info@earthjustice.org
Website: www.earthjustice.org

Founded: NA
Scope: Regional

Description: Earthlaw's mission is to protect and preserve the West's biodiversity, air, land and water by providing free legal services to grassroots environmental groups and by training new lawyers in the strategy, skills and value of public interest environmental work.

Publication(s): Publications on website

Keyword(s): Environmental Law

Contact(s):
Buck Parker, EXECUTIVE DIRECTOR

EARTH SHARE
3400 International Dr., NW, Suite 2K
Washington, DC 20008 USA
Phone: 202-537-7100   Fax: 202-537-7101
E-mail: www.earthshare.org
Website: www.earthshare.org

Founded: 1988
Membership: 50
Scope: National

Description: Earth Share is a nonprofit, federated fund-raising organization that represents nonprofit environmental and conservation organizations in workplace payroll deduction campaigns nationwide. Funds raised support these organizations' environmental and conservation programs and services. Earth Share also provides educational public service announcements about the environment.

Publication(s): Annual Newsletter, Tips

Keyword(s): Air Quality and Pollution, Environmental and Conservation Education, Protected Areas, Exotic species, Aquatic nuisance species, training

Contact(s):
Jay Feldman, CHAIRMAN
701 E St. SE, Washington, DC 20003
Phone: 202-543-5450
Fax: 202-543-4791
Kalman Stein, PRESIDENT
Chuck Paquette, SECRETARY
8925 Leesburg Pike, Vienna, VA 22184
Phone: 703-790-4016

EARTHJUSTICE
DENVER OFFICE
1631 Glenarm Pl.
Denver, CO 80202 USA
Phone: 303-623-9466   Fax: 303-623-8083
E-mail: eajusco@earthjustice.org
Website: www.earthjustice.org

Founded: NA
Membership: 12
Scope: Regional

Description: (formerly Sierra Club Legal Defense Fund, Inc.)

Contact(s):
Susan Daggett, MANAGING ATTORNEY
Jim Angell, STAFF ATTORNEY
Eric Hubert, STAFF ATTORNEY

EARTHJUSTICE
JUNEAU OFFICE
325 4th St.
Juneau, AK 99801 USA
Phone: 907-586-2751   Fax: 907-463-5891
E-mail: eajusak@earthjustice.org
Website: www.earthjustice.org

Founded: NA
Membership: 9
Scope: National

Description: (formerly Sierra Club Legal Defense Fund, Inc.)

Contact(s):
Eric Jorgensen, MANAGING ATTORNEY
Janis Searles, STAFF ATTORNEY
Thomas Waldo, STAFF ATTORNEY

EARTHJUSTICE
SEATTLE, WASHINGTON OFFICE
705 Second Ave., Suite 203
Seattle, WA 98104 USA
Phone: 206-343-7340   Fax: 206-343-1526
E-mail: eajuswa@earthjustice.org
Website: www.earthjustice.org

Founded: NA
Membership: 9
Scope: Regional

Description: (formerly Sierra Club Legal Defense Fund, Inc.)

Contact(s):
Patti Goldman, MANAGING ATTORNEY
pgoldman@earthjustice.org
Todd True, STAFF ATTORNEY
ttrue@earthjustice.org

EARTHJUSTICE
WASHINGTON, DC OFFICE
1625 Massachusetts Ave., NW, Suite 702
Washington, DC 20036 USA
Phone: 202-667-4500   Fax: 202-667-2356
E-mail: eajusdc@earthjustice.org
Website: www.earthjustice.org

Founded: NA
Membership: 68
Scope: Regional

Description: (formerly Sierra Club Legal Defense Fund, Inc.)

Contact(s):
Howard Fox, MANAGING ATTORNEY
David Baron, STAFF ATTORNEY

EARTHJUSTICE (FLORIDA OFFICE)
FLORIDA OFFICE
111 S. Martin Luther King Jr. Blvd., P.O. Box 1329
Tallahassee, FL 32302 USA
Phone: 850-681-0031   Fax: 850-681-0020
Website: www.earthjustice.org

Founded: NA
Membership: 6
Scope: Regional

Description: (formerly Sierra Club Legal Defense Fund, Inc.)

Contact(s):
David Guest, MANAGING ATTORNEY
Ansley Samson, STAFF ATTORNEY

EARTHJUSTICE LEGAL DEFENSE FUND
Headquarters, 180 Montgomery St., Suite 1400
San Francisco, CA 94104-4209 USA
Phone: 415-627-6700   Fax: 415-627-6740
E-mail: eajus@earthjustice.org
Website: www.earthjustice.org

Founded: NA
Scope: National

Description: (formerly Sierra Club Legal Defense Fund, Inc.) A nonprofit, tax-deductible public-interest law firm created to bring lawsuits on behalf of environmental and citizens' organizations to protect the environment. As such, provides staff lawyers to initiate legal action. Also engages in administrative proceedings before federal, state, and local agencies, and negotiates settlement agreements whenever possible.

Publication(s): In Brief (newsletter), Ka Palila (Hawaii newsletter), Annual Report

Keyword(s): Air Quality and Pollution, Environmental Law, Forests and Forestry, Public Lands, training

Contact(s):
Bruce Neighbor, SENIOR PRESIDENT FINANCE AND ADMINISTRATION
Tom Turner, DIR. OF PUBLICATIONS
Buck Parker, EXECUTIVE DIR.
Bill Curtiss, SENIOR DIR. PROGRAMS
Steve Katz, SENIOR DIRECTOR OF DEVELOPMENT

EARTHJUSTICE LEGAL DEFENSE FUND
National Office 180 Montgomery Street, Suite 1400
San Francisco, CA 94104-4809 USA
Phone: 415-627-6700   Fax: 415-627-6740
E-mail: eajus@earthjustice.org
Website: www.earthjustice.org

Founded: 1971
Scope: Regional

Description: Nonprofit law firm for the environment representing without charge—hundreds of public interest clients—large numbers and small. Played leading role in shaping the development of environmental laws.

Contact(s):
Monique Harden, COMMUNITY LIAISON
Nathalie Walker, MANAGING ATTORNEY

Eric Huber, STAFF ATTORNEY
Esther Boykin, STAFF ATTOURNEY

EARTHJUSTICE LEGAL DEFENSE FUND
CALIFORNIA OFFICE
180 Montgomery St., Suite 1400
San Francisco, CA 94104 USA
Phone: 415-627-6700   Fax: 415-627-6740
E-mail: eajus@earthjustice.org
Website: www.earthjustice.org

Founded: NA
Membership: 150
Scope: National

Description: (formerly Sierra Club Legal Defense Fund)

Publication(s): In Brief

Contact(s):
Deborah Reames, MANAGING ATTORNEY
Paul Beach, STAFF ATTORNEY
Mike Sherwood, STAFF ATTORNEY

EARTHJUSTICE LEGAL DEFENSE FUND
HAWAII OFFICE
223 S. King, 4th Fl.
Honolulu, HI 96813 USA
Phone: 808-599-2436   Fax: 808-521-6841
E-mail: eajushi@earthjustice.org
Website: www.earthjustice.org

Founded: NA
Scope: Regional

Description: (formerly Sierra Club Legal Defense Fund, Inc.)

Contact(s):
Paul Achitoff, MANAGING ATTORNEY
David Henkin, STAFF ATTORNEY

EARTHJUSTICE LEGAL DEFENSE FUND
MONTANA OFFICE
209 S. Willson Ave.
Bozeman, MT 59715 USA
Phone: 406-586-9699   Fax: 406-586-9695
E-mail: eajusmt@earthjustice.org
Website: www.earthjustice.org

Founded: NA
Membership: 8
Scope: Regional

Description: (formerly Sierra Club Legal Defense Fund, Inc.)

Publication(s): In Brief - monthly newsletter, Pamphlets

Contact(s):
Douglas Honnold, MANAGING ATTORNEY
dhonnold@earthjustice.org

EARTHSHARE OF GEORGIA
1447 Peachtree St., Suite 214
Atlanta, GA 30309 USA
Phone: 404-873-3173   Fax: 404-873-3135
E-mail: info@earthsharega.org
Website: www.earthsharega.org

Founded: 1993
Scope: Regional

Description: The Environmental Fund for Georgia is a federation of 22 leading environmental organizations raising funding and awareness through workplace giving campaigns.

Publication(s): E Share Quarterly Newsletter

Keyword(s): Environment, Environmental and Conservation Education, Funding

Contact(s):
Elicia Fritsch, CAMPAIGN COORDINATOR
Alice Rolls, EXECUTIVE DIRECTOR
alice@efg.org

EARTHSTEWARDS NETWORK
P.O. Box 10697
Bainbridge Island, WA 98110 USA
Phone: 206-842-7986   Fax: 206-842-8918
E-mail: office@earthstewards.org
Website: www.earthstewards.org

Founded: 1980
Membership: 500
Scope: International

Description: International network for global conflict resolution. Utilizes rainforest reforestation and urban forestry projects to bring together peoples of cultures-in-conflict to work to heal the environment.

Publication(s): Warriors of the Heart, Earthstewards Newsletter, Essence Book of Days - 2001, Essence Book of Meditations and Blessings, Earthstewards Handbook

Keyword(s): Communications, Urban Forestry, Youth Organizations

Contact(s):
Lynn Ellis, OUTREACH COORDINATOR
157 Granite St., Apt. 1A, Biddeford, ME 04005
Phone: 207-284-4132
Fax: 207-284-9921
outreach@maine.rr.com
Rosemary Jones, PUBLISHING
healingpgs@aol.com

EARTHTRUST
25 Kaneohe Bay Dr.
Kailua, HI 96734 USA
Phone: 808-254-2866   Fax: 808-254-6409
E-mail: earthtrust@aloha.net
Website: www.earthtrust.org

Founded: 1976
Scope: International

Description: Earthtrust is an international nonprofit wildlife conservation organization. It involves small groups of highly capable people, involved with innovative investigations and projects, in partnership with private industry, governments and other environmental groups. Earthtrust is aimed at resolving wildlife crisis situations. Earthtrust's focus is to expose the poaching of endangered species and the sale of endangered whale meat in Asian markets through the use of DNA analysis and the protection of dolphins world-wide.

Keyword(s): Endangered Species, Environmental Law, Marine Mammals, Whale, Dolphin, Seal, training, Dolphins

Contact(s):
Donald White, PRESIDENT

EARTHWATCH INSTITUTE
3 Clocktower Place, Suite 100, Box 75
Maynard, MA 01754 USA
Phone: 978-461-0081   Fax: 978-461-2332
E-mail: info@earthwatch.org
Website: www.earthwatch.org

Founded: 1971
Scope: International

Description: Earthwatch is a nonprofit organization which sponsors scientific field research worldwide. It recruits paying volunteers to help field scientists with their research. Volunteers go on short term expeditions to 50 countries and 25 US states.

Publication(s): Earthwatch Magazine, Earthwatch Expedition Guide

Keyword(s): Endangered Species, Biodiversity, Volunteering, Cultural Preservation, Environmental and Conservation Education, Geology, Grants, Flowers, Plants, and Trees, Internships, Sustainable Development, Sustainable Ecosystems, Zoology, Research

Contact(s):
John Walker, CHIEF FINANCIAL OFFICER
jwalker@earthwatch.org
Marie Studer, EXECUTIVE DIRECTOR OF CENTER FOR FIELD RESEARCH
mstuder@earthwatch.org
Roger Bergen, PRESIDENT
rbergen@earthwatch.org
Andrew Mitchell, VICE PRESIDENT
amitchell@earthwatch.org

EAST CENTRAL ILLINOIS FUR TAKERS
853 E. 1000 N. Rd
Onarga, IL 60955 USA
Phone: 217-394-2577

Founded: 1974
Membership: 50
Scope: Local

Description: State chapter of Fur Takers of America. Helps monitor furbearing wildlife populations in the state and helps conserve this renewable resource.

Contact(s):
Louis Krumwiede, PRESIDENT
Phone: 217-394-2577

EASTERN SHORE LAND CONSERVANCY
P.O. Box 169
Queenstown, MD 21658 USA
Phone: 410-827-9756   Fax: 410-827-5765
E-mail: INFO@ESLC.ORG
Website: www.eslc.org

Founded: 1990
Membership: 1000
Scope: Regional

Description: The Eastern Shore Land Conservancy preserves farms, forests and natural areas for future generations, utilizing a variety of voluntary land protection tools that are available to landowners.

Publication(s): Preserving Land for Our Future, Fact Sheets, Panarama

Keyword(s): Land Preservation, Natural Areas, National Parks, Open Space, Land Protection

Contact(s):
Robert Etgen, EXECUTIVE DIRECTOR

ECOLOGICAL SOCIETY OF AMERICA, THE
1707 H St., NW, Suite 400
Washington, DC 20006 USA
Phone: 202-833-8773   Fax: 202-833-8775
E-mail: esahq@esa.org
Website: www.esa.sdsc.edu

Founded: 1915
Membership: 3500
Scope: National

Description: The Ecological Society of America is the nation's premier professional society of ecologists. ESA promotes the responsible application of ecological principles to the solution of environmental problems through ESA reports, journals, and expert testimony to Congress. Each summer, ESA convenes a conference featuring the latest findings in ecological research.

Publication(s): Ecological Applications, Ecological Monographs, Bulletin of the Ecological Society of America, Issues in Ecology, Ecology

Keyword(s): Biodiversity, Ecology, Environment, Wildlife Rehabilitation, Professional Organization, Ecosystems, training, Ecological Education, Endangered Species, Internships

Contact(s):
Nadine Lymn, DIRECTOR FOR PUBLIC AFFAIRS
Katherine Mccarter, EXECUTIVE DIRECTOR
Pamela Matson, PRESIDENT

ECOLOGY CENTER
2530 San Pablo Ave.
Berkeley, CA 94702 USA
Phone: 510-548-2220   Fax: 510-548-2240
E-mail: info@ecologycenter.org
Website: www.ecologycenter.org

Founded: 1969
Membership: 1200
Scope: Local, Regional, Statewide

Description: A nonprofit organization working to develop a more responsible society by identifying environmentally destructive practices and demonstrating sound alternatives. Programs include an environmental information clearinghouse, library, classes, book and ecoproducts store, sponsorship of three weekly farmers' markets, and weekly residential curbside recycling service in the city of Berkeley, CA. Primary service area: Greater San Francisco Bay region.

Publication(s): Terrain

Keyword(s): Ancient Forests, EcoAction, Endangered Species, Environment, Environmental Justice, Environmental and Conservation Education, Internships, Gardening and Horticulture, Land Preservation, People of Color in the Environment, Pesticides, Renewable Resources, Women in the Environment, Public Lands, training

Contact(s):
Laird Townsend, EDITOR

ECOTOURISM SOCIETY, THE

P.O. Box 668
Burlington, VT 05402 USA
Phone: 802-651-9818   Fax: 802-651-9819
E-mail: ecomail@ecotourism.org
Website: www.ecotourism.org

Founded: 1990
Membership: 1,400
Scope: International

Description: The Ecotourism Society is an international nonprofit membership organization dedicated to finding the resources and building the expertise to make tourism a viable tool for conservation and sustainable development.

Publication(s): Ecotourism Bibliography, Ecotourism Guidelines for Nature Tour Operators, Ecotourism Society Quarterly Newsletter, Ecolodge Sourcebook for Planners and Developers, Ecotourism: A Guide for Planners and Managers Volume I&II

Keyword(s): Biodiversity, Outdoor Recreation, Sustainable Development, Developing Countries, Tourism, Travel, Environmental and Conservation Education, Ecotourism

Contact(s):
Nicole Riotte, MEMBERSHIP AND BOOK PROGRAM DIRECTOR
Megan Wood, PRESIDENT
P.O. Box 755, North Bennington, VT 05257
Phone: 802-447-2121
Fax: 802-447-2122
ecomail@ecotourism.org

EDUCATIONAL COMMUNICATIONS

P.O. Box 351419
Los Angeles, CA 90035 USA
Phone: 310-559-9160   Fax: 310-559-9160
E-mail: ecnp@aol.com
Website: www.ecoprojects.org

Founded: 1958
Scope: International

Description: EC creates and promotes educational and scientific projects and programs for the public, focusing on environmental concerns. It founded The Ecology Center of Southern California in 1972; since 1977 has sponsored the award-winning Environmental Directions, a weekly national and international radio series heard in 8 states and on shortwave and internet; and since 1984 has produced three-time Emmy-nominated ECONEWS, a weekly television series broadcast on over 100 cable and PBS outlets nationally. Educational Communications, Inc. is credited with over 400 award-winning programs including "Gem in the Heart of the City," "Wind: Energy for the '90's and Beyond" and "Population Crisis USA." In 1993, started Project Ecotourism, to promote responsible travel and donations overseas.

Publication(s): catologs of programs, Econews TV and Environmental Directions Radio, Directory of Environmental Organizations, The Compendium Newsletter: A Guide to Ecological Activism

Keyword(s): Ecotourism, Communications, Environmental and Conservation Education, Population Growth, training

Contact(s):
Leslie Lewis, ADMINISTRATIVE COORDINATOR
Anna Harlowe, ASSOCIATE DIRECTOR
Nancy Pearlman, EXECUTIVE PRODUCER AND DIRECTOR

ELM RESEARCH INSTITUTE

Elm St., P.O. Box 150
Westmoreland, NH 03467 USA
Phone: 603-358-6198   Fax: 603-358-6305
E-mail: libertyelm@webryders.com
Website: www.libertyelm.com

Founded: 1967
Membership: 3000
Scope: National

Description: A nonprofit organization which has funded over $1,000,000 in research for the treatment of Dutch elm disease and development of the disease-resistant American Liberty Elm, supplies equipment and information pertaining to elm care and treatment of Dutch elm disease, propagates the American Liberty Elm and distributes it under the auspices of the Johnny Elmseed Project with the assistance of local Boy Scouts and other nonprofit groups. Over 750 nurseries have been established since 1984. ERI provides a free ceremonial tree and commemorative plaque to the first 100 communities which agree to plant 10 more trees in 2000.

Publication(s): Elm Leaves, Data on Elm Injections, Specialized Elm Care Information

Keyword(s): Endangered Species, Flowers, Plants, and Trees, Historic Preservation, Urban Forestry

Contact(s):
Yvonne Spalthoff, ASSISTANT DIRECTOR
John Hansel, EXECUTIVE DIRECTOR

ENDANGERED SPECIES COALITION

1101 14th St., NW, Suite 1001
Washington, DC 20005 USA
Phone: 202-682-9400   Fax: 202-756-2804
Website: www.stopextinction.org

Founded: 1982
Scope: National

Description: The goal of the Coalition is to broaden and mobilize public support for protecting endangered species.

Publication(s): Newsletter, Activist Tools

Keyword(s): Biodiversity, Endangered Species, Environmental Law

Contact(s):
Ed Lytwak, COMMUNICATIONS DIRECTOR
Brock Evans, EXECUTIVE DIRECTOR

ENGENDERHEALTH

440 9th Ave,
New York, NY 10001 USA
Phone: 212-561-8000   Fax: 212-561-8067
E-mail: info@engenderhealth.org
Website: www.engenderhealth.org

Founded: 1943
Scope: International

Description: A nonprofit family planning and reproductive health organization. To ensure through education, research, and service, that men and women everywhere have access to reprroductive health care, including safe contraception. Special expertise in the areas of female sterilization, vasectomy, postpartum contraceptive care, postabortion contraceptive care, and reproductive health programs for men.

Publication(s): see publications on website, EngenderHealth News

Keyword(s): Health and Nutrition, Population Growth, Family Planning

Contact(s):
Carmela Cordero, (ACTING) MEDICAL DIRECTOR AND VP
Lyman Brainerd, CHAIR
Terrence Jezowski, CHIEF OF OPERATIONS AND VICE PRESIDENT
Lynn Bakamjian, DIRECTOR OF PROGRAMS
Amy Pollack, PRESIDENT

ENTOMOLOGICAL SOCIETY OF AMERICA

9301 Annapolis Rd.
Lanham, MD 20706-3115 USA
Phone: 301-731-4535   Fax: 301-731-4538
E-mail: esa@entsoc.org
Website: www.entsoc.org

Founded: 1889
Membership: 5000
Scope: International

Description: To promote the scientific study of insects and related arthropods. Specialty sections include systematic behavior, toxicology, biogenetics, plant protection, medical and veterinary, regulatory and extension, and related scientific disciplines.

Publication(s): Journal of Economic Entomology, Environmental Entomology, Journal of Medical Entomology, American Entomologist, ESA Newsletter, Arthropod Management Tests, Annals of the Entomological Society of America

Keyword(s): Agriculture, Wildlife Rehabilitation, Biotechnology, Insects and Butterflies, Zoology

Contact(s):
Paula Luttice, EXECUTIVE DIRECTOR
Phone: 301-731-4535
esa@entsoc.org
Sharron Quinsenberry, PAST PRESIDENT
Larry Larson, PRESIDENT

ENVIRONMENTAL EDUCATION ASSOCIATION OF WASHINGTON

2142 Cispus Rd.
Randle, WA 98377 USA
Phone: 360-497-7131   Fax: 360-497-7132
E-mail: eeaw@eeaw.org
Website: www.eeaw.org

Founded: NA
Scope: Statewide

Description: The EEAW promotes and stimulates the development of effective environmental education in our state's schools and communities. The organization successfully creates an environmentally literate citizenry who practice care and respect for our state's natural environments. EEAW is a strong and vital organization that has successfully positioned environmental education as a resource to improve student learning and achievement, enhance business practices and support sustainable communities.

Keyword(s): Environmental and Conservation Education, Education

Contact(s):
Marty Fortin, PAST-PRESIDENT
Phone: 360-497-7131
fortin@myhome.net
Robert Olson, PRESIDENT
Phone: 509-624-4884

ENVIRONMENT COUNCIL OF RHODE ISLAND

P.O. Box 9061
Providence, RI 02940 USA
Phone: 401-621-8048   Fax: 401-331-5266
E-mail: environ@studentweb.providence.edu

Founded: NA
Membership: 500
Scope: Statewide

Description: A representative statewide organization, affiliated with the National Wildlife Federation, dedicated to the protection and enhancement of wildlife and its habitat through public education and government interaction.

Keyword(s): Birds, Environmental and Conservation Education, education, Renewable Resources, training

Contact(s):
Laura Landen, ALTERNATE REPRESENTATIVE
Alicia Karpick, EDITOR
Jeff Kos, PRESIDENT
Paul Beaudette, REPRESENTATIVE
Nina Rooks, TREASURER

ENVIRONMENTAL ACTION FUND (EAF)

P.O. Box 22421
Nashville, TN 37202 USA
Phone: 615-385-4389
Website: www.civictrust.org.uk/eaf

Founded: 1976
Scope: Statewide

Description: A nonprofit, nonpartisan union of citizen groups joined to preserve and protect Tennessee's natural resources and environmental health. EAF works for strong environmental legislative programs and policies.

Contact(s):
Mark Manner, PRESIDENT
2424 Golf Club Ln., Nashville, TN 37215
Sandy Bivens, SECRETARY
3504 General Bates Dr., Nashville, TN 37204
Paul Davis, TREASURER
5462 Vanderbilt Rd., Old Hickory, TN 37138

ENVIRONMENTAL ADVOCATES

353 Hamilton St.
Albany, NY 12210 USA
Phone: 518-462-5526   Fax: 518-449-4937
E-mail: info@eany.org
Website: www.eany.org

Founded: 1969
Scope: Statewide

Description: A representative statewide organization, affiliated with the National Willdlife Federation, dedicated to the protection and enhancement of wildlife and its habitat through public education and government interaction.

Publication(s): Albany Report, The Greensheet, Voters Guide

Contact(s):
Charles Kruzansky, ALTERNATE REPRESENTATIVE & TREASURER
Jeff Jones, EDITOR
jjones@envadvocates.org
Val Washington, EXECUTIVE DIRECTOR & EDUCATION PROGRAMS CONTACT
vwash@envadvocates.org
Oakes Ames, PRESIDENT
Steve Allinger, REPRESENTATIVE

ENVIRONMENTAL AIR FORCE

22 Rittenhouse Rd.
Broomall, PA 19008 USA
Phone: 610-353-1535   Fax: 610-356-5814

Founded: 1989
Scope: National

Description: A nonprofit membership organization dedicated to providing free aviation services to environmental and conservation groups worldwide. These services are provided through a network of member pilots and include aerial surveys and photography, flying essential observers, etc.

Publication(s): Despatches

Keyword(s): Environmental and Conservation Education, Protected Areas, Land Preservation, training

Contact(s):
Alan Brecher, EXECUTIVE DIRECTOR

ENVIRONMENTAL AND ENERGY STUDY INSTITUTE (EESI)

122 C St., NW
Washington, DC 20001 USA
Phone: 202-628-1400   Fax: 202-628-1825
E-mail: eesi@eesi.org
Website: www.eesi.org

Founded: 1985
Membership: 8
Scope: National

Description: The EESI is dedicated to promoting environmentally sustainable societies. EESI produces credible, timely information, and innovative public policy initiatives that lead to transitions to social and economic patterns that sustain people, the environment, and the natural resources upon which present and future generations depend.

Publication(s): The ECO Newsletter, Briefing Summaries

Keyword(s): Energy, Greenhouse Effect/Global Warming, Transportation, Exotic species, Aquatic nuisance species

Contact(s):
Richard Ottinger, CHAIR
Carol Werner, DIRECTOR OF ENERGY AND CLIMATE CHANGE PROGRAM

ENVIRONMENTAL CAREER CENTER

100 Bridge St., Suite A1
Hampton, VA 23669 USA
Phone: 757-727-7895   Fax: 757-727-7904
E-mail: : eccinfo@environmentalcareer.com
Website: http://www.environmental-jobs.com/

Founded: 1980
Scope: National

Description: Dedicated to helping people help the environment through internships, career counseling, diversity achievement program, career research, newsletters, and career seminars. Environmental Partnership Program provides paid apprenticeships in natural resources management and environmental protection. Partnerships with environmental employers and universities. Conducts career seminars for professional societies, universities, and agencies.

Keyword(s): Environmental and Conservation Education, Internships, People of Color in the Environment, Training

Contact(s):
John Esson, PRESIDENT AND EXECUTIVE DIRECTOR
Barbara Routten, SECRETARY
Kesha Oliver, TREASURER
John Damron, VICE PRESIDENT

ENVIRONMENTAL CAREERS ORGANIZATION, INC., THE

179 South St., 3rd Fl.
Boston, MA 02111 USA
Phone: 617-426-4375   Fax: 617-423-0998
Website: www.eco.org

Founded: 1972
Scope: National

Description: ECO protects and enhances the environment through the development of professionals, the promotion of careers and the inspiration of individual action. This is accomplished through placement, career advisement, career products and research and consulting. ECO has three regional offices and an alumni network of over 6,500 individuals.

Publication(s): Beyond the Green, Complete Guide to Environmental Careers, The

Keyword(s): Internships, People of Color in the Environment, Careers

Contact(s):
John Cook, PRESIDENT
jcook@eco.org

ENVIRONMENTAL CONCERN INC.

201 Boundary Lane, P.O. Box P
St. Michaels, MD 21663 USA
Phone: 410-745-9620   Fax: 410-745-3517
E-mail: order@wetland.org
Website: www.wetland.org

Founded: 1972
Membership: 600
Scope: National

Description: A nonprofit corporation founded for the purpose of researching, developing, and applying the technology and methodologies used in creating, restoring, and preserving wetlands. Environmental Concern focuses on consulting services, supply native wetland plants, the construction of wetlands, wetland horticulture and professional and scholastic educational programs.

Publication(s): Wow! The Wonders of Wetlands, Wetland Planting Guide for the Northeastern United States, Evaluation for Planned Wetlands: A Procedure for Assessing Wetland Functions and a Guide to Functional Design, "A Comprehensive Review of Wetlands" Wetland Journal, POW—the planning of wetlands

Keyword(s): Environmental and Conservation Education, Gardening and Horticulture, Renewable Resources, Research, Wetlands

Contact(s):
Edgar Garbisch, PRESIDENT
Joanna Garbisch, VICE PRESIDENT, SECRETARY & TREASURER

ENVIRONMENTAL DEFENSE
257 Park Ave. South
New York City, NY 10010 USA
Phone: 212-505-2100   Fax: 212-505-2375
Website: www.environmentaldefense.org
Scope: National

Description: Environmental Defense is an advocacy and research organization made up of scientists, economists, engineers, and attorneys who seek practical solutions to a broad range of environmental and human health problems.

Contact(s):
Karen Kenyon, EXECUTIVE ASSISTANT

ENVIRONMENTAL DEFENSE
HEADQUARTERS
257 Park Ave. South
New York, NY 10010 USA
Phone: 212-505-2100   Fax: 212-505-2375
Website: www.environmentaldefense.org

Founded: 1967
Scope: National

Description: Environmental Defense is an advocacy and research organization made up of scientists, economists, engineers, and attorneys who seek practical solutions to a broad range of environmental and human health problems.

Publication(s): EDF Letter

Keyword(s): Air Quality and Pollution, Aquatic Habitats, Biodiversity, Biotechnology, Chemical Pollution Control, Wildlife, Energy, Pollution Prevention, Exotic species, Aquatic nuisance species, Water Quality, Transportation, Watersheds, Wetlands

Contact(s):
John Wilson, CHAIRMAN OF THE BOARD OF TRUSTEES
Ed Bailey, DEPUTY DIRECTOR OF OPERATIONS
Marcia Aronoff, DEPUTY DIRECTOR OF PROGRAMS
Joel Plagenz, EDITOR
Phone: 212-505-2100
Fred Krupp, EXECUTIVE DIRECTOR
Phone: 212-505-2100
James Tripp, GENERAL COUNSEL
Annie Petsonk, INTERNATIONAL COUNSEL
Elizabeth Thompson, LEGISLATIVE DIRECTOR
Steve Cochran, STRATEGIC COMMUNICATIONS DIRECTOR

ENVIRONMENTAL DEFENSE
ALLIANCE FOR ENVIRONMENTAL INNOVATION
6 North Market Bldg., Faneuil Hall Marketplace
Boston, MA 02109 USA
Phone: 617-723-2996   Fax: 617-723-2999
Website: www.envirnomentaldefense.org/alliance

Founded: NA
Scope: Regional

Description: The Alliance for Environmental Innovation is a joint project of EDF and The Pew Charitable Trusts.

Contact(s):
Gwen Ruta, DIRECTOR

ENVIRONMENTAL DEFENSE
CAPITAL OFFICE
1875 Connecticut Ave., NW Ste 1016
Washington, DC 20009 USA
Phone: 202-387-3500   Fax: 202-234-6049
E-mail: members@environmentaldefense.org
Website: www.environmentaldefense.org

Founded: NA
Membership: 200
Scope: National

Contact(s):
Fred Krupp, EXECUTIVE DIRECTOR
257 Park Ave S, New York, New York 10010
Phone: 212-505-2100
fkrupp@environmentaldefense.org

ENVIRONMENTAL DEFENSE
NORTH CAROLINA OFFICE
2500 Blue Ridge Rd., Suite 330
Raleigh, NC 27607 USA
Phone: 919-881-2601   Fax: 919-881-2607
Website: www.environmentaldefense.org

Founded: NA
Membership: 250
Scope: Regional

ENVIRONMENTAL DEFENSE
TEXAS OFFICE
44 East Ave.
Austin, TX 78701 USA
Phone: 512-478-5161   Fax: 512-478-8140
Website: www.environmentaldefense.org

Founded: NA
Membership: 300000
Scope: Regional

ENVIRONMENTAL DEFENSE CENTER INC.
906 Garden St.
Santa Barbara, CA 93101 USA
Phone: 805-963-1622   Fax: 805-962-3152
E-mail: edc@edcnet.org
Website: www.edcnet.org

Founded: 1977
Membership: 22
Scope: Regional

Description: A nonprofit, public-interest environmental law firm providing legal services to citizens' groups and environmental organizations on environmental issues facing California's central coast region since 1977. The Center focuses on a wide range of issues, including oil development, toxic wastes, air and water pollution, species and habitat protection, open space preservation, land use, and coastal access.

Keyword(s): Environmental Law, Land Use Planning, Exotic species, Aquatic nuisance species, training

Contact(s):
Linda Krop, EXECUTIVE DIRECTOR
John Buse, STAFF ATTORNEY

ENVIRONMENTAL DEFENSE INC.
ROCKY MOUNTAIN OFFICE
2334 N. Broadway
Boulder, CO 80304 USA
Phone: 303-440-4901   Fax: 303-440-8052
Website: www.environmentaldefense.org

Founded: NA
Membership: 7
Scope: International

Contact(s):
Fred Krupp, REGIONAL DIRECTOR
257 Park Ave S 17th Floor, New York New York, 10010
Phone: 212-505-2100

ENVIRONMENTAL DEFENSE INC.
WEST COAST OFFICE
5655 College Ave. Suite 304
Oakland, CA 94618 USA
Phone: 510-658-8008   Fax: 510-658-0630
Website: www.environmentaldefense.org

Founded: NA
Membership: 3,000,000
Scope: International

ENVIRONMENTAL EDUCATION ASSOCIATES
2929 Main Street
Buffalo, NY 14214 USA
Phone: 716-833-2929   Fax: 716-833-9292
E-mail: eea@EnvironmentalEducation.com
Website: http://www.environmentaleducation.com/

Founded: 1994
Scope: Statewide

Description: A non-profit organization working in public high schools to provide students with coursework in environmental law and policy with a focus on endangered species, environmental justice and water quality.

Publication(s): Endangered Species Act: The Case of the Yellow-backed Rat Skunk, Environmental Justice: A Planning Commission Hearing to Approve/Deny a Household Hazardous Wast Facility Plan

Keyword(s): Education, Environmental and Conservation Education, Endangered Species, Environmental Law, Environmental Justice, Water Quality

ENVIRONMENTAL EDUCATION ASSOCIATION OF ILLINOIS
26893 Carol Ln.
Ingleside, IL 60041 USA
Phone: 847-740-2590
Website: web.stclair.k12.il.us/eeai/

Founded: 1970
Scope: Statewide

Description: Environmental Education Association of Illinois is the only organization in Illinois that makes environmental literacy its primary goal as it strives to instill a sense of community between the native ecosystems and people.

Publication(s): Illinois Environmental Education UPDATE

Keyword(s): Aquatic Habitats, Environmental and Conservation Education, Flowers, Plants, and Trees, Solid Waste Management, training

Contact(s):
Curt Carter, MEMBERSHIP
S.I.U.E. Mail Code 6888, Carbondale, IL 62901
Phone: 618-453-1121
Deb Chapman, PRESIDENT
Karen Zuckerman, PRESIDENT-ELECT
Phone: 309-697-1325
Kim Petzing, SECRETARY
Phone: 217-384-4062
Dave Guritz, TREASURER
Phone: 847-428-2240

ENVIRONMENTAL EDUCATION ASSOCIATION OF INDIANA
Attn: President, Richardson, Wildlife Sanctuary, 64 West Rd., Dune Acres
Chesterton, IN 46304 USA
Phone: 219-787-8983   Fax: 219-787-1341
E-mail: jhtjr@niia.net
Website: http://www.hasti.org/groups/eeai.html

Founded: 1969
Scope: Statewide

Description: A statewide, nonprofit organization dedicated to the wise use and management of natural resources through environmental conservation education. Activities include an annual meeting, workshops, teaching materials, exhibits, and youth environmental summit.

Publication(s): CREED Newsletter

Keyword(s): Environmental and Conservation Education, Education

Contact(s):
Paul Steury, PRESIDENT ELECT
P.O. BOX 263, Wolf Lake, IN 46796
Phone: 219-779-5869
paulds@goshen.edu
Doug Waldman, TREASURER
11832 Kress Rd., Roanoke, IN 46783
Phone: 317-672-3842
Deborah Messenger, VICE PRESIDENT
402 W. Washington St. W-265, Indianapolis, IN 46204-2739
Phone: 317-233-3872
deborahmessenger@netzero.net

ENVIRONMENTAL EDUCATION COUNCIL OF OHIO
P.O. Box 2911
Akron, OH 44309-2911 USA
Phone: 330-761-0855   Fax: 330-761-0856
E-mail: director@eeco-online.org
Website: www.eeco-online.org

Founded: 1967
Membership: 600
Scope: State

Description: EECO is a statewide organization whose purpose is to promote environmental education which nurtures knowledge, attitudes, and behaviors that foster global stewardship. EECO brings together educators from many settings to provide opportunities to share ideas, materials, and techniques. Members include classroom teachers, naturalists, camp staff, teacher educators, youth leaders, and agency personnel.

Publication(s): Ohio Sampler: Outdoor and Environmental Education, Directory of Ohio Environmental Education Sites and Resources, Integrating Environmental Education and Science, EECO Newsletter

Keyword(s): Environmental and Conservation Education, Youth Organizations, Stewardship, Education

Contact(s):
Teresa Mourad, EXECUTIVE DIRECTOR
Phone: 330-761-0855
Dave Irvine, PRESIDENT
Jeanne Russell, SECRETARY
Phone: 614-265-6682
jeanne.russell@dnr.state.oh.us
Angela Manuszak, TREASURER
Charles McClaugherty, VICE PRESIDENT
Phone: 330-666-9200
irvine@crownpt.com

ENVIRONMENTAL EDUCATORS OF NORTH CAROLINA (EENC)

P.O. Box 4901
Chapel Hill, NC 27515-4901 USA
Phone: 919-250-1050
Website: www.eenc.org

Founded: 1990
Membership: 300
Scope: Statewide

Description: EENC advocates and supports the development and implementation of quality education which promotes responsible environmental decision-making and actions. Sponsors workshops and an annual conference.

Publication(s): EENC Networking Directory, EENC Newsletter, EENC Brochure

Keyword(s): Environmental and Conservation Education, Education

Contact(s):
Deborah Miller, ADVISOR
Phone: 919-541-5552
Aaryn Kay, PRESIDENT
Phone: 910-251-0191

ENVIRONMENTAL ENTERPRISES ASSISTANCE FUND, INC.

1655 N. Fort Meyer Dr., Fifth Fl.
Arlington, VA 22209 USA
Phone: 703-522-5928   Fax: 703-522-6450
E-mail: eeaf@igc.org
Website: www.eeaf.org

Founded: 1990
Membership: 18
Scope: National

Description: EEAF is a non-profit organization that operates as a venture capital fund; it provides long term risk capital and management asistance to environmentally beneficial businesses in developing countries, where such capital is otherwise unavailable.

Keyword(s): Agriculture, Biodiversity, Developing Countries, Energy, Wildlife, Energy Conservation, Pollution Prevention, Renewable Resources, Sustainable Development, Coral Reefs, Solar Energy

Contact(s):
Marion Heckclay, CHIEF FINANCIAL OFFICER
Brooks Browne, PRESIDENT
J. Doliner, VICE PRESIDENT

ENVIRONMENTAL LAW ALLIANCE WORLDWIDE (E-LAW)

U.S. Office: 1877 Garden Ave.
Eugene, OR 97403 USA
Phone: 541-687-8454   Fax: 541-687-0535
E-mail: elawus@elaw.org
Website: www.elaw.org

Founded: 1989
Membership: 9
Scope: International

Description: E-LAW is an international network of public-interest attorneys and scientists dedicated to using law to protect the environment. E-LAW's 24 offices, located around the world exchange vital legal and scientific information. Advocates in more than 50 countries are linked electronically and can call on E-LAW for information to support their environmental protection work.

Publication(s): E-LAW Advocate

Keyword(s): Environmental Law

Contact(s):
Karin Sheldon, CHAIR
School of Law, University of Oregon, Eugene, OR 97403
Bern Johnson, EXECUTIVE DIRECTOR
Penelope Foster, PRESIDENT
School of Law, University of Oregon, Eugene, OR 97403

ENVIRONMENTAL LAW AND POLICY CENTER OF THE MIDWEST

35 East Wacker Dr., Suite 1300
Chicago, IL 60601-2208 USA
Phone: 312-673-6500   Fax: 312-795-3730
E-mail: elpc@elpc.org
Website: www.elpc.org

Founded: 1993
Membership: 40
Scope: Regional

Description: A nonprofit public interest environmental advocacy organization working to implement sustainable energy strategies, promote innovative transportation approaches, expand and develop green markets and develop sound environmental management practices in Illinois, Indiana, Michigan, Minnesota, Ohio and Wisconsin.

Publication(s): Visions, Choosing a Future for Growing Communities, Lake County at the Crossroads No. 2, Repowering of the Midwest

Keyword(s): Transportation, Sustainable Energy, Sustainability, Environmental Living, Management Plans, Environmental Law

Contact(s):
Kevin Brubaker, DIRECTOR OF OPERATIONS
kbrubaker@elpc.org
Howard Learner, EXECUTIVE DIRECTOR
hlearner@elpc.org

ENVIRONMENTAL LAW INSTITUTE, THE
1616 P St., NW
Suite 200
Washington, DC 20036 USA
Phone: 202-939-3800   Fax: 202-939-3868
Website: www.eli.org

Founded: 1969
Scope: National

Description: The Environmental Law Institute advances environmental protection by improving law, policy and management. ELI researches pressing problems, educates professionals and citizens about the nature of these issues, and convenes all sectors in forging effective solutions.

Publication(s): National Wetlands Newsletter, Environmental Forum, The, ELR - Environmental Law Reporter, The

Keyword(s): Biodiversity, Conservation, Natural Resource Conservation, Forests and Forestry, Public Lands, Rivers, Water Pollution Management, Wilderness, International Environmental Law, Pollution Control, Federalism, Environmental Law, Wetlands, Sustainable Development

Contact(s):
Donald Stever, CHAIRMAN OF THE BOARD
Stephen Dujack, DIRECTOR OF COMMUNICATIONS
Lawrence Ross, HEAD LIBRARIAN
Phone: 202-328-5150
J. Futrell, PRESIDENT
Robert Percival, SECRETARY AND TREASURER

ENVIRONMENTAL LEAGUE OF MASSACHUSETTS
14 Beacon, Suite 714
Boston, MA 02108 USA
Phone: 617-742-2553   Fax: 617-742-9656
E-mail: elm@environmentalleague.org
Website: environmentalleague.org

Founded: 1898
Scope: Statewide

Description: A representative statewide organization, affiliated with the National Wildlife Federation, dedicated to the protection and enhancement of wildlife and its habitat through public education and government interaction.

Publication(s): ELM Bulletin, ELM Action Alerts

Keyword(s): Aquatic Habitats, Wildlife, Renewable Resources, Water Pollution Management, Wetlands

Contact(s):
Nancy Smith, ALTERNATE REPRESENTATIVE
Jeremy Marin, EDITOR & EDUCATION PROGRAMS CONTACT
James Gomes, EXECUTIVE DIRECTOR & PRESIDENT
Stephen Leonard, REPRESENTATIVE
John Cronin, TREASURER

ENVIRONMENTAL LEAGUE OF MASSACHUSETTS (ELM)
14 Beacon St.
Boston, MA 02108 USA
Phone: 617-742-2553   Fax: 617-742-9656
E-mail: elm@environmentalleague.org
Website: www.environmentalleague.org

Founded: 1898
Scope: Statewide

Description: Advocates for responsible environmental policy on the state level and the effective implementation of state programs dealing with issues such as land use, toxics use reduction, recycling, water resources protection and funding for environmental programs, in addition to being educating the public about the environmenta and environmental issues.

Publication(s): ELM Action Alerts, ELM Bulletin

Keyword(s): Environmental Law, Land Use Planning, Pollution Prevention, Open Space, Toxic Substances, Nuclear-free, Water quantity, Water export and diversion, Exotic species, Aquatic nuisance species, Air Quality and Pollution, Chemical Pollution Control, Environment, Land Preservation, Pesticides, Rivers, Coral Reefs, Watersheds, Sustainable Development

Contact(s):
Jessica Champness, BUSINESS MANAGER
Lauren Stiller-Rikleen, CHAIRMAN
Jeremy Marin, COMMUNICATIONS/ MKTG. DIRECTOR
Pamela Dibona, LEGISLATIVE DIRECTOR
James Gomes, PRESIDENT
Nancy Goodman, RESEARCHER DIRECTOR
John Cronin, TREASURER

ENVIRONMENTAL MEDIA ASSOCIATION
10780 Santa Monica Blvd., Suite 210
Los Angeles, CA 90025 USA
Phone: 310-446-6244   Fax: 310-446-6255
E-mail: ema@ema-online.org
Website: www.ema-online.org

Founded: 1989
Scope: National

Description: EMA works to mobilize the entertainment community in a global effort to educate people about environmental problems, and inspire them to act on those problems now.

Publication(s): Green Light

Keyword(s): Environmental Communication

Contact(s):
Patie Maloney, DIRECTOR OF SPECIAL EVENTS & PUBLIC RELATIONS
Debbie Levin, EXECUTIVE DIRECTOR
Jennifer Deperalta, PROGRAM & EVENTS MANAGER

ENVIRONMENTAL RESOURCE CENTER (ERC)
411 East Sixth St., P.O. Box 819
Ketchum, ID 83340 USA
Phone: 208-726-4333   Fax: 208-726-1531
E-mail: erc@ercsv.org
Website: www.ercsv.org

Founded: 1989
Membership: 600
Scope: Local, Regional

Description: The ERC is a nonprofit organization that provides resources and educational programs to the public about local, regional, and global environmental issues.

Publication(s): Local Dirt

Keyword(s): Environmental and Conservation Education, Ecology, Internships, Energy Conservation, education

Contact(s):
Craig Barry, EXECUTIVE DIRECTOR

ENVIROSOUTH, INC.
P.O. Box 11468
Montgomery, AL 36111 USA
Phone: 334-277-7050   Fax: 205-277-7080
E-mail: scrc@mindspring.com

Founded: 1975
Scope: Regional

Description: A private nonprofit organization specializing in recycling information and related services for the Southeast Recycling Market Council, and the annual Southeast Recycling Conference and Trade Show.

Keyword(s): Communications, Environmental and Conservation Education, education, Renewable Resources, Solid Waste Management

Contact(s):
Martha McInnis, PRESIDENT

E-P EDUCATION SERVICES, INC.
15 Brittany Ct.
Cheshire, CT 06410 USA
Phone: 203-271-2756   Fax: 203-271-2756
Website: http://www.nvwf.org/web/

Founded: 1972
Scope: Statewide

Description: A nonprofit group formed to promote environmental and population education in Connecticut and committed to assisting educators in the task of providing quality environmental education for the citizens of our state.

Keyword(s): Environmental and Conservation Education, Land Use Planning, Pesticides, Planning Management, Wetlands

Contact(s):
Larry Schaefer, PRESIDENT AND EXECUTIVE DIRECTOR
Lina Lawall, SECRETARY
J. Bouchard, TREASURER
Michael Schaefer, VICE PRESIDENT

EVERGLADES COORDINATING COUNCIL (ECC)
22951 Southwest 190 Ave.
Miami, FL 33170 USA
Phone: 305-248-9924   Fax: 305-248-9924
E-mail: evcoord@aol.com

Founded: 1970
Membership: 15
Scope: State

Description: ECC is an umbrella organization of south Florida sportspersons and conservation organizations united in a desire to protect wildlife habitat, assure sound wildlife management practices, and provide for properly regulated outdoor recreational activities.

Publication(s): Newsletter

Keyword(s): Hunting, Outdoor Recreation, Wetlands, training, Everglades

Contact(s):
Ralph Johnson, DIRECTOR
7901 W. 25th Ct., Hialeah, FL 33016
Phone: 305-825-4667
Fax: 305-362-7584
rj005@aol.com
Bishop Wright, PRESIDENT
Barbara Powell, SECRETARY
22951 SW 190th Ave., Miami, FL 33170
barjnpwll@aol.com
Dave Charland, TREASURER
3559 NW 52nd St., Ft. Lauderdale, FL 33309
Phone: 305-484-7777
Fax: 954-484-7834

F

FAIRFAX AUDUBON SOCIETY
4022 Hummer Rd.
Annandale, VA 22003-0128 USA
Phone: 703-256-6895   Fax: 703-256-2060
E-mail: fairfaxaud@erols.com
Website: www.fairfaxaudubon.org

Founded: 1980
Membership: 4800
Scope: Regional

Description: The Fairfax Audubon Society—a chapter of the National Audubon Society, is committed to the Audubon mission which is to conserve and restore natural ecosystems, focusing on birds and other wildlife, and their habitats.

Publication(s): Potomac Flier

Keyword(s): Birds, Biodiversity, Conservation, Ecology, Environmental and Conservation Education, Environmental Planning, Protected Areas, Environmental Protection, Natural Areas, Nongame Wildlife, Raptors, training

Contact(s):
Christine Winslow, PRESIDENT
cwinslow@birchdavis.com
Maynard Davis, TREASURER
Annandale, VA 22003
mdavis@tnc.org

FEDERAL CARTRIDGE COMPANY
900 Ehlen Dr.
Anoka, MN 55303 USA
Phone: 612-323-3827   Fax: 612-323-2506
E-mail: om
Website: www.federalcartridge.com

Founded: NA
Scope: National

Keyword(s): Hunting

Contact(s):
William Stevens, CONSERVATION MANAGER

FEDERAL WILDLIFE OFFICERS ASSOCIATION
4094 Majestic Ln. PMB 214
Fairfax, VA 22033 USA
Phone: 603-433-0502
Website: www.fwoa.org

Founded: NA
Scope: National

Description: The Federal Wildlife Officer's Association is an organization dedicated to the protection of wildlife and plants, the enforcement of federal wildlife law, the fostering of cooperation and communication among federal wildlife officers and the perpetuation, enhancement and defense of the wildlife officer profession.

Publication(s): The Federal Wildlife Officer Newsletter

Keyword(s): Law Enforcement, Professional Organization, Environmental and Conservation Education, Environmental Protection, Hunting, Mammals, Marine Mammals, Whale, Dolphin, Seal, Nongame Wildlife, Pesticides, Predators, Birds, Endangered Species, Environmental Justice, Raptors

Contact(s):
Timothy Santel, PRESIDENT
Phone: 217-793-9554
Fax: 217-793-2835
James Gale, SECRETARY AND TREASURER
Phone: 517-686-4578
Fax: 517-686-2837
Christopher Dowd, VICE PRESIDENT
Phone: 617-242-7874

FEDERATION OF ALBERTA NATURALISTS

Box 1472
Edmonton, Alberta T5J 2N5 Canada
Phone: 708-427-8124
Website: http://fanweb.ca

Founded: 1970
Scope: Statewide

Description: To increase Albertans' knowledge of natural history; foster creation of new natural history groups; promote natural areas; and provide a forum for discussion and means of taking action on environmental problems of concern to naturalists.

Publication(s): Alberta Naturalist

Keyword(s): education, Birds, Environmental and Conservation Education, National Parks, Public Lands

Contact(s):
Brian Parker, EDITOR
Glen Semenchuk, EXECUTIVE DIRECTOR
Derek Johnson, PRESIDENT
Pat Clayton, TREASURER

FEDERATION OF FLY FISHERS

P.O. Box 1595, 502 S. 19th, Suite 101
Bozeman, MT 59771 USA
Phone: 406-585-7592   Fax: 406-585-7596
E-mail: fffoffice@fedflyfishers.org
Website: www.fedflyfishers.org

Founded: 1965
Membership: 1000
Scope: International

Description: To promote international fly fishing as a most enjoyable and sportsmanlike method of fishing and to preserve all species of fish in all classes of waters through local stream and fisher restoration projects, conservation grants, audiovisual programs, public education, and international committees.

Publication(s): Osprey,The, Clubwire, The, Flyfisher, The

Keyword(s): Environmental and Conservation Education, Protected Areas, Wildlife, Sport Fishing

Contact(s):
Lauren Bagley, MANAGING DIRECTOR

FEDERATION OF NEW YORK STATE BIRD CLUBS, INC.

P.O. Box 440
Loch Sheldrake, NY 12759 USA
Website: birds/cornell.edu/fnysbc

Founded: 1947
Scope: Statewide

Description: To further the study of birdlife in New York state and to disseminate knowlege thereof, to educate the public on the need for conserving natural resources and to document the ornithology of the state.

Publication(s): New York Birders, Checklist of the Birds of New York State, Kingbird, The

Keyword(s): Birds, training

Contact(s):
Mary Koeneke, PRESIDENT
362 Nine Mile Point Rd., Oswego, NY 13126
Phone: 315-342-3402

FEDERATION OF ONTARIO NATURALISTS

355 Lesmill Rd.
Don Mills, Ontario M3B 2W8 Canada
Phone: 416-444-8419   Fax: 416-444-9866
E-mail: info@ontarionature.org
Website: www.ontarionature.org

Founded: 1931
Scope: State

Description: Committed to protecting and increasing awareness of Ontario's natural areas and wildlife, and exerts influence to protect our natural environment. Eighty-three federated clubs across Ontario.

Publication(s): Seasons

Keyword(s): Natural Areas

Contact(s):
Jean Labrecque, CHIEF ADMINISTRATIVE OFFICER
Gregory Beck, DIRECTOR OF CONSERVATION
Nancy Clarke, EDITOR
Ric Stymmes, EXECUTIVE DIRECTOR
Mark Dorfman, PRESIDENT

FEDERATION OF WESTERN OUTDOOR CLUBS

512 Boylston Ave. E., 106
Seattle, WA 98102 USA
Phone: 206-322-3041

Founded: 1932
Scope: Regional

Description: Established for mutual service and for the promotion of the proper use, enjoyment and protection of America's scenic, wilderness, and outdoor recreation resources. Forty-three affiliated clubs in Alaska, British Columbia, and the western states.

Publication(s): Outdoors West

Keyword(s): Environmental and Conservation Education, Forests and Forestry, Outdoor Recreation, training, Deserts

Contact(s):
Hazel Wolf, EDITOR
512 Boylston Ave., E, 106, Seattle, WA 98102
Brock Evans, PRESIDENT
5449 33rd Ave., NW, Washington, DC 20015
Nancy Kroening, SECRETARY
5615 40th Ave., W, Seattle, WA 98199
Martin Huebner, TREASURER
1995 McKinzie Dr., Idaho Falls, ID 83404
Winchell Hayward, VICE PRESIDENT
208 Willard N., San Fransico, CA 94118

FISH AND WILDLIFE REFERENCE SERVICE

5430 Grosvenor Ln.
Bethesda, MD 20814 USA
Phone: 800-582-3421   Fax: 301-564-4059
E-mail: fw9_fa_reference_service@fws.gov
Website: http://fa.r9.fws.gov/r9fwrs

Founded: 1965
Scope: National

Description: A computerized information retrieval system and clearinghouse, providing selected reports on fish and wildlife management. Operated under a contract with the U.S. Fish and Wildlife Service to provide access to reports produced by the Federal Aid in Fish and Wildlife Restoration Program, the Cooperative Fishery and Wildlife Research Units Program, the Endangered Species Grants Program, and the Anadromous sport Fish Conservation Program.

Publication(s): Fish and Wildlife Reference Service (Newsletter)

Keyword(s): Endangered Species, Wildlife, Nongame Wildlife, training

Contact(s):
Geoffrey Yeadon, EDITOR
Paul Wilson, PROJECT LEADER

FISH FOREVER

1271 Quaker Hill Dr.
Alexandria, VA 22314 USA
Phone: 703-461-9201   Fax: 703-461-9290

Founded: NA
Membership: 2000
Scope: National

Contact(s):
David Allison, PRESIDENT
dallison@msn.com

FLORIDA ASSOCIATION OF SOIL AND WATER CONSERVATION DISTRICTS

Attn: President, 16806 NW 40th PL
Newberry, FL 32669 USA
Phone: 352-472-5462   Fax: 352-472-4473

Founded: NA
Scope: Statewide

Keyword(s): Conservation Districts

Contact(s):
Tim Ford, PRESIDENT
Phone: 352-472-5462
Fax: 352-472-5435

FLORIDA B.A.S.S. CHAPTER FEDERATION

Attn: President
Cape Coral, FL 33915 USA
Phone: 863-763-9265
Website: www.floridabassfederation.com

Founded: NA
Membership: 1300
Scope: Statewide

Description: An organization of Bassmaster chapters, affiliated with the Bass Anglers Sportsman Society, organized to fight pollution, assist state and national conservation agencies in their efforts, and teach young people of our good conservation practices. Dedicated to the realistic conservation of our water resources.

Publication(s): see web site for pub listings

Contact(s):
Carroll Head, CONSERVATION DIRECTOR
2252 SW 22nd Circle., Okeechobee, FL 34974
Phone: 863-763-3568
Harvey Ford, PRESIDENT
Phone: 863-763-9265

FLORIDA DEFENDERS OF THE ENVIRONMENT, INC.

HOME OFFICE
4424 NW 13 St., Suite C-8
Gainesville, FL 32609 USA
Phone: 352-378-8465   Fax: 352-377-0869
E-mail: fde@bellsouth.net
Website: www.fladefenders.org

Founded: NA
Scope: Statewide

Description: FDE promotes conservation, restoration, and sustainable use of Florida's natural resources by providing the public and private sector with objective information and analysis developed through a statewide network of voluteer specialists. Guided by the motto "FDE gets the facts," the organization achieves realistic goals by targeting a limited number of complex environmental issues and providing expert scientific analysis, sustained tracking, advocacy, and litigation when necessary.

Publication(s): Monitor, The

Keyword(s): Watersheds, Rivers, Exotic species, Aquatic nuisance species, Water Quality, Environmental Law, training, Biodiversity, Land Use Planning

Contact(s):
Steve Leitman, COORDINATOR OF APALACHICOLA RIVER
David White, COORDINATOR OF OCKLAWAHA RIVER
Susan Vince, COORDINATOR OF SUWANNEE RIVER
Bob Simons, COORDINATOR OF SUWANNEE RIVER & PUBLIC LANDS COMMITTEE
Leslie Straub, EXECUTIVE DIRECTOR
Richard Hamann, PRESIDENT
Phone: 352-392-2237
Fax: 352-392-1457
Kristina Jackson, RESTORATION SPECIALIST
David Bruderly, SECRETARY
Frank Nordlie, TREASURER
Joe Siry, VICE PRESIDENT

FLORIDA EXOTIC PEST PLANT COUNCIL

P.O. Box 24680
West Palm Beach, FL 33416 USA
Phone: 305-242-7846
E-mail: tony_ternas@nps.gov
Website: www.fleppc.org

Founded: 1984
Membership: 350
Scope: Statewide

Description: FLEPPC goals are directed toward building public awareness about the serious threat invasive plants pose to native ecosystems, secure funding, and support for control and management of exotic plants, and developing integrated management and control methods.

Publication(s): Wildland Weeds, Florida Exotic Pest Plant Council Newsletter

Keyword(s): Conservation, Wildlife Rehabilitation, Plants, Protected Areas, Natural Areas

Contact(s):
Ken Langeland, CHAIRPERSON
Phone: 305-242-7846
tony_pernas@nps.gov
Amy Ferriter, EDITOR
Phone: 561-682-6097
aferriter@sfwmd.gov
Jackie Smith, SECRETARY
Phone: 561-791-4720
Dan Thayer, TREASURER
Phone: 561-682-6129
Fax: 561-681-6232

FLORIDA FORESTRY ASSOCIATION

P.O. Box 1696
Tallahassee, FL 32302 USA
Phone: 850-222-5646   Fax: 850-222-6179
E-mail: info@forestfla.org
Website: www.floridaforest.org

Founded: 1923
Membership: 2000
Scope: Statewide

Description: Nonprofit, trade-supported organization of industries, businesses, and individuals who encourage the promotion, development and protection of forestry in Florida.

Publication(s): Florida Forests Magazine, Pines and Needles

Keyword(s): Forests and Forestry

Contact(s):
Jeff Doran, EXECUTIVE VICE PRESIDENT AND EDITOR
Doyle Majors, PRESIDENT
Charles Thompson, SECRETARY & TREASURER

FLORIDA NATIVE PLANT SOCIETY

P.O. Box 690278
Vero Beach, FL 32969-0278 USA
Phone: 561-562-1598

Founded: 1980
Scope: Statewide

Description: Promotes preservation, conservation, and restoration of native plants and native plant communities of Florida, and provides information through publications, conferences, workshops and a statewide membership organized by local chapters.

Publication(s): Florida's Incredible Wild Edibles, Planning and Planting Your Native Plant Yard, Butterfly Gardening with Florida's Native Plants, Common Grasses of Florida and the Southeast, Big Trees: The Florida Register, Palmetto, The

Keyword(s): Native Plants, Biodiversity, Endangered Species, Environmental and Conservation Education, Flowers, Plants, and Trees, Natural Areas

Contact(s):
Candace Weller, PRESIDENT
1515 Country Club Rd. N., St. Petersburg, FL 33710
Phone: 727-345-4619
Robert Bareiss, TREASURER
10301 Bellwood Ave., New Port Richey, FL 34654
Phone: 727-842-3133
Don Spence, VICE PRESIDENT
P.O. Box 321, Roseland, FL 32957
Phone: 407-589-0319
Kim Zarillo, VICE PRESIDENT
760 Cajeput Cir., Melbourne Village, FL 32904
Phone: 407-727-1713

FLORIDA ORNITHOLOGICAL SOCIETY

1503 Wekewa Nene c/o Peter Merritt
Hobe Sound, FL 33455 USA
Phone: 850-942-2489   Fax: 561-546-2781
Website: www.FOSbirds.org

Founded: 1072
Membership: 500
Scope: Regional

Description: To engage in pursuits that advance ornithology in Florida; to facilitate education about birds in the wild; to unite amateurs and professionals on the study of birds in the wild; and to publish a scientific journal and other publications, relevant to the member's common interests.

Publication(s): Florida Ornithological Society Newsletter, Florida Field Naturalist

Keyword(s): Birds, Endangered Species, Environmental and Conservation Education, education, Nongame Wildlife

Contact(s):
Jerry Jackson, EDITOR
Whitaker Center 10501 Fl Coast University Blvd, Fort Myers, FL 33965-6565
Phone: 941-348-1468
Fax: 941-590-7200
Peter Merritt, PRESIDENT
Bob Henderson, SECRETARY
Dean Jue, TREASURER
dsjue@earthlinl.net
Ann Paul, VICE PRESIDENT
7217 N. Ola, Tampa, FL 33604
Phone: 941-643-2249
Phone: 813-623-4086
aschnapf@audubon.org

FLORIDA PANTHER PROJECT, INC., THE

P.O. Box 19866
Sarasota, FL 34276 USA
Phone: 941-379-2221

Founded: 1993
Scope: Statewide

Description: To assist in the sensible and responsible recovery of the Florida Panther in Florida, by raising funds to purchase environmentally sensitive panther habitat across Florida. Guest Speakers Available

Keyword(s): Endangered Species, Environmental and Conservation Education, Land Purchase, Nongame Wildlife, training

Contact(s):
Tim Mallon, ADVISORY BOARD
3715 Felda St., Cocoa, FL 32926
Phone: 407-633-4799
Judy Conda, BOARD OF DIRECTORS
6551 Gulfgate Pl., Sarasota, FL 34321
Phone: 941-921-7300
Bob Mills, EXECUTIVE DIRECTOR
Gifts and Fundraising 4269 Hearthstone Pl.,
Sarasota, FL 34238
Phone: 941-966-7765
William Samuels, PRESIDENT
P.O. Box 19866, Sarasota, FL 34276
Phone: 941-379-2221

FLORIDA PUBLIC INTEREST RESEARCH GROUP (FLORIDA PIRG)

704 West Madison St.
Tallahassee, FL 32304 USA
Phone: 850-224-3321   Fax: 850-224-1310
E-mail: floridapirg@pirg.org
Website: www.pirg.org/floridapirg

Founded: 1981
Scope: State

Description: Florida PIRG is a nonprofit organization committed to researching, educating, organizing, and advocating programs to protect Florida's environment. These programs include preventing offshore drilling, promoting recycling, and other vital issues.

Publication(s): Florida PIRG reports, Citizen Agenda

Keyword(s): Environmental Protection

Contact(s):
Mark Ferrulo, EXECUTIVE DIRECTOR

FLORIDA SPORTSMEN'S CONSERVATION ASSOCIATION

P.O. Box 20051
West Palm Beach, FL 33416-0051 USA
Phone: 561-478-5965

Founded: 1994
Scope: Statewide

Description: The Florida Sportsmen's Conservation Association promotes conservation, preservation, and propagation of all forms of game wildlife species, nongame wildlife species, and marine life. The Association stimulates a greater interest in any and all legitimate outdoor recreational activities, assures sportsmen that they may continue to use areas for legitimate outdoor recreational activities, works towards the opening of all lands and waters for legitimate outdoor recreational activities, where feasible, and promote the enactment of fair game and fisheries laws and to assist in the enforcement of the laws.

Keyword(s): Environmental and Conservation Education, Land Use Planning, Outdoor Recreation, Public Lands, training

Contact(s):
Robert Stossel, 1ST VICE PRESIDENT
14241-77th Pl. N., Loxahatchee, FL 33470
Phone: 561-753-7880
Mark Dombroski, 2ND VICE PRESIDENT
1842 Lynton Cir., Wellington, FL 33414
Phone: 561-793-7200
Bishop Wright, PRESIDENT
15439-94th St. N., West Palm Beach, FL 33412
Phone: 561-795-1375
Bruce Britt, PUBLICATION DIRECTOR
7407 Southern Blvd., West Palm Beach, FL 33413
Phone: 561-688-2553
Kevin Smith, SECRETARY
15856-93rd St. N., West Palm Beach, FL 33412
Phone: 561-795-4112
Richard Andrea, TREASURER
12334-77th Pl. N., West Palm Beach, FL 33412
Phone: 561-795-1136

FLORIDA TRAIL ASSOCIATION, INC.

5415 SW 13th St.
Gainesville, FL 32608 USA
Phone: 352-378-8823   Fax: 352-378-4550
E-mail: fta@florida-trail.org
Website: www.florida-trail.org

Founded: 1964
Membership: 5500
Scope: State

Description: This association was formed to instill in Floridians and in visitors to Florida an appreciation and a desire to conserve the natural beauty of Florida by all lawful means; to promote the creation of a hiking trail, to be called the Florida Trail, to run the length of the state; and to provide an opportunity for hiking and camping.

Publication(s): Footprint, The

Keyword(s): Environmental and Conservation Education, Outdoor Recreation, Natural Areas, Pedestrian Environment, Trail

Contact(s):
Eileen Wyand, 1ST VICE PRESIDENT FOR ADMINISTRATION
Sylvia Dunnam, 2ND VICE PRESIDENT FOR MEMBERSHIP
Joan Hobson, 3RD VICE PRESIDENT FOR TRAILS
Kevin Butler, 4TH VICE PRESIDENT FOR PUBLIC RELATIONS
Deborah Stewart-Kent, PRESIDENT
Mary Anne Freyer, SECRETARY
Ernest Baldini, TREASURER

 ## FLORIDA WILDLIFE FEDERATION

P.O. Box 6870
Tallahassee, FL 32314-6870 USA
Phone: 850-656-7113   Fax: 850-942-4431
E-mail: FWF@fwf@flawildlife.org
Website: http://flawildlife.org

Founded: 1937
Membership: 1100
Scope: State

Description: A representative statewide organization, affiliated with the National Wildlife Federation, dedicated to the protection and enhancement of wildlife and its habitat through public education and government interaction.

Publication(s): Florida Fish and Wildlife News

Contact(s):
  Patricia Pearson, BACKYARD WILDLIFE HABITATS
  CONTACT
  Richard Farren, EDITOR
  Diane Hines, EDUCATION PROGRAMS CONTACT
  Manley Fuller, EXECUTIVE DIRECTOR/PRESIDENT
  Bob Reid, NWF ALTERNATE REPRESENTATIVE
  Jenny Brock, NWF REPRESENTATIVE
  David White, PRESIDENT/CHAIR
  Mike Webster, TREASURER & VICE CHAIR OF RECORDS

FOREST FIRE LOOKOUT ASSOCIATION

374 Maple Ave., E., Suite 210
Vienna, VA 22180 USA
Phone: 703-255-2700   Fax: 703-281-9200
Website: www.firelookout.org

Founded: 1990
Scope: National

Description: A national organization devoted to forest protection
  through the inventory, maintenance and volunteer staffing of
  forest fire lookouts and fire towers in the 49 states that have
  them and throughout the world. Maintains a data base of
  designs and available lookout parts and salvage. Organized
  into 21 states and regional chapters.

Publication(s): Lookout Network (quarterly)

Keyword(s): Environmental Protection, Protected Areas, Forest
  Management, Forests and Forestry, Historic Preservation,
  Preservation and Protection, Museum

Contact(s):
  Keith Argow, CHAIRMAN OF THE BOARD
  Phone: 703-255-2700
  Mark Haughwout
  EASTERN DEPUTY CHAIR
  Phone: 802-476-8341
  Michael Pfeiffer, HISTORIAN
  Phone: 501-967-4167
  Joseph Higgins, LEGAL COUNSEL
  Phone: 201-391-1091
  Nancy Gabriel, NATIONAL HISTORIC LOOKOUT REGISTER
  Phone: 703-255-2700
  Henry Isenberg, RESTORATIONS
  Phone: 508-883-0834
  Ray Grimes Jr., SECRETARY
  Phone: 973-835-4487
  Shirley Goodrich, TREASURER
  Phone: 207-324-6537
  Gary Weber, WESTERN DEPUTY CHAIR
  Phone: 207-442-2405

FOREST HISTORY SOCIETY, INC.

701 William Vickers Ave
Durham, NC 27701 USA
Phone: 919-682-9319   Fax: 919-682-2349
Website: www.lib.duke.edu/forest/

Founded: 1946
Membership: 2000
Scope: National

Description: A nonprofit educational institution, the Forest History
  Society is dedicated to the advancement of historical understanding of human interaction with the forest environment—
  forest industries, forestry, conservation, and other forms of use

and appreciation. A membership organization, it sponsors
programs in research, publication, archives-library, and professional service.

Publication(s): Forest Time Line Newsletter, Environmental
History, Forest History Today

Keyword(s): Conservation, Environmental Law, Forests and
Forestry, Land Use Planning, Public Lands

Contact(s):
  William Bauthman, CHAIRMAN
  Summerville, SC 843-851-4653
  Adam Rome, EDITOR
  Pennsylvania State University,
  University Park, PA
  814-863-0184
  axr26@psu.edu
  Cheryl Oakes, LIBRARIAN
  Steve Anderson, PRESIDENT

FOREST LANDOWNERS ASSOCIATION, INC.

3776 Lavista Rd. Suite 250
Tucker, GA 30084 USA
Phone: 404-325-2954   Fax: 404-325-2955
E-mail: snswton@forestland.org
Website: www.forestland.org

Founded: 1941
Membership: 11000
Scope: National

Description: Nonprofit forestry organization of timberland owners
  large and small in 17 southern states seeking to give private
  timberland owners and related interests a greater voice in
  matters affecting their business.

Publication(s): Forest Landowner Manual, Forest Landowner
  Magazine

Keyword(s): Environmental and Conservation Education, Forests
  and Forestry, Renewable Resources, Urban Forestry

Contact(s):
  Paige Cash, EDITOR
  Steve Newton, EXECUTIVE VICE PRESIDENT
  Otis Ingram, GOVERNMENT AFFAIRS CHAIRMAN
  Carroll Cochran, REGIONAL VICE PRESIDENT
  Charles Tomlinson, REGIONAL VICE PRESIDENT
  Kirk Rodgers, REGIONAL VICE PRESIDENT
  Guerry Doolittle, REGIONAL VICE PRESIDENT
  Champion International, 9485 Regency Sq. Blvd,
  Jacksonville, FL 32225-8155
  C. Bush, REGIONAL VICE PRESIDENT
  6701 Carmel Road, Suite 404, Charlotte, NC 28226
  John Bowen, REGIONAL VICE PRESIDENT
  Box 159, Louisville, TN 37777
  L. Larson, REGIONAL VICE PRESIDENT
  P.O. Box 2143, Mobile, AL 36652

FOREST SERVICE EMPLOYEES FOR ENVIRONMENTAL ETHICS (FSEEE)

P.O. Box 11615
Eugene, OR 97440 USA
Phone: 541-484-2692   Fax: 541-484-3004
E-mail: fseee@fseee.org
Website: www.fseee.org

Founded: 1989
Membership: 1200
Scope: National

Description: A national nonprofit organization of Forest Service employees, retirees, other resource professionals, and concerned citizens working to change from within the Forest Service's basic management philosophy to a land ethic that ensures ecologically and economically sustainable management.

Publication(s): see publications on website, Forest Magazine

Keyword(s): Biodiversity, Environmental and Conservation Education, Forests and Forestry, Public Lands, Professional Organization, training

Contact(s):
Mark Blaine, EDITOR
Andy Stahl, EXECUTIVE DIRECTOR
Bob Dale, FIELD DIRECTOR
Dave Iverson, PRESIDENT
Cynthia Reichelt, SECRETARY AND TREASURER

FOREST SOCIETY OF MAINE
P.O. Box 775, 115 Franklin St.
Bangor, ME 04402 USA
Phone: 207-945-9200   Fax: 207-945-9229
E-mail: info@fsmaine.org

Founded: 1984
Scope: Regional

Description: A statewide land trust working with landowners on forest land conservation projects that maintain the environmental, recreational and economic values of Maine's forests, primarily through conservation easements.

Keyword(s): Conservation, Ecosystems, Biodiversity, Forest Management, Forest Stewardship, Forests and Forestry, Land Preservation, Land Purchase, Sustainable Ecosystems, Sustainable Resources

Contact(s):
Leslie Hudson, DEPUTY EXECUTIVE DIRECTOR
leslie@fsmaine.org
Alan Hutchinson, EXECUTIVE DIRECTOR

FOREST TRUST
P.O. Box 519
Santa Fe, NM 87504-0519 USA
Phone: 505-983-8992   Fax: 505-986-0798
E-mail: forest@theforesttrust.org
Website: www.theforesttrust.org

Founded: 1984
Scope: Regional, National

Description: A nonprofit organization dedicated to protecting the integrity of the forest ecosystem and improving the lives of people in rural communities. The Trust challenges conventional forest management philosophies and provides protection strategies to grassroots environmental organizations, rural communities, and public agencies. The Trust also provides land management services to owners of private lands with significant conservation values and serves as the institutional home for the Forest Stewards Guild.

Publication(s): Annual Report, Distant Thunder, Forest Trust Quarterly Report

Keyword(s): Protected Areas, Forests and Forestry, Land Use Planning, Rural Development, Sustainable Development

Contact(s):
Ron Dryden, ACCOUNTANT

Laura McCarthy, ASSISTANT DIRECTOR
Henry Carey, DIRECTOR
Steven Harrington, FOREST STEWARDS GUILD
Shirl Harrington, NATIONAL FOREST PROGRAM

FOREST WATCH
10 Langdon St., Suite 1
Montpelier, VT 05602 USA
Phone: 802-223-3216   Fax: 802-223-1363
E-mail: forestwatch@forestwatch.org
Website: www.forestwatch.org

Founded: 1994
Membership: 4000
Scope: Regional

Description: Forest Watch saves and recreates wild forests, reforms public land management, advocates ecological forestry and watches over the forests with a network of citizen volunteers.

Publication(s): Visions: A Wild-eyed Look at Forests, State of the Forest Report

Keyword(s): Ancient Forests, Conservation Biology, Endangered Species, Forest Management, Forest Stewardship, Forests and Forestry, Legal Advocacy, National Forests, Natural Areas, Public Lands, Reforestation, Sustainable Ecosystems, Trees, Urban Forestry, training

Contact(s):
Andrew Vota, ADVOCACY DIRECTOR
avota@forestwatch.org
Sue Higby, DEPUTY DIRECTOR
shigby@together.net
Jim Northup, EXECUTIVE DIRECTOR

FOSSIL RIM WILDLIFE CENTER
P.O. Box 2189
Glen Rose, TX 76043 USA
Phone: 254-897-2960   Fax: 254-897-3785
E-mail: visitor-services@fossilrim.org
Website: www.fossilrim.org

Founded: 1987
Scope: National

Description: A 2,100-acre wildlife preserve dedicated to the preservation of endangered and rare species with the ultimate goal of returning these species to the wild. Sixty animal species are represented, and Fossil Rim participates in 12 Species Survival Plan programs. Programs include public education, research into the management and propagation of endangered species, training of conservation professionals, and support for the creation of similar efforts around the world.

Keyword(s): Endangered Species, Environmental and Conservation Education, Internships, Sustainable Ecosystems

Contact(s):
M. Jurzykowski, CHAIRMAN OF THE BOARD
Jerry Millhon, CHIEF OPERATING OFFICER
Phone: 254-897-2960
Yola Carlough, DIRECTOR OF COMMUNICATIONS
Phone: 254-897-2960
Bruce Williams, VICE PRESIDENT OF CONSERVATION
Phone: 254-897-2960

FOUNDATION FOR NORTH AMERICAN BIG GAME

P.O. Box 2710
Woodbridge, VA 22193 USA
Phone: 703-878-2119   Fax: 703-878-2119

Founded: 1992
Membership: 500
Scope: National

Description: A nonprofit membership organization with major objectives of protection and encouragement of sport hunting in North America; education of the general public to the values of sport hunting, both direct and indirect; and the conservation and welfare of the big-game species of the continent.

Publication(s): North American Big Game (quarterly magazine)

Keyword(s): Environmental and Conservation Education, Hunting, Mammals, training

Contact(s):
Edward Nannini, BOARD MEMBER
Phone: 916-485-8111
Fax: 916-485-1709
William Nesbitt, EXECUTIVE DIRECTOR
Phone: 703-590-4449
Fax: 703-878-2119
Don Kirn, PRESIDENT
Phone: 816-761-4351
Fax: 816-761-8737
Don Morgan, SECRETARY
Phone: 352-473-2662
Fax: 352-473-2166
E. Pocius, TREASURER
Phone: 215-536-9616
Fax: 215-536-5815
Warren Parker, VICE PRESIDENT
Phone: 816-229-8899
Fax: 816-229-5933

FOUNDATION FOR NORTH AMERICAN WILD SHEEP

720 Allen Ave.
Cody, WY 82414 USA
Phone: 800-623-9657
E-mail: fnaws@fnaws.org
Website: http://fnaws.org

Founded: 1977
Scope: National

Description: A nonprofit organization whose purposes are to: promote the management of and safeguard against the extinction of all species of wild sheep native to the continent of North America; promote the protection of the remaining wild sheep populations and their habitat; and promote the re-establishment of wild sheep populations in suitable habitat. The Foundation funds wild sheep research, wildlife studies, improves habitat, finances sheep transplants, and supports hunting and game management policies based on sound, proven principles.

Publication(s): Wild Sheep

Keyword(s): Environmental and Conservation Education, training, standards, accreditation

Contact(s):
Matt Wolfe, FIRST VICE-PRESIDENT
P.O Box 309, Kasilof, AK 99610
Phone: 907-262-7058
Jeff Reynolds, PRESIDENT
P.O Box 146, Douglas, WY 82633
Phone: 307-358-3693
Fax: 307-358-3262
Curt Thompson, SECOND VICE-PRESIDENT
Box 9176, 29 Wann Rd., Whitehorse, Yukon Territory YIA4A2
Phone: 867-668-6564
Wayne Heimer, SECRETARY
1098 Chena Pump Rd., Fairbanks, AK 99709
Phone: 907-457-6847
weheimer@alaska.net
Ron Pomeroy, TREASURER
2969 Baseline Rd., Boulder, CO 80303
Phone: 303-444-4500
Fax: 303-449-1289

FRIENDS OF ACADIA

P.O. Box 45
Bar Harbor, ME 04609 USA
Phone: 207-288-3340   Fax: 207-288-8938
Website: www.friendsofacadia.org

Founded: 1986
Scope: State

Description: Friends of Acadia is a nonprofit organization providing citizen support in partnership with the National Park Service to preserve and protect Acadia National Park and the communities that surround it.

Publication(s): Friends of Acadia Journal

Keyword(s): Public Lands, Conservation, Outdoor Recreation

Contact(s):
H. Judd, BOARD OF DIRECTORS CHAIRMAN
Marla Major, DIRECTOR OF ASSOCIATION CONSERVATION
Stephanie Clement, DIRECTOR OF CONSERVATION
Diana McDowell, DIRECTOR OF OPERATIONS
W. Olson, PRESIDENT

FRIENDS OF ANIMALS INC.

777 Post Rd., Suite 205
Darien, CT 06820 USA
Phone: 203-656-1522   Fax: 203-656-0267
E-mail: foa@igc.org
Website: www.friendsofanimals.org

Founded: 1957
Scope: National

Description: An international animal protection organization that works to protect animals from cruelty, abuse, and institutionalized exploitation. FOA's efforts protect and preserve animals and their habitats around the world.

Publication(s): Act ionLine

Keyword(s): Endangered Species, Hunting, Marine Mammals, Whale, Dolphin, Seal, Trapping, training

Contact(s):
Priscilla Feral, PRESIDENT AND EDITOR
Sally LEINER-MALANGA, SECRETARY AND TREASURER

FRIENDS OF DISCOVERY PARK

P O Box 99662
Seattle, WA 98199-0662 USA
Phone: 206-283-8643
E-mail: info@discoveryparkfriends.org
Website: www.discoveryparkfirends.org

Founded: 1974
Membership: 650
Scope: Regional

Description: To create and protect an open space of quiet and tranquility where the works of man are minimized. A place which emphasizes its natural environment and to promote the development of Discovery Park according to a master plan responsive to these goals.

Publication(s): Explorer - Quarterly Newsletter

Keyword(s): Environmental Planning, Landscape Architecture, Nongame Wildlife, Sustainable Ecosystems, Open Space, Wetlands

Contact(s):
Valerie Cholvin, PRESIDENT
John Wooten, SECRETARY
Terry Mueller, TREASURER

FRIENDS OF THE BOUNDARY WATERS WILDERNESS

401 North 3rd St., Suite 290
Minneapolis, MN 55401 USA
Phone: 612-332-9630   Fax: 612-332-9624
E-mail: info@friends-bwca.org
Website: www.friends-bwca.org

Founded: 1976
Membership: 4000
Scope: International

Description: Established to protect and preserve the wilderness character of northeastern Minnesota's million-acre Boundary Waters Canoe Area (BWCA) Wilderness and the surrounding areas in the international Quetico-Superior Ecosystem. Works on wilderness protection, acid rain, forestry issues, water quality, rivers protection, wilderness management, and wildlife species, such as the wolf.

Publication(s): BWCA Wilderness News

Keyword(s): Biodiversity, Endangered Species, Forests and Forestry, Wilderness, Wildlands, Water Quality, Acid Rain

Contact(s):
Jon Nelson, CHAIRPERSON
Melissa Lindsay, EXECUTIVE DIRECTOR
Elizabeth Schuniesing, SECRETARY
Jeff Evans, TREASURER
Becky Rom, VICE-CHAIRPERSON

FRIENDS OF THE EARTH

The Global Bldg., 1025 Vermont Ave., NW, Suite 300
Washington, DC 20005 USA
Phone: 202-783-7400   Fax: 202-783-0444
E-mail: foe@foe.org
Website: www.foe.org

Founded: NA
Scope: National-International

Description: A global advocacy organization based in Washington, DC, with 61 international affiliates. Merged with Environmental Policy Institute and the Oceanic Society in 1990. Dedicated to protecting the planet from environmental disaster and preserving biological, cultural, and ethnic diversity. With strong ties to the grassroots in the U.S. and around the world, Friends of the Earth believes individuals and communities must have a voice in environmental policymaking that affects their lives. Fights ozone depletion, global warming, toxic chemical threats, nuclear hazards, groundwater contamination, environmentally unsound international lending policies, and irresponsible corporate practices. Produces the Earth Budget; works to change tax code to support environmentally sustainable policies. Budget is $2.7 million.

Publication(s): Anatomy of a Deal, The Technical and Economic Feasability of Replacing Methyl Bromide, River of Red Ink, Green Scissors Annual Report, Friends of the Earth Newsmagazine

Keyword(s): Air Quality and Pollution, Environmental Protection, Pesticides, Transportation, Water Quality

Contact(s):
Brent Blackwelder, PRESIDENT

FRIENDS OF THE EARTH

NORTHWEST REGIONAL OFFICE (WA, OR, ID)
1025 Vermont Ave. NW
Washington, D.C 20005 USA
Phone: 202-783-7400   Fax: 202-783-0444
E-mail: foe@foe.org
Website: foe.org

Founded: NA
Scope: National

Contact(s):
Shawn Cantrell, CONTACT

FRIENDS OF THE REEDY RIVER

P.O. Box 9351
Greenville, SC 29604 USA
E-mail: reedyriver@aol.com
Website: www.reedyriver.org

Founded: 1993
Scope: Regional

Description: FORR is a nonprofit river advocacy group committed to watershed protection and restoration of the Reedy River in upstate South Carolina.

Publication(s): Membership Brochure and Newsletter, NPS Brochure, Padding Guide

Keyword(s): Conservation Easements, Environmental Protection, Greenways, Habitat Conservation, Nonpoint Source Pollution, Riparian Restoration, Rivers, Runoff, Trail, Watersheds

Contact(s):
Dave Hargett, EXECUTIVE DIRECTOR
Phone: 864-297-3566
hargett@prodigy.net

FRIENDS OF THE RIVER

915 20th St.
Sacramento, CA 95814 USA
Phone: 916-442-3155   Fax: 916-442-3396
E-mail: chare@friendsoftheriver.org
Website: www.friendsoftheriver.org

Founded: 1973
Scope: Statewide

Description: A nonprofit membership organization dedicated to the preservation, protection, and restoration of rivers, streams, and watersheds through public education, citizen activist training and organizing, and expert advocacy to influence public policy.

Publication(s): Rivers Reborn: Removing Dams And Restoring Rivers in California, Rivers of Power: A Citizens Guide to hydropower and River Restoration, Headwaters

Keyword(s): Energy, Environmental and Conservation Education, Outdoor Recreation, Rivers, Exotic species, Aquatic nuisance species, Watersheds

Contact(s):
Jim Wheaton, BOARD OF DIRECTORS, CHAIR
Charlie Casey, COMMUNICATIONS DIRECTOR
Steve Evans, CONSERVATION DIRECTOR
Betsy Reifsnider, EXECUTIVE DIRECTOR

FRIENDS OF THE SAN JUANS

P.O. Box 1344
Friday Harbor, WA 98250 USA
Phone: 360-378-2319   Fax: 360-378-2324
E-mail: friends@sanjuans.org
Website: www.sanjuans.org

Founded: 1979
Membership: 1300
Scope: Regional

Description: To protect to the fullest extent possible the scenic, aesthetic, ecological and sociological qualities, and resources of the San Juan Islands and the Northwest straits marine ecosystem. Promoting long-range planning and monitoring development.

Publication(s): Friends of the San Juan Newsletter

Keyword(s): Aquatic Habitats, Coasts, Forest Management, Islands, Shorelines, Land Use Planning, Oceanography, Sea Grass, Marine Mammals, Whale, Dolphin, Seal, Sustainable Development

Contact(s):
Stephanie Buffem, EXECUTIVE DIRECTOR

FRIENDS OF THE SEA OTTER

2150 Garden Rd., A3
Monterey, CA 93940 USA
Phone: 831-373-2747   Fax: 831-373-2749
E-mail: info@seaotters.org
Website: www.seaotters.org

Founded: 1968
Membership: 4000
Scope: State

Description: A nonprofit organization dedicated to the protection and maintenance of a healthy population of southern sea otters, a threatened species, as well as sea otters throughout their north pacific range, and all sea otter habitat. Encourages research and public education to develop a sound conservation program.

Publication(s): Otter Raft, The

Keyword(s): Aquatic Habitats, Wildlife Rehabilitation, Biodiversity, Coasts, Endangered Species, Environment, Wildlife, Environmental and Conservation Education, Otters, Mammals, Coral Reefs

Contact(s):
Matt Rutishauser, SCIENCE DIRECTOR

FUND FOR ANIMALS INC., THE

200 W. 57th St.
New York, NY 10019 USA
Phone: 212-246-2096   Fax: 212-246-2633
E-mail: hdquarters@fund.org
Website: http://www.fund.org

Founded: 1967
Membership: 250,000
Scope: National

Description: National nonprofit animal-protection organization whose purpose is to preserve wildlife and promote humane treatment for all animals. Primarily serves as an advocacy group and information and education agency to help domestic and wild animals. The Fund operates four hands-on facilities.

Keyword(s): Endangered Species, Marine Mammals, Whale, Dolphin, Seal, Trapping, training, accreditation

Contact(s):
Christine Wolf, CHIEF OF LEGISLATIVE SERVICES
8121 Georgia Ave., Silver Spring, MD 20910
Michael Markarian, EXECUTIVE VICE PRESIDENT
8121 Georgia Ave., Silver Spring, MD 20190
Kimberly Sturla, FIELD OFFICER
808 Alamo Dr., Suite 306, Vacaville, CA 95688
Chuck Traisi, FIELD OFFICER, MANAGER OF ANIMAL TRUST SANCTUARY,
18740 Highland Valley Rd., Ramona, CA 92065
Phone: 760-789-2324
Doris Dixon, FIELD OFFICER
2841 Colony Rd., Ann Arbor, MI 48104
Phone: 313-971-4632
Caroline Gilbert, FIELD OFFICER
Rt. 2 Box 559, Simponsville, SC 29681
Fax: 803-963-4389
Laura Simon, FIELD OFFICER
30 Mountainview Rd., Bethany, CT 06524
Phone: 203-393-3669
Virginia Handley, FIELD OFFICER
Fort Mason Center, Rm. 3262, San Francisco, CA 94123
Phone: 415-474-4020
Andrea Reed, FIELD OFFICER
P.O. Box 11294, Jackson, WY 307-859-8840
Marion Stark, FIELD OFFICER
P.O. Box 9029, Albany, NY 12209
Chris Byrne, FIELD OFFICER, MANAGER OF BLACK BEAUTY RANCH,
P.O. Box 367, Murchison, TX 75778
Phone: 903-469-3011
Lia Albo, FIELD OFFICER, MANAGER OF HAVE A HEART CLINIC
335 W. 52nd St., New York, NY 10019
Phone: 212-977-6877
Edward Walsh, LEGAL COUNSEL
Vedder, Price, Kaufman, Kammholz, and Day, 805 3rd Ave., New York, NY 10022
Phone: 212-407-7740
Marian Probst, PRESIDENT
Phone: 212-246-2096
Fax: 212-246-2633

FUTURE FISHERMAN FOUNDATION

225 Reinekers Lane Ste 420
Alexandria, VA 22314 USA
Phone: 703-519-9691   Fax: 703-519-1872
E-mail: info@asafishing.org
Website: www.asafishing.org

Founded: NA
Membership: 3
Scope: National

Description: The educational arm of the American Sportfishing
Association, the Foundation is a nonprofit organization
dedicated to promoting participation and education in fishing as
well as enhancement and protection of aquatic resources.
Develops and coordinates the national program "Hooked On
Fishing - Not on Drugs". The Foundation is a national leader in
recreational fishing and aquatic resource education and offers
student and instructor educational materials.

Keyword(s): Youth Organizations, Environmental and
Conservation Education, Outdoor Recreation, Sport Fishing,
Coral Reefs, Exotic species, Aquatic nuisance species

Contact(s):
Anne Glick, EXECUTIVE DIRECTOR

G

GALIANO CONSERVANCY ASSOCIATION

R.R. 1, Sturdies Bay Rd.
Galiano Island, British Columbia V0N 1P0 Canada
Phone: 250-539-2424   Fax: 250-539-2424
E-mail: galiano_conservancy@gulfislands.com

Founded: 1989
Scope: Local

Description: The Galiano Conservancy Association is a
community-based, regionally-oriented conservation organization and land trust. Its purposes are to preserve, protect, and
enhance the quality of the human and natural environment of
the area through public education; management, and ownership
of conservatrion land; and research and restoration projects.

Publication(s): Newsletter, Bulletin, Archipelago

Keyword(s): Aquatic Habitats, Biodiversity, Coasts, Conservation,
Environmental and Conservation Education, Wildlife,
Environmental Planning, Forest Management, Islands, Land
Purchase, Natural Areas, Land Preservation, Land Use
Planning, Pesticides

Contact(s):
John Pritchard, CO-COORDINATOR
Phone: 250-539-2424
Ken Millard, COORDINATOR
Phone: 250-539-2424
Fax: 250-539-2424
Rose Longini, SECRETARY
Phone: 250-539-2424
Fax: 250-539-2424

GAME CONSERVANCY U.S.A.

P.O. Box 922
Darien, CT 06820-0922 USA
Phone: 203-662-0886   Fax: 203-662-9298
Website: www.gcusa.org

Founded: 1985
Membership: 525

Scope: National

Description: (formerly American Friends of the Game
Conservancy) Game Conservancy USA's primary functions are
to raise funds to support the research and educational
activities of The Game Conservancy Trust in the U.K., to
support antipoaching efforts in Tanzania, and to promote
beneficial programs that pertain to the conservation of wildlife
resources worldwide.

Publication(s): American Friends of the Game Conservancy
Newsletter

Keyword(s): Scholarships and Grants, training, standards,
accreditation

Contact(s):
Kimberly McMahon, ADMINISTRATIVE DIRECTOR
365-20th Ave, Vero Beach, FL 32962
Phone: 516-234-8718
Fax: 561-234-6081
George Morris, EXECUTIVE DIRECTOR
190 Camelia Ct. N., Vero Beach, FL 32963
Phone: 561-234-8718
Fax: 561-234-6081
James Long, LIBRARIAN
The Game Conservancy Trust,
Fordingbridge, Hampshire, SP61EF
011-44-1425-655848
F. Gillet, PRESIDENT
159 Via del Lago, Palm Beach, FL 33480
Phone: 407-655-6789
Fax: 407-832-1762
William Murray, SECRETARY
200 E. 61st ST. 4-G, New York, NY 10021
Phone: 212-832-2323
Fax: 212-319-5238

GAME CONSERVATION INTERNATIONAL (GAME COIN)

4600 Broad Ave.
Ft. Worth, TX 76107 USA
Phone: 817-738-5438   Fax: 817-737-2911

Founded: 1967
Scope: International

Description: A nonprofit organization dedicated to responsible
sustainable use of fish and wildlife and preserving the hunting
and fishing heritage for future generations. Supports strong
educational programs for classrooms and sponsors the state
and province Outstanding Hunter Education awards from
Mexico to Canada. Its National Junior Wildlife Artist
competition encourages high school youngsters to compete for
thousands of dollars in prizes under the theme: OUR
WILDLIFE HERITAGE: PASS IT ON!. Created the 4-H
Shooting Competition in Texas which now has been copied
nationwide.

Keyword(s): Endangered Species, training

Contact(s):
Harry Tennison, PRESIDENT
Phone: 817-738-5438

GARDEN CLUB OF AMERICA, THE

14 East 60th St.
New York, NY 10022 USA
Phone: 212-753-8287   Fax: 212-753-0134
E-mail: hq@gcamerica.org
Website: www.gcamerica.org

Founded: 1913
Membership: 17624
Scope: National

Description: A national nonprofit organization with member clubs from coast to coast and in Hawaii. Its purpose is to stimulate the knowledge and love of gardening, to share the advantages of association by means of educational meetings, conferences, correspondence and publications, and to restore, improve, and protect the quality of the environment through educational programs and action in the fields of conservation and civic improvement.

Keyword(s): Endangered Species, Environmental and Conservation Education, Flowers, Plants, and Trees, Gardening and Horticulture, Scholarships and Grants

Contact(s):
Peter Goedecke, CONSERVATION CHAIRMAN
Daniel Will III, CORRESPONDING SECRETARY
A. John Gregg, HORTICULTURE CHAIRMAN
John Murphy Jr, NATIONAL CHAIRMAN
Joseph Frierson, PRESIDENT

GENERAL FEDERATION OF WOMEN'S CLUBS

1734 N St., NW
Washington, DC 20036 USA
Phone: 202-347-3168   Fax: 202-835-0246
E-mail: gfwc@gfwc.org
Website: www.gfwc.org

Founded: 1890
Scope: National

Description: The General Foundation of Women's Clubs (GFWC) is an international organization of community-based volunteer women's clubs dedicates to community service since 1890. GFWC programs and projects encompass the major issues of our time including literacy, health, preservation of natural resources, abuse prevention, and solid waste management.

Publication(s): GFWC Clubwoman

Keyword(s): Energy, Environmental and Conservation Education, Protected Areas, National Parks, Solid Waste Management

Contact(s):
Judy Lutz, 1ST VICE PRESIDENT
Ernie Shriner, 2ND VICE PRESIDENT
Joyce Schaefer, BEAUTIFICATION PROGRAM CHAIR
Rd 4 Box 32, Seaford, DE 19973
Terri Wogan, CONSERVATION DEPARTMENT COORDINATOR
5401 E. Marilyn Rd., Scottsdale, AZ 85254
Maryanne Potter, DIRECTOR OF JUNIOR CLUBS
Sarah Wees, EDITOR
Shelby Hamlett, PRESIDENT
Judy Lutz, PRESIDENT-ELECT
Pat Nolan, PROGRAM DIRECTOR
Jacquelyn Pierce, RECORDING SECRETARY
Barbara Nunnari, RESOURCE CONSERVATION PROGRAM CHAIR
13200 Ridge Dr., Rockville, MD 20850
Rose Ditto, TREASURER

Norma Chesney, WATER QUALITY PROGRAM CHAIR
1331 Jill Terrace, Homewood, IL 60430

GEORGE MIKSCH SUTTON AVIAN RESEARCH CENTER INC.

P.O. Box 2007
Bartlesville, OK 74005 USA
Phone: 918-336-7778   Fax: 918-336-7783
E-mail: gmsarc@aol.com
Website: www.suttoncenter.org

Founded: 1983
Membership: 7
Scope: National & International

Description: The Sutton Research Center is a non-profit, tax-exempt organization conducting scientific studies, conservation projects and educational programs regarding avian species worldwide. Topics of particular interest include raptor population surveys and studies, bald eagle population monitoring, avian captive breeding and reintroductions, ecological studies of grassland birds including songbirds and gamebirds, public education projects and cooperative wildlife conservation efforts with landowners.

Publication(s): The Sutton Newsletter

Keyword(s): Birds, Endangered Species, Nongame Wildlife, Prairies, Biodiversity, Environment, Environmental and Conservation Education, Conservation, Protected Areas, Environmental Protection, National Parks, Predators, Raptors, Renewable Resources

Contact(s):
Stephen Adams, BOARD CHAIRMAN
Oklahoma Land & Cattle Co.,
1437 S. Boulder Ave., Suite 930, 74119
Phone: 918-585-5411
sadams@adamsaff.com
Steve Sherrod, EXECUTIVE DIRECTOR
Howard Burman, TREASURER

GEORGE WRIGHT SOCIETY, THE

P.O. Box 65
Hancock, MI 49930 USA
Phone: 906-487-9722   Fax: 906-487-9405
E-mail: info@georgewright.org
Website: www.georgewright.org

Founded: 1980
Membership: 750
Scope: International

Description: The George Wright Society is organized for the purposes of promoting the application of knowledge, fostering communication, improving resource management, and providing information to improve public understanding and appreciation of the basic purposes of natural and cultural parks and equivalent reserves.

Publication(s): George Wright Forum, The

Keyword(s): Protected Areas, Environmental Protection, Historic Preservation

Contact(s):
David Harmon, EXECUTIVE DIRECTOR
Hancock Office,
Bob Krumenaker, PRESIDENT
bob_krumenaker@nps.gov

GEORGIA ASSOCIATION OF CONSERVATION DISTRICT SUPERVISORS

P.O. Box 8024
Athens, GA 30603 USA
Phone: 706-542-9233
E-mail: info@gacds.org
Website: http://www.gacds.org

Founded: NA
Scope: Statewide

Keyword(s): Conservation Districts

GEORGIA B.A.S.S. CHAPTER FEDERATION

Attn: President, 11575 Northgate Trail
Roswell, GA 30075 USA
Phone: 770-993-6597
Website: www.gabassfed.org

Founded: NA
Scope: Statewide

Description: An organization of Bassmaster chapters, affiliated with the Bass Anglers Sportsman Society, organized to fight pollution, assist state and national conservation agencies in their efforts, and teach young people of our country's good conservation practices. Dedicated to the realistic conservation of our water resources.

Publication(s): Georgia Outdoor News, Georgia Federation Newsletter

Contact(s):
Scott Hendricks, CONSERVATION DIRECTOR
5131 Maner Rd., Smyrna, GA 30080
Phone: 404-799-2159
Larry Lewis, PRESIDENT
Phone: 770-993-6597

GEORGIA CONSERVANCY, INC., THE

1776 Peachtree Rd. NW, Suite 400, S.
Atlanta, GA 30309 USA
Phone: 404-876-2900   Fax: 404-872-9229
E-mail: mail@gaconservancy.org
Website: www.gaconservancy.org

Founded: 1967
Scope: State

Description: The Georgia Conservancy is a nonprofit organization of people dedicated to the responsible stewardship of Georgia's vital natural resources. We strive to balance the demands of social and economic progress with our commitment to protect the environment.

Publication(s): Highroad guide to the Georgia Coast and Okefenokee, Teching Conservation, The Hiking Trails of North Georgia, Wetlands: Georgia's Vanishing Treasure, Panorama: Highroad Guide to the North Georgia Mountains

Keyword(s): Air Quality and Pollution, Chemical Pollution Control, Development, Smart Growth, Environmental Protection, Forest Management, Land Use Planning, Renewable Resources, Open Space, Environment, Pollution Prevention, Urban Environment, Wetlands, Transportation, Exotic species, Aquatic nuisance species

Contact(s):
John Izard, CHAIRMAN OF THE BOARD
Patricia McIntosh, COASTAL PROGRAMS DIRECTOR
428 Bull St., Savannah, GA 31410

Phone: 912-447-5910
Fax: 912-447-5911
John Sibley, PRESIDENT
Betty Mori, SECRETARY
James Bostic, TREASURER
Susan Kidd, VICE PRESIDENT FOR EDUCATION AND ADVOCACY
Robert Smulian, VICE PRESIDENT FOR PLANNING AND DEVELOPMENT

GEORGIA ENVIRONMENTAL COUNCIL, INC.

P.O. Box 997
Suwanee, GA 300324 USA
Phone: 706-546-7507   Fax: 770-614-0593
Website: www.gecweb.org

Founded: NA
Membership: 60
Scope: Statewide

Description: A statewide umbrella for organizations interested in environmental protection that seeks to facilitate the exchange of information among member organizations, to provide a forum for discussion of environmental issues of interest to the members, and to monitor state government legislative activities having to do with the environment.

Publication(s): Directory of Environmental Groups in Georgia, Legislative Monitor, GEC Monitor

Keyword(s): Environmental and Conservation Education

Contact(s):
Carol Hassell, ADMINISTRATIVE DIRECTOR
Terrilyn Bayne, PRESIDENT
Patty McIntosh, SECRETARY
Hans Neuhauser, TREASURER

GEORGIA ENVIRONMENTAL ORGANIZATION, INC (GEO)

3185 Center St.
Smyrna, GA 30080-7039 USA
Phone: 404-605-0000   Fax: 404-350-9997
E-mail: gaenz@gaenz.org
Website: www.gaenz.org

Founded: 1991
Membership: 49
Scope: State

Description: GEO is a non-profit, citizen-oriented organization established to preserve and protect Georgia's environment through education, collaboration, research, planning, legislation, and grassroots organizing. The mission of GEO is to create an ecologically sound, sustainable society by developing and implementing cooperative, long-range policies, plans and programs and by carrying out hands-on projects within local communities.

Publication(s): Georgians on Sustainability

Keyword(s): Environmental and Conservation Education, Sustainable Development

Contact(s):
Trey Gibbs, EXECUTIVE DIRECTOR

GEORGIA ENVIRONMENTAL POLICY INSTITUTE

380 Meigs St.
Athens, GA 30601 USA
Phone: 706-546-7507   Fax: 706-613-7775
E-mail: gepi@ix.netcom.com
Website: www.gepinstitute.com

Founded: 1993
Scope: Local

Description: GEPI helps communities develop proactive strategies for a healthy environment through technical and legal services. The organization's primary focus is on land conservation.

Publication(s): see publications on website, A Landowner's Guide - Conservation Easements for Natural Resources Protection, A Summary of Takings Law, Right Whale News

Keyword(s): Conservation Easements, Conservation, Conservation Plannning, Environmental Law, Land Conservation, Land Management, Land Preservation, Land Protection, Land Purchase, Marine Conservation, Natural Areas, Stewardship, Sustainability, Watersheds, Wetlands

Contact(s):
Edwin Speir, CHAIR
Phone: 706-369-5650
Fax: 706-369-5792
lcarmon@negia.net
Hans Neuhauser, EXECUTIVE DIRECTOR

GEORGIA FEDERATION OF FOREST OWNERS

2402 Manchester Dr.
Waycross, GA 31501-7554 USA
Phone: 912-283-0871   Fax: 912-283-9141
E-mail: aemceuen@accessatc.net

Founded: 1974
Scope: Statewide

Description: A statewide organization affiliated with the National Woodland Owners Association to perpetuate good forest practices on private woodlands in Georgia including soil and water conservation, wildlife management, reforestation, and utilization of forest products.

Keyword(s): Forests and Forestry

Contact(s):
Patricia McCarthy, PRESIDENT
Phone: 912-283-0075
mccarthy5@aol.com
Archie McEuen, SECRETARY
aemceuen@accessatc.net

GEORGIA FORESTRY ASSOCIATION, INC.

505 Pinnacle Ct.
Norcross, GA 30071-3656 USA
Phone: 770-416-7621   Fax: 770-840-8961
E-mail: info@gfagrow.org
Website: www.gfagrow.org

Founded: 1907
Membership: 2500
Scope: Regional
Publication(s): GFA News, Legislative Bulletin, Tops
Keyword(s): Communications, Forests and Forestry, Transportation, training

Contact(s):
Blake Sullivan, PRESIDENT
Americus, GA
Dale Greene, TREASURER
Paul Mott, VICE PRESIDENT

GEORGIA TRAPPERS ASSOCIATION

P.O. Box 335
Doerun, GA 31744 USA
Phone: 912-782-5417
E-mail: tbehle@indy.tds.net
Website: www.geocities.com/yosemite/trails/5950/GA-app.html

Founded: 1979
Scope: Statewide

Description: An organization of Georgia trappers and friends of trappers, affiliated with Georgia Wildlife Federation and National Trappers Association, organized to protect the rights of trappers to trap, to coordinate a trappers education program with the Georgia Department of Natural Resources and to conserve and protect the natural resources of Georgia.

Contact(s):
Tommy Key, GENERAL ORGANIZER
Rt. 2, Newnan, GA 30623
Ralph Goodson, NTA DIRECTOR
P.O. Box 4398, Albany, GA 31706
Tom Ethridge, PRESIDENT
P.O. Box 335, Doerun, GA 31744
Grace Conder, SECRETARY AND TREASURER
P.O. Box 474, Brooklet, GA 30415

GEORGIA TRUST FOR HISTORIC PRESERVATION

1516 Peachtree St., NW
Atlanta, GA 30309-2916 USA
Phone: 404-881-9980   Fax: 404-875-2205
E-mail: info@georgiatrust.org
Website: www.georgiatrust.org

Founded: 1973
Scope: local

Description: The Georgia Trust for Historic Preservation promotes an appreciation of Georgia's diverse historic resources and provides for their protection and use to preserve, enhance and revitalize Georgia's communities.

Publication(s): The Rambler (bi-monthly newsletter)

Keyword(s): Cultural Preservation, Museum, Development, Historic Preservation, Internships, Open Spaces, Pedestrian Environment, Urban Environment, Smart Growth, Rural Development

Contact(s):
Gregory Paxton, PRESIDENT AND CEO
Phone: 404-885-7801
gpaxton@georgiatrust.org

GEORGIA WILDLIFE FEDERATION
11600 Hazelbrand Road
Covington, GA 30014 USA
Phone: 770-787-7887   Fax: 770-787-9229
E-mail: gwf@gwf.org
Website: www.gwf.org

Founded: 1936
Scope: Statewide

Description: A representative statewide organization, affiliated with the National Wildlife Federation, dedicated to the protection and enhancement of wildlife and its habitat through public education and government interaction.

Publication(s): Georgia Wildlife Magazine

Contact(s):
Joel Vinson, CHAIR
James Wilson, EDITOR
Laura Bryant, EDUCATION PROGRAMS CONTACT
Jerry McCollum, PRESIDENT AND CEO AND ALTERNATE REPRESENTATIVE
David Haire, REPRESENTATIVE
James Hayes, TREASURER

GEORGIANS FOR CLEAN ENERGY
427 Moreland Ave., Suite 100
Atlanta, GA 30307 USA
Phone: 404-659-5675   Fax: 770-234-3909
E-mail: georgia@cleanenergy.ws
Website: www.cleanenergy.ws

Founded: 1983
Membership: 300
Scope: Statewide

Description: CPG is a nonprofit statewide organization protects the environment and improves the economy by changing the way energy is produced and consumed in Georgia through public education and advocacy.

Publication(s): Technical Reports, Plugging In

Keyword(s): Air Quality and Pollution, Energy, Sustainable Development

Contact(s):
Amy Macklin, BOARD CHAIR
Amy Macklin, PRESIDENT
C. Copeland, SECRETARY
Miki Davis, TREASURER
Na`taki Osborne, VICE PRESIDENT

GET AMERICA WORKING!
1700 North Moore Street, Suite 2000
Arlington, VA 22209 USA
Phone: 703-527-5852   Fax: 703-527-8383
E-mail: info@getamericaworking.org
Website: www.getamericaworking.org

Founded: 1998
Membership: 6
Scope: National

Description: Get America Working! believes that a sustainable economy must reduce pollution and resource waste, while greatly expanding employment opportunities and improving the use of human capital. To this end, GAW! advocates a tax shift —reducing payroll taxes to encourage job creation, and instead taxing pollution, energy inefficiency, and resource waste. GAW! works with diverse constituencies including seniors, minorities, environmentalists, labor, women, and the disabled to build support for mutually beneficial policies.

Publication(s): Job Creation Tax Options: A Background Paper from Get America Working!, A Fresh Point of View: When Americans Get to Work, America Works!, Key Questions & Answers

Contact(s):
Susan Davis, ASSOCIATE CHAIR
William Drayton, CHAIRMAN

GIRL SCOUTS OF THE USA
420 5th Ave.
New York, NY 10018-2798 USA
Phone: 212-852-8000   Fax: 212-852-6509
Website: www.girlscouts.org

Founded: 1912
Membership: 500
Scope: National

Description: The national organization offers an informal education and recreation program designed to help each girl develop her own values and sense of worth as an individual. It provides opportunities for girls to experience, to discover, and to share planned activities that meet their interests. These activities encourage personal development through a wide variety of projects in social action, environmental action, wildlife values education, youth leadership, career exploration and community service.

Publication(s): Earth Matters, Investigaciones divertidas y faciles de la naturaleza y ciencia, Fun and Easy Activities with Nature and Science, Fun and Easy Nature and Science Investigations, Outdoor Education in Girl Scouting

Keyword(s): Environmental and Conservation Education, Outdoor Recreation, Urban Environment, Women in the Environment, Youth Organizations

Contact(s):
Sharon Woods-Hussey, NATIONAL DIRECTOR OF MEMBERSHIP AND PROGRAM CLUSTER
Phone: 212-852-8150
Fax: 212-852-6515
shussey@girlscouts.org
Marsha Evans, NATIONAL EXECUTIVE DIRECTOR
Phone: 212-852-5000
mevans@girlscouts.org
Connie Matsui, PRESIDENT
Phone: 212-852-5001
Fax: 212-852-6517
cmatsui@girlscouts.org
Laverne Alexander, WASHINGTON REPRESENTATIVE
1025 Connecticut Ave. NW, Suite 309, Washington, DC 20036-5405
Phone: 202-659-3780
Fax: 202-331-8065

GLACIER INSTITUTE, THE
P.O. Box 7457
Kalispell, MT 59904 USA
Phone: 406-755-1211   Fax: 406-755-7154
E-mail: glacinst@digisys.net
Website: www.glacierinstitute.org

Founded: 1983
Membership: 350
Scope: International

Description: The Glacier Institute serves students of all ages as an educational leader in the Crown of the Continent Ecosystem, emphasizing hands-on, field-based experiences promoting a balanced understanding of the science of ecology and human interaction with the environment.

Publication(s): Annual Course Catalog

Keyword(s): Aquatic Habitats, Conservation, Ecology, Geology, education, Internships, Sustainable Ecosystems, training

Contact(s):
Jim How, PRESIDENT
Jami Belt, PROGRAM DIRECTOR
R. Devitt, PROGRAM DIRECTOR
Sue Lind, SECRETARY
Doug Morehouse, VICE PRESIDENT

GLOBAL CITIES PROJECT, THE

2962 Fillmore St.
San Francisco, CA 94123 USA
Phone: 415-775-0791   Fax: 415-775-4159
E-mail: epc@globalcities.org
Website: http://www.globalcities.org

Founded: 1989
Scope: National

Description: The Global Cities Project provides local governments, businesses, and citizens with comprehensive, up-to-date information on local environmental policies and programs, promoting the development of environmentally and economically sound policy at the local level.

Publication(s): Case Studies, Building Sustainable Communities: A Guide for Local Government

Keyword(s): Energy, Energy Conservation, Environment, Air Quality and Pollution, Land Use Planning, Pollution Prevention, Solid Waste Management, Open Space, Sustainable Development, Transportation, Water Quality, Urban Forestry

Contact(s):
Colleen McCarty, CHIEF FINANCIAL OFFICER
2962 Fillmore St., San Francisco, CA 94123
Phone: 415-775-0791
Walter Mcguire, PRESIDENT

GLOBAL ENVIRONMENTAL MANAGEMENT INITIATIVE (GEMI)

1 Thomas Circle NW, 10th Fl.
Washington, DC 20005 USA
Phone: 202-296-7449   Fax: 202-296-7442
E-mail: gemi@worldweb.net
Website: www.gemi.org

Founded: 1990
Membership: 40
Scope: International

Description: The Global Environmental Management Initiative (GEMI), an industry-initiated coalition of domestic and multinational Fortune 500 companies, is dedicated to helping businesses achieve environmental, health, and safety excellence. Through the activities of its workgroups, it has generated and distributed concrete tools for industry use in a number of environmental management fields.

Publication(s): New Paths to Business Value; Strategic Sourcing -Environment, Health & Safety, Environment value to Business, Information Systems For Health Safety &

Environmental Management, www.bussinessandclimate.org, Fostering Environmental Prosperity Multinationals

Keyword(s): Communications, Environmental and Conservation Education, Planning Management, Sustainable Development

Contact(s):
Richard Guimond, CHAIRMAN
Amy Goldman, CONTACT
Steven Hellem, EXECUTIVE DIRECTOR

GOPHER TORTOISE COUNCIL

Florida Museum of Natural History, P.O. Box 117800, University of Florida
Gainesville, FL 32611 USA
Phone: 904-392-1721
Website: www.gophertortoisecouncl.org

Founded: 1978
Scope: Regional

Description: A nonprofit organization formed to assure the continued survival of viable populations of the gopher tortoise, Gopherus polyphemus, and its associated upland habitat in the southeastern United States.

Publication(s): Tortoise Burrow, The Bulletin

Keyword(s): Endangered Species, Nongame Wildlife, Reptiles and Amphibians, training

Contact(s):
George Heinrich, CO-CHAIR
Boyd Hill Nature Park, 1101 Country Club Way S.,
St. Petersburg, FL 33705
John Jensen, EDITOR
Georgia DNR, 116 Rum Creek Dr., Forsyth, GA 31029
Lora Smith, MEMBERSHIP SECRETARY
Terri Stilson, SECRETARY
2879 Thaxton Dr., 52, Palm Harbor, FL 34684
Christian Newman, TREASURER
3839 NW 67th Pl., Gainesville, FL 32653

GRAND CANYON TRUST

2601 N. Fort Valley Rd.
Flagstaff, AZ 86001 USA
Phone: 520-774-7488   Fax: 520-774-7570
E-mail: info@grandcanyontrust.org
Website: www.grandcanyontrust.org

Founded: 1985
Membership: 24
Scope: National

Description: The mission of the Grand Canyon is to protect and restore the canyon country of the Colorado Plateau—its spectacular landscapes, flowing rivers, clean air, diversity of plants and animals, and areas of beauty and solitude.

Publication(s): Annual Report, Action Alerts, Colorado Plateau Advocate

Keyword(s): National Parks

Contact(s):
Steele Wotkyns, COMMUNICATION MANAGER
Brad Ack, DIRECTOR OF CONSERVATION FIELD PROGRAMS
Katrina Rogers, DIRECTOR OF DEVELOPMENT
Tom Robinson, DIRECTOR OF GOVERMENT AFFAIRS
Darcy Allen, MEMBERSHIP MANAGER

Geoffrey Barnard, PRESIDENT
Phone: 520-774-7488
Bill Hedden, UTAH CONSERVATION DIRECTOR

GRAND CANYON TRUST

MOAB, UTAH OFFICE
2601 Fort Valley Rd
Flagstaff, AR 86001 USA
Phone: 435-259-5284   Fax: 928-774-7570
Website: www.grandcanyontrust.org

Founded: NA
Membership: 5000
Scope: Regional
Publication(s): Colorado Plateau Advocate

GRAND CANYON TRUST

ST. GEORGE, UT OFFICE
199 N. Main St.
St. George, UT 84770 USA
Phone: 435-673-8558   Fax: 435-673-8545
Website: www.grandcanyontrust.org

Founded: NA
Scope: Regional

Keyword(s): Land Preservation, Land Use Planning, Public
Lands, Sustainable Development, Exotic species, Aquatic
nuisance species

Contact(s):
Jim McMahon, DIRECTOR

GRASSLAND HERITAGE FOUNDATION

P.O. Box 394
Shawnee Mission, KS 66201 USA
Phone: 913-262-3506
E-mail: grasslandheritage@grapevine.net
Website: www.grasslandheritage.org

Founded: 1976
Membership: 200
Scope: Regional

Description: A tax-exempt, nonprofit organization dedicated to
prairie preservation and education. We encourage the preservation of all remaining prairies and work to increase public
awareness of our prairie heritage.

Publication(s): GHF News

Contact(s):
Sue Holcomb, OFFICE MANAGER
Phone: 913 8-29 -0037

GREAT BEAR FOUNDATION

P.O. Box No. 9383
Missoula, MT 59807 USA
Phone: 406-829-9378   Fax: 406-829-9379
E-mail: gbf@greatbear.org
Website: www.greatbear.org

Founded: 1982
Membership: 3000
Scope: International

Description: A membership-based organization dedicated to
protecting all eight species of bears and their habitat.
Programs range from supporting scientific research to
educational outreach in schools.

Publication(s): Bear News

Keyword(s): Endangered Species, Environmental and
Conservation Education, Public Lands, training

Contact(s):
Charles Jonkel, PRESIDENT

GREAT LAKES SPORT FISHING COUNCIL

P.O. Box 297
Elmhurst, IL 60126 USA
Phone: 630-941-1351   Fax: 630-941-1196
E-mail: imfo@great-lakes.org
Website: www.great-lakes.org

Founded: 1973
Membership: 350000
Scope: National

Description: A nonprofit confederation of organizations and
individuals throughout the Great Lakes states and provinces
whose members are concerned with the present and future of
sport fishing in the Great Lakes and adjoining waters. The
Council, which acts as a clearinghouse for the exchange of
information among members, also seeks to protect the Great
Lakes against pollution and exploitation by commercial,
individual, or other interests. over 100 U.S. and Canadian
organizations with a combined membership of more than
325,000 families.

Publication(s): Regional Government Reference Guide, Great
Lakes Basin Report

Keyword(s): Aquatic Habitats, Environmental and Conservation
Education, Wildlife, Sport Fishing, Wetlands

Contact(s):
Bob Schmidt, EDITOR
5016 West Argyle St., Chicago, IL 60630
Phone: 773-283-7871
rschmidt26@aol.com
Dan Thomas, PRESIDENT
Mel Both, SECRETARY
4633 Ridgecrest Dr., Racine, WI 53403
Phone: 262-598-8802
rodnreel@wi.net
Tom Couston, TREASURER
12 W. Schaumburg Rd, Schaumburg, IL 60194
Phone: 847-519-1711
tomdds@megsinet.net
Robert Mitchell, VICE PRESIDENT
6466 Parkview, Troy, MI 48098
Phone: 810-558-6547
Fax: 810-575-9713
emitchel@cecom.com
R. James, WEBMASTER
webmaster@great-lakes.org

GREAT LAKES UNITED

Headquarters, Buffalo State College, Cassety Hall,
1300 Elmwood Ave.
Buffalo, NY 14222 USA
Phone: 716-886-0142   Fax: 716-886-0303
E-mail: glu@glu.org
Website: www.glu.org

Founded: 1982
Membership: 600
Scope: International

Description: An international coalition of environmental, conservation, sports, labor, business, and community organizations, and individuals throughout the eight Great Lakes states, two Canadian provinces. GLU is dedicated to the protection and restoration of the Great Lakes-St. Lawrence River Basin ecosystem.

Publication(s): Great Lakes: Habitat Watch, Sustainable Waters Watch, and Toxic Watch, occasional reports, Great Lakes News

Keyword(s): Biodiversity, Communications, Conservation, Environmental Justice, Environmental Protection, Wildlife, Coral Reefs, Exotic species, Aquatic nuisance species, Watersheds, Wetlands, Pollution Prevention, Water Quality, Toxic Substances, Nuclear-free, Water quantity, Water export and diversion

Contact(s):
Jim Mahon, CANADIAN TREASURER, CANADIAN AUTO WORKERS, LOCAL 1520
120 Tufton Pl., London, Ontario N6C 4W9
Phone: 519-681-3680
Fax: 519-652-0586
jimahon@home.com
Alexandra McPherson, CLEAN PRODUCTION COORDINATOR
alex@glu.org
Stephane Gingras, COORDINATOR, QUEBEC
Phone: 514-396-3333
Fax: 514-396-0297
sgingras@glu.org
Margaret Wooster, EXECUTIVE DIRECTOR
Jennifer Nalbone, HABITAT AND BIODIVERSITY COORDINATOR
jen@glu.org
John Jackson, PAST PRESIDENT
Ed Michael, PRESIDENT, TROUT UNLIMITED
223 Barberry Rd., Highland Park, Illinois 60035
Phone: 847-831-4159
Phone: 847-831-1035
e1michael@cs.com
Patty Odonnell, SECRETARY, GRAND TRAVERSE BAND OF OTTAWA AND CHIPPEWA
2605 NW Bay Shore Dr., Suttons Bay, MI 49682
Phone: 231-271-7368
Fax: 231-271-3576
patty@freeway.net
Reg Gilbert, SENIOR COORDINATOR
reg@glu.org
Robin McClellan, UNITED STATES TREASURER, NYS CITIZENS ENVIRONMENTAL COALITION
2877 Gaines Basin Rd., Albion, NY 14411
Phone: 716-589-4695
robinm@eznet.net
Lynda Lukasik, VICE PRESIDENT, FRIENDS OF RED HILL VALLEY
148 Oakland Dr., Hamilton, Ontario L8E 1B6
Phone: 905-560-1177
lynda.lukasik@sympatico.ca

GREAT LAKES UNITED
4525 Derouen
Montreal, Quebec HiV IH1 Canada
Phone: 514-396-3333   Fax: 514-396-0297
Website: www.glu.org

Founded: NA

Scope: National

Publication(s): Newsletter-Great Lakes

Contact(s):
Daniel Green, DIRECTOR
Societe pour Vaincre la Pollution; C.P. 65 Place D'Armes, Montreal, Quebec H2Y 3E9
Phone: 514-844-5477
Fax: 514-844-1446
greentox@total.net
Julian Holenstein, DIRECTOR
Environment North; 427 Queen St., Thunder Bay, Ontario P7B 2K3
Phone: 807-345-7784
julian@tbaytel.net
Jane Wilkins, DIRECTOR
Sierra Club of Eastern Canada; 699 Bush St., Bel Fountain, Ontario L0N 1B0
Phone: 519-927-5924
Fax: 519-927-9828
Liliane Cotnoir, DIRECTOR
Front Common Quebecois pour une Gestion Ecologique des Dechets; 2025 A Masson 001, Montreal, Quebec H2H 2P7
Phone: 514-396-2286
Fax: 514-396-9041
cotnoirl@mlink.net
Jim Mahon, DIRECTOR
Canadian Auto Workers Local 1520; 120 Tufton Pl., London, Ontario N6C 4W9
Phone: 519-681-3680
Fax: 519-652-0586
jimahon@home.com
Margaret Wooster, EXECUTIVE DIRECTOR
Ed Michael, PRESIDENT

GREAT OUTDOORS CONSERVANCY, THE
4311 Manatee Ave., West, Suite 210
Bradenton, FL 34209-3948 USA
Phone: 941-708-3456   Fax: 941-708-3535
E-mail: conserve@TheGreatOutdoors.org
Website: www.thegreatoutdoors.org

Founded: 1998
Membership: 2500
Scope: National

Description: The conservancy is a nonprofit national marketing, fundraising, and educational organization for land conservation and expands wild, natural, scenic and recreational areas in the United States by the acquisition of land for the benefit of wildlife and the public's enjoyment for generations to come.

Publication(s): Partnerships in Preservation

Keyword(s): Conservation Easements, Environmental and Conservation Education, Land Conservation, Land Management, Land Preservation, Land Protection, Land Purchase, Mountain Ecosystems, Open Space, Outdoor Education, Public Access, Watersheds, Wilderness, training

Contact(s):
Bill Lamee, PRESIDENT
BlaMee@TheGreatOutdoors.org

GREAT PLAINS NATIVE PLANT SOCIETY
P. O. Box 461
Hot Springs, SD 57747 USA
Phone: 605-745-3397   Fax: 605-745-3397
E-mail: cascade@gwtc.net

Founded: 1984
Membership: 250
Scope: Regional

Description: Promotes the protection and study of native plants of the Great Plains through the formation of a Botanic Garden, an annual seed exchange, field trips and newsletter.

Publication(s): Plains Plants

Keyword(s): Botanical Gardens, Botany, Flowers, Plants, and Trees, Gardening and Horticulture, Grasslands, Great Plains, Native Plants, Wildflowers

Contact(s):
Cynthia Reed, PRESIDENT
Joe Lux, SECRETARY-TREASURER
Ronald Weedon, VICE-PRESIDENT

GREAT SMOKY MOUNTAINS INSTITUTE AT TREMONT

9275 Tremont Rd.
Townsend, TN 37882 USA
Phone: 865-448-6709   Fax: 865-448-9250
E-mail: mail@gsmit.org
Website: www.gsmit.org

Founded: 1969
Membership: 10
Scope: Statewide, National

Description: A residential environmental education center in the Great Smoky Mountains National Parks. Programs promote awareness, appreciation, and stewardship of national parks and are offered for children and adults.

Publication(s): Walker Valley Reflections (Newsletter), Connecting People and Nature

Keyword(s): Environmental and Conservation Education, education, National Parks

Contact(s):
Ken Voorhis, EXECUTIVE DIRECTOR
Phone: 423-448-6709
Fax: 423-448-9250
ken@smokiesnha.org
Bill Cobble, PRESIDENT
Phone: 800-721-6064
Phone: 423-982-6583
Norma Ogle, SECRETARY
Phone: 803-635-3561
Fax: 803-635-3561
Herb Handly, TREASURER
Phone: 423-974-1755
Fax: 423-974-4631
Bill Oliphant, VICE PRESIDENT

GREATER YELLOWSTONE COALITION

P.O. Box 1874, 13 S. Willson, Suite 2
Bozeman, MT 59771 USA
Phone: 406-586-1593   Fax: 406-586-0851
E-mail: gyc@greateryellowstone.org
Website: www.greateryellowstone.org

Founded: 1983
Membership: 10000
Scope: Regional, National

Description: A nonprofit, tax-exempt organization to preserve and protect the Greater Yellowstone Ecosystem and its unique quality of life by enhancing the ecosystem concept, raising the national public consciousness about the Greater Yellowstone Ecosystem, and combining the political effectiveness of the coalition's 7,500 individual members and more than 120 national and regional member organizations.

Publication(s): Annual Report, EcoAction Alerts, Greater Yellowstone Report

Keyword(s): Endangered Species, Forests and Forestry, Public Lands, training, Private Land Development

Contact(s):
Jon Catton, COMMUNICATIONS DIRECT
Mike Clark, EXECUTIVE DIRECTOR
Stephanie Kessler, PRESIDENT
Farrell Smith, SECRETARY AND TREASURER
Stephen Unfried, VICE PRESIDENT

GREEN (GLOBAL RIVERS ENVIRONMENTAL EDUCATION NETWORK)

1908 Mt. Vernon Ave., Second Fl.
Alexandria, VA 22301 USA
Phone: 703-299-9400   Fax: 703-299-9485
E-mail: green@earthforce.org
Website: www.earthforce.org

Founded: 1989
Scope: National

Description: The Global Rivers Environmental Education Network (GREEN) seeks to improve education through a global network that promotes watershed stewardship.

Publication(s): Investigating Streams and Rivers, Sourcebook for Watershed Education, Environmental Education for Empowerment, Field Manual for Water Quality Monitoring

Keyword(s): Wetlands, Watersheds, Water Quality, Exotic species, Aquatic nuisance species, Coral Reefs, Pollution Prevention, Environmental Protection, Rivers, Protected Areas, Environmental and Conservation Education, Wildlife Rehabilitation, Inquiry Based Education, Environment

Contact(s):
William Stapp, FOUNDER
University of Michigan, 2050 Delaware Ave.,
Ann Arbor, MI 48104
Thomas Martin, PRESIDENT
Phone: 703-519-6867
Vince Meldrum, VICE PRESIDENT OF NATIONAL PROGRAMS
Phone: 703-519-6864

GREEN MEDIA TOOLSHED

1320 18th Street, NW., Suite 500
Washington, D.C. 20036 USA
Phone: 202-223-2114   Fax: 202-463-6671
E-mail: info@greenmediatoolshed.org
Website: www.greenmediatoolshed.org

Founded: 2000
Membership: Limited to Environmental Nonprofit Organizations
Scope: National

Description: Green Media Toolshed is an online Internet portal designed to serve the environmental communications community by providing media-related tools and information for its members. It does not do advocacy or policy wrok or place stories in the media —it is a service bureau that provides the tools for other groups so that they can work more effectively.

Publication(s): Media Database, Image Managemant System,

Opinion Research Library, Calendar, tips, resources, advice for the communications staff of environmental nonprofit organizations

Contact(s):
Bobbi Russell, DIRECTOR OF MEDIA SERVICES AND MARKETING
bobbi@greenmediatooshed.org
Martin Kearns, EXECUTIVE DIRECTOR

GREEN MOUNTAIN CLUB INC., THE

4711 Waterbury-Stowe Rd.
Waterbury Center, VT 05677 USA
Phone: 802-244-7037   Fax: 802-244-5867
E-mail: gmc@greenmountainclub.org
Website: www.greenmountainclub.org

Founded: 1910
Membership: 9000
Scope: National

Description: The mission of the GMC is to make Vermont mountains play a larger part in the life of the people by protecting and maintaining the Long Trail System and fostering, through education, the stewardship of Vermont's hiking trails and mountains. The Club operates field programs and publishes guidebooks, maps, and educational materials in its efforts to maintain and protect the 440-mile Long Trail system. It is the advocate group for hiking in Vermont.

Publication(s): Day Hiker's Guide to Vermont, Long Trail End-to-Ender's Guide, Green Mountain Adventure, Vermont's Long Trail, Long Trail News, The Long Trail Guide

Keyword(s): Trail, Environmental and Conservation Education, Internships, education, Outdoor Recreation

Contact(s):
Ben Rose, EXECUTIVE DIRECTOR
Marty Lawthers, PRESIDENT
4 Hillside Dr., Peru, NY 12927
Richard Windish, SECRETARY
16 Forest St., Brattleboro, VT 05301
Walter Pomroy, TREASURER
Box 280, Johnson, VT 05606

GREEN SEAL

1001 Connecticut Ave., NW, Suite 827
Washington, DC 20036 USA
Phone: 202-872-6400   Fax: 202-872-4324
E-mail: greenseal@greenseal.org
Website: www.greeenseal.org

Founded: 1989
Membership: 12
Scope: National

Description: Green Seal helps organizations and individuals make environmentally responsible choices in their purchases. It develops environmental standards and tests products against these standards, identifing those products that are environmentally responsible through the award of an environmental "seal of approval.". The Environmental Partners Program helps businesses develop green procurement plans through buying guides and monthly reports on green products.

Publication(s): Catalog of Green Seal-Certified Products, Campus Green Buying Guide, Office Green Buying Guide, Monthly Choose Green Reports, Greening Your Property (A buying guide for hotels and motels), Environmental Criteria and Standards

Keyword(s): Air Quality and Pollution, Energy, Environmental and Conservation Education, Renewable Resources, Sustainable Development

Contact(s):
Bryan Thomlison, CHAIR OF THE BOARD
Phone: 609-737-8841
Arthur Weissman, PRESIDENT

GREEN TV

1125 Hayes St.
San Francisco, CA 94117 USA
Phone: 415-255-4797   Fax: 415-255-4664
Website: www.greentv.org

Founded: 1992
Scope: National

Description: A non-profit video production company specializing in television programming on subjects about human interaction with the natural world. Offers catalog of stock footage of California wildlife, endangered species habitats and the timber industry.

Contact(s):
Frank Green, OWNER/PRESIDENT

GREENPEACE, INC.

702 H St., NW. Suite 300
Washington, DC 20001 USA
Phone: 202-462-1177   Fax: 202-462-4507
E-mail: gp1@sharewest.com
Website: www.greenpeace.org

Founded: 1971
Scope: National

Description: A nonprofit organization dedicated to preserving the earth and the life it supports through nonviolent direct action, lobbying, public education, and research. Greenpeace seeks to protect biodiversity in all its forms; prevents pollution and abuse of the earth's ocean, land, air, and fresh water; end all nuclear threats; and promotes peace, global disarmament, and nonviolence.

Publication(s): Greenpeace Quarterly (magazine)

Keyword(s): Energy, Wildlife, Greenhouse Effect/Global Warming, Nuclear/Radiation, Toxic Substances, Nuclear-free, Water quantity, Water export and diversion

Contact(s):
Susan Sabella, BIODIVERSITY AND OCEAN ECOLOGY CAMPAIGN COORDINATOR
David Barre, DIRECTOR OF COMMUNICATIONS
Julie Crudele, DIRECTOR OF DEVELOPMENT
David Barre, EDITOR-IN-CHIEF
Lynn Thorp, NATIONAL CAMPAIGNS DIRECTOR

GROUNDWATER FOUNDATION, THE

P.O. Box 22558
Lincoln, NE 68542-2558 USA
Phone: 402-434-2740   Fax: 402-434-2742
E-mail: info@groundwater.org
Website: www.groundwater.org

Founded: 1985
Membership: 11
Scope: National

Description: The Groundwater Foundation is a nonprofit foundation dedicated to educating the public about conservation and management of groundwater. The Foundation is a clearinghouse for general groundwater information, sponsors the Nebraska Children Groundwater Festival, and coordinates the "Groundwater Guardian", a national community recognition program.

Publication(s): The Groundwater Catalog, The Aquifer

Keyword(s): Environmental and Conservation Education, Pollution Prevention, Exotic species, Aquatic nuisance species, Ground Water Protection

Contact(s):
Rachael Herpel, GROUNDWATER GUARDIAN PROGRAM DIRECTOR
Susan Seacrest, PRESIDENT
Chad Foust, YOUTH PROGRAMS DIRECTOR

GULF OF MEXICO FISHERY MANAGEMENT COUNCIL

The Commons at Rivergate, Suite 1000; 3018 U.S. Highway 301 North
Tampa, FL 33619-2266 USA
Phone: 813-228-2815   Fax: 813-225-7015
E-mail: gulfcouncil@gulfcouncil.org
Website: www.gulfcouncil.org

Founded: 1976
Membership: 22
Scope: Regional

Description: The Gulf Council is responsible for developing and monitoring fishery management plans to provide for the best use of the fishery resources in the federal waters of Gulf of Mexico.

Keyword(s): Wildlife

Contact(s):
Wayne Swingle, EXECUTIVE DIRECTOR

H

H. JOHN HEINZ III CENTER FOR SCIENCE, ECONOMICS, AND THE ENVIRONMENT

1001 Pennsylvania Ave., NW, Suite 735, South
Washington, DC 20004 USA
Phone: 202-737-6307   Fax: 202-737-6410
E-mail: info@heinzctr.org
Website: www.heinzctr.org

Founded: 1995
Membership: 17
Scope: National

Description: The H. John Heinz III Center is a nonprofit institution dedicated to improving the scientific and economic foundation of environmental policy. The Center's mission is to collaboratively identify emerging environmental issues, conduct related scientific research and economic analyses, and create and disseminate nonpartisan policy options for solving environmental problems.

Keyword(s): Coasts, Environmental Protection, Wildlife, Research

Contact(s):
G. William Miller, BOARD OF TRUSTEES CHAIR
William Merrell, SENIOR FELLOW AND PRESIDENT

Mary Katsouros, SENIOR FELLOW AND SENIOR VICE PRESIDENT
Robert Friedman, SENIOR FELLOW AND VICE PRESIDENT FOR RESEARCH

HARDWOOD FOREST FOUNDATION

P.O. Box 34518
Memphis, TN 38184-0518 USA
Phone: 901-377-1818   Fax: 901-382-6419
E-mail: nhla@natlhardwood.org
Website: www.natlhardwood.org

Founded: 1989
Scope: National

Description: The Hardwood Forest Foundation is a public, nonprofit organization dedicated to supporting research and education about North American hardwood forests. The main programs focus on distributing factual information about forest management and forest-related issues in the form of video tapes, computer programs, and publications to elementary and high schools.

Keyword(s): Forest Management

Contact(s):
Paul Houghland, EXECUTIVE MANAGER
Greg Kitchens, PRESIDENT

HAWAII NATURE CENTER

2131 Makiki Heights Dr.
Honolulu, HI 96822 USA
Phone: 808-955-0100   Fax: 808-955-0116
E-mail: hawaiinaturecenter@hawaii.rr.com

Founded: 1981
Scope: Statewide

Description: The Hawaii Nature Center promotes wise stewardship of the Islands through environmental education programs for school children and the general public. School programs focus on full-day, hands-on field adventures; community programs include adult interpretive hikes, family nature adventures and customized excursions for scouts, senior citizens and other special groups. The Iao Valley Interactive Science Arcade includes hands-on exhibits of native flora, fauna and streamlife for residents and visitors to Maui. Field sites on Oahu and Maui.

Publication(s): The Steward (Newsletter Quarterly)

Keyword(s): Nature Centers, Aquatic Habitats, Biodiversity, Birds, Ecology, Endangered Species, Environment, Environmental and Conservation Education, Insects and Butterflies, education, Terrestrial Habitats, Wetlands, Interpretive Center, Rainforests

Contact(s):
Diana King, EDUCATION DIRECTOR

HAWAII SOCIETY OF AMERICAN FORESTERS

5400 Grosvenor Ln
Honolulu, HI 96813 USA
Phone: 301-897-8720   Fax: 301-897-3690
E-mail: safweb@safnet.org

Founded: 1970
Scope: Statewide

Description: A nonprofit, tax-exempted organization working on environmental issues facing the state of Hawaii. Concerned with land use, native forests, water pollution, pesticides, coastal issues, and overdevelopment.

Keyword(s): Biodiversity, Endangered Species, Forests and Forestry, Islands, Sustainable Development

HAWAIIAN BOTANICAL SOCIETY

3190 Maile Way
Honolulu, HI 96822 USA
Phone: 808-956-8072   Fax: 808-956-3923

Founded: 1924
Scope: Statewide

Description: Objectives of society are: to advance the science of botany in all of its applications; To encourage research in botany in all of its phases; to promote the botanical welfare of its members; and to develop the spirit of good fellowship and cooperation in botanical matters. The Society is particularly interested in the preservation of the Hawaiian flora.

Publication(s): Newsletter of the Hawaiian Botanical Society

Contact(s):
Clifford Morden, EDITOR
Mindy Wilkinson, PRESIDENT
Leilani Durand, SECRETARY
Ron Fenstemacher, TREASURER
Alvin Yoshinaga, VICE PRESIDENT

HAWK MIGRATION ASSOCIATION OF NORTH AMERICA

Attn: Treasurer, R.R. 2 Box 301-A
New Ringgold, PA 17960 USA

Founded: 1974
Scope: National

Description: A nonprofit organization whose purpose is to advance the knowledge of bird-of-prey migration across the continent, to monitor raptor populations as an indicator of environmental health, to study further the behavior of raptors, and to contribute to greater public understanding of birds of prey.

Publication(s): Hawk Migration Studies

Keyword(s): Raptors

Contact(s):
William Barnard, CHAIR
Norwich University, Biology Department, Northfield, VT 05663
Phone: 802-485-2342
William Gallagher, SECRETARY
P.O. Box 822, Boonton, NJ 07005-0822
Phone: 973-335-0674
Douglas Wood, TREASURER
R.R. 2 Box 301-A, New Ringgold, PA 17960

HAWK MOUNTAIN SANCTUARY ASSOCIATION, VISITOR CENTER

1700 Hawk Mountain Rd.
Kempton, PA 19529 USA
Phone: 610-756-6961   Fax: 610-756-4468
Website: www.hawkmountain.org

Founded: 1934
Membership: 9000
Scope: International

Description: The Association is a nonprofit organization devoted to the conservation of birds of prey worldwide and a greater understanding of the central Appalachian environment. A full-time staff assisted by interns and volunteers, carries out coordinated programs in education, research, monitoring, and sanctuary management. A visitor center is open year-round, and the 2,400-acre Sanctuary is maintained as a high-quality natural area with trails open to the public.

Publication(s): Mountain and the Migration, The, Hawks Aloft, Raptor Watch, A Global Directory in Raptor Migration sites, Hawk Mountain News

Keyword(s): Birds, Mountain Ecosystems, Nongame Wildlife, Raptors, Migration

Contact(s):
Jeffrey Weil, CHAIRMAN
Cynthia Lenhart, EXECUTIVE DIRECTOR
Keith Bildstein, RESEARCH DIRECTOR
Harry Cerino, TREASURER

HAWKWATCH INTERNATIONAL, INC.

1800 South West Temple, Suite A226-1
Salt Lake City, UT 84115 USA
Phone: 801-484-6808   Fax: 801-484-6810
E-mail: hwi@hawkwatch.org
Website: www.hawkwatch.org

Founded: 1986
Membership: 1800
Scope: International

Description: HawkWatch International is a nonprofit organization promoting healthy and sustainable populations of hawks, eagles, falcons and other raptors through high quality science, education and conservation programs.

Publication(s): Raptorwatch

Keyword(s): Birds, Endangered Species, Environmental and Conservation Education, Raptors, training

Contact(s):
Dawn Sebesta, CHAIR
2466 Meadows Dr., Park City, UT 84060-7032
Phone: 435-649-3024
stoney@pcfastnet.com
Howard Gross, EXECUTIVE DIRECTOR
Phone: 801-484-6502
hgross@hawkwatch.org
Jeff Smith, SCIENCE DIRECTOR
Phone: 801-484-6758
jsmith@hawkwatch.org
Marlene Ford, SECRETARY
2798 East 2880 South, Salt Lake City, UT 84109-2029
Phone: 801-467-8057
tabby@slkc.uswest.net
Benita Pulins, TREASURER
C/O Pricewaterhouse Coopers LLP, 36 S. State St., Suite 1700, Salt Lake City, UT 84111
Phone: 801-537-5227
benita.r.pulins@us.pwcglobal.com

HEADLANDS INSTITUTE

Golden Gate National Recreation Area, Bldg. 1033
Sausalito, CA 94965 USA
Phone: 415-332-5771   Fax: 415-332-5784
E-mail: hi@yni.org
Website: www.yni.org

Founded: 1979
Membership: 50

Scope: National

Description: To create sustained global environmental stewardship through educational adventures in nature's classroom.

Keyword(s): Environmental and Conservation Education, education, Sustainable Development, Urban Environment

Contact(s):
Mike Lee, EXECUTIVE DIRECTOR

HENRY A. WALLACE INSTITUTE FOR ALTERNATIVE AGRICULTURE (HAWIAA)

9200 Edmonston Rd., Suite 117
Greenbelt, MD 20770-1551 USA
Phone: 301-441-8777   Fax: 301-220-0164
E-mail: hawiaa@access.digex.net
Website: igc.apc.org

Founded: 1983
Scope: National

Description: HAWIAA is a nonprofit, membership research and education organization established to encourage and facilitate adoption of resource-conserving, low-cost, environmentally sound, and economically viable farming systems.

Keyword(s): Agriculture, Research, Sustainable Agriculture

Contact(s):
Garth Youngberg, EXECUTIVE DIRECTOR
David Ervin, POLICY STUDIES PROGRAM DIRECTOR

HENRY STIFEL SCHRADER ENVIRONMENTAL EDUCATION CENTER

Oglebay Institute, Burton Center
Wheeling, WV 26003 USA
Phone: 304-242-6855   Fax: 304-242-5197
Website: www.oilonline.com

Founded: NA
Scope: Statewide

Description: Oglebay Institute operates a variety of programs: Resident nature summer camps for adults and children; Ecotourism Club; resident environmental education programs; children's day camping; special workshops and weekends; exhibits; school programs; and also the A.B. Brooks Environmental Education Center and Speidel Observatory.

Keyword(s): Aquatic Habitats, Wildlife, Renewable Resources, Water Pollution Management, Wetlands

Contact(s):
Cathy Gielty, ASSOCIATE DIRECTOR OF ENVIRONMENTAL EDUCATION
Greg Park, ASSOCIATE DIRECTOR OF ENVIRONMENTAL EDUCATION
Ford Parker, DIRECTOR OF NATURE AND ENVIRONMENTAL EDUCATION

HERP DIGEST

67-87 Booth Street
Forest Hills, NY 11375 USA
Phone: 718-275-2190   Fax: 718-275-3307
Website: www.herpdigest.org
Membership: 25000
Scope: International

Description: Free weekly newsletter delivered to your E-mail: address that covers the latest news in conservation and science on reptiles and Amphibians. Aimed to supply information usually not found in scientific journals in a timely manner. Newsletter prints items from all over the world. Past items have included new discoveries, new techniques for vets to use, reports and announcements on conferences and jobs, scientific abstracts from journals all over the world, and reports on the latest, conservation battles. Subscribers include herpetologists, zookeepers, conservationists, journalists, pet-owners, and just plain people who like herps.

Publication(s): Weekly Newsletter, current science and conservation, reptiles and amphibians, Herp Digest

Contact(s):
Allen Salzberg, CONTACT
asalzberg@nyc.rr.com

HIGH DESERT MUSEUM, THE

59800 S. Highway 97
Bend, OR 97702-7963 USA
Phone: 503-382-4754   Fax: 541-382-5256
E-mail: info@highdesert.org
Website: www.highdesert.org

Founded: 1974
Membership: 5,300
Scope: National

Description: Created to broaden the knowledge and understanding of the natural and cultural history and resources of the high desert country for the purpose of promoting thoughtful decision-making that will sustain the region's natural and cultural heritage. It is a "living," participation-oriented museum which focuses on the Intermountain West — portions of eight Western states and the Canadian province of British Columbia. Opened to the public in 1982.

Publication(s): Sagebrush Legacy, High Desert Quarterly

Keyword(s): Cultural Preservation, Deserts, Environmental and Conservation Education, education, training

Contact(s):
Sue McWilliams, EDUCATION MANAGER
SMcWilliams@highdesert.org
Sheila Timony, EXHIBITS MANAGER
Stimony@highdesert.org
Forrest Rodgers, PRESIDENT
Frodgers@highdesert.org
Kevin Britz, VICE PRESIDENT FOR PROGRAMS
Kbritz@highdesert.org
Kristi Jacobs, VOLUNTEER PROGRAM MANAGER
Kjacobs@highdesert.org
Becky Anderson, ZOOLOGICAL MANAGER
BAnderson@highdesert.org

HOLDEN ARBORETUM, THE

9500 Sperry Rd.
Kirtland, OH 44094 USA
Phone: 440-256-1110   Fax: 440-256-1655
E-mail: holden@holdenarb.org
Website: www.holdenarb.org

Founded: 1931
Membership: 7000
Scope: National

Description: The Holden Arboretum's mission is to acquire, maintain and display collections of plants and conserve natural areas for education, scientific inquiry and personal inspiration. The Arboretum develops and maintains documented biological collections, emphasizing woody plants of northeast Ohio; develops improved plants for the landscape through breeding and selection; studies, manages and conserves the natural environment, including the Holden Arboretum lands; and provides diverse learning experiences in horticulture, botany and natural history.

Publication(s): Environmental Thinking and Learning (ETAL), The Arboretum Class Schedule (quarterly class schedule), Arboretum Leaves, The

Keyword(s): Botanical Gardens, Ecology, Environmental and Conservation Education, Flowers, Plants, and Trees, Gardening and Horticulture, education

Contact(s):
Paul Spector, DIRECTOR OF EDUCATION
Robert Marquard, DIRECTOR OF RESEARCH
Nadia Aufderheide, HEAD LIBRARIAN
Warren H. Corning Library, 9500 Sperry Rd., Kirtland, OH 44094
Phone: 440-256-1110
Fax: 440-256-5836
Elaine Price, INTERIM EXECUTIVE DIRECTOR

HOLLY SOCIETY OF AMERICA, INC.
4738 Hale Haven Dr.
Ellicott City, MD 21043-6669 USA
Phone: 410-730-0243
E-mail: SECRETARY@HOLLYSOCAM.ORG
Website: hollysocam.org

Founded: 1947
Membership: 600
Scope: National

Description: National nonprofit organization dedicated to bringing together persons interested in any phase of holly culture. Collects and disseminates information about holly; studies methods of conservatively cutting and marketing holly; promotes research and hybridization; publishes research papers; and popularizes the use of holly as a landscape material.

Publication(s): H.S.A. Books, Holly Society Journal (including the proceedings)

Contact(s):
Ronald Solt, ADMINISTRATIVE VICE PRESIDENT
esolt79087@aol.com
Nancy Smith, EDITOR
Michael Pontti, EXECUTIVE VICE PRESIDENT
ponttim@gunet.georgetown.edu
Daniel Turner, PRESIDENT
Rondalyn Reeser, SECRETARY
Ruth Bradley, TREASURER

HOOD CANAL LAND TRUST
P.O. Box 861
Belfair, WA 98528 USA
Phone: 360-275-3505

Founded: NA
Scope: Local

Description: The Hood Canal Land Trust works to preserve shorelines, wetlands, forests and farmlands crucial to local wildlife, water quality and scenic splendor for future generations. The Land Trust sees preservation as a pathway to saving the special tranquility of life which has long enchanted residents and attracted visitors to the Hood Canal and nearby watersheds. The Land Trust holds 510 Acres under its protection.

Keyword(s): Protected Areas, Wetlands, Waterfowl, training, Water Quality, Land Preservation

Contact(s):
Joseph Testu, PRESIDENT

HOOSIER ENVIRONMENTAL COUNCIL
520 E. 12th St., Suite 14
Indianapolis, IN 46202 USA
Phone: 317-685-8800   Fax: 317-686-4794
E-mail: hec@hecweb.org
Website: www.hecweb.org

Founded: 1983
Membership: 25000
Scope: Statewide

Description: To encourage and promote more aggressive environmental regulation and enforcement in the state of Indiana. The Council objectives are as follows: Facilitation of communication between environmental groups and individuals; coordination of action on current environmental issues, educational programs and publications; and representation of the concerns of the membership before administrative officials and regulatory boards/agencies of the state and federal government.

Publication(s): Boardwatch, LDF Report, Monitor

Keyword(s): Air Quality and Pollution, Biodiversity, Environmental and Conservation Education, Solid Waste Management, Water Quality

Contact(s):
Denise Baker, EDITOR
Tim Maloney, EXECUTIVE DIRECTOR
520 E 12th St., Ste.14, Indianapolis, IN 46202
Phone: 317-685-8800
Fax: 317-686-4794
Jack Miller, PRESIDENT
520 E. 12th St., Suite 14, Indianapolis, IN 46202
Phone: 317-872-3516
Fax: 317-297-9271
Dona Young, SECRETARY
Alice Schloss, TREASURER
4525 N. Park Ave., Indianapolis, IN 46205

HUDSONIA LIMITED
Bard College Field Station P.O Box 5000
Annandale, NY 12504-0500 USA
Phone: 845-758-7053   Fax: 914-758-7033
Website: www.hudsonia.org

Founded: 1981
Scope: Regional

Description: Hudsonia Limited is a nonprofit, nonadvocacy institute for research, education, and technical assistance in the environmental sciences, focusing on the Hudson River Valley. There are over 25 research associates and other technical personnel. Hudsonia conducts pure and applied research on natural and social-sciences aspects of the environment, produces educational publications, and offers programs for environmental decision makers and natural history courses for a broader audience.

Publication(s): Guide to Biodiversity in the Hudson River Valley (forthcoming), News From Hudsonia (quarterly)

Keyword(s): Ecological Education, Aquatic Habitats, Endangered Species, Wildlife, Flowers, Plants, and Trees

Contact(s):
Erik Kiviat, SCIENCE DIRECTOR
kiviat@bard.edu
Gretchen Stevens, STAFF BOTANIST

HUMAN ECOLOGY ACTION LEAGUE, INC., THE (HEAL)

P.O. Box 29629
Atlanta, GA 30359-0629 USA
Phone: 404-248-1898   Fax: 404-248-0162
E-mail: healnatnl@aol.com
Website: www.members.aol.com/healNatnl/index.html

Founded: 1977
Scope: International

Description: A nonprofit volunteer organization of people affected by or concerned about environmental conditions that are hazardous to human health. It serves as an information clearinghouse on exposure-related illness; alerts the general public about the potential dangers of chemicals; and encourages healthy lifestyles that minimize potentially hazardous environmental exposures.

Publication(s): Bibliographies, Travel Directory, Environmental Consultant Directory, Frangrance and Health, Human Ecologist, The, Resource List

Keyword(s): Air Quality and Pollution, Health and Nutrition, Pesticides, Toxic Substances, Nuclear-free, Water quantity, Water export and diversion, Urban Environment

Contact(s):
Diane Thomas, EDITOR
Muriel Dando, PRESIDENT
Kenneth King, SECRETARY
Donald Jones, TREASURER AND BUSINESS MANAGER

HUMANE SOCIETY OF THE UNITED STATES, THE

2100 L St., NW
Washington, DC 20037 USA
Phone: 202-452-1100   Fax: 301-258-3077
Website: www.hsus.org

Founded: 1954
Scope: International

Description: A nonprofit organization dedicated to the protection of animals, both domestic and wild. Professional staff experienced in animal control, cruelty investigation, humane and environmental education, farm animals, federal and state legislative activities, wildlife and habitat protection, and laboratory animal welfare; offer resources to local organizations, government, media, and the general public.

Publication(s): Kind News, Shelter Sense, Kind Teacher, HSUS News

Keyword(s): Endangered Species, Hunting, Marine Mammals, Whale, Dolphin, Seal, Trapping, training

Contact(s):
G. Waite, CFO
O. Ramsey, CHAIRMAN OF THE BOARD
Richard Clugston, DIRECTOR OF THE CENTER FOR RESPECT OF LIFE AND ENVIRONMENT
Patricia Forkan, EXECUTIVE VICE PRESIDENT
Sharon Geiger, LIBRARY ASSISTANT
The Joyce Mertz Gilmore Library, HSUS Offices, 700 Professional Dr., Gaithersburg, MD 20879
Phone: 202-452-1100
Paul Irwin, PRESIDENT OF EARTHVOICE
Paul Irwin, PRESIDENT, CEO
Paul Irwin, PRESIDENT, HUMANE SOCIETY INTERNATIONAL
Amy Lee, SECRETARY
John Grandy, SENIOR VICE PRESIDENT OF WILDLIFE
David Wiebers, VICE CHAIRMAN
Roger Kindler, VICE PRESIDENT AND GENERAL COUNSEL

HUMBOLT FIELD RESEARCH INSTITUTE

P.O. Box 9
Steuben, ME 04680-0009 USA
Phone: 207-546-2821   Fax: 207-546-3042
E-mail: humboldt@nemaine.com
Website: www.maine.maine.edu/~taglehill

Founded: 1981
Scope: International

Description: A nonprofit educational and research organization providing advanced and professional training programs in all aspects of natural history (terrestrial, freshwater and marine) and encouraging similar pursuits. Classical natural history training programs are held in Maine and the American Tropics. Ecological restoration seminars are held in a number of cities across the United States and Canada.

Publication(s): Northeastern Naturalist (a quarterly, peer-reviewed scientific journal)

Keyword(s): Biodiversity, Environmental and Conservation Education, education, Wetlands, Restoration

Contact(s):
Joerg-Henner Lotze, DIRECTOR

HUMMINGBIRD SOCIETY, THE

P.O. Box 394
Newark, DE 19715 USA
Phone: 302-369-3699   Fax: 302-369-1816
E-mail: info@hummingbird.org
Website: www.hummingbird.org

Founded: 1996
Scope: International

Description: The Hummingbird Society is a nonprofit corporation dedicated solely to hummingbirds, through disseminating information, education, support of scientific research, and protection of habitat.

Publication(s): The Hummingbird Connection

Keyword(s): Birds, Conservation, Endangered Species, Environmental and Conservation Education, Research

Contact(s):
Robert Gell, DIRECTOR
Phone: 410-287-2988
H. Hawkins, PRESIDENT

Frances Oates, SECRETARY
Phone: 610-274-0551
hamilton@magpage.com
William Barry, TREASURER
Phone: 302-239-1797
billb@wserve.com
Gary Griffith, VICE PRESIDENT
Phone: 410-392-4491
garygriffith@mris.com

HUNTSMAN MARINE SCIENCE CENTRE

1 Lower Campus Rd.
St. Andrews, New Brunswick E5B 2L7 Canada
Phone: 506-529-1200   Fax: 506-529-1212
E-mail: huntsman@huntsmanmarine.ca
Website: www.huntsmarine.ca

Founded: 1969
Scope: National

Description: The HMSC is a nonprofit organization with a reputation for excellence in coastal and marine science research and education. It is supported by universities, corporations, federal and provincial government agencies, and the public. Located on one of the most biologically active bodies of water in the world, it provides information, research, education, and training opportunities for students, investigators, industry, government and the public.

Publication(s): Seawords, Sea Trek Bulletin, Atlantic Reference Centre Species Identification Series, Huntsman Marine Science News

Keyword(s): Coasts, Air Quality and Pollution, Aquariums, Aquatic Habitats, Biodiversity, Wildlife Rehabilitation, Ecology, Birds, Environment, Wildlife, Research, Training, Oceanography, Sustainable Ecosystems, Aquaculture

Contact(s):
Michael Burt, ASSOCIATE DIRECTOR
Phone: 506-529-1222
mburt@huntsmanmarine.ca
Paul Hebert, BOARD OF DIRECTORS CHAIR
Chair of Zoology Dept.; University of Guelph, Guelph, Ontario N1G 2W1
Mark Costello, EXECUTIVE DIRECTOR
Phone: 506-529-1200
costello@huntsmanmarine.ca

I

IDAHO ASSOCIATION OF SOIL CONSERVATION DISTRICTS

P O Box 2637
Boise, ID 83701 USA
Phone: 208-338-5900   Fax: 208-338-9537
E-mail: kfoster@agri.state.id.us
Website: www.iascd.state.id.us

Founded: NA
Scope: Statewide

Keyword(s): Conservation Districts

Contact(s):
David Ellsworth, BOARD MEMBER
Kevin Koester, DIRECTOR
Phone: 208-776-5382
Fax: 208-776-5043

Kent Foster, EXECUTIVE DIRECTOR
P.O. Box 2637, Boise, ID 83701
Phone: 208-338-5900
Fax: 208-338-9537
Alice Wallace, PRESIDENT
921 N. 5th Ave, Sandpoint, ID 83864
Phone: 208-263-0895
Fax: 208-265-8486
Roger Stutzman, SECRETARY
1937-B E. 4100 N, Buhl, ID 83316
Phone: 208-543-6824
Fax: 208-543-6824
Art Beal, TREASURER
Kyle Hawley, VICE PRESIDENT
1180 Lewis Rd., Moscow, ID 83843
Phone: 208-882-1290
Fax: 208-883-4239

IDAHO CONSERVATION LEAGUE

P.O. Box 844
Boise, ID 83701 USA
Phone: 208-345-6933   Fax: 208-344-0344
E-mail: icl@wildidaho.org
Website: www.wildidaho.org

Founded: 1973
Scope: Statewide

Description: The Idaho Conservation League is Idaho's largest statewide conservation organization. Based on grassroots activism and a professional staff, ICL works to preserve and protect Idaho's wild lands, water and wildlife.

Publication(s): Citizen's Guide to the Legislature, Idaho Conservationist

Keyword(s): Biodiversity, Protected Areas, Public Lands, Coral Reefs, Wilderness

Contact(s):
Linn Kincannon, CENTRAL IDAHO ASSOCIATE
John McCarthy, CONSERVATION DIRECTOR
Rick Johnson, EXECUTIVE DIRECTOR
Larry McLaud, NORTH IDAHO ASSOCIATE
Jerry Pavia, PRESIDENT
P.O. Box 912, Bonners Ferry, ID 83805
Phone: 208-267-7374
John Schmidt, SECRETARY
jschmidt@srv.net
Dallas Gudgell, STATE AFFAIRS DIRECTOR
Tom Pomeroy, TREASURER
P.O. Box 1765, Ketchum, ID 83340
Pat Ford, VICE-PRESIDENT
1511 N. 11th St., Boise, ID 83702
Phone: 208-345-9067

IDAHO ENVIRONMENTAL COUNCIL

1568 Lola St.
Idaho Falls, ID 83402 USA
Phone: 208-523-6692

Founded: NA
Scope: Statewide

Description: Founded to coordinate and stimulate the creative ideas, manpower, and financial resources of conservation-minded individuals and organizations; and to provide an increased understanding of modern man's impact upon his environment. Action, the objective, is based on information and research.

Publication(s): IEC Newsletter

Contact(s):
Jerry Jayne, EDITOR
Phone: 208-523-6692
Alan Hausrath, PRESIDENT
Phone: 208-336-4930
Dennis Baird, VICE PRESIDENT FOR NORTHERN IDAHO
Phone: 208-882-8289
Ralph Maughan, VICE PRESIDENT
FOR SOUTHEASTERN IDAHO
Phone: 208-233-7091

IDAHO FOREST OWNERS ASSOCIATION
P.O. Box 1257
Coeur d'Alene, ID 83816 USA
Phone: 208-762-9303

Founded: 1983
Membership: 300
Scope: Statewide

Description: A statewide organization, affiliated with the National Woodland Owners Association, of forest landowners dedicated to the management, use, and protection of private forest resources in Idaho.

Publication(s): Northwest Woodlands

Keyword(s): Forests and Forestry

Contact(s):
Lori Rasor, EDITOR
4033 SW Canyon Rd., Portland, OR 97221
Phone: 503-288-1367
Amy Gillette, EXECUTIVE DIRECTOR
Kennon McClintock, PRESIDENT
Phone: 208-267-7064
Kirk David, SECRETARY
Phone: 208-769-1525
Ozzie Osborn, TREASURER
Phone: 208-664-3889
Jim Thomas, VICE PRESIDENT
Phone: 208-245-2758

IDAHO STATE B.A.S.S. FEDERATION
16135 Gilenna Drive
Wilder, ID 83676 USA
Phone: 208-286-7138
Website: www.bassclubs.net/clubpages/idahofed

Founded: NA
Scope: Statewide

Description: An organization of Bassmaster chapters, affiliated with the Bass Anglers Sportsman Society, organized to fight pollution, assist state and national conservation agencies in their efforts, and teach the young people of our country good conservation practices. Dedicated to the realistic conservation of our water resources.

Contact(s):
Steve Spicklemier, CONSERVATION DIRECTOR
3766 S. Rush Creek Place, Boise, ID 83706
Phone: 208-342-5006
Allan Chandler, PRESIDENT
Phone: 208-286-7138
Steve Day, SECRETARY, TREASURER
J. Worthen, TOURNAMENT DIRECTOR
James Raitter, VICE PRESIDENT, COMMUNICATION
Larry Raganit, YOUTH

IDAHO TROUT LIMITED
57 Maxie Lane
Sandpoint, ID 83864 USA
Phone: 208-263-4433   Fax: 815-346-1400
Website: www.idahotrout.com

Founded: NA
Membership: 2000
Scope: Statewide

Description: A statewide council with eight active chapters dedicated to the protection and enhancement of the coldwater fishery resource.

Contact(s):
Robert Dunnagan, PRESIDENT
57 Maxie Ln., Sandpoint, ID 83864
Phone: 208-263-4433
Phone: 815-346-1400
rdunnagan@nidlink.com

IDAHO WILDLIFE FEDERATION
P.O. Box 6426
Boise, ID 83707-6426 USA
Phone: 208-342-7055   Fax: 208-342-7097
E-mail: iwfboise@micron.net

Founded: NA
Scope: Regional

Description: A representative statewide organization, affiliated with the National Wildlife Federation, dedicated to the protection and enhancement of wildlife and its habitat through public education and government interaction.

Publication(s): Idaho Wildlife News

Contact(s):
Corrine Fisher, ALTERNATE REPRESENTATIVE
Bill Goodnight, EDITOR
Jack Fisher, PRESIDENT AND REPRESENTATIVE

IL DEPARTMENT OF NATURAL RESOURCES
524 S. Second St.
Springfield, IL 62701 -1787 USA
Phone: 217-785-0075   Fax: 217-785-9236
Website: www.dnr.state.il.us
Scope: State

Contact(s):
Diane Giannone, CONTACT
Brent Manning, DIRECTOR

ILLINOIS ASSOCIATION OF CONSERVATION DISTRICTS
9313 Bull Valley Rd.
Woodstock, IL 60098 USA
Phone: 815-338-7664   Fax: 815-338-2773
E-mail: conserve1@aol.com

Founded: 1972
Membership: 30
Scope: State

Description: To promote the objectives and activities of the Conservation District of Illinois as set forth in the Illinois Conservation District Act and to cooperate with county, state, federal, and private agencies in resource management.

Keyword(s): Cultural Preservation, Endangered Species, Environmental and Conservation Education, Land Preservation, Sustainable Ecosystems

Contact(s):
Ken Fiske, ASSISTANT SECRETARY AND TREASURER
Phone: 815-338-7664
Ken Konsis, PRESIDENT
Phone: 217-442-1691
Dan Kane, SECRETARY/TREASURER
Kathy Merner, VICE PRESIDENT

ILLINOIS ASSOCIATION OF SOIL AND WATER CONSERVATION DISTRICTS

2520 Main St. State Fairgrounds, Emerson Bldg.
Springfield, IL 62702 USA
Phone: 217-744-3414   Fax: 217-744-3420
E-mail: aiswcdcs@aol.com
Website: www.aiswcd.org.com

Founded: NA
Membership: 4
Scope: Statewide
Keyword(s): Conservation Districts

Contact(s):
Kim Pate, ADMINISTRATIVE COORDINATOR
Terry Bogner, BOARD MEMBER
Rte. 1 Box 186, Henry, IL 61537
Phone: 309-364-3478
Fax: 309-364-3802
Chris Stone, EXECUTIVE DIRECTOR
Phone: 217-744-3414
Fax: 217-744-3420
Mark Besse, PRESIDENT
7341 Sand Rd., Erie, IL 61250
Phone: 309-659-7716
Fax: 309-659-7716
Virginia Hayter, TREASURER
2020 Hassell Rd. Apt. 101, Hoffman, IL 60195
Phone: 847-882-9100
Fax: 847-882-2621
virginia.hayter@hoffmanestate.org
Jerry Snodgrass, VICE PRESIDENT
13501 N. 1700th Ave., Geneseo, IL 61254
Phone: 309-944-2869
Fax: 309-937-2171
jerrypam@netexpress.net

ILLINOIS AUDUBON SOCIETY

425 B N. Gilbert St., P.O. Box 2418
Danville, IL 61834 USA
Phone: 217-446-5085   Fax: 217-446-6375
Website: www.illinoisaudubon.org

Founded: 1897
Membership: 2400
Scope: Statewide

Description: The Society is dedicated to the preservation and enjoyment of wildlife and their habitats.

Publication(s): Cardinal News, The, Illinois Audubon

Keyword(s): Birds, Endangered Species, Biodiversity, Land Preservation, training

Contact(s):
Debbie Newman, EDITOR, ILLINOIS AUDUBON
Marilyn Campbell, EXECUTIVE DIRECTOR AND EDITOR, CARDINAL NEWS
David Miller, PRESIDENT
813 N Cntr., McHenry, IL 60050

Mary Hoeffliger, VICE PRESIDENT
6752 E 2000th Ave., Shumway, IL 62461

ILLINOIS B.A.S.S. CHAPTER FEDERATION

Attn: President, 2425 Huntington Rd.
Springfield, IL 62703 USA
Phone: 217-529-8341
Website: www.ilbassfed.com

Founded: NA
Scope: Statewide

Description: An organization of Bassmaster chapters, affiliated with the Bass Anglers Sportsman Society, organized to fight pollution, assist state and national conservation agencies in their efforts, and to teach the young people of our country good conservation practices. Dedicated to the realistic conservation of our water resources.

Publication(s): Illinois B.A.S.S. Federation Newsletter inserted in "Midwest Outdoors" magazine

Contact(s):
John Gross, PRESIDENT
2425 Huntington Rd, Springfield, IL 62703, 217 5-29 -8341
president@ilbassfed.com

ILLINOIS CHAPTER OF THE SIERRA CLUB

ILLINOIS CHAPTER
200 N. Michigan Ave., Suite. 505
Chicago, IL 60601-5908 USA
Phone: 312-251-1680   Fax: 312-251-1780
E-mail: illinois.chapter@sierraclub.org
Website: www.sierraclub.org/chapters/il

Founded: NA
Membership: 22000
Scope: State
Publication(s): Lake and Prairie

Contact(s):
Jack Darin, CONTACT

ILLINOIS ENVIRONMENTAL COUNCIL

319 W. Cook St.
Springfield, IL 62704 USA
Phone: 217-544-5954   Fax: 217-544-5958
E-mail: iec@ilenviro.org
Website: www.ilenviro.org

Founded: 1975
Scope: State

Description: Statewide coalition committed to advocating for Illinois laws and policies that promote a healthful environment and conservation of resources. The Illinois Environmental Council Education Fund administers programs of education and outreach for the coalition.

Publication(s): Action Alerts, Environmental Voting Record, IEC Bulletin

Keyword(s): Air Quality and Pollution, Pesticides, Solid Waste, Exotic species, Aquatic nuisance species, training, Politics and Government

Contact(s):
Eleanor Roemer, BOARD OF DIRECTORS/PRESIDENT
Ellen Schmidt, DIRECTOR OF ADMINISTRATION
Lynne Padovan, EXECUTIVE DIRECTOR AND LEGISLATIVE DIRECTOR

ILLINOIS NATIVE PLANT SOCIETY

Forest Glen Preserve, 20301 E. 900 N. Rd.
Westville, IL 61883 USA
Phone: 217-662-2142
E-mail: ilnps@aol.com
Website: www.vccd.org

Founded: 1982
Membership: 500
Scope: Statewide

Description: Dedicated to the preservation, conservation, and study of the native plants and vegetation of Illinois.

Publication(s): Harbinger, Erigenia

Keyword(s): Native Plants, Aquatic Habitats, Biodiversity, Botanical Gardens, Conservation, Endangered Species, Protected Areas, Environmental Protection, Flowers, Plants, and Trees, Natural Areas, Prairies, Wetlands

Contact(s):
Ken Konsis, EXECUTIVE BOARD

ILLINOIS PRAIRIE PATH

P.O. Box 1086
Wheaton, IL 60189 USA
Phone: 630-752-0120
Website: www.ipp.org

Founded: 1963
Scope: Statewide

Description: To preserve natural areas and establish footpaths and other protected areas to be used for scientific, educational, and recreational purposes by the public. Adds trail amenities and promotes development of a 61-mile trail for bicyclists, hikers, and joggers on a former railroad right-of-way spanning DuPage County, extended Jan. 1972 into Kane County to the Fox River, and extended Dec. 1979 4 1/2 miles into Cook County. Incorporated 1965, in 1971 designated part of National Trails System.

Publication(s): Illinois Prairie Path, The, Trail Map, Newsletter

Keyword(s): Environmental and Conservation Education, Protected Areas, Outdoor Recreation, Public Lands, training, Trail, Bicycle

Contact(s):
Jean Mooring, EDITOR
295 Abbotsford Ct., Glen Ellyn, IL 60137
Phone: 630-469-4289
David Tate, PRESIDENT
Nancy Becker, SECRETARY
Paul Mooring, TREASURER

ILLINOIS WALNUT COUNCIL

Forest Glen Preserve, 20301 E. 900 N. Rd.
Westville, IL 61883 USA
Phone: 217-442-1691   Fax: 217-442-1695
E-mail: vccd@soltec.net
Website: www.vccd.org

Founded: NA
Scope: Regional

Description: To promote the growth and use of the black walnut (Juglans nigra), and the education of good forestry practices with concerns toward wildlife and soil erosion.

Publication(s): Juglans, Walnut Council Bulletin

Keyword(s): Trees, Forest Management, Pesticides, Renewable Resources, Research, Soil Conservation, Sustainable Development, Watersheds, training

Contact(s):
Bob Trimble, PRESIDENT
804 Tyler Court, Monticello, IL 61856
Steve Felt, SECRETARY
522 Roberts Ln., Sherrard, IL 61281
Wayne Wildy, VICE PRESIDENT
7718 Wildy Rd., New Athens, IL 62264

INDIAN CREEK NATURE CENTER

6665 Otis Rd., SE
Cedar Rapids, IA 52403 USA
Phone: 319-362-0664   Fax: 319-362-2876
E-mail: naturecenter@aol.com
Website: www.indiancreeknaturecenter.org

Founded: 1973
Membership: 800
Scope: Statewide

Description: The Indian Creek Nature Center is dedicated to fostering an appreciation of nature through environmental education and providing a natural facility for education and non-obtrusive recreation.

Publication(s): Indian Creek Currents

Keyword(s): Environmental and Conservation Education, Nature Study

Contact(s):
Rich Patterson, DIRECTOR
Dennis Redmond, PAST PRESIDENT
Phone: 319-366-2163
Fax: 319-366-7710
dredmond@rbjcpas.com
Leslie Smith, PRESIDENT
lsmith@berthel.com

INDIANA ASSOCIATION OF SOIL AND WATER CONSERVATION DISTRICTS, INC.

225 S. East St., Suite 740
Indianapolis, IN 46202 USA
Phone: 317-692-7374   Fax: 317-692-7363
E-mail: iaswcd@iaswcd.org
Website: www.iaswcd.org

Founded: 1968
Scope: Statewide

Description: We represent Indian's 92 soil and water conservation districts. We support the districts in their efforts to combat non-point source pollution.

Keyword(s): Conservation Districts

Contact(s):
Steve Graber, PRESIDENT
3850 Greenhurst Court, Auburn, IN 46706
Phone: 219-925-0676
Sherman Bryant, VICE-PRESIDENT
7343 N 650 E., N. Webster, IN 46555-9332
Phone: 219-834-2496

INDIANA AUDUBON SOCIETY, INC.

Mary Gray Bird Sanctuary, R.R. 6 Box 163
Connersville, IN 47331 USA
Phone: 765-825-9788
Website: www.indianaaudubon.org

Founded: 1898
Scope: Statewide

Description: Works for the conservation of wildlife, especially birds.

Keyword(s): Birds, Conservation, Ecology, Environmental and Conservation Education, Flowers, Plants, and Trees, Insects and Butterflies, Waterfowl

Contact(s):
Charles Keller, EDITOR
2505 E. Maynard Dr., Indianapolis, IN 46226
Phone: 317-786-5822
Mary Gough, EDITOR
901 Maplewood Dr., New Castle, IN 47362
Phone: 317-529-5225
Jane Miller, PRESIDENT
4020 S. Rural, Independence, IN 46227-3865
Deanna Barricklow, RESIDENT AGENT AND MANAGER OF SANCTUARY MANAGEMENT
3499 S. Bird Sanctuary Rd., Connersville, IN 47331-8721
Phone: 317-825-9788
Dan Leach, SECRETARY
2313 S. 30th St., Bedford, IN 47421-5415
Clare Oskay, TREASURER
551 Teton Trail, Indianapolis, IN 46217-3927
Larry Carter, VICE PRESIDENT
7496 N. Co. Rd. 2005, Ridgeville, IN 47380-9546

INDIANA B.A.S.S. CHAPTER FEDERATION

Attn: President, 1415 Cherokee Rd.
Ft. Wayne, IN 46808 USA
Phone: 219-483-0525
Website: www.ibfweb.com

Founded: NA
Scope: Statewide

Description: An organization of Bassmaster chapters, affiliated with the Bass Anglers Sportsman Society, organized to fight pollution, assist state and national conservation agencies in their efforts, and teach the young people of our country good conservation practices. Dedicated to the realistic conservation of our water resources.

Contact(s):
Dan Pardue, CONSERVATION DIRECTOR
7244 Holmestead Rd., Morgantown, IN 46160
Phone: 812-888-8788
Paul Hollabaugh, PRESIDENT
1415 Cherokee Rd., Ft. Wayne, IN 46808
Phone: 219-483-0525

INDIANA FORESTRY AND WOODLAND OWNERS ASSOCIATION

5578 South 500 W.
Atlanta, IN 46031-9363 USA
Phone: 317-758-4735

Founded: 1977
Scope: Statewide

Description: A statewide organization affiliated with the National Woodland Owners Association, providing leadership and programs to advance forestry in Indiana.

Publication(s): Leaves and Limbs

Keyword(s): Forests and Forestry

Contact(s):
Thomas Moehl, 1ST VICE PRESIDENT
Alan Bolenbaugh, 2ND VICE PRESIDENT
Jan Myers, EDITOR
Phone: 317-583-2422
Pete Halstead, FORESTRY EDUCATIONAL FOUNDATION
Robert Koenig, PRESIDENT
William Sigman, SECRETARY
Warren Baird, TREASURER

INDIANA NATIVE PLANT AND WILDFLOWER SOCIETY

6106 Kingsley Dr.
Indianapolis, IN 46220 USA
Phone: 317-253-3863
E-mail: rai38@aol.com
Website: www.inpaws.org

Founded: 1993
Scope: Statewide

Description: To promote the appreciation, preservation, conservation, utilization and scientific study of the flora native to Indiana; and to educate the public about the values, beauty, diversity, and environmental importance of indigenous vegetation.

Publication(s): Indiana Native Plant and Wildflower Society News

Keyword(s): Biodiversity, Flowers, Plants, and Trees, National Parks, Prairies, Sustainable Ecosystems

Contact(s):
Carolyn Bryson, PRESIDENT
quinnell@iquest.net
Ken Collins, VICE PRESIDENT
Phone: 317-891-9804

INDIANA STATE TRAPPERS ASSOCIATION, INC.

20941 Fir Road
Tippecanoe, IN 46570 USA
Phone: 219-498-6354
Website: krause.com/outdoors/tr/associations

Founded: 1961
Scope: Statewide

Description: A statewide organization dedicated to the conservation, restoration, and wise use of wildlife and other renewable natural resources. Provides public education concerning the role of trapping in the management of wildlife.

Contact(s):
Doyle Flory, PRESIDENT
Phone: 219 4-98 -6354
Richard McOlvanine, VICE PRESIDENT
Phone: 812-834-5514

INDIANA WILDLIFE FEDERATION
950 N. Rangeline Rd.,
Suite A
Carmel, IN 46032-1315 USA
Phone: 317-571-1220   Fax: 317-571-1223
E-mail: iwf@indy.net
Website: www.indianawildlife.org

Founded: NA
Membership: 5000
Scope: Statewide

Description: A representative statewide organization, affiliated with the National Wildlife Federation, dedicated to the protection and enhancement of wildlife and its habitat through public education and government interaction.

Publication(s): Hoosier Conservation

Contact(s):
Jack Dold, EDITOR & ALTERNATE REPRESENTATIVE
Becky Scheibelhut, EDUCATION PROGRAMS CONTACT
Paula Yeager, EXECUTIVE DIRECTOR
Dale Back, PRESIDENT
Dwight Shelton, REPRESENTATIVE
George Vargo, TREASURER

INFORM, INC.
120 Wall St., 16th Fl.
New York, NY 10005 USA
Phone: 212-361-2400   Fax: 212-361-2412
E-mail: brown@informinc.org
Website: www.informinc.org

Founded: 1973
Scope: National

Description: A nonprofit tax-exempt environmental research and education organization that identifies and reports on practical solutions for problems in municipal solid waste, chemical hazards, air quality, and alternative vehicle fuels, with an emphasis on pollution prevention and waste reduction.

Publication(s): China at the Crossroads, Gearing up for Hydrogen, Building for the Future, Tracking Toxic Chemicals, Rethinking Resources, INFORM Reports (Newsletter)

Keyword(s): Air Quality and Pollution, Energy, Solid Waste Management, Toxic Substances, Nuclear-free, Water quantity, Water export and diversion, Transportation

Contact(s):
Stephen Land, CHAIRMAN OF THE BOARD
Phone: 212-424-9018
sbland@linklaters.com
Samuel Arnoff, DIRECTOR OF OPERATION
arnoff@informinc.org
Joanna Underwood, DIRECTOR OF RESEARCH
Joanna Underwood, PRESIDENT
underwood@informinc.org

INITIATIVE FOR SOCIAL ACTION AND RENEWAL IN EURASIA
1601 Connecticut Ave., NW, Suite 301
Washington, DC 20009 USA
Phone: 202-387-3034   Fax: 202-667-3291
E-mail: postmaster@isar.org
Website: www.isar.org

Founded: 1983
Membership: 10

Scope: International

Description: ISAR promotes citizens participation and the development of the NGO sector in the former Soviet Union by supporting community activists and grassroots groups.

Publication(s): Give and Take, ISAR in Focus

Keyword(s): Women in the Environment, Environmental and Conservation Education, Grants, Training

Contact(s):
Eliza Klose, EXECUTIVE DIRECTOR
eliza@isar.org

INLAND BIRD BANDING ASSOCIATION
P.O. Box 832
Tiffin, OH 44883 USA
E-mail: MCGREEN@AOL.COM
Website: http://aves.net/inlandbba/ibbamain.htm

Founded: 1922
Scope: National

Description: Promotes cooperation among its members and other organizations, with state, federal, or other officials or individuals engaged in bird banding or other scientific work with birds; informs the public of the purposes and results secured by banding.

Publication(s): Inland Bird Banding Newsletter, North American Bird Bander

Contact(s):
Dan Kramer, EDITOR
3451 Co. Rd. 256, Victory, OH 43464
Wiletta Lueshen, EDITOR
R. 2 Box 26, Wisner, NE 68791
Al Valentine, MEMBERSHIP SECRETARY
17403 Oakington Ct., Dallas, TX 75252
Ruth Green, PRESIDENT
Carol Rudy, SECRETARY
W. 3866 Hwy. H, Chilton, WI 53084
C. Smith, TREASURER
6305 Cumberland Rd. SW, Sherrodsville, OH 44675

INSTITUTE AND SCHOOL FOR ENVIRONMENT AND NATURAL RESOURCES (IENR AND SENR)
UNIVERSITY OF WYOMING
P.O. Box 3971
Laramie, WY 82071 USA
Phone: 307-766-5080   Fax: 307-766-5099
E-mail: ienr@uwyo.edu
Website: www.uwyo.edu/enr/

Founded: 1994
Membership: 10
Scope: National

Description: Current projects include workshops on ESA and NEPA, reports on Brucellosis in the Greater Yellowstone Area, and an open spaces inititatives in Grand Teton National Park

Keyword(s): Open Spaces, Endangered Species

Contact(s):
Dianna Hulme, ASSISTANT DIRECTOR
Phone: 307-766-5354
Harold Bergman, DIRECTOR
Phone: 307-766-5150
bergman@uwyo.edu

INSTITUTE FOR CONSERVATION LEADERSHIP

6930 Carroll Ave. Suite 420
Takoma Park, MD 20912 USA
Phone: 301-270-2900   Fax: 301-270-0610
E-mail: icl@icl.org
Website: www.icl.org

Founded: NA
Scope: National

Description: The mission of the Institute is to train and empower volunteer leaders and to build volunteer institutions that protect and conserve the earth's environment. Services offered include training and technical assistance for nonprofit organizations and leaders in organizational development, fundraising, board development, volunteer recruitment, strategic planning, and related topics. Services also include meeting facilitation, coalition development, and network building.

Keyword(s): Communications, Environment, Pollution Prevention, Training

Contact(s):
Baird Straughan, ASSOCIATE DIRECTOR
Dianne Russell, EXECUTIVE DIRECTOR
Chiquita Edwards, OFFICE MANAGER
Peter Lane, PROJECT MANAGER

INSTITUTE FOR EARTH EDUCATION, THE

Cedar Cove
Greenville, WV 24945 USA
Phone: 304-832-6404   Fax: 304-832-6077
E-mail: iee1@aol.com
Website: www.eartheducation.org

Founded: 1974
Scope: International

Description: The Institute for Earth Education develops and disseminates focused educational programs to promote an understanding of, appreciation for, and harmony with the earth's natural systems and communities. The Institute conducts workshops, provides a seasonal journal, hosts an international conference, supports local and international branches, and publishes numerous books and program materials.

Publication(s): Sunship Earth, Earth Education Sourcebook, Earth Education: A New Beginning, Earthkeepers, Earth Speaks, The, Sunship III, Talking Leaves Journal

Keyword(s): Environmental and Conservation Education, Protected Areas, Internships, Environmental and Conservation Education, Ecology

Contact(s):
Steve Van Matre, CHAIR
Bill Weiler, EXECUTIVE STAFF CHAIR
Fran Bires, INTERNATIONAL INTERNSHIP COORDINATOR
Laurie Farber, INTERNATIONAL MEMBERSHIP SERVICES COORDINATOR
Bruce Johnson, INTERNATIONAL PROGRAM COORDINATOR
Mike Mayer, INTERNATIONAL TRAINING COORDINATOR

INSTITUTE OF ECOSYSTEM STUDIES

Mary Flagler Cary Arboretum, Box AB
Millbrook, NY 12545-0129 USA
Phone: 845-677-5343   Fax: 845-677-5976
Website: www.ecostudies.org

Founded: NA

Scope: International

Description: Devoted to the understanding of ecosystem structure and function. The program focus is on disturbance and recovery of northern temperate ecosystems. Education and research interests include wildlife management, biogeochemistry, landscape ecology, aquatic ecology, plant-animal interactions, microbial ecology, forest ecology, chemical ecology, and air and water quality.

Publication(s): Scientific journals, occasional publications, newsletter

Keyword(s): Acid Rain, Air Quality and Pollution, Environmental and Conservation Education, Ecology, Ecological Education, Environmental and Conservation Education

Contact(s):
Joseph Warner, ADMINISTRATOR
Richard Ostfeld, ANIMAL ECOLOGIST
Stuart Findlay, AQUATIC ECOLOGIST
Michael Pace, AQUATIC ECOLOGIST
Jonathan Cole, AQUATIC MICROBIOLOGIST
Nina Caraco, BIOGEOCHEMIST
Gene Likens, DIRECTOR
Clive Jones, ECOLOGIST
Charles Canham, FOREST ECOLOGIST
Kathleen Weathers, FOREST ECOLOGIST
David Strayer, FRESHWATER ECOLOGIST
Alan Berkowitz, HEAD OF EDUCATION
Phone: 845-677-5359
Chloe Keefer, LIBRARIAN
Peter Groffman, MICROBIAL ECOLOGIST
Gary Lovett, PLANT ECOLOGIST
Steward Pickett, PLANT ECOLOGIST
Raymond Winchcombe, WILDLIFE BIOLOGIST AND FIELD RESEARCH FACILITIES
Phone: 845-677-9818

INTERFAITH COUNCIL FOR THE PROTECTION OF ANIMALS AND NATURE INC. (ICPAN)

3691 Tuxedo Rd., NW
Atlanta, GA 30305 USA
Phone: 404-814-1371   Fax: 404-814-0440

Founded: 1980
Membership: 3000
Scope: National

Description: Composed of people of all faiths, ICPAN works to promote conservation and environmental and humane education, mainly within the religious community. We try to make religious leaders, institutions, and the general public aware of our moral spiritual obligations, as emphasized in the Bible, to protect animals and the natural environment.

Publication(s): Replenish the Earth: A Booklet on The Bibles Message of Conservation and Kindness to Animals, Cleaning up America the Poisoned: How to Survive our Polluted Society, Losing Paradise: The Growing Threat to Our Animals,Our Environment, and Ourself, Replenish the Earth: The Teachings of the Worlds Religions on Protecting Animals and Nature

Keyword(s): Endangered Species, Environmental and Conservation Education, Sustainable Development, training

Contact(s):
John Hoyt, CHAIRMAN
2100 L St. NW, Washington, DC 20037
Phone: 202-452-1100

Paul Irwin, DIRECTOR
2100 L St. NW, Washington, DC 20037
Phone: 202-452-1100
Lewis Regenstein, PRESIDENT
3691 Tuxedo Rd. NW, Atlanta, GA 30327
Phone: 404-814-1371

INTERNATIONAL ASSOCIATION FOR BEAR RESEARCH AND MANAGEMENT

UNIVERSITY OF TENNESEE
274 Ellington PSB
Knoxville, TN 37901-1071 USA
Phone: 250-837-7767   Fax: 865-974-3555
Website: www.bearbiology.com

Founded: 1968
Membership: 700
Scope: International

Description: A professional organization of biologists, animal or land managers, and private citizens with an interest or involvement in bear research and management. The Association encourages and reports research and management by various agencies or university research groups, sponsors the triannual International Conference on Bear Research and Management, publishes the proceedings of the conference, and sponsors or aids a world network of regional bear workshops, groups, and committees, and the IUCN Bear Group.

Publication(s): International Bear News (Quarterly Newsletter), Ursus, formerly Bears: Their Biology and Management (Conference Proceedings)

Keyword(s): Wildlife Rehabilitation, Endangered Species, Land Use Planning, Public Lands, Bears

Contact(s):
Harry Reynolds, PRESIDENT
1300 College Rd., Fairbanks, Alaska 99701
Phone: 907-459-7238
Joe Clark, SECRETARY
274 Ellington PSB, Knoxville, Tennessee 379011071
Phone: 865-974-4790
Frank Mannen, TREASURER
274 Ellington PSB, Knoxville, Tennessee 37901071
Phone: 865-974-0200
Sterling Miller, VICE PRESIDENT
240 N Higgins Ste 2, Massouli, Montana 59802
Phone: 406-721-6705

INTERNATIONAL ASSOCIATION FOR ENVIRONMENTAL HYDROLOGY (IAEH)

P.O. Box 35324
San Antonio, TX 78235 USA
Phone: 210-344-5418   Fax: 210-344-9941
E-mail: hydroweb@mail.org
Website: www.hydroweb.com

Founded: 1991
Scope: International

Description: IAEH works to foster a global interchange of ideas, approaches, and technologies for environmental cleanup and protection of fresh water resources and pollution prevention; to place special focus on approaches to cleanup, prevention, and protection that are practical in less affluent countries; to further the development of environmentally sound solutions that are realistic from the economic standpoint; to seek solutions to cleanup, pollution prevention, and environmental protection; and to place pollution cleanup and prevention in the context of broader water resource and environmental issues.

Publication(s): Environmental Hydrology Report, Journal of Environmental Hydrology

Keyword(s): Developing Countries, Environment, Environmental and Conservation Education, Pollution Prevention, Exotic species, Aquatic nuisance species, Coral Reefs

Contact(s):
Roger Peebles, PRESIDENT
308 Montfort Dr., San Antonio, TX 78216
Phone: 210-344-5418

INTERNATIONAL ASSOCIATION OF FISH AND WILDLIFE AGENCIES

444 North Capitol St., NW Suite 544
Washington, DC 20001 USA
Phone: 202-624-7890   Fax: 202-624-7891
E-mail: iafwa@sso.org
Website: www.iafwa.org

Founded: NA
Membership: 400
Scope: International

Description: Association of states or territories of the United States, provinces of Canada, the Commonwealth of Puerto Rico, the United States Government, the Dominion Government of Canada, and governments of countries located in the western hemisphere, as well as individual associate members whose principal objective is conservation, protection, and management of wildlife and related natural resources.

Publication(s): Annual Proceedings, Newsletter

Contact(s):
Wm. Nesbitt, ANNUAL PROCEEDINGS EDITOR
Edward Parker, CT DEPT OF ENVIRONMENTAL PROTECTION
79 Elm St, Hartford Ct, 06106
Phone: 860-424-3010
Fax: 860-424-4078
Robert McDowell, EXECUTIVE COMMITTEE CHAIR
Director, New Jersey Division of Fish, Game and Wildlife, P.O. Box 400, Trenton, NJ 08625-0400
G. Manning, EXECUTIVE COMMITTEE MEMBER
George Meyer, EXECUTIVE COMMITTEE MEMBER
Steven Williams, EXECUTIVE COMMITTEE MEMBER
John Baughman, EXECUTIVE COMMITTEE MEMBER
Allan Egbert, EXECUTIVE COMMITTEE MEMBER
Roger Holmes, EXECUTIVE COMMITTEE MEMBER (PAST PRESIDENT)
Arnold Boer, EXECUTIVE COMMITTEE VICE CHAIR
Executive Director of Fish and Wildlife, New Brunswick Department of Natural Resources and Energy, P.O. Box 6000, Fredericton, New Brunswick E3B 5H1
R. Peterson, EXECUTIVE VICE PRESIDENT
Samara Trusso, FUR RESOURCES COMMITTEE PROJECT COORDINATOR
Donald MacLauchlan, INTERNATIONAL RESOURCE DIRECTOR
Paul Lenzini, LEGAL COUNSEL
Gary Taylor, LEGISLATIVE DIRECTOR
Len Ugarenko, NAWMP COORDINATOR
Pat Graham, PRESIDENT
1420 E. 6th Helena, MT 59620
Phone: 406-444-3186
Fax: 406-444-4952

Bob Miles, RESOURCE DIRECTOR
C. Bennett, SECRETARY AND TREASURER
Kentucky Department of Fish and Wildlife Resources, One Game Farm Rd., Frankfort, KY 40601
Patrick Graham, VICE PRESIDENT
Director, Montana Department of Fish, Wildlife and Parks, 1420 E. 6th, P.O. Box 200701, Helena, MT 59620-0701
Naomi Edelson, WILDLIFE DIVERSITY DIRECTOR

INTERNATIONAL ASSOCIATION OF NATURAL RESOURCE PILOTS

200 Patrick St., SW
Vienna, VA 22180-6703 USA
Phone: 703-560-1271

Founded: NA
Membership: 200
Scope: International

Description: Performs aviation and aircrew conservation-related responsibilities for federal and state game and fish divisions and departments of natural resources throughout the U.S. and for their counterparts in the Canadian provinces. Additional membership includes a variety of aviation-oriented corporations and advanced technological suppliers of equipment used in the performance of the aviation missions.

Publication(s): Conservation Aviation

Keyword(s): Agriculture, Biodiversity, Birds, Chemical Pollution Control, Conservation, Environmental Law, Environmental Planning, Endangered Species, Wildlife, Forest Management, Lakes, Land Use Planning, Natural Areas, Hunting, Nongame Wildlife

Contact(s):
Val Judkins, NEWSLETTER EDITOR
Washington Fish and Wildlife, Olympia, WA 98501
Phone: 260-753-4717
vjudkins@aol.com
Paul Anderson, PRESIDENT
Phone: 701-328-6613
panderson@state.nd.us
Francis Satterlee, PUBLIC AFFAIRS OFFICER
200 Patrick St., SW, Vienna, VA 22180
Phone: 703-560-1271
Michael Jeffries, SECRETARY
Technical Representative AOS, Boise, ID 83705
Phone: 208-334-9310
michaeljeffries@ios.doi.gov
John Clem, TREASURER AND LIBRARIAN
Ohio Division of Wildlife, 9740 Briarwood Dr., Plain City, OH 43064
Phone: 614-873-4163
Fax: 6147834163- ce-ll
john_clem@compuserve.com

INTERNATIONAL ASSOCIATION OF WILDLAND FIRE

E. 8109 Bratt Rd.
Fairfield, WA 99012 USA
Phone: 509-523-4003   Fax: 509-523-5001
E-mail: greenlee@cet.com
Website: www.wildfiremagazine.com

Founded: 1991
Membership: 1,500
Scope: International

Description: (formerly Fire Research Institute) The International Association of Wildland Fire was organized to promote a fuller understanding of wildland fire. The Association is built on the belief that an understanding of this dynamic natural force is vital for natural resource management, firefighter safety, and harmonious interactions between people and their environment.

Publication(s): International Bibliography of Wildland Fire, International Journal of Wildland Fire, Wildfire Magazine, Current Titles in Wildland Fire, International Directory of Wildland Fire

Keyword(s): Ecology, Endangered Species, Forests and Forestry, Terrestrial Habitats, Wildlands

Contact(s):
Mike Weber, EDITOR
Phone: 403-435-7210
Jason Greenlee, EXECUTIVE DIRECTOR
Phone: 509-283-2397
Mike Degrosky, PRESIDENT
Phone: 307-543-0949

INTERNATIONAL BICYCLE FUND

4887 Columbia Dr. S.
Seattle, WA 98108-1919 USA
Phone: 206-767-0848
E-mail: ibike@ibike.org
Website: www.ibike.org

Founded: 1983
Scope: International

Description: The International Bicycle Fund's programs fall into the areas of transportation planning, sustainable economic development, safety education and promoting international understanding. Within these programs we address issues of the environment, energy policy, public health, appropriate technology, land use patterns, sustainable systems, resource conservation and employment generation. IBF coordinates and cooperates with organizations and individuals worldwide. IBF is a nonprofit organization.

Publication(s): IBF News, see publications on website

Keyword(s): Environmental and Conservation Education, Land Use Planning, Pedestrian Environment, Sustainable Development, Transportation, Bicycle

Contact(s):
David Mozer, PRESIDENT
etoole@irf.org

INTERNATIONAL CENTER FOR EARTH CONCERNS

2162 Baldwin Rd.
Ojai, CA 93023 USA
Phone: 805-649-3535   Fax: 805-649-1757
E-mail: information@earthconcerns.org
Website: www.earthconcerns.org

Founded: 1994
Membership: 200
Scope: Local

Description: The ICEC involves people with nature by fostering their appreciation of the natural world through environmental education and training.

Publication(s): Annual Newsletters, Brochure

Keyword(s): Botanical Gardens, Ecology, Environmental and Conservation Education, Environmental Protection

Contact(s):
John Taft, CHAIRMAN
Melody Taft, EXECUTIVE DIRECTOR
Paul Irwin, PRESIDENT

INTERNATIONAL CENTER FOR GIBBON STUDIES

P.O. Box 800249
Santa Clarita, CA 91380 USA
Phone: 661-296-2737   Fax: 661-296-1237
E-mail: gibboncntr@aol.com
Website: www.gibboncenter.org

Founded: 1977
Scope: International

Description: The International Center for Gibbon Studies ensures the preservation and propagation and a safe haven for all gibbon species living in the wild and in captivity; supports ongoing field conservation projects; and educates the public about the importance of this species and saving their natural habitat.

Publication(s): The Gibbon's Voice—yearly newsletter, brochures

Keyword(s): Environmental and Conservation Education, Zoology, Research, Training

Contact(s):
Bjorn Merker, ACTING DIRECTOR OF RESEARCH
Institute for Biomusicology, Mid Sweden, Ostersund, S-83125
Elaine Baker, ASSISTANT DIRECTOR OF RESEARCH
Department of Psychology, Marshall University, Huntington, WV 25755
Alan Mootnick, BOARD OF DIRECTORS PRESIDENT, FACILITY DIRECTOR, AND CHAIRMAN OF THE BOARD
P.O. Box 800249, Santa Clarita, CA 91380
Phone: 661-296-2737
Geri-Ann Galanti, BOARD OF DIRECTORS VICE PRESIDENT
2906 Ocean Ave., Venice, CA 90291
Phone: 310-827-0937
Lori Sheeran, DIRECTOR OF EDUCATION AND CONSERVATION
California State University at Fullerton
Phone: 714-773-2765

INTERNATIONAL CENTER FOR TROPICAL ECOLOGY

The University of Missouri at St. Louis, R224 Research Bldg., 8001 Natural Bridge Rd.
St. Louis, MO 63121-4499 USA
Phone: 314-516-5219   Fax: 314-516-6233
E-mail: icte@umsl.edu
Website: http://icte.umsl.edu

Founded: 1990
Membership: 100
Scope: International

Description: The ICTE is one of the premier institutes in the United States for the study of tropical biology and conservation. The Center's three primary missions include the training of graduate students in the vital areas of tropical ecology and conservation, the education of undergraduates about the importance of these areas, and involvement of the community in educational actvities with respect to issues related to conservation and biodiversity.

Keyword(s): Ecology, Biodiversity, Tropical Biodiversity and Conservation

Contact(s):
Bette Loiselle, DIRECTOR
Phone: 314-516-6224
Fax: 314-516-6233
loiselle@jinx.umsl.edu
Patrick Osborne, EXECUTIVE DIRECTOR
Phone: 314-516-5219
Fax: 314-516-6233
posborne@jinx.umsl.edu

INTERNATIONAL CRANE FOUNDATION

E-11376 Shady Ln. Rd., P.O. Box 447
Baraboo, WI 53913-0447 USA
Phone: 608-356-9462   Fax: 608-356-9465
E-mail: cranes@savingcranes.org
Website: www.savingcranes.org

Founded: 1973
Membership: 30
Scope: International

Description: Preservation of cranes through research, conservation, captive propagation, restocking, field ecology, and public education.

Publication(s): Proceedings of the 7th N. American Crane Workshop, 1997, Reflections: The Story of Cranes, ICF Bugle, The (Quarterly Magazine)

Keyword(s): Birds, Endangered Species, Environmental and Conservation Education, Wetlands, Birds

Contact(s):
Susan Finn, ASSISTANT TO THE PRESIDENT
Phone: 608-356-9462
sfinn@savingcranes.org
George Archibald, CO-FOUNDER
george@savingcranes.org
Claire Mirande, CONSERVATION COORDINATOR
Phone: 608-356-9462
mirande@savingcranes.com
Mike Putnam, CURATOR OF BIRDS
Robert Hallam, DEVELOPMENT
Phone: 608-356-9462
bhallam@savingcranes.org
C. Dietrich Schaaf, DIRECTOR OF EDUCATION
Phone: 608-356-9462
cdschaaf@savingcranes.org
Kate Fitzwilliams, DIRECTOR OF PUBLIC RELATIONS AND MARKETING
Phone: 608-356-9462
kate@savingcranes.org
Jeb Barzen, FIELD ECOLOGIST
Phone: 608-356-9462
jeb@savingcranes.org
Betsy Didrickson, LIBRARIAN
The Ron Sauey Memorial Library For Bird Conservation, Baraboo, WI 53913-0447
Phone: 608-356-9462
betsy@savingcranes.org
James Harris, PRESIDENT
Phone: 608-356-9462
harris@savingcranes.org

David Chesky, SITE MANAGER
Phone: 608-356-9462
dchesky@savingcranes.org
Peter Murray, VICE PRESIDENT OF FINANCE AND
ADMINISTRATION
Phone: 608-356-9462
pmurray@savingcranes.org

INTERNATIONAL ECOLOGY SOCIETY (IES)

1471 Barclay St.
St. Paul, MN 55106-1405 USA
Phone: 612-579-7008

Founded: 1975
Scope: International

Description: One hundred percent volunteer-staffed, nonprofit international organization dedicated to the protection of the environment and the encouragement of better understanding of all life forms.

Publication(s): Sunrise (neighborhood news), Action Alerts, Eco-Humane Letter

Keyword(s): Endangered Species, Marine Mammals, Whale, Dolphin, Seal, Trapping, training

Contact(s):
Maggie Warren, CONTACT (DAKOTA)
Hermosa, SD
Bina Robinson, NORTH EAST REPRESENTATIVE
Box 26, Swain, NY 14884-0026
R. Kramer, PRESIDENT AND PUBLISHER
George Johnson, VICE PRESIDENT

INTERNATIONAL EROSION CONTROL ASSOCIATION (IECA)

P.O. Box 774904
Steamboat Springs, CO 80477 USA
Phone: 970-879-3010   Fax: 970-879-8563
E-mail: ecinfo@ieca.org
Website: www.ieca.org

Founded: 1972
Membership: 3000
Scope: International

Description: To provide opportunities for the worldwide exchange of information and economic methods of erosion control.

Publication(s): Membership Directory, Products and Services Directory, Proceedings of Annual Conference

Keyword(s): Engineering, Environmental and Conservation Education, Soil Conservation, Coral Reefs

Contact(s):
Ben Northcutt, EXECUTIVE DIRECTOR

INTERNATIONAL FUND FOR ANIMAL WELFARE

411 Main St.
Yarmouth Port, MA 02675 USA
Phone: 508-362-4944   Fax: 508-744-2009
E-mail: info@ifaw.org
Website: www.ifaw.org

Founded: 1969
Membership: 500000
Scope: International

Description: An international nonprofit, tax-exempt organization in the U.S. dedicated to the protection of wild and domestic animals and their habitats. IFAW's goals are pursued through a strategic plan consisting of three distinct program areas: Commercial Expoitation and Trade of Wild Animals, Animals in Crisis and Distress, and Habitat for Animals.

Contact(s):
Fred O'Regan, CHIEF EXECUTIVE OFFICER
Aczedine Downes, CONTACT

INTERNATIONAL GAME FISH ASSOCIATION

300 Gulf Stream Way
Dania Beach, FL 33004 USA
Phone: 954-927-2628   Fax: 954-924-4299
E-mail: igfahq@aol.com
Website: www.igfa.org

Founded: 1939
Scope: International

Description: A nonprofit, tax-deductible organization which maintains and promotes ethical international angling regulations and compiles world game fish records for saltwater, freshwater, and fly fishing. Also represents and informs recreational fishermen regarding research, conservation, and legislative developments related to their sport. Encourages and supports game fish tagging programs and other scientific data collection efforts. There are over 250 international representatives and 1,000 affiliated fishing clubs.

Publication(s): International Angler, The, Rule Book for Freshwater, Saltwater and Fly Fishing, World Record Game Fishes

Keyword(s): Wildlife, Sport Fishing

Contact(s):
Michael Levitt, CHAIRMAN
Michael Leech, PRESIDENT
Roy Naftzger, SECRETARY
Pamela Basco, TREASURER
John Anderson, VICE CHAIRMAN

INTERNATIONAL HUNTER EDUCATION ASSOCIATION

P.O. Box 490
Wellington, CO 80549 USA
Phone: 970-568-7954   Fax: 970-568-7955
E-mail: ihea@frii.com
Website: www.ihea.com

Founded: NA
Scope: International

Description: To provide leadership and establish standards in the development of hunters to be safe, responsible, knowledgeable, and involved.

Publication(s): Hunter Education Student Guide, Hunter Education Journal

Keyword(s): Environmental and Conservation Education, Hunting, Outdoor Recreation, Training

Contact(s):
David Knotts, IHEA EXECUTIVE VICE PRESIDENT
Albert Ross, IHEA LEGAL COUNCEL
John Panio, INSTRUCTOR BOARD MEMBER
Bill Blackwell, INSTRUCTOR BOARD REPRESENTATIVE
Christopher Tymeson, INSTRUCTOR BOARD REPRESENTATIVE OF ZONE 2
Jan Morris, INSTRUCTOR BOARD REPRESENTATIVE OF ZONE 3

Tim Lawhern, PRESIDENT
Mac Lang, PRESIDENT -ELECT
Mark Birkhauser, SECRETARY
Joe Huggins, TREASURER
Robert Paddon, VICE PRESIDENT OF ZONE 1
Helen McCracken, VICE PRESIDENT OF ZONE 2
Keith Snyder, VICE PRESIDENT OF ZONE 3

INTERNATIONAL INSTITUTE FOR ENERGY CONSERVATION

750 1st St., Suite 940 NE
Washington, DC 20002 USA
Phone: 202-842-3388　　Fax: 202-842-1565
E-mail: iiec@iiec.org

Founded: 1984
Scope: International

Description: A nonprofit organization established to accelerate the global adoption of energy-efficiency policies, technologies, and practices to enable econimically and ecologically sustainable development.

Publication(s): Integrated Transport Management and Development, Opportunities for the U.S. Energy Efficiency Industry in Chile, Global Energy Efficiency Initiative Sustainable Energy Guide, E-Notes

Keyword(s): Energy, Greenhouse Effect/Global Warming, Sustainable Development, Transportation

Contact(s):
John Fox, CHAIRMAN OF THE BOARD
Steve Hall, DIRECTOR
Terry Oliver, DIRECTOR
Stewart Boyle, DIRECTOR
Russell Sturm, EXECUTIVE DIRECTOR AND PRESIDENT

INTERNATIONAL MARINE MAMMAL PROJECT, THE

EARTH ISLAND INSTITUTE
300 Broadway
Suite 28
San Francisco, CA 94133 USA
Phone: 415-788-3666　　Fax: 415-788-7324
Website: www.earthisland.org

Founded: 1982
Scope: International

Description: IMMP is a nonprofit research, education, and monitoring project of Earth Island Institute. IMMP is committed to ending dolphin mortality caused by the U.S. and international tuna industries, stopping the use of driftnets, and promoting sustainable fishing practices. In addition, IMMP aims to halt commercial whaling worldwide and ban live capture and display of marine mammals.

Publication(s): Ocean Alert, Earth Island Journal

Keyword(s): Endangered Species, Environmental and Conservation Education, Wildlife, Marine Mammals, Whale, Dolphin, Seal, Aquatic Habitats, Biodiversity, Aquariums, Environmental Law, Internships, Dolphins, National Parks, training, Mammals

Contact(s):
David Phillips, EXECUTIVE DIRECTOR

INTERNATIONAL OCEANOGRAPHIC FOUNDATION

University of Miami, Rosenstiel School of Marine & Atmosphere Science, 4600 Rickenbacker Causeway
Virginia Key
Miami, FL 33149 USA
Phone: 305-361-4061　　Fax: 305-361-4931
Website: www.rsmas.miami.edu/iof/UNDERRENOVATION

Founded: 1953
Scope: International

Description: Nonprofit foundation organized to encourage the extension of human knowledge by scientific study and exploration of the oceans in all their aspects and to acquaint and educate the general public concerning the vital role of the oceans to all life on this planet.

Keyword(s): Environmental and Conservation Education, Wildlife, Marine Mammals, Whale, Dolphin, Seal, Oceanography, Outdoor Recreation

Contact(s):
Edward Foote, PRESIDENT
Lourdes Lapaz, SECRETARY
400 SE 2nd Ave., 4th Fl., Miami, FL 33131
Phone: 305-375-8498
Fax: 305-375-9188
Diane Cook, TREASURER
Otis Brown, VICE PRESIDENT
David Lieberman, VICE PRESIDENT
Luis Glaser, VICE PRESIDENT

INTERNATIONAL OSPREY FOUNDATION INC., THE

P.O. Box 250
Sanibel, FL 33957 USA
Phone: 941-472-1862

Founded: 1981
Scope: International

Description: A nonprofit organization dedicated to studying the problem of restoring osprey numbers to a stable population, making recommendations to enhance the continued survival of the osprey and initiating educational programs. Yearly grant of up to $1000.00 given for graduate work. Work relating to all raptors is acceptable, but osprey study is given priority.

Publication(s): TIOF Newsletter - One International NL

Keyword(s): Birds, International Wildlife, Nongame Wildlife, Raptors

Contact(s):
David Loveland, PRESIDENT
Inge Glissman, SECRETARY AND TREASURER
Anne Mitchell, VICE PRESIDENT

INTERNATIONAL PLANT PROPAGATORS SOCIETY, INC.

Washington Park Arboretum, 2300 Arboretum Dr.
Seattle, WA 98112 USA
Phone: 206-543-8602　　Fax: 206-325-8893
E-mail: ippsint@aol.com
Website: www.ipps.org

Founded: 1950
Membership: 3000
Scope: International

Description: The Society was founded to seek and share information on plant propagation. The Sociey has nine regional chapters, three in USA and Canada, Australia, New Zealand, Great Britain and Ireland, Scandinavian, Japan and Southern Africa and holds area meetings in Latin America.

Publication(s): regional newsletters of meetings for members, Annual Proceedings of all regional meetings and papers

Keyword(s): Flowers, Plants, and Trees, Urban Environment, Urban Forestry, Plant Propagation

Contact(s):
John Wott, EXECUTIVE SECRETARY AND TREASURER

INTERNATIONAL PRIMATE PROTECTION LEAGUE

P.O. Box 766
Summerville, SC 29484 USA
Phone: 843-871-2280   Fax: 843-871-7988
E-mail: ippl@awod.com
Website: www.ippl.org/

Founded: 1973
Scope: International

Description: A nonprofit international organization devoted to the conservation and protection of nonhuman primates. There are branches in the United States, United Kingdom and field representatives in 32 countries.

Publication(s): International Primate Protection League News (IPPI News)

Keyword(s): Endangered Species, Forests and Forestry, Mammals, accreditation

Contact(s):
Shirley McGreal, CHAIRPERSON
Marjorie Doggett, SECRETARY
Diane Walters, TREASURER

INTERNATIONAL RIVERS NETWORK (IRN)

1847 Berkeley Way
Berkeley, CA 94703 USA
Phone: 510-848-1155   Fax: 510-848-1008
E-mail: irn@irn.org
Website: www.irn.org

Founded: 1986
Scope: International

Description: IRN supports local communities working to protect their rivers and watersheds. We work to halt destructive river development projects and encourage equitable and sustainable methods of meeting needs for water, energy and flood management. Members include environmentalists, engineers, hydrologists, human rights activists, and academics who are committed to the study and defense of rivers and riverine communities.

Publication(s): special briefings, action alerts, working papers, World Rivers Review

Keyword(s): Rivers, Flood Control, Development

Contact(s):
Lori Pottinger, AFRICA CAMPAIGNS & EDITOR
lori@irn.org
Annie Ducmanis, ASSISTANT TO EXECUTIVE DIRECTOR
annie@irn.org
Patrick McCully, CAMPAIGN DIRECTOR
patrick@irn.org

Juliette Majot, EXECUTIVE DIRECTOR
juliette@irn.org
Yvonne Cuellar, LIBRARY COORDINATOR
Glenn Switkes, SOUTH AMERICA CAMPAIGNS
glenn@altanet.com.br

INTERNATIONAL SNOW LEOPARD TRUST

4649 Sunnyside Ave., N., Suite 325
Seattle, WA 98103 USA
Phone: 206-632-2421   Fax: 206-632-3967
E-mail: info@snowleopard.org
Website: www.snowleopard.org

Founded: 1981
Membership: 700
Scope: International

Description: A nonprofit organization dedicated to the conservation of the endangered snow leopard and its mountain habitat through a balanced approach that considers the needs of the local people and the environment; and provides workshops, field training, equipment, publications, conservation education programs, and a centralized database for organizing and disseminating information.

Publication(s): Snow Leopard News, Snow Line

Keyword(s): Biodiversity, Endangered Species, Environmental and Conservation Education, Sustainable Ecosystems

Contact(s):
Tom McCarthy, CONSERVATION DIRECTOR
tmccarthy@snowleopard.org
Pricilla Allen, CONSERVATION PROGRAM OFFICE
Brad Rutherford, EXECUTIVE DIRECTOR
brad@snowleopard.org
Helen Freeman, FOUNDER
Charlie Morse, PRESIDENT
Steven Kearsley, TREASURER
Lewis Macfarlane, VICE PRESIDENT

INTERNATIONAL SOCIETY FOR ECOLOGICAL ECONOMICS (ISEE)

1313 Dolly Madison Blvd., Suite 402
McLean, VA 22101 USA
Phone: 703-790-1745   Fax: 703-790-2672
E-mail: isee@igc.com
Website: www.ecologicaleconomics.org

Founded: 1988
Scope: International

Description: ISEE actively encourages the integration of the study and the management of ecology and economics in order to achieve an ecologically and economically sustainable world.

Publication(s): Ecological Economics

Keyword(s): Biodiversity, Environmental Planning, Sustainable Development, Sustainable Ecosystems

Contact(s):
Richard Norgarrd, PRESIDENT OF BOARD OF DIRECTORS

INTERNATIONAL SOCIETY FOR ENDANGERED CATS (ISEC)

3070 Riverside Dr., Suite 160
Columbus, OH 43221 USA
Phone: 614-487-8760   Fax: 614-487-8769
E-mail: eduacation@asec.org
Website: www.isec.org

Founded: 1988
Scope: International

Description: ISEC's purpose is to raise awareness of the plight of endangered wild cats, and thereby prevent their extinction. ISEC offers conservation education programs, collects and disseminates information about wild cats, and supports specific conservation projects around the world. Member IUCN

Publication(s): Cat Tales

Keyword(s): Endangered Species, Environmental and Conservation Education, training, Wild Cats

Contact(s):
Patricia Currie, EXECUTIVE DIRECTOR
196 W. Central, Delaware, OH 43015
Phone: 740-369-9794
Bill Simpson, PRESIDENT
3070 Riverside Dr., Suite 160, Columbus, OH 43221

INTERNATIONAL SOCIETY FOR ENVIRONMENTAL ETHICS

Department of Philosophy, University of Windsor
Windsor, Ontario N9B 3P4 Canada
Phone: 519-253-3000   Fax: 519-971-3610
E-mail: philos@uwindsor.ca
Website: www.cep.unt.edu/isEE.html

Founded: 1990
Scope: International

Description: The International Society for Environmental Ethics' main purpose is to promote the critical analysis of ethical issues related to the natural environment, to further and support philosophical and scientific meetings and conferences nationally and internationally, and to provide material and media aids suitable for teaching environmental philosophy and environmental ethics.

Publication(s): International Society for Environmental Ethics Newsletter

Contact(s):
Mark Sagoff, PRESIDENT
Director of Institute for Philosophy and Public Policy,
University of Maryland, Baltimore, MD 20742
Laura Westra, SECRETARY
University of Windsor, Windsor, Ontario N9B 3P4
Phone: 519-253-4232
Edward Hettinger, TREASURER
College of Charleston, Charleston, SC 29424
J. Callicott, VICE PRESIDENT
Philosophy Department, University of Wisconsin at Stevens Point, Stevens Point, WI 54481

INTERNATIONAL SOCIETY FOR THE PRESERVATION OF THE TROPICAL RAINFOREST, THE

3931 Camino De La Cumbre
Sherman Oaks, CA 91423 USA
Phone: 818-788-2002   Fax: 818-990-3333
E-mail: forest@nwc.net
Website: isptr/pard.org

Founded: 1984
Membership: 8000
Scope: International

Description: The International Society for the Preservation of the Tropical Rainforest is dedicated to the global conservation of tropical forest resources through the promotion of park implementation, sustainable agriculture, and timber harvesting.

Publication(s): Amazon Hotline, Tropical Rainforest Our Most Valuable And Endangered Habitat With A Blueprint For Its Survival Into The Third Millennium

Keyword(s): Biodiversity, Cultural Preservation, Endangered Species, Protected Areas, Greenhouse Effect/Global Warming

Contact(s):
Roxanne Kremer, CO-DIRECTOR
3931 Camino De La Cumbre, Sherman Oaks, CA 91423
Phone: 626-572-0233
Fax: 626-572-9521
Edward Asner, CO-DIRECTOR
3931 Camino De La Cumbre, Sherman Oaks, CA 91423
Phone: 818-788-2002
Arnold Newman, CO-DIRECTOR
3931 Camino De La Cumbre, Sherman Oaks, CA 91423
Phone: 818-788-2002

INTERNATIONAL SOCIETY OF ARBORICULTURE

P.O. Box 3129
Champaign, IL 61826-3129 USA
Phone: 217-355-9411   Fax: 217-355-9516
E-mail: @isa-arbor.com
Website: www.isa-arbor.com

Founded: 1924
Membership: 15000
Scope: International

Description: To promote and improve the care and preservation of shade and ornamental trees through research and education.

Publication(s): Arborist News, Valuation of Landscape Trees, Shrubs, and Other Plants, Planting Tree and Shrubs, Publication listings to include "A Photographic Guide for Evaluation of Hazard Trees in Urban Areas", " Arborist Certificate Guide", Journal of Arboriculture

Contact(s):
Robert Miller, EDITOR
Phone: 715-346-4189
Fax: 715-346-3624
Paul Harter, EXECUTIVE DIRECTOR
Peggy Currid, MANAGING EDITOR
Kim Coder, PRESIDENT
Michael Neal, PRESIDENT-ELECT
Bailey Hudson, PRESIDENT-ELECT
Melinda Jones, VICE PRESIDENT
Lauren Lanphear, VICE PRESIDENT
Harvey Holt, VICE PRESIDENT
Lisle, IL

INTERNATIONAL SOCIETY OF TROPICAL FORESTERS, INC.

5400 Grosvenor Ln.
Bethesda, MD 20814 USA
Phone: 301-897-8720　　Fax: 301-897-3690
E-mail: istfi@igc.org
Website: www.cof.orst.edu/org/ists

Founded: 1950
Membership: 1500
Scope: International

Description: A nonprofit organization founded with the objective of providing an information exchange for members involved in the management, protection, and wise use of tropical forests.

Publication(s): ISTF Notices (Spanish), ISTF News

Keyword(s): Forest Management, Urban Forestry, Renewable Resources, Sustainable Ecosystems, Biodiversity

Contact(s):
Chum Lai, DIRECTOR AT LARGE
Philippines
Jeffery Burley, DIRECTOR AT LARGE
United Kingdom
John Fox, DIRECTOR AT LARGE
Australia
B. Tazl, DIRECTOR OF AFRICA
Gambia
Napoleon Vergata, DIRECTOR OF ASIA
Philippines
Rodolfo Salazar, DIRECTOR OF LATIN AMERICA
Costa Rica
Frank Wadsworth, EDITOR
Patricia Holmgren, OFFICE MANAGER
Warren Doolittle, PRESIDENT
USA
Napoleon Vergata, VICE PRESIDENT & DIRECTOR OF ASIA & PHILIPPINES

INTERNATIONAL SONORAN DESERT ALLIANCE

P.O. Box 687
Ajo, AZ 85321 USA
Phone: 520-387-6823　　Fax: 520-387-5626
E-mail: isda@tabletoptelephone.com
Website: www.isdanet.org

Founded: 1992
Membership: 1200
Scope: International

Description: The purpose of the International Sonoran Desert Alliance is to promote environmentally sustainable and culturally sound economic development while protecting the natural and tri-cultural heritage of the western Sonoran Desert.

Keyword(s): Cultural Preservation, Deserts, People of Color in the Environment, Sustainable Development

Contact(s):
Reynaldo Cantu, EXECUTIVE DIRECTOR
Carlos Nagel, PRESIDENT
closfree@aol.com
Isabel Granillo, SECRETARY
isabel@laruta.org
Sue Tout, TREASURER
Manuel Gonzales, VICE PRESIDENT
manuelg@email.arizona.edu

INTERNATIONAL UNION FOR CONSERVATION OF NATURE

REGIONAL OFFICE FOR MESO AMERICA
Phone: 506-236-2733　　Fax: 506-240-9934
E-mail: CORREO@ORMA.IUCN.ORG
Website: WWW.IUCN.ORG

Founded: 1988

Contact(s):
Enrique Lahmann, REGIONAL DIRECTOR

INTERNATIONAL UNION FOR CONSERVATION OF NATURE AND NATURAL RESOURCES (IUCN) THE WORLD CONSERVATION UNION

CANADA COUNTRY OFFICE
380 St. Antoine St. W., Office 3200
Montreal, Quebec H2Y 3X7 Canada
Phone: 1 514 287 97-04/-1830　　Fax: 1 514 287 -9057

Founded: NA
Scope: International

Contact(s):
Malcolm Mercer, HEAD

INTERNATIONAL UNION FOR CONSERVATION OF NATURE AND NATURAL RESOURCES (IUCN) THE WORLD CONSERVATION UNION

U.S. OFFICE, WASHINGTON, DC
1630 Connecticut Ave., NW
Washington, DC 20009 USA
Phone: 202-387-4826　　Fax: 202-387-4823
E-mail: postmaster@iucnus.org
Website: www.iucn.org

Founded: NA
Membership: 6
Scope: International
Publication(s): Tooth & Law-newsletter, Life At The Edge, Amman 2000

INTERNATIONAL WILD WATERFOWL ASSOCIATION

5614 River Styx Rd.
10114 54th Place N.E., Everett WA 98205
Medina, OH 44256 USA
Website: greatnorthern.net

Founded: 1958
Scope: International

Description: Works toward protection, conservation, and reproduction of any species of wild waterfowl considered in danger of eventual extinction; encourages breeding of well known and rare species in captivity. Established Avicultural Hall of Fame. Sponsors annual conference and gives grants in field.

Publication(s): IWWA Newsletter

Keyword(s): Endangered Species, Scholarships and Grants, Waterfowl, training

Contact(s):
Edward Asper, 1ST VICE PRESIDENT
Vice President of Sea World, 7007 Sea World Dr., Orlando, FL 32821
Phone: 407-351-3600

Paul Dye, 2ND VICE PRESIDENT
10114 54th Pl. NE, Everett, WA 98205
Phone: 425-334-8223
Fax: 425-397-8136
dye@greatnorthern.net
Walter Sturgeon, PRESIDENT
7 James Farm, Durham, NH 03824
Phone: 603-659-5442
Nancy Collins, SECRETARY
5614 River Styx Rd., Medina, OH 44256
Phone: 330-725-8782
William Lowe, TREASURER
3010 Shady Ln., Billings, MT 59102
Phone: 406-245-6119

INTERNATIONAL WILDERNESS LEADERSHIP (WILD) FOUNDATION

P.O. Box 1380
Ojai, CA 93024 USA
Phone: 805-640-0390　　　Fax: 805-640-0230
E-mail: info@wild.org
Website: www.wild.org

Founded: 1974
Scope: International

Description: The International Wilderness Leadership Foundation is dedicated to protecting wilderness and wildlife, providing environmental experience and training and promoting the correct use of wildlands worldwide.

Publication(s): Wilderness Management, For the Conservation of Earth, Wilderness, the Way Ahead, Arctic Wilderness, International Journal of Wilderness, Leaf Newsletter, The

Keyword(s): Environmental and Conservation Education, Wilderness, Wildlands

Contact(s):
I. Michael Sweatman, CHAIRMAN
350 Indiana St., Golden, CO 80401
Vance Martin, PRESIDENT
Phone: 805-640-0390
Michael Sweatman, TREASURER
Rte. 1, P.O. Box 659 1503, Waterbury, VT 05672
Phone: 802-244-8981
John Hendee, VICE PRESIDENT OF SCIENCE AND EDUCATION
College of Forestry, University of Idaho, Moscow, ID 83843
Phone: 208-885-2267

INTERNATIONAL WILDLIFE COALITION (IWC) AND THE WHALE ADOPTION PROJECT

70 E. Falmouth Highway
E. Falmouth, MA 02536 USA
Phone: 508-548-8328　　　Fax: 508-548-8542
Website: www.iwc.org

Founded: 1984
Scope: International

Description: IWC is a nonprofit, tax-exempt organization dedicated to preserving wildlife and their habitats. As an internationally recognized non-governmental organization, IWC's achievements have been accomplished through grassroots advocacy, activism, research, and education efforts. IWC's Whale Adoption Project protects and researches marine mammals.

Publication(s): Wildlife Watch, Whales of the World Teacher's Kit, Wildlife and You and What You Can Do To Help, WhaleWatch

Keyword(s): Endangered Species, Mammals, Marine Mammals, Whale, Dolphin, Seal, training, Whales

Contact(s):
Jose Palazzo, BRAZIL PROJECT COORDINATOR
P.O. Box 5087, Florianopolis, S.C. 88040
brazilian_wildlife@zaz.com.br
Ronald Orenstein, CANADA PROJECT DIRECTOR
130 Adelaide St. West, Suite 1940, Toronto, Ontario M5H 3P5
Phone: 905-820-7886
Fax: 905-569-0116
ornstn@inforamp.net
Daniel Morast, PRESIDENT
70 E. Falmouth Highway, E. Falmouth, MA 02536
Phone: 508-548-8328
Fax: 508-548-8542
dmorast@iwc.org
Charles Wartenberg, UNITED KINGDOM DIRECTOR
141A, High St., Edenbridge, Kent TN85AX

INTERNATIONAL WILDLIFE REHABILITATION COUNCIL (IWRC)

4437 Central Pl., Suite B-4
Suisun, CA 94585-1633 USA
Phone: 707-864-1761　　　Fax: 707-864-3106
E-mail: iwrc@inreach.com
Website: www.iwrc-online.org

Founded: 1972
Membership: 2000
Scope: International

Description: An organization dedicated to conserving and protecting wildlife and habitat through wildlife rehabilitation

Publication(s): Basic Wildlife Rehabilitation, Minimum Standards and Accreditation, other publications and catalog available. IWRC Literature Catalog, Journal of Wildlife Rehabilitation

Keyword(s): Wildlife Rehabilitation, education, training, standards, accreditation

Contact(s):
Edward Clark, PRESIDENT
Lee Theisen-Watt, SECRETARY
Dody Wyman, TREASURER
Penny Elliston, VICE PRESIDENT

INTERNATIONAL WOLF CENTER

1396 Highway 169
Ely, MN 55731 USA
Phone: 218-365-4695　　　Fax: 218-365-3318
E-mail: wolfinfo@wolf.org
Website: www.wolf.org

Founded: 1985
Membership: 9000
Scope: International

Description: The International Wolf Center supports the survival of the wolf around the world by teaching about its life, its associations with other species and its dynamic relationship to humans.

Publication(s): Guidelines for Gray Wolf Management, various educational pamphlets, International Wolf Magazine

Keyword(s): Endangered Species, Environmental and Conservation Education, training, Wolves, Mammals

Contact(s):
Nancy Tubbs, CHAIR
Walter Medwid, EXECUTIVE DIRECTOR

Liz Harper, INFORMATION SPECIALIST
Rolf Peterson, SECRETARY
Paul Anderson, TREASURER
L. Mech, VICE CHAIR

INTERNATIONAL WOLF CENTER
ADMINISTRATIVE OFFICES
3300 Bath Lake Rd. 202
Minneapolis, MN 55429 USA
Phone: 763-560-7374   Fax: 763-560-7368
E-mail: mplspac@wolf.org
Website: www.wolf.org

Founded: NA
Membership: 26
Scope: International
Publication(s): International Wolf
Contact(s):
George Knoter, DEVELOPMENT DIRECTOR
3300 Bass Lake Rd. Suite 202, Minneapolis, Minnesota 55429
Walter Medwid, EXECUTIVE DIRECTOR
3300 Bass Lake Rd, 55429

INTERPRETATION CANADA
c/o Kerry Wood Nature Centre, 6300-45 Ave.
Red Deer, Ontario T4N 3M4 Canada
Phone:
E-mail: webmaster@interpcan.ca
Website: interpcan.cae http://interpcan.cae

Founded: 1973
Scope: National

Description: Interpretation Canada is dedicated to raising public awareness, understanding, and appreciation for Canada's natural and cultural heritage, provides training, networking, and advocacy for interpretors, and promotes the role of interpretation in fields such as conservation, education, recreation, and tourism.

Publication(s): regional newsletters from Northwest Territories, British Columbia, Alberta, Ontario, and Atlantic region, annual membership directory, Interpscan - national journal

Keyword(s): Culture, Interpretation, Training
Contact(s):
Ann Finlayson, CHAIRPERSON
782 E. Kings Rd., North Vancouver, BC V1K 1E3
Phone: 604-987-8653
Fax: 604-987-8629
annfinlayson@bc.sympatico.ca
Jim Robertson, MEMBER SERVICES
Kerrywood Nature Centre, 6000 45 Ave., Red Deer, Alberta T4N 3M4
Phone: 403-346-2010
Fax: 403-347-2590
kwnc@supernet.ab.ca

INTERTRIBAL BISON COOPERATIVE (ITBC)
1560 Concourse Dr.
Rapid City, SD 57703 USA
Phone: 605-394-9730   Fax: 605-394-7742
E-mail: itbc@enetis.net
Website: www.intertribalbison.com

Founded: 1991
Membership: 50
Scope: National

Description: The ITBC is dedicated to the restoration of buffalo to Indian lands in a manner which is compatible with the cultural practices and spiritual beliefs of the respective tribes.

Publication(s): Bison, A Living Story, educational CD rom, Buffalo Tracks, Annual Report

Keyword(s): Cultural Preservation, Land Preservation, Prairies, Sustainable Development, training

Contact(s):
Donald Lake, INTERIM EXECUTIVE DIRECTOR
Fred Dubray, PRESIDENT
Winnebago Tribe of NE, Rte 1, Box 15, Winnebago, NE 68071
Phone: 402-878-2711
Carl Tsosie, SECRETARY
Picuris Pueblo, P.O. Box 127, Penasco, NM 87553
Phone: 505-587-2519
Michael Fox, TREASURER
Taos Pueblo, P.O. Box 3164, Taos, NM 87571
Phone: 505-758-3883
Ervin Carlton, VICE PRESIDENT

IOWA ACADEMY OF SCIENCE
University of Northern Iowa 175 Baker Hall
Cedar Falls, IA 50614-0508 USA
Phone: 319-273-2021   Fax: 319-273-2807
Website: www.iren.net/iaf/

Founded: 1875
Membership: 1000
Scope: Statewide

Description: To further the work of scientists, facilitate cooperation among them, and increase public understanding and appreciation of the importance and promise of the methods of science in human progress. A conservation section meets each year as part of an annual convention submitting papers dealing with all conservation happenings.

Publication(s): Iowa Science Teachers Newsletter, Journal of the Iowa Academy of Science, IAS Bulletin

Keyword(s): Environmental and Conservation Education
Contact(s):
David McCalley, EXECUTIVE DIRECTOR
Phone: 319-273-2021
Fax: 319-273-2807
davidmccaulley@uni.edu
David McCalley, EXECUTIVE DIRECTOR AND MANAGER EDITOR
Charlie Martinson, PAST PRESIDENT
Phone: 515-294-1062
Fax: 515-294-9420
cmartins@iastate.edu
Lynn Brant, PRESIDENT
50011
Phone: 319-273-6160
Fax: 319-273-7124
lynnbrant@uni.edu
Raymond Anderson, PRESIDENT ELECT
Phone: 319-335-1575
Fax: 319-335-2754
randerson@igsb.uiowa.edu

IOWA ASSOCIATION OF NATURALISTS

CONSERVATION EDUCATION CENTER
2473 160th Rd.
Guthrie Center, IA 50115 USA
Phone: 641-747-8383　　　Fax: 641-747-3951
E-mail: ajay.winter@dnr.state.ia.us

Founded: 1978
Membership: 120
Scope: State

Description: Organization of persons interested in promoting the development of skills and education within the art of interpreting the natural and cultural environment. Members representing county, state, federal, and private conservation education agencies, organizations, and facilities.

Keyword(s): Environmental and Conservation Education, Interpretation

Contact(s):
Steve Martin, PRESIDENT
Phone: 319-278-1130
srmartin@ins.net
Patrice Peterson-Keys, SECRETARY OF THE BOARD
Phone: 515-999-2557
Fax: 515-999-2709
Brad Freidhof, TREASURER OF THE BOARD
IA 712-335-4395
Phone: 712-335-3606
pokyccb@mcn.net
Ann Burns, VICE PRESIDENT
Phone: 319-652-3783
jc-cons@co.jackson.ia.us
Kelly McKeown, WORKSHOP COORDINATOR
Phone: 712-263-3409
kmac@frontier.net

IOWA ASSOCIATION OF SOIL AND WATER CONSERVATION DISTRICT COMMISSIONERS

USA
Phone: 515-289-2331

Founded: NA
Scope: Statewide

Keyword(s): Conservation Districts

Contact(s):
Jill Knapp, EXECUTIVE DIRECTOR
3829 71st St., Ste. A, Urbandale, IA 50322
Phone: 515-278-5362
Fax: 515-278-5362
Dan Bruene, PRESIDENT, BOARD MEMBER
Phone: 515-473-2338
Fax: 515-473-2455
Bernie Bolton, SECRETARY
38995 Honeysuckle Rd., Oakland, IA 52560-9686
Phone: 712-482-3386
Fax: 712-482-3386
Art Ralston, TREASURER
2629 200th St., Moville, IA 51039-8036
Phone: 712-873-3719
William Bennett, VICE PRESIDENT
4331 Dove Rd., Elgin, IA 82141-9526
Phone: 319-426-5695
Fax: 319-426-5695

IOWA AUDUBON

P.O. Box 71174
Grinnell, IA 50325 USA
Phone: 515-727-4271
E-mail: p2eph@audubon.org

Founded: NA
Scope: Statewide

Description: The state office of the National Audubon Society supporting the 12 Audubon groups in Iowa. Iowa Audubon's mission is to promote the enjoyment, protection and restoration of Iowa's natural ecosystems with a focus on birds, other wildlife and their habitats.

Keyword(s): Environmental and Conservation Education, Restoration, Biodiversity, Birds, Environmental Protection, training

Contact(s):
Paul Zeph, EXECUTIVE DIRECTOR

IOWA B.A.S.S. CHAPTER FEDERATION

Attn: President, 3282 Midway
Marion, IA 52302 USA
Phone: 319-393-1481
E-mail: tbowler1@go.com

Founded: NA
Scope: Statewide

Description: An organization of Bassmaster chapters, affiliated with the Bass Anglers Sportsman Society, organized to fight pollution, assist state and national conservation agencies in their efforts, and teach the young people of our country good conservation practices. Dedicated to the realistic conservation of our water resources.

Contact(s):
Russell Engelbart, CONSERVATION DIRECTOR
12565 Amber Rd., Highway X44, Anamosa, IA 52205
engelbartbass101@uswest.net
Tom Bowler, PRESIDENT
Phone: 319-393-1481

IOWA ENVIRONMENTAL COUNCIL

711 E. Locust St.
Des Moines, IA 50309 USA
Phone: 515-244-1194　　　Fax: 515-244-7856
E-mail: iecmail@earthweshare.org
Website: www.earthweshare.org

Founded: 1994
Membership: 4
Scope: Statewide

Description: The Iowa Environmental Council is an alliance of diverse organizations and individuals working with all Iowans to protect our natural environment. We seek a sustainable future through shaping public policy, research and education, coalition-building, and advocacy.

Publication(s): Iowa Environmental Quarterly, News Bulletin, Legislative Action

Keyword(s): Agriculture, Biodiversity, Pesticides, Water Quality

Contact(s):
Susan Heathcote, DIRECTOR (ACTING)
David Hurd, PRESIDENT
Ray Heinicke, SECRETARY
Mark Ackelson, TREASURER
Debbie Neustadt, VICE PRESIDENT

IOWA NATIVE PLANT SOCIETY

DEPT OF BOTANY
Botany Department, 341A Bessey Hall, Iowa State
University
Ames, IA 50011-1020 USA
Phone: 515-294-9499   Fax: 515-294-1337

Founded: 1995
Scope: Statewide

Description: Iowa Native Plant Society is an organization of
amateurs and professionals who are interested in the scientific,
educational, cultural aspects, preservation, and conservation
of Iowa's native plants.

Publication(s): Iowa Native Plant Society Newsletter

Keyword(s): Native Plants, Biodiversity, Conservation,
Endangered Species, Environmental and Conservation
Education, Protected Areas, Wetlands, Prairies, Flowers,
Plants, and Trees

Contact(s):
Tom Rosburg, PRESIDENT
Drake University, Olin Hall, Des Moines, IA 50311
Phone: 515-377-2930
thomas.rosburg@drake.edu
Diana Horton, SECRETARY AND TREASURER
720 Sandusky Drive, Iowa City, IA 52240
Phone: 319-337-5430
diana-horton@uiowa.edu
Mary Hatfield, VICE PRESIDENT
2505 Tullamori Lane, Ames, IA 50010
Phone: 515-232-7555

IOWA NATURAL HERITAGE FOUNDATION

Attn: Communications Coordinator, Insurance Exchange
Bldg., Suite 444, 505 Fifth Ave.
Des Moines, IA 50309 USA
Phone: 515-288-1846   Fax: 515-288-0137
E-mail: info@inhf.org
Website: www.inhf.org

Founded: 1979
Scope: Statewide

Description: An independent, statewide, nonprofit organization
founded by business and community leaders to involve the
private sector in protecting Iowa's natural resources. Program
emphasis on land protection, landowner education, resource
planning, wetland restoration and rail-trail development in
Iowa.

Publication(s): The Landowner's Options, Iowa Natural Heritage,
Enjoy Iowa's Recreation Trails Guidebook

Keyword(s): Conservation, Endangered Species, Flowers,
Plants, and Trees, Internships, Land Preservation, Natural
Areas, National Parks, Rivers, Sustainable Development,
Wetlands, training, Prairies, Wildlands, Outdoor Recreation,
Trail

Contact(s):
Barb MacGregor, CHAIRMAN
Cathy Engstrom, COMMUNICATIONS COORDINATOR
Brenda Binder, DIRECTOR OF ADMINISTRATION
Anita O'Gara, DIRECTOR OF DEVELOPMENT AND
COMMUNICATIONS
Bruce Mountain, DIRECTOR OF LAND PROJECTS
Joe McGovern, DIRECTOR OF LAND STEWARDSHIP
Lisa Hein, DIRECTOR OF TRAILS AND GREENWAYS
Laura McVay, FINANCE MANAGER
Cheri Grauer, GIFT PLANNER

Perry Thostenson, LAND PROTECTION PROGRAM
DIRECTOR
Mark Ackelson, PRESIDENT
Richard Ramsay, SECRETARY
Michael Riley, TREASURER
Mike Lamair, VICE CHAIRMAN

IOWA PRAIRIE NETWORK

6736 Laural
Omaha, NE 68104 USA
Phone: 402-571-6230   Fax: 402-571-6230
E-mail: pollockg@top.net
Website: www.IowaPrairieNetwork.org

Founded: 1990
Scope: Statewide

Description: The Iowa Prairie Network is dedicated to protecting
Iowa prairie heritage.

Publication(s): IPN News, Native Prairie Management Guide, A
Prairie Bioliography

Keyword(s): Prairies

Contact(s):
David Hansen, DIRECTOR
Phone: 515-357-3665
Glenn Pollock, PRESIDENT
Carole Kern, TREASURER
Phone: 319-273-2813
Cindy Hildebrand, VICE PRESIDENT

IOWA TRAILS COUNCIL

P.O. Box 131
Center Point, IA 52213-0131 USA
Phone: 319-849-1844   Fax: 3198491844@-aol-.com

Founded: 1983
Membership: 1200
Scope: National

Description: A membership nonprofit organization primarily active
in the Midwest, but with membership in over one-half the states
and in several foreign countries. Primary purpose is to acquire
and convert former railroad rights-of-way into recreational
trails.

Publication(s): Bicycle Trails of Iowa (128 page color illustrated
book), Trails Advocate (bi-monthly mini magazine)

Keyword(s): Bicycle, Protected Areas, Trail, Land Preservation,
Land Use Planning, National Parks, Outdoor Recreation,
Transportation

Contact(s):
Eldon Colton, CHAIRMAN
1008 Bowier St., Hiawatha, IA 52233
Phone: 319-378-8971
Fax: 319-294-1914
Tom Neenan, SECRETARY/TREASURER AND EXECUTIVE
DIRECTOR
P.O. Box 131, Center Point, IA 52213-0131
Phone: 319-849-1844
tomneenan1@aol.com
David Lyon, VICE CHAIRMAN
116 10th Ave., S., Mt. Vernon, IA 52314
Phone: 319-895-8240

IOWA TRAPPERS ASSOCIATION, INC.

c/o Anna Marie Scalf, 123 N. Madison Ave.
Ottumwa, IA 52501 USA
Phone: 641-682-3937   Fax: 641-682-9092
E-mail: iantadtr@lisco.com

Founded: 1950
Scope: State

Description: A nonprofit organization that works to continue the wise use and harvest of Iowa's renewable resource of furbearing animals. Cooperates with all recognized conservation agencies, law enforcement agencies, and legislative committees, and provides input on the benefits and necessity of trapping.

Publication(s): Trapper and Predator Caller, The

Keyword(s): Environmental and Conservation Education, Outdoor Recreation, Trapping, training

Contact(s):
Paul Wait, EDITOR
700 E. State St., Iola, WI 54990
Phone: 715-445-2214
Tom Walters, PRESIDENT
1723 20th St., Bettendorf, IA 52722-3829
Phone: 319-359-6949
Chris Grillot, SECRETARY
2769 110th Ave., Wheatland, IA 52777
Phone: 319-374-1074
Anna Scalf, TREASURER
123 N. Madison Ave., Ottumwa, IA 52501
Phone: 515-682-3937
James Stauffer, VICE PRESIDENT
29602 202nd St., Clarksville, IA 50619-9801
Phone: 319-278-4004

IOWA WILDLIFE FEDERATION

P.O. Box 3332
Des Moines, IA 50316-0332 USA
Phone: 319-624-3107   Fax: 319-644-3213

Founded: NA
Scope: Statewide

Description: A representative statewide organization, affiliated with the National Wildlife Federation, dedicated to the protection and enhancement of wildlife and its habitat through public education and government interaction.

Contact(s):
John Zietlow, ALTERNATE REPRESENTATIVE
Mike Hodges, EDITOR
Joe Wilkinson, PRESIDENT & EDUCATION PROGRAM CONTACT
Doug Thompson, REPRESENTATIVE
Kevin Thomasson, TREASURER

IOWA WILDLIFE REHABILITATORS ASSOCIATION

1005 Harken Hill Dr., P.O. Box 217
Osceola, IA 50213 USA
Phone: 641-342-2783

Founded: 1986
Membership: 1
Scope: Statewide

Description: A nonprofit organization established to disseminate information pertaining to wildlife rehabilitation and medicine to veterinarians, rehabilitators, naturalists and others; to communicate and cooperate with environmental/conservation organizations; and to encourage the public to become more aware of the earth and its wild creatures. This is done through newsletters, educational material, state and regional conferences, and presentations.

Publication(s): Newsletters, educational materials

Keyword(s): Accreditation

Contact(s):
Marlene Ehresman, PRESIDENT
Phone: 515-296-2995
Wendy Dewalle, SECRETARY
Phone: 641-964-9592
Beth Brown, TREASURER
Phone: 641-342-2783
Heather Blevins, VICE PRESIDENT
Phone: 641-277-7745

IOWA WOMEN IN NATURAL RESOURCES

Des Moines, IA 50320-0083 USA
Phone: 319-872-5495
Website: www.hometown.aol.com

Founded: 1988
Membership: 1
Scope: Statewide

Description: A nonprofit organiaztion dedicated to providing professional development to individuals interested in all natural resource careers by promoting communication among professionals, encouraging girls and women to consider natural resource careers, conducting outdoor skills workshops, providing networking and support systems for women working in natural resources and providing career enhancement training.

Publication(s): IWINR Membership Directory, IWINR News

Keyword(s): Women in the Environment, Training, Environment

Contact(s):
Theresa Blackburn, EDUCATION CHAIR
Phone: 563-872-5495
Fax: 563-872-5659
theresa_blackburn@usgs.gov
Theresa Minaya, TREASURER
Phone: 712-258-0838

ISLAND CONSERVATION EFFORT

90 Edgewater Dr. 901
Coral Gables, FL 33133 USA
Phone: 305-666-5381   Fax: 305-663-9941
E-mail: tropbird@unspoiedqueen.com

Founded: 1988
Scope: International

Description: Island Conservation Effort is dedicated to the preservation of island natural resources, fauna, and habitats on which their preservation depends. We promote conservation, education, and research to obtain necessary data to support conservation measures.

Keyword(s): Birds, Endangered Species, Environmental and Conservation Education, Research, Islands

Contact(s):
Martha Walsh-McGehee, PRESIDENT
90 Edgewater Dr., Suite 901, Coral Gables, FL 33133
Phone: 305-666-5381
tropbird@gate.net

Rosemarie Gnam, SECRETARY AND TREASURER
1872 Stanhope St., Ridgewood, NY 11385
Michelle Pugh, VICE PRESIDENT
P.O. Box 4254, Christinsted, St. Croix 00820
Phone: 340-773-7030
divexp@viaccess.net

ISLAND INSTITUTE, THE

P.O. Box 648
Rockland, ME 04841 USA
Phone: 207-594-9209　　　Fax: 207-594-9314
E-mail: inquiry@islandinstitute,org
Website: www.islandinstitute.org

Founded: 1983
Scope: National

Description: Private, nonprofit organization dedicated to sustaining island and coastal communities through community initiatives, publications, resource management, science and marine research.

Publication(s): Working Waterfront, Gulf of Maine Environmental Atlas, Islands in Time, Island Journal

Keyword(s): Cultural Preservation, Wildlife, Islands, Sustainable Development

Contact(s):
Horace Hildreth, CHAIRMAN
Phone: 207-774-5981
Josee Shelley, CHIEF OPERATING OFFICER
Phone: 207-594-9209
David Platt, MANAGING EDITOR
Phone: 207-594-9209
Leslie Fuller, MARINE RESOURCES DIRECTOR
Phone: 207-594-9209
Philip Conkling, PRESIDENT
Phone: 207-594-9209
Peter Ralston, VICE PRESIDENT
Phone: 207-594-9209

ISLAND RESOURCES FOUNDATION

1718 P St., NW, Suite T-4
Washington, DC 20036 USA
Phone: 202-265-9712　　　Fax: 202-232-0748
E-mail: irf@irf.org
Website: www.irf.org

Founded: 1972
Membership: 220
Scope: International

Description: An independent center for the study of island systems, dedicated to improved resources management, comprehensive development planning, the conservation of cultural, physical, and natural resources of islands.

Publication(s): Publications on website

Keyword(s): Coasts, Islands, Land Use Planning, Exotic species, Aquatic nuisance species

Contact(s):
Edward Towle, CHAIRMAN
etowle@irf.org
Bruce Potter, PRESIDENT
Charles Consolvo, SECRETARY
Judith Towle, TREASURER
jtowle@irf.org
Henry Wheatley, VICE CHAIRMAN

ISSAQUAH ALPS TRAILS CLUB (I.A.T.C.)

P.O. Box 351
Issaquah, WA 98027 USA
Phone: 206-328-0480
E-mail: LATCDrew@aol.com
Website: www.issaquahalps.org

Founded: 1979
Scope: Local

Description: A nonprofit membership organization established to preserve and promote trails and open space in the area east of Seattle along the I-90 highway corridor from Lake Washington to the Cascades, primarily in the area known as the "Issaquah Alps."

Publication(s): Targeting Tomorrow - Washington's Economy Adjusts to the 90's, Washington State We're In The, Speaking of Ground Water, Washington State Public Port Districts

Keyword(s): Air Quality and Pollution, Energy, Planning Management, Solid Waste, Wetlands

Contact(s):
Harvey Manning, FOUNDER
Steve Drew, PRESIDENT
Kitty Gross, SECRETARY
Steve Drew, TREASURER-ACTING
Barbara Johnson, VICE PRESIDENT OF OPERATIONS

IZAAK WALTON LEAGUE OF AMERICA ENDOWMENT

3185 Dubuque St., NE
Iowa City, IA 52240 USA
Phone: 319-351-7037　　　Fax: 319-351-7037
Website: www.iwla.org

Founded: 1943
Scope: National

Description: Organized to help rebuild Outdoor America by the acquisition for governmental agencies of unique natural areas for the use of future generations. Members of the Izaak Walton League of America.

Keyword(s): Acid Rain, Environmental and Conservation Education, Wildlife, Hunting, Outdoor Recreation, Natural Areas

Contact(s):
Robert Russell, EXECUTIVE SECRETARY
Howard White, HONORARY PRESIDENT
P.O. Box 527, Havana, IL 62644
Phone: 309-543-4391
Wendell Haley, PRESIDENT
1040 NE 92nd Ave., Portland, OR 97220
Phone: 503-253-9749
Charles Eldridge, SECRETARY
2008 74th St., Des Moines, IA 50322
Phone: 515-244-0932
William Weber, TREASURER
6357 W. Encantado Ct., Rockford, MI 49341
Phone: 616-456-8691
Fax: 616-456-1915
Larry Smith, VICE PRESIDENT
1611 Alderman Dr., Greensboro, NC 27408
Phone: 336-834-0018

IZAAK WALTON LEAGUE OF AMERICA, INC., THE

Headquarters, 707 Conservation Ln.
Gaithersburg, MD 20878-2983 USA
Phone: 301-548-0150   Fax: 301-548-0146
Website: www.iwla.org

Founded: NA
Membership: 50000
Scope: National

Description: Promotes means and opportunities for educating the public to conserve, maintain, protect, and restore the soil, forest, water, air, and other natural resources of the U.S. and promotes the enjoyment and wholesome utilization of those resources.

Publication(s): Outdoor America

Keyword(s): Environmental Protection, Hunting, Public Lands, Water Quality, Outdoor Ethics, Sustainability

Contact(s):
Jim Mosher, CONSERVATION DIRECTOR
Jason McGarvey, EDITOR AND MEDIA DIRECTOR
Paul Hansen, EXECUTIVE DIRECTOR
Stan Adams, NATIONAL PRESIDENT
Georgia Townsend, SECRETARY
William West, TREASURER

IZAAK WALTON LEAGUE OF AMERICA, INC., THE

c/o President, 3826 Lane Rd.
Cazenovia, NY 13035 USA
Phone: 315-655-3375
E-mail: cheneyweb@aol.com

Founded: NA
Membership: 500
Scope: Regional
Publication(s): Periodical Outdoor America

Contact(s):
Matt Webber, PRESIDENT
Phone: 315-655-3375
Les Monostory, SECRETARY
NY 315-435-6600
hllmomo@health.ongov.net

IZAAK WALTON LEAGUE OF AMERICA, INC., THE

ALASKA DIVISION
P.O. Box 670650
Chugiak, AK 99567 USA
Phone: 907-333-0243

Founded: NA
Scope: Statewide

Contact(s):
Thomas Carter, PRESIDENT
Phone: 907-333-0243

IZAAK WALTON LEAGUE OF AMERICA, INC., THE

CALIFORNIA DIVISION
700 E. Lake Drive 86
Orange, CA 92866 USA
Phone: 714-516-9483

Founded: 1938
Scope: Statewide

Contact(s):
Peter Hillebrecht, PRESIDENT
Phone: 310-791-0793

IZAAK WALTON LEAGUE OF AMERICA, INC., THE

COLORADO DIVISION
12175 West Ohio Place
Lakewood, CO 80228 USA
Phone: 303-986-1747
Website: www.iwla.org

Founded: NA
Membership: 1000
Scope: Statewide
Publication(s): Outdoor America-Magazine

Contact(s):
Nelson Burton, NATIONAL DIRECTOR
Phone: 719-473-0700
Leah Whellan, PRESIDENT
Amy Miller, SECRETARY AND NATIONAL DIRECTOR
513 Strachan Dr., Fort Collins, CO 80525-2130
Phone: 970-223-5379

IZAAK WALTON LEAGUE OF AMERICA, INC., THE

ILLINOIS DIVISION
Attn: President, 314 Townhall Rd., RR-2
Metamora, IL 61548 USA
Phone: 309-383-4203

Founded: NA
Scope: Statewide

Publication(s): Illini Ike (Newsletter)

Contact(s):
Jim Tyas, PRESIDENT
Phone: 309-383-4203
Marsha Johnson, SECRETARY
1512 45th St., Moline, IL 61265-3544
Phone: 309-797-8255

IZAAK WALTON LEAGUE OF AMERICA, INC., THE

INDIANA DIVISION
Attn: President, 2173 Pennsylvania St.
Portage, IN 46368-2444 USA
Phone: 219-762-4876
Website: www.in-iwla.org

Founded: NA
Membership: 5000
Scope: National
Publication(s): Hoosier Waltonian, The

Contact(s):
James Daniels, EDITOR/VICE PRES.
1808 Ravenswood Dr., Evansville, IN 47717
Phone: 812-477-7250
jimdaniels3@juno.com
Charles Siar, PRESIDENT
Phone: 219-762-4876
Bobby Schroader, SECRETARY
418 N. Harvey St., Griffith, IN 46319-2116
Phone: 219-924-2343
Emil Garcia, TREASURER
3420 W 40Th Pl., Gary, IN 46408
Phone: 219-980-2612
elgarcia@earthlink.net

IZAAK WALTON LEAGUE OF AMERICA, INC., THE
IOWA DIVISION
707 Conservation Lane
Gaithersburg, MD 20878 USA
Phone: 301-548-0150   Fax: 301-548-0146
Website: iwla.org

Founded: NA
Scope: Statewide

Contact(s):
Doyle Adams, PRESIDENT
Phone: 707 N. 7th St., Indianola, IA 50125-1430
Phone: 515-961-6004

IZAAK WALTON LEAGUE OF AMERICA, INC., THE
MARYLAND DIVISION
703 Conservation Lane
Gaithersburg, MD 20871 USA
Phone: 301-972-1627

Founded: NA
Membership: 4000
Scope: Local Regional

Contact(s):
Bill Gorman, EXECUTIVE SECRETARY
Georgia Townsend, PRESIDENT
406 Leighton Ave., Silver Spring, MD 20901
Phone: 301-588-8335

IZAAK WALTON LEAGUE OF AMERICA, INC., THE
MICHIGAN DIVISION
c/o President, 6260 Blythefield NE
Rockford, MI 49341 USA
Phone: 616-866-8475
Website: www.mi-iwla.org

Founded: NA
Membership: 580
Scope: Statewide

Contact(s):
E. John Trimberger, PRESIDENT
Robert Stegmier, SECRETARY
5285 Windmill Dr. NE, Rockford, MI 49341-9311
Phone: 616-866-4769

IZAAK WALTON LEAGUE OF AMERICA, INC., THE
OREGON DIVISION
15056 Quall Rd.
Silverton, OR 97381 USA
Phone: 503-873-2681
E-mail:

Founded: 1930
Scope: Statewide

Description: To protect, perpetuate, and strive for renewal of
Oregon's natural resources, including the air, soil, woods,
waters, and wildlife; to promote means and opportunities for
education of the public in respect to such resources and the
enjoyment and utilization thereof.

Contact(s):
Jeanne Norton, PRESIDENT
Phone: 503-235-7634
Coral Torley, SECRETARY
1820 NW Woodland Dr., Corvallis, OR 97330-1019
Phone: 541-752-0114

IZAAK WALTON LEAGUE OF AMERICA, INC., THE
PENNSYLVANIA DIVISION
460 New Salem Rd.
Uniontown, PA 17313 USA
Website: www.iwla.org

Founded: NA
Scope: Statewide

Keyword(s): Land Preservation, Land Purchase, Land Use
Planning, Natural Areas

Contact(s):
Raymond Kossler, PRESIDENT
460 New Salem Rd., Uniontown, PA 15401-9013
Phone: 724-437-5356
Martha Shaffer, SECRETARY
P.O. Box 35, Loganville, PA 17342-0035
Phone: 717-428-2883

IZAAK WALTON LEAGUE OF AMERICA, INC., THE
SOUTH DAKOTA DIVISION
Attn: President, 798 11th St., SW
Watertown, SD 57350-3060 USA
Phone: 605-352-2598
E-mail: clayton@santel.net
Website: http:// itc-web.com/sdikes

Founded: NA
Scope: Statewide

Contact(s):
Charles Clayton, PRESIDENT
Phone: 605-352-2598

IZAAK WALTON LEAGUE OF AMERICA, INC., THE
VIRGINIA DIVISION
Attn: President, 5235 Richardson Dr.
Fairfax, VA 22032-3930 USA
Phone: 703-361-5729

Founded: NA
Scope: Statewide

Publication(s): Periodic books and special reports., PEC
Newsreporter

Contact(s):
Birtrun Kidwell, PRESIDENT
Phone: 703-232-6563
Jeanne Kling, SECRETARY
6110 Occoquan Forest Drive, Manassas, VA 20112-3018

IZAAK WALTON LEAGUE OF AMERICA, INC., THE
WASHINGTON DIVISION
Attn: Bruce McGlenn, 2031 Franklin Ave E, 304
Seattle, WA 98102 USA
Phone: 425-455-1986   Fax: 425-453-9629
Website: www.seattleikes.org

Founded: NA
Scope: Statewide

Contact(s):
Ronni McGlenn, PRESIDENT, WASHINGTON STATE
DIVISION
ronnimc@juno.com

IZAAK WALTON LEAGUE OF AMERICA, INC., THE
WEST VIRGINIA DIVISION
79 E. Main Street
Richwood, WV 26261 USA
Phone: 304-846-6818
Website: www.izaakwaltonleague.com

Founded: NA
Scope: Statewide

Contact(s):
Don McClung, PRESIDENT
Phone: 304-876-2457

IZAAK WALTON LEAGUE OF AMERICA, INC., THE
WYOMING DIVISION
Attn: President, 1072 Empinado
Laramie, WY 82070 USA
Phone: 307-742-2785
E-mail:

Founded: NA
Scope: Statewide

Keyword(s): Agriculture, Energy, Environment, Solid Waste
Management, Exotic species, Aquatic nuisance species

Contact(s):
Raymond Jacquot, PRESIDENT
Phone: 307-742-2785

IZAAK WALTON LEAGUE OF AMERICA, INC., THE
YORK CHAPTER 57
Attn: William Shaffer
P.O. Box 35
Loganville, PA 17342
Phone: 717-428-2883
E-mail: reg6govike@aol.com
Website: www.iwla.org

Founded: 1926

Description: The strive for the purity of water, the clarity of air, the
wise stewartship of the land and its resources; to know the
beauty and understanding of nature and the value of wildlife,
woodlands, and open space; to the preservation of this
heritage and to man's sharing in it.

MISSION STATEMENT

We're a diverse group of 50,000 men and women dedicated to
protecting our nation's soil, air, woods, waters and wildlife. Our
strength lies in our grassroots, commonsense approach to
solving local, regional and national conservation issues. Our
interests span the spectrum of outdoor recreation and conservation activities, from angling and birding to stream monitoring,
wildlife photography and hunting. But we all share one major
goal: to protect and use sustainably America's rich resources
to ensure a high quality of life for all people, now and in the
future.

Publication(s): Waltonian News Monthly, Outdoor America

Contact(s):
William Shaffer, CORRESPONDING SECRETARY
P.O. Box 35, Loganville, PA 17342

IZAAK WALTON LEAGUE OF AMERICA, INC., THE
OHIO DIVISION
Attn: Secretary, 953 Greenwood Ave.
Hamilton, OH 45011-1817 USA
Phone: 513-697-6100
E-mail: kflowers@fuse.net
Website: www.iwla.org

Founded: 1922
Membership: 2800
Scope: Statewide
Publication(s): Tri annual newsletter - Buckeye Ike Line,
quarterly magazine - Outdoor America

Contact(s):
Kevin Flowers, ENVIRONMENTAL DIRECTOR
6793 Midnight Sun Dr., Mainville, OH 45039
Phone: 513-697-6100
Bill Ashbaugh, PRESIDENT
Yvonne Hayes, SECRETARY
Phone: 513-863-8018

IZAAK WALTON LEAGUE OF AMERICA, INC., THE
NEBRASKA DIVISION
Attn: President, 3017 Midway Rd.
Grand Island, NE 68803-2436 USA
Phone: 308-384-0656

Founded: NA
Scope: Statewide

Contact(s):
Roger Mettenbrink, PRESIDENT
Phone: 308-384-0656
Lurlie Campbell, SECRETARY
17125 Sodtown Rd., Ravenna, NE 68869
Phone: 308-452-3800

IZAAK WALTON LEAGUE OF AMERICA, INC., THE
FLORIDA DIVISION
P.O. Box 97
Estero, FL 33928 USA
Phone: 941-992-2184   Fax: 941-495-0201
E-mail: koreshanfound@mindspring.com
Website: www.iwla.org

Founded: NA
Membership: 93
Scope: Regional

Contact(s):
Michael Chenoweth, NATIONAL DIRECTOR
P.O. Box 236, Homestead, FL 33090-0236
Phone: 305-451-0993
Fax: 305-451-3627
michael.chenoweth@mail.com
Charles Dauray, PRESIDENT
Sarah Bergquist, SECRETARY

J

J.N. (DING) DARLING FOUNDATION
3333 Sanibel-Captiva Road, P.O. Box 482
Sanibel, FL 33957 USA
Phone: 941-472-2329   Fax: 305-361-9789
E-mail: kipkoss@hotmail.com
Website: www.dingdarling@dingdarling.org

Founded: 1962

Scope: National

Description: A nonprofit organization formed to continue the ideals and work of pioneer conservationist "Ding" Darling, with an emphasis on conservation education. The Foundation has no paid staff. With all services, including legal and accounting, provided by its trustees, the Foundation is able to funnel 100% of contributed funds into selected projects.

Keyword(s): Environmental and Conservation Education, Waterfowl, Exotic species, Aquatic nuisance species, training

Contact(s):
Kristie Anders, EXECUTIVE DIRECTOR
P O Box 482, Sanibel, FL 33957
Phone: 941-472-2329
Fax: 941-472-6421
kanders@sccf.org
Christopher Koss, PRESIDENT OF BOARD OF TRUSTEES AND CHAIRMAN OF EXECUTIVE COMMITTEE

JACK H. BERRYMAN INSTITUTE FOR WILDLIFE DAMAGE MANAGEMENT

DEPT. OF FISHERIES AND WILDLIFE
Utah State University
Logan, UT 84322-5210 USA
Phone: 435-797-2436   Fax: 435-797-1871
Website: www.berymaninstitute.org

Founded: NA
Membership: 20
Scope: National

Description: The Jack H. Berryman Institute is a national non-profit organization which is centered at Utah State University. It engages in research, education, and extension activities aimed at resolving human and wildlife conflicts, enhancing the positive aspects of wildlife, and increasing human tolerance of wildlife problems.

JACK MINER MIGRATORY BIRD FOUNDATION, INC.

P.O. Box 39
Kingsville, Ontario N9Y 2E8 Canada
Phone: 519-733-4034
E-mail: info@jackminer.com
Website: www.jackminer.com

Founded: 1904
Scope: International

Description: A nonprofit (501(c)3) foundation inc. in both the U.S. and Canada. This sanctuary and its founder, Jack Miner, have become internationally known as one of the earliest efforts in waterfowl conservation. Often referred to as "The Father of Conservation", Jack Miner pioneered the tagging of waterfowl in 1909. The sanctuary is open year round to the public with no admission fee.

Keyword(s): Nature Centers, Birds, Waterfowl

Contact(s):
Kirk Miner, PRESIDENT AND TREASURER
Marilyn Hageniers, SECRETARY
Edna Miner, VICE PRESIDENT

JACKSON HOLE CONSERVATION ALLIANCE

P.O. Box 2728
Jackson, WY 83001 USA
Phone: 307-733-9417   Fax: 307-733-9008
E-mail: jhca@wyoming.com
Website: www.jhalliance.org

Founded: 1979
Membership: 1800
Scope: Local, Regional

Description: The Alliance is a nonprofit organization dedicated to responsible land stewardship in Jackson Hole, Wyoming, to ensure that human activities are in harmony with the area's irreplaceable wildlife, scenic and other natural resources.

Publication(s): Alliance News, The, Fiscal Impacts of Growth in Teton County, Mosquito Abatement Program in Teton County, Wyoming, The, Welcome to the Neighborhood

Keyword(s): Biodiversity, Endangered Species, Forest Management, Land Use Planning, Conservation, Wildlife, Environment, Open Space, training, Wilderness

Contact(s):
Lisa Robertson, CO-VICE PRESIDENT
Marcia Kunstel, CO-VICE PRESIDENT
Franz Camenzind, EXECUTIVE DIRECTOR
franz@jhalliance.com
David Hardie, PRESIDENT
Pamela Lichtman, PROGRAM DIRECTOR
pam@jhalliance.com
Jean Barash, SECRETARY
Phone: 307-739-8669
Fax: 307-739-9691
jbarash@wyoming.com
John Carney, TREASURER

JACKSON HOLE LAND TRUST

P.O. Box 2897
Jackson, WY 83001 USA
Phone: 307-733-4707   Fax: 307-733-4144
E-mail: info@jhlandtrust.org
Website: www.jhlandtrust.org

Founded: 1980
Membership: 11
Scope: Local

Description: A private, nonprofit land conservation organization which works to preserve open space and the scenic, ranching, and wildlife values of Jackson Hole by assisting landowners who wish to protect their land in perpetuity. Not a membership organization.

Publication(s): Land Trust Newsletter

Keyword(s): Agriculture, Land Preservation, Land Purchase, Open Space, training

Contact(s):
Leslie Mattson, EXECUTIVE DIRECTOR
Scott Pierson, PRESIDENT
Michael Caruso, SECOND VICE PRESIDENT
Richard Vangyeek, SECRETARY
Mia Jensen, TREASURER

JACKSON HOLE PRESERVE, INC.
30 Rockefeller Plaza, Rm. 5600
New York, NY 10112 USA
Phone: 212-649-5819   Fax: 212-649-5729

Founded: 1940
Scope: National

Description: Nonprofit, charitable, and educational organization, established to conserve areas of outstanding primitive grandeur and natural beauty and to provide facilities for their use and enjoyment by the public.

Contact(s):
C.W. Frye, CHAIRMAN OF THE BOARD
Antonia Grumbach, SECRETARY
Carmen Reyes, TREASURER

JANE GOODALL INSTITUTE, THE
P.O. Box 14890
Silver Spring, MD 20911 USA
Phone: 301-565-0086   Fax: 301-565-3188
E-mail: jgiinformation@janegoodall.org
Website: www.janegoodall.org

Founded: 1977
Membership: 10
Scope: International

Description: The Jane Goodall Institute is an international organization dedicated to the conservation and understanding of wildlife, particularly chimpanzees, and to promoting environmental education, reforestation, and humanitarianism worldwide.

Publication(s): Annual JGI World Report, Semi-annual Roots and Shoots Network, ChimpanZOO Newsletter

Keyword(s): training, Environmental and Humanitarian Education, Wildlife Research, Reforestation, Animal Welfare, Chimpanzees

Contact(s):
Gary North, DEPUTY DIRECTOR OF MERCHANDISE
gwnjhu@aol.com
Jeanne McCarty, DIRECTOR OF ROOTS AND SHOOTS
j.mccarty@janegoodall.org

JOHN INSKEEP ENVIRONMENTAL LEARNING CENTER
19600 S. Molalla Ave.
Oregon City, OR 97045 USA
Phone: 503-657-6958   Fax: 503-650-6669
E-mail: elc@clackamas.cc.or.us
Website: www.clackamas.cc.or.us

Founded: 1972
Scope: Statewide

Description: A source of teacher training and community education on environmental education topics, focusing on urban watershed issues. Located on a restored industrial site featuring buildings made from salvaged and recycled materials.

Keyword(s): Environmental and Conservation Education, Watersheds, Recycling

Contact(s):
John Lecavalier, DIRECTOR

K

KANSAS ACADEMY OF SCIENCE
DIVISION OF BIOLOGICAL SCIENCES, EMPORIA STATE UNIVERSITY
P.O. Box 65
Baldwin City, KS 66006
Emporia State University
Emporia, KS 66901 USA
Phone: 785-594-6451   Fax: 785-594-6721
E-mail: cink@harvey.bekeru.edu
Website: washburn.edu

Founded: 1868
Membership: 500
Scope: Statewide

Description: A nonprofit organization to increase, diffuse, and promote knowledge in various departments of science; interest young people in science and encourage them to consider science as their profession; aid the improvement of science teaching; and aid in development of the state's economic growth.

Publication(s): Transactions of the Kansas Academy of Science

Keyword(s): Wildlife Rehabilitation, Chemistry, Geography, Geology, education

Contact(s):
Dan Merriam, EDITOR
Kansas Geological Survey University of Kansas, Lawrence, KS 66047
Phone: 913-864-4991
Karen Debres, PRESIDENT
Department of Geography Kansas State University, Manhattan, KS 66506
Phone: 785-532-6727
David Saunders, PRESIDENT-ELECT
Division of Biological Sciences Emporia State University, Emporia, KS 66901
Phone: 316-341-5610
Pieter Berendsen, SECRETARY
Kansas Geological Survey University of Kansas, Lawrence, KS 66047
Phone: 785-864-4991

KANSAS ADVISORY COUNCIL FOR ENVIRONMENTAL EDUCATION
Attn: President, 1005 Merchants Tower
2610 Claflin Rd, Manhattan, KS 66502
Topeka, KS 66612 USA
Phone: 785-532-3314   Fax: 785-532-3305
E-mail: jstrickl@oz.oznet.ksu.edu
Website: uwsp.edu

Founded: 1969
Scope: Statewide

Description: Organized to promote and support effective environmental education in order to enhance awareness, knowledge, and concern about the environment among the citizens of Kansas. The advisory council is made up of representatives of over 180 public and private organizations, institutions, business organizations, and individuals.

Publication(s): KACEE News

Contact(s):
Clark Duffy, PRESIDENT AND COUNCIL MEMBER
1005 Merchants Tower, Topeka, KS 66612
Phone: 913-234-0589

Connie Elders, SECRETARY
455 N. Main 11th Fl., Wichita, KS 67202
Phone: 316-264-8323
Ruth Gennrich, TREASURER
Museum of Natural History University of Kansas, Lawrence,
KS 66045-2454
Phone: 913-864-4173
Carol Williamson, VICE PRESIDENT
1209 Willow Dr., Olathe, KS 66061
Phone: 913-764-6036

KANSAS ASSOCIATION FOR CONSERVATION AND ENVIRONMENTAL EDUCATION
2610 Claflin Rd.
Manhattan, KS 66502-2743 USA
Phone: 785-532-3322   Fax: 785-532-3305
Website: http://www.kacee.org

Founded: 1969
Scope: Statewide

Description: Kansas Association for Conservation and Environmenal Education was organized to promote and support effective conservation and environmental education in Kansas. The Association is made up of over 200 public and private organizations and individuals.

Publication(s): KACEE NEWS, Annual Report, Workshop Brochure

Keyword(s): Environment, Environmental and Conservation Education

Contact(s):
Laura Downey, EXECUTIVE DIRECTOR
Phone: 785-532-3322
Fax: 785-532-3305
ldowney@oznet.ksu.edu
Brad Loveless, PRESIDENT
Phone: 785-575-8115
Fax: 785-575-8039
brad_loveless@wstnres.com
Clark Duffy, TREASURER
Phone: 785-296-3185
Fax: 785-296-0878
Kate Grover, VICE PRESIDENT
Phone: 785-368-3801
Fax: 785-368-3806
kgover@topeka.org

KANSAS ASSOCIATION OF CONSERVATION DISTRICTS
Attn: President, Rt. 1 Box 110
Glen Elder, KS 67446 USA
Phone. 785-475-2342   Fax: 785-475-3886

Founded: NA
Scope: Statewide

Keyword(s): Conservation Districts

Contact(s):
Don Paxson, BOARD MEMBER
P.O. Box 487, Penokee, KS 67659
Phone: 785-421-2480
Fax: 785-421-5662
Richard Jones, EXECUTIVE DIRECTOR
522 Winn Rd., Salina, KS 67401-3668
Phone: 785-827-5847
Fax: 785-827-7784

Carl Jordan, PRESIDENT, ALTERNATE BOARD MEMBER
Phone: 785-545-3361
Fax: 785-545-3659
Don Rezac, SECRETARY-TREASURER
12350 Ranch Rd., Emmett, KS 66422
Phone: 785-535-2961
Fax: 785-457-2868
Sandra Jones, VICE PRESIDENT
5160 E Rd 17, Johnson, KS 67855
Phone: 316-492-6495
Fax: 316-492-2772

KANSAS B.A.S.S. CHAPTER FEDERATION
Attn: President, P.O. Box 330
Alba, MO 64830 USA
Phone: 417-525-4940
E-mail: onemorefish@ckt.net
Website: www.kbcf.com

Founded: NA
Scope: Statewide

Description: An organization of Bassmaster chapters, affiliated with the Bass Anglers Sportsman Society, organized to fight pollution, assist state and national conservation agencies in their efforts, and teach the young people of our country good conservation practices. Dedicated to the realistic conservation of our water resources.

Publication(s): KBCF News and Views (Quarterly Newsletter)

Contact(s):
Greg Clark, CONSERVATION DIRECTOR
9320 E. Osie Apt 2104, Wichita, KS 67207
Phone: 316-681-1887
Jon Stewart, PRESIDENT
Phone: 417-525-4940

KANSAS HERPETOLOGICAL SOCIETY
University of Kansas Natural History Museum, Dyche Hall
Lawrence, KS 66045 USA

Founded: 1974
Scope: Statewide

Description: The Kansas Herpetological Society is a nonprofit organization designed to encourage education and dissemination of scientific information through the facilities of the Society; and to encourage conservation of wildlife in general and of amphibians and reptiles in Kansas in particular.

Publication(s): Kansas Herpetological Society Newsletter

Keyword(s): Ecology, Environmental and Conservation Education, education, Nongame Wildlife, Reptiles and Amphibians

Contact(s):
Eric Rundquist, EDITOR
Animal Care Unit, B054 Malott, University of Kansas,
Lawrence, KS 66045
Karen Toepfer, TREASURER
303 W. 39th St., Hays, KS 67601
Phone: 785-628-1437

KANSAS NATURAL RESOURCE COUNCIL
Topeka, KS 66601 USA
Phone: 785-746-8885
Website: www.knrc.ws

Founded: 1981
Membership: 500
Scope: Statewide

Description: Environmental advocacy including public education, lobbying, and litigation.

Publication(s): Weekly Legislative Updates, KNRC Journal

Keyword(s): Agriculture, Energy, Prairies, Rivers, Exotic species, Aquatic nuisance species, Environmental Legislation

Contact(s):
John Barnes, EXECUTIVE DIRECTOR
Joan Vibert, PRESIDENT
1981 Indiana, Ottawa, KS 66067
Phone: 785-746-8885
joan@windwalker-farm.com

KANSAS ORNITHOLOGICAL SOCIETY

14207 Robin Rd.
Leavenworth, KS 66048 USA
Phone: 913-651-2565
Website: http://ksbirds.org/kos

Founded: 1949
Scope: Statewide

Description: Formed to promote the study of ornithology, to advance the members in ornithological science, to promote conservation, and the appreciation of birds by the general public.

Publication(s): Horned Lark, The, K.O.S. Bulletin

Keyword(s): Birds, Endangered Species, Environmental and Conservation Education, Raptors, training

Contact(s):
John Schukman, PRESIDENT
Leavenworth, KS 66048
schuksaya@aol.com
Gene Young, VICE PRESIDENT
P.O. Box 1147 Natural Science Dept, Arkansas City, KS 67005
youngg@cowley.cc.ks.us

KANSAS WILDFLOWER SOCIETY

R.L. McGregor Herbarium, 2045 Constant Ave.
Lawrence, KS 66047-3729 USA
Phone: 785-864-3453   Fax: 785-864-5093

Founded: 1978
Scope: Statewide

Description: The Society provides educational materials and sponsors activities to promote the conservation and cultivation of the native plants of Kansas.

Publication(s): KWS Newsletter

Keyword(s): Conservation, Prairies, Flowers, Plants, and Trees, Native Plants

Contact(s):
Craig Freeman, AGENT
Dwight Platt, PRESIDENT
Phone: 316-283-2500
Fax: 316-284-5286
Cynthia Ford, SECRETARY
Phone: 316-235-4726
Patricia Stanley, TREASURER
Phone: 316-689-4070
wichitacsj@feist.com

KANSAS WILDLIFE FEDERATION

P.O. Box 8237
Wichita, KS 67208-0237 USA
Phone: 785-526-7466   Fax: 785-658-2466

Founded: NA
Scope: Statewide

Description: A representative statewide organization, affiliated with the National Wildlife Federation, dedicated to the protection and enhancement of wildlife and its habitat through public education and government interaction.

Publication(s): Kansas Wildlife Federation: The Voice of Outdoor Kansas (newsletter)

Contact(s):
Velma Miller, ALTERNATE REPRESENTATIVE & EDITOR
Steve Montgomery, EDUCATION PROGRAMS CONTACT
Tommie Berger, PRESIDENT & REPRESENTATIVE
Phone: 785-658-2465
Roger Brooner, TREASURER
Phone: 316-768-3827

KANSAS WILDSCAPE FOUNDATION

P.O. Box 4029
Lawrence, KS 66046 USA
Phone: 785-843-9453   Fax: 785-843-6379
Website: kansaswildscape.com

Founded: 1991
Scope: Statewide

Description: The Kansas Wildscape Foundation is dedicated to conserving and perpetuating the land, wild species, and the rich beauty of Kansas for the use and enjoyment of all. Wildscape is a public/private partnership with the Kansas Department of Wildlife and Parks.

Keyword(s): Land Preservation, Wetlands, training, Youth Organizations

Contact(s):
Bob Beachy, CHAIRMAN
P.O. Box 134446, Kansas City, MO 64199-3446
Harland Priddle, EXECUTIVE DIRECTOR
Jim Huntington, PRESIDENT
Charlie Becker, TREASURER

KEEP AMERICA BEAUTIFUL, INC.

1010 Washington Blvd., 7th Fl.
Stamford, CT 06901 USA
Phone: 203-323-8987   Fax: 203-325-9199
Website: www.kab.org

Founded: 1953
Scope: National

Description: A national nonprofit public education organization dedicated to litter prevention and improved waste handling practices in American communities. Keep America Beautiful trains and certifies communities into the Keep America Beautiful System, a behavior-based approach to improved waste handling.

Publication(s): Network News

Keyword(s): Communications, Environmental and Conservation Education, Protected Areas, Public Lands, Solid Waste Management

Contact(s):
Thomas Tomoney, CHAIRMAN OF THE BOARD
G. Empson, PRESIDENT

Susanne Woods, SENIOR VICE PRESIDENT OF
DEVELOPMENT AND ENVIRONMENTAL PROGRAMMING
John Bard, VICE CHAIRMAN OF THE BOARD

KEEP FLORIDA BEAUTIFUL, INC.
2615 N. Monroe St., Suite 200
Tallahassee, FL 32303-4027 USA
Phone: 850-385-1528   Fax: 850-385-4020
Website: www.keepfloridabeutiful.org

Founded: 1991
Membership: 3
Scope: Statewide

Description: KFB's mission is to empower individuals to take greater responsibility for their community environment.

Keyword(s): Aquatic Habitats, Communications, Protected Areas, Renewable Resources, Solid Waste Management

Contact(s):
Shane McIntosh, CHAIRMAN, BOARD OF DIRECTORS

KENTUCKY ACADEMY OF SCIENCE
Attn: President, Dept. of Biology, Berea College
Berea, KY 40404 USA
Phone: 859-985-3351   Fax: 859-985-3303
Website: http://kas.wku.edu/kas/

Founded: 1914
Membership: 10
Scope: Regional

Description: To encourage scientific research, promote the diffusion of scientific knowledge, and unify the scientific interests of Kentucky.

Publication(s): Journal of the Kentucky Academy of Science

Keyword(s): Research

Contact(s):
John Thieret, EDITOR
Dept. of Biological Sciences, Northern Kentucky University, Highland Heights, KY 41099
Phone: 606-572-6390
Fax: 606-572-5639
Blaine Ferrell, PAST PRESIDENT
Dept. of Biology, Western Kentucky Univeristy, Bowling Green, KY 42101
Phone: 270-745-5999
Ron Rosen, PRESIDENT
Dept. of Biology, Berea College, Berea, KY 40403
Phone: 859-986-9341
Fax: 859-986-4506
Jerry Warner, PRESIDENT-ELECT
Dept. of Biological Science, Northern Kentucky University, Highland Heights, KY 41076
Phone: 606-572-5277
Stephanie Dew, SECRETARY
Dept. of Biology, Centre College, Danville, KY 40422
Phone: 859-238-5316
Kenneth Crawford, TREASURER
Dept. of Biology, Western Kentucky University, Bowling Green, KY 42103
Phone: 270-745-6005
Fax: 270-745-6856
kenneth.crawford@wku.edu
Robert Barney, VICE-PRESIDENT
Atwood Research Facility, Kentucky State University, Frankfort, KY 40601
Phone: 502-227-6178

KENTUCKY ASSOCIATION FOR ENVIRONMENTAL EDUCATION (KAEE)
P.O. Box 176055
Covington, KY 41017 USA
Phone: 859-578-3012
Website: ww.kaee.org

Founded: NA
Scope: Statewide

Description: Organized to promote and support formal and nonformal environmental education programs throughout the state. Promotes information sharing, research, and development of EE programs and activities. Annually sponsors a three-day conference.

Publication(s): E.E. Resource Guide, Earth Day Handbook, Newsletter

Keyword(s): Environmental and Conservation Education

Contact(s):
Karen Reagor, EXECUTIVE DIRECTOR
P.O. Box 176055, Covington, KY 41017
Phone: 606-578-0312
KPReagor@aol.com
Joe Baust, PRESIDENT
Center for Environmental Education, Murray State University, Murray, KY 42071
Phone: 270-762-2595
joe.baust@coe.murraystate.edu

KENTUCKY ASSOCIATION OF CONSERVATION DISTRICTS
Attn: President, 1299 Lillies Ferry Rd.
Winchester, KY 40391 USA
Phone: 786-836-2272

Founded: NA
Scope: Statewide

Keyword(s): Conservation Districts

Contact(s):
James Lacy, BOARD MEMBER
300 Sanfield Rd., Campton, KY 41301
Phone: 606-662-4161
Fax: 606-668-7033
John Chism, PRESIDENT, ALTERNATE BOARD MEMBER
Phone: 606-744-8909
Phone: 502-564-9195
Kevin Jeffries, SECRETARY-TREASURER
1503 E. Hwy. 22, Crestwood, KY 40014
Phone: 502-222-9877
Fax: 502-222-0046
Patrick Henderson, VICE PRESIDENT
Rt. 1 Box 146, Irvington, KY 40146
Phone: 502-547-6206
Fax: 502-564-9195

KENTUCKY AUDUBON COUNCIL
Attn: President, 306 Hoover Hill Rd.
Hartford, KY 42347 USA
Phone: 270-298-4237
E-mail: bandb3@minspring.com
Website: www.audubon.wku.edu

Founded: 1971
Membership: 3500
Scope: Regional

Description: A statewide Audubon Council for the seven key chapters of the National Audubon Society. Works to promote, foster, and encourage the conservation and preservation of all wildlife, plants, soils, water, air, and other natural resources for the benefit of all people.

Publication(s): Kentucky's Cause

Contact(s):
Jeff Frank, PAST-PRESIDENT
16509 Bradbe Rd., Fisherville, KY 40023
Phone: 502-266-7181
Bill Little, PRESIDENT
Phone: 270-298-4237
Maggie Selvidge, SECRETARY
904 North Dr., Hopkinsville, KY 42240
Phone: 502-886-8078
Bertha Timmel, TREASURER
3604 Graham Rd., Louisville, KY 40207
Phone: 502-893-5601

KENTUCKY B.A.S.S. CHAPTER FEDERATION
Attn: President, 562 N. Saint Gregory Church
Coxs Creek, KY 40013 USA
Phone: 502-331-9656
Website: www.kybassfed.com

Founded: NA
Scope: Statewide

Description: An organization of Bassmaster chapters, affiliated with the Bass Anglers Sportsman Society, organized to fight pollution, assist state and national conservation agencies in their efforts, and teach the young people of our country good conservation practices. Dedicated to the realistic conservation of our water resources.

Contact(s):
John Romans, CONSERVATION DIRECTOR
209 Park Ave., Carrollton, KY 41008
john.romans@dowcorning.com
Steven Taylor, PRESIDENT
Phone: 502-331-9656
steve@wareinc.com
Donnie Keeton, VICE PRESIDENT
P.O. Box 71, Warsaw, KY 41095
Phone: 859-567-2885
donkee311@earthlink.net

KENTUCKY RESOURCES COUNCIL
P.O. Box 1070
Frankfort, KY 40602 USA
Phone: 502-875-2428   Fax: 502-875-2845
E-mail: fitzkrc@aol.com

Founded: NA
Scope: Statewide

Description: The KRC is a nonprofit, membership-based statewide organization dedicated to the conservation and prudent use of Kentucky's natural resources. The Council is comprised of Kentuckians from all walks of life: Urban dwellers and rural residents, farmers, river recreationists, and conservationists. This broad-based membership shares a common concern with the impact of mineral extraction, natural resource development, and economic development on our homes, health, and quality of life.

Keyword(s): Environmental Law, Environmental Protection, Toxic Substances, Nuclear-free, Water quantity, Water export and diversion, Water Pollution Management

Contact(s):
Tom Fitzgerald, DIRECTOR

KENTUCKY WOODLAND OWNERS ASSOCIATION
433 Chestnut St.
Berea, KY 40403 USA
Phone: 859-986-2373   Fax: 859-986-1299

Founded: 1991
Scope: Statewide

Description: A statewide nonprofit organization, affiliated with the National Woodland Owners Association, organized to promote good forest stewardship, circulate information on timber marketing, and encourage private property responsibility among woodland owners throughout the Commonwealth of Kentucky.

Publication(s): Kentucky Woodlands

Keyword(s): Forests and Forestry

Contact(s):
Bridget Abernathy, EXECUTIVE SECRETARY
Don Girton, PRESIDENT AND EDITOR
Paschal Phillips, SECRETARY
Herb Lloyd, TREASURER

KEYSTONE CENTER, THE
1628 Saints John Rd.
Keystone, CO 80435 USA
Phone: 970-513-5800   Fax: 970-262-0152
E-mail: tkcspp@keystone.org
Website: www.keystone.org

Founded: 1975
Membership: 25+
Scope: National

Description: A nonprofit center for environmental dispute resolution, mediation, and facilitation. Conducts national policy dialogues on environmental, energy, natural resources, health, and science/technology issues; assists in environmental decisionmaking and regulatory negotiations; provides environmental mediation services; provides training and organizational development services in environmental conflict resolution.

Publication(s): Discovery, Consensus

Keyword(s): Biotechnology, Energy, Environmental and Conservation Education, Health and Nutrition, Natural Resource Conservation, Forests and Forestry, Public Lands, Rivers, Water Pollution Management, Wilderness

Contact(s):
Tom Grumbly, PRESIDENT

KIDS FOR SAVING EARTH WORLDWIDE
P.O. Box 421118
Minneapolis, MN 55442 USA
Phone: 763-559-1234   Fax: 763-559-6980
E-mail: kseww@aol.com
Website: www.kidsforsavingearth.org/

Founded: 1989
Membership: 3,000
Scope: National

Description: KSEW's mission is to educate and empower children to help to protect the Earth's environment by providing free educational materials to kids, schools, and organizations through the KSE Network. Curriculum guides are also available.

Publication(s): Action Guide, Earth Works Central Notebook, KSE Action Guide, Rock the World Concert Kit, KSE News, Free Membership, List available upon request., Travel of Earth Book, Toxic waste site poster

Keyword(s): Environmental and Conservation Education, Protected Areas, Youth Organizations, Air Quality and Pollution, Conservation, Endangered Species, Aquatic Habitats, Energy Conservation, Environment, Flowers, Plants, and Trees, Land Preservation, National Parks, Environmental Protection, Pollution Prevention

Contact(s):
Tessa Hill, PRESIDENT AND DIRECTOR

KODIAK BROWN BEAR TRUST

11930 Circle Dr.
Anchorage, AK 99516 USA
Phone: 907-345-2939   Fax: 907-348-0450

Founded: 1981
Scope: National

Description: The Kodiak Brown Bear Trust is an Alaska-based nonprofit wildlife conservation trust whose mission is to support conservation of the majestic Kodiak brown bear through funding of habitat protection, research and public education.

Publication(s): Exxon Valdez conservation saga in Alaska's Kodiak archipelago

Keyword(s): Wildlife Protection, training, Research

Contact(s):
Dave Cline, CHAIRMAN
Tim Richardson, EXECUTIVE DIRECTOR
6707 Old Stage Rd., North Bethesda, MD 20852-4329
Phone: 301-770-6496

L

LADY BIRD JOHNSON WILDFLOWER CENTER

4801 La Crosse Ave.
Austin, TX 78739 USA
Phone: 512-292-4200   Fax: 512-292-4627
Website: www.wildflower.org

Founded: 1982
Scope: National

Description: The Lady Bird Johnson Wildflower Center's purpose is to educate people about the environmental necessity, economic value, and natural beauty of native plants. The Wildflower Center, a nonprofit organization, serves North America by promoting the preservation and use of native plants through education programs, information dissemination, and by example.

Publication(s): Wild Ideas The Store Catalog, Native Plants magazine

Keyword(s): Wildflowers, Flowers, Plants, and Trees, Trees, Environmental and Conservation Education, Landscape Architecture, Endangered Species

Contact(s):
Flo Oxley, ACTING DIRECTOR OF EDUCATION/SENIOR BOTANIST
Lady Bird Johnson, CO-FOUNDER
Helen Hayes, CO-FOUNDER
Karen Stevenson, COMMUNICATIONS DIRECTOR

Leslie Lewis, CONTACT
Denise Delaney, DIRECTOR OF HORTICULTURE
Robert Breunig, EXECUTIVE DIRECTORY

LAKE ERIE CLEAN-UP COMMITTEE, INC.

Attn: President, 29789 Fort Rd.
Rockwood, MI 48173 USA
Phone: 313-379-3891

Founded: 1959
Scope: National

Description: The LECC's mission is to stop pollution of Lake Erie and of all freshwater lakes and streams; to inform the public of the need for greater pollution controls; to prevent the return to the methods of the past; and to encourage industry to do more research. Our Great Lakes are a fragile part of our ecosystem and we must continue to protect them. Membership includes representatives of Michigan and Ohio citizen groups.

Keyword(s): Coasts, Water Pollution Management, Exotic species, Aquatic nuisance species, Waterfowl, Wetlands, Remedial Action Plans

Contact(s):
Leonard Mannausa, PRESIDENT
29789 Fort Rd., Rockwood, MI 48173
Phone: 313-379-3891
Richard Micka, SECRETARY
47 E. Elm, Monroe, MI 48162
Phone: 313-242-0909
Jerome Falwell, TREASURER
30251 Worth, Gibraltor, MI 48173

LAKE MICHIGAN FEDERATION

220 S. State St., Suite 1900
Chicago, IL 60604 USA
Phone: 312-939-0838   Fax: 312-939-2708
E-mail: Info@lakemichigan.org
Website: www.lakemichigan.org

Founded: 1970
Membership: 1000
Scope: Regional

Description: A coalition of citizens and citizen organizations in Wisconsin, Illinois, Indiana, and Michigan dedicated to protecting Lake Michigan through community action and research. Supported by foundation and corporate grants, membership and contributions.

Publication(s): A Citizen's Action Guide, A Citizen's Guide to Cleaning Up Contaminated Sediments (book), Wetlands and Water Quality: A Citizen's Guide., Lake Michigan Monitor

Keyword(s): Environment, Lakes, EcoAction, Habitat Conservation, Pollution Prevention, Coral Reefs, Protecting Special Places, Watersheds

Contact(s):
Sophia Twichell, BOARD OF DIRECTORS
Cameron Owens, EXECUTIVE DIRECTOR

LAKE SUPERIOR GREENS

P.O. Box 1144
Superior, WI 54880 USA
Phone: 715-392-5782

Founded: 1991
Membership: 50

Scope: National

Description: Lake Superior Greens is a grassroots group joined to other Green groups in our dedication to a more sustainable lifestyle and a healthy planet. We are active locally as well as on a state, national, and international basis, recognizing that all issues are interrelated.

Publication(s): monthly newsletter

Keyword(s): Water Quality, Sustainability, Pesticides, Toxic Substances, Nuclear-free, Water quantity, Water export and diversion, Wetlands

Contact(s):
John Schraufnagel, CONTACT
1506 N. 19th, Superior, WI 54880
Phone: 715-394-6660
Rosie Seymour, CONTACT
1606 N. 18th St., Superior, WI 54880
Phone: 715-395-0494
Bob Browne, CONTACT
422 Fisher Ogden, Superior, WI 54880
Phone: 715-394-6235
Jan Conley
STEERING COMMITTEE
2406 Hughitt, Superior, WI 54880
Phone: 715-392-5782

LAND BETWEEN THE LAKES ASSOCIATION
Golden Pond, KY 42211-9001 USA
Phone: 270-924-2000   Fax: 270-924-2119
Website: www.lbl.org

Founded: 1983
Membership: 3000
Scope: National

Description: A private nonprofit membership organization supporting and promoting Tennessee Valley Authority's Land Between The Lakes, a 170,000-acre national demonstration in natural resource management, environmental education, and recreation.

Keyword(s): Environmental and Conservation Education, Prairies, Public Lands, Natural Areas, Lakes, Cultural Preservation

Contact(s):
Ramay Winchester, CHAIRMAN
Gaye Lueber, DIRECTOR
Loran Wagoner, PRESIDENT

LAND TRUST ALLIANCE, THE
1331 H St., NW, 4th Fl.
Washington, DC 20005 USA
Phone: 202-638-4725   Fax: 202-638-4730
E-mail: lta@lta.org
Website: www.lta.org

Founded: 1982
Scope: National

Description: Provides training, technical assistance and publications for local and regional land trusts to increase their skills and strengthen the land trust movement; fosters public policies that further land trusts' goals; sponsors the National Land Trust Rally; and builds awareness among a broad constituency of the consequences of diminishing land resources and the role of land trusts in saving land.

Publication(s): Conservation Easement Handbook, Appraising Easements, Federal Tax Law of Conservation Easements, Starting a Land Trust, National Directory of Conservation Land Trusts, Conservation Easement Stewardship Guide, Exchange

Keyword(s): Environmental Law, Environmental and Conservation Education, Protected Areas, Land Purchase, Land Use Planning, Land Conservation, Land Protection, Conservation Easements

Contact(s):
James Espy, CHAIRMAN
Jean Hocker, PRESIDENT
Constance Best, SECRETARY
David Hartwell, TREASURER
John Turner, VICE CHAIR
Andrew Zepp, VICE PRESIDENT FOR PROGRAMS
John Chappell, VICE PRESIDENT OF DEVELOPMENT
Phil Jones, VICE PRESIDENT OF OPERATIONS

LANDOWNER PLANNING CENTER
P.O. Box 2242
Boston, MA 02101 USA

Founded: NA

Publication(s): Preserving Family Land, Book 1, Preserving Family Lands, Book 2

Contact(s):
Connie Small

LANDWATCH MONTEREY COUNTY
Salinas, CA 93902 USA
Phone: 831-422-9390   Fax: 831-422-9391
E-mail: landwatch@mclw.org
Website: www.landwatch.org

Founded: 1997
Membership: 1000
Scope: Regional

Description: Land Watch is committed to fundamental land use reform and works to build public support for better land use policies at the local, regional, and state level.

Publication(s): State of Monterey County, Update

Keyword(s): Environment, Environmental and Conservation Education, Environmental Planning, Protected Areas, Environmental Protection, Land Management, Land Preservation, Land Protection, Land Use Planning, Protecting Special Places, Transportation, Urban Environment

Contact(s):
Arianne Tucker, ADMINISTRATIVE DIRECTOR
atucker@mclw.org
Chris Fitz, DEPUTY DIRECTOR
Gary Patton, EXECUTIVE DIRECTOR
Phone: 831-375-7396
gapatton@mclw.org

LEAGUE OF CONSERVATION VOTERS
1920 L St., NW, Suite 800
Washington, DC 20036 USA
Phone: 202-785-8683   Fax: 202-835-0491
E-mail: lcv@lcv.org
Website: www.lcv.org

Founded: 1970
Membership: 40

Scope: National

Description: The LCV is the national, bipartisan political action arm of the environmental movement. LCV works to elect pro-environment candidates to Congress; publishes the National Environmental Scorecard, which rates members of Congress on key environmental votes; raises funds for campaigns through its Political Action Committee and Earthlist; and is governed by a Board of Directors made up of leaders from major national environmental organizations.

Publication(s): Presidential Scorecard, LCV Insider Newsletter., National Environmental Scorecard

Keyword(s): Politics and Government, Conservation

Contact(s):
Theodore Roosevelt, CHAIR
Anne Saer, CHIEF FINANCIAL OFFICER
Beth Sullivan, EXECUTIVE DIRECTOR, EDUCATIONAL FUND
Deb Callahan, PRESIDENT
Wade Greene, SECRETARY
Winsome McIntosh, TREASURER

LEAGUE OF KENTUCKY SPORTSMEN, INC.

P.O. Box 8527
Lexington, KY 40533 USA
Phone: 859-276-3518
E-mail: office@kentuckysportsmen.com
Website: kentuckysportsmen.com

Founded: 1935
Scope: Statewide

Description: A representative statewide organization, affiliated with the National Wildlife Federation, dedicated to the protection and enhancement of wildlife and its habitat through public education and government interaction.

Publication(s): Kentucky Sportsman, The

Contact(s):
Jim Thompson, ALTERNATE REPRESENTATIVE
Ben Hall, EDITOR
Rowland Beers, PRESIDENT AND EDUCATION PROGRAMS CONTACT
Linda Saunders, REPRESENTATIVE
Don York, TREASURER

LEAGUE OF OHIO SPORTSMEN

3953 Indianola Ave.
Columbus, OH 43214 USA
Phone: 614-268-9924   Fax: 614-268-9924
E-mail: info@leagueofohiosportsmen.org
Website: www.leagueofohiosportsmen.org

Founded: NA
Scope: Statewide

Description: A representative statewide organization, affiliated with the National Wildlife Federation, dedicated to the protection and enhancement of wildlife and its habitat through public education and government interaction.

Publication(s): Ohio Out Of Doors Magazine

Contact(s):
George Lynch, ALTERNATE REPRESENTATIVE
Marilyn Lieb, ALTERNATE REPRESENTATIVE, TREASURER & EDUCATION PROGRAMS CONTACT
Larry Mitchell, EDITOR & EXECUTIVE DIRECTOR

Larry Mitchell, PRESIDENT & REPRESENTATIVE
Pat Agner, SECRETARY

LEAGUE OF WOMEN VOTERS OF IOWA

P.O. Box 93775
Des Moines, IA 50393-3775 USA
Phone: 641-777-9739
E-mail: vote@iwvia.org

Founded: 1920
Membership: 750
Scope: Statewide

Description: A nonpartisan organization of local chapters and members-at-large, affiliated with the League of Women Voters of the U.S., whose purpose is to promote political responsibility through informed and active participation of citizens in government and to act on selected governmental issues. We promote and support management, preservation, and conservation of our natural resources.

Publication(s): Legislative Newsletter, Iowa Voter

Keyword(s): Air Quality and Pollution, Chemical Pollution Control, Energy Conservation, Environmental Protection, Land Use Planning, Pollution Prevention, Soil Conservation, Solid Waste Management, Water Quality, Politics and Government

Contact(s):
Judie Hoffman, ENVIRONMENTAL COORDINATOR
Phone: 515-292-2660
Gail Quinn, PRESIDENT
Phone: 515-684-8094
Myrna Loehrlein, VICE PRESIDENT
Phone: 319-365-3199

LEAGUE OF WOMEN VOTERS OF THE U.S.

1730 M St., NW
Washington, DC 20036 USA
Phone: 202-429-1965   Fax: 202-429-0854
E-mail: lwv@lwv.org
Website: www.lwv.org

Founded: 1920
Scope: National

Description: Nonpartisan organization of 100,000 members located in all 50 states, the District of Columbia, Hong Kong, and the Virgin Islands, working to promote political responsibility through informed and active participation of citizens in government. Takes political action on water and air quality, solid and hazardous waste management, land use, and energy. The League of Women Voters Education Fund carries out educational projects, publishes materials, and arranges conferences on water and energy issues.

Publication(s): National Voter, The

Keyword(s): Communications, Environmental and Conservation Education, Energy, Exotic species, Aquatic nuisance species

Contact(s):
Bob Adams, EDITOR
Nancy Tate, EXECUTIVE DIRECTOR
Carolyn Jenkins, PRESIDENT

LEAGUE OF WOMEN VOTERS OF WASHINGTON

4710 D. Wilsonian, 4710 University Way NE
Seattle, WA 98105 USA
Phone: 206-622-8961   Fax: 206-622-4908
E-mail: lwvwa@lwvwa.org
Website: www.lwvwa.org

Founded: NA
Membership: 2200
Scope: Statewide

Description: The League of Women Voters is a nonpartisan political organization that encourages the informed and active participation of citizens in government and influences public policy through education and advocacy. Any citizen over 18 may become a voting member.

Publication(s): Washington, County Finance, Evaluation of Major Election Methods & Selected Laws, Washingtons Dynamic Forest 1&2, Targeting Tomorrow - Washington's Economy Adjusts to the 90's, Speaking of Ground Water, Washington State Public Port Districts, Public Assistance as Social Policy, Higher Education in Washington State, Gun Control in Washington, The State We're In

Keyword(s): Forests and Forestry, Outdoor Recreation, Public Lands, Rivers, training

Contact(s):
Jean Wells, FIRST VICE PRESIDENT
Judy Hedden, PRESIDENT
Elizabeth Davis, SECOND VICE PRESIDENT
Betsy Greene, SECRETARY
Myra Howrey, TREASURER

LEAGUE TO SAVE LAKE TAHOE

955 Emerald Bay Rd.
South Lake Tahoe, CA 96150 USA
Phone: 530-541-5388   Fax: 530-541-5454
E-mail: info@keeptahoeblue.org
Website: www.keeptahoeblue.org

Founded: 1957
Scope: National

Description: A private, nonprofit corporation dedicated to preserving the environmental balance, scenic beauty, and recreational opportunities of the Lake Tahoe Basin.

Publication(s): Keep Tahoe Blue

Keyword(s): Environmental and Conservation Education, Protected Areas, Lakes, Land Use Planning, Coral Reefs

Contact(s):
Rochelle Nason, EXECUTIVE DIRECTOR
Tom Mertens, PRESIDENT
Steffi Mooers, SECRETARY
William Callender, TREASURER
Tom Bates, VICE PRESIDENT
Adolphus Andrews, VICE PRESIDENT
John Jorgenson, VICE PRESIDENT

LEARNING FOR ENVIRONMENTAL ACTION PROGRAMME (LEAP)

OISE/UT, Rm. 7-115, 252 Bloor St. W.
Toronto, Ontario M5S 1V6 Canada
Phone: 416-923-6641   Fax: 416-926-4749

Founded: 1990
Scope: International

Description: LEAP is an international program which develops and supports environmental adult education through research, community programs, workshops, seminars and policy work.

Publication(s): Case Studies in Environmental Adult and Popular Education, The Nature of Transformation: Environmental Adult Education, second ediition, Pachamara (International Newsletter), Convergence on Environmental Adult Education, 2000, Convergence on Environmental Adult Education, 1995 -

Keyword(s): Education, Environmental Communication, Environmental and Conservation Education, Environmental Justice, Environmental Literacy, Networking, Women in the Environment, Public Participation, Research, Overconsumption

Contact(s):
Darlene Clover, INTERNATIONAL COORDINATOR
dclover@oise.utoronto.ca

LEGACY INTERNATIONAL

1020 Legacy Drive
Bedford, VA 24523 USA
Phone: 540-297-5982   Fax: 540-297-1860
E-mail: mail@legacyintl.org
Website: www.legacyintl.org

Founded: 1979
Membership: 10
Scope: National

Description: Legacy International is a nonprofit, private voluntary organization serving public and private organizations facing the need to manage change. Legacy's expertise and experience bring about practical designs. Creative and innovative projects use interdisciplinary teams, extended networks, public and private partnerships, and citizen exchanges. Legacy has achieved international recognition for its accomplishments in environmentally sound development, conflict resolution, curriculum design, and leadership training.

Publication(s): Global Youth Village, Organization Brochure

Keyword(s): Environmental and Conservation Education, Environmental Planning, Planning Management, Sustainable Development, Youth Organizations

Contact(s):
Mary Helmig, DIRECTOR
1020 Legacy Drive, Bedford, VA 24523
Phone: 703-297-5982

LEGAL ENVIRONMENTAL ASSISTANCE FOUNDATION INC. (LEAF)

1114 Thomasville Rd., Suite E
Tallahassee, FL 32303-6290 USA
Phone: 850-681-2591   Fax: 850-224-1275
Website: www.leaf-envirolaw.org

Founded: 1979
Scope: Regional

Description: LEAF is a charitable public-interest environmental law firm that protects human health from pollution. We provide legal and technical assistance to citizens and grassroots organizations in Florida, Georgia, and Alabama. LEAF is a membership organization and provides assistance and services free of charge.

Publication(s): various educational documents. Citizens may write for a publications list., LEAF BRIEFS quarterly newsletter

Keyword(s): Energy, Environmental Law, Energy Conservation, Coral Reefs, Solar Energy, Environmental Protection, Pesticides, Environmental Justice, Pollution Prevention, Rivers, Solar Energy

Contact(s):
Jim Presswood, ENERGY PROGRAM
jpresswood@leaf-enviro-law.org
Deb Swim, ENERGY PROGRAM ATTORNEY
dswim@leaf-enviro-law.org
David Ludder, GENERAL COUNSEL
dludder@leaf-enviro-law.org
B. Ruhl, PRESIDENT
Aliki Moncrief, STAFF ATTORNEY
amoncrief@leaf-enviro-law.org
Jeanne Zokovitch, STAFF ATTORNEY
jzokovitch@leaf-enviro-law.org
Cynthia Valencic, VICE PRESIDENT
cvalencic@leaf-enviro-law.org
Larry Thompson, VICE PRESIDENT
lthompson@leaf-enviro-law.org

LIFE OF THE LAND
76 North King St., Suite 203
Honolulu, HI 96817 USA
Phone: 808-533-3454   Fax: 808-533-0993

Founded: NA
Membership: 1,000
Scope: Statewide

Description: To preserve and protect the life of the land through promoting sustainable energy and land use policies and to promote open government through research, education, advocacy and, when necessary, litigation.

Keyword(s): Sustainable Energy, Land Use Planning

LIGHTHAWK
The Presidio of San Francisco, P.O. Box 29231
San Francisco, CA 94129-0231 USA
Phone: 415-561-6250   Fax: 415-561-6251
E-mail: sfo@lighthawk.org
Website: www.lighthawk.org

Founded: 1979
Scope: International

Description: LightHawk's mission is to use aerial education and advocacy, harnessing the power of flight, to defend the environment. LightHawk's staff and volunteer pilots conduct aerial missions with key decision-makers, media representatives, community leaders and conservation groups, illuminating critical environmental concerns by flying over and into lands otherwise inaccessible. LightHawk operates regional programs in the Pacific Northwest, British Columbia, California, the Rocky Mountains, Mexico and Central America.

Publication(s): Lighthawk Newsletter

Keyword(s): Aquatic Habitats, Ancient Forests, Biodiversity, Birds, Coasts, Endangered Species, Environment, Conservation, Environmental and Conservation Education, Environmental Planning, Environmental Protection, Forest Management, Protected Areas, Geography

Contact(s):
Marty Fujita, EXECUTIVE DIRECTOR
Michael Azeez, PRESIDENT
Blaine Townsend, SECRETARY
Patricia Farrar, TREASURER

LIGHTHAWK
NORTHERN ROCKY MOUNTAIN FIELD OFFICE
31845 Frontage Rd.
Bozeman, MT 59715 USA
Phone: 406-586-8572   Fax: 406-585-7835
E-mail: sarahd@lighthawk.org

Founded: NA
Scope: Regional

Contact(s):
Sarah Deopscine, PROGRAM COORDINATOR

LIGHTHAWK
NORTHWEST FIELD OFFICE
2915 E. Madison St., Suite 306
Seattle, WA 98112 USA
Phone: 360-344-3550   Fax: 360-301-4253
E-mail: susen@lighthawk.org
Website: www.lighthawk.org

Founded: NA
Scope: Regional

Contact(s):
Susen Seth, PROGRAM MANAGER
susen@lighthawk.org

LIGHTHAWK
SOUTHERN ROCKY MOUNTAIN FIELD OFFICE
311 Unit K AABC
Aspen, CO 81611 USA
Phone: 970-925-6987   Fax: 970-925-2701
E-mail: micheleg@lighthawk.org
Website: www.lighthawk.org

Founded: 1979
Scope: International

Contact(s):
Bruce Gordon, ASSOCIATE EXECUTIVE DIRECTOR

LONG LIVE THE KINGS
19435 184th Pl., NE
Woodinville, WA 98072 USA
Phone: 206-382-9555
Website: www.longlivethekings.org/home.html

Founded: 1985
Scope: Regional

Description: To rebuild wild salmon populations in specific Northwest Rivers and to enhance their habitat. We are supported by foundations, corporations, individuals, Indian tribes, and fishing and environmental organizations. We are not a membership group.

Publication(s): Long Live the Kings Newsletter

Keyword(s): Aquatic Habitats, Endangered Species, Wildlife, Rivers, Exotic species, Aquatic nuisance species

Contact(s):
Jim Youngren, CHAIRMAN OF THE BOARD
John Sayre, EXECUTIVE DIRECTOR

LOUISIANA ASSOCIATION OF CONSERVATION DISTRICTS

Attn: President, 663 Holmes Rd.
Keatchie, LA 71046 USA
Phone: 318-933-5375

Founded: NA
Scope: Statewide

Keyword(s): Conservation Districts

Contact(s):
John Compton, BOARD MEMBER
6267 Moss Side Ln., Baton Rouge, LA 70808
Phone: 225-766-7979
Jerry Holmes, PRESIDENT
663 Holmes Rd., Keatchie, LA 71046
Phone: 318-933-5375
Fax: 318-872-3178
Charles Dupuy, SECRETARY/TREASURER AND BOARD MEMBER
313 N. Monroe St, Ste 4, Marksville, LA 71351
Phone: 318-253-7603
Fax: 318-253-8890
John Woodward, VICE PRESIDENT, BOARD MEMBER
1902 Savanne Rd., Houma, LA 70360
Phone: 504-879-3528
Fax: 504-876-5267

LOUISIANA AUDUBON COUNCIL

355 Napoleon St.
Baton Rouge, LA 70802-5955 USA
Phone: 225-346-8761   Fax: 225-338-9806

Founded: 1989
Membership: 15
Scope: Statewide

Description: To implement the Audubon cause in Louisiana on issues of statewide concern; coordinate activities among the Audubon chapters in Louisiana; and advocate on behalf of birds, wildlife and their habitat.

Keyword(s): Birds, Endangered Species, Nongame Wildlife, Wetlands, training

Contact(s):
Doris Falkenheiner, PRESIDENT
Esther Boykin, SECRETARY
Clyde Mattison, TREASURER
Donna Lafleur, VICE PRESIDENT

LOUISIANA B.A.S.S. CHAPTER FEDERATION

Attn: President, 603 Terri Dr.
Luling, LA 70070 USA
Phone: 504-785-9069
E-mail: kevgobear@aol.com

Founded: NA
Scope: Statewide

Description: An organization of Bassmaster chapters, affiliated with the Bass Anglers Sportsman Society, organized to fight pollution, assist state and national conservation agencies in

their efforts, and teach the young people of our country good conservation practices. Dedicated to the realistic conservation of our water resources.

Contact(s):
Will Courtney, CONSERVATION DIRECTOR
4548 Chelsea Dr., Baton Rouge, LA 70809
Phone: 225-923-1908
Kevin Gaubert, PRESIDENT
Phone: 504-785-9069

LOUISIANA FORESTRY ASSOCIATION

P.O. Drawer 5067
Alexandria, LA 71307 USA
Phone: 318-443-2558   Fax: 318-443-1713
E-mail: lfa@laforestry.com
Website: www.laforestry.com

Founded: 1947
Membership: 3000
Scope: Statewide

Description: Trade-supported association whose purpose is the conservation of the state's forest land and the promotion of the products and services derived therefrom.

Publication(s): Louisana Logger, Forests and People

Keyword(s): Forests and Forestry, Research, Scholarships and Grants, training

Contact(s):
Charles Vandersteen, EXECUTIVE DIRECTOR
Clyde Todd, STAFF FORESTER

LOUISIANA WILDLIFE FEDERATION, INC.

P.O. Box 65239
Baton Rouge, LA 70896-5239 USA
Phone: 225-344-6762   Fax: 225-344-6707
E-mail: lawildfed@aol.com

Founded: NA
Scope: Statewide

Description: A representative statewide organization, affiliated with the National Wildlife Federation, dedicated to the protection and enhancement of wildlife and its habitat through public education and government interaction.

Publication(s): Louisiana Wildlife Federation Magazine

Contact(s):
Kathy Wascom, EDUCATION PROGRAMS CONTACT
Randy Lanctot, EXECUTIVE DIRECTOR AND EDITOR
Keith Saucier, PRESIDENT AND ALTERNATIVE REPRESENTATIVE
Edgar Veillon, REPRESENTATIVE
Eugene Dauzat, TREASURER

LOWER MISSISSIPPI RIVER CONSERVATION COMMITTEE

2524 S. Frontage Rd., Suite C
Vicksburg, MS 39180-5269 USA
Phone: 601-629-6602   Fax: 601-636-9541
Website: www.lmrcc.org

Founded: NA
Scope: Regional

Description: The Committee provides an organizational structure and forum for coordinating and facilitating cooperative activities involving the natural resources of the Lower Mississippi River. Also encourages sustainable use of Lower Mississippi River natural resources for long-term environmental, social, and economic benefits.

Publication(s): LMRCC Newsletter, The

Keyword(s): Wildlife, Land Use Planning, Rivers, Sustainable Ecosystems, Water Quality

Contact(s):
Dugan Sabins, CHAIRMAN
P.O. Box 82178, Baton Rouge, Louisiana 70884
Phone: 225-765-0246
Fax: 225-765-0617
dugans@deq.state.la.us
Ron Nassar, COORDINATOR
2524 S. Frontage Rd., Ste. C, Vicksburg, MS 39180-5269
Phone: 601-629-6602

M

MACBRIDE RAPTOR PROJECT

W.H., KCC, 6301 Kirkwood Blvd., SW
Cedar Rapids, IA 52406 USA
Phone: 319-398-5495   Fax: 319-398-5611
E-mail: iaraptor@avalon.net
Website: www.ai-design.com/stargig/raptor/ai/main.html

Founded: 1985
Membership: 500
Scope: Statewide

Description: The Macbride Raptor Project is devoted to the preservation of Iowa's birds of prey and their natural habitats through rehabilitation of sick or injured raptors, education of the public to the role of raptors in our environment, and research on various aspects of raptor biology.

Publication(s): Raptor Review

Keyword(s): Raptors, accreditation

Contact(s):
Gail Dawson, ASSISTANT DIRECTOR
Phone: 319-398-4979
Jodeane Cancilla, DIRECTOR
Eric Burrough, VETERINARIAN
Phone: 319-398-4979
Mary Ebert, VETERNARIAN
Phone: 319-398-5495
Gail Dawson, VOLUNTEER COORDINATOR

MAGIC

P.O. Box 15894
Stanford, CA 94309 USA
Phone: 650-323-7333   Fax: 650-323-4233
E-mail: magic@ecomagic.org
Website: www.ecomagic.org

Founded: 1979
Scope: Regional

Description: Magic's programs apply methods and principles of ecology to clarify values, improve health, increase cooperation, and steward the environment. Activities include lectures and seminars about the nature of value; life-planning workshops, swim, run, and hatha yoga instruction; mentoring, community organizing, habitat enhancement, water and land, resource planning; neighborhood design and publishing.

Publication(s): Liveable City, Oak Regeneration on Stanford Lands, Human Ecology, A Science for Living Well

Keyword(s): Ecology, Environmental and Conservation Education, EcoAction, Health and Nutrition, Sustainable Ecosystems, Exotic species, Aquatic nuisance species, Urban Environment, training

Contact(s):
Robin Bayer, PRESIDENT
robin@ecomagic.org
David Schrom, TREASURER
david@ecomagic.org

MAINE ASSOCIATION OF CONSERVATION COMMISSIONS (MACC)

P.O. Box 702
Bath, ME 04330 USA
Phone: 207-443-2925   Fax: 207-443-6913
E-mail: macc@clinic.net

Founded: 1969
Membership: 80 commissions
Scope: Statewide

Description: A membership organization whose objectives are twofold: to assist Maine municipalities in establishing conservation commissions; to assist the existing 200+ conservation commissions through technical assistance and educational programs.

Publication(s): Grass Roots

Keyword(s): Environmental and Conservation Education, Environmental Planning, Environmental Protection, Solid Waste, Exotic species, Aquatic nuisance species

Contact(s):
Bob Cummings, EXECUTIVE DIRECTOR
Mike Cline, PRESIDENT

MAINE ASSOCIATION OF CONSERVATION DISTRICTS

Attn: President, 2467 Exeter Rd.
Exeter, ME 04435-3107 USA
Phone: 207-622-7589

Founded: NA
Scope: Statewide

Keyword(s): Conservation Districts

Contact(s):
Raymond Harris, BOARD MEMBER
Rt. 1 Box 8396, Washburn, ME 04786
Phone: 207-764-3217
William Dell, EXECUTIVE DIRECTOR
P.O. Box 228, Augusta, ME 04330
Phone: 207-622-4443
Fax: 207-623-3748
newengag@mint.net
John Hemond, PRESIDENT
46 N. Verreill Rd, Minot, ME 04258
Phone: 207-345-5333
Neil Crane, PRESIDENT, ALTERNATE BOARD MEMBER
Phone: 207-379-2641
Fax: 207-379-2644
Larry Macdonald, SECRETARY
Box 1187, Greenville, ME 04441
Phone: 207-695-2639

Fred Hardy, TREASURER
879 Weeks Mill Rd., New Sharon, ME 04955
Phone: 207-778-4320
Bruce Roope, VICE PRESIDENT

MAINE AUDUBON SOCIETY
20 Gilsland Farm Rd.
Falmouth, ME 04105 USA
Phone: 207-781-2330   Fax: 207-781-6185
E-mail: maineaudubon@maineaudubon.org
Website: www.maineaudubon.org

Founded: 1843
Scope: Statewide

Description: Dedicated to the protection, conservation, and enhancement of Maine's ecosystems through the promotion of individual understanding and actions. Programs focusing on forest conservation, endangered and threatened species protection, wildlife and wildlife habitats, grassroots activism, environmental education, and school curriculum enhancement. Nature day camp, field trip and world tour program, store, and 13 sanctuaries.

Publication(s): Habitat: Journal of the Maine Audubon Society

Keyword(s): Coasts, Endangered Species, Environmental and Conservation Education, Forests and Forestry, education

Contact(s):
Carol Hammond, EDITOR
chammond@maineaudubon.org

MAINE B.A.S.S. CHAPTER FEDERATION
Attn: President, 39 Vokes Drive
Trenton, ME 04605 USA
Phone: 207-266-6914

Founded: NA
Scope: Statewide

Description: An organization of Bassmaster chapters, affiliated with the Bass Anglers Sportsman Society, organized to fight pollution, assist state and national conservation agencies in their efforts, and teach young people of our country good conservation practices. Dedicated to the realistic conservation of our water resources.

Publication(s): Federation Guide

Contact(s):
Norm Moulton, PRESIDENT
Phone: 207-266-6914

MAINE COAST HERITAGE TRUST
1 Main St.
Topsham, ME 04086 USA
Phone: 207-729-7366   Fax: 207-729-6863
E-mail: info@mcht.org
Website: www.mcht.org

Founded: 1970
Membership: 29
Scope: Statewide

Description: To protect land that is essential to the character of Maine, in particular its coastline and islands. Provides free advisory services on open-space protection to landowners, town officials, state and federal agencies, land trusts, and other private conservation organizations.

Publication(s): Directory of Maine Land Conservation Trusts, Conservation Options, A Guide For Maine Landowners, Annual Report, Maine Heritage, Technical Bulletins

Keyword(s): Coasts, Islands, Land Preservation, Natural Areas, Open Space

Contact(s):
Harold Woodsum, CHAIRMAN
Chris Hamilton, EDITOR
chamiltion@mcht.org
James Espy, PRESIDENT
jespy.mcht.org
John Robinson, TREASURER

MAINE ENVIRONMENTAL EDUCATION ASSOCIATION
485 Chewonki Neck Rd.
Wiscasset, ME 04578 USA
Phone: 207-882-7323   Fax: 207-882-4074

Founded: 1981
Membership: 150
Scope: Statewide

Description: The Maine Environmental Education Association (MEEA) Facilitates and promotes environmental education in Maine through the sharing of ideas, resources, information and cooperative programs among educators, organizations and concerned individuals. MEEA offers a newsletter, annual conference, Environmental Educator of the Year award and Teacher Mine Grants.

Publication(s): Connections, New England Journal of Environmental Education

Keyword(s): Environmental and Conservation Education, education

Contact(s):
Dot Lamson, PRESIDENT
Phone: 207-882-7323
dlamson@chewonki.org

MANASOTA-88
P.O. Box 14119
Bradenton, FL 34280 USA
Phone: 941-966-6256

Founded: 1968
Scope: Statewide

Keyword(s): Biodiversity, Environmental Law, Nuclear/Radiation, Toxic Substances, Nuclear-free, Water quantity, Water export and diversion, Sustainable Ecosystems

Contact(s):
Glenn Compton, CHAIRMAN
419 Reubens Drive, Nokomis, FL 34275
Phone: 941-966-6256
Rebecca Eger, DIRECTOR
Phone: 941-366-1765
Laurence Quy, DIRECTOR
1619 Palma Sola Blvd., Bradenton, FL 34209
Phone: 941-792-5509
Glenn Compton, EDITOR

MANITOBA NATURALISTS SOCIETY

401-63 Albert St.
Winnipeg, Manitoba R3B 1G4 Canada
Phone: 204-943-9029   Fax: 204-943-9029
E-mail: mns@escape.ca
Website: www.manitobanature.ca

Founded: 1920
Membership: 1500
Scope: Regional

Description: Fosters an awareness and appreciation of the natural environment and an understanding of humanity's place therein; and sponsors lectures, workshops, field trips on natural history topics, and recreational outings that are environmentally friendly.

Publication(s): Manitoba's Tall Grass Prairie: A Field Guide to an Endangered Space, Bulletin, The Birds of Southeastern Manitoba, Wings Along Winnipeg, The Wild Plants of Birds Hill Park, The Wild Plants of the Great Plains

Keyword(s): education, Outdoor Recreation

Contact(s):
Gordon Fardoe, EXECUTIVE DIRECTOR
Phone: 204-943-9029
Larry De March, PRESIDENT
Phone: 204-943-9029

MANITOBA WILDLIFE FEDERATION

70 Stevenson Rd.
Winnipeg, Manitoba R3H 0W7 Canada
Phone: 204-633-5967   Fax: 204-632-5200
E-mail: mwf@mb.sympatico.ca
Website: www.mwf.mb.ca

Founded: 1944
Scope: Statewide

Description: Promotes conservation, safety, and good sportsmanship. Manages the Habitat Trust Fund which secures critical land to ensure habitat for wildlife. Protects the interests of anglers and hunters.

Publication(s): Wildlife Crusader/Outdoor Edge

Keyword(s): training

Contact(s):
Randy Walker, PAST PRESIDENT
Lloyd Lintott, PRESIDENT
Darlene Garnham, SECRETARY

MANOMET CENTER FOR CONSERVATION SCIENCES

P.O. Box 1770
Manomet, MA 02345-1770 USA
Phone: 508-224-6521   Fax: 508-224-9220
E-mail: info@manomet.org
Website: www.manomet.org

Founded: 1969
Scope: International

Description: Manomet is a non-profit conservation research institute dedicated to promoting informed conservation policy and natural resource management through applied research. At study sites throughout the Americas, Manomet scientists and volunteers monitor migrant songbird and shorebird populations, identify critical wetlands habitats, design fisheries conservation and management strategies, and develop plans for sustainable management of temperate and tropical forest ecosystems. Manomet's environmental education program serves as the interpretive link between our research work and the information needs of the public. Manomet is membership supported.

Publication(s): Conservation Sciences, Various articles and books

Keyword(s): Birds, Endangered Species, Wildlife, Land Use Planning, Research, Coasts, Ecology, Biodiversity, Environment, Forest Management, Internships, Natural Areas, Nongame Wildlife, Pesticides

Contact(s):
Jeptha Wade, CHAIR
Jennie Robbins, OFFICE MANAGER
jrobbins@manomet.org
Linda Leddy, PRESIDENT AND DIRECTOR
lleddy@manomet.org

MARIN CONSERVATION LEAGUE

55 Mitchell Blvd., Suite 21
San Rafael, CA 94903 USA
Phone: 415-472-6170   Fax: 415-472-1404
E-mail: mcl@conservationleague.org
Website: www.conservationleague.org

Founded: 1934
Scope: Statewide

Description: The Marin Conservation League has worked to preserve and protect the natural assets of Marin County. The league works on all issues affecting the county environment, seeking partnerships with diverse groups to influence public policy and educate citizens and decisionmakers in understanding critical issues and options.

Publication(s): MCL News

Keyword(s): Agriculture, Environmental and Conservation Education, Protected Areas, Land Use Planning, Public Lands

Contact(s):
Jim Goodwin, 1ST VICE PRESIDENT
Jana Haehl, 2ND VICE PRESIDENT
Tom Hinman, EXECUTIVE DIRECTOR
Susan Stompe, PRESIDENT
Robert Berner, SECRETARY
Kenneth Drexler, TREASURER

MARINE CONSERVATION BIOLOGY INSTITUTE

15806 NE 47th Ct.
Redmond, WA 98052-5208 USA
Phone: 425-883-8914   Fax: 425-883-8017
Website: www.mcbi.org

Founded: 1996
Membership: 5
Scope: National

Description: MCBI is a nonprofit, non-partisan, tax-exampt organization dedicated to advancing the multidisciplinary science of marine conservation biology. MCBI helps scientists to generate information that arms people with knowledge crucial for informed decision-making.

Keyword(s): Biodiversity, Marine Protected Areas, Seabed Disturbance

Contact(s):
Elliott Norse, PRESIDENT
William Chandler, VICE PRESIDENT
Phone: 702-465-5959
bill@mcbi.org

MARINE ENVIRONMENTAL RESEARCH INSTITUTE (MERI)

772 W. End Ave.
New York, NY 10025 USA
Phone: 212-864-6285   Fax: 212-864-1470
E-mail: meri@downeast.net
Website: www.merireserch.org

Founded: 1990
Membership: 400
Scope: National

Description: MERI is a nonprofit organization dedicated to protecting the health and biodiversity of the marine environment. MERI's programs are international in scope and include direct field research, environmental and conservation education, training, and collaboration with the world's scientific community. MERI strives to address the problems of global marine pollution, endangered species and habitat degradation, and environmental emergencies affecting marine life.

Publication(s): MERI Resource Center News, research publications, MERI News

Keyword(s): Biodiversity, Endangered Species, Environmental and Conservation Education, Marine Mammals, Whale, Dolphin, Seal

Contact(s):
Lemuel Evans, CHAIRMAN
3536 Paintwater Pl., Las Vegas, NV 89129-7338
Susan Shaw, PRESIDENT
P.O. Box 179, Brooklin, ME 04616
Elizabeth Petterson, RESOURCE CENTER DIRECTOR
MERI Resource Center, Main St., P.O. Box 300, Brooklin, ME 04616
Phone: 207-359-8078
Fax: 207-359-8079
meri@downeast.net
Joan Koven, SECRETARY
Astrolabe Inc., 4812 V St. NW, Washington, DC 20007
Pamela Stacey, TREASURER
Suzanne Hopkins, VICE PRESIDENT
15200 Old York Road, Monkton, MD 21111

MARINE MAMMAL CENTER, THE

Marin Headlands 1065 Ft. Cronkhite
Sausalito, CA 94965 USA
Phone: 415-289-7325   Fax: 415-289-7333
E-mail: com@tmmc.org
Website: www.tmmc.org

Founded: 1975
Membership: 35000
Scope: National

Description: The Marine Mammal Center is a nonprofit organization licensed to rescue and rehabilitate sick, injured, and orphaned marine mammals that strand along the northern and central California coast. Information derived from routine medical treatment is shared with scientists worldwide. Through education programs, the Center promotes public awareness of the ocean environment among over 100,000 visitors annually.

Publication(s): Annual Report, various scientific papers., Release

Keyword(s): Environmental and Conservation Education, Marine Mammals, Whale, Dolphin, Seal, accreditation, standards

Contact(s):
Dennis Didomenico, CHAIR
Phone: 415-289-7335
B.J. Griffin, EXECUTIVE DIRECTOR
Sheldon Wolfe, TREASURER
Steefel, Levitt, & Weiss, One Embarcadero Center, 30th Fl., San Francisco, CA 94111-3784
Phone: 415-788-0900

MARINE TECHNOLOGY SOCIETY

1828 L St., NW, Suite 906
Washington, DC 20036-5104 USA
Phone: 202-775-5966   Fax: 202-429-9417
E-mail: mtspubs@aol.com
Website: www.mtsociety.org

Founded: 1963
Scope: International

Description: An ocean-oriented, multidisciplinary, international professional society, formed to encourage the development of the technology, education, operational expertise, and public awareness needed to advance man's capability to work effectively in all ocean areas and depths.

Publication(s): MTS Newsletter Currents, various proceedings, Marine Technology Society Journal

Keyword(s): Engineering, Oceanography, Exotic species, Aquatic nuisance species

Contact(s):
Judith Krauthamer, EXECUTIVE DIRECTOR

MARYLAND ASSOCIATION OF CONSERVATION DISTRICTS

53 Falma Rd
Edgewater, MD 21037 USA
Phone: 410-956-5771   Fax: 410-956-0161

Founded: NA
Membership: 2
Scope: Local
Keyword(s): Conservation Districts

Contact(s):
Lynne Hoot, EXECUTIVE DIRECTOR
53 Slama Rd., Edgewater, MD 21037
Phone: 410-956-5771
Fax: 410-956-0161
Robert Wilson, PRESIDENT AND ALTERNATIVE BOARD MEMBER
Sharon Mariaca, SECRETARY
Donald Spickler, TREASURER AND COUNCIL MEMBER
14854 Hicksville Rd., Clear Spring, MD 21722
Phone: 301-842-2534
Fax: 301-842-2534
dspick@erols.com
Robert Fitzgerald, VICE PRESIDENT
27570 Fitzgerald Rd., Princess Anne, MD 21853
Phone: 410-651-3701

MARYLAND B.A.S.S. CHAPTER FEDERATION

Attn: President, 1106 West Washington St.
Hagerstown, MD 21740 USA
Phone: 301-791-3724
Website: www.mdbass.com

Founded: NA
Scope: Statewide

Description: An organization of Bassmaster chapters, affiliated with the Bass Anglers Sportsman Society, organized to fight pollution, assist state and national conservation agencies in their efforts, and teach the young people of our country good conservation practices. Dedicated to the realistic conservation of our water resources.

Publication(s): Maryland State Federation Update

Contact(s):
Ken Penrod, CONSERVATION DIRECTOR
4708 Sellman Rd, Beltsville, MD 20705
Phone: 301-937-0010
Jim Kline, PRESIDENT
Phone: 301-791-3724

MARYLAND CHAPTER, SIERRA CLUB

MARYLAND CHAPTER
7338 Baltimore Ave., Suite 101A
College Park, MD 20740-3211 USA
Phone: 301-277-7111   Fax: 301-277-6699
Website: www.sierraclub.org/chapters/md/

Founded: NA
Membership: 1500
Scope: State, National, Intl
Publication(s): see publication web site

Contact(s):
Jon Robinson, CHAPTER CHAIRMAN

MARYLAND FORESTS ASSOCIATION

P.O. Box 599
Grantsville, MD 21536 USA
Phone: 301-895-5369   Fax: 301-895-5369
E-mail: mfa@hereintown.net
Website: www.mdforests.org

Founded: NA
Membership: 550
Scope: Statewide

Description: A nonprofit 501c (3) citizens organization for people interested in trees, forests, related natural resources, and forestry. To promote the maintenance of a healthy and productive forestland base to enhance the economic, environmental, and social well-being of all who live in the state.

Publication(s): MFA Legislative Update, Crosscut, The

Keyword(s): Environment, Environmental and Conservation Education, Forests and Forestry, Renewable Resources, Sustainable Ecosystems

Contact(s):
Karin Miller, EXECUTIVE DIRECTOR
Peter Alexander, PRESIDENT
Richard Stanfield, SECRETARY AND TREASURER
Kevin Simpson, VICE PRESIDENT
Tony Dipaolo, VICE PRESIDENT

MARYLAND NATIVE PLANT SOCIETY

P.O. Box 4877
Silver Spring, MD 20914 USA
Phone: 410-286-2928
Website: www.mdflora.org/index.html

Founded: 1990
Scope: State, National

Description: MNPS is a nonprofit organization that uses education, research, and community service to foster awareness and appreciation for Maryland's native flora and habitats, leading to their conservation.

Publication(s): Marilandica (Journal/Newsletter)

Keyword(s): Biodiversity

Contact(s):
Marc Imlay, DIRECTOR
Phone: 301-283-0808
Karyn Molinas, PRESIDENT
Samuel Jones, SECRETARY
Phone: 410-838-7950
Joseph Metzger Jr., TREASURER
Roderick Simmons, VICE PRESIDENT
Louis Aronica, VICE PRESIDENT
Phone: 202-722-1081

MARYLAND ORNITHOLOGICAL SOCIETY, INC.

Cylburn Mansion, 4915 Greenspring Ave.
Baltimore, MD 21209 USA
Phone: 800-823-0050
Website: www.mdbirds.org http://www.mdbirds.org

Founded: 1945
Membership: 2,400
Scope: Statewide

Description: Nonprofit statewide organization of 16 chapters. Aims to promote the knowledge, protection, and conservation of wildlife and natural resources; to foster appreciation of the natural environment; to establish educational and scientific projects to inform and enrich the public; and to record, evaluate, and publish observations of birdlife in Maryland.

Publication(s): Maryland Yellowthroat, The, Maryland Birdlife

Keyword(s): Birds, Environmental and Conservation Education, education, training

Contact(s):
Chandler Robbins, EDITOR
Patuxent Wildlife Research Center, Laurel, MD 20811
Phone: 301-498-0281
Robert Rineer, PRESIDENT
8326 Philadelphia Rd., Baltimore, MD 21237
Phone: 410-391-8499
Sybil Williams, SECRETARY
2000 Baltimore Rd. A24, Rockville, MD 20851
Phone: 301-762-0560
Jeff Metter, TREASURER
1301 N. Rolling Rd., Catonsville, MD 21228
Phone: 410-788-4877
Norm Saunders, VICE PRESIDENT
1261 Cavendish Rd., Colesville, MD 20905
Phone: 301-989-9035

MASSACHUSETTS ASSOCIATION OF CONSERVATION COMMISSIONS (MACC)

10 Juniper Rd.
Belmont, MA 02478 USA
Phone: 617-489-3930   Fax: 617-489-3935
E-mail: staff@maccweb.org
Website: www.maccweb.org

Founded: 1961
Membership: 3000
Scope: Statewide

Description: Protects wetlands and open space through education and advocacy.

Publication(s): Environmental Handbook for Massachusetts Conservation Commissioners 2000 revised, Newsletter of the Association for members of conservation commissions, government agencies, educational institutions, and environmental organizations, Environmental Handbook for Massachusetts Conservation Commissioners, 1991 Edition

Keyword(s): Environmental and Conservation Education, Protected Areas, Land Purchase, Public Lands, Wetlands

Contact(s):
Ingeborg Hegeman, 1ST VICE PRESIDENT
ingeborg@maccweb.org
Helen Bethell, 2ND VICE PRESIDENT
helen@maccweb.org
Sally Zielinski, EXECUTIVE DIRECTOR
sally@maccweb.org
George Hall, PRESIDENT
Patrick Gerner, SERETARY

MASSACHUSETTS ASSOCIATION OF CONSERVATION DISTRICTS

Attn: President, 25 Shore Rd.
Bourne, MA 02532 USA
Phone: 508-759-4363   Fax: 508-759-4363

Founded: NA
Scope: Statewide
Contact(s):
Thomas Quink, BOARD MEMBER
67 Church St., Gilbertville, MA 01031-9864
Phone: 413-477-8870
Fax: 413-477-8870
Peggy Pacheco, PRESIDENT, ALTERNATE BOARD MEMBER
25 Shore Rd., Bourne, MA 02532
Phone: 508-759-4363
Fax: 508-759-4363
Anne Merriam, SECRETARY
157 State Rd E, Westminster, MA 01473
Phone: 978-874-2432
Donald Lambert, TREASURER
178 Moulton Hill Rd, Monson, MA 01057
Phone: 413-267-4837
Ed Himlan, VICE PRESIDENT
P.O. Box 577, Leaminister, MA 01453
Phone: 978-534-0379
Fax: 978-534-1329

MASSACHUSETTS AUDUBON SOCIETY, INC.

208 S. Great Rd.
Lincoln, MA 01773 USA
Phone: 781-259-9500   Fax: 781-259-8899
E-mail: info@massaudubon.org
Website: www.massaudubon.org

Founded: 1896
Membership: 65000
Scope: Statewide

Description: A nonprofit organization committted to the protection of the environment for people and wildlife. One of the oldest conservation organizations in the world and the largest in New England. Owns and protects 28,000 acres with 13 wildlife sanctuaries across Massachusetts. Programming priorities: conservation, education and advocacy.

Publication(s): Sanctuary

Keyword(s): Birds, Biodiversity, Ecology, Environment, training, Environmental Protection, Land Preservation, education

Contact(s):
John Mitchell, EDITOR
jmitchell@massaudubon.org
Laura Johnson, PRESIDENT
ljohnson@massaudubon.org
Eleanor Pansar, SECRETARY
epansar@massaudubon.org

MASSACHUSETTS B.A.S.S. CHAPTER FEDERATION

Attn: President, 15A Bolton St.
Waltham, MA 02453 USA
Phone: 781-647-5288
Website: www.massbass.com

Founded: NA
Scope: Statewide

Description: An organization of Bassmaster chapters, affiliated with the Bass Anglers Sportsman Society, organized to fight pollution, assist state and national conservation agencies in their efforts, and teach young people of our country good conservation practices. Dedicated to the realistic conservation of our water resources.

Contact(s):
Dean Percival, CONSERVATION DIRECTOR
396 Green St., Northboro, MA 01532
Phone: 508-366-2030
mail@whiznet.com
Joe Mckinnon, PRESIDENT
Phone: 781-647-5288

MASSACHUSETTS DIVISION OF FISHERIES AND WILDLIFE

251 Causeway Street, Suite 400
Boston, MA 02114 USA
Phone: 617-626-1590   Fax: 617-626-1517
E-mail: mass.wildlife@state.ma.us
Website: www.masswildlife.org
Scope: Regional

Contact(s):
Jack Buckley, DEPUTY DIRECTOR
Wayne MacCullum, DIRECTOR

MASSACHUSETTS FORESTRY ASSOCIATION

P.O. Box 1096
Belchertown, MA 01007-1096 USA
Phone: 413-323-7326

Founded: 1970
Scope: Statewide

Description: A voluntary nonprofit association, affiliated with the National Woodland Owners Association; dedicated to conservation, stewardship, and advocacy of the forestland of Massachusetts. An educational organization offering information, workshops, conferences, publications, and professional assistance. Begun in 1970 as the Massachusetts Land League, changed to present name in 1986.

Publication(s): Woodland Steward, The

Keyword(s): Forests and Forestry

Contact(s):
Gregory Cox, EXECUTIVE DIRECTOR, EDITOR
Mary Lees, PRESIDENT
Tim Fowler, SECRETARY AND TREASURER
Hugh Putnam, VICE PRESIDENT

MASSACHUSETTS TRAPPERS ASSOCIATION, INC.

741 Pulaski Blvd.
Bellingham, MA 02019 USA
Phone: 508-883-4214

Founded: 1950
Scope: Statewide

Description: Objectives are to develop leadership for the advancement of the interests of the trapper and the fur industry, and to promote sound management for the conservation of furbearing animals.

Publication(s): Fur Ever

Keyword(s): Mammals, Trapping, Wetlands, training

Contact(s):
David Black, PRESIDENT
741 Pulaski Blvd, Bellingham, MA 02019
Phone: 508-883-4214
Tom Hayes, PUBLIC RELATIONS
155 Williams Rd., Concord, MA 01742
Phone: 508-369-5065
Irene Hayes, SECRETARY
155 Williams Rd., Concord, MA 01742
Phone: 978-369-5065
Debra Benedetto, TREASURER
P.O. Box 60, Wakefield, MA 01880
Phone: 781-246-2136
Frederick Frazier, VICE PRESIDENT EAST
111 Newport Rd., Hull, MA 02045
Phone: 781-925-5841

MATLOCK AND HOLTORF WILDLIFE PARK

226 Tower View Lane
Ozark, MO 65721 USA

MATTS (MID-ATLANTIC TURTLE AND TORTOISE SOCIETY, INC.)

2914 E. Joppa Rd.
Baltimore, MD 21234-3031 USA
Phone: 410-882-2769   Fax: 410-882-0839
Website: www.matts.herptiles.com

Founded: 1997
Scope: National

Description: A nonprofit organization dedicated to promoting the study of Mid-Atlantic chelonian natural history, responsible herpetoculture, and conservation of habitat.

Publication(s): Terrapin Tales, The

Keyword(s): Conservation, education, Zoology, Herpetoculture, Herpetology, Turtles, Tortoises

Contact(s):
Gregory Pokrywka, PRESIDENT
Donald Keefer, SECRETARY
Phone: 410-561-1668
keefercham@aol.com
Brian McLaren, VICE PRESIDENT
Phone: 301-384-7444
briancrcc@aol.com

MAX MCGRAW WILDLIFE FOUNDATION

P.O. Box 9
Dundee, IL 60118 USA
Phone: 847-741-8000   Fax: 847-741-8157
E-mail: mcgrawwild@aol.com

Founded: 1962
Scope: National

Description: Conducts wildlife and fisheries research and management and conservation education projects; cooperates with other conservation agencies and institutions.

Publication(s): Wildlife Management Notes Series, Annual Research Report, Descriptive brochure

Keyword(s): Biodiversity, Wildlife Rehabilitation, Birds, Conservation Tillage, Endangered Species, Protected Areas, Wildlife, Environmental and Conservation Education, Hunting, Internships, Mammals, Natural Areas, National Parks, Land Use Planning, Nongame Wildlife

Contact(s):
John Thompson, DIRECTOR OF RESEARCH

MERCK FOREST AND FARMLAND CENTER

Rte. 315, Box 86
Rupert, VT 05768 USA
Phone: 802-394-7836   Fax: 802-394-2519
E-mail: merck@vermontel.net
Website: www.merckforest.org

Founded: NA
Membership: 2000
Scope: Statewide

Description: Over 3,100 acres of field, farm, and forest open year-round to the public in the heart of the Taconic range in southwestern Vermont. Outdoor and environmental education experiences for individuals, families, and organized groups. Over 28 miles of trails, 65-acre organic demonstration farm, camping cabins and sites, a solar-powered visitor center and sustainable forestry information.

Publication(s): Ridgeline-newsletter

Keyword(s): Solar Energy, Environmental and Conservation Education, Forests and Forestry

Contact(s):
Ken Smith, DIRECTOR
Alan Calfee, PRESIDENT

MICHIGAN ASSOCIATION OF CONSERVATION DISTRICTS
Attn: President, 14302, OP Ave. E.
Climax, MI 49034 USA
201 N Mitchell St., Ste. 301, Cadillac, MI 49601
Phone: 231-876-0348   Fax: 231-871-0372

Founded: NA
Scope: Statewide

Keyword(s): Conservation Districts

Contact(s):
Carol Bogard, ADMINISTRATIVE ASSISTANT
101 S. Main, P.O. Box 539, Lake City, MI 49651
Phone: 616-839-3360
Fax: 616-839-3361
Marilyn Shy, EXECUTIVE DIRECTOR
101 S. Main P.O. Box 539, Lake City, MI 49651
Phone: 616-839-3360
Fax: 616-839-3361
mdistricts@aol.com
Larry Leach, PRESIDENT AND BOARD MEMBER
Phone: 616-746-4648
Fax: 616-746-4393
Rodney Dragicevich, SECRETARY AND TREASURER,
ALTERNATE BOARD MEMBER
29396 Heritage Lane, Paw Paw, MI 49079
Phone: 616-375-3005
Joe Slater, VICE PRESIDENT
6780 Brunswick Rd, Holton, MI 49425
Phone: 616-821-2843

MICHIGAN AUDUBON SOCIETY
6011 W. St. Joseph, Suite 403, P.O. Box 80527
Lansing, MI 48908-0527 USA
Phone: 517-886-9144   Fax: 517-886-9466
E-mail: mas@michiganaudubon.org
Website: www.mas.mi.audubon.org

Founded: 1904
Membership: 10000
Scope: Statewide

Description: The Michigan Audubon Society works to protect the Great Lakes ecosystem for people and wildlife. The society conducts scientific research, educates, and advocates for the protection of species and habitats through five major centers, three affiliate organizations, forty-six chapters, and sanctuaries totalling over 5,000 acres of land.

Publication(s): Michigan Birds & Natural History, Jack-Pine Warbler

Keyword(s): Birds, Environmental and Conservation Education, Land Preservation, EcoAction

Contact(s):
Loretta Gold, 1ST VICE PRESIDENT
143 Lillie Ave., Battle Creek, MI 49015
Harold Prowse, 2ND VICE PRESIDENT
P.O. Box 336, Metamora, MI 48455
Eileen Scamehorn, BUSINESS MANAGER

David Worthington, EDITOR-IN-CHIEF
Julie Craves, EDITOR-IN-CHIEF
Gary Siegrist, PRESIDENT
11772 Trist Rd., Grass Lake, MI 49240
Larry Uhrie, SECRETARY
19057 12 Mile Rd., Battle Creek, MI 49014
Charles Macdonald, TREASURER
945 Tihart, Okemos, MI 48864

MICHIGAN B.A.S.S. CHAPTER FEDERATION
Attn: President, 41970 Jason Drive
Clinton Township, MI 48038 USA
Phone: 810-286-3523   Fax: 810-286-3588
Website: www.michiganbass.org

Founded: NA
Membership: 1000
Scope: Statewide

Description: An organization of Bassmaster chapters, affiliated with the Bass Anglers Sportsman Society, organized to fight pollution, assist state and national conservation agencies in their efforts, and teach the young people of our country good conservation practice. Dedicated to the realistic conservation of our water resources.

Publication(s): Bass Lines

Contact(s):
Ron Spitler, CONSERVATION DIRECTOR
2710 Browning Dr., Lake Orion, MI 48360
Phone: 248-391-4393
Dennis Beltz, PRESIDENT
Phone: 810-286-3523

MICHIGAN ENVIRONMENTAL COUNCIL
119 Pere Marquette
Lansing, MI 48912 USA
Phone: 517-487-9539   Fax: 917-487-9541
E-mail: mlenvcouncil@igc.apc.org

Founded: 1980
Scope: Statewide

Description: A statewide coalition of more than 50 environmental, public health and faith-based organizations with a collective membership of over 175,000 residents. In addition to serving as a clearinghouse of environmental information, MEC develops public policy, educates state officials and the public, and provides technical assistance and support to member organizations.

Publication(s): Land: Michigan's Promise, Michigan's Future, Groundwater at Risk: A Citizen's Guide, Michigan Environmental Report

Keyword(s): Communications, Energy, Land Use Planning, Pollution Prevention, Urban Environment

Contact(s):
Elizabeth Harris, CHAIRMAN
E. Michigan Environmental Action Council 21220 W. 14 Mile Rd., Bloomfield Township, MI 48301-4000
Phone: 313-258-5188
Carol Misseldine, EXECUTIVE DIRECTOR
Lana Pollack, PRESIDENT
Phone: 517-487-9539
Brian Imus, SECRETARY
Phone: 734-662-6597
Alice Austin, VICE CHAIR
Phone: 517-663-2400
Alison Horton, VICE CHAIR

MICHIGAN FORESTS ASSOCIATION

1558 Barrington St.
Ann Arbor, MI 48103-5603 USA
Phone: 734-665-8279   Fax: 734-913-9167
E-mail: mfa@i-star.com
Website: www.mfa.nu

Founded: 1951
Membership: 675
Scope: Statewide

Description: A statewide organization affiliated with the National Woodland Owners Association, with concern for the full spectrum of forest activity, enterprise, development, and conservation in Michigan.

Publication(s): Michigan Forests, Leaves -Newsletter, Green Gold:Michigan Forrest History

Keyword(s): Forests and Forestry

Contact(s):
Don Ingle, EDITOR
P.O. Box 78, Baldwin, MI 49304-0078
McClain Smith, EXECUTIVE DIRECTOR
Gordon Terry, PRESIDENT
Allan Kerton, TREASURER
Collin Burnett, VICE PRESIDENT

MICHIGAN LAND USE INSTITUTE

P.O. Box 228
Benzonia, MI 49016 USA
Phone: 231-882-4723   Fax: 213-882-7350
Website: www.mlui.org

Founded: 1995
Membership: 2200
Scope: Statewide

Description: Michigan Land Use Institute is a nonprofit environmental economic policy research organization focused on reforming land use policy and curbing sprawl.

Publication(s): Great Lakes Bulletin, Rivers at Risk, Benzie County Wetlands - A Resource Worth Protecting

Keyword(s): Environment, Environmental and Conservation Education, Environmental Protection

Contact(s):
Hans Voss, EXECUTIVE DIRECTOR
Keith Schneider, PROGRAM DIRECTOR
Richard Hitchingham, TREASURER

MICHIGAN NATURAL AREAS COUNCIL

University of Michigan, Botanical Gardens, 1800 N. Dixboro Rd.
Ann Arbor, MI 48105 USA
Phone: 313-461-9390
E-mail: mnac@cyberspace.org
Website: www.cyberspace.org

Founded: 1947
Scope: Statewide

Description: The Michigan Natural Areas Council promotes the preservation of outstanding natural areas, prepares reports based on field investigations, and serves as an informed-citizens advisory on such matters.

Publication(s): Michigan Natural Areas News and Views (natural areas, endangered species, and relevant conservation news

Keyword(s): Endangered Species, Flowers, Plants, and Trees, Islands, National Parks, Prairies, Wilderness

Contact(s):
Sylvia Taylor, CHAIR
10353 Judd Rd., Willis, MI 48191
Phone: 313-461-9390
smtaylot@umich.edu
Robert Grese, EDITOR
Christopher Graham, TREASURER
725 Peninsula Ct., Ann Arbor, MI 48105
kfdh64@prodigy.com
Robert Grese, VICE CHAIR
1512 Carlton, Ann Arbor, MI 48103
bgrese@umich.edu

MICHIGAN NATURE ASSOCIATION

P.O. Box 102
Avoca, MI 48006-0102 USA
Phone: 810-387-3771
E-mail: mna@greatlakes.net

Founded: 1952
Scope: Statewide

Description: Purpose is to acquire and maintain nature sanctuaries that contain examples of Michigan's original flora and fauna. Holds title to 149 properties totaling 7,702 acres in 52 counties of Michigan. MNA lands contain 206 of Michigan's endangered, threatened, and of special concern species. Available to public for nature education and appreciation.

Publication(s): MNA Nature Sanctuary Guidebook 7th edition, 1993, MNA—In Retrospect: A Celebration of 28 Years of Preserving Michigan's Wild and Rare Natural Lands 1960-1988, Walking Paths in Keweenaw, In Our Trust (1990-91, 30-minute Wildlife Video, Members' Newsletter

Keyword(s): Endangered Species, Land Preservation, Land Purchase, Birds, training, Mammals, Reptiles and Amphibians, Flowers, Plants, and Trees, Insects and Butterflies, Natural Areas, education, Nongame Wildlife, Prairies, National Parks, Wetlands

Contact(s):
Bertha Daubendek, EDITOR
Box 102, Avoca, MI 48006
Bertha Daubendek, EXECUTIVE SECRETARY AND TREASURER
Phone: 810-324-2626
Karen Weingarden, PRESIDENT
Phone: 810-546-5429

MICHIGAN STATE UNIVERSITY DEPT OF FISHERIES & WILDLIFE

EDUCATION SECTION
13 Natural Resources Bldg.
East Lansing, MI 48824 USA
Phone: 517-353-3373   Fax: 517-432-1699
E-mail: WEBMASTER@PERM3.FW.MSU.EDU
Website: www.fw.msu.edu

Founded: NA
Membership: 7500
Scope: Statewide
Publication(s): Publications on website

Contact(s):
Thomas Coon, DEPT. CHAIRPERSON
coontg@msu.edu

MICHIGAN UNITED CONSERVATION CLUBS, INC.

2101 Wood St.
Lansing, MI 48912-3728 USA
Phone: 517-371-1041   Fax: 517-371-1505
E-mail: mucc@mucc.org
Website: www.mucc.org

Founded: 1937
Membership: 100000
Scope: Statewide

Description: A representative statewide organization, affiliated with the National Wildlife Federation, dedicated to the protection and enhancement of wildlife and its habitat through public education and government interaction.

Publication(s): Michingan Out-of-Doors Magazine

Contact(s):
William Whippen, ALTERNATE REPRESENTATIVE
Dennis Knickerbocker, EDITOR
Kevin Frailey, EDUCATION PROGRAMS CONTACT
Jim Goodheart, EXECUTIVE DIRECTOR
Dan Delisle, PRESIDENT
James Campbell, REPRESENTATIVE
Michael Leach, TREASURER

MICHIGAN WILDLIFE HABITAT FOUNDATION

6380 Drumheller Road, P.O. Box 393
Bath, MI 48808 USA
Phone: 517-641-7677   Fax: 517-641-7877
E-mail: wildlife@mwhf.org
Website: www.mwhf.org

Founded: 1982
Membership: 2,500
Scope: Statewide

Description: The Michigan Wildlife Habitat Foundation is a nonprofit membership organization, which restores and improves wildlife habitat through cost-effective projects. We want future generations to enjoy the same world of natural experiences we do today.

Publication(s): Wildlife Volunteer, The

Keyword(s): Aquatic Habitats, Renewable Resources, Terrestrial Habitats, Wetlands, training

Contact(s):
Michael Depolo, CHAIRMAN
Dennis Fijalkowski, EXECUTIVE DIRECTOR
Keith Groty, PRESIDENT

MID-ATLANTIC COUNCIL OF WATERSHED ASSOCIATIONS

12 Morris Rd.
Ambler, PA 19002 USA
Phone: 215-372-3916

Founded: NA
Scope: National

Description: Promotes exchange of ideas on citizen watershed association activities and advises any group wishing to start a new watershed association.

Keyword(s): Environmental and Conservation Education, Land Use Planning, Rivers, Coral Reefs, Exotic species, Aquatic nuisance species

MID-ATLANTIC FISHERY MANAGEMENT COUNCIL

300 S. New St., Rm. 2115
Dover, DE 19904 USA
Phone: 302-674-2331   Fax: 302-674-5399
Website: www.mafmc.org

Founded: 1976
Membership: 11
Scope: National

Description: Mid-Atlantic Fishery Management Council is one of eight regional fishery management councils established to carry out provisions of Magnuson-Stevens Fishery Conservation and Management Act. The Council is charged with responsibility to prepare fishery management plans and amendments to such plans for implmentation by the Secretary of Commerce.

Publication(s): Newsletter

Keyword(s): Marine Fisheries, Wildlife, Marine Mammals, Whale, Dolphin, Seal, Wetlands

Contact(s):
Daniel Furlong, EXECUTIVE DIRECTOR
dfurlong@mafmc.org
Marla Trollan, PUBLIC AFFAIRS
mtrollan@mafmc.org

MINERAL POLICY CENTER

1612 K St., NW, Suite 808
Washington, DC 20006 USA
Phone: 202-887-1872   Fax: 202-887-1875
E-mail: mpc@mineralpolicy.org
Website: www.mineralpolicy.org

Founded: 1988
Membership: 5000
Scope: International

Description: MPC is a national environmental membership organization. The Center is a research, education, and advocacy organization dedicated to cleaning up and preventing pollution from mining. The Center works for common sense environmental reform of mineral policy. The Center produces educational materials on mining impact, offers training for and works closely with, citizens groups affected by mining damage.

Publication(s): Canary Calls / not on web, Mine Wire, Golden Dreams, Poison Streams, MPC News

Keyword(s): Environmental Law, Public Lands, Solid Waste, Toxic Substances, Nuclear-free, Water quantity, Water export and diversion, Coral Reefs, Mining, Mineral Resources

Contact(s):
Alan Septoff, RESEARCH DIRECTOR

MINNESOTA ASSOCIATION OF SOIL AND WATER CONSERVATION DISTRICTS

790 Cleaveland Ave. S.
Ste. 216
St. Paul, MN 55116 USA
Phone: 651-690-9028   Fax: 651-690-9065
E-mail: maswcd@maswcd.org

Founded: NA
Scope: Statewide

Keyword(s): Conservation Districts

Contact(s):
Richard Zupp, PRESIDENT
417 136th St., Pipestone, MN 56164
Phone: 507-825-3024
Fax: 507-825-2855
rlzupp@svtv.com
Leann Buck, EXECUTIVE DIRECTOR
790 Cleveland Ave. S. Suite 216, St. Paul, MN 55116
Phone: 612-690-9028
Fax: 612-690-9065
lbuck@pioneerplanet.infi.net
Scott Hoese, VICE PRESIDENT
5520 Polk Ave., Mayer, MN 55360
Phone: 952-657-2223
sfhoese@aol.com

MINNESOTA B.A.S.S. CHAPTER FEDERATION

Attn: President, P.O. Box 225
Howard Lake, MN 55349 USA
Phone: 612-339-5609
Website: www.mnbf.org

Founded: NA
Membership: 550
Scope: Regional

Description: An organization of Bassmaster chapters, affiliated
with the Bass Anglers Sportsman Society, organized to fight
pollution, assist state and national conservation agencies in
their efforts, and teach the young people of our country good
conservation practices. Dedicated to the realistic conservation
of our water resources.

Contact(s):
Jay Green, PRESIDENT
Phone: 612-339-5609

MINNESOTA CENTER FOR ENVIRONMENTAL ADVOCACY (MCEA)

26 E. Exchange St., Suite 206
St. Paul, MN 55101-2264 USA
Phone: 651-223-5969   Fax: 651-223-5967
E-mail: mcea@mncenter.org
Website: www.mncenter.org

Founded: 1974
Membership: 1550
Scope: Statewide

Description: The Minnesota Center for Environmental Advocacy
is a nonprofit organization that uses law, science, and research
to protect Minnesota's natural resources, wildlife, and the
health of its people.

Publication(s): Advocacy Update

Keyword(s): Feedlots and Pollution, Water Quality, Air Quality and
Pollution, Toxic Reduction, Legal Advocacy, Pesticides

Contact(s):
Steven Thorne, CHAIR
Peter Bachman, EXECUTIVE DIRECTOR

MINNESOTA CONSERVATION FEDERATION

551 S. Snelling Avenue South, Suite B
St. Paul, MN 55116-1525 USA
Phone: 651-690-3077
E-mail: mncf@mtn.org
Website: www.mncf.org

Founded: 1935
Scope: Statewide

Description: A representative statewide organization, affiliated
with the National Wildlife Federation, dedicated to the
protection and enhancement of wildlife and its habitat through
public education and government interaction.

Publication(s): Minnesota Out-of-Doors and Walk on The
Wildside

Contact(s):
Chris Vokaty, ALTERNATE REPRESENTATIVE
Barb Prindle, EDUCATION PROGRAMS CONTACT
Gordy Meyer, PRESIDENT AND REPRESENTATIVE
gmeyer9330@aol.com
Joan Moore, TREASURER

MINNESOTA DIVISION OF THE IZAAK WALTON LEAGUE OF AMERICA

555 Park St., Suite140
St. Paul, MN 55103 USA
Phone: 651-221-0215   Fax: 651-221-0215
E-mail: mn-ikes@mtn.org
Website: www.mtn.org/~mn-ikes

Founded: NA
Membership: 1400
Scope: Statewide
Publication(s): Available on website

Contact(s):
Lee Barthel, PRESIDENT

MINNESOTA FORESTRY ASSOCIATION

P.O. Box 496
Grand Rapids, MN 55744 USA
Phone: 218-326-3000   Fax: 218-326-3224
E-mail: info@mnforest.com
Website: www.mnforest.com

Founded: NA
Membership: 800
Scope: Statewide

Description: A nonprofit organization, affiliated with the National
Woodland Owners Association, dedicated to promoting the
high potential advantages of intensive scientific management
of forests, woodlots, and other renewable resources.

Publication(s): Minnesota Forest newsletter, Minnesota Better
Forests

Keyword(s): Forests and Forestry

Contact(s):
Stephanie Kessler, EXECUTIVE DIRECTOR
Phone: 218-326-3000
mfakessler@yahoo.com
James Lemmerman, PRESIDENT
6316 Nashua St., Duluth, MN 55807
Phone: 218-624-3847

Richard Holter, TREASURER
Phone: 218-328-5173
Culver Adams, VICE PRESIDENT
Phone: 612-823-2618

MINNESOTA GROUND WATER ASSOCIATION

4779 126th St., N.
White Bear Lake, MN 55110-5910 USA
Phone: 651-296-7822   Fax: 651-297-8676

Founded: 1981
Membership: 500
Scope: Statewide

Description: MGWA's mission is to advocate the wise use and protection of gound water, and to provide education to the users of Minnesota's ground water.

Publication(s): Minnesota Ground Water Association Directory, Minnesota Ground Water Association Newsletter

Keyword(s): Environmental and Conservation Education, Environmental Planning, Protected Areas, Ground Water Protection

Contact(s):
Leigh Harrod, ADVERTISING MANAGER
Phone: 651-474-8678
mn_homebase@worldnet.att.net
Jeanette Leete, BUSINESS MANAGER
Phone: 651-426-6122
Fax: 651-426-5449
Tom Clark, EDITOR
Phone: 651-296-8580
Fax: 651-297-7709
tom.p.clark@pca.state.mn.us
James Piegat, PAST PRESIDENT
Phone: 612-470-6075
James Lundy, PRESIDENT
Phone: 651-296-7822
Fax: 651-297-8676
jm.lundy@pca.state.mn.us
Jan Falteisek, SECRETARY AND MEMBERSHIP
Phone: 651-296-3877
Fax: 651-296-0445
jan.falteisek@dnr.state.mn.us
Lee Trotta, TREASURER
Phone: 651-638-3160
Fax: 651-638-3226
trottaLC@usfilter.com

MINNESOTA HERPETOLOGICAL SOCIETY

JAMES FORD BELL MUSEUM OF NATURAL HISTORY
10 Church St., SE, University of Minnesota
Minneapolis, MN 55455-0104 USA
Phone: 612-624-7065

Founded: 1981
Scope: Statewide

Description: A nonprofit organization chartered for the conservation and preservation of reptiles and amphibians, through the education of members and the public.

Publication(s): MHS Newsletter

Keyword(s): Endangered Species, Environmental and Conservation Education, Reptiles and Amphibians, National Parks, Nongame Wildlife

Contact(s):
Bill Moss, PRESIDENT
mngatorguy@qwest.net

MINNESOTA NATIVE PLANT SOCIETY

220 Biological Sciences Center, 1445 Gortner Ave.,
University of Minnesota
St. Paul, MN 55108 USA
Phone: 507-867-4692
E-mail: david.johnson@usfamily.net
Website: www.stolaf.edu/depts/biology/mnps

Founded: 1982
Membership: 306
Scope: Statewide

Description: A nonprofit organization dedicated to education about native Minnesota flora and to its preservation and conservation. Activities include monthly meetings, summer field trips, sponsorship of symposia and publication of a regular newsletter.

Publication(s): Minnesota Plant Press

Keyword(s): Biodiversity, Endangered Species, Environmental and Conservation Education, Protected Areas, Prairies, Native Plants

MINNESOTA ORNITHOLOGISTS' UNION

James Ford Bell Museum of Natural History, 10 Church
St. SE, University of Minnesota
Minneapolis, MN 55455 USA
Phone: 763-780-8890
E-mail: mou@biosci.umn.edu

Founded: 1937
Membership: 1,400
Scope: Statewide

Description: Statewide organization contributing to scientific knowledge through bird observations; stimulating public interest in birds; and working to preserve bird life and bird habitat.

Publication(s): Minnesota Birder, Loon, The

Keyword(s): Wildlife Rehabilitation, Birds, Geography, Nongame Wildlife, Waterfowl

Contact(s):
Jim Williams, EDITOR
5239 Cranberry Lane, Webster, WI 54893
Anthony Hertzel, EDITOR
8461 Pleasant View Dr., Mounds View, MN 55112
Elizabeth Bell, MEMBERSHIP SECRETARY
5868 Pioneer Rd., St. Paul Park, MN 55071
Ann Kessen, PRESIDENT
31145 Genesis Ave., Stacy, MN 55079
Al Batt, RECORDING SECRETARY
RR 1, Box 56A, Hartland, MN 56042
Mark Citsay, TREASURER
210 Mariner Way, Bayport, MN 55003

MINNESOTA PARKS AND TRAILS COUNCIL

275 E. 4th St. 642
St. Paul, MN 55107 USA
Phone: 651-726-2457   Fax: 651-726-2458
E-mail: info@parksandtrails.org
Website: www.mnptc.org

Founded: 1954
Membership: 1,100
Scope: Statewide

Description: The mission of the Council is to further the establishment, development, and enhancement of parks and trails within the state of Minnesota, and to encourage their prudent use and protection.

Publication(s): Newsletter

Keyword(s): Trail, Environmental and Conservation Education, Protected Areas, Land Purchase, Outdoor Recreation, Public Lands

Contact(s):
Dorian Grilley, EXECUTIVE DIRECTOR
Eleanor Winston, PRESIDENT
Alan Ruvelson, SECRETARY
Michael Prichard, TREASURER
Mark Strobel, VICE PRESIDENT
Jeff Olson, VICE PRESIDENT
John Leinen, VICE PRESIDENT

MINNESOTA WILDLIFE HERITAGE FOUNDATION, INC.

5701 Normandale Rd., Suite 325
Minneapolis, MN 55424 USA
Phone: 952-925-1923   Fax: 952-925-3487

Founded: NA
Membership: 400
Scope: Statewide

Description: Formed to promote the idea of charitable giving for conservation purposes and to assist people in making charitable donations of property for wildlife habitat.

Contact(s):
James Mady, PRESIDENT
7338 Frontier Trail, Chanhassen, MN 55317
Laurence Koll, SECRETARY AND LEGAL COUNSEL
633 Sunset Ln., Mendota Heights, MN 55118
Phone: 612-291-9155
Hugh Price, VICE PRESIDENT AND DIRECTOR
5707 Knox Ave S, Minneapolis, MN 55419
Phone: 612-925-2486

MINNESOTA WINGS SOCIETY, INC.

P.O. Box 11323
Minneapolis, MN 55411 USA
Phone: 612-588-2966

Founded: 1978
Scope: Statewide

Description: To present a program to high school students called "Sight and Save Wildlife Management." This program helps students and enables them to improve wildlife habitat around some of the openings they use every day.

Publication(s): Wings (newsletter)

Keyword(s): Birds, Environmental and Conservation Education, Land Preservation, training

Contact(s):
Thurman Tucker, PRESIDENT
1321 N. Irving Ave., Minneapolis, MN 55411
Phone: 612-588-2466
Martin Hanson, SECRETARY
1530 Quinlan Ave., So., St. Croix Beach, MN 55043
Phone: 612-436-8242
Jim McLellan, TREASURER
10273 Yellow Cir. Dr., Minnetonka, MN 55343
Phone: 612-933-2263

David Donna, VICE PRESIDENT
4200 IDS Center, 80 S. 8th St., Minneapolis, MN 55402
Phone: 612-371-3211

MISSISSIPPI ASSOCIATION OF CONSERVATION DISTRICTS, INC.

Jackson, MS 39225-3005 USA

Founded: NA
Scope: Statewide

Keyword(s): Conservation Districts

Contact(s):
Marc Curtis, 1ST VICE PRESIDENT
P.O. Box 958, Leland, MS 38756
Phone: 601-686-2321
Jack Winstead, 2ND VICE PRESIDENT
5337 Lawrence Rd., Lawrence, MS 39336
Daryl Burney, BOARD MEMBER
P.O. Box 603, Coffeeville, MS 38922
Phone: 601-675-2703
Fax: 601-675-2786
Benny Goff, PRESIDENT
Phone: 228-769-3070
Fax: 228-769-3005
Gale Martin, SECRETARY AND TREASURER
P.O. Box 23005, Jackson, MS 39225-3005
Phone: 601-354-7645
Fax: 601-354-6628

MISSISSIPPI B.A.S.S. CHAPTER FEDERATION

295 Country Rd. 4701
Meridian, MS 39301 USA
Website: www.msbass.com

Founded: NA
Scope: Statewide

Description: An organization of Bassmaster chapters, affiliated with the Bass Anglers Sportsman Society, organized to fight pollution, assist state and national conservation agencies in their efforts, and teach the young people of our country good conservation practices. Dedicated to the realistic conservation of our water resources.

Contact(s):
John Hamilton, CONSERVATION DIRECTOR
404 Meadow Lane, Aberdeen, MS 39730
Phone: 662-369-8290

MISSISSIPPI INTERSTATE COOPERATIVE RESOURCE ASSOCIATION

P.O. Box 774
Bettendorf, IA 52722-0774 USA
Phone: 309-793-5811
Website: www.aux.cerc.cr.usgs.gov/MICRA

Founded: 1989
Membership: 200
Scope: Regional

Description: An interstate organization of 28 state departments of conservation and natural resources working in collaboration with federal agencies, Native American tribes, and other interests to improve the conservation, development, management, and utilization of interjurisdictional fishery resources in the Mississippi River basin through improved coordination and communication among the responsible management entities.

Publication(s): Other Periodic Reports, River Crossings

Keyword(s): Wildlife, Rivers, Sport Fishing, Endangered Species, Exotic species, Aquatic nuisance species

Contact(s):
Bill Reeves, CHAIRMAN
Phone: 615-781-6575
Fax: 615-781-6667
breeves@mail.state.tn.us
Norm Stucky, VICE CHAIRMAN
Phone: 573-781-4115
Fax: 573-526-4047
stuckyn@mail.conservation.state.mo.us

MISSISSIPPI NATIVE PLANT SOCIETY

c/o Mississippi Museum Natural Science,
2148 Riverside Dr.
Jackson, MS 39202 USA
Phone: 601-354-7303   Fax: 601-354-7227
Website: www.mdwfp.state.ms.us/museum

Founded: 1981
Scope: Statewide

Description: The Mississippi Native Plant Society promotes the study and use of native and naturalized species of Mississippi, their use in landscaping, the appreciation of natural ecological communities of the state, and the conservation or preservation of these species, habitats and plant associations, using the principles of conservation biology and ecosystem management.

Publication(s): Mississippi Native Plants

Keyword(s): Environmental and Conservation Education, Flowers, Plants, and Trees, Landscape Architecture, education, Native Plants

Contact(s):
Debora Mann, SECRETARY AND TREASURER
Millsaps College, 1701 North State St, Jackson, MS 39210
Phone: 601-974-1415
Fax: 601-974-1401
manndl@millsap.edu

MISSISSIPPI RIVER BASIN ALLIANCE

708 N. First St. Ste 238
Minneapolis, MN 55401 USA
Phone: 612-334-9460   Fax: 612-340-1632
E-mail: mrbaoffice@mrba.org
Website: www.mrba.org

Founded: 1992
Membership: 150
Scope: Regional

Description: To protect and restore the ecological, economic, cultural, historical, and recreational resources in the Basin, and to eliminate barriers of race, class, and economic status which divide us in the quest to achieve these purposes.

Publication(s): Mississippi River Basin Directory, Alliance Newsletter

Keyword(s): People of Color in the Environment, Rivers, Water Quality, Wetlands, Greenways, Navigation

Contact(s):
James Falvey, ASSISTANT DIRECTOR
Tim Sullivan, EXEC. DIR.

MISSISSIPPI WILDLIFE FEDERATION

855 South Pear Orchard Road, Suite 500
Ridgeland, MS 39157-5138 USA
Phone: 601-206-5703   Fax: 601-206-5705
E-mail: mboyd@mswf.org
Website: www.mswildlife.org

Founded: 1946
Membership: 4000
Scope: State

Description: A representative statewide organization, affiliated with the National Wildlife Federation, dedicated to the protection and enhancement of wildlife and its habitat through public education and government interaction.

Publication(s): Mississippi Wildlife Magazine

Contact(s):
Bob Fairbank, ALTERNATE REPRESENTATIVE
Martha Boyd, EDUCATION PROGRAMS CONTACT
Cathy Shropshire, EXECUTIVE DIRECTOR
Marty Brunson, PRESIDENT
Jimmy Bullock, REPRESENTATIVE
Johnny McArthur, TREASURER

MISSOURI ASSOCIATION OF SOIL AND WATER CONSERVATION DISTRICTS

19050 State Hwy O
Tarkio, MO 64491 USA
Phone: 660-736-4368

Founded: NA
Scope: Statewide

Keyword(s): Conservation Districts

Contact(s):
Peggy Lemons, EXECUTIVE SECRETARY
1209 Biscayne Dr., Jefferson City, MO 65109
Phone: 573-893-5188
Fax: 573-893-7328
peggy@mojefferso.fsc.usda.gov
Steve Hopper, PRESIDENT AND BOARD MEMBER
Phone: 660-639-2575
David Dix, TREASURER
P.O. Box 756, Eminence, MO 65466
Phone: 573-226-3787

MISSOURI B.A.S.S. CHAPTER FEDERATION

Attn: President, 220 W. 6th Street
Sedalia, MO 65301 USA
Phone: 660-826-5251
Website: www.MOBASS.COM

Founded: NA
Scope: Statewide

Description: An organization of Bassmaster chapters, affiliated with the Bass Anglers Sportsman Society, organized to fight pollution, assist state and national conservation agencies in their efforts, and teach the young people of our country good conservation practices. Dedicated to the realistic conservation of our water resources.

MISSOURI FOREST PRODUCTS ASSOCIATION

611 E. Capitol Ave., Suite One
Jefferson City, MO 65101 USA
Phone: 573-634-3252   Fax: 573-636-2591
E-mail: moforest@moforest.org
Website: www.moforest.org

Founded: 1970
Membership: 7
Scope: Statewide

Description: The Missouri Forest Products Association is a nonprofit organization committed to promoting closer working relationships among the wood products industry and the conservation and wise use of natural resources.

Publication(s): MFPA News, Professional Timber Harvester

Keyword(s): Conservation Planning, Urban Forestry, Trees,
Training, Sustainable Resources, Forest Management, Forest
Stewardship, Forests and Forestry, Internships, Environmental
Ethics, Conservation

Contact(s):
Cory Ridenhour, EXECUTIVE DIRECTOR
cory@mfpanews.org

MISSOURI NATIVE PLANT SOCIETY

P.O. Box 20073
St. Louis, MO 63144-0073 USA
Phone: 314-894-9021
Website: www.missouri.edu/~umo_herb/monps

Founded: 1979
Scope: Statewide

Description: To promote the enjoyment, preservation, conservation, restoration, and study of the flora native to Missouri; to
educate the public about the values of the beauty, diversity and
environmental importance of indigenous vegetation; and to
publish related information.

Publication(s): Petal Pusher, Missouriensis

Keyword(s): Native Plants, Biodiversity, Endangered Species,
Environment, Flowers, Plants, and Trees, Protected Areas,
Forest Management, Gardening and Horticulture,
Environmental Protection, Natural Areas, education, Public
Lands, Wildlands, National Parks, Prairies

Contact(s):
Pat Harris, EDITOR
Phone: 314-894-9021
pharris@stlnet.com
George Yatskievych, EDITOR
Phone: 314-577-9522
gyatskievych@rschctr.mobot.org
Jack Harris, PRESIDENT
Donna Kennedy, TREASURER
Phone: 636-256-7578
fishn2@primary.net
Sue Hollis, VICE PRESIDENT
Phone: 816-561-9419
serngro@worldnet.att.net

MISSOURI PRAIRIE FOUNDATION

P.O. Box 200
Columbia, MO 65205 USA
Phone: 417-537-4412
E-mail: gfreeman@mail.com.missouri.edu

Founded: 1966
Scope: Statewide

Description: A nonprofit citizens' group organized to ensure the
preservation of native prairie along with associated plant and
animal life by acquisition, management protection, control, and
perpetuation of the prairie; to carry on educational programs;
and to provide scientific research relative to native prairie.

Publication(s): Missouri Prairie Journal

Keyword(s): Prairies, Land Preservation, Land Purchase,
Conservation, Environmental and Conservation Education,
Biodiversity, Endangered Species, Native Plants

Contact(s):
D. Christisen, ADVISOR TO THE BOARD
Carol Davit, EDITOR
c/o MDC, P.O. Box 180, Jefferson City, MO 65102
Phone: 573-751-4115
George Nichols, PRESIDENT
Phone: 417-682-8768
Russell Runge, SECRETARY
Phone: 573-581-8754
John Cline, TREASURER
Phone: 314-581-6566
Elizabeth Garrett, VICE PRESIDENT
Phone: 573-446-3778

MONITOR INTERNATIONAL

300 State St.
Annapolis, MD 21403 USA
Phone: 410-268-5155   Fax: 410-268-8788
E-mail: info@monitorinternational.org
Website: www.monitorinternational.org

Founded: 1978
Scope: International

Description: Monitor International is a nonprofit organization,
conserves biological diversity and cultural heritage, and
promotes environmentally sustainable development of marine
and freshwater ecosystems throughout the world.

Publication(s): Sustainable Development, Success Stories, Lake
Toba-Lake Champlain Exchange

Keyword(s): Lakes, Coasts, Developing Countries, Sustainable
Development

Contact(s):
Milton Kaufmann, BOARD OF TRUSTEES CHAIRMAN
David Barker, PRESIDENT
John Dolan, SECRETARY
Richard Tobin, TREASURER
Lisa Borre, VICE PRESIDENT

MONO LAKE COMMITTEE

P.O. Box 29
Lee Vining, CA 93541 USA
Phone: 760-647-6595   Fax: 760-647-6377
E-mail: info@monolake.org
Website: www.monolake.org

Founded: 1978
Membership: 1700

Scope: National

Description: The Mono Lake Committee is a nonprofit citizens' group dedicated to protecting and restoring the Mono Basin ecosystem; educating the public about Mono Lake and the impacts on the environment of excessive water use; and promoting cooperative solutions that protect Mono Lake and meet real water needs without transferring environmental problems to other areas.

Publication(s): Mono Lake Guidebook, Mono Lake Newsletter, Plants of the Mono Basin, Geology of the Mono Basin, South Tufa: A Self-guided Walking Tour of Mono Lake

Keyword(s): Birds, Sustainable Development, Lakes, Exotic species, Aquatic nuisance species, Wetlands

Contact(s):
Ed Manning, CHAIR
Sally Gaines, CHAIR
Arya Degenhardt, EDITOR
Geoffrey McQuilkin, EXECUTIVE DIRECTOR-OPERATIONS
Francis Spivy-Weber, EXECUTIVE DIRECTOR-POLICY
Phone: 310-316-0041
Tom Soto, SECRETARY

MONTANA ASSOCIATION OF CONSERVATION DISTRICTS
501 N. Sanders, Suite 2
Helena, MT 59601 USA
Phone: 406-443-5711   Fax: 406-443-0174
E-mail: mail@macdnet.org
Website: www.macdnet.org

Founded: NA
Membership: 58
Scope: Statewide

Publication(s): Conservation Conversation - Monthly Newsletter

Keyword(s): Conservation Districts

Contact(s):
Jan Fontaine, ADMINISTRATIVE ASSISTANT
501 N. Sanders, Suite 2, Helena, MT 59601
Phone: 406-443-5711
Fax: 406-443-0174
Mike Wendland, PRESIDENT
Dale Marxer, TREASURER
Bob Fossum, VICE PRESIDENT

MONTANA AUDUBON
P.O. Box 595
Helena, MT 59624 USA
Phone: 406-443-3949   Fax: 406-443-7144
E-mail: mtaudubon@montana.com
Website: www.mtaudubon.org

Founded: 1976
Membership: 3500
Scope: Statewide

Description: The statewide orgnization of the nine National Audubon Society chapters in Montana. Montana Audubon is involved in education, research, conservation, and public advocacy on issues affecting Montana's birds, wildlife, and other natural heritage. We enable Montana's 3,000 Audubon members to work together for the Audubon cause.

Publication(s): Montana Bird Distribution

Keyword(s): Biodiversity, Birds, Land Use Planning, Wetlands

Contact(s):
Ray Johnson, EXECUTIVE DIRECTOR
Dorothy Poulsen, PRESIDENT
MT 406-727-7516
Chuck Carlson, SECRETARY
P.O. Box 227, Ft. Peck, MT 59223
Phone: 406-526-3245
Bill Ballard, TREASURER
5120 Larch Av., Missoula, MT 59802
Phone: 406-549-5097

MONTANA B.A.S.S. CHAPTER FEDERATION
Attention: President, P.O. Box 4952
Missoula, MT 59808 USA
Phone: 406-549-1303
E-mail: riska@montana.com

Founded: NA
Scope: Statewide, Regional

Description: An organization of Bassmaster chapters, affiliated with the Bass Anglers Sportsman Society, organized to fight pollution, assist state and national conservation agencies in their efforts, and teach the young people of our country good conservation practice. Dedicated to the realistic conservation of our water resources.

Contact(s):
Tony Quinnell, CONSERVATION DIRECTOR
1535 Trumbel Creek Rd., Kallispell, MT 59901
Phone: 406-755-7867
Mike Riska, PRESIDENT
Phone: 406-549-1303

MONTANA ENVIRONMENTAL INFORMATION CENTER
P.O. Box 1184
Helena, MT 59624 USA
Phone: 406-443-2520   Fax: 406-443-2507
E-mail: meic@meic.org
Website: www.meic.org

Founded: 1973
Membership: 4000
Scope: Statewide

Description: Overall purpose is to protect and restore Montana's natural environment. Educates and mobilizes citizens on Montana environmental issues to press for wise decisions at local, state, and federal levels. Priority issues include: Water quality, solid waste, hardrock mining, hazardous waste, environmental policy, air quality, land use planning, toxic chemicals, and energy conservation.

Publication(s): Montana Environment, Capitol Monitor, Down to Earth

Keyword(s): Air Quality, Energy, Land Use Planning, Toxic Substances, Nuclear-free, Water quantity, Water export and diversion, Water Quality, Mining, Solid Waste Management

Contact(s):
John Ray, BOARD OF DIRECTORS PRESIDENT
Harvey Bjornlie, BOARD OF DIRECTORS VICE PRESIDENT
Jim Jensen, EXECUTIVE DIRECTOR

MONTANA FOREST OWNERS ASSOCIATION

17975 Ryan's Ln.
Evaro, MT 59802 USA
Phone: 406-726-3787   Fax: 406-549-2287

Founded: NA
Scope: Statewide

Description: A statewide organization affiliated with the National Woodland Owners Association, dedicated to the careful use and active enjoyment of private forest lands in Montana. Goals are achieved through active forestry education programs, public communications, networking, and political advocacy.

Publication(s): Big Sky NIPF-TY Notes

Keyword(s): Forests and Forestry

Contact(s):
Thorn Liechty, PRESIDENT
Karen Liechty, SECRETARY
Jim Haviland, TREASURER
Peter Kolb, VICE PRESIDENT
Tom Castles, VICE PRESIDENT

MONTANA LAND RELIANCE

P.O. Box 355
Helena, MT 59624-0355 USA
Phone: 406-443-7027   Fax: 406-443-7061
Website: www.mtlandreliance.org

Founded: NA
Membership: 18
Scope: Statewide

Description: A private nonprofit land trust protecting and conserving ecologically and agriculturally significant land in Montana, as well as sharing knowledge of voluntary, private-sector land conservation techniques. Pioneering ways to assure a legacy of responsibly managed private land.

Publication(s): Montana Spaces, Annual Report, The Montana Land Reliance - spring & fall newsletter, Better Trout Habitat, Tax Implications of Donated Conservation Easements: An Introduction to Conservation Easements, A Guide to Planned Giving: Creation of a Conservation Legacy

Keyword(s): Agriculture, Environment, Environmental Protection, Wildlife, Forest Management, Natural Areas, National Parks, Land Preservation, Open Space, Rivers, Wetlands, Exotic species, Aquatic nuisance species, Watersheds, training

Contact(s):
Christopher Montague, EASTERN MANAGER
P.O. Box 171, Billings, MT 59103-0171
Phone: 406-259-1382
mli@mcn.net
Amy Eaton, GLACIER/FLATHEAD REGIONAL OFFICE
P.O. Box 460, Bigfork, MT 59911-0460
Phone: 406-837-2178
mlrnw@digisys.net
Roy O'Connor, PRESIDENT
5015 Larch Ave., Missoula, MT 59802
rsocmt@bigsky.net
George Olsen, SECRETARY AND TREASURER
Galusha, Higgens & Galusha, Box 1699, Helena, MT 59624-1699
Jerry Townsend, VICE PRESIDENT
Elk Run Ranch, Highwood, MT elkrun@3rivers.net

MONTANA WILDERNESS ASSOCIATION

P.O. Box 635
Helena, MT 59624 USA
Phone: 406-443-7350   Fax: 406-443-0750
E-mail: mwa@wildmontana.org
Website: www.wildmontana.org

Founded: 1958
Membership: 4500
Scope: Statewide

Description: A nonprofit membership organization dedicated to the preservation and proper management of Montana's wild lands, including designated and de facto wilderness areas, national parks, national forests, wildlife refuges, and BLM lands in Montana. The Montana Wilderness Association has five chapter affiliates and four field offices.

Publication(s): Wilderness Walks Program, Wild Montana

Keyword(s): Wilderness, Biodiversity, Forests and Forestry, Public Lands, Sustainable Ecosystems, training

Contact(s):
Bob Decker, DIRECTOR
Susan Miles, DIRECTOR OF MEMBERSHIP SERVICES
Ross Rogers, PRESIDENT
John Gatchell, PROGRAM DIRECTOR
Gerry Jennings, VICE PRESIDENT

MONTANA WILDLIFE FEDERATION

P.O. Box 1175
Helena, MT 59624-1175 USA
Phone: 406-458-0227   Fax: 406-458-0373
E-mail: mwf@mtwf.org
Website: www.montanawildlife.com

Founded: 1935
Scope: Statewide

Description: A representative statewide organization, affiliated with the National Wildlife Federation, dedicated to the protection and enhancement of wildlife and its habitat through public education and government interaction.

Publication(s): Montana Wildlife

Contact(s):
Stan Frasier, ALTERNATE REPRESENTATIVE
Craig Sharpe, EDITOR & EXECUTIVE DIRECTOR
Brian Logan, EDUCATION PROGRAMS CONTACT
John Gibson, PRESIDENT
jcgibson@imt.net
Kathy Hadley, REPRESENTATIVE
khadley@ncat.org
Bill Orsollo, TREASURER

MOTE MARINE LABORATORY

1600 Ken Thompson Parkway
Sarasota, FL 34236 USA
Phone: 941-388-4441   Fax: 941-388-4312
E-mail: info@mote.org
Website: www.mote.org

Founded: 1955
Membership: 195
Scope: Regional

Description: MML is an independent, nonprofit research organization dedicated to excellence in marine and environmental sciences. Since its inception, the laboratory's primary missions have been the pursuit of excellence in scientific research and the dissemination of information to the scientific community as well to the general public. MML operates a public aquarium, the Arthur Vining Davis Library and a marine science education and long distance learning program.

Publication(s): Views from Mote, Mote Marine Laboratory Collected Papers (list upon request), Mote News

Keyword(s): Aquaculture, Aquariums, Wildlife Rehabilitation, Chemistry, Endangered Species, Environmental and Conservation Education, Wildlife, Dolphins, Environmental and Conservation Education, Manatees, Sea Turtles

MOUNT GRACE LAND CONSERVATION TRUST

137 N. Main St.
New Salem, MA 01355 USA
Phone: 978-544-7170   Fax: 978-544-3877
E-mail: mtgrace@shaysnet.com
Website: www.mtgrace.org

Founded: 1987
Membership: 650
Scope: Regional

Description: Mount Grace Land Conservation Trust is dedicated to the protection of forests, agricultural land, and other open space in North Central Massachusetts. In 12 years, Mount Grace Land Conservation Trust has permanently protected 11,000 acres in 90 separate projects.

Publication(s): Views From Mount Grace Quarterly

Keyword(s): Agriculture, Conservation, Forest Management, National Parks, Open Space, training, Stewardship

Contact(s):
Alain Peteroy, CONSERVATION COORDINATOR
Leigh Youngblood, DIRECTOR OF LAND PROTECTION
Dan Leahy, PRESIDENT

MOUNT SHASTA AREA AUDUBON SOCIETY

P.O. Box 530
Mount Shasta, CA 96067 USA
Phone: 916-842-2537

Founded: 1971
Membership: 124
Scope: Statewide

Description: Protects, enhances, and enjoys the natural environment of the Mount Shasta region, including its birds, wildlife, forests, mountains, meadows, and waters.

Publication(s): Endeavor

Contact(s):
Mike Hauptman, EDITOR
Phone: 916-842-2537
Bette Koerner, PRESIDENT
Willo Balfrey, SECRETARY
2934 Nighthawk Ln., Weed, CA 96094
Phone: 916-938-2342

MOUNT CONSERVATION TRUST OF GEORGIA, INC.

104 N. Main St., Suite B3
Jasper, GA 30143 USA
Phone: 706-692-4077   Fax: 706-692-4077
E-mail: mctg@mindspring.com

Founded: 1994
Membership: 360
Scope: Local Region

Description: Dedicated to the permanent conservation of the natural resources and senic beauty of the mountains and foothills of north Georgia thorugh land protection, partnerships and education.

Keyword(s): Environmental and Conservation Education, Land Protection, Watersheds, training

Contact(s):
Barbara Decker, EXECUTIVE DIRECTOR
Gary Reece, PRESIDENT
Phone: 706-692-2424

MOUNTAIN LION FOUNDATION

P.O. Box 1896
Sacramento, CA 95812 USA
Phone: 916-442-2666   Fax: 916-442-2871
E-mail: mlf@mountainlion.org
Website: www.mountainlion.org

Founded: 1986
Membership: 10000
Scope: Statewide

Description: The Mountain Lion Foundation is a nonprofit conservation and education organization dedicated to protecting wildlife and their habitat throughout California.

Publication(s): Cougar: The American Lion, Mountain Lion Update, Preserving Cougar Country, Crimes Against the Wild: Poaching in California

Keyword(s): Biodiversity, Conservation, Endangered Species, Environmental Law, Hunting, Mammals, Nongame Wildlife, Land Preservation, Predators, training

Contact(s):
Michelle Cullens, DIRECTOR OF CONSERVATION POGRAMS
Lynn Sadler, EXECUTIVE DIRECTOR
Phone: 916-442-2666
Kathy Fletcher, PRESIDENT
Sharon Cavallo, SECRETARY
Toby Cooper, TREASURER
Joseph Hurwitz, VICE PRESIDENT

MOUNTAINEERS, THE

CONSERVATION DIVISION
300 3rd Ave., W.
Seattle, WA 98119 USA
Phone: 206-284-6310   Fax: 206-284-4977
E-mail: clubmail@mountaineers.org
Website: www.mountaineers.org

Founded: 1906
Membership: 1500
Scope: Regional

Description: The Mountaineers provides opportunities for outdoor recreation and training to its members and strives to protect the environment through community outreach, education, and political action.

Publication(s): see publications on website, numerous titles published by Mountaineers Books

Keyword(s): Coasts, Public Lands, Rivers, Wilderness, Outdoor Recreation, training

Contact(s):
Kelly McCaffrey, CONSERVATION COORDINATOR
Steve Costie, EXECUTIVE DIRECTOR
Ed Henderson, PRESIDENT

MULE DEER FOUNDATION, THE
1005 Terminal Way, Suite 170
Reno, NV 89502 USA
Phone: 775-322-6558   Fax: 775-322-3421
E-mail: tmdfreno@aol.com
Website: www.muledeer.org

Founded: 1988
Scope: National

Description: The Mule Deer Foundation is a wildlife conservation organization that focuses on mule deer and their subspecies for habitat improvement.

Publication(s): Mule Deer Chronicle, Mule Deer Magazine

Keyword(s): Environmental and Conservation Education, Hunting, Outdoor Recreation, training

MUSKIES, INC.
P.O. Box 120870
New Brighton, MN 58112 USA
Phone: 701-239-9540
E-mail: info@muskiesinc.org
Website: www.muskiesinc.org

Founded: 1966
Scope: National

Description: A nonprofit organization dedicated to establishing hatcheries and introducing the Muskellunge into suitable waters, abating water pollution, promoting a high quality muskellunge sport fishery, supporting selected conservation practices, promoting muskellunge research, disseminating muskellunge information, maintaining records of habits, growth, and range, and promoting good fellowship and sportsmanship.

Publication(s): Muskie

Keyword(s): Lakes, Outdoor Recreation, Sport Fishing, Exotic species, Aquatic nuisance species

N

NATIONAL 4-H COUNCIL
7100 Connecticut Ave.
Chevy Chase, MD 20815-4999 USA
Phone: 301-961-2800
Website: www.fourhcouncil.edu

Founded: 1976
Scope: National

Description: The National 4-H Council is a youth development organization fostering innovation and shared learning. An Environmental Stewardship program engages youth and adults to work as partners in developing creative, community-based solutions to environmental challenges. We also provide science-based educational materials that promote critical thinking skills and youth action grants.

Publication(s): environmental education materials on biotechnology, energy, transportation, food issues, endangered species and water quality, Monthly E-mail: update

Keyword(s): Energy, Environmental and Conservation Education, Transportation, Biotechnology, Pesticides, Sustainable Development, Renewable Resources, training, Youth Organizations

Contact(s):
David Carrier, PROJECT COORDINATOR
Phone: 301-961-2894
carrier@fourhcouncil.edu
Kashyap Choksi, PROJECT DIRECTOR
Phone: 301-961-2894
choksi@fourhcouncil.edu

NATIONAL ARBOR DAY FOUNDATION
100 Arbor Ave.
Nebraska City, NE 68410 USA
Phone: 402-474-5655   Fax: 402-474-0820
E-mail: info@arborday.org
Website: www.arborday.org

Founded: 1971
Membership: 1000000
Scope: National

Description: A nonprofit, membership organization, sponsors Trees for America, Arbor Day, Tree City USA, Conservation Trees and Rain Forest Rescue educational programs. The Foundation publishes "Arbor Day National Poster Contest" and other instructional units for grade schools.

Publication(s): Tree City USA Bulletin, Conservation Trees (booklet), Celebrate Arbor Day (booklet), Library of Trees, ALL PUBLICATIONS ON WEBSITE, Arbor Day

Keyword(s): Environmental and Conservation Education, Flowers, Plants, and Trees, Forests and Forestry, Soil Conservation, Urban Forestry, Trees

Contact(s):
Tony Dorrell, CHAIR
James Fazio, EDITOR
Stewart Udall, HONORARY TRUSTEE AND CHAIRMAN
Gary Brienzo, INFORMATION DIRECTOR
John Rosenow, PRESIDENT
Mary Yager, PROGRAM DIRECTOR
211 N. 12th St., Lincoln, NE 68508
Preston Cole, VICE CHAIR
211 N. 12th St., Lincoln, NE 68508

NATIONAL ASSOCIATION FOR INTERPRETATION
P.O. Box 2246
Fort Collins, CO 80522 USA
Phone: 970-484-8283   Fax: 970-484-8179
E-mail: membership@interpnet.com
Website: ww.interpnet.com

Founded: 1954
Membership: 4500
Scope: International

Description: A nonprofit professional organization, employed by agencies and organizations concerned with natural and cultural resources, conservation, management and with the interpretation of the natural and historical environment.

Publication(s): Centers Directory, The, Interp News, Jobs in Interpretation, Journal of Interpretation Research, El Intérprete and Investigacíones en Interpretación, Legacy

Keyword(s): Communications, Environmental and Conservation Education, Interpretation, education, Internships, Professional Organization

Contact(s):
Nancy Nichols, EDITOR
Communication Director of NAI, P.O. Box 2246, Fort Collins, CO 80522
Phone: 970-484-8283
naicom@aol.com
Tim Merriman, EXECUTIVE DIRECTOR
P.O. Box 2246, Fort Collins, CO 80522
Phone: 970-484-8283
Fax: 970-484-8179
naiexec@aol.com
Heather Manier, MEMBERSHIP MANAGER
Sarah Blodgett, PRESIDENT
cbasman@siu.edu

NATIONAL ASSOCIATION OF BIOLOGY TEACHERS

11250 Roger Bacon Dr., 19
Reston, VA 20190-5202 USA
Phone: 703-471-1134   Fax: 703-435-5582
E-mail: nabter@aol.com
Website: www.nabt.org

Founded: 1938
Membership: 7,500
Scope: National

Description: The only national association specifically organized to assist teachers in the improvement of biology/life science teaching. NABT offers teachers an opportunity to develop professionally through its journal, annual convention, summer workshops, and other publication programs.

Publication(s): News and Views, The Monograph Series, American Biology Teacher, The

Keyword(s): Wildlife Rehabilitation, Biotechnology, Environmental and Conservation Education, Zoology

Contact(s):
Randy Moore, EDITOR
College of Arts & Sciences, University of Louisville, Louisville, KY 40292
Phone: 502-852-6490
r0moor01@homer.louisville.edu
Wayne Carley, EXECUTIVE DIRECTOR
NABT, 11250 Roger Bacon Drive, 19, Reston, VA 20190-5202
Phone: 703-471-1134
nabt31@bellatlantic.net
Christine Chantry, MANAGING EDITOR
NABT, 11250 Roger Bacon Drive, 19, Reston, VA 20190-5202
Phone: 703-471-1134
Vivian Ward, PAST PRESIDENT
Access Excellence-Genentech, Inc., Mail Stop 16B, 460 Point San Bruno Blvd., South San Francisco, CA 94080
Phone: 650-225-8750
vlward@gene.com
Richard Storey, PRESIDENT
Chair, Dept. of Biology, The Colorado College, Colorado Springs, CO 80903
Phone: 719-389-6406
rstorey@coloradocollege.edu
Catherine Wilcoxson, SECRETARY AND TREASURER
2833 Douglas Dr., Fremont, NE 68025
catherine.wilcoxson@nau.edu

NATIONAL ASSOCIATION OF CONSERVATION DISTRICTS

509 Capitol Ct., NE
Washington, DC 20002 USA
Phone: 202-547-6223   Fax: 202-547-6450
E-mail: info@nacdnet.org
Website: www.nacdnet.org

Founded: 1946
Membership: 3000
Scope: National

Description: A nonprofit organization serving as the national instrument of its membership - 3,000 local districts and 54 state and territorial associations. Conservation districts, local subdivisions of state government, work to promote the conservation, wise use and orderly development of land, water, forests, wildlife, and related natural resources.

Publication(s): Tuesday Letter, District Leader, The, America's Conservation Districts, Guide to Conservation Careers, Environmental Film Service Catalogue, Forestry Notes, Buffer Notes

Keyword(s): Agriculture, Environmental and Conservation Education, Soil Conservation, Coral Reefs, Urban Environment, Conservation Districts

Contact(s):
Gary Mast, 2ND VICE PRESIDENT
6055 CR 203 Rte 4, Millersburg, OH 44654
Phone: 330-674-6278
Fax: 330-674-3690
Donna Smith, ADMINISTRATIVE ASSISTANT
donna-smith@nacdnet.org
Ernest Shea, CHIEF EXECUTIVE OFFICER
Phone: 202-547-6223
Fax: 202-547-6450
ernie-shea@nacdnet.org
David Gagner, DIRECTOR OF GOVERNMENT AFFAIRS
david-gagner@nacdnet.org
Debra Bogar, DIRECTOR OF LEADERSHIP SERVICES, NORTH
9150 W. Jewell Ave Ste 111, Lakewood, CO 80232-6469
Phone: 303-988-1893
Fax: 303-988-1896
Robert Toole, DIRECTOR OF LEADERSHIP SERVICES, SOUTH
4617 Cahaan Creek Rd., Edmond, OK 73034
Phone: 405-359-9011
Fax: 405-359-9047
Ray Ledgerwood, DIRECTOR OF LEADERSHIP SERVICES, WEST
NE 1615 Eastgate Blvd., Suite B, Pullman, WA 99163
Phone: 509-334-1823
Fax: 509-334-3453
Bill Horvath, DIRECTOR OF NORTH CENTRAL PROGRAM OFFICE
1052 Main St. Ste. 204, Stevens Point, Wi 54481-2895
Phone: 715-341-1022
Fax: 715-341-1023
Robert Doucette, DIRECTOR OF OPERATIONS
bob-doucette@nacdnet.org
Eugene Lamb, DIRECTOR OF PROGRAMS
eugene-lamb@nacdnet.org
Ron Francis, DIRECTOR OF PUBLIC AFFAIRS
408 East Main Street, League City, TX 77574
Phone: 281-332-3402
Fax: 281-332-5259
ron-francis@nacdnet.org

Laura McNichol, GOVERNMENT AFFAIRS/
COMMUNICATIONS SPECIALIST
laura-mcnichol@nacdnet.org
Linda Neel, MEETING SERVICES MANAGER
9150 W. Jewell Ave. Suite 102, Lakewood, CO 80232-6469
Phone: 303-988-1810
Fax: 303-988-1896
J. Smith, PRESIDENT
11751 Lancaster Rd., St. John, WA 99171-9723
Phone: 509-648-3922
Fax: 509-648-3293
read-smith@nacdnet.org
Billy Wilson, SECOND VICE PRESIDENT
Phone: 918-768-3542
bwilson@cwis.net
Tim Reich, SECRETARY/TREASURER
1007 Kingsbury St., Belle Fourche, SD 57717
Phone: 605-892-4366

NATIONAL ASSOCIATION OF CONSERVATION DISTRICTS

509 Capitol Ct., NE
Washington, DC 2002-4946 USA
Phone: 202-547-6223   Fax: 202-547-6450
Website: www.nacdnet.org

Founded: 1962
Membership: 3000
Scope: National

Description: Directed by the Conservation Districts Foundation,
Inc., an adjunct of the National Association of Conservation
Districts. Collects various conservation and environmental
education materials. Dedicated to the memory of Waters S.
Davis, Past President of NACD.

Publication(s): News and Views, eNotes

Keyword(s): Libraries, Environmental and Conservation
Education, Soil Conservation, Exotic species, Aquatic
nuisance species, Wetlands

Contact(s):
Ernest Shea, CHIEF EXECUTIVE OFFICER
NACD, 509 Capitol Ct., NE, Washington, DC 20002
Phone: 202-547-6223
Ronald Francis, DIRECTOR OF OFFICE OF PUBLIC
AFFAIRS
Phone: 281-332-3402
ron-francis@nacdnet.org
Read Smith, PRESIDENT
11751 Lancaster Rd, St John, Washington 99171
Phone: 509-648-3922
read-smith@nacdnet.org

NATIONAL ASSOCIATION OF CONSERVATION DISTRICTS

LEAGUE CITY OFFICE
P.O. Box 855
League City, TX 77574 USA
Phone: 281-332-3402   Fax: 281-332-5259
Website: www.nacd.net.org

Founded: NA
Membership: 3000
Scope: National

Description: tf

Publication(s): NACD News & Views - newsletter

Contact(s):
Ronald Francis, OFFICE OF PUBLIC AFFAIRS DIRECTOR
ron-francis@nacdnet.org
Maxine Mathis, SERVICE CENTER PRODUCTION
MANAGER
maxine-mathis@nacdnet.org

NATIONAL ASSOCIATION OF ENVIRONMENTAL PROFESSIONALS, THE

NATIONAL OFFICE
P.O. Box 2086
Bowie, MD 20718 USA
Phone: 888-251-9902   Fax: 301-860-1141
E-mail: office@naep.org
Website: www.naep.org

Founded: 1975
Membership: 2,000
Scope: National

Description: NAEP is the professional association of the environmental professions, dedicated to the promotion of ethical
practice, technical competency, and professional standards in
the environmental field and recognition of the environmental
profession since 1975.

Keyword(s): Air Quality and Pollution, Energy, Environmental
Planning, Solid Waste Management, Exotic species, Aquatic
nuisance species, Professional Organization

NATIONAL ASSOCIATION OF RECREATION RESOURCE PLANNERS

c/o Tim Hogsett, Treasurer, Texas Parks & Wildlife Dept.,
4200 Smith School Rd.
Austin, TX 78744-3291 USA
Phone: 512-912-7109   Fax: 512-707-2742
E-mail: rec.grants@tpwd.state.tx.us
Website: www.tpwd.state.tx.us/park/grants

Founded: NA
Membership: 14
Scope: Statewide

Description: A nonprofit organization involved in the exchange of
recreation resource planning information among fedreal, state
and regional agencies. Participates in national recreation
concerns, promotes improvements in the state-of-the-art of
recreation planning and professionalism among its members
and acts as an advocate for conservation and recreation
opportunities for the future.

Publication(s): NARRP Newsletter

Keyword(s): Land Use Planning, Open Space, Outdoor
Recreation, Planning Management, Public Lands

Contact(s):
Gordon Kimball, PRESIDENT
Minnesota,
Robert Sammon, VICE PRESIDENT
New Hampshire,

NATIONAL ASSOCIATION OF SERVICE & CONSERVATION CORPS

666 11th St., N.W.,
Suite 1000
Washington, DC 20001-4542 USA
Phone: 202-737-6272
Website: www.nascc.org/

Contact(s):
T. Jarvis
destryjarvis@earthlink.net

NATIONAL ASSOCIATION OF SERVICE AND CONSERVATION CORPS (NASCC)

666 11th St., NW, Suite 1000
Washington, DC 20001 USA
Phone: 202-737-6272   Fax: 202-737-6277
E-mail: nascc@nascc.org
Website: www.nascc.org

Founded: 1985
Scope: National

Description: NASCC unites and supports youth corps as a preminent strategy for achieving the nation's youth development, community service, and environmental restoration goals. NASCC serves as an advocate, central reference point, and source of assistance for the growing number of state and local youth corps around the country.

Publication(s): Youth Corps Resource Book, Corpsmember Wellness Guide, Urban Waterways Restoration Training Manual, Youth Corps Profiles

Keyword(s): Environmental and Conservation Education, People of Color in the Environment, Training, Urban Environment, Youth Organizations, Riparian Restoration, Trail

Contact(s):
Leslie Wilkoff, DIRECTOR FOR MEMBER SERVICES
lwilkoff@nascc.org
Kathleen Selz, PRESIDENT
kselz@nascc.org
Harry Bruell, VICE PRESIDENT, FIELD SERVICES
hbruell@nascc.org
Andrew Moore, VICE PRESIDENT, GOVERNMENT RELATIONS AND PUBLIC AFFAIRS
amoore@nascc.org

NATIONAL ASSOCIATION OF STATE DEPARTMENTS OF AGRICULTURE

1156 15th St., NW, Suite 1020
Washington, DC 20005 USA
Phone: 202-296-9680   Fax: 202-296-9686
E-mail: nasda@patriot.net
Website: www.nasda-hq.org

Founded: NA
Scope: National

Description: The National Association of State Departments of Agriculture (NASDA) is a nonprofit, nonpartisan association of public officials comprised of the executive heads of the fifty State Departments of Agriculture and those from territories of Puerto Rico, Guam, American Samoa, and the Virgin Islands. NASDA's mission is to support and promote the American agriculture industry, while protecting consumers and the environment, through the development, implementation, and communication of sound policy and programs.

Publication(s): Ag In Perspective (quarterly), NASDA News (weekly)

Keyword(s): Agriculture, Biotechnology, Conservation, Conservation Tillage, Environmental Protection, Pesticides, Precision Farming, Health and Nutrition, Public Farming, Public Lands, Coral Reefs, Exotic species, Aquatic nuisance species, Water Quality, Rural Development, Wetlands

Contact(s):
Richard Kirchhoff, CHIEF EXECUTIVE OFFICER
rick@nasda-hq.org

NATIONAL ASSOCIATION OF STATE FORESTERS

444 N. Capitol St., NW, Suite 540
Washington, DC 20001 USA
Phone: 202-624-5415   Fax: 202-624-5207
E-mail: nasf@sso.org
Website: www.stateforesters.org

Founded: NA
Membership: 58
Scope: National

Description: Members are state foresters or equivalent officials whose agencies are the legally-constituted authorities for public forestry work within the states. In cooperation with federal agencies, private organizations, and individuals, NASF promotes sound forest management on public and private lands.

Publication(s): NASF Washington Update

Keyword(s): Forests and Forestry, Public Lands, Soil Conservation, Urban Forestry, Exotic species, Aquatic nuisance species, Professional Organization

Contact(s):
Gerald Thiedy
NORTHEASTERN AREA REGIONAL REPRESENTATIVE
P.O. Box 30452, Lansing, MI 48909
Phone: 517-335-4225
Fax: 517-373-2443
Connie Motyka, PRESIDENT
VT Dept. of Forest, Parks and Recreation, 103 S. Main St., Waterbury, VT 05671-0601
Phone: 802-241-3670
Jim Hull, SOUTHERN GROUP REGIONAL REPRESENTATIVE
301 Tarrow, Suite 364, College Station, TX 77840
Phone: 979-458-6600
Fax: 979-548-6610
James Sledge, TREASURER
Frankfort, KY 40601
Phone: 502-564-4496
Larry Kotchman, VICE PRESIDENT
ND Forest Service, 307 First St., Bottineau, ND 58318-1100
Phone: 701-228-5422
Tom Ostermann, WESTERN COUNCIL REGIONAL REPRESENTATIVE
Wyoming State 4th Street Division 1100 West, 22 N, Cheyenne, WY 82002
Phone: 307-777-7586

NATIONAL ASSOCIATION OF STATE PARK DIRECTORS

9894 E. Holden Pl.
Tucson, AZ 85748 USA
Phone: 520-298-4924   Fax: 520-298-6515
Website: www.naspd.org

Founded: 1962
Membership: 53
Scope: National

Description: Works to unite the states on a common ground for the development of park systems to meet the intensive public demand for out-of-doors recreational opportunities; to promote the exchange of ideas regarding the development of state park systems; to encourage and develop professional leadership; and to expand and improve park policies and practices.

Publication(s): NASPD, Annual Information Exchange, NASPD, The Directory

Keyword(s): Environmental and Conservation Education, Land Purchase, Land Use Planning, Outdoor Recreation, Public Lands, Professional Organization

Contact(s):
Glen Alexander, EXECUTIVE DIRECTOR
9894 E. Holden Pl., Tucson, AZ 85748
Phil McKnelly, PRESIDENT

NATIONAL ASSOCIATION OF STATE PARK DIRECTORS

9894 E. Holden Place
Tucson, AZ 85748 USA
Phone: 520-298-4924   Fax: 520-298-6515
E-mail: naspdglen@home.com
Website: www.indiana.edu/~naspd/
http://www.indiana.edu/~naspd/

Founded: 1967
Scope: National

Description: An organization of 56 gubernatorial-appointed state and territorial officials working with the National Park Service and the Department of the Interior to strengthen the nation's total outdoor recreation program. Represents state and local interests in administration of the Land and Water Conservation Fund Program, which provides money for acquisition and development of recreation land and facilities.

Keyword(s): Land Purchase, Outdoor Recreation, Planning Management

Contact(s):
Glen Alexander, EXECUTIVE DIRECTOR
9894 E. Holden Place, Tucson, AZ 85748
Phone: 520-298-4924
Fax: 520-298-6515
naspdglen@home.com

NATIONAL ASSOCIATION OF UNIVERSITY FISHERIES AND WILDLIFE PROGRAMS

Attn: President, Department of Animal Ecology, Iowa State University
Ames, IA 50011-3221 USA
Phone: 515-294-6148   Fax: 515-294-7874
Website: www.naufwp.iastate.edu

Founded: 1991
Membership: 65
Scope: National

Description: Meets annually at the North American Wildlife and Natural Resources Conference. The purpose is to foster improved communications among members and between other agencies, organizations, and the general public in order to provide a unified voice for academic fisheries and wildlife programs.

Keyword(s): Environmental and Conservation Education, Wildlife, training

Contact(s):
Bruce Menzel, PRESIDENT
Department of Animal Ecology, Iowa State University, Ames, IA 50011-3221
Phone: 515-294-7419
Bruce Menzel, PRESIDENT AND CHAIR
Daniel Pletscher, PRESIDENT-ELECT
Wildlife Biology Program, School of Forestry at the University of Montana, Missoula, Montana 59812
Phone: 406-243-6364
Erik Fritzell, SECRETARY AND TREASURER
Department of Fisheries and Wildlife, Oregon State University, Corvallis, OR 97331
Phone: 541-737-5906

NATIONAL AUDUBON SOCIETY

Headquarters, 700 Broadway
New York, NY 10003-9501 USA
Phone: 212-979-3000   Fax: 212-979-3188
Website: www.audubon.org

Founded: 1905
Membership: 550000
Scope: National

Description: Solid Science, policy research, lobbying, citizen science and action, and education — these are the tools used by the Audubon Society to protect the land and habitat that are critical to our health and the health of the planet. With the support of 550,000 members (in addition to the 500,000 elementary school students in the Audubon Adventures Program) and an extensive chapter network in the United States and Latin America, Audubon draws on the enthusiasm and power of the grassroots to save our threatened ecosystems.

Publication(s): Audubon Field Notes, Audubon Adventures, Audubon

Keyword(s): Birds

Contact(s):
Donal O'Brien, CHAIRMAN OF THE BOARD
Phone: 212-353-0377
Daniel Beard, CHIEF OPERATING OFFICER
David Seideman, EDITOR-IN-CHIEF AUDUBON MAGAZINE
John Flicker, PRESIDENT AND CEO
Carol May, SENIOR VICE PRESIDENT OF DEVELOPMENT
Glenn Olson, SENIOR VICE PRESIDENT OF FIELD OPERATIONS AND SANCTUARIES
James Cunningham, SENIOR VICE PRESIDENT OF OPERATIONS
Frank Gill, SENIOR VICE PRESIDENT OF SCIENCE
Alan Bayersdorfer, VICE PRESIDENT OF MEMBERSHIP
Patrick Downes, VICE PRESIDENT OF PUBLISHING
Carole McNamara, VICE PRESIDENT/CONTROLLER

NATIONAL AUDUBON SOCIETY
PROJECT PUFFIN
159 Sapsucker Woods Rd.
700 Broadway, NYC-NY 10003
Ithaca, NY 14850 USA
Phone: 212-979-3000   Fax: 212-979-3188

Founded: NA
Scope: Regional

Contact(s):
Stephen Kress, DIRECTOR

NATIONAL AUDUBON SOCIETY
SCULLY SCIENCE CENTER
National Audubon Society 700 Broadway
New York, NY 10003 USA
Phone: 212-979-3000   Fax: 212-979-3188
E-mail: education@audubon.org
Website: www.audubon.org

Founded: NA
Scope: Regional

Contact(s):
Carl Safina, DIRECTOR

NATIONAL AUDUBON SOCIETY
TAVERNIER SCIENCE CENTER
115 Indian Mound Tr.
Tavernier, FL 33070 USA
Phone: 305-852-5318   Fax: 305-852-8012

Founded: NA
Membership: 5
Scope: Regional

Contact(s):
Jerry Lorenz, DIRECTOR OF RESEARCH

NATIONAL AUDUBON SOCIETY
WASHINGTON, D.C. OFFICE
P.O. Box 15726 Washington DC 20003-5726
Suite 1100
Washington, DC 20006 USA
Phone: 202-861-2242   Fax: 202-861-4290
E-mail: judyschaefer@attglobal.net
Website: www.dcaudubon.org

Founded: NA
Scope: National

Publication(s): Audubon Magazine

Contact(s):
Dan Beard, SENIOR VICE PRESIDENT
dbeard@audubon.org

NATIONAL AUDUBON SOCIETY
LIVING OCEANS PROGRAM
550 South Bay Ave.
Islip, NY 11751 USA
Phone:
E-mail: livingoceans@audubon.org
Website: www.audubon.org/campaign/lo
http://www.audubon.org/campaign/lo

Founded: 1993
Scope: National

Description: Living Oceans is the marine conservation program of National Audubon Society. The program is dedicated to reversing the mismanagement of marine fisheries which has led to the depletion of marine wildlife, and to restore the health of our marine environment and coastal habitats.

Publication(s): Living Oceans News, Audubon Guide to Seafood

Keyword(s): Ocean Conservation, Marine Conservation

Contact(s):
Mercedes Lee, ASSISTANT DIRECTOR
Phone: 516-224-3669
Carl Safina, DIRECTOR
Merry Camhi, STAFF SCIENTIST
Phone: 516-581-2927

NATIONAL AVIARY
Allegheny Commons West
Pittsburgh, PA 15212-5248 USA
Phone: 412-323-7235   Fax: 412-321-4364
E-mail: info@aviary.org
Website: www.aviary.org

Founded: 1952
Membership: 1800
Scope: National

Description: The National Aviary works to inspire respect for nature through the appreciation of birds. Travel the world at the National Aviary and visit over 500 exotic and endangered birds in natural habitats.

Publication(s): Bird Calls

Keyword(s): Birds, Raptors, Biodiversity, Conservation, Endangered Species, Zoology, Environmental and Conservation Education, training, Botanical Gardens, Flowers, Plants, and Trees

Contact(s):
Dayton Baker, EXECUTIVE DIRECTOR
dayton.baker@aviary.org

NATIONAL BIRD-FEEDING SOCIETY
P.O. Box 23
Northbrook, IL 60065-0023 USA
Phone: 847-272-0135   Fax: 773-404-0923
E-mail: feedbirds@aol.com
Website: www.birdfeeding.org

Founded: 1989
Scope: National

Description: The National Bird-Feeding Society works to make feeding of wild birds better for birds as well as for people. The Society also works to encourage suitable feeding and nesting habitats for backyard birds to thrive; sponsors education and research on backyard birds; and helps people learn more ways to attract and care for birds, to share bird-feeding experiences, observations, and to help the environment through the backyard.

Publication(s): The Bird -Eye reView

Keyword(s): Birds, Conservation, Environment, Flowers, Plants, and Trees, Birds, Research

Contact(s):
Donald Stokes, CHAIRMAN EMERITUS
Sue Wells, EXECUTIVE DIRECTOR

NATIONAL BISON ASSOCIATION

4701 Marion St.,
Denver, CO 80216 USA
Phone: 303-292-2833   Fax: 303-292-2564
E-mail: info@bisoncentral.com
Website: www.bisoncentral.com

Founded: NA
Membership: 2400
Scope: National

Description: The National Bison Association represents public and private bison herds through educational programs, public education and scholarships.

Publication(s): Bison World (quarterly)

NATIONAL BOATING FEDERATION

P.O. Box 4111
Annapolis, MD 21403 USA

Founded: 1966
Scope: National

Description: National, all-volunteer, non-profit boating organization consisting principally of regional or special interest boating organizations and yacht clubs. Activities include monitoring federal legislation and rule-making as they affect recreational boating. The NBF promotes the interests of those who use boats for cruising, water skiing, fishing, and other water sports.

Publication(s): Recreational Boating News and Legislative Issues, The LOOKOUT

Keyword(s): Coasts, Lakes, Outdoor Recreation, Rivers, Coral Reefs

Contact(s):
William Mitchelson, PAST PRESIDENT
9483 N. Fairway Cr., Milwaukee, WI 53217-1316
Phone: 414-352-0967
Robert David, PRESIDENT
70 Garfield Lane, West Dennis, MA 02670-2321
Phone: 508-394-5670
Fax: 508-394-7236
Jill Andrick, PUBLIC RELATIONS AND EDITOR
P.O. Box 211, Salem, OR 97308-0211
Phone: 503-580-0769
William Heider, SECRETARY AND TREASURER
1114 Appletree Lane, Erie, PA 16509-3917
Phone: 814-825-3011
Fax: 814-825-5284
Roger Brown, VICE-PRESIDENT
111 S. View Court, Shore Acres, NJ 08723-7520
Phone: 732-236-3516
Fax: 702-477-0071

NATIONAL COALITION FOR MARINE CONSERVATION

3 N. King St.
Leesburg, VA 20176 USA
Phone: 703-777-0037   Fax: 703-777-1107
Website: www.savethefish.org

Founded: 1973
Membership: 1500
Scope: National

Description: A nonprofit, privately-supported organization devoted exclusively to the conservation of ocean fish and the protection of their environment. Promotes public awareness of marine conservation issues and stimulates the formulation of responsible public policy.

Publication(s): Marine Bulletin

Keyword(s): Coasts, Environmental and Conservation Education, Wildlife, Sport Fishing, Wetlands

Contact(s):
Christine Wilkins, DIRECTOR OF COMMUNICATION AND DEVELOPMENT
Tim Hobbs, FISHERIES PROJECT DIRECTOR
Ken Hinman, PRESIDENT AND EDITOR

NATIONAL COUNCIL FOR GEOGRAPHIC EDUCATION

16A Leonard Hall, Indiana University of Pennsylvania
Indiana, PA 15705 USA
Phone: 412-357-6290
Website: http://www.oneonta.edu

Founded: 1915
Membership: 4,000
Scope: National

Description: To promote and advance geographic and environmental education in the public schools and colleges of the U.S. and Canada.

Publication(s): Water In the Global Environment, list of other publications available upon request., Journal of Geography

Keyword(s): Environmental and Conservation Education

Contact(s):
Jonathan Leib, EDITOR
Department of Geography, Florida State University, Tallahassee, FL 32306-4016
Ruth Shirey, EXECUTIVE DIRECTOR
James Peterson, PRESIDENT
Sandra Mather, SECRETARY
Robert Bednarz, VICE PRESIDENT OF CURRICULUM AND INSTRUCTION
Celeste Fraser, VICE PRESIDENT OF FINANCE
Gary Elbow, VICE PRESIDENT OF PUBLICATIONS AND PRODUCTS

NATIONAL EDUCATION ASSOCIATION

1201 16th St., NW
Washington, DC 20036 USA
Phone: 202-833-4000
Website: www.nea.org

Founded: 1857
Membership: 2600000
Scope: National

Description: Works to elevate the character and advance the interests of the teaching profession and to promote the cause of education in the U.S.

Publication(s): newletter- NEA Now, newsletter- NEA Today

Contact(s):
John Wilson, EXECUTIVE DIRECTOR
Robert Chase, PRESIDENT
Dennis Roekel, SECRETARY AND TREASURER
Reg Weaver, VICE PRESIDENT

NATIONAL ENVIRONMENTAL HEALTH ASSOCIATION

720 S. Colorado Blvd., S. Tower, Suite 970
Denver, CO 80246-1925 USA
Phone: 303-756-9090   Fax: 303-691-9490
E-mail: staff@neha.org
Website: www.neha.org

Founded: 1937
Membership: 5300
Scope: National

Description: NEHA is a member nonprofit organization that offers a wide variety of educational credentialing and advancement opportunities for people involved or interested in environmental health issues. It is the largest society of environmental health practitioners in the nation today, numbering almost 5,000 members and growing.

Publication(s): Self Paced Learning Modules, various books, manuals, etc., Journal of Environmental Health

Keyword(s): Air Quality and Pollution, Solid Waste, Toxic Substances, Nuclear-free, Water quantity, Water export and diversion, Coral Reefs, Exotic species, Aquatic nuisance species, Environmental Health, Food Safety

Contact(s):
Nelson Fabian, EXECUTIVE DIRECTOR

NATIONAL FARMERS UNION

11900 E. Cornell Ave.
Aurora, CO 80014-3194 USA
Phone: 303-337-5500   Fax: 303-368-1390
Website: www.nfu.org

Founded: 1902
Scope: National

Description: Believes that the soil, water, forest and other natural resources of the nation should be used and conserved in a manner to pass these resources on undiminished to future generations and that publicly and privately owned land and resources should be administered in the interest of all the public.

Publication(s): National Farmers Union News

Keyword(s): Agriculture, Environmental Law, Health and Nutrition, Renewable Resources, Soil Conservation

Contact(s):
Rae Price, EDITOR
11900 E. Cornell Ave., Aurora, CO 80014-3194
Phone: 303-337-5500
Leland Swenson, PRESIDENT
David Carter, TREASURER/SECRETARY
Charles Nash, VICE PRESIDENT
Larry Mitchell, VICE PRESIDENT OF LEGISLATIVE SERVICES
400 Virginia Ave. SW, Suite 710, Washington, DC 20024
Phone: 202-554-1600

NATIONAL FFA ORGANIZATION

P.O. Box 68960, 6060 FFA Drive
Indianapolis, IN 46268-0960 USA
Phone: 317-802-6060   Fax: 317-802-6061

Founded: 1928
Scope: National

Description: The FFA is a national organization of high school agriculture students in public secondary schools. Congress granted the organization a federal charter in 1950, making it an integral part of the high school agriculture program. Major aims are to provide activities that will stimulate students to higher achievement in the study of production agriculture, agriscience, agribusiness, and agrimarketing, and give them opportunities through student-planned programs for leadership and self-development.

Publication(s): FFA Advisor Publication, Update Newsletter, FFA New Horizons Magazine, The

Keyword(s): Agriculture, Biotechnology, Gardening and Horticulture, Renewable Resources, Youth Organizations

Contact(s):
Larry Case, ADVISOR
Phone: 703-360-3600
C. Harris, EXECUTIVE SECRETARY
Phone: 703-360-3600

NATIONAL FIELD ARCHERY ASSOCIATION

31407 Outer I-10
Redlands, CA 92373 USA
Phone: 909-794-2133   Fax: 909-794-8512
E-mail: nfaarchery@aol.com
Website: www.nfaa-archery.org

Founded: NA
Membership: 20,000
Scope: National

Description: A nonprofit national membership headquarters for all archers.

Publication(s): Archery Magazine

Keyword(s): Hunting, Scholarships and Grants, Sport Fishing, Youth Organizations

Contact(s):
Tim Atwood, BOWHUNTING COMMITTEE CHAIRMAN
3175 Racine, Riverside, CA 92503
Marihelen Rogers, EDITOR
Marihelen Rogers, EXECUTIVE SECRETARY
Walter Rueger, PRESIDENT
122 Stanton Ave., Ripon, WI 54971

NATIONAL FISH AND WILDLIFE FOUNDATION

1120 Connecticut Ave., NW, Suite 900
Washington, DC 20036 USA
Phone: 202-857-0166   Fax: 202-857-0162
E-mail: info@nfwf.org
Website: ww.nfwf.org

Founded: 1984
Scope: National

Description: A national nonprofit grant-making and grant-seeking organization dedicated to the conservation of natural resources — fish, wildlife, and plants. NFWF was established by Congress to leverage federally, appropriated funds by forging public and private partnerships which result in conservation activities that pinpoint and solve root causes of environmental problems.

Keyword(s): Birds, Endangered Species, Environmental and Conservation Education, Wildlife, training

Contact(s):
Steve Peet, CHAIRMAN OF THE BOARD

Ginette Ring, CHIEF FINANCIAL OFFICER
Jerry Clark, DEPUTY DIRECTOR REGIONAL PROGRAMS
clark@nfwf.org
Tom Kelsch, DIRECTOR OF CONSERVATION EDUCATION
INITIATIVE
Lorraine Howerton, DIRECTOR OF CONSERVATION
POLICY
Whitney Tilt, DIRECTOR OF CONSERVATION PROGRAMS
Gary Guinn, DIRECTOR OF DEVELOPMENT AND
MARKETING
guinnn@nfwf.org
Gary Kania, DIRECTOR OF WILDLIFE AND HABITAT
INITIATIVE
John Berry, EXECUTIVE DIRECTOR
berry@nfwf.org
Peter Stangel, REGIONAL DIRECTOR OF SOUTHEAST
stangel@nfwf.org
Alex Echols, SPECIAL ASSISTANT

NATIONAL FLYWAY COUNCIL
South Dakota Game, Fish and Parks, 523 E. Capitol
Pierre, SD 57501 USA
Phone: 605-773-4192   Fax: 605-773-6245
E-mail: george.vandel@state.sd.us

Founded: NA
Scope: National

Contact(s):
George Vandel, CHAIRMAN
South Dakota Game, Fish, and Parks Department, 523 E.
Capitol, Pierre, SD 57501

NATIONAL FLYWAY COUNCIL
ATLANTIC FLYWAY OFFICE
Forest, Wildlife and Heritage Administration
Tawes State Office Building
580 Taylor Ave.
Annapolis, MD 21401 USA
Phone: 410-260-8534   Fax: 410-260-8595
Website: www.dec.state.ny.us

Founded: NA
Scope: Regional

Contact(s):
Gerald Barnhart, CHAIRMAN

NATIONAL FLYWAY COUNCIL
CENTRAL FLYWAY OFFICE
WY Game & Fish Dept.
5400 Bishop Blvd.
Cheyenne, WY 82006 USA
Phone: 307-777-4501
E-mail: bwiche@missc.state.wy.us

Founded: NA
Scope: Regional

Contact(s):
William Wichers, CHAIRMAN

NATIONAL FLYWAY COUNCIL
MISSISSIPPI FLYWAY OFFICE
IA DEPT. OF NATURAL RESOURCES
Wallace State Office Bldg.
502 E-9th St.
Des Moines, IA 50036 USA
E-mail: rbishop@max.state.ia.us

Founded: NA
Scope: Regional

Contact(s):
Richard Bishop, CHAIRMAN

NATIONAL FOREST FOUNDATION
2715 M St., NW
Washington, DC 20007 USA
Phone: 202-496-4963   Fax: 202-822-0632
E-mail: www.nationalforest.org
Website: www.nationalforest.org

Founded: 1993
Membership: 7500
Scope: National

Description: As the independent, not-for-profit partner organization of the U.S. Forest Service, the NFF seeks to build relationships that result in measurable improvements in the health, productivity, and diversity of our National Forests and Grasslands for present and future generations.

Publication(s): Mosaic

Keyword(s): Environmental and Conservation Education, Forests and Forestry, Outdoor Recreation, Public Lands, training

Contact(s):
Laura Dunleavy, CONSERVATION PROGRAMS MANAGER
William Possiel, PRESIDENT
Mathew Esserman, PROGRAM DEVELOPMENT OFFICER

NATIONAL FOREST FOUNDATION
MONTANA OFFICE
32 South Ewing, Suite 324
Helena, MT 59601 USA
Phone: 406-495-8312   Fax: 703-495-8312
Website: www.natlforests.org/donate.html

Description: As the independent, not-for-profit partner organization of the U.S. Forest Service, the NFF seeks to build relationships that result in measurable improvements in the health, productivity, and diversity of our National Forests and Grasslands for present and future generations.

NATIONAL GARDEN CLUBS INC
4401 Magnolia Ave.
St. Louis, MO 63110 USA
Phone: 314-776-7574   Fax: 314-776-5108
E-mail: headquarters@gardenclub.org
Website: www.gardenclub.org

Founded: 1929
Membership: 243597
Scope: International

Description: Coordinates and furthers the interests and activities of the State Federations of Garden Clubs and aids in the protection and conservation of natural resources; protects civic beauty and encourages the improvement of roadsides and parks; encourages and assists in establishing and maintaining botanical gardens and horticultural centers; and advances the arts of gardening and landscape design, and study of horticulture.

Publication(s): National Gardener, The

Keyword(s): Natural Resource Conservation, Forests and Forestry, Public Lands, Rivers, Water Pollution Management, Wilderness, Gardening and Horticulture

Contact(s):
June Wood, 1ST VICE PRESIDENT
7000 Seminole Rd., NE, Albuquerque, NM 87110-2739
Jan Blair, CONSERVATION/NATURAL RESOURCES WILDLIFE/ENDANGERED SPECIES CHAIRPERSON AND "BACKYARD HABITATS FOR WILDLIFE"
Louis Rd, Irvington, NY 914-591-6959
envirojb@aol.com
Katrina Vollmer, CORRESPONDING SECRETARY
3134 N Greenbriar, Nashville, IN 47448
Phone: 812-988-0063
katrina@bigfoot.com
Susan Davidson, EDITOR
102 S. Elm St., St. Louis, MO 63119
Phone: 314-968-1664
susand4@juno.com
Fran Mantler, EXECUTIVE DIRECTOR
Lois Shuster, PRESIDENT
77 Water Wheel Drive, Champion, PA 15622
Phone: 814-352-7777
lois@shalkl.com
Susan Slivken, TREASURER
4613 37th Ave., Rock Island, IL 61201-7108

NATIONAL GARDENING ASSOCIATION

1100 Dorset St.
South Burlington, VT 05403 USA
Phone: 800-538-7476   Fax: 802-864-6889
Website: www.kidsgardening.com

Founded: 1972
Membership: 16
Scope: National

Description: The mission of the National Gardening Association is to sustain the essential values of life and community, renewing the fundamental links between people, plants, and the earth. Through gardening, we promote environmental responsibility, advance multidisciplinary learning and scientific literacy, and create partnerships that restore and enhance communities.

Publication(s): Guide to Kids Gardening, Community Garden Book, The, National Gardening Survey, Gardening, Gardening Video Series, GrowLab: A Complete Guide to Gardening in the Classroom, Ruth Page's Gardening Journal, GROWLAB: Activities for

Keyword(s): Environmental and Conservation Education, Flowers, Plants, and Trees, Gardening and Horticulture, Youth Organizations

Contact(s):
Amy Gifford, ADMINISTRATVE COORDINATOR OF GROWLAB
William Vandeventer, ASSISTANT TREASURER AND CFO
billv@kidsgardening.com
Eve Pranis, ASSOCIATE DIRECTOR OF EDUCATION
evep@kidsgardening.com

Larry Sommers, DIRECTOR OF ADVERTISING
180 Flynn Ave., Burlington, VT 05401
Phone: 802-863-1308
Fax: 802-863-5962
l.sommers@nationalgardening.com
Charlie Nardozzi, HORTICULTURIST
180 Flynn Ave., Burlington, VT 802-863-1308
Phone: 802-863-5962
c.nardozzi.nationalgardening.com
Bruce Butterfield, MARKET RESEARCH
bruceb@gardenresearch.com
Valerie Kelsey, PRESIDENT
valeriek@kidsgardening.com

NATIONAL GEOGRAPHIC SOCIETY

1145 17th St., NW
Washington, DC 20036 USA
Phone: 800-647-5463
E-mail: askngs@nationalgeographic.com
Website: www.nationalgeographic.com

Founded: 1888
Scope: International

Description: For the increase and diffusion of geographic knowledge.

Publication(s): National Geographic World Magazine (for children), National Geographic Traveler, Books/Maps/Atlases, Globes, Filmstrips, Documentary Films, Classroom Materials, National Geographic Adventure Magazine, National Geographic Channel, National Geographic

Keyword(s): Environmental and Conservation Education, Geography, education, Oceanography, Public Lands

NATIONAL GRANGE, THE

1616 H St., NW
Washington, DC 20006-4999 USA
Phone: 202-628-3507   Fax: 202-347-1091
Website: www.nationalgrange.org

Founded: 1867
Scope: National

Description: Rural family service organization with special interests in community service and agriculture.

Publication(s): Grange Today, View From The Hill

Keyword(s): Agriculture, Environmental Law, Land Use Planning, Soil Conservation, Transportation

Contact(s):
Robert Clark, EXECUTIVE COMMITTEE CHAIRMAN
Phone: 360-683-4431
A. Henninger, EXECUTIVE COMMITTEE SECRETARY
Phone: 815-544-4522
Leroy Watson, LEGISLATIVE DIRECTOR
Washington D.C. Office,
Kermit Richardson, MASTER
Washington DC Office,
Shirley Lawson, SECRETARY
120 Wilson Ave., Rumford, RI 02916
Phone: 401-434-1491
Fax: 401-434-6772

NATIONAL GROUND WATER ASSOCIATION, THE

601 Dempsey Rd.
Westerville, OH 43081 USA
Phone: 614-898-7791   Fax: 614-898-7786
E-mail: ngwa@ngwa.org
Website: www.ngwa.org

Founded: 1948
Membership: 16500
Scope: International

Description: The NGWA is the world's leading organization committed to the study of the occurrence, development, and protection of ground water. The Association annually sponsors educational programs dealing with a wide variety of water issues, including toxic substances, solid waste, and water pollution. Operates on-line data bases at Web Site.

Publication(s): Ground Water Monitoring and Remediation, Journal of Ground Water, Water Well Journal

Keyword(s): Environmental and Conservation Education, Geology, Coral Reefs, Exotic species, Aquatic nuisance species

Contact(s):
Kevin McCray, EXECUTIVE DIRECTOR
Sandy Masters, INFORMATION
National Ground Water Information Center, 601 Dempsey Dr., Westerville, OH 43081

NATIONAL HUNTERS ASSOCIATION, INC.

P.O. Box 820
Knightdale, NC 27545 USA
Phone: 919-365-7157   Fax: 919-366-2142
E-mail: nhadvs@worldnet.att.net
Website: www.nationalhunters.com

Founded: 1976
Scope: National

Description: The National Hunters Association, Inc. was incorporated under the laws of NC to protect your hunting rights in the U.S. and around the world. Dedicated to hunter safety, the preservation of the rights of the individual sportsman to pursue the sport of hunting and the preservation of an adequate supply of game for the sportsman to hunt — now and in the future.

Publication(s): NHA Newsletter

Keyword(s): Environmental and Conservation Education, Hunting, Trapping, Youth Organizations

NATIONAL MILITARY FISH AND WILDLIFE ASSOCIATION

12428 Pinecrest Ln.
Newburg, MD 20664 USA
Fax: 619-545-5225

Founded: 1983
Membership: 700
Scope: National

Description: A nonprofit organization established to promote professional natural resources management on over 25.5 million acres of United States Department of Defense lands worldwide. Membership is comprised primarily of professional Department of Defense natural resources personnel.

Publication(s): Fish and Wildlife News

Keyword(s): Biodiversity, Wildlife, Land Use Planning, Public Lands, training, Professional Organization

Contact(s):
Jim Bailey, AT-LARGE DIRECTOR
USAGAPG,STEAP-SH-ER, Aberdeen Proving Ground, MD 21005-5001
Phone: 410-278-6748
Fax: 410-278-6779
jbailey@dshe.apg.army.mil//
Pat Walsh, AT-LARGE DIRECTOR
347WG, DET 1, OLA/CEVN, Avon Park AF Range, FL 33825
Phone: 941-452-4254
Fax: 941-452-4221
Patrick.Walsh@avonpark.macdill.af.mil
Mike Passmore, NEWSLETTER EDITOR
Environmental Laboratory, US Army Engineer Research & Development Center, 3909 Halls Ferry Rd., ATTN: CEERD-EN-S, Vicksburg, MS 39180-6199
Phone: 601-634-4862
Fax: 601-634-3726
passmom@wes.army.mil//end
Dave Tazik, PRESIDENT
Environmental Lab, US Army Engineer Research & Development Center, ATTN: CEERD-EN, 3909 Halls Ferry Rd., Vicksburg, MS 39180-6199
Phone: 601-634-2610
Fax: 601-634-3726
tazikd@mail.wes.army.
Don Pitts, PRESIDENT-ELECT
ENGINEERING RESEARCH AND DEVEL CCTR
ATTN:CEERD-CD-N, Champaign, IL 61825-9005
Phone: 800-usa—cer
Fax: 217-373-7266
donald.pitts@erdc.usace.army.mil
Scott Smith, REGIONAL DIRECTORS EAST
Dare County AF Range, P.O. Box 2480, Mantelo, NC 27954
Phone: 919-722-1011
Fax: 919-722-0494
scott.smith@seymourjohnson.af.mil
John Joyce, REGIONAL DIRECTORS EAST
Naval Engineering Station, Code 8000EB5-2, Route 547, Lakehurst, NJ 08733-5065
Phone: 732-323-2911
Fax: 732-323-2718
joycejg@navair.navy.mil//
Coralie Cobb, REGIONAL DIRECTORS WEST
10954 Creekbridge Place, San Diego, CA 92131
Phone: 858-748-1719
cobb@aznet.net
Rhys Evans, REGIONAL DIRECTORS WEST
NREA Division, Bldg. 1451, Box 788110, MCAGCC, Twenty-nine Palms, CA 92278-8110
Phone: 760-830-7396
Fax: 760-830-5718
evansrm@29palms.usmc.mil//
Tammy Conkle, SECRETARY/TREASURER
Natural Resources Office, P.O. Box 357088 (Code N4515TC), NAS North Island (Bldg.3), San Diego, CA 92135-7088
Phone: 619-545-3703
Fax: 619-545-3489
conkle.tamara@ni.cnrsw.navy.mil//
Glenn Wampler, VICE PRESIDENT
ATTN: ATZR-BN, Fort Sill, OK 73503
Phone: 580-442-4324
Fax: 580-442-7207
wamplerg@sill.army.mil//

NATIONAL NETWORK OF FOREST PRACTITIONERS

305 S Main St.
Providence, RI 02903
Phone: 401-273-6507　　Fax: 401-273-6508
E-mail: info@nnfp.org
Website: www.nnfp.org

Founded: 1990
Membership: 500
Scope: National

Description: The National Network of Forest Practitioners is a grassroots alliance of rural people, organizations and businesses finding practical ways to integrate economic development, environmental protection and social justice.

Publication(s): Practitioner (newsletter), Directory, Engaging Communities in the Research Process

Keyword(s): Environmental Justice, Environmental Protection, Forest Stewardship, Forests and Forestry, Indigenous People, Natural Resource Conservation, Public Lands, Rivers, Water Pollution Management, Wilderness, Reforestation, Research, Rural Development, Sustainable Development

Contact(s):
Thomas Bendler, EXECUTIVE DIRECTOR
29 Temple Pl., 2nd Fl., Boston, MA 02111
Phone: 617-338-7821
Fax: 617-422-0881
tbendler@igc.org

NATIONAL ORGANIZATION FOR RIVERS (NORS)

212 W. Cheyenne MT Blvd.
Colorado Springs, CO 80906 USA
Phone: 719-579-8759　　Fax: 719-576-6238
E-mail: nors@rml.net
Website: www.nationalriver.org

Founded: 1979
Scope: National

Description: A nonprofit organization dedicated to education about whitewater river sports, including kayaking, rafting, and canoeing; to preserving rivers; and to protecting river access rights of the general public.

Publication(s): Currents

Contact(s):
Earl Perry, BOARD MEMBER
Fletcher Anderson, BOARD MEMBER
Gary Lacy, PRESIDENT
Eric Leaper, SECRETARY, TREASURER, AND EXECUTIVE DIRECTOR
Ben Harding, VICE PRESIDENT

NATIONAL PARK TRUST

415 2nd St., NE, Suite 210
Washington, DC 20002 USA
Phone: 202-548-0500　　Fax: 202-548-0595
E-mail: legacy@parktrust.org
Website: www.parktrust.org

Founded: 1983
Scope: National

Description: The private nonprofit land conservancy dedicated exclusively to protecting resources within and around parklands and other natural and historic properties. The Trust is the only private citizen group recognized by Congress to own and manage, in cooperation with the National Park Service, a unit of the National Park System, The Tallgrass Prairie National Preserve, established in 1996. The Trust has acquired land in over 40 other parks units and finished 4 parks.

Publication(s): Legacy Report, NPT Legacy News

Keyword(s): Historic Preservation, Land Purchase, Public Lands, training, National Parks

Contact(s):
Stephan Miller, CHAIRMAN OF THE BOARD OF TRUSTEE
Paul Pritchard, PRESIDENT
paul@parktrust.org
William Brownell, SECRETARY
Barry Schimel, TREASURER
Davinder Khanna, VICE PRESIDENT FOR OPERATIONS
dkhanna@aol.com
Susan Hawley, VICE PRESIDENT FOR PROGRAMS
susan@parktrust.org

NATIONAL PARKS AND CONSERVATION ASSOCIATION (NPCA)

Headquarters, 1300 19th St NW Ste. 300
Washington, DC 20036 USA
Phone: 800-628-7275
E-mail: npca@npca.org
Website: www.npca.org

Founded: 1919
Scope: National

Description: A private nonprofit citizen organization, dedicated solely to preserving, protecting, and enhancing the U.S. National Park System. As a "watchdog" group, NPCA has been an advocate as well as a constructive critic of the National Park Service. NPCA has focused on the health of the entire system, from specific sites and programs to the processes of planning, management, and evaluation.

Publication(s): ParkWatcher, National Parks

Keyword(s): Cultural Preservation, Land Preservation, Public Lands, National Parks

Contact(s):
Jerome Uher, DIRECTOR OF COMMUNICATIONS
Jennifer Robertson, SECRETARY
Carol Aten, VICE PRESIDENT
Jessie Brinkley, VICE PRESIDENT OF DEVELOPMENT
William Chandler, VICE PRESIDENT OF CONSERVATION POLICY
Terry Vines, VICE PRESIDENT OF MEMBERSHIP

NATIONAL PARKS AND CONSERVATION ASSOCIATION (NPCA)

HEARTLAND REGIONAL OFFICE
P.O. Box 25354
Woodbury, MN 55125-5354 USA
Phone: 612-735-8008
E-mail: npca@npca.org
Website: npca.org

Founded: NA
Scope: Regional

Contact(s):
Lori Nelson, DIRECTOR

NATIONAL PARKS AND CONSERVATION ASSOCIATION (NPCA)

NORTHEAST REGIONAL OFFICE
41 Winter Street, Suite 403
Boston, MA 02108 USA
Phone: 617-338-0126   Fax: 617-338-0232
E-mail: northeast@npca.org
Website: http://www.npca.org

Founded: NA
Scope: Regional

Contact(s):
Eileen Woodford, DIRECTOR

NATIONAL PARKS AND CONSERVATION ASSOCIATION (NPCA)

SOUTHEAST REGIONAL OFFICE
101 South Main Street, Suite 322
Clinton, TN 37716 USA
Phone: 865-457-7775   Fax: 865-457-6499
E-mail: southeast@npca.org
Website: http://www.npca.org/

Founded: NA
Scope: Regional

Contact(s):
Don Barger, DIRECTOR

NATIONAL PARKS CONSERVATION ASSOCIATION (NPCA)

PACIFIC REGIONAL OFFICE
P.O. Box 1289
Oakland, CA 94604 USA
Phone: 510-839-9922   Fax: 510-839-9926
E-mail: pacific@npca.org
Website: www.npca.org

Founded: NA
Membership: 4
Scope: Regional
Publication(s): Pacific Park News - biweekly email newsletter

Contact(s):
Christophe Foubert, ADMIN. ASST.
Courtney Cuff, REGIONAL DIRECTOR
Elizabeth North, REGIONAL DIRECTOR OF DEVELOPMENT
enorth@npc.org

NATIONAL PARKS CONSERVATION ASSOCIATION (NPCA)

ROCKY MOUNTAIN REGIONAL OFFICE
P.O. Box 737
Fort Collins, CO 80521 USA
Phone: 970-493-2545   Fax: 970-493-9164
Website: www.npca.org

Founded: NA
Membership: 400,000
Scope: National
Contact(s):
Mark Peterson, DIRECTOR FOR STATE OF THE PARKS PROGRAM

NATIONAL PARKS CONSERVATION ASSOCIATION (NPCA)

SOUTHWEST REGIONAL OFFICE
823 Gold Ave., SW
Albuquerque, NM 87102 USA
Phone: 505-247-1221   Fax: 505-247-1222
E-mail: southwest@npca.org
Website: www.npca.org

Founded: NA
Membership: 2
Scope: Regional

Publication(s): NPCA Policy Papers, Defending the Dessert, NPCA Policy Paper, Park Policy Agenda, NPCA Policy Paper, Vanishing Night Skies, NPAC Policy Paper, Ten Parks in Jeopardy, 2000 Annual Report, National Park Activist Guide, A Manual for Citizen Action, Guide To National Parks Southwest Region, Guide to National Parks Pacific Northwest Region, Guide to National Parks Rocky Mountian Region, Guide to National Parks South East Region, Guide to National Parks Heartland Region, Refer to Publication Website for further listings, Guide to National Parks North East Region, Guide to National Parks Pacific Region

Contact(s):
David Simon, DIRECTOR

NATIONAL PARKS CONSERVATION ASSOCIATION ALASKA REGION OFFICE

ALASKA REGIONAL OFFICE
750 West 2nd Ave.
Anchorage, AK 99501 USA
Phone: 907-277-6722   Fax: 907-277-6723
E-mail: AKRO@NPCA.ORG
Website: www.eparks.org

Founded: NA
Membership: 425000
Scope: Regional
Publication(s): National Parks Magazine

Contact(s):
Chip Dennerlein, DIRECTOR

NATIONAL PARKS FOUNDATION

1101 17th St., NW, Suite 1102
Washington, DC 20036 USA
Phone: 202-785-4500   Fax: 202-785-3539
Website: www.nationalparks.org

Founded: 1967
Scope: National

Description: The National Park Foundation is the official nonprofit partner of the National Park Service. Created by Congress in 1967, the Foundation raises support from corporations, foundations, and individuals to preserve and enhance America's National Parks. Over the past five years, the National Park Foundation has raised more than $42 million in direct support for the National Parks.

Publication(s): Complete Guide to America's National Parks, The

Keyword(s): Cultural Preservation, Environmental and Conservation Education, Historic Preservation, Outdoor Recreation, Public Lands, National Parks

Contact(s):
Bruce Babbitt, CHAIRMAN
Jill Nicoll, EXECUTIVE VICE PRESIDENT
James Maddy, PRESIDENT

Robert Stanton, SECRETARY
Claudia Schechter, TREASURER
B. West, VICE CHAIRMAN

NATIONAL RECREATION AND PARK ASSOCIATION

22377 Belmont Ridge Rd.
Ashburn, VA 20148 USA
Phone: 703-858-0784   Fax: 703-858-0794
Website: www.nrpa.org

Founded: NA
Membership: 23500
Scope: National

Description: A national nonprofit service, education, and research organization dedicated to the improvement of park and recreation leadership, programs, and facilities. The Association attempts to build public understanding that leisure programs and environments are indispensable to the well-being of a nation and its citizens.

Publication(s): Journal of Leisure Research, Therapeutic Recreation Journal, Recreation and Parks Law Reporter, Dateline, Parks and Recreation Magazine

Contact(s):
Rip Wilkenson, CHAIRMAN
Baton Rouge, LA
rwillkenson@nrpa.org
T Jarvis, CONTACT
tjarvis@nrpa.org
Barry Tindall, DIRECTOR OF PUBLIC POLICY
btindall@nrpa.org
Kathy Spangler, NATIONAL PROGRAMS DIRECTOR
Recreation and Park Assoc., 22377 Belmont Ridge Road, Ashburn, VA 20148
Phone: 703-858-0784
kspangler@nrpa.org
Pamela Earle, PACIFIC REGIONAL DIRECTOR
350 S. 333rd St., 103, Federal Way, WA 98003
Phone: 206-661-2265
pearle@nrpa.org
Alice Conkey, PRESIDENT
aconkey@nrpa.org
Larry Zehnder, SOUTHEAST REGIONAL DIRECTOR
1285 Parker Rd., Conyers, GA 30207
Phone: 404-760-1668
lzehnder@nrpa.org
Suzanne Mathis, TRUSTEE LIAISON & COORDINATOR OF FRIENDS OF PARKS AND RECREATION
smathis@nrpa.org
Maria Stamats, WESTERN REGIONAL DIRECTOR
CO 719-632-7031
mstamats@nrpa.org

NATIONAL RESEARCH COUNCIL

2101 Constitution Ave., NW
Washington, DC 20418 USA
Phone: 202-334-2000
Website: www.nas.edu

Founded: 1916
Scope: National

Description: An independent advisor to the federal government on scientific and technical questions of national importance. Jointly administered by the National Academies of Sciences and Engineering and the Institute of Medicine.

Publication(s): Catalogue available upon request.

Contact(s):
Bruce Alberts, CHAIRMAN
Suzanne Woolsey, CHIEF OPERATING OFFICER
Susan Vines, DIRECTOR OF OFFICE OF NEWS AND PUBLIC INFORMATION

NATIONAL RIFLE ASSOCIATION OF AMERICA

11250 Waples Mill Rd.
Fairfax, VA 22030 USA
Phone: 703-267-1000   Fax: 703-267-3909
E-mail: nra.contact@nra.org
Website: www.nra.org

Founded: 1871
Scope: National

Description: A nonprofit organization dedicated to protect and defend the Constitution of the United States, especially the right to possess and use firearms for recreation and personal protection; to promote public safety, law and order, and the national defense; and to train members of law enforcement agencies, the military, and private citizens of good repute in marksmanship and the safe handling and efficient use of small arms.

Publication(s): American Hunter, Insights, America's First Freedom, American Rifleman

Keyword(s): Hunting, Outdoor Recreation, training, Youth Organizations

Contact(s):
Kayne Robinson, 1ST VICE PRESIDENT
Sandra Froman, 2ND VICE PRESIDENT
Janice Taylor, ASSISTANT MANAGER OF HUNTER SERVICES
Phone: 703-267-1523
Susan Lamson, DIRECTOR OF CONSERVATION, WILDLIFE AND NATURAL RESOURCES DIVISION
Phone: 703-267-1541
William Poole, DIRECTOR OF EDUCATION AND TRAINING DIVISION
Phone: 703-267-1414
Craig Sandler, EXECUTIVE DIRECTOR OF GENERAL OPERATIONS
Wayne Lapierre, EXECUTIVE VICE PRESIDENT
Howard Moody, INSTRUCTOR AND COACH TRAINER FOR TRAINING
Phone: 703-267-1401
Charles Mitchell, MANAGER FOR TRAINING
Phone: 703-267-1431
Matthew Szramoski, MANAGER FOR YOUTH PROGRAMS
Phone: 703-267-1596
Robert Davis, MANAGER OF HUNTER SERVICES
Phone: 703-267-1522
Charlton Heston, PRESIDENT
Montey Embrey, PROGRAM ASSISTANT OF HUNTER SERVICES
Phone: 703-267-1503
Britt Ford, PROGRAM COORDINATOR OF HUNTER SERVICES
Phone: 703-267-1516
Edward Land, SECRETARY
Wilson Phillips, TREASURER
Billy Templeton, WILDLIFE MANAGEMENT SPECIALIST OF HUNTER SERVICES (ECHO)
Phone: 703-267-1501

NATIONAL SCIENCE TEACHERS ASSOCIATION

1840 Wilson Blvd.
Arlington, VA 22201 USA
Phone: 703-243-7100   Fax: 703-243-7177
Website: www.nsta.org

Founded: 1944
Membership: 50000
Scope: National

Description: NSTA is the world's largest organization committed to improving science education at all levels - preschool through college. NSTA's membership includes science teachers, science supervisors, administrators, scientists, business and industry representatives, and others involved in science education.

Publication(s): Science Scope, The Science Teacher, Journal of College Science Teaching, NSTA Reports!, Quantum, Science and Children

Keyword(s): Environmental and Conservation Education

Contact(s):
Shelley Carey, EDITOR
Phone: 703-312-9238
scarey@nsta.org
Gerald Wheeler, EXECUTIVE DIRECTOR
Phone: 703-312-9254
gwheeler@nsta.org

NATIONAL SHOOTING SPORTS FOUNDATION, INC.

Flintlock Ridge Office Center, 11 Mile Hill Rd.
Newtown, CT 06470-2359 USA
Phone: 203-426-1320   Fax: 203-426-1087
E-mail: info@nssf.org
Website: www.nssf.org

Founded: 1960
Membership: 30
Scope: National

Description: Nonprofit educational, trade-supported association sponsors a wide variety of programs to create a better understanding of and a more active participation in the shooting sports and in practical conservation.

Keyword(s): Environmental and Conservation Education, Hunting, Outdoor Recreation, Renewable Resources, training

Contact(s):
Larry Ference, DIRECTOR OF RESEARCH AND INFORMATION SERVICES
Bill Brassard, EDITORIAL DIRECTOR
Douglas Painter, EXECUTIVE DIRECTOR
Jodi DiCamillo, NATIONAL COORDINATOR OF STEP OUTSIDE
Robert Delfay, PRESIDENT AND CEO
Nancy Coburn, VICE PRESIDENT OF FINANCE AND ADMINISTRATION

NATIONAL SPELEOLOGICAL SOCIETY, INC.

2813 Cave Ave.
Huntsville, AL 35810-4431 USA
Phone: 256-852-1300   Fax: 256-851-9241
E-mail: nss@caves.org
Website: www.caves.org

Founded: 1941
Membership: 12000

Scope: National

Description: A nonprofit membership organization dedicated to the exploration, study, and conservation of America's caves and caverns, related features, and the ecology of caves.

Publication(s): Journal of Cave and Karst Studies, publishers of speleological books, NSS News

Keyword(s): Wildlife Rehabilitation, Endangered Species, Environmental and Conservation Education, Geology

Contact(s):
David Jagnow, CONSERVATION CHAIRMAN
1300 Iris St., Apt. 103, Los Alamos, NM 87544-3140
Phone: 505-662-0553
djagnow@roadrunner.com
John Moses, INTERNATIONAL SECRETARY
15807 River Roads Dr., Houston, TX 77079-5041
Phone: 281-597-1494
Michael Hood, PRESIDENT

NATIONAL TRAPPERS ASSOCIATION, INC.

P.O. Box 550
New Martinsville, WV 26155 USA
Phone: 304-455-2656   Fax: 309-829-7615
E-mail: trappers@aol.com
Website: www.nationaltrappers.com

Founded: 1959
Scope: National, International

Description: A national trappers organization dedicated to promoting sound conservation legislation; to conserving the nation's natural resources; to helping implement environmental education programs; and to promoting a continued annual furbearer harvest as a necessary wildlife management tool.

Publication(s): American Trapper

Keyword(s): Renewable Resources, Sustainable Development, Trapping, training

Contact(s):
Robert Colona, CONSERVATION DIRECTOR
5539 Sharptown Rd., Rhodesdale, MD 21659
Phone: 410-883-2607
Fax: 410-376-3916
Steve Greene, DIRECTOR OF COMMUNICATION
St Marys, KS 66536
Phone: 785-437-2930
Fax: 785-437-2940
Scott Hartman, DIRECTOR OF NATIONAL AND INTERNATIONAL AFFAIRS
New Martinsville, WV 26155
Phone: 304-455-4865
Fax: 304-455-6736
Tom Krause, EDITOR AND ADVERTISING MANAGER
P.O. Box 513, Riverton, WY 82501
Phone: 307-856-3830
Fax: 307-857-2993
tkrause@wyoming.com
Royl Schoonover, GENERAL ORGANIZER
P.O. Box 308, Westminster, VT 05158-0308
Phone: 802-722-9062
Fax: 802-722-9062
Dave Sollman, PRESIDENT
RR 1, Box 391-1, Heltonsville, IN 47436
Phone: 812-834-5334
Fax: 812-834-5334

Al Barton, VICE PRESIDENT
P O Box 435, Maypearl, TX 76064
Phone: 972-435-3071

NATIONAL TREE TRUST

1120 G St., NW, Suite 770
Washington, DC 20005 USA
Phone: 202-628-8733   Fax: 202-628-8735
E-mail: info@nationaltreetrust.org
Website: www.nationaltreetrust.org

Founded: 1990
Membership: 17
Scope: National

Description: The National Tree Trust serves as a catalyst for local
volunteer and community service groups in growing, planting,
and maintaining trees in rural communities, urban areas, and
along the nation's highways. NTT mobilizes volunteer groups,
promotes public awareness, provides educational and tree
planting grants, and unites civic and corporate institutions in
support of public land tree plantings.

Publication(s): National Tree Trust News, The

Keyword(s): Environmental and Conservation Education, Forests
and Forestry, Public Lands, Urban Forestry, Environment

Contact(s):
Cindy Zimar, ASSISTANT EXECUTIVE DIRECTOR
George Cates, EXECUTIVE DIRECTOR
Major General, USMC (Ret.),

NATIONAL TRUST FOR HISTORIC PRESERVATION

1785 Massachusetts Ave., NW
Washington, DC 20036 USA
Phone: 202-588-6000   Fax: 202-588-6038
E-mail: feedback@nthp.org
Website: www.nthp.org

Founded: 1949
Membership: 250,000
Scope: National

Description: Private nonprofit membership organization
chartered by Congress to encourage the public to participate in
the preservation of America's historic and cultural heritage
through advocacy, education, technical assistance, financial
aid to nonprofit groups, and demonstration programs.

Publication(s): Preservation Magazine, Preservation Law
Reporter, Historic Preservation Forum

Keyword(s): Environmental and Conservation Education, Historic
Preservation, Land Use Planning, Renewable Resources,
Urban Environment

Contact(s):
Richard Moe, PRESIDENT
Phone: 200-258-8860

NATIONAL TRUST FOR HISTORIC PRESERVATION

MID ATLANTIC
One Penn Center at Suburban Station, Suite 1520,
1617 John F. Kennedy Blvd.
Philadelphia, PA 19144 USA
Phone: 215-568-8162

Founded: NA
Scope: Regional

Contact(s):
Patricia Wilson, DIRECTOR

NATIONAL TRUST FOR HISTORIC PRESERVATION

MIDWEST OFFICE
53 W. Jackson Blvd., Suite 350
Chicago, IL 60604 USA
Phone: 312-939-5547   Fax: 312-939-5651
E-mail: mwronthp.org

Founded: NA
Scope: Regional

Publication(s): Preservation Magazine

Contact(s):
James Mann, DIRECTOR

NATIONAL TRUST FOR HISTORIC PRESERVATION

MOUNTAINS - PLAINS OFFICE
910 16th St., Ste .1100
Denver, CO 80202 USA
Phone: 303-623-1504   Fax: 303-623-1508
E-mail: mpro@nthp.org
Website: www.nthp.org

Founded: NA
Scope: Regional

Contact(s):
Barbara Pahl, DIRECTOR

NATIONAL TRUST FOR HISTORIC PRESERVATION

NORTHEAST OFFICE
7 Faneuil Hall Marketplace, 4th Fl.
Boston, MA 02109 USA
Phone: 617-523-0885   Fax: 617-523-1199
E-mail: nero@nthp.org
Website: www.nationaltrust.org

Founded: NA
Membership: 62,000
Scope: National

Contact(s):
Tina White, ADMINISTRATIVE ASSISTANT
7 Faneuil Hall Marketplace, Boston, MA 02109
Wendy Nicholas, DIRECTOR

NATIONAL TRUST FOR HISTORIC PRESERVATION

SOUTHERN OFFICE
456 King St.
Charleston, SC 29403 USA
Phone: 803-722-8552   Fax: 843-722-8652
E-mail: soro@hthp.org
Website: www.nationaltrust.org

Founded: 1949
Scope: Regional

Contact(s):
John Hildreth, DIRECTOR

NATIONAL TRUST FOR HISTORIC PRESERVATION

WESTERN
One Sutter St., Suite 707
San Francisco, CA 94104 USA
Phone: 415-956-0610   Fax: 415-956-0837
E-mail: wro@nthp.org

Founded: NA
Scope: Regional

Contact(s):
Kathryn Burns, DIRECTOR

NATIONAL TRUST FOR HISTORIC PRESERVATION

SOUTHWEST OFFICE
500 Main St., Suite 1030
Fort Worth, TX 76102 USA
Phone: 817-332-4398   Fax: 817-332-4512
E-mail: swo@nthp.org
Website: www.nthp.org

Founded: NA
Scope: Regional

Contact(s):
Daniel Carey, DIRECTOR

NATIONAL WATER RESOURCES ASSOCIATION

3800 N. Fairfax Dr., Suite 4
Arlington, VA 22203 USA
Phone: 703-524-1544   Fax: 703-524-1548
E-mail: nwra@nwra.org
Website: www.nwra.org

Founded: NA
Membership: 5000
Scope: National

Description: Promotes development, conservation and management of the water resources of 17 western state associations, including cities, counties, conservation districts, and individual members.

Publication(s): Water Report, National Waterline

Keyword(s): Agriculture, Endangered Species, Coral Reefs, Exotic species, Aquatic nuisance species, Wetlands, Nonpoint Source Pollution

Contact(s):
Thomas Donnelly, EXECUTIVE VICE PRESIDENT
David Sprynczynatyk, PRESIDENT

NATIONAL WATERSHED COALITION

9304 Lundy Ct.
Burke, VA 22015 USA
Phone: 703-455-6886   Fax: 703-455-6888

Founded: 1989
Membership: 7000
Scope: National

Description: The NWC is a nonprofit coalition made up of national, regional, state, and local organizations, associations, and individuals, that advocate dealing with natural resources problems and issues using the watershed as the planning and implementation unit.

Publication(s): Watershed Newsletter & Conference Proceedings, Watershed News

Contact(s):
Carry Smith, CHAIR
John Peterson, EXECUTIVE DIRECTOR
D. Sebert, SECRETARY, TREASURER
Oklahoma City, OK
Dan Lowranch, VICE CHAIR
Alma, OK

NATIONAL WATERWAYS CONFERENCE INC.

1130 17th St., NW
Washington, DC 20036-4676 USA
Phone: 202-296-4415   Fax: 202-835-3861
E-mail: information@waterways.org
Website: www.waterways.org

Founded: 1960
Scope: National

Description: To promote a better understanding of the public value of water resource and water transportation programs and to show their importance to the total environment.

Publication(s): Washington Watch

Keyword(s): Coasts, Energy, Rivers, Exotic species, Aquatic nuisance species

Contact(s):
Harry Cook, PRESIDENT AND EDITOR
Michael Toohey, SECRETARY
Director of Federal Government Relations, Ashland, Inc., 601 Pennsylvania Ave., NW, 540-N, Washington, DC 20004
Phone: 202-223-8290
Scott Robinson, TREASURER
Port Director of Muskogee City-County Port Authority, 4901 Harold Scoggins Dr., Muskogee, OK 74403
Phone: 918-682-7886
Fred Raskin, VICE CHAIRMAN
Pres. and CEO of Eastern Enterprises, Inc., 9 Riverside Rd., Weston, MA 02493-2214
Phone: 781-647-2300

NATIONAL WHISTLEBLOWER CENTER

National Whistleblower Legal Defense and Education Fund, 3238 P. St., NW
Washington, DC 20007 USA
Phone: 202-342-1902   Fax: 202-342-1904
E-mail: whistle@whistleblowers.org
Website: www.whistleblowers.org

Founded: 1988
Scope: National

Description: The Fund is the only public insterest law firm dedicated to enforcing and enhancing the legal protections of employees who blow the whistle on significant violations of law, environmental protection, nuclear safety, and first amendment rights. The Fund provides legal advice, resources, and referrals for counsel to whistleblowers nationwide and conducts seminars and other outreach activities.

Publication(s): Law Reporter, Whistleblower News

Keyword(s): Environmental Ethics, Environmental Justice, Environmental Law, Training

Contact(s):
Stephen Kohn, CHAIRPERSON
Phone: 202-342-6980

NATIONAL WILD TURKEY FEDERATION, CANADA, INC., THE

c/o Kevin Townsend, Regional Director
Wroxeter, Ontario N0G 2X0 Canada
Phone: 519-335-6893   Fax: 519-335-6050
E-mail: ontrdkt@wcl.on.ca

Founded: 1998
Scope: International

Description: A non-profit organization dedicated to the wise conservation and management of the North American Wild Turkey
and protecting the turkey hunting tradition. Supports annual
research grants program.

Publication(s): Turkey Caller, Turkey Call Television

Keyword(s): Hunting, Research Grants, Conservation

Contact(s):
Rob Keck, CEO/EXECUTIVE VICE-PRESIDENT
Russ Davies, PRESIDENT
Jack Playne, SECRETARY-TREASURER
Randy Roloson, VICE-PRESIDENT
James Kennamer, VICE-PRESIDENT FOR CONSERVATION
PROGRAMS

NATIONAL WILD TURKEY FEDERATION, INC., THE

770 Augusta Rd., P.O. Box 530
Edgefield, SC 29824-0530 USA
Phone: 803-637-3106   Fax: 803-637-0034
E-mail: nwtf@nwtf.net
Website: www.nwtf.org

Founded: 1973
Membership: 390,000
Scope: International

Description: A nonprofit organization dedicated to the wise conservation and management of the North American Wild Turkey
and protecting the turkey hunting tradition. Comprised of 1,300
state and local affiliates. Supports annual research grants
program.

Publication(s): Caller, The, JAKES Magazine, Women in the
Outdoors Magazine, Turkey Call Magazine

Keyword(s): Women in the Environment, Research Grants,
Forests and Forestry, Hunting, Public Lands

Contact(s):
Rob Keck, CEO/EXECUTIVE VICE PRESIDENT
James Sparks, CHIEF FINANCIAL OFFICER
Carl Brown, CHIEF OPERATING OFFICER
Tammy Bristow, DIRECTOR OF COMMUNICATIONS
Donna Leggett, DIRECTOR OF DEVELOPMENT
James Kennamer, VICE PRESIDENT FOR CONSERVATION
PROGRAMS

NATIONAL WILDLIFE FEDERATION
ALASKA PROJECT OFFICE
750 W. Second Ave.
Anchorage, AK 99501 USA
Phone: 907-258-8480   Fax: 907-258-4811
Website: www.nwf.org

Founded: NA
Scope: Regional

Description: National Wildlife Federation's Alaska office was
established in 1988 and specializes in wetlands issues. Our
work ranges from educational programs in public schools to
federal court lawsuits designed to influence national wetlands
policy. Increasingly, the office builds and leads conservation
coalitions. NWF currently coordinates the Cooper River Delta
Coalition and Prince William Sound Alliance. We also manage
Alaska Women's Environmental Network and Alaska Youth for
Environmental Action, and are key players in the campaign to
protect the Arctic National Wildlife Refuge.

Contact(s):
Tony Turrini, DIRECTOR
Turrini@nwf.org

NATIONAL WILDLIFE FEDERATION
EVERGLADES PROJECT OFFICE
2590 Golden Gate Parkway, Suite 109
Naples, FL 34105 USA
Phone: 941-643-4111   Fax: 941-643-5130
Website: www.nwf.org

Founded: NA
Scope: Regional

Description: The Everglades Project Office specializes in
conserving and protecting the Big Cypress Watershed region
of the Everglades. Advocating conservation-based land use
planning, staff engages policy and decision-makers at the
federal, state, and local level, and uses litigation and
educational programs to influence management, legal, and
legislative issues effecting the Everglades.

Publication(s): Everglades Newsletter

Keyword(s): Everglades

NATIONAL WILDLIFE FEDERATION
FEDERAL & CONGRESSIONAL AFFAIRS
1400 16th St., NW, Suite 501
Washington, DC 20036 USA
Phone: 202-797-6800   Fax: 202-797-6646
Website: www.nwf.org

Founded: NA
Scope: National

Description: Advocates NWF's position on important issues
affecting the national community. It provides technical,
scientific, and legal support to state affiliates and field offices
on selected issues. Advocates the Federation's position before
all three branches of government and works cooperatively with
the private sector to achieve mutually desired goals.

Contact(s):
Jim Lyon, LEGISLATIVE DIRECTOR
Lyon@nwf.org

NATIONAL WILDLIFE FEDERATION
GREAT LAKES NATURAL RESOURCE
CENTER
213 W. Liberty, Ste. 200
Ann Arbor, MI 48104-2210 USA
Phone: 734-769-3351   Fax: 734-769-1449
E-mail: greatlakes@nwf.org
Website: www.nwf.org

Founded: NA
Scope: Regional

Description: The National Wildlife Federation's Great Lakes
Natural Resource Center unites people throughout the eight-
state Great Lakes region, the U.S. and Canada to protect the

world's greatest freshwater seas and the surrounding ecosystem. The Center's staff of scientists, educators, lawyers, and organizers work with citizens and activists to end the toxic pollution and habitat destruction that threaten the health of wildlife, fish, and people in the Great Lakes region.

Contact(s):
Guy Williams, DIRECTOR
williams@nwf.org

NATIONAL WILDLIFE FEDERATION
GULF STATES NATURAL RESOURCE CENTER
44 East Ave., Suite 200
Austin, TX 78701 USA
Phone: 512-476-9805   Fax: 512-476-9810
Website: www.nwf.org

Founded: NA
Scope: Regional

Description: The Gulf States Natural Resource Center is working to protect threatened rivers and important wetlands in the region and to restore polluted watersheds. The Center also promotes NWF's educational programs by working with schools and other organizations.

Publication(s): Publication on website

Contact(s):
Susan Kaderka, DIRECTOR
Phone: 512-476-9818
Fax: 512-476-9810
kaderka@nwf.org

NATIONAL WILDLIFE FEDERATION
HEADQUARTERS
11100 Wildlife Center Drive.
Reston, VA 20190-5362 USA
Phone: 703-438-6000   Fax: 703-442-7332
Website: www.nwf.org

Founded: 1936
Membership: 4,000,000+
Scope: National, International

Description: A nonprofit organization whose mission is to educate, inspire, and assist individuals and organizations of diverse cultures to conserve wildlife and other natural resources and to protect the Earth's environment in order to achieve a peaceful, equitable, and sustainable future. NOTE. Any correspondence for a member of the Board of Directors of the National Wildlife Federation should be directed to the National Wildlife Federation mailing address or fax number.

Publication(s): International Wildlife, Ranger Rick, Conservation Directory, Your Big Backyard, EnviroAction, NatureScope, National Wildlife Week, EarthSavers, Wild Animal Baby, National Wildlife

Keyword(s): Endangered Species, Land Preservation, Sustainable Development, Water Quality, Wetlands, Everglades

Contact(s):
Bryan Pritchett, BOARD OF DIRECTORS, CHAIR
Rebecca Scheibelhut, BOARD OF DIRECTORS, VICE CHAIR
Edward Clark, BOARD OF DIRECTORS, VICE CHAIR
Eileen Johnson, GENERAL COUNSEL
johnsone@nwf.org
Mark Van Putten, PRESIDENT AND CHIEF EXECUTIVE OFFICER
vanputten@nwf.org
Christopher Palmer, PRESIDENT AND CHIEF EXECUTIVE OFFICER OF NATIONAL WILDLIFE PRODUCTIONS
Phone: 703-790-4078
Fax: 703-790-4076
palmer@nwf.org
Doug Inkley, SENIOR ADVISOR
inkley@nwf.org
Tom Dougherty, SENIOR ADVISOR
dougherty@nwf.org
Barbara Bramble, SENIOR ADVISOR
bramble@nwf.org
Jamie Clark, SENIOR VICE PRESIDENT, CONSERVATION PROGRAMS
clark@nwf.org
Natalie Waugh, SENIOR VICE PRESIDENT, CONSTITUENT PROGRAMS
Phone: 703-790-4010
Fax: 703-790-4039
waugh@nwf.org
Lawrence Amon, SR. VICE PRESIDENT AND CFO
amon@nwf.org
Wayne Schmidt, STAFF DIRECTOR
schmidt@nwf.org
Dan Chu, VICE PRESIDENT OF AFFILIATE RELATIONS
chu@nwf.org
Philip Kavits, VICE PRESIDENT OF COMMUNICATIONS
kavitz@nwf.org
Jessie Brinkley, VICE PRESIDENT OF DEVELOPMENT
brinkley@nwf.org
James Stofan, VICE PRESIDENT OF EDUCATION
stofan@nwf.org
Robert Ertter, VICE PRESIDENT OF HUMAN RESOURCES
ertter@nwf.org
Jaime Matyas, VICE PRESIDENT OF INTERNET AND CAUSE RELATED MARKETING
matyas@nwf.org
Thomas McGuire, VICE PRESIDENT OF MEMBERSHIP
mcguire@nwf.org
Carole Fox, VICE PRESIDENT OF OPERATIONS, WINCHESTER FACILITY
fox@nwf.org
Susan Rieff, POLICY DIRECTOR
Phone: 512-476-9805
rieff@nwf.org
R. Fischer, POLICY DIRECTOR
Phone: 802-229-0650
fischer@nwf.org

NATIONAL WILDLIFE FEDERATION
INTERNATIONAL AFFAIRS
1400 16th St., NW, Suite 501
Washington, DC 20036 USA
Phone: 202-797-6603
Website: www.nwf.org

Founded: NA
Scope: International

Description: The international affairs team, also based in Washington D.C. office, works to advance the conservation agenda, recognizing the borderless reality of ecosystems and migratory species in North America and beyond. Staff members work with state affiliates and field offices to educate and to build a constituency for U.S. leadership on key international conservation issues and with other stakeholders and like-minded organizations, domestic and foreign, to address international conservation priorities. Dialogue and advocacy are conducted before both domestic and multilateral government agencies and forums. NWF's international affairs programs seek to ensure that the rules and institutions relating to the economic and social forces of globalization are shaped ion ways that safeguard wildlife and other natural resources to achieve a peaceful, equitable, and sustainable future.

Contact(s):
Paul Joffe, ACTING DIRECTOR
Joffe@nwf.org

NATIONAL WILDLIFE FEDERATION
NORTHEAST NATURAL RESOURCE CENTER
58 State St., Suite 1
Montpelier, VT 05602 USA
Phone: 802-229-0650   Fax: 802-229-4532
Website: www.nws.org

Founded: NA
Scope: Regional

Description: The Northeast Natural Resource Center of the National Wildlife Federation works on a range of conservation and natural resource issues across the six-state New England region and often in close coordinate with our state-based affiliate conservation groups. We also collaborate with a variety of other like-minded organizations to focus on the protection and restoration of our unique "woods, water, and wildlife" across with the region. Applying common-sense programs in education, advocacy, and research, NWF's overarching goals are to provide conservation leadership and protection for wildlife and their habitat for generations to come.

Contact(s):
Eric Palola, DIRECTOR
palola@nwf.org

NATIONAL WILDLIFE FEDERATION
NORTHERN ROCKIES PROJECT OFFICE
240 N. Higgins, Suite 2
Missoula, MT 59802 USA
Phone: 406-721-6705   Fax: 406-721-6714
Website: ww.nwf.org

Founded: NA
Scope: Regional

Description: The Northern Rockies Project Office focuses on endangered species recovery as a key to protecting not only the species themselves, but many other fish and wildlife populations as well. Through its work on wolf and grizzly bear recovery, the Northern Rockies office works at a landscape level to ensure the conservation of these species across millions of acres of important wildlife habitat. Across the prairie and the great grassland basins of the intermountain West, NWF's work on sage grouse, black tailed prairie dogs and black-footed ferrets helps ensure quality habitat for wildlife and clean water an healthy riparian areas for fish.

Contact(s):
Thomas France, DIRECTOR
france@nwf.org

NATIONAL WILDLIFE FEDERATION
NORTHWESTERN NATURAL RESOURCE CENTER
418 First Ave. West
Seattle, WA 98119 USA
Phone: 206-285-8707   Fax: 206-285-8698
E-mail: salmon@nwf.org
Website: www.nwf.org

Founded: NA
Scope: Regional

Description: Located in Seattle, Washington, the Northwestern Natural Resource Center focuses on the issues critical to this region, such as wild salmon, water quality, smart growth, and wolf recovery. NWF provides the tools, the expertise, and the grassroots clout to make a difference for wildlife and wild places in the Northwest, the nation, and the world.

Contact(s):
Thomas France, ACTING DIRECTOR

NATIONAL WILDLIFE FEDERATION
ROCKY MOUNTAIN NATURAL RESOURCE CENTER
2260 Baseline Rd., Suite 100
Boulder, CO 80302 USA
Phone: 303-786-8001   Fax: 303-786-8911
Website: www.nwf.org

Founded: NA
Scope: Regional, National

Description: The Rocky Mountain Natural Resource Center is dedicated to the conservation of wildlife and natural resources on public and private lands throughout the Rocky Mountains and Plains States. Center staff are working to restore biological diversity on millions of acres of native grasslands, to promote the restoration of water quality in our streams and rivers, and to promote the restoration of bison populations residing in Yellowstone National Park to lands throughout their historic range.

Contact(s):
Catherine Johnson, DIRECTOR
Johnsonc@nwf.org

NATIONAL WILDLIFE FEDERATION
SOUTHEASTERN NATIONAL RESOURCE CENTER
1330 West Peachtree St., Suite 475
Atlanta, GA 30309 USA
Phone: 404-876-8733   Fax: 404-892-1744
Website: www.nwf.org

Founded: NA
Scope: National

Description: The Southeastern Natural Resource Center in Atlanta forges links between people and the environment, promoting sustainable practices to enhance the quality of life in our communities for people and wildlife.

The SE Natural Resource Center was established in 1986 to address the water issues that affect both people and animals in the southeastern United States. In fact, the southeastern states are home to some of the highest levels of freshwater and upland diversity in the world. Our work focuses on improving water quality, protecting water supplies and improving and improving wildlife habitat in the Chattahoochee River basin and the Greater Okefenokee Ecosystem in southern Georgia and northern Florida.

Contact(s):
Andrew Schock, DIRECTOR
schock@nwf.org

NATIONAL WILDLIFE FEDERATION
WESTERN NATURAL RESOURCE CENTER
3500 5th Avenue, Suite 101
San Diego, CA 92103 USA
Phone: 619-296-8353   Fax: 619-296-8355
E-mail: wnrc@nwf.org
Website: www.nwf.org

Founded: 2000
Scope: Regional

Description: The mission of the Western Natural Resource Center of National Wildlife Federation is to educate, inspire and empower people from all walks of life to conserve wildlife and other natural resources throughout California, Nevada, and within the Mexico/U.S. border region. Specific efforts include working with diverse partners to promote smart growth development in the biodiversity-rich areas of South California.

Contact(s):
David Younkman, DIRECTOR
Phone: 619-296-8353
younkman@nwf.org

NATIONAL WILDLIFE FEDERATION ENDOWMENT, INC.
11100 Wildlife Center Drive
Reston, VA 20190-5362 USA
Phone: 703-438-6000   Fax: 703-438-6060

Founded: NA
Scope: National

Description: Established to support the conservation education and resource management programs of the National Wildlife Federation. Gifts and bequests are invested, and income is transferred to the National Wildlife Federation.

Contact(s):
Raymond Golden, BOARD OF TRUSTEE
Mary Harris, BOARD OF TRUSTEE
Allen Guisinger, BOARD OF TRUSTEE VICE CHAIR
John Rainey, CHAIRMAN AND TRUSTEE
Eileen Johnson, SECRETARY
Lawrence Amon, TREASURER

NATIONAL WILDLIFE PRODUCTIONS, INC.
11100 Wildlife Center Dr.
Reston, VA 20190-5362 USA
Phone: 703-438-6077   Fax: 703-438-6076
E-mail: palmer@nwf.org
Website: www.nwf.org

Founded: NA
Scope: National

Description: National Wildlife Productions is the television, film, and multimedia arm of the National Wildlife Federation (NWF). The goal of NWP is to fulfill NWF's conservation mission by creating and producing television and mass-media projects, including children's television programs, documentaries, large format films for IMAX(r) theaters, feature films, TV movies, and interactive multimedia programs.

Contact(s):
Christopher Palmer, PRESIDENT AND CEO

NATIONAL WILDLIFE REFUGE ASSOCIATION
1010 Wisconsin Avenue NW, Suite 200
Washington, DC 20007 USA
Phone: 202-333-9075   Fax: 202-333-9077
E-mail: nwra@refugenet.org
Website: www.refugenet.org

Founded: 1975
Scope: National

Description: The NWRA is the only national membership, nonprofit organization dedicated solely to protecting and preserving the National Wildlife Refuge System and to increasing public understanding and appreciation of it. Our mission is to preserve and enhance the integrity of the nation's largest network of lands and waters set aside primarily for the benefit of wildlife. Membership open to interested persons, organizations, institutions.

Publication(s): Friends Flyer (Quarterly), Taking Flight, Flyer (Quarterly)

Keyword(s): training, Public Lands, Conservation, Environmental Protection, Protected Areas

Contact(s):
Michael Moran, ADIMINISTRATION
Kate Malleck, DIRECTOR, FIELD SERVICES AND COMMUNICATIONS
nwra_kate@refugenet.org
Beverly Heinze-Lacey, FRIENDS INITIATIVE CONSULTANT
c/o Parker River NWR, 261 Northern Blvd., Newburyport, MA 01950
Phone: 978-465-4178
Phone: 978-465-2807
bevlacey@aol.com
David Tobin, PRESIDENT
nwra_david@refugenet.org
Ellen Croteau, VICE PRESIDENT OF DEVELOPMENT AND MEMBERSHIP
nwra_ellen@refugenet.org

NATIONAL WILDLIFE REHABILITATORS ASSOCIATION

14 N. 7th Ave.
St. Cloud, MN 56303 USA
Phone: 320-259-4086
E-mail: nwra@nwrawildlife.org
Website: www.nwrawildlife.org

Founded: 1982
Membership: 2000
Scope: National

Description: A nonprofit membership organization committed to promoting and improving the integrity and professionalism of wildlife rehabilitation and contributing to the preservation of natural ecoystems. The Organization disseminates information, provides training, and encourages networking through a quarterly journal, an annual membership directory, reviewed publications, annual symposia, and active committees for standards, wildlife medicine, education, awards, and grants.

Publication(s): Principles of Wildlife Rehabilitation, NWRA Quick Reference Guide, Training Opportunities in Wildlife Rehabilitation, Minimum Standards for Wildlife Rehabilitation, Wildlife Rehabilitation Bulletin - comes out twice a yr, Wildlife Rehabilitation annual volumes

Keyword(s): Birds, Environmental and Conservation Education, Mammals, training, accreditation

Contact(s):
Daneil Ludwig, EDITOR
P.O. Box 2339, Glen Ellyn, IL 60138
Elaine Thrune, PRESIDENT
Phone: 320-255-4911
Erica Miller, SECRETARY
John Huckabee, TREASURER
Phone: 425-787-2500
Michael Cox, VICE PRESIDENT
Diane Nickerson, VICE PRESIDENT
Phone: 609-883-6606

NATIONAL WOODLAND OWNERS ASSOCIATION

374 Maple Ave., E., Suite 310
Vienna, VA 22180 USA
Phone: 703-255-2700   Fax: 703-281-9200
E-mail: nwoa@mindspring.com
Website: www.nationalwoodlands.org

Founded: 1983
Scope: National

Description: A nationwide association of woodland owners united to foster good stewardship of their nonindustrial private forest lands. Working together with cooperating and affiliated state woodland owners and forestry associations, the Association is a voice for private landowners on forestry, wildlife, and resource conservation issues. Sponsors the American Federation of Forest and Woodland Owner Associations.

Publication(s): National Woodlands Magazine, Woodland Report (newsletter)

Keyword(s): Forests and Forestry

Contact(s):
Peter Demarsh, CANADIAN FEDERATION OF WOODLOT OWNERS
Phone: 506-459-2990
Fax: 506-459-3515

Eric Johnson, EDITOR, NATIONAL WOODLANDS
Phone: 315-369-3078
Keith Argow, EDITOR, WOODLAND REPORT
Bert Udell, EXECUTIVE COMMITTEE CHAIR
Phone: 541-258-6643
Keith Argow, PRESIDENT
Warren Baird, VICE-PRESIDENT, MIDWEST REGION
Phone: 317-758-4735
lonice_barrett.dnr.state.us
Jill Cornell, VICE-PRESIDENT, NORTHEAST REGION
Phone: 518-753-4336
Don Girton, VICE-PRESIDENT, SOUTH REGION
Phone: 859-635-7826
Nels Hanson, VICE-PRESIDENT, WESTERN REGION
Phone: 360-943-3875

NATIVE AMERICAN FISH AND WILDLIFE SOCIETY (NAFWS)

750 Burbank St.
Broomfield, CO 80020 USA
Phone: 303-466-1725   Fax: 303-466-5414
Website: www.nafws.org

Founded: 1982
Membership: 1500
Scope: National

Description: The Native American Fish and Wildlife Society is a nonprofit organization serving the needs of fish, wildlife, and natural resources on Tribal lands across the United States, including Alaska. The Society membership is comprised of approximately 1,500 professional and technical personnel associated with Native American natural resource programs. Two hundred sixteen federally-recognized tribes represent the Society, and many federal agencies rely on the Society's expertise and established network.

Publication(s): From the Eagles Nest

Keyword(s): Cultural Preservation, Environmental Planning, Wildlife, Forests and Forestry

Contact(s):
Ken Poynter, EXECUTIVE DIRECTOR
Phone: 303-466-1725
Matthew Vanderhoop, PRESIDENT
Phone: 508-645-9265
natres@vineyard.net
McGruther Faith, SECRETARY AND TREASURER
Phone: 906-632-0043
cotfma@up.net
Mike Fox, TECHNICAL SERVICES DIRECTOR
Harris Teresa, VICE PRESIDENT
Phone: 803-366-4792
harristeresa@yahoo.com

NATIVE PLANT SOCIETY OF NORTHEASTERN OHIO

2651 Kerwick
University Heights, OH 44118 USA
Phone: 216-371-4454

Founded: 1982
Scope: Regional

Description: The Native Plant Society of Northeastern Ohio educates about, views, protects and preserves native plants, trees, and shrubs.

Publication(s): On the Fringe

Keyword(s): Flowers, Plants, and Trees, Conservation, Aquatic Habitats, Native Plants, Ancient Forests, Botanical Gardens, Prairies, Endangered Species, Wetlands

Contact(s):
Tom Sampliner, BOARD MEMBER
Phone: 216-371-4454
Jean Roche, PRESIDENT
Brian Gilbert, SECRETARY
Phone: 216-486-8765
Judy Bradt-Hart, TREASURER
Phone: 440-548-2414
George Wilder, VICE PRESIDENT
Phone: 216-932-3351

NATIVE PLANT SOCIETY OF OREGON
P.O. Box 902
Eugene, OR 97440 USA
Phone: 541-343-2364   Fax: 541-341-1752
Website: www.npsoregon.org/

Founded: 1961
Scope: Statewide

Description: The Native Plant Society of Oregon is a nonprofit statewide organization. The Society is dedicated to the enjoyment, conservation, and study of Oregon's native vegetation.

Publication(s): Biography of Louis F Henderson, Conservation and Management of Native Plants and Fungi, Atlas of Oregon Carex, Proceedings from a Conference of the Native Plant of Oregon, KALMIOPSIS: Journal of the Native Plant Society of Oregon, Bulletin of the Native Plant Society of Oregon

Keyword(s): Endangered Species, Conservation, Biodiversity, Environmental Protection, Native Plants

Contact(s):
Bruce Newhouse, PRESIDENT
Phone: 541-343-2364
Kelli Van Norman, SECRETARY
Eric Wold, TREASURER
Michael McKeag, VICE-PRESIDENT
Phone: 503-642-3965
vice_president@npsoregon.org

NATIVE PLANT SOCIETY OF TEXAS
P.O. Box 891
Georgetown, TX 78627 USA
Phone: 512-238-0695   Fax: 512-238-0703
E-mail: dtucker@io.com
Website: npsot.org

Founded: 1980
Membership: 2,000
Scope: Statewide

Description: A nonprofit organization dedicated to the education and promotion of conservation, preservation, and utilization of the native plants and the plant habitats of Texas.

Publication(s): Native Plant Society of Texas News

Keyword(s): Native Plants

NATIVE PRAIRIES ASSOCIATION OF TEXAS
P.O. Box 210
Georgetown, TX 78627 USA
Phone: 512-339-0618
E-mail: prairie65@aol.com
Website: www.texasprairie.org

Founded: 1986
Membership: 200
Scope: Statewide

Description: Native Prairies Association of Texas is dedicated to conservation and restoration of native prairies, through education, research, public awareness, agency cooperation, management, restoration, and acquisitions.

Publication(s): Prairie Dog, The

Keyword(s): Prairies, Restoration, Conservation Easements

Contact(s):
Lee Stone
Phone: 512 5-81 -9822
Gene Heinemann, PRESIDENT
Clint Josey, VICE PRESIDENT

NATURAL LAND INSTITUTE
320 S. 3rd St.
Rockford, IL 61104 USA
Phone: 815-964-6666   Fax: 815-964-6661
E-mail: nli@aol.com
Website: www.naturalland.org

Founded: 1982
Membership: 700
Scope: Regional

Description: A nonprofit organization to protect Illinois's native flora and fauna, and to encourage wise stewardship of the natural resources that affect them.

Publication(s): Land and Nature (Quarterly Newsletter), Flora of Winnebago County, Pecatonica River Water Shed

Keyword(s): Native Plants, Endangered Species, Protected Areas, Biodiversity

Contact(s):
Jill Kennay, ASSISTANT DIRECTOR
Jerry Paulson, EXECUTIVE DIRECTOR
Rebecca Olson, LAND PRESERVATION SPECIALIST
Randall Vincent, PRESIDENT
Gary McIntyre, VICE PRESIDENT

NATURAL AREAS ASSOCIATION
Bend, OR 97709 USA
Phone: 541-317-0199   Fax: 541-317-0140
E-mail: naa@natareas.org
Website: www.natareas.org

Founded: 1980
Membership: 6000
Scope: International

Description: A nonprofit organization of professional and active volunteers in natural area identification, preservation, protection, management, and research. Provides a medium of exchange and coordination to advance the understanding and appreciation of natural areas and natural diversity.

Publication(s): Natural Areas Journal, Natural Area News

Keyword(s): Biodiversity, Conservation, Natural Areas, Endangered Species, Land Preservation, Nongame Wildlife, Wildlands, National Parks

Contact(s):
Reid Schuller, EXECUTIVE DIRECTOR
P.O. Box 1504, Bend, OR 97709
Phone: 541-317-0199
Fax: 541-317-0140
naa@natareas.org
Chuck Williams, JOURNAL EDITOR
Biology Dept., Clarion University, Clarion, PA 16214
Phone: 814-226-1936
Fax: 814-226-2731
cwilliams@mail.clarion.edu
Harry Tyler, PRESIDENT
Maine State Planning Office, 184 State St., State House Station 38, Augusta, ME 04333
Phone: 207-287-1489
Fax: 207-287-7379
hank.tyler@state.me.us
Peg Kohring, SECRETARY
P.O. Box 506, Sawyer, MI 49125
Phone: 312-913-9459
Fax: 312-913-9523
Pkohring@aol.com
Sam Pearsall, TREASURER
1307 Chaney Rd, Raleigh, NC 27606
Phone: 919-403-8558
Fax: 919-403-0379
spearsall@tnc.org
Carl Becker, VICE PRESIDENT
Natural Heritage Division, Illinois Dept. of Natural Resources, 524 Second St., Springfield, IL 62701-1787
Phone: 217-785-8774
Fax: 217-785-8277
cbecker@dnrmail.state.il.us

NATURAL HISTORY SOCIETY OF MARYLAND, INC., THE

2643 N. Charles St.
Baltimore, MD 21218-4590 USA
Phone: 410-235-6116
Website: www.naturalhistory.org

Founded: 1929
Membership: 4
Scope: National

Description: A nonprofit membership organization formed to promote the appreciation of natural history through education, research, and publication—thereby fostering stewardship of natural and cultural resources. The Society maintains a museum of Maryland objects and a library, and bestows the Edmund B. Fladung Award to recognize persons exemplifying the society's goals.

Publication(s): The Maryland Naturalist, News and Views, Bulletin of the Maryland Herpetological Society

Keyword(s): Birds, Flowers, Plants, and Trees, Insects and Butterflies, Mammals, education, Reptiles and Amphibians

Contact(s):
Charles Davis, CHAIRMAN OF THE BOARD

NATURAL LAND INSTITUTE

320 S. 3rd St.
Rockford, IL 61104 USA
Phone: 815-964-6666   Fax: 815-964-6661
E-mail: nli@aol.com
Website: www.naturalland.org

Founded: 1958
Scope: Statewide

Description: A nonprofit organization dedicated to preserving natural areas and natural diversity through a comprehensive program of land protection, stewardship, research, education, and advocacy.

Publication(s): Boone and Winnebago Regional Greenways Plan, Land and Nature, Flora of Winnebago County

Keyword(s): Biodiversity, Natural Areas, Land Preservation, Watersheds, Birds, Conservation, Ecology, Aquatic Habitats, Endangered Species, Environmental Planning, Land Preservation, Land Use Planning, Protected Areas, Environmental Protection, Land Purchase

Contact(s):
Jerry Paulson, EXECUTIVE DIRECTOR

NATURAL LANDS TRUST, INC.

Hildacy Farm, 1031 Palmers Mill Rd.
Media, PA 19063 USA
Phone: 610-353-5587   Fax: 610-353-0517
Website: www.natlands.org

Founded: 1961
Scope: Regional

Description: Natural Lands Trust is a regional land trust dedicated to working with people to conserve land in the Philadelphia metropolitan region and other nearby areas of environmental concern by acquiring and managing preserve properties, accepting conservation easements, and encouraging and supporting the conservation efforts of landowners, communities, government agencies, and other nonprofit organizations.

Publication(s): Natural Lands

NATURAL RESOURCES COUNCIL OF AMERICA

1025 Thomas Jefferson St., NW, Suite 109
Washington, DC 20007-5291 USA
Phone: 202-333-0411   Fax: 202-333-0412
E-mail: nrca@naturalresourcescouncil.org
Website: http://www.naturalresourcescouncil.org

Founded: 1946
Scope: National

Description: An association of nonprofit environmental and conservation organizations dedicated to the protection, conservation, and responsible management of the nation's natural resources. The Council coordinates cooperative efforts between its members, government agencies, private citizens, and businesses. The Council also administers the Conservation Round Table Luncheon series, an annual Conservation Community Banquet and Awards program, and publishes a bimonthly newsletter.

Publication(s): The Conservation voice, NEP

Keyword(s): Wildlife, Forests and Forestry, Public Lands, Exotic species, Aquatic nuisance species, training

Contact(s):
Melissa Bondi, ASSISTANT DIRECTOR
melissa@nationalresourcescouncil.org
Andrea Yank, EXECUTIVE DIRECTOR
andrea@nationalresourcescouncil.org

NATURAL RESOURCES COUNCIL OF MAINE

3 Wade St.
Augusta, ME 04330-6351 USA
Phone: 207-622-3101　　　　　　Fax: 207-622-4343
E-mail: nrcm@nrcm.org
Website: maineenvironment.org

Founded: NA
Membership: 5000
Scope: Statewide

Description: A representative statewide organization, affiliated with the National Wildlife Federation, dedicated to the protection and enhancement of wildlife and its habitat through public education and government interaction.

Publication(s): other available on website, Maine Environment

Contact(s):
Patty Renaud, EDITOR & EDUCATION PROGRAMS CONTACT
Brownie Carson, EXECUTIVE DIRECTOR AND ALTERNATE REPRESENTATIVE
Ellen Baum, PRESIDENT
Paul Liebow, REPRESENTATIVE
Mac Deford, TREASURER

NATURAL RESOURCES DEFENSE COUNCIL

6310 San Vicente Blvd.
Los Angeles, CA 90048 USA
Phone: 323-934-6900　　　Fax: 323-934-1210
Website: www.nrdc.org

Founded: NA
Membership: 500000
Scope: National

Contact(s):
Gayle Petersen, OFFICE MANAGER

NATURAL RESOURCES DEFENSE COUNCIL, INC.

Headquarters, 40 W. 20th St.
New York, NY 10011 USA
Phone: 212-727-2700　　　Fax: 212-727-1773
E-mail: nrdcinfo@nrdc.org
Website: www.nrdc.org

Founded: 1970
Membership: 520,000
Scope: National

Description: Nonprofit membership organization dedicated to protecting America's endangered natural resources and to improving the quality of the human environment. Combines interdisciplinary legal and scientific approach in crafting innovative solutions, monitoring government agencies, bringing legal action, and disseminating citizen information. Areas of concentration: air and water pollution, global warming, nuclear safety, land use, urban environment, pollution prevention, ecosystem management, wilderness and wildlife protection, international environment, Alaska, energy efficiency, forestry, and ocean and fisheries protection.

Publication(s): A complete list of NRDCs books and reports is available upon request., Amicus Journal, The

Contact(s):
Frederick Schwarz, CHAIRMAN OF THE BOARD
Phone: 212-727-2700
Kathrin Lassila, EDITOR
Frances Beinecke, EXECUTIVE DIRECTOR
John Adams, PRESIDENT
Phone: 212-727-2700

NATURAL RESOURCES DEFENSE COUNCIL, INC.

SAN FRANCISCO, CALIFORNIA OFFICE
71 Stevenson St.,　1825
San Francisco, CA 94105 USA
Phone: 415-777-0220　　　Fax: 415-495-5996
Website: www.nrdc.org

Founded: NA
Scope: National

Contact(s):
Gwen Thomas, CLERICAL ASSISTANT
Linda Ward, OFFICE ADMINISTRATOR

NATURAL RESOURCES DEFENSE COUNCIL, INC.

WASHINGTON, D.C. OFFICE
1200 New York Ave., NW, Suite 400
Washington, DC 20005 USA
Phone: 202-289-6868　　　Fax: 202-289-1060
E-mail: nrdcinfo@nrdc.org
Website: www.nrdc.org

Founded: NA
Membership: 50
Scope: Regional
Publication(s): The Amicus Journal, see website for publications

Contact(s):
John Adams, PRESIDENT

NATURAL RESOURCES INFORMATION COUNCIL

Anne Hedrich Science & Technology Library, 3100 Old Main Hill
Logan, UT 84322-3100 USA
Phone: 435-797-2165　　　Fax: 435-797-7475
E-mail: annhed@cc.usu.edu
Website: www.quinneylibrary.usu.edu/NRIC/Index.htm

Founded: 1991
Scope: International

Description: Federal, state, provincal, academic, and special research librarians and information specialists from U.S. and Canada who facilitate the exchange of information on sustainable natural resources. Goals are to build a network of resource people to collect and disseminate information on sustainable natural resources and to provide continuing education.

Publication(s): Fish and Game Natural Resource Library Survey, Annual Newsletter (Yearly)

Keyword(s): education, Natural Science, Sustainability, Sustainable Resources, Librarians/Information Professionals

Contact(s):
Anne Hedrich, MEMBERSHIP
Phone: 435-797-2165
Fax: 435-797-7475
annhed@cc.usu.edu
Barbara Voeltz, TREASURER
bvoeltz@ngpc.state.ne.us

NATURAL SCIENCE FOR YOUTH FOUNDATION

130 Azalea Dr.
Roswell, GA 30075 USA
Phone: 770-594-9367   Fax: 770-594-7738

Founded: 1952
Scope: National

Description: Provides counseling to community groups in the planning and development of environmental and natural science centers, museums, and native animal parks which are designed particularly to meet the needs and interests of children and young people. Conducts an annual conference as part of its widespread effort to promote professional excellence in environmental and natural science centers and museums.

Publication(s): Directory of Natural Science Centers

Keyword(s): Environmental and Conservation Education, education, Outdoor Recreation, accreditation, Youth Organizations

Contact(s):
John Forbes, FOUNDER AND PRESIDENT EMERITUS
Joe Witley, PRESIDENT
Owen Winters, PUBLICATIONS DIRECTOR
Georgine Pindar, SECRETARY
John Hammaker, TREASURER

NATURE CONSERVANCY OF CANADA, THE

110 Eglinton Ave. W., Suite 400
Toronto, Ontario M4R 1A3 Canada
Phone: 416-932-3202   Fax: 416-932-3208
E-mail: nature@natureconservancy.ca
Website: www.natureconservancy.ca

Founded: 1962
Scope: National

Description: The Nature Conservancy of Canada is the only national charity dedicated to preserving ecologically significant areas, places of special beauty and education interest thorugh outright purchase, donations and conservation agreements.

Publication(s): The Ark, Annual Report

Keyword(s): Land Purchase, Land Preservation

Contact(s):
Ted Boswell, CHAIRMAN
Lynn Gran, DIRECTOR OF DEVELOPMENT
Ext. 284
lynn.gran@natureconservancy.ca
John Lounds, EXECUTIVE DIRECTOR

NATURE CONSERVANCY, THE

4245 North Fairfax Dr.
Arlington, VA 22208 USA
Phone: 703-841-5300   Fax: 703-841-1283
E-mail: unknown
Website: www.nature.org

Founded: 1951
Membership: 1,000,000

Scope: International

Description: International nonprofit membership organization committed to preserving biological diversity by protecting natural lands, and the life they harbor; cooperates with educational institutions, public and private conservation agencies. Works with states through "natural heritage programs" to identify ecologically significant natural areas. Manages a system of over 1,600 nature sanctuaries nationwide.

Publication(s): The Nature Conservancy Magazine

Keyword(s): Biodiversity, Endangered Species, Land Purchase, training

Contact(s):
Ray Culter, ADMINISTRATION
Phone: 703-841-5300
Bob Reynolds, CHAIRMAN OF THE BOARD
Phone: 703-841-5300
Bill Weeks, CHIEF CONSERVATION OFFICER
Phone: 703-841-5300
Steve Howell, CHIEF OPERATIONS OFFICER
Phone: 703-841-5300
David Williamson, COMMUNICATIONS
Phone: 703-841-5300
Deborah Jensen, CONSERVATION SCIENCE
Phone: 703-841-5300
Grace Vance, DEVELOPMENT AND MARKETING
Phone: 703-841-5300
Greg Row, DOMESTIC CONSERVATION
Phone: 703-841-5300
Ron Geatz, EDITOR-IN-CHIEF
Phone: 703-841-5300
Mike Dennis, GENERAL COUNSEL
Phone: 703-841-5300
Maggie Coon, GOVERNMENT RELATIONS
Phone: 703-841-5300
mcoon@tnc.org
Diane Gosting, HUMAN RESOURCES BUSINESS SERVICES
Phone: 703-841-5300
Joy Gaddy, HUMAN RESOURCES FIELD SERVICES
Phone: 703-841-5300
Alexander Watson, INTERNATIONAL CONSERVATION
Phone: 703-841-5300
Donna Cherel, MEMBERSHIP
Phone: 703-841-5300
Steve McCormick, PRESIDENT
Phone: 703-841-5300

NATURE CONSERVANCY, THE

700 SW Jackson St., Suite 804
Topeka, KS 66603 USA
Phone: 785-233-4400   Fax: 785-233-2022
E-mail: rpalmer@tnc.org
Website: www.nature.org

Founded: NA
Membership: 13
Scope: Statewide

Contact(s):
Alan Pollom, VICE PRESIDENT

NATURE CONSERVANCY, THE
P.O. Box 4125
Baton Rouge, LA 70821 USA
Phone: 225-338-1040   Fax: 225-338-0103
E-mail: lafo@tnc.org
Website: www.louisiananature.org

Founded: NA
Membership: 5600
Scope: Statewide

Contact(s):
    Keith Ouchley, STATE DIRECTOR

NATURE CONSERVANCY, THE
ALABAMA CHAPTER
2821 C 2nd Ave.
Birmingham, AL 35233 USA
Phone: 205-251-1155   Fax: 205-251-4444
E-mail: cwilborn@tnc.org
Website: www.nature.org

Founded: NA
Membership: 13
Scope: Statewide

Contact(s):
    Kathy Stiles Freeland, STATE DIRECTOR

NATURE CONSERVANCY, THE
ALASKA CHAPTER
421 W. First Ave.,
Anchorage, AK 99501 USA
Phone: 907-276-3133   Fax: 907-276-2584
E-mail: alaska@tnc.org
Website: www.nature .org

Founded: 1988
Membership: 1800
Scope: Statewide
Publication(s): Nature Conservancy Of Alaska

Contact(s):
    Erin Dovichin, DIRECTOR OF COMMUNICATIONS
    alaska@tnc.org

NATURE CONSERVANCY, THE
ARIZONA CHAPTER
300 E. University Blvd., 230
Tuscon, AZ 85705 USA
Phone: 520-622-3861
Website: http://nature.org

Founded: 1951
Scope: Statewide

NATURE CONSERVANCY, THE
ARKANSAS FIELD OFFICE
601 N. University Ave.
Little Rock, AR 72205 USA
Phone: 501-663-6699   Fax: 501-663-8332

Founded: NA
Scope: Statewide

Contact(s):
    Nancy Delamar, STATE DIRECTOR
    cbornemeier@tnc.org

NATURE CONSERVANCY, THE
ASIA/PACIFIC PROGRAM
1116 Smith St., 201
Honolulu, HI 96817 USA
Phone: 808-537-4508

Founded: NA
Scope: International

NATURE CONSERVANCY, THE
CALIFORNIA CHAPTER
201 Mission St. 4th Fl.
San Francisco, CA 94105 USA
Phone: 415-777-0487

Founded: NA
Scope: Statewide

Contact(s):
    Steve McCormick, CONTACT

NATURE CONSERVANCY, THE
COLORADO CHAPTER
1881 Ninth St.
Suite 200
Boulder, CO 80302 USA
Phone: 303-444-2950   Fax: 303-444-2986
Website: www.nature.org /colorado

Founded: NA
Membership: 26,500
Scope: Statewide
Publication(s): Land Mark, The

Contact(s):
    Linda Giandinoto, OFFICE COORDINATOR

NATURE CONSERVANCY, THE
CONNECTICUT CHAPTER
55 High St.
Middletown, CT 06457 USA
Phone: 860-344-0716   Fax: 860-344-1334
Website: www.nature.org

Founded: NA
Membership: 40
Scope: International

Contact(s):
    Dennis McGrath, DIRECTOR

NATURE CONSERVANCY, THE
DELAWARE CHAPTER
100 W. 10th St 1107
Wilmington, DE 19801 USA
Phone: 302-369-4144   Fax: 302-654-4708
E-mail: delaware@tnc.org
Website: www.nature.org

Founded: NA
Membership: 3800
Scope: Statewide

Contact(s):
    Roger Jones, STATE DIRECTOR

NATURE CONSERVANCY, THE
EASTERN NEW YORK CHAPTER
200 Broadway, 3rd Floor
Troy, NY 12180 USA
Phone: 518-272-0195   Fax: 518-272-0298
Website: www.tnc.org

Founded: NA
Membership: 18000
Scope: Local/Regional
Publication(s): Available on website

Contact(s):
Kieley Michasiow, OFFICE MANAGER
Phone: 518-272-0195

NATURE CONSERVANCY, THE
EASTERN NEW YORK CHAPTER
19 N. Moger
Mt. Kisco, NY 10549 USA
Phone: 914-244-3271   Fax: 914-244-3275

Founded: NA
Scope: Local Region

Contact(s):
Kathleen Moser, EXECUTIVE DIRECTOR

NATURE CONSERVANCY, THE
FLORIDA CHAPTER
222 S. Westmonte Dr. Suite 300
Altamonte Springs, FL 32714 USA
Phone: 407-682-3664   Fax: 407-682-3077
E-mail: mcantillo@tnc.org
Website: http://nature.org/

Founded: NA
Scope: Statewide

NATURE CONSERVANCY, THE
GEORGIA CHAPTER
1330 W. Peachtree St., Ste.410
Atlanta, GA 30309-2904 USA
Phone: 404-873-6946   Fax: 404-873-6984
Website: www.nature.org

Founded: 1987
Membership: 20000
Scope: Statewide
Publication(s): Quarterly newsletter

Contact(s):
Tavia McCuean, VICE PRESIDENT AND STATE DIRECTOR

NATURE CONSERVANCY, THE
GREAT PLAINS DIVISION
1313 5th St., SE, Suite 320
Minneapolis, MN 55414 USA
Phone: 612-331-0750   Fax: 612-331-0770
Website: www.nature.org

Founded: NA
Scope: Regional

Publication(s): Annual Report, Newsletter, National Magazine
Contact(s):
Robert McKim, STATE DIRECTOR

NATURE CONSERVANCY, THE
HAWAII CHAPTER
923 Nu'uanu Avenue
Honolulu, HI 96817 USA
Phone: 808-537-4508   Fax: 808-545-2019
E-mail: mwaits@tnc.org
Website: http://nature.org

Founded: NA
Scope: Statewide

NATURE CONSERVANCY, THE
IDAHO CHAPTER
Sun Valley, ID 83353 USA
Phone: 208-726-3007   Fax: 208-726-1258
Website: www.nature.org

Founded: NA
Scope: Statewide

Contact(s):
Geoff Pampush, STATE DIRECTOR

NATURE CONSERVANCY, THE
ILLINOIS CHAPTER
8 S. Michigan Ave.
Chicago, IL 60603 USA
Phone: 312-346-8166   Fax: 312-346-5606

Founded: NA
Scope: Statewide

Contact(s):
Bruce Boyd, STATE DIRECTOR

NATURE CONSERVANCY, THE
INDIANA CHAPTER
1505 N. Delaware St.
Indianapolis, IN 46202 USA
Phone: 317-923-7547   Fax: 317-917-2478
E-mail: csutton@tnc.org
Website: www.nature.org

Founded: NA
Membership: 1700
Scope: Regional
Publication(s): Chapter Newsletter (twice yearly)

Contact(s):
Mary McConnell, STATE DIRECTOR

NATURE CONSERVANCY, THE
IOWA CHAPTER
108 3rd St.,
Des Moines, IA 50309-4758 USA
Phone: 515-244-5044   Fax: 515-244-8890
E-mail: iowa@tnc.org
Website: www.nature.org/iowa

Founded: NA
Membership: 9000
Scope: Statewide
Publication(s): Iowa Field Notes, Nature Conservancy

Contact(s):
Ann Robinson, DIRECTOR OF DEVELOPMENT
Dave Degeus, DIRECTOR OF PROTECTION

NATURE CONSERVANCY, THE
KENTUCKY CHAPTER
642 W. Main St.
Lexington, KY 40508 USA
Phone: 859-259-9655   Fax: 859-259-9678
E-mail: nature@mis.net
Website: www.nature.org

Founded: NA
Membership: 90000
Scope: Regional
Contact(s):
James Aldrich, STATE DIRECTOR

NATURE CONSERVANCY, THE
MAINE CHAPTER
14 Maine St., 401
Brunswick, ME 04011 USA
Phone: 207-729-5181

Founded: NA
Scope: Statewide

NATURE CONSERVANCY, THE
MASSACHUSETTS CHAPTER
70 Milk St., #300
Boston, MA 02109
Phone: 617-227-7017   Fax: 617-227-7688
E-mail: mmail@tnc.org
Website: www.nature.org

Founded: NA
Scope: Statewide

NATURE CONSERVANCY, THE
MICHIGAN CHAPTER
2840 E. Grand River Ave., Suite 5
East Lansing, MI 48823 USA
Phone: 517-332-1741   Fax: 517-332-8382

Founded: NA
Scope: Statewide
Contact(s):
Helen Taylor, STATE DIRECTOR

NATURE CONSERVANCY, THE
MID-ATLANTIC DIVISION OFFICE
4705 University Dr., Suite 290
Durham, NC 27707 USA
Phone: 919-403-8558   Fax: 919-403-0379

Founded: NA
Scope: Regional
Contact(s):
Katherine Skinner, VICE PRESIDENT

NATURE CONSERVANCY, THE
MINNESOTA CHAPTER
1313 Fifth St., SE, 320
Minneapolis, MN 55414 USA
Phone: 612-331-0750   Fax: 612-331-0770

Founded: NA
Scope: Statewide
Contact(s):
Rob McKim, STATE DIRECTOR

NATURE CONSERVANCY, THE
MISSISSIPPI CHAPTER
6400 Lakeover Rd., Suite C
Jackson, MS 39213 USA
Phone: 601-713-3355   Fax: 601-982-9499
Website: nature.org/Mississippi

Founded: NA
Scope: National
Contact(s):
Robbie Fisher, STATE DIRECTOR

NATURE CONSERVANCY, THE
MISSOURI CHAPTER
2800 S. Brentwood Blvd.
St. Louis, MO 63144 USA
Phone: 314-968-1105   Fax: 314-968-3659
E-mail: missouri@tnc.org
Website: http://nature.org

Founded: NA
Membership: 12
Scope: Statewide

NATURE CONSERVANCY, THE
MONTANA CHAPTER
32 South Ewing
Helena, MT 59601 USA
Phone: 406-443-0303   Fax: 406-443-8311
Website: www.nature.org

Founded: NA
Membership: 5000
Scope: Statewide
Contact(s):
Jamie Williams, STATE DIRECTOR

NATURE CONSERVANCY, THE
NEBRASKA CHAPTER
1019 Leaveniworth St.
Omaha, NE 68102 USA
Phone: 402-342-0282   Fax: 402-342-0474
E-mail: nebraska@tnc.org
Website: www.nature.org

Founded: NA
Membership: 4000
Scope: International
Contact(s):
Vince Shay, STATE DIRECTOR

NATURE CONSERVANCY, THE
NEVADA CHAPTER
1771 E. Flamingo,
Las Vegas, NV 89119 USA
Phone: 702-737-8744   Fax: 702-737-5787
E-mail: swainscott@tnc.org
Website: www.nature.org

Founded: NA
Membership: 6000
Scope: Statewide
Contact(s):
Ame Hellman, STATE DIRECTOR

NATURE CONSERVANCY, THE
NEW HAMPSHIRE CHAPTER
2 1/2 Beacon St., 6
Concord, NH 03301 USA
Phone: 603-224-5853   Fax: 603-228-2459

Founded: NA
Scope: Statewide

Contact(s):
   Daryl Burtnett, STATE DIRECTOR

NATURE CONSERVANCY, THE
NEW JERSEY CHAPTER
200 Pottersville Rd.
Chester, NJ 07930 USA
Phone: 908-879-7262   Fax: 908-879-2172
Website: www.nature.org

Founded: NA
Membership: 31000
Scope: Local
Publication(s): Oak Leaf Quarterly Newsletter

Contact(s):
   Michael Catania, STATE DIRECTOR

NATURE CONSERVANCY, THE
NEW MEXICO CHAPTER
212 E. Marcy, 200
Santa Fe, NM 87501 USA
Phone: 505-988-3867
E-mail: smacfarland@tnc.org
Website: www.nature.org

Founded: NA
Scope: Statewide

Contact(s):
   Bill Waldman, STATE DIRECTOR

NATURE CONSERVANCY, THE
NEW YORK CENTRAL/WESTERN CHAPTER
339 East Ave.
Rochester, NY 14604 USA
Phone: 716-546-8030   Fax: 706-546-7825
Website: www.nature.org

Founded: NA
Membership: 13
Scope: Local Region
Publication(s): Quarterly newsletter

NATURE CONSERVANCY, THE
NEW YORK CITY CHAPTER
570 Seventh Ave.,
New York, NY 10018 USA
Phone: 212-997-1880   Fax: 212-997-8451
Website: www.nature.org

Founded: NA
Scope: International

Publication(s): Nature Conservancy of New York

NATURE CONSERVANCY, THE
NEW YORK LONG ISLAND CHAPTER
250 Lawrence Hill Rd.
Cold Spring Harbor, NY 11724 USA
Phone: 631-367-3225   Fax: 631-367-4775
E-mail: cgordon@tnc.org
Website: http://nature.org/

Founded: NA
Scope: Local Region

Description: Works cooperatively with other conservation groups,
businesses and many levels of government. Two conservancy
chapters on Long Island have protected 40,000 acres and own
approximately 5,000 acres in preserves.

Contact(s):
   John Turner, DIRECTOR OF CONSERVATION
   Paul Rabinovitch, EXECUTIVE DIRECTOR

NATURE CONSERVANCY, THE
NEW YORK SOUTH FORK/ SHELTER ISLAND
CHAPTER
P.O. Box 5125
E. Hampton, NY 11937 USA
Phone: 631-329-7689   Fax: 631-329-0215

Founded: NA
Scope: Local Region

Contact(s):
   Nancy Kelley, DIRECTOR

NATURE CONSERVANCY, THE
NORTH CAROLINA CHAPTER
4705 Univeristy Dr., 290
Durham, NC 27707 USA
Phone: 919-403-8558   Fax: 919-403-0379

Founded: NA
Scope: Statewide

Contact(s):
   Katherine Skinner, EXECUTIVE DIRECTOR

NATURE CONSERVANCY, THE
NORTH DAKOTA
1256 N. Parkview Dr.
Bismarck, ND 58501 USA
Phone: 701-222-8464   Fax: 701-222-8061
E-mail: nature.org
Website: www.nature.org

Founded: NA
Membership: 1800
Scope: Statewide

Contact(s):
   Gerald Reichert, FIELD REPRESENTATIVE

NATURE CONSERVANCY, THE
NORTHEAST DIVISION OFFICE
159 Waterman St.
Providence, RI 02906 USA
Phone: 401-751-2521   Fax: 401-751-7596
Website: www.tnc.org

Founded: NA
Scope: Regional

Publication(s): Nature Conservancy - Bi-monthly magazine

Contact(s):
John Cook, VICE PRESIDENT

NATURE CONSERVANCY, THE
NORTHWEST & HAWAII DIVISION OFFICE
217 Pine St., Suite 1100
Seattle, WA 98101 USA
Phone: 206-343-4344

Founded: NA
Scope: Regional

Contact(s):
Elliot Marks, VICE PRESIDENT

NATURE CONSERVANCY, THE
OHIO CHAPTER
6375 Riverside Dr.
Dublin, OH 43017 USA
Phone: 614-717-2770   Fax: 614-717-2777
Website: www.nature.org

Founded: NA
Membership: 45
Scope: Nationwide

Contact(s):
Rich Shank, STATE DIRECTOR

NATURE CONSERVANCY, THE
OHIO CHAPTER
6375 Riverside Dr., Suite 50
Dublin, OH 43017 USA
Phone: 614-717-2770   Fax: 614-717-2777

Founded: NA
Scope: Regional

Contact(s):
David Weekes, VICE PRESIDENT

NATURE CONSERVANCY, THE
OKLAHOMA CHAPTER
2727 East 21st
Tulsa, OK 74114 USA
Phone: 918-585-1117   Fax: 918-585-2383
Website: www.nature.org

Founded: NA
Scope: Statewide

Contact(s):
Mary Collins, STATE DIRECTOR

NATURE CONSERVANCY, THE
OREGON CHAPTER
821 SE 14th Ave.
Portland, OR 97214 USA
Phone: 503-230-1221   Fax: 503-230-9639
Website: www.nature.org

Founded: NA
Membership: 29000
Scope: Statewide

Contact(s):
Russell Hoeflich, STATE DIRECTOR

NATURE CONSERVANCY, THE
PENNSYLVANIA CHAPTER
1100 E. Hector St.
Conshohocken, PA 19428 USA
Phone: 610-834-1323   Fax: 610-834-6533
Website: www.nature.org

Founded: NA
Membership: 36,000
Scope: Statewide
Publication(s): Penns Woods-Bi-annually

Contact(s):
P. Gray, STATE DIRECTOR

NATURE CONSERVANCY, THE
RHODE ISLAND FIELD OFFICE
159 Waterman St.
Providence, RI 02906 USA
Phone: 401-331-7110   Fax: 401-273-4902

Founded: NA
Scope: Statewide

Contact(s):
Doug Parker, STATE DIRECTOR

NATURE CONSERVANCY, THE
ROCKY MOUNTAIN DIVISION OFFICE
117 E. Mountain Ave.
Fort Collins, CO 80524 USA
Phone: 970-484-2886   Fax: 970-498-0225
Website: www.nature.org

Founded: NA
Scope: National

Contact(s):
Bruce Runnels, VICE PRESIDENT

NATURE CONSERVANCY, THE
SOUTH CAROLINA CHAPTER
P.O. Box 5475
Columbia, SC 29250 USA
Phone: 803-254-9049   Fax: 803-252-7134

Founded: NA
Scope: Statewide

Contact(s):
Mark Robertson, STATE DIRECTOR

NATURE CONSERVANCY, THE
SOUTH CENTRAL DIVISION OFFICE
P.O. Box 1440
San Antonio, TX 78295-1440 USA
Phone: 210-224-8774   Fax: 210-228-9805
E-mail: nmcdaniel@tnc.org
Website: www.nature.org

Founded: NA
Scope: International

Contact(s):
Robert Potts, VICE PRESIDENT

NATURE CONSERVANCY, THE
SOUTH DAKOTA CHAPTER
1000 West Ave. N., Suite 100
Sioux Falls, SD 57104 USA
Phone: 605-331-0619   Fax: 605-874-8518

Founded: NA
Scope: Statewide

Contact(s):
Clint Miller, FIELD REP.

NATURE CONSERVANCY, THE
SOUTHEAST DIVISION OFFICE
222 S. Westmonte Dr., Suite 300
Altamonte Springs, FL 32714 USA
Phone: 407-682-3664   Fax: 407-682-3077

Founded: NA
Scope: Regional

Contact(s):
Robert Benedick, VICE-PRESIDENT

NATURE CONSERVANCY, THE
TENNESSEE CHAPTER
2021 21st. Ave. S.
Nashville, TN 37212 USA
Phone: 615-383-9909   Fax: 615-383-9717
Website: www.nature.org/tennessee

Founded: NA
Membership: 12000
Scope: Statewide
Publication(s): Newsletter - Bi-annual

Contact(s):
Scott Davis, STATE DIRECTOR

NATURE CONSERVANCY, THE
TEXAS CHAPTER
P.O. Box 1440
San Antonio, TX 78295-1440 USA
Phone: 210-224-8774   Fax: 210-228-9805

Founded: NA
Scope: Statewide

Contact(s):
Robert Potts, STATE DIRECTOR

NATURE CONSERVANCY, THE
UTAH CHAPTER
559 E. South Temple
Salt Lake City, UT 84102 USA
Phone: 801-531-0999   Fax: 801-531-1003
Website: www.nature.org

Founded: NA
Membership: 8600
Scope: Statewide

Contact(s):
David Livermore, STATE DIRECTOR

NATURE CONSERVANCY, THE
VERMONT CHAPTER
27 State. St.
Montpelier, VT 05602 USA
Phone: 802-229-4425   Fax: 802-229-1347

Founded: NA
Scope: Statewide

Contact(s):
Robert Klein, STATE DIRECTOR

NATURE CONSERVANCY, THE
VIRGIN ISLANDS CHAPTER
14B Norre Gade, 2nd Fl.
Charlotte Amalie, VI 00802 USA
Phone: 340-774-7633   Fax: 340-774-7736
E-mail: c.philyaw@att.net
Website: www.nature.org

Founded: NA
Scope: Statewide

Publication(s): available on web

Contact(s):
Robert Weary, DIRECTOR

NATURE CONSERVANCY, THE
VIRGINIA CHAPTER
TESTING
490 Westfield Rd.
Charlottesville, VA 22901 USA
Phone: 804-295-6106   Fax: 804-979-0370
E-mail: dwhite@tnc.org
Website: http://nature.org

Founded: NA
Scope: Statewide

NATURE CONSERVANCY, THE
WASHINGTON CHAPTER
217 Pine St., 1100
Seattle, WA 98101 USA
Phone: 206-343-4344   Fax: 206-343-5608

Founded: NA
Membership: 35000
Scope: Statewide

Contact(s):
David Weekes, STATE DIRECTOR
Elliot Marks, VICE PRESIDENT

NATURE CONSERVANCY, THE
WEST VIRGINIA CHAPTER
723 Kanawha Blvd. East, 500
Charleston, WV 25301 USA
Phone: 304-345-4350   Fax: 304-345-4351
Website: www.nature.org

Founded: NA
Membership: 4000
Scope: Statewide

Contact(s):
Paul Trianosky, STATE DIRECTOR

NATURE CONSERVANCY, THE
WISCONSIN CHAPTER
633 W. Main St.
Madison, WI 53703 USA
Phone: 608-251-8140   Fax: 608-251-8535
Website: www.nature.org

Founded: NA
Membership: 60
Scope: Statewide
Publication(s): The Nature Conservancy

Contact(s):
    Anne Sayers, OFFICE ADMINISTRATOR

NATURE CONSERVANCY, THE
WYOMING CHAPTER
258 Main St., 200
Lander, WY 82520 USA
Phone: 307-332-2971   Fax: 307-332-2974
E-mail: wyoming@tnc.org
Website: http://nature.org

Founded: NA
Scope: Statewide

Contact(s):
    Dave Neary, STATE DIRECTOR

NATURE CONSERVANCY, THE
MARYLAND/DC CHAPTER
4510 Grosvenor Lane
Bethesda, MD 20814 USA
Phone: 301-897-8570   Fax: 301-897-0858
Website: www.nature.org

Founded: NA
Scope: Statewide

Contact(s):
    Nat Williams, STATE DIRECTOR & VICE PRESIDENT

NATURE CONSERVANCY, THE
NEW YORK ADIRONDACK CHAPTER
Route 73
Keene Valley, NY 12943 USA
Phone: 518-576-2082   Fax: 518-576-4203

Founded: NA
Membership: 15
Scope: Local Region

Contact(s):
    Michael Carr, DIRECTOR

NATURE CONSERVANCY, THE OF CALIFORNIA
WESTERN DIVISION OFFICE
201 Mission St., 4th Fl.
San Francisco, CA 94105 USA
Phone: 415-777-0487   Fax: 415-777-0244
Website: www.tnccalifornia.org

Founded: NA
Membership: 150
Scope: Regional

Contact(s):
    Williams Jody, PROGRAM MANAGER

NATURE SASKATCHEWAN
Attn: Administration Coordinator, 206-1860 Lorne St.
Regina, Saskatchewan S4P 2L7 Canada
Phone: 306-780-9273   Fax: 306-780-9263
E-mail: nature.sask@unibase.com
Website: www.nature.sask.com

Founded: 1947
Scope: Provincewide

Description: Nature Saskatchewan is the largest non-profit nature organization in the province, committed to preserving our natural environment. We operate a nature bookshop, offer ecological tours, own nature sanctuaries, and support conservation and research activities and nature education.

Publication(s): Nature Views, special publication, Blue Jay

Keyword(s): Aquatic Habitats, Biodiversity, Birds, Conservation, Ecology, Environment, Endangered Species, Environmental Law, Flowers, Plants, and Trees, Land Preservation, Natural Areas, Insects and Butterflies, Mammals, education

NEBRASKA ASSOCIATION OF RESOURCE DISTRICTS
601 South 12th St.
Lincoln, NE 68508 USA
Phone: 402-471-7670   Fax: 402-471-7677
E-mail: nard@nrd.net.org
Website: www.nrd.net.org

Founded: NA
Membership: 5
Scope: Regional
Keyword(s): Conservation Districts, Natural Resource Conservation, Forests and Forestry, Public Lands, Rivers, Water Pollution Management, Wilderness

Contact(s):
    Dean Edson, EXECUTIVE DIRECTOR
    601 S. 12th St., Suite 201, Lincoln, NE 68508
    Phone: 402-471-7674
    Fax: 402-471-7677
    Clint Johannes, PAST PRESIDENT
    326 Road K, Richland, NE 68601
    Phone: 402-352-5640
    Fax: 402-563-4272
    Pete Rubin, PRESIDENT
    416 Bellvue North, Bellvue, NE 68005
    Phone: 40-273-3369
    Orval Gigstad, VICE PRESIDENT
    RR 1 Box 54, Syracuse, NE 68446
    Phone: 402-269-3267

NEBRASKA B.A.S.S. CHAPTER FEDERATION
Attn: President, 1518 Kozy Dr.
Columbus, NE 68601 USA
Phone: 402-563-2297
E-mail: jlcitta@nppd.com

Founded: NA
Scope: Statewide

Description: An organization of Bassmaster chapters, affiliated with the Bass Anglers Sportsman Society, organized to fight pollution, assist state and national conservation agencies in their efforts, and teach the young people of our country good conservation practices. Dedicated to the realistic conservation of our water resources.

Contact(s):
Tom Boyd, CONSERVATION DIRECTOR
1610 S. Blaine, Grand Island, NE 68803
Phone: 308-382-8357
Joe Citta, PRESIDENT
Phone: 402-563-2297

NEBRASKA ORNITHOLOGISTS UNION, INC.
W436 Nebraska Hall
Lincoln, NE 68588-0514 USA
Phone: 402-472-8366

Founded: 1899
Scope: Statewide

Description: To promote the study of ornithology in Nebraska by both professionals and amateurs; to publish the results of independent studies; and to promote the passage and enforcement of judicious laws for bird protection.

Publication(s): Nebraska Bird Review, The

Keyword(s): Birds

Contact(s):
Jan Uttecht, BOARD OF DIRECTORS
NE
Mark Brogie, DIRECTOR
508 Seeley, Box 316, Creighton, NE 68729
Phone: 402-358-5675
Mary Pritchard, LIBRARIAN
6325 O St. 515, Lincoln, NE 68510-2246
Clem Klaphake, PRESIDENT
Mitzi Fox, SECRETARY
NE 68840-9654
Janice Paseka, VICE PRESIDENT
NE

NEBRASKA SIERRA CLUB
NEBRASKA CHAPTER
5106 Western Ave.
Omaha, NE 68132 USA
Phone: 402-556-1830
E-mail: nebraska.chapter@sierraclub.org
Website: www.sierraclub.org/chapters/ne/

Founded: NA
Membership: 2
Scope: Statewide

Publication(s): The Nebraska Sierran (Newsletter)

Contact(s):
Mary Green, CHAIR.
Pat Knapp, STATE COORDINATOR
1614 N. 31st. St., Lincoln NE. 68503
Phone: 402-464-8537
patanap@alltel.net

NEBRASKA WILDLIFE FEDERATION, INC.
P.O. Box 81437
Lincoln, NE 68501-1437 USA
Phone: 402-476-9081　　　Fax: 402-994-2021
E-mail: nebraskawildlife@alltel.net
Website: http:// www.omaha.org/newf

Founded: 1970
Scope: Statewide

Description: A representative statewide organization, affiliated with the National Wildlife Federation, dedicated to the protection and enhancement of wildlife and its habitat through public education and government interaction.

Publication(s): Stream Conservation - quarterly newsletter, Prairie Blade

Contact(s):
Galen Wray, ALTERNATE REPRESENTATIVE & TREASURER
Duane Hovorka, EDITOR & EXECUTIVE DIRECTOR
Mike Coe, EDUCATION PROGRAMS CONTACT
Gene Oglesby, PRESIDENT
David Koukol, REPRESENTATIVE

NEVADA ASSOCIATION OF CONSERVATION DISTRICTS
Attn: President, HC 65 Box 11
Carlin, NV 89822-9701 USA
Phone: 775-738-8431　　　Fax: 775-738-7229

Founded: NA
Membership: 150
Scope: Local
Keyword(s): Conservation Districts

Contact(s):
Joe Sicking, BOARD MEMBER
1550 Cushman Rd., Fallon, NV 89406
Phone: 775-423-5216
Fax: 775-738-7229
Eleanor Odonnell, EXECUTIVE DIRECTOR
Phone: 775-738-8431
Fax: 775-738-7229
Patsy Tomera, PRESIDENT, ALTERNATIVE BOARD MEMBER
Phone: 775-754-2333

NEVADA WILDLIFE FEDERATION, INC.
P.O. Box 71238
Reno, NV 89570 USA
Phone: 775-885-0405　　　Fax: 775-885-0405
E-mail: nvwf@nvwf.org
Website: www.nvwf.org

Founded: NA
Scope: Statewide

Description: A representative statewide organization, affiliated with the National Wildlife Federation, dedicated to the protection and enhancement of wildlife and its habitat through public education and government interaction.

Publication(s): Nevada Wildlife

NEW BRUNSWICK WILDLIFE FEDERATION
Fredericton, New Brunswick E3B 7A2 Canada
Phone: 888-272-6411　　　Fax: 506-458-9941
E-mail: nbwf1@nbnet.nb.ca
Website: http://www.wildlife.nb.ca/

Founded: 1924
Scope: Statewide

Description: Promotes the wise use of renewable natural resources, with prime emphasis on education of the young. Affiliated with the Canadian Wildlife Federation.

Keyword(s): Environmental and Conservation Education

Contact(s):
Sharon Kingston Ellridge, PRESIDENT

NEW ENGLAND ASSOCIATION OF ENVIRONMENTAL BIOLOGISTS (NEAEB)

60 Westview St.
Lexington, MA 02173 USA
Phone: 617-860-4300

Founded: 1976
Scope: Regional

Description: A professional society of environmental scientists, engineers and planners from industry and state and federal agencies in the northeast, working to coordinate and enhance environmental programs in each state. The organization advances technical information on environmental research, planning and management and evaluates the effectiveness of environmental regulations for protection of water quality.

Keyword(s): Professional Organization

Contact(s):
Ernest Pizzuto, EXECUTIVE COMMITTEE
David McDonald, INFORMATION OFFICER
EPA 60 Westview St., Lexington, MA 02421
Phone: 781-860-4609

NEW ENGLAND COALITION FOR SUSTAINABLE POPULATION (NECSP)

P.O. Box 194
Sullivan, NH 03445 USA
Phone: 603-847-9798
E-mail: d9cat@cheshire.net
Website: http://cheshire.net/~d9cat/necsp.html

Founded: 1996
Scope: Regional

Description: NECSP is a network of organizations and individuals committed to achieving a sustainable human population at the local, state, regional, national and global levels.

Publication(s): NECSP News (quarterly newsletter)

Keyword(s): Family Planning, Habitat Conservation, Networking, Population Growth, Overconsumption, Reproductive Rights, Sprawl, Sustainability

Contact(s):
Annie Faulkner, COORDINATOR

NEW ENGLAND NATURAL RESOURCES CENTER

Box 44
Wayland, MA 01778 USA
Phone: 508-358-2261　　Fax: 508-358-2261
E-mail: hagenstein@aol.com

Founded: 1970
Membership: 13
Scope: Regional

Description: A nonprofit trust organized to provide a focal point for discussion and resolution of regional natural resource and environmental issues.

Keyword(s): Communications, Renewable Resources

Contact(s):
Perry Hagenstein, CHAIRMAN
Box 44, Wayland, MA 01778
Robert Eisenmenger, TREASURER
Russell Brenneman, VICE CHAIRMAN
Murtha, Cullina, Richter & Pinney, 101 Pearl St.,
Hartford, CT 06102

NEW ENGLAND WILD FLOWER SOCIETY, INC.

180 Hemeway Rd.
Framingham, MA 01701-2699 USA
Phone: 508-877-7630　　Fax: 508-877-3658
E-mail: newfs@newfs.org
Website: www.newfs.org

Founded: 1900
Scope: National

Description: The New England Wild Flower Society is a nonprofit organization that promotes the conservation of temperate North American plants through conservation and research, education, horticulture and habitat preservation.

Publication(s): New England Wild Flower - Journal and Program Events, New England Wild Flower - Conservation Notes of New England Wild Flower Society, Garden in the Woods Trail guide, Annual Seed and book Catalogue, Botanical Clubs and Native Plant Societies of the U.S.

Keyword(s): Native Plants, Botanical Gardens, Conservation, Endangered Species, Environmental and Conservation Education, Gardening and Horticulture, education

NEW HAMPSHIRE ASSOCIATION OF CONSERVATION COMMISSIONS

54 Portsmouth St.
Concord, NH 03301 USA
Phone: 603-224-7867　　Fax: 603-228-0423
E-mail: info@nhacc.org
Website: www.nhacc.org

Founded: 1970
Scope: State

Description: A nonprofit organization whose purpose is to foster conservation and wise use of NH's natural resources and to assist and facilitate communication and cooperation among the state's 200 conservation commissions.

Publication(s): Handbook for Municipal Conservation Commissions in New Hampshire, Legislative Updates, NH Commission News

Keyword(s): Environmental Law, Protected Areas, Land Use Planning, Exotic species, Aquatic nuisance species, Wetlands, Conservation

Contact(s):
Marjory Swope, EXECUTIVE DIRECTOR AND EDITOR
Katherine Metzger, PRESIDENT
Bob Boynton, SECRETARY
James Meiklejohn, TREASURER
Mason Westfall, VICE PRESIDENT

NEW HAMPSHIRE ASSOCIATION OF CONSERVATION DISTRICTS

Attn: President, 357 Prospect Hill Rd.
Rumney, NH 03266 USA
Phone: 603-786-9601

Founded: NA
Scope: Statewide

Keyword(s): Conservation Districts

Contact(s):
Calvin Perkins, PRESIDENT
68 Isaac Perkins Rd., Lyme, NH 03768
Phone: 603-795-2584
Fax: 603-743-3477

Francesca Latawiec, 1ST VICE PRESIDENT
595 New Rd., Center Barnstead, NH 03225
Phone: 603-271-2155
Mark Perry, 2ND VICE PRESIDENT
767 Salmon Falls Rd., Rochester, NH 03868
Phone: 603-335-0798
Joan Richardson, ADMINISTRATOR
73 Main St., P.O. Box 533, Conway, NH 03818-0533
Phone: 603-447-2771
Fax: 603-447-8945
John Hodsdon, BOARD MEMBER
85 Daniel Webster Hwy, Meredith, NH 03253
Phone: 603-279-6126
Fax: 603-528-8783
Robert Ward, EXECUTIVE DIRECTOR
P.O. Box 404, Sunapee, NH 03782
Phone: 603-763-5425
Fax: 603-763-4194
Stanely Grimes, SECRETARY AND TREASURER
529 Buck St., Pembroke, NH 03275
Phone: 603-485-9326
Fax: 603-233-6030

NEW HAMPSHIRE B.A.S.S. CHAPTER FEDERATION

Attn: President, P.O. Box 282
Wolfeboro, NH 03894 USA
Phone: 603-569-6035
Website: www.nhbassfederation.com

Founded: NA
Scope: Statewide

Description: An organization of Bassmaster chapters, affiliated with the Bass Anglers Sportsman Society, organized to fight pollution, assist state and national conservation agencies in their efforts, and teach the young people of our country good conservation practices. Dedicated to the realistic conservation of our water resources.

Contact(s):
A. Disilva, CONSERVATION DIRECTOR
P.O. Box 923, North Conway, NH 03860
Phone: 603-356-2220
Doug Plasencia, PRESIDENT
Phone: 603-569-6035

NEW HAMPSHIRE FISH AND GAME DEPT.

2 Hazen Dr
Concord, NH 03301 USA
Phone: 603-271-3511   Fax: 603-271-1438
E-mail: info@wildlife.state.nh.us
Website: www.wildlife.state.nh.us

Founded: NA
Membership: 205
Scope: Regional

Description: Consists of heads of fish and game agencies in 11 northeastern states, Canadian Maritime Provinces, Newfoundland, Ontario, and Quebec. Meets at least yearly to review progress, consider mutual problems, coordinate programs on a regional basis, and promote sound fish and game management programs.

Publication(s): bi monthly - Wildlife Journal

Contact(s):
Wayne Vetter, DIRECTOR OF NEW HAMPSHIRE
director@wildlife.state.nh.us

NEW HAMPSHIRE LAKES ASSOCIATION

5 South State St
Concord, NH 03301 USA
Phone: 603-226-0299   Fax: 603-224-9442
E-mail: info@nhlakes.org
Website: www.nhlakes.org

Founded: 1992
Membership: 800
Scope: Local

Description: The New Hampshire Lakes Association is a nonprofit education and advocacy organization dedicated to protecting and preserving New Hampshire's lakes for the responsible and equitable enjoyment of everyone. The NHLA provides assistance to individuals and lake associations throughout New Hampshire.

Publication(s): Lakeside (quarterly magazine), Educational Brochures

Keyword(s): Lakes, Watersheds, Water Quality, Aquatic Habitats, Environment, Environmental Planning, Land Use Planning, Environmental Protection, Environmental and Conservation Education, Outdoor Recreation, Open Space

Contact(s):
Nancy Christie, EXECUTIVE DIRECTOR
nchristie@nhlakes.org

NEW HAMPSHIRE TIMBERLAND OWNERS ASSOCIATION

54 Portsmouth St.
Concord, NH 03301 USA
Phone: 603-224-9699   Fax: 603-225-5898
Website: www.nhtoa.org

Founded: 1911
Membership: 1500
Scope: National, Local

Description: A statewide organization affiliated with the National Woodland Owners Association, dedicated to the promotion of wise forest management and the protection of forestry interests in New Hampshire.

Publication(s): The Timber Crier - quarterly newsletters, The Forest Fax - biweekly for members only

Keyword(s): Forests and Forestry

Contact(s):
Jasen Stock, EXECUTIVE DIRECTOR
jstock@nhtoa.org
Bruce Jacobs, PRESIDENT
Hunter Carbee, PROGRAM DIRECTOR
hcarbee@nhtoa.org
Don Winsor, SECRETARY
Tim Frizzell, TREASURER

NEW HAMPSHIRE WILDLIFE FEDERATION

54 Portsmouth St.
Concord, NH 03301 USA
Phone: 603-228-0423
E-mail: nhwf@aol.com
Website: www.nhwf.org

Founded: NA
Membership: 7000
Scope: Statewide

Description: A representative statewide organization, affiliated with the National Wildlife Federation, dedicated to the protection and enhancement of wildlife and its habitat through public education and government interaction.

Publication(s): New Hampshire Wildlife

Contact(s):
Russ Kott, ALTERNATE REPRESENTATIVE
Margaret Lane, EDITOR
Mary Brown, EXECUTIVE DIRECTOR
Sharon Guaraldi, PRESIDENT
John Monson, REPRESENTATIVE

NEW JERSEY AGRICULTURAL SOCIETY
Trenton, NJ 08625 USA
Phone: 609-394-7766   Fax: 609-292-3978
Website: www.state.nj.us/agriculture/agsociety

Founded: 1781
Membership: 800
Scope: Regional

Description: The Society has continually worked to educate the public and promote agriculture in New Jersey. The Society is charitable, nonprofit, and conducts numerous educational programs about agriculture's vital role in the economy of New Jersey.

Publication(s): Garden View, Harbinger

Keyword(s): Agriculture

Contact(s):
Terry Haaf, ASSISTANT SECRETARY AND TREASURER
Joni Elliott, EDITOR
Pam Mount, PRESIDENT
Arthur Brown, SECRETARY AND TREASURER
Richard Nieuwenhuis, VICE PRESIDENT

NEW JERSEY ASSOCIATION OF CONSERVATION DISTRICTS
POB 330
Trenton, NJ 08625 USA
Phone: 973-398-2511
E-mail: clifford-lundin@nj.nacdnet.org

Founded: NA
Scope: Statewide

Keyword(s): Conservation Districts

Contact(s):
Kenneth Marsh, 1ST VICE PRESIDENT
534 Hanford Place, Westfield, NJ 07909
Phone: 908-233-4528
Kenneth Roehrich, BOARD MEMBER
451 Schooley's Mountain Rd., Hackettstown, NJ 07840
Phone: 908-852-5787
Jay Kandle, PAST PRESIDENT
Phone: 609-589-7916
Clifford Lundin, PRESIDENT
8 Skytop Rd., Andover, NJ 07821
Phone: 973-398-2511
Edward Dipolvere, SECRETARY, ALTERNATIVE BOARD MEMBER
53 Cubberly Rd., Trenton, NJ 08690
Phone: 609-586-2684
Allen Carter, TREASURER
P.O. Box 403, Tuckahoe, NJ 08250
Phone: 609-628-2466

NEW JERSEY AUDUBON SOCIETY
P.O. Box 126, 9 Hardscrabble Rd.
Bernardsville, NJ 07924 USA
Phone: 908-204-8998   Fax: 908-204-8960
E-mail: hq@njaudubon.org
Website: www.njaudubon.org

Founded: 1897
Membership: 17,000
Scope: Statewide

Description: Fosters environmental awareness and a conservation ethic among New Jersey citizens; protects New Jersey's birds, mammals, other animals, and plants, especially endangered and threatened species; and promotes preservation of New Jersey's valuable natural habitats.

Publication(s): New Jersey Audubon, Records of New Jersey Birds, Birds of New Jersey, Bridges to the Natural World

Keyword(s): Environmental and Conservation Education, Land Preservation, education

Contact(s):
Jean Clark, BOARD CHAIRPERSON
Gordon Schultze, DIRECTOR OF LORRIMER SANCTUARY
P.O. Box 125, 790 Ewing Ave., Franklin Lakes, NJ 07417
Phone: 201-891-2185
Gretchen Ferrante, DIRECTOR OF NATURE CENTER OF CAPE MAY
1600 Delaware Ave., Cape May, NJ 08204
Phone: 609-898-8848
Pete Bacinski, DIRECTOR OF OWL HAVEN NATURE CENTER
250 Route 522, P.O. Box 26, Tennent, NJ 07763
Phone: 732-780-7007
Karl Anderson, DIRECTOR OF RANCOCAS NATURE CENTER
794 Rancocas Rd., Mt. Holly, NJ 08060
Phone: 609-261-2495
Karla Risdon, DIRECTOR OF WEIS ECOLOGY CENTER
150 Snake Den Rd., Ringwood, NJ 07456
Phone: 973-835-2160
Thomas Gilmore, PRESIDENT AND CEO
Phone: 908-204-8998
Fax: 908-204-8960
Richard Kane, VICE PRESIDENT FOR CONSERVATION
Scherman—Hoffman Sanctuaries
P.O. Box 693, Bernardsville, NJ 07924
Phone: 908-766-5787
Patricia Kane, VICE PRESIDENT FOR EDUCATION
Sherman—Hoffman Sanctuaries, 11 Hardscrabble Rd., P.O. Box 693, Bernardsville, NJ 07924
Phone: 908-766-5787
Peter Dunne, VICE PRESIDENT OF NATURAL HISTORY INFORMATION
Cape May Bird Observatory, Center for Research and Education, 600 Rt. 47 N., Cape May Court House, NJ 08210
Phone: 609-861-0700

NEW JERSEY B.A.S.S. CHAPTER FEDERATION
Attn: President, 77 Kenvil Ave.
Succasunna, NJ 07876 USA
Phone: 973-584-9387
E-mail: amgolng@bellatlantic.net
Website: www.njbassfed.org

Founded: NA
Membership: 900
Scope: International

Description: An organization of Bassmaster chapters, affiliated with the Bass Anglers Sportsman Society, organized to fight pollution, assist state and national conservation agencies in their efforts, and teach the young people of our country good conservation practices. Dedicated to the realistic conservation of our water resources.

Publication(s): Fishing Line, Reel to Reel

Contact(s):
John Carlone, CONSERVATION DIRECTOR
402 Ames Rd., Highland Lakes, NJ 07422
Phone: 973-764-9723
Tony Going, PRESIDENT
Phone: 973-584-9387

NEW JERSEY CONSERVATION FOUNDATION
170 Longview Rd.
Far Hills, NJ 07931 USA
Phone: 908-234-1225   Fax: 908-234-1189
E-mail: info@njconservation.org
Website: www.njconservation.org

Founded: 1960
Membership: 3000
Scope: Statewide

Description: A nonprofit membership organization concerned with environmental education issues related to land use and open-space acquisition and preservation through use of its revolving land fund. Formed from the Great Swamp Committee of the North American Wildlife Foundation.

Publication(s): New Jersey Highlands, The: Treasures at Risk, New Jersey Conservation, Greenways to the Arthurkill, Charting a Course for the Delaware Bay Watershed

Keyword(s): Agriculture, Biodiversity, Environmental and Conservation Education, Environmental Planning, Land Preservation, Land Use Planning, Natural Areas, Land Purchase, Open Space, Public Lands, Urban Environment, National Parks, Sustainable Development, Sustainable Ecosystems, Watersheds

Contact(s):
Samuel Lambert, PRESIDENT OF BOARD OF TRUSTEES

NEW JERSEY ENVIRONMENTAL LOBBY
204 W. State St.
Trenton, NJ 08608 USA
Phone: 609-396-3774   Fax: 609-396-4521
Website: http://www.njenvironment.org

Founded: 1969
Scope: State

Description: To advocate for legislation and regulation that is protective and preservative of both the natural and the built environment with a view, always, of protecting human health for all citizens and future generations. Conversely, we oppose those laws and regulations that are detrimental to the above.

Publication(s): periodic special reports, NJ Environmental Lobby News

Keyword(s): Air Quality and Pollution, Energy, Land Use Planning, Sustainable Development, Transportation

Contact(s):
Anne Poole, PRESIDENT
43 Four Mile Rd., Pemberton, NJ 08068
Phone: 609-894-4113
newpoole@bellatlantic.net

Mark Herzberg, TREASURER
24 Clinton Pl., Metuchen, NJ 08840
Phone: 908-494-4883
mherzb8468@aol.com
Eileen Hogan, VICE PRESIDENT
96 Briarcliff Rd, Mountain Lakes, NJ 07046
Phone: 973-267-6100

NEW JERSEY FORESTRY ASSOCIATION
1628 Prospect St.
Trenton, NJ 08638 USA
Phone: 609-771-8301
Website: http://loki.stockton.edu/~forestry/

Founded: 1975
Scope: Statewide

Description: A statewide organization affiliated with the National Woodland Owners Association. Formed to encourage the scientific management and perpetuation of woodlands in New Jersey.

Publication(s): New Jersey Woodlands

Keyword(s): Forests and Forestry

Contact(s):
Richard West, EDITOR
Ron Sheay, EXECUTIVE SECRETARY AND EDITOR
Thomas Bullock, PRESIDENT
Phone: 609-696-5300
George Pierson, VICE PRESIDENT
Phone: 609-737-0489

NEW MEXICO ASSOCIATION OF CONSERVATION DISTRICTS
163 Trail Canyon Rd.
Carlsbad, NM 88220 USA
Phone: 505-981-2400   Fax: 505-981-2422
E-mail: nmacd@dellcity.com
Website: www.nm.nacdnet.org

Founded: NA
Membership: 47
Scope: Statewide

Keyword(s): Conservation Districts

Contact(s):
Debbie Hughes, EXECUTIVE DIRECTOR
163 Trail Canyon Rd., Carlsbad, NM 88220
Phone: 505-981-2400
Fax: 505-981-2422
Brian Greene, PRESIDENT
Rt. 1 Box 22, Mountainair, NM 87036
Phone: 505-849-1080
Fax: 505-847-0615
Eddie Vigil, SECRETARY TREASURER
Leedrue Hyatt, VICE PRESIDENT
8410 Flying U Rd., NE, Deming, NM 88030
Phone: 505-546-9694
Fax: 505-546-3265

NEW MEXICO B.A.S.S. CHAPTER FEDERATION

Attn: President, 164 Monte Rey S.
Los Alamos, NM 87544 USA
Phone: 505-672-3536
Website: www.becksfishing.com/federation.htm

Founded: NA
Scope: Statewide

Description: An organization of Bassmaster chapters, affiliated with the Bass Anglers Sportsman Society, organized to fight pollution, assist state and national conservation agencies in their efforts, and teach the young people of our nation good conservation practices. Dedicated to the realistic conservation of our water resources.

Publication(s): New Mexico B.A.S.S. Federation Newsletter "BigMouth"

Contact(s):
Ron Gilworth, CONSERVATION DIRECTOR
P.O. Box 717, Socorro, NM 87801
Phone: 505-835-1200
Larry Knecht, PRESIDENT
Phone: 505-672-3536

NEW MEXICO ENVIRONMENTAL LAW CENTER

1405 Luisa St., Suite 5
Santa Fe, NM 87505 USA
Phone: 505-989-9022   Fax: 505-989-3769
E-mail: nmelc@nmelc.org

Founded: 1987
Membership: 350
Scope: Regional

Description: The New Mexico Environmental Law Center is a nonprofit, public interest law firm. The Law Center is the only New Mexico organization that provides free legal services for the preservation of the state's natural resources and protection of its citizens against environmental hazards. The Law Center represents grassroots organizations, individuals, and other environmental groups in site-specific efforts; participates in statewide and federal legislative advocacy; and provides public education.

Publication(s): The Green Fire Report

Keyword(s): Environmental Law

Contact(s):
Douglas Meiklejohn, EXECUTIVE DIRECTOR

NEW MEXICO WILDLIFE FEDERATION, INC.

3240-A Juan Tabo NE, Suite 204
Albuquerque, NM 87111 USA
Phone: 505-299-5404

Founded: NA
Scope: Statewide

Description: A representative statewide organization, affiliated with the National Wildlife Federation, dedicated to the protection and enhancement of wildlife and its habitat through public education and government interaction.

Publication(s): Outdoor Reporter

Contact(s):
Cliff Mendel, EDITOR
Ed Machin, PRESIDENT AND ALTERNATE REPRESENTATIVE
Mary Reed, REPRESENTATIVE AND EDUCATION PROGRAMS CONTACT

NEW YORK ASSOCIATION OF CONSERVATION DISTRICTS, INC.

Attn: President,104 Edwards Ave.
Calberton, NY 11933 USA
Phone: 631-727-3777   Fax: 631-727-3721
Website: www.agmkt.stats.ny.us/soilwater/intro.asp

Founded: NA
Scope: Statewide

Keyword(s): Conservation Districts

Contact(s):
Anita Cartin, EXECUTIVE VICE PRESIDENT
1 Winners Cir., Albany, NY 12235
Phone: 518-457-7229
Fax: 518-457-2716
Joe Gerdela, PRESIDENT
P.O. Box 341, Center Moriches, NY 11934
Phone: 631-727-3777
Carl Seymour, SECRETARY
242 Grange Hall Rd., Schuylerville, NY 12871
Phone: 518-695-9249
William Chamberlain, TREASURER
Box 487, Henderson Harbor, NY 13651
Phone: 315-938-7106

NEW YORK B.A.S.S. CHAPTER FEDERATION

Attn: President, 177 Barmore Rd.
LaGrangeville, NY 12540 USA
Website: www.nybassfed.com

Founded: NA
Scope: Statewide

Description: An organization of Bassmaster chapters, affiliated with the Bass Anglers Sportsman Society, organized to fight pollution, assist state and national conservation agencies in their efforts, and teach the young people of our country good conservation practices. Dedicated to the realistic conservation of our water resources.

Publication(s): Fishlines

Contact(s):
Bernie Haney, CONSERVATION DIRECTOR
826 C Tamarack Dr., West Carthage, NY 13619
Phone: 315-493-2356
bernie@nybassfed.com
Wayne Tomassi, PRESIDENT
wayne@nybassfed.com

NEW YORK FOREST OWNERS ASSOCIATION, INC.

P.O. Box 180
Fairport, NY 14450 USA
Phone: 716-377-6060   Fax: 716-388-7592
E-mail: nyfoainc@excite.com
Website: www.nyfoa.org

Founded: 1962
Membership: 2000
Scope: Statewide

Description: A statewide organization, affiliated with the National Woodland Owners Associaiton, organized to unite the 500,000 owners of 11 million acres of forest land in New York in encouraging the wise management of private woodland resources in New York State by promoting, protecting, representing, and serving the interests of woodland owners.

Publication(s): Forest Owner

Keyword(s): Forests and Forestry

Contact(s):
Deborah Gill, ADMINISTRATOR
Phone: 716-377-6060
Mary Malmsheimer, EDITOR
Desktop Solutions, Cazenovia, NY 13035
Phone: 315-655-4110
Fax: 315-655-9694
Ronald Pederson, PRESIDENT
Phone: 518-785-6061
Jerry Michael, TREASURER
Phone: 315-733-7391

NEW YORK PUBLIC INTEREST RESEARCH GROUP (NYPIRG)

Main Office, 9 Murray St., 3rd Fl.
New York, NY 10007 USA
Phone: 212-349-6460   Fax: 212-349-7474
E-mail: nypirg@nypirg.org
Website: www.nypirg.org

Founded: 1973
Scope: Statewide

Description: The NYPIRG is a nonprofit, nonpartisan research group established and directed by New York state college students. Staff lawyers, researchers, and advocates work with students and other citizens developing citizenship skills and shaping public policy on environmental preservation, good government, and consumer issues.

Publication(s): NYPIRG Agenda, NYC CouncilWatch, Get the Lead Out

Keyword(s): Environmental Justice, Pesticides, Solid Waste Management, Toxic Substances, Nuclear-free, Water quantity, Water export and diversion, Transportation

NEW YORK STATE DEPARTMENT OF ENVIRONMENTAL CONSERVATION

Bureau of Wildlife, Division of Fish, Wildlife and Marine Resources, 50 Wolf Rd., Rm. 562
Albany, NY 12233-4754 USA
Phone: 518-457-3730   Fax: 518-457-0691

Founded: 1983
Scope: National

Description: Members include the wildlife management program administrators of the northeast U.S. state and eastern Canadian provincial fish and wildlife agencies. The goal of NEWAA is to coordinate and facilitate the work of regional technical wildlife management committees to promote sound wildlife management programs and to exchange ideas, methods, and approaches to administrative and operational problems in wildlife management.

Keyword(s): Professional Organization, Endangered Species, Hunting, Nongame Wildlife, training

Contact(s):
John Major, CHAIRMAN

NEW YORK TURTLE AND TORTOISE SOCIETY

P.O. Box 878
Orange, NJ 07051-0878 USA
Phone: 212-459-4803
E-mail: info@nytts.org
Website: www.nytts.org/

Founded: 1970
Scope: National

Description: The Society is dedicated to the conservation and preservation of habitat, and the promotion of proper husbandry and captive propagation of turtles. Education of members and the public is a key goal. Events held in the NYC area include a seminar, field trips, and show.

Publication(s): Journal of the New York Turtle and Tortoise Society, NYTTS, NewsNotes, Plastron Papers

Keyword(s): Environmental and Conservation Education, Nongame Wildlife, Reptiles and Amphibians, accreditation, Zoology

Contact(s):
Jim Van Abbemg, EDITOR OF PROCEEDINGS
Joan Frumkies, MEMBERSHIP
Suzanne Dohm, PRESIDENT
Rita Devine, SECRETARY
Lori Craner, TREASURER OF WILDLIFE REHABILITATION
Allen Foust, VICE PRESIDENT

NEW YORK-NEW JERSEY TRAIL CONFERENCE INC.

156 Ramapo Valley Road
Mahurah, NJ 07430 USA
Phone: 212-685-9699   Fax: 212-779-8102
E-mail: office@nynjtc.org
Website: www.nynjtc.org

Founded: 1920
Scope: Regional

Description: A nonprofit organization which coordinates the efforts of hiking and outdoor groups in New York and New Jersey to build and maintain over 1,300 miles of foot trails and whose purpose is to protect and conserve open space, wildlife, and places of natural beauty and interest.

Publication(s): New York Walk Book, New Jersey Walk Book, Guide to the Appalachian Trail in New York and New Jersey, Guide To The Long Path, Catskill Trails Map Set, Hiking the Catskills, Iron Mine Trails, Delaware Water Gap National Recreation Area Hiking, Trail Walker

Keyword(s): Land Preservation, Open Space, Outdoor Recreation, Trail, Natural Areas

Contact(s):
Gary Haugland, CHAIRMAN OF TRAILS COUNCIL
Gary Haugland, PRESIDENT
Daniel Chazin, SECRETARY
Jane Daniels, VICE PRESIDENT

NEWFOUNDLAND LABRADOR WILDLIFE FEDERATION

Attn: President, 67 Commonwealth Ave
Mount Pearl, Newfoundland A1N 1W7 Canada
Phone: 709-364-8415   Fax: 709-753-4709

Founded: NA
Membership: 24000
Scope: International

Keyword(s): Acid Rain, Air Quality and Pollution, Aquatic Habitats, Ancient Forests, Biodiversity, Ecology, Birds, Environmental Ethics, Forest Management, training, Lakes, Solid Waste Management, Mammals, Trapping, Nongame Wildlife

Contact(s):

Clifford Head, SECRETARY
49 Harnum Crescent, Mt Pearl, Newfoundland A1N 1W7
Phone: 709-335-2226
Gordon Cooper, VICE PRESIDENT
67 Commonwealth Ave., Mount Pearl, Newfoundland A1N 1W7
Phone: 709-368-6180

NORTH AMERICAN ASSOCIATION FOR ENVIRONMENTAL EDUCATION

410 Tarvin Rd.
Rock Spring, GA 30739 USA
Phone: 703-764-2926   Fax: 706-764-2094
E-mail: email@naaee.org
Website: www.naaee.org

Founded: 1971
Scope: National

Description: NAAEE is dedicated to promoting environmental education and supporting the work of environmental educators in North America and around the world. NAAEE is made up of students and professionals who have thought seriously about how individuals become literate concerning environmental issues and about how to prepare people to work together towards resolving environmental problems.

Publication(s): Environmental Communicator, Annual Conference Proceedings, NAAEE Directory of Environmental Educators, and others

Keyword(s): Communications, Environmental and Conservation Education, Training, Urban Environment, Environment, Professional Development, Environmental Literacy, Networking

Contact(s):

Bonnie Shelton, EXECUTIVE DIRECTOR
Judy Braus, PRESIDENT
James Elder, TREASURER

NORTH AMERICAN ASSOCIATION FOR ENVIRONMENTAL EDUCATION

CONFERENCE, PUBLICATIONS AND MEMBERSHIP OFFICE
410 Tarvin Rd.
Rock Spring, GA 30739 USA
Phone: 706-764-2926   Fax: 706-764-2094
Website: www.naaee.org

Founded: NA
Membership: 2400
Scope: International

Contact(s):

Barbara Eager, ACTING DEPUTY DIRECTOR
Sarah Gray, CONFERENCE & PUBLICATIONS ADMINISTRATIVE ASSISTANT
Connie Smith, MEMBERSHIP DEVELOPMENT SERVICES MANAGER

NORTH AMERICAN BEAR FEDERATION

3503 Hwy. 89
South Livingston, MT 59047 USA
Phone: 406-333-4414   Fax: 406-333-9733
E-mail: nabear@nabear.org
Website: www.nabear.org
Membership: 1600
Scope: National

Description: -Bear and bear habitat conservation

-Bear hunting and advocacy

-Non-profit organization

Publication(s): The Bear Facts - magazine comes out opposite month, Bear Bulletin - every other month

Contact(s):

Carl Brooke, PRESIDENT, CO-FOUNDER
Phone: 406-333-4414

NORTH AMERICAN BENTHOLOGICAL SOCIETY

c/o Allen Marketing and Management, P.O. Box 1897
Lawrence, KS 66044-8897 USA
Website: benthos.org http://benthos.org

Founded: 1953
Scope: National

Description: The Society is an international scientific organization whose purpose is to promote better understanding of the biotic communities of lake and stream bottoms and their role in aquatic ecosystems. The Society provides media for disseminating results of scientific investigations and other information to aquatic biologists and to the scientific community at large.

Publication(s): Journal of the North American Benthological Society, Bulletin of the North American Benthological Society, Current and Selected Bibliography of Benthic Biology

Keyword(s): Wildlife Rehabilitation, Aquatic Habitats, Lakes, Streams

Contact(s):

Steve Canton, EDITOR
Chadwick and Associates Inc., 5575 S. Sycamore St., Suite 101, Littleton, CO 80120
David Rosenberg, EDITOR
Freshwater Institute, 501 University Crescent, Winnipeg, MB R3T 2N6
Donald Webb, EDITOR
Illinois Natural History Survey, 607 E. Peabody St., Champaign, IL 61820
Nancy Grimm, PRESIDENT
Phone: 602-965-4735
Donna Giberson, SECRETARY
Phone: 902-566-0797
Kim Haag, TREASURER
Phone: 813-243-5800

NORTH AMERICAN BLUEBIRD SOCIETY

Darlington, WI 53530-0074 USA
Phone: 608-329-6403   Fax: 608-329-7057
E-mail: info@nabluebirdsociety.org
Website: www.nabluebirdsociety.org

Founded: 1978
Membership: 3
Scope: International

Description: The North American Bluebird Society, a non-profit conservation, education and research organization, promotes the recovery of bluebirds and other native, cavity-nesting species. On-going research, educational material development, including two slide programs, outreach initiatives through the NABS Speakers Bureau and a comprehensive website on bluebirding, address issues related to bluebirds and other native cavity-nesting bird species.

Publication(s): Bluebird Magazine (formerly Sialia), Stokes Bluebird Basics "10 Minute Introductory Video", Educational Bluebird posters and Educational Slide Program, Bluebird Educators Packet, Transcontinental Bluebird Trail Program (network of bluebird trails across North America)

Keyword(s): Birds

Contact(s):
Lisa Kivirist, CO-EXECUTIVE DIRECTOR
John Ivanko, CO-EXECUTIVE DIRECTOR
Jim Williams, EDITOR
Doug Levasseru, PRESIDENT
Arlene Ripley, RECORDING SECRETARY
Bob Martin, TREASURER
Carol McDaniel, VICE PRESIDENT OF COMMUNITY RELATIONS

NORTH AMERICAN BUTTERFLY ASSOCIATION
4 Delaware Rd.
Morristown, NJ 07960 USA
Phone: 973-285-0907   Fax: 973-285-0936
E-mail: naba@naba.org
Website: www.naba.org

Founded: 1993
Membership: 4500
Scope: International

Description: NABA promotes public enjoyment and conservation of butterflies, encouraging non-consumption activities such as butterfly watching, gardening and photography.

Publication(s): Naba Checklist and English Names of North America Butterflies, American Butterflies, Butterfly Gardener, NABA 4th of July Butterfly Count Report

Keyword(s): Biodiversity, Conservation, Ecotourism, Endangered Species, Environmental and Conservation Education, Insects and Butterflies, Gardening and Horticulture, Interpretive Center, Outdoor Recreation

Contact(s):
Jeffrey Glassberg, PRESIDENT
Ann Swengel, VICE-PRESIDENT
Jim Springer, WEBMASTER
springer@naba.org

NORTH AMERICAN COALITION ON RELIGION AND ECOLOGY (NACRE)
5 Thomas Cir., NW
Washington, DC 20005 USA
Phone: 202-462-2591   Fax: 202-462-6534
E-mail: nacre@earthlink.net
Website: caringforcreation.net

Founded: 1989
Membership: 6,300
Scope: National

Description: NACRE is an ecumenical and interfaith environmental organization designed to help the North American religious community enter into the environmental movement in the 1990s and to help environmental organizations and the wider society become aware and act upon these same ethical values.

Publication(s): ECO-Letter

Keyword(s): Biodiversity, Renewable Resources, Urban Environment

Contact(s):
Bruce Anderson, CHAIRMAN OF THE BOARD
Donald Conroy, PRESIDENT AND CEO
Carolyn Gutowski, SECRETARY AND TREASURER

NORTH AMERICAN CRANE WORKING GROUP
341 W. Olymipic Pl. Suite 300
Seattle, WA 98119-3719 USA
Phone: 206-286-8607
E-mail: thoffmann@hoffmannf.com
Website: www.portup.com/~nacwg

Founded: 1988
Scope: International

Description: An organization of professional biologists, aviculturists, land managers, and other interested individuals dedicated to the conservation of cranes and their habitats in North America.

Publication(s): Unison Call, The

Keyword(s): Biodiversity, Birds, Endangered Species, Wetlands, training

Contact(s):
Jane Nicolich, EDITOR
R.R. 2 Box 264A, Laurel, MD 20708
Scott Hereford, PRESIDENT
Mississippi Crane Refuge, 7200 Crane Ln., Gauthier, MS 39553
Stephen Nesbitt, SECRETARY
4005 S. Main St., Gainesville, FL 32601
Thomas Hoffmann, TREASURER
Wendy Brown, VICE PRESIDENT
1208 Claire Ct. NW, Albuquerque, NM 87104

NORTH AMERICAN FALCONERS ASSOCIATION
Ellinwood, KS 67526-9801 USA
Website: n-a-f-a.org http://n-a-f-a.org

Founded: 1962
Scope: National

Description: A nonprofit fraternal organization with the following purposes: improve and encourage competency in the practice of falconry; urge recognition of falconry as a legal field sport; and promote scientific study, conservation, and welfare of birds of prey with an appreciation of their value in nature.

Publication(s): Hawk Chalk (quarterly), Journal (annual)

Keyword(s): Conservation, Hunting, Raptors

Contact(s):
Robert Glass, CORRESPONDING SECRETARY
33 Tanager Ln., Robbinville, NJ 08691
Dan Cecchini, EDITOR
7220 Burgess Rd., Colorado Springs, CO 80908
Fax: 303-495-4506
Williston Shor, EDITOR
318 Montford Ave., Mill Valley, CA 94941
Brian Millsap, PRESIDENT
Sue Cecchini, TREASURER
7220 Burgess Rd., Colorado Springs, CO 80908

NORTH AMERICAN GAMEBIRD ASSOCIATION, INC.

1214 Brooks Ave.
Raleigh, NC 27607 USA
Phone: 919-782-6758   Fax: 919-515-7070
E-mail: gamebird@naga.org
Website: www.naga.org

Founded: 1932
Scope: National

Description: To promote educational work and develop interest in game bird breeding and hunting preserves (nonprofit); to afford a means of cooperation with the federal and state governments in all matters of concern to the industry; and to encourage study of the sciences connected with the live production, preparation for markets, and marketing of game bird eggs and game birds.

Publication(s): Game Bird Propagation Book, List of Hunting Resort Members., Wildlife Harvest Magazine, Membership Directory

Keyword(s): Birds, Hunting, Waterfowl, training

Contact(s):
Gary Davis, EXECUTIVE DIRECTOR
1214 Brooks Ave., Raleigh, NC 27607
Phone: 919-782-6758
Royd Hatt, PRESIDENT
Box 134, Green River, UT 84525
Phone: 801-564-3224

NORTH AMERICAN LOON FUND

6 Lily Pond Rd.
Gilford, NH 03246 USA
Phone:
Website:
charlotte.uwn.edu/biology/nalf/aNALFhomepage.html

Founded: 1979
Scope: National

Description: A nonprofit organization established to sponsor loon conservation, public education, and scientific research projects across the U.S. and Canada. Sponsors annual grant program, and organizes annual research conference.

Publication(s): Loon Call Newsletter, Educational poster and resource directory, Annotated Bibliography of the Loons, Gaviidae

Keyword(s): Endangered Species, Lakes, Nongame Wildlife, Scholarships and Grants, training

Contact(s):
Jordan Prouty, CHAIRMAN
Guy Swenson, CLERK
Linda Obara, EXECUTIVE DIRECTOR
Ellen Barth, TREASURER

NORTH AMERICAN MEMBERSHIP GROUP

12301 Whitewater Dr.
Minnetonka, MN 55343 USA
Phone: 952-936-9333   Fax: 952-936-9755
Website: www.naog.com

Founded: 1978
Membership: + million
Scope: International

Description: The North American Fishing Club is a membership organization dedicated to enhancing the fishing skills and enjoyment of anglers. The NAFC is the largest association of multi-species anglers in North America.

Publication(s): North American Fisherman

Keyword(s): Aquatic Habitats, Wildlife, Lakes, Sport Fishing

Contact(s):
Steve Pennaz, EXECUTIVE DIRECTOR
Nancy Evensen, PRESIDENT

NORTH AMERICAN NATIVE FISHES ASSOCIATION

123 W. Mt. Airy Ave.
Philadelphia, PA 19119 USA
Phone:
E-mail: nanfa@att.net
Website: nanfa.org

Founded: 1972
Scope: National

Description: Membership includes ichthyologists, students, sportsmen, amateur naturalists, and aquarists who seek to promote the study, research, and conservation of North American native fishes. Goals are to promote the restoration and protection of habitat and to distribute information about native fishes.

Publication(s): American Currents Magazine

Keyword(s): Aquatic Habitats, Endangered Species, Wildlife, Nongame Wildlife

Contact(s):
Bruce Stallsmith, PRESIDENT
801 Wells Ave., Huntsville, IL 35801
Phone: 256 519 -2099
fundulus@hotmail.com
D. Martin Moore, SECRETARY
155 David Henderson Rd., Penahatchie, MS 39145
Phone: 601 5-46 -2320
archimed@netdoor.com
Stephanie Brough, TREASURER
1107 Argonne Dr., Baltimore, MD 21218
Phone: 410 2-43 -9050
ichthos@charm.net
Mark Binkley, VICE PRESIDENT
21 Orchard Dr., Worthington, OH 43085
Phone: 614 8-44 -6042
mbinkley@columbus.rr.com

NORTH AMERICAN WILDLIFE PARK FOUNDATION, INC.

Wolf Park 4004 E 800 N.
Battle Ground, IN 47920 USA
Phone: 765-567-2265   Fax: 765-567-4299
E-mail: wolfpark@wolfpark.org
Website: www.wolfpark.org

Founded: 1972
Membership: 1600
Scope: International

Description: A nonprofit organization which operates a Wolf Park; provides continuous behavior research programs; offers lectures and a teaching program, as well as four Wolf Behavior seminars per year; monitors legislation on predators; and provides research opportunities for scientists and students.

Publication(s): Wolf Park News, Wolf ! Magazine

Contact(s):
Erich Klinghammer, DIRECTOR

NORTH AMERICAN WOLF ASSOCIATION

23214 Tree Bright Lane
Spring, TX 77373 USA
Phone: 281-821-4884   Fax: 281-821-4417
E-mail: nawa@nawa.org
Website: www.nawa.org

Founded: 1973
Membership: 1000+
Scope: International

Description: Nonprofit, tax-exempt volunteer organization dedicated to the wise stewardship of the wolf and other wild canids found in North America. Produces educational materials appropriate to all age levels, as well as an annual summary of wolf recovery and management activities in North America.

Publication(s): NAWA NEWS (newsletter), publications on web, yearly summaries

Keyword(s): Endangered Species, Environmental and Conservation Education, training

Contact(s):
Rae Evening Earth Ott, DIRECTOR
Eric Schweig
INTERNATIONAL CELEBRITY SPOKESPERSON

NORTH CAROLINA ASSOCIATION OF SOIL AND WATER CONSERVATION DISTRICTS

512 N. Salisbury St.
1614 Mail Service Center, Raleigh, NC 27699
Raleigh, NC 27604 USA
Phone: 919-733-2302   Fax: 919-715-3559
Website: enr.state.nc.us

Founded: NA
Scope: Statewide

Keyword(s): Conservation Districts

Contact(s):
David Vogel, DIRECTOR
Phone: 919 7-15 -6097
david.vogel@ncmail.net

NORTH CAROLINA B.A.S.S. CHAPTER FEDERATION

Attn: President, 403 Red Wood Ct.
Lenoir, NC 28645 USA
Phone: 828-728-8550   Fax: 828-728-8549

Founded: NA
Membership: 2300
Scope: Statewide

Description: An organization of Bassmaster chapters, affiliated with the Bass Anglers Sportsman Society, organized to fight pollution, assist state and national conservation agencies in their efforts and teach the young people of our country good conservation practices. Dedicated to the realistic conservation of our water resources.

Contact(s):
Randy Lee, CONSERVATION DIRECTOR
1730 Allens Crossroads Road, Four Oaks, NC 27524
SmTrp19@aol.com

Ed Cannon, PRESIDENT
Phone: 828-728-8550

NORTH CAROLINA BEACH BUGGY ASSOCIATION, INC.

Box 940
Manteo, NC 27954 USA
Phone: 252-473-4880
Website: www.ncbba.org

Founded: NA
Scope: National

Description: NCBBA is an organization dedicated to preserving natural resources and coastal areas of North Carolina. Its purpose is to unite in an organization all persons interested in the natural beach resources of the Outer Banks of North Carolina and elsewhere, and establish a Code of Ethics of beach behavior to which each member must subscribe to uphold.

Publication(s): NCBBA News, The

Keyword(s): Coasts, Endangered Species, Wildlife, Outdoor Recreation, Sport Fishing

Contact(s):
John Newbold, EDITOR
W. Keene, PRESIDENT
23134 Homestead Lane, Franklin, VA 23851
Phone: 757-562-2554
Sharon Newbold, SECRETARY
600 S. Memorial Drive, Kill Devil Hills, NC 27948
Phone: 252-480-2453
Brenda Outlaw, TREASURER
P.O. Box 940, Manteo, NC 27954
Phone: 252-473-4880
brendaoutlaw@hotm-ail-.com
Tom Burke, VICE PRESIDENT
2512 S. Virginia Dare Trl., Nags Head, NC 27959

NORTH CAROLINA COASTAL FEDERATION, INC.

Attn: Alicia Kramer 3609 Highway 24 (Ocean)
Newport, NC 28570 USA
Phone: 252-393-8185   Fax: 252-393-7508
Website: www.nccoast.org

Founded: 1982
Membership: 6000
Scope: Regional

Description: The Coastal Federation focuses on the twenty coastal counties in North Carolina, with citizens working together for a healthy coast..

Publication(s): State of Coast Report, Sound Advice, Alternative to shoreline erosion control magazine, Coastal Review

Keyword(s): Water Quality, Estuaries, Land Use Planning, Environmental and Conservation Education

Contact(s):
Sally Steele, DIRECTOR OF DEVELOPMENT
sallys@nccoast.org
Jessica Kester, DIRECTOR OF EDUCATION
Todd Miller, EXECUTIVE DIRECTOR
toddm@nccoast.org

NORTH CAROLINA FORESTRY ASSOCIATION

1600 Glenwood Ave., Suite I
Raleigh, NC 27608-2355 USA
Phone: 919-834-3943   Fax: 919-832-6188
Website: www.ncforestry.org

Founded: 1911
Membership: 2500
Scope: Regional

Description: The North Carolina Forestry Association promotes and protects the long-term health and productivity of the forest ecosystem to enhance the environment and economy of North Carolina.

Publication(s): TreeLine

Keyword(s): Communications, Environmental and Conservation Education, Forests and Forestry, Renewable Resources

Contact(s):
Robert Slocum, EXECUTIVE VICE PRESIDENT
1600 Glenwood Ave., Suite I, Raleigh, NC 27608-2355
Phone: 919-834-3943
Dan Owens, PRESIDENT
P O Box 1536, Wendell, NC 27591
Phone: 919-365-1003

NORTH CAROLINA MUSEUM OF NATURAL SCIENCES

11 W Jones
Raleigh, NC USA
Phone: 877-462-8724
Website: naturalsciences.org

Founded: 1978
Scope: Statewide

Description: A nonprofit group formed to promote an interest in and to educate members and the general public concerning the ecological importance and conservation of reptiles and amphibians.

Publication(s): NC HERPS

Keyword(s): Reptiles and Amphibians

Contact(s):
Alvin Braswell, ADVISOR
1208 Buffaloe Rd., Garner, NC 27529
Phone: 919-733-7450
alvin_braswell@mail.enr.state.nc.us
Jeff Beane, EDITOR
4433 Graham Newton Rd., Raleigh, NC 27606
Phone: 919-733-7450
jeff_beane@mail.enr.state.nc.us
Tom Thorp, PRESIDENT
Phone: 004-201-0200
tt-threelakes@juno.com
Joe Zawadowski, SECRETARY
503 Valley Dr., Durham, NC 27704
Phone: 919-684-6062
Joe_Zawadowski@bba.mc.duke.edu
Dan Dombrowski, TREASURER
NC State Museum of Natural Sciences, P.O. Box 29555, Raleigh, NC 27626-0555
Phone: 919-733-7450
dan_dombrowski@mail.enr.state.nc.us
Dan Lockwood, VICE PRESIDENT
112 E. Skyhawk Dr., Cary, NC 27513
Phone: 919-460-3504
ddlockwood@email.msn.com

NORTH CAROLINA RECREATION AND PARK SOCIETY, INC.

883 Washington St.
Raleigh, NC 27605 USA
Phone: 919-832-5868   Fax: 919-832-3323
E-mail: ncrps@bellsouth.net
Website: www.ncrps.org

Founded: 1944
Scope: State

Description: A nonprofit organization formed to promote the wise use of leisure and intelligent development of the state's recreation resources. An affiliate of The National Recreation and Park Association.

Publication(s): NCRPS News, North Carolina Recreation and Park Review

Keyword(s): Protected Areas, Open Space, Outdoor Recreation, Youth Organizations

Contact(s):
Paul Herbert, EDITOR
Cornelius Park Recreation, Cornelius, NC 28031
Phone: 704-892-6031
Fax: 704-892-2462
pherbert@cornelius.org
Mike Waters, EXECUTIVE DIRECTOR
883 Washington St., Raleigh, NC 27605
Phil Rea, PRESIDENT
NC State University, Raleigh, NC 27695
Phone: 919-515-3675
Fax: 919-515-3687
phil_rea@ncsu.edu

NORTH CAROLINA WATERSHED COALITION, INC.

P.O. Box 337
Colfax, NC 27035 USA
Phone: 336-992-8734
E-mail: newcom@ynet.net
Website: www.ncwatershedcoalition.org

Founded: 1998
Scope: Statewide

Description: The coalition promotes conservation, protection and enhancement of watersheds and rivers; encourages founding and growth of local organizations devoted to those ends through information sharing and cooperation.

Keyword(s): Air Quality and Pollution, Conservation, Dams, Drinking Water Protection, Resource Law Enforcement, Rivers, Runoff, Streams, Upstream Flood Prevention, Urban and Rural Development, Water and Air Quality, Coral Reefs, Watersheds, training

NORTH CAROLINA WILD FLOWER PRESERVATION SOCIETY

c/o NC Botanical Garden, CB 3375, Totten Center, UNC-CH
Chapel Hill, NC 27599-3375 USA
Fax: 336-370-8172
E-mail: alice@ncwildflower.org
Website: www.ncwildflower.org/

Founded: 1951
Scope: Statewide

Description: A non-profit organization dedicated to the enjoyment and conservation of native plants and their habitats through education, protection and propagation.

Publication(s): NC Wild Flower Preservation Society Newsletter, NC Native Plant Propagation Handbook

Keyword(s): Botanical Gardens, Botany, Endangered Species, Habitat Conservation, Native Plants, Plant Propagation, Wildflowers, Research Grants, Stewardship, Sustainable Ecosystems

Contact(s):
Carla Oldham, CORRESPONDING SECRETARY
Phone: 919-932-3311
Craig Moretz, EDITOR
Phone: 919-563-1795
Charlotte Patterson, PRESIDENT
Phone: 336-643-4656
Evelyn Caldwell, RECORDING SECRETARY
Phone: 919-851-5780
Alice Zawadzki, VICE-PRESIDENT
Phone: 919-834-4172
alice_zawadzki@mail.agr.state.nc.us

NORTH CAROLINA WILDLIFE FEDERATION
P.O. Box 10626
Raleigh, NC 27605 USA
Phone: 919-833-1923   Fax: 919-829-1192
E-mail: ncwf-chuck@mindspring.com
Website: www.ncwildlifefed.org

Founded: 1945
Membership: 2500
Scope: Statewide

Description: A representative statewide organization, affiliated with the National Wildlife Federation, dedicated to the protection and enhancement of wildlife and its habitat through public education and government interaction.

Publication(s): Friend Of Wildlife Magazine

Contact(s):
Eddie Nickens, EDITOR
Lisa West, EDUCATION PROGRAMS CONTACT
Chuck Rice, EXECUTIVE DIRECTOR
Gary Shull, PRESIDENT AND ALTERNATE REPRESENTATIVE
Richard Mode, REPRESENTATIVE
Stan Warlen, TREASURER

NORTH CASCADES CONSERVATION COUNCIL
P.O. Box 95980
Seattle, WA 98145-2980 USA
Phone: 206-282-1644   Fax: 206-684-1379
Website: www.n3c-wedb1

Founded: 1957
Membership: 1000
Scope: Regional

Description: The Council seeks to protect and perserve the North Cascades' scenic, scientific, recreational, educational, wildlife, and wilderness values from the Columbia River to the US-Canadian border in the state of Washington.

Publication(s): Wild Cascades, The

Keyword(s): Ancient Forests, Biodiversity, Conservation, Protected Areas, Environmental Protection, Forest Management, Wildlife, Lakes, Land Preservation, Mining, National Parks, Land Use Planning, Natural Areas, Outdoor Recreation

Contact(s):
Patrick Goldsworthy, CHAIRMAN
Phone: 206-282-1644
Marc Bardsley, PRESIDENT
steveb@premier1.net
Phil Zalesky, SECRETARY
Thomas Brucker, TREASURER
Charles Ehlert, VICE PRESIDENT

NORTH DAKOTA ASSOCIATION OF SOIL CONSERVATION DISTRICTS
P.O. Box 1601
Bismarck, ND 58502 USA
Phone: 701-223-8518   Fax: 701-223-1291
E-mail: lincoln@tic.bisman.com
Website: www.lincolnoakes.com

Founded: NA
Membership: 260
Scope: Statewide
Keyword(s): Conservation Districts

Contact(s):
Gary Puppe, EXECUTIVE VICE PRESIDENT
P.O. Box 1601, 3310 University Dr.,
Bismarck, ND 58502-1601
Phone: 701-223-8518
Fax: 701-223-1291
Rodney Hickle, PRESIDENT
1631 28th Ave., SW, Center, ND 58530
Phone: 701-794-3342
Dale Tinjum, VICE PRESIDENT

NORTH DAKOTA NATURAL SCIENCE SOCIETY
Dept. of Biological Sciences, Box 4050
Emporia, KS 66801 USA
Phone: 316-341-5623
Website: www.emporia.edu/s/www/biosci/pn/pn.htm

Founded: 1967
Scope: Regional

Description: Dedicated to the observation, recording, study, and preservation of all aspects of the natural history of the Great Plains.

Publication(s): Prairie Naturalist, The

Keyword(s): Wildlife, education, Nongame Wildlife, Prairies, training, Great Plains

Contact(s):
Elmer Finck, EDITOR
Div. of Biological Sciences, Box 4050, Emporia, KS 66801
Jane Austin, PRESIDENT
Northern Prairie Wildlife Research Center,
Jamestown, ND 58401

NORTH DAKOTA WILDLIFE FEDERATION

P.O.Box 7248
Bismarck, ND 58507-7248 USA
Phone: 701-222-2557   Fax: 701-222-0334
E-mail: ndwf@gcentral.com
Website: www.ndwf.org

Founded: 1935
Scope: Statewide

Description: A representative statewide organization, affiliated with the National Wildlife Federation, dedicated to the protection and enhancement of wildlife and its habitat through public education and government interaction.

Publication(s): Flickertales

Contact(s):
Paula Mielke, EDUCATION PROGRAMS CONTACT
Carolyn Merbach, EXECUTIVE DIRECTOR
Art Mielke, PRESIDENT AND ALTERNATE REPRESENTATIVE
James Boley, REPRESENTATIVE
Conrad Carlson, TREASURER
Dick McCabe, VICE-PRESIDENT

NORTHCOAST ENVIRONMENTAL CENTER

575 H Street
Arcata, CA 95521 USA
Phone: 707-822-6918   Fax: 707-822-0827
E-mail: nec@igc.org
Website: www.necandeconews.to

Founded: 1971
Membership: 4500
Scope: Regional

Description: A tax-exempt educational organization dedicated to illuminating the relationships between humankind and the biosphere. The Center provides environmental information and referral services for northwestern California and southwestern Oregon and operates a library open to the public.

Publication(s): Econews

Keyword(s): Libraries, Ancient Forests, Endangered Species, Environmental Protection, Forests and Forestry, Biodiversity, Watersheds, Wilderness, Public Lands

Contact(s):
Tim McKay, DIRECTOR
Sid Dominitz, EDITOR
Gail Sellstrom, LIBRARIAN
Connie Stewart, OFFICE MANAGER

NORTHEAST CONSERVATION LAW ENFORCEMENT CHIEFS' ASSOCIATION (CLECA)

Attn: Secretary, RI Dept. of Environmental Management, 83 Park St.
Providence, RI 02908 USA
Phone: 401-277-2284

Founded: NA
Scope: Regional

Keyword(s): Professional Organization, Law Enforcement

Contact(s):
Ronald Alie, PRESIDENT
Chief, New Hampshire Fish and Game Dept., Law Enforcement Division, 2 Hazen Dr., Concord, NH 033301
Phone: 603-271-3127
Fax: 603-271-1438
Thomas Greene, SECRETARY AND TREASURER
Thomas Kamerzel, VICE PRESIDENTS
Dir., Bureau of Law Enforcement, Pennsylvania Fish and Boat Commission, P.O. Box 67000, Harrisburg, PA 17106-7000
Phone: 717-567-4542
Fax: 717-657-4033

NORTHEAST SUSTAINABLE ENERGY ASSOCIATION

50 Miles St.
Greenfield, MA 01301 USA
Phone: 413-774-6051   Fax: 413-774-6053
E-mail: nesea@nesea.org
Website: www.nesea.org

Founded: 1974
Membership: 1500
Scope: Regional

Description: The Northeast Sustainable Energy Association (NESEA) aims to strengthen the economy and lessen our impact on the environment by bringing sustainable energy into everyday use. Through its programs and activities, NESEA offers an alternative vision of responsible energy use and works with policymakers, industry, educators, students, and the general public to make this vision a reality.

Publication(s): Getting Around Without Gasoline, Totally Tree-mendous Activities, Northeast Sun

Keyword(s): Sustainable Energy, Transportation, Sustainable Buildings, Solar Energy, Environmental and Conservation Education

Contact(s):
Nancy Hazard, ASSOCIATE DIRECTOR
nhazard@nesea.org
Peter Taggert, BOARD OF DIRECTORS CHAIR
Chris Mason, DIRECTOR OF EDUCATION
cmason@nesea.org
Sandy Thomas, DIRECTOR OF ENERGY PARK
sthomas@nesea.org
Nancy Hazard, DIRECTOR OF TRANSPORTATION PROGRAMS
nhazard@nesea.org
Warren Leon, EXECUTIVE DIRECTOR
wleon@nesea.org
Jonathon Tauer, MANAGER OF BUILDING PROGRAM
jtauer@nesea.org
Michael Tennis, SECRETARY

NORTHERN ALASKA ENVIRONMENTAL CENTER

830 College Rd
Fairbanks, AK 99701-2806 USA
Phone: 907-452-5021   Fax: 907-452-3100
E-mail: info@northern.org
Website: www.northern.org

Founded: 1971
Membership: 1200
Scope: State, Loc, Reg, Nat, Intl

Description: The Northern Alaska Environmental Center works to protect some of the wildest country left in North America — the vast Interior and Arctic regions of Alaska — through education, advocacy, grassroots organizing, and sheer perseverance. The Northern Center's priorities are to protect Alaska's Arctic from oil drilling; promote ecologically sound, sustainable management of Alaska's boreal forests; defend wild rivers from mining pollution and roads; and encourage better environmental understanding, particularly by our youth.

Publication(s): Camp Habitat-email list, Arctic Action, Boreal Briefs, Conservation Abstracts, Northern Line, The, Mining Memos - email list

Keyword(s): Wildlands, Public Lands, Wilderness, Water Quality, Rivers, Natural Areas, Sustainable Ecosystems, Forest Management, Environmental and Conservation Education, Arctic, Tundra, Taiga

Contact(s):
Deb Moore, ARCTIC COORDINATOR
deb@northern.org
Nancy Fresco, BOREAL FOREST CAMPAIGN COORDINATOR
nancy@northern.org
Paul Ollid, COMMUNICATIONS AND MEMBERSHIP DIRECTOR
Arthur Hussey, EXECUTIVE DIRECTOR
arthur@northern.org
Clarice Dukeminier, PRESIDENT
Margaret Eagleton, TREASURER

NORTHERN PLAINS RESOURCE COUNCIL

2401 Montana Ave., Suite 200
Billings, MT 59101-2336 USA
Phone: 406-248-1154   Fax: 406-248-2110
E-mail: info@nprcmt.org
Website: www.nprcmt.org

Founded: 1972
Scope: National

Description: NPRC is a grassroots citizens' organization of farmers, ranchers, townspeople, and other conservationists. NPRC works on natural resource and agricultural issues to promote sustainable economic development, and to maintain Montana's unique rural quality of life. NPRC is dedicated to family agriculture and to stewardship of air, land, and water.

Publication(s): Reclaiming the Wealth (A Citizens' Guide to Hard Rock Mining in Montana), Legislative Bulletin, Plains Truth, The

Keyword(s): Agriculture, Air Quality and Pollution, Energy, Solid Waste Management, Water Quality, Mining

Contact(s):
Denna Hoff, CHAIR
Amy Frykman, RESEARCH COORDINATOR
Jeanie Alderson, SECRETARY
Teresa Erickson, STAFF DIRECTOR
Dan Teigen, TREASURER
Mary Fitzpatrick, VICE CHAIR

NORTHWEST ATLANTIC FISHERIES ORGANIZATION (NAFO)

Dartmouth, Nova Scotia B2Y 3Y9 Canada
Phone: 902-468-5590   Fax: 902-468-5538
E-mail: info@nafo.ca
Website: www.nafo.ca

Founded: 1979

Membership: 17
Scope: International

Description: Works for the optimum utilization, rational management, and conservation of the fishery resources of the convention area in the Northwest Atlantic. Contracting Parties: Bulgaria, Canada, Cuba, Denmark for the Faroes and Greenland, Estonia, European Union, France (for St. Pierre and Miquelon), Iceland, Japan, Korea, Latvia, Lithuania, Norway, Poland, Romania, Russia and the USA.

Publication(s): All publications available on web, Journal of Northwest Atlantic Fishery Science, Statistical Bulletin, Scientific Council Studies, Sampling Yearbook, Index of Meeting Documents, Annual Report

Keyword(s): Wildlife Rehabilitation, Wildlife, Oceanography

Contact(s):
T. Amaratunga, ASSISTANT EXECUTIVE SECRETARY
Leonard Chepel, EXECUTIVE SECRETARY
P. Gullestad, FISHERIES COMMISSION CHAIRMAN
D. Swanson, FISHERIES COMMISSION VICE CHAIRMAN
P. Chamut, GENERAL COUNCIL, VICE CHAIRMAN
E. Otuski, PRESIDENT OF NAFO AND CHAIRMAN OF GENERAL COUNCIL
W. Brodie, SCIENTIFIC COUNCIL CHAIRMAN
R. Mayo, SCIENTIFIC COUNCIL VICE CHAIRMAN

NORTHWEST COALITION FOR ALTERNATIVES TO PESTICIDES

P.O. Box 1393
Eugene, OR 97440 USA
Phone: 541-344-5044   Fax: 541-344-6923
E-mail: info@pesticide.org
Website: www.pesticide.org

Founded: 1977
Membership: 1950
Scope: Regional

Description: The Northwest Coalition for Alternatives to Pesticides works to protect people and the environment by advancing healthy solutions to pest problems.

Publication(s): Journal of Pesticide Reform, Farmer Exhange, The

Keyword(s): Agriculture, Alternative Agriculture, Pest Management, Pesticides, Toxicology

Contact(s):
Becky Long, DEVELOPMENT DIRECTOR
Caroline Cox, EDITOR/STAFF SCIENTIST
ccox@pesticide.org
Norma Grier, EXECUTIVE DIRECTOR
ngrier@pesticide.org
Kay Rumsey, LIBRARIAN
Edward Winter, OFFICE COORDINATOR/ FINANCIAL COORDINATOR
Megan Kemple, PUBLIC EDUCATION COORDINATOR
Aimee Code, RIGHT-TO-KNOW COORDINATOR
Pollyanna Lind, SALMON AND WATER QUALITY

NORTHWEST ECOSYSTEM ALLIANCE

1421 Cornwall Ave., Suite 201
Bellingham, WA 98225 USA
Phone: 360-671-9950   Fax: 360-671-8429
E-mail: nwea@ecosystem.org
Website: www.ecosystem.org

Founded: 1989
Membership: 11000
Scope: Regional

Description: The Northwest Ecosystem Alliance protects and restores wildlands in the Pacific Northwest and supports such efforts in British Columbia. The Alliance bridges science and advocacy, working with activists, policymakers, and the public to conserve our natural heritage.

Publication(s): Wild Salmon and Trout Action Plan: Of Wolves and Washington, Cascadia Wild: Protecting an International Ecosystem, Conservation Biology and National Forest Management in the Inland Northwest: A Handbook for Act, Northwest Conservation: News and Priorities

Keyword(s): Endangered Species, Environment, Forest Management, Biodiversity, Ecology, Reptiles and Amphibians, National Parks, Sustainable Ecosystems, training, Trapping, Ancient Forests, Conservation, Raptors, Internships, Predators

Contact(s):
Mary Humphries, BUSINESS MANAGER AND DEVELOPMENT DIRECTOR
meh@ecosystem.org
Lisa McShane, COMMUNITY OUTREACH DIRECTOR
lmcshane@ecosystem.org
Joe Scott, CONSERVATION DIRECTOR
jscott@ecosystem.org
Mitch Friedman, EXECUTIVE DIRECTOR

NORTHWEST ENVIRONMENT WATCH
1402 3rd Ave., Suite 500
Seattle, WA 98101 USA
Phone: 206-447-1880   Fax: 206-447-2270
E-mail: new@northwestwatch.org
Website: www.northwestwatch.org

Founded: 1993
Scope: Regional

Description: NEW's mission is to foster sustainability in the Pacific Northwest. NEW provides citizens with intelligence reports on the latest findings from the natural and social sciences, and guides them in creating a sustainable economy.

Publication(s): Over our Heads, This Place on Earth, Hazardous Handouts, Green Collared Jobs, Tax Shift, The Car and The City, Stuff, Misplaced Blame, State of the Northwest

Keyword(s): Population Growth, Sustainable Development, Transportation

Contact(s):
Alan Durning, EXECUTIVE DIRECTOR
Rhea Connors, OFFICE MANAGER

NORTHWEST INTERPRETIVE ASSOCIATION
909 1st Ave., Suite 630
Seattle, WA 98104-3627 USA
Phone: 206-220-4140   Fax: 206-220-4143
Website: www.nwpubliclands.com

Founded: 1974
Scope: Regional

Description: The Association supports interpretation and education on public lands administered by the National Park Service, U.S. Forest Service, and other agencies in the Pacific Northwest. Proceeds from the sale of interpretive publications are donated to these agencies to educate visitors in the area's natural and cultural history.

Keyword(s): Cultural Preservation, Environmental and Conservation Education, education, Outdoor Recreation, Wilderness

Contact(s):
Jim Torrence, CHAIRMAN OF THE BOARD
Mary Quackenbush, EXECUTIVE DIRECTOR
Tom Scribner, VICE CHAIRMAN

NORTHWEST RESOURCE INFORMATION CENTER
2811 W. State St., P.O. Box 427
Eagle, ID 83616 USA
Phone: 208-939-0714   Fax: 208-939-8731
Website: www.nwric.org

Founded: 1976
Membership: 3
Scope: National, International

Description: NRIC promotes through research, public education, technology transfer, and litigation the concept that ecological diversity and environmental quality are synonymous with long-term economic productivity and quality of life.

Contact(s):
Ed Chaney, EXECUTIVE DIRECTOR
2811 W. State St., Eagle, ID 83616
Phone: 208-939-8731

NOVA SCOTIA FEDERATION OF ANGLERS AND HUNTERS
P.O. Box 654
Halifax, Nova Scotia B3J 2T3 Canada
E-mail: tr.NSWF@chebucto.ns.ca

Founded: 1930
Membership: 6,500
Scope: Statewide

Description: Affiliated with the Canadian Wildlife Federation and the National Coalition of Provincial and Territorial Wildlife Federations. Aims to unite all conservation organizations in Nova Scotia, fosters appreciation of wildlife and habitat, promotes fish and game management, seeks enactment and enforcement of laws necessary for environmental controls as well as conservation of wildlife resources.

Contact(s):
A. Rodgers, EXECUTIVE DIRECTOR
Rannie Gillis, PRESIDENT

NOVA SCOTIA FORESTRY ASSOCIATION
P.O. Box 1113
Truro, Nova Scotia D2N 5O9 Canada
Phone: 902-893-4653   Fax: 902-893-1197
Website: www.nsfa.ca

Founded: 1959
Scope: international

Description: The Nova Scotia Forestry Association is a nonprofit, charitable organization dedicated to promoting the wise use and management of our forest resources through education programs for youth. Programs emphasize the importance of being good stewards of our natural resources.

Publication(s): Annual Teachers Guide, for National Forest Week

Keyword(s): Forest Management, Environmental and Conservation Education, Pollution Prevention

Contact(s):
Debbie Totten, EXECUTIVE DIRECTOR
Russ Waycott, PRESIDENT

NW ENERGY COALITION
219 First Ave., S.
Seattle, WA 98104 USA
Phone: 206-621-0094   Fax: 206-621-0097
E-mail: nwec@nwenergy.org
Website: www.nwenergy.org

Founded: 1981
Scope: Regional

Description: NWEC is a regionwide coalition of public interest groups and progressive utilities. The energy coalition works for a clean, affordable energy policy for the Pacific Northwest and British Columbia. The Coalition advocates for energy conservation and renewable resources, wild salmon, and low income/consumer protection.

Publication(s): Energy Activist, Plugging People into Power, NW Energy Coalition Report

Keyword(s): Energy, Wildlife, Salmon Recovery, Renewable Resources, Solar Energy

Contact(s):
Mark Glyde, COMMUNICATIONS DIRECTOR
Sara Patton, DIRECTOR
Rob Gala, OUTREACH DIRECTOR
Nancy Hirsh, POLICY DIRECTOR

O

OCEAN VOICE INTERNATIONAL
3332 McCarthy Rd.
Ottawa, Ontario K1V 0W0 Canada
Phone: 613-721-4541   Fax: 613-721-4562
E-mail: oceans@superaje.com
Website: www.ovi.ca

Founded: 1987
Membership: 600
Scope: National

Description: To conserve the diversity of marine life, protect and restore marine ecosystems and ecological services, enhance the quality of life and equity of benefits for coastal peoples, and promote ecologically sustainable harvest of marine resources.

Publication(s): Save Our Coral Reefs, How Green is Your School?, Green School Biodiversity Booklet, The Status of the World Ocean and its Biodiversity, Global Freshwater Biodiversity, Striving for the Integrity of Freshwater Ecosystems, Sea Wind

Keyword(s): Aquatic Habitats, Biodiversity, Developing Countries

Contact(s):
Jaime Baquero, PRESIDENT
Phone: 819-243-1334

OFFICE OF PROTECTIVE RESOURCES
NATIONAL MARINE FISHERIES SERVICE
NOAA/NMFS F/HP4 Off., Hab. Prot., 1315 East-West Highway
Silver Spring, MD 20910-3282 USA
Phone: 301-713-2319   Fax: 301-713-0376
Website: www.noaa.gov

Founded: NA
Membership: 56
Scope: National

Contact(s):
Don Knowles, DIRECTOR
don.knowles@noaa.org

OFFICE OF THE SECRETARY OF DEFENSE
INSTALLATIONS
3E1074 Defense Pentagon
Washington, DC 20321 USA
Phone: 703-697-8080

Description: The Installations office of Acquisition and Technology in the Office of the Secretary of Defense is responsible for U.S. and worldwide policy and Services coordination regarding land management, energy policy and energy purchasing policy, and base closure issues.

OHIO ACADEMY OF SCIENCE, THE
1500 W. 3rd Ave., Suite 223
Columbus, OH 43212-2817 USA
Phone: 614-488-2228   Fax: 614-488-7629
E-mail: oas@iwaynet.net
Website: www.ohiosci.org

Founded: 1891
Membership: 2000
Scope: Regional

Description: A nonprofit organization designed to stimulate interest in the sciences, to promote research, to improve instruction in the sciences, to disseminate scientific knowledge, and to recognize high achievement in attaining these objectives.

Publication(s): Ohio Academy of Science Newsletter, Ohio Journal of Science, The

Keyword(s): Biotechnology, Communications, Environmental and Conservation Education, Youth Organizations, Ecological Education

Contact(s):
Thomas Schmidlin, EDITOR
Kent State University, Department of Geography, Kent, OH 44242
Phone: 330-672-4632
Lynn Elfner, EXECUTIVE OFFICER
William Hummon, PRESIDENT
Lynn Elfner, SECRETARY
Michael Herschler, TREASURER

OHIO ALLIANCE FOR THE ENVIRONMENT
1500 West Third Ave.
Columbus, OH 43212 USA
Phone: 614-487-9957   Fax: 614-487-9957
Website: www.ohioalliance.org

Founded: 1977
Membership: 500
Scope: Statewide

Description: A statewide nonprofit organization established to provide leadership in resolving environmental conflicts and to promote and support adult environmental education in Ohio. The Alliance is a nonadvocacy organization working to establish channels of communication among diverse groups with environmental concerns and to provide balanced environmental information through conferences, publications, roundtable discussions, and seminars.

Publication(s): OAE Newsletter, Focus on the Issue

Keyword(s): Communications, Environmental and Conservation Education

Contact(s):
Jane Haynes, EDITOR
Natural Resources Specialist, League of Women Voters of Ohio Education Fund, 17 South High St.., Columbus, OH 43215
Phone: 614-469-1505
Fax: 614-469-7918
sjhaynes@juno.com
Irene Probasco, EXECUTIVE DIRECTOR
probasco@ohioalliance.org
Jim Swaney, PAST PRESIDENT
Dept. of Economics, Wright State Univ., 3640 Colonel Glenn Hwy., Dayton, OH 45435
Phone: 937-775-2769
Fax: 937-775-3545
jswaney@nova.wright.edu
Lisa Novosat-Gradert, PRESIDENT
Community Outreach Manager, Lafarge Corporation, P.O. Box 160, Paulding, OH 45879-0160
Phone: 419-399-4861
Fax: 419-399-2459
amfdjt@bright.net
Kenneth Schaublin, SECRETARY
Facilitator and Trainer, 58 Olentangy St., P.O. Box 82249, Columbus, OH 43202-2249
Phone: 614-571-8705
Fax: 614-261-0163
scovey@iwaynet.net
Mary Wiard, TREASURER
Brouse and McDowell, 500 First National Tower, Akron, OH 44308-1471
Phone: 330-535-5711
Fax: 330-253-8601
lngradert@browse.com

OHIO AUDUBON COUNCIL, INC.

121 Larchmont Rd.
Springfield, OH 45503 USA
Phone: 614-224-3303   Fax: 614-224-3305

Founded: 1969
Membership: 20,000
Scope: Statewide

Description: Associated with the National Audubon Society. Works to promote, foster, and encourage the conservation and preservation of all wildlife, plants, soil, water, air, and other natural resources for the benefit of all citizens. 20 chapters, two affiliates.

Keyword(s): Birds, Endangered Species, Environmental and Conservation Education, Wetlands, training

Contact(s):
Stephen Sedam, EXECUTIVE DIRECTOR
692 N. High St., Suite 208, Columbus, OH 43215
Phone: 614 2-24 -3303

OHIO B.A.S.S. CHAPTER FEDERATION

Attn: President, 376 N. Dorset
Troy, OH 45373 USA
E-mail: dbecker@erinet.com
Website: www.ohiobass.org

Founded: NA
Scope: Statewide

Description: An organization of Bassmaster chapters, affiliated with the Bass Anglers Sportsman Society, organized to fight pollution, assist state and national conservation agencies in their efforts, and teach the young people of our country good conservation practices. Dedicated to the realistic conservation of our water resources.

Contact(s):
Jim Doss, CONSERVATION DIRECTOR
43 Portsmouth Rd., Gallipolis, OH 45631
Phone: 740-446-9810
JSDoss@zoomnet.net
Dennis Becker, PRESIDENT
376 North Dorset, Troy, OH 45373
Phone: 937-335-2078
dbecker@erinet.com

OHIO BIOLOGICAL SURVEY

1315 Kinnear Rd.
Columbus, OH 43212-1192 USA
Phone: 614-292-9645   Fax: 614-688-4322
E-mail: obsinfo01@postbox.acs.ohio-state.edu
Website: www.obs.biosci.ohio-state.edu

Founded: 1912
Membership: 200
Scope: Regional

Description: An inter-institutional organization of 106 colleges, universities, museums and other organizations in Ohio, ten other states and the province of Ontario. Produces and disseminates scientific and technical information concerning the flora and fauna of the Ohio environment, and larger areas of which Ohio is an integral part.

Publication(s): publications available on website, Miscellaneous Publications, Notes, Informative Publications, In Ohio's Backyard Series, Bulletins

Keyword(s): Biodiversity, Endangered Species, Environmental and Conservation Education, education, National Parks, Flowers, Plants, and Trees

Contact(s):
Terry Keiser, CHAIRMAN OF THE ADVISORY BOARD
Ohio Northern University,
Brian Armitage, EXECUTIVE DIRECTOR
armitage.7@osu.edu

OHIO ENERGY PROJECT

640 Enterprise Dr.
Suite A
Lewis Center, OH 43035 USA
Phone: 614-785-1717   Fax: 614-785-1731
E-mail: oep@ohioenergy.org
Website: www.ohioenergy.org

Founded: 1984
Membership: 500
Scope: Statewide

Description: A nonprofit organization promoting energy education, efficiency and conservation, youth leadership development, using a fun hands on multi-dimensional, Inter-discipinary approach and a "kids teaching kids" philosophy.

Publication(s): Curriculum Materials

Keyword(s): Energy, Energy Conservation, Environmental and Conservation Education, Scholarships, Youth Organizations, Youth Leadership

Contact(s):
Rich Smith, DIRECTOR
Melissa Fu, EDUCATION COORDIANTOR
Mike Stranges, EDUCATION COORDINATOR
Shauni Nix, EXECUTIVE DIRECTOR
Annie Rasor, PROGRAMS MANAGER
Mary McCarron, STATEWIDE COORDINATOR

OHIO ENVIRONMENTAL COUNCIL, INC.
Suite 201, 1207 Grandview Ave.
Columbus, OH 43212 USA
Phone: 614-487-7506   Fax: 614-487-7510
Website: www.theoec.org

Founded: 1969
Membership: 1500
Scope: Regional

Description: The Ohio Environmental Council is a statewide organization providing resources for local environmental organizations across Ohio. The OEC promotes improved envi-ronmental quality in the state through advocacy, research, education and collaborative efforts.

Publication(s): Publications available on web, Green Pages (listing of Ohio environmental groups and resources, variety of other publications on critical environmental issues for Ohio, Ohio Environmental Report Newsletter

Keyword(s): Air Quality and Pollution, Energy, Environmental and Conservation Education, Environmental Justice, Watersheds, Factory Farms, Pollution Prevention, Energy, Lakes, Sustainable Development, Coral Reefs, Water Quality, Sustainable Ecosystems, Toxic Substances, Nuclear-free, Water quantity, Water export and diversion, Exotic species, Aquatic nuisance species

Contact(s):
Vicki Deisner, EXECUTIVE DIRECTOR
Daniel Bender, PRESIDENT
817 S Remmington Rd., Bexle, OH 43209
David Rynbolt, SECRETARY
Bruce Cornett, TREASURER
Green Environmental Coalition, Box 266, Yellow Springs, OH 45387
Phone: 937-767-5000
Fax: 93776785-87/-9041
bcornett@greenlink.org
Guy Denny, VICE PRESIDENT

OHIO FEDERATION OF SOIL AND WATER CONSERVATION DISTRICTS
4383 Faintain Sq. Ct. Building B-3
Columbus, OH 43224 USA
Phone: 614-265-6610
Website: dnr.state.oh.us/odnr/soil+water

Founded: NA
Scope: Statewide

Description: To provide leadership and services that enable Ohioans to conserve, protect and enhance soil, water and land resources

Keyword(s): Conservation Districts

Contact(s):
Brad Ross, ADMINISTRATOR
Phone: 614 2-65 -6616

OHIO FORESTRY ASSOCIATION, INC., THE
Grove City, OH 43123 USA
Phone: 614-497-9580   Fax: 614-497-9581
E-mail: nichoel@ohioforest.org
Website: www.ohioforest.org

Founded: 1903
Membership: 900+
Scope: Regional

Description: A statewide organization, affiliated with the National Woodland Owners Association, organized to promote the welfare of the people and private enterprise of Ohio by improving, through education, the wise management of Ohio's forest resource. Sponsors annual forestry camp for youths 14-19; assists schools' conservation activities and education; and coordinates American Tree Farm Program in Ohio.

Publication(s): Bark and Bunk, Ohio Woodlands

Keyword(s): Forests and Forestry, Camp

Contact(s):
C. Wayne Lashbrook
1ST VICE PRESIDENT
Jim Doll, 2ND VICE PRESIDENT
Roy Palmer, 3RD VICE PRESIDENT
Karl Gebhardt, INTERIM
Melvin Yoder, PRESIDENT
Robert B. Redett, TREASURER

OHIO NATIVE PLANT SOCIETY
6 Louise Dr.
Chagrin Falls, OH 44022 USA
Phone: 440-338-6622
Website: http:/dir.gardenweb.com/directory/onps1

Founded: 1982
Scope: Statewide

Description: The Ohio Native Plant Society is dedicated to preservation, conservation and education concerning all native plants of Ohio.

Keyword(s): Botany, Botanical Gardens, Native Plants, Wildflowers, Flowers, Plants, and Trees, Environmental and Conservation Education

Contact(s):
A. Malmquist, EXECUTIVE SECRETARY

OKLAHOMA ACADEMY OF SCIENCE
P.O. Box 701915
Tulsa, OK 74170-1915 USA
Phone: 918-495-6944
Website: http://bmb-fs1.biochem.okstate.edu/OAS/

Founded: 1909
Membership: 705 members, 60 libraries
Scope: Statewide

Description: To stimulate scientific research; to promote fraternal relationship among those engaged in scientific work in

Oklahoma; to diffuse among the citizens of the state a knowledge of the various departments of science; and to investigate and make known the material, education, and other resources of the state.

Publication(s): Annals-Oklahoma Academy of Science, Transactions-Oklahoma Junior Academy of Science, CAS Newsletter, Proceedings-Oklahoma Academy of Science

Keyword(s): Environmental and Conservation Education, education, Research, training

OKLAHOMA ASSOCIATION OF CONSERVATION DISTRICTS

Attn: President, P.O. Box 107
Chelsea, OK 74016-0107 USA
Phone: 918-696-7612

Founded: NA
Scope: Statewide

Keyword(s): Conservation

Contact(s):
Billy Wilson, BOARD MEMBER
P.O. Box 208, Kinta, OK 74552-0208
Phone: 918-768-3542
bwilson@cwis.net
George Fraley, PAST PRESIDENT
P.O. Box 107, Chelsea, OK 74016-0107
Phone: 918-789-2511
Fax: 918-789-2835
Carol Gaunt, PRESIDENT, ALTERNATE BOARD MEMBER
Rt 5 Box 244, Weatherford, OK 73096-8815
Phone: 405-772-5107
Christy Kimble, SECRETARY
Oklahoma County CD, 1120 NW 63rd STE G101, Oklahoma City, OK 73116
Phone: 405-848-1933
Fax: 405-842-8744
Wayne Smith, TREASURER
506 N. Pennsylvania, Mangum, OK 73554-3036
Phone: 405-782-3575
Fax: 405-782-3581
Matt Gard, VICE PRESIDENT
Rt. 1, Box 16, Fairview, OK 73737-9621
Phone: 580-438-2320
Rick Jeans, VICE PRESIDENT
Rt 1 Box 184, Tonkawa, OK 74653
Phone: 405-628-2223
Mark Moehle, VICE PRESIDENT
1601 Shadow Court, Edmond, OK 73013-2683
Phone: 405-340-8884
Fax: 405-048-0744

OKLAHOMA AUDUBON COUNCIL

P.O. Box 2476
Tulsa, OK 74101 USA
Phone: 918-592-1614
E-mail: info@tulsaaudubon.org

Founded: 1987
Membership: 4,000
Scope: Statewide

Description: A statewide council of representatives of the eight National Audubon Society chapters in Oklahoma. The council coordinates the efforts of the chapters on statewide environmental issues, and advocates protection, preservation, and wise use of soil, water, plants, and wildlife.

OKLAHOMA B.A.S.S. CHAPTER FEDERATION

Attn: President, 2300 E. Coleman Rd.
Ponca City, OK 74604 USA
Phone: 580-765-0165
E-mail: bigc@ponca.net
Website: www.okbass.org

Founded: NA
Scope: Statewide

Description: An organization of Bassmaster chapters, affiliated the with Bass Anglers Sportsman Society, organized to fight pollution, assist state and national conservation agencies in their efforts, and teach the young people of our country good conservation practices. Dedicated to the realistic conservation of our water resources.

Publication(s): Bulletin of the Oklahoma Ornithological Society, The, Oklahoma B.A.S.S. Federation Newsletter, Scissortail, The

Contact(s):
Don Linder, CONSERVATION DIRECTOR
2409 Cardinal, Ponca City, OK 74604
Phone: 580 7-62 -3301
dlinder@horizon.hit.net
Robert Cartlidge, PRESIDENT
2300 E. Coleman Road, Ponca City, OK 74604
Phone: 580 7-65 -0165
R. Kitterman, SECRETARY, EDITOR
411 W. 6th, Dewey, OK 74029
Phone: 918 5-34 -1720
James Hardage, TREASURER
1352 Bridle Path Lane, Lindale, TX 75771
Phone: 903 8-82 -8652

OKLAHOMA DEPT. OF WILDLIFE & CONSERVATION

OKLAHOMA CHAPTER
75-B Rte. 1
Porter, OK 74454 USA
Phone: 918-683-1031	Fax: 918-683-9406
E-mail: trodwc@oknet1.net
Website: www.wildlifedepartment.com

Founded: 1968
Membership: 9
Scope: Statewide

Contact(s):
Paul Balkenbush, REGIONAL SUPERVISOR

OKLAHOMA NATIVE PLANT SOCIETY

c/o Tulsa Garden Center, 2436 S. Peoria
Tulsa, OK 74114 USA
Phone: 405-872-9652	Fax: 405-872-8361
E-mail: cox.chadwick@worldnet.http.net
Website: www.telepath.com/chadcox/onps.html

Founded: 1986
Membership: 400
Scope: State

Description: Oklahoma Native Plant Society encourages the study, protection, propagation, appreciation, and use of Oklahoma's native plants.

Publication(s): Native Plant Selection Guide for Oklahoma Woody Plants, Gaillardia, The

Keyword(s): Native Plants, Biodiversity, Conservation, Ecology,

Endangered Species, Flowers, Plants, and Trees, Environmental Protection, Natural Areas, Prairies, Public Lands, Rural Development, Wilderness, Rivers, Terrestrial Habitats, Wetlands

Contact(s):
Patricia Folley, PRESIDENT
Phone: 405-872-8361
pfolley7@juno.com
Maurita Nations, SECRETARY
nationsokc@juno.com
Mary Korthase, TREASURER
Phone: 918-743-2743
mkorthase@webzone.net
Chad Cox, VICE PRESIDENT
Phone: 405-598-6742

OKLAHOMA ORNITHOLOGICAL SOCIETY

Attn: Business Manager, 1701 W. Will Rogers
Claremore, OK 74017 USA
Phone: 918-343-7706   Fax: 918-343-7563
E-mail: kwmartin@rsu.edu

Founded: 1950
Membership: 450
Scope: Statewide

Description: Affiliated with National Audubon Society and the Oklahoma Wildlife Federation. Dedicated to the observation, study, and conservation of birds in Oklahoma.

Publication(s): Bulletin of the Oklahoma Ornithological Society

Contact(s):
Jo Loyd, BUSINESS MANAGER
6736 E. 28th St., Tulsa, OK 74129
Phone: 918-835-2946
Richard Stewart, EDITOR
birdbander@aol.com
Charles Brown, EDITOR
Biology Dept. University of Tulsa, Tulsa, OK 74104
Phone: 918-631-3943
Keith Martin, PRESIDENT
Michael Bay, SECRETARY
Dept. Biology, East Central State Univ., Ada, OK 74820
Marty Kamp, TREASURER
6422 S. Indianapolis Pl., Tulsa, OK 74136

OKLAHOMA WILDLIFE FEDERATION

P.O. Box 60126
Oklahoma City, OK 73146-0126 USA
Phone: 405-524-7009   Fax: 405-521-9270
E-mail: owf@nstar.net
Website: okwildlife.org

Founded: 1963
Scope: Statewide

Description: A representative statewide organization, affiliated with the National Wildlife Federation, dedicated to the protection and enhancement of wildlife and its habitat through public education and government interaction.

Publication(s): Urban Landscaping, Outdoor News

Contact(s):
Mich Entz, ALTERNATE REPRESENTATIVE
Lance Meek, EDITOR
Margaret Ruff, EDUCATION PROGRAMS CONTACT & EXECUTIVE DIRECTOR
Royce Meek, PRESIDENT AND REPRESENTATIVE

Dick Gunn, SECRETARY
James Menzer, TREASURER

OLYMPIC PARK ASSOCIATES

13245 40th Ave., NE
Seattle, WA 98125-4617 USA
Phone: 206-364-3933

Founded: 1948
Scope: Local

Description: Dedicated to preservation of the integrity and wilderness of Olympic National Park and surrounding areas, as well as supporting wild land and wildlife habitat protection elsewhere in the nation. Currently working on restoration of the Elwha River ecosystem, elimination of exotic species from Olympic National Park and restoration and protection of salmon spawning areas.

Publication(s): Voice of the Wild Olympics

Keyword(s): National Parks, Salmon Recovery, Native Plants, Wilderness

Contact(s):
Sally Soest, EDITOR
Phone: 206-860-2865
Polly Dyer, PRESIDENT
Phone: 206-364-3933
Philip Zalesky, SECRETARY
Phone: 425-337-2479
John Anderson, TREASURER
Phone: 206-523-5043
Tim McNulty, VICE PRESIDENT
Phone: 360-681-2480

OLYMPIC PARK INSTITUTE

111 Barnes Point Rd.
Port Angeles, WA 98363 USA
Phone: 360-928-3720   Fax: 360-928-3046
E-mail: opi@yni.org
Website: www.yni.org/opi

Founded: 1987
Scope: National

Description: Mission is to inspire personal connection to the natural world and responsible actions to sustain it. OPI provides residential field science programs in Olympic National Park for adults, families and K-12 classrooms. Elderhostel and field seminar programs are also available. Programs introduce themes of ecology, sustainability and stewardship.

Publication(s): Publications on website

Keyword(s): Environmental Education Curriculum, Ancient Forests, National Parks, Biodiversity, Wildlife Rehabilitation, Birds, Conservation, Coasts, Cultural Preservation, Ecology, Endangered Species, Environment, Environmental Protection, Energy, Wildlife

ONTARIO FEDERATION OF ANGLERS AND HUNTERS, INC., THE

Box 2800
Peterborough, Ontario K9J 8L5 Canada
Phone: 705-748-6324   Fax: 705-748-9577
E-mail: ofah@ofah.org
Website: www.ofah.org

Founded: 1928

Scope: Regional

Description: Purpose is to conserve Ontario's natural resources and promote ethical angling and hunting practices.

Publication(s): Hunter Education News, Canadian Fishing and Hunting Trade News, Angler and Hunter Hotline, Call of the Loon

Contact(s):
Mark Holmes, EDITOR
mark_holmes@ofah.org
Mike Reader, EXECUTIVE DIRECTOR
mike_reader@ofah.org

ONTARIO FORESTRY ASSOCIATION
307—200 Consumers Rd.
Toronto, Ontario M2J 4R4 Canada
Phone: 416-493-4565   Fax: 416-493-4608
E-mail: forestry@oforest.on.ca
Website: www.oforest.on.ca

Founded: 1949
Scope: International

Description: Ontario Forestry Association works to raise awareness and understanding of all aspects of Ontario's forests and to develop commitment to stewardship of forest ecosystems. Programs include: Envirothon, Community Woodland Steward Initiative, Consultant Registry, Land use and forest management, forestry publicity and historical data.

Publication(s): Re:View, Ontario Forest Products Marketing Bulletin

Keyword(s): Forests and Forestry, Forest Management, Forestry

Contact(s):
Alex Rigsby, 1ST VICE PRESIDENT
James Farrell, 2ND VICE PRESIDENT
Erik Turk, EXECUTIVE DIRECTOR
Anne Koven, PRESIDENT

OPENLANDS PROJECT
25 E. Washington St., Suite 1650
Chicago, IL 60602 USA
Phone: 312-427-4256   Fax: 312-427-6251
E-mail: info@openlands.org
Website: www.openlands.org

Founded: NA
Scope: Local

Description: A private, nonprofit organization, Openlands Project was founded in 1963 to protect, enhance and expand land and water—to provide a healthy environment and a more livable place for all people of the region. Openlands preserves open space through land acquisition, greenways, watershed planning and restoration, urban greening initiates, advocacy and technical assistance.

Publication(s): Annual Report, Under Pressure: Land Consumption in the Chicago Region, Openlander

Keyword(s): Protected Areas, Land Purchase, Land Use Planning, Urban Environment, Watersheds

Contact(s):
Gerald Adelmann, EXECUTIVE DIRECTOR
Ders Anderson, GREENWAYS DIRECTOR
Tony Dean, PRESIDENT
Charles Saltzman, SECRETARY

J. Ritchie, TREASURER
Glenda Daniel, URBAN GREENING DIRECTOR

OREGON ASSOCIATION OF CONSERVATION DISTRICTS
c/o Yamhill SWCD
P.O. Box 2200 W. 2nd McMinnville 97128
Salem, OR 97393 USA

Founded: NA
Scope: Statewide

Keyword(s): Conservation Districts

Contact(s):
Don Lucas, 1ST VICE PRESIDENT, ALTERNATE BOARD MEMBER
HC 60 Box 4000, Lakeview, OR 97630
Phone: 541-947-2482
John McDonald, 2ND VICE PRESIDENT
30730 SW Simpson Rd, Cornelius, OR 97113
Phone: 503-640-2841
Fax: 541-947-5854
donald_lucas@yahoo.com
Mike Barlow, PRESIDENT AND BOARD MEMBER
2524 Mitchell Butte Rd., Nyssa, OR 97913
Phone: 541-372-2886
Fax: 541-889-4304
Stan Christensen, SECRETARY/TREASURER
16350 Delashmutt Ln., McMinnville, OR 97128
Phone: 503-472-6307

OREGON B.A.S.S. CHAPTER FEDERATION
Attn: President, 1601 S. Dogwood St.
Cornelius, OR 97113 USA
Phone: 503-357-4798
Website: http://orbass.oregonbass.net

Founded: NA
Scope: Statewide

Description: An organization of Bassmaster chapters, affiliated with Bass Anglers Sportsman Society, organized to fight pollution, assist state and national conservation agencies in their efforts, and teach the young people of our country good conservation practices. Dedicated to the realistic conservation of our water resources.

Publication(s): EarthWatch Oregon

Keyword(s): Air Quality and Pollution, Land Use Planning, Transportation, Exotic species, Aquatic nuisance species, Water Quality

Contact(s):
Chuck Lang, CONSERVATION DIRECTOR
4775 Gardner Rd., SE, Salem, OR 97302
Phone: 503-588-1920
charleslang@home.com
Orville Alleman, PRESIDENT

OREGON ENVIRONMENTAL COUNCIL
520 SW 6th.
Portland, OR 97204-1535 USA
Phone: 503-222-1963   Fax: 503-222-1405
E-mail: oec@orcouncil.org
Website: www.orcouncil.org

Founded: 1968
Membership: 2000

Scope: Statewide

Description: Oregon Environmental Council is a nonprofit organization whose mission is to restore and protect Oregon's clean water and air now and for future generations. OEC brings Oregonians together to create and promote socially just and economically sound environmental policies.

Publication(s): EarthWatch Oregon

Keyword(s): Air Quality and Pollution, Environmental Justice, Pesticides, Pollution Prevention, Rivers, Coral Reefs, Water Quality, Transportation

Contact(s):
Jeff Allen, EXECUTIVE DIRECTOR
Jesse Reeder, PRESIDENT
Dorothy Fisher Atwood, SECRETARY
James Whitty, VICE PRESIDENT

OREGON NATURAL RESOURCES COUNCIL

5825 N. Greeley Avenue
Portland, OR 97217 USA
Phone: 503-283-6343   Fax: 503-283-0756
E-mail: sr@onrc.org
Website: www.onrc.org

Founded: 1972
Membership: 5000
Scope: Statewide

Description: A nonprofit, state-wide organization to aggressively protect and restore Oregon's wildlands, wildlife, and waters as an enduring legacy, and dedicated to permanently protect public forest lands and protect and restore critical habitat for native species.

Publication(s): Wild Oregon

Keyword(s): Ancient Forests, Conservation, Environmental Protection, Protected Areas, Natural Areas, Watersheds, Wilderness, Public Lands, Wildlands

Contact(s):
Tim Lillebo, ADVOCACY DIRECTOR
16 NW Kansas, Bend, OR 97701
Phone: 541-382-2616
Fax: 503-385-3370
tl@onrc.org
Ken Rait, CONSERVATION DIRECTOR
kr@onrc.org
Jacki Richey, DIRECTOR OF FINANCE
Regna Merritt, EXECUTIVE DIRECTOR
rm@onrc.org
Alex Brown, GRASSROOTS COORDINATOR
em@onrc.org
Pat Clancy, PRESIDENT
Wendell Wood, SOUTHERN OREGON FIELD REP.
943 Lakeshore Drive, Klamath Falls, OR 97601
Phone: 541-885-4886
Fax: 541-885-4887
ww@onrc.org
Doug Heiken, WESTERN OREGON FIELD REP.
P.O. Box 11648, Eugene, OR 97440
Phone: 541-344-0675
Fax: 541-343-0996
onrcdoug@efn.org

OREGON SMALL WOODLANDS ASSOCIATION

1775 32nd Place, NE, Suite C
Salem, OR 97303 USA
Phone: 503-588-1813   Fax: 503-588-1970
E-mail: oswa@oswa.org
Website: www.oswa.org

Founded: 1967
Membership: 2300
Scope: Statewide

Description: A statewide organization affiliated with the National Woodland Owners Association, dedicated to the protection, management, use and enhancement of Oregon's forest resources.

Publication(s): The Update, Northwest Woodlands

Keyword(s): Environmental and Conservation Education, Forests and Forestry, Renewable Resources

Contact(s):
Bill Arsenault, 1ST VICE-PRESIDENT
Phone: 541-584-2272
Ken Faulk, 2ND VICE-PRESIDENT
Phone: 541-447-6762
Lori Rasor, EDITOR
Phone: 503-228-3624
Denny Miles, EXECUTIVE DIRECTOR
John Poppino, PRESIDENT
Phone: 541-447-1342

OREGON TROUT

117 SW Naito Parkway
Portland, OR 97204-3595 USA
Phone: 503-222-9091   Fax: 503-222-9187
E-mail: info@ortrout.org
Website: www.ortrout.org

Founded: NA
Membership: 1500
Scope: Regionally

Description: An Oregon-based organization focused on the protection and restoration of native fish and their ecosystems.

Publication(s): Riverkeeper

Keyword(s): Native Fish, Habitat Conservation, training

Contact(s):
Jim Myron, CONSERVATION DIRECTOR
Joe Whitworth, EXECUTIVE DIRECTOR

OREGON WILDLIFE HERITAGE FOUNDATION

P.O. Box 30406
Portland, OR 97294-3406 USA
Phone: 503-255-6059   Fax: 503-255-6467
E-mail: owhf@aol.com

Founded: 1981
Membership: 400
Scope: Statewide

Description: The Oregon Wildlife Heritage Foundation is a nonprofit, tax-exempt foundation incorporated under the laws of the state of Oregon. It has a 501(c(3 determination under the I.R.S. code. It receives grants and contributions to be used to fund selected projects beneficial to the fish and wildlife resources of Oregon and the people who enjoy them.

Keyword(s): training

Contact(s):
Rod Brobeck, EXECUTIVE DIRECTOR
Kim MacColl, PRESIDENT
Charles Lilley, SECRETARY
Nelson Rutherford, TREASURER
Marcia Hartman, VICE PRESIDENT

ORGANIZATION FOR BAT CONSERVATION
1553 Haslett Rd.
Haslett, MI 48840 USA
Phone: 517-339-5200   Fax: 517-339-5618
E-mail: obcbats@aol.com
Website: www.batconservation.org

Founded: 1990
Scope: International

Description: One of the only international non-profit organizations dedicated to bat conservation. This mission is fulfilled through education and habitat preservation.

Publication(s): Understanding Bats, Bats: The True Story (video), Simple Guide to Bat House Designs

Keyword(s): Conservation, Endangered Species, Habitat Conservation, Interpretation, Rehabilitation, Zoological Parks, Bats

Contact(s):
Rob Mies, DIRECTOR
Denise Tomlinson, DIRECTOR OF OPERATIONS
Phone: 517-339-5481
Kim Williams, EXECUTIVE DIRECTOR
Phone: 517-339-5200

ORGANIZATION OF WILDLIFE PLANNERS
1900 Kanawha Blvd.
East Charleston, WV 25305 USA
Phone: 304-558-2771   Fax: 304-558-3147
E-mail: wildlife@dnr.state.wv.us
Website: www.dnr.state.wv.us

Founded: 1978
Membership: 28
Scope: Statewide

Description: A nonprofit, tax-exempt organization comprised of professional state and federal fish and wildlife resource planners, natural resources educators, professional conservationists, and associated interests dedicated to improving, through education and training, the quality of state-level resources management and planning. The focus of the organization is on developing the necessary tools and skills to conduct effective planned management systems.

Publication(s): Tomorrow's Management, Newsletter

Keyword(s): Planning Management

Contact(s):
Paul Johnasen, ASSISTANT CHIEF, GAME MANAGEMENT

ORION SOCIETY
189 Main St
Great Barrington, MA 01230 USA
Phone: 413-528-4422   Fax: 413-528-0676
E-mail: orion@orionsociety.org
Website: www.oriononline.org

Founded: NA
Membership: 8000

Scope: National

Description: The Orion Society is an award-winning publisher, an environmental eduation organization, and a communication support network for grassroots environmental and community organizations across North America. It is a nonprofit member organization with 8000 members, individual and organization representing all fifty states and thirty-one countries.

Publication(s): Orion Magazine, Orion Afield

ORNITHOLOGICAL COUNCIL
THE NATIONAL COUNCIL FOR SCIENCE &
THE ENVIRONMENT
1725 K St., NW, Suite 212
Washington, DC 20006-1401 USA
Phone: 202-530-5810   Fax: 202-628-4311
Website: www.ncsonline.org

Founded: 1992
Scope: International

Description: The Council provides impartial scientific information about birds for sound decisions, policies, or management actions; links the scientific community with public and private decision-makers; informs ornithologists of actions that affect birds or the study of birds; and speaks for scientific ornithology when the study of birds might be affected. Website contains links to ornithological scientific socities.

Keyword(s): Wildlife Rehabilitation, Birds, Environmental and Conservation Education, Raptors, Waterfowl, Research Grants, Birds

Contact(s):
David Blockstein, CHAIR
Peter Saundry, EXECUTIVE DIRECTOR
Phone: 301-986-8568
Fax: 301-986-5205
peter@ncsonline.org

OUTDOOR CIRCLE, THE
1314 S. King
Honolulu, HI 96814 USA
Phone: 808-593-0300   Fax: 808-593-0525
E-mail: mail@outdoorcircle.org
Website: www.outdoorcircle.org

Founded: 1912
Membership: 3500
Scope: Local

Description: A nonprofit organization whose purpose is to work for and develop a more beautiful state, freeing it from disfigurement, conserving and developing its natural beauty, and cooperating in educational and other efforts towards preservation of open spaces, parklands, recycling, and antilitter.

Publication(s): The Greenleaf Newsletter, Pua Nani: Hawaii is a Garden, Majesty: Exceptional Trees of Hawaii, Majesty II, Exceptional Trees of Hawaii, The, Trees and Flowers of the Hawaiian Islands, Our Familiar Island Trees, Keep Hawaii Green

Keyword(s): Land Preservation, Natural Areas, Open Space, Public Lands, Urban Forestry, Trees, Beautification

Contact(s):
Mary Steiner, CHIEF EXECUTIVE OFFICER
Chris Snyder, LANDSCAPE AND PLANTING AND PROJECT MANAGER
chris@outdoorcircle.org

OUTDOOR RECREATION COUNCIL OF BRITISH COLUMBIA

334 - 1367 W. Broadway
Vancouver, British Columbia V6H 4A9 Canada
Phone: 604-737-3058   Fax: 604-737-3666
E-mail: orc@intergate.ca
Website: www.orcbc.ca

Founded: 1976
Scope: Statewide

Description: The ORC of BC is a nonprofit society formed to serve as a mechanism independent from government, through which the interests and activities of groups organized on a provincial basis concerned with outdoor recreation, education, and conservation can be coordinated and represented to government, industry, and the public.

Publication(s): Outdoor Report

Keyword(s): Conservation, Land Use Planning, Outdoor Recreation, Research, Rivers, Wilderness

OUTDOOR WRITERS ASSOCIATION OF AMERICA, INC.

121 Hickory St., Suite 1
Missoula, MT 59801 USA
Phone: 406-728-7434   Fax: 406-728-7445
E-mail: owaa@montana.com
Website: www.owaa.org

Founded: 1927
Membership: 1900
Scope: National

Description: We strive to improve ourselves in the art and media of our craft and to increase our knowledge and understanding in supporting the conservation of our natural resources. To this end we pledge ourselves to maintain the highest ethical standards in the exercise of our craft.

Publication(s): Outdoors Unlimited

Keyword(s): Communications, Environmental and Conservation Education, Outdoor Recreation, training

Contact(s):
Laurie Dovey, 1ST VICE PRESIDENT
160 White Pines Dr., Alpharetta, GA 30004
Phone: 770-475-3793
Fax: 770-772-9925
lldovey@webimages.net
Jim Casada, CHAIRMAN OF THE BOARD
1250 Yorkdale Dr., Rock Hill, SC 29730
Phone: 803-329-4354
Fax: 803-329-2420
jcasada@jounalist.com
Kevin Rhoades, EDITOR
oueditor@montana.com
Steve Wagner, EXECUTIVE DIRECTOR
Bill Monroe, PRESIDENT
14292 S. Forsythe Rd., Oregon City, OR 97045
Phone: 888-222-8231
Phone: 503-221-8168
billmonroe@news.oregonian.com

OZARK SOCIETY, THE

P.O. Box 2914
Little Rock, AR 72203 USA
Phone: 501-666-2989   Fax: 501-666-2989
E-mail: steward810@aol.com
Website: ozarksociety.net

Founded: 1962
Membership: 1,000
Scope: Statewide

Description: To promote the knowledge and enjoyment of the scenic and scientific resources, particularly free-flowing streams, wilderness areas, and unique natural areas of the Ozark-Ouachita mountain region, and to help protect those resources for present and future generations.

Publication(s): Pack and Paddle, The

Keyword(s): Wilderness, Natural Areas

Contact(s):
Stewart Noland, PRESIDENT
5210 Sherwood Rd., Little Rock, AR 72207
Phone: 501-666-2989

OZARKS RESOURCE CENTER

P.O. Box 3
Brixey, MO 65618 USA
Phone: 417-679-4773
E-mail: jlorrain@goin.missouri.org
Website: csf.colorado.edu/sustainability/community/ozarks

Founded: 1978
Membership: 1,750
Scope: National

Description: The Center provides research, education, technical assistance, and dissemination of information on renewable resources-based technolog, sustainable agriculture, environmentally responsible practices, sustainable community economic development, and self-reliance for the family, farm, community, Ozarks, and other bio-regions.

Publication(s): Talking Oak Leaves (newsletter), Broadcaster, The (newsletter)

Keyword(s): Environmental and Conservation Education, Rural Development, Sustainable Development, Protected Areas, Cultural Preservation

Contact(s):
Janice Lorrain, EXECUTIVE DIRECTOR
Rt. 1 Box 393, Ava, MO 65608
Phone: 417-683-5049
Donna Jones, PRESIDENT
RR 1 Box 68A-1, Dora, MO 65637
Phone: 417-261-2518
Kathi Trantham, SECRETARY
7969 Covnty Rd. 3010, West Plains, MO 65775
Phone: 417-256-6518
Denise Vaughn, TREASURER
Rt. 3 Box 200, Mtn. View, MO 65548
Phone: 417-256-6518
Corliss Schaffer, VICE PRESIDENT
HCR 64 Box 221, West Plains, MO 65775
Phone: 417-257-0670

OZONE ACTION
1700 Connecticut Ave., NW, 3rd Fl.
Washington, DC 20009 USA
Phone: 202-265-6738   Fax: 202-986-6041
E-mail: ozone_action@ozone.org

Founded: 1992
Scope: National

Description: Ozone Action educates the public about threats from ozone depletion and human-induced climate change. Ozone Action investigates and publicizes attempts to weaken our global environmental protections and exposes attempts by industry to distort public debate.

Publication(s): Ties That Blind, Black Market CFC Reports, Climate Change Current Effects Summaries, Ozone Action News

Keyword(s): Greenhouse Effect/Global Warming, Ozone Depletion

Contact(s):
Kevin Sweeney, BOARD CHAIR

P

PACIFIC FISHERY MANAGEMENT COUNCIL
2130 SW 5th Ave., Suite 224
Portland, OR 97201 USA
Phone: 503-326-6352   Fax: 503-326-6831
Website: www.pcouncil.org

Founded: NA
Scope: National

Description: Nonprofit organization established by the Magnuson-Stevens Fishery Conservation and Management Act of 1976. Develops management plans for fisheries off the coasts of Washington, Oregon, and California. Fourteen voting and five nonvoting members, of which nine are appointed by the Secretary of Commerce. Members include state and federal fishery agency managers, knowledgable citizens, and a tribal representative.

Keyword(s): Aquatic Habitats, Wildlife, Renewable Resources, Sport Fishing

Contact(s):
James Seger, ECONOMIC ANALYSIS COORDINATOR
Donald McIsaac, EXECUTIVE DIRECTOR
James Glock, FISHERY MANAGEMENT COORDINATOR, MARINE
John Coon, FISHERY MANAGEMENT COORDINATOR, SALMON

PACIFIC INSTITUTE FOR STUDIES IN DEVELOPMENT, ENVIRONMENT, AND SECURITY
654 13th St.
Oakland, CA 94612 USA
Phone: 510-251-1600   Fax: 510-251-2203
E-mail: pistaff@mindstring.com
Website: www.pacinst.org

Founded: 1987
Membership: 20
Scope: International

Description: The institute is a policy research organization that focuses on the interface of security, development and environmental protection issues. Areas of focus include climate change and water.

Publication(s): Pacific Institute Report, Global Change, Occasional papers and reports on an ongoing basis.

Keyword(s): Asia Water Environment, Dams, Developing Countries, Development, Drinking Water Protection, Environmental and Conservation Education, Greenhouse Effect/Global Warming, Ground Water Protection, Internships, Local Resource Conservation, Ozone Depletion, Sustainable Development, Sustainable Ecosystems, Water Conservation

Contact(s):
William Burns, DIRECTOR OF COMMUNICATIONS
wburns1@mindstring.com
Peter Gleick, EXECUTIVE DIRECTOR
pgleick@pipeline.com

PACIFIC NORTHWEST TRAIL ASSOCIATION
13595 Avon Allen Rd.
Mount Vernon, WA 98273 USA
Phone: 306-424-0407
E-mail: JDMEL@NWLINK.COM
Website: www.pnt.org

Founded: 1977
Membership: 300
Scope: National

Description: The PNTA was formed to promote the development of a continuous foot and horse trail from the Continental Divide at Glacier National Park to the Pacific Ocean at Olympic National Park. The PNTA encourages land use and conservation education through exposure to the historic and natural diversity of the Pacific Northwest.

Publication(s): Pacific Northwest Trail, The, a guidebook to the 1,100 mile PNT (available on diskette), Blanchard Hill and Chuckanut Mountain Map, Nor'wester

Keyword(s): Environmental and Conservation Education, Internships, Outdoor Recreation, Pedestrian Environment, Public Lands

Contact(s):
Duane Melcher, CHAIR
Phone: 306-424-0407

PACIFIC RIVERS COUNCIL
P.O. Box 10798
Eugene, OR 97440 USA
Phone: 541-345-0119   Fax: 541-345-0710
E-mail: info@pacrivers.org
Website: www.pacrivers.org

Founded: 1987
Scope: National

Description: The purpose of the Pacific Rivers Council is to protect and restore rivers, their watersheds and native aquatic species.

Publication(s): Freeflow, various briefing books and reports, Entering the Watershed

Keyword(s): Biodiversity, Wildlife, Public Lands, Rivers, Native Fish, Watersheds

Contact(s):
David Bayles, CONSERVATION DIRECTOR
Tryg Sletteland, EXECUTIVE DIRECTOR

PACIFIC SEABIRD GROUP, UNIVERSITY OF VICTORIA BIOLOGY DEPARTMENT

UNIVERSITY OF VICTORIA BIOLOGY DEPT.
Box 179, 4505 University Way, NE
Seattle, WA 98105 USA
Website: www.nmnh.sl.edu/BIRDNET/PacBirds/

Founded: 1972
Scope: International

Description: An international organization to promote the knowledge, study and conservation of Pacific seabirds.

Publication(s): Pacific Seabird Group Bulletin

Keyword(s): Birds, Endangered Species, Environmental and Conservation Education, training, Zoology

Contact(s):
Bill Sydeman, CHAIR
Point Reyes Bird Observatory 4990 Shoreline Hwy, Stinson, CA 94970
Phone: 415-868-1221
Fax: 415-868-1946
wjsydeman@prbo.org
Lisa Ballance, CHAIR-ELECT
Noah, MNFS Southwest Fisheries Science Center, La Jolla, CA 92037
Phone: 858-546-7173
Fax: 858-546-7003
lisa@caliban.ucsd.edu
Vivian Mendenhall, EDITOR
4600 Rabbit Creek Rd., Anchorage, AK 99516
Phone: 907-345-7124
Fax: 907-345-0686
fasgadair@worldnet.att.net
Lora Leschner, SECRETARY
Washington Dept. of Fish & Wildlife, Mill Creek, WA 98102
Phone: 425-776-1311
Fax: 425-338-1066
leschlll@dfw.wa.gov
Breck Tyler, TREASURER
Long Marine Laboratory 100 Shaffer Rd, Santa Cruz, CA 95060
Phone: 831-426-5740
ospr@cats.ucsc.edu

PACIFIC WHALE FOUNDATION

101 N. Kihei Rd.
Kihei, HI 96753 USA
Phone: 808-879-8860   Fax: 808-879-2615
E-mail: info@pacificwhale.org
Website: www.pacificwhale.org

Founded: 1980
Scope: National

Description: A nonprofit tax-exempt 501(c)(3) organization dedicated to saving whales, dolphins and their ocean habitats through marine research, public education and marine conservation. Pacific Whale Foundation actively studies whales, dolphins and coral reef communities throughout the Pacific to ensure their survival and recovery. The award-winning Ocean Outreach Program educates people about marine conservation through Research Internships, school programs, an Adopt-a-Whale program, and an Adopt-a-Dolphin program.

Publication(s): Soundings, Whalewatch News, Maui Outdoor Adventure, Fin and Fluke

Keyword(s): Aquatic Habitats, Endangered Species, Environmental and Conservation Education, Internships, Marine Mammals, Whale, Dolphin, Seal, Dolphins, Whales

Contact(s):
Anne Rillero, EDITOR
Gregory Kaufman, PRESIDENT
Dixie Bongolan, SECRETARY AND TREASURER
Paul Forestell, VICE PRESIDENT

PANOS INSTITUTE, THE

1701 K St., NW, 11th Fl.
Washington, DC 20006 USA
Phone: 202-223-7949   Fax: 202-223-7947
E-mail: panos@cais.com
Website: www.panosinst.org

Founded: 1986
Scope: International

Description: The Panos Institute consists of three autonomous nonprofit, nongovernmental organizations located in London, Paris, and Washington, DC, working to raise public understanding of sustainable development issues. The Washington, DC institute focuses its work on Latin America, the Caribbean, and the United States.

Publication(s): SIDAmerica, Eco-Reports, From Information to Education, We Speak For Ourselves

Keyword(s): Communications, Environmental and Conservation Education, Internships, Population Growth, Sustainable Development

Contact(s):
John Kramer, ACTING CHAIRMAN
Gretchen Maynes, EXECUTIVE DIRECTOR/ SECRETARY
George Woodring, TREASURER
Michael McDowell, VICE CHAIRMAN

PARTNERS IN AMPHIBIAN AND REPTILE CONSERVATION (PARC)

PO Drawer E
Aiken, SC 29802 USA
Phone: 803-725-2473   Fax: 803-725-3309
E-mail: forrest@srel.edu
Website: www.parcplace.org

Founded: 1998
Scope: International

Description: PARC is an international organization and a multi-sector partnership dedicated to the conservation of the herepetofauna (amphibians and reptiles) and their habitats.

PARTNERS IN PARKS

4916 Butterworth Pl. NW
Washington, DC 20016 USA
Phone: 202-364-7244   Fax: 202-364-7246
E-mail: partpark@cqi.com

Founded: 1988
Scope: National

Description: A nonprofit organization that encourages, promotes, and establishes professional level partnerships between national park and other public land managers and those who would contribute their time and skills to studying, protecting, and interpreting natural and cultural features.

Keyword(s): Research, Natural Areas, Conservation, Cultural Preservation, Public Lands, National Parks

Contact(s):
Sarah Bishop, PRESIDENT
David Kikel, SECRETARY AND TREASURER

PAWS/OLYMPIC WILDLIFE RESCUE
1393 Mox-Chehaus Rd.
McCleary, WA 98557 USA
Phone: 360-495-3337   Fax: 360-495-4285
E-mail: info@paws.org
Website: www.paws.org

Founded: NA
Scope: Statewide

Description: Olympic Wildlife Rescue is a nonprofit, tax-exempt organization dedicated to the care of injured, orphaned, and ill northwest wildlife. We operate a licensed wildlife rehabilitation center in McCleary, Washington (about 20 miles west of Olympia).

Publication(s): Ptarmigan Ptales

Keyword(s): Environmental and Conservation Education, Outdoor Recreation, Renewable Resources, Wilderness, training, accreditation

Contact(s):
Jeanne Wasserman, DIRECTOR
Gary Bankers, PRESIDENT
Pat Pringle, SECRETARY
Carol Chatwood, TREASURER
Shawn Newman, VICE PRESIDENT

PENNSYLVANIA ASSOCIATION OF CONSERVATION DISTRICTS INC.
4999 Jonestown Rd.
Harrisburg, PA 17109 USA
Phone: 717-545-8878   Fax: 717-545-8850
E-mail: pacd@pacd.org
Website: www.pacd.org

Founded: NA
Membership: 800
Scope: Statewide

Contact(s):
Franklin Long, 1ST VICE PRESIDENT
Rd. 1 Box 441, Tyrone, PA 16886
Phone: 814-648-4838
Robert Wagner, BOARD MEMBER
373 Scott Rd., Quarryville, PA 17566
Phone: 717-529-2831
Fax: 717-529-2155
Susan Fox, EXECUTIVE DIRECTOR
225 Pine St., Harrisburg, PA 17101
Phone: 717-230-1000
Fax: 717-236-6410
Murray Laite, LEGISLATIVE DIRECTOR
512 Electric Ave, Lewistown, PA 17044
Phone: 717-248-6733
Ron Rohall, PRESIDENT, ALTERNATE BOARD MEMBER
534 Kennedy Rd., Airville, PA 17032
Phone: 717-862-3486
Fax: 717-862-3486
Ronald Rohall, SECRETARY
P.O. Box 27, Rector, PA 15677
Phone: 412-238-4973
Cloyd Brenneman, TREASURER
103 Courthouse, Mercer, PA 16137
Phone: 412-662-3800

PENNSYLVANIA B.A.S.S. CHAPTER FEDERATION, INC.
Attn: President, 769 N. Cottage Rd.
Mercer, PA 16137 USA
E-mail: 2dunks@infonline.net
Website: www.pabass.com

Founded: NA
Scope: Statewide

Description: An organization of Bassmaster chapters, affiliated with the Bass Anglers Sportsman Society, organized to fight pollution, assist state and national conservation agencies in their efforts, and teach the young people of our country good conservation practices. Dedicated to the realistic conservation of our water resources.

Contact(s):
Bill Reichert, CONSERVATION DIRECTOR
51 N. 4th St., Cressona, PA 17929
Phone: 570-385-2122
breichert@losch.net
Mike Dunkerley, PRESIDENT
Phone: 412-475-2422

PENNSYLVANIA CITIZENS ADVISORY COUNCIL TO DEPARTMENT OF ENVIRONMENTAL PROTECTION
13th Fl., RCSOB P.O. Box 8459
Harrisburg, PA 17105-8459 USA
Phone: 717-787-4527   Fax: 717-787-2878
E-mail: suswilson@state.pa.us
Website: www.cacdep.state.pa.us

Founded: NA
Membership: 18
Scope: Regional

Description: Created by Act 275 of the PA General Assembly, 1971.

Publication(s): Annual Report, Regional Report, Advisory

Keyword(s): Exotic species, Aquatic nuisance species, Air Resources, Waste Resources, Public Participation

Contact(s):
Stephanie Mioff, ADMINISTRATIVE ASSISTANT
Dave Strong, CHAIRPERSON
Susan Wilson, EXECUTIVE DIRECTOR

PENNSYLVANIA ENVIRONMENTAL COUNCIL, INC. (PEC)
117 S. 17th St., 23rd Fl.
Philadelphia, PA 19103-5022 USA
Phone: 215-563-0250   Fax: 215-563-0528
Website: www.pecpa.org

Founded: 1969
Scope: Statewide

Description: Private nonprofit statewide membership organization devoted to the protection and improvement of Pennsylvania's environment through education, advocacy and consensus-building. PEC brings together nonprofits, government agencies, businesses and citizens to develop environmental policy, take action on environmental issues and work for effective environmental legislation, regulation and enforcement.

Publication(s): PA Legislative Updates, Guiding Growth, Building Better Communities and Preserving our Countryside, Environmental Advisory Council Handbook, Transit-Oriented Development Handbook, Urban Vacant Land Handbook, Environmental A, Environmental Forum Newsletter

Keyword(s): Watersheds, Environmental Legislation, Sustainable Development, Greenways

Contact(s):
Andrew McElwaine, PRESIDENT AND CEO
Anna Brienich, REGIONAL DIRECTOR OF COMMUNITY PLANNING WESTERN PA OFFICE
Ellen Alaimo, REGIONAL DIRECTOR OF NORTHEASTERN PA OFFICE
Patrick Starr, REGIONAL DIRECTOR OF SOUTHEASTERN OFFICE
Davitt Woodwell, REGIONAL DIRECTOR OF WESTERN PA OFFICE
Brian Hill, VICE PRESIDENT AND DIRECTOR OF FRENCH CREEK PROJECT

PENNSYLVANIA FEDERATION OF SPORTSMENS CLUBS
2426 N. Second St.
Harrisburg, PA 17110 USA
Phone: 717-232-3480   Fax: 717-231-3524
E-mail: pawild@aol.com
Website: pfsc.org

Founded: 1932
Scope: Statewide

Description: A representative statewide organization, affiliated with the National Wildlife Federation, dedicated to the protection and enhancement of wildlife and its habitat through public education and government interaction.

Publication(s): On Target

Contact(s):
Ed Zygmut, ALTERNATE REPRESENTATIVE
Linda Steiner, EDITOR
Melody Zullinger, EXECUTIVE DIRECTOR
Ted Onufrak, PRESIDENT
Mark Henry, REPRESENTATIVE
John Riley, TREASURER

PENNSYLVANIA FISH AND BOAT COMMISSION: REGION 5
Northcentral Region P.O. Box 5306
Pleasant Gap, PA 16823-5306 USA
Phone: 814-359-5250   Fax: 814-359-5254
E-mail: fbncregion@state.pa.us
Website: www.fish.state.pa.us
Membership: 17
Scope: State

Contact(s):
William Hartle, REGIONAL MANAGER
Phone: 814-359-5250

PENNSYLVANIA FORESTRY ASSOCIATION, THE
56 E. Main St.
Mechanicsburg, PA 17055 USA
Phone: 717-766-5371
E-mail: thepfa@juno.com
Website: www.pfa.cas.psu.edu

Founded: 1886
Membership: 1200
Scope: Statewide

Description: An independent nonprofit conservation organization, affiliated with the National Woodland Owners Association dedicated to environmental improvement and wise use of natural resources in Pennsylvania. Membership includes a cross section of all groups and individuals interested in true conservation.

Publication(s): Pennsylvania Forest Magazine - Quarterly

Keyword(s): Environmental and Conservation Education, Forests and Forestry, Outdoor Recreation, Public Lands, Youth Organizations

Contact(s):
Jan Zinn, EDITOR
Phone: 717-632-1648
Roy Seifert, PRESIDENT
Phone: 717-225-4711
William Corlett, SECRETARY
Phone: 717-737-7118
William Cook, TREASURER
Phone: 717-787-2039
Lloyd Casey, VICE-PRESIDENT

PENNSYLVANIA RECREATION AND PARK SOCIETY, INC.
1315 W. College Ave., Suite 200
State College, PA 16801-2776 USA
Phone: 814-234-4272   Fax: 814-234-5276
E-mail: prps@vicon.net
Website: www.prps.org

Founded: 1935
Membership: 1600
Scope: Statewide

Description: To promote quality recreation and park opportunities for all the citizens of the Commonwealth of Pennsylvania by actively involving professionals and citizens in recreation, park, and conservation programs, by fostering and maintaining high standards of professional qualifications and ethics, and by providing quality educational opportunities.

Publication(s): Pennsylvania Recreation and Parks, The PRPS Update (Monthly)

Keyword(s): Environmental and Conservation Education, Flowers, Plants, and Trees, Pesticides, Solid Waste, Coral Reefs

Contact(s):
Vanyla Tierney, EDITOR
Recreation Planner, DCNR-Bureau of State Parks, P.O. Box 1519, Mechanicsburg, PA 17055-9019
Phone: 717-783-2654
Robert Griffith, EXECUTIVE DIRECTOR
R. Mcfate, PRESIDENT-ELECT
Park Manager, Caledonia State Park, 40 Rocky Mtn. Rd., Fayetteville, PA 17222-9610
Phone: 717-352-2161
Steven Landes, SECRETARY
William Rosevear, TREASURER
Park Manager, PA Bureau of State Parks, 1100 Pine Grove Rd., Gardners, PA 17324-7174
Phone: 717-486-7174
Fax: 717-486-4961

PENNSYLVANIA RESOURCES COUNCIL, INC.

3606 Providence Rd.
Newtown Square, PA 19073 USA
Phone: 610-353-1555   Fax: 610-353-6257
Website: www.prc.org

Founded: 1939
Scope: Regional

Description: (formerly PA Roadside Council) The Pennsylvania Resources Council (PRC) is recognized nationally for its expertise in recycling, waste reduction, and litter control. PRC produces educational materials, as well as seminars and conferences for citizens, municipalities, civic groups and corporations. PRC also sponsors the nation's only environmental shopping hotline (1-800-Go-To-PRC). Open to the public, PRC's environmental living center has exhibits and workshops to show the impact of lifestyle choices on the environment.

Publication(s): Recyclers Roundup, Environmental Living Magazine

Keyword(s): Beautification, Healthy Home, Environmental and Conservation Education, education, Outdoor Recreation, training, Nature Study, Recycling, Waste Management, Litter, Environmental and Conservation Education, Pollution Prevention, Environmental Living

Contact(s):
Howard Wein, PRESIDENT
Marcia Weller
REGIONAL DIRECTOR PHILADELPHIA AREA
David Mazza
REGIONAL DIRECTOR PITTSBURGH AREA
Phone: 412-488-7490

PENNSYLVANIA TROUT, A COUNCIL OF TROUT UNLIMITED

RD4 Box 140 AA
Greensburg, PA 15601 USA
Phone: 814-863-7585   Fax: 814-865-9131
Website: www.patrout.org

Founded: NA
Membership: 11000
Scope: Regional

Description: A statewide council with 56 active chapters working for the protection and enhancement of the coldwater fishery resource.

Publication(s): Pennsylvania Trout

Keyword(s): Environmental and Conservation Education, Land Use Planning, Watersheds, Water Quality, accreditation

Contact(s):
Ken Undercoffer, CHAIRMAN

PEOPLE FOR PUGET SOUND

911 Western Ave.
Seattle, WA 98104 USA
Phone: 206-382-7007   Fax: 206-382-7006
E-mail: people@pugetsound.org
Website: www.pugetsound.org

Founded: 1991
Membership: 5000
Scope: Regional

Description: People for Puget Sound works to protect and restore the imperiled marine and estuarine ecosystems of Puget Sound and the Northwest Straits.

Publication(s): Sound & Straits (quarterly newsletter), Orcas in the Balance (documentary video), Habifacts (newsletter), Kids Sound (quarterly newsletter for children)

Keyword(s): Estuaries, Marine Protected Areas, Marine Conservation, Environmental and Conservation Education, Oil Spill Response, Salmon Recovery, Water Quality, Volunteering

Contact(s):
Kayleen Dunson, COMMUNICATIONS DIRECTOR
Stephanie Raymond, EDUCATION COORDINATOR
Kathy Fletcher, EXECUTIVE DIRECTOR
Pam Johnson, FIELD DIRECTOR
Jacques White, HABITAT DIRECTOR
Kate Janeway, PRESIDENT OF THE BOARD OF DIRECTORS

PEOPLE FOR PUGET SOUND

NORTH SOUND OFFICE
407 Main St. Ste 201
Mt. Vernon, WA 98273 USA
Phone: 306-336-1931   Fax: 360-336-5422
E-mail: northsound@pugetsound.org
Website: www.pugetsound.org

Founded: NA
Scope: Local Region

Contact(s):
Mike Sato, DIRECTOR

PEOPLE FOR PUGET SOUND

SOUTH SOUND OFFICE
1063 Capitol Way S., Suite 206
Olympia, WA 98501 USA
Phone: 360-754-9177   Fax: 360-534-9371
E-mail: southsound@pugetsound.org
Website: www.pugetsound.org

Founded: NA
Membership: 5000
Scope: Regional
Publication(s): Sound and Straits

Contact(s):
Bruce Wishart, DIRECTOR
bwishart@pugetsound.org

PEREGRINE FUND, THE

5666 W. Flying Hawk Ln.
Boise, ID 83709 USA
Phone: 208-362-3716   Fax: 208-362-2376
E-mail: tpf@peregrinefund.org
Website: www.peregrinefund.org

Founded: 1970
Scope: National, Internationl

Description: The Peregrine Fund works nationally and internationally to conserve biological diversity and enhance environmental health by working with birds through management and conservation of species and their habitat, and through education and scientific investigation. Although best known nationally for species restoration, they have assisted on conservation projects in over 40 countries.

Publication(s): Annual Report, Operation Report, progress reports, The Peregrine Fund Newsletter

Keyword(s): Birds, Endangered Species, Raptors

Contact(s):
Henry Paulson, Jr, CHAIRMAN OF BOARD
William Burnham, PRESIDENT AND CEO
Ronald Yanke, SECRETARY
Paxson Offield, TREASURER
D. Nelson, VICE CHAIRMAN OF THE BOARD
J. Peter Jenny, VICE PRESIDENT
Jeffrey Cilek, VICE PRESIDENT

PHEASANTS FOREVER, INC.

1783 Buerkle Circle
St. Paul, MN 55110 USA
Phone: 651-773-2000   Fax: 651-773-5500
E-mail: pf@pheasantsforever.org
Website: www.pheasantsforever.org

Founded: 1982
Membership: 92,000 members and 550 chapters
Scope: National

Description: Pheasants Forever, Inc. is a nonprofit conservation organization formed in response to the continued decline of ring-necked pheasants. The mission of Pheasants Forever is to protect and to enhance pheasant and other wildlife populations throughout North America through public awareness and education, habitat restoration, development and maintenance, and improvements in land and water management policies.

Publication(s): Pheasants Forever

Keyword(s): Agriculture, Birds, Environmental and Conservation Education, Hunting, training

Contact(s):
George Wilson, CHAIRMAN
Howard Vincent, CHIEF EXECUTIVE OFFICER
Phone: 651-773-2000
hvincent@pheasantsforever.org
Peter Berthelsen, DIRECTOR OF CONSERVATION-
NEBRASKA
1101 Alexander Ave., Elba, NE 68835
Phone: 308-754-5339
phasianus@aol.com
Mark Herwig, EDITOR
Phone: 651-773-2000
herwig@pheasantsforever.org
Matthew O'Connor
REGIONAL FIELD REPRESENTATIVE
2880 Thunder Rd., Hopkinton, IA 52237
Phone: 319-926-2357
niapfmatt@n-connect.net
Jeff Gaska, REGIONAL REPRESENATIVE
W. 9947 Ghost Hill Rd., Beaver Dam, WI 53916
Phone: 920-927-3579
jgaska@pheasantsforever.org
Mike Pruss, REGIONAL REPRESENTATIVE
HC 67 Box 104A, Mifflin, PA 17058
Phone: 717-436-0005
mpruss@pheasantsforever.org
Keith Brus, REGIONAL REPRESENTATIVE
5995 W. Little Portage Rd., Pt. Clinton, OH 43452
Phone: 419-732-7149
kbrus@cros.net

Dan Hare, REGIONAL REPRESENTATIVE
315 Tucson Ave, Bismarck, ND 58504
Phone: 701-250-9921
danhare@home.com
Mark Heckenlaible, REGIONAL REPRESENTATIVE
2103 County Rd, 23 Lyons, NE 68038
Phone: 402-687-2004
mh52934@navix.net
Eric Henning, REGIONAL REPRESENTATIVE
29570 Camp Adair Rd, Monmouth, OR 97361
Phone: 541-745-5363
henning_erk@hotmail.com
Aaron McCormick, REGIONAL REPRESENTATIVE
P.O. Box 537
408 S. Said St., Caird, NE 68824
Phone: 308-485-0154
pfbiologist@nebi.com
Tom Schwartz, REGIONAL REPRESENTATIVE
40 Crater Lake Dr., Springfield, IL 62707
Phone: 214-498-7558
tschwartz@pheasantsforever.org
Thomas Kirschenmann, REGIONAL REPRESENTATIVE
600 W. Beck St., Worthing, SD 57077
Phone: 605-372-2037
tkirschenmann@pheasantsforever.org
Barth Crouch, REGIONAL REPRESENTATIVE
1625 E. Beloit Ave., Salina, KS 67401
Phone: 785-823-0240
vcrouch@juno.com
Matthew Holland, REGIONAL REPRESENTATIVE
679 W. River Dr., New London, MN 56273
Phone: 320-354-4377
ringneck@tds.net
Walt Bodie, REGIONAL REPRESENTATIVE
2909 Navajo Drive, Nampa, ID 83686
Phone: 208-461-7350
wbodie@pheasantsforever.org
Mike Parker, REGIONAL REPTRESENATIVE
117 Wilson St., De Witt, MI 48820
Phone: 517-668-1033
mparkerpf@aol.com
Robert Larson, SECRETARY
James Wooley, SENIOR REGIONAL WILDLIFE BIOLOGIST
1205 Ilion Ave., Chariton, IA 50049
Phone: 641-774-2238
jwooley@pheasantsforever.org
Bruce Hertzke, TREASURER
Joseph Duggan, VICE PRESIDENT OF DEVELOPMENT
AND PUBLIC AFFAIRS
Phone: 651-773-2000
jduggan@pheasantsforever.org
Rick Young, VICE PRESIDENT OF FIELD OPERATIONS
Phone: 651-773-2000
ryoung@pheasantsforever.org
David Nomsen, VICE PRESIDENT OF GOVERNMENTAL
AFFAIRS
2101 Ridgewood Dr., Alexandria, MN 56308
Phone: 320-763-6103
pfnomsen@rea-alp.com

PHYSICIANS FOR SOCIAL RESPONSIBILITY

1875 Connecticut Avenue NW, Suite 1012
Washington, DC 20009 USA
Phone: 202-667-4260   Fax: 202-667-4201
E-mail: psrnatl@psr.org
Website: www.psr.org

Founded: 1961
Scope: National

Description: Promotes arms reduction, international cooperation to protect the environment, and education and programs aimed at reducing violence.

Publication(s): PSR Reports, Monitor

Keyword(s): Environmental Health, Nuclear Abolition, Gun Violence Prevention, Nuclear/Radiation, Pesticides, Toxic Substances, Nuclear-free, Water quantity, Water export and diversion

Contact(s):
Robert Musil, EXECUTIVE DIRECTOR
Peter Wilk, PRESIDENT

PIEDMONT ENVIRONMENTAL COUNCIL

P.O. Box 460
Warrenton, VA 20188 USA
Phone: 540-347-2334   Fax: 540-349-9003
E-mail: pecva@pecva.org
Website: www.pecva.org

Founded: 1972
Membership: 3000
Scope: Statewide

Description: A nonprofit organization formed to conserve natural resources and the pastoral landscape of a nine-county region of the Northern Virginia Piedmont. Public education and services to public officials and citizens, covering: land use; farmland retention; open space conservation; historic preservation; rural transportation policy; and rural planning legislation. Active statewide and federally on rural conservation issues.

Publication(s): AsPECts- newsletter

Keyword(s): Protected Areas, Land Purchase, Outdoor Recreation, Pedestrian Environment, Public Lands

Contact(s):
Eve Fout, CHAIRMAN
Christopher Miller, PRESIDENT
John Birdsall, VICE CHAIRMAN
Douglas Larson, VICE PRESIDENT

PIEDMONT ENVIRONMENTAL COUNCIL

45 Horner St.
Warrenton, VA 20186 USA
Phone: 540-347-2334   Fax: 540-349-9003
E-mail: pec@pecva.org
Website: www.pecva.org
Scope: Regional
Publication(s): AsPECts

Contact(s):
Christopher Miller, PRESIDENT

PINCHOT INSTITUTE FOR CONSERVATION

1616 P St., NW, Suite 100
Washington, DC 20036 USA
Phone: 202-797-6580   Fax: 202-797-6583
E-mail: pinchot@pinchot.org
Website: www.pinchot.org

Founded: 1963
Scope: National

Description: The Pinchot Institute for Conservation is an independent nonprofit organization established to advance forest conservation thought, policy, and action. Serves as a bridge between the scientific and policymaking communities, providing timely, objective policy research, facilitation, leadership, training, and environmental education on issues relating to the protection and sustainable management of forests.

Publication(s): Grey Towers Press books, numerous policy reports, discussion papers and lecture series., The Pinchot Letter

Keyword(s): Forests and Forestry, Planning Management, Sustainable Ecosystems, Public Lands, Research

Contact(s):
Dennis Lemaster, CHAIR
1159 Forestry Building W Lafayette Purdue University,
IN 47907
dclmstr@fnr.purdue.edu
Edgar Brannon, DIRECTOR OF GREY TOWERS NATIONAL HISTORIC LANDMARK
P.O. Box 188, Milford, PA 18337
Phone: 570-296-9630
Fax: 570-296-9675
V. Sample, PRESIDENT
Ann Hanus, SECRETARY
775 Summer St. NE, Salem, OR 97301
Hugh Miller, TREASURER
2629 W Grace St, Richmond, VA 23220-1945
Peter Pinchot, VICE CHAIR
225 Moose Hill Rd, Guilford, CT 06437

PITTSBURGH HERPETOLOGICAL SOCIETY, THE

c/o The Pittsburgh Zoo & Aquarium, One Wild Pl.
Pittsburgh, PA 15206 USA
Phone: 412-361-0835   Fax: 412-361-2718
E-mail: phs@trfn.clpgh.org
Website: www.trfn.clpgh.org/phs

Founded: 1993
Membership: 120
Scope: National

Description: The society is dedicated to fostering an appreciation of all reptiles and amphibians through husbandry, conservation, and education.

Publication(s): Pittsburgh Herpetological Society Newsletter, The

Keyword(s): Conservation, Conservation Planning, Outdoor Education, Reptiles and Amphibians, Streams, Water Quality

Contact(s):
Dolly Ellerbrock, CO-FOUNDER
diguana@bellatlantic.net

PLANNED PARENTHOOD FEDERATION OF AMERICA, INC.

810 Seventh Ave.
New York, NY 10019 USA
Phone: 212-541-7800   Fax: 212-245-1845
E-mail: communications@ppfa.org
Website: www.plannedparenthood.org

Founded: 1916
Membership: 200
Scope: National

Description: Planned Parenthood Federation of America (PPFA) is a federation of 132 not-for-profit affiliates operating nearly

900 medically-supervised health centers nationwide. Planned Parenthood centers provide a wide range of services—including family planning counseling, contraception, prenatal care, adoption referrals, abortion services, cancer screening, testing and treatment for HIV/AIDS, and other sexually transmitted infections, and sexuality education—to nearly five million men and women each year. For the Planned Parenthood center nearest you, call 1-800-230-PLAN.

Keyword(s): Health and Nutrition, Population Growth, Family Planning, Reproductive Rights

Contact(s):
Mary Shallenberger, CHAIRPERSON
John Romo, CHIEF OPERATING OFFICER
Susan Pichler, LIBRARIAN
Katherine Dexter McCormick Library, 810 Seventh Ave., New York, NY 10019
Phone: 212-261-4637
Gloria Feldt, PRESIDENT
Alfred Poindexter, SECRETARY
Barbara Singhaus, TREASURER
Alfredo Vigil, VICE CHAIRPERSON

PLANNING AND CONSERVATION LEAGUE
926 J St.,
Suite 612
Sacramento, CA 95814 USA
Phone: 916-444-8726   Fax: 916-448-1789
E-mail: pclmail@pcl.org
Website: www.pcl.org

Founded: 1965
Scope: Regional

Description: A representative statewide organization, affiliated with the National Wildlife Federation, dedicated to the protection and enhancement of wildlife and its habitat through public education and government interaction.

Publication(s): California Today

Contact(s):
Gerald Meral, EXECUTIVE DIRECTOR, EDITOR AND ALTERNATE REPRESENTATIVE, EDUCATION PROGRAMS CONTACT
Sage Sweetwood, PRESIDENT
Dan Frost, REPRESENTATIVE
William Yeates, TREASURER

PLANTIFICACION DE RECURSOS INTEGRL
NATURALES Y AMBIENTALES DE PUERTO RICO, PUERTO RICO

Contact(s):
Carlos Bibiloni

POCONO ENVIRONMENTAL EDUCATION CENTER
R.R. 2 Box 1010
Dingmans Ferry, PA 18328 USA
Phone: 570-828-2319   Fax: 570-828-9695
E-mail: peec@ptd.net
Website: www.peec.org

Founded: 1986
Scope: Statewide

Description: The Pocono Environmental Education Center (PEEC) advances environmental awareness, knowledge, and skills through education, in order that those who inhabit and will inherit the planet may better understand the complexities of natural and human-designed environments.

Contact(s):
Thomas Shimalla, ASSISTANT DIRECTOR
Paul Brandwein, ASSOCIATE DIRECTOR
Florence Mauro, DIRECTOR
John Padalino, PRESIDENT
jack.peec@oal.com

POLLUTION PROBE FOUNDATION
625 Church St., Suite 402
Toronto, Ontario M4Y 2G1 Canada
Phone: 416-926-1907   Fax: 416-926-1601
E-mail: pprobe@pollutionprobe.org
Website: www.pollutionprobe.org

Founded: 1969
Scope: National

Description: Pollution Probe is a Canadian nonprofit organization that exists to define environmental problems through research; to promote understanding through education; and to press for practical solutions through advocacy.

Publication(s): see publications on website, Canadian Junior Green Guide, The, Kitchen Handbook, An Environmental Guide, Canadian Green Consumer Guide, The

Contact(s):
Edward Babin, CHAIRMAN OF POLLUTION PROBE FOUNDATION
Ken Ogilvie, EXECUTIVE DIRECTOR
Phone: 416-926-1907

POPE AND YOUNG CLUB
273 Mill Creek Rd, P.O. Box 548
Chatfield, MN 55923 USA
Phone: 507-867-4144   Fax: 507-867-4144
E-mail: pyclub@isl.net
Website: www.pope-young.org

Founded: 1961
Scope: National

Description: A North American bowhunting and wildlife conservation organization dedicated to the promotion and protection of our bowhunting heritage and North America's wildlife.

Publication(s): Bow Hunting Record Book "Bow Hunting Big Game Records of North America"

Contact(s):
Kevin Hisey, EXECUTIVE SECRETARY
C. Randall Byers, FIRST VICE PRESIDENT
G. Asbell, PRESIDENT
Donald Morgan, TREASURER

POPULATION ACTION INTERNATIONAL
1300 19th St., NW, 2nd Fl.
Washington, DC 20036 USA
Phone: 202-557-3400   Fax: 202-728-4177
E-mail: pai@popact.org
Website: www.populationaction.org

Founded: 1965
Membership: 42
Scope: International

Description: Develops worldwide support for international population and voluntary family planning programs through public education, policy analysis, and liaison with international leaders and organizations.

Publication(s): population on funding country status re, legislative and policy updates, annual report of activities, studies of population-environment linkages, Population & reproductive health

Keyword(s): Environment, Health and Nutrition, Population Growth, Renewable Resources, Sustainable Development

Contact(s):
Amy Coen, PRESIDENT
Phyllis Piotrow, SECRETARY
Scott Spangler, TREASURER
Sally Ethelston, VICE PRESIDENT, COMMUNICATIONS
Carol Wall, VICE PRESIDENT, DEVELOPMENT
Terri Bartlett, VICE PRESIDENT, PUBLIC POLICY
Robert Engelman, VICE PRESIDENT, RESEARCH

POPULATION COMMUNICATIONS INTERNATIONAL

777 United Nations Plaza 5th Fl.
New York, NY 10017 USA
Phone: 212-687-3366   Fax: 212-661-4188
E-mail: pciny@poulation.org
Website: www.poulation.org

Founded: 1985
Scope: International

Description: PCI works through mass media and nongovernmental organizations to promote elevation of women's status, use of family planning, and small family norms. PCI's social-content soap operas in developing countries are locally researched and produced and weave social themes into long-term script development for radio and television dramas. In the United States PCI also works with broadcasters and NGOs. Currently PCI is collaborating with NWF to develop an environmental/human sexuality soap opera for U.S. audiences.

Publication(s): see publications on website, Global Intersections (Monthly Electronic Newsletter), On Air (quarterly newsletter)

Keyword(s): Communications, Population Growth, Environmental Communication

Contact(s):
David Andrews, PRESIDENT

POPULATION INSTITUTE, THE

107 Second St., NE
Washington, DC 20002 USA
Phone: 202-544-3300   Fax: 202-544-0068
E-mail: web@populationinstitute.org
Website: www.populationinstitute.org

Founded: 1969
Membership: 30000
Scope: International

Description: To enlist and motivate key leadership groups to participate in the effort to bring population growth into balance with resources by means consistent with human dignity and freedom. Works in communications and with mass membership organizations, educational leaders, and policy leaders.

Publication(s): Annual Report, Towards The 21st Century (Monograph Series), POPLINE - World Population News Service

Keyword(s): Population Growth, Developing Countries, Air Quality and Pollution, Internships, Exotic species, Aquatic nuisance species, Environment

Contact(s):
Suzanne Kellerman, CHAIRPERSON OF THE BOARD
St. Petersburg, FL 727-525-0989
Phone: 727-586-1400
Werner Fornos, PRESIDENT
Phone: 202-544-3300
Jyoti Singh, SECRETARY
Joyce Cramer, TREASURER
Marilyn Hempel, VICE CHAIRPERSON

POPULATION REFERENCE BUREAU, INC.

1875 Connecticut Ave., NW, Suite 520
Washington, DC 20009 USA
Phone: 202-483-1100   Fax: 202-328-3937
E-mail: popref@prb.org
Website: www.prb.org

Founded: 1929
Scope: National

Description: PRB is a nonprofit educational organization which provides timely and objective information on U.S. and international population trends and their implications.

Publication(s): PRB Reports on America, Population Today, World and U.S. Population Data Sheets, teaching kits on population topics, list available on request., Population Bulletin

Keyword(s): Communications, Environmental and Conservation Education, Geography, Population Growth, Sustainable Development

Contact(s):
Michael Bentzmen, CHAIR OF THE BOARD
Ellen Carnevale, DIRECTOR OF COMMUNICATIONS
Zuali Malsawma, LIBRARIAN
Peter Donaldson, PRESIDENT
Carl Haub, SENIOR DEMOGRAPHER

POPULATION-ENVIRONMENT BALANCE, INC.

2000 P St., NW, Suite 600
Washington, DC 20036-5915 USA
Phone: 202-955-5700   Fax: 202-955-6161
E-mail: uspop@balance.org
Website: www.balance.org

Founded: 1973
Membership: 10,000
Scope: National

Description: Population-Environment Balance is a grassroots membership organization dedicated to public education regarding the adverse effects of population growth on the environment. "BALANCE" advocates measures that would encourage population stabilization in the U.S.; encourages a responsible immigration policy for the U.S.; and promotes increased funding for contraceptive research and availability. Activities include public education, advocacy, media campaigns, and publications.

Publication(s): Balance Data, Action Alerts, Balance Activist

Keyword(s): Air Quality and Pollution, Environment, Population Growth

Contact(s):
Virginia Abernethy, CHAIRMAN OF THE BOARD

POTOMAC APPALACHIAN TRAIL CLUB
118 Park St., SE
Vienna, VA 22180 USA
Phone: 703-242-0693   Fax: 703-242-0968
Website: www.patc.net

Founded: 1927
Membership: 7000
Scope: Regional

Description: Maintains 240 miles of the Appalachian Trail from Rock Fish Gap in Virginia to Pine Grove Furnace State Park in Pennsylvania Also maintains an additional 750 miles of trails. Activities include publication of maps and guidebooks, outdoor recreation leadership, construction and maintenance of shelters and cabins and conservation of trail lands through purchase or easements.

Publication(s): newsletter Monthly- The Potomac Appalachian

Keyword(s): Environmental and Conservation Education, Trail, Outdoor Recreation

Contact(s):
Linda Shannon Beaver, CHIEF EDITOR
PA@patc.net
Wilson Riley, DIRECTOR OF ADMINISTRATION
Warren Sharp, GENERAL SECRETARY
Walter Smith, PRESIDENT
Kerry Snow, SUPERVISOR OF TRAILS
Gerhard Salinger, TREASURER

POULSBO MARINE SCIENCE CENTER
18743 Front St., NE, P.O. Box 2079
Poulsbo, WA 98370 USA
Phone: 360-779-5549   Fax: 360-779-8960
E-mail: info@poulsbomsc.org
Website: www.polsbomsc.org

Founded: 1968
Scope: Regional

Description: The Marine Science Center works to meet the science education needs of public citizens and of students and teachers nationally, regionally, and locally. Hands-on environmental education is at the heart of its mission and is reflected in direct instruction from its facility on Washington's State's Liberty Bay, part of the Puget Sound.

Keyword(s): Marine Conservation, Nature Centers, Oceanography

Contact(s):
Michelle Benedict, DIRECTOR
Cindy Rathbone, MARINE SOCIETY OF THE PACIFIC NORTHWEST, PRESIDENT

POWDER RIVER BASIN RESOURCE COUNCIL
P.O. Box 1178
Douglas, WY 82633 USA
Phone: 307-358-5002   Fax: 307-358-6771
E-mail: doprbrc@coffey.com

Founded: 1973
Scope: Statewide

Description: Nonprofit grassroots organization whose major purpose is to help Wyoming people work to prevent and alleviate environmental and rural problems. Major issues include: Coal mining, water development, toxics, wastes, energy conservation, agriculture, and accountable government.

Publication(s): The Powder River Breaks - quarterly newsletter

Keyword(s): Mining, Toxic Substances, Nuclear-free, Water quantity, Water export and diversion, Exotic species, Aquatic nuisance species

Contact(s):
Pennie Vance, CHAIR
xxrobin@wavecom.net
Kevin Lind, DIRECTOR
23 N. Scott, Sheridan, WY 82801
Phone: 307-672-5809
Fax: 307-672-5800
prbrc@wavecom.net
Chessie Lee, TREASURER
chessie@coffey.com

PRAIRIE CLUB, THE
533 W. North Ave Suite 10
Elmhurst, IL 60626 USA
Phone: 630-516-1277   Fax: 630-516-1278
E-mail: prairieclb@aol.com
Website: www.prairieclb.org

Founded: 1911
Membership: 820
Scope: National

Description: Organized for the promotion of outdoor recreation in the form of walks, outings, camping, and canoeing; the establishment and maintenance of permanent and temporary camps; and the encouragement of the love of nature.

Publication(s): The Bulletin

Keyword(s): Biodiversity, Protected Areas, Land Purchase, Outdoor Recreation, training

Contact(s):
Leo Nelson, 1ST VICE PRESIDENT
Linda Mullikin, 2ND VICE PRESIDENT
Milt Davies, CHAIRMAN OF CONSERVATION COMMITTEE
Susan Messer, EDITOR
Loretta Davies, EXECUTIVE DIRECTOR
Lloyd Anderson, PRESIDENT
Glenn Krus, SECRETARY
Tom Meyers, TREASURER

PRAIRIE GROUSE TECHNICAL COUNCIL
WILDLIFE AND FISHERIES SCIENCES DEPT
Wildlife and Fisheries Sciences Department 2258 TAMU
College Station, TX 77843-2258 USA
Phone: 979-845-5777   Fax: 979-845-3786
Website: wfscnet.tamu.edu

Founded: NA
Scope: National

Description: Comprises federal, state, and private agency biologists or administrators concerned with the status, research, and management of the prairie-chicken and sharp-tailed grouse in North America.

Publication(s): Newsletter and Proceedings

Keyword(s): Birds, Endangered Species, Prairies, training

Contact(s):
Kenneth Giesen, EXECUTIVE COMMITTEE
317 W. Prospect, Ft. Collins, CO 80526

PRAIRIE RIVERS NETWORK
809 S. Fifth St.
Champaign, IL 61820 USA
Phone: 217-344-2371   Fax: 217-344-2381
E-mail: info@prairierivers.org
Website: www.prairierivers.org

Founded: 1968
Scope: Statewide

Description: (Formerly Central States Education Center) A statewide river conservation organization working to protect the rivers and streams of illinois. Organizational and technical assistance is provided to persons and organizations in related activities.

Publication(s): Rivers Directory, NPDES Permit Handbook, Dirty Water, Dirty Business

Keyword(s): Watersheds, River Conservation, Water Quality, Exotic species, Aquatic nuisance species, Rivers

Contact(s):
Robert Moore, EXECUTIVE DIRECTOR
robmoore@prairierivers.org
Bruce Hannon, PRESIDENT
Phone: 217-352-3646
John McNussen, TREASURER
Phone: 217-398-8531
Marc Miller, WATERSHED ORGANIZER
Phone: 217-344-2371
mmiller@prairierivers.org

PRAIRIE RIVERS NETWORK
809 South 5th St
Champaign, IL 61820 USA
Phone: 217-344-2371   Fax: 217-344-2381
E-mail: info@prairierivers.org

Founded: NA
Scope: Statewide

Description: Prairie River Networks strives to protect the rivers & streams of IL and to promote the lasting health & beauty of watershed communities

Publication(s): Prairie River Notes, Prairie Rivers Directory

Keyword(s): Water Quality, Coral Reefs, Watershed Protection, River Conservation

Contact(s):
Rob , EXECUTIVE DIRECTOR
Mark , WATERSHED ORGANIZER

PREDATOR CONSERVATION ALLIANCE
P.O. Box 6733
Bozeman, MT 59771 USA
Phone: 406-587-3389   Fax: 406-587-3178
E-mail: pca@predatorconservation.org
Website: www.predatorconservation.org

Founded: 1991
Scope: National

Description: Since its inception in 1991, Predator Project has worked to conserve and restore ecosystem integrity by protecting predators and their habitats—saving a place for America's predators. We advocate on behalf of over 12 species including: the grizzly bear, wolf, lynx, wolverine, fisher, marten, black-footed ferret, swift fox, burrowing owl, coyote, northern gas hawk, mountain lion, and black bear. Of these, nine are imperiled. We also work on behalf of important prey species, such as prairie dogs, because these animals provide predators with essential food and/or habitat.

Publication(s): Conservation of Prairie Dog Ecosystems: Learning From the Past to Insure the Prairie Dog's Future, The Wild Bunch, Motorizing Montana's National Forest Trails, At a Crossroads: The wolf and its place in the Norther, The Home Range (quarrerly newsletter)

Keyword(s): Endangered Species, Mammals, Public Lands, training, Predators, Environmental Protection, Biodiversity, Nongame Wildlife, Prairies, Predators

Contact(s):
Tom Skeele, EXECUTIVE DIRECTOR
Sharon Negri, PRESIDENT
Steve Forrest, SECRETARY/TREASURER

PRIORITIES INSTITUTE, THE
P.O. Box 89
Pine, CO 80470-0089 USA
Phone: 303-447-3792   Fax: 303-838-8105
E-mail: mail@priorities.org
Website: www.priorities.org

Founded: 1996
Scope: International

Description: Non-partisan, non-profit research organization focusing on sustainable land use planning, designing car-free eco-cities, international communities, holistic indexing and moral evolution.

Publication(s): Livable Cities, Perspectives and Priorities

Keyword(s): Environmental and Conservation Education, Environmental Planning, Land Use Planning, Pedestrian Environment, Preservation and Protection, Sustainable Development, Transportation, Urban and Rural Development

Contact(s):
Melissa Moon, ASSISTANT
Logan Perkins, DIRECTOR/FOUNDER
logan@priorities.org

PROFESSIONAL BOWHUNTERS SOCIETY
P.O. Box 246
Terrell, NC 28682 USA
Phone: 704-664-2534   Fax: 704-664-7471
E-mail: bowhunters@worldnet.att.net
Website: www.bowsite.com/pbs

Founded: 1963
Membership: 2500
Scope: National

Description: Created as an organization of dedicated bowhunters interested in promoting a high level of ethics in the taking of wild game with bow and arrow. To provide training for others in safety, shooting skill, and hunting techniques. To practice and promote the wise use of our natural resources and conservation of wildlife.

Publication(s): Professional Bowhunter Magazine, The

Contact(s):
Jack Smith, EDITOR
Brenda Kisner, PBS OFFICE
Phone: 704-664-2534
Fax: 704-664-7471
Wayne Capp, PRESIDENT
Jack Smith, SECRETARY AND TREASURER
P.O. Box 246, Terrell, NC 28682
Phone: 704-664-2534
Fax: 704-664-7471
Louie Adams, SENIOR COUNCILMAN
Larry Fischer, VICE PRESIDENT

PROVINCE OF QUEBEC SOCIETY FOR THE PROTECTION OF BIRDS, INC.

Station B
Montreal, Quebec H2B 3J5 Canada
Phone: 514-637-2141
Website: www.minet.ca/~pqspb

Founded: 1917
Membership: 500
Scope: Regional
Publication(s): Field Check List of Birds in the Montreal Area, Birdfinding in the Montreal Area, The Song Sparrow (monthly newsletter), Tchebec (Annual Review)
Keyword(s): Birds

Contact(s):
Sheila Arthur, EDITOR
3804 Royal Ave., Montreal, Quebec H4E 1P1
Phone: 514-487-4047
Kyra Emo, HON. SECRETARY
140 Irvine Ave., Westmount, Quebec H3Z 2K2
Phone: 514-939-9666
Kenneth Thorpe, HON. TREASURER
5615 Eldridge, Cote St-Luc, Quebec H4W 2C9
Phone: 514-483-5031
Bess Muhlstock, PRESIDENT
4807 Jeanne Mance, Montreal, Quebec H2V 4J6
Phone: 514-274-3810
Betsy McFarlane, VICE PRESIDENT

PTARMIGANS, THE

P.O. Box 1821
Vancouver, WA 98668 USA
Phone: 206-687-2436   Fax: 206-695-5385
Website: www.ptarmigans.org

Founded: 1960
Scope: Statewide

Description: The Ptarmigans were established for the purpose of conducting mountaineering activities in the northwest and in promoting the preservation of the northwest forests, wilderness lands, and mountain scenery. The only membership requirement is a love of the outdoors. Offer a variety of outdoor activities led in an ecologically-minded manner. Help maintain several local trail systems and conduct an annual Basic Climbing School. Visit and observe changes proposed in our northwest forests and mountains. The conservation committee studies issues that affect our outdoor recreation area.

Publication(s): A Place in the Islands, Landowner's Guide, Mom's Marsh and Other Fine Places (video), Preserve Farmlands

Keyword(s): Birds, Environmental and Conservation Education, Protected Areas, Islands, training

Contact(s):
Ruth Rowland, 1ST VICE PRESIDENT
Jon Bell, 2ND VICE PRESIDENT
Tim Kutscha, PRESIDENT
Don Spencer, SECRETARY
Jon Yrjanson, TREASURER

PUBLIC EMPLOYEES FOR ENVIRONMENTAL RESPONSIBILITY (PEER)

2001 S St., NW, Suite 570
Washington, DC 20009 USA
Phone: 202-265-7337   Fax: 202-265-4192
E-mail: info@peer.org
Website: http://www.peer.org

Founded: 1993
Scope: National

Description: PEER is an alliance of land managers, scientists, biologists, law enforcement officials, and other government professionals dedicated to the protection of the nation's environment. PEER advocates the responsible management of natural resources and promotes environmental ethics, professional integrity and accountability within local, state and federal agencies.

Publication(s): various employee-authored white papers, PEEReview

Keyword(s): Environmental Ethics, Government Accountability, Environmental Protection, Forests and Forestry, Public Lands, Exotic species, Aquatic nuisance species

Contact(s):
Danielle Lawson, ADMINISTRATIVE DIRECTOR
Howard Wilshire, BOARD CHAIR
Amanda Carufel, COMMUNICATIONS DIRECTOR
Dennis McKinney, DEVELOPMENT DIRECTOR
Jeffrey Ruch, EXECUTIVE DIRECTOR
Dan Meyer, GENERAL COUNSEL
Eric Wingerter, NATIONAL FIELD DIRECTOR
Mark Davis, OUTREACH MEMBERSHIP COORDINATOR

PUBLIC EMPLOYEES FOR ENVIRONMENTAL RESPONSIBILITY (PEER)

WEST COAST OFFICE
P.O. Box 30
Hood River, OR 97031 USA
Phone: 530-333-1106   Fax: 541-387-4783
Website: www.peer.org

Founded: 1992
Scope: National

Description: Organize a broad base of support among employees within local, state, and federal resource management agencies. Monitor natural resource management agencies by serving as a "watch dog" for the public interest. Inform the administration, congress, state officials, media and the public about substantive environmental issues of concern to public employees who speak out about issues concerning natural resource management and environmental protection. Provide free legal assistance if and when necessary.

PUBLIC LANDS FOUNDATION

P.O. Box 7226
Arlington, VA 22207 USA
Phone: 703-790-1988   Fax: 703-821-3490
E-mail: leaplf@erols.com
Website: www.publicland.org

Founded: 1987
Membership: 1000
Scope: National

Description: A national, nonprofit, independent advocate to keep
the public lands public and for the proper use and protection of
the public lands administered by the Bureau of Land
Management; implementation of the Federal Land Policy and
Management Act (FLPMA); and for professional land
management by professional employees.

Publication(s): Public Lands Monitor, The

Keyword(s): Environmental and Conservation Education, Land
Use Planning, Public Lands, Renewable Resources, Soil
Conservation

Contact(s):
George Lea, PRESIDENT/ EDITOR
Phone: 703-790-1988
Bill Leavell, VICE PRESIDENT
Phone: 301-881-0964

PUERTO RICO ASSOCIATION OF SOIL AND WATER CONSERVATION DISTRICTS

Attn: President, P.O. Box 91
Orocovis, PR 00720 USA

Founded: NA
Scope: Statewide

Publication(s): Puerto Rico Environmental Laws, Verde Luz
Newsletter, Puerto Rico Conservation Directory, Puerto Rican
Parrot Teacher's Kit, Pablo y Marisol van a la Playa

Keyword(s): Biodiversity, Coasts, Endangered Species,
Environmental and Conservation Education, Land Purchase,
Wetlands, Conservation Districts

Contact(s):
Migdalia Rodriguez, ADMINISTRATIVE SECRETARY
P.O. Box 1225, Casguas, PR 00726
Phone: 787-258-0490
Fax: 787-258-0490
Pedro Fuentes, PRESIDENT
P.O. Box 91, Orocovis, PR 00720
Phone: 787-867-4707
Hilda Bonilla, SECRETARY
HC 1 Box 8162, Luquillo, PR 00773
Phone: 787-860-0045
Norberto Colon, TREASURER
P.O. Box 175, Barranquitas, PR 00794
Phone: 787-857-7965
Carlos Mantras, VICE PRESIDENT, BOARD MEMBER
1722 Pastemark St., Urb Purple Tree, San Juan, PR 00926
Phone: 787-761-6247

PUERTO RICO CONSERVATION FOUNDATION, THE (PRCF)

527 Avenue Andalucia, Suite 75
San Juan, PR 00920-4131 USA
Phone: 787-763-9875   Fax: 787-772-4645
E-mail: fconserv@tld.net
Website: www.prcf.org

Founded: 1987
Membership: 30
Scope: Local

Description: The PRCF is a private nonprofit tax-exempt 501(c)3
organization, dedicated to protecting Puerto Rico's biological
biodiversity, focusing on the conservation of threatened and
endangered species and other keystone and lesser known
species.

Publication(s): Verde Luz Newsletter, Puerto Rico NGOs
Directory, Puerto Rican Parrot Teacher's Kit, Pablo y Marisol
van Playa, Sea Turtle Kit and various brochures

Keyword(s): Biodiversity, Endangered Species, Environmental
and Conservation Education, Research, Conservation

Contact(s):
Miguel Iturregui, DIRECTOR
Esther Rojas, EXECUTIVE DIRECTOR
Roberto Biaggi, PRESIDENT
Juan Ricart, SECRETARY
Jose Dueno, TREASURER

PUGET SOUNDKEEPER ALLIANCE

1415 W. Dravus
Seattle, WA 98119 USA
Phone: 206-286-1309   Fax: 206-286-1082
E-mail: volunteer@pugetsoundkeeper.org
Website: www.pugetsoundkeeper.org

Founded: 1984
Scope: National

Description: A nonprofit organization whose mission is to serve
as stewards for the protection and enhancement of Puget
Sound through education, advocacy, monitoring, and
celebration.

Publication(s): Sounder

Keyword(s): Coasts, Environmental Law, Environmental and
Conservation Education, Coral Reefs, Water Quality

Contact(s):
Sue Joerger, EXECUTIVE DIRECTOR / SOUNDKEEPER
Tom Diller, PRESIDENT
Seattle, WA
Tobey Watkins, TREASURER AND SECRETARY

PURPLE MARTIN CONSERVATION ASSOCIATION

Edinboro University of Pennsylvania
Edinboro, PA 16444 USA
Phone: 814-734-4420   Fax: 814-734-5803
E-mail: pmca@edinboro.edu
Website: www.purplemartin.org

Founded: 1987
Membership: 8
Scope: International

Description: An international tax-exempt, nonprofit organization
dedicated to the conservation of the Purple Martin (Progne
subis) species of bird through scientific research, state-of-the-
art wildlife management techniques, and public education. The
PMCA's scientific staff conducts research on all aspects of
martin biology throughout the bird's North, South, and Middle
American breeding, wintering, and migratory ranges. The
organization functions as a centralized data-gathering and
information source on the species, serving both the scientist
and the martin enthusiast. Its major mission is educating martin

enthusiasts in the proper techniques for managing this human-dependent species.

Publication(s): Purple Martin Update

Keyword(s): Birds, Endangered Species, Environmental and Conservation Education, Nongame Wildlife

Contact(s):
James Hill, DIRECTOR AND EDITOR

Q

QUAIL UNLIMITED, INC.
31 Quail Run, P.O. Box 610
Edgefield, SC 29824-0610 USA
Phone: 803-637-5731   Fax: 803-637-0037
E-mail: national@qu.org
Website: www.qu.org

Founded: 1981
Scope: National

Description: A national nonprofit conservation organization dedicated to improving quail and upland game bird populations through habitat management and research. Organized to re-establish and manage suitable upland game habitat, both public and private lands across the country, and to educate the public to the needs for wildlife habitat management.

Publication(s): Quail Unlimited Magazine Bi-Monthly

Keyword(s): training, Birds

Contact(s):
Jerry Allen, ADMINISTRATIVE VICE PRESIDENT
1884 Highway 23 West, Edgefield, SC 29824
Phone: 803-637-5877
national@qu.org
David Howell, DIRECTOR OF AGRICULTURAL WILDLIFE SERVICES
10364 S. 950 E., Stendal, IN 47585
Phone: 812-536-2272
dhowell@psci.net
Tommy Dean, DIRECTOR OF CHAPTER DEVELOPMENT
815 Shawnee Dr., N. Augusta, SC 29841
Phone: 803-637-5731
chapterdev@qu.org
D. Kogon, EDITOR
qumag@qu.org
Joseph Evans, EXECUTIVE VICE PRESIDENT
3012 Sussex Rd., Augusta, GA 30909
Phone: 706-738-0692
national@qu.org
Jeff Hodges, GREAT PLAINS REGIONAL DIRECTOR
382 NW Hwy 18, Clinton, MO 64735
Phone: 660-885-7057
Fax: 660-885-7152
Wade Teague, MID-ATLANTIC REGIONAL DIRECTOR
271 Stevens Church Rd, Goldsboro, NC 27530
Phone: 919-689-3884
Fax: 919-689-2726
wadequ@earthlink.net
Chris Wolkonowski
MIDWEST REGIONAL DIRECTOR
HCR 76 Box 645 Hwy. 108, Gruetli-Laager, TN 37339
Phone: 931-779-4868
Phone: 812-536-2272
cwolk@qu.org

Roger Wells, NATIONAL HABITAT COORDINATOR
868 Road 290, Americus, KS 316-443-5834
rwells@americusks.net
Mike Newell, OKLAHOMA REGIONAL DIRECTOR
2607 NW Columbia, Lawton, OK 73505
Phone: 580-357-0619
mnewell@qu.org
Steve Mcghee, PRESIDENT
100 Patterson Circle, Oliver Springs, TN 37840
Phone: 423-574-3685
Harvey Bray, ROCKY MOUNTAIN REGIONAL DIRECTOR
13 Archway Lane, Pueblo, CO 81005
Phone: 719-561-3825
Fax: 719-561-8977
Randy Guthrie, SOUTH CENTRAL REGIONAL DIRECTOR
2061 Crow Mt. Rd, Russellville, AR 72801
Phone: 501-967-2200
Fax: 501-767-2716
rguthrie@cswnet.com
Yale Leiden, SOUTHEAST REGIONAL DIRECTOR
Chip Martin, SOUTHWEST REGIONAL DIRECTOR
3320 FM 3326, Anson, TX 79501
Phone: 915-823-3347
Fax: 915-823-3340
Dick Haldeman, WESTERN REGIONAL DIRECTOR
39455 Black Oak Rd., Temecula, CA 92592
Phone: 909-767-3435
Fax: 909-767-2716
quwest@pe.net

QUEBEC WILDLIFE FEDERATION
6780 1st Ave., Bureau 109
Charlesbourg, Quebec G1H 2W8 Canada
Phone: 418-626-6858   Fax: 418-622-6168
E-mail: fede@fqf.qc.ca
Website: www.fqf.qc.ca

Founded: 1945
Scope: Providence

Publication(s): INFO-FQF

Keyword(s): Hunting, Outdoor Recreation, Public Lands, Renewable Resources, Sport Fishing, Training, Sustainable Development, Waterfowl, training

Contact(s):
Gaetan Roy, BIOLOGIST
Annie Guertin, COMMUNICATION COORDINATOR
Patrick Filiatrault, FOREST ENGINEER
Alain Cossette, GENERAL DIRECTOR
Aurele Blais, PRESIDENT
Bruno Paradis, TREASURER
Alain Gagnon, VICE PRESIDENT
Michel Savard, VICE PRESIDENT
Rodolphe Lasalle, VICE PRESIDENT
Alain Bisson, VICE PRESIDENT AND SECRETARY

R

RACHEL CARSON COUNCIL, INC.
8940 Jones Mill Rd.
Chevy Chase, MD 20815 USA
Phone: 301-652-1877   Fax: 301-587-3863
E-mail: rccouncil@aol.com
Website: http://members.aol.com/rccouncil/ourpage

Founded: 1965
Membership: 7

Scope: National

Description: (formerly Rachel Carson Trust for the Living Environment Inc.) An international clearinghouse for information on toxic substances, particularly pesticides for both scientists and laymen, distributed by means of publications, conferences, and response to specific questions. Rachel Carson Council is devoted to fostering a sense of wonder and respect toward nature and helping society realize Rachel Carson's vision of a healthy and diverse environment.

Publication(s): Rachel Carson Council News, books, pamphlets and sheets on specific alternative pest control methods and on pesticides effects, list of current publications available on request., Basic Guide to Pesticides

Keyword(s): Agriculture, Environmental and Conservation Education, Pesticides, Toxic Substances, Nuclear-free, Water quantity, Water export and diversion, Coral Reefs

Contact(s):
Diana Post, EXECUTIVE DIRECTOR AND SECRETARY
Aaron Blair, LIAISON TO THE BOARD
David McGarth, TREASURER
Martha Talbot, VICE PRESIDENT

RAILS-TO-TRAILS CONSERVANCY
P.O. Box 61091
Durham, NC 27715 USA
Phone: 919-233-8444   Fax: 202-331-9680
Website: http://www.ncrail-trails.org/trtc/INDEX.HTM

Founded: 1986
Scope: National

Description: The mission of the Rails-to-Trails Conservancy is to enhance America's communities and countrysides by converting thousands of miles of abandoned rail corridors, and connecting open space into a nationwide network of public trails.

Publication(s): Seven-Hundred Great Rail-Trails: A Directory of Multi-Use Paths Created from Abandoned Railroads, Secrets of Successful Rail-Trails: An Acquisition and Organizing Manual for Converting Rails into Trails, Enhancing America's Communities: A Nat, Trailblazer

Keyword(s): Environment, Research, Outdoor Recreation, Land Preservation, Internships

Contact(s):
Mark Ackelson, BOARD OF DIRECTORS CHAIR
Richard Angle, BOARD OF DIRECTORS TREASURER
Mark Wolf-Armstrong, EXECUTIVE VICE PRESIDENT
David Burwell, PRESIDENT

RAINBOW PUSH COALITION
1002 Wisconsin Ave., NW
Washington, DC 20007 USA
Phone: 202-333-5270   Fax: 202-728-1192
E-mail: info@rainbowpush.org
Website: www.rainbowpush.org

Founded: 1984
Scope: International

Description: RPC is a national progressive membership organization committed to public education, empowerment, economic and social justice, and gender and racial equality. The Rainbow Push Coalition has state chapters and a national membership base. The Rainbow Push Coalition addresses such issues as education, voter registration, economic justice, civil rights, environment, labor, and working people's rights.

Publication(s): various issue papers, speeches, and briefings, Rainbow Push Magazine or The Rainbow JaxFax, Rainbow Newsletter

Keyword(s): Environmental and Conservation Education, Sustainable Development, Toxic Substances, Nuclear-free, Water quantity, Water export and diversion, Urban Environment

Contact(s):
Willie Barrow, CO-CHAIRMAN OF THE BOARD
Dennis Rivera, CO-CHAIRMAN OF THE BOARD
Jesse Jackson, PRESIDENT AND FOUNDER

RAINFOREST ACTION NETWORK
221 Pine St.,
San Francisco, CA 94104 USA
Phone: 415-398-4404   Fax: 415-398-2732
E-mail: rainforest@ran.org
Website: www.ran.org

Founded: 1985
Scope: National

Description: RAN works nationally and internationally on major campaigns to protect rainforests and defend the rights of indigenous people, using non-violent direct action such as: letter-writing campaigns, boycotts, and demonstrations against corporations and lending agencies contributing to rainforest destruction. RAN also produces educational materials, a teachers' packet, and fact sheets for community organizers.

Publication(s): Action Alert, World Rainforest Week Organizers Manual

Keyword(s): Rainforests, Biodiversity, Environmental and Conservation Education, Forests and Forestry

Contact(s):
Jim Gollin, BOARD CHAIR
Randall Hayes, BOARD SECRETARY, PRESIDENT
Scott Price, BOARD TREASURER
Sara Riggs, COMMUNICATIONS DIRECTOR
Christopher Hatch, EXECUTIVE DIRECTOR
Randy Hayes, PRESIDENT
Laura Fauth, PUBLICATIONS EDITOR

RAINFOREST ALLIANCE
65 Bleecker St.
New York, NY 10012 USA
Phone: 212-677-1900   Fax: 212-677-2187
E-mail: canopy@ra.org

Founded: 1986
Scope: International

Description: The Rainforest Alliance is an international nonprofit organization dedicated to the conservation of tropical forests for the benefit of the global community. Its primary mission is to develop and promote economically viable and socially desirable alternatives to the destruction of tropical forests.

Publication(s): Floods of Fortune, Tales from the Jungle, So Fruitful a Fish, Catfish Connection, The, CANOPY, The

Keyword(s): Biodiversity, Forests and Forestry, Grants, Sustainable Development, Rainforests

Contact(s):
Karin Kreider, ASSOCIATE DIRECTOR

Daniel Katz, EXECUTIVE DIRECTOR
Tensie Whealan, PRESIDENT

RAINFOREST RELIEF

P.O. Box 150566
Brooklyn, NY 11215 USA
Phone: 718-398-3760   Fax: 718-398-3760
E-mail: relief@igc.org
Website: www.enviroweb.org/rainrelief

Founded: 1989
Membership: 200
Scope: National

Description: Rainforest Relief, a nonprofit 501(c)3 organization, works through education and non-violent direct action to end the loss of tropical and temperate rain forests by reducing the demand for products and materials for which rainforests are destroyed. These materials include tropical hardwoods, paper, oil, metals and agricultural products such as bananas, beef, coffee and chocolate.

Publication(s): Raindrops, Rainforest Relief Reports, Roots

Keyword(s): Rainforests, Tropical Biodiversity and Conservation, Biodiversity, Cultural Preservation, Environmental and Conservation Education, Forests and Forestry, Overconsumption

Contact(s):
Tim Keating, PRESIDENT AND DIRECTOR
Jeffrey Lockwood, VICE PRESIDENT AND PORTLAND OREGON CHAPTER, DIRECTOR
P.O. Box 14232, Portland, OR 97293
Phone: 503-236-3031
rainrelief@hotmail.com

RAINFOREST TRUST, INC., THE

6001 SW 63rd Ave.
Miami, FL 33143 USA
Phone: 305-667-2779   Fax: 305-665-0691
E-mail: rforest@rainforesttrust.com
Website: www.rainforesttrust.com

Founded: 1995
Scope: International

Description: The Trust has established a jaguar sanctuary and rainforest preserve in Belize. It also promotes eco-tourism and alternative rainforest sustainable agriculture as economically viable alternatives to deforestation, particularly in Jamaica; and promotes programmes to educate public school children and farmers about conservation of eco-systems, organic agriculture, and wildlife. The trust has established a second wildlife sanctuary and rainforest preserve in Jamaica.

Keyword(s): Agriculture, Endangered Species, Environmental and Conservation Education, Sustainable Ecosystems, training, Conservation, Ancient Forests, Environmental Protection, Historic Preservation, Rainforests

Contact(s):
T. Hawkins, CHAIRMAN
Phone: 305-666-2158
Brett Ashmeade-Hawkins, PRESIDENT
Mark Ashmeade-Hawkins, SECRETARY AND TREASURER

RAPTOR CENTER, THE

University of Minnesota, 1920 Fitch Ave.
St. Paul, MN 55108 USA
Phone: 612-624-4745   Fax: 612-624-8740
E-mail: raptor@umn.edu
Website: www.raptor.cvm.umn.edu

Founded: 1974
Scope: International

Description: To preserve biological diversity among raptors and other avian species through medical treatment, scientific investigation, education, and management of wild populations.

Publication(s): Medical Management of Birds of Prey, Care and Management of Captive Raptors, Raptor Release, The

Keyword(s): Birds, Endangered Species, Environmental and Conservation Education, Raptors, accreditation

Contact(s):
Ron Osterbauer, ASSOCIATE DIRECTOR
Michael Gabriel, BOARD CHAIR
Patrick Redig, DIRECTOR

RAPTOR EDUCATION FOUNDATION, INC.

P.O. Box 200400
Denver, CO 80220 USA
Phone: 303-680-8500   Fax: 303-680-8502
E-mail: raptor2@usaref.org
Website: www.usaref.org

Founded: 1980
Scope: National

Description: A nonprofit, charitable educational organization utilizing nonreleasable raptors to promote environmental literacy. Lecturers travel nationwide.

Publication(s): Talon Supplement, Castings (volunteer newsletter), Talon

Keyword(s): Raptors, Environmental and Conservation Education, Environmental Protection, Protected Areas

Contact(s):
Anne Price, CURATOR OF RAPTORS
Patrick Duran, EXECUTIVE DIRECTOR
Shellie Sage, MEWS MANAGER
Peter Reshetniak, PRESIDENT AND EDITOR
Anne Price, SECRETARY

RAPTOR RESEARCH FOUNDATION, INC.

USGS Forest and Rangeland Ecosystem Science Center, Snake River Field Station, 970 Lusk St.
Boise, ID 83706 USA
Phone: 208-426-5201
Website: biology.boisestate.edu/raptor

Founded: 1966
Scope: National

Description: A nonprofit corporation formed to stimulate and coordinate the dissemination of information on the biology and management of birds of prey and their habitats. Areas of particular interest include: raptor banding, behavior, captive breeding, conservation, ecology, research techniques, management, migration, population monitoring, pathology, and rehabilitation.

Publication(s): Wingspan, Journal of Raptor Research

Keyword(s): Endangered Species, Environmental and Conservation Education, Nongame Wildlife, Raptors, training

Contact(s):
James Bednarz, EDITOR-IN-CHIEF
Department of Biology, Boise State University,
Michael Kochert, PRESIDENT
Phone: 208-426-5201
Fax: 208-426-5210
mkochert@eagle.idbsu.edu
Patricia Hall, SECRETARY
436 David Dr. E, Flagstaff, AZ 86001
Phone: 520-526-6222
pah@alpine.for.nau.edu
Jim Fitzpatrick, TREASURER AND MEMBERSHIP INFORMATION
Carpenter, St. Croix Valley Nature Center, 12805 St. Croix Tr., Hastings, MN 55033
Phone: 651-437-4359
jim@cncstcroix.com
Keith Bildstein, VICE PRESIDENT
Hawk Mountain Sanctuary, Route 2, Box 191, Kempton, PA 19529
Phone: 910-756-6961
bildstein@hawkmountain.org

RARE CENTER FOR TROPICAL CONSERVATION
1840 Wilson Blvd, Ste 402
Arlington, VA 22201 USA
Phone: 703-522-5070   Fax: 703-522-5027
E-mail: rare@rarecenter.org
Website: rarecenter.org

Founded: 1973
Membership: 2,400
Scope: International

Description: RARE Center is a small organization doing innovative work to preserve threatened habitats and ecosystems in Latin America, the Caribbean, and the Pacific. Our programs focus on education and training, habitat protection, and research.

Keyword(s): Environmental and Conservation Education, Protected Areas

Contact(s):
Roger Pasquier, CHAIRMAN
Paul Butler, DIRECTOR, CONSERVATION EDUCATION PROGRAM
Brett Jenks, DIRECTOR, ECOTOURISM AND COMMUNITY DEVELOPMENT PROGRAM
Brett Jenks, PRESIDENT (ACTING)
Emilie Pryor, SECRETARY
Sumner Pingree, TREASURER
Susan Babcock, VICE CHAIRMAN

REEF RELIEF
P.O. Box 430
Key West, FL 33041 USA
Phone: 305-294-3100   Fax: 305-293-9515
E-mail: reef@bellsouth.net
Website: www.reefrelief.org

Founded: 1989
Membership: 3000
Scope: National

Description: Reef Relief is a nonprofit membership organization dedicated to preserve and protect Living Coral Reef Ecosystems through local, regional, and global efforts.

Publication(s): Florida's Coral Reef Ecosystem, Booklets, Flyers, Reef Line

Keyword(s): Biodiversity, Environmental and Conservation Education, Sustainable Ecosystems, Water Pollution Management

Contact(s):
Joel Biddle, EDUCATIONAL COORDINATOR
Deevon Quirolo, EXECUTIVE DIRECTOR
Craig Quirolo, FOUNDER AND DIRECTOR OF MARINE PROJECTS
Marci Rose, PRESIDENT
Michael Blades, PROJECT COORDINATOR
David Ethridge, TREASURER

REEFKEEPER INTERNATIONAL
2809 Bird Ave., PMB 162
Miami, FL 33133 USA
Phone: 305-358-4600   Fax: 305-358-3030
E-mail: reefkeeper@reefkeeper.org
Website: www.reefkeeper.org

Founded: NA
Scope: International

Publication(s): ReefKeeper Report, Reef Alert, Reef Dispatch, Reef Monitor Update

Contact(s):
Alexander Stone, DIRECTOR

RENEW THE EARTH
1200 18th St., NW, Suite 1100
Washington, DC 20036 USA
Phone: 202-721-1545   Fax: 202-467-5780
E-mail: renew@renewtheearth.org
Website: www.renewtheearth.org

Founded: 1978
Scope: International

Description: Renew America specializes in broad-based environmental program identification, verification, and promotion of positive models for change. By seeking out, promoting and awarding exemplary programs among all sectors, we offer innovative, constructive models to inspire communities and businesses to meet environmental challenges.

Publication(s): Environmental Embassador Index

Keyword(s): Sustainability, Environmental and Conservation Education

Contact(s):
Katy Moran, EXECUTIVE DIRECTOR
Phone: 202-721-1545
katymoran@renewtheearth.org

RENEWABLE ENERGY POLICY PROJECT (REPP)
1612 K St., NW, Suite 202
Washington, DC 20006 USA
Phone: 202-293-2898   Fax: 202-293-5857
Website: www.repp.org

Founded: 1995
Scope: International

Description: The Renewable Energy Policy Project (REPP) investigates the emerging relationships among policies, markets and public demand for renewable energy technologies. REPP's mission is to accelerate growth of the renewable energy industry and maximize deployment of renewable energy technology, by providing credible information, insightful analysis and innovative strategies.

Publication(s): Rural Electrification with Solar Energy as a Climate Protection Strategy (Jan 2000), Federal Energy Subsidaries:Not all Technologies are Created Equal (July 2000), Renewable Energy Policy Outside the United States (Oct !999), Clean Government: Options for Governments to Buy Renewable Energy (May 1999), Natural Gas: Bridge to a Renewable Energy Future (May 1997), Wind Clusters: Expanding the Market Appeal of Wind Energy Systems (Nov 1996)

Keyword(s): Energy, Renewable Resources

Contact(s):
Karl Rabago, CHAIRMAN OF THE BOARD OF DIRECTORS
Mary Campbell, INTERNET AND PUBLICATIONS DIRECTOR
Virinder Singh, RESEARCH DIRECTOR
Fred Beck, RESEARCH MANAGER

RENEWABLE NATURAL RESOURCES FOUNDATION

5430 Grosvenor Ln.
Bethesda, MD 20814-2193 USA
Phone: 301-493-9101   Fax: 301-493-6148
E-mail: info@rnrf.org
Website: www.rnrf.org

Founded: 1972
Membership: 4
Scope: National

Description: A public, nonprofit, operating foundation. Conducts conferences and symposia on renewable natural resource subjects and public policy alternatives. Developer of the Renewable Natural Resources Center, an office-park complex for natural resource organizations.

Publication(s): Renewable Resources Journal

Keyword(s): Environmental and Conservation Education, Land Use Planning, Renewable Resources, Urban Environment

Contact(s):
David Moody, CHAIRMAN OF BOARD OF DIRECTORS
Chandru Krishna, DIRECTOR OF ADMINISTRATION AND FINANCE
Ryan Colker, DIRECTOR OF PROGRAMS
Robert Day, EXECUTIVE DIRECTOR
Albert Grant, VICE CHAIRMAN OF THE BOARD

REP AMERICA/REPUBLICANS FOR ENVIRONMENTAL PROTECTION

Deerfield, IL 60015 USA
Phone: 847-940-0320   Fax: 847-940-0320
Website: www.repamerica.org

Founded: 1995
Scope: National

Description: REP aims to resurrect the GOP's conservation tradition and restore natural resource conservation and sound environmental protection as fundamental elements of the Republican Party.

Publication(s): The Green Elephant

Keyword(s): Environmental Protection, Environmental Legislation, Environmental Ethics, Politics and Government, Natural Resource Conservation, Forests and Forestry, Public Lands, Rivers, Water Pollution Management, Wilderness, Conservation, Endangered Species, Biodiversity, Forest Stewardship, Pollution Prevention

Contact(s):
Jim Scarantino, EXECUTIVE DIRECTOR
Martha Marks, PRESIDENT
martha@repamerica.org
Vince Williams, SECRETARY
Anthony Cobb, TREASURER
Aurie Kryzuda, VICE-PRESIDENT

RESOURCE RENEWAL INSTITUTE, THE

Fort Mason Center Building A
San Francisco, CA 94123 USA
Phone: 415-928-3774   Fax: 415-928-6529
E-mail: info@rri.org
Website: www.rri.org

Founded: 1983
Scope: National

Description: The RRI is a national, nonprofit organization advocating state and national comprehensive, integrated environmental strategies (known as Green Plans), modeled on those of the Netherlands and New Zealand. RRI has set up a Global Green Plan Center to act as a clearinghouse for information on Green Plans. For more information, use RRI's E-mail: address.

Publication(s): The International Green Planner, Green Plans: Greenprint for Sustainability, Saving Cities, Saving Money

Keyword(s): Environmental and Conservation Education, Environmental Planning, Renewable Resources, Sustainable Development, Sustainable Ecosystems, Planning Management

Contact(s):
Danielle Kraaijvanger, EXECUTIVE DIRECTOR
Huey Johnson, PRESIDENT

RESOURCES FOR THE FUTURE

1616 P St., NW
Washington, DC 20036 USA
Phone: 202-328-5000   Fax: 202-939-3460
Website: www.rff.org.com

Founded: 1952
Scope: National

Description: An independent nonprofit organization that works to advance research and education in the development, conservation, and use of environmental and natural resources. Staff is comprised primarily of economists and policy analysts who research a variety of environmental and natural resource issues.

Publication(s): Resources

Keyword(s): Air Quality and Pollution, Energy, Renewable Resources, Sustainable Development, Transportation, Developing Countries, Biodiversity, Endangered Species, Energy Conservation, Environment, Forest Management, Land Use Planning, Environmental Protection, Mining

Contact(s):

J. Davies, CENTER FOR RISK MANAGEMENT DIRECTOR
Phone: 202-328-5093
davies@rff.org.com
Lesli Creedon, DIRECTOR OF DEVELOPMENT
creedon@rff.org.com
Michael Toman, ENERGY AND NATURAL RESOURCES
DIVISION DIRECTOR
Phone: 202-328-5091
toman@rff.org.com
Christopher Clotworthy, LIBRARIAN
Phone: 202-328-5089
clotworthy@rff.org.com
Dan Quinn, MANAGER OF PUBLIC AFFAIRS
Phone: 202-328-5019
quinn@rff.org.com
Paul Portney, PRESIDENT
Phone: 202-328-5000
portney@rff.org.com
Alan Krupnick, QUALITY OF THE ENVIRONMENT DIVISION
DIRECTOR
Phone: 202-328-5059
krupnick@rff.org.com
Edward Hand, VICE PRESIDENT OF FINANCE AND
ADMINISTRATION
Phone: 202-328-5029
hand@rff.org.com

RESOURCE-USE EDUCATION COUNCIL

P.O. Box 11104
Richmond, VA 23230 USA
Phone: 804-698-4442   Fax: 804-698-4533
E-mail: sgilley@dgif.state.va.us
Website: www.vruec.org/

Founded: 1952
Membership: 30
Scope: Regional

Description: A volunteer, nonprofit organization, composed of
members of the state and federal government, colleges, and
private industry, working to promote the broad principle of environmental education. The Council offers conservation
education workshops for educators across Virginia, and is staff
to the Governor's EE initiative, Virginia Naturally 2000
(www.vanaturally.com)

Publication(s): The Virginia Natural Resources Education Guide,
www.vanaturally.com

Contact(s):

Ann Regn, CHAIRMAN
Department of Environmental Quality, P.O. Box 10009,
Richmond, VA 23240-0009
Phone: 804-698-4442
Susan Gilley, SECRETARY AND TREASURER
Department of Game and Inland Fisheries, Box 11104,
Richmond, VA 23230
Phone: 804-367-1000

RESPONSIVE MANAGEMENT

130 Franklin St.
Harrisonburg, VA 22801 USA
Phone: 540-432-1888   Fax: 540-432-1892
Website: www.responsivemanagement.com

Founded: 1986
Scope: National

Description: Developed to help fish and wildlife organizations
understand and work with their constituents. Responsive
Management conducts focus group research, telephone and
mail surveys, public opinion and attitude research, literature
reviews, demographic analysis, and workshops in public
opinion polling, marketing, change, communications, dispute
resolution, and human dimensions of natural resource
management.

Publication(s): Responsive Management Report

Keyword(s): Communications, Research, Training

Contact(s):

Mark Duda, EXECUTIVE DIRECTOR
Phone: 540-432-1888

RETURNED PEACE CORPS VOLUNTEER FOR ENVIRONMENT AND DEVELOPMENT (RPCV-ED)

P.O. Box 102
Iowa City, IA 52246-0102 USA
Phone: 319-351-3375

Founded: 1991
Scope: National

Description: The RPCVs-ED was formed to serve as a focal point
for action on environment and development issues by Peace
Corps alumni and friends.

Keyword(s): Developing Countries, Environment, Sustainable
Development

Contact(s):

Katy Hansen, CO-CHAIR
Mary Steinmaus, CO-CHAIR
Susan Singh, EDITOR
1762 E. 60th St., Tulsa, OK 74105
Phone: 918-749-7004

RHODE ISLAND B.A.S.S. CHAPTER FEDERATION

10 Ridgeway Drive
Warren, RI 02885 USA
Phone: 401-245-6264
E-mail: ribassfed@aol.com
Website: www3.edgenet.net/ebbm/state/main.html

Founded: NA
Scope: Statewide

Description: An organization of Bassmaster chapters, affiliated
with the Bass Anglers Sportsman Society, organized to fight
pollution, assist state and national conservation agencies in
their efforts, and teach the young people of our country good
conservation practices. Dedicated to the realistic conservation
of our water resources.

Publication(s): Forest Conservationist, The

Keyword(s): Forests and Forestry

Contact(s):

Bill Weikert, CONSERVATION DIRECTOR
156 Ridgewood Rd., Middletown, RI 02842
Phone: 401-846-0512
riscnrdbil@aol.com
Roger Pray, PRESIDENT

RHODE ISLAND FOREST CONSERVATORS ASSOCIATION

P.O. Box 40328
Providence, RI 02908 USA
Phone: 401-273-8037

Founded: 1989
Scope: Statewide

Description: A statewide organization affiliated with the National Woodland Owners Association organized to promote stewardship of Rhode Island's wooded lands and watersheds and protect their heritage for future generations.

Publication(s): (newsletter)

Contact(s):
Marc Tremblay, PRESIDENT AND EDITOR
Milton Schumacher, SECRETARY
Virginia Warrender, TREASURER
Alex Merrimian, VICE PRESIDENT

RHODE ISLAND STATE CONSERVATION COMMITTEE

Chair, Sosnowski Farm, P.O. Box 722
W. Kingston, RI 02892 USA

Founded: NA

Contact(s):
Susna Sosnowski, CHAIR
Sosnowski Farm, P.O. Box 722, W. Kingston, RI 02892
Phone: 401-783-7704
senmike@uriacc.uri.edu

RHODE ISLAND WILD PLANT SOCIETY

Box 114
Peace Dale, RI 02883-0114 USA
Phone: 410-783-5895

Founded: 1987
Membership: 600
Scope: Statewide

Description: The Rhode Island Wild Plant Society is a nonprofit conservation organization dedicated to the preservation and protection of Rhode Island's native plants and their habitats. Activities include talks, inventories of local flora, native plant restoration projects and a spring flower show and garden exhibit.

Publication(s): RIWPS Newsletter (biannual newsletter)

Keyword(s): Native Plants, Wildflowers, Environmental and Conservation Education, Protected Areas, Flowers, Plants, and Trees

Contact(s):
Deborah Poor, EXECUTIVE DIRECTOR
Jules Cohen, PRESIDENT
85 Scrabbletown Rd., N. Kingstown, RI 02852

RIVER ALLIANCE OF WISCONSIN

306 East Wilson, Suite 2W
Madison, WI 53703 USA
Phone: 608-257-2424   Fax: 608-260-9799
E-mail: wisrivers@wisconsinrivers.org
Website: www.wisconsinrivers.org

Founded: 1993
Membership: 1700

Scope: Statewide

Description: The River Alliance is a nonprofit, nonpartisan citizen advocacy organization for rivers. Our mission is to lead the growing statewide effort to protect, enhance and restore Wisconsin's rivers and watersheds for their ecological, recreational, aesthetic, and cultural values. Recent program work includes education and information about the impacts of dams on river system health, minimizing ecosystem damage through federal relicensing of hydro dams, and advocacy for selective removal of uneconomical dams for purpose of river restoration.

Publication(s): Periodic Action Alerts, Wisconsin Rivers (Quarterly), News Bulletins and fact Sheets, Canoe (E-mail: newsletter), Dam Removal "A citizens guide to restoring rivers", Local Groups Directory for Wisconsin, Small Groups Building Tool Kit

Keyword(s): Aquatic Habitats, Wildlife, Rivers, Watersheds, Dams, Coral Reefs, Hydropower Relicensing

Contact(s):
Todd Ambs, EXECUTIVE DIRECTOR

RIVER NETWORK

520 SW 6th Ave., Suite 1130
Portland, OR 97204 USA
Phone: 503-241-3506   Fax: 503-241-9256
E-mail: info@rivernetwork.org
Website: www.rivernetwork.org

Founded: 1988
Scope: National

Description: River Network supports river advocates at the grassroots, state and regional levels; helps them build effective organizations; and links them together in a national movement to protect and restore America's rivers and watersheds. River Network also works with river conservationists to acquire and conserve riverlands and riparian areas critical for wildlife, fisheries, drinking water, flood plain management, and recreation.

Publication(s): How to save a River, Living Waters, a book on monitoring a river, River Voices, 2000-2001 River and Watershed Conservation Directory, River Fundraising Alert, Starting Up: A Handbook for New River and Watershed Organizati, The Watershed Innovators Workshop: Proceedings and The Swift River Principles

Keyword(s): Watersheds, Land Purchase, Rivers

Contact(s):
Lindy Walsh, ADMINISTRATIVE DIRECTOR
Peter Kirsch, CHAIRMAN OF BOARD OF DIRECTORS
Thalia Zepatos, COMMUNICATIONS DIRECTOR
Cary Schaye, DIRECTOR OF DEVELOPMENT
Don Elder, DIRECTOR OF WATERSHED PROGRAMS
Sue Doroff, NORTHWEST (RLC)
Ken Margolis, PRESIDENT

RIVER NETWORK

EASTERN OFFICE
4000 Albemarle St., NW, Suite 303
Washington, DC 20016 USA
Phone: 202-364-2550   Fax: 202-364-2520
E-mail: dc@rivernetwork.org
Website: www.rivernetwork.org

Founded: NA
Membership: 600

Scope: National

Publication(s): The Clean Water Act Owners Manual, How to Save a River

Contact(s):
Robin Chanay, WATERSHED PROGRAM ASSOCIATE
rchanay@rivernetwork.org

RIVER NETWORK

NORTHERN ROCKIES OFFICE: RIVER CONSERVANCY FIELD OFFICE
44 North Last Chance Gulch, 4
Helena, MT 59601 USA
Phone: 406-442-4777   Fax: 406-442-8883
E-mail: montanazac@aol.com
Website: www.rivernetwork.org

Founded: NA
Scope: National

Contact(s):
Hugh Zackheim, CONTACT

RIVER OTTER ALLIANCE, THE

6733 S. Locust Ct.
Englewood, CO 80112 USA
Phone: 303-773-2749   Fax: 303-722-7703

Founded: 1990
Membership: 150
Scope: National

Description: The River Otter Alliance promotes the survival of the North American River Otter through education, research, and habitat protection. We support current research and reintroduction programs, monitor abundance and distribution in the United States, and educate through our newsletter on the need to restore and sustain river otter populations.

Publication(s): River Otter Journal, The

Keyword(s): Endangered Species, Rivers, Trapping, accreditation, Water Quality, Otters

Contact(s):
Tracy Johnstone, PRESIDENT
John Mulvihill, TREASURER
6733 S. Locust Ct., Englewood, CO 80112-1007
Carol Peterson, VICE PRESIDENT

RIVERS COUNCIL OF WASHINGTON

509 10th Ave., East, Suite 200
Seattle, WA 98102 USA
Phone: 206-568-1380   Fax: 206-568-1381
E-mail: riverswa@brigadoon.com
Website: www.riverscouncilofwa.org

Founded: 1984
Scope: Statewide

Description: (formerly Northwest Rivers Council) The mission of the Rivers Council of Washington is to lead an expanding grassroots effort to preserve, enhance, and restore rivers and their watersheds in Washington state for their natural, recreational, and cultural values, and support compatible efforts of other organizations in the Pacific Northwest.

Publication(s): Washington Rivers

Keyword(s): Watersheds, Environmental and Conservation Education, Outdoor Recreation, Rivers, Sustainable Ecosystems, Exotic species, Aquatic nuisance species

Contact(s):
Scott Andrews, EXECUTIVE DIRECTOR
Doug North, PRESIDENT AND CHAIR OF TRUSTEES
Andy Held, SECRETARY
Matt Scobel, TREASURER
Kate Sullivan, VICE PRESIDENT

ROBERT ROADS ILLINOIS DEPARTMENT OF NATURAL RESOURCES

524 S Second St
Springfield, IL 62701 USA
Phone: 217-782-1329   Fax: 217-782-9599
E-mail: BROADS@DNRMAIL.IL.STATE.US
Website: www.conservation.state.mo.us/engineering/ace/

Founded: 1961
Scope: National

Description: To encourage and broaden the educational, social, and economic interests of conservation engineering practices; to promote recognition of the importance of sound engineering practices in fish, wildlife, and recreation development; to enable each member to take advantage of the experience of other states.

Publication(s): Handbook, Conference Proceedings, Informational Brochure, A.C.E. Newsletter, Membership Directory

Keyword(s): Engineering, Environmental and Conservation Education, Outdoor Recreation, Exotic species, Aquatic nuisance species, training

Contact(s):
Robert Roads, PRESIDENT
Phone: 217-782-2605
Norval Olson, SECRETARY AND TREASURER

ROCKY MOUNTAIN BIGHORN SOCIETY

P O Box 8320
Denver, CO 80201 USA
Phone: 303-697-4896   Fax: 303-697-2921
Website: www.bighornsheep.org

Founded: 1975
Membership: 1200
Scope: National

Description: The purpose of the Society is to support the sound management of the Rocky Mountain bighorn sheep and its habitat and to promote the advancement and knowledge of the bighorn.

Publication(s): The Bighorn

Keyword(s): Hunting, Trapping, training, accreditation

Contact(s):
Dennis Gardner, PRESIDENT
19114 Silver Ranch Rd., Conifer, CO 80433
Phone: 303-697-4896
Todd Brickell, SECRETARY
8181 Cooper River Dr., Colorado Springs, CO 80920
Kevin Wilson, TREASURER
P.O. Box 485, Conifer, CO 80433
Victor Lauer, VICE PRESIDENT
P. O. Box 1811, Woodland, CO 80866

ROCKY MOUNTAIN ELK FOUNDATION

P.O. Box 8249
Missoula, MT 59807-8249 USA
Phone: 406-523-4500   Fax: 406-523-4550
E-mail: rmef@rmef.org
Website: www.elkfoundation.org

Founded: 1984
Membership: 130000
Scope: International

Description: The Foundation's mission is to ensure the future of elk, other wildlife, and their habitat. Projects funded by RMEF include: land protection, habitat enhancement, management, research, conservation education, and hunting heritage.

Publication(s): Wild Outdoor World (W.O.W.: Kids Conservation Magazine), BUGLE: Journal of Elk Country

Keyword(s): Environmental and Conservation Education, Hunting, Land Purchase, training, Open Space, Hunting

Contact(s):
John Fasel, CHAIRMAN BOARD OF DIRECTORS
Gary Wolfe, PRESIDENT & CEO
Lance Schelvan, SENIOR VICE PRESIDENT OF COMMUNICATION/PUBLIC RELATIONS
Ron Marcoux, SENIOR VICE PRESIDENT, OPERATIONS
George Bettas, VICE CHAIRMAN, BOARD OF DIRECTORS
Alan Christensen, VICE PRESIDENT OF LANCE
Grant Parker, VICE PRESIDENT, LEGAL AFFAIRS/GENERAL COUNSEL
David Wesley, VICE PRESIDENT, MARKETING AND PLANNED GIVING

ROCKY MOUNTAIN ELK FOUNDATION

INTERMOUNTAIN REGION OFFICE
P O Box 290
Coaldale, CO 81222 USA
Phone: 719-547-8212   Fax: 719-942-5505
Website: www.elkfounation.org

Founded: NA
Membership: 1
Scope: Regional

Contact(s):
Marty Holmes, DIRECTOR OF FIELD OPERATIONS
Phone: 719-547-8212
Tom Brown, INTERMOUNTAIN REGIONAL DEV. DIRECTOR
Phone: 303-279-7974
tjbrown@rmi.net
Joe Coker, REG. DIR OF KENTUCKY & TENNESSEE
Lance Schul, REGIONAL DIRECTOR CO.
Phone: 303-984-4456
Bill Christensen, REGIONAL DIRECTOR OF UT.
Phone: 801-254-1922
Blake Henning, REGIONAL DIRECTOR WY
Phone: 307-634-4099
Doug Robinson, SOUTHERN ROCKIES REGIONAL LANDS MNGR.
Phone: 303-216-1953
drobinson@rmef.org

ROCKY MOUNTAIN ELK FOUNDATION

INTERNATIONAL HEADQUARTERS
2291 W. Broadway
P.O. Box 8249
Missoula, MT 59807 USA
E-mail: info@elkfoundation.org

Founded: NA
Scope: Regional

Contact(s):
Lyle Button, CONTACT FOR AZ
Phone: 520-714-1774
Mike Ford, CONTACT FOR NORTHERN CA
Phone: 530-842-2021
Tony Kavalok, CONTACT FOR NV
Phone: 775-971-9000
Bill Harris, CONTACT FOR SOUTHERN CA
Phone: 619-486-5601
Dakota Livesay, REGIONAL DIRECTOR
Phone: 520-286-1833

ROCKY MOUNTAIN ELK FOUNDATION

NORTH-CENTRAL REGION OFFICE
1320 Canal St.
Custer, SD 57730 USA
Phone: 605-644-2396

Founded: NA
Scope: Regional

Contact(s):
Ralph Cinfio, CONTACT FOR MN AND IA
Phone: 320-203-0932
Larry Baesler, CONTACT FOR ND
Phone: 605-673-2396
Mike Mueller, CONTACT FOR SD
Phone: 605-644-2396
Bill Hunyadi, CONTACT FOR WI
Phone: 715-769-3559
Mike Mueller, REGIONAL DIRECTOR
Phone: 605-673-2396

ROCKY MOUNTAIN ELK FOUNDATION

NORTHEAST REGION OFFICE
198 Bennett Rd.
Julian, PA 16844 USA
Phone: 814-353-1667   Fax: 814-353-2963

Founded: NA
Scope: Regional, International

Publication(s): Bugle

Contact(s):
Dave Messics, DIRECTOR OF NOTHEAST FIELD OPERATIONS
Phone: 814-353-1667
Bruce Wojcik, MICHIGAN REGIONAL DIRECTOR
Phone: 517-668-0613
David Kelner, REGIONAL DIRECTOR FOR NY, CT, RI, MA, VT, NH, ME.
Phone: 315-633-0308
Dave Ragantesi, REGIONAL DIRECTOR OF EASTERN PA, NJ, MD, DE.
Phone: 570-756-3867
Dennis McGraw, REGIONAL DIRECTOR WESTERN PA & OH
Phone: 814-357-3799

ROCKY MOUNTAIN ELK FOUNDATION

NORTHWEST REGION OFFICE
10400 Duck Ln.
Nampa, ID 83686 USA
Phone: 208-466-0204　　　　Fax: 208-466-0204
Website: www.atalsmarema.org

Founded: NA
Membership: 135000
Scope: Regional

Publication(s): Bugle-Magazine, Wapiti-Newsletter

Contact(s):
Lloran Johnson, CONTACT FOR AK, AND WESTERN WA
Phone: 253-539-7651
Kevin Brown, CONTACT FOR EASTERN WA
Kirk Murphy, CONTACT FOR MT
Phone: 406-883-1147
Todd Bastain, CONTACT FOR OR
Art Talsma, CONTACT FOR SOUTHERN ID
Phone: 208-466-0204
Pat Cudmore, CONTACT FOR SW, MT AND EASTERN ID
Steve , EASTERN MONTANA
Art Talsma, REGIONAL DIRECTOR
Phone: 208-466-0204

ROCKY MOUNTAIN ELK FOUNDATION

SOUTH-CENTRAL REGION OFFICE
1480 8th NE, 186th St.
Holt, MO 64048 USA
Phone: 816-320-2681　　　　Fax: 816-320-2682
E-mail: tcloutier@rmef.org
Website: www.rmef.org

Founded: NA
Membership: 135000
Scope: Regional

Publication(s): WOW Magazine, Bugle Magazine

Contact(s):
Terry Cloutier, DIRECTOR OF SOUTH CENTRAL OPERATIONS
Phone: 816-320-2681
Don Blakely, REGIONAL DIRECTOR FOR IL
Phone: 618-893-4142
Wayne Bivans, REGIONAL DIRECTOR FOR INDIANA
4541 Greenthread Ct., Zionsville, IN 46077
Phone: 317-769-2163
wbivans@rmef.org
Dale Miller, REGIONAL DIRECTOR FOR MO
311 Salt Lake Circle, Napoleon, MO 64074
Phone: 816-240-2846
dmiller@rmef.org
Randy Porterfield, REGIONAL DIRECTOR FOR OK AND EAST TX
Phone: 903-677-5740

ROCKY MOUNTAIN ELK FOUNDATION

SOUTHEAST REGION OFFICE
320 Fallen Oak Cir.
Seymore, TN 37865 USA
Fax: 865-609-9190
Website: www.rmef.org

Founded: 1984
Scope: Regional

Description: International non-profit wildlife conservation organization, who's mission is to ensure the future of Elk, other wildlife and their habitat.

Contact(s):
Kevin Brown, REGIONAL DIRECTOR AL, AR, LA, MS
Phone: 877-286-3016
Dale Nolan, REGIONAL DIRECTOR GA, FL
Phone: 828-389-1692
John Mechler, REGIONAL DIRECTOR KY, TN
Phone: 865-609-9593
Tom Jones, REGIONAL DIRECTOR VA, WV, NC, SC
Phone: 540-382-0022

ROGER TORY PETERSON INSTITUTE OF NATURAL HISTORY

311 Curtis St.
Jamestown, NY 14701 USA
Phone: 716-665-2473　　　　Fax: 716-665-3794
E-mail: webmaster@rtpi.org
Website: www.rtpi.org

Founded: 1984
Membership: 2800
Scope: National

Description: The mission of the Roger Tory Peterson Institute is to create a passion for and knowledge of the natural world in the hearts and minds of children by guiding and inspiring the study of nature in our schools and communities.

Publication(s): Quarterly publication for members

Keyword(s): Environmental and Conservation Education, Environment, education, Urban and Rural Development, Nature Study

Contact(s):
Mike Lyons, DIRECTOR OF DEVELOPMENT
Phone: 716-665-2473
Fax: 716-665-3794
mike@rtpi.org
Mark Baldwin, DIRECTOR OF EDUCATION
Phone: 716-665-2473
Fax: 716-665-3794
mark@rtpi.org
Jim Berry, PRESIDENT
Phone: 716-665-2473
Fax: 716-665-3794
jim@rtpi.org

RUFFED GROUSE SOCIETY, THE

451 McCormick Rd.
Coraopolis, PA 15108 USA
Phone: 412-262-4044　　　　Fax: 412-262-9207
E-mail: rgshg@aol.com
Website: www.ruffedgrousesociety.org

Founded: 1961
Membership: 23000
Scope: National

Description: Nonprofit conservation organization dedicated to improving the environment for ruffed grouse, woodcock, and other forest wildlife through maintenance, improvement, and expansion of their habitat. Assists private, industrial, county, state, and federal landholders in forest wildlife habitat improvement programs.

Publication(s): RGS

Keyword(s): Birds, Environmental and Conservation Education, Forests and Forestry, training

Contact(s):
Samuel Pursglove, EXECUTIVE DIRECTOR PHD.
S. Mellon, EXECUTIVE VICE PRESIDENT
Ronald Burkert, GROUP DIRECTOR, ADMINISTRATION
AND INFORMATION SYSTEMS
William Goudy, GROUP DIRECTOR, FIELD SERVICES
Phone: 570-745-7313
Brenda Osborn, GROUP DIRECTOR, HEADQUARTERS
SERVICES
Paul Carson, GROUP DIRECTOR, PUBLICATIONS AND
COMMUNICATIONS
Joe Irwin, PRESIDENT
Edwin Gott, PRESIDENT
Paul Karczmarczyk, REGIONAL BIOLOGIST
Phone: 802-325-2114
rgskarz@prodigy.net
Mark Banker, REGIONAL BIOLOGIST (MID-ATLANTIC)
Phone: 616-829-4797
Rick Horton, REGIONAL BIOLOGIST (MN)
Adam Bump, REGIONAL BIOLOGIST(LOWER MI, OH, IN)
Gary Zimmer, REGIONAL BIOLOGIST(WI,UPPER MI, IA, IL)
William Klein, REGIONAL DIRECTOR (MD, OH, PA,
WASHINGTON, DC, VA, SOUTHEASTERN STATES)
Phone: 724-938-3705
Mark Fouts, REGIONAL DIRECTOR (MN, ND)
Phone: 715-399-2270
Donnie Nickerson, REGIONAL DIRECTOR (NJ, NY, NEW
ENGLAND)
Phone: 607-638-5456
Walt Noa, REGIONAL DIRECTOR (WI)
Phone: 906-384-6366
Louis George, REGIONAL SUPERVISOR OF WESTERN
STATES
Phone: 608-788-1786
Jim Abbey, RGS-CANADA EXECUTIVE DIRECTOR (ALL
CANADA)
Phone: 519-842-6027
jimabbey@oxford.net
James Jurries, SECRETARY
Dan Dessecker, SR. BIOLOGIST (ALL AREAS)
Phone: 715-234-8302
David Sandstron, TREASURER
Stephen Quill, VICE PRESIDENT

S

SAFARI CLUB INTERNATIONAL
4800 W. Gates Pass Rd.
Tucson, AZ 85745 USA
Phone: 520-620-1220   Fax: 520-622-1205
Website: www.safariclub.org

Founded: 1971
Membership: 42000
Scope: International

Description: A world-wide charitable organization of hunter-con-
servationists dedicated to the conservation of wildlife,
education of people, service to people in need and the
protection of hunters' rights. Sponsors wildlife management
research, field projects and works with national and international agencies and governments to promote conservation
programs worldwide. Operates two education facilities: the
International Wildlife Museum at headquarters and the
American Wilderness Leadership School in Jackson, WY.

Publication(s): Safari Times, Safari Times Africa, Safari Club,
Worldwide Hunting Annual, Record Book of Trophy Animals.
Safari Magazine

Keyword(s): Endangered Species, Environmental and
Conservation Education, Hunting, training, Museum

Contact(s):
Pat Johnson, CONVENTIONS DIRECTOR
Ronald Simmons, CORPORATE SECRETARY
Richard Parsons, DIRECTOR OF GOVERNMENTAL AND
CONSERVATION AFFAIRS
Donald Brown, EDUCATION DIRECTOR
George Banks, PRESIDENT
James Brown, PUBLIC RELATIONS DIRECTOR
Steve Comus, PUBLICATIONS DIRECTOR

SAFARI CLUB INTERNATIONAL
INTERNATIONAL HEADQUARTERS
4800 West Gates Pass Rd.
Tucson, AZ 85745 USA
Phone: 520-620-1220   Fax: 520-622-1205
Website: safariclub.org

Founded: 1973
Membership: 200
Scope: Statewide

Description: Goals are to educate the public, especially children,
in the realm of conservation as it relates to the hunter.
Especially concerned with environmental respect and the
preservation of any endangered species.

Contact(s):
Betty Dykstra, EDUCATION
Phone: 616-676-9704
Terry Blauwkamp, PRESIDENT
Phone: 616-669-1610
Len Vining, PRESIDENT-ELECT
Phone: 616-698-7440
Lance Norris, PROGRAM
Phone: 616-798-3410
Denny Anastor, SECRETARY AND TREASURER
Phone: 616-942-8877

SAFARI CLUB INTERNATIONAL
WASHINGTON, DC OFFICE
501 2nd St., NE
Washington, DC 20007 USA
Phone: 703-709-2293   Fax: 703-709-2296
Website: www.sci-dc.org

Founded: NA
Scope: International

Publication(s): Safari Times, Safari Magazine

SAFE ENERGY COMMUNICATION COUNCIL
1717 Massachusetts Ave., NW, Suite 106
Washington, DC 20036 USA
Phone: 202-483-8491   Fax: 202-234-9194
E-mail: safeenergy@erols.com
Website: www.safeenergy.org

Founded: 1980
Scope: National

Description: A coalition of 10 national environmental, safe energy,
and public interest media groups. SECC produces broadcast
and print ads, studies, commentaries, and graphic editorial
services to promote sustainable energy policies and respond to
nuclear industry and utility campaigns and helps groups
develop media skills.

Publication(s): MYTHBusters Series, Viewpoint, Enfacts, polls and various energy reports, Power Boosters

Keyword(s): Energy, Energy Conservation, Environment, Nuclear Energy, Renewable Resources, Training

Contact(s):
Amy Ringold, ADMINISTRATIVE DIRECTOR
aringold@erols.com
Linda Gunter, COMMUNICATIONS DIRECTOR
lcpentz@erols.com
Scott Denman, EXECUTIVE DIRECTOR
sdenman@erols.com
Andrew Schwartzman, PRESIDENT
Christopher Sherry, RESEARCH DIRECTOR
csherry@erols.com

SAN JUAN PRESERVATION TRUST, THE

Lopez Island, WA 98261 USA
Phone: 360-468-3202   Fax: 360-468-3509
E-mail: sjptrust@rockisland.com
Website: www.rockisland.com/~sjptrust/

Founded: 1979
Scope: Regional

Description: The trust is supported by voluntary contributions from members who support the preservation of wildlife, scenery, and natural heritage of the San Juan Islands of Washington state.

Publication(s): A Place in the Islands, Preserve Farmlands, Landowner's Guide, Mom's Marsh and Other Fine Places (video)

Keyword(s): Islands, training

Contact(s):
Judy Moody, PRESIDENT
Bob Dittmer, SECRETARY
Morris Dalton, TREASURER
Dale Hazen, VICE PRESIDENT

SANIBEL-CAPTIVA CONSERVATION FOUNDATION, INC.

P.O. Box 839, 3333 Sanibel-Captiva Rd.
Sanibel, FL 33957 USA
Phone: 941-472-2329   Fax: 941-472-6421
E-mail: sccf@sccf.org
Website: www.sccf.org

Founded: 1967
Membership: 2500
Scope: Regional

Description: The Sanibel-Captiva Conservation Foundation is a not-for-profit organization dedicated to the preservation of natural resources and wildlife habitat on and around Sanibel and Captiva Islands. Community programs include: Land acquisition, habitat management, landscaping for wildlife, research, education, wildlife assistance, and sea turtle conservation program. The Foundation utilizes the time and talent of over 250 dedicated volunteers.

Publication(s): Walk in the Wetlands, Bighorn, The, Conservation Update, Growing Native, Stewardship Update

Keyword(s): Environmental and Conservation Education, Islands, Land Protection, Wetlands, training, Habitat Conservation, Land Management

Contact(s):
Jean Laswell, BUSINESS MANAGER
Kristie Anders, EDUCATION DIRECTOR
Erick Lindblad, EXECUTIVE DIRECTOR
Kathy Boone, NATIVE PLANT NURSERY MANAGER
David Ceilley, RESTORATION ECOLOGIST

SASKATCHEWAN WILDLIFE FEDERATION

444 River St., W.
Moose Jaw, Saskatchewan S6H 6J6 Canada
Phone: 306-692-8812   Fax: 306-692-4370
E-mail: sask.wildlife@sk.sympatico.ca
Website: www.swf.sk.ca

Founded: 1929
Membership: 26000
Scope: Statewide

Description: Affiliated with the Canadian Wildlife Federation. A nonprofit, citizens' conservation group established for the protection and enhancement of fish and wildlife habitat. One hundred and thirty-seven local branches representing 32,000 members. Includes the Habitat Trust Fund holding title to 15,000 purchased and donated acres, and the Wildlife Tomorrow Program with 400,000 acres under free easement.

Publication(s): Outdoor Edge

Contact(s):
James Kroshus, LAND COORDINATOR
444 River St. W., Moose Jaw, Saskatchewan S6H 6J6
Phone: 306-693-9022
Sandra Dewald, OFFICE MANAGER
444 River St. W., Moose Jaw, Saskatchewan S6H 6J6
Phone: 306-692-8812
Joe Schemenauer, PRESIDENT

SAVE AMERICAS FORESTS

4 Library Ct., SE
Washington, DC 20003 USA
Phone: 202-544-9219
E-mail: info@saveamericaforest.org
Website: www.saveamericasforests.org

Founded: 1990
Scope: National

Description: A nationwide coalition of grassroots regional and national environmental groups, public interest groups, responsible businesses, and individuals working together to pass strong forest protection legislation in the U.S. Congress.

Publication(s): Save America's Forests Magazine

Keyword(s): Biodiversity, Forests and Forestry, Public Lands, Sustainable Development, Sustainable Ecosystems, National Forests

Contact(s):
Carl Ross, DIRECTOR

SAVE OUR RIVERS, INC.

P.O. Box 122
Franklin, NC 28744 USA
Phone: 828-369-7877   Fax: 828-369-7877
E-mail: rivers@dnet.net

Founded: 1990
Scope: Statewide

Description: Committed to facilitating active public involvement in decisions concerning our rivers by providing information, initiating programs, encouraging public awareness, promoting coordination of services, activities, resources and opportunities.

Publication(s): The Current

Keyword(s): Conservation, Cultural Preservation, Habitat Conservation, Litter, Watersheds, Riparian Restoration, Rivers, Runoff, Streams, Coral Reefs

Contact(s):
Peg Jones, PRESIDENT

SAVE SAN FRANCISCO BAY ASSOCIATION
1600 Broadway
Oakland, CA 94612 USA
Phone: 510-452-9261   Fax: 510-452-9266
E-mail: savebay@savesfbay.org
Website: www.savesfbay.org

Founded: 1961
Membership: 30000
Scope: Local

Description: Member-supported, non-profit environmental organization dedicated to restoring and protecting San Francisco Bay. We work for the improvement of water quality, adequate fresh water inflow and protection of the Bay's plant, wildlife, fish and human populations and their habitats. Our efforts are focused on public education, collaboration with other organizations, coalition-building, litigation, the monitoring of regulatory agencies and input into the legislative process.

Publication(s): information fact sheets, Watershed

Keyword(s): Environmental and Conservation Education, Coral Reefs, Exotic species, Aquatic nuisance species, Wetlands, training

Contact(s):
Paul Revier, EDITOR
David Lewis, EXECUTIVE DIRECTOR
Ralph Benson, PRESIDENT
Joe Engbeck, VICE PRESIDENT

SAVE THE BAY, INC.
434 Smith St.
Providence, RI 02908-3770 USA
Phone: 401-272-3540   Fax: 401-273-7153
E-mail: savebay@savebay.org
Website: www.savebay.org

Founded: 1970
Membership: 25000
Scope: Statewide

Description: Save The Bay is dedicated to restoring and protecting Narragansett Bay—a designated estuary of national significance. As a nonprofit, member-supported environmental organization, Save The Bay works to ensure that the environmental quality of Narragansett Bay and its watershed is restored and protected from the harmful effects of human activity.

Publication(s): Backyards on the Bay: A Yard Care Guide for the Coastal Home Owner, Bay Bulletin, The Uncommon Guide to common life of Narragansett Bay, Coastal Property and Landscape Management Guidebook

Keyword(s): Watersheds, Water Pollution Management

Contact(s):
Christopher Hamblett, DIRECTOR OF ADVOCACY

Christy Law-Blanchard, EDITOR / COMMUNICATIONS DIRECTOR
H. Spalding, EXECUTIVE DIRECTOR
Timothy Burns, PRESIDENT

SAVE THE DUNES CONSERVATION FUND
SAVE THE DUNES CONSERVATION FUND
444 Barker Rd.
Michigan City, IN 46360 USA
Phone: 219-879-3564   Fax: 219-872-4875
Website: www.savedunes.org

Founded: NA
Membership: 1
Scope: Local

Description: Educational and non-lobbying 501(c)(3) fund to support the goals of the Save the Dunes Council.

Publication(s): Save the Dunes (newsletter)

Keyword(s): Watersheds, Environmental and Conservation Education, Research, Streams

Contact(s):
Thomas Anderson, EXECUTIVE DIRECTOR
Sandra Wilmore, FUND DIRECTOR

SAVE THE DUNES COUNCIL
444 Barker Rd.
Michigan City, IN 46360 USA
Phone: 219-879-3937   Fax: 219-872-4875
E-mail: std@savedunes.org
Website: www.savedunes.org

Founded: 1952
Membership: 500
Scope: National

Description: Dedicated to the preservation of the Indiana Dunes National Lakeshore for public use and enjoyment. Concerned with protecting the ecological values of the dunes region, preserving Lake Michigan, and combating air, water, and hazardous waste pollution. Established by Dorothy Buell.

Publication(s): Newsletter

Keyword(s): Air Quality and Pollution, Land Preservation, Outdoor Recreation, Public Lands, Coral Reefs

Contact(s):
Dorothy Potucek, 1ST VICE PRESIDENT
1608 Parkview Ave., Whiting, IN 46394
Christine Livingston, ADMINISTRATIVE ASSISTANT
Charlotte Read, ASSISTANT DIRECTOR
Thomas Anderson, EXECUTIVE DIRECTOR
Thomas Serynek, PRESIDENT
1000 N. Warrick, Gary, IN 46403
Phone: 219-938-5410
Sandra Wilmore, PROGRAM DIRECTOR
Mark Mihalo, TREASURER
8 Diana Road, Ogden Dunes, Portage, IN 46368
Phone: 219-763-4871

SAVE THE HARBOR/SAVE THE BAY
59 Temple Pl., Suite 304
Boston, MA 02111 USA
Phone: 617-451-2860   Fax: 617-451-0496
E-mail: wolfe@savetheharbor.org
Website: www.savetheharbor.org

Founded: 1986

Membership: 2000
Scope: Statewide

Description: Save the Harbor/Save the Bay is a nonprofit organization whose mission is to foster a positive vision of Boston Harbor and Massachusetts Bay, and to build a broad-based constituency to promote the restoration and protection of these valuable resources. Services include narrated boat tours of Boston Harbor, discussions of harbor pollution, cleanup projects, history, celebratory events, summer youth program, and a Baywatch Program.

Publication(s): Splash (newsletter)

Keyword(s): Environment, Coral Reefs, Exotic species, Aquatic nuisance species, Water Quality, Watersheds

Contact(s):
Bruce Berman, BAYWATCH & COMMUNICATIONS DIRECTOR
Beth Nicholson, CHAIRPERSON & PRES.
Matt Wolfe, EVENTS COORDINATOR
Patricia Foley, EXEC. DIRECTOR
Lisa Mantoni, PROGRAM COORDINATOR
mantoni@savetheharbor.org

SAVE THE MANATEE CLUB

500 N. Maitland Ave.
Maitland, FL 32751 USA
Phone: 407-539-0990   Fax: 407-539-0871
E-mail: education@savethemanatee.org
Website: www.savethemanatee.org

Founded: 1981
Membership: 40000
Scope: International

Description: A national nonprofit organization founded by Governor Bob Graham and singer and songwriter Jimmy Buffett. Objectives are public awareness and education; funding research, rescue, rehabilitation and advocacy and appropriate legal action for the endangered West Indian manatee and its habitat. Funded primarily by the club's Adopt-A-Manatee program.

Keyword(s): Endangered Species, Environmental and Conservation Education, Marine Mammals, Whale, Dolphin, Seal, Mammals

Contact(s):
Jimmy Buffett, CO-CHAIRMAN
Helen Spivey, CO-CHAIRMAN
Nancy Sadusky, COMMUNICATIONS DIRECTOR
Judith Vallee, EXECUTIVE DIRECTOR

SAVE THE SOUND, INC.

185 Magee Ave.
Stamford, CT 06902 USA
Phone: 203-327-9786   Fax: 203-967-2677
E-mail: savethesound@savethesound.or
Website: www.savethesound.org

Founded: 1972
Scope: Regional National

Description: Save the Sound, Inc. is devoted to protecting, restoring, and appreciating Long Island Sound and its watershed. With a staff of nine full-time employees and additional seasonal employees, STS operates year-round programs in education, research, and advocacy, including Sea Camp, Soundshore Ecology, water quality monitoring, habitat restoration, beach cleanups, and an extensive library.

Publication(s): Water Quality Monitoring: A Guide for Concerned Citizens, Save the Sound News (Quarterly Newsletter), Harbor Watch Report (1986-1992), The Sound Connection Curriculum Guide, Long Island Sound Municipal Report Cards

Keyword(s): Coasts, Environmental and Conservation Education, Protected Areas, Coral Reefs, Wetlands, Restoration

Contact(s):
Bill Jessup, CHAIRMAN
John Atkin, PRESIDENT
Larry Mcgaughey, SECRETARY
William Jessup, TREASURER
Larry Bingaman, VICE CHAIRMAN

SAVE THE SOUND, INC. AT GARVIES POINT MUSEUM

50 Barry Dr.
Glen Cove, NY 11542 USA
Phone: 516-759-2165   Fax: 516-759-0644
E-mail: savethesound@savethesound.org
Website: www.savethesound,org

Founded: NA
Membership: 5000
Scope: Regional
Publication(s): Sound Bites, Long Island Sound Conservation Blueprint, Sound Connection

Contact(s):
John Atkins, PRESIDENT

SAVE WETLANDS AND BAYS

24353 Thorneby Trace
Millsboro, DE 19966 USA
Phone: 302-945-1317   Fax: 302-945-1317

Founded: 1989
Membership: 220
Scope: Local

Description: To protect Delaware's inland bays and fringing marshes from perceived threats.

Contact(s):
Til Purnell, EXECUTIVE DIRECTOR
purnell@ce.net

SAVE-THE-REDWOODS LEAGUE

114 Sansome St., Suite 1200
San Francisco, CA 94104 USA
Phone: 415-362-2352   Fax: 415-362-7017
E-mail: info@savetheredwoods.org
Website: www.savetheredwoods.org

Founded: 1918
Membership: 4500
Scope: National

Description: Established to rescue from destruction representative areas of our primeval forests; to cooperate with the California Department of Parks and Recreation, the National Park Service, and other agencies in establishing redwood parks and other parks and reservations; to purchase redwood groves by private subscription; to support reforestation and conservation of our forest areas.

Publication(s): Trees, Shrubs and Flowers of the Redwood Region, Redwood Forest, Redwoods of the Past, Story Told by a Fallen Redwood, California Redwood Parks and Preserves

Keyword(s): Forests and Forestry, Land Purchase, Public Lands

Contact(s):
Bruce Howard, CHAIRMAN OF THE BOARD
Meg Reilly, GENERAL COUNSEL
Richard Otter, PRESIDENT
Kate Anderton, SECRETARY AND EXECUTIVE DIRECTOR
Frank Westworth, TREASURER
P.O. Box 44614, San Francisco, CA 94144-0001

SCENIC AMERICA
801 Pennsylvania Ave., SE, Suite 300
Washington, DC 20003 USA
Phone: 202-543-6200   Fax: 202-543-9130
E-mail: scenic@scenic.org
Website: www.scenic.org

Founded: 1978
Membership: 7
Scope: National

Description: National membership organization dedicated to preserving and enhancing the scenic character of America's communities and countryside. Provides information and technical assistance on billboard and sign control, scenic byways, tree preservation, highway design, cellular tower siting, and other scenic conservation issues.

Publication(s): The Grassroots Advocate, Fighting Billboard Blight: An Action Guide for Citizens and Elected Officials, series of videos and technical bulletins, Viewpoints

Keyword(s): Environmental and Conservation Education, Land Preservation, Land Use Planning, Landscape Architecture, Urban Environment, education

Contact(s):
Kathryn Whitmire, CHAIRMAN
Meg Maguire, PRESIDENT

SCENIC AMERICA
SCENIC CALIFORNIA
c/o LSA Associates, 2215 Fifth Street
Berkeley, CA 94710 USA
Phone: 510-540-7331
E-mail: sceniccalifornia@earthlink.net
Website: http://www.sceniccalifornia.org

Founded: 1998
Scope: Statewide

Description: Scenic California is dedicated to protecting natural beauty in the environment, preserving and enhancing landscapes and streetscapes, protecting historical and cultural resources, promoting the enhancement of scenic approaches and settings of cities and towns, improving community appearance and fostering the establishment and preservation of scenic roads and viewsheds.

Contact(s):
Sheila Brady, MANAGER

SCENIC AMERICA
SCENIC MICHIGAN
445 E. Mitchell St.
Petoskey, MI 49770 USA
Phone: 231-347-1171   Fax: 231-347-1185
E-mail: info@scenicmichigan.org
Website: www.mucc.org/scenic

Founded: NA

Scope: Statewide

Description: Scenic Michigan started under the aegis of Michigan United Conservation Clubs, the largest nonprofit conservation organization in the United States, as a billboard control task force in 1989. The mission of Scenic Michigan is to protect and enhance the appearance and scenic character of Michigan's communities and countryside.

Publication(s): Scenic Michigan (Newsletter)

Contact(s):
Debbie Rohe, PRESIDENT
Bethany Goodman, SECRETARY
Mary Tanton, TREASURER
Pam Frucci, VICE PRESIDENT

SCENIC HUDSON, INC.
9 Vassar St.
Poughkeepsie, NY 12601 USA
Phone: 845-473-4440   Fax: 845-473-2648
E-mail: info@scenichudson.org
Website: www.scenichudson.org

Founded: 1963
Scope: Statewide

Description: A nonprofit conservation and environmental organization dedicated to protecting and enhancing the scenic, natural, recreational and historic treasures of the Hudson River Valley. Speakers' Bureau available to the public.

Publication(s): Signs of the Times (Creative Ideas for Signage), Dealing with Airport Growth, Cooling Tower Report, Understanding Traffic and its Impacts, PCB Dredging Report, Cell Tower Report, Adventure Guide, Scenic Hudson News

Keyword(s): Air Quality and Pollution, Protected Areas, Land Preservation, Rivers, Transportation, education, Sprawl, Land Protection, Open Space

Contact(s):
Marjorie Hart, CHAIRMAN
Michelle Terwilliger, EDITOR
Mona Burkhart, EXECUTIVE ASSISTANT
Edward Sullivan, EXECUTIVE DIRECTOR
Rudolph Rauch, SECRETARY
Anne Impellizzeri, TREASURER

SCENIC MISSOURI
SCENIC MISSOURI
5650A South Sinclair
Columbia, MO 65203 USA
Phone: 573-446-3129   Fax: 573-443-3748
E-mail: scenicmo@tranquility.net
Website: www.scenicmissouri.org

Founded: 1993
Membership: 1000
Scope: Statewide

Description: Scenic Missouri was founded because of a growing concern about the loss of Missouri's scenic heritage. Its mission is to preserve and enhance the scenic beauty of Missouri.

Publication(s): Scenic Views

Contact(s):
Karl Kruse, EXECUTIVE DIRECTOR

SCENIC NORTH CAROLINA INC.

P.O. Box 628
Raleigh, NC 27602 USA
Phone: 919-832-3687   Fax: 919-832-3299
E-mail: scenic.nc@worldnet.att.net
Website: www.scenicnc.org

Founded: NA
Scope: Statewide

Description: Our sense of identity as North Carolinians is tied to special places and buildings and views. Each type of landscape has its own kind of beauty, character and uniqueness. Scenic North Carolina is dedicated to preserving and enhancing scenic resources and community appearance in North Carolina.

Publication(s): Scenic North Carolina News

Contact(s):
Dale McKeel, EXECUTIVE DIRECTOR

SCENIC TEXAS, INC

SCENIC TEXAS
3015 Richmond Suite 220
Houston, TX 77098 USA
Phone: 713-629-0481   Fax: 713-629-0485
E-mail: scenic@scenictexas.org
Website: www.scenictexas.org

Founded: NA
Membership: 4000
Scope: Statewide

Description: The mission of Scenic Texas is to preserve and enhance the scenic character of the visual environment. Scenic Texas has chapters in Houston, Austin, San Antonio, Fort Worth, and Galveston.

Publication(s): Scenic Views

Contact(s):
Cece Fowler, EXECUTIVE DIRECTOR

SCIENTISTS CENTER FOR ANIMAL WELFARE

7833 Walker Dr., Suite 410
Greenbelt, MD 20770 USA
Phone: 301-345-3500   Fax: 301-345-3503
E-mail: info@scaw.com
Website: www.scaw.com

Founded: NA
Membership: 900
Scope: National, International

Description: A nonprofit educational organization that promotes the belief that high standards of animal welfare complement the quality of scientific results. SCAW publishes educational material about current issues of animal use in research, testing, and teaching.

Publication(s): see publications on website, SCAW (Newsletter)

Keyword(s): Agriculture, Aquariums, Wildlife Rehabilitation, Biotechnology, Birds, Environmental Ethics, Endangered Species, Wildlife, Health and Nutrition, Mammals, Nongame Wildlife, Reptiles and Amphibians, Marine Mammals, Whale, Dolphin, Seal, Public Health Protection, Research

Contact(s):
Lee Krulisch, EXECUTIVE DIRECTOR

SEA SHEPHERD CONSERVATION SOCIETY

P.O. Box 2616
Friday Harbor, WA 98250 USA
Phone: 360-370-5500   Fax: 360-370-5506
Website: www.seashepherd.org

Founded: 1977
Scope: International

Description: An international direct action marine mammal conservation organization involved in stopping marine mammal slaughters. Special projects include: campaigns against drift net fishing, whaling, the faeroese pilot whale slaughter, and sealing. The Society owns and operates three ships, the Ocean Warrior, the submarine Mirage and the Edward Abbey.

Publication(s): Sea Shepherd Log Quarterly

Keyword(s): Endangered Species, Environmental and Conservation Education, Wildlife, Marine Mammals, Whale, Dolphin, Seal

Contact(s):
Paul Watson, EDITOR
Paul Watson, FOUNDER
Andrew Christie, INFORMATION DIRECTOR
andrew@seashepherd.org

SEA SHEPHERD CONSERVATION SOCIETY

USA OFFICE
22774 Pacific Coast Hwy.
Malibu, CA 90265 USA
Phone: 310-456-1141   Fax: 310-456-2488
E-mail: seashepherd@seashpherd.org
Website: www.seashepherd.org

Founded: 1979
Membership: 4000+
Scope: National, International

Description: The Sea Shepherd Conservation Society has been actively engaged in marine wildlife conservation for over 20 years. A non-profit organization SSCS is involved with the investigation and documentation of violations of international laws, regulations and treaties that maintain global biodiversity through the protection of marine wildlife species, and utilizes many innovative approaches to marine conservation. Sea Shepherd publishes a quarterly newsletter which contains reports and updates of SSCS conservation activities, educational information, and updates on current laws and legistlations.

Publication(s): The Log

Contact(s):
Angela Navarro, ACCOUNTS MANAGER
Jose Fernandez, EXECUTIVE DIRECTOR
Capt. Paul Watson, FOUNDER AND PRESIDENT
Andrew Christie, INFORMATION DIRECTOR

SEA SHEPHERD CONSERVATION SOCIETY

CANADA OFFICE
P.O. Box 2670
Malibu, CA V7X 1A2 Canada
Phone: 310-456-1141   Fax: 310-456-2248
E-mail: seashepherd@cshepherd.org
Website: www.seashepherd.org

Founded: NA
Membership: 70,000
Scope: International

Publication(s): Earth Force, Captain Log, Annual Report, Ocean Warrior

Contact(s):
Pepper Fernandez, EXECUTIVE DIRECTOR

SEACAMP ASSOCIATION, INC.
1300 Big Pine Ave.
Big Pine Key, FL 33043-3336 USA
Phone: 305-872-2331　　　Fax: 305-872-2555
E-mail: info@seacamp.org
Website: www.seacamp.org

Founded: 1964
Scope: International

Description: Non-profit organization encompassing two marine education organizations in the Florida Keys, a summer camp and the school program Newfound Harbor Marine Institute (NHMI). Strong international program with Russia. Member NAAEE.

Keyword(s): Camp, Inquiry Based Education, Environmental and Conservation Education, Mangrove Habitats, Marine Conservation, Nature Study, Outdoor Education, Outdoor Recreation, Ocean Conservation, Internships

Contact(s):
Grace Upshaw, CAMP DIRECTOR
Mary Hensel, DEVELOPMENT DIRECTOR
Elena Istoma, DIRECTOR OF INTERNATIONAL PROGRAMS
Russel Bachert, DIRECTOR OF SPECIAL PROJECTS
Irene Hooper, EXECUTIVE DIRECTOR
Chuck Brand, INSTITUTE DIRECTOR, NHMI
John Booker, PROGRAM DIRECTOR

SEACOAST ANTI-POLLUTION LEAGUE
P.O. Box 1136
Portsmouth, NH 03802 USA
Phone: 603-431-5089

Founded: 1969
Scope: Statewide

Description: To promote the wise use of natural resources of the seacoast region, and to alert and educate the community and relevant government agencies of threats to the environment. SAPL works to prevent ecological, economic and public health damage from the Seabrook nuclear reactor, the Portsmouth Naval Shipyard and over-development.

Keyword(s): Nuclear/Radiation, Development

Contact(s):
Steve Haberman, EXECUTIVE DIRECTOR
Davie Hills, PRESIDENT
Johanna Lyons, SECRETARY
Peter Vandermark, TAG COORDINATOR
Jim Horrigan, TREASURER
Mary Metcalf, VICE PRESIDENT

SEAPLANE PILOTS ASSOCIATION
4315 Highland Park Blvd. Suite C
Lakeland, FL 33813 USA
Phone: 301-695-2083
Website: www.seaplanes.org

Founded: 1972
Membership: 7500+
Scope: National

Description: A unit formed to provide seaplane services to agencies and environmental groups involved in forest fire detection, search and rescue, wildlife surveys, pollution patrols, and other related environmental and ecological projects.

Publication(s): Water Flying Annual, Water Landing Directory, Water Flying

Keyword(s): Coasts, Lakes, Outdoor Recreation, Rivers, Exotic species, Aquatic nuisance species

Contact(s):
Michael Volk, EXECUTIVE DIRECTOR
J. Frey, PRESIDENT
Jerry Potter, SECRETARY
Walter Windus, VICE PRESIDENT

SHELBURNE FARMS
1611 Harbor Rd.
Shelburne, VT 05482 USA
Phone: 802-985-8686　　　Fax: 802-985-8123
Website: www.shelburnefarms.org/

Founded: NA
Scope: National

Description: Shelburne Farms is a 1,400 acre working farm, national historic site and non-profit environmental education center. Our mission is to cultivate a conservation ethic by teaching and demonstrating stewardship of our natural and agricultural resources.

Publication(s): Project Seasons, This Lake Alive

Keyword(s): Agriculture, Environmental and Conservation Education, Environmental Education Curriculum, Forest Stewardship, Historic Preservation, Land Protection, Natural Systems, Stewardship, Sustainable Agriculture, Wetlands

Contact(s):
Alexander Webb, PRESIDENT
awebb@shelburnefarms.org
Linda Wellings, SCHOOL PROGRAMS DIRECTOR
jelson@shelburnefarms.org
Megan Camp, VP AND PROGRAM DIRECTOR
mcamp@shelburnefarms.org

SIERRA CLUB
85 2nd St.
San Francisco, CA 94105-3459 USA
Phone: 415-977-5500　　　Fax: 415-977-5799
E-mail: information@sierraclub.org
Website: www.sierraclub.org/

Founded: 1892
Membership: 700000
Scope: National

Description: To explore, enjoy, and protect the wild places of the earth; to practice and promote the responsible use of the earth's ecosystems and resources; to educate and enlist humanity to protect and restore the quality of the natural and human environment; and to use all lawful means to carry out these objectives. With 65 chapters and 396 groups in North America, the Club's nonprofit program work includes legislation, litigation, public information, publishing, wilderness outings, and conferences. Founded by John Muir.

Publication(s): Planet, The, Chapter and group newsletters, Sierra

Keyword(s): Air Quality and Pollution, Energy, Public Lands, Toxic Substances, Nuclear-free, Water quantity, Water export and diversion

Contact(s):

Bruce Hamilton, ASSOCIATE EXECUTIVE DIRECTOR OF CONSERVATION AND COMMUNICATIONS
Kim Haddow, COMMUNICATION DIRECTOR
Bob Bingaman, DIRECTOR OF CONSERVATION FIELD SERVICES
Joan Hamilton, EDITOR-IN-CHIEF OF SIERRA
Carl Pope, EXECUTIVE DIRECTOR
Jennifer Ferenstein, PRESIDENT
jennifer.ferenstein@sierraclub.org
Helen Sweetland, PUBLISHER OF BOOKS
Gene Coan, SENIOR ADVISOR TO THE EXECUTIVE DIRECTOR
Charlie Ogle, VICE PRESIDENT

SIERRA CLUB

54 Chicon St.
Austin, TX 78702-5461 USA
Phone: 512-472-9094   Fax: 512-472-8710
E-mail: txar@earthlink.net
Website: www.sierraclub.org

Founded: NA
Membership: 2400
Scope: Regional

Contact(s):

Ayelet Hines, END COMMERCIAL LOGGING ON PUBLIC LAND ORGANIZER
Nicole Holt, GLOBAL WARMING & ENERGY PROGRAM ORGANIZER
Larry Freilich, REPRESENTATIVE

SIERRA CLUB

ALASKA OFFICE
201 Barrow St.
Anchorage, AK 99501 USA
Phone: 907-276-4048   Fax: 907-258-6807
E-mail: nw-ak.field@sierraclub.org
Website: www.sierraclubalaska.org

Founded: NA
Membership: 1600+
Scope: Regional

Contact(s):

Sara Callaghan-Chapell, ALASKA REPRESENTATIVE
Kevin Myers, CONSERVATION ORGANIZER
Jack Hession, SENIOR ALASKA SPECIALIST

SIERRA CLUB

ALASKA RAINFOREST FIELD OFFICE
201 Barrow St. Field Office
Anchorage, AK 99501 USA
Phone: 907-276-4048   Fax: 907-258-6807
E-mail: nw-ak.field@sierraclub.org

Founded: NA
Scope: Regional

Keyword(s): Rainforests

SIERRA CLUB

APPALACHIAN FIELD OFFICE
200 N. Glebe Rd., Suite 905
Arlington, VA 22203-3728 USA
Phone: 703-312-0533   Fax: 703-312-0508
E-mail: ap-va.field@sierraclub.org
Website: www.sierraclub.org

Founded: NA
Scope: Regional

Description: DC, DE, GA, MD, NC, SC, TN, VA, WV
Contact(s):

Joy Oakes, APPALACHIAN FIELD DIRECTOR

SIERRA CLUB

ATLANTIC COAST OFFICE
P.O. Box 160
Nassau, DE 19969 USA
Phone: 902-422-5091   Fax: 302-644-9712
Website: www.sierraclub.org

Founded: NA
Scope: Regional

SIERRA CLUB

CALIFORNIA/NEVADA/HAWAII OFFICE AND CALIFORNIA LEGISLATIVE OFFICE
1414 K St., Suite 300
Sacramento, CA 95814-3929 USA
Phone: 916-557-1100   Fax: 916-557-9669

Founded: NA
Membership: 650
Scope: Regional

Contact(s):

Barbara Boyle, STAFF DIRECTOR
Bill Craven, STATE DIRECTOR

SIERRA CLUB

CLEVELAND OFFICE
2460 Fairmont Blvd., Suite C
Cleveland, OH 44106-3125 USA
Phone: 216-791-9110   Fax: 216-791-9138
E-mail: mw-oh.field@sierraclub.org
Website: www.sierraclub.org

Founded: NA
Membership: 700,000
Scope: National

Contact(s):

Glenn Landers, CLEAN AIR SPECIALIST

SIERRA CLUB

COLORADO FIELD OFFICE
2260 Baseline Rd., Suite 105
Boulder, CO 80302-7737 USA
Phone: 303-449-5595   Fax: 303-449-6520
E-mail: sw-co.field@sierraclub.org
Website: www.sierraclub.org

Founded: NA
Membership: 19000
Scope: Regional

SIERRA CLUB
COLUMBIA BASIN OFFICE
85 Second St., Second Floor
San Francisco, CA 94105 USA
Phone: 415-977-5500   Fax: 41-597-7579
E-mail: nw-cb.field@sierraclub.org
Website: sierraclub.org

Founded: NA
Scope: Regional

SIERRA CLUB
EASTERN CANADA CHAPTER
24 Mercer St
Toronto, Ontario M5V 1H3 Canada
Phone: 416-960-9606   Fax: 416-960-0020
E-mail: eastern.canada.chapter@sierraclub.org
Website: http;//eastern.sierraclub.ca

Founded: NA
Membership: 1000
Scope: Regional
Publication(s): Sierra, Newsletters

Contact(s):
    Kim Neill, CHAPTER COORDINATOR
    Phone: 416-960-9606

SIERRA CLUB
FLORIDA FIELD OFFICE
475 Central Ave., Suite M-1 (Florida Field Office)
St. Petersburg, FL 33701 USA
Phone: 727-824-8813   Fax: 727-824-0936
Website: www.sierraclub.org

Founded: NA
Membership: 228000
Scope: Regional

SIERRA CLUB
FLORIDA-MIAMI FIELD OFFICE
2700 SW 3rd Ave.,
Suite 2F
Miami, FL 33129 USA
Phone: 305-860-9888   Fax: 305-860-9862
E-mail: information@sierraclub.org
Website: www.seirraclub.org

Founded: NA
Membership: 700000
Scope: National, Local, Regional

SIERRA CLUB
GEORGIA FIELD OFFICE/LOUISIANA AND ALABAMA
FIELD OFFICE
1447 Peachtree St., NE, Suite 305
Atlanta, GA 30309-3034 USA
Phone: 404-888-9778   Fax: 404-876-5260
E-mail: ap-ga.field@sierraclub.org

Founded: NA
Scope: Regional

SIERRA CLUB
MONTANA FIELD OFFICE
P.O. Box 1290
Bozeman, MT 59771 USA
Phone: 406-582-1281   Fax: 406-582-9417
E-mail: wildgriz@aol.com
Website: www.sierraclub.org

Founded: NA
Membership: 350
Scope: Regional
Publication(s): Newsletter

Contact(s):
    Kathryn Hohmann, SR. REGIONAL REPRESENTATIVE

SIERRA CLUB
NEW YORK CITY OFFICE
116 John St., 31st Fl.
New York, NY 10038 USA
Phone: 212-791-9291   Fax: 212-791-0839
E-mail: ne-nyc.field@sierraclub.org
Website: www.sierraclub.org

Founded: NA
Scope: Regional

Contact(s):
    Emma McGregor, ADMINISTRATIVE ASSISTANT

SIERRA CLUB
NORTHEAST REGION OFFICE
85 Washington St.
Saratoga Springs, NY 12866 USA
Phone: 518-587-9166   Fax: 518-583-9062
E-mail: ne-ny.field@sierraclub.org
Website: www.sierraclub.org

Founded: NA
Membership: 600,000
Scope: Regional

Description: CT, MA, ME, NH, NJ, NY, PA, RI, VT

Publication(s): Magazine- Sierra

Contact(s):
    Baret Pinyoun, ASSOCIATE REGIONAL REPRESENTATIVE
    Mark Bettinger, STAFF DIRECTOR

SIERRA CLUB
NORTHERN PLAINS REGION
23 N. Scott St., Suite 27
Sheridan, WY 82801 USA
Phone: 307-672-0425   Fax: 307-674-6187
E-mail: np-wy.field@sierraclub.org
Website: www.sierraclub.org

Founded: NA
Membership: 9800
Scope: Regional

Description: KS, MT, NE, ND, SD, WY

Publication(s): National Sierra newsletter

Contact(s):
    Liz Howell, CONSERVATION ORGANIZER
    Steve Thomas, DEPUTY FIELD DIRECTOR
    larry.mehlhaff@sierra.org
    Kirk Koepsel, REGIONAL REP
    kirk.koepsel@sierra.org

SIERRA CLUB
NORTHERN ROCKIES CHAPTER
(IDAHO/WASHINGTON)
P.O. Box 552
Boise, ID 83701-0552 USA
Phone: 208-384-1023   Fax: 208-384-0239
E-mail: northern.rockies.chapter@sierraclub.org
Website: www.sierraclub.org/chapters/id/

Founded: NA
Membership: 4500
Scope: Regional

Contact(s):
Roger Singer, CHAPTER DIRECTOR
roger.singer@sierraclub.org

SIERRA CLUB
NORTHWEST OFFICE
180 Nickerson Ave. Suite 207
Seattle, WA 98109 USA
Phone: 206-378-0114   Fax: 206-378-0034
E-mail: nw-wa.field@sierraclub.org
Website: www.sierraclub.org

Founded: NA
Membership: 500,000
Scope: Regional

Description: AK, ID, OR, WA

Publication(s): Sierra Magazine

Contact(s):
Jim Young, ASSOCIATE REPRESENTATIVE
Bill Arthur, STAFF DIRECTOR

SIERRA CLUB
OZARK CHAPTER (MISSOURI)
1007 North College Ave., Suite 1
Columbia, MO 65201-4725 USA
Phone: 573-815-9250   Fax: 573-442-7051
E-mail: ozark.chapter@sierraclub.org
Website: www.sierraclub.org/chapters/mo/

Founded: NA
Membership: 4
Scope: State
Publication(s): Ozark Sierran, 3 other local group publications

Contact(s):
Ken Midkiff, DIRECTOR OF THE MISSOURI SIERRA CLUB
Phone: 573-815-9250
ozarj.chapter@sierraclub.org

SIERRA CLUB
ROCKY MOUNTAIN CHAPTER (COLORADO)
1410 Grant St., Suite B303
Denver, CO 80203-1848 USA
Phone: 303-861-8819   Fax: 303-861-2436
Website: www.rmc.sierraclub.org/

Founded: NA
Membership: 20000
Scope: State
Publication(s): The Peak & Prairie - bi-monthly newsletter, The Sierra - bimonthly magazine

Contact(s):
Susan Lefever, PROGRAM DIRECTOR
slefever@rmi.net

SIERRA CLUB
SAN GORGONIO CHAPTER (SOUTHERN CALIFORNIA)
4079 Mission Inn Ave.
Riverside, CA 92501-3204 USA
Phone: 909-684-6203   Fax: 909-684-6172
Website: http://sangorgonio.sierraclub.org/

Founded: NA
Scope: Local, Regional

SIERRA CLUB
SOUTHERN CALIFORNIA/NEVADA FIELD OFFICE
3435 Wilshire Blvd., Suite 320
Los Angeles, CA 90010 USA
Phone: 213-387-6528   Fax: 213-387-5383
E-mail: ca-sc.field@sierraclub.org
Website: www.sierraclub.org/field/southerncal/

Founded: NA
Scope: Regional

Publication(s): Publications on website

Contact(s):
Jim Blomquist, SR. REGIONAL REPRESENTATIVE

SIERRA CLUB
SOUTHWEST OFFICE
812 N. 3rd St.
Phoenix, AZ 85004 USA
Phone: 602-254-9330   Fax: 602-258-6533
E-mail: sw-az.field@sierraclub.org
Website: www.sierraclub.org

Founded: NA
Membership: 50,000
Scope: Regional

Description: AZ, CO, NM, OK, UT

Contact(s):
Philip Church, ADMINISTRATIVE COORDINATOR, SOUTHWEST REGION
Rob Smith, STAFF DIRECTOR

SIERRA CLUB
UTAH FIELD OFFICE
2273 S. Highland Dr., Suite 2-D
Salt Lake City, UT 84106-2832 USA
Phone: 801-467-9294   Fax: 801-467-9296
E-mail: sw-ut.field@sierraclub.org
Website: www.sierraclub.org

Founded: NA
Membership: 5000
Scope: Regional

SIERRA CLUB
WASHINGTON, DC OFFICE
408 C St., NE
Washington, DC 20002 USA
Phone: 202-547-1141   Fax: 202-547-6009
Website: www.sierraclub.org

Founded: NA
Membership: 600,000
Scope: National
Publication(s): Sierra Magazine

Contact(s):
Bob Bingaman, FIELD DIRECTOR
Debbie Sease, LEGISLATIVE DIRECTOR

SIERRA CLUB
WYOMING CHAPTER
Jackson, WY 83001-0263 USA
Phone: 307-734-0441   Fax: 307-734-0443
E-mail: wyoming.chapter@sierraclub.org
Website: www.sierraclub.org/chapters/wy/

Founded: NA
Membership: 750
Scope: State
Contact(s):
Liz Howell, WYOMING CHAPTER STAFF
Phone: 307-672-0425
liz.howell@sierraclub.org

SIERRA CLUB BAY AREA FIELD OFFICE
CALIFORNIA, NEVADA, HAWAII FIELD OFFICE
827 Broadway
Oakland, CA 94607 USA
Phone: 510-622-0290   Fax: 510-622-0278
E-mail: ca-oa.field@sierraclub.org
Website: www.california.sierraclub.org

Founded: NA
Scope: Regional

SIERRA CLUB
BRITISH COLUMBIA CHAPTER
576 Johnson St.
Victoria, British Columbia V8W 1M3 Canada
Phone: 250-386-5255   Fax: 250-386-4453
E-mail: info@sierraclubbc.org
Website: www.sierraclub.ca/bc/

Founded: NA
Scope: Regional

Publication(s): The Sierra Report - magazine
Contact(s):
Bill Warham, EXECUTIVE DIRECTOR

SIERRA CLUB FOUNDATION, THE
85 Second St., Suite 750
San Francisco, CA 94105 USA
Phone: 415-995-1780   Fax: 415-995-1791
E-mail: sierraclub.foundation@sierraclub.org
Website: www.tscf.org

Founded: 1960
Scope: National

Description: A nonprofit, tax-deductible, public foundation established to finance the educational, literary, and scientific projects of citizen-based groups working on national and international environmental problems. Manages assets in excess of $25 million and over 600 regional or special interest funds principally for charitable conservation purposes. Also manages charitable remainder unitrusts and a pooled income fund with assets over $7.5 million.

Keyword(s): Environmental Law, Environmental and Conservation Education, Land Use Planning, Public Lands, Urban Environment

Contact(s):
Iqbal Parupia, CONTROLLER
iqbal.parupia.sierraclub.org
Mary McGarrahan, DIRECTOR OF ADMINISTRATION
Nancy Thomas, DIRECTOR OF FINANCE
nancy.thomas@sierraclub.org
John Decook, EXECUTIVE DIRECTOR
john.decook@sierraclub.org
Michael Loeb, FIFTH OFFICER
michael.loeb@sierraclub.org
Andrea Manion, GRANTS MANAGER
Michael Loeb, PRESIDENT
Roger Hershey, SECRETARY
Richard Cellarius, TREASURER
richard.cellarius@sierraclub.org
Marlene Fluharty, VICE PRESIDENT
marlene.fluharty@sierraclub.org

SIERRA CLUB JOHN MUIR CHAPTER
JOHN MUIR CHAPTER
222 S. Hamilton St., Suite 1
Madison, WI 53703-3201 USA
Phone: 608-256-0565   Fax: 608-256-4562
E-mail: john.muir.chapter@sierraclub.org
Website: www.sierraclub.org/chapters/wi/

Founded: NA
Membership: 12000
Scope: Regional
Publication(s): The Muir View

SIERRA CLUB MAINE CHAPTER
MAINE CHAPTER
One Pleasant St.
Portland, ME 04101-3936 USA
Phone: 207-761-5616   Fax: 207-773-6690
E-mail: maine.sierra@prodigy.net
Website: www.sierraclub.org/chapters/me/

Founded: NA
Membership: 3000
Scope: Statewide
Publication(s): Mainely Sierran Bimonthly Newsletter

Contact(s):
Karen Woodsum, MAINE WOODS REGIONAL
Phone: 207-791-2821
maine.woods@prodigy.net

SIERRA CLUB OF CANADA
#1 Nicholas St., Suite 412
Ottawa, Ontario K1N 7B7 Canada
Phone: 613-241-4611   Fax: 613-241-2292
Website: www.sierraclub.ca/national

Founded: NA
Scope: International

Publication(s): Sierra Magazine, SCAN
Contact(s):
Elizabeth May, EXECUTIVE DIRECTOR

SIERRA CLUB SAN FRANCISCO BAY CHAPTER
SAN FRANCISCO BAY CHAPTER
2530 San Pablo Ave., Suite 1
Berkeley, CA 94702-2000 USA
Phone: 510-848-0800   Fax: 510-848-3383
E-mail: info@sfasc.org
Website:
www.sierraclub.org/chapters/sanfranciscobay/nindex.html

Founded: NA
Scope: Local

Publication(s): SF Bay chapter schedule of activities - subscription item 3 times yearly

Contact(s):
Michael Bornstein, CHAPTER DIRECTOR
michael.bornstein@sierraclub.org

SIERRA CLUB SOUTHEAST OFFICE
SOUTHEAST OFFICE
1330 21st Way South, Suite 100
Birmingham, AL 35205 USA
Phone: 205-933-9111   Fax: 205-939-1020
E-mail: jim.price@sierraclub.org
Website: www.sierraclub.org

Founded: NA
Scope: National

Description: AL, AR, FL, LA, MS, TX

Contact(s):
Jim Price, SENIOR REGIONAL STAFF DIRECTOR
jimprice@sierraclub.org

SIERRA CLUB,
921 N. Congress St.
Jackson, MS 39202-2554 USA
Phone: 601-352-1026
Website: http://mississippi.sierraclub.org/

Founded: NA
Scope: Statewide

SIERRA CLUB,
13114 W. 125th Terrace
Overland Park, KS 66213-2463 USA
Phone: 913-402-9244   Fax: 913-402-7244
E-mail: wildlife1@aol.com
Website: www.kssierra.org/

Founded: NA
Scope: State

Contact(s):
Scott Smith, 9844 Georgia Avenue, Kansas City, KS 66109
Phone: 785-539-1973
wizard1@kscable.com

SIERRA CLUB,
1330 21st Way South, Suite 110
Birmingham, AL 35205 USA
Phone: 205-972-0252
Website: http://alabama.sierraclub.org/

Founded: NA
Scope: Statewide

Contact(s):
Jay Hudson, CHAIR
Phone: 205-972-0252
jayhudson@mindspring.com

SIERRA CLUB,
ALASKA CHAPTER
P.O Box 103441
Anchorage, AK 99501-3441 USA
Phone: 907-276-4048   Fax: 907-276-4048
E-mail: nw-ak.field@sierraclub.org
Website: www.sierraclub.org/chapters/ak/

Founded: NA
Scope: Statewide

SIERRA CLUB,
ANGELES CHAPTER
3435 Wiltshire Blvd., Suite 320
Los Angeles, CA 90010-1904 USA
Phone: 213-387-4287   Fax: 213-387-5383
E-mail: info@angeleschapter.org
Website: www.angeleschapter.sierraclub.org

Founded: NA
Scope: Local, Regional

Publication(s): The Schedule of Activities (quarterly), The Southern Sierran

Contact(s):
Martin Schlageter, CONSERVATION COORDINATOR
Bill Corcoran, PUBLIC LANDS COORDINATOR

SIERRA CLUB,
ARKANSAS CHAPTER
Little Rock, AR 72221-2446 USA
Phone: 501-224-2582
E-mail: davisvh@aristotle.net
Website: www.aristotle.net/~sierra/

Founded: NA
Membership: 600+
Scope: Statewide
Publication(s): Arkansas Sierran -Bi-Annual

Contact(s):
Randy Zurcher, STAFF PERSON WORKING ON CONSERVATION

SIERRA CLUB,
ATLANTIC CHAPTER
116 John St.
New York, NY 10038-3401 USA
Phone: 212-791-2400   Fax: 212-791-0839
E-mail: atlantic.chapter@sierraclub.org
Website: www.atlantic.sierraclub.org

Founded: NA
Membership: 10,000
Scope: State
Publication(s): Sierra Atlantic

SIERRA CLUB,
CASCADE CHAPTER
8511 15th Ave. NE, Rm. 201
Seattle, WA 98115-3101 USA
Phone: 206-523-2147   Fax: 206-729-2468
E-mail: cascade.chapter@sierraclub.org
Website: www.cascadechapter.org/

Founded: NA
Membership: 24,000
Scope: Statewide

Publication(s): The Cascade Crest

Contact(s):
Roy Goodman, CHAPTER COORDINATOR

SIERRA CLUB,
CONNECTICUT CHAPTER
118 Oak St.
Hartford, CT 06106-1514 USA
Phone: 860-525-2500
Website: www.sierraclub.org/chapters/ct/

Founded: NA
Scope: Statewide

SIERRA CLUB,
CUMBERLAND CHAPTER
259 W. Short St.
Lexington, KY 40507-1226 USA
Phone: 806-299-4410
Website: www.sierraclub.org/chapters/ky/

Founded: 1968
Scope: Statewide

SIERRA CLUB,
DAKOTA CHAPTER
311 E. Thayer
Bismarck, ND 58501 USA
Phone: 701-594-4275   Fax: 701-530-9290
Website: www.sierraclub.org/chapters/nd/

Founded: NA
Membership: 500
Scope: Statewide
Publication(s): Dakota Prairie, newsletter

Contact(s):
Todd Leake, CHAIR

SIERRA CLUB,
DELAWARE CHAPTER
1304 N. Rodney St.
Wilmington, DE 19806 USA
Phone: 302-425-4911
E-mail: delaware.chapter@sierraclub.org
Website: delaware.sierraclub.org
http://delaware.sierraclub.org

Founded: NA
Scope: Statewide

Contact(s):
Jim Steffens, CHAIR
Phone: 302-239-9601
jjsteff@magpage.com
Matt Urban, VICE CHAIR
Phone: 302-661-2050
Matt@mobiusnm.com

SIERRA CLUB,
DELTA CHAPTER
P.O. Box 19469
New Orleans, LA 70179-0469 USA
Phone: 504-836-3062
E-mail: delta.chapter@sierraclub.org
Website: www.sierraclub.org/chapters/la/

Founded: NA

Scope: Statewide

SIERRA CLUB,
FLORIDA CHAPTER
475 Central Ave., Suite M1
St. Petersburg, FL 33701-3817 USA
Phone: 813-824-8813   Fax: 813-824-0936
E-mail: geraldine.swormstead@sierraclub.org
Website: http://florida.sierraclub.org/

Founded: NA
Scope: Statewide

Contact(s):
Geraldine Swormstead, CHAIR

SIERRA CLUB,
GEORGIA CHAPTER
1447 Peachtree St., NE,
Atlanta, GA 30309-3034 USA
Phone: 404-607-1262   Fax: 404-876-5260
E-mail: georgia.chapter@sierraclub.org
Website: www.georgia.sierraclub.org

Founded: NA
Membership: 14,700
Scope: Statewide
Publication(s): Georgia Sierran (newsletter)

Contact(s):
Karen Austin, CHAPTER COORDINATOR
Bryan Hager, CONSERVATION ORGANIZER

SIERRA CLUB,
GRAND CANYON CHAPTER
812 N. 3rd St. Grand Canyon
Phoenix, AZ 85004-2020 USA
Phone: 602-253-8633   Fax: 602-258-6533
E-mail: grandcanyon@qwest.net
Website: www.sierraclub.org/chapters/az/

Founded: NA
Membership: 12,000
Scope: Statewide
Publication(s): Canyon Echo - monthy newsletter

SIERRA CLUB,
HAWAII CHAPTER
P.O. Box 2577
Honolulu, HI 96803-2577 USA
Phone: 808-538-6616   Fax: 808-537-9019
Website: http://hawaii.sierraclub.org/

Founded: NA
Scope: Statewide

SIERRA CLUB,
HOOSIER CHAPTER
6224 N. College Ave
Indianapolis, IN 46220 USA
Phone: 317-466-9992
E-mail: sierra@netdirect.net
Website: hoosier.sierraclub.org/

Founded: NA
Scope: Statewide

SIERRA CLUB,
IOWA CHAPTER
Thoreau Center, 3500 Kingman Blvd.
Des Moines, IA 50311-3798 USA
Phone: 515-277-8868
E-mail: iowa.chapter@sierraclub.org
Website: http://iowa.sierraclub.org

Founded: NA
Scope: Statewide

Contact(s):
   Charlie Winterwood, CHAIR
   Phone: 319-588-2783

SIERRA CLUB,
KERN-KAWEAH CHAPTER
P.O. Box 3357
Bakersfield, CA 93385-3357 USA
Phone: 661-323-5569
E-mail: kern-kaweah.chapter@sierraclub.org
Website: www.sierraclub.org/chapters/kernkaweah/

Founded: NA
Membership: 1400
Scope: Local, Regional
Publication(s): The Road Runner

Contact(s):
   Lorraine Unger, EXECUTIVE OFFICER

SIERRA CLUB,
LONE STAR CHAPTER
54 Chicon St.
Austin, TX 78702-5451 USA
Phone: 512-477-1729   Fax: 512-477-8526
E-mail: lonestar.chapter@sierraclub.org
Website: www.texas.sierraclub.org

Founded: NA
Membership: 23000
Scope: Statewide
Publication(s): State Capital Report, The Lone Star Sierran

Contact(s):
   Jennifer Walker, ADMINSTRATIVE ASSISTANT
   Ken Kramer, CHAPTER DIRECTOR
   Neil Carman, CLEAN AIR PROGRAM DIRECTOR
   Fred Richardson, COMMUNICATIONS DIRECTOR
   fred.richardsom@sierraclub.org
   Tracy Arambula, ENVIRONMENTAL JUSTICE DIRECTOR
   tracy.arambula@sierraclub.org
   Erin Rogers, GRASS ROOTS COORDINATOR
   erin.rogers@sierraclub.org
   Brian Sybert, NATURAL RESOURCES

SIERRA CLUB,
LOS PADRES CHAPTER
P.O. Box 90924
Santa Barbara, CA 93190-0924 USA
Phone: 805-966-6622
Website: http://lospadres.sierraclub.org/

Founded: NA
Scope: Statewide

Contact(s):
   Rick Skillin, CHAPTER CHAIR
   Phone: 805-735-4190
   rick.skillin@sierraclub.org

SIERRA CLUB,
MASSACHUSETTS CHAPTER
100 Boylston St., Suite 760
Boston, MA 02116-4610 USA
Phone: 617-423-5775   Fax: 617-423-5858
E-mail: office@sierraclubmass.org
Website: www.sierraclubmass.org

Founded: NA
Scope: Statewide

Publication(s): Massachusetts Sierran

Contact(s):
   James McCaffrey, DIRECTOR
   director@sierraclubmass.org

SIERRA CLUB,
MONTANA CHAPTER
P.O. Box 7312
Missoula, MT 59807 USA
Phone: 406-549-6031
E-mail: accipiter4@juno.com
Website: www.sierraclub.org

Founded: NA
Scope: Regional

Publication(s): Big Sky Sierrian

Contact(s):
   Kathryn Hohmann, ASSOCIATE FIELD REP.
   kathryn.holmann@sierraclub.org
   Christine Phillips, CONTACT
   Bozeman, MT 59771-1290
   Phone: 406-582-1281
   magpie@mcn.net

SIERRA CLUB,
MOTHER LODE CHAPTER
1414 K St., Suite 300
Sacramento, CA 95814-3929 USA
Phone: 916-557-1100   Fax: 916-557-9669
E-mail: motherload@mcsweb1.com
Website: www.motherload.sierraclub.org

Founded: NA
Scope: Regional

Publication(s): The Bonanza, bimonthly newsletter
Contact(s):
   Julie Parker, ADMINISTRATIVE ASSISTANT

SIERRA CLUB,
NEW COLUMBIA CHAPTER
1416 33rd St., NW
Washington, DC 20007 USA
Phone: 202-333-5424   Fax: 202-965-3769
Website: newcolumbia.sierraclub.org

Founded: NA
Scope: State

Contact(s):
   Danilo Pelletiere, CONSERVATION CHAIR
   Phone: 202-543-7791
   dpelleti@gmu.edu
   Mark Wenzler, VICE CHAIR
   Phone: 202-547-3410

SIERRA CLUB,
NEW HAMPSHIRE CHAPTER
Three Bicentennial Sq.
Concord, NH 03301-4058 USA
Phone: 603-224-8222   Fax: 603-224-4719
Website: www.sierraclub.org/chapters/nh/

Founded: NA
Scope: State

Publication(s): The New Hampshire Sierran

SIERRA CLUB,
NORTH CAROLINA CHAPTER
112 S. Blount St.
Raleigh, NC 27601 USA
Phone: 919-833-8467   Fax: 919-833-8460
E-mail: info@sierraclub-nc.org
Website: www.sierraclub-nc.org

Founded: NA
Membership: 17,000
Scope: Statewide
Publication(s): Footnotes

Contact(s):
  David Knight, LOBBYIST
  John Hudson, RESOURCES DEVELOPMENT
  COORDINATOR
  Molly Diggins, STATE DIRECTOR

SIERRA CLUB,
NORTH STAR CHAPTER (MINNESOTA)
1313 5th St., SE, Suite 323
Minneapolis, MN 55414-4504 USA
Phone: 612-379-3853   Fax: 612-379-3855
E-mail: north.star.chapter@sierraclub.org
Website: www.northstar.sierraclub.org/

Founded: NA
Membership: 20000
Scope: Regional
Publication(s): The North Star Journal - bimonthly newsletter

Contact(s):
  Scott Elkins, STATE DIRECTOR
  selkins@igc.org

SIERRA CLUB,
OHIO CHAPTER
36 W Gayn St., Suite 314
Columbus, OH 43215-3006 USA
Phone: 606-255-1946   Fax: 606-233-4099
E-mail: ogeralds@lexkylaw.com
Website: http://kentucky.sierraclub.com

Founded: NA
Scope: Statewide

Publication(s): Ohio Sierran Chapter Newsletter

Contact(s):
  Marc Conte, LEGISLATIVE COORDINATOR
  Shannon Harps, TRANSPORTATION POLICY SPECIALIST

SIERRA CLUB,
OKLAHOMA CHAPTER
P.O. Box 60644
85 Second St. Second Floor,
San Francisco, CA 94105-3441
Oklahoma City, OK 73146-0644 USA
Phone: 415-977-5500   Fax: 415-977-5799
E-mail: oklahoma.chapter@sierraclub.org
Website: www.sierraclub.org/chapters/ok/

Founded: NA
Scope: Statewide

SIERRA CLUB,
PRAIRIE CHAPTER (AB, MB, SK)
10511 Saskatchewan Dr.
Edmonton, Alberta T6E 4S1 Canada
Phone: 780-439-1160   Fax: 780-437-3932
E-mail: sierraclub@connect.ab.ca
Website: www.sierraclub.ca/prairie

Founded: NA
Scope: Regional

SIERRA CLUB,
REDWOOD CHAPTER (NORTHERN CALIFORNIA)
P.O. Box 466
Santa Rosa, CA 95402 USA
Phone: 707-544-7651   Fax: 707-544-9861
E-mail: heyneedles@aol.com
Website: redwood.sierraclub.org
http://redwood.sierraclub.org

Founded: NA
Scope: Statewide

Publication(s): Redwood Needles

Contact(s):
  Margaret Pennington, CHAPTER CHAIR
  penningt@sonic.net

SIERRA CLUB,
RIO GRANDE CHAPTER (NEW MEXICO/WEST TEXAS)
207 Ricardo Road
Santa Fe, NM 87501 USA
Phone: 505-988-5760
E-mail: jhanna505@aol.com
Website: http://riogrande.sierraclub.org/

Founded: NA
Scope: Regional

Contact(s):
  Jennifer De Garmo, STAFF MEMBER
  202 Central Avenue SE, Albuquerque, NM 87102
  Phone: 505-243-7767
  nmex.field1@prodigy.net

SIERRA CLUB,
SAN DIEGO CHAPTER (SOUTHERN CALIFORNIA)
3820 Ray St.
San Diego, CA 92104-3623 USA
Phone: 619-299-1743   Fax: 619-299-1742
E-mail: san-diego.chapter@sierraclub.org
Website: http://sandiego.sierraclub.org/home/index.asp

Founded: 1948
Scope: Local Region

Contact(s):
Cheryl Reiff, OFFICE ADMINISTRATOR

SIERRA CLUB,
SANTA LUCIA CHAPTER
P.O. Box 15755
San Luis Obispo, CA 93406-5755 USA
Phone: 805-543-8717
E-mail: gfelsman@thegrid.net.
Website: http://santalucia.sierraclub.org/

Founded: NA
Scope: Local Region

SIERRA CLUB,
SOUTH DAKOTA CHAPTER
P.O. Box 1624
Rapid City, SD 57709-1624 USA
Phone: 605-348-1345   Fax: 605-348-1344
E-mail: brademey@rapidnet.com
Website: www.sierraclub.org/chapters/sd/

Founded: NA
Membership: 7,000,000
Scope: Local, Regional, National
Publication(s): Sierra Magazine, Newsletter Quarterly, Pines & Prairie

Contact(s):
Heather Morijah, CONSERVATION ORGANIZER
1101 E, Phildelphia St., Rapid City, SD 57701
Phone: 605-342-2244
Fax: 605-342-2255
heather.morijah@sierraclub.org
Sam Clauson, S. D. CHAPTER CHAIR

SIERRA CLUB,
TEHIPITE CHAPTER (NORTHERN CALIFORNIA)
P.O. Box 5396
Fresno, CA 93755-5396 USA
Phone: 559-271-0652
E-mail: Tehipite.Chapter@sierraclub.org
Website: http://tehipite.sierraclub.org/

Founded: NA
Scope: Local Region

SIERRA CLUB,
TENNESSEE CHAPTER
4641 Villa Green Dr.
Nashville, TN 37215-4331 USA
Phone: 615-665-1010
E-mail: tennessee.chapter@sierraclub.org
Website: www.sierraclub.org/chapters/tn/

Founded: NA
Scope: Statewide

SIERRA CLUB,
TOIYABE CHAPTER (NEVADA/EASTERN CALIFORNIA)
P.O. Box 8096
Reno, NV 89507-8096 USA
Phone: 775-323-3162
Website: http://nevada.sierraclub.org/

Founded: NA
Scope: Statewide

SIERRA CLUB,
UTAH CHAPTER
2273 S. Highland Dr., Suite 2D
Salt Lake City, UT 84106-2832 USA
Phone: 801-467-9297
E-mail: utah.chapter@sierraclub.org
Website: http://utah.sierraclub.org/

Founded: NA
Scope: Statewide

Contact(s):
Nina Dougherty, CHAPTER CHAIR
Dan Schroeder, SECRETARY/TREASURER
dschroeder@weber.edu
Tony Guay, VICE-CHAIR
tpguay@hotmail.com

SIERRA CLUB,
VENTANA CHAPTER (NORTHERN CALIFORNIA)
P.O. Box 5667
Carmel, CA 93921-5667 USA
Phone: 831-624-8032
E-mail: ventana@mbay.net
Website: www.ventana.org/

Founded: NA
Scope: Local, Regional

SIERRA CLUB,
VERMONT CHAPTER
P.O. Box 3154
Burlington, VT 05401-0031 USA
Phone: 802-651-0169   Fax: 888-729-4109
Website: http://vermont.sierraclub.org/

Founded: NA
Scope: Statewide

Publication(s): Vermont Serrian (Quarterly newsletter)

SIERRA CLUB,
WEST VIRGINIA
P.O. Box 4142
Morgantown, WV 26504-4142 USA
Phone: 304-363-4006
E-mail: shalom.tazewell@sierraclub.org
Website: www.wvsierra.org/

Founded: NA
Membership: 300
Scope: Statewide

Publication(s): Chapter Newsletter - Mountain State Sierran
Contact(s):
Paul Potter, CHAPTER CHAIR
Phone: 304-363-4006
paul_wilson@fws.gov

SIERRA CLUB, (RHODE ISLAND CHAPTER)
RHODE ISLAND CHAPTER
21 Meeting St. Garden Entrance
Providence, RI 02903-1000 USA
Phone: 401-521-4734   Fax: 401-521-4001
E-mail: contactus@sierraclubri.org
Website: www.sierraclubri.org

Founded: NA

Membership: 2700
Scope: Statewide

Publication(s): Nasty Nine Sprawl Report, Transit Users Survey 2000, Coastlines (newsletter), Transportation Reform Alliance

SIERRA CLUB, LOMA PRIETA CHAPTER
LOMA PRIETA CHAPTER
3921 E. Bayshore Rd., Suite. 204
Palo Alto, CA 94303-4303 USA
Phone: 650-390-8411   Fax: 650-390-8497
E-mail: ctsierraclub@aol.com
Website: http://connecticut.sierraclub.org

Founded: NA
Membership: 24,000
Scope: Local/Regional

Publication(s): Loma Prietan

Contact(s):
Dan Kalb, DIRECTOR
loma.prieta.director@sierraclub.org

SIERRA CLUB, MACKINAC CHAPTER
MACKINAC CHAPTER
109 E. Grand River
Lansing, MI 48906 USA
Phone: 517-484-2372   Fax: 517-484-3108
E-mail: mackinac.chapter@sierraclub.org
Website: www.michigan.sierraclub.org

Founded: NA
Membership: 19,000
Scope: Statewide

Publication(s): Mackinac, The, Quarterly

Contact(s):
Anne Woiwode, DIRECTOR

SIERRA CLUB, MIDWEST OFFICE
MIDWEST OFFICE
214 N. Henry St., Suite 203
Madison, WI 53703 USA
Phone: 608-257-4994   Fax: 608-257-3513
E-mail: mw-wi.field@sierraclub.org
Website: www.sierraclub.org

Founded: NA
Membership: 6
Scope: Regional

Description: IA, IL, IN, KY, MI, MN, MO, OH, WI

Publication(s): Mississippi Times

Contact(s):
Emily Green, GREAT LAKES PROGRAM DIRECTOR
emily.green@sierraclub.org
Bill Redding, MIDWEST REGIONAL REPRESENTATIVE
bill.redding@sierraclub.org
Brett Hulsey, SENIOR REGIONAL REPRESENTATIVE
brett.hulsey@sierraclub.org

SIERRA CLUB, NEW JERSEY CHAPTER
NEW JERSEY CHAPTER
57 Mountain Ave.
Princeton, NJ 08540-2611 USA
Phone: 609-924-3141   Fax: 609-924-8799
Website: www.sierraactivist.org

Founded: NA
Membership: 27,000
Scope: Local

Publication(s): The Sierran

Contact(s):
Jeff Tiddle, DIRECTOR

SIERRA CLUB, PENNSYLVANIA CHAPTER
PENNSYLVANIA CHAPTER
Box 663
Harrisburg, PA 17108 USA
Phone: 717-232-0101   Fax: 717-238-6330
E-mail: sierraclub.pa@paonline.com
Website: www.sierraclub.org/chapters/pa/

Founded: NA
Membership: 26,000
Scope: Local, State

Publication(s): Sierra Club PA Club Newsletter Sylvanian

Contact(s):
Jeff Schmidt, GOVERNMENTAL LIAISON

SIERRA CLUB, SOUTH CAROLINA CHAPTER
SOUTH CAROLINA CHAPTER
P.O. Box 2388, 1314 Lincoln St., Suite 211
Columbia, SC 29202 USA
Phone: 803-256-8487   Fax: 803-256-8448
E-mail: scsierra@conterra.com
Website: www.sierraclub.org/chapters/sc/

Founded: 1978
Membership: 4900
Scope: Statewide

Publication(s): The Congaree Chronical - bi monthy newsletter

Contact(s):
Dell Isham, CHAPTER DIRECTOR
Phone: 803-256-8487
scsierra@conterra.com

SIERRA CLUB VIRGINIA CHAPTER
VIRGINIA CHAPTER
Six N. 6th St., Suite 401
Richmond, VA 23219-2419 USA
Phone: 804-225-9113   Fax: 804-225-9114
Website: www.sierraclubva.org

Founded: NA
Membership: 13,000
Scope: Statewide

Publication(s): Old Dominion Sierran

Contact(s):
Pat Dezern, CONSERVATION ORGANIZER

SIERRA STUDENT COALITION
SIERRA STUDENT COALITION
P.O. Box 2402
Providence, RI 02906-0402 USA
Phone: 401-861-6012   Fax: 401-861-6241
Website: www.ssc.org

Founded: NA
Membership: 5
Scope: National

Publication(s): Generation E

Contact(s):
Myke Bybee, DIRECTOR

SINAPU
2260 Baseline Rd., Suite 212
Boulder, CO 80302 USA
Phone: 303-447-8655　　　Fax: 303-447-8612
E-mail: sinapu@sinapu.org
Website: www.sinapu.org

Founded: 1991
Membership: 600
Scope: Regional

Description: Sinapu, named after the Ute word for wolves, is dedicated to the recovery of native carnivores in the Southern Rocky Mountains and to the restoration of the wild habitat in which all species flourish.

Publication(s): Wild Again

Keyword(s): Biodiversity, Endangered Species, Predators, Public Lands, training, EcoAction, Ecology, Ancient Forests, Forest Management, Wildlands, Wolves

Contact(s):
Kimberly Riggs, EXECUTIVE DIRECTOR
kim@sinapu.org
Wendy Keefover-Ring, PROGRAM STAFF
wendy@sinapu.org
Rob Edward, PROGRAM DIRECTOR

SISKIYOU PROJECT
P.O. Box 220
Cave Junction, OR 97523 USA
Phone: 541-592-4459

Description: OUR STATEMENT OF PURPOSE: We believe in the power of place and of biological cycles, and in modeling our lives, actions, and community on the ideals of wholeness and being-of-a-place. For us the ultimate model is the wild, and we are reaching for the wild inside ourselves as well as "out there". We see all life forms as interconnected and inseparable, and realizing that we are in a time of crisis, we feel urgency in effecting change in the way human industrial culture deals with wild nature. We base our relationships with each other and with the natural world on knowledge, and we are here because we love to live and work in this place.

Contact(s):
Barry Snitkin, COMMUNITY OUTREACH
Barbara Ulliam, CONSERVATION DIRECTOR
Kelpie Wilson, DEVELOPMENT DIRECTOR
Tim Dimilio, EAST SISKIYOU CONSERVATION COORDINATOR
Steve Marsden, EXECUTIVE DIRECTOR
Jim "Jake" McBride, FINANCE DIRECTOR
Marjorie Reynolds, MAIL AND DATE PROCESSOR
Linda Serrano, NETWORK COORDINATOR
Kindi Fahrnkopf, OFFICE MANAGER
David Johns, PRESIDENT
Romain Cooper, PROGRAM DIRECTOR
Steven Jessup, SECRETARY
Rich Nawa, SENIOR ECOLOGIST
Vicky Rummel, SFI ADMIN. COORDINATOR
Jennifer Marsden, SFI DIRECTOR
Lori Cooper, STAFF ATTORNEY
Lou Gold, STORYTELLER EMERITUS

Erik Jules, TREASURER
Dave Willis, VICE PRESIDENT
Julie Norman, VIDEO PROJECT COORDINATOR

SMALL WOODLAND OWNERS ASSOCIATION OF MAINE
153 Hospital St., P.O. Box 836
Augusta, ME 04332 USA
Phone: 207-626-0005　　　Fax: 207-626-7992
E-mail: swoam@mint.net
Website: www.swoam.com

Founded: 1975
Scope: Statewide

Description: A statewide nonprofit organization, affiliated with the National Woodland Owners Association, which pursues better understandings, skills, and directions in small woodland ownership/management under integrated use objectives.

Publication(s): SWOAM News

Keyword(s): Forests and Forestry

SMITHSONIAN INSTITUTE NATIONAL ZOOLOGICAL PARK
3001 Connecticut Ave. NW
Washington, DC 20008 USA
Phone: 202-673-4717
Website: www.nationalzoologicalpark.com

Founded: NA
Membership: 300
Scope: National

Description: Research concentrates on a better understanding of animal behavior and health, particularly endangered species. Through the operation of the zoo's Conservation and Research Center in Front Royal, VA, the NZP is developing a program of animal propagation which will aid in the survival of threatened and endangered species. Undertakes a number of programs overseas to develop new methodology and increase knowledge of species in the wild. The Migratory Bird Center is located at the zoo.

Contact(s):
Lucy Spelman, DIRECTOR
Phone: 202-673-4721

SMITHSONIAN INSTITUTION
1000 Jefferson Dr., SW
Washington, DC 20560 USA
Phone: 202-357-2700
Website: www.si.edu

Founded: 1846
Scope: National

Description: An education, museum, and research complex as well as an independent trust instrumentality of the United States, established for the increase and diffusion of knowledge. Mission accomplished by: field investigations; national collections development in arts, history, and science, and their preservation for study, reference, and exhibition; scientific research and publications; programs of national and international cooperative research, conservation, education, and training; answering inquiries from the general public and educational and scientific organizations; long-term loan of selected objects; and sharing of research and educational material on the World Wide Web.

Publication(s): The Smithsonian Magazine, Smithsonian Institution Press

Keyword(s): Aquatic Habitats, Ancient Forests, Biodiversity, Birds, Cultural Preservation, Culture, Endangered Species, Gardening and Horticulture, Historic Preservation, Mammals, education, Insects and Butterflies, Marine Mammals, Whale, Dolphin, Seal, Reptiles and Amphibians

Contact(s):
Lawrence Small, SECRETARY
Phone: 202-357-1300
J. O'Connor, UNDER SECRETARY OF SCIENCE
Phone: 202-357-2903

SMITHSONIAN INSTITUTION
NATIONAL MUSEUM OF NATURAL HISTORY
10th St. and Constitution Ave., NW
Washington, DC 20560 USA
Phone: 202-357-2700   Fax: 202-357-1729
E-mail: info@infor.si.edu
Website: www.si.edu

Founded: NA
Scope: National

Description: A center for the study of humans, plants, animals, fossil organisms, terrestrial and extraterrestrial rocks, and minerals as well as other fields of scientific investigation.

Contact(s):
David Correll, CHIEF SCIENTIST OF ENVIRONMENTAL RESEARCH CENTER
Smithsonian Environmental Research Center, P.O. Box 28, Edgewater, MD 21037
Phone: 301-261-4190
Robert Fri, DIRECTOR

SMITHSONIAN INSTITUTION
OFFICE OF FELLOWSHIPS AND GRANTS
Victor Bldg, 750 9th St. N. W.
Suite 9300
Washington, DC 20560 USA
Phone: 202-357-2700   Fax: 202-275-0489
Website: www.si.edu/research+study

Founded: NA
Membership: 9
Scope: International

Description: Oversees all Smithsonian fellowships and supports a wide range of research activities. It also provides program and administrative assistance for cooperative teaching arrangements between the Institution and local universities in American history, museum studies, and other areas.

Contact(s):
Roberta Rubinoff, DIRECTOR

SMITHSONIAN INSTITUTION
OFFICE OF INTERNATIONAL RELATIONS
Smithsonian Institution, 1100 Jefferson Dr., SW,
Washington, DC 20560 USA
Phone: 202-357-4795   Fax: 202-786-2557
Website: www.prism.edu

Founded: NA
Membership: 10
Scope: National

Description: The Foreign Currency Program supports the research activities of American institutions of higher learning through grants in U.S.-owned local currencies.

Contact(s):
Francine Berkowitz, DIRECTOR

SMITHSONIAN INSTITUTION
SMITHSONIAN PRESS/SMITHSONIAN PRODUCTIONS
470 L'Enfant Plaza, Suite 7100
Washington, DC 20560 USA
Phone: 202-287-3738
Website: http://www.si.edu/sipress/

Founded: NA
Scope: National

Description: Information on history, art, and science research is presented in non-technical style in Smithsonian Institution Research Reports issued four times a year by the Office of Public Affairs (202-357-2627). Smithsonian, the official magazine of the Institution, presents general interest feature articles each month in every subject area of the Smithsonian museums: art, culture, history, science, and technology.

Publication(s): Research in various fields is reported in a continuing series of publications by the Smithsonian Institution Press

Contact(s):
Daniel Goodwin, DIRECTOR
David Umansky, DIRECTOR OF COMMUNICATIONS
Arts and Industries Bldg., 900 Jefferson Dr. SW, Rm. 4210, Washington, DC 20560
Phone: 202-357-2627
Don Moser, EDITOR
Smithsonian Magazine, Arts and Industries Bldg., 900 Jefferson Dr. SW, Rm. 1310C, Washington, DC 20560

SMITHSONIAN INSTITUTION
SMITHSONIAN TROPICAL RESEARCH INSTITUTE
APO AA, FL 34002-0948 USA
Phone: 202-357-2700
Scope: National

SMITHSONIAN MARINE STATION AT FORT PIERCE
701 Seaway Dr.
Fort Pierce, FL 34949 USA
Phone: 561-465-6630   Fax: 561-461-8154
Website: www.sms.si.edu

Founded: NA
Scope: International

Description: Marine studies aim at understanding the ecological function of inland waterways and their relationship to land use policy.

Contact(s):
Mary Rice, DIRECTOR
rice@sms.si.edu

SOCIETY FOR ANIMAL PROTECTIVE LEGISLATION
P.O. Box 3719, Georgetown Station
Washington, DC 20007 USA
Phone: 202-337-2334   Fax: 202-338-9478
E-mail: sapl@saplonline.org
Website: www.saplonline.org

Founded: 1955
Scope: National

Description: Nonprofit organization which keeps its 7,000 correspondents apprised of current developments in legislation for the protection of animals. Has been instrumental in obtaining enactment of 14 federal laws.

Keyword(s): Endangered Species, Mammals, Marine Mammals, Whale, Dolphin, Seal, Trapping, training

Contact(s):
John Gleiber, EXECUTIVE SECRETARY
Phone: 202-337-2334
John Kullberg, PRESIDENT
Christine Stevens, SECRETARY
Phone: 202-337-2334

SOCIETY FOR CONSERVATION BIOLOGY

Attn: Executive Coordinator, Univ. of Washington, Box 351800
Seattle, WA 98195-1800 USA
Phone: 206-616-4054 Fax: 206-543-3041
E-mail: conbio@u.washington.edu
Website: http://conbio.net.scb

Founded: 1985
Scope: International, National

Description: A professional society dedicated to providing the scientific information and expertise required to protect the world's biological diversity. Incorporated as a tax-exempt scientific organization, the Society has a board composed of scholars, government personnel, and members of both national and international scientific and conservation organizations.

Publication(s): Conservation Biology In Practice - quarterly magazine, Society for conservation biology - quarterly newsletter, Conservation Biology bi-monthly scientific journal

Keyword(s): Biodiversity, Endangered Species, Environmental and Conservation Education, Sustainable Development

Contact(s):
Gary Meffe, EDITOR OF CONSERVATION BIOLOGY
Phone: 352-846-0557
Fax: 352-846-2823
conbio@gnv.ifas.ufl.edu
Kathy Kohm, EDITOR OF CONSERVATION BIOLOGY IN PRACTICE
Phone: 206-685-4724
Fax: 206-221-7839
kkohm@u.washington.edu
Alice Blandin, EXECUTIVE COORDINATOR
Phone: 206-616-4054
Fax: 206-543-3041
conbio@u.washington.edu
Reed Noss, PRESIDENT
Phone: 541-757-0687
Fax: 541-758-3454
Sarah Reichard, SECRETARY
Phone: 206-616-5020
Fax: 206-685-2692
reichard@u.washington.edu
Stephen Humphrey, TREASURER
Phone: 352-392-9230
Fax: 352-392-9748
humphrey@ufl.edu

SOCIETY FOR ECOLOGICAL RESTORATION

1207 Seminole Highway, Suite B
Madison, WI 53711 USA
Phone: 608-262-9547 Fax: 608-265-8557

Founded: 1989
Membership: 2,600
Scope: International

Description: Created to promote the development of ecological restoration both as a discipline and as a model for a healthy relationship with nature, and to raise awareness of the value and limitations of restoration as a conservation strategy.

Publication(s): SER News, Restoration Ecology, Proceedings from the Seventh SER Conference, 1995, Ecological Restoration

Keyword(s): Conservation, Ecology, Environmental and Conservation Education, Renewable Resources, Restoration

Contact(s):
Edith Read, CHAIR
Phone: 714-751-7373
William Niering, EDITOR
Phone: 203-447-1911
Donald Falk, EXECUTIVE DIRECTOR
Phone: 520-626-7201
Eric Higgs, SECRETARY
Phone: 403-492-5469
William Halvorson, TREASURER
Phone: 520-670-6885
George Gann, VICE CHAIR
Phone: 305-245-6547

SOCIETY FOR INTEGRATIVE AND COMPARATIVE BIOLOGY

1313 Dolley Madison Blvd. Ste. 402
McLean, VA 22101 USA
Phone: 703-790-1745 Fax: 703-790-2672
E-mail: sicb@burkinc.com
Website: www.sicb.org

Founded: 1890
Membership: 2,200
Scope: National

Description: (formerly AMERICAN SOCIETY OF ZOOLOGISTS) The Society for Integrative and Comparative Biology (SICB) is one of the largest and most prestigious professional associations of its kind. SICB is dedicated to promoting the pursuit and public dissemination of important information relating to comparative biology.

Publication(s): American Zoologist, The

Keyword(s): Wildlife Rehabilitation, Insects and Butterflies, Mammals, Reptiles and Amphibians, Zoology

Contact(s):
Marquesa Mills, BUSINESS MANAGER
104 Sirius Cir., Thousand Oaks, CA 91360
Phone: 805-492-3585
Fax: 805-492-0370
Mary Adams-Wiley, EXECUTIVE OFFICER
104 Sirius Cir., Thousand Oaks, CA 91360
Phone: 805-492-3585
Fax: 805-492-0370

Milton Fingerman, MANAGING EDITOR
Department of Biology, Tulane University, New Orleans,
LA 70118
Phone: 504-865-5546
Albert Bennett, PRESIDENT
School of Biological Sciences, University of California+I365,
Irvine, CA 92717
Phone: 714-856-6930
Fax: 714-725-2181
Mary Ottinger, SECRETARY
Department of Poultry Science, University of Maryland,
College Park, MD 20742
Phone: 301-405-5780
Fax: 301-314-9557
Marjorie Reaka, TREASURER
Department of Zoology, University of Maryland, College Park,
MD 20742
Phone: 301-454-0259

SOCIETY FOR MARINE MAMMALOGY, THE
BIOLOGICAL SCIENCES AND CENTER FOR MARINE
SCIENCE RESEARCH
CORPORATE ZOOLOGICAL OPERATIONS
SEAWORLD, INC.
Busch Entertainment Corporation, 7007 Seaworld Dr.,
Orlando, FL 32821
Phone: 407-363-2662   Fax: 407-345-5397
E-mail: dan.odell@seaworld.com

Founded: 1981
Scope: National

Description: To promote the educational, scientific, and
managerial advancement of marine mammal science; gather
and disseminate scientific, technical, and management
information, through publications and meetings to members of
the society, the public, and public and private institutions; and
promote the wise conservation and management of marine
mammal resources.

Publication(s): Marine Mammal Science

Keyword(s): Endangered Species, Marine Mammals, Whale,
Dolphin, Seal, education

Contact(s):
Carol Fairfield, AWARDS & SCHOLARSHIP COMMITTEE
NOAA/NMFS/SEFSC, 1002 Forest Dr., Arnold, MD 21012
Phone: 410-757-7224
carol.fairfield@noaa.gov
Steven Swartz
COMMITTEE OF SCIENTIFIC ADVISORS
National Marine Fisheries Service, 75 Virginia Beach Dr.,
Miami, FL 33149
Phone: 305-361-4487
Fax: 305-361-4478
Steven.Swartz@noaa.gov
William Perrin, EDITOR
Southwest Fisheries Science Center, NMFS, P.O. Box 271,
LaJolla, CA 92109
Phone: 619-546-7093
Fax: 619-546-7003
wperrin@ucsd.edu
Edward Keith, EDUCATION COMMITTEE
Oceanographic Center, Nova Southeastern University,
8000 N. Ocean Dr., Dania, FL 33004
Phone: 954-262-8322
Fax: 954-921-7764
edwardok@hpd.nova.edu

Glenn Vanblaricom, MEMBERSHIP COMMITTEE
WA Cooperative Fish & Wildlife Research Unit, Box 357980,
University of Washington, Seattle, WA 98195
Phone: 206-543-6475
Fax: 206-616-9012
Daniel Odell, PRESIDENT-ELECT
Sea World, Inc., 7007 Sea World Dr., Orlando,
FL 32821-8097
Paul Nachtigall, SCIENTIFIC PROGRAM COMMITTEE
Marine Mammal Research Program, Hawaii Institute of
Marine Science, University of Hawaii, P.O. Box 1106, Kailua,
HI 96734
Phone: 808-247-5297
Fax: 808-247-5831
nachtiga@hawaii.edu
D. Pabst, SECRETARY
Phone: 910-962-7266
Fax: 910-962-4066
pbasta@uncwil.edu

SOCIETY FOR RANGE MANAGEMENT
445 Union Blvd Suite 230
Lakewood, CO 80228 USA
Phone: 303-986-3309   Fax: 303-986-3892
E-mail: srmden@ix.netcom.com
Website: www.srm.org

Founded: 1948
Membership: 4000
Scope: National

Description: Professional society which promotes understanding
of rangeland ecosystems and their management and use for
tangible products and intangible values; reports new findings
and techniques in range science; promotes public appreciation
of rangelands and benefits derived from them; promotes professional development of members.

Publication(s): Rangelands, Journal of Range Management

Keyword(s): Agriculture, Land Use Planning, Prairies, Public
Lands, Renewable Resources, Exotic species, Aquatic
nuisance species, Sustainable Resources, Water Quality,
Watersheds, Wetlands, Ecology, Environment, Professional
Organization

Contact(s):
Samuel Albrecht, EXECUTIVE VICE-PRESIDENT
sam-albrecht@ix.netcom.com

SOCIETY FOR THE PRESERVATION OF BIRDS OF PREY
P.O. Box 66070, Mar Vista Station
Los Angeles, CA 90066-0070 USA
Phone: 310-840-2322

Founded: 1966
Scope: National

Description: A private charity, non-membership, national
association which advocates the strictest possible protection
for birds of prey; educates the public about the role of raptors
in the ecosystem; opposes lenient harvesting practices and the
sale of birds of prey for profit; endorses captive raptor breeding
as a conservation technique; and supports the largest
collection of literature on birds of prey at any public university
or facility. The Society is the only and oldest raptor organization
which places emphasis on birds of prey occuring naturally in
the wild.

Publication(s): Leaflet Series, Raptor Report, The

Keyword(s): Raptors, Birds, Falconry, Endangered Species

Contact(s):
J. Hilton, PRESIDENT AND EDITOR
Phone: 310-636-0072

SOCIETY FOR THE PROTECTION OF NEW HAMPSHIRE FORESTS
THE FOREST SOCIETY
54 Portsmouth St.
Concord, NH 03301-5400 USA
Phone: 603-224-9945   Fax: 603-228-0423
E-mail: info@spnhf.org
Website: www.spnhf.org

Founded: 1901
Membership: 10,000
Scope: Regional

Description: A voluntary nonprofit organization promoting balanced conservation of New Hampshire's renewable natural resources through land protection, education, advocacy, and forestry.

Publication(s): Forest Notes

Keyword(s): Environmental and Conservation Education, Forests and Forestry, Land Purchase, training

Contact(s):
Jane Difley, PRESIDENT/ FORESTER
Paul Doscher, SENIOR DIRECTOR OF LAND CONSERVATION

SOCIETY OF AMERICAN FORESTERS
5400 Grosvenor Ln.
Bethesda, MD 20814 USA
Phone: 301-897-8720   Fax: 301-897-3690
E-mail: safweb@safnet.org

Founded: 1900
Scope: National

Description: The national organization representing all segments of the forestry profession and the accreditation authority for professional forestry education in the U.S. Objectives are to advance the science, technology, education, and practice of professional forestry and to use the knowledge and skills of the profession to benefit society.

Publication(s): Journal of Forestry, Forest Science, Southern Journal of Applied Forestry, Northern Journal of Applied Forestry, Western Journal of Applied Forestry, Forestry Source, The

Keyword(s): Environmental Law, Environmental and Conservation Education, Forests and Forestry, Public Lands, Renewable Resources

Contact(s):
Lori Gardner, DIRECTOR OF COMMUNICATIONS AND MARKETING SERVICES
Diane Perl, DIRECTOR OF CONVENTIONS AND MEETINGS
Charles Jackson, DIRECTOR OF FINANCE AND ADMINISTRATION
Lawrence Hill, DIRECTOR OF RESOURCE POLICY
P. Smith, DIRECTOR OF SCIENCE AND EDUCATION
Rebecca Staebler, EDITORIAL DIRECTOR AND DIRECTOR OF PUBLICATIONS

William Banzhaf, EXECUTIVE VICE PRESIDENT
Robert Bosworth, PAST PRESIDENT
Harry Wiant, PAST PRESIDENT
Karl Wenger, PRESIDENT
James Coufal, VICE PRESIDENT

SOCIETY OF AMERICAN FORESTERS
NORTH WEST OFFICE
4033 SW Canyon Rd.
Portland, OR 97221 USA
Phone: 503-224-8046   Fax: 503-226-2515
Website: www.forestry.org

Founded: 1900
Membership: 1925
Scope: Regional

Description: The mission of the society is to advance the science, education, technology, and practice of forestry; enhance its members' competency and professionalism; and use the knowledge and skills of the profession to benefit society.

Publication(s): Western Forester

Contact(s):
Lori Rasor, MANAGER AND EDITOR
4033 SW Canyon Rd., Portland, OR 97221
Phone: 503-224-8046

SOCIETY OF TYMPANUCHUS CUPIDO PINNATUS LTD.
Stone Ridge II, Suite 280, N 14 W23777 Stone Ridge Dr.
Waukesha, WI 53188-1188 USA
E-mail: mihal@execpc.com

Founded: 1961
Scope: National

Description: Nonprofit organization dedicated to the preservation of the prairie chicken for all future generations in Wisconsin and all threatened and endangered species native to the state of Wisconsin.

Publication(s): Boom

Keyword(s): Endangered Species, Prairies, training

Contact(s):
Russell Schallert, PRESIDENT
Kurt Remus, Jr., SECRETARY
3860 N. Port Washington Rd., Milwaukee, WI 53217
Glenn Goergen, TREASURER
Deloitte and Touche, 250 E. Wisconsin Ave., Milwaukee, WI 53202
Lawrence Deleers, Jr., VICE PRESIDENT
4065 Highway Y, Saukville, WI 53080
Phone: 773-373-3366
William Emory, VICE PRESIDENT
Klug and Smith Company, 4425 W. Mitchell, Milwaukee, WI 53214
Gregory Septon, VICE PRESIDENT
Milwaukee Public Museum; 800 W. Wells Street, Milwaukee, WI 53233

SOCIETY OF WETLAND SCIENTISTS
P.O. Box 7060
Lawrence, KS 66044-7060 USA
Phone: 785-843-1235   Fax: 785-843-1274
E-mail: sws@allenpress.com
Website: www.sws.org

Founded: 1979
Membership: 5000
Scope: International

Description: International nonprofit education and charitable society of persons interested in wetland science, technology, and related fields. Encourages educational, scientific, and technological development and advancement in all fields of wetland science. Encourages protection, restoration, and stewardship of wetlands. Student memberships and scholarships.

Publication(s): SWS Bulletin, Wetlands

Keyword(s): Aquatic Habitats, Environmental and Conservation Education, Sustainable Ecosystems, Exotic species, Aquatic nuisance species, Wetlands

Contact(s):
Virginia Carter, PAST PRESIDENT
Barry Warner, PRESIDENT
Glenn Guntenspergen, SECRETARY
Phone: 301-497-5523
Mary Kentula, TREASURER
Phone: 541-754-4478

SOIL AND WATER CONSERVATION SOCIETY

Attn: Deb Happe-vonArb, 7515 NE Ankeny Rd.
Ankeny, IA 50021-9764 USA
Phone: 515-289-2331   Fax: 515-289-1227
E-mail: swcs@swcs.org
Website: www.swcs.org

Founded: 1945
Scope: National

Description: (formerly Soil Conservation Society of America) The Soil and Water Conservation Society is a multidisciplinary membership organization advocating protection, enhancement, and wise use of soil, water, and related natural resources. SWCS programs emphasize the interdependence of natural resources through education, publications, and a network of local chapters throughout the U.S. and Canada. SWCS also manages the World Association of Soil and Water Conservation.

Publication(s): Conservation Voices, Journal of Soil and Water Conservation

Keyword(s): Agriculture, Environmental and Conservation Education, Soil Conservation, training, Wetlands, Environmental Protection, Conservation Tillage, Renewable Resources, Research, Sustainable Ecosystems, Water Quality, Sustainable Development, Training, Watersheds

Contact(s):
Deb Happe-vonArb, COMMUNICATIONS DIRECTOR/EDITOR
deb@swcsswcs.org
Craig Cox, EXECUTIVE VICE PRESIDENT
craigcox@swcs.org
Charles Persinger, MEMBER SERVICES DIRECTOR
charliep@swcs.org
James Bruce, OTTAWA, CANADA REPRESENTATIVE
Phone: 613-731-5929
Fax: 613-731-3509

Dana Chapman, PRESIDENT
Ag Consulting Service 1634 Monroe Ave, Rochester, NY 14618
Phone: 716-341-5312
Phone: 315-255-9266
dcc@acsoffice.com
Norman Berg, WASHINGTON, DC REPRESENTATIVE

SONORAN INSTITUTE

7630 E. Broadway Blvd., Suite 203
Tucson, AZ 85710 USA
Phone: 520-290-0828   Fax: 520-290-0969
E-mail: sonoran@sonoran.org
Website: www.sonoran.org

Founded: 1990
Scope: National

Description: The mission of the Sonoran Institute is to promote community-based conservation strategies that preserve the ecological integrity of protected lands and at the same time meet the economic aspirations of adjoining landowners and communities. Underlying this mission is the conviction that locally-driven and inclusive approaches to conservation produce the most effective results.

Keyword(s): Land Use Planning, Public Lands, Sustainable Development, Riparian Restoration, Stewardship, Community Conservation

Contact(s):
John Shepard, ASSOCIATE DIRECTOR
Joaquin Murriet, ASSOCIATE DIRECTOR OF BORDERLANDS PROGRAM
Frank Gregg, CHAIRMAN
Steve Cornelius, DIRECTOR OF BORDERLANDS PROGRAM
Lara Schmit, DIRECTOR OF COMMUNICATIONS
Lee Nellis, DIRECTOR OF LAND USE POLICY
Mark Briggs, DIRECTOR OF RESEARCH
Luther Propst, EXECUTIVE DIRECTOR
Josh Schachter, PROGRAM ASSOCIATE
Susan Culp, RESEARCH ASSOCIATE
Jake Kittle, SECRETARY/TREASURER
Fred Bosselman, VICE CHAIR

SONORAN INSTITUTE

NORTHWEST OFFICE
201 South Wallace Ave.
Bozeman, MT 59715 USA
Phone: 406-587-7331   Fax: 406-587-2027
Website: www.sonoran.org

Founded: 1997
Membership: 30
Scope: International

Description: The Sonoran Institute works with communities to consume and restore important natural landscapes in western North America, including the wildlife and cultural values of these lands.

Publication(s): Measuring Change in World Communities In Economics Workbook For Northern Canada, Landscape Wildlife and People Community Workbook For Habitat Conservation

Contact(s):
Ray Rasker, DIRECTOR
Barb Cestero, DIRECTOR OF THE YELLOWSTONE TO YUKON PROGRAM

Ben Alexander, DIRECTOR OF WORKING LANDSCAPE PROGRAM

SOUND EXPERIENCE

2310 Washington St.
Port Townsend, WA 98368 USA
Phone: 360-379-0438   Fax: 360-379-0439
E-mail: soundexp@olypen.com
Website: www.soundexp.org

Founded: 1989
Scope: National

Description: Sound Experience involves participants in exploration of Puget Sound from the decks of a traditional sailing ship (the 101' Schooner Adventuress). Our mission is protecting Puget Sound through education.

Publication(s): publications available on website

Keyword(s): Environmental and Conservation Education, Historic Preservation, Outdoor Recreation, Youth Organizations

Contact(s):

Jenell Dematteo, EXECUTIVE DIRECTOR
Nick Worden, PRESIDENT
Jan Vulk, VICE PRESIDENT

SOUTH ATLANTIC FISHERY MANAGEMENT COUNCIL

One Southpark Cir.
Charleston, SC 29407-4699 USA
Phone: 843-571-4366
E-mail: safmc.net
Website: www. safmc.net

Founded: 1976
Membership: 12
Scope: Regional

Description: Responsible for the conservation and management of fish stocks within the 200-mile limit (federal waters) of the Atlantic off the coasts of North Carolina, South Carolina, Georgia, and Florida.

Publication(s): Fishery Management Plans, South Atlantic Update

Keyword(s): Marine Fisheries, Wildlife

Contact(s):

Fulton Love, CHAIRMAN
Robert Mahood, EXECUTIVE DIRECTOR

SOUTH CAROLINA ASSOCIATION OF CONSERVATION DISTRICTS

1835 Assembly St., Rm. 950 Strom Thurmond Federal Building
Columbia, SC 29201 USA
Phone: 803-755-0319   Fax: 803-253-3670

Founded: NA
Membership: 330
Scope: Statewide

Keyword(s): Conservation Districts

Contact(s):

Larry Nates, PRESIDENT
112 Luther Dr., Gaston, SC 29053
Phone: 803-755-0319
Linda Tansill, EXECUTIVE DIRECTOR

Amanda Bauknight, SECRETARY
1967 Burles Ridge Rd., Easley, SC 23640
Phone: 86-485-9338
Diane Edwins, TREASURER
4169 State Rd., Ridgeville, SC 29472
Phone: 843-688-5461
Ed McAllister, VICE PRESIDENT

SOUTH CAROLINA B.A.S.S. CHAPTER FEDERATION

Attn: President, 1469 Schurlknight Rd.
St. Stephen, SC 29479-3627 USA
Phone: 803-567-4680
E-mail: tonybennett@dycon.com
Website: www.scbass.com

Founded: NA
Scope: Statewide

Description: An organization of Bassmaster chapters, affiliated with the Bass Anglers Sportsman Society, organized to fight pollution, assist state and national conservation agencies in their efforts, and teach the young people of our country good conservation practices. Dedicated to the realistic conservation of our water resources.

Publication(s): South Carolina B.A.S.S. Federation, Inc. Newsletter, South Carolina Forestry Journal

Keyword(s): Forests and Forestry, Land Use Planning, Transportation, Wetlands, training

Contact(s):

Tom Hueble, CONSERVATION DIRECTOR
446 Baker Rd., Whitmire, SC 29178
Phone: 803-694-3602
hueblefamily@mindsprings.com
Tony Bennett, PRESIDENT
Phone: 803-567-4680

SOUTH CAROLINA COASTAL CONSERVATION LEAGUE

P.O. Box 1765
Charleston, SC 29402 USA
Phone: 843-723-8035   Fax: 843-723-8308
E-mail: scccl@charleston.net
Website: www.scccl.org

Founded: 1989
Scope: Local

Description: SCCCL works to protect our state's coastal resources through programs in land use, forestry, water quality and public education.

Publication(s): Conservation League Newsletter

Keyword(s): Beaches, Biodiversity, Birds, Coastal Construction and Erosion, Environmental and Conservation Education, Environmental Planning, Protected Areas, Greenways, Habitat Conservation, Land Management, Land Use Planning, Nonpoint Source Pollution, Urban and Rural Development, Water Quality, Wetlands

Contact(s):

Jane Lareau, CONTACT FOR FORESTRY
janel@sccoast.net
Nancy Vinson, CONTACT FOR H2O
nancyv@scccl.org
Sam Passmore, CONTACT FOR LAND USE
samp@scccl.org

Dana Beach, EXECUTIVE DIRECTOR
danabeach@scccl.org

SOUTH CAROLINA ENVIRONMENTAL LAW PROJECT

P.O. Box 1380
Pawleys Island, SC 29585 USA
Phone: 843-527-0078   Fax: 843-527-0540
Website: www.scelp.org

Founded: 1987
Membership: 2
Scope: Regional

Description: SCELP is a nonprofit organization whose mission is to protect the natural environment of South Carolina by providing legal services and advice to environmental organizations and concerned citizens, and by improving the state's system of environmental regulation.

Publication(s): Mountains and Marshes

Keyword(s): Environmental Law, Protected Areas, Environmental Protection, Legal Advocacy, Natural Resource Conservation, Forests and Forestry, Public Lands, Rivers, Water Pollution Management, Wilderness, Water Quality, Wetlands

Contact(s):
James Chandler, PRESIDENT AND GENERAL COUNSEL

SOUTH CAROLINA FORESTRY ASSOCIATION

4901 Broad River Rd., P.O. Box 21303
Columbia, SC 29221 USA
Phone: 803-798-4170   Fax: 803-798-2340
E-mail: fcfa@scforestry.org
Website: www.scforestry.org

Founded: 1968
Scope: State

Description: A nonprofit educational organization with a membership of timberland owners, wood dealers, wood-using industries, equipment suppliers, and individuals interested in forest conservation and wise use of natural resources.

Contact(s):
Sam Coker, CHAIRMAN OF THE BOARD
SC Pole at Piling Company P.O. Box 3309,
Leesville, SC 29070
Phone: 803-532-5806
Robert Scott, PRESIDENT

SOUTH CAROLINA NATIVE PLANT SOCIETY

P.O. Box 759
Pickens, SC 29671 USA
Phone: 864-868-7798
Website: www.clemson.edu/scnativeplants

Founded: 1996
Scope: Statewide

Description: Promotes native plants and plant communities through an education-based agenda. The Society sponsors field trips, symposiums, workshops and lectures. The Society also works with government agencies to assist in seed collection and management.

Publication(s): Newsletter, Brochure

Keyword(s): Native Plants, Flowers, Plants, and Trees

Contact(s):
Rick Huffman, PRESIDENT
Phone: 864-868-7798
rhuffman@innova.net
Bill Stringer, VICE-PRESIDENT
Phone: 864-656-3527

SOUTH CAROLINA WILDLIFE FEDERATION

2711 Middleburg Dr., Suite 104
Columbia, SC 29204 USA
Phone: 803-256-0670   Fax: 803-256-0690
E-mail: mail@scwf.org
Website: www.scwf.org

Founded: 1931
Membership: 5000
Scope: Statewide

Description: A representative statewide organization, affiliated with the National Wildlife Federation, dedicated to the protection and enhancement of wildlife and its habitat through public education and government interaction.

Publication(s): Working for Wildlife, Out of Doors

Keyword(s): Aquatic Habitats, Wildlife, Renewable Resources, Water Pollution Management, Wetlands

Contact(s):
John Helms, ALTERNATE REPRESENTATIVE
Joyce Peters, EDITOR
Sara Theiben, EDUCATION PROGRAMS CONTACT
Angela Viney, EXECUTIVE DIRECTOR
Dave Hargett, REPRESENTATIVE
R. Patten Watson III, TREASURER

SOUTH DAKOTA ASSOCIATION OF CONSERVATION DISTRICTS

FDACD
14321 465th Ave
Marvin, SD 57251 USA
Phone: 605-938-4579   Fax: 605-895-9424
E-mail: conserve@wcenet.com
Website: www.fdacd.com

Founded: NA
Membership: 340
Scope: Regional
Keyword(s): Conservation Districts

Contact(s):
Gene Williams, BOARD MEMBER NACD
P.O. Box 2, Interior, SD 57750-0002
Phone: 605-433-5469
Fax: 605-433-5470
gsw111@gwtc.net
Angela Ehlers, EXECUTIVE SECRETARY
116 N. Euclid, P.O. Box 275, Pierre, SD 57501-0275
Phone: 605-224-0361
Fax: 605-773-4531
Gerald Thaden, PRESIDENT
14321 465th Ave., Marvin, SD 57251-9720
Phone: 605-938-4579
John Majeres, SECRETARY/TREASURER
RR2 Box 122, Dell Rapids, SD 57022-0208
Phone: 605-428-3090
Fax: 605-988-5773

Lynn Denke, VICE PRESIDENT
19580 224th St., Creighton, SD 57729-9747
Phone: 605-279-2633
ldenke@gwtc.net

SOUTH DAKOTA B.A.S.S. CHAPTER FEDERATION

Attn: President, P.O. Box 377
Brandon, SD 57005 USA
Phone: 605-582-2309   Fax: 605-582-2309
Website: www.sdbassfederation.com

Founded: NA
Membership: 250
Scope: Statewide

Description: An organization of Bassmaster chapters, affiliated with the Bass Anglers Sportsman Society, organized to fight pollution, assist state and national conservation agencies in their efforts, and teach young people of our country good conservation practices. Dedicated to the realistic conservation of our water resources.

Publication(s): Birds of South Dakota, 1991, The South Dakota Breeding Bird Atlas, 1995, B.A.S.S. Federation Newsletter "Dakota Bassin", South Dakota Bird Notes

Keyword(s): Birds, Endangered Species, education, Raptors, Waterfowl

Contact(s):
Phillip Risnes, CONSERVATION DIRECTOR
26643 461st. Ave., Hartford, SD 57033
Phone: 605-526-4339
philrisnes@aol.com
Chuck Doom, PRESIDENT

SOUTH DAKOTA ORNITHOLOGISTS UNION

Dept. of Biology, University of South Dakota
Vermillion, SD 57069 USA
Phone: 605-677-6175   Fax: 605-677-6557

Founded: 1949
Membership: 300
Scope: Statewide

Description: To encourage the study of birds in South Dakota and to promote the study of ornithology by more closely uniting the students of this branch of natural science.

Publication(s): Birds of South Dakota (1991), South Dakota Breeding Bird Atlas, The (1995), South Dakota Bird Notes

Keyword(s): Agriculture, Air Quality and Pollution, Energy, Solid Waste Management, Exotic species, Aquatic nuisance species

Contact(s):
Dan Tallman, EDITOR
Box 740, Northern State University, Aberdeen, SD 57401
Phone: 605-226-2255
Jeffrey Palmer, PAST PRESIDENT
821 NW Fifth St., Madison, SD 57041
Phone: 605-256-9745
Robb Schenck, PRESIDENT
422 N. Linwood Ct., Sioux Falls, SD 57103
David Swanson, SECRETARY
Biology Department, University of South Dakota, Vermillion, SD 57069
Phone: 605-624-0203
Nelda Holden, TREASURER
1620 Elmwood Dr., Brookings, SD 57006
Phone: 605-692-8278

SOUTH DAKOTA RESOURCES COALITION

P.O. Box 66
Brookings, SD 57006 USA
Phone: 605-697-6675
E-mail: sdrc@brookings.net
Website: www.geocities.com/sdrc2/

Founded: 1972
Scope: Statewide

Description: Seeks to promote the survival and integrity of water, energy, land, wildlife, and air resources, along with justice in their allocation.

Publication(s): ECO-Forum

Keyword(s): Environmental Protection

Contact(s):
Loanne Nappon, CHAIR
Phone: 605-693-4893
Jeff Cornforth, EDITOR
Phone: 605-692-8579
Lawrence Novotny, SECRETARY
Phone: 605-688-6172
Robert Roby, TREASURER
4512 Belmont St., Sioux Falls, SD 57102
Phone: 605-371-0743
Kaye Hunt, VICE CHAIR
P.O. Box 309, Garretson, SD 57030
Phone: 605-594-3558

SOUTH DAKOTA WILDLIFE FEDERATION

P.O. Box 7075
Pierre, SD 57501-7075 USA
Phone: 605-224-7524   Fax: 605-224-7524
E-mail: sdwf@sbtc.net
Website: http://sdwf.org

Founded: 1945
Membership: 4500
Scope: National

Description: A representative statewide organization, affiliated with the National Wildlife Federation, dedicated to the protection and enhancement of wildlife and its habitat through public education and government interaction.

Publication(s): Out of Doors

Keyword(s): Aquatic Habitats, Wildlife, Renewable Resources, Water Pollution Management, Wetlands

Contact(s):
Chuck Clayton, ALTERNATE REPRESENTATIVE
Chris Hesla, EXECUTIVE DIRECTOR AND EDITOR & EDUCATION PROGRAMS CONTACT
Mike Larsen, PRESIDENT AND REPRESENTATIVE
Robert Jacobson, TREASURER

SOUTH OAHU SOIL AND WATER CONSERVATION DISTRICT

938 Kamiloniu Place,
Honolulu, HI 96825 USA
E-mail: swcd@soswcd.org
Website: http://sos.wcd.org

Founded: 1939
Membership: 2,200
Scope: Statewide

Description: For better understanding, appreciation, and conservation of Hawaii's native wildlife resources, especially its unique and endangered bird species and their associated ecosystems.

Keyword(s): Conservation Districts

Contact(s):
Joloyce Kaia, 1ST VICE PRESIDENT
P.O. Box 404, Hana, HI 96713
Phone: 808-248-7725
Ted Inouye, ALTERNATE BOARD MEMBER
P.O. Box 278, Hanamaulu, HI 96715
Phone: 808-245-3027
Valerie Mendes, ALTERNATE BOARD MEMBER
1100 Alakea St. 1200, Honolulu, HI 96813
Phone: 808-531-8181
Mike Tulang, EXECUTIVE DIRECTOR
919 Ala Moana Blvd. Rm. 309, Honolulu, HI 96814
Phone: 808-586-4389
Fax: 808-586-4300
David Norbriga, PRESIDENT
Phone: 808-244-7951
Fax: 808-244-4108

SOUTHEAST ALASKA CONSERVATION COUNCIL (SEACC)

419 6th St., Suite 200
Juneau, AK 99801 USA
Phone: 907-586-6942   Fax: 907-463-3312
E-mail: info@seacc.org
Website: www.seacc.org

Founded: 1969
Membership: 1300
Scope: Regional

Description: SEACC is a coalition of 17 local conservation groups, dedicated to preserving the integrity of Southeast Alaska's magnificent natural environment. Protection of the region's pristine coastal rainforests, abundant fish and wildlife, and outstanding scenery. Provides for a sustainable approach to economic stability, subsistence use areas, recreational opportunities, and a unique way of life.

Publication(s): Action Alerts, RAVENCALL

Keyword(s): Conservation, Environment, Wildlife, Forest Management, Land Use Planning, Outdoor Recreation, Internships, Public Lands, Rivers, Coral Reefs, Sustainable Development, Water Quality, Watersheds, Rainforests

Contact(s):
Paige Else, Bart Koehler, ASSOCIATE DIRECTOR
P.O. Box 1620, Durango, CO 81302
Buck Lindekugel, CONSERVATION DIRECTOR AND STAFF ATTORNEY
Katya Kirsch, EXECUTIVE DIRECTOR
Wayne Weihing, PRESIDENT
P.O. Box 1193, Wardcove, AK 99928
Dana Owen, TREASURER
949 Goldbelt, Juneau, AK 99801
Bruce Baker, VICE PRESIDENT
P.O. Box 211384, Auke Bay, AK 99821

SOUTHEASTERN ASSOCIATION OF FISH AND WILDLIFE AGENCIES

8005 Freshwater Farms Rd.
Tallahassee, FL 32308 USA
Phone: 850-893-1204   Fax: 850-893-6204
E-mail: seafwa@aol.com
Website: www.seafwa.org

Founded: 1938
Scope: Regional

Description: To protect the best interests of the southeastern states by maintaining their right of jurisdiction over their wildlife resources on public and private lands, by supporting or opposing state and federal wildlife legislation, and by making recommendations on federal programs involving aid. Conducts annual conference for the exchange of ideas and research and land management information concerning wildlife and native and fresh water fisheries.

Keyword(s): Wildlife Rehabilitation, training, Forest Management, Wildlife

Contact(s):
Robert Brantly, EXECUTIVE SECRETARY
8005 Freshwater Farms Rd., Tallahassee, FL 32308
Phone: 850-893-1204
Bill Woodfin, PRESIDENT
Director, Virginia Dept. of Game and Inland Fisheries, 4010 W. Broad St., Richmond, VA 23230
Phone: 804-367-1000
William Woodfin, PRESIDENT
Phone: 804-367-9231

SOUTHEASTERN COOPERATIVE WILDLIFE DISEASE STUDY

College of Veterinary Medicine,University of Georgia
Athens, GA 30602 USA
Phone: 706-542-1741   Fax: 706-542-5865
Website: www.acwds.org

Founded: 1957
Membership: 25
Scope: Regional

Description: The first regional diagnostic and research service in the U.S. for the specific purpose of investigating wildlife diseases. This joint-state organization currently is sponsored by the Southeastern Association of Fish and Wildlife Agencies; Veterinary Services of APHIS, USDA; and the Biological Resources Division of USDI. Participating states: AL, AR, FL, GA, KY, LA, MD, MO, MS, NC, PR, SC, TN, VA, WV.

Publication(s): SCWDS BRIEFS Newsletter

Keyword(s): Agriculture, Environmental and Conservation Education, Health and Nutrition, standards

Contact(s):
John Fischer, DIRECTOR

SOUTHEASTERN FISHES COUNCIL

c/o Stephen T. Ross, Dept. of Biological Studies, University of Southern Mississippi
Hattiesburg, MS 39406-5018 USA
E-mail: STEPHENROSS@USM.EDU
Website: flmnh.ufl.edu/fish/organization

Founded: NA
Scope: National

Description: Objectives are to provide for the pursuit and transmittal of information on the status and protection of southeastern fishes and their habitats, and to promote the perpetuation of rich natural assemblages of fishes and their habitats, as well as the localized unique forms and their habitats.

Publication(s): Proceedings of the Southeastern Fishes Council

Keyword(s): Biodiversity, Aquatic Habitats, Conservation, Endangered Species, Wildlife, Rivers, education, Watersheds

Contact(s):
Stephen Ross, CHAIR
Frank Pezold, EDITOR
Department of Biology, Northeast Louisiana State University, Monroe, LA 71209
Melvin Warren, PAST CHAIR
Gerry Dinkins, SECRETARY
3D International Environmental Group, 7039 Maynardville Highway, Knoxville, TN 37830-7976
Peggy Shute, TREASURER
Tennessee Valley Authority, Natural Heritage Program, Norris, TN 37820

SOUTHERN APPALACHIAN BOTANICAL SOCIETY

Biology Department, 2100 College St., Newberry College
Newberry, SC 29108 USA
Phone: 803-321-5257   Fax: 803-321-5636

Founded: 1936
Membership: 750
Scope: Regional

Description: A nonprofit organization to desseminate information on the native plants of eastern North America through meetings and publications.

Publication(s): Castanea Journal, Chinquapin

Keyword(s): Botany, Ecology, Ecosystems, Endangered Species, National Forests, Native Plants, education, Plants, Research, State Parks, Trees, Wildflowers

Contact(s):
Joe Winstead, PRESIDENT
Phone: 870-235-4289
jewinstead@saumag.edu
Charles Horn, SECRETARY AND TREASURER
Phone: 803-321-5257
Fax: 803-321-5232
chorn@newberry.edu

SOUTHERN ENVIRONMENTAL LAW CENTER

201 W. Main St., Suite 14
Charlottesville, VA 22902-5065 USA
Phone: 804-977-4090   Fax: 434-977-1483
E-mail: selcva@selcva.org
Website: www.southernenvironment.org

Founded: 1985
Membership: 45
Scope: Regional

Description: A regional nonprofit public interest advocacy organization committed to protecting the natural resources of the southeast through direct advocacy in court and before regulatory agencies; through assistance to state and local environmental groups in the region; and through providing regional leadership on key southeastern environmental issues.

Publication(s): Energy Choices: A Primer on Electric Utility Industry Restructuring, Citizen's Guides to Protecting Wetlands in South Carolina, Alabama, and Georgia, Visual Pollution and Sign Control, Energy 2000: A Blueprint for an Energy Efficient Virginia

Keyword(s): Energy, Environmental Law, Forests and Forestry, Water Pollution Management, Wetlands, Air Quality and Pollution, Acid Rain, Coasts, Energy Conservation, Land Preservation, Open Space, Public Lands, Natural Areas, Planning Management, Transportation

Contact(s):
Deaderick Montague, BOARD OFFICER AND CHAIRMAN OF THE BOARD OF TRUSTEES
Cathryn McCue, COMMUNICATIONS COORDINATOR
Phone: 804-977-4090
Lark Hayes, DIRECTOR OF CAROLINAS OFFICE
Phone: 919-967-1450
Frederick Middleton, EXECUTIVE DIRECTOR
Phone: 804-977-4090
Mary Rice, SECRETARY
Frederick Middleton, TREASURER
Terence Sieg, VICE PRESIDENT

SOUTHERN ENVIRONMENTAL LAW CENTER

NORTH CAROLINA OFFICE
200 W. Franklin St., Suite 330
Chapel Hill, NC 27516-2520 USA
Phone: 919-967-1450   Fax: 919-929-9421
E-mail: selcnc@selcnc.org
Website: www.southernenvironment.org

Founded: NA
Membership: 50
Scope: Regional
Publication(s): Southern Resources

SOUTHERN NEW ENGLAND FOREST CONSORTIUM, INC. (SNEFCI)

P.O. Box 760
Chepachet, RI 02816 USA
Phone: 401-568-1610   Fax: 401-568-7874
E-mail: sneforest@efortress.com

Founded: 1985
Membership: 3
Scope: Regional

Description: SNEFCI promotes wise conservation practices in Southern New England. Our goals are to reduce forest fragmentation, promote stewardship of forest resources, and enhance urban and community forst resources

Publication(s): Your Family Lands: Legacy or Memory: Commonly Asked Questions, Cost of Community Services in Southern New England, Foresters and The Care of Your Land, Land Conservation Development and Property Taxes in Rhode Island, Threatened and Endangered Species Field Guide in New England, Forest Land Conversion, Fragmentation, and Partialization, Your Family Land: Legacy or Memory, Preferencial Property Tax Treatment of Open Space Land in New England

Keyword(s): Forest Management, Urban Forestry, Planning Management

Contact(s):
Christopher Modisette, EXECUTIVE DIRECTOR
Thomas Dupree, PRESIDENT

Phone: 401-277-1414
Fax: 401-647-3590
riforestry@edgenet.net
Hans Bergey, TREASURER
Phone: 401-821-8746
Fax: 401-821-8746
hberg16@aol.com
Donald Smith, VICE PRESIDENT
Phone: 860-424-3630
Fax: 860-424-4070
don.smith@po.state.ct.us

SOUTHERN RHODE ISLAND STATE ASSOCIATION OF CONSERVATION DISTRICTS

60 Quaker Ln, Suite 46
Warwick, RI 02866-0114 USA
Phone: 401-822-8832   Fax: 401-828-0433
Website: ri.nacdnet.org/sricd_web/index.htm

Founded: NA
Scope: Statewide

Publication(s): Cultivation Notes 1-16, Bi-Annual Newsletter, Wild Plants! (some basic information, other resources and fact sheets)

Keyword(s): Endangered Species, Environmental and Conservation Education, Flowers, Plants, and Trees, National Parks, Terrestrial Habitats, Conservation Districts

Contact(s):
  Robert Swanson, CHAIR
  39 Shannock Hill Rd., Carolina, RI 02812
  Phone: 401-364-4069
  Jesse Carpenter, PRESIDENT, BOARD MEMBER
  Phone: 401-762-7346
  Fahma1@aol.com
  John Devany, TREASURER
  Emerson Wildes, VICE PRESIDENT
  Whimshaw Farm, Shaw Rd., Little Compton, RI 02837
  Phone: 401-635-2935

SOUTHERN UTAH WILDERNESS ALLIANCE

MOAB OFFICE
P.O.Box 968
Moab, UT 84532-0968 USA
Phone: 435-259-5440   Fax: 435-259-9151
E-mail: suwa.org
Website: www.suwa.org

Founded: NA
Scope: Statewide

SOUTHERN UTAH WILDERNESS ALLIANCE

ST. GEORGE OFFICE
P.O. Box 1726
Cedar City, UT 84721 USA
Phone: 801-486-3161

Founded: NA
Scope: Statewide

SOUTHERN UTAH WILDERNESS ALLIANCE

WASHINGTON, DC OFFICE
122 C St., NW
Washington, DC 20001 USA
Phone: 202-546-2215   Fax: 202-544-5197
Website: www.suwa.org

Founded: NA
Scope: Statewide

Publication(s): Redrock Wilderness (Quarterly Newsletter)
Contact(s):
  Keith Hammond

SOUTHERN UTAH WILDERNESS ALLIANCE (SUWA)

Headquarters, 1471 S. 1100 E.
Salt Lake City, UT 84105-2423 USA
Phone: 801-486-3161   Fax: 801-486-4233
E-mail: suwa@suwa.org
Website: www.suwa.org

Founded: 1983
Membership: 16000
Scope: International

Description: SUWA advocates wilderness preservation for qualifying federal public lands in Utah's incomparable canyon country. Through the allied efforts of SUWA's staff, Utah activists, and concerned citizens across the United States, SUWA seeks to give its members and the general public a voice in deciding the fate of America's redrock wilderness.

Publication(s): bulletins, America's Redrock Wilderness—quarterly newsletter

Contact(s):
  Hansjorg Wyss, CHAIRMAN
  Larry Young, EXECUTIVE DIRECTOR
  Greg Miner, SECRETARY
  Mark Ristow, TREASURER
  Ted Wilson, VICE CHAIRMAN

SOUTHFACE ENERGY INSTITUTE

SOUTHFACE ENERGY & ENVIRONMENTAL RESOURCE CENTER
241 Pine St.
Atlanta, GA 30308 USA
Phone: 404-872-3549   Fax: 404-872-5009
E-mail: info@southface.org
Website: www.southface.org

Founded: 1978
Membership: 1000
Scope: International

Description: The Southface is a nonprofit organization that promotes the use of sustainable energy and environmental technologies and policies in the building sciences through education, research, and technical assistance.

Publication(s): Sustainable Design, Construction & Land Development Guidelines for the Southeast, A Builder's Guide to Energy Efficient Homes in Georgia, Home Energy Projects, The Southface Journal of Sustainable Building

Keyword(s): Energy Conservation, Internships, Sustainable Development, Environment

Contact(s):
  Dennis Creech, EXECUTIVE DIRECTOR
  dcreech@southface.org
  Aziza Cooper, OUTREACH COORDINATOR
  aziza@southface.org
  David Dimling, PRESIDENT

SOUTHWEST RESEARCH AND INFORMATION CENTER

105 Standford SE
P.O Box 4524
Albuquerque, NM 87106 USA
Phone: 505-262-1862   Fax: 505-262-1864
E-mail: sricdon@earthlink.net
Website: www.sric.org

Founded: 1971
Scope: National

Description: SRIC is a nonprofit organization founded to provide timely, accurate information to the public on a broad range of issues related to the environment, human, and natural resources. SRIC's twin objectives are to promote citizen participation and environmental justice, and to protect natural resources.

Publication(s): Workbook, The

Keyword(s): Environmental Justice, Mining, Coral Reefs

Contact(s):
Don Hancock, ADMINISTRATOR
Lalora Charles, SECRETARY
Wilfred Rael, TREASURER
Anne Albrink, VICE PRESIDENT

SOUTHWESTERN HERPETOLOGISTS SOCIETY

P.O. Box 7469
Van Nuys, CA 91409 USA
Phone: 818-503-2052
Website: www.swhs.org

Founded: 1954

Description: A California non profit corporation dedicated to the education of its members and the public concerning the roles of reptiles and amphibians in the natural world, the conservation of all wildlife, in particular reptiles and amphibians, and the cooperation between amateur and professional herpetologists, the hobbyist and the academician, for the promotion of the study of lizards, snakes, turtles, tortoises, geckos, skinks, monitors, frogs, toads, and all other reptiles and amphibians.

Contact(s):
Jarron Lucas, PRESIDENT

SPORTSMANS ALLIANCE OF MAINE

R.R. 12 , Church Hill Rd.
Augusta, ME 04330-9749 USA
Phone: 207-622-5503   Fax: 207-622-5596
Website: www.samcef.org

Founded: 1975
Membership: 1500
Scope: Statewide

Description: SAM is a statewide nonprofit organization of sportsmen and women dedicated to hunting, fishing, trapping, protection of wildlife habitat, and conservation. Lobbies and works with state agencies on behalf of Maine sportsmen.

Publication(s): SAM News

Contact(s):
James Hilly, 1ST VICE PRESIDENT
Robert Cram, 2ND VICE PRESIDENT
George Smith, EXECUTIVE DIRECTOR AND EDITOR
Edye Cronk, PRESIDENT
Herbert Morse, SECRETARY AND CLERK
Richard Paradis, TREASURER

ST. REGIS MOHAWK TRIBE

Environment Division 412 State Rt. 37
Hogansburg, NY 13655 USA
Phone: 518-358-5937   Fax: 518-358-6252
E-mail: earth@northnet.org
Website: www.northnet.org/earth

Founded: NA
Membership: 18
Scope: National

Description: To monitor, maintain, and protect the environment of the St. Regis Mohawk Tribe for the prevention of disease and injury to body, mind, and spirit. Participation in hazardous waste remediation, Superfund site cleanups, reservation environmental protection, and air and water quality.

Publication(s): Iroquois Environmental Newsletter

Keyword(s): Environmental Protection

Contact(s):
Carol White, AKWESASNE LIBRARY/CULTURAL CENTER LIBRARIAN
Les Benedict, ASSISTANT DIRECTOR OF ENVIRONMENT DIVISION
earth-benlbenedic@northnet.org
Ken Jock, DIRECTOR OF ENVIRONMENT DIVISION

STANFORD ENVIRONMENTAL LAW SOCIETY

Stanford Law School-559 Nathan Abbott Way
Stanford, CA 94305-8610 USA
Phone: 650-723-4421   Fax: 650-723-0501
Website: www.els.stanford.edu

Founded: 1969
Scope: Local

Description: The Stanford Environmental Law Society is the oldest student organization of its kind in the United States. Its primary function is sponsorship of original research in developing areas of environmental law. The Society relies on contributions, grants, and proceeds from the sale of publications.

Publication(s): Strategies for Environmental Law Enforcement, Endangered Species Act, Handbook, Hazardous Waste Disposal Sites, Who Runs the Rivers?, Stanford Environmental Law Journal

Keyword(s): Environmental Law, Research

Contact(s):
Louise Warren, BUSINESS MANAGER
Janelle Smith, CO-PRESIDENT
Katherine Wannamaker, CO-PRESIDENT

STATE AND TERRITORIAL AIR POLLUTION PROGRAM ADMINISTRATORS AND THE ASSOCIATION OF LOCAL AIR POLLUTION CONTROL OFFICIALS (STAPPA AND ALAPCO)

444 N. Capitol St., NW, Suite 307
Washington, DC 20001 USA
Phone: 202-624-7864   Fax: 202-624-7863
E-mail: 4cleanair@sso.org
Website: www.4cleanair.org

Founded: 1980
Membership: 200
Scope: National

Description: The national associations of air pollution control agencies in the states, territories, and major metropolitan areas. The associations' members have primary responsibility for ensuring healthy air quality and represent the technical expertise behind the implementation of our nation's air pollution control laws and regulations.

Publication(s): Controlling Nitrogen Oxides Under The Clean Air Act: A Menu of Options (1994), Controlling Particulate Matter Under The Clean Air Act: A Menu of Options (1995), Meeting the 15% Rate of Progress Requirement Under The Clean Air Act: A Menu of Options (1993)

Keyword(s): Acid Rain, Air Quality and Pollution, Environment, Greenhouse Effect/Global Warming, Pollution Prevention

Contact(s):
S. Becker, EXECUTIVE DIRECTOR

STATE IOWA WOODLANDS ASSOCIATIONS
2735 14th Ave.
Marion, IA 52302-1848 USA
Phone: 515-233-1161

Founded: 1987
Scope: Statewide

Description: A statewide organization affiliated with the National Woodland Owners Association, organized to advance good forestry on the 1.5 million acres of timberland owned by 28,000 nonindustrial private landowners in Iowa.

Publication(s): Timber Talk

Keyword(s): Forests and Forestry

Contact(s):
Al Manning, PRESIDENT
E. Frye, SECRETARY AND EDITOR
Phone: 319-377-2540
Joanne Mensinger, TREASURER
Phone: 319-259-1160
Tom Woodruff, VICE PRESIDENT

STATEWIDE PROGRAM OF ACTION TO CONSERVE OUR ENVIRONMENT (SPACE)
N.H. Current Use Coalition, 54 Portsmouth St.
Concord, NH 03301 USA
Phone: 603-224-3306   Fax: 603-228-0423
E-mail: space@conknet.com
Website: www.nhspace.org

Founded: 1966
Scope: State

Description: A private, not-for-profit advocacy coalition of groups dedcated to conserving open space land. S.P.A.C.E.'s work includes advocacy, education, supporting research and working with the state, towns, and individuals on the administration and monitoring of the current use program.

Publication(s): SPACE Newsletter

Keyword(s): Land Preservation

STEAMBOATERS, THE
P.O. Box 176
Idleyld Park, OR 97447 USA
Phone: 503-496-3003
Website: http://www.steamboaters.org

Founded: 1966

Membership: 400
Scope: National

Description: Formed to preserve, promote, and restore the natural production of wild fish populations, the habitat which sustains them, and the unique aesthetic values of the North Umpqua River for present and future generations.

Keyword(s): Biodiversity, Endangered Species, Environmental Law, Environmental and Conservation Education, Planning Management

Contact(s):
Jim Watson, PRESIDENT
Phone: 541-496-3512
samnjim@rosenet.net
Charlie Spooner, SECRETARY
Paul Moore, TREASURER
Len Janssen, VICE PRESIDENT
Phone: 541-440-9375

STOP
230-651 Notre Dame West
Montreal, Quebec H3C 1H9 Canada
Phone: 514-393-9559   Fax: 514-393-9588

Founded: 1970
Scope: State

Description: Devoted to preserving and improving the quality of the physical and human environment, and to promoting rational utilization of natural resources.

Publication(s): Stop Press

Keyword(s): Acid Rain, Air Quality and Pollution, Chemical Pollution Control, Environmental and Conservation Education, Solid Waste Management, Transportation, Toxic Substances, Nuclear-free, Water quantity, Water export and diversion, Urban Environment, Coral Reefs

STRIPERS UNLIMITED, INC.
P.O. Box 3045
S. Attleboro, MA 02703 USA
Phone: 508-226-4007   Fax: 508-226-2031

Founded: 1965
Membership: 500+
Scope: National

Description: Nonprofit organization formed to promote, conserve, and protect striped bass and to protect and restore its environment. Promotes and encourages research on striped bass in order to preserve it as a natural resource and to increase its areas of reproduction. Members in 21 states and two Canadian provinces.

Publication(s): Newsletter in North East Woods & Waters, monthly

Keyword(s): Environmental and Conservation Education, Wildlife, Sport Fishing, Toxic Substances, Nuclear-free, Water quantity, Water export and diversion, Coral Reefs

Contact(s):
Robert Pond, EXECUTIVE DIRECTOR AND TREASURER
P.O. Box 3045, S. Attleboro, MA 02703
Phone: 508-226-4007
Avis Boyd, EXECUTIVE SECRETARY AND EDITOR
P.O. Box 3166, S. Attleboro, MA 02703
Phone: 508-761-4627
Carleen Proulx, MEMBERSHIP SECRETARY

Norman Whitten, PRESIDENT
24 Ellsworth Terr., Lynn, MA 01904

STROUD WATER RESEARCH CENTER

970 Spencer Rd.
Avondale, PA 19311 USA
Phone: 610-268-2153   Fax: 610-268-0490
E-mail: webmaster@stroudcenter.org
Website: www.stroudcenter.org

Founded: 1967
Membership: 35
Scope: International

Description: The mission of the Stroud Center is to advance the knowledge of river and stream ecosystems through research and education.

Publication(s): Upstream (newsletter)

Keyword(s): Research, Riparian Restoration, Rivers, Coral Reefs, Water Quality, Exotic species, Aquatic nuisance species, Watersheds, Fieldwork, Outdoor Education, Environmental and Conservation Education, Wildlife Rehabilitation, Chemistry

Contact(s):
Claire Birney, DEVELOPEMENT DIRECTOR
clairebirney@stroudcenter.org
Bernard Sweeney, DIRECTOR
sweeney@stroudcenter.org
James Mcgonigle, EDUCATION DIRECTOR
jmcgonigle@stroudcenter.org

STUDENT CONSERVATION ASSOCIATION, CAPITAL REGION OFFICE

CAPITAL REGION OFFICE
1800 N. Kent St., Suite 102
Arlington, VA 22209 USA
Phone: 703-524-2441   Fax: 703-524-2451
E-mail: info@sca-inc.org
Website: www.sca-inc.org

Founded: NA
Scope: Regional, National

Contact(s):
Theresa Drakeford, OFFICE MANAGER

STUDENT CONSERVATION ASSOCIATION, INC.

P.O. Box 550
Charlestown, NH 03603 USA
Phone: 603-543-1700   Fax: 603-543-1828
Website: www.sca-inc.org

Founded: 1957
Membership: 110
Scope: National

Description: SCA is guided by two premises: high school and college-age volunteers can accomplish a variety of conservation tasks vital to the protection of America's natural resources and the experience gained by these volunteers can significantly impact their education, personal development and career goals.

Keyword(s): Environmental and Conservation Education, Internships, Public Lands, Youth Organizations, Careers, Cultural Preservation, Environment, People of Color in the Environment, Renewable Resources

Contact(s):
Edmund Bartlett, CHAIR OF THE BOARD
Mark Bodin, CHIEF FINANCIAL OFFICER
Kevin Hamilton, COMMUNICATIONS
Elizabeth Titus, FOUNDING PRESIDENT
Dale Penny, PRESIDENT
Robert Holley, VICE PRESIDENT OF DEVELOPMENT
Scott Weaver, VICE PRESIDENT OF PROGRAMS

STUDENT CONSERVATION ASSOCIATION, INC.

CALIFORNIA SOUTHWEST REGIONAL OFFICE
655 13th St., Suite 304
Oakland, CA 94612 USA
Phone: 510-832-1966   Fax: 510-832-4726
Website: www.sca-inc.org

Founded: NA
Scope: Regional, National

Publication(s): The Volunteer - newsletter

Contact(s):
Rick Covington, DIRECTOR OF REGIONAL PROGRAMS
rick@sca-inc.org
Bob Coates, REGIONAL VICE PRESIDENT

STUDENT CONSERVATION ASSOCIATION, INC.

NEWARK OFFICE
689 River Rd.
Charlestown, NJ 03603 USA
Phone: 603-543-1700   Fax: 603-543-1828
Website: www.sca-inc.org

Founded: NA
Scope: National

STUDENT CONSERVATION ASSOCIATION, INC.

NORTHWEST OFFICE
1265 S. Main St., Suite 210
Seattle, WA 98144 USA
Phone: 206-324-4649   Fax: 206-324-4998
E-mail: susan@sea-inc.org
Website: www.sca-inc.org

Founded: NA
Scope: Regional, National, International

Publication(s): The Volunteer - quarterly newletters

Contact(s):
Jay Satz, VICE PRESIDENT NATIONAL FIELD OPERATIONS NW REGIONAL OFFICER

STUDENT ENVIRONMENTAL ACTION COALITION (SEAC)

P.O. Box 31909
Philadelphia, PA 19104 USA
Phone: 215-222-4711   Fax: 215-222-2896
E-mail: seac@seac.org
Website: www.seac.org

Founded: 1988
Membership: 1000
Scope: National

Description: SEAC is a national student-run and student-led network of progressive organizations and individuals whose aim is to uproot environmental injustices through action and education. We define the environment to include the physical, economical, political, and cultural conditions in which we live. By challenging the power structure which threatens these environmental conditions, SEAC works to create progressive social change on both the local and global levels.

Publication(s): book- The Student Environmental Organizing Guide, Threshold

Keyword(s): Solid Waste Management, Wilderness, Youth Organizations, Environmental Justice, Social Justice

Contact(s):
Kelly Nagy, NATIONAL COUNSEL COORDINATOR
Phone: 215-222-4711
ncc@seac.org

STUDENT PUGWASH USA
815 15th St., NW, Suite 814
Washington, DC 20005 USA
Phone: 202-393-6555   Fax: 202-393-6550
E-mail: spusa@spusa.org
Website: www.spusa.org

Founded: NA
Scope: National

Description: The mission of Student Pugwash USA is to promote the socially-responsible application of science and technology in the 21st century. As a student organization, Student Pugwash USA encourages young people to examine the ethical, social, and global implications of science and technology, and to make these concerns a guiding focus of their academic and professional endeavors.

Publication(s): Pugwatch, Mindfull: a brainsnack for future leaders with ethical appetites, Global Issues Guidebook, Jobs You Can Live With: Working at the Crossroads of Science, Technology, and Society

Keyword(s): Biotechnology, Communications, Energy, Environmental and Conservation Education, Nuclear/Radiation

Contact(s):
Eric Roberts, EXECUTIVE COMMITTEE CHAIRMAN
Susan Veres, EXECUTIVE DIRECTOR

SUNCOAST SEABIRD SANCTUARY INC.
18328 Gulf Blvd.
Indian Shores, FL 33785 USA
Phone: 727-391-6211   Fax: 727-399-2923
Website: www.webcoast.com/seabird

Founded: 1972
Membership: 30
Scope: National

Description: A private, nonprofit, membership organization dedicated to the rescue, repair, recuperation, and release of healed sick and injured wild birds. The Sanctuary treats and releases over 9,000 birds each year. It provides a safe home for over 600 permanently injured avian species. The Sanctuary is open for visitation every day from 9:00 a.m. till dusk. Guided tours available - free admission.

Publication(s): S.S.S. Brochure (Blue), If You Find A Baby Bird Book, Suncoast Seabird Sanctuary Newsletter

Keyword(s): Birds, Endangered Species, Environmental and Conservation Education, Waterfowl, accreditation

Contact(s):
Ralph Heath Jr., FOUNDER AND DIRECTOR
18323 Sunset Blvd., Redington Shores, FL 33708
Phone: 727-391-6211

Helen Heath, TREASURER
18323 Sunset Blvd., Redington Shores, FL 33708
Phone: 727-393-0933

SUSTAINABLE ENERGY INSTITUTE
P.O. Box 4347
Arcata, CA 95518-4347 USA
Phone: 707-826-7775
E-mail: info@culturechange.org
Website: www.culturechange.org

Founded: 1988
Scope: International

Description: As the founder of the Alliance for a Paving Moratorium, Fossil Fuels Action directs road-fighting and education to address loss of farmland, wilderness, and community. APM unites over 150 groups and businesses in a call for an end to new road construction while promoting alternative transportation and the car-free lifestyle. We envision sustainability, embracing nature and sharing, rejecting the technofix designed to perpetuate status quo economics.

Publication(s): Culture Change

Keyword(s): Road Construction, Air Quality and Pollution, Environmental Planning, Protected Areas, Population Growth, Sustainable Development, Land Use Planning, Transportation, Urban Environment, Wilderness, NAFTA Superhighway, Energy, Watersheds

Contact(s):
Pincas Jawetz, BOARD MEMBER
Richard Register, BOARD OF DIRECTORS MEMBER
Lonnie Maxfield, BOARD OF DIRECTORS MEMBER
Debbie Lukas, BOARD OF DIRECTORS MEMBER
Jan Lundberg, BOARD OF DIRECTORS PRESIDENT
Eve Gilmore, VICE PRESIDENT, BOARD MEMBER AND SECRETARY

SUSTAINABLE ENERGY INSTITUTE
1062 G Street, Suite K
P.O. Box 4347
Arcata, CA 95518 USA
Phone: 707-826-7775   Fax: 707-822-7007
E-mail: info@culturechange.org
Website: culturechange.org

Founded: NA
Scope: International

Contact(s):
Raul Riutor, CONTACT

T

TAHOE REGIONAL PLANNING AGENCY
308 Dorla Ct., P.O. Box 1038
Zephyr Cove, NV 89448-1038 USA
Phone: 775-588-4547   Fax: 775-588-4527
E-mail: trpa@trpa.org
Website: www.trpa.org

Founded: 1969
Membership: 75
Scope: Regional

Description: To establish and implement land use and environmental plans and regulations in the Lake Tahoe Region. Established by Public Law No. 91-148, December 1969, amended by Public Law No. 96-551, December 1980.

Keyword(s): Air Quality and Pollution, Lakes, Land Use Planning, Coral Reefs, Wetlands

Contact(s):
Juan Palma, EXECUTIVE DIRECTOR
John Marshall, LEGAL COUNSEL
Jordan Kahn, LEGAL COUNSEL

TALL TIMBERS RESEARCH STATION

13093 Henry Beadel Dr.
Tallahassee, FL 32312-9712 USA
Phone: 850-893-4153   Fax: 850-668-7781
Website: www.talltimbers.org

Founded: 1958
Membership: 2500
Scope: International

Description: A nonprofit, tax-exempt scientific and educational organization with a focus on land management, conservation, ecological research, and fire ecology. Information is exchanged in print and on the Internet. Tall Timbers provides publications, seminars, conferences and training programs for land owners and managers, scholars, research scientists, students and concerned citizens.

Publication(s): Newsletters, annual reports, proceedings, technical reports, and informational bulletins.

Keyword(s): Endangered Species, Environmental and Conservation Education, Flowers, Plants, and Trees, training

Contact(s):
Kate Ireland, CHAIRMAN
Lane Green, EXECUTIVE DIRECTOR
Ann Bruce, LIBRARIAN
Tall Timbers Library, 850-668-7781
Leonard Brennan, RESEARCH DIRECTOR
Lawton Langford, TREASURER
Walter Sedgwick, VICE CHAIRPERSON

TALLAHASSEE MUSEUM OF HISTORY AND NATURAL SCIENCE

3945 Museum Dr.
Tallahassee, FL 32310 USA
Phone: 850-575-8684   Fax: 850-574-8243
Website: www.tallahasseemuseum.org

Founded: 1957
Membership: 4200
Scope: Local

Description: To educate residents and visitors of Tallahassee and the Big Bend area about the region's natural and cultural history, from the beginning of the 19th-century until the present. For this purpose, the museum collects, preserves, and exhibits artifacts and historic buildings, maintains native animals in natural habitats, and operates a 19th century farmstead.

Publication(s): The Newsletter of The Tallahassee Museum of History and Natural Science

Keyword(s): Zoology, Zoological Parks, Biodiversity, Culture, Museum, Environment, Environmental and Conservation Education, Endangered Species, Historic Preservation, Mammals, Nongame Wildlife, Natural Areas, education

Contact(s):
Mike Jones, CURATOR OF ANIMALS
Linda Deaton, CURATOR OF COLLECTIONS AND EXHIBITS
Jennifer Golden, DIRECTOR OF EDUCATION

Paula Moyer, DIRECTOR OF INSTITUTIONAL ADVANCEMENT
Russell Daws, EXECUTIVE DIRECTOR/CEO
daws@tallhasseemuseum.org

TEENS FOR RECREATION AND ENVIRONMENTAL CONSERVATION (TREC)

Seattle Department of Parks and Recreation,
100 Dexter Ave., N.
Seattle, WA 98109-5199 USA
Phone: 206-684-7097   Fax: 206-684-7025
Website: www.seattletrec.org

Founded: 1992
Scope: Statewide

Description: TREC is an outdoor expedition-level program designed to expose multi-ethnic teens to environmental education, urban conservation, and stewardship, while creating an environment for community leadership and empowerment.

Contact(s):
Robert Warner, CONTACT

TENNESSEE ASSOCIATION OF CONSERVATION DISTRICTS

Attn: President, 2205 Armour Dr., Rte. 2, Box 3712
Somerville, TN 38068 USA
Phone: 901-465-9684

Founded: NA
Scope: Statewide

Keyword(s): Conservation Districts

Contact(s):
John Wilson, BOARD MEMBER
560 Orr Rd., Arlington, TN 38002
Phone: 901-867-8289
Phil Cherry, EXECUTIVE DIRECTOR
144 Southeast Parkway, Suite 210, Franklin, TN 37064
Phone: 615-595-9978
Fax: 615-595-9982
Harris Armour, PRESIDENT
Phone: 901-465-9684
Fax: 901-465-5608
Barry Lake, SECRETARY-TREASURER
P.O. Box 107, Hickory Valley, TN 38042
Phone: 901-764-2909
Roy Gills, VICE PRESIDENT
419 Nofattie Rd., Limestone, TN 37615
Phone: 423-257-2305

TENNESSEE B.A.S.S. CHAPTER FEDERATION

2597 Ogden Rd.
McEwen, TN 37101 USA
Phone: 931-296-4428
Website: www.tnbass.com

Founded: NA
Scope: Statewide

Description: An organization of Bassmaster chapters, affiliated with the Bass Anglers Sportsman Society, organized to fight pollution, assist state and national conservation agencies in their efforts, and teach the young people of our country good conservation practices. Dedicated to the realistic conservation of our water resources.

Publication(s): Chapter Newsletter

Keyword(s): Aquatic Habitats, Wildlife

Contact(s):
Chuck Harger, CONSERVATION DIRECTOR
731 Oakland Dr., New Johnsonville, TN 37134
Phone: 931-535-2209
Charles Mitchell, PRESIDENT
Phone: 931-296-4428

TENNESSEE CITIZENS FOR WILDERNESS PLANNING

130 Tabor Rd.
Oak Ridge, TN 37830 USA
Phone: 432-481-0286
E-mail: tcwp@korrnet.org
Website: www.korrnet.org/tcwp/

Founded: 1966
Scope: State

Description: Dedicated to achieving and perpetuating protection of natural lands and waters by means of public ownership, legislation, or cooperation with the private sector. Our first focus is the Cumberland and Appalachian regions of East Tennessee, but efforts may extend to the rest of the state and the nation.

Publication(s): TCWP Newsletter

Keyword(s): Wilderness, Natural Areas, Public Lands, Rivers, Watersheds, Biodiversity, Nongame Wildlife, Exotic species, Aquatic nuisance species, Forest Management

Contact(s):
Jimmy Groton, PRESIDENT
87 Outer Dr., Oak Ridge, TN 37830
Phone: 423-482-5799
Mary Lynn Dobson, SECRETARY
209 Cove Point Road, Rockwood, TN 37854
Phone: 423-354-4924
Eric Hirst, VICE PRESIDENT
106 Capital Cir., Oak Ridge, TN 37830
Phone: 423-483-1289

TENNESSEE CONSERVATION LEAGUE

300 Orlando Ave.
Nashville, TN 37209-3257 USA
Phone: 615-353-1133   Fax: 615-353-0083
E-mail: tcl@conservetn.com
Website: conservetn.com

Founded: 1946
Membership: 2500
Scope: Statewide

Description: A representative statewide organization, affiliated with the National Wildlife Federation, dedicated to the protection and enhancement of wildlife and its habitat through public education and government interaction.

Publication(s): Tennessee Out-of-Doors

Contact(s):
Bruce Newport, DIRECTOR OF DEVELOPMENT AND COMMUNICATIONS
Marty Marina, EXECUTIVE DIRECTOR, ALTERNATE REPRESENTATIVE
Greer Tidwell, PRESIDENT
Rick Murphree, REPRESENTATIVE

Steve Wise, TREASURER
David Jackson, VICE-PRESIDENT

TENNESSEE ENVIRONMENTAL COUNCIL

1700 Hayes St., Suite 101
Nashville, TN 37203 USA
Phone: 615-321-5075   Fax: 615-321-5082
E-mail: tec@nol.com

Founded: 1970
Membership: 44 organizations, 2,200 individuals
Scope: Statewide

Description: A nonprofit coalition working to protect and improve Tennessee's public health, quality of life,, and natural heritage. TEC is a 28-year old organization, focused on carrying out the state's environmental policies and regulations on behalf of Tennessee citizens.

Publication(s): ProTECt

Keyword(s): Air Quality and Pollution, Chemical Pollution Control, Environment, Environmental Justice, Environmental Protection, Planning Management, Public Health Protection, Forest Management, Solid Waste Management, Sustainable Development, Coral Reefs, Toxic Substances, Nuclear-free, Water quantity, Water export and diversion

Contact(s):
Will Callaway, EXECUTIVE DIRECTOR
will@tectn.org
Robert Diehl, PRESIDENT
Sandi Kurtz, SECRETARY
Phone: 423-892-4403
William Miller, TREASURER
Phone: 931-486-9504

TENNESSEE FORESTRY ASSOCIATION

Nashville, TN 37229 USA
Phone: 615-883-3832   Fax: 615-883-0515

Founded: 1951
Scope: State

Description: A nonprofit conservation group of about 1300 woodland owners, public and private foresters, educators, and wood using companies, as well as individual citizens and allied businesses, encouraging the development and wise use of Tennessee's forest resources.

Publication(s): Treeline Newsletter

Keyword(s): Forestry

Contact(s):
Steve Bond, 1ST VICE PRESIDENT
Larry Pitts, PRESIDENT

TENNESSEE WOODLAND OWNERS ASSOCIATION

P.O. Box 1400
Crossville, TN 38557 USA
Phone: 615-484-5535

Founded: 1976
Membership: 80
Scope: Statewide

Description: A statewide organization affiliated with the National Woodland Owners Association to focus on the special needs and concerns of non-industrial private forest owners throughout Tennessee and to promote responsible resource stewardship.

Keyword(s): Forests and Forestry

Contact(s):
Robert Harrison, SECRETARY/TREASURER

TERRENE INSTITUTE, THE
4 Herbert St.
Alexandria, VA 22305 USA
Phone: 703-548-5473   Fax: 703-548-6299
E-mail: terrinst@aol.com
Website: www.terrene.org

Founded: 1990
Scope: National

Description: The Terrene Institute is a nonprofit organization that works with corporate, environmental, and government partners to promote innovative and economical solutions for improving environmental quality. Terrene is an environmental education organization that develops conferences, issues forums, and publications that serve both corporate and nonprofit audiences. Terrene's principal work is water issues, including the coordination of American Wetlands Month activities, and nonpoint source water quality. A catlaog of Terrene publications is available.

Keyword(s): Runoff, Watersheds, Wetlands, Exotic species, Aquatic nuisance species

Contact(s):
Judy Taggart, EXECUTIVE VICE PRESIDENT
4 Herbert St., Alexandria, VA 22305
Phone: 703-548-5473

TEXAS ASSOCIATION OF SOIL AND WATER CONSERVATION DISTRICTS
Atttn: President P.O. Box 13
Temple, TX 77653 USA
Phone: 254-778-8741   Fax: 254-773-3311

Founded: NA
Scope: Statewide

Publication(s): Official Park Newspaper, Big Bend seminars and sales catalog available on request., Big Bend Paisano, The

Keyword(s): Conservation Districts

Contact(s):
Jose Dodier, PRESIDENT
P.O. Box 13, Zapata, TX 78076
Phone: 956-936-2007
Beatrice White, SECRETARY
P.O. Box 658, Temple, TX 76503
Phone: 254-778-8741
Fax: 254-770-0011
Dayton Elam, SECRETARY-TREASURER
600 SW 21st St., Seminole, TX 79360
Phone: 915-758-3504
Aubrey Russell, VICE PRESIDENT

TEXAS B.A.S.S. CHAPTER FEDERATION
Attn: President, 2221 Apache Dr.
Harker Heights, TX 76548 USA
Phone: 254-698-2015
Website: www.texas-bass.com

Founded: NA
Scope: Statewide

Description: An organization of Bassmaster chapters, affiliated with the Bass Anglers Sportsman Society, organized to fight pollution, assist state and national conservation agencies in their efforts, and teach the young people of our country good conservation practices. Dedicated to the realistic conservation of our water resources.

Publication(s): Texas Tightline Magazine, SCOT Sportsmen Conservation of Texas

Keyword(s): Environmental and Conservation Education, Forests and Forestry, Rivers, Exotic species, Aquatic nuisance species, Wilderness

Contact(s):
Alan Allen, CONSERVATION DIRECTOR
807 Brazos, Suite 311, Austin, TX 78701
Phone: 512-472-2267
AlanAllen-SCOT@att.net
Stacy Twiggs, PRESIDENT

TEXAS COMMITTEE ON NATURAL RESOURCES
1301 South IH-35, Suite 301
Austin, TX 78741 USA
Phone: 512-441-1122   Fax: 512-328-3399
E-mail: tconr@texas.net
Website: http://tconr.home.texas.net

Founded: 1968
Membership: 1200
Scope: Regional

Description: A representative statewide organization, affiliated with the National Wildlife Federation, dedicated to the protection and enhancement of wildlife and its habitat through public education and government interaction.

Publication(s): Conservation Progress

Contact(s):
David Gray, CHAIR/PRESIDENT
Edward Fritz, EDITOR
Janice Bezanson, EXECUTIVE DIRECTOR AND ALTERNATE REPRESENTATIVE & EDUCATION PROGRAMS CONTACT
Susan Petersen, REPRESENTATIVE
Claude Albritton III, TREASURER

TEXAS FORESTRY ASSOCIATION
P.O. 1488
Lufkin, TX 75902-1488 USA
Phone: 936-632-8733   Fax: 936-632-9461
E-mail: tfa@lcc.net
Website: www.texasforestry.org

Founded: 1914
Scope: Statewide

Description: Private nonprofit statewide organization promoting the conservation, fullest economic development, and utilization of forest and related resources.

Publication(s): Texas Forestry

Keyword(s): Biodiversity, Endangered Species, Environmental and Conservation Education, Reforestation, Scholarships and Grants, Sustainability, Forests and Forestry

Contact(s):
Ronald Hufford, EXECUTIVE VICE PRESIDENT
Frank Shockley, PRESIDENT

TEXAS ORGANIZATION FOR ENDANGERED SPECIES

P.O. Box 12773
Austin, TX 78711-2773 USA
Website: rice.edu

Founded: 1972
Membership: 275
Scope: Statewide

Description: A nonprofit statewide organization dedicated to the conservation of endangered, threatened, or rare species and biotic communities of Texas.

Publication(s): TOES News & Notes

Keyword(s): Hunting, Outdoor Recreation, Sustainable Development, training

Contact(s):
Jason Singhurst, CHAIR OF CONSERVATION COMMITTEE
Phone: 512-912-7011
Peggy Homer, CHAIR OF NATURAL RESOURCES
Phone: 512-912-7047
David Lemke, EDITOR
Phone: 512-245-2178
Bob Murphy, EDUCATION CHAIRMAN
Lee Ann Linam, PAST-PRESIDENT
Phone: 512-847-9480
Gary Valentine, PRESIDENT
Phone: 254-297-1291
Deborah Holle, SECRETARY
Phone: 512-482-5700
C. Sherrod, TREASURER
Phone: 512-328-2430

TEXAS PARKS & WILDLIFE

TEXAS CHAPTER
Attn: President, c/o Texas Parks and Wildlife Division, 4200 Smith School Rd.
Austin, TX 78744 USA
Phone: 512-389-4800
Website: www.tpwd.state.tx.us

Founded: 1976
Membership: 400
Scope: Statewide

Contact(s):
Paul Hammerschmidt, COASTAL FISHERIES REGULATIONS MANAGER
Phone: 512-389-4650
Fax: 512-389-4388
paul.hammerschmidt@tpwd.state.tx.us

TEXAS WILDLIFE ASSOCIATION

401 Isom Rd. Ste. 237
San Antonio, TX 78216 USA
Phone: 210-826-2904   Fax: 210-826-4933
E-mail: twa@texas-wildlife.org
Website: www.texas-wildlife.org

Founded: 1985
Scope: Statewide

Description: Texas Wildlife Association is a nonprofit corporation, formed to protect and promote the rights of Texas' wildlife managers, land owners, sportsmen, and the state's wildlife resources—especially on private lands.

Publication(s): Texas Wildlife

THE CONSERVATION EDUCATION CENTER

2473 160th Rd.
Guthrie Center, IA 50115 USA
Phone: 641-747-8383   Fax: 641-747-3951
Website: www.consed@nef.net

Founded: 1958
Scope: Statewide

Description: To encourage and lead the development and practice of a widespread and effective conservation education program in Iowa.

Contact(s):
A. Jay Winter, TRAINING OFFICER
ajwinter@pionet.net
Don Sievers, TRAINING OFFICER
dsievers@pionet.net

THE FLORIDA PANTHER SOCIETY, INC.

Route 1, P.O. Box 1895
White Springs, FL 32096 USA
Phone: 904-397-2945   Fax: 386-397-2945
E-mail: oldflorida@atlantic.net
Website: www.atlantic.net/~oldfla/panther/panther.html
Membership: 150
Scope: International

Publication(s): Quarterly News Letter Cat Track

Contact(s):
Stephen Williams, PRESIDENT
oldfla@atlantic.net
Karen Hill, VICE PRESIDENT
coolcat@atlantic.net

THE IZAAK WALTON LEAGUE OF AMERICA, INC.

WISCONSIN DIVISION
Attn: President, 5316 Forest Cir., N
Stevens Point, WI 54481-5605 USA
Phone: 715-344-1803
Website: www.iwa.org

Founded: NA
Membership: 1300
Scope: Statewide

Publication(s): Wisconsin Waltonian Newsletter

Keyword(s): Environmental and Conservation Education, Rivers, Water Quality, Exotic species, Aquatic nuisance species, Watersheds

Contact(s):
Robert Elliker, PRESIDENT
Phone: 715-344-1803
Gerald Ernst, SECRETARY
811 4th St., Plover, WI 54467-2253
Phone: 715-344-4668

THE NATIONAL COUNCIL FOR SCIENCE AND THE ENVIRONMENT

1725 K St., NW, Suite 212
Washington, DC 20006 USA
Phone: 202-530-5810   Fax: 202-628-4311
E-mail: cnie@cnie.org
Website: www.cnie.org

Founded: 1990
Scope: National, International

Description: The CNIE is a nonprofit organization of scientists, policymakers, environmentalists, business representatives and other stakeholders in environmental policy. The CNIE's mission is to improve the scientific basis for making decisions on environmental issues through the creation and successful operation of a National Institute for the Environment. The proposed entity, which would be affiliated with the National Science Foundation, would reform the relationship between science and environmental decision-making by establishing programs for environmental research, assessment, education and training, and a National Library for the Environment.

Publication(s): Federal Environmental Research and Development Programs, over 800 Congressional Service Reports and many other resources available through on-line library.

Keyword(s): Communications, Environmental and Conservation Education, Sustainable Development, Sustainable Ecosystems, training, Libraries

Contact(s):
Stephen Hubbell, CHAIR
Professor of Botany, University of Georgia, Dept. of Botany, Athens, GA 30602
Phone: 706-583-0393
Fax: 706-542-1805
Peter Saundry, EXECUTIVE DIRECTOR
Richard Benedick, PRESIDENT
A. Ahmed, SECRETARY AND TREASURER
President, Global Children's Health and Environmental Fund
Phone: 202-789-1201
Fax: 202-789-1206

THE NATIONAL WOODLAND OWNERS ASSOCIATION

374 Maple Ave., E., Suite 310
Vienna, VA 22180 USA
Phone: 703-255-2700   Fax: 703-281-9200
E-mail: cam@nwoa.net
Website: www.nationalforestry.net

Founded: 1983
Scope: National

Description: A nationwide advocate of sustainable forestry on private and public lands. Programs include the National Forestry Network, Green Tag Forestry, the American Hardwood Management Advisory Board and the National Historic Lookout Register.

Publication(s): Forestry Advantage, The

Keyword(s): Renewable Resources, Conservation, Sustainable Resources, Forest Management, Forests and Forestry, Green Certification

Contact(s):
Eric Johnson, EDITOR
Phone: 315-369-3078
Arlyn Perkey, HARDWOOD MANAGEMENT ADVISORY BOARD
Phone: 304-285-1523
Nancy Gabriel, NATIONAL HISTORIC LOOKOUT REGISTER
Keith Argow, PRESIDENT

THE OCEAN CONSERVANCY

1725 DeSales St., NW, Suite 600
Washington, DC 20036 USA
Phone: 202-429-5609   Fax: 202-872-0619
Website: www.oceanconservancy.org

Founded: 1972
Membership: 127,000+
Scope: National

Description: A nonprofit, scientific organization dedicated to protecting marine wildlife and its habitats, and to conserving coastal and ocean resources. The center's programs are conducted in five major areas: Fisheries and Wildlife Conservation, Ecosystem Protection, Biodiversity Conservation, International Initiatives, Citizen Monitoring, and Outreach. Program efforts focus on research, policy analysis, education, and public information and involvement.

Publication(s): Blue Planet Quarterly, Coastal Connection, International Coastal Cleanup Report, list of additional publications on request.

Keyword(s): Biodiversity, Cleanup, Endangered Species, Wildlife, Coral Reefs, Environmental and Conservation Education, Conservation, Ecology, Pollution Prevention, Sustainable Ecosystems, Watersheds, Marine Mammals, Whale, Dolphin, Seal

Contact(s):
John Bierwirth, CHAIRMAN OF THE BOARD
David Dixon, DIRECTOR OF CONSTITUENCY DEVELOPMENT
Phone: 202-429-5609
Jack Sobel, DIRECTOR OF ECOSYSTEM PROGRAMS
Phone: 202-429-5609
Nina Young, DIRECTOR OF MARINE CONSERVATION WILDLIFE
Phone: 202-429-5609
Fax: 202-872-0619
Kris Balliet, DIRECTOR OF THE ALASKA REGION
425 G. St., Anchorage, AK 907-258-9922
Linda Sheehan, DIRECTOR OF THE PACIFIC COAST REGION
580 Market St., Suite 550, San Francisco, CA 94104
Phone: 415-391-6204
David White, DIRECTOR OF THE SOUTHEAST ATLANTIC & GULF OF MEXICO REGIONAL OFFICE
One Beach Dr. SE, 304, St. Petersburg, FL 33701
Phone: 727-895-2188
Nicole Sandberg
INTERNATIONAL COASTAL CLEANUP MANAGER
Phone: 202-429-5609
Fax: 202-872-0619
Roger Rufe, PRESIDENT
Phone: 202-429-5609
Stephanie Drea, VICE PRESIDENT FOR COMMUNICATIONS AND MARKETING
Phone: 202-429-5609
David Guggenheim, VICE PRESIDENT FOR CONSERVATION POLICY
Phone: 202-429-5609
Elliot Gruber, VICE PRESIDENT FOR DEVELOPMENT AND MEMBERSHIP
Phone: 202-429-5609
Peter Jones, VICE PRESIDENT FOR FINANCE AND ADMINISTRATION

David Hoskins, VICE PRESIDENT FOR GOVERNMENT AFFAIRS AND GENERAL COUNSEL
Phone: 202-429-5609
Warner Chabot, VICE PRESIDENT FOR REGIONS
580 Market St., Suite 550, San Francisco, CA 94104
Phone: 415-391-6204

THE SHEEPSCOT VALLEY CONSERVATION ASSOCIATION

MAINE COUNCIL
P.O. Box 125
Alna, ME 04535 USA
Phone: 207-586-5616   Fax: 207-586-6442
E-mail: svaca@lincoln.midcoast.com
Website: www.lincoln.midcoast.com/~svca

Founded: NA
Membership: 225
Scope: Local

Description: A statewide council with six active chapters working for the protection and enhancement of coldwater fishery resources.

Contact(s):
Sam Merrill, EXECUTIVE DIRECTOR

THE SIERRA CLUB OREGON CHAPTER

OREGON CHAPTER
2950 SE Stark St., Suite 110
Portland, OR 97214 USA
Phone: 503-238-0442   Fax: 503-238-6281
E-mail: oregon.chapter@sierraclub.org
Website: www.oregon.sierraclub.org

Founded: NA
Membership: 19,000
Scope: Statewide

Publication(s): The Conifer -Bi-Monthly Newsletter

Contact(s):
Mari Margil, CONSERVATION COORDINATOR
Phone: 503-232-1723
mari.margil@sierraclub.org

THE SONORAN/RINCON INSTITUTES

7650 E. Broadway Blvd., Suite 203
Tucson, AZ 85710 USA
Phone: 520-290-0828   Fax: 520-290-0969
E-mail: sonoran@sonoran.org
Website: www.sonoran.org

Founded: 1990
Scope: International

Description: The Rincon Institute is a nonprofit organization dedicated to protecting the natural resources of the Rincon Mountain District of Saguaro National Park and the surrounding lands. This goal is accomplished primarily through ecological research, environmental education, natural area protection, private land stewardship and the development of cooperative approaches to resolving land use conflicts.

Keyword(s): Environmental and Conservation Education, Sustainable Development, Sustainable Ecosystems, Terrestrial Habitats, National Parks, Land Use Planning, Land Preservation, Research, Conservation, Ecology, Natural Areas, Riparian Restoration

Contact(s):
Nancy Laney, CHAIRMAN OF THE BOARD
pbtradings@aol.com
Mary Vint, COMMUNITY OUTREACH COORDINATOR
mary@sonoran.org
Mark Briggs, DIRECTOR OF SCIENCE
mark@sonoran.org
Luther Propst, EXECUTIVE DIRECTOR
luther@sonoran.org
Josh Schachter, PROGRAM ASSOCIATE
josh@sonoran.org

THE WILDERNESS SOCIETY

1615 M Street, NW
Washington, DC 20036 USA
Phone: 202-833-2300   Fax: 202-429-3958
Website: http://www.wilderness.org/

Contact(s):
Fran Hunt, DEPUTY VICE PRESIDENT OF REGIONAL CONSERVATION
Dave Alberswerth, DIRECTOR OF BLM PROGRAM
Sue Gunn, DIRECTOR OF NATIONAL PARKS PROGRAM & LWCF ISSUES
Greg Aplet, DIRECTOR, CENTER FOR LANDSCAPE ANALYSIS
7475 Dakin Street, Suite 410, Denver, CO 80221
Phone: 303-650-5818
Donald Barry, EXECUTIVE VICE PRESIDENT
Julie Wormser, REGIONAL DIRECTOR
45 Bromfield Street, Suite 1109, Boston, MA 02108
Phone: 817-350-8866
Robert Friemark, REGIONAL DIRECTOR
1424 Fourth Ave., Suite 816, Seattle, WA 98101
Phone: 206-624-6430
Craig Gehrke, REGIONAL DIRECTOR
2600 Rose Hill, Suite 201, Boise, ID 83705
Phone: 208-343-8153
Jerry Greensberg, VICE PRESIDENT OF COMMUNICATIONS
G. Bancroft, VICE PRESIDENT OF ECOLOGY AND ECONOMIC RESEARCH
Elizabeth Colt, VICE PRESIDENT OF MEMBERSHIP AND DEVELOPMENT
Linda Lance, VICE PRESIDENT OF PUBLIC POLICY

THE OKLAHOMA WOODLAND OWNERS ASSOCIATION

2657 S. Trenton
Tulsa, OK 74114-2727 USA
Phone: 918-569-4287   Fax: 918-743-6941

Founded: NA
Scope: Statewide

Description: A statewide organization affiliated with the National Woodland Owners Association to advance forest management skills of Oklahoma woodland owners. Other objectives are to promote education and networking, provide timber marketing, and to monitor and act upon legislation.

Publication(s): OWOA bulletin

Keyword(s): Forests and Forestry

Contact(s):
John Ahern, 1ST VICE PRESIDENT
Sue Paschell, BOARD MEMBER
Miles Schulze, BOARD MEMBER
Rick Hutchinson, BOARD MEMBER

Patt Nelson, PRESIDENT AND EDITOR
Tom Kee, SECRETARY
Sue Paschall, TREASURER

THEODORE ROOSEVELT CONSERVATION ALLIANCE

27 Ft. Missoula Rd.
Missoula, MT 59804 USA
Phone: 406-541-9975   Fax: 406-549-7402
E-mail: info@trca.org
Website: www.trca.org

Founded: 1999
Membership: 45000
Scope: National

Description: The Alliance's mission is to inform and engage Americans to foster our conservation legacy while working to nurture, enhance and protect our fish, wildlife and habitat resources on America's Public Lands.

Publication(s): Square Dealer-quarterly newsletter

Contact(s):
Kristen Wagner, EXECUTIVE ASSISTANT
Phone: 406-549-0101
kwagner@trca.org

THORNE ECOLOGICAL INSTITUTE

P.O. Box 19107
Boulder, CO 80308-2107 USA
Phone: 303-447-1769   Fax: 303-499-8340
E-mail: info@thorne-eco.org
Website: www.thorne-eco.org

Founded: 1954
Scope: State

Description: A nonprofit educational institute creating inovative outdoor learning experiences and other educational opportunities that teach stewardship of the earth to children and adults. The Institute offers a variety of environmental education classes to children from the Boulder, Denver area with science-based activities and exciting hands-on lessons.

Keyword(s): Camp, Environmental and Conservation Education, Internships, Wetlands, training, Ecology, Watersheds, Natural Science, Fieldwork

Contact(s):
Susan Peterson, CHAIRMAN
susankae@aol.com
Oakleigh Thorne, FOUNDER AND HONORARY PRESIDENT

THREE CIRCLES CENTER FOR MULTICULTURAL ENVIRONMENTAL EDUCATION

P.O. Box 1946
Sausalito, CA 94965 USA
Phone: 415-331-4540
E-mail: circlecenter@igc.apc.org

Founded: 1990
Scope: National

Description: Three Circles Center introduces, encourages, and cultivates multicultural perspectives and values in environmental and outdoor education, recreation, and interpretation.

Publication(s): Perspectives, Monographs, Research Papers, Journal of Multicultural Environmental Education

Keyword(s): Culture, Environment, Environmental and Conservation Education, Training, Justice

Contact(s):
Running-Grass, EXECUTIVE DIRECTOR
P.O. Box 1946, Sausalito, CA 94965
Phone: 415-331-4540

THRESHOLD, INC.

International Center for Environmental Renewal,
Drawer CU
Bisbee, AZ 85603 USA
Phone: 602-432-5814
E-mail: sactedway.com
Website: www.sactedway.com

Founded: 1972
Scope: International

Description: An international center seeking to improve mankind's understanding of and relationship to the environment at five levels: Individual and home, neighborhood, city, bioregion, national, and international. Projects involve environmental research, case studies, planning, education, communication, conferencing, and demonstration activities. Primary focus is on wilderness retreats, vision quests, and awareness training in nature.

Keyword(s): Wilderness, Vision Quest, Environmental and Conservation Education, Internships, Land Use Planning, training

Contact(s):
John Milton, CHAIRMAN
Phone: 520-432-5814
John Diamante, DIRECTOR OF SAN FRANCISCO OFFICE
Phone: 415-986-0999
Sarah Sher, DIRECTOR OF TUCSON, ARIZONA OFFICE
Phone: 520-432-7353
Bud Wilson, SECRETARY, TREASURER AND ADMINISTRATIVE OFFICER OF ARIZONA OFFICE
Phone: 800-294-4795
George Binney, VICE CHAIRMAN
Phone: 520-398-9163

TOGETHER FOUNDATION, THE

113 East 64th St
2nd floor
NYC, NY 10021 USA
Phone: 212-879-9334   Fax: 212-879-9440
E-mail: info@together.org

Founded: 1989
Scope: National

Description: To facilitate positive global change by establishing communications and information systems that inventory and integrate the resources and needs of people, projects, and organizations working on environment, sustainable development, and human rights.

Publication(s): The Together Foundation Newsletter

Keyword(s): Communications, Environmental and Conservation Education, Environmental Planning, Sustainable Development

Contact(s):
Ella Cisneros, PRESIDENT
Robin Rugg, SECRETARY
James MacIntyre, VICE PRESIDENT

TRAFFIC NORTH AMERICA

c/o World Wildlife Fund, 1250 24th St., NW
Washington, DC 20037 USA
Phone: 202-293-4800   Fax: 202-775-8287
E-mail: tna@wwfus.org
Website: www.traffic.org

Founded: NA
Scope: International

Description: Trade Records Analysis of Flora and Fauna in Commerce is a scientific, information-gathering program that monitors the trade in wild animals and plants and the products made from them. It is a program of World Wildlife Fund and is a part of an international network of TRAFFIC offices.

Publication(s): Special Reports, Traffic North America

Keyword(s): Birds, Endangered Species, Mammals

Contact(s):
Simon Habel, DIRECTOR
simon.habel@wwfus.org
Holly Reed, PROGRAM ASSISTANT
holly.reed@wwfus.org
Chris Robbins, PROGRAM OFFICER
chris.robbins@wwfus.org
Craig Hoover, SENIOR PROGRAM OFFICER
craig.hoover@wwfus.org

TREAD LIGHTLY! INC

298 24th St., Suite 325
Ogden, UT 84401 USA
Phone: 801-627-0077   Fax: 801-621-8633
E-mail: tlinc@xmission.org
Website: www.treadlightly.org

Founded: 1990
Membership: 4
Scope: National

Description: Tread Lightly!, Inc. is a nonprofit organization that is an ethical and educational force among outdoor enthusiasts and the industries that serve them. Tread Lightly annually carries out programs designed to instill a proactive, low impact message among enthusiasts, manufacturers, advertising agencies, the media and children of all ages.

Publication(s): Guide to Responsible Four-Wheeling, Guide To Responsible Snowmobiling, Guide to Responsible Trail Biking, Guide to Responsible ATV Riding, Guide to Responsible Mountain Biking, Guide to Responsible Personal Watercraft Use, Tread Lightly! Trails Newsletter

Keyword(s): Environmental and Conservation Education, Protected Areas, Outdoor Recreation

Contact(s):
Philip Milburn, CHAIRMAN
1 Olympic Plaza, Colorodo Springs, CO 80909
pmilburn@usacycling.org
Lori Davis, EXECUTIVE DIRECTOR
lori@treadlightly.org
Andrew Clurman, SECRETARY AND TREASURER
929 Pearl St., Suite 200, Boulder, CO 80302
aclurman@skinet.com
Scott Heath, VICE CHAIRMAN
9 Calle Catrina, Rancho Santa Margarita, CA 92688
Phone: 949-713-6574
Fax: 949-713-6579
scottheath4@home.com

TREEPEOPLE

12601 Mulholland Dr.
Beverly Hills, CA 90210 USA
Phone: 818-753-4600   Fax: 818-753-4635
E-mail: info@treepeople.org
Website: www.treepeople.org

Founded: 1973
Scope: Regional

Description: Andy Lipkis and his teenage friends became known as the "TreePeople" when they began planting trees to restore a dying forest. Through innovative education and training programs, TreePeople has involved thousands of students and volunteers in neighborhood renewal and community service throughout southern California. Today, TreePeople is at the forefront of the urban forestry movement, offering sustainable solutions for the urban ecosystem.

Publication(s): The Simple Act of Planting a Tree, Healing Your Neighborhood, Your City and Your World, Second Nature: Adapting L.A.'s Landscape for Sustainable Living, Seedling News (member newsletter)

Keyword(s): Environmental and Conservation Education, Forests and Forestry, Urban Environment, Trees, Urban Forestry, Youth Organizations

Contact(s):
Jeff Hohensee, DIRECTOR OF EDUCATION PROGRAM
Jim Summers, DIRECTOR OF FORESTRY PROGRAMS
Andy Lipkis, PRESIDENT

TREES ATLANTA

96 Poplar St., NW
Atlanta, GA 30303 USA
Phone: 404-522-4097   Fax: 404-522-6855
E-mail: info@treesatlanta.org
Website: www.treesatlanta.org

Founded: 1985
Membership: 4000
Scope: Regional

Description: Trees Atlanta is a citizens organization that plants, maintains, and conserves trees in metro Atlanta area and educates the public about the importance of trees.

Publication(s): Atlanta Treebune, Tree Walk Brochure

Keyword(s): Ecology, Ecosystems, Environment, Environmental Protection, Greenhouse Effect/Global Warming, Plants, Stewardship, Trees, Urban Environment, Urban Forestry

Contact(s):
Cheryl Bramblett, DIRECTOR OF COMMUNICATIONS
Marcia Bansley, EXECUTIVE DIRECTOR
Greg Levine, VOLUNTEER COORDINATOR

TREES FOR THE FUTURE, INC.

9000 16th St., P.O. Box 7027
Silver Spring, MD 20907 USA
Phone: 301-565-0630   Fax: 301-565-5012
E-mail: info@treesftf.org
Website: www.treesftf.org

Founded: 1989
Membership: 4800
Scope: International

Description: Trees for the Future offers multi-purpose tree seeds, training materials, and technical assistance for requesting communities, institutions and individuals throughout the developing regions of the world. In addition, Trees for the Future creates awareness of the potential threat of Global Climatic Change and the simple, cost-effective solutions of tree-planting to counter the "Global Warming" effect.

Publication(s): Technical Papers, Johnny Appleseed News

Keyword(s): Agriculture, Air Quality and Pollution, Chemical Pollution Control, Developing Countries, Energy Conservation, Environmental Planning, Environment, Protected Areas, Flowers, Plants, and Trees, Greenhouse Effect/Global Warming, Renewable Resources, Rural Development, Sustainable Development

Contact(s):
Scott Bode, AFRICA PROGRAM COORDINATOR
John Moore, CHAIRMAN
Joseph Permetti, DIRECTOR OF DEVELOPMENT
Bill Ligon, EXECUTIVE DIRECTOR
Julio Navarro-Monzo, EXECUTIVE DIRECTOR
Amy Martin Burns, OFFICE MANAGER
Dave Deppner, PRESIDENT EMERITUS
Patricia Aiken, SECRETARY
Celso Maatac, TREASURER

TREES FOR TOMORROW, INC., NATURAL RESOURCES EDUCATION CENTER
P.O. Box 609
Eagle River, WI 54521 USA
Phone: 715-479-6456   Fax: 715-479-2318
E-mail: trees@nnex.net
Website: www.treesfortomorrow.com

Founded: 1944
Membership: 15
Scope: Regional

Description: Original purpose to help reforest northern Wisconsin and provide land management assistance; currently an education center which conducts multi-day workshops on the sensible management and use of natural resources. Also sells tree seedlings.

Publication(s): Northbound- natural resources journal, Tree Tips - biweekly newsletter

Keyword(s): Trees, Environmental and Conservation Education, Land Management

Contact(s):
Jim Holperin, DIRECTOR
Fred Souba, PRESIDENT
Stora Enso North America, Rapids, WI 54495
Phone: 715-422-3669
Fax: 715-422-3620
fred.souba@conpapers.com

TREES, WATER, & PEOPLE
633 S. College Ave.
Fort Collins, CO 80524 USA
Phone: 970-484-3678   Fax: 970-224-0126
E-mail: twp@treeswaterpeople.org
Website: www.treeswaterpeople.org

Founded: 1998
Membership: 6000
Scope: International, Regional

Description: Trees, Water, and People is a nonprofit conservation organization established in 1998 with the mission of working collaboratively with communties to establish sustainable forests and watersheds while improving people's lives. We currently have programs in four countries in Central America, and in Colorado, and Wyoming in the U.S.

Publication(s): Watershed Currents, Forests Forever

Contact(s):
Tempra Board, DIRECTOR OF DEVELOPMENT
twp@treeswaterpeople.org

TRI-STATE BIRD RESCUE AND RESEARCH, INC.
110 Possum Hollow Rd.
Newark, DE 19711 USA
Phone: 302-737-9543   Fax: 302-737-9562
Website: www.tristatebird.org

Founded: 1976
Scope: International

Description: To study and promote healthy populations of native wildlife by rehabilitation of oiled birds; rehabilitation of injured, diseased, and orphaned birds for release back into the wild; conducting training and education programs for colleagues, peers, and the general public; and conducting medical and biological research consistent with goals of providing for the general well-being of native wildlife.

Publication(s): Medical Notes for Rehabilitators, Oiled Bird Rehabilitation, Wildlife and Oil Spills Bulletin, Wildlife and Oil Spills: Rehabilitation, Research, and Contingency Planning, Effects of Oil on Wildlife, The

Keyword(s): Birds, Endangered Species, Environmental and Conservation Education, standards, accreditation, Oil Spill Response

Contact(s):
Chris Motoyoshi, EXECUTIVE DIRECTOR, CONTACT FOR HUMAN RESOURCES
David Mooberry, PRESIDENT
106 Spottswood Ln., Kennett Square, PA 19348
Phone: 610-444-5495
Barbara Druding, SECRETARY
110 Possum Hollow Rd., Newark, DE 19711
Phone: 302-737-9543
Cindy Peterson, TREASURER
John Frink, VICE PRESIDENT
400 Milton Dr., Wilmington, DE 19802

TROUT UNLIMITED
National Headquarters, 1500 Wilson Blvd., Suite 310
Arlington, VA 22209-2404 USA
Phone: 703-522-0200   Fax: 703-284-9400
E-mail: trout@tu.org
Website: http://tu.org

Founded: 1959
Membership: 125,000
Scope: National

Description: A nonprofit, tax-deductible international coldwater fisheries organization dedicated to the conservation, protection, and restoration of coldwater fisheries and their watersheds. Affiliates in Canada, New Zealand, and Australia.

Publication(s): Lines To Leaders, Trout Magazine

Keyword(s): Acid Rain, Aquatic Habitats, Endangered Species, Environmental and Conservation Education, Rivers

Contact(s):
Stephen Born, CHAIRMAN OF NATIONAL RESOURCE BOARD
424 Washburn Pl., Madison, WI 53403
Phone: 608-257-6625
Oakleigh Thorne, CHAIRMAN OF THE BOARD
Kenneth Mendez, CHIEF OPERATING AND FINANCIAL OFFICER
Don Duff, COORDINATOR OF TU AND FS
Whit Fosburgh, DIRECTOR OF DEVELOPMENT
Joseph Mcgurrin, DIRECTOR OF RESOURCES
Sarah Johnson, DIRECTOR, NATIONAL VOLUNTEER OPERATIONS
Phone: 608-250-2757
Fax: 608-255-1326
johnson@tu.org
Christine Arena, EDITOR
Wendy Reed, MANAGER OF MEMBERSHIP SERVICES
Charles Gauvin, PRESIDENT AND CHIEF EXECUTIVE OFFICER
David Nickum, REGIONAL DIRECTOR OF SOUTHERN ROCKIES
1900 13th St., Ste. 101, Boulder, CO 80302
Phone: 303-440-2937
Fax: 303-440-7933
Mike Brock, REGIONAL VICE PRESIDENT OF GREAT LAKES
23410 Beech Rd., Southfield, MI 48034-3482
Phone: 248-356-8195
Phone: 810-592-6098
mikebrock@medidone.net
Lou Schmidt, REGIONAL VICE PRESIDENT OF MID-ATLANTIC
Rt. 1 Box 109-A, Bristol, WV 26332-9801
Phone: 304-367-2724
Fax: 304-367-2727
lschmidt@lolina.net
Ray Smith, REGIONAL VICE PRESIDENT OF MIDWEST
70 N. College Ave., Suite 11, Fayetteville, AR 72701-5337
Phone: 501-521-7011
Fax: 501-443-4333
rsmith7011@aol.com
David Bowie, REGIONAL VICE PRESIDENT OF NEW ENGLAND
540 Duck Pond Rd., Westbrook, ME 04092-2510
Phone: 207-854-9978
Fax: 207-770-1211
usunmz6m@ibmmail.com
Paul Maciejewski, REGIONAL VICE PRESIDENT OF NORTHEAST
47 Flintlock Dr., Long Valley, NJ 07835-0320
Phone: 973-765-6673
Fax: 973-705-5974
72077.1327@compuserve.com
Lorine Albright, REGIONAL VICE PRESIDENT OF NORTHERN ROCKIES
P.O. Box 1525, Great Falls, MT 59403-1525
Phone: 406-454-1384
Fax: 406-761-2610
evenson@mch.net
K. Johnson, REGIONAL VICE PRESIDENT OF PACIFIC NORTHWEST
14727 SE 145th Pl., Renton, WA 98059-7336
Phone: 425-865-2201
Fax: 425-271-6378
kbob@halcyon.com

Kirk Otey, REGIONAL VICE PRESIDENT OF SOUTHEAST
1308 Lexington Ave., Charlotte, NC 28203-4837
Phone: 704-334-3060
Fax: 704-334-0768
kskotey@mindspring.com
Fred Rasmussen, REGIONAL VICE PRESIDENT OF SOUTHERN ROCKIES
225 County Road 516, Ignacio, CO 81137-9728
Phone: 790-563-6517
Fax: 790-563-9599
engelbj@compuserve.com
Stan Griffin, REGIONAL VICE PRESIDENT OF SOUTHWEST
27 Dorset Ln., Mill Valley, CA 94941-5203
Phone: 510-528-5390
Fax: 510-528-7880
tucalif@ziplink.net
Steven Moyer, VICE PRESIDENT OF CONSERVATION PROGRAMS
Kathy Buchner, WYOMING TU OFFICE ADMINISTRATOR
P.O. Box 4069, Jackson, WY 83001
Phone: 307-733-6991
Fax: 307-733-9678
kbuchner@wyoming.com

TROUT UNLIMITED
Attn: Chairman, 75 Valley Rd.
Louisville, KY 40204-1516 USA
Phone: 502-562-0115
Website: www.tu.org

Founded: NA
Membership: 800+
Scope: Statewide

Description: A statewide council with three active chapters working for the protection and enhancement of coldwater fishery resources.

Publication(s): see publications on website, Trout Magazine

Contact(s):
Stephen Woodring, CHAIRMAN

TROUT UNLIMITED
820 Old Crystal Bay Rd.
Wayzata, MN 55391-9365 USA
Phone: 612-341-9360   Fax: 612-341-9363

Founded: NA
Scope: Statewide

Description: A statewide council with nine active chapters working for the protection and enhancement of coldwater fishery resources.

Contact(s):
George Hust, CHAIRMAN
Phone: 952-475-2054

TROUT UNLIMITED
1966 13th St., Suite LL60
Boulder, CO 80302 USA
Phone: 303-440-2937   Fax: 303-440-7933
E-mail: dnickum@tu.org
Website: www.cotrout.org

Founded: NA
Scope: Statewide

Description: A statewide council with 28 active chapters working for the protection and enhancement of coldwater fishery resources.

Publication(s): Rocky Mountain Streamside, Currents

Contact(s):
David Nickum, EXECUTIVE DIRECTOR
Tom Krol, STATE CHAIRMAN

TROUT UNLIMITED
784 Texas Way
Fayetteville, AR 72701-4449 USA
Phone: 501-521-0837

Founded: NA
Scope: Statewide

Description: A statewide council with 3 active chapters working for the protection and enhancement of coldwater fishery resources.

Contact(s):
John Hehr, CHAIRMAN

TROUT UNLIMITED
ALASKA COUNCIL
P.O. Box 876675, Wasilla, AK 99687-6675
Homer, AK 99603-3324 USA
Phone: 907-376-1666   Fax: 907-376-1666
E-mail: tuakcoun@alaska.net

Founded: 1988
Membership: 530+
Scope: Statewide

Description: A statewide council with ten active chapters dedicated to the protection and enhancement of the cold water fishery resource.

Contact(s):
Jack Willis, CHAIRMAN
Phone: 907-235-3860

TROUT UNLIMITED
ARIZONA COUNCIL
Arizona Council TU, 77 E. Columbus Ave 200
Phoenix, AZ 85012 USA
Phone: 602-264-5840
E-mail: carm.moehle@azbar.org

Founded: NA
Scope: Statewide

Description: A statewide council with four active chapters working for the protection and enhancement of coldwater fishery resources.

Contact(s):
Carm Moehle, CHAIRMAN

TROUT UNLIMITED
CALIFORNIA COUNCIL
State Office, 828 San Pablo Ave.,
Suite 208
Albany, CA 94706 USA
Phone: 510-528-5390   Fax: 510-528-7880
E-mail: tucalif@earthlink.net

Founded: NA
Scope: Statewide

Description: A statewide council with ten active chapters working for the protection and enhancement of coldwater fishery resources.

Contact(s):
Stan Griffin, OFFICE DIRECTOR

TROUT UNLIMITED
CONNECTICUT COUNCIL
654 Cyprus Rd.
Newington, CT 06111-5612 USA
Phone: 860-667-2515
E-mail: fraa@fraa.org

Founded: NA
Scope: Statewide

Description: A statewide council with nine active chapters working for the protection and enhancement of coldwater fishery resources.

Contact(s):
Steve Lewis, CHAIRMAN

TROUT UNLIMITED
GEORGIA COUNCIL
108 Sycamore St. Apt. B
Rome, GA 30165-4036 USA
Phone: 706-234-5310   Fax: 706-234-9135
E-mail: gmoran@redmax.com

Founded: NA
Scope: Statewide

Description: A statewide council with 14 active chapters working for the protection and enhancement of coldwater fishery resources.

TROUT UNLIMITED
ILLINOIS COUNCIL
Attn: Chairman, P O Box 1280
Oak Brook, IL 60522-1280 USA
Phone: 312-751-4730   Fax: 219-756-7735
Website: www.tu.org

Founded: NA
Scope: Regional

Description: A statewide council with four active chapters working for the protection and enhancement of coldwater fishery resources.

Contact(s):
Walter Bock, CHAIRMAN
wjbock1@home.com

TROUT UNLIMITED
MARYLAND COUNCIL, MID-ATLANTIC
Attn: President, 3509 Pleasant Plains Dr.
Reisterstown, MD 21136-4417 USA
Phone: 410-239-8468   Fax: 410-374-5719
E-mail: tedgodfrey@erols.com
Website: www.tu.org/

Founded: NA
Scope: Statewide

Description: A statewide council with seven active chapters working for the protection and enhancement of coldwater fishery resources.

Contact(s):
Ted Godfrey, CHAIRMAN

TROUT UNLIMITED
MICHIGAN COUNCIL
7 Trowbridge NE
Grand Rapids, MI 49503 USA
Phone: 616-460-0477

Founded: NA
Scope: Statewide

Description: A statewide council with 21 active chapters working for the protection and enhancement of coldwater fishery resources.

Contact(s):
Richard Rowman, EXECUTIVE DIRECTOR

TROUT UNLIMITED
MONTANA COUNCIL
P.O. Box 7186
Missoula, MT 59807 USA
Phone: 406-543-0054   Fax: 406-543-0054
E-mail: emontrout@montana.com
Website: www.montanatu.org

Founded: NA
Membership: Regular $15; Family $35; 3-Year Regular $80; 3-Year Family $90; Sponsor $75; Cuntury $150; Business
Scope: Statewide

Description: A statewide council with 12 chapters working for the protection and enhancement of coldwater fishery resources.

Contact(s):
Bruce Farling, EXECUTIVE DIRECTOR

TROUT UNLIMITED
NEVADA COUNCIL
474 South Blakeland Drive
Spring Creek, NV 89815 USA
Phone: 775-778-3159   Fax: 775-778-8199
E-mail: nvtu@rabbitbrush.com
Website: http://www.rabbitbrush.com/nvtu/

Founded: NA
Scope: Statewide

Description: A statewide council with three active chapters, dedicated to the protection and enhancement of coldwater fishing resources.

Contact(s):
Matt Holford, EXECUTIVE DIRECTOR

TROUT UNLIMITED
NEW HAMPSHIRE COUNCIL
Attn: Chairman, 9 Sirod Rd.
Windham, NH 03087-1401 USA
Phone: 603-896-2236

Founded: NA
Scope: Statewide

Description: A statewide council with six active chapters working for the protection and enhancement of coldwater fishery resources.

Contact(s):
James Norton, CHAIRMAN

TROUT UNLIMITED
NEW YORK COUNCIL
Attn: Chairman, 111 High Point Mountain Rd.
West Shokan, NY 12494-5337 USA
Phone: 914-892-8630
E-mail: karwac@ibm.org
Website: http://nyscounciltu.homestead.com/HOME.html

Founded: NA
Scope: Statewide

Description: A statewide council with 37 active chapters working for the protection and enhancement of coldwater fishery resources.

Publication(s): Long Casts

Contact(s):
Chester Karwatowski, CHAIRMAN

TROUT UNLIMITED
NORTH CAROLINA
Attn: Chairman, 438 Armfield St.
Statesville, NC 28677-5702 USA
Phone: 704-878-3560   Fax: 704-878-3464
E-mail: davidstewart@usiway.net

Founded: NA
Scope: Statewide

Description: Dedicated to the protection and enhancement of coldwater fishing resources.

Contact(s):
David Stewart, CHAIRMAN

TROUT UNLIMITED
NORTH CAROLINA COUNCIL
Attn: Chairman, 135 Tacoma Cir.
Asheville, NC 28801-1625 USA
Phone: 704-684-5178   Fax: 704-687-1689
E-mail: grant@vnet.net
Website: http://www.nctu.org

Founded: NA
Scope: Statewide

Description: A statewide council with 18 active chapters working for the protection and enhancement of coldwater fishery resource.

Contact(s):
Kirk Otey, CHAIRMAN
1308 Lexington Ave., Charlotte, NC 28203
Phone: 800-432-6268
Phone: 704-375-6425
kskotey@mindspring.com

TROUT UNLIMITED
OHIO COUNCIL
Attn: Chairman, 1487 New Way Dr.
Beaver Creek, OH 45434-6925 USA
Phone: 937-426-5757
E-mail: mark.blauvelt@stdreg.com

Founded: NA
Scope: Statewide

Description: Dedicated to the protection and enhancement of the coldwater fishery resource.

Contact(s):
Mark Blauvelt, PRESIDENT

TROUT UNLIMITED
OREGON COUNCIL
213 SW Ashe, Ste 205
Portland, OR 97024 USA
Phone: 503-827-5700   Fax: 503-827-5672
Website: www.teleport.com/~tuwest

Founded: NA
Scope: Statewide

Description: A statewide council with seven active chapters working for the protection and enhancement of the coldwater fishery resource.

Contact(s):
Alan Moore, WESTERN COMMUNICATIONS COORDINATOR
dmoore@tu.org
Jeff Curtis, WESTERN CONSERVATION DIRECTOR
jcurtis@tu.org
Scott Yates, WESTERN LEGAL AND POLICY COORDINATOR
syates@tu.org

TROUT UNLIMITED
OZARKS COUNCIL, MISSOURI
2010 Daisy Ln.
Jefferson, MO 65109-1810 USA
Phone: 573-751-1039   Fax: 573-634-3096
Website: www.agron.missouri.edu/lyfishing/mmtu.html

Founded: NA
Membership: 1600
Scope: Statewide

Description: A statewide council with four active chapters working for the protection and enhancement of coldwater fishery resources.

Contact(s):
John Wenzlick, CHAIRMAN

TROUT UNLIMITED
TENNESSEE COUNCIL
Attn: Chairman, 1326 Lipscomb Dr.
Brentwood, TN 37027 USA
Phone: 615-371-9211
E-mail: johns@bwood.com

Founded: NA
Scope: Statewide

Description: A statewide council with 10 active chapters working for the protection and enhancement of the coldwater fishery resource.

Contact(s):
John Smitherman, CHAIRMAN

TROUT UNLIMITED
UTAH COUNCIL
Attn: Chairman, 1471 E. Canyon Dr.
South Weber, UT 84405-9629 USA
Phone: 801-538-7353   Fax: 801-538-7278
E-mail: rwj@utw.com

Founded: NA

Scope: Statewide

Description: Dedicated to the protection and enhancement of the coldwater fishery resource.

Contact(s):
Wes Johnson, CHAIRMAN

TROUT UNLIMITED
VIRGINIA COUNCIL
Attn: Chairman, 202 Deerfield Ln.
Lynchburg, VA 24502-3122 USA
Phone: 804-239-1017
E-mail: dorffly@aol.com

Founded: NA
Scope: Statewide

Description: A statewide organization with 16 active chapters working for the protection and enhancement of the coldwater fishery resource.

Contact(s):
Thomas Reisdorf, CHAIRMAN

TROUT UNLIMITED
WASHINGTON COUNCIL
Attn: Chairman, 2701 NE 148th Ave.
Vancouver, WA 98684-7877 USA
Phone: 360-896-6967
E-mail: ohio12@uswest.net

Founded: NA
Scope: Statewide

Description: A statewide council with 31 active chapters working for the protection and enhancement of coldwater fishery resources.

Publication(s): Trout and Salmon Leader

Contact(s):
James Derry, CHAIRMAN

TROUT UNLIMITED
WEST VIRGINIA COUNCIL
Att. Dave Bott, 124 Ohio Ave
Westover, WV 26501 USA
Phone: 304-937-2214
E-mail: dwbott@wstdm1.westco.net
Website: www.tu.org

Founded: NA
Membership: 1400
Scope: Statewide

Description: A statewide council with 0 active chapters working for the protection and enhancement of coldwater fishery resources.

Publication(s): Vpale of the Pool

Contact(s):
Dave Bott, CHAIRMAN OF WEST VIRGINIA COUNCIL

TROUT UNLIMITED
WYOMING COUNCIL
Attn: Chairman, P.O. Box 1022
Jackson, WY 83001 USA
Phone: 307-733-4944   Fax: 413-383-1125

Scope: Statewide

Description: A statewide council with 16 active chapters dedicated to the protection and enhancement of the coldwater fishery resource.

Contact(s):
Jay Buchner, CHAIRMAN
Kathy Buchner, DIRECTOR
kbuchner@wyoming.com

TROUT UNLIMITED
SC COUNCIL
115 Conrad Cir.
Columbia, SC 29212-2619 USA
Phone: 803-777-7652   Fax: 803-777-4760

Founded: NA
Membership: 12
Scope: Statewide

Description: A statewide council with four active chapters working for the protection and enhancement of coldwater fishery resources.

Contact(s):
Malcolm Leaphart, CHAIRMAN

TRUMPETER SWAN SOCIETY, THE
3800 County Rd. 24
Maple Plain, MN 55359 USA
Phone: 612-476-4663   Fax: 612-476-1514
E-mail: ttss@hennepinparks.org

Founded: 1968
Scope: International

Description: International scientific and educational organization dedicated to assuring the vitality and welfare of wild Trumpeter Swan populations in North America, and to restoring the species to its original range. The Society promotes research into Trumpeter ecology and management, and provides a framework for exchange of knowledge about the species.

Publication(s): Trumpetings, Proceedings and Papers of the 14th Trumpeter Swan Society Conference, Proceedings and Papers of the 15th Trumpeter Swan Society Conference, Proceedings and Papers of the 16th Trumpeter Swan Society Conference, North American Swans, Bulletin of the Trumpeter Swan Society

Keyword(s): Birds, Nongame Wildlife, Waterfowl, Wetlands, training

Contact(s):
Jane West, EDITOR
Madeleine Linck, EDITOR
Ruth Shea, EXECUTIVE DIRECTOR
3346 East 200 N., Rigby, ID 83442
Harvey Nelson, PRESIDENT AND EDITOR
10515 Kell Ave., Bloomington, MN 55437
Larry Gillette, VICE PRESIDENT

TRUST FOR PUBLIC LAND, THE
National Office, 116 New Montgomery St. 4th Fl.
San Francisco, CA 94105 USA
Phone: 415-495-4014   Fax: 415-495-4103
E-mail: info@tpl.org
Website: www.tpl.org

Founded: 1972
Membership: 1500
Scope: National

Description: TPL specializes in conservation real estate, applying its expertise in negotiations, public finance and law to protect land for public use and enjoyment. TPL has helped protect more than 1 million acres nationwide and has recently launched its Greenprint for Growth campaign to help sprawl-threatened communities protect land as a way to guide development and sustain a healthy economy and a high quality of life.

Publication(s): Regional newsletters, Land and People

Contact(s):
Susan Ives, EDITOR, VICE PRESIDENT, DIRECTOR OF PUBLIC AFFAIRS
Ralph Benson, EXECUTIVE VICE PRESIDENT
Will Rogers, PRESIDENT
Ernest Cook, SENIOR VICE PRESIDENT
33 Union St., 4th Fl., Boston, MA 02108
Phone: 617-367-6200
Stephen Thompson, SENIOR VICE PRESIDENT
418 Montezuma, Santa Fe, NM 87501
Phone: 505-988-5922
Bowen Blair, SENIOR VICE PRESIDENT / NATIONAL DIRECTOR OF PROJECTS
Smith Tower, Suite 1510, 506 2nd Ave., Seattle, WA 98104
Phone: 206-587-2447
Robert McIntyre, SENIOR VICE PRESIDENT AND CFO
Nelson Lee, SENIOR VICE PRESIDENT, GENERAL COUNSEL
Rose Harvey, SENIOR VICE PRESIDENT, MID-ATLANTIC REGIONAL DIRECTOR
666 Broadway, New York, NY 10012
Phone: 212-677-7171
Alan Front, SENIOR VICE PRESIDENT, DIRECTOR OF FEDERAL AFFAIRS
Ted Harrison
SOUTHWEST REGIONAL MANAGER
418 Montezuma, Santa Fe, NM 87501
Phone: 505-988-5922
Kathy Blaha, VICE PRESIDENT
666 Pennsylvania Ave. SE, Washington, DC 20003
Phone: 202-543-7552
W. Allen, VICE PRESIDENT AND SOUTHEAST REGIONAL MANAGER
306 N. Monroe St., Tallahassee, FL 32301-7635
Phone: 904-222-7911
Reed Holderman, VICE PRESIDENT, WESTERN REGIONAL DIRECTOR
116 New Montgomery, 3rd Fl., San Francisco, CA 94105
Phone: 415-495-5660
Whitney Hatch, VICE PRESIDENT, REGIONAL DIRECTOR
33 Union St., 4th Fl., Boston, MA 02108
Phone: 617-367-6200
Rand Wentworth, VICE PRESIDENT, ASSOC. REGIONAL DIRECTOR
1447 Peachtree St., NE, Suite 601, Atlanta, GA 30309
Phone: 404-873-7306

TRUST FOR PUBLIC LAND, THE
Headquarters, National Office, Attn: Public Affairs Assistant, 116 New Montgomery St. 4th Fl.
San Francisco, CA 94105 USA
Phone: 415-495-4014   Fax: 415-495-4103
E-mail: info@tpl.org
Website: www.tpl.org

Founded: 1972
Scope: National

Description: Since its founding in 1972, TPL has worked with public agencies, landowners and citizen groups to protect more than 1.2 million acres in 45 states nationwide and has recently launched its Greenprint for Growth campaign to help sprawl-threatened communities protect land as a way to guide development and sustain a healthy economy and a high quality of life.

Publication(s): Regional newsletters, Land and People

Keyword(s): Land Purchase, Open Space, Smart Growth, Outdoor Recreation, Sprawl, Watersheds, Wilderness

Contact(s):
Becky Mitchell, DIRECTOR OF EXTERNAL AFFAIRS
Ralph Benson, EXECUTIVE VICE PRESIDENT
Will Rogers, PRESIDENT
Ernest Cook, SENIOR VICE PRESIDENT
33 Union St., 4th Fl., Boston, MA 02108
Phone: 617-367-6200
Stephen Thompson, SENIOR VICE PRESIDENT
418 Montezuma, Santa Fe, NM 87501
Phone: 505-988-5922
Alan Front, SENIOR VICE PRESIDENT & DIRECTOR OF FEDERAL AFFAIRS
Kathy Blaha, SENIOR VICE PRESIDENT & GREEN CITIES INITIATIVE PROGRAM DIRECTOR
666 Pennsylvania Ave. SE, Washington, DC 20003
Phone: 202-543-7552
Rose Harvey, SENIOR VICE PRESIDENT & MID-ATLANTIC REGIONAL DIRECTOR
666 Broadway, New York, NY 10012
Phone: 212-677-7171
Bowen Blair, SENIOR VICE PRESIDENT & NATIONAL DIRECTOR OF PROJECTS
1211 SW Sixth Ave., Portland, OR 97204-1001
Phone: 503-228-6620
W. Allen, SENIOR VICE PRESIDENT & REGIONAL DIRECTOR
306 N. Monroe St., Tallahassee, FL 32301-7635
Phone: 904-222-7911
Ted Harrison, SENIOR VICE PRESIDENT & SOUTHWEST REGIONAL DIRECTOR
418 Montezuma, Santa Fe, NM 87501
Phone: 505-988-5922
Robert McIntyre, SENIOR VICE PRESIDENT AND CFO
Nelson Lee, SENIOR VICE PRESIDENT AND GENERAL COUNSEL
Rand Wentworth, V.P. & ASSOCIATE REGIONAL DIRECTOR
1447 Peachtree St., NE, Suite 601, Atlanta, GA 30309
Phone: 404-873-7306
Tod Dobratz, V.P. & ASST. CFO
Lesley Kane-Synal, V.P. & DIRECTOR FEDERAL LEGISLATIVE OFFICE
666 Pennsylvania Ave. SE Suite 401, Washington, D.C. 20003-4334
Phone: 202-543-7552
Luara Brehm, V.P. & DIRECTOR OF DEVELOPMENT
Cynthia Whiteford, V.P. & REGIONAL DIRECTOR
2610 University Ave. 300,
Reed Holderman, V.P. & WESTERN REGIONAL DIRECTOR
116 New Montgomery, 3rd Fl., San Francisco, CA 94105
Phone: 415-495-5660
Susan Ives, V.P./DIRECTOR OF PUBLIC AFFAIRS AND EDITOR OF LAND & PEOPLE MAGAZINE

Whitney Hatch, VICE PRESIDENT & NEW ENGLAND REGIONAL DIRECTOR
33 Union St., 4th Fl., Boston, MA 02108
Phone: 617-367-6200

TRUST FOR WILDLIFE, INC.
127 Ehrich Rd.
Shaftsbury, VT 05262 USA
Phone: 802-447-0746   Fax: 802-447-0746
E-mail: trustforwildlife@net0.net
Website: www.neotropicalbirds.net

Founded: 1983
Membership: 20
Scope: International

Description: Dedicated to wildlife conservation and education with a focus on wildlife habitats, international partnerships with a focus on Russia, and wildlife rehabilitation with an emphasis on public education.

Keyword(s): Environmental and Conservation Education, education, training, accreditation, Land Preservation, Biodiversity, Youth Organizations, Communications, National Parks

Contact(s):
Les Line, DIRECTOR
P.O. Box 323, Amenia, NY 12501
Phone: 914-373-9135
Ed Metcalfe, DIRECTOR
6373 Vermont Rt 100, Whitingham, VT 05361
Phone: 802-464-0048
Marshal Case, PRESIDENT
Gregory Sharp, SECRETARY
225 Reeds Gap Rd., East Northford, CT 6472
Phone: 203-240-6046

TRUSTEES FOR ALASKA
1026 West 4th Ave., Suite 201
Anchorage, AK 99501-2101 USA
Phone: 907-276-4244   Fax: 907-276-7110
E-mail: ecolaw@trustees.org
Website: www.trustees.org

Founded: 1974
Scope: Statewide

Description: Nonprofit, public-interest, environmental law firm working to ensure that Alaska's unique environmental values are not lost to future generations through irresponsible development or irrational public management. Primarily concerned with protection of Alaska's environment and the wise management of Alaska's resources.

Publication(s): The Environmental Advocate (Quarterly Newsletter), Under The Influence "Oil and The Industrialization of America's Arctic"

Keyword(s): Environmental Law

Contact(s):
Ken Robertson, CHAIR
Ann Rothe, EXECUTIVE DIRECTOR
Joanna Parker, LEGAL SUPPORT STAFF
Peter van Tuyn, LITIGATION DIRECTOR
Chris Rose, SECRETARY
Michael Frank, STAFF ATTORNEY
Lisa Taylor, TREASURER
Mary Mcburney, VICE CHAIR

TRUSTEES OF RESERVATIONS, THE

572 Essex St.
Beverly, MA 01915-1530 USA
Phone: 978-921-1944   Fax: 978-921-1948
E-mail: information@ttor.org
Website: www.thetrustees.org

Founded: 1891
Membership: 30000
Scope: Statewide

Description: The Trustees of Reservations is the nation's oldest statewide non-profit conservation and preservation organization. TToR is dedicated to preserving, for public use and enjoyment, properties of exceptional scenic, historic and ecological value in addition to other special places throughout Massachusetts.

Publication(s): Land of Commonwealth, Annual Report, Land Conservation Options: A Guide For Massachusetts Landowners, Newsletter, Reservations Guidebook, Conserving our Common Wealth: A Vision for the Massachusetts Landscape

Keyword(s): Gardening and Horticulture, Historic Preservation, Land Conservation, National Parks, Open Space, Environmental Protection

Contact(s):
Elliot Surkin, CHAIRMAN
Michael Triff, DIRECTOR OF COMMUNICATIONS AND MARKETING
Ann Powell, DIRECTOR OF DEVELOPMENT
John McCrae, DIRECTOR OF FINANCE AND ADMINISTRATION
Wesley Ward, DIRECTOR OF LAND CONSERVATION
John Bradley, DIRECTOR OF MEMBERSHIP
Sarah Carothers, DIRECTOR OF PLANNED GIVING
Richard Howe, DIRECTOR OF PROPERTY MANAGEMENT
Andy Kendall, EXECUTIVE DIRECTOR
Janice Hunt, PRESIDENT
F. Smithers, SECRETARY
Charles Kane, TREASURER

TUG HILL TOMORROW LAND TRUST

P.O. Box 6063
Watertown, NY 13601 USA
Phone: 315-785-2382   Fax: 315-785-2574
E-mail: thtomorro@northnet.org
Website: www.tughilltomorrowlandtrust.org

Founded: 1990
Membership: 100
Scope: State

Description: Tug Hill Tomorrow is a private nonprofit corporation that works to help retain the forests, farms, recreational, and wild lands of the Tug Hill region through education, research, and voluntary land protection.

Publication(s): Tug Hill Recreation Guide: A Guide to Cross-Country Skiing, Hiking, Biking, and Fishing, Tug Hill Working Lands, Tug Hill Resource Guide to Educational Programs, Tug Hill Natural History Field Guide, Greenings

Keyword(s): Environmental and Conservation Education, Forests and Forestry, Land Preservation, Natural Areas, Open Space

Contact(s):
Robert Boice, CHAIR
Greg Gardner, TREASURER

TURTLE CREEK WATERSHED ASSOCIATION, INC.

325 Commerce Street, Suite 204
Wilmerding, PA 15148 USA
Phone: 412-829-2817
E-mail: GOODFISH@helicon.net
Website: http://trfn.clpgh.org/tcwa/

Founded: 1969
Scope: Local Region

Description: The objective of the Turtle Creek Watershed Association, Inc. is to preserve and protect natural resources; educate the community about important environmental issues; monitor and improve water quality; and work with responsible agencies to encourage wise land use planning in the Turtle Creek Watershed.

Publication(s): Conserve, Business Associate, TCWA Report

Keyword(s): Endangered Species, Environmental Protection, Land Purchase, Rivers, training, Water Quality, Environmental and Conservation Education, Watersheds

Contact(s):
James Tempero, DIRECTOR EMERITUS
Edward Fischer, EXECUTIVE DIRECTOR (TEMPORARY)
Robert Mazik, PRESIDENT
Edward Fischer, SECRETARY
A.B. Carl, TREASURER
Henry Hoffman, VICE PRESIDENT

U

U.S. PUBLIC INTEREST RESEARCH GROUP

218 D St., SE
Washington, DC 20003 USA
Phone: 202-546-9707   Fax: 202-546-2461
E-mail: uspirg@pirg.org
Website: www.pirg.org

Founded: 1983
Scope: National

Description: U.S. PIRG is the national lobbying office for state PIRGs around the country, representing more than one million members. We conduct independent research and lobby for national environmental and consumer protections.

Publication(s): A publication list available upon request., Citizen Agenda

Keyword(s): Air Quality and Pollution, Chemical Pollution Control, Conservation, Endangered Species, Energy, Environment, Solid Waste Management, Coral Reefs, Wetlands, Environmental Protection, Energy Conservation, Pollution Prevention

Contact(s):
Elizabeth Hitchcock, COMMUNICATIONS DIRECTOR
Gene Karpinski, DIRECTOR

UNION OF CONCERNED SCIENTISTS

Two Brattle Square
Cambridge, MA 02238 USA
Phone: 617-547-5552   Fax: 617-864-9405
E-mail: ucs@ucsusa.org
Website: www.ucsusa.org

Founded: 1969
Membership: 50000

Scope: National

Description: The Union of Concerned Scientists is a national non-profit working for a cleaner environment and a safer world. UCS is working to encourage preservation of life-sustaining resources, promote energy technologies that are renewable, safe, and cost-effective, promote advanced transportation technologies, encourage sustainable agriculture, and curtail weapons proliferation.

Publication(s): Earthwise, The Consumer's Guide to Effective Environmental Choices, Powerful Solutions: Seven Ways to Switch America to Renewable Electricity, Now or Never: Serious New Plans to Save a Natural Pest Control, Newsletter: NUCLEUS

Keyword(s): Biotechnology, Energy, Environmental and Conservation Education, Renewable Resources, Sustainable Development, Biodiversity, Agriculture, Environment, Greenhouse Effect/Global Warming, Solar Energy, Nuclear Energy, Transportation

Contact(s):
Kurt Gottfried, CHAIRMAN
Howard Ris, PRESIDENT

UNITED NATIONS ENVIRONMENT PROGRAMME
NEW YORK LIAISON OFFICE
2 United Nations Plaza, Rm. DC2-0803
New York, NY 10017 USA
Phone: 212-963-8210   Fax: 212-963-7341
E-mail: info@nyo.unep.org
Website: www.nyo.unep.org

Founded: NA
Membership: 12
Scope: International
Publication(s): Our Planet, The Global Environment Outlook
Contact(s):
Adnan Amin, DIRECTOR
Phone: 212-963-8138
adnan.amin@nyo.unep.org
James Sniffen, INFORMATION OFFICER
Phone: 212-963-8094
sniffenj@nyo.unep.org

UNITED STATES COMMITTEE FOR THE UNITED NATIONS ENVIRONMENT PROGRAMME, THE (U.S. AND UNEP)
2013 Q St., NW
Washington, DC 20009 USA
Phone: 202-234-3600   Fax: 202-332-3221

Founded: NA
Scope: National

Description: A nonprofit support group for the U.N. Environment Programme, U.S. and UNEP generates public awareness of global environmental issues, including ozone layer depletion, the greenhouse effect, and the transport of hazardous chemicals, and UNEP's response to these issues. The organization links UNEP to environmental groups across the U.S.

Publication(s): US - UNEP NEWS

Keyword(s): Agriculture, Endangered Species, Environmental and Conservation Education, Greenhouse Effect/Global Warming

Contact(s):
Richard Hellman, PRESIDENT

UNITED STATES TOURIST COUNCIL
Drawer 1875
Washington, DC 20013-1875 USA

Founded: NA
Scope: National

Description: A nonprofit association of conservation-concerned individuals, industries, and institutions who travel or cater to the traveler. Emphasis is on historic and scenic preservation, wilderness and roadside development, ecology through sound planning and education, and support of scientific studies of natural wilderness.

Keyword(s): Aquatic Habitats, Forests and Forestry, Historic Preservation, Wetlands

Contact(s):
Stanford West, CHAIRMAN OF BOARD OF TRUSTEES AND EXECUTIVE DIRECTOR

UNIVERSITY OF KENTUCKY DEPT. OF FORESTRY
UK Thomas Poe Cooper Bldg
Lexington, KY 40546-0073 USA
Phone: 859-257-5994   Fax: 859-323-1031
Website: www.uky.edu-agriculture-forestry-forestry.html

Founded: NA
Membership: 45
Scope: National, Regional

Description: K-T SAF is the Kentucky-Tennessee section of the Society of American Foresters, and carries out the policies and programs of SAF within these two states. See the Society of American Foresters listing for more information.

Keyword(s): Renewable Resources, Urban Forestry, Environmental Planning

Contact(s):
Jim Ringe, DIRECTOR OF UNDERGRADUATE STUDIES
Phone: 859-257-7594
jringe@uky.edu
Jeff Stringer, EXTENSION COORDINATOR
Dept. of Forestry, University of Ky., Lexington, KY 40546-0073
Phone: 859-257-5994

UPPER CHATTAHOOCHEE RIVERKEEPER
1900 Emery St., Suite 450
Atlanta, GA 30318 USA
Phone: 404-352-9828   Fax: 404-352-8676
E-mail: bbolton@usriverkeeper.org
Website: www.chattahoochee.org

Founded: 1994
Membership: 2500
Scope: Regional

Description: To advocate snd secure the protection and stewardship of the Chattahoochee River, its tributaries and watershed using education, research, communication, cooperation, monitoring and legal actions.

Publication(s): River Chat, Stream Chat

Keyword(s): Rivers, Watershed Protection, Exotic species, Aquatic nuisance species, Environmental and Conservation Education, Research

Contact(s):
Darcie Doden, DIRECTOR OF HEADWATER CONSERVATION
Sally Bethea, EXECUTIVE DIRECTOR
sbethea@mindspring.com
Michelle Fried, GENERAL COUNSEL
mfriverkeeper@mindspring.com
Matt Kales, PROGRAM MANAGER FOR RIVER BASIN PROTECTION
mknverkeeper@mindspring.com
Alice Chamipagne, WATERSHED PROTECTION SPECIALIST

UPPER MISSISSIPPI RIVER CONSERVATION COMMITTEE

4469 - 48th Avenue Ct.
Rock Island, IL 61201 USA
Phone: 309-793-5800   Fax: 309-739-5804
E-mail: umrcc@mississippi-river.com
Website: www.mississippi-river.com/umrcc

Founded: 1943
Scope: Regional

Description: Promotes preservation, development, and wise use of the natural and recreational resources of the Upper Mississippi River and formulates policies, plans, and programs for conducting cooperative studies. Members: state conservation departments of Illinois, Iowa, Minnesota, Missouri, and Wisconsin.

Publication(s): Newsletter, and miscellaneous technical reports, Annual Proceedings

Keyword(s): Wildlife, Rivers, Aquatic Habitats, Exotic species, Aquatic nuisance species, Wildlife Rehabilitation, Water Quality, Mussels, Sustainable Development

Contact(s):
Jon Duyvejonck, COORDINATOR
Phone: 309-793-5800
Ken Brummett, SECRETARY AND TREASURER
Missouri Dept. of Conservation, 653 Clinic Rd., Hannibal, MO 63401
Phone: 309-582-5611

URBAN HABITAT PROGRAM

P.O. Box 29908, Presidio Station
San Francisco, CA 94129 USA
Phone: 415-561-3333   Fax: 415-561-3334
E-mail: contact@urbanhabitatprogram.org
Website: www.urbanhabitatproram.org

Founded: 1989
Scope: National

Description: The Urban Habitat Program is a project of Tides Center. Its mission is to build multi-cultural urban environmental leadership for socially-just and sustainable communities in the San Francisco Bay area. Our project areas include transportation, regional land use and social justice, land recycling and brown fields, leadership institute for sustainability and justice, and the goal of ecological literacy, all from an ecological and social justice perspective.

Publication(s): Race, Poverty & the Environment

Keyword(s): Cultural Preservation, Land Use Planning, Transportation, Urban Environment, People of Color in the Environment, Brown Fields, Environmental Justice

Contact(s):
Carl Anthony, DIRECTOR

URBAN WILDLIFE RESOURCES

5130 W. Running Brook Rd.
Columbia, MD 21044 USA
Phone: 410-997-7161   Fax: 410-997-6849
Website: www.erols.com/urbanwildlife

Founded: 1995
Scope: International

Description: Urban Wildlife Resources works to facilitate interaction and cooperation among land managers and planners, biologists, landscape architects, and others in achieving better management of natural resources in urban and urbanizing areas.

Publication(s): Urban Open Space Manager, The

Keyword(s): Conservation, Open Space, Urban and Rural Development, Urban Environment, Urban Forestry, training

Contact(s):
Lowell Adams, PRESIDENT

US GEOLOGICAL SURVEY

WESTERN REGION
345 Middlefield Road
Menlo Park, CA 94025 USA
Phone: 650-853-8300
Website: http://www.usgs.gov/

Contact(s):
John Buffington, REGIONAL DIRECTOR

UTAH ASSOCIATION OF CONSERVATION DISTRICTS

1860 N.100 East
Logan, UT 84341 USA
Phone: 435-753-6029   Fax: 435-755-2117
Website: www.uacd.org

Founded: NA
Membership: 230
Scope: Statewide
Publication(s): newsletter - The Leader, book - Citizen Planners Guide to Sub Division Development, book - Study on low impact street design
Keyword(s): Conservation Districts

Contact(s):
Gordon Younker, EXECUTIVE VICE PRESIDENT
1860 N. 100 E, North Logan, UT 84341-2215
Phone: 435-753-6029
Fax: 435-755-2117
gordon-yonker@ut.nacdnet.org
William Rigby, PAST PRESIDENT
Randy Greenhalgh, PRESIDENT
Phone: 435-623-0845
Fax: 435-623-0845
Richard Saunders, SECRETARY AND TREASURER
4083 W 12680 S., Payson, UT 84651
Phone: 801-465-2777

Larry Johnson, VICE PRESIDENT /BOARD MEMBER
P.O. Box 177, Randolph, UT 84064
Phone: 435-793-5625
Fax: 435-793-5625

UTAH B.A.S.S. CHAPTER FEDERATION
Attn: President, 3460 Scott Cir.
Salt Lake City, UT 84115 USA
Phone: 801-487-8711
E-mail: gjlables@aol.com
Website: www.utahbassfederation.org

Founded: NA
Scope: Statewide

Description: An organization of Bassmaster chapters, affiliated with the Bass Anglers Sportsman Society, organized to fight pollution, assist state and national conservation agencies in their efforts, and teach young people of our country good conservation practices. Dedicated to the realistic conservation of our water resources.

Publication(s): Nature News Notes

Keyword(s): Environmental and Conservation Education, Natural Areas, training, National Parks, Nongame Wildlife

Contact(s):
Walter Maldonado, CONSERVATION DIRECTOR
P.O. Box 482, Green River, UT 84525-0482
Phone: 435-564-8147
viper@etv.net
George Sommer, PRESIDENT
Phone: 801-487-8711

UTAH NATIVE PLANT SOCIETY
P.O. Box 520041
Salt Lake City, UT 84152-0041 USA
Phone: 801-272-3275
E-mail: unps@xmission.com
Website: www.unps.org

Founded: 1978
Membership: 430
Scope: Local

Description: Our orgaization is a charitable, non-profit dedicated to the understanding, preservation, enjoyment, and responsible use of the Utah native plants. We wish to foster public recognition of the diverse flora of the state.

Publication(s): Heritage Garden Native Plant Propagation Workshop booklet, UNPS.org website, Sego Lily

Contact(s):
Susan Meyer, CHAIRMAN
Phone: 1-356-3636
smeyer@sisna.com
Janett Warner, CENTRAL UTAH CHAPTER
janettw@hubwest.com
Mindy Wheeler, SALT LAKE CHAPTER PRESIDENT
Phone: 801-561-0779
mindywheeler@usa.net
Therese Meyer, SECRETARY
Phone: 801-272-3275
tmeyer@xmission.com

UTAH NATURE STUDY SOCIETY
Attn: President Utah Nature Study Society,
2853 S. 23rd East
Salt Lake City, UT 84109 USA

Founded: 1954
Scope: Statewide

Description: Promotes conservation and nature education through workshops and field trips for members; publicizes conservation problems and issues through meetings and its newsletter. Member of Utah Associated Garden Clubs.

Publication(s): UWA Review

Keyword(s): Biodiversity, Environmental and Conservation Education, Public Lands, Wilderness

Contact(s):
Catherine Quinn, EDITOR
1383 S. 300 East, Salt Lake City, UT 84115
Jean White, EXECUTIVE SECRETARY
377 E. 5300 S., Murray, UT 84107-6019
Dorothy Platt, PRESIDENT
2853 S. 23rd East, Salt Lake City, UT 84109
Maria Dickerson, SECRETARY
323 S. 2nd W., Tooele, UT 84074

UTAH WILDERNESS COALITION
P.O. Box 520974
Salt Lake City, UT 84152-0974 USA
Phone: 801-486-2872   Fax: 801-485-5572
E-mail: wildutah@xmission.com
Website: http://www.uwcoalition.org

Founded: 1985
Scope: National

Description: Promote and coordinate the preservation of U.S. BLM wildlands in southern and western Utah through public education and the passage of American's Redrock Wilderness Act. The goal includes protection of the remaining wilderness quality public lands under the National Wilderness Preservation System.

Keyword(s): Outdoor Recreation, Public Lands, Wilderness

UTAH WILDLIFE FEDERATION
P.O. Box 526367
Salt Lake City, UT 84152-6367 USA
Phone: 801-487-1946   Fax: 801-773-0412
E-mail: uwfhal@xmission.com

Founded: NA
Scope: Statewide

Description: A representative statewide organization, affiliated with the National Wildlife Federation, dedicated to the protection and enhancement of wildlife and its habitat through public education and government interaction.

Publication(s): Utah Wildlife News

UTAH WOODLAND OWNERS COUNCIL
2829 Sleep Hollow Dr.
Salt Lake City, UT 84117 USA
Phone: 801-277-1615

Founded: 1997
Membership: 7
Scope: Statewide

Description: A statewide organization affiliated with the National Woodland Owners Association and associated with Utah Farms Bureau, that is working for good forest management practices on the private forest and ranch land in Utah.

Keyword(s): Forests and Forestry

Contact(s):
Richard Oldroyd, CHAIRMAN

V

VERMONT ASSOCIATION OF CONSERVATION DISTRICTS
487 Rowell Hill Rd.
Berlin, VT 05602 USA
Phone: 802-229-9250

Founded: NA
Scope: Statewide

Publication(s): Annual Reports

Keyword(s): Conservation Districts

Contact(s):
Rita Visson, TREASURER
240 Vermont Rt. 100, Orange, VT 05641
Phone: 802-479-9538
Claire Ayer, VICE PRESIDENT
Phone: 802-545-2142

VERMONT AUDUBON COUNCIL
65 Millet St.
Richmond, VT 05477 USA
Phone: 802-434-4300
E-mail: vermont@audubon.org
Website: audubon.org

Founded: NA
Scope: Statewide

Description: A statewide council consisting of eight chapters of the National Audubon Society formed to protect birds, other wildlife and their habitats by promoting a culture of conservation through research, education and advocacy.

Keyword(s): Environmental and Conservation Education, Forests and Forestry, Nongame Wildlife, Planning Management, Wetlands, Birds, training

Contact(s):
Warren King, PRESIDENT
P.O. Box 77, Ripton, VT 05766
Suzanna Liepmann, TREASURER
P.O. Box 112, South Strafford, VT 05070
Seward Weber, VICE PRESIDENT
R.D. 2, Box 390, Plainfield, VT 05667

VERMONT B.A.S.S. CHAPTER FEDERATION
Attn: President, 19 Pinewood Rd.
Montpelier, VT 05602 USA
Phone: 802-223-7793
E-mail: nsk@together.net
Website: http://www.vermontbass.com

Founded: NA
Scope: Statewide

Description: An organization of Bassmaster chapters, affiliated with the Bass Anglers Sportman Society, organized to fight pollution, assist state and national conservation agencies in their efforts, and teach the young people of our country good conservation practices. Dedicated to the realistic conservation of our water resources.

Keyword(s): Birds, Endangered Species, Environmental and Conservation Education, education, Raptors

VERMONT INSTITUTE OF NATURAL SCIENCE
27023 Church Hill Rd.
Woodstock, VT 05091 USA
Phone: 802-457-2779   Fax: 802-457-1053
E-mail: info@vinsweb.org
Website: www.vinsweb.org

Founded: 1972
Scope: State

Description: The mission of VINS is to protect Vermont's natural heritage through environmental education and research. VINS Raptor Center, Living Museum of birds of prey, on VINS nature preserve.

Publication(s): Vermont Institute of Natural Science, Records of Vermont Birds, Hands on Nature

Keyword(s): Agriculture, Environmental and Conservation Education, Forests and Forestry, Land Preservation

Contact(s):
Marsha Whitney, EDUCATION DIRECTOR
Sherman Kent, EXECUTIVE DIRECTOR
Deborah Granquist, PRESIDENT BOARD OF DIRECTORS
Christopher Rimmer, RESEARCH DIRECTOR
Jenepher Linglebach, VICE PRESIDENT

VERMONT LAND TRUST
8 Bailey Ave.
Montpelier, VT 05602 USA
Phone: 802-223-5234   Fax: 802-223-4223
Website: www.vlt.org

Founded: 1977
Scope: Regional

Description: Conserving the productive, recreational, and scenic lands that help give Vermont and its communities their distinctive rural character.

Publication(s): Tri-annual newsletters, Annual Report

Keyword(s): Land Conservation

Contact(s):
Darby Bradley, PRESIDENT
8 Bailey Ave., Montpelier, VT 05602
Phone: 802-223-5234
darby@vlt.org
Gil Livingston, VICE PRESIDENT FOR LAND CONSERVATION
8 Bailey Ave., Montpelier, VT 05602
Phone: 802-223-5234
Barbara Wagner, VICE PRESIDENT OF OPERATIONS
8 Bailey Ave., Montpelier, VT 05602
Phone: 802-223-5234

VERMONT NATURAL RESOURCES COUNCIL

9 Bailey Ave.
Montpelier, VT 05602 USA
Phone: 802-223-2328   Fax: 802-223-0287
E-mail: info@vnrc.org
Website: www.vnrc.org

Founded: NA
Membership: 9
Scope: Statewide

Description: A representative statewide organization, affiliated with the National Wildlife Federation, dedicated to the protection and enhancement of wildlife and its habitat through public education and government interaction.

Publication(s): Vermont Environmental Directory, The Legislative Bulletin, Vermont Environmental Report

Contact(s):
Stephan Holmes, ALTERNATE REPRESENTATIVE
Mark Naud, CHAIR
Stephanie Mueller, EDITOR
Patrick Berry, EDUCATION PROGRAMS CONTACT
Elizabeth Courtney, EXECUTIVE DIRECTOR
Leonard Wilson, REPRESENTATIVE
Bill Amberg, TREASURER

VERMONT STATE-WIDE ENVIRONMENTAL EDUCATION PROGRAMS (SWEEP)

c/o Vermont Natural Resources Council, 9 Bailey Ave.
Montpelier, VT 05602 USA
Phone: 802-223-2328

Founded: 1973
Scope: Statewide

Description: SWEEP is a coalition of individuals and organizations promoting environmental education in Vermont. SWEEP's purpose is to foster environmental appreciation and understanding in order to enable Vermonters to make responsible decisions affecting the environment.

Publication(s): SWEEP Newsletter

Keyword(s): Environmental and Conservation Education

Contact(s):
Susan Clark, CHAIR
165-A Wood Rd., N Middlesex, VT 05682
Phone: 802-223-5824
Marie Caduto, CHAIR
198 Kerwin Hill Rd., Norwich, VT 05055
Phone: 802-763-8303
Linda Wellings, SECRETARY
Shelburne Farms, Shelburne, VT 05482
Phone: 802-985-8686

VERMONT WOODLANDS ASSOCIATION

664 North County Rd.
Groton, VT 05046 USA
Phone: 802-584-3333
E-mail: vtwoods@together.net

Founded: NA
Scope: State

Description: A statewide organization, affiliated with the National Woodland Owners Association, organized to promote sound forest management throughout Vermont.

Publication(s): Vermont Woodlands, forestry and caring for Vermont forests

Keyword(s): Forests and Forestry

Contact(s):
Harry Chandler, EXECUTIVE DIRECTOR
Robert Darrow, IMMEDIATE PAST PRESIDENT
Phone: 802-773-7144
Putnam W. Blodgett, PRESIDENT
putblodgett@valley.net
Stanley James, VICE PRESIDENT
822 Lemon Fair Rd, Weybridge VT, 05753,
John Hemenway, VICE PRESIDENT
Phone: 802-765-4324
jthemenway@aol.com

VIRGIN ISLANDS CONSERVATION DISTRICT

Attn: President, P.O. Box 1576
Fredericksted, VI 00841 USA

Founded: NA
Scope: Statewide

Publication(s): Federation Record, The

Keyword(s): Conservation Districts

Contact(s):
Cedrick Lewis, ALTERNATE BOARD MEMBER
P.O. Box 303142, St. Thomas, VI 00803
Phone: 340-775-7393
Hans Lawaetz, PRESIDENT AND BOARD MEMBER
P.O. Box 1576, Fredericksted, VI 00841
Phone: 340-788-2229
Fax: 340-778-0270
Enrico Gasperi, SECRETARY AND TREASURER
P.O. Box 895, Christiansted, VI 00824
Phone: 340-773-2386
Joseph Samuel, VICE PRESIDENT
P.O. Box 241, Fredericksted, St. Croix, VI 00841
Phone: 340-772-3168

VIRGIN ISLANDS CONSERVATION SOCIETY, INC.

Arawak Bldg., Suite 3, Gallows Bay
Christiansted, VI 00820 USA
Phone: 340-773-1989   Fax: 340-773-7545
E-mail: sea@viaccess.net

Founded: 1968
Membership: 750
Scope: Regional

Description: A representative statewide organization, affiliated with the National Wildlife Federation, dedicated to the protection and enhancement of wildlife and its habitat through public education and government interaction.

Keyword(s): Agriculture, Environmental and Conservation Education, Pollution Prevention, Soil Conservation

Contact(s):
Emy Thomas, EDITOR
fhenry@vvi.edu
Tysha Jules, EDUCATION PROGRAMS CONTACT
Carlos Tesitor, PRESIDENT
Carla Joseph, REPRESENTATIVE
Stevie Ketcham, TREASURER

VIRGINIA ASSOCIATION FOR PARKS

5616 Bloomfield Drive 103
Alexandria, VA 22312 USA
Phone: 703-941-1350   Fax: 202-548-0595
Website: www.parksonline.org/vap

Membership: 75
Scope: Statewide

Contact(s):
Robert Williams, CONTACT
Phone: 540-972-9954
Davinder Khanna, CONTACT
davinder@parksonline.org

VIRGINIA ASSOCIATION OF CONSERVATION DISTRICTS

7293 Hanover Green Dr., Suite B-101
Mechanicsville, VA 23111 USA
Phone: 804-559-0324

Founded: NA
Scope: Statewide

Keyword(s): Conservation Districts

Contact(s):
Jay Gilliam, 1ST VICE PRESIDENT
Raphine, VA 24472
Phone: 540-377-6179
strmiwla@cfw.com
Greg Evans, 2ND VICE PRESIDENT
8400 Oakford Dr., Springfield, VA 22152
Phone: 703-644-1227
soilandh2o@aol.com
Stephanie Martin, EXECUTIVE DIRECTOR
7293 Hanover Green Dr., Suite B101, Mechanicsville,
VA 23111
Phone: 804-559-0324
Fax: 804-559-0325
John Dixon, PAST PRESIDENT,
1228 Rendezous Ln., Bedford, VA 24523
Phone: 540-586-8969
Daphne Jamison, PRESIDENT
Phone: 540-721-2361
rjam229@aol.com
James Byrne, SECRETARY/TREASURER
Rt. 1 Box 351, Reva, VA 22735
Phone: 540-547-2932
tohisplace1@juno.com

VIRGINIA B.A.S.S. CHAPTER FEDERATION

Attn: President, 28447 Cabin Point Rd.
Disputanta, VA 23842 USA
Phone: 757-428-4280   Fax: 804-834-8198
Website: www.vabass.com

Founded: NA
Membership: 2400
Scope: Statewide

Description: An organization of Bassmaster chapters, affiliated with the Bass Anglers Sportsman Society, organized to fight pollution, assist state and national conservation agencies in their efforts, and teach the young people of our country good conservation practices. Dedicated to the realistic conservation of our water resources.

Publication(s): Virginia B.A.S.S. Federation Newsletter "Tightlines"

Keyword(s): Air Quality and Pollution, Land Use Planning, Water Quality

Contact(s):
Mitchell Perkins, ACTING CONSERVATION DIRECTOR
12003 Bourne Road, Glen Allen, VA 23059
Phone: 804-264-1124
HUNTNBASS1@aol.com
Roger Fitchett, PRESIDENT

VIRGINIA CONSERVATION NETWORK

1001 E. Broad St., Suite LL 35-C
Richmond, VA 23219 USA
Phone: 804-644-0283   Fax: 804-644-0286
E-mail: ellenshepard@yahoo.com
Website: www.vcnva.org

Founded: 1969
Scope: Statewide

Description: The Conservation Council of Virginia merged with the Virginia Environmental Network in 1994 to form the Virginia Conservation Network. The Network's member organizations are devoted to advancing a common, environmentally-sound vision for Virginia.

Publication(s): 1999 Voting Summary

Keyword(s): Environmental Legislation

Contact(s):
Katherine Slaughter, PRESIDENT
kslaughter@selcva.org
Michael Nelson, SECRETARY
mike@audubonnaturalist.org
Martha Wingfield, TREASURER
marlridge@aol.com
Anne Marshall, VICE PRESIDENT
aamvirginia@hotmail.com

VIRGINIA FORESTRY ASSOCIATION

8810B Patterson Ave.
Richmond, VA 23229 USA
Phone: 804-741-0836   Fax: 804-741-0838
E-mail: vafa@erols.com
Website: www.vaforestry.org

Founded: 1943
Scope: Statewide

Description: An association of landowners and forest industry that promotes stewardship and wise use of forest resources for the economic and environmental benefits of all Virginians.

Publication(s): All publications on web, Virginia Forest Magazine, News and Notes - newsletter, chapter newsletters, wildflower conservation guidelines, checklists, nursery source list, fact sheets, brochure, invasive alien plant list, Bulletin

Keyword(s): Endangered Species, Environmental and Conservation Education, Flowers, Plants, and Trees, Gardening and Horticulture, Natural Areas

Contact(s):
Paul Howe, EXECUTIVE VICE PRESIDENT & EDITOR
Dave Froggatt, TREASURER

VIRGINIA NATIVE PLANT SOCIETY

Blandy Experimental Farm, 400 Blandy Farm Lane—
Unit 2
Boyce, VA 22620 USA
Phone: 540-837-1600   Fax: 540-837-1523
E-mail: vnpsofc@shentel.net
Website: www.vnps.org

Founded: 1982
Membership: 2000
Scope: Statewide

Description: The VNPS and nine chapters throughout Virginia seek further appreciation and conservation of Virginia's wild plants and habitats. Programs emphasize public education, protection of endangered species, habitat preservation, control of invasive alien plants and encouragement of appropriate landscape use of native plants. Includes both amateurs and professionals.

Publication(s): Virginia Wildflower of the Year, Nursery Sources to Native Plants, List of Invasive Alien Plants for Virginia, Factsheets on Invasive Alien Plants, List of Recommended Native Plants for Landscaping and Restoration, Chapter Newsletters, BULLETIN

Keyword(s): Flowers, Plants, and Trees, Botany, Habitat Conservation, Conservation, Landscape Architecture, Public Lands, Prairies, Natural Areas, Endangered Species, Environment, Forest Management, Urban Forestry, Wetlands, Watersheds, Native Plants

Contact(s):

Butch Kelly, 1ST VICE PRESIDENT
8564 Grave Hill Road, Catawba, VA 24070
Phone: 540-562-3772
rkelly@rcs.k12.va.us
Shirely Gay, 2ND VICE PRESIDENT
210 South Abington St., Arlington, VA 22204
cgay1153@aol.com
Julie Alexander, BLUE RIDGE WILDFLOWER SOCIETY CHAPTER PRESIDENT
628 Walnut Avenue, Roanoke, VA 24014
John Magee, CORRESPONDING SECRETARY
2716 West Ox Road, Herndon, VA 20171
euphorbia@aol.com
Allen Bellden, DIRECTOR
1202 W 45th St., Richmond, VA 23225
Phone: 804-786-7951
ajb@dcr.state.va.us
Faith Campbell, DIRECTOR
8208 Dabney Ave., Springfield, VA 22152
phytodoer@aol.com
Nancy Hugo, DIRECTOR
11208 Gwathmey Church Rd, Ashland, VA 23005
nancyhugo@aol.com
Cole Burrell, DIRECTOR
P.O. Box 76, Free Union, VA 22940
Phone: 804-975-2859
nldr@aol.com
Jim Bruce, DIRECTOR
20042 Sterling Creek Ln., Rockville, VA 23146
Phone: 804-749-4304
jgbruce@erols.com
Pat Baldwin, DIRECTOR
430 Yale Dr., Hampton, VA 23666
Phone: 757-874-0892
Fax: 757-874-3037

Stanwyn Shetler, DIRECTOR OF BOTANY
142 E. Meadowland Lane, Sterling, VA 20164-1144
Phone: 202-786-2996
Fax: 202-786-2563
shetler.stanwyn@nmnh.si.edu
Jessica Strother, DIRECTOR OF CONSERVATION
6004 Windward Dr., Burke, VA 703-324-1795
sylvantica9@juno.com
Charles Smith, DIRECTOR OF FUND RAISING
8407 Sunset Dr., Manassas, VA 201112
chrissmith@juno.com
Deanne Eversmyer, DIRECTOR OF HORTICULTURE
1918 Leonard Road, Falls Church, VA 22043
d.eversmeyer@progidy.net
Mary Painter, DIRECTOR OF MEMBERSHIP CHAIR
P.O. Box D, Hume, VA 22639
vanatvs@erols.com
Pam Weiringo, DIRECTOR OF PUBLICATION
2740 Derwent Dr., SW, Roanoke, VA 24015
Phone: 540-772-3660
Bruce Jones, DIRECTOR OF PUBLICITY
601 Long Mountain Rd, Washington, VA 22747
bsjonz@aol.com
Boleyn Dale, DIRECTOR OF REGISTRY
P.O. Box 85, Rt. 1006, Moon, VA 23119
bkd@visi.net
Pat Willis, JEFFERSON CHAPTER PRESIDENT
1611 Hamilton Rd., Louisa, VA 23093
Michael Sawyer, JOHN CLAYTON CHAPTER PRESIDENT
P.O. Box 369, Toano, VA 23168-0369
Phone: 804-262-9887
Waterbone@aol.com
Jody Lyon, PIEDMONT PRESIDENT
19106 Black Oak Rd., Purcellville, VA 20132
rlyon@megapipe.net
Richard Moss, POCAHONTAS PRESIDENT
12565 Brook Lane, Chester, VA 23831
rmoss@richmond.infi.net
Marianne Mooney, POTOWMACK PRESIDENT
1112 N. Powhatan St., Arlington, VA 22205
e-mail-moosfy@webtv.net
Nicky Staunton, PRESIDENT
8815 Fort Drive, Manassas, VA 20110
Phone: 703-368-9803
nstaunton@earthlink.net
Nancy Vehrs, PRINCE WILLIAM WILDFLOWER SOCIETY PRESIDENT
8318 Highland St., Manassas, VA 20110-3671
nvehrs@attglobal.net
June Griffin, RECORDING SECRETARY
1622 Bruce Avenue, Charlottesville, VA 22903
Carol Gardner, SHENANDOAH PRESIDENT
3858 Wayfarers Trail, Bridgewater, VA 22812
w-cgardner@rica.net
Karen Renda, SOUTH HAMPTON ROADS PRESIDENT
4433 Revere Drive, Virginia Beach, VA 23456
karenrenda@hotmail.com
Ellie Leonard, TREASURER
6168 Carters Run Road, Marshall, VA 20115
elleonard@aol.com

VIRGINIA SOCIETY OF ORNITHOLOGY

7451 Little River Turnpike, 202
Annandale, VA 22003 USA
Phone: 703-305-7381

Founded: 1929

Scope: Statewide

Description: Dedicated to all aspects of the birds of Virginia, including conservation, field research, education of any interested person or group, and dissemination of all types of information. The VSO coordinates with state agencies and with other private organizations in this mission.

Publication(s): VSO Newsletter, Raven, The

Keyword(s): Birds

Contact(s):
Larry Lynch, PRESIDENT
Lauren Scott, SECRETARY
Barbara Thrasher, TREASURER
120 Woodbine Dr., Lynchburg, VA 24502
Phone: 804-239-5850
Larry Lynch, VICE PRESIDENT
9430 Tuxford Rd., Richmond, VA 23236
Phone: 804-272-8582

W

WASHINGTON ASSOCIATION OF CONSERVATION DISTRICTS
Attn: Executive Director
Colfax, WA 99111 USA
Phone: 253-473-4999
Website: www.wacd.org

Founded: NA
Scope: Statewide

Keyword(s): Conservation Districts

Contact(s):
Pat McGregor, EXECUTIVE DIRECTOR
3911 S. K St., Tacoma, WA 98418
Phone: 253-473-4999
Fax: 253-473-7246
Bob Haberman, NATIONAL DIRECTOR
771 Hungry Junction Rd., Ellensburg, WA 98926
Phone: 509-925-1713
Fax: 509-925-7730
bobhaber@eburg.net
Colin Bennett, PRESIDENT
185 Beebe Rd., Goldendale, WA 98620
Phone: 509-773-5065
Fax: 509-773-5600
cbennett@gorge.net
Monte Marti, SECRETARY AND TREASURER
11605 33rd Ct., NE, Lake Stevens, WA 98258
Phone: 425-261-6678
Fax: 425-258-4839
Wade Troutman, VICE PRESIDENT
Bridgeport, WA 98813
Phone: 509-686-2061

WASHINGTON B.A.S.S. CHAPTER FEDERATION
1721 South Methow St.
Wenatchee, WA 98801 USA
Phone: 425-251-3214
E-mail: joe.arballo@wabass.org
Website: www.wabass.org

Founded: NA
Membership: 500
Scope: Statewide

Description: An organization of Bassmaster chapters, affiliated with the Bass Anglers Sportsman Society, organized to fight pollution, assist state and national conservation agencies in their efforts, and teach the young people of our country good conservation practices. Dedicated to the realistic conservation of our water resources.

Publication(s): The Washington State B.A.S.S (Quarterly Newsletter)

Keyword(s): Energy, Environmental and Conservation Education, Protected Areas, Renewable Resources, Solid Waste

Contact(s):
Martin Bixby, CONSERVATION DIRECTOR
427 West 18th Ave., Kennewick, WA 99337
Phone: 509-582-7239
martin.bixby@wabass.org
Joe Arballo, PRESIDENT

WASHINGTON ENVIRONMENTAL COUNCIL
615 2nd Avenue,
Seattle, WA 98104 USA
Phone: 206-622-8103   Fax: 206-622-8113
E-mail: wec@wecprotects.org
Website: www.wecprotects.org

Founded: 1967
Membership: 2700
Scope: Statewide

Description: A statewide coalition of 88 member groups and over 3,000 individuals working to protect Washington's forests and wildlife, water and fish, open spaces and quality of life. NEC advocates for the environment at the state legislature, develops environmental policy, educates and involves the public and takes legal action.

Publication(s): State Legislative Briefing Book Annual, Forest Fish & Wildlife, WEC Voices

Keyword(s): Exotic species, Aquatic nuisance species, Fish Wildlife Management, Sustainable Development, Forests and Forestry

Contact(s):
Tom Geiger, EDITOR
Joan Crooks, EXECUTIVE DIRECTOR
Josh Baldi, POLICY DIRECTOR
Jay Manning, PRESIDENT
John Anderson, TREASURER

WASHINGTON FARM FORESTRY ASSOCIATION
Olympia, WA 98507 USA
Phone: 360-459-0984   Fax: 360-570-1537
E-mail: wafarmforestry.com
Website: www.wafarmforestry.com

Founded: 1944
Membership: 1500
Scope: Statewide

Description: A statewide organization affiliated with the National Woodland Owners Association, founded to help small woodland owners acquire information on better management of small timber tracts.

Publication(s): Landowner News, Northwest Woodlands

Keyword(s): Environmental Law, Environmental and Conservation Education, education, Scholarships and Grants, training, Forests and Forestry

Contact(s):
Lori Rasor, EDITOR, NORTHWEST WOODLANDS
4033 SW Canyon Rd., Portland, OR 97221
Phone: 502-228-1367
Nels Hanson, EXECUTIVE DIRECTOR AND EDITOR,
LANDOWNER NEWS
Phone: 360-943-3875
nelswh@home.com
Sherry Fox, PRESIDENT
Phone: 360-978-6448
tmp@i-link-2.net
Erin Woods, SECRETARY
Bill Woods, SECRETARY
Norma Green, TREASURER
nfgreen@reachone.com

WASHINGTON FOUNDATION FOR THE ENVIRONMENT

P.O. Box 2123
Seattle, WA 98111 USA
Phone: 360-866-9204
E-mail: info@wffe.org
Website: http://www.wffe.org

Founded: 1979
Scope: Statewide

Description: Dedicated to preserving and enhancing the environmental heritage of Washington state by making small grants to support educational and innovative projects in both the public and private sectors. For grant guidelines, send a message to JudyTurpin@aol.com

Keyword(s): Grants, Environmental and Conservation Education, Protected Areas, Environmental Protection

Contact(s):
Melanie Rowland, GRANTS CHAIR
13742 41st. Ave. NE, Seattle, WA 98125
Phone: 206-526-6537
Konrad Liegel, VICE PRESIDENT
1103 18th Ave. E., Seattle, WA 98112
Phone: 206-320-8582

WASHINGTON NATIVE PLANT SOCIETY

7400 Sand Point Way NE
Seattle, WA 98115 USA
Phone: 206-527-3210
E-mail: wnps@wnps.org
Website: www.wnps.org

Founded: 1976
Membership: 1,800
Scope: Statewide

Description: To promote the appreciation and conservation of Washington's native plants and their habitats through study, education, and advocacy.

Publication(s): Syllabus

Keyword(s): Native Plants, Outdoor Recreation, Public Lands, Urban Forestry, Wetlands, Youth Organizations

Contact(s):
Dottie Knecht, PAST PRESIDENT
P.O. Box 48, Peshatin, WA 98847
Phone: 509 5-48 -7393

Joan Frazee, PRESIDENT
P.O. Box 1082, Leavenworth, WA 98826
Phone: 509 5-48 -2166
Tom Johnson, SECRETARY
7742 32nd Ave., Seattle, WA 98115
Phone: 206 5-25 -3176
Richard Easterly, TREASURER
P.O. Box 4027, Tenino, WA 98589
Phone: 360 2-64 -5644
Richard Robohm, VICE PRESIDENT
963 N. Motor Pl. 4, Seattle, WA 98103
Phone: 206 5-45 -1823

WASHINGTON RECREATION AND PARK ASSOCIATION

350 S. 333rd St., Suite 103
Federal Way, WA 98003 USA
Phone: 253-874-1283   Fax: 253-661-3929
E-mail: wrpa@wrpatoday.org
Website: www.wrpatoday.org

Founded: 1947
Membership: 1200
Scope: Statewide

Description: Dedicated to enhancing and promoting parks, recreation, and leisure pursuits in Washington state, and plays a vital role in promoting public support for parks and recreation.

Publication(s): Syllabus

Keyword(s): Outdoor Recreation

Contact(s):
Lynn Devoir, EXECUTIVE DIRECTOR
Larry Otos, PRESIDENT
1717 S. 13th St., Mt. Vernon, WA 98274
Phone: 360-336-6215
larryo@ci.mt-vernon.wa.us
Don Williams, PRESIDENT-ELECT
11609 164th St., Puyallup, WA 98374
Phone: 253-848-9587
Tracy Thomas, SECRETARY
1000 S. 220th St., Des Moines, WA 98198
Phone: 206-870-6528
tthomas@cityofdesmoines.com

WASHINGTON SOCIETY OF AMERICAN FORESTERS

NORTHWEST
4033 SW Canyon Rd.
Portland, OR 97221 USA
Phone: 503-224-8046   Fax: 503-226-2515

Founded: 1900
Scope: Statewide

Description: Represents the forestry profession in advancing the science, technology, education, and practice of forestry for the benefit of forests, forest managers, and the public.

Contact(s):
Jocko Burks, CHAIR
3302 Sounview Dr. West, University Place, WA 98466
Art Schick, CHAIR-ELECT ('01)
2585 NE Ortis Rd., Poulsbo, WA 98370
Lori Rasor, MANAGER/EDITOR
4033 SW Canyon Rd., Portland, OR 97221

WASHINGTON TOXICS COALITION

4649 Sunnyside Ave., N.
Seattle, WA 98103 USA
Phone: 206-632-1545   Fax: 206-632-8661
E-mail: info@watoxics.org
Website: www.watoxics.org

Founded: 1981
Membership: 1100
Scope: Statewide

Description: Works to reduce society's reliance on toxic chemicals through research, education, advocacy, organizing and litigation.

Publication(s): Grow Smart, Grow Safe: A Consumer Guide to Lawn and Garden Products, Trubbling Bubbles: The Case for Replacing Alkylphenol Ethoxylate Surfacants, No Place For Poisons: Reducing Pesticides in School, Home Safe Home (fact sheets), Fact sheets, Alternatives

Keyword(s): Cultural Preservation, Environmental and Conservation Education, Outdoor Recreation, Toxic Substances, Nuclear-free, Water quantity, Water export and diversion, Toxic Reduction, Public Lands, Transportation

Contact(s):
Gregg Small, EXECUTIVE DIRECTOR
David Stitzhal, PRESIDENT
Dave Coffman, SECRETARY
Jennifer Dold, SECRETARY
Don Bollinger, TREASURER

WASHINGTON TRAILS ASSOCIATION

1305 4th Ave., Suite 512
Seattle, WA 98101-2401 USA
Phone: 206-625-1367   Fax: 206-625-9249
Website: www.wta.org

Founded: 1973
Membership: 5000
Scope: Statewide

Description: Washington Trails Association works to protect and enhance hiking opportunities in Washington state through education, volunteer trail maintenance, advocacy and cooperation with other trail users.

Publication(s): Washington Trails - Monthly Magazine

Keyword(s): Trail, Volunteering, Environmental and Conservation Education, Natural Areas, Open Space, Outdoor Recreation, training

Contact(s):
Elizabeth Lunney, EXECUTIVE DIRECTOR
Susan Elderkin, PRESIDENT

WASHINGTON WILDERNESS COALITION

4649 Sunnyside Ave., N., Suite 520
Seattle, WA 98103 USA
Phone: 206-633-1992   Fax: 206-633-1996
E-mail: info@wawild.org
Website: www.wawild.org

Founded: 1979
Membership: 15000
Scope: Statewide

Description: WWC is a statewide organization of individuals and groups dedicated to preserving wilderness and biodiversity for the benefit of future generations. WWC works to protect and restore wildlands and waters in Washington State through outreach, public education, organizing, and support of grassroots conservation groups.

Publication(s): Washington Wildfire

Keyword(s): Biodiversity, Wilderness

Contact(s):
Jon Owen, CAMPAIGN DIRECTOR
Kristen Tremoulet, CANVASS DIRECTOR
John Leary, EXECUTIVE DIRECTOR
Martin Loesch, PRESIDENT
Cyndi Lewis, SECRETARY
Michelle Kinsch, TREASURER
Mike Peterson, VICE PRESIDENT

WASHINGTON WILDLIFE AND RECREATION COALITION

811 First Avenue, Suite 262
Seattle, WA 98104 USA
Phone: 206-748-0082   Fax: 206-748-0580
Website: http://www.wildliferecreation.org

Founded: 1989
Scope: Statewide

Description: The Washington Wildlife and Recreation Coalition was formed in 1989 to promote the acquisition of land for wildlife and outdoor recreation through public education, support of appropriate legislation, and research into outdoor recreation and conservation needs in Washington state. Our long-range goal is to secure state funding for a $450 million, 8-10 year program to acquire and develop parks, trails, water access, wildlife habitat, and natural areas. To date, the state legislature has allocated over $270 million towards that goal.

Publication(s): Land News

Keyword(s): Land Preservation, Land Purchase, National Parks, Public Lands, training

Contact(s):
Joanna Grist, EXECUTIVE DIRECTOR
Shamra Harrison, OUTREACH DIRECTOR
Peter Scholes, PRESIDENT
Karen Munro, SECRETARY
Craig Lee, TREASURER
David Veley, VICE PRESIDENT

WASHINGTON WILDLIFE FEDERATION

P.O. Box 1966
Olympia, WA 98507-1966 USA
Phone: 360-705-1903
E-mail: washingtonwildlife.org
Website: www.washingtonwildlife.org

Founded: NA
Scope: Statewide

Description: A representative statewide organization, affiliated with the National Wildlife Federation, dedicated to the protection and enhancement of wildlife and its habitat through public education and government interaction.

Publication(s): Washington Wildlife News

Keyword(s): Environmental and Conservation Education, Protected Areas, Land Purchase, Water Pollution Management, Wetlands

Contact(s):
Ken Hilton, PRESIDENT

WASHINGTON WILDLIFE HERITAGE FOUNDATION

32610 Pacific Highway South
Federal Way, WA 98003 USA

Founded: NA
Scope: Statewide

Description: The Foundation including Heritage Land Trust, is dedicated to fish and wildlife/water and land conservation through resource management, and enhancing habitat and public education, with the involvement, support, and cooperation of both the public and private sector.

Publication(s): Land News

Keyword(s): training, Land Management, Outdoor Recreation

Contact(s):
Bruce Stuwe, CHAIRMAN
4727 Crisman Ct., SE, Olympia, WA 98501
Phone: 360-491-9195
Robert Gribble, SECRETARY
26442 164th SE, Kent, WA 98042
Phone: 253-631-9244

WATER ENVIRONMENT FEDERATION

601 Wythe St.
Alexandria, VA 22314-1994 USA
Phone: 703-684-2400　　Fax: 703-684-2492
Website: www.wef.org

Founded: 1928
Scope: International

Description: A nonprofit technical and educational organization with the mission to preserve and enhance the global water environment. Federation members are water quality specialists from around the world, including environmental, civil and chemical engineers, biologists, government officials, treatment plant managers and operators, laboratory technicians, college professors, students, and equipment manufacturers and distributors.

Publication(s): Water Environment and Technology, WEF Highlights, Water Environment Regulation Watch, WEF Industrial Wastewater, Watershed and Wet Weather Technical Bulletin, other titles available on request, Water Environment Research

Keyword(s): Engineering, Environmental and Conservation Education, Toxic Substances, Nuclear-free, Water quantity, Water export and diversion, Water Pollution Management, Exotic species, Aquatic nuisance species, Asia Water Environment

Contact(s):
Quincalee Brown, EXECUTIVE DIRECTOR
Joe Stowe, PRESIDENT
James Clark, VICE-PRESIDENT
Joe Stowe, PRESIDENT-ELECT
Prad Khare, TREASURER

WATER RESOURCES ASSOCIATION OF THE DELAWARE RIVER BASIN

P.O. Box 867
Valley Forge, PA 19482-0867 USA
Phone: 610-917-0090　　Fax: 610-917-0091
E-mail: wradrb@aol.com
Website: www.wrabrb.org

Founded: 1959
Scope: Regional

Description: Nonprofit federation of businesses, industries, academia, government, environmental, and citizen organizations which serves to advise of and advocate the need for adequate water supplies through the orderly conservation, development, and equitable use and reuse of the water and related land resources of the Delaware River Basin.

Publication(s): newsletter

Keyword(s): Environmental and Conservation Education, Rivers, Coral Reefs, Exotic species, Aquatic nuisance species, Wetlands

Contact(s):
William McElroy, CHAIR
William Palmer, EXECUTIVE DIRECTOR

WATERLOO-WELLINGTON WILDFLOWER SOCIETY

Botany Dept., University of Guelph
Guelph, Ontario N1G 2W1 Canada
Phone: 519-824-4120　　Fax: 519-767-1991
Website: www.uoguelph.ca/~botcal/

Founded: 1990
Scope: Local

Description: (Formerly the Dogtooth Group) A non-profit organization based in Guelph, Ontario dedicated to the use and protection of native plants in parks, gardens and other open spaces.

Publication(s): Dogtooth (monthly newsletter)

Keyword(s): Native Plants, Gardening and Horticulture, Environmental and Conservation Education, Flowers, Plants, and Trees, National Parks

Contact(s):
Carole Lacroix, PRESIDENT
Phone: 519-824-4120
botcal@uoguelph.ca

WATERSHED MANAGEMENT COUNCIL

P.O. Box 1090
Mammoth Lakes, CA 93546 USA
E-mail: WMC@watershed.org
Website: http://www.watershed.org/

Founded: 1986
Scope: Regional

Description: The Watershed Management Council is a nonprofit, educational organization dedicated to advancing the art and science of watershed management, with an emphasis on the Western region.

Publication(s): Networker, The, Proceedings

Keyword(s): Land Management, Nonpoint Source Pollution, Rivers, Streams, Water Pollution Management, Watersheds

WELDER WILDLIFE FOUNDATION

P.O. Box 1400
Sinton, TX 78387 USA
Phone: 361-364-2643   Fax: 361-364-2650
E-mail: welderwf@aol.com
Website: www.hometown.aol.com/welderwf/welderweb.html

Founded: 1954
Scope: National

Description: Established by the will of the late Rob Welder, the Foundation is dedicated to the cause of conservation through research and education in wildlife ecology and management and closely related fields. Operates through a small staff, with research fellowships to graduate students only.

Keyword(s): Birds, Mammals, Scholarships and Grants, training

Contact(s):
Selma Glasscock, ASSISTANT DIRECTOR/
CONSERVATION EDUCATOR
Terry Blankenship, ASSISTANT DIRECTOR/WILDLIFE
BIOLOGIST
D. Drawe, DIRECTOR

WEST MICHIGAN ENVIRONMENTAL ACTION COUNCIL

1514 Wealthy SE, Suite 280
Grand Rapids, MI 49506-2755 USA
Phone: 616-451-3051   Fax: 616-451-3054
Website: www.wseac.org

Founded: 1968
Scope: Local

Description: Provide leadership in environmental protection and preservation in west Michigan and throughout Michigan on issues such as water quality, land use planning and sustainable business. Through the involvement of concerned volunteers, WMEAC has helped landmark environmental legislation and assured application of existing laws.

Publication(s): see publications on website, Action Issue

Keyword(s): Air Quality and Pollution, Environmental and Conservation Education, Land Use Planning, Exotic species, Aquatic nuisance species, Conservation, Waste Management, Sustainability

Contact(s):
Tom Leonard, EXECUTIVE DIRECTOR
Karel Rogers, PRESIDENT

WEST VIRGINIA ASSOCIATION OF CONSERVATION DISTRICT SUPERVISORS ASSOCIATION, INC.

Attn: President, P.O. Box 711
Gallipolis Ferry, WV 25515 USA

Founded: NA
Scope: Statewide

Keyword(s): Conservation Districts

WEST VIRGINIA B.A.S.S. CHAPTER FEDERATION

Attn: President
Buckhannon, WV 26201 USA
Phone: 304-472-3600
E-mail: jburdette@neumedia.net
Website: www.wvbass.com

Founded: NA
Membership: 828
Scope: Statewide

Description: An organization of Bassmaster chapters, affiliated with the Bass Anglers Sportsman Society, organized to fight pollution, assist state and national conservation agencies in their efforts, and teach the young people of our country good conservation practices. Dedicated to the realistic conservation of our water resources.

Publication(s): Highlands Voice, The, Monongahela National Forest Hiking Guide, The

Keyword(s): Forests and Forestry, Public Lands, Rivers, Water Pollution Management, Wilderness

Contact(s):
Jim Summers, CONSERVATION DIRECTOR
Rte. 1 Box 205, Worthington, WV 26591
Phone: 304-287-7700
JSummers8@compuserve.com
John Burdette, PRESIDENT

WEST VIRGINIA HIGHLANDS CONSERVANCY

P.O. Box 306
Charleston, WV 25321 USA

Founded: 1967
Membership: about 1,000
Scope: Statewide

Description: An organization devoted to the conservation and wise management of West Virginia's natural and historic resources. Active in wilderness preservation, river conservation, public lands management, forestry, mining, air and water quality, water resources management and a wide variety of other environmental and conservation issues.

Publication(s): The Highlands Voice, The Monongahela National Forest Hiking Guide

Keyword(s): Natural Resource Conservation, Forests and Forestry, Public Lands, Rivers, Water Pollution Management, Wilderness, Water and Air Quality, Mining, Exotic species, Aquatic nuisance species, Forestry

Contact(s):
Bill Reed, EDITOR
350 Bucks Branch, Beckley, WV 25801
Phone: 304-934-5828
Dave Saville, MEMBERSHIP SECRETARY
P.O. Box 569, Morgantown, WV 26507
Phone: 304-284-9548
Frank Young, PRESIDENT
Rt. 1 Box 108, Ripley, WV 25271
Phone: 304-372-9329
Andrew Maier, SECRETARY
Rt. 1 Box 27, Hinton, WV 25952
Phone: 304-466-3864
Judy Rodd, SENIOR VICE PRESIDENT
Rt. 1, Box 178, Moatsville, WV 26405
Phone: 304-265-0018

Jacqueline Hallinan, TREASURER
1120 Swan Rd., Charleston, WV 25314
Phone: 304-345-3718
Norm Steenstra, VICE PRESIDENT OF STATE AFFAIRS
1001 Valley Rd., Charleston, WV 25302
Phone: 304-346-5891

 **WEST VIRGINIA WILDLIFE FEDERATION, INC.**
P.O. Box 275
Paden City, WV 26159 USA
Phone: 304-782-3685
E-mail: pleinbach@aol.com
Website: www.wvwf.org

Founded: NA
Scope: Statewide

Description: A representative statewide organization, affiliated with the National Wildlife Federation, dedicated to the protection and enhancement of wildlife and its habitat through public education and government interaction.

Publication(s): West Virginia Wildlife Notes

Contact(s):
William Mullins, PRESIDENT AND ALTERNATE REPRESENTATIVE

WEST VIRGINIA, WOODLAND OWNERS ASSOCIATION OF
P.O. Box 13695
Sissonville, WV 25360 USA
Phone: 304-594-3648   Fax: 304-594-3648

Founded: 1991
Scope: State

Description: A statewide organization affiliated with the National Woodland Owners Association that promotes good forestry and sustainable management by non-industrial private owners in West Virginia.

Publication(s): West Virginia Woods

Keyword(s): Forests and Forestry

Contact(s):
Clay Smith, EDITOR
HC 64 Box 50, Parsons, WV 26287-9709
Phone: 304-478-2104
Bob Whipkey, FORESTRY ADVISOR
Phone: 304-558-2788
Mark Burke, PRESIDENT
dadobourke@aol.com
Edward Murriner, SECRETARY
Rt. 3. Box 186D, Hurricane, WV 25526
Phone: 304-727-5591
Fax: 304-558-0143
emmurin@gwmail.state.wv.us
Mark Metz, TREASURER
1017 Mt. Vernon Circle, Barboursville, WV 25504
Phone: 304-733-1043
themetzs@gateway.net
Russ Richardson, VICE-PRESIDENT
P.O. Box 206, Weston, WV 26452
Phone: 304-269-3862
Phone: 304-269-3964
forestruss@aol.com

WESTERN ASSOCIATION OF FISH AND WILDLIFE AGENCIES
5400 Bishop Blvd.
Cheyenne, WY 82006 USA
Phone: 307-777-4569   Fax: 307-777-4699

Founded: NA
Scope: Regional, National

Description: A regional organization including 18 fish and wildlife agencies of 15 states and three Canadian provinces. Meets annually to consider mutual problems and provide a forum for the exchange of information at both administrative and technical levels.

Publication(s): Western Proceedings

Contact(s):
Larry Bell, 1ST VICE PRESIDENT
P.O. Box 25112, Santa Fe, NM 87504
Phone: 505-827-7899
Jeff Koenings, 2ND VICE PRESIDENT
Olympia, WA 98504
John Kimball, PRESIDENT
Salt Lake City, UT 84114-6301
Phone: 801-538-4703
Larry Kruckenberg, SECRETARY AND TREASURER
Game and Fish Department, 5400 Bishop Blvd., Cheyenne, WY 82006
Phone: 307-777-4569

WESTERN ENVIRONMENTAL LAW CENTER
1216 Lincoln St.
Eugene, OR 97401 USA
Phone: 541-485-2471   Fax: 541-485-2457
E-mail: eugene@westernlaw.org
Website: www.westernlaw.org

Founded: 1993
Scope: Statewide

Description: WELC specializes in environmental law enforcement, working with grassroots citizen groups and native American tribes to implement our nation's environmental laws. WELC has offices in Eugene, Oregon and Taos, New Mexico.

Publication(s): Defending the West, Biannual Report

Keyword(s): Environmental Law, Air Quality and Pollution, Mining, Toxic Substances, Nuclear-free, Water quantity, Water export and diversion, Water Quality, Pesticides, Cultural Preservation, training, Forests and Forestry

Contact(s):
Grove Burnett, DIRECTOR OF TAOS OFFICE
Phone: 505-751-0351
Fax: 505-751-1775
law@welctaos.org
Peter Frost, EXECUTIVE DIRECTOR
Michael Axline, LITIGATION DIRECTOR
Corrie Yackulic, PRESIDENT
yackulic@schroeter-goldmark.com
Mary Wood, VICE PRESIDENT, SECRETARY AND TREASURER
mwood@law.uoregon.edu

WESTERN FORESTRY AND CONSERVATION ASSOCIATION

4033 SW Canyon Rd.
Portland, OR 97221 USA
Phone: 503-226-4562   Fax: 503-226-2515
Website: www.westernforestry.org

Founded: 1909
Scope: Regional

Description: The mission of the WFCA is to promote forest stewardship in western North America. The Association's objectives are to promote the science and practice of forestry, promote the dissemination of forestry research and technical information, and foster cooperation between federal, state, provincial, and private forest agencies.

Keyword(s): Forest Management

Contact(s):
Richard Zabel, PRESIDENT
Blair Holman, TREASURER

WESTERN HEMISPHERE SHOREBIRD RESERVE NETWORK (WHSRN)

c/o Manomet Center for Conservation Services, 81 Stage Point Rd., P.O. Box 1770
Manomet, MA 02345 USA
Phone: 508-224-6521   Fax: 508-224-9220
Website: www.manomet.org/whsrn.htm

Founded: 1985
Membership: 3500
Scope: International

Description: As a partnership of Manomet and Wetlands International: the Americas, WHSRN is a voluntary nonregulatory network of 40 critical wetland sites in seven countries which have joined together to study, manage, and promote the sustainable conservation of shorebirds and their habitats for the benefit of the ecosystems and people. WHSRN's strategy promotes a multiple species ecosystem approach to protection of over nine million acres of habitats that are critical staging, nesting, and nonbreeding sites of migratory shorebirds, throughout North and South America.

Publication(s): Conservation Sciences-Quarterly, Shorebird Migrations: Fundamentals for Land Managers, Important Shorebird Staging Sites Meeting WHSRN Criteria in the U.S., The Amazing Migration of Shorebirds (video), Save Our Migratory Shorebirds (curriculum guide), Shorebird Atlas, WHSRNews

Keyword(s): Shorebirds, Birds, Sustainable Ecosystems, Wetlands

Contact(s):
Linda Leddy, DIRECTOR

WESTERN PACIFIC REGIONAL FISHERY MANAGEMENT COUNCIL

1164 Bishop St., Suite 1400
Honolulu, HI 96813 USA
Phone: 808-522-8220   Fax: 808-522-8226
Website: www.wpcouncil.org

Founded: 1977
Membership: 16
Scope: National

Description: The Council is the policy-making organization for the management of fisheries in and around the EEZs of American Samoa, Guam, Hawaii, and the Northern Mariana Islands, and U.S. possessions in the Pacific Ocean. Council members and members of its advisory bodies: Scientific and Statistical Committee, Plan Teams, and Advisory Panels represent the fishing community, government agencies, and national international fisheries management organizations throughout the region.

Publication(s): Pacific Islands Fishery News

Keyword(s): Management Plans, Commercial Sport Fishing, Wildlife, Highly Migratory Species, Islands, Fish Wildlife Management

Contact(s):
Judith Guthertz, CHAIRMAN
Kitty Simonds, EXECUTIVE DIRECTOR

WESTERN PENNSYLVANIA CONSERVANCY

209 4th Ave.
Pittsburgh, PA 15222 USA
Phone: 412-288-2777   Fax: 412-281-1792
E-mail: wpc@paconserve.org
Website: www.paconserve.org/

Founded: 1932
Scope: Local

Description: The Western Pennsylvania Conservancy, working together to save the places we care about, protects natural lands, promotes healthy and attractive communities and preserves Frank Lloyd Wright's masterwork Fallingwater. The Conservancy fosters the integration of ecological protection with economic and social needs while building on the core values of the community and has protected more than 200,000 acres of natural lands in Pennsylvania.

Publication(s): Annual Calendar, Conserve

Keyword(s): Conservation, Endangered Species, Gardening and Horticulture, Land Preservation, Sustainable Ecosystems, Urban Environment

Contact(s):
Mike Boyle, CHAIRMAN
Julie Lalo, EDITOR & VICE PRES. PUBLIC AFFAIRS
Cynthia Carrow, EXECUTIVE VICE PRESIDENT AND COO
Larry Schweiger, PRESIDENT AND CEO
Lynda Waggoner, VICE PRESIDENT AND DIRECTOR OF FALLINGWATER
Jacquelyn Bonomo, VICE PRESIDENT OF CONSERVATION PROGRAMS

WETLAND HABITAT ALLIANCE OF TEXAS

118 E. Hospital, Suite 208
Nacogdoches, TX 75961 USA
Phone: 936-569-9428   Fax: 936-569-6349
E-mail: whatduck@txucom.net
Website: www.whatduck.org

Founded: 1984
Membership: 8000
Scope: Regional

Description: A nonprofit organization of conservationists, dedicated to preserving, reclaiming, and enhancing Texas wetland habitat, that promotes the wise use of our natural resources and the progress of our society. Constructs habitat improvement projects on public and private lands, promotes educational programs, performs priority wetland research, and supports legislative conservation efforts.

Publication(s): Texas Wetlands

Keyword(s): Wetlands

Contact(s):
Bruce Klingman, CHAIRMAN
John Frasier, EXECUTIVE DIRECTOR
Neal Jenkins, TREASURER
John Gardere, VICE PRESIDENT

WHITE CLAY WATERSHED ASSOCIATION
760 Chambers Rock Rd.
Landenberg, PA 19350 USA
Phone: 610-274-8499
Website: http://home.ccil.org/~wcwa/

Founded: 1965
Scope: Regional

Description: The White Clay Watershed Association is a nonprofit organization devoted to protection and improvement of the environmental quality of the White Clay Creek and valley. The Association works to improve water quality in local streams, conserve open space, woodlands, wetlands and geological features; aid in the preservation of cultural, historical and archaeological sites; increase outdoor recreation opportunities; and conduct educational programs relating to the environment.

Keyword(s): Cultural Preservation, National Parks, Environmental Protection, Water Quality, Watersheds

Contact(s):
John Murray, PRESIDENT
Carol Catanese, SECRETARY
Donna Bush, TREASURER
Robert Stark, VICE PRESIDENT

WHITETAILS UNLIMITED, INC.
P.O. Box 720, 1715 Rhode Island St.
Sturgeon Bay, WI 54235 USA
Phone: 920-743-6777   Fax: 920-743-4658
E-mail: wtu@itol.com
Website: www.whitetailsunlimited.com

Founded: 1982
Membership: 23
Scope: National

Description: Whitetails Unlimited is a national, nonprofit conservation organization. Its purpose is to raise funds in support of education, habitat enhancement, and the preservation of the hunting tradition for the direct benefit of the white-tailed deer and other wildlife species.

Publication(s): Whitetails Unlimited Magazine

Keyword(s): Environmental and Conservation Education, Hunting, training

Contact(s):
Kim McKinney, EVENT PROGRAM MANAGER
Peter Gerl, EXECUTIVE DIRECTOR AND PRODUCTION MANAGER
William Gerl, EXECUTIVE VICE PRESIDENT
Peter Schoonmaker, FIELD EDITOR
Kevin Naze, FIELD EDITOR
Arlene Peterson, INVENTORY SHIPMENT COORDINATOR
Kevin Devault, MANAGER OF CONSERVATION FUNDING

Eric Carper, MANAGER OF MERCHANDISE AND ADVERTISING
Cheryl Uecker, MEMBERSHIP SERVICES COORDINATOR
Janet Gerl, OFFICE MANAGER
Jeffrey Schinkten, PRESIDENT
Denise Dubick, PRODUCTION/DESIGN
David Hawkey, VICE PRESIDENT OF FIELD OPERATIONS

WHOOPING CRANE CONSERVATION ASSOCIATION INC.
1393 Henderson Highway
Breaux Bridge, LA 70517 USA
Phone: 337-228-7563   Fax: 337-228-7424
E-mail: wcca@excelonline.com
Website: www.whoopingcrane.com

Founded: 1961
Membership: unkown
Scope: National, International

Description: A scientific and educational organization, international in scope, working to prevent the extinction of the whooping crane and save wetland habitats.

Publication(s): The Whooping Crane: North America's Symbol of Conservation by Jerome J. Pratt, Grus Americana

Keyword(s): Endangered Species, Birds, Waterfowl, Wetland Habitat

Contact(s):
Marie Maltese, EDITOR
Baltimore, MD
Jerome Pratt, PAST EDITOR & COMMUNICATION COORDINATOR
Mary Courville, SECRETARY AND TREASURER

WILD CANID SURVIVAL AND RESEARCH CENTER
P.O. Box 760
Eureka, MO 63025 USA
Phone: 636-938-5900   Fax: 636-938-6490
E-mail: edu@wolfsantuary.org
Website: www.wolfsanctuary.org

Founded: 1971
Membership: 900
Scope: National

Description: A nonprofit, conservation organization dedicated to the preservation of wolves and other wild canids through education, research, and captive breeding.

Publication(s): Wolf Pack Press, Wild Canid Center Review

Keyword(s): Endangered Species, Predators, Environmental and Conservation Education, Research, Reintroduction, Wolves

Contact(s):
William Sadler, CHAIRMAN
Sue Lindsey, EXECUTIVE DIRECTOR
Patricia Biggerstaff, TREASURER
Margaret Ratz, VICE-CHAIR

WILD DOG FOUNDATION, THE
P.O. Box 1603
Mineola, NY 11501-0901 USA
Phone: 516-746-0005   Fax: 516-746-0005
E-mail: savewilddogs@hotmail.com
Website: www.wilddog.org

Founded: 1996
Membership: 30
Scope: International

Description: The foundation is a conservation and educational group. The foundation promotes wolf restoration to the Adirondack State Park in New York and the Northeast, and deals with less popular predators, mostly wild canines and hyenas. Its flagship species are the african Wild Dog, and coyote.

Publication(s): Wild, The

Keyword(s): Conservation, Endangered Species, Environmental and Conservation Education, International Wildlife, Mammals, Predators, Preservation and Protection, Wolves, Wild Dogs, Hyenas

Contact(s):
Frank Vincenti, PRESIDENT
Phone: 516-746-0005
savewilddogs@hotmail.com
Lew Egol, VICE PRESDENT
Hope Ryden, VICE PRESIDENT
Peggy Weinberg, VICE PRESIDENT
Pat Traub, VICE PRESIDENT
Robert Berghaier, VICE PRESIDENT

WILD HORSE ORGANIZED ASSISTANCE, INC. (WHOA)

P.O. Box 555
Reno, NV 89504 USA
Phone: 702-851-4817
Website: http://www.ipt.com/htmlpub/jpi/whoa.htm

Founded: 1971
Membership: 12,000
Scope: National

Description: Directs efforts toward the welfare of wild horses and burros; implementation of federal efforts in carrying out terms of the management, protection, and control program for their welfare; student projects pertaining to all phases of our heritage.

Keyword(s): Nongame Wildlife

Contact(s):
Dawn Lappin, EXECUTIVE DIRECTOR AND CHAIRMAN OF THE BOARD
Phone: 702-851-4817
Bert Lappin, SECRETARY
Phone: 702-851-4817
Leslie Johnson, TREASURER
Phone: 702-851-4817
Russell Johnson, VICE CHAIRMAN
Phone: 702-786-7600

WILD ONES - NATURAL LANDSCAPERS, LTD

Headquarters, P.O. Box 1274
Appleton, WI 54912-1274 USA
Phone: 877-394-9453   Fax: 920-730-8654
E-mail: woresource@aol.com
Website: www.for-wild.org

Founded: 1977
Membership: 2,500 households
Scope: National

Description: Wild Ones is a nonprofit organization seeking to educate and inform members and the public at the plants-roots level and to promote biodiversity and environmental sound practices, thru natural landscaping using native species in developing plant communities.

Publication(s): Wild Ones Handbook, Wild Ones Journal

Keyword(s): Native Plants, Ancient Forests, Birds, Conservation, Endangered Species, Protected Areas, Environmental Ethics, Flowers, Plants, and Trees, Gardening and Horticulture, Grants, Land Preservation, National Parks, Insects and Butterflies, Natural Areas

Contact(s):
Donna Vanbuecken, ADMINISTRATIVE DIRECTOR
Bret Rappaport, PRESIDENT
Joe Powelka, SECRETARY
Klaus Wisiol, TREASURER
Mandy Ploch, VICE PRESIDENT

WILDERNESS EDUCATION ASSOCIATION

900 East 7th Street
Bloomington, IN 47405 USA
Phone: 812-855-4095   Fax: 812-855-8697
E-mail: wea@indiana.edu
Website: http://ebl.org/wea/

Founded: 1977
Membership: 2,500
Scope: National

Description: WEA is a nonprofit membership organization. It promotes national wilderness education and preservation programs by providing for-credit, expedition-based wilderness leadership training programs, developing and publishing state-of-the-art wilderness education publications and training manuals, promoting scholarly research programs, establishing and maintaining national outdoor leadership certification standards, providing support to wildland management agencies to promote wilderness education, and help foster a preservationist land ethic.

Publication(s): Trustees and Affiliates Briefing System (TABS), The Backcountry Classroom, WEA Affiliate Handbook, New Wilderness Handbook, Wilderness Educator, WEA Legend

Keyword(s): Environmental and Conservation Education, Protected Areas, Internships, Outdoor Recreation, Wilderness

Contact(s):
Darla Deruiter, EXECUTIVE DIRECTOR
WEA Department of Natural Resource Recreation and Tourism, Colorado State University, Fort Collins, CO 80523
Phone: 970-223-6252
Fax: 970-223-6252
David Cockrell, PRESIDENT
Department of Human Performance and Leisure Studies, University of Southern Colorado, 2200 Bonforte Blvd., Pueblo, CO 81001-4901
Phone: 719-549-2775
Fax: 719-549-2732
W. Norton, PUBLISHER
Jeff Olson, SECRETARY
Confidence Learning Center,
6260 Mary Fawcett Memorial Dr., Brainerd, MN 56401
Phone: 218-828-2344
William Forgey, TREASURER
One Tower Plaza, 109 E. 89th Ave., Merrillville, IN 46410
Phone: 219-769-6055
Fax: 219-769-6035

Mitchell Sakofs, VICE PRESIDENT
Outward Bound USA, Rt. 9, R. D. 2, Box 280,
Garrison, NY 10524-9757
Phone: 914-424-4000

WILDERNESS LAND TRUST, THE

4060 Post Canyon Dr.
Hood River, OR 97031 USA
Phone: 541-386-9546   Fax: 541-386-9547
Website: www.wildernesstrust,.org

Founded: 1992
Scope: Regional

Description: To facilitate public acquisition of private lands
(inholdings) within units of the National Wilderness
Preservation System to fulfill the promise of Congress made in
The Wilderness Act of 1964 that all generations of Americans
will enjoy an enduring resource of wilderness.

Publication(s): Wilderness Heritage Newsletter

Keyword(s): Conservation, Environmental Protection, Land
Preservation, Land Purchase, Wilderness

Contact(s):
John Fielder, CHAIRMAN
P.O. Box 1261, Englewood, CO 80150
Phone: 303-935-0900
Jon Mulford, PRESIDENT
Andy Wiessner, SECRETARY AND TREASURER
811 Potato Patch Dr., Vail, CO 81657
Phone: 303-715-3570

WILDERNESS SOCIETY, THE

1615 M St., NW
Washington, DC 2036 USA
Phone: 202-833-2300   Fax: 202-429-3945
Website: www.wilderness.org
Scope: National

Description: Works to protect America's wilderness and to
develop a nationwide network of wild lands through public
education, scientific analysis and advocacy

Publication(s): Annual: Wilderness America, Newsletter,
Wilderness Year, The

Contact(s):
Donald Barry, EXECUTIVE VICE PRESIDENT
1615 M St. NW #100, Washington, DC 20036
Phone: 202-429-8458
Fax: 202-429-3958
don_barry@tws.org
William Meadows, PRESIDENT
1615 M St. NW, Washington, DC 20036
Phone: 202-429-2007
Fax: 202-429-3958
william_meadows@tws.org

WILDERNESS SOCIETY, THE

1615 M Street, NW
Washington, DC 20036 USA
Phone: 202-833-2300   Fax: 202-429-3958
E-mail: members@tws.org
Website: http://www.wilderness.org

Founded: 1935
Membership: 300,000
Scope: National

Description: A nonprofit membership organization devoted to
preserving wilderness and wildlife, protecting America's prime
forests, parks, rivers, and shorelands, and fostering an
American land ethic. The Society welcomes membership
inquiries, contributions, and bequests.

Publication(s): Annual: Wilderness America, Newsletter, The
Wilderness Year

Keyword(s): Biodiversity, Forests and Forestry, Public Lands,
Sustainable Ecosystems, Wilderness

Contact(s):
Bert Fingerhut, CHAIR
Gaylord Nelson, COUNSELOR
Sue Gunn, DIRECTOR OF LWCF
Michael Francis, DIRECTOR OF NATIONAL FOREST
ISSUES
Rose Fennell, DIRECTOR OF NATIONAL PARKS AND
ALASKA ISSUES
Jim Waltman, DIRECTOR OF NATIONAL WILDLIFE
REFUGES AND ENDANGERED SPECIES ISSUES
William Meadows, PRESIDENT
Fran Hunt, PROGRAM DIRECTOR OF BLM ISSUES
Craig Gehrke, REGIONAL AND STATE DIRECTOR
413 W. Idaho St., Suite 102, Boise, ID 83702
Phone: 208-343-8153
Jay Watson, REGIONAL DIRECTOR
Presidio Bldg. 1016, P.O. Box 29241, San Francisco, CA
94129
Phone: 415-561-6641
Pamela Eaton, REGIONAL DIRECTOR
7475 Dakin St., Suite 410, Denver, CO 80221
Phone: 303-650-5818
Steve Whitney, REGIONAL DIRECTOR
1424 Fourth Ave., Suite 816, Seattle, WA 98101
Phone: 206-624-6430
Robert Perschel, REGIONAL DIRECTOR
45 Bromfield St., Suite 1101, Boston, MA 02108
Phone: 617-350-8866
Robert Ekey, REGIONAL DIRECTOR
105 W. Main St., Suite E, Bozeman, MT 59715
Phone: 406-586-1600
Allen Smith, REGIONAL DIRECTOR
430 W. 7th Ave., 210, Anchorage, AK 99501
Phone: 907-272-9453
Sue Lomenzo, VICE PRESIDENT OF COMMUNICATIONS
Thomas Bancroft, VICE PRESIDENT OF ECOLOGY AND
ECONOMICS RESEARCH
Rindy O'Brien, VICE PRESIDENT OF PUBLIC POLICY
Darrell Knuffke, VICE PRESIDENT OF REGIONAL
CONSERVATION
Steve Howard, VICE PRESIDENT OF RESOURCE
DEVELOPMENT (ACTING)

WILDERNESS WATCH

P.O. Box 9175
Missoula, MT 59807 USA
Phone: 406-542-2048   Fax: 406-542-7714
E-mail: wild@wildernesswatch.org
Website: www.wildernesswatch.org

Founded: NA
Membership: 1100
Scope: National

Description: Wilderness Watch is a national, nonprofit, citizen
organization dedicated solely to the protection and proper
administration of lands within the National Wilderness

Preservation System and Wild and Scenic Rivers System. We achieve our goals through the efforts of citizen activists, local chapters, wilderness "adopters", and by working with other local organizations concerned about wilderness and wild river issues.

Publication(s): Wilderness Watcher

Keyword(s): Wilderness

Contact(s):
George Nickas, EXECUTIVE DIRECTOR

WILDFLOWER ASSOCIATION OF MICHIGAN
3853 Farrell Rd.
Hastings, MI 49058 USA
Phone: 616-948-2496   Fax: 616-948-2957
E-mail: wam@iserv.net
Website: www.wildflowersmich.org

Founded: 1986
Scope: Regional

Description: The Wildflower Association of Michigan promotes, coordinates, and participates in education, enjoyment, science, and stewardship of native wildflowers and their habitats.

Publication(s): Wildflowers quarterly newsletter, Annual Conference Program

Keyword(s): Beautification, Botany, Conservation, Endangered Species, Environmental and Conservation Education, Grants, Habitat Conservation, Native Plants, Natural Areas, Outdoor Education, Plant Propagation, Prairies, Restoration, Wildflowers

Contact(s):
Marji Fuller, EDITOR
marjif@iserv.net
Marilyn Case, MEMBERSHIP COORDINATOR
Phone: 616-781-8470
mcase15300@aol.com
Stephan Keto, PRESIDENT
Phone: 616-343-1669
Fax: 616-343-0768
Esther Durnwald, SECRETARY
Phone: 517-647-6010
Fax: 517-647-6072
wildflowers@voyager.net

WILDFOWL TRUST OF NORTH AMERICA, INC., THE
P.O. Box 519, Discovery Ln.
Grasonville, MD 21638 USA
Phone: 410-827-6694   Fax: 410-827-6713
E-mail: horschead@wildfowltrust.org
Website: www.wildfowltrust.org

Founded: 1979
Membership: 700
Scope: International

Description: A nonprofit, tax-exempt organization dedicated to the preservation of wildlife and wetlands through education, conservation, and research. The Trust operates The Horsehead Wetlands Center on its, 500-acre wetland refuge on the Chesapeake Bay's Eastern Shore. The Center provides environmental education programs, a collection of resident waterfowl and raptors in natural habitat settings, a Visitor's Center, trails, and observation blinds and towers. Canoes are available to members.

Publication(s): On The Wing-Newsletter

Keyword(s): Biodiversity, Environmental and Conservation Education, Research, Wetlands, training, Birds

Contact(s):
Edward Delaney, EXECUTIVE DIRECTOR
Torrey Brown, PRESIDENT
William Stott, VICE PRESIDENT

WILDLANDS CONSERVANCY
3701 Orchid Pl.
Emmaus, PA 18049-1637 USA
Phone: 610-965-4397   Fax: 610-965-7223
E-mail: wildlands@aol.com
Website: www.wildlandspa.org

Founded: 1973
Membership: 1500
Scope: Local

Description: A nonprofit, member-supported organization serving eastern Pennsylvania, involved in land and river preservation and environmental education. Wildlands has preserved over 30,000 acres of open space, much of it in cooperation with the Pennsylvania Game Commission. Some of the activities involve the operation of nature preserves and sanctuaries; developing and implementing plans for river conservation and other preservation projects; and the creation of a curriculum (K-College) that teaches responsible watershed stewardship.

Publication(s): Wildlands quarterly newsletter

Keyword(s): Rivers, Nature Centers, Land Preservation, Wildlands, Watersheds

Contact(s):
David Kepler, CHAIRMAN
702 Hamilton Mall, Allentown, PA 18101

WILDLANDS PROJECT, THE
52 Bridge St. Fl. 2
Richmond, VT 05477 USA
Phone: 520-884-0875   Fax: 802-434-5980
E-mail: wildlands@twp.org
Website: www.twp.org

Founded: 1992
Membership: 4500
Scope: International

Description: The mission of The Wildlands Project is to protect and restore the natural heritage of North America, through the establishment of a connected system of wildlands. TWP coordinates the efforts of regional organizations and individuals in the development of reserve design proposals for a continental vision.

Publication(s): Wildlands Project, The: First Thousands Days of the Next Thousand Years, Wild Earth - special edition

Keyword(s): Wildlands, Biodiversity, Conservation, Wilderness

Contact(s):
Dave Foreman, CHAIRMAN
Leanne Klyza Linck, EXECUTIVE DIRECTOR
Bob Howard, PRESIDENT
David Johns, SECRETARY AND TREASURER

WILDLIFE ACTION, INC.

P.O. Box 866
Mullins, SC 29574 USA
Phone: 843-464-8473   Fax: 843-464-8859
Website: www.wildlifeaction.com

Founded: 1977
Scope: National

Description: Wildlife Action is a private nonprofit 501(c)3 tax-exempt organization dedicated to the appreciation and enjoyment of our wildlife heritage and to educating the public in the value of protection, restoration, enhancement, and wise use of our natural resources.

Publication(s): Wild Things - Our Resource Education Center, Wildlife Pride

Keyword(s): Environmental and Conservation Education, Environmental Ethics, Protected Areas, Outdoor Recreation, Youth Organizations

Contact(s):
M. Beeson, PRESIDENT AND CEO
Sandra Bane, SECRETARY
Ted Williams, TREASURER
Tommy Simpson, VICE PRESIDENT

WILDLIFE CENTER OF VIRGINIA, THE

P.O. Box 1557
Waynesboro, VA 22980-1414 USA
Phone: 540-942-9453   Fax: 540-943-9453
E-mail: wildlife@wildlifecenter.org
Website: www.wildlifecenter.org

Founded: 1982
Membership: 25
Scope: International

Description: A nonprofit organization that operates the nation's largest professionally-staffed veterinary teaching and research hospital for native wildlife. Study and documentation of environmental factors that cause injuries, especially pesticide poisoning, are used to monitor environmental and wildlife health trends and support public policy positions. The Center also trains students and professionals from the fields of veterinary medicine, wildlife management and wildlife rehabilitation.

Publication(s): The Wildlife Center Teacher's Packet, Annual and Mid-year reports, reprints of articles and papers on various topics., Handbook of Wildlife Medicine

Keyword(s): Environmental and Conservation Education, training, Raptors, standards, accreditation, Nongame Wildlife, Endangered Species, Mammals, Birds

Contact(s):
Erwin Bohmfalk, CHAIRMAN OF THE BOARD
Lisa Briskey, DIRECTOR OF ENVIRONMENTAL EDUCATION
briskey@wildlifecenter.org
Jonathan Sleeman, DIRECTOR OF VETERINARY SERVICES
Edward Clark, PRESIDENT
eclark@wildlifecenter.org
Lisa Briskey, VICE PRESIDENT

WILDLIFE CONSERVATION SOCIETY

2300 Southern Blvd.
Bronx, NY 10460-1099 USA
Phone: 718-220-5100   Fax: 718-220-2685
E-mail: feedback@wcs.org
Website: www.wcs.org

Founded: 1895
Scope: National, International

Description: A nonprofit organization which operates an international wildlife and wildlands conservation program with a full-time staff of wildlife biologists conducting field research and training programs around the world. Headquartered in New York City, the Society operates the Bronx Zoo; The New York Aquarium, Central Park Wildlife Center and Tisch Childhood Zoo, Queens Wildlife Center, Prospect Park Wildlife Center, St. Catherines Island Wildlife Survival Center and a zoological photo library. Education Department offers training programs and cirriculum materials globally to teachers K-12.

Publication(s): Annual Report, Wildlife Conservation

Keyword(s): Aquariums, Endangered Species, training, Environmental Education Curriculum, Environmental and Conservation Education, Libraries, Zoological Parks

Contact(s):
David Schiff, CHAIRMAN
George Amato, DIRECTOR OF SCIENCE RESOURCE CENTER
Joan Downs, EDITOR-IN-CHIEF
Phone: 718-220-5897
Fax: 718-584-2625
jdowns@wcf.org
Steve Johnson, LIBRARIAN
Phone: 718-220-6874
James Large, PRESIDENT AND CEO
Jennifer Herring, PUBLIC AFFAIRS AND DEVELOPMENT
Richard Lattis, SENIOR VICE-PRESIDENT FOR ZOOS AND AQUARIUMS
Phone: 718-220-6526
Annette Berkovits, SENIOR VICE-PRESIDENT OF EDUCATION
Phone: 718-220-5131
aberkovits@wcs.org
John Robinson, SENIOR VICE-PRESIDENT OF INTERNATIONAL CONSERVATION
Bonnie Koeppel, VICE-PRESIDENT — STRATEGIC OPERATIONS
Robert Cook, VICE-PRESIDENT FOR WILDLIFE HEALTH SCIENCE
John Hoare, VICE-PRESIDENT & COMPTROLLER
Louis Garibaldi, VICE-PRESIDENT, DIR. OF AQUARIUM SCIENCE
John Hoare, VICE-PRESIDENT—FINANCIAL SERVICES
W. McKeown, VICE-PRESIDENT—GENERAL COUNSEL

WILDLIFE DAMAGE REVIEW (WDR)

P.O. Box 85218
Tucson, AZ 85754 USA
Phone:
E-mail: wdr@azstarnet.com
Website: www.Azstarnet.com/~WDR

Founded: 1991
Membership: 3,500
Scope: National

Description: Wildlife Damage Review's mission is to bring much needed public attention to the USDA's Animal Damage Control (ADC) program, renamed Wildlife Services in 1997. This taxpayer supported program traps, snares, poisons, and aerial guns 1-2 million of America's wildlife yearly for private interests. WDR's ultimate goal is to place wildlife management into the hands of those agencies whose vested interest is protection of native diversity and banish management guided by predator prejudice.

Publication(s): Special Edition Update, Waste, Fraud, Abuse in the U.S. Animal Damage Control Program, Wildlife Damage Review, Audit of the USDA Animal Damage Control Program, The War on Wildlife (audio CD), Investigating J.F.K. International Airport Gull Hazard Reduction

Keyword(s): Biodiversity, Agriculture, Environmental and Conservation Education, Predators, Birds, Endangered Species, Mammals, Chemical Pollution Control, Pesticides, Toxic Substances, Nuclear-free, Water quantity, Water export and diversion, Trapping

Contact(s):
Nancy Zierenberg, EXECUTIVE DIRECTOR

WILDLIFE DISEASE ASSOCIATION

P.O. Box 1897
Lawrence, KS 66044-8897 USA
Phone: 785-843-1221   Fax: 785-843-1274
Website: wildlifedisease.org

Founded: 1951
Membership: 1,300
Scope: International

Description: An international nonprofit organization of scientists interested in advancing knowledge of the effects of infectious, parasitic, toxic, genetic, and physiologic diseases and environmental factors upon the health and survival of free-living and captive wild animals, and upon their relationships to humans.

Publication(s): Newsletter, Journal of Wildlife Diseases

Keyword(s): Wildlife Rehabilitation, Health and Nutrition, standards

Contact(s):
Irwin Polls, BUSINESS MANAGER
Daniel Pence, EDITOR
Robert McLean, PRESIDENT
Elizabeth Howerth, SECRETARY
Leslie Uhazy, TREASURER

WILDLIFE EDUCATION PROGRAM AND DESIGN

44781 Bittner Point Rd
Bovey, MN 55709 USA
Phone: 218-245-3049

Founded: NA
Membership: 1
Scope: International

Description: A non-profit education organization with slide lectures, Wolf Display, education programs and teachers workshops with "Wolves and Humans" curriculum and a Wolf and Wetland Learning Stations box of environmental education materials. Nationwide programs available. No jobs available.

Contact(s):
Karlyn Berg, DIRECTOR
karlyn@uslink.net

WILDLIFE FEDERATION OF ALASKA

1120 E. Huffman Road, 216
Anchorage, AK 99515-3516 USA
Phone: 907-274-3388   Fax: 907-258-4811
E-mail: wfa@micronet.net
Website: www.micronet/users/~wfa/default.html

Founded: 1985
Scope: Statewide

Description: A representative statewide organization, affiliated with the National Wildlife Federation, dedicated to the protection and enhancement of wildlife ands its habitat through public education and government interaction.

Publication(s): Tracks

Contact(s):
Laurie Fairchild, EDITOR & ALTERNATE REPRESENTATIVE
Rosa Meehan, EDUCATION PROGRAMS CONTACT & TREASURER
Tracy Shafer, PRESIDENT AND REPRESENTATIVE

WILDLIFE FOREVER

10365 West 70th St.
Eden Prairie, MN 55344 USA
Phone: 952-833-1522   Fax: 952-833-0804
E-mail: info@wildlifeforever.org
Website: www.wildlifeforever.org

Founded: 1987
Membership: 70000
Scope: National

Description: Wildlife Forever is a non-profit conservation organization dedicated to conserving America's wildlife heritage through the preservation of habitat, the management of fish and wildlife, and conservation education.

Publication(s): Sports Fish Pocket Guide Series, Wildlife Forever, Wildlife Forever CD-ROM Curriculum, Wildlife Forever Critter Pocket Guide Series, Wildlife Forever Fish On, Annual Report, Cry of the Wild

Keyword(s): Environmental and Conservation Education, Aquatic Habitats, Environment, Wildlife, Flowers, Plants, and Trees, Land Purchase, Hunting, Mammals, Nongame Wildlife, Outdoor Recreation, Public Lands, Waterfowl, Raptors, Sport Fishing

Contact(s):
James Gallagher, ACCOUNTANT
Ann McCarthy, DIRECTOR OF EDUCATION
Pete Wuebker, DIRECTOR OF MARKETING
David Fredrick, GRANTS COORDINATOR
Mark Petersen, MERCHANDISE MANAGER
Douglas Grann, PRESIDENT, CEO

WILDLIFE FOUNDATION OF FLORIDA, INC.

Tallahassee, FL 32302 USA
Phone: 850-487-3796   Fax: 850-488-6988
Website: www.wildlifefoundationofflorida.com

Founded: 1994
Membership: 6
Scope: Statewide

Description: The mission of the Wildlife Foundation of Florida, Inc. is to provide assistance, funding, and promotional support for the Florida Game and Wildlife Conservation Commission, and in so doing, contribute to the health and well-being of Florida's fish and wildlife resources and their habitats.

Keyword(s): Wildlife, Hunting, Sport Fishing, training

Contact(s):
William Blake, BOARD OF DIRECTORS
George Matthews, BOARD OF DIRECTORS
Robert Brantly, BOARD OF DIRECTORS
Allan Egbert, BOARD OF DIRECTORS
William Bostick, BOARD OF DIRECTORS
Linda Bremer, BOARD OF DIRECTORS
C. Rainey, BOARD OF DIRECTORS

WILDLIFE HABITAT CANADA

7 Hinton Ave., North,
Ottawa, Ontario K1Y 4P1 Canada
Phone: 613-722-2090   Fax: 613-722-3318
E-mail: reception@whc.org
Website: www.whc.org

Founded: 1984
Scope: National

Description: Wildlife Habitat Canada is a national non-profit organization dedicated to working with private citizens, governments, non-government organizations, and industry to conserve the great variety of wildlife habitats across Canada. The organization develops and implements its own conservation initiatives, such as the Forest Biodiversity Program, but also provides grants for conservation, research, communication and education projects and has a graduate scholarship program.

Publication(s): (publications are available upon request.), Annual Reports (a list of free publications is contained there)

Keyword(s): Stewardship, Research Grants, training, Wetlands, Scholarships and Grants, Forest Management, Biodiversity, Wildlife, Agriculture

Contact(s):
Doug Wolthausen, DIRECTOR OF PROGRAMS
Jean Cinq-Mars, EXECUTIVE DIRECTOR

WILDLIFE HABITAT COUNCIL

1010 Wayne Ave., Suite 920
Silver Spring, MD 20910 USA
Phone: 301-588-8994   Fax: 301-588-4629
E-mail: whc@wildlifehc.org
Website: www.wildlifehc.org

Founded: 1987
Membership: 20
Scope: National, International

Description: A joint effort between the conservation and corporate communities, WHC is an international, nonprofit organization formed to assist corporations in enhancing their lands for the benefit of wildlife. WHC's program includes technical assistance in establishing and maintaining responsible corporate wildlife management practices, environmental mediation, habitat certification, information sharing, employee involvement, and community outreach.

Publication(s): Registry of Certified and Internationally Accredited Corporate Wildlife Habitat Programs, Corporate Homes for Wildlife annual desk calender, Wildlife Habitat

Keyword(s): training, Sustainable Development, Environmental and Conservation Education, Biodiversity

Contact(s):
Lawrence Selzer, BOARD OF DIRECTORS VICE CHAIR
Robert Fenech, CHAIRMAN OF THE BOARD

Laurie Coran, CONTROLLER
Robert Johnson, EXECUTIVE VICE PRESIDENT
Hugh Dillingham III, PAST DIRECTOR
William Howard, PRESIDENT
Steven Elbert, SECRETARY

WILDLIFE HERITAGE FOUNDATION OF WYOMING (WHFW)

5400 Bishop Blvd., P.O. Box 20088
Cheyenne, WY 82003-7002 USA
Phone: 307-777-4693   Fax: 307-777-4699
Scope: National

Description: The Wildlife Heritage Foundation of Wyoming has been established as an independent, public benefit, nonprofit corporation to provide private support for the game and fish department in its stewardship of the state's terrestrial and aquatic wildlife resrouces.

Contact(s):
Marlene Brown
FOUNDATION LIAISON

WILDLIFE INFORMATION CENTER, INC.

P.O. Box 198
Slatington, PA 18080 USA
Phone: 610-760-8889
E-mail: wildlife@fast.net
Website: www.wildlifeinfo.org

Founded: 1986
Scope: National

Description: A nonprofit, member-supported organization whose purpose is to secure and disseminate wildlife conservation, education, recreation, and scientific research information. Programs include: The Kittatinny Raptor Corridor Project and long-term hawk migration field studies at Bake Oven Knob, PA.; in-service teacher training courses and public education; research and preparation of conservation-education papers and reports; maintaining wildlife libraries, photographs, computer databases; and advocating wildlife observation and wildlife tourism. The Center currently is raising funds for the purchase of land for its own wildlife refuge and headquarters.

Publication(s): Wildlife Activist, American Hawkwatcher, Wildlife Conservation Reports

Keyword(s): Biodiversity, Birds, Environmental and Conservation Education, Raptors, training

Contact(s):
Dan Kunkle, EDITOR
Dan Kunkle, PRESIDENT
Kathie Romano, SECRETARY
Margaret Libonati, TREASURER

WILDLIFE LEGISLATIVE FUND OF AMERICA, THE, AND WILDLIFE CONSERVATION FUND OF AMERICA, THE

801 Kingsmill Parkway
Columbus, OH 43229-1137 USA
Phone: 614-888-4868   Fax: 614-888-0326
E-mail: info@wlfa.org
Website: www.wlfa.org

Founded: 1978
Membership: 13
Scope: National

Description: Companion nonprofit organizations established to protect America's hunting, trapping, and fishing heritage, and the scientific wildlife management practices which support it. The WLFA is the legislative arm. The WCFA is the legal defense, public education, and research arm.

Publication(s): Update

Keyword(s): Hunting, Sport Fishing, Trapping, training

Contact(s):
Vincent Shiel, CHAIRMAN OF THE BOARD
Doug Jeanneret, DIRECTOR OF COMMUNCATIONS
William Horn, DIRECTOR OF NATIONAL AND
INTERNATIONAL AFFAIRS AND
WASHINGTON, D.C. COUNSEL
Robert Sexton, DIRECTOR OF STATE SERVICES
Walter Pidgeon, PRESIDENT & CEO
Gilbert Humphrey, TREASURER
F. Maddox, VICE CHAIRMAN
Rick Story, VICE PRESIDENT

WILDLIFE MANAGEMENT INSTITUTE
1101 14th St., NW
Washington, DC 20005 USA
Phone: 202-371-1808   Fax: 202-408-5059
Website: www.wildlifemgt.org/wmi

Founded: NA
Membership: 11
Scope: National

Description: International nonprofit scientific and educational private membership organization, supported by industries, groups, and individuals, promoting improved professional management of wildlife and other natural resources for the benefit of those resources and North American, including its people.

Publication(s): Transactions North American Wildlife and Natural Resources Conference, books and booklets, Outdoor News Bulletin

Contact(s):
Ronald Helinski, CONSERVATION POLICY SPECIALIST
Terry Riley, DIRECTOR OF CONSERVATION
Carol Peddicord, FINANCE MANAGER
Rob Manas, MIDWEST FIELD REPRESENTATIVE
Scot Williamson, NORTHEAST FIELD REPRESENTATIVE
R.R. 1, Box 587, Spur Rd., North Stratford, NH 03590
Phone: 603-636-9846
Fax: 603-636-9853
wmisw@together.net
Robert Davison, NORTHWESTERN FIELD
REPRESENTATIVE
Kathryn Reis, PARTNERS NETWORK COORDINATOR
Rollin Sparrowe, PRESIDENT
James Woehr, SENIOR SCIENTIST
Donald McKenzie, SOUTHEAST FIELD REPRESENTATIVE
Len Carpenter, SOUTHWEST FIELD REPRESENTATIVE
4015 Cheney Dr., Fort Collins, CO 80526
Phone: 970-223-1099
Fax: 970-204-9198
lenc@verinet.com
Richard McCabe, VICE PRESIDENT
Robert Byrne, WILDLIFE PROGRAM COORDINATOR

WILDLIFE PRESERVATION TRUST INTERNATIONAL, INC.
1520 Locust St., Suite 704
Philadelphia, PA 19102 USA
Phone: 215-731-9770   Fax: 215-731-9766
E-mail: homeoffice@wpti.org
Website: www.wpti.org

Founded: 1971
Membership: 3,000
Scope: National

Description: Wildlife Preservation Trust International conserves threatened wild species and their habitats in partnership with local scientists and educators around the world.

Publication(s): Dodo, The, Dodo Dispatch, Wild Times, The, Annual Report, On the Edge

Keyword(s): standards, Endangered Species, Environmental and Conservation Education, Grants, training

Contact(s):
Joanne Gullifer, DIRECTOR OF ADMINISTRATION
Fred Koontz, DIRECTOR OF CONSERVATION PROGRAM
Peter Wilmerding, DIRECTOR OF DEVELOPMENT
Mary Pearl, EXECUTIVE DIRECTOR
A. Aguirre, INTERNATIONAL FIELD VETERINARIAN
Thomas McHenry, PRESIDENT
Victoria Mars, SECRETARY
John Tuten, TREASURER
Virginia Mars, VICE PRESIDENT
Allen Model, VICE PRESIDENT

WILDLIFE SOCIETY
ALABAMA CHAPTER
Attn: President, 331 Funchess Hall, Auburn University
Auburn, AL 36830 USA

Founded: NA
Scope: Statewide

Contact(s):
Tommy Counts, PAST-PRESIDENT
P.O. Box 278, Double Springs, AL 35553
Phone: 205-489-5111
Fax: 205-489-3427
tom.counts@al.usda.gov
James Armstrong, PRESIDENT
Auburn University
108 M. White Smith Hall, Auburn, AL 36830
Phone: 334-844-9233
Fax: 334-844-9234
jarmstro@acesag.auburn.edu
Jeff Makemson, SECRETARY-TREASURER
11481 Colonial Dr., Duncanville, AL 35456
Phone: 205-345-3807
Fax: 205-333-2900

WILDLIFE SOCIETY
ALASKA CHAPTER
Attn: President, P.O. Box 72962
Fairbanks, AK 99707 USA
Phone: 907-235-8191

Founded: NA
Scope: Statewide

Contact(s):
  Theron Schenck, NEWSLETTER EDITOR
  1913 Parkview Cir., Anchorage, AK 99501-5753
  Phone: 907-271-2839
  tschenck/r10_chugach@fs.fed.us
  Roger Post, PAST-PRESIDENT
  P.O. Box 72962, Fairbanks, AK 99707
  Phone: 907-455-6583
  Fax: 907-455-6583
  rpost@mosquitonet.com
  Gino Del Frate, PRESIDENT
  AK Dept. of Fish and Game
  3298 Douglas Place, Homer, AK 99603-8027
  Phone: 907-235-8191
  Fax: 907-235-2448
  gino_delfrate@fishgame.state.ak.us
  Doug Larsen, PRESIDENT-ELECT
  AK Dept. of Fish and Game
  1910 Glacier Ave., Juneau, AK 99801-7802
  Phone: 907-465-5277
  Fax: 907-465-6142
  doug_larsen@fishgame.state.ak.us
  Theron Schenck, SECRETARY-TREASURER
  1913 Parkview Cir., Anchorage, AK 99501-5753
  Phone: 907-271-2839
  tschenck/r10_chugach@fs.fed.us

WILDLIFE SOCIETY
ALBERTA CHAPTER
Attn: President, Rural Route 4
Sherwood Park, Alberta T8A 3K4 Canada
Phone: 780-778-7116
Website: http://www.rr.ualberta.ca/wildlifesociety/

Founded: NA
Scope: Statewide

Contact(s):
  Troy Sorensen, NEWSLETTER EDITOR
  Suite 203, 111-54 St., Edson, AB T7E 1T2
  Phone: 780-723-8244
  Fax: 780-723-8502
  troy.sorensen@gov.ab.ca
  Michael Dorrance, PAST-PRESIDENT
  RR4, Sherwood Park, AB T8A 3K4
  Phone: 780-467-4396
  Fax: 780-436-9540
  mathdorr@telusplanet.net
  Elston Dzus, PRESIDENT
  AB Pacific Forest Ind. Inc.
  P.O. Box 8000, Boyle, AB T0A 0M0
  Phone: 780-525-8393
  Fax: 780-525-8007
  dzusel@alpac.ca
  Arlen Todd, PRESIDENT-ELECT
  1263 Berkley Dr. NW, Calgary, AB T3K 1T1
  Phone: 403-297-7349
  Fax: 403-297-3362
  arlen.todd@gov.ab.ca
  Ronald Mumme, SECRETARY-TREASURER
  Dept. Bio.
  Allegheny College, Meadville, PA 16335
  Phone: 814-332-2382
  rmumme@alleg.edu

WILDLIFE SOCIETY
ARIZONA CHAPTER
Attn: President
130 W. Calle Melendez
Green Valley, AZ 85614 USA
Fax: 520-648-6556
E-mail: hrs@u.arizona.edu

Description: Student organization

WILDLIFE SOCIETY
ARKANSAS CHAPTER
Attn: President, P.O. Box 279 - Arkansas Tech University
Altus, AR 72821-0279
Scope: Statewide

Contact(s):
  Kendall Moles, PRESIDENT
  P.O. Box 599, State University, AR 72467
  Phone: 870-972-3082

WILDLIFE SOCIETY
CALIFORNIA CENTRAL COAST CHAPTER
Attn: President, USDA Forest Service 6144 Calle Real
Goleta, CA 93117 USA
Website: wildlife.org

Founded: NA
Scope: Statewide

Contact(s):
  Maeton Freel, PRESIDENT
  273 Santa Barbara Shore Dr., Goleta, CA 93117
  Phone: 805-681-2764
  Justin Vreeland, SECRETARY
  UCCE
  425 Waupelani Dr. 508, State College, PA 16801
  Phone: 814-237-8567
  jkv104@psu.edu
  Michael Hanson, TREASURER
  1203 Madonna, San Luis Obispo, CA 93405
  Phone: 805-541-0272
  Fax: 805-756-1419
  mthanson@calpoly.edu
  Kevin Cooper, VICE-PRESIDENT
  452 Lawrence Dr., San Luis Obispo, CA 93401
  Phone: 805-925-9538
  Fax: 805-681-2781
  lecoop@juno.com

WILDLIFE SOCIETY
CALIFORNIA NORTH COAST CHAPTER
Attn: President, Simpson Timber Co., P.O. Box 08
Korbel, CA 95550 USA
Website: wildlife.org

Founded: NA
Scope: Statewide

Contact(s):
  Sandra Arb, PRESIDENT
  P.O. Box 532, Scotia, CA 95565-0532
  Phone: 707-764-4488
  Fax: 707-764-4118
  vonarb@scopac.com

WILDLIFE SOCIETY
COLORADO CHAPTER
Attn: President, 0772 S. Rd. 1E
Monta Vista, CO 81144 USA
E-mail: cws-fmp@cws.cnchost.com
Website: wildlife.org

Founded: NA
Scope: Statewide

Description: Student organization

Contact(s):
Francie Pusateri, PRESIDENT
317 Prospect Rd., Ft. Collins, CO 82526
Phone: 970 4-72 -4336
franciep@concentric.net

WILDLIFE SOCIETY
FLORIDA CHAPTER
Attn: President, USDA Forest Service, P.O. Box 579
Bristol, FL 32321 USA
Fax: 352-955-2230
E-mail: molerp@fwc.state.fl.us
Website: wildlife.org

Founded: NA
Scope: Statewide

Contact(s):
Sara Schuweitzer, PRESIDENT
School of Forest Resources, University of Georgia, Athens,
GA 30602

WILDLIFE SOCIETY
GEORGIA CHAPTER
Attn: President, D.B. Warnell School of Forest Resources,
University of Georgia
Athens, GA 30602-2152 USA

Founded: NA
Scope: Statewide

Contact(s):
Chuck Waters, NEWSLETTER EDITOR
2150 Dawsonville Highway, Gainesville, GA 30501
Phone: 770-535-5700
Fax: 770-535-5953
chuck_waters@mail.dnr.state.ga.us
Douglass Hall, PAST-PRESIDENT
1161 Crooked Creek Rd., Watkinsville, GA 30677
Phone: 706-546-2020
Fax: 706-546-2004
douglas.i.hall@usda.gov
Sara Schweiter, PRESIDENT
U. of GA
D.B. Warnell Sch. Forest Res., Athens, GA 30602-2152
Phone: 706-542-1150
Fax: 706-542-8356
schweitz@smokey.forest.uga.edu
Mark Whitney, PRESIDENT-ELECT
1057 Plantation Way SE, Conyers, GA 30094
Phone: 770-761-1697
Fax: 706-557-3042
mark_whitney@mail.dnr.state.ga.us
Douglass Hoffman, SECRETARY-TREASURER
120 Diamond Dr., Athens, GA 30605
Phone: 706-546-2020
Fax: 706-546-2004
douglas.m.hoffman@usda.gov

WILDLIFE SOCIETY
HAWAII CHAPTER
40 Kunihi Ln., 221
Kahului, HI 96732 USA

Founded: NA
Scope: Statewide

Contact(s):
Cathleen Hodges, NEWSLETTER EDITOR
20 Kumano Dr., Pukalani, HI 96768
cathleen_hodges@nps.gov
Cathleen Hodges, PAST-PRESIDENT
20 Kumano Dr., Pukalani, HI 96768
cathleen_hodges@nps.gov
Carrie Haurez, PRESIDENT
40 Kunihi Ln. 221, Kahului, HI 96732
Phone: 808-877-1455
Fern Duvall, PRESIDENT-ELECT
211 Ulana St., Makawao, HI 96768-8034
Phone: 808-873-3502
Fax: 808-873-3505
mawildl@aloha.net
Joy Tamayose, SECRETARY
c/o Haleakala National Park
P.O. Box 369, Makawao, HI 96822
Phone: 808-572-4492
Fax: 808-572-4498
ulukitty@aol.com
Dan McNulty-Huffman, TREASURER
85 Haele Place, Makawao, HI 96768-8053
Phone: 808-572-4485
dmh@t-link.net

WILDLIFE SOCIETY
IDAHO CHAPTER
Attn: President, College of Forestry, Wildlife & Range
Sciences, University of Idaho
Moscow, ID 83843-1136 USA
E-mail: tws@uidaho.edu
Website: uidaho.edu/student-orgs/wlfsoc
http://uidaho.edu/student-orgs/wlfsoc

Founded: NA
Scope: Statewide

Contact(s):
Robyn Januszewski, PRESIDENT
Phone: 208-324-1160
mcommons@idfg.state.id.us

WILDLIFE SOCIETY
ILLINOIS CHAPTER
Attn: President, Max McGraw Wildlife Foundation,
P.O. Box 9
Dundee, IL 60118 USA
Fax: 618-453-6944
Website: siu.edu/-siuctws

Founded: NA
Scope: Statewide

Contact(s):
Tim Van Deelen, PRESIDENT
607 E. Peabody, Champaign, IL 61820
Phone: 217-333-6856
Phone: 618-453-2806
feldhamer@zoology.siu.edu

WILDLIFE SOCIETY
INDIANA CHAPTER
Department of Forestry and Wildlife Resources 1159
Forestry Building
West Lafayette, IN 47907-1159 USA
Phone: 765-494-3567   Fax: 765-496-2422
Website: www.fnr.purdue.edu

Founded: NA
Scope: Statewide

Contact(s):
Linda Byer, NEWSLETTER EDITOR
6615 S. 875 E., Monterey, IN 46960
Phone: 219-896-3522
Fax: 219-896-3038
byer@pwtc.com
James Gerbracht, PAST-PRESIDENT
IN Dept. Nat. Resources, State Parks & Reservoirs 402 W
Washington St W298, Indianapolis, IN 46204-2745
Phone: 317-232-4124
Fax: 317-232-4132
jim_gerbracht_at_dnrian@ima.isd.state.in.us
Mark Pochon, PRESIDENT
2045 Graddy Rd., Mount Vernon, IN 47620
Phone: 812-838-2927
Fax: 812-838-5473
bfhovey@evansville.net
Phil Seng, PRESIDENT-ELECT
1010 Yeardley Ln., Mishawaka, IN 46544-6766
Phone: 219-258-0100
Fax: 219-258-0189
phil@djcase.com
Linnea Floyd, SECRETARY-TREASURER
1266 Farley Dr., Indianapolis, IN 46214
Phone: 317-233-6527
Fax: 317-232-8150
lfloyd@dnr.state.in.us

WILDLIFE SOCIETY
KANSAS CHAPTER

Founded: NA
Scope: Statewide

Contact(s):
M. McCord, NEWSLETTER EDITOR
4413 Ponderosa Ln., Temple, TX 76502
Phone: 254-742-9812
Fax: 254-742-9848
brad.mccord@tx.usda.gov
Michael McFadden, PAST-PRESIDENT
1110 North 900 Road, Lawrence, KS 66047
Phone: 785-206-2630
Fax: 785-295-7630
miketmks@aol.com
Elmer Finck, PRESIDENT
Emporia State U.
Div. of Bio. Sci. Box 4050, Emporia, KS 66801
Phone: 620-341-5623
Fax: 620-341-5607
finckelm@emporia.edu
M. McCord, PRESIDENT-ELECT
4413 Ponderosa Ln., Temple, TX 76502
Phone: 254-742-9812
Fax: 254-742-9848
brad.mccord@tx.usda.gov
Charles Lee, SECRETARY-TREASURER

Kansas State University
127 Call Hall, Manhattan, KS 66505-1600
Phone: 758-532-5734
Fax: 785-532-5681
clee@oz.oznet.ksu.edu

WILDLIFE SOCIETY
KENTUCKY CHAPTER
Attn: President, KY Dept. Fish & Wildlife,
#1 Game Farm Rd.
Frankfort, KY 40601 USA

Founded: NA
Scope: Statewide

Contact(s):
Mark Cramer, NEWSLETTER EDITOR
KY Dept F & W Resources
#1 Game Farm Road, Frankfort, KY 40601
Phone: 502-564-4404
Fax: 502-564-6508
roy.grimes@mail.state.ky.us
Robert Morton, PAST-PRESIDENT
8407 US 41 A, Henderson, KY 42420-9637
Phone: 502-827-2673
mmorton@apex.net
Roy Grimes, PRESIDENT
#1 Game Farm Road, Frankfort, KY 40601
Phone: 502-564-4404
Fax: 502-564-6508
roy.grimes@mail.state.ky.us
Charles Elliot, PRESIDENT-ELECT
Dept. of Biology
EKU 521 Lancaster Ave., Richmond, KY 40475-3102
Phone: 606-622-1538
Fax: 606-622-1020
bioelliott@acs.eku.edu
Dan Figert, SECRETARY-TREASURER
KY Dept. Fish & Wildlife
#1 Game Farm Rd., Frankfort, KY 40601
Phone: 800-858-1549
Phone: 502-564-4859
dan.figert@mail.state.ky.us

WILDLIFE SOCIETY
LOUISIANA CHAPTER
Attn: President, 5492 Grand Chenier Hwy
Grand Chenier, LA 70643 USA

Founded: NA
Scope: Statewide

Contact(s):
Frank Hohwer, NEWSLETTER EDITOR
Louisiana State U.
Forestry, Wldlfe, & Fisheries, Baton Rouge, LA 70803
Phone: 504-388-4131
Fax: 504-388-4227
frohwer@lsu.edu
Edmond Mouton, PAST-PRESIDENT
LDWF
2415 Darnall Rd, New Iberia, LA 70560
Phone: 318-373-0032
Fax: 318-373-0181
mouton_ec@wlf.state.la.us

Martin Floyd, PRESIDENT
2044 Bayou Road, Cheneyville, LA 71325
Phone: 318-473-7690
Fax: 318-473-7747
marty_floyd@la.usda.gov
Mike Olinde, PRESIDENT-ELECT
2130 Terrace Ave., Baton Rouge, LA 70806
Phone: 225-765-2353
olinde_mw@wlf.state.la.us
Virginia Rettig, SECRETARY
USFWS
1010 Gause Blvd. Bldg. 936, Slidell, LA 70458
Phone: 540-646-7555
Fax: 504-646-7588
virginia_rettig@fws.gov
John Pitre, TREASURER
263 White Oak Blvd., Boyce, LA 71409
Phone: 318-473-7809
Fax: 318-473-7616
john.pitre@la.usda.gov

WILDLIFE SOCIETY
MAINE CHAPTER
Attn: Secretary, ME Dept. Inland Fish & Wildlife,
P.O. Box 416
Ashland, ME 04732 USA

Founded: NA
Scope: Statewide

Contact(s):
James Nelson, EXECUTIVE COMMITTEE
34 Cates Rd., Thorndike, ME 04986
Phone: 207-948-3131
Fax: 207-948-6277
jnelson@unity.unity.edu
Mitschka Hartley, PAST-PRESIDENT
U of ME
Natl. Audubon Soc.
230 E. Lake Rd., DeRuyter, NY 13052
Phone: 315-662-7900
mhartley@audubon.org
James Ecker, PRESIDENT
58 Canterbury Rd., Brewer, ME 04412
Phone: 207-827-6191
Fax: 207-827-8441
jestump@aol.com
Joseph Wiley, SECRETARY-TREASURER
Bureau of Parks and Lands
22 State House Station, Augusta, ME 04333
Phone: 207-287-4921
Fax: 207-287-8111
joe.wiley@state.me.us

WILDLIFE SOCIETY
MANITOBA CHAPTER
Attn: President, Dillion Consulting Ltd., 6 Donald St. S.
Winnipeg, Manitoba R3L 0K6 Canada
Website: http://twsmb.tripod.com

Founded: NA
Scope: Statewide

Contact(s):
Marc Schuster, NEWSLETTER EDITOR
242 Hartford Ave., Winnipeg, MB R2V 0W1
Phone: 204-269-2184
Fax: 204-983-5248
marc.schuster@ec.gc.ca

Rhiannon Christie, PAST-PRESIDENT
3-395 River Ave., Winnipeg, MB R3L 0C5
Phone: 204-632-2938
Fax: 204-693-9673
rchristie@mb.sympatico.ca
Cory Lindgren, PRESIDENT
One Hammock Marsh
Box 1160, Stonewall, MB R0C 2Z0
Fax: 204-437-3000
c_lindgren@ducks.ca
Tanys Uhmann, SECRETARY AND TREASURER
1017 Kilkenny Dr., Winnipeg, MB R3T 4K5
Phone: 204-261-2184
Fax: 204-261-0038
umuhmann@cc.umanitoba.ca
Neil Mochnacz, STUDENT REPRESENTATIVE
468 Clelsea Avenue, Winnipeg, MB R2K 1A1
Phone: 204-984-2425
Fax: 204-983-2403
mochnaczn@dfg-mpo.gc.ca

WILDLIFE SOCIETY
MARYLAND-DELAWARE CHAPTER
Attn: President, 1053 Hampton Dr.
Crownsville, MD 21032-1315 USA

Founded: NA
Scope: Statewide

Contact(s):
Brenda Belensky, NEWSLETTER EDITOR
25 Montrose Manor Ct. H, Baltimore, MD 21228
Phone: 410-313-4724
Fax: 410-313-4660
bbelensky@co.ho.md.us
Carol Bernstein, PAST-PRESIDENT
1053 Hampton Drive, Crownsville, MD 21032-1315
Phone: 410-962-3208
Fax: 410-962-4698
carol.l.bernstein@usace.army.mil
Philip Norman, PRESIDENT
723 Roland Ave., Bel Air, MD 21014
Phone: 410-313-1675
Fax: 410-313-4660
pnorman@co.ho.md.us
Edward Morgereth, PRESIDENT-ELECT
Biohabitat, Inc. 602
15 W. Aylesbury Rd., Timonium, MD 21093
Phone: 410-337-3659
Fax: 410-583-5678
edward@biohabitat.com
Amy Deller-Jacobs, SECRETARY
P.O. Box 1455, Cambridge, MD 21613-5455
Phone: 410-330-3911
Donald Rohrback, TREASURER
Reg. Wldlf. Manager Indian Springs WMA
14038 Blairs Valley Rd., Clear Spring, MD 21722
Phone: 301-842-3355

WILDLIFE SOCIETY
MICHIGAN CHAPTER
Attn: President, 5525 Hayes Tower Rd.
Gaylord, MI 49735 USA

Founded: NA
Scope: Statewide

Contact(s):
Craig Albright, NEWSLETTER EDITOR
Hwy. 2, 41, & M-35, Gladestone, MI 49837
Phone: 906-786-2351
Fax: 906-786-1300
albrighc@state.mi.us
Larry Caldwell, PAST-PRESIDENT
Central MI Univ.
Biology Dept., Mt. Pleasant, MI 48859
Phone: 517-774-3387
Fax: 517-774-3462
larry.caldwell@cmich.edu
Craig Albright, PRESIDENT
Hwy. 2, 41, & M-35, Gladestone, MI 49837
Phone: 906-786-2351
Fax: 906-786-1300
albrighc@state.mi.us
Henry Campa, PRESIDENT-ELECT
MI State University
Dept. of Fish & Wldlfe., East Lansing, MI 48824
Phone: 517-353-2042
Fax: 517-432-1699
campa@pilot.msu.edu
Kelly Millenbah, SECRETARY-TREASURER
MI State U.
13 Natural Resources Bldg., East Lansing, MI 48824-1222
Phone: 517-353-4802
Fax: 517-432-1699
millenba@pilot.msu.edu

WILDLIFE SOCIETY
MINNESOTA CHAPTER
Attn: President, 24201 County Rd.10
Bovey, MN 55709 USA
Website: www.crk.edu/tws/mn

Founded: NA
Scope: Statewide
Contact(s):
Margaret Anderson, MEMBERSHIP
1823 Robin Hood Drive, Thief River Falls, MN 56701
Phone: 218-449-4115
margaret_anderson@mail.fws.gov
Joel Huener, NEWSLETTER EDITOR
HCR 3 Box 17, Middle River, MN 56737
Phone: 218-222-3747
Fax: 218-222-3746
joel.huener@dnr.state.mn.us
Janet Boe, PAST-PRESIDENT
24201 City. Rd. 10, Bovey, MN 55709
Phone: 218-755-4028
Fax: 218-327-1617
janet.boe@dnr.state.mn.us
Gary Huschle, PRESIDENT
RR 1 Box 114, Thief Rvr. Fls., MN 56701-9739
Phone: 218-449-4115
Fax: 218-449-3241
gary_huschle@fws.gov
Martha Minchak, PRESIDENT
RR 1 Box 51, St. Hilaire, MN 56754-9725
Phone: 218-783-6861
Fax: 218-783-6832
gmehmel@wiktel.com

Gretchen Mehmel, SECRETARY-TREASURER
P.O. Box 100, Roosevelt, MN 56673-0100
Phone: 218-783-6861
Fax: 218-783-6832
gmehmel@witkel.com

WILDLIFE SOCIETY
MISSISSIPPI CHAPTER
Attn: President, P.O. Box 451
Jackson, MS 39205 USA
Website: www.cfr.msstate.edu/mstws

Founded: NA
Scope: Statewide

Contact(s):
Darren Miller, NEWSLETTER EDITOR
Weyerhaeuser Co.
Southern Forestry Res., Box 2288, Columbus, MS 39704-2288
Phone: 662-245-5249
Fax: 662-245-5228
darren.miller@weyerhaeuser.com
Marcus Spencer, PAST-PRESIDENT
104 Jess Dean Dr., Brandon, MS 39047-9539
Phone: 601-364-2229
Fax: 601-364-2209
randys@mdwfp.state.ms.us
Kristina Godwin, PRESIDENT
610 Hospital Rd., Starkville, MS 39759
Phone: 662-325-3014
Fax: 662-325-3690
kris.godwin@usda.gov
K. Godwin, PRESIDENT-ELECT
610 Hospital Rd., Starkville, MS 39759
Phone: 662-325-5119
Fax: 662-325-8726
dgodwin@cfr.msstate.edu
Julie Marcy, SECRETARY-TREASURER
P.O. Box 820161, Vicksburg, MS 39182
Phone: 601-631-5302
Fax: 601-631-7133
julie.b.marcy@usace.army.mil

WILDLIFE SOCIETY
MISSOURI CHAPTER
Attn: President, 21999 Hwy. B
Maitland, MO 64466 USA

Founded: NA
Scope: Statewide

Contact(s):
Donald Martin, NEWSLETTER EDITOR
2207 Oak Cliff Dr., Columbia, MO 65203
Phone: 573-751-4115
martind@mail.conservation.state.mo.us
R. Jackson, PAST-PRESIDENT
21999 Hwy. B, Maitland, MO 64466
Phone: 660-446-3371
jacksr@mail.conservation.state.mo.us
John Schulz, PRESIDENT
Fish & Wldlfe. Research Cntr.
MO Dept. of Cons.
1110 College Ave., Columbia, MO 65201
Phone: 573-882-9880
schulj@mail.conservation.state.mo.us

Donald Martin, PRESIDENT-ELECT
2207 Oak Cliff Dr., Columbia, MO 65203
Phone: 573-751-4115
martind@mail.conservation.state.mo.us
Dennis Browning, SECRETARY
1811 Eastview Dr., Trenton, MO 64683
Phone: 816-675-2205
Fax: 816-675-2221
brownd@mail.conservation.state.mo.us
Phil Rockers, TREASURER
P.O. Box 248, Sullivan, MO 63080
Phone: 573-468-3335
Fax: 573-468-5434
rockep@mail.conservation.state.mo.us

WILDLIFE SOCIETY
MONTANA CHAPTER
Attn: President, 107 Mark Jensen Ln.
Polson, MT 59860 USA
Website: www.montanatws.org

Founded: NA
Scope: Statewide

Contact(s):
Daniel Young, NEWSLETTER EDITOR
Box 916, Eureka, MT 59917
Phone: 406-296-2536
Fax: 406-296-2588
lyoung@fs.fed.us
Frank Pickett, PRESIDENT
45 Basin Creek Road, Butte, MT 59701
PHONE: 406-533-3445
FAX:406-533-6000
EMAIL: fjpickett@pplmt.com
Bev Dickerson, PRESIDENT-ELECT
3710 Fallon, Suite C Bozeman, MT 59718
Phone: 406-522-2541
Fax: 406-522-2528
Email: bdixon@fs.fed.us
Marion Cherry, SECRETARY-TREASURER
518 Fieldstone Dr., Bozeman, MT 59715
Phone: 406-587-6257
mcherry@fs.fed.us

WILDLIFE SOCIETY
NATIONAL CAPITAL CHAPTER
USA

Founded: NA
Scope: Statewide

Contact(s):
Douglas Hobbs, NEWSLETTER EDITOR
5807 Blaine Dr., Alexandria, VA 22303-1914
Phone: 703-960-4271
doug_hobbs@fws.gov
Stephanie Hussey, NEWSLETTER EDITOR
208 N. Trenton St. 4, Arlington, VA 22203
Phone: 703-526-0272
saoffice1@pipeline.com
Kristen La Vine, PAST-PRESIDENT
1912 N. Rhodes Street, Arlington, VA 22201
Phone: 703-519-0013
Fax: 703-519-9565
kp_lavine@yahoo.com

Douglas Hobbs, PRESIDENT
5807 Blaine Dr., Alexandria, VA 22303-1914
Phone: 703-960-4271
doug_hobbs@fws.gov
Stephanie Hussey, SECRETARY-TREASURER
208 N. Trenton St. 4, Arlington, VA 22203
Phone: 703-526-0272
saoffice1@pipeline.com

WILDLIFE SOCIETY
NEBRASKA CHAPTER
Grand Island, NE 68801 USA
Website: wildlifeconsult.com/netws

Founded: NA
Scope: Statewide

Contact(s):
Mark Humpert, NEWSLETTER EDITOR
45090 Elm Island Rd., Gibbon, NE 68840
Phone: 308-865-5308
Fax: 308-865-5309
mhumpert@ngpun.ngpc.state.ne.us
Mark Czaplewski, PAST-PRESIDENT
NE Game & Parks Commision
1617 First Avenue, Kearney, NE 68847
Phone: 308-385-6282
Fax: 308-385-6285
czaplews@linux3.nrc.state.ne.us
Garry Steinauer, PRESIDENT
Nebraska Game and Parks Commission, 1703 L Street
Aurora, NE 68818
Phone:402-694-2498
Mark Humpert, PRESIDENT-ELECT
45090 Elm Island Rd., Gibbon, NE 68840
Phone: 308-865-5308
Fax: 308-865-5309
mhumpert@ngpun.ngpc.state.ne.us
Laurel Badura, SECRETARY
306 E. 29th, Kearney, NE 68847
Phone: 308-865-5332
Fax: 308-865-5309
lbadura@ngpc.state.ne.us
Jeanine Lackey, TREASURER
18909 N. 84th St., Ceresco, NE 68017-4209
jlackey2@unl.edu

WILDLIFE SOCIETY
NEVADA CHAPTER
Attn: President, 4321 Jody Ave.
Las Vegas, NV 89120 USA

Founded: NA
Scope: Statewide

Contact(s):
James Jeffress, PRESIDENT
2085 Skyland Blvd., Winnemucca, NV 89445
Phone: 702-623-4959
David Pulliam, PRESIDENT-ELECT
8003 Moss Creek Dr., Reno, NV 89506
Phone: 775-688-1561
dpulliam@govmail.state.nv.us
Alan Jenne, SECRETARY AND TREASURER
4080 Bluewing Ln., Carson City, NV 89704
Phone: 775-888-7689

WILDLIFE SOCIETY
NEW ENGLAND CHAPTER
USA

Founded: NA
Scope: National

Contact(s):
Robert Gilmore, NEWSLETTER EDITOR
P.O. Box 121, West Simsbury, CT 06092
Phone: 860-424-3866
Fax: 860-424-4075
Paul Rego, PAST-PRESIDENT
Sessions Woods WMA
P.O. Box 1550, Burlington, CT 06013
Phone: 860-675-8130
Fax: 860-675-8141
paul.rego@po.state.ct.us
John McDonald, PRESIDENT
MA Div. Fish and Wildlife
Field HQ, Westborough, MA 01581
Phone: 508-792-7270
Fax: 508-792-7275
john.mcdonald@state.us
Jenny Dickson, PRESIDENT-ELECT
391 Jackson St., Thomaston, CT 06787-2016
Phone: 860-675-8130
Phone: 860-675-8141
jenny.dickson@po.state.ct.us
Susan Langlois, SECRETARY-TREASURER
MA Div. Fish and Wildlife
Field Headquarters, Westboro, MA 01581
Phone: 508-792-7270
Fax: 508-792-7275
sue.langlois@state.ma.us

WILDLIFE SOCIETY
NEW JERSEY CHAPTER
Attn: President, P.O. Box 34
Oceanville, NJ 08231-0034 USA

Founded: NA
Scope: Statewide

Contact(s):
Tracy Casselman, PRESIDENT
Phone: 609-652-1665
Laurance Torok, SECRETARY
139 George St., Lambertville, NJ 08530-1611
Phone: 609-633-6755
James Sciascia, TREASURER
4667 McDermott Rd., Bangor, PA 18013
Phone: 908-735-8975

WILDLIFE SOCIETY
NEW MEXICO CHAPTER
USA

Founded: NA
Scope: Statewide

Contact(s):
James Biggs, PAST PRESIDENT
M887ESH-20, Albuquerque, NM 87545
Phone: 505-665-5714
Fax: 505-667-0731
biggsj@lanl.gov

Gail Tunberg, PAST-PRESIDENT
331 Camino de la Tierra, Corrales, NM 87048-8554
Phone: 505-842-3151
Fax: 505-842-3457
gtunberg@fs.fed.us
Eric Rominger, PRESIDENT
141 Sereno Dr., Santa Fe, NM 87501
Phone: 505-992-8651
e_rominger@gmfsh.state.nm.us

WILDLIFE SOCIETY
NEW YORK CHAPTER
NY USA
Website: cobleskill.edu/nychaptws

Founded: NA
Scope: Statewide

Contact(s):
Nancy Heaslip, NEWSLETTER EDITOR
3750 Skyline Dr., Schenectady, NY 12306
Phone: 518-357-2156
Fax: 518-357-2460
nxheasil@gw.dec.state.ny.us
Mark Lowery, PAST-PRESIDENT
325 Randall Road, Ridge, NY 11961
Phone: 516-444-0350
Fax: 516-444-0349
mdlowery@gw.dec.state.ny.us
Michael Matthews, PRESIDENT
NYS DEC
108 Game Farm Road, Delmar, NY 12054
Phone: 518-457-3720
mjmatthe@gw.dec.state.ny.us
George Mattfeld, PRESIDENT ELECT
Richard Chipman, SECRETARY
USDA/APHIS/WS, 1930 Route 9, Castleton, NY 12033-9653
Phone: 518-477-4837
Fax: 518-477-4899
richard.b.chipman@usda.gov
James Daley, TREASURER
9 Dunbar Rd., Westerlo, NY 12193-2505
Phone: 518-783-5733
jgdaley@gw.dec.state.ny.us
Chuck Dente, VICE-PRESIDENT
14 Marvin Ave., Demar, NY 12054
Phone: 518-478-3009
Fax: 518-478-3004
cxdente@gw.dec.state.ny.us

WILDLIFE SOCIETY
NORTH CAROLINA CHAPTER
NC USA
Website: main.nc.us/nctws/

Founded: NA
Scope: Statewide

Contact(s):
Bethany Olmstead, NEWSLETTER EDITOR
2279-13 Lystra Road, Chapel Hill, NC 27514
Phone: 919-968-1575
awolmstre@mindsprings.com
Gordon Warburton, PAST-PRESIDENT
245 Deep Woods Dr., Marion, NC 28752
Phone: 828-659-8352
Fax: 828-652-8170
warburg@wnclink.com

Gary Marshall, PRESIDENT
marshgd@co.mecklenburg.nc.us
Randall Wilson, PRESIDENT-ELECT
200 Sandy Run, Knightdale, NC 27545
Phone: 919-661-4872
Fax: 919-661-4878
wilsonrc@mail.wildlife.state.nc.us
Stephen Brown, SECRETARY
933 Mudham Rd., Wendell, NC 27591
Phone: 919-846-9332
Fax: 919-846-1261
steve.brown@usace.army.mil
Christopher McGrath, TREASURER
315 Morgan Branch Rd., Leicester, NC 28748
Phone: 828-683-0671
mcgrathc@ncdial.net

WILDLIFE SOCIETY
NORTH DAKOTA CHAPTER
Attn: President, USFWS, 1500 E. Capital Ave.
Bismarck, ND 58501 USA
Website: ndctws.homestead.com/ndctws_home.html

Founded: NA
Scope: Statewide

Contact(s):
Alicia Waters, NEWSLETTER EDITOR
6721 Valley Vista Lane, Bismarck, ND 58501
Phone: 701-250-4242
Fax: 701-250-4590
Tim Phalen, PRESIDENT-ELECT
ND 701-439-2007
phalen@rrt.net
John Schulz, PRESIDENT-PAST
7928 45th St. NE, Devil's Lake, ND 58301-8501
Phone: 701-662-3617
Fax: 701-662-3618
jwschulz@state.nd.us
Greg Hiemenz, SECRETARY-TREASURER
830 N. 34th Street, Bismacrk, ND 58501
Phone: 701-250-4242
Fax: 701-250-4590
ghiemenz@gp.usbr.gov

WILDLIFE SOCIETY
OHIO CHAPTER
OH USA

Founded: NA
Scope: Statewide

Contact(s):
Edward Smith, NEWSLETTER EDITOR
OSU Ext. E. District Ofc.
16714 SR 215, Caldwell, OH 43724
Phone: 740-732-2381
Fax: 740-732-5992
smith.25@osu.edu
Scott Butterworth, PRESIDENT
952 Lima Ave., Findley, OH 45840
Phone: 419-424-5000
Kendra Wecker, SECRETARY
107 Glenmont Ave., Columbus, OH 43214
Phone: 614-265-7043
Fax: 614-262-1143
kendra.wecker@dnr.state.oh.us

Tim Plageman, TREASURER
7403 Twp. Rd. 32, Jenera, OH 45841
Phone: 419-424-5000
Fax: 419-422-4875
tim.plagman@dnr.state.oh.us

WILDLIFE SOCIETY
OKLAHOMA CHAPTER
OK USA

Founded: NA
Scope: Statewide

Contact(s):
Eric Jorgensen, NEWSLETTER EDITOR
Environmental Protection Agency
Robt. S. Kerr Env. Res. Center
919 Kerr Research Drive, Ada, OK 74820
Phone: 580-436-8545
Fax: 580-436-8703
jorgensen.eric@epamail.epa.gov
James Shaw, PAST-PRESIDENT
OK State U.
Dept. of Zoology, Stillwater, OK 74078
Phone: 405-744-9668
Fax: 405-744-7824
shawjh@okstate.edu
Michael Porter, PRESIDENT
Noble Foundation
P.O. Box 2180, Ardmore, OK 73402-2180
Phone: 580-221-7272
Fax: 580-221-7320
mdporter@noble.org
John Skeen, PRESIDENT-ELECT
OK Dept. Wldlf. Cons.
HCR 75 Box 308-12, Broken Bow, OK 74728-9020
Phone: 580-241-7875
okwild@pine-net.com
Jullianne Whitaker-Hoagland, SECRETARY
OK Dept. of Wildlife. Cons.
1801 N. Lincoln Blvd.
P.O. Box 53465, Oklahoma City, OK 73152
Phone: 405-522-0189
Fax: 405-521-6235
jhoagland@odwc.state.ok.us
Jerry Brabander, TREASURER
U.S. Fish and Wildlife Services
10960 S. 241st West Ave., Sapulpa, OK 74066
Phone: 918-581-7458
Fax: 918-581-7467
jerry_brabander@fws.gov

WILDLIFE SOCIETY
OREGON CHAPTER
OR USA
Website: orst.edu/dept/fish_wild/tws

Founded: NA
Scope: Statewide

Publication(s): On Target

Contact(s):
Laura Todd, PAST-PRESIDENT
P.O. Box 50, Rhododendron, OR 97049
Phone: 503-231-6179
Fax: 503-231-6195

Laura_Todd@fws.gov
Jim Thraikill, PRESIDENT
OR Coop. Wldlf. Res. Unit
McKenzie Ecological
45304 Goodpasture Rd., Vida, OR 97488
Phone: 541-687-9076
Fax: 541-687-1065
jimt@pond.net
Cheryl Friesen, PRESIDENT-ELECT
45304 Goodpasture Rd., Vida, OR 97488
Phone: 541-822-7232
Fax: 541-822-7254
cafriesen@msh.com
Katherine Beal, SECRETARY
P.O. Box 429, Lowell, OR 97452
Phone: 541-937-2131
Fax: 541-937-3401
kat.beal@usace.army.mil
Edward Arnett, TREASURER
Weyerhaeuser Co.
OR State U.
321 Richardson Hall, Corvallis, OR 97331
Phone: 541-737-8469
Fax: 541-737-1393
ed.arnett@orst.edu

WILDLIFE SOCIETY
PENNSYLVANIA CHAPTER
Attn: President, 415 E. McCormick Ave.
State College, PA 16801 USA

Founded: NA
Scope: Statewide

Contact(s):
Carolyn Mahan, NEWSLETTER EDITOR
Penn State Altoona
Dept. Bio.
205 Force Bldg., Altoona, PA 16601
Phone: 814-949-5530
Fax: 814-865-3725
cgm2@psu.edu
Michelle Cohen, PAST-PRESIDENT
3490 North Third Street, Harrisburg, PA 17110
Phone: 717-232-0593
Fax: 717-232-0593
mcohen@skellyloy.com
Shayne Hoachlander, PRESIDENT
RD 2 Box 140 Factory Rd., Corry, PA 16407
Phone: 814-664-8867
shoachlander@tbscc.com
Michelle Cohen, PRESIDENT-ELECT
3100 North Third Street, Harrisburg, PA 17110
Phone: 717-232-0593
Fax: 717-232-0593
mcohen@skellyloy.com
J. Benner, SECRETARY
RD 1 Box 87, Liverpool, PA 17045
Phone: 717-787-3706
Fax: 717-783-5109
mbenner@dcnr.state.pa.us
Thomas Hardisky, TREASURER
2621 E. Winter Rd., Loganton, PA 17747
Phone: 570-725-2287
Fax: 570-725-2287
disky@cub.kcnet.org

WILDLIFE SOCIETY
SACRAMENTO-SHASTA CHAPTER
Attn: President, W.M. Beaty & Associates,
P.O. Box 990898
Redding, CA 96099-0898 USA
Website: www.tws-west.org/sac-shastal/index.html

Founded: NA
Scope: Statewide

Contact(s):
Debra Hawk, NEWSLETTER EDITOR
P.O. Box 610, Mammoth Lakes, CA 93546-0610
Phone: 760-872-1134
dhawk@dfg.ca.gov
Michael Bradbury, PAST-PRESIDENT
3251 S Street, Sacramento, CA 95816
Phone: 916-227-7527
Fax: 916-227-7554
mbradbur@water.ca.gov
Robert Carey, PRESIDENT
W.M. Beaty & Associates
P.O. Box 990898, Redding, CA 96099-0898
Phone: 530-243-2783
Fax: 530-243-2900
bobc@sunset.net
Craig Bailey, PRESIDENT-ELECT
1313 Shadowglen Rd., Sacramento, CA 95864-2723
Phone: 916-331-8810
Fax: 916-331-8755
craig_bailey73@hotmail.com
Thomas Boullion, SECRETARY-TREASURER
18005 Willow Dr., Cottonwood, CA 96022
Phone: 530-244-8600
Fax: 530-244-7656
boullion@shasta.com

WILDLIFE SOCIETY
SAN FRANCISCO BAY AREA CHAPTER
CA USA
Website: www.tws-west.org/bayarea/index.html

Founded: NA
Scope: Statewide

Contact(s):
Steven Bobzien, PAST-PRESIDENT
2950 Peralta Oaks Ct.
P.O. Box 5381, Oakland, CA 94605-0381
Phone: 510-635-0138
Fax: 510-635-3478
sbobzien@ebparks.org
David Cook, PRESIDENT
Sonoma County Water Agency, Oakland, CA 94605-0381
Phone: 707-547-1944
dcook@scwa.ca.gov.
John Baas, PRESIDENT-ELECT
210 MacCalvey Dr., Martinez, CA 94553
Phone: 510-335-9778
Fax: 510-335-9778
karthikl@value.net
Jessica Martini-Lamb, SECRETARY-TREASURER
2643 Diablo Street, Napa, CA 94558
Phone: 707-547-1903
Fax: 707-524-3782
jesmartini@hotmail.com

WILDLIFE SOCIETY
SAN JOAQUIN VALLEY CHAPTER
Attn: President, P.O. Box 9622
Bakersfield, CA 93389 USA

Founded: NA
Scope: Statewide

Contact(s):
Brian Cypher, NEWSLETTER EDITOR
Endangered Species Rec. Prog.
P.O. Box 9622, Bakersfield, CA 93389-9622
Phone: 661-398-2201
Fax: 661-398-0549
bcypher@tcsn.net
Brian Cypher, PAST-PRESIDENT
Endangered Species Rec. Prog.
P.O. Box 9622, Bakersfield, CA 93389-9622
Phone: 661-398-2201
Fax: 661-398-0549
bcypher@tcsn.net
Scott Frazer, PRESIDENT
1017 Jefferson Avenue, Los Banos, CA 93635
Phone: 209-826-3508
Fax: 209-826-1445
scott_frazer@fws.gov
Marcia Wolfe, PRESIDENT-ELECT
P.O. Box 10254, Bakersfield, CA 93389
Phone: 661-837-1169
Fax: 661-837-8467
yakimapark@aol.com
Michelle Selmon, SECRETARY
628 W. Euclid Ave., Clovis, CA 93612
mselmon@esrp.org
Christine Horn Job, TREASURER
3517 Sedona Way, Bakersfield, CA 93309
Phone: 661-834-6781
cvanjob@aol.com

WILDLIFE SOCIETY
SOUTH CAROLINA CHAPTER
SC USA
Website: www.tws-west.org/sjvc

Founded: NA
Scope: Statewide

Contact(s):
Karen Dulik, NEWSLETTER EDITOR
kdulik@water.ca.gov
Benjamin Miller, PAST-PRESIDENT
Mulberry Plantation
1904 N. Mulberry Dr., Moncks Corner, SC 29461
Phone: 843-761-5220
Fax: 843-761-5292
Kevin O'Conner, PRESIDENT
Columbia, SC
koconnor@dfg.ca.gov
William Baughman, PRESIDENT-ELECT
P.O. Box 1950, Summerville, SC 29485
Phone: 843-851-4629
Fax: 843-873-2654
wmbaugh@westvaco.com
Paul Jones, SECRETARY-TREASURER
2441 Williston Rd., Aiken, SC 29803
Phone: 803-725-5337
Fax: 803-725-3309
johns@srel.edu

WILDLIFE SOCIETY
SOUTH DAKOTA CHAPTER
SD USA
Website: http://wfs.sdstate.edu/sdtws.htm

Founded: NA
Scope: Statewide

Contact(s):
Carl Madsen, PAST PRESIDENT
2205 North Shore Dr., Brookings, SD 57006
Phone: 605-697-2500
Fax: 605-697-2505
Carl_Madsen@fws.gov
Daniel Hubbard, PRESIDENT
SDSU
P.O. Box 2140B, Brookings, SD 57007
Phone: 605-688-6121
Fax: 605-688-4515
Daniel_Hubbard@sdstate.edu
Paul Coughlin, PRESIDENT ELECT
SD Game, Fish & Parks, Box 218, DeSmet, SD 57231-0218
Phone: 605-773-3658
paul.coughlin@state.sd.us

WILDLIFE SOCIETY
SOUTHERN CALIFORNIA CHAPTER
CA USA
Website: tws-west.org/social/index.html

Founded: NA
Scope: Statewide

Contact(s):
Mari Schroeder, PRESIDENT
mschroeder@chambersgroupine.com
Anna Schroeder, PRESIDENT-PAST
12551 Hinton Way, Santa Ana, CA 92705
Phone: 949-261-5414
Fax: 949-261-8950
mschroeder@chambersgroupinc.com
Kathleen Keane, SECRETARY
5546 E. Parkcrest St., Long Beach, CA 90808
Phone: 310-425-6842
John Stephenson, TREASURER
199 Via Del Cerrito, Encinitas, CA 92024
Phone: 619-436-8340
jstephen/r5_cleveland@fs.fed.us
Brad Blood, VICE-PRESIDENT
12702 Cowley Ave., Downey, CA 90242
Phone: 626-683-3547
Fax: 626-683-3548
pizonyx@aol.com

WILDLIFE SOCIETY
TENNESSEE CHAPTER
TN USA
Website: www.utm.edu/department/gr/agnatres/tn-tws/tn-tws.html

Founded: NA
Scope: Statewide

Contact(s):
Edward Warr, NEWSLETTER EDITOR
Tennessee Wldlf. Res. Agency
Wildlife Division
P.O. Box 40747, Nashville, TN 37204

Phone: 615-781-6613
Fax: 615-781-6654
ewarr@mail.state.tn
David Buehler, PAST-PRESIDENT
U. of Tennessee
Dept. For., Wldlf. & Fish
P.O. Box 1071, Knoxville, TN 37901
Phone: 423-974-7992
Fax: 423-974-4714
dbuehler@utk.edu
Eric Pelren, PRESIDENT
Dept. Agri. & Nat. Res.
114 Brehm Hall, Martin, TN 38238
Phone: 901-587-7263
Fax: 901-587-7968
epelren@utm.edu
Lisa Muller, SECRETARY-TREASURER
Dept. For., Wldlf., & Fish.
P.O. Box 1071, Knoxville, TN 37901
Phone: 423-974-7981
Fax: 423-974-4714
lmuller@utk.edu

WILDLIFE SOCIETY
TEXAS CHAPTER
NM USA
Website: http://www.tctws.org

Founded: NA
Scope: Statewide
Contact(s):
Jim Gallagher, NEWSLETTER EDITOR
183 CR 4604, Dilley, TX 78017
Phone: 830-676-3413
Fax: 830-676-3493
gal4028@vsta.com
Penny Bartnicki, PAST-PRESIDENT
1810 Coulter, Rio Rancho, NM 87124
Phone: 505-248-7465
Fax: 505-248-7471
p_l_bartnicki@yahoo.com
Kirby Brown, PRESIDENT ELECT
4200 Smith School Road, Austin, TX 78744
Phone: 512-389-4395
Fax: 512-389-4398
kirby.brown@tpwd.state.tx.us
Clark Adams, PRESIDENT
TX A&M U.
College of Agriculture
Wildlife & Fish Sci., College Station, TX 77843-2258
Phone: 409-845-8824
Fax: 409-845-3786
cadams@wfscgate.tamu.edu
Tamara Trail, SECRETARY
7887 U.S. Hwy 87N., San Angelo, TX 76901-9782
Phone: 915-653-4576
Fax: 915-655-7791
t-trail@tamu.edu
Donald Davis, TREASURER
TX A&M U.
Dept. Vet. Pathology, College Station, TX 77843
Phone: 409-845-5174
Fax: 409-862-1088
ddavis@cvm.tamu.edu

Scott Henke, VICE PRESIDENT
TX A&M U.-Kingsville
Campus Box 218, Kingsville, TX 78363-8202
Phone: 361-593-3689
Fax: 361-593-3788
kfseh00@tamuk.edu

WILDLIFE SOCIETY
UTAH CHAPTER
Attn: President, Bureau of Land Management,
318 N. 100 E.
Kanab, UT 84741 USA

Founded: NA
Scope: Statewide

Contact(s):
Stanley Beckstrom
NEWSLETTER EDITOR
4311 S. 4625 W., Salt Lake City, UT 84120-4931
Phone: 435-865-6112
nrdwr.sbeckstr@state.ut.us
Harry Barber, PAST-PRESIDENT
BLM
318 N. 100 E., Kanab, UT 84741
Phone: 435-644-4311
Fax: 435-644-2672
hbarber@ut.blm.gov
Kathleen Paulin, PRESIDENT
Vernal Ranger District
Ashley National Forest
355 North Vernal Avenue, Vernal, UT 84078
Phone: 435-781-5160
kpaulin@fs.fed.us
Lisa Church, PRESIDENT-ELECT
1381 S. Ford, Kanab, UT 84741
Phone: 435-644-4600
Fax: 435-644-4620
lchurch@ut.blm.gov
Stanley Beckstrom, SECRETARY
4311 S. 4625 W., Salt Lake City, UT 84120-4931
Phone: 435-865-6112
nrdwr.sbeckstr@state.ut.us
Randall Thacker, TREASURER
P.O. Box 337, Altamont, UT 84001
Phone: 435-454-3081
mrdwr/rtjacler@state.ut.us

WILDLIFE SOCIETY
VIRGINIA CHAPTER
VA USA

Founded: NA
Scope: Statewide

Contact(s):
Jack Gwynn, NEWSLETTER EDITOR
2503 Brunswick Rd., Charlottesville, VA 22903
Phone: 804-295-4681
Fax: 804-975-1005
jackgwynn@aol.com
Jesse Overcash, PAST-PRESIDENT
110 Southpark Dr., Blacksburg, VA 24060
Phone: 540-522-4641
Fax: 540-552-4376
jovercas@vt.edu

Bruce Lemmert, PRESIDENT
21 S. Church St., Lovettsville, VA 22080
Phone: 540-822-4219
blemmert@dgif.state.va.us
Lisa Sausville, SECRETARY
966 Rt. 17 W., Addison, VT 05491
Ralph Keel, TREASURER
1232 Geranium Crescent, Virginia Beach, VA 23456
Phone: 757-986-3706
Fax: 757-986-2353
r5rw_gdsnwr@mail.fws.gov
Jefferson Waldon, VICE-PRESIDENT
VPI & SU, Fish & Wildlife Info Exchange, 203 W. Roanoke
St., Blacksburg, VA 24061
Phone: 540-231-7348
Fax: 540-231-7019
fwiexchg@vt.edu

WILDLIFE SOCIETY
WASHINGTON CHAPTER
WA USA
Website:
http://www.washingtonwildlifesoc.org/ns/default_ns.htm

Founded: 1966
Membership: 200
Scope: Statewide

Contact(s):
Kenneth Bevis, NEWSLETTER EDITOR
Yakima Indian Nation
Wildlife
P.O. Box 151, Toppenish, WA 98948-0151
beviskrb@dfw.wa.gov
Paul Fielder, PRESIDENT
1633 Concord Place, Wenatchee, WA 98801
Phone: 509-663-8121
Fax: 509-664-2338
paul@chelanpud.org
Con Utzinger, PRESIDENT ELECT
44691 Baker Lake Road, Concrete, WA 98237
Phone: 360-853-7806
Fax: 360-853-7806
utzinger@fidalgo.net
John Lehmkuhl, PRESIDENT-PAST
1133 N. Western Ave., Wenatchee, WA 98801
Phone: 509-662-4315
jlehmkhul/r6pnw_wenatchee@fs.fed.us
Ann Sprague, SECRETARY
Box 188, Twisp, WA 98856
Phone: 509-997-2131
Fax: 509-997-9770
sprague@nethow.com
Catherine Raley, TREASURER
Forestry Sciences Lab
3625 93rd Ave. SW, Olympia, WA 98502
Phone: 360-753-7686
craley@fs.fed.us

WILDLIFE SOCIETY
WEST VIRGINIA CHAPTER
WV USA

Founded: NA
Scope: Statewide

Contact(s):
Shawn Head, PAST-PRESIDENT
WV DNR
P.O. Box 67, Elkins, WV 26241
Phone: 304-637-0245
Fax: 304 637-0250
Jim Fregonara, PRESIDENT
210 Boundary Ave., Elkins, WV 26241
Phone: 304-637-0245
Fax: 304-637-0250
James Anderson, SECRETARY-TREASURER
WV DNR-Wildlife
2006 Robert C. Byrd Drive, Beckley, WV 25801
Phone: 304-293-2941
Fax: 304-293-2441
jander25@wvu.edu
Christopher Ryan, VICE-PRESIDENT
P.O. Box 73, Shirley, WV 26434
Phone: 304-758-2681

WILDLIFE SOCIETY
WISCONSIN CHAPTER
WI USA

Founded: NA
Scope: Statewide

Publication(s): Wisconsin Association for Environmental
Education Bulletin

Contact(s):
Alan Crossley, NEWSLETTER EDITOR
459 Sidney St., Madison, WI 53703
Phone: 608-275-3242
Fax: 608-275-3338
crossa@dnr.state.wi.us
Gerald Bartelt, PAST PRESIDENT
6315 Clovernook Rd., Middletown, WI 53562-3824
Phone: 608-221-6344
Fax: 608-221-6353
barteg@dnr.state.wi.us
Jonathan Gilbert, PRESIDENT
25350 Fischer Rd., Ashland, WI 54806
Phone: 715-682-6619
Fax: 715-682-9294
jgilbert@glifwc.org
Gary Zimmer, SECRETARY-TREASURER
P.O. Box 116, Laona, WI 54541
Phone: 715-674-4481
Fax: 715-276-3594
gzimmer@fs.fed.us

WILDLIFE SOCIETY
WYOMING CHAPTER
Attn: President, 260 Buena Vista
Lander, WY 82520 USA
Website: http://www.wyotws.org

Founded: NA
Scope: Statewide

Contact(s):
Lori Hunter, NEWSLETTER EDITOR
P.O. Box 64, Elk Mtn, WY 82324
Phone: 307-348-7464
lhunter@union-tel.com

Mark Hinschberger, PRESIDENT
U.S. Forest Service
P.O. Box 186, Dubois, WY 82513
Phone: 307-455-2466
mhinschb/r2_shashone@fs.fed.us
Stan Anderson, PRESIDENT ELECT
Wyoming Cooperative Fish and Wildlife Research Unit,
Dept. of Zoology & Physiology, Biological Sciences Bldg.,
Room 419, Laramie, WY 82071
Phone: 307-455-2466
anderson@uwyo.edu
Tom Thorne, PRESIDENT-PAST
WY Game & Fish Dept., 5400 Bishop Blvd., Cheyenne,
WY 82006
Phone: 307-777-4591
Fax: 307-777-4602
tthorn1@state.wy.us
Vicki Herren, SECRETARY
1338 Kimberly Ave., Rock Springs, WY 82901
Phone: 307-352-0236
Fax: 307-352-0329
vicki_herren@blm.gov
Timothy Thomas, TREASURER
WY Game & Fish Dept., P.O. Box 6249, Sheridan, WY 82801
Phone: 307-672-8003
Fax: 307-767-0594
tthoma@state.wy.us

WILDLIFE SOCIETY, THE

5410 Grosvenor Ln.
Bethesda, MD 20814-2197 USA
Phone: 301-897-9770　　Fax: 301-530-2471
E-mail: tws@wildlife.org
Website: www.wildlife.org

Founded: 1937
Membership: 9,000
Scope: National

Description: International scientific and educational organization
of professionals and students engaged in wildlife research,
management, education, and administration. Dedicated to
sound stewardship of wildlife resources and the environments
upon which wildlife and humans depend; undertakes an active
role in preventing human-induced environmental degradation;
increases awareness and appreciation of wildlife values; and
seeks the highest standards in all activities of the wildlife
profession.

Publication(s): Wildlife Monographs, Wildlife Society Bulletin,
The Wildlifer, The Journal of Wildlife Management

Keyword(s): Nongame Wildlife, Renewable Resources, training

Contact(s):

Gerald Kobriger, CENTRAL MOUNTAINS & PLAINS
SECTION REPRESENTATIVE
225 30th Ave., SW, Dickinson, ND 58601
Phone: 701-227-7431
gkobrige@state.nd.us
Harry Hodgdon, EXECUTIVE DIRECTOR
Gary Potts, NORTH CENTRAL SECTION
REPRESENTATIVE
Northwest Experimental Station, University of Minnesota,
ARC Building, Crookston, MN 56716
Phone: 218-281-8129
dsvedars@mail.crk.umn.edu

John Organ, NORTHEAST SECTION REPRESENTATIVE
P.O. Box 45, Buckland, MA 01338-0045
Phone: 413-253-8501
john_organ@mail.fws.gov
Winifred Kessler, NORTHWEST SECTION
REPRESENTATIVE
Dept. of Fisheries & Wildlife, Oregon State University, 104
Nash Hall, Corvallis, OR 97331-3803
Phone: 541-737-1953
edgew@ucs.orst.edu
Len Carpenter, PAST PRESIDENT
Wildlife Management Institute, 4015 Cheney Dr., Fort Collins,
CO 80526
Phone: 970-223-1099
lenc@verinet.com
Diana Hallett, PRESIDENT
MO Dept. of Conservation, 1110 College Ave.,
Columbia, MO 65201
Phone: 573-882-9880
halled@mail.conservation.state.mo.us
Robert Warren, PRESIDENT ELECT
Warnell School of Forest Resources, University of Georgia,
Athens, GA 30602-2152
Phone: 706-542-6474
warren@smokey.forestry.uga.edu
Sandra Staples-Bortner
PROGRAM DIRECTOR
18214 NE 125th Way, Brush Prairie, WA 98606
Phone: 360-253-4611
twsssb@aol.com
Richard Lancia, SOUTHEAST SECTION REPRESENTATIVE
Robert Brown, SOUTHWEST SECTION REPRESENTATIVE
Dept. of Wildlife & Fisheries Sciences, 210 Nagle Hall, Texas
A&M University, College Station, TX 77843-2258
Phone: 409-845-1261
rdbrown@tamu.edu
Daniel Decker, VICE PRESIDENT
Marti Kie, WESTERN SECTION REPRESENTATIVE
CA Dept. of Forestry, P.O. Box 670, Santa Rosa, CA 95401
Phone: 707-576-2937
brad_valentine@fire.ca.gov
Thomas Franklin, WILDLIFE POLICY DIRECTOR

WILDLIFE SOCIETY

IOWA CHAPTER
124 Science Hall II, Iowa State University
Ames, IA 50011 USA
Phone: 515-294-7429　　Fax: 515-294-7874
E-mail: jlpease@iastate.edu
Website: www.wildlife.org

Founded: NA
Scope: Statewide

Contact(s):

Jamie Edwards, NEWSLETTER EDITOR
3011 Nelson Ct. SE., 55094, MN 55904
Phone: 507-280-5070
Fax: 507-285-7144
noodlerun@aol.com
James Pease, PAST-PRESIDENT
Iowa State University
124 Science II, Ames, IA 50011
Phone: 515-294-7429
Fax: 515-294-7874

Donald Sievers, PRESIDENT
109 W. Wilcoxway, Jefferson, IA 50129
Phone: 515-747-8383
Fax: 515-747-3951
dsiever@pionet.net
Donald Pheiffer, PRESIDENT-ELECT
110 Lake Darling Road, Brighton, IA
Phone: 319-653-4912
Todd Bogenschutz, SECRETARY-TREASURER
Wildlife Res. Sta.
1436 255 St., Boone, IA 50036
Phone: 515-432-2823
Fax: 515-432-2835

WILDLIFE WAYSTATION

14831 Little Tujunga Canyon Rd.
Angeles National Forest, CA 91342-5999 USA
Phone: 818-899-5201   Fax: 818-890-1107
Website: www.waystation.org

Founded: 1969
Scope: National

Description: A southern California nonprofit refuge providing medical care, refuge, rehabilitation, and placement services for over 4,000 wild and exotic animals annually. Public tours and educational programs available.

Publication(s): Wild Proofing the Human Habitat (brochure), Wildlife Waystation (newsletter)

Keyword(s): Endangered Species, Environmental and Conservation Education, Mammals, Raptors, accreditation

Contact(s):
Martine Colette, FOUNDER AND PRESIDENT

WILSON ORNITHOLOGICAL SOCIETY

Wilson Ornithological Society
Museum of Zoology
University of Michigan
Ann Arbor, MI 48109 USA
Phone: 508-543-8988   Fax: 508-286-8278
E-mail: wedavis@bu.edu
Website: http://www.ummz.lsa.umich.edu/birds/wos.html

Founded: 1888
Membership: 7
Scope: National

Description: To advance the science of ornithology and to secure cooperation in measures tending to this end.

Publication(s): Wilson Bulletin, The

Keyword(s): Biodiversity, Birds, training, Research, education, Grants, Ecology, Environment, Zoology, Wildlife Rehabilitation

Contact(s):
Robert Beason, EDITOR
Dept. of Biology, State University of New York, 1 College Circle, Geneseo, NY 14454
Phone: 716-245-5310
Fax: 716-245-5007
wilsonbull@uno.cc.geneseo.edu
William Davis, FIRST VICE PRESIDENT
College of General Studies, 871 Commonwealth Avenue, Boston University, Boston, MA 02215
Phone: 617-353-2886
Fax: 617-353-5868
wedavis@bu.edu

John Kricher, PRESIDENT
Dept. of Biology, Wheaton College, Norton, MA 02766
Phone: 508-285-8200
jkricher@wheatonma.edu
Charles Blem, SECOND VICE PRESIDENT
Dept. of Biology, 816 Park Ave./P.O. Box 842012; Virginia Commonwealth University, Richmond, VA 23284-2012
Phone: 804-828-1562
Fax: 804-828-0503
cblem@cabell.vcu.edu
John Smallwood, SECRETARY
Dept. of Biology, Montclair State University, Upper Montclair, NJ 07043
Phone: 973-655-5345
Fax: 973-655-7047
smallwood@saturn.montclair.edu
Doris Watt, TREASURER
Dept. of Biology, Saint Mary's College, Notre Dame, IN 46556-5001
Phone: 219-284-4668
Fax: 219-284-4716
dwatt@jade.saintmarys.edu

WINCHESTER NILO FARMS

Olin Corporation, 427 N. Shamrock
E. Alton, IL 62024 USA
Phone: 618-258-3133   Fax: 618-258-2370

Founded: NA
Membership: 10
Scope: National

Contact(s):
Roger Jones, MANAGER
Phone: 618-466-0613

WINDSTAR FOUNDATION, THE

7 Avenida Vista Grande 304
Santa Fe, NM 87505 USA
Fax: 970-963-1463
E-mail: windstar@rof.net
Website: www.wstar.org

Founded: 1976
Scope: National

Description: A nonprofit organization co-founded by John Denver and Tom Crum. Windstar works to inspire individuals to make responsible choices and take direct action to achieve a peaceful and environmentally sustainable future.

Keyword(s): Environmental and Conservation Education, Sustainability

Contact(s):
Cheryl Charles, CHAIRMAN OF BOARD OF TRUSTEES
Jeanie Tomlinson, LIAISON
Phone: 970-963-5534
Beth Miller, SECRETARY AND TREASURER

WISCONSIN ASSOCIATION FOR ENVIRONMENTAL EDUCATION, INC.

233 Nelson Hall, UWSP
Stevens Point, WI 54481 USA
Phone: 715-346-2796   Fax: 715-346-3819
E-mail: waee@uwsp.edu
Website: www.uwsp.edu/waee

Founded: 1974

Membership: 500
Scope: Statewide

Description: Promotes environmental education in schools and other institutions and organizations in Wisconsin.

Publication(s): EE News

Keyword(s): Environmental Law, Environmental and Conservation Education, Lakes, Water Pollution Management, Exotic species, Aquatic nuisance species, Education

Contact(s):
Christy Allar, ADMINISTRATIVE ASSISTANT
Geoffrey Bishop, CHAIR
Paul Denowski, CO-CHAIRMAN

WISCONSIN ASSOCIATION OF LAKES (WAL)

P.O. Box 126
Stevens Point, WI 54481-0126 USA
Phone: 608-662-0923   Fax: 715-346-3624
E-mail: info@wisconsinlakes.org
Website: www.wisconsinlakes.org

Founded: 1980
Scope: Statewide

Description: WAL is a coalition of 287 lake management organizations, as well as hundreds of individual members. The organization is dedicated to the protection of lake ecosystems in Wisconsin. WAL works closely with the Wisconsin Department of Natural Resources and University Extension in the Wisconsin Lakes Partnership.

Publication(s): Lake Connection, The

Keyword(s): Aquatic Habitats, Protected Areas, Environmental Protection, Lakes, Land Use Planning, Sustainable Ecosystems, Watersheds, Pollution Prevention, Wetlands

Contact(s):
Hal Krueger, COMMUNICATIONS AND DEVELOPMENT COORDINATOR
Donna Sefton, EXECUTIVE DIRECTOR
Debra Sweeney, MEMBERSHIP COORDINATOR
Jim Burgess, PRESIDENT
Phone: 608-257-4443
jeburg@aol.com
Judy Jooss, SECRETARY
Phone: 414-877-9301
jjooss@techheadnet.com
John Seibel, TREASURER
Phone: 715-479-4714
jpsmis@nnex.net
Susan Tesarik, WATER CLASSIFICATION OUTREACH COORDINATOR

WISCONSIN B.A.S.S. CHAPTER FEDERATION

Attn: President, 6503 Lani Ln.
McFarland, WI 53558 USA
Phone: 608-838-3040   Fax: 608-838-3040
Website: www.swiftsite.com/wsbf

Founded: NA
Scope: Statewide

Description: An organization of Bassmaster chapters, affiliated with the Bass Anglers Sportsman Society, organized to fight pollution, assist state and national conservation agencies in their efforts, and teach young people good conservation practices. Dedicated to the realistic conservation of our water resources.

Publication(s): Wisconsin Bass News

Contact(s):
Kevin Fassbind, CONSERVATION DIRECTOR
12 Bel Aire, Madison, WI 53713
Phone: 608-224-0029
fassbind@hotmail.com
Chuck Rolfsmeyer, PRESIDENT

WISCONSIN LAND AND WATER CONSERVATION ASSOCIATION

One Point Place, Suite 101
Madision, WI 53719 USA
Phone: 608-833-1833   Fax: 608-833-7179
E-mail: wlwca@exectc.com
Website: www.execpc.com/~wlwca

Founded: NA
Membership: 72
Scope: Regional

Description: Wisconsin Land and Water Conservation Association is a 501(c) (3) non-profit organization representing Wisconsin's 72 county land conservation committees and departments, assisting them with the protection, enhancement and sustainable use of Wisconsin's natural resources, and representing them through education and government interaction.

Publication(s): Thursday Note

Keyword(s): Communications, Acid Rain, Public Lands, Air Quality, Biodiversity, Birds, Chemical Pollution Control, Biotechnology, Coasts, Conservation, EcoAction, Ecology, Energy, Conservation Tillage, Endangered Species

Contact(s):
Roger Hahn, BOARD MEMBER
705 Pease St., Augusta, WI 54722
Phone: 715-286-5343
Rebecca Baumann, EXECUTIVE DIRECTOR
One Point Place, Ste 101, Madison, WI 53719-2809
Phone: 608-833-1833
Fax: 608-833-7179
wlwca3@execpc.com
Marvin Fox, PRESIDENT
N2538 Cty Road, J, Kaukauna, WI 54130
Phone: 414-766-3242
Robert Washkuhn, VICE PRESIDENT
W8225 Sand Rd, Shell Lake, WI 54871
Phone: 715-468-7657

WISCONSIN PARK AND RECREATION ASSOCIATION

0001-C Northway
Greendale, WI 53129 USA
Phone: 414-423-1210   Fax: 414-423-1296
E-mail: wpra@execpc.com
Website: www.nrta.org/member/wpra/

Founded: NA
Membership: 1200
Scope: Statewide

Description: A nonprofit organization, affiliated with the National Recreation and Park Association, working with other groups and organizations to achieve the best in park services and recreational opportunities.

Publication(s): P.R. Monthly Newsletter, Impact Magazine

Contact(s):
Steve Thompson, EDITOR
7000 Greenway, Suite 201, Greendale, WI 53129
Phone: 414-423-1210
Roger Kist, PRESIDENT
Washington County

WISCONSIN SOCIETY FOR ORNITHOLOGY, INC., THE

5188 Bittersweet Ln.
Oshkosh, WI 54901 USA
Phone: 920-233-1973
Website: www.uwgb.edu\birds\wso\

Founded: 1939
Membership: 1500
Scope: Regional

Description: To stimulate interest in and promote the study of birds in Wisconsin for a better understanding of their biology and basis for their preservation.

Publication(s): Badger Birder, Passenger Pigeon

Keyword(s): Environmental Law, Environmental and Conservation Education, Waterfowl, Wetlands, training, Birds

Contact(s):
Mary Uttech, EDITOR
Phone: 262-675-6482
muttech@asq.org
R. Highsmith, EDITOR
702 Schiller Ct., Madison, WI 53704
Phone: 608-242-1168
William Brooks, PRESIDENT
Ripon College Dept. of Biology, Ripon, WI 54971
Bettie Harriman, PUBLICITY CHAIR
5188 Bittersweet Ln., Oshkosh, WI 54901
Phone: 920-233-1973
Jane Dennis, SECRETARY
138 S. Franklin Ave., Madison, WI 53705-5248
Phone: 608-231-1741
Alex Kailing, TREASURER
W330 N8275 W. Shore Dr., Hartland, WI 53029
Phone: 414-966-1072
Daryl Christenson, VICE PRESIDENT

WISCONSIN WATERFOWL ASSOCIATION, INC.

78th Enterprise Rd. Ste. A
Delafield, WI 53018-0496 USA
Phone: 262-646-5926   Fax: 262-646-5949
E-mail: h2ofowl@powercom.net
Website: www.wisducks.org

Founded: 1983
Membership: 7000
Scope: Statewide

Description: A statewide nonprofit environmental/educational organization that establishes, promotes, assists, and contributes to conservation, restoration, and management of Wisconsin wetlands to perpetuate waterfowl and wildlife. Represents waterfowl enthusiasts via a unified statewide voice on Wisconsin migratory bird hunting regulations and conservation legislation benefiting the protection of wetlands. Educational programs and waterfowl hunting seminars.

Publication(s): Wisconsin Waterfowl

Keyword(s): Wetlands, Conservation, Waterfowl

Contact(s):
Kelcy McCarthy, ADMINISTRATOR
Jeff Bord, EXECUTIVE DIRECTOR
jbord@powercom.net
Dennis Tetzlaff, PRESIDENT
508 E. South St., Beaverdam, WI 53916
Phone: 920-887-2686
Fax: 920-885-9399

 ## WISCONSIN WILDLIFE FEDERATION

2036 W. 9th Street
Oshkosh, WI 54904 USA
Phone: 920-235-9136   Fax: 920-235-6030
E-mail: wiwf@execpc.com
Website: execpc.com/-wiwf

Founded: NA
Membership: 5000
Scope: Statewide

Description: A representative statewide organization, affiliated with the National Wildlife Federation, dedicated to the protection and enhancement of wildlife and its habitat through public education and government interaction.

Publication(s): Wisconservation

Keyword(s): Aquatic Habitats, Wildlife, Renewable Resources, Water Pollution Management, Wetlands

Contact(s):
Daniel Gries, EDITOR
Ruth Lee, EDUCATION PROGRAMS CONTACT
James Weishan, PRESIDENT AND ALTERNATE REPRESENTATIVE
Martha Kilishek, REPRESENTATIVE
Russell Hitz, TREASURER

WISCONSIN WOODLAND OWNERS ASSOCIATION

P.O. Box 285
Stevens Point, WI 54481-0285 USA
Phone: 715-346-4798   Fax: 715-346-4821
Website: www.wisconsinwoodlands.org

Founded: 1979
Membership: 2000
Scope: Statewide

Description: A statewide organization affiliated with the National Woodland Owners Association, established to advance the interests of woodland owners and the cause of forestry in Wisconsin.

Publication(s): Woodland Management, WWOA Seedlings

Keyword(s): Forests and Forestry

Contact(s):
Timothy Eisele, EDITOR
Phone: 608-233-2904
Nancy Bozek, EXECUTIVE DIRECTOR
Virgil Kopitske, PRESIDENT
Phone: 715-878-4331
Beverly Schendel, SECRETARY
Phone: 612-881-7610
Dale Lightfuss, TREASURER
Phone: 920-244-7668
Marv Meier, VICE PRESIDENT
Phone: 608-271-9718

WISCONSIN DEPARTMENT OF NATURAL RESOURCES

101 S. Webster St. P.O. Box 7921
Madison, WI 53707-7921 USA
Phone: 608-266-2121   Fax: -267-9380
Website: www.dnr.state.wi.us
Scope: National

Contact(s):
Darrell Bazzell, SECRETARY

WOLF EDUCATION AND RESEARCH CENTER

P. O. Box 217
Winchester, ID 83555 USA
Phone: 208-924-6960   Fax: 208-924-6959
E-mail: wolfsta@mce.net
Website: www.wolfcenter.org

Founded: 1992
Membership: 5000
Scope: National

Description: The Wolf Education and Research Center is dedicated to providing public information, education, and research concerning endangered species, with an emphasis on the gray wolf, its habitat and ecosystem in the Northern Rocky Mountain region. Our efforts seek to improve public awareness of endangered and threatened species in the area and to develop, in concert with regional cultures and residents, ways to coexist with these species.

Publication(s): Wild Wolves Sponsorship Newsletter, Educational Track of Wolf - monthly pub, Wolf Education and Research Center Membership Newsletter, Sawtooth Pack Sponsorship Newsletter

Keyword(s): Endangered Species, Predators, Cultural Preservation, Environmental and Conservation Education, Research, Outdoor Recreation, Biodiversity, Training, Nongame Wildlife, National Parks, Wolves

Contact(s):
Douglass Christensen, PRESIDENT
Phone: 208-726-1982
1dmc@sunvalley.net
Sally Farrar, SECRETARY
Phone: 208-336-6562
Fax: 208-384-0540
Roy Farrar, VICE PRESIDENT
Phone: 208-384-0540
wolfsta@mce.net

WOLF HAVEN INTERNATIONAL

3111 Offut Lake Rd.
Tenino, WA 98589 USA
Phone: 360-264-4695   Fax: 360-264-4639
E-mail: info@wolfhaven.org
Website: www.wolfhaven.org

Founded: NA
Membership: 3500
Scope: International, National

Description: The organization's mission is "Working for Wolf Conservation". With the intent on making a difference, all of Wolf Haven's activities have been undergoing changes that seek to strenghten our involvement in research and education.

Publication(s): Wolf Tracks

Keyword(s): Endangered Species

Contact(s):
Julie Palmquist, COMMUNICATIONS DIRECTOR
julie@wolfhaven.org
Rick Castellano, EXECUTIVE DIRECTOR
Rick Schaefer, PRESIDENT
Dana Maher, TREASURER

WOMEN'S ENVIRONMENT AND DEVELOPMENT ORGANIZATION (WEDO)

355 Lexington Avenue, 3rd Floor
New York, NY 10017 USA
Phone: 212-973-0325   Fax: 212-973-0335
E-mail: wedo@iwedo.org
Website: www.wedo.org

Founded: 1990
Scope: National

Description: On January 27, 1995, Women USA Fund, Inc. changed its name to WEDO. The organization is an international advocacy network actively working to transform society to achieve social, political, economic, and environmental justice for all through the empowerment of women, in all their diversity, and through their equal participation with men in decision-making from grassroots to global arenas.

Publication(s): News and Views (contact WEDO for a comprehensive list)

Keyword(s): Biotechnology, Environmental and Conservation Education, Population Growth, Renewable Resources

Contact(s):
June Zeitlin, EXECUTIVE DIRECTOR
Jocelyn Dow, PRESIDENT
jocelyndow@hotmail.com
Elizabeth Calvin, SECRETARY
Brownie Ledbetter, TREASURER
Thais Corral, VICE PRESIDENT
Bisi Ogunleye, VICE PRESIDENT

WOMEN'S SHOOTING SPORTS FOUNDATION

4620 Edison Ave., Suite C
Colorado Springs, CO 80915 USA
Phone: 719-638-1299   Fax: 719-638-1271
E-mail: wssf@worldnet.att.net
Website: http://www.wssf.org/

Founded: 1993
Membership: 3,500
Scope: National

Description: The Women's Shooting Sports Foundation is a national, nonprofit membership organization offering an ongoing series of programs to expand shooting opportunities for women.

Publication(s): Women's Resource List, The, Outdoors for Women

Keyword(s): Hunting, Training, Waterfowl, Women in the Environment

Contact(s):
Shari Legate, EXECUTIVE DIRECTOR

WORLD BIRD SANCTUARY

Box 270270
St. Louis, MO 63127 USA
Phone: 636-938-6193   Fax: 636-938-9464
E-mail: info@worldbirdsanctuary.org
Website: www.worldbirdsanctuary.org

Founded: 1977
Membership: 2000
Scope: International

Description: (formerly The Raptor Rehabilitation and Propagation Project Inc.) The WBS was established by Walter C. Crawford, Jr. near St. Louis, Missouri. It is a nonprofit, tax-exempt organization whose mission is to preserve the earth's biological diversity and to secure the future of threatened bird species in their natural environments. We work to fullfill that mission through education, propagation, and rehabilitation. We also have a hands-on internship program.

Publication(s): Methods of Feather Replacement in Birds of Prey, Techniques for Artificial Incubation and Hand-rearing of Raptors, Stress in Captive Birds of Prey, Mews News

Keyword(s): accreditation, Endangered Species, Biodiversity, Birds, Environment, Raptors

Contact(s):
Marion Ernst, EDITOR
Walter Crawford, EXECUTIVE DIRECTOR
Phone: 636-938-6193
Susan Poling, PRESIDENT
Mary Roth, SECRETARY
Dennis Breite, TREASURER
Tom Rollins, VICE PRESIDENT

WORLD FORESTRY CENTER

4033 SW Canyon Rd.
Portland, OR 97221 USA
Phone: 503-228-1367   Fax: 503-228-4608
Website: www.worldforrest.org

Founded: 1966
Scope: National

Description: The World Forestry Center is a nonprofit organization promoting a greater appreciation and understanding of the world's forests and related natural resources. The Center operates a forestry museum adjacent to the Hoyt Arboretum, conference facilities, an international institute, and an 80-acre demonstration forest and outdoor education site. Public tours and classes, school programs, exhibits and special events, conferences, curriculum materials, and publications are available.

Publication(s): Branching Out Newsletter, Forest Education Program Guide

Keyword(s): Museum, Environmental and Conservation Education, Forests and Forestry, education, Renewable Resources

Contact(s):
Rick Zenn, EDUCATION DIRECTOR
Dennis Dykstra, PRESIDENT

WORLD PAL (WORLD POPULATION ALLOCATION LIMITED INC.)

52 Stevens St., Suite 1100
White Plains, NY 10606 USA
Phone: 914-684-6539   Fax: 914-684-9607
Website: www.worldpal.org

Founded: NA
Membership: 25
Scope: International

Description: World Pal is concerned with increasing the awareness that young people have for their environment. World Pal conducts educational awareness programs on the decks of a newly-built, four-masted barquentine. The programs are offered in many ports throughout Latin America, from the Rio Grande to Patagonia. While on board, passengers are encouraged to participate in the educational program.

Publication(s): Information available upon request

Keyword(s): Environmental Protection, Environment, Population Growth, Sustainable Development, Ecosystems

Contact(s):
D. Anderson, EXECUTIVE DIRECTOR

WORLD PARKS ENDOWMENT INC.

1616 P St., NW, 200
Washington, DC 20036 USA
Phone: 202-939-3803
E-mail: worldparks@juno.com
Website: www.worldparks.org

Founded: 1988
Scope: International

Description: World Parks Endowment, Inc. is a unique organization which acquires land in the rain forest and other critical sites for biological diversity. It provides funds for park management of tropical rain forests and other ecosystems of great conservation importance, and has developed projects in over 12 countries, including the Sierra de las Minas Biosphere Reserve in Guatemala and the Bilsa Reserve in Ecuador.

Publication(s): Annual Report

Keyword(s): Land Preservation, Land Purchase, Biodiversity, Tropical Biodiversity and Conservation, Rainforests

Contact(s):
Daniel Katz, CHAIRMAN AND PRESIDENT
Byron Swift, SECRETARY AND EXECUTIVE DIRECTOR
Roger Pasquier, VICE PRESIDENT AND TREASURER

WORLD RESOURCES INSTITUTE

10 G St., NE, Suite 800
Washington, DC 20002 USA
Phone: 202-729-7600   Fax: 202-729-7610
E-mail: front@wri.org
Website: www.wri.org

Founded: 1982
Membership: 135
Scope: International

Description: A policy research center created with funding from the John D. and Catherine T. MacArthur Foundation and others, to help governments, international organizations, the private sector, and others address vital issues of environmental integrity, natural resource management, economic growth, and international security.

Publication(s): Research Report Series, World Resources Report, Policy Studies Series

Keyword(s): Air Quality and Pollution, Forests and Forestry, Greenhouse Effect/Global Warming, Renewable Resources, Sustainable Development

Contact(s):
Jonathan Lash, PRESIDENT
jlash@wri.org
Matthew Arnold, SENIOR VICE PRESIDENT/COO
Marjorie Beane, V.P. FOR ADMINISTRATION AND CFO

WORLD SOCIETY FOR THE PROTECTION OF ANIMALS (WSPA)

34 Deloss Street
Framingham, MA 01702 USA
Phone: 508-879-8350   Fax: 508-620-0786
E-mail: wspa@wspausa.com
Website: http://www.wspa-americas.org

Founded: 1981
Scope: National

Description: The World Society for the Protection of Animals (WSPA) aims to promote the protection of animals, to prevent cruelty to animals, and to relieve animal suffering in every part of the world. For decades, our tools have been hands-on field work, along with humane education and legislative action as we strive for the humans treatment and safety of animals. WSPA has over 350 members societies in 75 nations that provide support for our many animal protection initiatives.

Publication(s): WSPA World, WSPA Campaign News, Annual Report, By-Laws, Policy Statement, Animals International

Keyword(s): Endangered Species, Mammals, Marine Mammals, Whale, Dolphin, Seal, Nongame Wildlife, accreditation

Contact(s):
Andrew Dickson, CHIEF EXECUTIVE
John Walsh, INTERNATIONAL PROJECTS DIRECTOR
Paul Irwin, PRESIDENT
Murdaugh Madden, SECRETARY
Robert Cummings, TREASURER
Peter Davies, VICE PRESIDENT
Hans Haering, VICE PRESIDENT

WORLD WILDLIFE FUND

1250 24th St., NW
Washington, DC 20037 USA
Phone: 202-243-4800   Fax: 202-293-9211
E-mail: archer@wwfus.org
Website: www.worldwildlife.org

Founded: 1061
Scope: International

Description: WWF is the largest private U.S. organization working worldwide to protect wildlife and wildlands—especially in the tropical forests of Latin America, Asia, and Africa. WWF has helped create and protect more than 450 national parks and nature reserves; supports scientific investigations; monitors international trade in wildlife; promotes ecologically-sound development; assists local groups to take the lead in needed conservation projects; and seeks to influence public opinion and the policies of governments and private institutions to promote conservation of the earth's living resources.

Publication(s): FOCUS

Keyword(s): Endangered Species, Urban Environment, training

Contact(s):
Bruce Bunting, ASIA, CONSERVATION FINANCE AND SPECIES CONSERVATION VICE PRESIDENT
Roger Sant, CHAIRMAN OF BOARD
Edward Bass, CHAIRMAN OF EXECUTIVE COMMITTEE
Margaret Ackerly, GENERAL COUNSEL
Lou Ann Dietz, LAC SENIOR PROGRAM OFFICER
Phone: 202-778-9657
dietz@wwfus.org
Twig Johnson, LATIN AMERICA AND CARIBBEAN VICE PRESIDENT
Deborah Hechinger, MANAGING VICE PRESIDENT FOR OPERATIONS
David Evanich, MARKETING, MEMBERSHIP AND COMMUNICATIONS VICE PRESIDENT
Kathryn Fuller, PRESIDENT
James Leape, SENIOR VICE PRESIDENT
William Eichbaum, U.S. CONSERVATION AND GLOBAL THREATS VICE PRESIDENT
Diane Wood, VICE PRESIDENT OF RESEARCH AND DEVELOPMENT

WORLDWATCH INSTITUTE

1776 Massachusetts Ave., NW
Washington, DC 20036-1904 USA
Phone: 202-452-1999   Fax: 202-296-7365
E-mail: worldwatch@worldwatch.org
Website: www.worldwatch.org

Founded: 1974
Scope: International

Description: A nonprofit research organization designed to inform policymakers and the public about emerging global problems and trends and the complex links between the world economy and its environmental support systems. Recent studies have covered issues such as global warming, world water shortages, soil erosion, and the decline in food production compared to population growth, renewable energy, deforestation, transportation, oceans, fisheries, carrying capacity and environmental refugees.

Publication(s): State of the World, World Watch Magazine, Vital Signs, Environmental Book series, Database Diskette, www.worldwatch.org, Worldwatch papers

Keyword(s): Agriculture, Energy, Sustainable Development, Population Growth, Renewable Resources, Exotic species, Aquatic nuisance species

Contact(s):
Reah Kauffman, ASSISTANT TO THE CHAIRMAN, DIRECTOR OF INTL PUBLICATIONS
rjkauffman@worldwatch.org
Lester Brown, CHAIRMAN
rjkauffman@worldwatch.org
Lori Brown, LIBRARIAN
lorib@worldwatch.org
Christopher Flavin, PRESIDENT
cflavin@worldwatch.org
Hilary French, VICE PRESIDENT FOR RESEARCH
Richard Bell, VICE PRESIDENT OF COMMUNICATIONS
dbell@worldwatch.org
Ed Ayres, WORLDWATCH MAGAZINE EDITOR
edayres@worldwatch.org

WYOMING ASSOCIATION OF CONSERVATION DISTRICTS

2304 E 13Th St.
Cheyenne, WY 82001 USA
Phone: 307-632-5716   Fax: 307-638-4099
E-mail: waocd@trib.com
Website: www.conservewy.com

Founded: NA
Membership: 3
Scope: Statewide
Keyword(s): Conservation Districts
Contact(s):
Tracy Renner, BOARD MEMBER
P.O. Box 271, Meeteetse, WY 82433
Phone: 307-868-2355
Fax: 307-868-2470
Bobbie Frank, DIRECTOR
waocdt-rib-.com
Olin Sims, PRESIDENT
Phone: 307-632-5716
Fax: 307-632-5716
Veronica Canfield, VICE PRESIDENT
P.O. Box 952, Sundance, WY 82729
Phone: 307-283-2062
Fax: 307-283-2170

WYOMING B.A.S.S. CHAPTER FEDERATION

Attn: President, 1008 Rosewood Dr.
Rock Springs, WY 82901 USA
Phone: 307-362-5863
E-mail: jweber@wyoming.com

Founded: NA
Scope: Statewide

Description: An organization of Bassmaster chapters, affiliated with the Bass Anglers Sportsman Society, organized to fight pollution, assist state and national conservation agencies in their efforts, and teach the young people of our country good conservation practices. Dedicated to the realistic conservation of our water resources.

Publication(s): Wyoming B.A.S.S. Frederation (Newsletter)
Contact(s):
Leonard Nichols, CONSERVATION DIRECTOR
421 Sage Avenue, Kemmer, WY 83101
Phone: 307-877-3629
nichols@hamsfork.net
John Weber, PRESIDENT

WYOMING NATIVE PLANT SOCIETY

P.O. Box 3452
Laramie, WY 82071 USA
Phone: 307-766-3020
Website:
www.uwadmnweb.uwyo.edu/wyndd/wnps/wnps_home.htm

Founded: 1981
Membership: 200
Scope: Statewide

Description: The Wyoming Native Plant Society promotes the use and appreciation of the state's native flora through education and supporting research.

Publication(s): Castilleja, Landscaping with Wildflowers and Native Plants

Keyword(s): Native Plants, Endangered Species, Flowers, Plants, and Trees, Gardening and Horticulture, education, Wildlands, Natural Areas
Contact(s):
Joy Handley, PRESIDENT
Walter Fertig, SECRETARY AND TREASURER AND EDITOR
Nina Haas, VICE PRESIDENT

WYOMING OUTDOOR COUNCIL

262 Lincoln St.
Lander, WY 82520 USA
Phone: 307-332-7031   Fax: 307-332-6899
E-mail: woc@wyomingoutdoorcouncil.org
Website: www.wyomingoutdoorcouncil.org

Founded: 1967
Membership: 1000
Scope: Statewide

Description: A statewide membership organization dedicated to the conservation of Wyoming's natural resources. Promotes sound environmental policy and education of the public for wise decisionmaking. Serves as an active citizen lobby for environmental policies, conducts research, and monitors state and federal agencies.

Publication(s): State Legislative Analysis, Frontline Report (newsletter), various reports and alerts

Keyword(s): Politics and Government, Environmental and Conservation Education, Public Lands, Solid Waste, Toxic Substances, Nuclear-free, Water quantity, Water export and diversion, training
Contact(s):
Dan Heilig, EXECUTIVE DIRECTOR
Joyce Evans, PRESIDENT
Barbara Oakleaf, SECRETARY
Cherry Landen, TREASURER
Nancy Debevoise, VICE PRESIDENT

WYOMING WILDLIFE FEDERATION

P.O. Box 106
Cheyenne, WY 82003 USA
Phone: 307-637-5433   Fax: 307-637-6629
E-mail: admin@wyomingwildlife.org
Website: www.wyomingwildlife.org

Founded: 1937
Membership: 6000
Scope: Statewide

Description: A representative statewide organization, affiliated with the National Wildlife Federation, dedicated to the protection and enhancement of wildlife and its habitat through public education and government interaction.

Publication(s): The Pronghorn News
Contact(s):
Jim Narva, ALTERNATE REPRESENTATIVE
Vicky Urbanek, EDITOR & EDUCATION PROGRAMS CONTACT
Mark Winland, REPRESENTATIVE
Larry Durante, TREASURER

X

XERCES SOCIETY, THE
4828 SE Hawthorne Blvd.
Portland, OR 97215 USA
Phone: 503-232-6639   Fax: 503-233-6794
E-mail: xerces@teleport.com
Website: www.xerces.org

Founded: 1971
Membership: 5500
Scope: National

Description: An international nonprofit organization dedicated to invertebrates and the preservation of critical biosystems worldwide. The Society is committed to protecting invertebrates as major components of biological diversity. Emphasis: aquatic invertebrate monitoring to assist in conservation of Pacific Northwest watersheds, butterfly farming in NE Costa Rica, enhancing wild pollinator populations in out-of-play areas of selected Columbia Plateau golf courses, and education through publications.

Publication(s): Butterfly Gardening: Creating Summer Magic in Your Garden, Common Names of North American Butterfies, The, Wings: Essays on Invertebrate Conservation (membership magazine)

Keyword(s): Biodiversity, Endangered Species, Insects and Butterflies

Contact(s):
Matthew Shepherd, EDITOR
Scott Hoffman Black, EXECUTIVE DIRECTOR
Katherine Janeway, LEGAL ADVISOR
1932 First Avenue, Suite 510, Seattle, WA 98101
Phone: 206-583-8304
Thomas Eisner, PRESIDENT
Cornell University, Neurobiology and Behavior, W347 Mudd Hall, Ithaca, NY 14853-2702
Phone: 607-255-4464
Ed Grosswiler, SECRETARY, TREASURER
Kathy Parker, VICE PRESIDENT

Y

YELLOWSTONE GRIZZLY FOUNDATION (YGF)
P.O. Box 12769
Jackson Hole, CO 83002 USA
Phone: 800-996-4638

Founded: 1986
Membership: 500
Scope: National

Description: The YGF is a nonprofit organization dedicated to the conservation of the threatened grizzly bear and habitat preservation in the Greater Yellowstone Ecosystem. It conducts independent research and draws from that to produce educational materials and programs for both professional and general public audiences.

Publication(s): Yellowtone Grizzly Journal

Keyword(s): Endangered Species, Environmental and Conservation Education, Mammals, training

Contact(s):
Marilyn French, PRESIDENT
6675 Upper Cascade Dr., Jackson, WY 83001
Phone: 307-733-8630

Timothy Floyd, SECRETARY
P.O. Box 3229, Hailey, ID 83333
Phone: 208-726-3968
Steven French, TREASURER
Karen Shirley, VICE PRESIDENT
P.O. Box 3468, Jackson, WY 83001
Phone: 307-733-5311

YOSEMITE RESTORATION TRUST
1212 Broadway, Suite 810
Oakland, CA 94612 USA
Phone: 510-763-1403   Fax: 510-208-4435
Website: yosemitetrust.org

Founded: 1990
Scope: National

Description: To ensure protection of the natural, scenic, and historic resources of Yosemite National Park and its ecosystems, and to ensure that visitors have the highest quality experience of the park's natural environment.

Publication(s): special reports on regional transportation, day-use reservations, and housing., Yosemite Viewpoints (Newsletter)

Keyword(s): National Parks, Outdoor Recreation, Transportation, Wilderness, Rural Development

Contact(s):
Janet Cobb, PRESIDENT
Thomas Gwyn, TREASURER
Walter Kieser, VICE PRESIDENT
Hal Browder, VICE PRESIDENT

YOUNG ENTOMOLOGISTS SOCIETY, INC.
6907 W. Grand River Ave.
Lansing, MI 48906-9131 USA
Phone: 517-886-0630   Fax: 517-886-0630
E-mail: yesbugs@aol.com

Founded: 1965
Membership: 650
Scope: National

Description: An international nonprofit organization educates and serves youth and amateur entomologists via publications, programs and the minibeast Zooseum and Education center; assists in information, talent, scientific literature, and insect specimen exchanges (informational networks); distributes and develops resource materials; and promotes awareness of arthropod importance and the contributions youth and amateur entomologists make to the science of entomology. Founded as Teen International Entomology Group.

Publication(s): Insect World, Project B.U.G.S., Insect Indentification Guide, Caring for Insect Livestock

Keyword(s): Environmental and Conservation Education, Insects and Butterflies, Outdoor Recreation, Youth Organizations, Zoology

Contact(s):
Gary Dunn, DIRECTOR OF EDUCATION
Dianna Dunn, EXECUTIVE DIRECTOR
Phone: 517-887-0499

YUKON FISH AND GAME ASSOCIATION

P.O. Box 4434
Whitehorse, Yukon Territory Y1A 3T5
Phone: 403-667-2843
E-mail:
Website:

Founded: NA
Scope: Statewide

Description: Affiliated with the Canadian Wildlife Federation.

Z

ZERO POPULATION GROWTH, INC.

1400 16th St., NW.
Washington, DC 20036 USA
Phone: 202-332-2200   Fax: 202-332-2302
E-mail: info@zpg.org
Website: www.zpg.org

Founded: 1968
Membership: 75,000
Scope: National

Description: ZPG is a national nonprofit membership organization that works to educate and motivate Americans to help
meet the global population challenge. ZPG mobilizes
grassroots support for the adoption of policies and programs
necessary to stabilize global population growth.

Publication(s): Teachers' PET Term Paper, Action Alerts,
Factsheets, Backgrounders, Media Targets, reports, and
promotional brochures, ZPG Reporter

Keyword(s): Environmental and Conservation Education,
Population Growth, Public Health Protection, Renewable
Resources, Sustainable Development

Contact(s):
Tim Cline, DIRECTOR OF COMMUNICATIONS
tim@zpg.org
Brian Dixon, DIRECTOR OF GOVERNMENT RELATIONS
brian@zpg.org
Elizabeth Borg, DIRECTOR OF MEMBERSHIP AND
DEVELOPMENT
liz@zpg.org
Pamela Wasserman, DIRECTOR OF POPULATION
EDUCATION
pam@zpg.org
John Seager, EXECUTIVE DIRECTOR
john@zpg.org
Jay Keller, FIELD DIRECTOR
jay@zpg.org
Peter Kostmayer, POLICY COUNSELOR
peter@zpg.org

NON-GOVERNMENTAL FOR-PROFIT ORGANIZATIONS

A

ABSEARCH, INC.
121 Sweet Ave
Moscow, ID 83843 USA
Phone: 800-867-1877   Fax: 208-885-3803
Website: www.absearch.com

Founded: NA

Description: Produces ABSEARCH databases which include thousands of abstracts and citations from professional research journals in the area of natural resources. The databases are updated as new research becomes available. Format: CD-ROM and online subscription.

Contact(s):
Leah Lipar, OFFICE MANAGER

AIZA BIBY
P.O. Box 701
East Setauket, NY 11733 USA
Phone: 516-658-6871
E-mail: biby@biby.org
Website: http://www.biby.org

Founded: NA

Contact(s):
Rebecca Grella, EXECUTIVE DIRECTOR
Phone: 631-744-7668

AMERICAN CHEMICAL SOCIETY
1155 16th St. NW
Washington, D.C. 20036 USA
Phone: 202-872-4582   Fax: 202-872-4403
E-mail: est@acs.org
Website: www.acs.org

Founded: NA
Scope: National
Publication(s): Environmental Science and Technology

Contact(s):
Sally Pecor, KEY CONTACT

ANIMALS AGENDA
P.O. Box 25881
Baltimore, MD 21224 USA
Phone: 410-675-4566   Fax: 410-675-0066
E-mail: office@animalsagenda.org
Website: www.animalsagenda.org

Founded: NA
Scope: International
Publication(s): The Animals' Agenda

Contact(s):
Kim Stallwood, EDITOR IN CHIEF

ATLANTA AUDUBON SOCIETY
1447 Peachtree St.
Suite 214
Atlanta, GA 30309 USA
Phone: 404-873-3034   Fax: 404-873-3135
Website: www.atlantaaudubon.org

Founded: NA
Scope: Local
Publication(s): Wingbars

Contact(s):
Jim Wilson, IBA COORDINATOR

B

BERLET FILMS AND VIDEOS
1646 West Kimmel Rd
Jackson, MI 49201 USA
Phone: 517-784-6969   Fax: 517-796-2646
E-mail: mark@berletfilms-video.com
Website: www.berletfilms-video.com

Founded: NA
Scope: International

Description: Produces environmental/nature audio-video resources. Free catalogues upon request.

Contact(s):
Mark Snedekeder, GENERAL MANAGER

BIOSIS
2100 Market St.,Suite 700
Philadelphia, PA 19103-7095 USA
Phone: 215-587-4800   Fax: 215-587-2016
E-mail: info@mail.biosis.org
Website: www.biosis.org

Founded: NA

Publication(s): Biosis Previews, Toxline

Contact(s):
Marisa Westcott

BRAUER PRODUCTIONS
530 S. Union St
Traverse City, MI 49684 USA
Phone: 231-941-0850   Fax: 231-941-0947
E-mail: brauer@brauer.com
Website: www.brauer.com

Founded: 1978
Scope: National

Description: Full Service film and video production company serving Traverse City and clients throughout Michigan and the N.W. since 1978.

Contact(s):
Susan McQuaid, CONTACT
Richard Brauer, PRESIDENT

BRITISH COLUMBIA CONSERVATION DATA CENTRE; MINISTRY OF SUSTAINABLE RESOURCE MANAGEMENT
MINISTRY OF ENVIRONMENT, LAND & PARKS
P.O. Box 9344
Station Provincial Gov.
Victoria, British Columbia V8W 9M1 Canada
Phone: 250-356-0928   Fax: 250-387-2733
E-mail: elpcdcdata@victoria1.gov.bc.ca
Website: www.elp.gov.bc.ca/rib/cbs/cdc

Founded: NA
Scope: International

Description: Providing information on rare organisms and ecosystems.

Publication(s): see publications on website

Contact(s):
Andrew Harcombe, COORDINATOR
andrewharcombe@gems2.gov.bc.ca

BUILDING GREEN, INC.
122 Birge St.
Brattleboro, VT 05301 USA
Phone: 802-257-7300   Fax: 802-257-7304
E-mail: ebn@buildinggreen.com
Website: www.buildinggreen.com

Founded: NA
Scope: International

Publication(s): Environmental Building News and Green Spec
Directory

Contact(s):
Dan Woodbury, PUBLISHER

BULLFROG FILMS
Olney, PA 19547 USA
Phone: 800-543-3764   Fax: 610-370-1978
E-mail: video@bullfrogfilms.com
Website: www.bullfrogfilms.com

Founded: NA

Description: Films and videos rented and sold worldwide to
educational institutions. Most programs come with study guide,
with suggested activities, research topics, debate subjects, and
bibliographies. Free catalogues available.

Contact(s):
Sieglinde Abromaitis

BUSINESS PUBLISHERS, INC.
8737 Colesville Rd., Suite 1100
Silver Spring, MD 20910 USA
Phone: 301-589-5103   Fax: 301-587-4530
E-mail: bpinews@bpinews.com
Website: www.bpinews.com

Founded: NA

Contact(s):
Beth Early

C

C&O CANAL NATIONAL HISTORICAL PARK
NATURAL RESOURCES DIVISION
P.O. Box 4
Sharpsburg, MD 21782 USA
Phone: 301-739-4200   Fax: 301-739-5275
E-mail: CHOH_Superintendent@nps.gov
Website: http://www.nps.gov/choh/

Founded: NA

C.A.R.E (CITIZENS AGAINST RACCOON EXTERMINATION)
5125 Paso Venado
Carmel, CA 93923 USA
Phone: 831-647-8400   Fax: 831-373-8531
Website: www.drpasten.com
Scope: Local

Description: The goal of C.A.R.E is to save the raccoons on the
Central Coast.

Contact(s):
Mari Fuentes-Jones, ADMINISTRATOR

CALIFORNIA INSTITUTE OF PUBLIC AFFAIRS
P.O. Box 189040
Sacramento, CA 95818 USA
Phone: 916-442-2472   Fax: 916-442-2478
E-mail: cipa@cipahq.org
Website: www.cipahq.org

Founded: NA

Publication(s): California Environmental Directory, World
Directory of Environmental Organizations 6th edition 2001, A
Sustainable World

CHATAUQUA WATERSHED CONSERVANCY
413 Main Street
Jamestown, NY 14701-1618 USA
Phone: 716-664-2166   Fax: 716-483-3524
E-mail: chautwsh@netsync.net
Website: www.chatauquawatershed.org

Founded: NA
Scope: Local

Description: The Chautauqua Watershed Conservancy is a
county-wide organization with a mission to preserve and
enhance the water quality, scenic beauty and ecological health
of the lakes, streams and watersheds of the Chautauqua
region.

Publication(s): The Shed Sheet

Contact(s):
John Jablonski, EXECUTIVE DIRECTOR

CJE ASSOCIATES
237 Gretna Green Ct.
Alexandria, VA 22304 USA
Phone: 703-823-0662   Fax: 703-823-5923
E-mail: eeiserer@netscape.net

Founded: NA
Scope: National
Publication(s): Ecology USA

Contact(s):
Elaine Eiserer, EDITOR

COASTAL RESOURCES CENTER
U.R.I. Narragansett Bay Campus, South Ferry Rd.
Naragansett, RI 02882 USA
Phone: 401-874-6224   Fax: 401-789-4670
E-mail: cyoung@gso.uri.edu
Website: http://crc.uri.edu

Founded: 1971
Scope: International-National

Description: CRC is active in the U.S. and world advancing
coastal management through field projects, education and
training, research and learning and sharing lessons learned
throughout the coastal community.

Publication(s): Intercoast Network, a manual for assessing
progress in coastal management, and Aquidneck Island: our
shared vision

Contact(s):
Chip Young, COMMUNITY LIAISON

Phone: 401-874-6630
Fax: 401-789-4670

CONGRESSIONAL GREEN SHEETS, INC.

406 E St., SE
Washington, D.C. 20003 USA
Phone: 202-546-2220   Fax: 202-546-7490
E-mail: wb@greensheets.com
Website: www.greensheets.com

Founded: NA
Scope: International

Publication(s): Environment and Energy weekly bulletin. Greensheets Express and News Room.

Contact(s):
John Dineen, EDITOR

CONNECTICUT CARIBOU CLAN

P.O. Box 9344
Bolton, CT 06043 USA
Phone: 860-643-2948
E-mail: captundra@aol.com
Website: http://hometown.aol.com/captundra/index.html

Founded: NA

Description: Tundra Talk Newsletter focuses on energy, transportation, and the environment, with special emphasis on permanently preserving the Arctic Refuge.

Contact(s):
Rodney Parlee, EDITOR/PUBLISHER

CROWNPOINT INSTITUTE OF TECHNOLOGY

NATURAL RESOURCE DEPARTMENT
P.O. Box 849
Crownpoint, NM 87313
Phone: 505-786-4100
E-mail: mattie@cit.cc.nm.us
Website: www.cit.cc.nm.us

CUTTER INFORMATION CORPORATION

37 Broadway
Arlington, MA 02474 USA
Phone: 781-641-9876
E-mail: cdoucette@cutter.com
Website: www.cutter.com

Founded: NA
Scope: International
Publication(s): Business & The Environment ISO 140 update, Air Quality Global Environmental Change Report, Environmental Design Update

Contact(s):
Christine Doucette

D

DIVISION DE PATRIMONIO NATURAL AREA DE PLANIFICATION DE RECUSOS INTEGRAL

DEPARTMENTO DE RECURSOS NATURALES Y AMBIENTALES DE PUERTO RICO
P.O. Box 9066600
Puerta de Tierra, 00906-6600 Puerto Rico
Phone: 787-724-8774   Fax: 787-723-4255

Founded: NA

Contact(s):
Aida Martinez, DIVISION DIRECTOR

E

ENVIRONMENT AND ENERGY PUBLISHING, LLC

122 C St., NW
Washington, D.C. 20001 USA
Phone: 202-628-6500   Fax: 202-737-5299
E-mail: pubs@eenews.net
Website: www.eenews.net

Founded: NA
Scope: International

Publication(s): Land Letter (The Newsletter for Natural Resource Professionals), Greenwire and Environmental and Energy Daily.

Contact(s):
Kevin Braun, MANAGING EDITOR
Drew Gagliano, MARKETING DIRECTOR

ENVIRONMENTAL ALLIANCE FOR SENIOR INVOLVEMENT (EASI)

P.O. Box 250, 9292 Old Dumfries Rd
Catlett, VA 20119 USA
Phone: 540-788-3274   Fax: 540-788-9301
E-mail: easi@easi.org
Website: www.easi.org

Founded: NA
Scope: International

ENVIRONMENTAL CAREERS WORLD

100 Bridge St.
Bldg. C
Hampton, VA 23669 USA
Phone: 757-727-7895   Fax: 757-727-7904
E-mail: eccinfo@environmentalcareer.com
Website: http://environmentalcareer.com

Founded: NA
Scope: International
Publication(s): National Environmental Employment Report

Contact(s):
John Esson, DIRECTOR
Debbie Gunn, OFFICE MANAGER

ENVIRONMENTAL MEDIA CORPORATION

1008 Paris Ave.
Port Royal, SC 29935-2410 USA
Phone: 843-986-9034   Fax: 843-986-9093
E-mail: bpendergraft@envmedia.com
Website: www.envmedia.com

Founded: 1989
Scope: National

Description: Environmental Media designs, produces, and distributes media to support environmental education. Programs are curriculum-based and most are accompanied by teaching guides Free catalogue available and other educational material.

Publication(s): see publications on website

Contact(s):
Bill Pendergraft, OWNER
Eileen Newton, RESELLER MANAGER

F

FISH AND WILDLIFE REFERENCE SERVICE

5430 Grosvenor Ln.
Suite 110
Bethesda, MD 20814 USA
Phone: 301-492-6403   Fax: 301-564-4059
Website: Fa.r9.fws.gov/r9fwrs/

Founded: NA
Scope: International

Description: Produces "Fish and Wildlife Reference Service Database", a bibliographic database of primarily state fish and wildlife agency research reports; some USFWS publications; COOP Unit Theses and dissertations and some USNBS publications.

Contact(s):
Paul Wilson, PROJECT MANAGER

FLORIDA NATURAL AREAS INVENTORY

1018 Thomasville Rd., Suite 200-C
Tallahassee, FL 32303 USA
Phone: 850-224-8207   Fax: 850-681-9364
E-mail: joetting@fnai.org
Website: www.fnai.org

Founded: 1981

Description: Information is collected on the status and distribution of natural communities, rare and endangered species of plants and animals and other natural features, then analyzed through an integrated data management system.

G

GECKO PRODUCTIONS, INC.

Attn: Director Nathalie Ward
P.O. Box 573
Woods Hole, MA 02543 USA
Phone: 508-548-3317   Fax: 508-548-3317
E-mail: nward@mbl.edu

Founded: 1995

Description: Designs conservation education materials and workshops about marine endangered species and marine protected areas, with emphasis on bringing environmental awareness and cultural understanding to the Caribbean, United States and beyond.

Contact(s):
Nathalie Ward, DIRECTOR

GLOBAL INFORMATION NETWORK

146 West 29th St., # 7E
New York, NY 10001 USA
Phone: 212-244-3123   Fax: 212-244-3522
E-mail: ipsgin@igc.org
Website: www.globalinfo.org

Founded: NA
Scope: International

Description: GIN is the distributor of Inter Press Service and other news wires from developing countries with special features on the environment.

Contact(s):
Katherine Stapp

GREEN MOUNTAIN POST FILMS

Turners Falls, MA 01376 USA
Phone: 413-863-4754   Fax: 413-863-8248
E-mail: gmfilms.com
Website: www.gmpfilms.com

Founded: 1975
Scope: International

Description: A film/video production and distribution company that specializes in media concerning environmental issues.

Contact(s):
Charles Light, BUSINESS MANAGER

GREENWIRE

ENVIRONMENT AND ENERGY PUBLISHING, LLC
122 C St., NW
Suite 722
Washington, D.C. 20001 USA
Phone: 202-628-6500   Fax: 202-737-5299
E-mail: pubs@eenews.net
Website: www.eenews.net

Founded: NA

Description: An online daily publication providing comprehensive coverage of environmental, energy and natural resources issues, politics, developments and policy action. Greenwire tracks and reports on the White House, federal agencies, states, court decisions and the stories being reported on by the media nationwide.

Publication(s): The Environment and Energy Daily, Greenwire, and Land Letter

Contact(s):
Drew Gagliano, MARKETING DIRECTOR

I

ILLINOIS STUDENT ENVIRONMENTAL NETWORK

110 S. Race St., Suite 202
Urbana, IL 61801 USA
Phone: 217-384-0830   Fax: 217-278-2105
E-mail: isen@isenonline.org
Website: www.isenonline.org

Founded: 1996
Scope: Regional

Description: Through educational programs such as newsletters, information alerts, speakers and conferences, training, and how-to manuals, ISEN keeps over 2,500 student environmentalists in 114 different student groups on over 80 college campuses across Illinois up-to-date on environmental issues that affect them. ISEN also provides students with timely, straightforward information on finding great environmental jobs and internships, what other student environmental groups around Illinois are up to, and gives them ideas for projects, fund-raisers, and membership drives their groups can undertake.

Publication(s): Tips for the Environmental Job Search, Advocating for Campus Waste Reduction

Contact(s):
Laura Huth, EXECUTIVE DIRECTOR AND FOUNDER

INSTITUTE FOR GLOBAL COMMUNICATIONS

P.O. Box 29904
San Francisco, CA 94129-0904 USA
Phone: 415-561-6100   Fax: 415-561-6101
E-mail: econet@igc.apc.org
Website: www.igc.org

Founded: 1984
Scope: International

Description: Creates and manages "EcoNet". Through the development of communication and information sharing systems, EcoNet seeks to increase collaboration and cooperation between organizations seeking environmental sustainability.

Contact(s):
Debra Farrell, EXECUTIVE DIRECTOR

INTERNATIONAL ACADEMY

ENVIRONMENTAL
Santa Barbara, CA 93140 USA
Phone: 805-964-0790   Fax: 805-564-4634
E-mail: info@iasb.org
Website: www.iasb.org

Founded: NA
Scope: International

Description: Produces "Environmental Bibliography", bibliographic database covering more than 400 scientific and popular journals in social, political and philosophical issues, air, energy, land and water resources, nutrition and health. Author Abstracts, 1997 forward.

Publication(s): Environmental Knowledge Base By Subscription, CD Rom Issued Quarterly Online

Contact(s):
Eric Boehm, CHAIRMAN
Steven Popps, DIRECTOR OF SALES AND CUSTOMER SERVICE
Lauren Everett, EDITOR
Amy Rushing, EDITOR
Joann St. John, PRESIDENT

INTERNATIONAL RESEARCH AND EVALUATION

21098 IRE Control Center
Eagan, MN 55121 USA
Fax: 952-888-9124

Founded: NA

Description: Environmental library for the application of knowledge, methods and means.

Publication(s): Waste Management Information Database, World Environment Report, World Environmental Directory

Contact(s):
R. Danford

JERE MOSSIER PRODUCTIONS/ UNDERWATER IMAGES

P.O. Box 1415
Hayden, ID 83835 USA
Phone: 208-683-8112
Website: www.jeremossier.com

Founded: NA
Scope: International

Description: Licensing of underwater and wildlife stock footage and professional video production of fisheries, underwater programs, wildlife, and natural history videos.

Publication(s): Videos, Underwater Exploration

Contact(s):
Jere Mossier

JONES & STOKES

2600 V Street
Sacramento, CA 95818 USA
Phone: 916-737-3000
Website: www.jonesandstokes.com

Founded: 1970

Description: Jones and Stokes was quickly positioned as industry leaders by providing their clients with scientifically accurate, innovative, and practical solutions to their environmental challenges. Today, their reputation for providing clients with incomparable quality and the broadest diversity of expertise in the industry remains unsurpassed. Their unique teams of knowledgeable, experienced professionals embrace a multi-disciplinary, problem-solving philosophy and approach that benefits their clients, communities, and the environment.

Contact(s):
Julie Jessen
julieb@jsanet.com

KEEPING TRACK

Huntington, VT 05472 USA
Phone: 802-434-7000   Fax: 802-434-5383
E-mail: keeptrak@together.net
Website: www.keeptracking.org

Founded: 1994
Scope: Reginal National

Description: The purpose of Keeping Track, Inc. is to protect wildlfe habitat through conservation education, planning and field research. This is achieved by teaching adults and children to observe, interpret and record evidence of wildlife, and encouraging broad-based volunteer tracking groups to use the information obtained to affect land use planning their towns.

Publication(s): Keep Tracking-Quarterly newsletter

Contact(s):
Lars Botzojoras, EXECUTIVE DIRECTOR
Susan Morse, FOUNDER
Monica Mac, OFFICE MANAGER

L

LAND AND WATER FUND OF THE ROCKIES

2260 Baseline Rd., Suite 200
Boulder, CO 80302 USA
Phone: 303-444-1188

Founded: 1991
Scope: Regional

Description: Founded in 1991, the Land and Water Fund of the
Rockies (the LAW FUND) is an environmental law and policy
center serving the Rocky Mountain and Desert Southwest
region. The LAW Fund promotes policy reform and provides
legal and strategic support to environmental and community
groups across an eight-state area. While our areas of expertise
are energy, water, public lands, growth and sprawl, and environmental justice, the LAW Fund serves as a clearinghouse for
information and resources on the full range of environmental
issues confronting the Intermountain West.

Contact(s):
Pam Hathaway, PROJECT DIRECTOR

LEXISNEXIS ACADEMIC & LIBRARY SOLUTIONS

4520 East-West Hwy., Suite 800
Bethesda, MD 20814-3389 USA
Phone: 301-654-1550   Fax: 301-657-3203
E-mail: academicinfo@lexis-nexis.com
Website: www.lexisnexis.com/academic

Founded: NA
Scope: International

Description: Publishes indexes, electronic databases and
microform collections that provide access to information
published by government, private and international sources.

Publication(s): Enviroline, Environment Abstracts

Contact(s):
Marcy Taylor, CONTACT
Henry Stoever, MARKETING DIRECTOR

LUMMI ISLAND HERITAGE TRUST

P.O. Box 158
Lummi Island, WA 98262-0158 USA
Phone: 360-758-7997   Fax: 360-758-7001
E-mail: heritagetrust@nas.com
Website: www.nas.com/heritagetrust

Founded: NA
Scope: National

Contact(s):
Dave Kershner

M

MANITOBA CONSERVATION DATA CENTRE

DEPARTMENT OF CONSERVATION
Box 24, 200 Saulteaux Crescent
Winnipeg, Manitoba R3J 3W3 Canada
Phone: 204-645-7743   Fax: 204-945-3077
E-mail: fblovun@nr.gov.mb.ca
Website: www2.gov.mb.ca/natres/db/

Founded: NA

MARINE FISH CONSERVATION NETWORK

660 Pennsylvania Ave., SE, Suite 302B
Washington, D.C. 20003 USA
Phone: 202-543-5509   Fax: 204-543-5774
E-mail: network@conservefish.org
Website: www.conservefish.org

Founded: NA
Scope: National

Description: The Marine Fish Conservation Network is a coalition
of national and regional environmental organizations,
commerical and recreational fishing associations, aquariums,
and marine science groups dedicated to promoting the long-
term sustainability of marine fisheries.

Publication(s): Network News

Contact(s):
Lee Crockett, EXECUTIVE DIRECTOR

MINNESOTA LAND TRUST

2356 University Ave. West
Saint Paul, MN 55114 USA
Phone: 651-647-9590   Fax: 651-647-9769
E-mail: mnland@mnland.org
Website: www.mnland.org

Founded: NA
Scope: Regional

Contact(s):
Susan McCallurm

N

NATIONAL AGRICULTURAL LIBRARY

INFORMATION SYSTEMS DIVISON
10301 Baltimore Ave
Beltsville, MD 20705 USA
Phone: 301-504-6813   Fax: 301-504-7473
E-mail: agref@nal.usda.gov
Website: www.nal.usda.gov

Founded: NA
Scope: National

Description: Produces "Agricola", a database of bibliographic
citations covering all aspects of agricultural and food sciences,
including natural resources, animal welfare, pollution,
pesticides and land and water management. The database has
over 3 million records.

Contact(s):
David Goldberg, CONTACT PERSON

NATIONAL GROUND WATER INFORMATION CENTER

601 Dempsey Rd.
Westerville, OH 43081 USA
Phone: 614-898-7791   Fax: 614-898-7786
E-mail: smaste@ngwa.org
Website: www.ngwa.org

Founded: NA

Description: Produces "Ground Water Network", a fee-based
information service conducting literature searches and
document delivery. Maintains six databases with over 82,000
abstracts related to ground water, water treatability, NGWA
Certified Ground Water Contractors and U.S. Census on
housing and water source information.

Contact(s):
Sandy Masters, DIRECTOR OF THE INFORMATION CENTER

NATIONAL INFORMATION SERVICES CORPORATION
Wyman Towers, 3100 St. Paul St.
Baltimore, MD 21218 USA
Phone: 410-243-0797   Fax: 410-243-0982
E-mail: sales@nisc.com
Website: www.nisc.com

Founded: NA
Scope: International
Publication(s): Fish & Fisheries Worldwide, Wildlife Worldwide

Contact(s):
Debbie Durr, SALES & MARKETING MANAGER

NAVAJO NATURAL HERITAGE PROGRAM
NAVAJO FISH & WILDLIFE DEPARTMENT
P.O. Box 1480
Window Rock, AZ 86515 USA
Phone: 520-871-7068   Fax: 520-871-7069
Website: www.navajofishandwildlife.org

Founded: NA
Scope: Regional
Publication(s): Newsletter- Proclamation (Annually)

Contact(s):
Jeff Cole, COORDINATOR

NEAL COMMUNICATIONS
1220 Bald Eagle Rd.
Kingston, TN 37082 USA
Phone: 615-662-1946   Fax: 615-952-4522
E-mail: nealcynthiap@yahoo.com

Founded: NA
Scope: International

Description: Produce outreach videos supporting sustainable ecosystems and development, conservation of natural resources, wildlife, and cultural integrity. Specialists in translating complex issues into compelling, motivating communications.

P

PEW WILDERNESS CENTER
122 C St.
P.O. Box 37
Boulder, CO 80306
Phone: 202-544-3691
Website: www.pewwildernesscenter,org

Founded: NA
Scope: Statewide

Description: The Pew Wilderness Center works to educate the general public about the concept of wilderness and to rejuvenate interest in protecting our nation & rescue remaining wild places. Whether as open space, places for a variety of recreational pursuits, or habitat for wildlife, wilderness is an important part of our natural heritage, and is quickly diminishing. Public recognition in regard to the value and extent of remaining wilderness will lead to active involvement in efforts to protect these landscapes as part of the National Wilderness Preservation System for our families today and for future generations.

Contact(s):
Mike Matz, EXECUTIVE DIRECTOR

PROPERTY CARETAKING OPPORTUNITIES WORLDWIDE
P.O. Box 540
River Falls, WI 54022 USA
Phone: 480-488-1970
Website: www.caretaker.org

Founded: NA
Scope: International
Publication(s): The Caretaker—Bi-monthly newsletter

Contact(s):
Gary Dunn, OWNER
Phone: 715-426-5500
caretaker@caretaker.org

PUBLIC LANDS INTERPRETIVE ASSOCIATION
SOUTHWEST NATURAL & CULTURAL HERITAGE ASSN.
6501 Fourth NW
Suite I
Albuquerque, NM 87107 USA
Phone: 505-345-9498   Fax: 505-344-1543
Website: www.publiclandsinfo.org

Founded: 1981
Scope: Regional

Description: Educational and interpretive not for profit operating bookstores in federal visitor centers as well as publisher of local interpretive booklets.

Publication(s): Wild & Scenic Rio Grande, Merrit Island National Wildlife Refuge, Pecos Wilderness Trail Guide, Bosque Del Apache National Wildllife Refuge

Contact(s):
Lisa Madsen, CEO
Stephen Maurer, DIRECTOR OF PUBLICATIONS
Ted Peay, PRESIDENT

S

SASKATCHEWAN CONSERVATION DATA CENTRE
SK ENVIRONMENT & RESOURCE MANAGEMENT
FISH AND WILDLIFE BRANCH
3211 Albert St
Regina, Saskatchewan S4S 5W6 Canada
Phone: 306-787-7196   Fax: 306-787-3913
E-mail: jkeith@serm.gov.sk.ca
Website: www.biodiversity.sk.ca

Founded: NA
Scope: Provincial, Local

Contact(s):
Jeff Keith, INFORMATION MANAGER

SEAWEB
1731 Connecticut Ave., 4th Floor
Washington, D.C. 20009 USA
Phone: 202-483-9570   Fax: 202-483-9354
Website: http://www.seaweb.org

Description: SeaWeb was launched in 1996 to raise awareness about the growing threat to the ocean and its living resources. SeaWeb's goal is to make ocean protection a high environmental priority in the U.S. and around the world. SeaWeb provides science-based information from a variety of sources to a variety of media outlets. With the help of scientists, educators, researchers and communications specialists, SeaWeb has become a respected independent resource for journalists, government officials and concerned citizens.

Contact(s):
Jessica Brown

SHARING NATURE FOUNDATION
14618 Tyler Foote Road
Nevada City, CA 95959 USA
Phone: 530-478-7650   Fax: 530-478-7650
E-mail: info@sharingnature.com
Website: www.sharingnature.com

Founded: 1979
Scope: International

Description: Established in 1979 by naturalist and author, Joseph Cornell, the Sharing Nature Foundation uses creative nature activities to give people joyful and inspiring experiences of nature. We believe it's only by uplifting people's consciousness that we truly change their way of looking at, and relating to the world around them. To do this we use Flow Learning (tm), a playful and inspirational teaching strategy that works with people where they are and gently brings them to a deeper, more profound experience of nature.

Publication(s): Sharing Nature Worldwide Journal (online journal), Sharing Nature with Children I & II, Listening to Nature, John Muir: My Life with Nature, With Beauty Before Me, Journey to the Heart of Nature

Contact(s):
Colleen Heater, PROGRAM COORDINATOR

SPORTSMAN NETWORK,INC.
501 S. Kentucky Ave., P.O. Box 427
Corbin, KY 40702-0427 USA
Phone: 606-528-1800   Fax: 859-824-0556
E-mail: sportsman@sportsmansnetwork.org
Website: www.sportsmansnetwork.org

Scope: Regional

Contact(s):
Peter Samples, PRESIDENT AND CEO
Phone: 859-824-7585
Elmer Chavies, Jr., REGIONAL DIRECTOR

SPORTSMANS NETWORK, INC., THE
501 S. Kentucky Ave
Corbin, KY 40702 USA
Phone: 859-824-6526   Fax: 606-528-2287
E-mail: sportsmen@sportsmansnetwork.org
Website: www.sportsmansnetwork.org

Founded: 1991
Scope: Statewide

Description: The Sportsman's Network is an incorporated statewide nonprofit conservation organization dedicated to educating the public and raising awareness of wildlife conservation through programs which promote controlled hunting, fishing, and other related activities. Also produces "A Moment in Conservation" radio program.

Keyword(s) Hunting, Land Preservation, National Parks, Rivers, Sport Fishing, Trapping, Training, Wetlands, Wildlands, training, Rehabilitation,

Contact(s):
Ernie Samples, EXECUTIVE DIRECTOR
Elmer Chavies, Jr., REGIONAL DIRECTOR
Paul Cookendorfer, SECRETARY
Peter Samples, STATE CHAIRMAN
Ken Hale, TREASURER
Keith Fullwood, VICE PRESIDENT

T

THE CENTER FOR A NEW AMERICAN DREAM
6930 Carroll Ave
Takoma Park, MD 20912 USA
Phone: 301-891-3683   Fax: 301-891-3684
E-mail: newdream@newdream.org
Website: www.newdream.org

Founded: October 1996
Scope: International

Description: The center for a New American Dream helps individuals and institutions reduce and shift consumption to enhance quality of life and protect the natural environment.

Publication(s): Enough!, Simply the Holidays, and Tips for Parenting in a Commerical Culture.

Contact(s):
Eric Brown, COMMUNICATIONS DIRECTOR
Eric@newdream.org
Monique Tilford, DEVELOPMENT DIRECTOR
Monique@newdream.org
Nancy Smith, DIRECTOR OF ADMINISTRATION
Sean Sheehan, DIRECTOR OF NETWORK SERVICES AND INTERNET PROGRAMS
Sean@newdream.org
Betsy Taylor, EXECUTIVE DIRECTOR
Betsy@newdream.org

THE MARIE SELBY BOTANICAL GARDENS
811 South Palm Ave.
Sarasota, FL 34236 USA
Phone: 941-955-7553   Fax: 941-366-9807
E-mail: contactus@selby.org
Website: www.selby.org

Founded: NA
Scope: International
Publication(s): Selbyana-Journal, Shelbyana Bulletin-Newsletter

Contact(s):
Barry Walsh, STAFF EDITOR

THE VIDEO PROJECT
San Francisco, CA 94107 USA
Phone: 800-475-2638
E-mail: video@videoproject.net
Website: www.videoproject.net/

Founded: NA

Description: "Media for a safe and sustainable world". Affordable films and videos on environmental and related issues. The project now offers over 600 programs for sale.

TURNER ENDANGERED SPECIES FUND

1123 Research Dr.
Bozeman, MT 59718 USA
Phone: 406-556-8500   Fax: 406-556-8501
E-mail: tesf@montana.net
Website: www.tesf.org#http://www.tesf.org#

Founded: NA

Contact(s):
    Kyran Kunkel

<hr>

U

U.S. FISH AND WILDLIFE SERVICE

SOUTHWEST REGION
P.O. Box 1306
500 Gold Ave. SW
Room 8526
Albuquerque, NM 87102 USA
Phone: 505-248-6282   Fax: 505-248-6845

Founded: NA

UNIVERSITY OF MARYLAND EASTERN SHORE

MARYLAND COOPERATIVE FISH & WILDLIFE
RESEARCH UNIT
1120 Trigg Hall
Princess Anne, MD 21853 USA
Phone: 410-651-7663   Fax: 410-651-7662

Founded: 1994
Scope: National

Description: The unit is sponsored by the Biological Resources
Division, US Geological Survey, Maryland Department of
Natural Resources, US Fish & Wildlife Service, University of
Maryland Eastern Shore and the Wildlife Management
Institute. Fish and Wildlife research, graduate education, and
technical assistance are the unit's primary purposes.

Contact(s):
    Dr. Steven Hughes, ASSISTANT UNIT LEADER OF
    FISHERIES
    Phone: 410-651-7664
    sghughes@mail.umes.edu
    Dr. Dixie Bounds, ASSISTANT UNIT LEADER OF WILDLIFE
    Phone: 410-651-6913
    dlbounds@mail.umes.edu
    James Wiley, UNIT LEADER
    Phone: 410-651-7654
    jwwiley@mail.umes.edu

UNIVERSITY OF TULSA

600 S. College.101 Harwell (Petroleum Abstracts)
Tulsa, OK 74104-3189 USA
Phone: 918-631-2295   Fax: 918-599-9361
E-mail: dbrown@utulsa.edu
Website: www.pa.utulsas.edu

Founded: NA
Scope: Statewide

Description: Information on ecology and pollution related to
petroleum exploration, production and transportation, plus
environmental, health and safety topics. TULSA, includes
50,000 environmentally related entries updated weekly.

Contact(s):
    David Brown, MARKETING MANAGER

V

VA POLYTECHNIC INSTITUTE

FISH AND WILDLIFE INFORMATION EXCHANGE
DPT. OF FISHERIES AND WILDLIFE
203 W. Roanoke St.
Blacksburg, VA 24060 USA
Phone: 540-231-7348   Fax: 540-231-7019
E-mail: fwiexchg@vt.edu
Website: fwie.fw.vt.edu

Founded: NA

Description: Produces "The Master Species File", an archive of
species accounts compiled by state and federal fish and
wildlife agencies in North America.

Contact(s):
    Sheila Ratcliff

VIDEO PROJECT

P.O. Box 77188
San Francisco, CA 94107 USA

Contact(s):
    Quinn Kanaly
    quinn@videoproject.net

W

WALKABOUT PRODUCTIONS, INC.

45 Old Solomons Island Rd.
Annapolis, MD 21401 USA
Phone: 410-573-1228   Fax: 410-573-9521
E-mail: info@walkaboutinc.com
Website: www.walkaboutinc.com

Founded: 1975
Scope: International

Description: Walkabout Productions, Inc., is an EMMY award
winning production team that focuses on environment, wildlife,
and science documentaries. Filmography is available.

Contact(s):
    Allison Nichols, PRODUCER

WEST VIRGINIA RAPTOR REHABILITATION CENTER

P.O. Box 333
Morgantown, WV 26505 USA
Phone: 304-366-9286
E-mail: raptor@wvrrc.org
Website: www.wvrrc.org

Founded: NA
Scope: Regional
Publication(s): The Falcon

WYOMING NATURAL DIVERSITY DATABASE

P.O. Box 3381
Laramie, WY 82071-3381 USA
Phone: 307-766-3023   Fax: 307-766-3026
E-mail: wndd@uwyo.edu
Website: www.uwyo.edu/wyndd

Founded: NA
Scope: Regional
Contact(s):
    Gary Beauvais

EDUCATIONAL INSTITUTIONS

A

ACADIA UNIVERSITY
24 University Ave., Patterson Hall
Wolfville, Nova Scotia B0P 1X0 Canada
Phone: 902-542-2201   Fax: 902-585-1059
E-mail: biology@acadiau.ca
Website: www.acadiau.ca

Founded: 1838
Membership: 18
Scope: National, International

Description: Primarily an undergraduate university, emphasizing a liberal education in a balanced blend of arts, science, and professional studies. Masters degrees are offered in biology, chemistry, computer science, education, English, geology, political science, physiology, and sociology.

Publication(s): See publication web site

Contact(s):
P. Taylor
Phone: 902-585-1287
Fax: 902-585-1059
philip.taylor@arcadiau.ca
Don Stewart, ASSISTANT PROFESSOR, BIOLOGY DEPARTMENT
Phone: 902-585-1391
Fax: 902-585-1059
don.stewart@arcadiau.ca
David Stiles, ENVIRONMENTAL CHEMISTRY, (B.SC., HON., M.SC.)
Phone: 902-585-1325
Fax: 902-585-1114
david.stiles@acadiau.ca
Robert Raeside, ENVIRONMENTAL GEOLOGY, (B.SC., HON., M.SC.)
Phone: 902-585-1323
robert.raeside@arcadiau.ca
Tom Herman, HEAD OF BIOLOGY DEPARTMENT
Phone: 902-585-1469
tom.herman@arcadiau.ca
Glyn Bissex, RECREATION MANAGEMENT, (B.R.M, HON., BKIN, HON.)
glyn.bissex@arcadiau.ca
Soren Bondrup-Niselsen, WILDLIFE, FISHERIES, AQUATIC BIOLOGY, MARINE ECOLOGY, MAMMALOGY, ANIMAL BEHAVIOR, ORNITHOLOGY, CONSERVATION BIOLOGY, MOLECULAR ECOLOGY, LANDSCAPE ECOLOGY
Phone: 902-585-1424
Fax: 902-585-1059
soren.bondrup-nielsen@aradiau.ca

ALFRED UNIVERSITY
DIVISION OF ENVIRONMENTAL STUDIES
Saxon Dr.
Alfred, NY 14802-1205 USA
Phone: 607-871-2634   Fax: 607-871-2697
E-mail: ens@alfred.edu
Website: www.ens@alfred.edu

Founded: 1971
Scope: Statewide

Description: The program offers an undergraduate degree in multidisciplinary environmental studies in a liberal arts setting. Students can focus on either natural or social sciences, and many take a second major in biology, geology, political science, economics, etc. The project-oriented program is supervised by fifteen faculty members from different disciplines.

Contact(s):
Diana Sinton, ASSISTANT PROFESSOR OF GEOGRAPHY AND ENVIRONMENTAL STUDIES
ens@alfred.edu
Michele Hluchy, CHAIR OF DIVISION ENS AND PROFESSOR OF GEOLOGY AND ENVIRONMENTAL STUDIES
Phone: 507-871-2634
ens@alfred.edu

ANTIOCH COLLEGE
795 Livermore St
Yellow Springs, OH 45387 USA
Phone: 937-767-7331   Fax: 93-776-7331
E-mail: admissions@anitoch-college.edu
Website: www.antioch-college.edu/

Founded: NA
Membership: 650
Scope: National
Publication(s): Colleges that Changes Lives

Contact(s):
Jill Yager, BIOLOGY
jyager@antioch-college.edu
Peter Townsend, GEOLOGY
ptownsend@antioch-college.edu
Charles Taylor, PHYSICS AND SOLAR ENERGY/ALTERNATIVE TECHNOLOGY
ctaylor@antioch-college.edu

ANTIOCH NEW ENGLAND GRADUATE SCHOOL, ENVIRONMENTAL STUDIES
40 Avon St.
Keene, NH 03431 USA
Phone: 603-357-3122   Fax: 603-357-0718
E-mail: admissions@antiochne.edu
Website: www.antiochne.edu

Founded: NA
Scope: State

Description: Antioch New England Graduate School offers professional training for effective, reflective, environmental leadership. The M.S. Degree in Environmental Studies is a field-oriented program that stresses professional preparation in environmental biology, teaching, communication, administration, policy, and environmental education. Biology and general science teacher certifications are available. The M.S. degree in Resource Management and Administration integrates environmental science, natural resources policy, and administration. The Ph.D Program in Environmental Studies offers an interdisciplinary approach to research in Environmental Education and Environmental Policy.

ANTIOCH UNIVERSITY SEATTLE
ENVIRONMENT AND COMMUNITY PROGRAM
2326 Sixth Ave.
Seattle, WA 98121-1814 USA
Phone: 206-441-5352   Fax: 206-441-3307
Website: www.seattleantioch.edu

Founded: est. in 1852; in Seattle since 1975
Scope: Local, Regional

Description: The E&C program approaches environmental challenges with natural science literacy for professionals in environmental or community development fields. Students gain a clear understanding of the scientific, economic and institutional dimensions of environmental issues. The program publishes the Environment and Community Newsletter.

Contact(s):
Don Comstock, DIRECTOR
Phone: 206-441-5352
Courtney Putnam, PROGRAM ASSISTANT
Phone: 206-441-5352
cputnam@antiochsea.edu
Jonathan Scherch, PROGRAM COORDINATOR
Phone: 206-441-5352
scherch@antiochsea.edu

APPALACHIAN STATE UNIVERSITY
426 Sanford Hall
Boone, NC 28608 USA
Phone: 828-262-2000   Fax: 828-262-6472
Website: www.appstate.edu

Founded: NA
Membership: 2000
Scope: Statewide

Contact(s):
Francis Borkowski, CHANCELLOR
Kim Siegenthaler, RECREATION MANAGEMENT PROGRAM, DIRECTOR
Phone: 828-262-2540
siiegenthalkl@appstate.edu
Jeff Boyer, SUSTAINABLE DEVELOPMENT MINOR, DIRECTOR
426 Sanford Hall, Boone, NC 28608
boyerjc@appstate.edu

ARIZONA STATE UNIVERSITY; CENTER FOR ENVIRONMENTAL STUDIES
CENTER FOR ENVIRONMENTAL STUDIES
Box 873211, Arizona State University
Tempe, AZ 85287-3211 USA
Phone: 480-965-2975   Fax: 480-965-8087
Website: www.asu.edu/ces or caplter.asu.edu

Founded: NA
Scope: State

Description: The Center is involved in the Central Arizona-Phoenix Long Term Ecological Research (CAP LTER) project at Arizona State University, funded by the National Science Foundation and is one of the first urban sites in the LTER network. CAP LTER provides a unique addition to CAP TER research by focusing upon an arid-land ecosystem profoundly influenced, even defined by the presence and activities of humans. Investigations of land-use and ecological consequences in an urban environmental also involves community partners and K-12 schools. Our aim is to understand the changing urban fabric of our arid urban ecosystems and to offer applications to arid cities across the globe.

Contact(s):
Nancy Grimm, CO-PROJECT DIRECTOR
Charles Redman, CO-PROJECT DIRECTOR

ARKANSAS STATE UNIVERSITY
DEPARTMENT OF BIOLOGICAL SCIENCE
P. O. Box 1030
State University, AR 72467 USA
Phone: 870-972-3082   Fax: 870-972-2638
Website: www.csm.astate.edu/~biology/biology.html

Founded: NA
Scope: State

Contact(s):
David Harding, DIRECTOR OF ENVIRONMENTAL SCIENCES PROGRAM
Phone: 870-972-2007
Fax: 870-972-2638
envirsci@navajo.astate.edu

ARKANSAS TECH UNIVERSITY
BIOLOGY DEPT.
1701 North Boulder Ave.
Russellville, AR 72801 USA
Phone: 501-964-0852   Fax: 501-964-0837
Website: www.atu.edu/fish&wildlife

Founded: 1909
Scope: Statewide

Description: Recreation and Park Adminstration offers five areas of emphasis: Recreation Administration, Therapeutic Recreation, Park Administration, Turf Management and Interpretive Naturalist

Contact(s):
Joseph Stoeckel, FISHERIES AND WILDLIFE BIOLOGY, DIRECTOR
Phone: 501-964-0852
joe.stoeckel@mail.atu.edu
Charlie Gagen, HEAD OF BIOLOGICAL SCIENCES
Phone: 501-964-0814
charlie.gagen@mail.atu.edu
Theresa Herrick, RECREATION AND PARK ADMINISTRATION, DIRECTOR
Williamson Hall, Room 100, Russellville, AR 72081
Phone: 501-968-0378
theresa.herrick@mail.atu.edu

AUBURN UNIVERSITY
COLLEGE OF AGRICULTURE
DEPT. OF FISHERIES AND ALLIED AQUACULTURES
Swingle Hall
Auburn University, AL 36849 USA
Phone: 334-844-4786   Fax: 334-844-9208
Website: www.ag.auburn.edu/dept/faa/

Founded: NA
Membership: 150
Scope: Statewide

Description: The department sponsors the Southeastern Cooperative Fish Disease Project, providing a fish-kill diagnostic service, training in fish diseases and research on fish diseases to the cooperating member states.

Publication(s): Publications on website

Contact(s):
John Grizzle, ASSOCIATE PROJECT DIRECTOR
John Jensen, DEPARTMENT HEAD
jjensen@acesag.auburn.edu

B. Duncan, DIRECTOR, INTERNATIONAL CENTER FOR AQUACULTURE AND AQUATIC ENVIRONMENTS
bduncan@acesag.auburn.edu

AUBURN UNIVERSITY

COLLEGE OF SCIENCES AND MATHEMATICS
DEPT. OF BIOLOGICAL SCIENCES
59 Duggar Dr., Extension Cottage
Auburn University, AL 36849 USA
Phone: 334-844-4830   Fax: 334-844-5748
Website: www.auburn.edu/cosam

Founded: NA
Scope: Statewide

Description: Newly formed from merger of Zoology and Botany departments.

Contact(s):
Alfred Brown, DEPARTMENT CO-HEAD
101 Life Science Bldg., Dept Of Biological Sciences, Auburn University, AL 36849
Phone: 334-844-1661
Fax: 334-844-1645

AUBURN UNIVERSITY SCHOOL OF FORESTRY & WILDLIFE SCIENCES

SCHOOL OF FORESTRY AND WILDLIFE SCIENCES
108 M. White Smith Hall
Auburn University, AL 36849-5418 USA
Phone: 334-844-1007   Fax: 334-844-1084
Website: www.forestry.auburn.edu/

Founded: NA
Membership: 100
Scope: International

Contact(s):
Richard Brinker, DEAN

B

BALL STATE UNIVERSITY, DEPARTMENT OF NATURAL RESOURCES & ENVIRONMENTAL MANAGEMENT

DEPARTMENT OF NATURAL RESOURCES AND ENVIRONMENTAL MANAGEMENT
Muncie, IN 47306 USA
Phone: 765-285-5780   Fax: 765-285-2606
E-mail: nrem@bsu.edu
Website: www.bsu.edu/nrem

Founded: NA
Membership: 8
Scope: Statewide

Contact(s):
John Pichtel, DEPARTMENTAL MINOR IN ENVIRONMENTAL MANAGEMENT, ADVISOR
Phone: 765-285-2182
jpichtel@bsu.edu
Amy Sheaffer, ENVIRONMENTAL INTERPRETATION AND OUTDOOR RECREATION MANAGEMENT OPTION, ADVISOR
Phone: 765-285-5781
asheaffer@bsu.edu
Fred Siewert, ENVIRONMENTAL PROTECTION OPTION, ADVISOR

James Eflin, INTERDEPARTMENTAL MINORS IN ENERGY RESOURCES AND ENVIRONMENTAL POLICY, ADVISOR
Phone: 765-285-2327
jeflin1@bsu.edu
Hugh Brown, NATURAL RESOURCES AND ENVIRONMENTAL MANAGAMENT, (B.S., B.A., M.A., M.S.), CHAIR
Phone: 765-285-5788
hbrown@bsu.edu
Paul Chandle, NATURAL RESOURCES STUDIES OPTION, ADVISOR
Phone: 765-285-5788
pchandle@bsu.edu
Thad Godish, OCCUPATIONAL AND INDUSTRIAL HYGIENE OPTION, ADVISOR
Phone: 765-285-5782
tgodish@bsu.edu
Timothy Lyon, TEACHING MINOR IN ENVIRONMENTAL STUDIES, ADVISOR
Phone: 765-285-5783
tlyon@bsu.edu

BARD COLLEGE

DEBARD CENTER FOR ENVIRONMENTAL POLICY
P.O. Box 5000
Annandale-on-Hudson, NY 12504-5000 USA
Phone: 845-758-7071   Fax: 845-758-7636
E-mail: cep@bard.edu
Website: www.bard.edu/cep

Founded: NA
Membership: 20
Scope: Statewide

Description: Bard College offers an intensive graduate degree program leading to a Master of Science in Environmental Studies. Students develop an understanding of key ecological and natural concepts and the ability to become effective environmental professionals. Coursework is offered during the summer in two four-week sessions. Students can complete degree requirements, including course and thesis, in three summers. In the year 2001, Bard is launching a master's degree program during the academic year as part of its new Center for Environmental Policy (CEP). The center is dedicated to teaching, research and public service.

Publication(s): Open Forum Report

Contact(s):
Marie Beichert, ASSISTANT DIRECTOR
Phone: 845-758-7071
Kris Feder, ASSOCIATE DIRECTOR
Joanne Fox-Pski, DIRECTOR

BEMIDJI STATE UNIVERSITY

CENTER FOR ENVIRONMENTAL, EARTH AND SPACE STUDIES
P.O. Box 27, 1500 Birchmont Dr., NE
Bemidji, MN 56601 USA
Phone: 218-755-2910   Fax: 218-755-4107
Website: www.bemidji.msus.edu/

Founded: 1968
Scope: International

Description: The Center for Environmental Studies is a research and teaching unit directed towards understanding our physical, biological, and social environment, and preventing its deterioration. The center conducts laboratory and field studies, both internally and externally funded, and offers baccalaureate and master's degree programs.

Contact(s):
Patrick Welle, DIRECTOR

BOWLING GREEN STATE UNIVERSITY

Center for Environmental Programs,153 College Park Office Building
Bowling Green, OH 43403 USA
Phone: 419-372-8207   Fax: 419-372-7243
E-mail: envs@bgnet.bgsu.edu)
Website: www.bgsu.edu/department/envp

Founded: NA
Membership: 9
Scope: Statewide

Description: Offer undergraduate environmental degree programs in Arts and Sciences, Environmental Policy, and Analysis/Environmental Science)/Health and Human Services.

Contact(s):
Holly Myers-Jones, DIRECTOR

BRADLEY UNIVERSITY OF LIBERAL ARTS SCIENCE BIOLOGY DEPT.

BIOLOGY DEPT.
1501 W. Bradley Ave.
Peoria, IL 61625 USA
Phone: 309-677-3020   Fax: 309-677-3558
Website: www.bradley.edu/academics/las/bio/

Founded: NA
Membership: 10
Scope: International

Contact(s):
Janet Gehring, PLANT BIOLOGY
Phone: 309-677-3017
jgehring@bradley.edu
Kelly McConnaughay, PLANT ECOLOGY
Phone: 309-677-3018
kdm@bradley.edu

BROWN UNIVERSITY

CENTER FOR ENVIRONMENTAL STUDIES
Box 1943
Providence, RI 02912 USA
Phone: 401-863-3449   Fax: 401-863-3503
E-mail: envstu@brown.edu
Website: www.envstudies.brown.edu-dept/

Founded: 1978
Membership: 70
Scope: Regional

Description: The Center for Environmental Studies offers three interdisciplinary degrees (A.B., Sc.B., and M.A.) in environmental problem-solving; coordinates and facilitates environmental efforts within the university community; and collaborates with both state government agencies and community-based groups on projects to improve environmental quality for all Rhode Island residents. All programs aim to integrate teaching, scholarship, and service.

Contact(s):
Patti Caton, ADMINISTRATIVE MANAGER
patti_caton@brown.edu
Harold Ward, DIRECTOR

harold_ward@brown.edu
Kurt Teichert, ENVIRONMENTAL COORDINATOR

C

CALIFORNIA POLYTECHNIC STATE UNIVERSITY; COLLEGE OF ARCHITECTURE AND ENVIRONMENTAL DESIGN

COLLEGE OF ARCHITECTURE AND ENVIRONMENTAL DESIGN
One Grand Ave.
San Luis Obispo, CA 93407 USA
Phone: 805-756-1321   Fax: 805-756-5986
E-mail: caed@polymail.calpoly.edu
Website: www.calpoly.edu/~caed/

Founded: NA
Scope: Statewide
Publication(s): Publications on website

Contact(s):
William Siembieda, DEPARTMENT HEAD, CITY AND REGIONAL PLANNING
Phone: 805-756-1315
wsiembie@calpoly.edu
Walter Bremer, DEPARTMENT HEAD, LANDSCAPE ARCHITECTURE
wbremer@calpoly.edu

CALIFORNIA STATE UNIVERSITY AT CHICO DEPT OF RECREATION & PARKS MANAGEMENT

Dept of Recreation and Parks Management
Chico, CA 95929-0560 USA
Phone: 530-898-6408   Fax: 530-898-6557
E-mail: recr@csuchico.edu
Website: www.csuchico.edu/recr

Founded: NA
Membership: 15
Scope: National

Description: Areas of study include environmental education and interpretation, recreation and natural resource management, parks maintenance and operations, and planning and design.

Contact(s):
Jon Hooper, COORDINATOR, PARKS AND NATURAL RESOURCES MANAGEMENT OPTION
Phone: 530-898-5811
Emilyn Sheffield, DEPARTMENT CHAIR
Phone: 530-898-4855
esheffield@csuchico.edu

CALIFORNIA STATE UNIVERSITY AT FULLERTON

SCHOOL OF HUMANITIES AND SOCIAL SCIENCES
ENVIRONMENTAL STUDIES PROGRAM
Humanities H-420A
Fullerton, CA 92834 USA
Phone: 714-278-4373
E-mail: mhogarth@fullerton.edu
Website: http://hss.fullerton.edu/envstud/index.html

Founded: 1970
Scope: Statewide

Description: Interdisciplinary graduate program leading to master's degree in environmental sciences, environmental policy and planning, or environmental education and communication.

Contact(s):
Robert Voeks, PROGRAM DIRECTOR
Phone: 714-278-3361

CALIFORNIA STATE UNIVERSITY AT SACRAMENTO

ENVIRONMENTAL STUDIES DEPARTMENT
6000 J St.
Sacramento, CA 95819 USA
Phone: 916-278-6620   Fax: 916-278-7582
E-mail: infodesk@csus.edu
Website: www.csus.edu\index.stm

Founded: NA
Membership: 2
Scope: Statewide

Description: Biology department offers a concentration in Biological Conservation. Interdisciplinary Environmental Studies program offers a B.A.. Recreation and Leisure Studies program offers a B.S. or B.A. in Park and Recreation Resource Management.

Contact(s):
C. Vanicek, ADVISOR, CONSERVATION BIOLOGY
Phone: 916-278-6569
Laurel Heffernan, CHAIR, DEPT. OF BIOLOGICAL SCIENCES
Phone: 916-278-6535
Fax: 916-278-6993
Steven Gray, CHAIR, RECREATION AND LEISURE STUDIES
graysw@csus.edu
Tom Krabacher, ENVIRONMENTAL STUDIES
Phone: 916-278-6620
Fax: 916-278-7582
wrighta@csus.edu
Cary Goulard, GRADUATE COORDINATOR, RECREATION AND LEISURE STUDIES
goulardc@hhsserver.hhs.csus.edu

CALIFORNIA UNIVERSITY OF PENNSYLVANIA

BIOLOGICAL AND ENVIRONMENTAL SCIENCES DEPT.
250 University Ave.
California, PA 15419-1394 USA
Phone: 724-938-4200   Fax: 724-938-1514
Website: www.cup.edu

Founded: NA
Scope: Statewide

Contact(s):
David Boehm, BIOLOGY, (B.S., M.S.OPTION), CHAIR
Phone: 724-938-4200
Thomas Moon, ENVIRONMENTAL CONSERVATION (OPTION)
Phone: 724-938-4204
moon@cup.edu
William Kimmel, ENVIRONMENTAL POLLUTION CONTROL (OPTION)
Phone: 724-938-4213
kimmel@cup.edu

Allan Miller, ENVIRONMENTAL STUDIES PROGRAM COORDINATOR
Phone: 724-938-4462
miller@cup.edu
David Argent, WILDLIFE BIOLOGY (OPTION)
Phone: 724-938-1529
argent@cup.edu

CENTRAL MICHIGAN UNIVERSITY

Department of Biology, 184 Brooks Hall
Mt. Pleasant, MI 48859 USA
Fax: 987-774-4000
Website: www.cmich.edu/

Founded: NA

Contact(s):
Michael Hamas, CONSERVATION BIOLOGY (B.S., M.S.), CONTACT
Phone: 517-774-3185
Douglas Peterson, FISHERIES (B.S., M.S) CONTACT
Phone: 517-774-3377
Scott McNaught, WATER RESOURCES (B.S., M.S.), CONTACT
Phone: 517-774-1335
John Krull, WILDLIFE, (B.S., M.S.), CONTACT
Phone: 517-774-3412

CITY UNIVERSITY OF NEW YORK

COLLEGE OF STATEN ISLAND
ENVIRONMENTAL SCIENCE MASTERS PROGRAM
6S-310, 2800 Victory Blvd.
Staten Island, NY 10314 USA
Phone: 718-982-2000   Fax: 718-982-3923
E-mail: gerstle@postbox.csi.cuny.edu
Website:
www.library.csi.cuny.edu/dept/as/ces/escpgm.htm

Founded: NA
Scope: Local

Description: The interdisciplinary masters program in Environmental Science includes ecology, geology, chemistry, environmental engineering, and computer modeling. The objective of the masters program is to expose the students to the scientific principles underlying environmental problems. Research is carried out on wetlands, park planning, air, water and soil pollution, waste disposal, aquatic toxics, environmental epidemiology and risk analysis. Courses are offered in the evenings for full and part time students.

Contact(s):
Alfred Levine, DIRECTOR

CITY UNIVERSITY OF NEW YORK

HUNTER COLLEGE
695 Park Ave.
New York, NY 10021 USA
Phone: 212-772-4490
Website: www.hunter.cuny.edu

Founded: NA

Contact(s):
Charles Heatwole, DEPARTMENT OF GEOGRAPHY, AFFILIATED WITH CITY UNIVERSITY OF NEW YORK PH.D PROGRAM IN EARTH AND ENVIRONMENTAL SCIENCE, CHAIRMAN
Phone: 212-772-5265
Fax: 212-772-5268

Jeffery Osleeb, ENERGY AND ENVIRONMENTAL POLICY STUDIES PROGRAM
Phone: 212-772-5413
Fax: 212-772-5268
Louise Sherby, WEXLER LIBRARY CHIEF LIBRARIAN
Phone: 212-772-4146
Fax: 212-772-4142

CLARK UNIVERSITY INTERNATIONAL DEVELOPMENT, COMMUNITY PLANNING AND ENVIRONMENT

INTERNATIONAL DEVELOPMENT, COMMUNITY PLANNING AND ENVIRONMENT
950 Main St.
Worcester, MA 01610 USA
Phone: 508-793-7201   Fax: 508-793-8820
E-mail: idce@clark.edu
Website: www.clarku.edu

Founded: 1972
Membership: 25
Scope: International

Description: The International Development Program uses a multidisciplinary approach in research and teaching to analyze issues of underdevelopment in Asia, Africa, and Latin America. It draws on faculty from the fields of geography (including GIS), environmental studies, management, anthropology, economics, politics, and history, and serves both U.S. and international students. (B.A. and M.A. degree offered)

Publication(s): Introduction to PRA, PRA Handbook, Implementing PRA, A Manual for Socio-Economic and Gender Analysis, Tools of Gender Analysis

Contact(s):
Richard Ford, CENTER FOR COMMUNITY-BASED DEVELOPMENT, DIRECTOR
Phone: 508-793-7691
rford@clarku.edu
William Fisher, INTERNATIONAL DEVELOPMENT COMMUNITY PLANNING & ENVIRONMENT
Phone: 508-421-3765
wfisher@clarku.edu
Barbara THOMAS-SLAYTER, INTERNATIONAL DEVELOPMENT PROGRAM
Phone: 508-793-7454
bslayer@clarku.edu

CLEMSON UNIVERSITY

AQUACULTURE, FISHERIES AND WILDLIFE
G08 Lehotsky Hall
Clemson, SC 29634 USA
Phone: 864-656-3117   Fax: 864-656-5332
Website: www.virtual.clemson.edu/groups/AFW/

Founded: NA
Membership: 350
Scope: Statewide

Description: The curriculum leading to a B.S. degree provides a solid foundation in basic and applied science, social science, and communication skills. Emphasis areas permit students to broaden their technical knowledge in their chosen career path. Those interested in pursuing a graduate degree program in aquaculture, fisheries, or wildlife management shoud have sound undergraduate training in the biological or related sciences. Programs of study are designed to emphasize relationships between wild animals and their changing environments, or production of aquatic organisms. The graduate program in wildlife biology is accredited by the Southeastern Section of the Wildlife Society.

Publication(s): see publication web site

Contact(s):
John Sweeney, CHAIR
Phone: 864-656-5333
jrswny@clemson.edu
Robert Barkley, DIRECTOR OF ADMISSIONS
Phone: 864-656-2287
Phone: 864-656-2464

CLEMSON UNIVERSITY

SCHOOL OF THE ENVIRONMENT
342 Computer Court Rich Lab, Research Park
Anderson, SC 29625 USA
Phone: 864-656-5567   Fax: 864-656-0672
Website: www.ces.clemson.edu/ees./

Founded: NA
Membership: 20
Scope: State

Description: Made up of the Environmental Engineering and Science Dept., the Environmental Toxicology Dept., and the Geological Sciences Dept. Administers university-wide Environmental Science and Policy Program.

Contact(s):
Alan Elzerman, CHAIR OF ENV. ENGINEERING AND SCIENCE; DIR. OF THE SCHOOL OF THE ENVIRONMENT; PROGRAM COORDINATOR OF THE ENVIRONMENTAL SCIENCE AND POLICY
Phone: 864-656-5568
awlzrmn@clemson.edu
John Rodgers, ENVIRONMENTAL TOXICOLOGY, (M.S., PH.D.), CHAIR
Phone: 864-646-2239
Alan Elzerman, GEOLOGICAL SCIENCE, CHAIR
Phone: 864-656-5568
awlzrmn@clemson.edu
Pam Fjeld, STUDENT SERVICES COORDINATOR
Phone: 864-656-1010
hpamela@clemson.edu

COLLEGE OF THE ATLANTIC

HUMAN ECOLOGY
105 Eden St.
Bar Harbor, ME 04609 USA
Phone: 207-288-5015   Fax: 207-288-2328
E-mail: inquiry@ecology.coa.edu
Website: www.coa.edu

Founded: NA
Membership: 75
Scope: National, International

Description: The College of the Atlantic is a fully accredited four-year residential college. Students are attracted to its excellent programs in marine biology, environmental studies and ecology, environmental design, public policy, education, and selected humanities. Over 250 students. Awards a B.A. and M. PH. in human ecology. Summer programs in field studies for teachers.

Contact(s):
Steven Katona, PRESIDENT

COLLEGE OF WILLIAM AND MARY

VIRGINIA INSTITUTE OF MARINE SCIENCE/SCHOOL
OF MARINE SCIENCE
P.O. Box 1346
Gloucester Point, VA 23062 USA
Phone: 804-684-7000   Fax: 804-684-7097
E-mail: http://www.vims.edu/mailman
Website: www.vims.edu/

Founded: 1940

Description: A state institution founded for providing research,
advisory services, and education for the public and for state
and federal agencies responsible for managing marine
resources.

Contact(s):
Richard Wetzel, BIOLOGICAL SCIENCES, CHAIR
Phone: 804-684-7381
Gene Silberhorn, COASTAL AND OCEAN POLICY, CHAIR
Phone: 804-684-7382
L. Wright, DEAN AND DIRECTOR
Phone: 804-684-7103
E. Burreson, DIRECTOR OF RESEARCH AND ADVISORY
SERVICES
Phone: 804-684-7108
M. Roberts, ENVIRONMENTAL SCIENCES, CHAIR
Phone: 804-684-7260
J. Graves, FISHERIES SCIENCES, CHAIR
Phone: 804-684-7352
William Dupaul, HEAD OF MARINE ADVISORY SERVICES
Phone: 804-684-7164
S. Kuehl, PHYSICAL SCIENCES, CHAIR
Phone: 804-684-7118

COLORADO MOUNTAIN COLLEGE

TIMBERLINE CAMPUS
901 S. Hwy. 24
Leadville, CO 80461 USA
Phone: 719-486-2015   Fax: 719-486-3212
Website: www.coloradomtn.edu

Founded: NA
Membership: 19
Scope: Local

Description: Colorado Mountain College is a public two-year
community college. The College is fully accredited by the North
Central Accrediting Association.

Contact(s):
Nancy Cain, ASSISTANT PROFESSOR OF BIOLOGY
Phone: 719-486-4241
ncain@coloradomtn.edu
Jessica Clement, CONTACT
Phone: 719-486-4209
jclement@coloradomtn.edu
Kent Clement, PROFESSOR OF OUTDOOR
RECREATIONAL LEADERSHIP

COLORADO STATE UNIVERSITY

COLLEGE OF NATURAL RESOURCES
101 Metro Resources Bldg
Fort Collins, CO 80523 USA
Phone: 970-491-6675   Fax: 970-491-0279
E-mail: webadmin@cnr.colstate.edu
Website: www.cnr.colostate.edu

Founded: NA

Scope: Statewide
Contact(s):
Joyce Berry, ASSISTANT DEAN
Phone: 970-491-5405
Fax: 970-491-0279
joyceb@cnr.colostate.edu
David Anderson, COOPERATIVE FISH AND WILDLIFE
RESEARCH UNIT, LEADER
Phone: 970-491-1414
A. Dyer, DEAN
Phone: 970-491-4997
Judith Hannah, EARTH RESOURCES (B.S., M.S., PH.D.),
HEAD
Phone: 970-491-5662
Randall Robinette, FISHERY AND WILDLIFE BIOLOGY
(B.S., M.S., PH.D.), HEAD
Phone: 970-491-5020
Susan Stafford, FOREST SCIENCES (B.S., M.S., M.F.,
PH.D.), HEAD
Phone: 970-491-6911
Fax: 970-491-6754
stafford@cnr.colostate.edu
Diana Wall, NATURAL RESOURCES ECOLOGY
LABORATORY, CONTACT
Phone: 970-491-2504
Michael Manfredo, NATURAL RESOURCES RECREATION
AND TOURISM (B.S., M.S., PH.D.), HEAD
Phone: 970-491-0474
Fax: 970-491-2255
manfredo.cnr.colostate.edu
Dennis Child, RANGELAND ECOSYSTEM SCIENCE
(B.S., M.S., PH.D.), HEAD
Phone: 970-491-4994
Fax: 970-491-2339
dennisc@cnr.colostate.edu

COLORADO STATE UNIVERSITY

DEPARTMENT OF POLITICAL SCIENCE
ENVIRONMENTAL POLITICS AND POLICY
Clark Building C-346,
Fort Collins, CO 80523 USA
Phone: 970-491-5156   Fax: 970-491-2490
Website: www.colostate.edu/depts/polisci/grad2.html

Founded: 1975
Scope: Statewide

Description: All Ph.D. students in the program choose
Environmental Politics and Policy as one of three subfields in
political science offered in preparation for their degree. The
program prepares doctoral students for university positions
and a wide variety of private and public sector careers related
to environmental politics and policy.

Contact(s):
Wayne Peak, CHAIR & UNDERGRADUATE COORDINATOR
wayne.peak@colostate.edu
Dimitris Stevis, GRADUATE COORDINATOR
sksmith@lamar.colostate.edu

CONNECTICUT COLLEGE

270 Mohegan Ave.
New London, CT 06320 USA
Phone: 860-439-5021   Fax: 860-439-2519
Website: www.conncoll.edu

Founded: NA

Scope: International

Description: Environmental Studies has a long and successful history at Connecticut College beginning in 1931 with the establishment of the Connecticut College Arboretum. Since then, a common theme in the program has been to understand the structure and functioning of both natural and managed ecosystems.

Contact(s):

T. Owen, BOTANY DEPARTMENT, CHAIR
Phone: 860-439-2147
tpowe@conncoll.edu
Glenn Dreyer, CENTER FOR CONSERVATION BIOLOGY AND ENVIRONMENTAL STUDIES, DIRECTOR
Phone: 860-439-2144
Fax: 860-439-5482
gddre@conncoll.edu
Peter Siver, DIRECTOR, ENVIRONMENTAL STUDIES PROGRAM
Phone: 860-439-2160
Fax: 860-439-2519
pasiv@conncoll.edu
Phillip Barnes, ZOOLOGY DEPARTMENT, CHAIR
Phone: 860-439-2148
Fax: 860-439-2519
ptbar@conncoll.edu

CONWAY SCHOOL OF LANDSCAPE DESIGN

46 Delabarre Ave.
Conway, MA 01341 USA
Phone: 413-369-4044   Fax: 413-369-4032
E-mail: info@csld.edu
Website: www.csld.edu

Founded: 1972
Membership: 4
Scope: National

Description: CSLD is a ten-month graduate program in environmentally sound site design and land use planning. The degree offered is a M.A. degree in Landscape Design. The curriculum is structured around professional level work for residential clients, municipal agencies, and non-profit organizations. Through these projects, students produce the drawings and reports characteristic of the designer/planner while learning technical skills and developing intellectual abilities. Integrated throughout is a strong emphasis on communication skills and ecological processes.

Publication(s): Con text (annual newsletter)

Contact(s):

Nancy Braxton, ADMINISTRATIVE DIRECTOR
Donald Walker, DIRECTOR

CORNELL UNIVERSITY, COLLEGE OF AGRICULTURE & LIFE SCIENCES

COLLEGE OF AGRICULTURAL AND LIFE SCIENCES
DEPT. OF NATURAL RESOURCES
118 Fernow Hall
Ithaca, NY 14853 USA
Phone: 607-255-2821   Fax: 607-255-0349
Website: www.dnr.cornell.edu

Founded: NA
Membership: 50
Scope: International

Contact(s):

James Lassoie, CHAIR
jpl4@cornell.edu
Barbara Knuth, CO-LEADER, HUMAN DIMENSIONS RESEARCH UNIT
Phone: 607-255-2822
bak3@cornell.edu
Edward Mills, CORNELL BIOLOGICAL FIELD STATION
900 Shackelton Point Rd., Bridgeport, NY 13030-9750
Phone: 315-633-9243
Fax: 315-633-2358
elm5@cornell.edu
Richard Baer, ENVIRONMENTAL ETHICS
Phone: 607-255-7797
rab12@cornell.edu
Timothy Fahey, FOREST SCIENCE
Phone: 607-255-5470
tjf5@cornell.edu
Charles Smith, PLANT AND WILDLIFE INVENTORY
Phone: 607-255-3219
crs6@cornell.edu
Marian Hovencamp, UNDERGRADUATE PROGRAM, ASSISTANT
Phone: 607-255-2809
mth6@cornell.edu

DALHOUSIE UNIVERSITY

SCHOOL FOR RESOURCE AND ENVIRONMENTAL STUDIES (SRES)
1312 Robie St.
Halifax, Nova Scotia B3H 3E2 Canada
Phone: 902-494-3632   Fax: 902-494-3728
E-mail: sres@is.dal.ca
Website: www.mgmt.dal.ca/sres/

Founded: 1975
Scope: Regional

Description: Graduate school within the Faculty of Management of Dalhousie University, offering a master of environmental studies (M.E.S.) degree, through a two year programme (thesis required). Emphasis of programme is on policy and management aspects.

Contact(s):

Peter Duinker, DIRECTOR

DARTMOUTH COLLEGE

ENVIRONMENTAL STUDIES PROGRAM
6182 Steele Hall, Rm. 113
Hanover, NH 03755-3577 USA
Phone: 603-646-2838   Fax: 603-646-1682
Website: www.dartmouth.edu/

Founded: 1970
Scope: International

Description: Interdisciplinary academic program providing students with the opportunity to assess the seriousness and complexity of environmental problems and to understand how to search for solutions. Faculty research interests include biological conservation, ecosystem ecology, air pollution, economics, and international environmental governance.

Contact(s):

Andrew Friedland, CHAIRMAN

DEPAUL UNIVERSITY
BIOLOGICAL SCIENCES
McGowan Center
2325 North Clifton Ave.
Chicago, IL 60614-3207 USA
Phone: 773-325-7595   Fax: 773-325-7596
Website: www.dupaul.edu/`biology
Membership: 17
Scope: State

Contact(s):
Stan Cohn, CONTACT

DEPAUL UNIVERSITY
ENVIRONMENTAL SCIENCES
2325 N. Clifton Ave.
Chicago, IL 60614-3207 USA
Phone: 773-325-7422   Fax: 773-325-7448
Website: http://www.depaul.edu/~envirsci/

Contact(s):
Thomas Murphy, CHAIRMAN

DONALD BREN SCHOOL OF ENVIRONMENTAL SCIENCE & MANAGEMENT
4670 Physical Sciences North
Santa Barbara, CA 93106 USA
Phone: 805-893-7611   Fax: 805-893-7612
E-mail: jricharson@bren.ucsb.edu
Website: http://www.esm.ucsb.edu

Description: We are a graduate program in environmental science and management, offering a specialization in conservation biology.

Contact(s):
Jill Richardson, OUTREACH COORDINATOR
Phone: 805-893-7980
jrichardson@bren.ucsb.edu

DREXEL UNIVERSITY
SCHOOL OF ENVIRONMENTAL SCIENCE, ENGINEERING, AND POLICY
32nd and Chestnut St.
Philadelphia, PA 19104 USA
Phone: 215-895-2266   Fax: 215-895-2267
E-mail: sesep@drexel.edu
Website: www.drexel.edu/sesep/

Founded: NA
Scope: Internationally

Description: Environmental Engineering and Science undergraduate and graduate study is offered by the School of Environmental Science, Engineering, and Policy at Drexel University. Over 25 faculty participate in SESEP programs. Degrees available with specializations in air pollution, environmental assessment, environmental biotechnology, environmental chemistry, environmental health, hazardous and solid waste, subsurface contaminant hydrology, water and wastewater treatment, water resources, environmental risk management and ecology. Programs are offered on a full or part-time basis.

Contact(s):
Claire Welty, ASSOCIATE DIRECTOR
Phone: 215-895-2281
weltyc@drexel.edu

DUKE UNIVERSITY (NICHOLAS SCHOOL OF EXTERNAL AFFAIRS)
NICHOLAS SCHOOL OF THE ENVIRONMENT
Box 90328
Durham, NC 27708-0328 USA
Phone: 919-613-8000   Fax: 919-684-8741
E-mail: scottee@duke.edu
Website: www.env.duke.edu

Founded: NA
Scope: Statewide

Contact(s):
Curtis Richardson, CHAIR OF THE DIVISION OF THE ENVIRONMENT
Michael Orbach, COASTAL ENVIRONMENTAL MANAGEMENT AND DUKE UNIVERSITY MARINE LABORATORY, DIRECTOR
mko@mail.duke.edu
Norman Christensen, DEAN
normc@duke.edu
Jeffrey Karson, EARTH AND OCEAN SCIENCES
jkarson@geo.duke.edu
Richard Di Giulio, ENVIRONMENTAL TOXICOLOGY AND CHEMISTRY
richd@duke.edu
Daniel Richter, FOREST RESOURCE MANAGEMENT AND RESOURCE ECOLOGY
drichter@duke.edu
Randall Kramer, PROGRAM CHAIR OF RESOURCE ECONOMICS AND POLICY
Phone: 919-613-8072
Fax: 919-684-8741
Kenneth Reckhow, WATER & AIR RESOUCES
reckhow@duke.edu

DUKE UNIVERSITY TROPICAL STUDIES
ORGANIZATION FOR TROPICAL STUDIES
Durham, NC 27708-0381 USA
Phone: 919-684-5774   Fax: 919-684-5661
E-mail: nao@duke.edu
Website: www.ots.duke.edu/

Founded: 1988
Membership: 20
Scope: International

Description: The goal of the Center is to contribute to the alleviation of the world environmental crisis, particularly as it affects the developing countries of the tropics. The CTC works toward this goal through interdisciplinary research into issues of environmental policy relevance, training in environmental management, and dissemination of information.

Publication(s): see publication web site

Contact(s):
Nora Bynum, ACADEMIC DIRECTOR
elb@duke.edu

E

EASTERN ILLINOIS UNIVERSITY
Charleston, IL 61920 USA
Phone: 217-581-3126   Fax: 217-581-7141
Website: www.eiu.edu/~biology/

Founded: NA
Membership: 35

Scope: Statewide

Description: Eastern Illinois University offer an undergraduate degree in Biology, with an three options: Teacher Certification, Environmental Biology and Biological Sciences. Within the Biological Sciences, students choose from among 4 concentrations: Biology, Botanical Sciences, Ecology and Systematics, and Cell and Functional Biology. Emphasis is placed upon a fundamental understanding of biology and environmental concerns.

Contact(s):

Charles Costa, BIOLOGICAL SCIENCES (M.S.), CONTACT
Phone: 217-581-2520
cfcjc@eiu.edu
James McGaughey, BIOLOGY TEACHER CERTIFICATE OPTION COORDINATOR
Phone: 217-581-2928
cfjam@eiu.edu
Kipp Kruse, BIOLOGY, CHAIR
Phone: 217-581-3126
cfkck@eiu.edu
Robert Fischer, ENVIRONMENTAL BIOLOGY OPTION COORDINATOR
Phone: 217-581-2817
cfruf@eiu.edu

EASTERN KENTUCKY UNIVERSITY

Biological Sciences Department, 521 Lancaster Ave.
Richmond, KY 40475-3102 USA
Phone: 859-622-1531   Fax: 859-622-1399
Website: www.eku.edu

Founded: NA
Membership: 22
Scope: National
Publication(s): Newsletter

Contact(s):

Barbara Ramey, APPLIED ECOLOGY (M.S.), CONTACT
Phone: 606-622-1531
bioramey@acs.eku.edu
Guenter Shuster, BIOLOGY—AQUATIC OPTION (B.S.)
bioschus@acs.eku.edu
Ross Clark, BIOLOGY—BOTANY OPTION (B.S.)
bioclark@acs.eku.edu
Charles Elliott, ENVIRONMENTAL STUDIES (B.S.), CONTACT
Phone: 859-622-1531
bioelliott@acs.eku.edu
William Martin, LILLIE WOODS RESEARCH NATURAL AREA
narmartin@acs.eku.edu
Robert Frederick, WILDLIFE MANAGEMENT (B.S.) CONTACT
Phone: 859-622-1531
biofred@acs.eku.edu

EASTERN MICHIGAN UNIVERSITY

316 Mark Jefferson
Ypsilanti, MI 48197 USA
Phone: 734-487-4242   Fax: 734-487-9235
Website: www.emich.edu

Founded: NA
Scope: National, International
Contact(s):
Robert Neely, BIOLOGY, (B.S., M.S.), HEAD
316 Mark Jefferson, EMU, Ypsilanti, MI 48197

Phone: 734-487-4242
bio_neely@online.emich.edu
Catherine Bach, CONSERVATION RESOURCE USE, CONTACT
Phone: 734-487-0212
bio_bach@online.emich.edu
Michael Kasenow, GEOGRAPHY AND GEOLOGY, (B.S.,M.S.), HEAD
203 Strong Hall, EMU, Ypsilanti, MI 48197
Phone: 734-487-0218
geo_kasenow@online.emich.edu
Ben Czinski, KRESGE ENVIRONMENTAL EDUCATION CENTER, DIRECTOR
2816 Fish Lake Rd., Lapeer, MI 48446
Phone: 810-667-2350
bio_czinski@online.emich.edu

EMORY UNIVERSITY

BIOLOGY DEPARTMENT
Rollins Research Center, Emory University
Atlanta, GA 30322 USA
Phone: 404-727-6048   Fax: 404-727-2880
Website: www.emory.edu/biology/

Founded: NA
Membership: 26
Scope: Statewide

Contact(s):

John Lucchesi, BIOLOGY DEPARTMENT, CHAIR
Chris Beck, ECOLOGY AND EVOLUTION, PROFESSOR

EMPORIA STATE UNIVERSITY

BIOLOGICAL SCIENCES
Biological Sciences Campus Box 4050
Emporia, KS 66801 USA
Phone: 316-341-5311   Fax: 316-341-5607
E-mail: mooredwi@emporia.edu
Website: www.emporia.edu/biosci/biology

Founded: NA
Scope: Statewide
Publication(s): see publication web site

Contact(s):

Marshall Sundberg, DEPT. OF BIOLOGICAL SCIENCES, ECOLOGY AND WILDLIFE BIOLOGY (B.S., M.S.), CONTACT

F

FAU PINE JOG ENVIRONMENTAL EDUCATION CENTER

6301 Summit Blvd.
West Palm Beach, FL 33415 USA
Phone: 561-686-6600   Fax: 561-687-4968
Website: www.pinejog.org

Founded: 1960
Membership: 22
Scope: Regional, Local

Description: Pine Jog is an environmental education center within the College of Education of Florida Atlantic University. The purpose of the Center is to provide environmental education programs which foster an awareness and appreciation of the natural world, promote an understanding of ecological concepts, and instill a sense of stewardship towards the earth and all of its inhabitants.

Contact(s):
Donald Mathis, CHAIR, BOARD OF DIRECTORS
Sartory, Mathis, & Beedle, 5840 Corporate Way, West Palm Beach, FL 33407
Phone: 561-683-7500
Patricia Welch, EXECUTIVE DIRECTOR
Phone: 561-686-6600

FERRIS STATE UNIVERSITY
COLLEGE OF ALLIED HEALTH SCIENCES
200 Ferris Dr.
Big Rapids, MI 49307-2740 USA
Phone: 231-591-2313   Fax: 231-591-3788
Website: www.ferris.edu/htmls/colleges/alliedhe/

Founded: 1964
Membership: 3
Scope: Statewide

Description: Educational institution offering B.S. in industrial and environmental health management with options in general environmental health, hazardous materials management, industrial hygiene, and industrial safety.

Contact(s):
Ellen Haneline, HEALTH MANAGEMENT DEPARTMENT, HEAD
Phone: 231-591-2313
ellen_j_haneline@ferris.edu

FERRUM COLLEGE
DEPT. OF FORESTRY & WILDLIFE
Ferrum, VA 24088 USA
Phone: 540-365-2121
E-mail: webmaster@ferrum.edu
Website: www.ferrum.edu

Founded: NA
Scope: National
Publication(s): The Iron Blade, The Chrysalis, Ferrum Alumni Magazine

Contact(s):
Ron Stephens, AGRICULTURE (B.S.)
Phone: 540-365-4360
rstephens@ferrum.edu
Bob Pohlad, BIOLOGY (B.S.)
Phone: 540-365-4367
bpohlad@ferrum.edu
David Johnson, CHEMISTRY (B.S.)
Phone: 540-365-4364
djohnson@ferrum.edu
Joseph Stogner, ENVIRONMENTAL STUDIES (B.S.)
Phone: 540-365-4369
jstogner@ferrum.edu
Michael Mengak, FORESTRY AND WILDLIFE
Phone: 540-365-4373
mmengak@ferrum.edu
Kathy Mengak, LEISURE SERVICES (B.S., B.A.) AND RECREATION
Phone: 540-365-4387
kmengak@ferrum.edu

FLORIDA STATE UNIVERSITY
UNIVERSITY RELATIONS
216 Westcott Bldg
Tallahassee, FL 32306 USA
Phone: 850-644-2525   Fax: 850-644-3612
Website: www.fsu.edu

Founded: NA
Membership: 30
Scope: Regional

Contact(s):
Bruce Grindal, ANTHROPOLOGY (B.S., M.S., PH.D.), CHAIRMAN
Bellamy G-24, Tallahassee, FL 32306-2150
Phone: 850-644-8147
Fax: 850-644-4283
bgrindal@mailer.fsu.edu
Thomas Roberts, BIOLOGICAL SCIENCE (B.S., M.S., PH.D.), CHAIRMAN
P.O. Box 4340, Tallahassee, FL 32306-4340
Phone: 850-644-3700
Fax: 850-644-9829
Patrick O'Sullivan, GEOGRAPHY (POLITICAL GEOGRAPHY AND ENVIRONMENTAL STUDIES) (B.S., M.S., PH.D.), CHAIRMAN
P.O. Box 2190, Tallahassee, FL 32306-2190
Phone: 850-644-7175
Fax: 850-644-5913
kmcclell@mailer.fsu.edu
J Tull, GEOLOGY (B.S., M.S., PH.D.), CHAIRMAN
Carraway Bldg., Tallahassee, FL 32306-4100
Phone: 904-644-1448
Fax: 904-644-4214
tull@gly.fsu.edu
David Stuart, METEOROLOGY (B.S., M.S., PH.D.), CHAIRMAN
404 Love Building, Tallahassee, FL 32306-4520
Phone: 850-644-6205
Fax: 850-644-9642
stuart@met.fsu.edu
Wilton Sturges, OCEANOGRAPHY (M.S., PH.D.), CHAIRMAN
329 OSB, West Call Street, Tallahassee, FL 32306-4320
Phone: 850-644-6700
Fax: 850-644-2581
sturges@ocean.fsu.edu

FROSTBURG STATE UNIVERSITY (UNIVERSITY OF MARYLAND)
DEPARTMENT OF BIOLOGY
101 Braddock Rd.
Frostburg, MD 21532 USA
Phone: 301-687-4166   Fax: 301-687-3034
Website: www.fsu.umd.edu

Founded: NA
Scope: Statewide

Description: Wildlife and Fisheries Program (B.A, M.A., Ph.D.), Wildlife/Fisheries Biology (M.S.), Applied Ecology; Conservation Biology (M.S.), Biology (B.S., Ph.D.)

Publication(s): see publication web site

Contact(s):
David Morton, DEPT. CHAIR
Phone: 301-687-4355
dmorton@frostburg.edu.

G

GEORGE WASHINGTON UNIVERSITY

2121 I Street , NW
Washington, DC 20052 USA
Phone: 202-994-1000
Website: www.gwu.edu/

Founded: NA
Scope: Statewide

Contact(s):
Henry Merchant, ENVIRONMENTAL AND RESOURCE
POLICY (M.A., PH.D.), DIRECTOR
Phone: 202-994-7123
Fax: 202-994-6100
Theodore Toridis, ENVIRONMENTAL ENGINEERING (B.S.,
M.S., D.SC.), ACTING CHAIR
801 22nd St., Washington, DC 20052
Phone: 202-994-6749
Fax: 202-944-0238
toridis@seas.gwu.edu
Henry Merchant, ENVIRONMENTAL STUDIES (B.A., B.S.,
M.S.), DIRECTOR
Phone: 202-994-7118

GEORGE WASHINGTON UNIVERSITY

LAW SCHOOL
2000 H. St., NW
Washington, DC 20052 USA
Phone: 202-994-6260
Website: www.law.gwu.edu/

Founded: 1865
Scope: International

Description: Nation's largest graduate and undergraduate environmental law program. Twenty-two environmental courses for J.D. and LL.M. students in addition to land use and other related topics. Emphasizes a practical approach.

Contact(s):
Laurent Hourcle, CO-DIRECTOR
Phone: 202-994-4823
lhourcle@main.nlc.gwu.edu

GEORGETOWN UNIVERSITY

LAW CENTER
600 New Jersey Ave., NW
Washington, DC 20001 USA
Phone: 202-662-9000   Fax: 202-662-9444
E-mail: admis@law.georgetown.edu
Website: www.law.georgetown.edu/

Founded: NA
Membership: 300
Scope: National

Contact(s):
Judith Areen, DEAN

GEORGETOWN COLLEGE

ENVIRONMENTAL SCIENCE PROGRAM
400 E. College St.
Georgetown, KY 40324 USA
Phone: 502-863-8088   Fax: 502-868-7744
Website: www.georgetowncollege.edu

Founded: NA

Scope: Statewide

Description: Environmental Science Degree with tracks in Chemical Science, Biological Science, Chemical-Biological Science and Environmental Policy.

Contact(s):
Rick Kopp, PROGRAM COORDINATOR
Phone: 502-868-7744
rkopp@georgetowncollege.edu

GEORGIA INSTITUTE OF TECHNOLOGY

GEORGIA WATER INSTITUTE
School of Civil & Environmental Engineering Georgia
Institute of Technology
Atlanta, GA 30332-0335 USA
Phone: 404-894-3776   Fax: 404-894-3828
Website: www.gatech.edu

Founded: NA
Scope: Statewide

Contact(s):
Aris Georgakokos, DIRECTOR, GEORGIA WATER
INSTITUTE
Phone: 404-894-2240

H

HAMLINE UNIVERSITY

CENTER FOR GLOBAL ENVIRONMENTAL EDUCATION
1536 Hewitt Ave.
St. Paul, MN 55104-1284 USA
Phone: 651-523-2480   Fax: 651-523-2987
E-mail: cgee@hamline.edu
Website: cgee.hamline.edu

Founded: 1990
Scope: Statewide

Description: CGEE was founded to nurture greater understanding of the interconnectedness of local and global environments among educators, students, scientists, and citizens.

Publication(s): Publications on website

Contact(s):
Tracy Fredin, DIRECTOR
Phone: 651-523-3105
Fax: 651-523-2987
tfredin@gw.hamline.edu

HOCKING COLLEGE

SCHOOL OF NATURAL RESOURCES
3301 Hocking Parkway
Nelsonville, OH 45764 USA
Phone: 740-753-3591   Fax: 740-753-2021
E-mail: admissions@hocking.edu
Website: www.hocking.edu

Founded: 1969
Scope: State

Description: The mission of our School of Natural Resources is to prepare individuals for careers as technicians in the natural resources profession. Emphasis is placed on basic theory, developing a sustained postive work ethic and the practical application of the development of the competencies required for entry-level positions in recreation, wildlife, forestry, timber harvesting/tree care and a wide variety of land management technology fields.

Contact(s):
  Russell Tippett, DEAN, 740-753-3591
  tippet_r@hocking.edu
  Lloyd Wright, FISHERIES, BIOLOGIST
  wright_ll@hocking.edu
  Albert Lecount, WILDLIFE, BIOLOGIST
  lecount_a@hocking.edu

HUMBOLDT STATE UNIVERSITY

1 Harpst St.
Arcata, CA 95521-8299 USA
Phone: 707-826-3256   Fax: 707-826-3562
E-mail: cnrs@humboldt.edu
Website: www.humboldt.edu/~cnrs/

Founded: NA
Scope: Statewide

Contact(s):
  Milton Boyd, CHAIRMAN, BIOLOGICAL SCIENCES
  Phone: 707-826-3246
  David Hankin, CHAIRMAN, FISHERIES
  Phone: 707-825-5645
  Susan Bicknell, CHAIRMAN, FORESTRY
  Phone: 707-826-4243
  Steven Carlson, CHAIRMAN, NATURAL RESOURCES
  PLANNING AND INTERPRETATION
  Phone: 707-826-4147
  Marie Deanglis, CHAIRMAN, OCEANOGRAPHY
  Phone: 707-826-4147
  Steven Carlson, CHAIRMAN, RANGELAND RESOURCES
  AND WILDLAND SOILS
  Phone: 707-826-4147
  Mark Colwell, CHAIRMAN, WILDLIFE
  Phone: 707-826-3723
  John Howard, DEAN
  Russel Boham, DIRECTOR, INDIAN NATURAL
  RESOURCES, SCIENCES, AND ENGINEERING PROGRAM
  Phone: 707-826-4994
  Walter Duffy, LEADER, COOPERATIVE FISHERY
  RESEARCH UNIT
  Phone: 707-826-3268
  Robert Ziemer, PROJECT LEADER, EXPERIMENT
  STATION, PACIFIC SOUTHWEST FOREST AND RANGE
  STATION
  Phone: 707-825-2936

I

IDAHO STATE UNIVERSITY

DEPARTMENT OF BIOLOGICAL SCIENCES
Box 8007
Pocatello, ID 83209 USA
Phone: 208-282-3765   Fax: 208-236-4570
E-mail: bios@isu.edu
Website: www.isu.edu/departments/bios/

Founded: NA

Description: The Department of Biological Sciences at Idaho
  State University has high quality degree programs in ecology.
  Strong basic coursework and original investigations are
  emphasized at the undergraduate and graduate levels.
  Habitats available for study range from cold sagebrush deserts
  to heavily forested areas and includes streams and riparian
  areas in the Snake River Canyon to its headwaters in
  Yellowstone National Park.

Contact(s):
  Mary Watwood, ASSOCIATE PROFESSOR
  Phone: 208 2-36 -3090
  watwmari@isu.edu
  Rod Seeley, ECOLOGY (B.S., M.S., PH.D.)
  Phone: 208 2-82 -2181
  seelrodn@isu.edu

ILLINOIS STATE UNIVERSITY

ENVIRONMENTAL HEALTH PROGRAM, DEPARTMENT
OF HEALTH SCIENCES
Campus Box 5220
Normal, IL 61790-5220 USA
Phone: 309-438-8329   Fax: 309-438-2450
Website: www.ilstu.edu/

Founded: 1974
Membership: 30
Scope: Statewide

Description: Undergraduate education for B.S. in environmental
  health. Five faculty persons and 165 enrolled students. Four-
  year undergraduate curriculum accredited by National
  Environmental Health Science and Protection Accreditation
  Council. Graduate Education for M.S. in Environmental Health
  and Safety.

Contact(s):
  Thomas Bierma, MASTERS PROGRAM COORDINATOR
  Marilyn Morrow, PROGRAM DIRECTOR (ACTING)

INDIANA STATE UNIVERSITY

Science Bldg., Rm. 256
Terre Haute, IN 47809 USA
Phone: 812-237-2400   Fax: 812-237-4480
Website: www.biology.indstate.edu/dls/

Founded: NA
Membership: 27
Scope: Statewide
Publication(s): see publication web site

Contact(s):
  Marion Jackson, ECOLOGY AND WILDLIFE
  (B.S., M.S., PH.D)
  lsmjack@scifac.indstate.edu

INDIANA UNIVERSITY

SCHOOL OF PUBLIC AND ENVIRONMENTAL AFFAIRS
1315 E 10th St., 4th Fl.
Bloomington, IN 47405 USA
Phone: 812-855-2457   Fax: 812-855-7802
E-mail: speainfo@indiana.edu
Website: www.spea.indiana.edu

Founded: 1972
Membership: 200
Scope: Statewide

Description: The School of Public Environmental Affairs brings an
  interdisciplinary approach to the study of the environmental
  sciences. The focus of the academic programs is to teach
  techniques that will help graduates preserve and protect the
  quality of natural resources, identify environmental hazards,
  and significantly contribute to solutions to enhance quality of
  life in the world's communities.

Contact(s):
  Robert Agranoff, ASSOCIATE DEAN
  Astri Merget, DEAN

David Jones, DIRECTOR OF GRADUATE PROGRAMS
J. Randolph, DIRECTOR OF PH.D PROGRAMS IN
ENVIRONMENTAL SCIENCE
Roger Parks, DIRECTOR OF PH.D. PROGRAMS IN PUBLIC
POLICY AND PUBLIC AFFAIRS
Frank Vilardo, DIRECTOR OF UNDERGRADUATE
PROGRAMS

IOWA STATE UNIVERSITY
COLLEGE OF AGRICULTURE COMMUNICATIONS
OFFICE
304 Curtis Hall
Ames, IA 50011-1050 USA
Phone: 515-294-5616   Fax: 515-294-8662
E-mail: edadcock@iastate.edu
Website: www.ag.iastate.edu/

Founded: NA
Membership: 7
Scope: Regional

Contact(s):
Bruce Menzel, ANIMAL ECOLOGY, FISHERIES AND
WILDLIFE BIOLOGY (B.S., M.S., PH.D.), CHAIRMAN
Phone: 515-294-6148
bmenzel@iastate.edu
J. Kelly, FORESTRY (B.S., M.S., PH.D.), CHAIRMAN
Dept. of Forestry, Ames, IA 50011-1021
Phone: 515-294-1166
jmkelly@iastate.edu
Jeff Iles, HORTICULTURE (B.S., M.S., PH.D.), HEAD
Rm. 106B Horticulture Hall, 50011-1100
Phone: 515-294-5893
Fax: 515-294-0730
iles@ia.state.edu

IOWA STATE UNIVERSITY
COLLEGE OF DESIGN
134 College of Design, Iowa State University
Ames, IA 50011 USA
Phone: 515-294-7428
Website: www.design.iastate.edu/

Founded: NA
Scope: Statewide
Publication(s): Design News

Contact(s):
Riad Mahayni, COMMUNITY AND REGIONAL PLANNING
(B.S., AND GRADUATE DEGREES), CHAIR
Phone: 515-294-8958
Fax: 515-294-4015
rmahayni@iastate.edu
J. Keller, LANDSCAPE ARCHITECTURE (B.L.A., M.L.A.),
CHAIRMAN
Phone: 515-294-5676
tkeller@iastate.edu

JOHNS HOPKINS UNIVERSITY
SCHOOL OF PUBLIC HEALTH
PEW ENVIRONMENTAL HEALTH COMMISSION
111 Market Pl., Suite 850
Baltimore, MD 21202 USA
Phone: 410-659-2690   Fax: 410-659-2699
E-mail: cllee@jhsph.edu
Website: pewenvirohealth.jhsph.edu/

Founded: NA
Scope: National

Description: The Pew Environmental Health Commission works
to strengthen the country's public health system to protect
against sickness and disease caused by environmental
threats.

Publication(s): Healthly From the Start, Attack Asthma, America's
Environmental Health Gap

Contact(s):
Paul Locke, DEPUTY DIRECTOR
Shelley Hearn, EXECUTIVE DIRECTOR

JOHNS HOPKINS UNIVERSITY
CENTER FOR A LIVABLE FUTURE
SCHOOL OF HYGIENE AND PUBLIC HEALTH
615 N. Wolfe St., Suite 8503
Baltimore, MD 21205 USA
Phone: 410-502-7576   Fax: 410-502-7579
E-mail: clf@jhsph.edu.
Website: www.jhsph.edu/environment/

Founded: NA
Scope: International

Description: The mission of the Center for a Livable Future is to
establish a global resource to develop and disseminate
information and to promote policies for the protection of health,
the global environment, and our ability to sustain life for future
generations.

Contact(s):
Polly Walker, ASSOCIATE DIRECTOR
Phone: 410-502-7578
pwalker@jhsph.edu
George Jakab, ASSOCIATE DIRECTOR FOR SCIENCE
gjakab@jhsph.edu
Robert Lawrence, DIRECTOR
Phone: 410-614-4590
rlawrenc@jhsph.edu
David Brubaker, PROGRAM DIRECTOR, HENRY
SPIRA/GRACE PROJECT ON INDUSTRIAL ANIMAL
PRODUCTION
Phone: 410-223-1722
dbrubake@jhsph.edu
Leo Horrigan, PROJECT COORDINATOR, URBAN
AGRICULTURE
Phone: 410-502-7575
lhorriga@jhsph.edu

JOHNS HOPKINS UNIVERSITY
DEPARTMENT OF GEOGRAPHY
313 Ames Hall, 3400 North Charles St.
Baltimore, MD 21218 USA
Phone: 410-516-7092   Fax: 410-516-8996
E-mail: dogee@jhu.edu
Website: www.jhu.edu/~dogee

Founded: NA
Membership: 25
Scope: Regional

Contact(s):
Grace Brush, ECOLOGY (M.A., PH.D.), CONTACT
Phone: 410-516-7107
gbrush@jhu.edu
Alan Stone, ENVIRONMENTAL CHEMISTRY (M.A., M.S.,
PH.D.), CONTACT
Phone: 410-516-8476
astone@jhu.edu
Edward Bouwer, ENVIRONMENTAL ENGINEERING (M.S.E.,
PH.D.), CONTACT
Phone: 410-516-7437
bouwer@jhu.edu
M. Wolman, NATURAL RESOURCES (M.A., M.S., PH.D.),
CONTACT
Phone: 410-516-7090
wolman@jhu.edu

JOHNSON STATE COLLEGE
DEPT. OF BIOLOGY AND ENVIRONMENTAL STUDIES
AND DEPT. OF HEALTH SCIENCES AND OUTDOOR
EDUCATION
337 College Hill
Johnson, VT 05656-9464 USA
Phone: 1 800 635 -2356   Fax: 802-635-1230
E-mail: jscapply@badger.jsc.vsc.edu
Website: www.jsc.vsc.edu/

Founded: NA

Contact(s):
John Wrazen, BABCOCK NATURE PRESERVE, DIRECTOR
Robert Genter, BIOLOGY
John Wrazen, ECOLOGY
Margaret Ottum, ENVIRONMENTAL SCIENCE AND
NATURAL RESOURCES
Martin Walker, GREEN CHEMISTRY
Karen Uhlendorf, OUTDOOR EDUCATION

K

KANSAS STATE UNIVERSITY
COLLEGE OF AGRICULTURE
117 Waters Hall College of Agriculture
Manhattan, KS 66506-5506 USA
Phone: 785-532-6151   Fax: 785-532-6897
E-mail: jax1@ksu.edu
Website: www.ag.ksu.edu/

Founded: NA
Membership: 3
Scope: International

Contact(s):
David Mengel, DEPT. OF AGRONOMY, HEAD
2004 Throckmorton Plant Science Center, Manhattan,
KS 66506
Phone: 785-532-6101
Fax: 785-532-6094
dmengel@bear.agron.ksu.edu
Thomas Warner, HORTICULTURE, FORESTRY AND
RECREATION RESOURCES, DEPT. HEAD
2021 Throckmorton Plant Science Center, Manhattan,
KS 66506
Phone: 785-532-6170
twarner@oznet.ksu.edu

Ted Cable, NATURAL RESOURCE MANAGEMENT
Phone: 785-532-1408
tcable@oznet.ksu.edu
Michel Ransom, SOIL AND WATER CONSERVATION
mdransom@ksu.edu

KANSAS STATE UNIVERSITY
DEPARTMENT OF LANDSCAPE ARCHITECTURE /
REGIONAL & COMMUNITY PLANNING
302 Seaton Hall
Manhattan, KS 66506-2909 USA
Phone: 785-532-5961   Fax: 785-532-6722
E-mail: la-rcp@ksu.edu
Website: www.aalto.arch.ksu.edu/lar/

Founded: NA
Scope: State, International

Contact(s):
Dan Donelin, DEPT. HEAD
dandon@ksu.edu
C. Keithley, REGIONAL AND COMMUNITY PLANNING,
DIRECTOR
Phone: 785-532-2440
cak@ksu.edu

KANSAS STATE UNIVERSITY
DIVISION OF BIOLOGY
232 Ackert Hall
Manhattan, KS 66506 USA
Phone: 785-532-6615   Fax: 785-532-6653
Website: www.ksu.edu/biology/

Founded: NA
Scope: Statewide

Contact(s):
Brian Spooner, DIRECTOR OF BIOLOGY
spoon1@ksu.edu
David Hartnett, DIRECTOR, KONZA PRAIRIE RESEARCH
NATURAL AREA
Phone: 785-532-5925
dchart@ksu.edu

KANSAS STATE UNIVERSITY
KANSAS COOPERATIVE FISH WILDLIFE UNIT
US GEOLOGICAL SURVEY
205 Leisure Hall
Manhattan, KS 66506-3501 USA
Phone: 785-532-6070   Fax: 785-532-7159
E-mail: kscfwru@ksu.edu
Website: www.ksu.edu

Founded: NA
Membership: 15
Scope: Local

Description: Provides graduate training and research in fisheries
and wildlife biology, research, management, ecology,
population dynamics, genetics, and related areas. Supported
cooperatively by Kansas State University, The Kansas
Department of Wildlife and Parks, the National Biological
Service and the Wildlife Management Institute.

Publication(s): Annual Report of Activities

Contact(s):
Christopher Guy, ASSISTANT LEADER OF FISHERIES
Jack Cully, ASSISTANT LEADER OF WILDLIFE
Philip Gipson, LEADER

KEENE STATE COLLEGE

DEPARTMENT OF ENVIRONMENTAL STUDIES
229 Main St.
Keene, NH 03435 USA
Phone: 603-352-1909 Fax: 603-358-2897
Website: www.keene.edu/

Founded: 1909
Scope: Statewide

Description: A multipurpose, predominantly undergraduate college with a central focus in the liberal arts and sciences. B.S. in Environmental Studies with options in Environmental Policy and Environmental Science and specializations in Environmental Biology, Environmental Chemistry and Environmental Geology

Contact(s):
Tim Allen, PROGRAM COORDINATOR
Phone: 603-358-2571
tallen@keene.edu

L

LAKE SUPERIOR STATE UNIVERSITY

SCHOOL OF NATURAL SCIENCES
650 W. Easterday Ave.
Sault Ste. Marie, MI 49783 USA
Phone: 906-635-2267 Fax: 906-635-2266
Website: www.lssu.edu

Founded: NA
Scope: Statewide

Description: Degrees offered in Biological science, conservation law enforcement, environmental chemistry, environmental science, fisheries/wildlife management and natural resources technology (A.D.).

Contact(s):
Gregory Zimmerman, BIOLOGY, CHAIR
Phone: 906-635-2470
David Myton, CHEMISTRY AND ENVIRONMENTAL SCIENCE, CHAIR (ACTING DEAN)
Phone: 906-635-2431
dmyton@gw.lssu.edu

LAKEHEAD UNIVERSITY

FORESTRY AND FOREST ENVIRONMENT
955 Oliver Rd.
Thunder Bay, Ontario P7B 5E1 Canada
Phone: 807-343-8507 Fax: 807-343-8116
E-mail: ulf.runesson@lakeheadu.ca
Website: www.lakeheadu.ca/~forwww/forestry.html

Founded: NA
Scope: Statewide

Description: B.S. in Forestry or Environmental Studies and M.S. in Forestry

Contact(s):
William Parker, ENVIRONMENTAL STUDIES
Phone: 807-343-8484
whparker@flash.lakeheadu.ca
K. Brown, GRADUATE FORESTRY PROGRAMS
Phone: 807-343-8114
ed.setliff@lakeheadu.ca
L. Meyer, UNDERGRADUATE FORESTRY PROGRAMS
Phone: 807-343-8445
gary.murchison@lakeheadu.ca

LEWIS AND CLARK COLLEGE

COLLEGE OF ARTS AND SCIENCES
0615 S.W. Palatine Hill Road
Portland, OR 97219 USA
Phone: 208-792-5272
Website: www.lclark.edu

Founded: NA

Description: Undergraduate major in Environmental Studies. Interdisciplinary with participting faculty drawn from all divisions of the college.

Contact(s):
Evan Williams, DIRECTOR OF ENVIRONMENTAL STUDIES
Phone: 503-768-7699
Fax: 503-768-7369
etw@lclark.edu

LEWIS AND CLARK COLLEGE

LAW SCHOOL
10015 S.W. Terwilliger Blvd .
Portland, OR 97219 USA
Phone: 503-768-6613 Fax: 503-768-6850
E-mail: lawadmss@lclark.edu
Website: www.lclark.edu

Founded: NA
Scope: International

Description: Strong environmental law training program (Environmental Law Certificate at J.D. level and specialized LL.M. in Environmental and Natural Resources Law); publish journal of Environmental Law; research program in Natural Resources Law Institute (newsletter: NRLI News); conferences and workshops through continuing education program; internships in natural resources; and environmental clinical opportunities.

Publication(s): 2002-2003 Catalog, Brochures

LOUISIANA STATE UNIVERSITY SCHOOL OF FORESTRY, WILDLIFE & FISHERIES

SCHOOL OF FORESTRY, WILDLIFE AND FISHERIES
Forestry Building, Rm. 124
Baton Rouge, LA 70803 USA
Phone: 225-388-4184 Fax: 578-388-4144
Website: www.coa.lsu.edu/fores/fores.html

Founded: NA
Membership: 30
Scope: Statewide

Contact(s):
Megan Lapayoro, ASSISTANT LEADER FISHERIES
Charles Bryan, COOPERATIVE FISH AND WILDLIFE RESEARCH UNIT, LEADER
Mary Ehrett, SECRETARY
mehrett@lsu.edu

LOUISIANA TECH UNIVERSITY

SCHOOL OF FORESTRY
P.O. Box 10138
Ruston, LA 71272 USA
Phone: 318-257-4985 Fax: 318-257-5061
Website: www.ans.latech.edu/forestry-index.html

Founded: NA
Membership: 13
Scope: Statewide

Educational Institutions

Description: Located in Louisiana's major forest region, the School of Forestry offers Bachelor of Science degrees in Forestry and Wildlife Conservation. Teaching facilities include a GIS/Remote Sensing Laboratory, a highly trained and diverse faculty and a successful placement record.

Contact(s):
James Dickson, COORDINATOR OF WILDLIFE PROGRAM
jdickson@ans.latech.edu
Mark Gibson, INTERIM DIRECTOR
mark.gibson@ans.atech.edu

M

MANCHESTER COLLEGE
KOINONIA ENVIRONMENTAL AND RETREAT CENTER
604 College Ave.
North Manchester, IN 46962 USA
Phone: 219-982-5010   Fax: 219-982-5043
Website: www.ares.manchester.edu/academic/koin.html

Founded: 1974
Scope: Regional

Description: The100-acre facility is used extensively to provide hands-on environmental science education for area students in grades, K-12. A two-story building houses the nature center with many educational displays. The retreat facility will accommodate 32 persons.

Contact(s):
Barbara Ehrhardt, DIRECTOR

MCGILL UNIVERSITY
DEPT. OF NATURAL RESOURCE SCIENCES
AVIAN SCIENCE AND CONSERVATION CENTRE
(ASCC)
21,111 Lakeshore Rd.
Ste. Anne de Bellevue, Quebec H9X 3V9 Canada
Phone: 514-398-7760   Fax: 514-398-7990
Website: www.nrs.mcgill.ca/ascc

Founded: 1974
Membership: 3
Scope: International

Description: To promote the study of birds and their conservation, we conduct pure and applied research in the field and laboratory; breed, release and manage endangered species; and train students and interns from all over the world. The Centre publishes a semi-annual newsletter, The Talon.

Publication(s): The Talon

Contact(s):
Rodger Titman, ASSOCIATE DIRECTOR
Ian Ritchie, CURATOR
Phone: 514-398-7932
Fax: 514-398-7540
ritchie@nrs.mcgill.ca
David Bird, DIRECTOR

MCNEESE STATE UNIVERSITY
Lake Charles, LA 70609 USA
Phone: 318-475-5690
Website: www.mcneese.edu/

Founded: NA
Scope: Statewide

Contact(s):
Billy Delany, DEPARTMENT OF AGRICULTURE, WILDLIFE MANGEMENT PROFESSOR
Phone: 318-475-5690

MIAMI UNIVERSITY
Boyd Hall
Oxford, OH 45056 USA
Phone: 513-529-5811   Fax: 513-529-5814
E-mail: havener@muohio.edu
Website: www.muohio.edu/

Founded: NA
Membership: 3
Scope: Regional

Contact(s):
Robert Benson, ARCHITECTURE (A.B., M.A.), CHAIRMAN
David Francko, BOTANY (A.B., B.S., M.A., PH.D.), CHAIRMAN
Michael Novak, CHEMISTRY (A.B., B.S., M.S., PH.D.), CHAIRMAN
James Rubenstein, GEOGRAPHY (A.B., M.A.) CHAIRMAN
Gene Willeke, INSTITUTE OF ENVIRONMENTAL SCIENCES (M.EN.), DIRECTOR
willekge@muohio.edu
Douglas Meikle, ZOOLOGY (A.B., B.S., M.A., M.S., PH.D.), CHAIRMAN

MICHIGAN STATE UNIVERSITY
COLLEGE OF AGRICULTURE AND NATURAL RESOURCES
13 Natural Resource Bldg.
East Lansing, MI 48824-1222 USA
Phone: 517-355-4478   Fax: 517-432-1699
E-mail: webmaster@perm3.sw.msu.edu
Website: www.fw.msu.edu

Founded: NA
Scope: Statewide

Contact(s):
Scott Witter, DEPARTMENT OF RESOURCES DEVELOPMENT (B.S., M.S., PH.D.), CHAIR
Phone: 517-355-3421
Fax: 517-353-8994
witter@msu.edu
Thomas Coon, FISHERIES AND WILDLIFE (B.S., M.S., PH.D.), CHAIR (ACTING)
Phone: 517-355-4478
Fax: 517-432-1699
Daniel Keathley, FORESTRY (B.S., M.S., PH.D.), CHAIRPERSON
Phone: 517-355-0093
Fax: 517-432-1143
keathley@msu.edu
Joseph Fridgen, PARK AND RECREATION RESOURCES (B.S., M.S., PH.D.), CHAIRPERSON
Phone: 517-353-5190
Fax: 517-432-3597

MICHIGAN TECHNOLOGICAL UNIVERSITY; SCHOOL OF FORESTRY AND WOOD PRODUCTS

SCHOOL OF FORESTRY AND WOOD PRODUCTS
1400 Townsend Dr.
Houghton, MI 49931 USA
Phone: 906-487-2454   Fax: 906-487-2915
E-mail: forestry@mtu.edu
Website: www.forestry.mtu.edu/

Founded: NA
Scope: Statewide

Description: Undergraduate concentrations and graduate programs in forest management science, forest biology and ecology, wildlife biology and ecology, and wood science and technology.

Publication(s): see publication web site

Contact(s):
Glenn Morz, FORESTRY, DEAN
Phone: 906-487-6303

MIDDLE TENNESSEE STATE UNIVERSITY ENVIRONMENTAL EDUCATION SYSTEM

CENTER FOR ENVIRONMENT EDUCATION
BIOLOGY DEPARTMENT
Box 60
Murfreesboro, TN 37132 USA
Phone: 615-898-5449   Fax: 615-898-5920
E-mail: jpkelly@mtsu.edu
Website: www.mtsu.edu/~biol/cee/ceehmpg.html

Founded: NA
Scope: Statewide

Description: The Center for Environment Education, an arm of the Biology Dept. offers a wide variety of offerings on topics related to the environment, including waste reduction and recycling. Workshops for school teachers are offered twice a semester. The center consults with teachers, youth leaders and education organizations in the areas of environmental education, curriculum, teacher training and hands-on learning.

Contact(s):
Padgett Kelly
Phone: 615-898-5615
jpkelly@mtsu.edu
Cindi Smith-Walters
Phone: 615-898-5449
csmithwa@mtsu.edu

MISSISSIPPI STATE UNIVERSITY

COLLEGE OF FOREST RESOURCES
Box 9820
Mississippi State, MS 39762 USA
Phone: 662-325-8530   Fax: 601-325-8726
Website: www.cfr.msstate.edu/

Founded: NA
Scope: Statewide

Contact(s):
Bob Karr, ASSICIATE DEAN
Box 9680, Mississippi State, MS 39762
Phone: 601-325-2793
bkarr@cfr.msstate.edu

Warren Thompson, EMERITUS
Box 9680, Mississippi State, MS 39762
Phone: 601-325-2952
wthompson@cfr.msstate.edu
Cynthia West, FOREST PRODUCTS (B.S., M.S.), HEAD
Phone: 601-325-2119
cwest@cfr.msstate.edu
Douglas Richards, FORESTRY (B.S., M.S.), HEAD
Box 9681, Mississippi State, MS 39762
Phone: 601-325-2949
drichards@cfr.msstate.edu
Bruce Leopold, WILDLIFE AND FISHERIES (B.S., M.S.), HEAD
Mississippi State, MS 39762
Phone: 601-325-2619
bleopold@cfr.msstate.edu

MONTANA STATE UNIVERSITY

COLLEGE OF AGRICULTURE
202 Linfield Hall
Bozeman, MT 59717 USA
Phone: 406-994-5744   Fax: 406-994-6579
E-mail: agweb@montana.edu
Website: www.montana.edu/agriculture/College/

Founded: NA
Scope: Statewide

Contact(s):
Myles Watts, AG ECONOMICS & ECONOMICS
Pete Burfemy, ANIMAL & RANGE SCIENCES
Greg Johnson, ANTOMOLOGY
Jeffery Jacobson, DEPT. OF LAND RESOURCES AND ENVIRONMENTAL SCIENCES (B.S., M.S., PH.D.), HEAD
Leon Johnson Hall, P.O. Box 173120, Bozeman, MT 59717
Phone: 406-994-7060
Fax: 406-994-3933
jefj@montana.edu, general information: kathyj@montana.edu
Norman Weeden, DEPT. OF PLANT SCIENCES AND HORTICULTURE, HEAD
119 AgBioScience Building, Bozeman, MT 59717-3150
Phone: 406-994-4832
Fax: 406-994-7600
nweeden@montana.edu, for general information: plantsciences@montana.edu
D. Harmsen, VETERINARY MOLECULAR BIOLOGY

MONTANA STATE UNIVERSITY DEPT. OF ECOLOGY

DEPARTMENT OF ECOLOGY
Lewis Hall
Bozeman, MT 59717 USA
Phone: 406-994-4548   Fax: 406-994-3190
E-mail: ecology@montana.edu
Website: www.montana.edu/ecology/

Founded: NA
Membership: 13
Scope: Statewide

Description: Biology Dept. offers B.S. in Biology, Biology Teaching, Biomedical Sciences or Fish and Wildlife Management and M.S. and Ph.D. programs with a concentration in Ecology, Conservation Biology, Plant Biology or Neurobiology.

Contact(s):
Jay Rotella, DEPARTMENT HEAD
Phone: 406-994-5676
rotella@montana.edu
Lynn Irby, FISH AND WILDLIFE MANAGEMENT PROGRAM,
COORDINATOR
Phone: 406-994-3252
ubili@montana.edu

MONTCLAIR STATE UNIVERSITY
COLLEGE OF SCIENCE AND MATHMATICS
One Normal Ave.
Upper Montclair, NJ 07043 USA
Phone: 973-655-4448
Website: csam.montclair.edu/

Founded: NA

Contact(s):
Bonnie Lustigman, BIOLOGY, CHAIR
lustigman@saturn.montclair.edu
Robert Taylor, EARTH AND ENVIRONMENTAL SCIENCE,
PROFESSOR
taylorr@saturn.montclair.edu

MOREHEAD STATE UNIVERSITY
DEPT. OF BIOLOGICAL & ENVIRONMENTAL
SCIENCES
123 Lappin Hall
Morehead, KY 40351 USA
Phone: 606-783-2944   Fax: 606-783-5002
Website: www.morehead-st.edu

Founded: NA
Membership: 20
Scope: Statewide
Publication(s): see publication web site

Contact(s):
David Magrane, BIOLOGY
A major in biology (organismal, genetics, microbiology, cell
biology, physiology, ecology) is offered (39 semester hours); a
minor is also available.
d.magrane@morehead-st.edu

MURRAY STATE UNIVERSITY
WILDLIFE
334 Blackburn Science
Murray, KY 42071-3346 USA
Phone: 270-762-2786   Fax: 270-762-2788
Website: www.murraystate.edu

Founded: NA
Membership: 4
Scope: Statewide
Publication(s): see publication web site

Contact(s):
David White, CENTER FOR RESERVOIR RESEARCH,
CONTACT
Phone: 270-474-2272
david.white@murraystate.edu
Tom Timmons, DEPARTMENT OF BIOLOGICAL SCIENCES,
FISHERIES, CHAIRMAN
Phone: 270-762-6754
tom.timmons@murraystate.edu
Stephen White, WILDLIFE (B.S., M.S.), CONTACT
Phone: 270-762-6298
steve.white@murraystate.edu

NEW MEXICO STATE UNIVERSITY
COLLEGE OF AGRICULTURE AND HOME
ECONOMICS
DEPT. OF FISHERY AND WILDLIFE SCIENCES
P.O. Box 30003, Dept. 4901
Las Cruces, NM 88003 USA
Phone: 505-646-1544   Fax: 505-646-1281
E-mail: natres@nmsu.edu
Website: www.leopold.nmsu.edu

Founded: NA
Scope: Statewide

Contact(s):
Donald Caccamise, DEPARTMENT HEAD

NEW MEXICO STATE UNIVERSITY
COLLEGE OF AGRICUTURE AND HOME ECONOMICS
DEPT. OF ANIMAL AND RANGE SCIENCES
Box 30003 Agriculture & Home Economics
Las Cruces, NM 88003 USA
Phone: 505-646-0111   Fax: 505-646-5975
Website: www.nmsu.edu/~dars

Founded: NA
Scope: Statewide

Description: In the Department of Animal and Range Sciences,
students can major in animal or range science. The
Department also offers pre-veterinary studies. In addition to
undergraduate degrees, the Department offers graduate
degrees at the Master of Science and Doctor of Philosophy
levels. The M.S. or Ph.D. in Animal Science can emphasize
nutrition or physiology, and the M.S. or Ph.D. in Range Science
students have the option to study in areas including, but not
exclusive to, range ecology and watershed management.

Contact(s):
Jerry Schickedanz, DEAN

NORTH CAROLINA STATE UNIVERSITY
COLLEGE OF AGRICULTURE & LIFE SCIENCES
Box 7642 115
Raleigh, NC 27695-7642 USA
Phone: 919-515-2614   Fax: 919-515-5266
Website: www.cals.ncsu.edu/

Founded: NA
Scope: Statewide

Contact(s):
James Young, BIOLOGICAL AND AGRICULTURAL
ENGINEERING (B.S., M.S., M.B.A.E., PH.D.), HEAD
Phone: 919-515-2694
Fax: 919-515-6772
jim_young@ncsu.edu
Gerald Van Dyke, BOTANY (B.S., M.S., M.L.S., PH.D.),
HEAD
Phone: 919-515-2222
William Grant, DIRECTOR OF UNDERGRADUATE
BIOLOGY PROGRAMS
Phone: 919-515-3341
Samuel Mozley, ECOLOGY (M.S.); ENVIRONMENTAL
SCIENCES (B.S.), COORDINATOR
Phone: 919-515-1981

Gerald Leblanc, ENVIRONMENTAL AND MOLECULAR TOXICOLOGY, HEAD
Phone: 919-515-7404
Fax: 919-515-7169
gal@unity.ncsu.edu
John Havlin, SOIL SCIENCE (B.S., M.S., M.N.R.A., PH.D.), HEAD
Phone: 919-515-2655
john_havlin@ncsu.edu
James Gilliam, ZOOLOGY (B.S., M.S., M.L.S., M.W.B., PH.D.), HEAD
Phone: 919-515-5978

NORTH DAKOTA STATE UNIVERSITY

Stevens Hall
Fargo, ND 58105 USA
Phone: 701-231-7087   Fax: 701-231-7149
Website: www.ndsu.nodak.edu/zoology/

Founded: NA
Scope: Local

Description: Wildlife and Fisheries Biology Option in Zoology

Contact(s):
Craig Stockwell, ASSISTANT PROFESSOR
Conservation Biology,
W. Bleier, DEPT. CHAIR
Gary Nuechterlein, PROFESSOR
Behavorial Ecology,
M. Butler, PROFESSOR
Aquatic Ecology, -213-7398

NORTHEASTERN UNIVERSITY

BIOLOGY DEPARTMENT
414 Mugar Life Sciences, 360 Huntington Ave.
Boston, MA 02115 USA
Phone: 617-373-2260   Fax: 617-373-3724
Website: www.dac.neu.edu/biology

Founded: NA
Membership: 24
Scope: Statewide

Contact(s):
Joseph Ayers, MARINE SCIENCE CENTER/MARINE BIOLOGY, (B.S., M.S., PH.D.), DIRECTOR
East Point Nahant, MA 01908
Phone: 617-581-7370
lobster@neu.edu
Gwilym Jones, VERTEBRATE SYSTEMATICS AND ECOLOGY (B.S., M.S., PH.D.)
Phone: 617-373-2851
g.jones@nunet.neu.edu

NORTHERN ARIZONA UNIVERSITY

COLLEGE OF ARTS AND SCIENCES
NAU Box 5640
Flagstaff, AZ 86011-5621 USA
Phone: 520-523-2381   Fax: 520-523-7500
E-mail: biology@nau.edu
Website: www.nau.edu/

Founded: NA
Scope: Statewide

Description: Dept. of Biology offers an emphasis in the areas of: Applied Plant Science, Aquatic Biology, Ecology, Cellular and Molecular Biology and Fish and Wildlife Management. Available emphasis areas for Environmental Science are: Biology, Chemistry, Applied Geology, Applied Mathematics, Microbiology, Environmental Administration and Policy, Environmental Communications and Environmental Management.

Contact(s):
Lee Drickamer, CHAIR, DEPT. OF BIOLOGICAL SCIENCES
Phone: 520-523-7501
Fax: 520-523-7500
lee.drickamer@nau.edu

NORTHERN ARIZONA UNIVERSITY

COLLEGE OF ECOSYSTEM SCIENCE AND MANAGEMENT
Box 15018
Flagstaff, AZ 86011-5018 USA
Phone: 520-523-3031   Fax: 520-523-1080
E-mail: esm.info@nau.edu
Website: www.cesm.nau.edu

Founded: NA
Scope: State

Description: Norhtern Arizona University is in an ideal location for the study of both forestry and recreation. Near Flagstaff are the largest ponderosa pine forest in America, five life zones within fifty miles, recreation and aesthetic areas, and extensive wildlife, grazing and watershed areas.

Contact(s):
Donald Arganbright, INTERIM DEAN
donald.g.arganbright@nau.edu

NORTHERN ARIZONA UNIVERSITY

COLLEGE OF ECOSYSTEM SCIENCE AND MANAGEMENT
DEPARTMENT OF GEOGRAPHY AND PUBLIC PLANNING
Box 15016
Flagstaff, AZ 86011-5016 USA
Phone: 520-523-2650   Fax: 520-523-1080
Website: www.geog.nau.edu/

Founded: NA
Scope: State

Description: For Public Planning, choice of emphasis in Land Use Planning or Environmental Planning

Contact(s):
Robert Clark, CHAIR, DEPARTMENT OF GEOGRAPHY AND PUBLIC PLANNING
roc@alpine.for.nau.edu

NORTHERN ARIZONA UNIVERSITY

NORTHERN ARIZONA ENVIRONMENTAL EDUCATION RESOURCES CENTER
CENTER FOR ENVIRONMENTAL SCIENCES AND EDUCATION
S. San Francisco St. 860011
Flagstaff, AZ 86011 USA
Phone: 928-523-9011   Fax: 520-523-5441
E-mail: paul.rowland@nau.edu
Website: http://www.nau.edu/~envsci/naeerc/index.html

Founded: 1994

Contact(s):
Paul Rowland, ASSOCIATE DIRECTOR
Phone: 520-523-5853

NORTHERN MICHIGAN UNIVERSITY

1401 Presque Isle Ave.
Marquette, MI 49855 USA
Phone: 906-227-2700   Fax: 906-227-2703
E-mail: artssci@nmu.edu
Website: www.nmu.edu

Founded: NA
Scope: Statewide
Publication(s): Update (Newsletter)

Contact(s):
Neil Cumberlidge, DEPARTMENT OF BIOLOGY (B.A., B.S., M.S.), HEAD
Luther S. West Science Bldg., Rm. 277, Marquette, MI 49855
Phone: 906-227-2310
Fax: 906-227-1063
ncumberl@nmu.edu
Michael Broadway, DEPARTMENT OF GEOGRAPHY, EARTH SCIENCE, CONSERVATION, AND PLANNING (B.A., B.S.), HEAD
Luther S. West Science Bldg., Rm. 213, Marquette, MI 49855
Phone: 906-227-2500
Fax: 906-227-1621
mbroadwa@nmu.edu

NORTHLAND COLLEGE

SIGURD OLSON ENVIRONMENTAL INSTITUTE
1411 Ellis Ave
Ashland, WI 54806 USA
Phone: 715-682-1223   Fax: 715-682-1218
E-mail: www.northland.edu/soei
Website: www.northland.edu/soei

Founded: 1972
Membership: 12
Scope: National

Description: The Sigurd Olson Environmental Institute was founded at Northland College in 1972 to increase public understanding of the complex relationships between natural and cultural environments in the Lake Superior region and to assist in developing workable solutions to regional environmental problems. The institute seeks to carry out Sigurd Olson's vision by fostering environmental citizenship and educating citizens for a sustainable future.

Contact(s):
Eileen Long, ADVISORY BOARD CHAIR
Mike Gardner, ASSISTANT DIRECTOR
Jerri Ridlon, COMMUNICATIONS SPECIALIST
Kenneth Bro, EXECUTIVE DIRECTOR
Jane Silberstein, LAKE SUPERIOR PROGRAM COORDINATOR
Carolyn Hanna, OFFICER MANAGER
Pam Troxell, TIMBER WOLF ALLIANCE COORDINATOR

NORTHWESTERN STATE UNIVERSITY OF LOUISIANA

WILDLIFE PROGRAM
Biology Dept.
Natchitoches, LA 71497 USA
Phone: 318-357-5323   Fax: 318-357-4518
Website: www.nsula.edu.

Founded: NA
Membership: 1
Scope: Local

Description: To educate students in principles and science of wildlife management; to prepare students for management of natural resources at the professional entry levels; and to provide an emphasis on biodiversity and ecosystems; to orient students toward interpersonal communication

Contact(s):
Dick Stalling, BIOLOGY SCIENCES, HEAD
Steven Gabrey, BIOLOGY/WILDLIFE MANAGEMENT (B.S.), ADVISOR
Phone: 318-357-5375
steveng@alpha.nsula.edu

O

OBERLIN COLLEGE

ADAM JOSEPH LEWIS CENTER
ENVIRONMENTAL STUDIES PROGRAM
122 Elm Street
Oberlin, OH 44074-1095 USA
Phone: 440-775-8747   Fax: 440-775-8946
E-mail: bev.burgess@oberlin.edu
Website: www.oberlin.edu/~envs/

Founded: NA

Description: An interdisciplinary program which includes 30+ courses across ten departments. Students are required to do significant academic work that spans the sciences, social sciences, and the humanities. The program offers significant off-campus opportunities for students through a Watershed Education Program, a local initiative in sustainable agriculture, and work with the city on energy and development issues.

Contact(s):
David Orr, PROGRAM CHAIR

OHIO STATE UNIVERSITY SCHOOL OF NATURAL RESOURCES

SCHOOL OF NATURAL RESOURCES
2021 Coffey Rd.
Columbus, OH 43210-1085 USA
Phone: 614-292-2265   Fax: 614-292-7432
Website: www.snr.osu.edu

Founded: NA
Membership: 40
Scope: Statewide

Contact(s):
Gary Mullins, DIRECTOR (B.S., M.S., PH.D.)
mullins.2@osu.edu

OKLAHOMA STATE UNIVERSITY

COOPERATIVE FISH AND WILDLIFE RESEARCH UNIT
Fishery & Wildlife
Stillwater, OK 74078 USA
Phone: 405-744-5000   Fax: 405-744-5006
E-mail: coopunit@okstate.edu
Website: www.okstate.edu

Founded: NA
Scope: Regional

Contact(s):
David Leslie, COOPERATIVE FISH AND WILDLIFE RESEARCH UNIT, LEADER
404 Life Sciences West, 405-744-6342

Becky Johnson, DEPARTMENT OF BOTANY (B.S., M.S., PH.D.), HEAD
Phone: 405-744-5559
Craig McKinley, DEPARTMENT OF FORESTRY (B.S., M.S.), HEAD
Phone: 405-744-5437
James Shaw, DEPARTMENT OF WILDLIFE/FISHERIES/ECOLOGY/ZOOLOGY (B.S., M.S., PH.D.), HEAD
Phone: 405-744-5555
Edward Knobbe, ENVIRONMENTAL SCIENCE (M.S., PH.D), PROGRAM COORDINATOR
Phone: 405-744-9229
David Engle, RANGE MANAGEMENT (B.S., M.S., PH.D.), PROGRAM COORDINATOR
477 Ag Hall, Stillwater, OK 74078
Phone: 405-744-6410

OREGON STATE UNIVERSITY DEPT OF FISHERIES & WILDLIFE

DEPARTMENT OF FISHERIES AND WILDLIFE
104 Nash
Corvallis, OR 97331 USA
Phone: 541-737-4531   Fax: 541-737-3590
Website: www.osu.orst.edu

Founded: NA
Membership: 300
Scope: Statewide

Contact(s):
Michael Unsworth, CENTER FOR ANALYSIS OF ENVIRONMENTAL CHANGE, DIRECTOR
George Stankey, CONSORTIUM ON SOCIAL VALUES OF NATURAL RESOURCES, COORDINATOR OF OUTDOOR RECREATION
Carl Schreck, COOPERATIVE FISHERY AND WILDLIFE RESEARCH UNIT, LEADER
Robert Anthony, COOPERATIVE FISHERY AND WILDLIFE RESEARCH UNIT, LEADER
Daniel Edge, DEPT. HEAD
Daniel Edge, FISHERIES AND WILDLIFE (B.S., M.S., PH.D.), HEAD OF WILDLIFE, INTERIM
Harold Salwasser, FORESTRY AND FOREST RECREATION (B.S., M.S., PH.D.), DEAN
Steven Radosevich, SUSTAINABLE FORESTRY PROGRAM, LEADER OF FORESTRY

P

PENNSYLVANIA STATE UNIVERSITY

SCHOOL OF FOREST RESOURCES
113 Ferguson Bldg., Schl. Forest Resources
University Park, PA 16802 USA
Phone: 814-863-7093   Fax: 814-865-3725
Website: www.sfr.cas.psu.edu

Founded: NA
Membership: 150
Scope: State

Contact(s):
Charles Strauss, DIRECTOR INTERIM
Robert Carline, FISHERIES AND WILDLIFE COOPERATIVE RESEARCH UNIT, LEADER
113A Merkle Bldg., Univeristy Park, PA 16802
Phone: 814-865-4511
Fax: 814-863-4710
f7u@psu.edu

M. Brittingham, WILDLIFE AND FISHERIES (B.S., M.S., PH.D.), CONTACT
320 Forest Resources Lab, University Park, PA 16802
Phone: 814-863-8442
Fax: 814-863-7193
mxb21@psu.edu
John Janowiak, WOOD PRODUCTS (B.S., M.S., PH.D.), PROFESSOR
307 Forest Resources Laboratory, University Park, PA 16802
Phone: 814-865-5722
Fax: 814-863-7193
jjj2@psu.edu

PINES ROWAN UNIVERSITY

120-13 Whitesbog Rd.
Browns Mills, NJ 08015 USA
Phone: 609-893-1765   Fax: 609-893-8297
Website: none

Founded: NA
Scope: Statewide

Contact(s):
Gary Patterson, DIRECTOR
Maria Peter, PROGRAM COORDINATOR

POLYTECHNIC UNIVERSITY OF NEW YORK

CIVIL AND ENVIRONMENTAL ENGINEERING DEPARTMENT
6 Metro Tech Center
Brooklyn, NY 11201 USA
Phone: 718-260-3220   Fax: 718-260-3433
E-mail: cee@poly.edu
Website: www.poly.edu/cee

Founded: NA
Membership: 15
Scope: International

Description: The Department is engaged in teaching and research in several areas of environmental science and engineering. Masters degrees with environmental focus are offered in civil engineering, environmental engineering, and environmental health science. The Ph.D. is also offered.

Contact(s):
F. (Bud) Grisses, DEPT. CHAIR OF CIVIL ENGINEERING
David Chang, PRESIDENT

PORTLAND STATE UNIVERSITY

ENVIRONMENTAL SCIENCES AND RESOURCES
P.O. Box 751
Portland, OR 97207-0751 USA
Phone: 503-725-4980   Fax: 503-725-3888
E-mail: envir@pdx.edu
Website: www.esr.pdx.edu

Founded: NA
Scope: Statewide

Description: The focus of the program is research on the problems of the environment and resources. The program offers Ph.D. degrees in cooperation with the departments of biology, chemistry, civil engineering, economics, geography, geology, and physics. Master programs include M.S., M.E.M. (Master of Environmental Management), and M.S.T (Master of Science in Teaching). Bachelor's programs (B.A., B.S.) include tracks in environmental science and environmental policy and management.

Contact(s):
Roy Koch, DIRECTOR
Phone: 503-725-8038
kochr@mail.pdx.edu

POULSBO MARINE SCIENCE CENTER
18743 Front St., NE,
Poulsbo, WA 98370 USA
Phone: 360-779-5549   Fax: 360-779-8960
E-mail: info@poulsbo.msc.org
Website: www.poulsbo,msc.org

Founded: 1968
Scope: Regional

Description: The Marine Science Center works to meet the science education needs of public citizens and of students and teachers nationally, regionally, and locally. Hands-on environmental education is at the heart of its mission and is reflected in direct instruction from its facility on Washington's State's Liberty Bay, part of the Puget Sound.

Publication(s): Seasquirt-newsletter, Wonders of Puget Sound

Contact(s):
Michelle Benedict, DIRECTOR
Cindy Ratheone, SOCIETY PRESIDENT

PRESCOTT COLLEGE, LIBERAL ARTS AND THE ENVIRONMENTAL STUDIES PROGRAM
ENVIRONMENTAL STUDIES PROGRAM
220 Grove Ave.
Prescott, AZ 86301 USA
Phone: 520-778-2090   Fax: 928-776-5137
Website: www.prescott.edu/rdp/rdp_es.html

Founded: NA
Scope: Regional

Description: The Environmental Studies Program is one of four programs within Prescott College. The program emphasizes experiential and interdisciplinary learning, and focuses on the interrelationaships between the human and nonhuman worlds and the reciprocal influences each has on the other.

Publication(s): Transitions, brochure, Alligator Juniper. Poems, stories and photographs, yearly softcover book, Wolfberry Sun, semi-annual newsletter

Contact(s):
Lisa Floyd-Hanna, PROGRAM COORDINATOR

PURDUE UNIVERSITY
DEPARTMENT OF FORESTRY AND NATURAL RESOURCES
1159 Forestry Bldg.
West Lafayette, IN 47907-1159 USA
Phone: 765-494-3591   Fax: 765-496-2422
Website: www.fnr.purdue.edu/

Founded: NA
Scope: Statewide

Contact(s):
Dennis Lemaster, DEPARTMENT OF FORESTRY AND NATURAL RESOURCES, HEAD
Phone: 765-494-3590
Fax: 765-496-2422

Robert Swihart, GRADUATE PROGRAM, DIRECTOR OF GRADUATE STUDIES
Phone: 765-494-3621
Fax: 765-496-2422
W. Mills, UNDERGRADUATE PROGRAMS, DIRECTOR OF STUDENT SERVICES
Phone: 765-494-3575
Fax: 765-496-2422

RENSSELAER POLYTECHNIC INSTITUTE
DEPT. OF EARTH AND ENVIRONMENTAL SCIENCES
Jonsson-Rowland Science Center, Rm. 1C25
Troy, NY 12180-3590 USA
Phone: 518-276-6474   Fax: 518-276-6680
E-mail: ees@rpi.edu
Website: www.rpi.edu/dept/geo/

Founded: NA
Scope: International

Contact(s):
Frank Spear, CHAIR
spearf@rpi.edu

RENSSELAER POLYTECHNIC INSTITUTE
ENVIRONMENTAL MANAGEMENT AND POLICY PROGRAM
110 8th St., Pittbsurgh Building
Troy, NY 12180-3590 USA
Phone: 518-276-6565   Fax: 518-276-2665
E-mail: emap@rpi.edu
Website: http://emat.mgmt.rip.edu/

Founded: NA
Scope: International

Description: Rensselaer's EMP program educates students at the masters of science level to undertake a professional role in companies, governmental agencies, and other organizations dealing with environmental and energy matters from a base of technical and managerial knowledge and understanding.

Publication(s): Corporate Environmental Strategy

Contact(s):
Frank Mendelson, DIRECTOR (ACTING)

RHODE ISLAND SCHOOL OF DESIGN
DEPARTMENT OF LANDSCAPE ARCHITECTURE
Two College St.
Providence, RI 02903 USA
Phone: 401-454-6282   Fax: 401-454-6299
E-mail: ldardept@risd.edu
Website: www.risd.edu

Founded: NA
Membership: 37
Scope: Statewide

Description: Graduates depart RISD with the necessary training to work from an informed position, with an environmental ethic, a personal philosophy, their interpretive abilities honed and with creative vision.

Contact(s):
Elizabeth Hermann, ASSOCIATE PROFESSOR
edherman@risd.edu
Leonard Newcomb, ASSOCIATE PROFESSOR
Phone: 401-454-6282

Miekyoung Kim, ASSOCIATE PROFESSOR
Phone: 401-454-6286
info@mikyoungkim.com
Colgate Searle, HEAD DEPT OF LANDSCAPE
ARCHITECTURE
csearle@risd.edu
Derek Bradford, PROFESSOR
Phone: 401-454-6292
dbradfor@risd.edu

RICE UNIVERSITY

Architecture MS 50, 6100 Main Street
Houston, TX 77005 USA
Phone: 713-348-4864   Fax: 713-348-5277
E-mail: arch@rice.edu
Website: www.arch.rice.edu/

Founded: NA
Membership: 25
Scope: Regional

Description: Master of Architecture in Urban Design for individuals who already hold a professional degree qualifying them for registration as architects or landscape architects

Publication(s): see publication web site

Contact(s):
Lars Lerup, DEAN
lars@rice.edu

RICE UNIVERSITY

6100 Main St.
Houston, TX 77251 USA
Phone: 713-527-8101
Website: www.rice.edu

Founded: 1912

Contact(s):
Frank Fisher, DIRECTOR OF WETLAND STUDIES, ECOLOGY DEPT.
Houston, 713-527-5917
fisher@rice.edu
Ronald Sass, ECOLOGY AND EVOLUTIONARY BIOLOGY DEPT. (B.S., M.A., PH.D.), CHAIR
6100 Main St., MS 170, Houston, TX 77005
Phone: 713-527-4919
Fax: 713-285-5232
eeb@rice.edu or sass@ruf.rice.edu
C. Ward, ENERGY AND ENVIRONMENTAL SYSTEMS INSTITUTE (B.A., M.E.S., M.E.E., M.S., PH.D.), DIRECTORY
6100 Main St., MS 316, Houston, TX 77005
eesi@rice.edu

RICHARD STOCKTON COLLEGE

DIVISION OF NATURAL SCIENCES AND MATHEMATICS
P.O. Box 195
Pomona, NJ 08240 USA
Phone: 609-652-1776   Fax: 609-748-5515
E-mail: iaprod573f@vax003.stockton.edu
Website: www.stockton.edu/

Founded: 1969
Scope: Statewide
Publication(s): NANS Safety

Contact(s):
Peter Straub, BIOLOGY COORDINATOR

Edward Paul, CHEMISTRY COORDINATOR
Dennis Weiss, DEAN
George Zimmerman, ENVIRONMENTAL SCIENCE & GEOLOGY
Gordan Grguric, MARINE SCIENCE
Charlie Helands, MATHEMATICS COORDINATOR

ROGER WILLIAMS UNIVERSITY

DEPT. OF BIOLOGY, MARINE BIOLOGY, CHEMISTRY & ENVIRONMENTAL SCIENCE
One Old Ferry Rd.
Bristol, RI 02809 USA
Phone: 401-254-3108   Fax: 401-254-3310
Website: www.rwuonline.cc/

Founded: 1969
Membership: 61
Scope: International

Contact(s):
Delia Anderson, ASSISTANT DEAN
danderson@rwu.edu

RUTGERS UNIVERSITY, COOK COLLEGE

COOK COLLEGE
P.O. Box 231
New Brunswick, NJ 08903 USA
Phone: 732-932-4636
Website: www.rutgers.edu/

Founded: NA
Scope: Statewide

Contact(s):
Adesoji Adelaja, AGRICULTURAL, FOOD, AND RESOURCE ECONOMICS DEPT., CHAIR
Cook Office Bldg., 55 Dudley Road, New Brunswick, NJ 08901
Phone: 732-932-9155
Fax: 732-932-8887
adelaja@aesop.rutgers.edu
Roni Avissar, ENVIRONMENTAL SCIENCES DEPT., CHAIR
14 College Farm Road, New Brunswick, NJ 08901-8551
Phone: 732-932-9185
Fax: 732-932-8644
CHAIR@envsci.rutgers.edu
Peter Parks, GRADUATE DIRECTOR
parks@aesop.rutgers.edu
Peter Strom, GRADUATE PROGRAM DIRECTOR
Phone: 732-932-8078
strom@envsci.rutgers.edu
Maurice Hartley, UNDERGRADUATE DIRECTOR
hartley@aesop.rutgers.edu
Robert Tate, UNDERGRADUATE PROGRAM COORDINATOR
Phone: 732-932-9810
tate@envsci.rutgers.edu

S

SAN FRANCISCO STATE UNIVERSITY
WILDLANDS STUDIES PROGRAM
3 Mosswood Circle
Cazadero, CA 95421 USA
Phone: 707-632-5665   Fax: 707-632-5665
E-mail: wildlnds@sonic.net
Website: wildlandsstudies.com/ws

Founded: 1979
Scope: Regional, National, International

Description: Wildlands Studies offers a year-round series of field study programs in North American and international wilderness locations. Participants join backcountry research teams in a search for answers to important environmental problems concerning wildlife populations and/or wildlands habitats. Participants can earn 3-14 units of university credit.

Publication(s): Course Catalog

Contact(s):
Crandall Bay, DIRECTOR
Phone: 707-632-5665

SAN JOSE STATE UNIVERSITY
DEPARTMENT OF ENVIRONMENTAL STUDIES
One Washington Sq.
San Jose, CA 95192-0115 USA
Phone: 408-924-5450   Fax: 408-924-5477
E-mail: envstdys@email.sjsu.edu
Website: www.sjsu.edu/depts/envstudies/

Founded: 1970
Scope: Statewide

Description: Special interests of the faculty include habitat restoration, environmental empact assessment, energy, water, and forest resource management, human ecology, international development, coastal resource management, solid waste management, and environmental education for teachers. Credit is given for beyond-the-classroom experiences for appropriate Peace Corps Service, Internships programs, Center for Development of Recycling (CDR) and Environmental Resource Center (ERC).

Contact(s):
Lester Rowntree, DEPARTMENT CHAIRPERSON

SCHOOL FOR FIELD STUDIES (BOSTON UNIVERSITY)
16 Broadway
Beverly, MA 01915-4499 USA
Phone: 800-989-4453   Fax: 978-927-5127
E-mail: admissions@fieldstudies.org
Website: www.fieldstudies.org

Founded: 1980

Description: The mission of The School for Field Studies is to provide highly motivated young people from the U.S. and abroad with an excellent practical education in environmental studies, in order that tomorrow's leaders may become more environmentally literate/aware as well as make immediate and future contributions toward the sustainable management of natural resources.

Contact(s):
Terry Andreas, PRESIDENT

SHAWNEE STATE UNIVERSITY
DEPARTMENT OF NATURAL SCIENCES
940 Second St.
Portsmouth, OH 45662 USA
Phone: 740-354-3205
Website: www.shawnee.edu

Founded: NA
Membership: 12
Scope: State

Description: B.S. in Natural Science field with minor or certificate in Environmental Studies

Contact(s):
Jeffrey Bauer, ENVIRONMENTAL CERTIFICATE ADVISOR
Phone: 740-355-2421
Fax: 740-355-2416
jbauer@shawnee.edu

SHEPHERD COLLEGE
INSTITUTE FOR ENVIRONMENTAL STUDIES
P.O. Box 3210
Byrd Center
Shepherdstown, WV 25443 USA
Phone: 304-876-5227   Fax: 304-876-5028
Website: www.shepherd.wvnet.edu/iesweb/

Founded: NA
Membership: 50
Scope: Statewide

Description: B.S. in Environmental Studies with focus in physical and biological sciences or resource management.

Contact(s):
Ed Snyder, DIRECTOR
Phone: 304-876-5227
Fax: 304-876-5028
iesweb@shepherd.edu

SLIPPERY ROCK UNIVERSITY
101 Eisenberg Bldg. SRU
Slippery Rock, PA 16057 USA
Phone: 724-738-2068   Fax: -724-7382
Website: www.sru.edu/

Founded: NA

Contact(s):
Dan Dziubek, ENVIRONMENTAL EDUCATION (B.S., M.ED.), COORDINATOR
Phone: 724-738-2958
Michael Stapleton, ENVIRONMENTAL SCIENCE (B.S.), PROGRAM COORDINATOR
Phone: 724-738-2495
Beverly Buchert, ENVIRONMENTAL STUDIES (B.S.), COORDINATOR
Phone: 724-738-2389
Dan Dziubek, INSTITUTE FOR THE ENVIRONMENT EXECUTIVE COMMITTEE CHAIR
Phone: 724-738-2958
Bruce Boliver, PARK AND RESOURCE MANAGEMENT (B.S., M.S.), CHAIRMAN
Phone: 724-738-2068
Paulette Johnson, PENNSYLVANIA CENTER FOR ENVIRONMENTAL EDUCATION, DIRECTOR
Phone: 724-738-4555
Karen Kainer, SUSTAINABLE SYSTEMS (M.S.), COORDINATOR
Phone: 724-738-2622

SOLAR AND ENERGY CONVERSION LABORATORIES

SOLAR ENERGY AND ENERGY CONVERSION
LABORATORIES
237 MEB, Box 116300
Gainesville, FL 32611 USA
Phone: 352-392-0812   Fax: 352-392-1071
E-mail: solar@cimar.me.ufl.edu
Website: www.me.ufl.edu/SOLAR/

Founded: 1954
Scope: International

Publication(s): Solar Touch Newsletter, Advances in Solar
Energy (Published every other year), Principles of Solar
Engineering (textbook)

Contact(s):
D. Goswami, DIRECTOR

SONOMA STATE UNIVERSITY

DEPARTMENT OF ENVIRONMENTAL STUDIES AND
PLANNING
1801 E. Cotati Ave.
Rohnert Park, CA 94928 USA
Phone: 707-664-2306   Fax: 707-664-4202
E-mail: ensp@sonoma.edu
Website: www.sonoma.edu/ensp/

Founded: NA
Scope: National

Description: Interdisciplinary academic program with B.S. and
B.A. degrees. Study tracks in environmental education, energy
management and design, city and regional planning, water
quality, hazardous materials management, and environmental
conservation and restoration

Contact(s):
Steve Orlick, DEPARTMENT CHAIR
Phone: 707-664-2414
steve.orlick@sonoma.edu

SONOMA STATE UNIVERSITY

EARTH LAB
Earth Lab Environmental Studies and Planning, Sonoma
State University
Rohnert Park, CA 94928 USA
Phone: 707-664-2577   Fax: 707-664-3920
E-mail: EarthLab@sonoma.edu
Website: www.sonoma.edu/ensp/earthlab.html

Founded: NA

Description: The Earthlab is an on-campus demonstration,
education, and research center which serves the campus and
surrounding communities through programs in environmental
education, professional training, teacher workshops, demonstration projects, and scientific research.

SOUTH DAKOTA STATE UNIVERSITY

DEPARTMENT OF WILDLIFE, FISHERIES SCIENCES
P.O. Box 2140B
Brookings, SD 57007-1696 USA
Phone: 605-688-6121   Fax: 605-688-4515
Website: wfs.sdstate.edu/wfsci.htm

Founded: NA
Scope: State

Description: Fish and wildlife research, education, and services
with emphasis on prairie pothole ecology, fisheries
management, wildlife management, and fisheries and wildlife
ecology.

Contact(s):
Charles Berry, COOPERATIVE FISH AND WILDLIFE
RESEARCH UNIT, LEADER
Charles Scalet, HEAD

SOUTHERN CONNECTICUT STATE UNIVERSITY

CENTER FOR THE ENVIRONMENT
501 Crescent St., Jennings Hall, Rm. 342
New Haven, CT 06515 USA
Phone: 203-392-6600   Fax: 203-392-6614
Website: www.scsu.ctstcteu.edu

Founded: NA
Membership: 8
Scope: International

Description: The Center for the Environment is an academic
center granting graduate and undergraduate degrees in environmental areas, conducting research, and developing episte-
mological models. An active field study program includes
experiences in Costa Rica, South Africa, Madagascar, Ecuador
(including the Galapagos Islands) and various sites in the U.S.

Publication(s): SEED Newsletter

Contact(s):
Susan Hageman, CHAIR OF SCIENCE EDUCATION AND
ENVIRONMENTAL STUDIES
Vincent Breslin, PROFESSOR OF ENVIRONMENTAL
STUDIES

SOUTHERN ILLINOIS UNIVERSITY CARBONDALE

DEPT. OF FORESTRY
Southern Illinois University, Carbondale 4411
Carbondale, IL 62901-6899 USA
Phone: 618-453-2121   Fax: 618-453-7475
Website: www.siu.edu/`forestry.com

Founded: NA
Membership: 10
Scope: Statewide

Contact(s):
John Phelps, ENVIRONMENTAL STUDIES PROGRAM,
DIRECTOR
Phone: 618-453-3341
jphelps@siu.edu
John Phelps, FORESTRY (B.S., M.S.), CHAIR
Mailstop 4411, SIU, Carbondale, IL 62001
Phone: 618-453-7464
jphelps@siu.edu

SOUTHERN OREGON UNIVERSITY

ENVIRONMENTAL EDUCATION PROGRAM
BIOLOGY DEPARTMENT
1250 Siskiyou Blvd .
Ashland, OR 97520 USA
Phone: 541-552-6797   Fax: 541-552-6415
Website: www.sou.edu/biology/enved/mainpage.htm

Founded: 1990
Membership: 13
Scope: Statewide

Description: This graduate program grants a Master of Science degree, and provides hands-on learning experiences in conservation biology, interpretive practices, field interpretation, and field studies in southwestern Oregon and elsewhere in the state for students committed to careers in environmental education. Studies are also required in biology and related disciplines of choice to complete the program.

Contact(s):
Stewart Janes, CONTACT

ST. CLOUD STATE UNIVERSITY
720 4th Ave., S
St. Cloud, MN 56301 USA
Phone: 320-255-3235　　Fax: 320-654-5122
E-mail: ets@condor.stcloudstate.edu
Website: www.stcloudstate.edu

Founded: NA
Scope: Statewide

Description: Linking the human and natural world with programs designed to foster environmental and technological literacy and prepare students who can integrate the interconnections of science, technology, society and the environment through research and assessment.

Contact(s):
Michael Karian, ASSOCIATE PROFESSOR
Phone: 320-255-3966
Charles Rose, DIRECTOR OF ENVIRONMENTAL

ST. LAWRENCE UNIVERSITY
ENVIRONMENTAL STUDIES PROGRAM
Canton, NY 13617 USA
Phone: 315-229-5814　　Fax: 315-229-5802
Website: web.stlawu.edu/envstudies

Founded: 1856
Scope: Statewide

Description: St. Lawrence University is a liberal arts and sciences institution. The institution offers one of the oldest environmental studies programs in the nation. The university is committed to environmentally responsible management practices and comprehensive outdoor education programs.

Contact(s):
Glenn Harris, DIRECTOR

ST. NORBERT COLLEGE
CENTER FOR INTERNATIONAL EDUCATION
100 Grant St.
De Pere, WI 54115-2099 USA
Phone: 920-403-3100　　Fax: 920-403-4083
E-mail: mediarel@mail.snc.edu
Website: www.snc.edu/

Founded: 1990
Scope: State

Description: The Center conducts an Annual Global Ecology Series on themes such as the Great Lakes as an endangered resource of North America, Africa and women, population, and international policy-making. Instructional resources and in-service programs are provided for K-1

Contact(s):
Joseph Tullbane, ASSOCIATE DEAN FOR INTERNATIONAL STUDIES
Phone: 920-403-3378
Fax: 920-403-4083
tulljd@mail.snc.edu

STANFORD UNIVERSITY
DEPARTMENT OF BIOLOGICAL SCIENCES
CENTER FOR CONSERVATION BIOLOGY
Herrin Labs, 385 Sierra Mall
Stanford, CA 94305-5020 USA
Phone: 650-723-5924　　Fax: 650-723-5920
E-mail: consbio@bing.stanford.edu
Website: www.stanford.edu/group/ccb/

Founded: 1984
Scope: National

Description: To develop the science of conservation biology, including its application to solutions for critical conservation problems. The Center conducts scientific and policy research that is building a sound basis for the conservation, management, and restoration of biotic diversity around the world. The overall goal is to develop ways and means for protecting Earth's life support systems and thus enhancing future human well-being.

Publication(s): see publication web site

Contact(s):
Paul Ehrlich, PRESIDENT

STANFORD UNIVERSITY
MORRISON INSTITUTE FOR POPULATION AND RESOURCE STUDIES
371 Sierra Mall, MC 5020
Stanford, CA 94305-5020 USA
Phone: 415-723-7518　　Fax: 415-725-8244
E-mail: morrinst@stanford.edu
Website: www.stanford.edu/group/morrinst/

Founded: 1986

Description: To support research and education in the interconnected global issues of population growth, its effects on the environment, the pressure on natural resources, and the capacity of many nations to achieve sustainable socioeconomic development. Issues are approached through interdisciplinary perspectives of population biology, economics, and social and medical sciences.

Contact(s):
Marcus Feldman, DIRECTOR
Phone: 650-725-1867

STATE UNIVERSITY OF NEW YORK AT CORTLAND
GEOLOGY DEPARTMENT
P.O. Box 2000
Cortland, NY 13045 USA
Phone: 607-753-2011　　Fax: 607-753-2927
E-mail: stouts@cortland.edu
Website: www.cortland.edu/

Founded: NA
Scope: Statewide

Contact(s):
Jack Sheltmire, ENVIRONMENTAL AND OUTDOOOR EDUCATION, COORDINATOR
Phone: 607-753-5488
sheltmirej@cortland.edu

Christopher Cirmo, ENVIRONMENTAL GEOLOGY AND ENVIRONMENTAL SCIENCES (CONCENTRATION FOR MAJORS IN BIOLOGY, GEOLOGY, CHEMISTRY AND PHYSICS), COORDINATOR
Phone: 607-753-2924
cirmoc@cortland.edu

STATE UNIVERSITY OF NEW YORK AT STONY BROOK

MARINE SCIENCES RESEARCH CENTER
Stony Brook, NY 11794 USA
Phone: 631-632-8700   Fax: 631-632-8820
Website: www.msrc.sunysb.edu/

Founded: NA

Description: University-wide center to develop marine and atmospheric research, instructional programs and facilities for the State University of New York. Ongoing research projects are directed toward coastal oceanographic processes, marine environmental problems and management, atmospheric sciences and resources management. Among the Center's organized units are the Living marine Resources Institute, the Waste Reduction and Management Institute, the Coastal Ocean Action Strategies Institute, the Institute for Urban Ports and Harbors, the Institute for Planetary and Terrestrial Atmospheres and the Flax Pond Laboratory. Publications of the center include Technical Report Series, Working Report Series; Special Report Series; and a newsletter.

Contact(s):
Nicholas Fisher, ASSOCIATE DEAN
Phone: 516-632-8649
nfisher@notes.cc.sunysb.edu
W. Wise, ASSOCIATE DIRECTOR
Phone: 516-632-8656
Marvin Geller, DEAN AND DIRECTOR
Phone: 516-632-8701
mgeller@notes.cc.sunysb.edu
Glenn Lopez, GRADUATE PROGRAMS
Phone: 516-632-8660
glopez@notes.cc.sunysb.edu

STATE UNIVERSITY OF NEW YORK COLLEGE OF ENVIRONMENTAL SCIENCE AND FORESTRY

1 Forestry Dr.
Syracuse, NY 13210-2778 USA
Phone: 315-470-6500   Fax: 315-470-6953
E-mail: esfinfo@esf.edu
Website: www.esf.edu

Founded: 1911
Scope: Statewide

Description: Research has been a hallmark of ESF since its inception. Recent wildlife studies have aimed toward reintroducing lynx and moose to the Adirondack Park; application of molecular biology techniques to identify migrant bird populations; restoration of muskellunge and sturgeon in the St. Lawrence River system; analysis of Flamingo population dynamics in Mexico; tailoring black cherry clones for fast growth and straight limbs; researching willow plantations as a source of biomass energy; cooperating with NASA to analyze polarized-light photographs of Earth; and detailing the effects of acid precipitation on forest ecosystems.

Contact(s):
William Porter, ADIRONDACK ECOLOGICAL CENTER, DIRECTOR
Phone: 315-470-6798
William Winter, CELLULOSE RESEARCH INSTITUTE, ACTING DIRECTOR
Phone: 315-470-6855
H. Underwood, COOPERATIVE PARK STUDIES UNIT, DIRECTOR
Phone: 315-470-6820
Hannu Makkonen, EMPIRE STATE PAPER RESEARCH INSTITUTE, DIRECTOR
Phone: 315-470-6900
Richard Smardon, ENVIRONMENTAL INSTITUTE, RANDOLF G. PACK, DIRECTOR
Phone: 315-470-6636
John Hassett, FACULTY OF CHEMISTRY (B.S., M.S., PH.D.), CHAIR
Phone: 315-470-6855
George Kyanka, FACULTY OF CONSTRUCTION MANAGEMENT AND WOOD PRODUCTS ENGINEERING (B.S., M.S., PH.D.), CHAIR
Phone: 315-470-6880
Neil Ringler, FACULTY OF ENVIRONMENTAL AND FOREST BIOLOGY (B.S., M.S., PH.D.), CHAIR
Phone: 315-470-6743
James Hassett, FACULTY OF ENVIRONMENTAL RESOURCES AND FOREST ENGINEERING (B.S., M.S., PH.D.), CHAIR
Phone: 315-470-6633
Richard Smardon, FACULTY OF ENVIRONMENTAL STUDIES (B.S., M.S., PH.D.), CHAIR
Phone: 315-470-6636
William Bentley, FACULTY OF FORESTRY (B.S., M.S., PH.D.), CHAIR
Phone: 315-470-6536
Thomas Amidon, FACULTY OF PAPER SCIENCE AND ENGINEERING (B.S., M.S., PH.D.), CHAIR
Phone: 315-470-6502
Christopher Westbrook, FOREST TECHNICIAN PROGRAM (A.A.S.), DIRECTOR
Phone: 315-848-2566
Richard Smardon, GREAT LAKES RESEARCH CONSORTIUM, CO-DIRECTOR
Phone: 315-470-6816
Robert Hanna, N.C. BROWN LABORATORY FOR ULTRA-STRUCTURE STUDIES, DIRECTOR
Phone: 315-470-6880
Israel Cabasso, POLYMER RESEARCH INSTITUTE, DIRECTOR
Phone: 315-470-4767
Neil Ringler, ROOSEVELT WILDLIFE STATION, DIRECTOR
Phone: 315-470-6770
Wayne Zipperer, U.S. FOREST SERVICE UNIT, DEPUTY PROJECT LEADER
Phone: 315-448-3201

STEPHEN F. AUSTIN STATE UNIVERSITY ARTHUR TEMPLE COLLEGE OF FORESTRY

ARTHUR TEMPLE COLLEGE OF FORESTRY
P.O. Box 6109
Nacogdoches, TX 75962-6109 USA
Phone: 936-468-3301   Fax: 936-468-2489
Website: www.sfasu.edu

Founded: NA

Membership: 42
Scope: Statewide
Contact(s):
R. Beasley, DEAN
Phone: 936-468-2164
sbeasley@sfasu.edu
Brian Oswald, FIRE MANAGEMENT AND SILVICULTURE
Phone: 936-468-2275
boswald@sfasu.edu
Hans Williams, FOREST ECO-PHYSIOLOGY
Phone: 936-468-2127
hwilliams@sfasu.edu
Gary Kronrad, FOREST ECONOMICS
Phone: 936-468-2473
gdkronrad@sfasu.edu
David Kulhavy, FOREST ENTOMOLOGY
Phone: 936-468-2141
dkulhavy@sfasu.edu
Mingteh Chang, FOREST HYDROLOGY
Phone: 936-468-2195
mchang@sfasu.edu
Edward Dougal, FOREST PRODUCTS
Phone: 936-468-2006
edougal@sfasu.edu
Michael Legg, FOREST RECREATION MANAGEMENT
Phone: 936-468-2246
mlegg@sfasu.edu
Jeffery Duguay, FOREST RESOURCES/WILDLIFE MANAGEMENT
Phone: 936-468-2196
jduguay@sfasu.edu
R. Whiting, FOREST WILDLIFE MANAGEMENT
Phone: 936-468-2125
mwhiting@sfasu.edu
James Kroll, FOREST WILDLIFE MANAGEMENT
Phone: 936-468-1198
jkroll@sfasu.edu
Peter Siska, GIS/REMOTE SENSING
Phone: 936-468-1347
siska@sfasu.edu
Paul Risk, INTERPRETATION/CONFLICT RESOLUTION
Phone: 936-468-2492
prisk@sfasu.edu
David Kulhavy, LANDSCAPE ECOLOGY
Shiyou Li, MEDICINAL PLANTS
Phone: 936-468-2071
lis@sfasu.edu
Daniel Unger, REMOTE SENSING/MENSURATION
Phone: 936-468-2234
unger@sfasu.edu
Michael Fountain, SILVICULTURE/FOREST ECOLOGY
Phone: 936-468-2313
mfountain@sfasu.edu
Kenneth Farrish, SOIL SCIENCE
Phone: 936-468-2475
kfarrish@sfasu.edu
Hans Williams, URBAN FORESTRY
R. Beasley, WATER QUALITY/FOREST HYDROLOGY

STERLING COLLEGE

Attn: Director of Admissions, P.O. Box 72
Craftsbury Common, VT 05827-0072 USA
Phone: 802-586-7711　　　Fax: 802-586-2596
E-mail: admissions@sterlingcollege.edu
Website: www.sterlingcollegeedu

Founded: 1958
Scope: Regional

Description: Sterling College offers a bachelor of arts degree with concentrations in outdoor education and leadership, sustainable agriculture, and wildlands ecology, and management.

Publication(s): Common Voice

Contact(s):
Chris Monz, DEAN
cmonz@sterlingcollege.edu
John Zaber, DIRECTOR OF ADMISSIONS
John Williamson, PRESIDENT

TEMPLE UNIVERSITY

309 Gladfelter Hall
Philadelphia, PA 19122 USA
Phone: 215-204-5918　　　Fax: 215-204-7833
Website: www.temple.edu/emv/stud

Founded: NA
Scope: Statewide

Description: This is a listing for an education program—undergraduate

Contact(s):
Robert Mason, DIRECTOR

TENNESSEE TECHNOLOGICAL UNIVERSITY

DEPT. OF BIOLOGY
Department of Biology
Box 5063
TTU
Cookeville, TN 38505 USA
Phone: 931-372-3134
E-mail: dlcombs@tntech.edu
Website: www.tntech.edu/www/acad/biol/

Founded: NA

Contact(s):
Dale Ensor, ENVIRONMENTAL SCIENCE (PH.D.)
P.O. Box 5055, Cookeville, TN 38505
Phone: 931-372-3493
densor@tntech.edu
Daniel Combs, WILDLIFE AND FISHERIES SCIENCE (B.S., M.S.)
dlcombs@tntech.edu

TEXAS A & M UNIVERSITY AT COMMERCE

DEPARTMENT OF AGRICULTURAL SCIENCES
2600 S. Neal St.
Commerce, TX 75429-3011 USA
Phone: 903-886-5358　　　Fax: 903-886-5990
Website: www.tamu-commerce.edu

Founded: 1889
Membership: 11
Scope: State

Description: Educational institution offering B.S. and M.S. degrees in agricultural fields and pre-wildlife management programs.

Contact(s):
Robert Williams, INTERIM, DEPT. HEAD

David Crenshaw, PRE-WILDLIFE MANAGEMENT, ADVISOR
Phone: 903-886-5329
david_crenshaw@tamu-commerce.edu

TEXAS A&M UNIVERSITY AT COLLEGE STATION

COLLEGE OF AGRICULTURE AND LIFE SCIENCES
113 Administration Bldg.
College Station, TX 77843-2142 USA
Phone: 979-845-4747   Fax: 979-845-9938
E-mail: agprogram@tamu.edu
Website: www.agprogram.tamu.edu

Founded: NA
Membership: 200
Scope: Statewide

Contact(s):
Tat Smith, FOREST SCIENCE (B.S., M.S., M.AGR., PH.D.), HEAD
Phone: 979-845-5000
Bob Brown, INSTITUTE FOR RENEWABLE NATURAL RESOURCES (B.S., M.AGR.), DIRECTOR
Phone: 979-845-5777
Bob Whitson, RANGELAND ECOLOGY AND MANAGEMENT (B.S., M.S., M.AGR., PH.D.), HEAD
Phone: 979-845-5579
Peter Witt, RECREATION PARKS, AND TOURISM SCIENCES (B.S., M.S., M.AGR., PH.D.), HEAD
Phone: 979-845-7324
Bob Brown, WILDLIFE AND FISHERIES SCIENCES (B.S., M.S., M.AGR., PH.D.), HEAD
Phone: 979-845-5777

TEXAS A&M UNIVERSITY AT KINGSVILLE

CAESAR KLEBERG WILDLIFE RESEARCH INSTITUTE
MSC 218
Kingsville, TX 78363 USA
Phone: 361-593-3922   Fax: 361-593-3924
Website: ckwri.tamuk.edu/

Founded: 1981

Description: A nonprofit institute that emphasizes research on wildlife and range management in Texas. Some work also done in Mexico and Canada. Research specialties include deer, quail, waterfowl, endangered cats, and nongame wildlife.

Contact(s):
Fred Bryant, DIRECTOR
Phone: 361-593-4025

TEXAS CHRISTIAN UNIVERSITY

ENVIRONMENTAL SCIENCE PROGRAM
2800 South University Drive
Fort Worth, TX 76109 USA
Phone: 817-257-7000
Website: www.ensc.tcu.edu/

Founded: NA

Contact(s):
Leo Newland, DIRECTOR
L.Newland@tcu.edu

TEXAS TECH UNIVERSITY

DEPARTMENT OF RANGE WILDLIFE & FISHERIES
P.O.Box 42125
Lubbock, TX 79409-2125 USA
Phone: 806-742-2841   Fax: 806-742-2280
Website: www.rw.ttu.edu/dept/

Founded: NA
Membership: 25
Scope: State
Publication(s): see publication web site

Contact(s):
Ernest Fish, WILDLIFE SCIENCE AND FISHERIES SCIENCE (M.S., PH.D.), CHAIRMAN
fish@water.rw.ttu.edu

TREASURE VALLEY COMMUNITY COLLEGE

DEPT. OF NATURAL RESOURCES
650 College Blvd.
Ontario, OR 97914 USA
Phone: 541-881-8822
Website: www.tvcc.cc.or.us/NatRes/

Founded: NA

Description: Offers Associate of Applied Science focusing on Forestry, range management or wildland fire management.

Contact(s):
John Russell, PROFESSOR
John_Russell@mailman.tvcc.cc.or.us

TUFTS UNIVERSITY CIVIL ENGINEERING

Anderson Hall
Medford, MA 02155 USA
Phone: 617-627-3211   Fax: 617-627-3994
Website: www.tufts.edu/

Founded: NA
Scope: National, International

Contact(s):
Linfield Brown, ENVIRONMENTAL ENGINEERING, CONTACT
Phone: 617-627-2273
lbrown1@tufts.edu
Christopher Swan, ENVIRONMENTAL GEOTECHNOLOGY AND GEOTECHNICAL ENGINEERING, CONTACT
Phone: 617-627-2212
cswan@emerald.tufts.edu
John Durant, HAZARDOUS MATERIAL MANAGEMENT, CONTACT
Phone: 617-627-5489
jdurant@emerald.tufts.edu
Richard Vogel, WATER RESOURCES, CONTACT
Phone: 617-627-4260
rvogel@tufts.edu

TULANE ENVIRONMENTAL LAW CLINIC

ENVIRONMENTAL LAW CLINIC
6329 Freret St.
New Orleans, LA 70118-6231 USA
Phone: 504-865-5789   Fax: 504-862-8721
Website: www.tulane.edu/~telc

Founded: 1989
Membership: 6
Scope: State

Description: Provides free legal assistance through its student attorneys to community organizations and indigent persons seeking to protect public health and the environment.

Contact(s):
Adam Babich, DIRECTOR

TULANE UNIVERSITY

6823 St. Charles Ave. Biology Dept.
New Orleans, LA 70118 USA
Phone: 504-865-5191   Fax: 504-862-8706
Website: www.tulane.edu

Founded: NA
Membership: 50
Scope: Statewide

Contact(s):
David Heins, ENVIRONMENTAL BIOLOGY (ECOLOGY AND SYSTEMATICS), CHAIRMAN

TULANE UNIVERSITY, ENVIRONMENTAL LAW PROGRAMS, TULANE LAW SCHOOL

LAW SCHOOL
ENVIRONMENTAL LAW PROGRAM
Weinmann Hall, Suite 255
New Orleans, LA 70118 USA
Phone: 504-865-5946   Fax: 504-862-8855
Website: www.law.tulane.edu/

Founded: 1981
Scope: Local, State, Regional, National

Description: Environmental law education, research, and advocacy through faculty, staff, JD and graduate student body.

Publication(s): see publication web site

Contact(s):
Gunthe Handl, CHAIR, INTERNATIONAL ENVIRONMENTAL LAW
Oliver Houck, DIRECTOR, ohouck@law.tulane.edu
Adam Babich, DIRECTOR, ENVIRONMENTAL LAW CLINIC
Eric Dannenmaier, DIRECTOR, INSTITUTE OF ENVIRONMENTAL LAW AND POLICY

U

UNITY COLLEGE

90 Quacker Hill Rd
Unity, ME 04988 USA
Phone: 207-948-3131   Fax: 207-948-6277
Website: www.unity.edu/

Founded: 1966
Scope: National

Description: Unity College is a small, liberal arts college in rural Maine with degree programs specializing in natural resource management and wilderness-based recreation.

Contact(s):
A. Chacko, AQUACULTURE
Doug Fox, ARBORICULTURE
Ed Beals, BOTANY AND ECOLOGY
Larry Farnsworth, CONSERVATION LAW ENFORCEMENT
David Oakes, ENVIRONMENTAL EDUCATION
Dave Potter, FISHERIES
Tom Mullins, PARK MANAGEMENT,
Jim Nelson, WILDLIFE

UNIVERSITE LAVAL

Cite Universitaire
Quebec, Canada G1K 7P4 Canada
Phone: 418-656-3333   Fax: 418-656-2809
E-mail: sg@sg.ulaval.ca
Website: www.ulaval.ca/

Founded: NA

Contact(s):
Andre Gosselin, ARGICULTURE (M.S., PH.D.), CONTACT
Phone: 418-656-7234
Fax: 416-656-7856
Michel Dessureault, FORESTRY, WOOD, AND FOREST SCIENCES DEPARTMENT (B.S., M.S., PH.D.), CONTACT
Phone: 418-656-7128
Fax: 418-656-3177
Michel.Dessureault@ffg.ulaval.ca

UNIVERSITY OF MIAMI

ROSENSTIEL SCHOOL OF MARINE AND ATMOSPHERIC SCIENCE
4600 Rickenbacker Causeway
Miami, FL 33149 USA
Phone: 305-361-4000   Fax: 305-361-9306
E-mail: libcirc@rsmas.miami.edu
Website: www.rsmas.miami.edu/

Founded: NA
Scope: Statewide

Contact(s):
Otis Brown, DEAN

UNIVERSITY OF AKRON

CENTER FOR ENVIRONMENTAL STUDIES
215 Crouse Hall
Akron, OH 44325-4102 USA
Phone: 330-972-5389   Fax: 330-972-7611
Website: www.uakron.edu/envstudies/

Founded: NA
Membership: 45
Scope: National

Contact(s):
Ira Sasowsky, DIRECTOR

UNIVERSITY OF ALASKA AT FAIRBANKS

COLLEGE OF SCIENCE, ENGINEERING AND MATHEMATICS
DEPT. OF BIOLOGY AND WILDLIFE
211 Irving
Fairbanks, AK 99775 USA
Phone: 907-474-7671   Fax: 907-474-5101
Website: www.uaf.edu/csem

Founded: NA
Scope: Statewide

Contact(s):
Joe Margraf, ALASKA COOPERATIVE FISH AND WILDLIFE RESEARCH UNIT, LEADER
209 Irving, UAF, Fairbanks, AK 99775-7020
Phone: 907-474-7661
Fax: 907-474-6716
ffjfm1@uaf.edu
David Woodall, DEAN OF COLLEGE OF SCIENCE, ENGINEERING & MATHMATICS

Brian Barnes, INSTITUTE OF ARCTIC BIOLOGY, INTERIM DIRECTOR
Phone: 907-474-7648

UNIVERSITY OF ALASKA FAIRBANKS
SCHOOL OF FISHERIES AND OCEAN SCIENCES
245 O'Neill Building, P.O. Box 757220
Fairbanks, AK 99775-7220 USA
Phone: 907-474-7824   Fax: 907-474-7204
E-mail: fysfos@uaf.edu
Website: www.sfos.uaf.edu:

Founded: NA
Membership: 300
Scope: International Regional
Publication(s): SFOS Today

Contact(s):
Vera Alexander, DEAN

UNIVERSITY OF ALBERTA
FACULTY OF AGRICULTURE, FORESTRY, AND HOME ECONOMICS
2-14 Agriculture Forestry Centre
Edmonton, Alberta T6G 2PS Canada
Phone: 780-492-4933   Fax: 780-492-0097
Website: www.afhe.ualberta.ca/

Founded: NA
Scope: National

Description: Undergraduate Degree Programs: B.Sc. in Agricultural and Food Business Management; Agriculture; Environmental and Conservation Sciences; Forest Business Management; Forestry; Human Ecology; Nutrition and Food Sciences; and Human Ecology/Bachelor of Education. Graduate Degree Programs: M.Sc., M. Ag., M.Eng., Ph.D. in Agricultural Food and Nutritional Science; M.A., MSc., and Ph.D. in Human Ecology; M.Sc., M. Ag., M.F., Ph. D., MBA/MF in Renewable Resources; MSc., MAg., Ph.D., MBA/MAg in Rural Economy.

Contact(s):
John Kennelly, AGRICULTURAL, FOOD, AND NUTRITIONAL SCIENCE, CHAIR
Phone: 780-492-3239
Fax: 780-492-4265
Nancy Gibson, HUMAN ECOLOGY, CHAIR
Phone: 780-492-3883
Fax: 780-492-4821
John Spence, RENEWABLE RESOURCES, CHAIR
Michele Veeman, RURAL ECONOMY, CHAIR
Phone: 780-492-4225
Fax: 780-492-0268

UNIVERSITY OF ARIZONA
SCHOOL OF RENEWABLE NATURAL RESOURCES
325 Biological Sciences East, P.O. Box 210043
Tucson, AZ 85721-0043 USA
Phone: 520-621-2543   Fax: 520-626-7401
Website: www.srnr.arizona.edu/

Founded: NA
Scope: Statewide

Description: The School of Renewable Natural Resources provides instruction, research, and extension in a range of disciplines. The specific academic programs of landscape resources, rangeland and forest resources, watershed resources, and wildlife and fisheries resources provide undergraduate and graduate education. Physical and biological sciences are integrated with socioeconomic and political factors necessary for the conservation, protection, and management of renewable natural resources.

Contact(s):
Malcom Zwolinski, ASSOCIATE DIRECTOR
Phone: 520-621-1432
Fax: 520-621-8801
mjz@ag.arizona.edu
Scott Bonar, COOPERATIVE FISH AND WILDLIFE RESEARCH UNIT (U.S. DEPARTMENT OF INTERIOR), LEADER
Phone: 520-626-8535
Fax: 520-621-8801
sbonar@ag.arizona.edu
William Halvorson, COOPERATIVE NATIONAL PARK RESOURCES STUDIES UNIT (U.S. DEPARTMENT OF INTERIOR), LEADER
Phone: 520-621-1174
Fax: 520-621-8801
halvor@srnr.arizona.edu
Michael Johnson, COOPERATIVE SOCIAL SCIENCES INSTITUTE (USDA NATURAL RESOURCES CONSERVATION SERVICE) (M.L.A.), CONTACT
Phone: 520-626-4685
Fax: 520-621-8801
mdjnrcs@ag.arizona.edu
C.P. Reid, DIRECTOR
Phone: 520-621-7257
Fax: 520-621-8801
cppr@ag.arizona.edu
Carl Edminster, FOREST SERVICE COOPERATIVE RESEARCH UNIT (USDA, ROCKY MOUNTAIN RESEARCH STATION), LEADER
Phone: 520-556-2177
D. Guertin, LANDSCAPE STUDIES (B.S., M.S., PH.D.)
Phone: 520-621-1723
Fax: 520-621-8801
phil@srnr.arizona.edu
George Ruyle, RANGELAND AND FOREST RESOURCES (B.S., M.S., PH.D.)
Phone: 520-621-1384
Fax: 520-621-8801
gruyle@ag.arizona.edu
Mitchel McClaren, RENEWABLE NATURAL RESOURCES STUDIES (M.S., PH.D.)
Phone: 520-621-1673
Fax: 520-621-8801
Ed De Steiguer, WATERSHED RESOURCES SCIENCE AND MANAGEMENT (B.S., M.S., PH.D.)
Phone: 520-621-3341
Fax: 520-621-8801
jedes@ag.arizona.edu
William Shaw, WILDLIFE AND FISHERIES RESOURCES (B.S., M.S., PH.D.)
Phone: 520-621-7265
Fax: 520-621-8801
wshaw@ag.arizona.edu

UNIVERSITY OF ARIZONA DEPT. OF HYDROLOGY & WATER RESOURCES
DEPARTMENT OF HYDROLOGY AND WATER
RESOURCES
P.O. Box 210011
Tucson, AZ 85721-0011 USA
Phone: 520-621-5082   Fax: 520-621-1422
E-mail: programs@hwr.arizona.edu
Website: www.hwr.arizona.edu

Founded: NA
Membership: 150
Scope: Statewide

Description: The mission of the department is to provide education,
research, and service in the fields of hydrology and water
resources and to engage in basic and applied research. The
department offers comprehensive programs in all areas of
surface and subsurface hydrology, water quality, and water
resources systems (management, administration, engineering).
The department is home to the NSF Science and Technology
Center on the Sustainability of Water Resources in Arid Regions,
the NASA Southwest Regional Earth Sciences Application Center
and the Arizona Research Laboratory for Riparian Studies.

Publication(s): Arizona HWR report, publications website
www.hwr.arizona.edu/pubs.html

Contact(s):
Terrie Thompson, ACADEMIC ADVISING COORDINATOR
Phone: 520-621-3131
Fax: 520-621-1422
terrie@hwr.arizona.edu
Victor Baker, DEPARTMENT HEAD
Dept. of Hydrology, 520-621-7120
Phone: 520-621-1422
baker@hwr.arizona.edu
Carla Stoffle, MAIN LIBRARY
P.O.Box 210055, 520-621-7440
Phone: 520-621-9733

UNIVERSITY OF ARKANSAS AT LITTLE ROCK
DEPARTMENT OF BIOLOGY
2801 S. University Ave.
Little Rock, AR 72204 USA
Phone: 501-569-3000
E-mail: webmaster@valer.edu
Website: www.ualr.edu

Founded: NA

Description: The Environmental Health Sciences Program
(www.ualr.edu/~ehsp/) curriculum consists of a common core
and a choice from four areas of concentrated study:
Environmental quality management; occupational safety and
health; environmental planning; and environmental/public
health sciences. Fish and Wildlife Management Program
(www.ualr.edu/~biology/programs/wild/) prepares students for
conservation biology research and management positions.
Meets certification requirements of American Fisheries Society
and the Wildlife Society. GIS Applications Laboratory available.

Contact(s):
Carl Stapleton, DIRECTOR, ENVIRONMENTAL HEALTH
SCIENCES PROGRAM
Phone: 501-569-3501
crstapleton@ualr.edu

Gary Heidt, DIRECTOR, FISHERIES & WILDLIFE
MANAGEMENT PROGRAM
Phone: 501-569-3511
gaheidt@ualr.edu

UNIVERSITY OF ARKANSAS AT LITTLE ROCK
DEPT. OF BIOLOGY
AR USA
Website: www.ualr.edu/~biology/programs/wild/

Founded: NA

UNIVERSITY OF ARKANSAS AT MONTICELLO
SCHOOL OF FOREST RESOURCES/ARKANSAS
FOREST RESOURCES CENTER
P.O.Box 3468, Forestry & Wildlife
Monticello, AR 71656 USA
Phone: 870-460-1052   Fax: 870-460-1092
Website: www.afrc.uamont.edu/sfr/index.htm

Founded: NA
Membership: 17
Scope: Statewide

Contact(s):
Richard Klunder

UNIVERSITY OF BRITISH COLUMBIA
ENVIRONMENTAL PROGRAMS
2075 Wesbrook Mall
Vancover, British Columbia V6T 1Z1 Canada
Phone: 604-822-8111   Fax: 604-822-1637
Website: www.hse.uvc.ca

Founded: NA
Membership: 30
Scope: Local

Contact(s):
Moura Quayle, AGRICULTRAL SCIENCES, DEAN
248-2357 Main Mall University Campus, Vancouver, British
Columbia V6T 1Z4
D. Shackleton, ANIMAL SCIENCES DEPARTMENT,
CONTACT
248-2357 Main Mall University Campus, Vancouver, British
Columbia V6T 1Z4
M. Isaacson, CIVIL ENGINEERING DEPARTMENT, HEAD
2324 Main Mall, Vancouver, British Columbia V6T 1Z4
J. McLean, FORESTRY, DEAN (ACTING)
2424 Main Mall, Vancouver, British Columbia V6T 1Z4
G. Wynn, GEOGRAPHY DEPARTMENT, HEAD
1984 West Mall, Vancouver, British Columbia V6T 1Z5
L. Lavkulich, INSTITUTE FOR RESOURCES AND
ENVIRONMENT, DIRECTOR
Rm. 436E, 2206 E. Mall, Vancouver, British Columbia
V6T 1Z3
A. Lewis, OCEANOGRAPHY DEPARTMENT, HEAD
6270 University Blvd., Vancouver, British Columbia V6T 1Z2
M. Healey, WESTWATER RESEARCH CENTRE, DIRECTOR
1933 W. Mall Annex, Rm. 200, Vancouver, British Columbia
V6T 1Z2
J. Berger, ZOOLOGY DEPARTMENT, HEAD
6270 University Blvd., Vancouver, British Columbia V6T 1Z4

UNIVERSITY OF CALIFORNIA AT DAVIS
COLLEGE OF AGRICULTURE AND ENVIRONMENTAL
SCIENCE
One Shields Ave.
Davis, CA 95616 - 8571 USA
Phone: 530-752-6586   Fax: 530-752-4154
Website: www.wscb.ucdavis.edu

Founded: NA
Membership: 20
Scope: Statewide

Description: Agricultural research programs and 21 departments.

Contact(s):
Colin Carter, CHAIR, AGRICULTURAL AND RESOURCE
ECONOMICS
Phone: 530-752-1517
Fax: 530-752-5614
cacarter@ucdavis.edu
Michael Parrella, CHAIR, ENTOMOLOGY
Phone: 530-752-0492
mpparrella@ucdavis.edu
Jo Stabb, CHAIR, ENVIRONMENTAL DESIGN
Phone: 530-752-6809
jcstabb@ucdavis.edu
D. Burger, CHAIR, ENVIRONMENTAL HORTICULTURE
Phone: 530-752-0130
Fax: 530-752-1819
dwburger@ucdavis.edu
Gary Polis, CHAIR, ENVIRONMENTAL SCIENCE AND
POLICY
Phone: 530-754-8994
Fax: 530-752-3350
gapolis@ucdavis.edu
Marion Miller, CHAIR, ENVIRONMENTAL TOXICOLOGY
Phone: 530-752-4526
mgmiller@ucdavis.edu
Larry Harper, CHAIR, HUMAN AND COMMUNITY
DEVELOPMENT
Phone: 530-752-3624
lharper@ucdavis.edu
Dennis Rolston, CHAIR, LAND, AIR AND WATER
RESOURCES
Phone: 530-752-2113
Fax: 530-752-1552
derolston@ucdavis.edu
Dean MacCannell, CHAIR, LANDSCAPE ARCHITECTURE
Phone: 530-752-6437
edmaccannell@ucdavis.edu
Harry Kaya, CHAIR, NEMATOLOGY
Phone: 530-752-1051
Fax: 530-752-5809
hkkaya@ucdavis.edu
Jim Macdonald, CHAIR, PLANT PATHOLOGY
Phone: 530-752-6897
Fax: 530-752-5674
jdmacdonald@ucdavis.edu
Arnold Bloom, CHAIR, VEGETABLE CROPS PROGRAM
Phone: 530-752-1743
Fax: 530-752-9659
ajbloom@ucdavis.edu
Deborah Elliot-Fisk, CHAIR, WILDLIFE, FISH, &
CONSERVATION BIOLOGY
Phone: 530-752-6586
Fax: 530-752-4514
dlelliottfisk@ucdavis.edu

Ruth Reck, DIRECTOR, NATIONAL INSTITUTE FOR
GLOBAL ENVIRONMENTAL CHANGE
Phone: 530-757-3401
Fax: 530-756-6499
rareck@ucdavis.edu
Roger Shaw, VICE-CHAIR, LAND, AIR AND WATER
RESOURCES
Phone: 530-752-1822
Fax: 530-752-1552
rhshaw@ucdavis.edu

UNIVERSITY OF CALIFORNIA AT DAVIS
HERBARIUM
One Shields Ave., Herbarium Plant Biology, University of
California
Davis, CA 95616 USA
Phone: 530-752-1091   Fax: 530-752-5410
Website: www.herbarium.ucdavis.edu

Founded: 1923
Scope: Statewide

Description: The UC Davis Herbarium is the center for research
in plant systematics at the University of California, Davis. The
Herbarium, of worldwide scope, includes 200,000 specimens.
Holdings from California include documentation for many rare
and endangered species.

Contact(s):
Ellen Dean, DIRECTOR AND CURATOR

UNIVERSITY OF CALIFORNIA AT LOS ANGELES
COLLEGE LETTERS & SCIENCE
Box 951361
Los Angeles, CA 90095 USA
Phone: 310-794-9766   Fax: 310-825-9368
Website: www.college.ucla.edu/

Founded: NA
Scope: National

Contact(s):
Roger Wakimoto, CHAIR, DEPARTMENT OF
ATMOSPHERIC SCIENCES
Phone: 310-825-1751

UNIVERSITY OF CALIFORNIA AT LOS ANGELES
SCHOOL OF ENGINEERING AND APPLIED SCIENCE
CIVIL AND ENVIRONMENTAL ENGINEERING
DEPARTMENT
5731 Boelter Hall, P.O. Box 951593
Los Angeles, CA 90095-1593 USA
Phone: 310-825-1346   Fax: 310-206-2222
E-mail: deeona@ea.ucla.edu
Website: www.cee.ucla.edu/

Founded: NA
Membership: 25
Scope: International/Local

Contact(s):
Jiann-Wen Ju, CHAIR
Phone: 310-206-1751
Fax: 310-206-2222
juj@seas.ucla.edu

UNIVERSITY OF CALIFORNIA AT RIVERSIDE

GRADUATE SCHOOL OF ENVIRONMENTAL SCIENCE
AND ENGINEERING
2217 Geology, University of California
Riverside, CA 92521 USA
E-mail: karenh@mail.ucr.edu
Website: ese.ucr.edu/

Founded: NA

UNIVERSITY OF CALIFORNIA AT RIVERSIDE ENVIRONMENTAL DEPT

DEPARTMENT OF ENVIRONMENTAL SCIENCE
Riverside, CA 92521 USA
Phone: 909-787-1012
Website: envisci.ucr.edu/

Founded: 1971
Scope: State

Description: The Environmental Sciences Program offers four curriculum tracks: Natural Science, Social Science, Environmental Toxicology and Soil Science. Opportunities are available for students to conduct research and to engage in environmental internships. Graduate Degrees available in Soil and Water Science.

Contact(s):
Walt Farmer, CHAIR OF ENVIRONMENTAL SCIENCES
Phone: 909-787-5116

UNIVERSITY OF CALIFORNIA AT SAN DIEGO

SCRIPPS INSTITUTION OF OCEANOGRAPHY
9500 Gilman Dr.
La Jolla, CA 92037 USA
Phone: 858-534-3206   Fax: 858-534-7889
E-mail: siodept@sio.ucsd.edu
Website: www-sio.ucsd.edu/

Founded: 1903
Membership: 200
Scope: International

Description: A part of the University of California, San Diego, the Scripps Institution of Oceanography is one of the oldest, largest, and most important centers for marine science research and graduate training in the world. The Birch Aquarium serves as the public education center for the institution.

Publication(s): Explorations

Contact(s):
Myrl Hendershott, CHAIR OF THE GRADUATE DEPARTMENT
Charles Kennel, DIRECTOR AND VICE CHANCELLOR FOR MARINE SCIENCES
ckennel@ucsd.edu

UNIVERSITY OF CALIFORNIA AT SANTA BARBARA

ENVIRONMENTAL STUDIES PROGRAM
Environmental Studies Program @ University of California
Santa Barbara, CA 93106-4170 USA
Phone: 805-893-2968   Fax: 805-893-8686
E-mail: envst_info@envst.ucsb.edu
Website: www.es.ucsb.edu

Founded: NA

Scope: State

Description: The Environmental Studies Program at UCSB remains one of the strongest in terms of student demand and national reputation. The Environmental Studies curriculum is designed to provide students with the scholarly background and intellectual skills necessary to understand complex environmental problems and formulate decisions that are environmentally sound. While the E.S. Program offers both a B.S. and B.A. degree, both majors recognize and stress the importance of understanding the interrelationships between the hamanities, social sciences, and natural science disciplines within the environment.

Publication(s): see publication web site

Contact(s):
Eric Zimmerman, ACADEMIC ADVISOR
Jo-Ann Shelton, PROGRAM CHAIR

UNIVERSITY OF CALIFORNIA AT SANTA CRUZ

ENVIRONMENTAL STUDIES
Santa Cruz, CA 95064 USA
Phone: 831-459-2634
E-mail: studies@zzyx.ucsc.edu
Website: http://zzyx.ucsc.edu/ES/es.html

Founded: NA

Contact(s):
David Goodman, CHAIRPERSON

UNIVERSITY OF CALIFORNIA, BERKELEY

DEPARTMENT OF ENVIRONMENTAL SCIENCE, POLICY AND MANAGEMENT
145 Mulford Hall
Berkeley, CA 94720-3114 USA
Phone: 510-642-6730   Fax: 510-642-4034
E-mail: undergraduate.espmug@nature.berkeley.edu
graduate
Website: www.cnr.berkeley.edu/departments/espm/

Founded: NA
Scope: International

Description: The Department has a strong undergraduate program awarding the B.S. degree in Forestry, Resource Management, Molecular Environmental Biology and Conservation and Resource Studies. The graduate degree program (M.S., Ph.D.) integrates the biological, social and physical sciences to provide advanced education in basic and applied environmental sciences, develops critical analytical abilities and fosters the capacity to conduct research on the structure and function of ecosystems through ecosystem levels and their interlocked human social systesms. A Master of Forestry (M.F.) and an M.S. in Range Management are also available.

Contact(s):
Sue Jennison, DIRECTOR OF STUDENT SERVICES
Phone: 510-642-6410
susan@nature.berkeley.edu

UNIVERSITY OF COLORADO

SCHOOL OF LAW
NATURAL RESOURCES LAW CENTER
Campus Box 401
Boulder, CO 80309-0401 USA
Phone: 303-492-1286   Fax: 303-492-1297
E-mail: nrlc@spot.colorado.edu
Website: www.colorado.edu/law/nrlc

Founded: NA
Membership: 5
Scope: National, International

Description: Conducts research on environmental and natural resources law and policy, including water, public lands, minerals, Indian law, etc. Sponsors conferences and workshops and hosts visiting scholars. Publishes books, research papers, and Resource Law Notes newsletter.

Contact(s):
Gary Bryner, DIRECTOR

UNIVERSITY OF COLORADO AT BOULDER

ENVIRONMENTAL CENTER
Campus Box 207
Boulder, CO 80309 USA
Phone: 303-492-8308   Fax: 303-492-1897
E-mail: ecenter@stripe.colorado.edu
Website: www.colorado.edu/ecenter

Founded: 1970
Membership: 250
Scope: Local

Description: The CU Environmental Center is the nation's largest student-run environmental resource center. With over 40 student staff and interns, five permanent staff and 100 volunteers, it is the focal point for efforts to make the Boulder campus more environmentally responsible. Besides giving students applied experience in interdisciplinary environmental problem solving, the center provides direct services to the University community, including award-winning recycling and student bus pass programs and a comprehensive library of environmental books, periodicals and video tapes.

Publication(s): Finding A New Way, Blueprint for a Green Campus

Contact(s):
Will Toor, DIRECTOR
Phone: 303-492-8309
toor@spot.colorado.edu

UNIVERSITY OF CONNECTICUT

WBY, Room 308, 1376 Storrs Road, Unit 4087
Storrs, CT 06269-4087 USA
Phone: 860-486-2840   Fax: 860-486-5408
Website: www.canr.uconn.edu/nrme/

Founded: NA
Scope: State

Description: The Dept. offers degrees in natural resources with emphasis in forestry, fisheries, wildlife, biometeorology, watershed hydrology, remote sensing, soil and water conservation and natural resources engineering.

Contact(s):
David Schroeder, DEPARTMENT HEAD
dschroed@canr.uconn.edu

UNIVERSITY OF DELAWARE

COLLEGE OF AGRICULTURE AND NATURAL RESOURCES
531 S College Ave., Townsend Hall
Newark, DE 19717 USA
Phone: 302-831-2501   Fax: 302-831-6758
Website: www.ag.udel.edu

Founded: NA
Membership: 50
Scope: State, Regional

Description: B.S. in wildlife conservation; M.S. and Ph.D. in entomology and applied ecology

Contact(s):
Judith Hough-Stein, DEPARTMENT OF ENTOMOLOGY AND APPLIED ECOLOGY, CHAIRPERSON
Phone: 302-831-8889
Fax: 302-831-3651
jhough@udel.edu
Roland Roth, PROFESSOR ENTOMOLOGY AND APPLIED ECOLOGY
rroth@udel.edu

UNIVERSITY OF FLORIDA

SCHOOL OF FOREST RESOURCES AND CONSERVATION
Gainesville, FL 32611-0410 USA
Phone: 352-846-0850   Fax: 352-392-1707
E-mail: sfrc@gnv.ifas.ufl.edu
Website: www.sfrc.ufl.edu

Founded: 1937
Scope: Statewide

Description: The school seeks to advance understanding of forests: interactions of their components and environment, their relationships with other ecosystems, and appropriate management practices and conservation strategies. Communicating this knowledge to students, public and other professionals is central to this mission; working together with other disciplines is essential to meeting its challenges.

Publication(s): Publications available on web

Contact(s):
Wayne Smith, DIRECTOR
whsmith@ufl.edu
Henry Gholz, GRADUATE PROGRAMS COORDINATOR
George Blakeslee, UNDERGRADUATE PROGRAMS COORDINATOR
gb4stree@usl.edu

UNIVERSITY OF GEORGIA

UGA Marine Institute
Sapelo Island, GA 31327 USA
Phone: 912-485-2221   Fax: 912-485-2133
Website: www.uga.edu/ugami/

Founded: 1953
Membership: 10
Scope: Regional

Description: Concerned with research into the system-ecology, biology, chemistry, and geology of the salt marshes, barrier islands, and nearshore zone of the Georgia coast.

Publication(s): University of Georgia Marine Institute Collected Reprints

Contact(s):
Jon Garbisch, EDUCATION PROGRAM SPECIALIST
jgarbisch@peachnet.campuscwix.net

UNIVERSITY OF GEORGIA SAVANNAH RIVER ECOLOGY LAB

SAVANNNAH RIVER ECOLOGY LABORATORY
Aiken, SC 29802-1030 USA
Phone: 803-725-2472   Fax: 803-725-3309
E-mail: forrest@srel.edu
Website: www.uga.edu/srel/

Founded: NA
Membership: 160
Scope: International, Regional

Description: Learning and communicating ecological processes and principles is the mission of the University of Georgia's Savannah River Ecology Laboratory. The Lab accomplishes its mission through research, outreach and education, and service. Research is conducted in wetlands ecology, wildlife ecology and toxicology, and biogeochemical ecology, including radioecology. Outreach and education activities reach more than 120,000 people annually in Georgia and South Carolina.

Publication(s): Most publications on website, EcoLines

Contact(s):
Carl Strojan, ASSOCIATE DIRECTOR
Savannah River Ecology Lab, Aiken, SC 29802
Phone: 803-725-8217
strojan@serl.edu
Paul Bertsch, DIRECTOR
Whit Gibbons, OUTREACH AND EDUCATION DIRECTOR
Rosemary Forrest, PUBLIC RELATIONS

UNIVERSITY OF GEORGIA WARNELL SCHOOL OF FOREST RESOURCES

DANIEL B. WARNELL SCHOOL OF FOREST RESOURCES
Athens, GA 30602-2152 USA
Phone: 706-542-2686   Fax: 706-542-8356
Website: www.uga.edu/wsfr/

Founded: NA
Membership: 200
Scope: Statewide

Description: The undergraduate degree (B.S.F.R) offers majors in Forestry, Wildlife, Fisheries and Aquaculture and Forest Environmental Resources. Graduate programs (M.S.,M.F.R., Ph.D.) offers a focus in Wildlife Ecology and Management, Fisheries and Aquaculture and a variety of Forest biology and management fields.

Contact(s):
Scott Merkle, GRADUATE PROGRAM COORDINATOR
Phone: 706-542-1183
Fax: 706-542-8356
Arnett Mace, DEAN

UNIVERSITY OF GUELPH

ONTARIO AGRICULTURAL COLLEGE
OAC Deans Office
Guelph, Ontario N1G 2W1 Canada
Phone: 519-824-4120   Fax: 519-766-1423
E-mail: oacinfo@oac.uoguelph.ca
Website: www.oac.uoguelph.ca

Founded: NA
Scope: International

Contact(s):
Alan Watson, ARBORETUM DIRECTOR
Phone: 519-824-4120
awatson@uoguelph.ca
Stu Hilts, CHAIR OF LAND RESOURCE SCIENCE
Mark Sears, ENVIRONMENTAL BIOLOGY (B.S., M.S., PH.D.), DEPT. CHAIR
msears@evbhort.uoguelph.ca
John Fitzgibbon, EXECUTIVE DIRECTOR, COLLEGE FACULTY OF ENVIRONMENTAL DESIGN & RURAL DEVELOPMENT
jfitzgib@rpd.uoguelph.ca
Nathan Perkins, UNDERGRADUATE PROGRAM COORDINATOR, LA
nperkins@la.uoguelph.ca
S. Marshall, UNIVERSITY OF GUELPH INSECT COLLECTION, CURATOR
smarshal@evbhort.uoguelph.ca

UNIVERSITY OF HAWAII

COOPERATIVE FISHERY RESEARCH UNIT
2538 The Mall
Honolulu, HI 96822 USA
Phone: 808-956-8350   Fax: 808-956-6438
E-mail: parrishj@hawaii.edu

Founded: NA
Scope: Local, State, Regional, National, International

Contact(s):
Charles Birkeland, ASSISTANT LEADER
James Parrish, LEADER

UNIVERSITY OF HOUSTON

DEPARTMENT OF CIVIL AND ENVIRONMENTAL ENGINEERING
4800 Calhoun Rd.
Houston, TX 77204-4003 USA
Phone: 713-743-4250   Fax: 713-743-4260
Website: www.egr.uh.edu/cive/

Founded: NA
Membership: 6
Scope: International

Contact(s):
Theodore Cleveland, ENVIRONMENTAL ENGINEERING PROGRAM (B.S., M.S.) DIRECTOR
Phone: 713-743-4250
cleveland@uh.edu

UNIVERSITY OF IDAHO

COLLEGE OF NATURAL RESOURCES
P.O. Box 441136
Moscow, ID 83844-1136 USA
Phone: 208-885-6434   Fax: 208-885-6226
Website: www.its.uidaho.edu/cnr

Founded: NA
Scope: Local, State, Regional

Contact(s):
J. Scott, COOPERATIVE FISH AND WILDLIFE RESEARCH UNIT, LEADER
Phone: 208-885-6336
Kerry Reese, PROFESSOR OF WILDLIFE RESOURCES

UNIVERSITY OF IDAHO
WOMEN IN NATURAL RESOURCES
P.O.Box 441114
Moscow, ID 83844-1114 USA
Phone: 208-885-6754   Fax: 208-885-5878
Website: www.its.uidaho.edu\winr

Founded: 1968
Membership: 500
Scope: International

Publication(s): Women in Natural Resources (Quarterly), Jobs Flyer (Monthly)

Contact(s):
Sandra Martin, EDITOR
winr@uidaho.edu

UNIVERSITY OF ILLINOIS AT URBANA-CHAMPAIGN
205 N. Mathews Ave., 1114 Newmark Civil Rng Lab
Urbana, IL 61801 USA
Phone: 217-333-1000
E-mail: consult@uiuc.edu
Website: www.uiuc.edu/

Founded: NA

Contact(s):
Scott Robinson, ANIMAL BIOLOGY (B.A., PH.D.), HEAD
David Daniel, CIVIL ENGINEERING (B.S., M.S., PH.D.), HEAD
Colin Thorn, GEOGRAPHY (B.S., M.S, PH.D.), HEAD
Vincent Bellafiore, LANDSCAPE ARCHITECTURE (B.L.A., M.L.A.), HEAD
Patrick Brown, NATURAL HISTORY SURVEY PROFESSOR
Gary Rolfe, NATURAL RESOURCES AND ENVIRONMENTAL SCIENCES (B.S., M.S., PH.D.), HEAD
Christopher Silver, URBAN AND REGIONAL PLANNING (B.A., M.U.P., PH.D.), HEAD

UNIVERSITY OF ILLNOIS EXTENSION
214 Mumford Hall (MC-710), 1301 W. Gregory Dr.
Urbana, IL 61801 USA
Phone: 217-333-5900   Fax: 217-244-5403
E-mail: www.web_extension@aces.uiuc.edu
Website: www.extension.uiuc.edu/welcome.html

Founded: NA
Scope: Local

Contact(s):
Richard Warner, ASSISTANT DEAN OFFICE OF RESOURCE
211 Mumford Hall (MC-710), 1301 W Gregory Dr, Urbana, IL 61801
Phone: 217-333-5199
Fax: 217-244-3219
Patricia Buchanan, ASSISTANT DEAN, EXTENSION OPERATIONS
buchananp@mail.aces.uiuc.edu
John Van Es, ASSISTANT DEAN, EXTENSION PROGRAM COORDINATION
e-van1@uiuc.edu
Dennis Campion, ASSOCIATE DEAN
dcampion@uiuc.edu

UNIVERSITY OF IOWA
2700 Steindler Bldg.
Iowa City, IA 52242 USA
Phone: 319-335-9627   Fax: 319-335-9200
Website: www.public-health.uiowa.edu

Founded: NA
Scope: Statewide

Contact(s):
Robert Ettema, CIVIL AND ENVIRONMENTAL ENGINEERING PROGRAM (M.S., PH.D.), DEPT. CHAIR
Dept. Office: 2130 Seamans Center, Iowa City, IA 52242
Phone: 319-335-5647
Fax: 319-335-5660
cee@engineering.uiowa.edu
James Merchant, ENVIRONMENTAL HEALTH SCIENCES RESEARCH CENTER, DIRECTOR
2707 Steindler Bldg., Iowa City, IA 52242
Phone: 319-335-9833
james-merchant@uiowa.edu

UNIVERSITY OF KANSAS
DEPARTMENT OF ENVIRONMENTAL STUDIES
517 W. 14th St., Bldg. 138
Lawrence, KS 66045 USA
Phone: 785-842-2059
E-mail: env-studies@ku.edu
Website: www.ku.edu/~kuesp

Founded: NA
Membership: 30
Scope: Statewide

Description: Environmental Studies Program offers options in ecology and field biology, environmental policy, environmental impact analysis, environmental health, geology and meterology, water resources, and environmental land-use analysis

Contact(s):
Stanford Loeb, DIRECTOR
Phone: 785-842-2059
Deborah Snyder, SECRETARY

UNIVERSITY OF KANSAS FIELD STATIONS
Kansas Biological Survey, 2021 Constant Ave
Lawrence, KS 66047 USA
Phone: 785-864-7720   Fax: 785-864-5093
Website: www.ker.ukans.edu

Founded: NA
Scope: Statewide

Contact(s):
Ed Martinko, DIRECTOR, KANSAS APPLIED REMOTE SENSING PROGRAM WITH EMPHASIS IN NATURAL RESOURCES MANAGEMENT
Phone: 785-864-7770
Thomas Taylor, ECOLOGY AND EVOLUTIONARY BIOLOGY (B.S., M.A., PH.D.), CHAIR
Phone: 785-864-3625
Ed Martinko, EXPERIMENTAL AND APPLIED ECOLOGY, WITH EMPHASIS IN AQUATIC ECOLOGY AND POPULATION BIOLOGY OF SMALL MAMMALS AND PLANTS (M.A., PH.D.), DIRECTOR
Phone: 785-864-4375
edmartinko@ku.edu

UNIVERSITY OF KENTUCKY
COLLEGE OF AGRICULTURE
Lexington, KY 40546 USA
Phone: 606-257-7596
Website: www.ca.uky.edu

Founded: NA

Contact(s):
Donald Graves, FORESTRY (B.S., M.S., M.S.F.), CHAIRMAN
Dewayne Ingram, HORTICULTURE (B.S., M.S.), CHAIR
Phone: 606-257-1758
dingram@ca.uky.edu
Karen Goodlet, LANDSCAPE ARCHITECTURE (B.S.)
Phone: 606-257-7295
kgoodlet@ca.uky.edu

UNIVERSITY OF LOUISVILLE
UNIVERSITY OF LOUISVILLE BIOLOGY
Belknap Campus
Louisville, KY 40292 USA
Phone: 502-852-6771   Fax: 502-852-0725
Website: www.louisville.edu/a-s/biology.edu

Founded: NA
Membership: 30
Scope: Local, Regional

Description: The Large River Laboratory was established in 1992 to conduct research on river and freshwater systems in Kentucky and surrounding states. Community and population studies of rivers and smaller streams constitute the primary focus of the laboratory.

Contact(s):
Jeff Jack, DEPT. OF BIOLOGY
Phone: 502-852-5940
jdjack01@gwise.louisville.edu
William Pearson, PROFESSOR
Phone: 502-852-3727
wdpear01@gwise.louisville.edu

UNIVERSITY OF MAINE
COLLEGE OF NATURAL SCIENCES, FORESTRY AND AGRICULTURE
5782 Winslow Hall, Suite 105
Orono, ME 04469-5782 USA
Phone: 207-581-3202   Fax: 207-581-3207
Website: www.umaine.edu/

Founded: NA
Scope: Statewide

Contact(s):
Charles Wallace, CHAIRMAN FOR ANIMAL & HORTICULTURAL SCIENCES
Phone: 207-581-2770
William Krohn, COOPERATIVE FISH AND WILDLIFE RESEARCH UNIT, LEADER
Phone: 207-581-2870
Bruce Wiersma, DEAN
Phone: 207-581-3202
John Singer, DEPARTMENT OF BIOCHEMISTRY, MICROBIOLOGY AND MOLECULAR BIOLOGY (B.A., B.S., M.S., M.P.S., PH.D.), CHAIRMAN
Phone: 207-581-2810
Christopher Campbell, DEPARTMENT OF BIOLOGICAL SCIENCES (B.A., B.S., M.S., PH.D.), CHAIRMAN
Phone: 207-581-2551
Rodney Bushway, DEPARTMENT OF FOOD SCIENCE AND HUMAN NUTRITION, FOOD AND NUTRITION SCIENCES (B.S., M.S., PH.D.), CHAIR
Phone: 207-581-1621
William Livingston, DEPARTMENT OF FOREST ECOSYSTEM SCIENCE, CHAIRMAN
Phone: 207-581-2884
David Field, DEPARTMENT OF FOREST MANAGEMENT (B.S., M.F., M.S., PH.D.), CHAIRMAN
Phone: 207-581-2856
Daniel Belknap, DEPARTMENT OF GEOLOGICAL SCIENCES (B.A., B.S., M.S., PH.D.), CHAIR
Phone: 207-581-2152
Ivan Fernandez, DEPARTMENT OF PLANT, SOIL, AND ENVIRONMENTAL SCIENCE (B.S., M.S., PH.D.), CHAIRMAN
Phone: 207-581-2932
George Criner, DEPARTMENT OF RESOURCE ECONOMICS AND POLICY (B.S., M.S.), CHAIRMAN
Phone: 207-581-3150
Dan Harrison, DEPARTMENT OF WILDLIFE ECOLOGY
Mark Anderson, PROGRAM DIRECTOR
Dave Townsend, SCHOOL OF MARINE SCIENCES, DIRECTOR

UNIVERSITY OF MAINE AT FORT KENT
25 Pleasant St.
Fort Kent, ME 04743 USA
Phone: 207-834-7617   Fax: 207-834-7503
E-mail: sselva@maine.main.edu
Website: www.umfk.maine.edu/

Founded: NA
Scope: Regional

Description: Located in the heart of Maine's Acadian forest region. Our Bachelor of Science in Environmental Studies degree program provides a solid experiential and academic background to students preparing for careers in education, industry, and public service.

Publication(s): University Catalog and Brochures

Contact(s):
Don Zillman, PRESIDENT (INTERIM)
Steven Selva, PROFESSOR OF BIOLOGY AND ENVIRONMENTAL STUDIES
sselva@maine.main.edu

UNIVERSITY OF MAINE AT ORONO
SCHOOL OF MARINE SCIENCES
5741 Libby Hall
Orono, ME 04469 USA
Phone: 207-581-4381   Fax: 207-581-4388
E-mail: marine@maine.edu
Website: www.umaine.edu/

Founded: NA
Scope: Statewide

Contact(s):
Bruce Sidell, CHAIR
Phone: 207-581-4381

UNIVERSITY OF MARYLAND AT COLLEGE PARK COLLEGE OF AGRICULTURE & NATURAL RESOURCES

COLLEGE OF AGRICULTURE AND NATURAL RESOURCES
0107 Symons Hall
College Park, MD 20742 USA
Phone: 301-405-7761   Fax: 301-405-8570
Website: www.agnr.umd.edu

Founded: NA
Scope: Statewide

Contact(s):
Thomas Fretz, DEAN
John Doerr, DEPT. OF ANIMAL AND AVIAN SCIENCES (B.S.), UNDERGRADUATE COORDINATOR
Animal Sciences Center, College Park, MD 20742
Phone: 301-405-1373
Fax: 301-314-9059
Richard Weismiller, DEPT. OF NATURAL RESOURCE SCIENCES AND LANDSCAPE ARCHITECTURE (B.S., B.L.A., M.S., PH.D.), CHAIR
Room 2104, Plant Sciences Bldg., College Park, MD 20742
Phone: 301-405-1306
Fax: 301-314-9308
rw22@umail.umd.edu
Mark Varner, DIRECTOR OF GRADUATE PROGRAM & AVIAN SCIENCES
Animal Sciences Center, College Park, MD 20742
Phone: 301-405-1396
varner@umd5.umd.edu

UNIVERSITY OF MARYLAND AT EASTERN SHORE

DEPT. OF NATURAL SCIENCES
Carver Hall
Princess Anne, MD 21853 USA
Phone: 301-651-2200
Website: hawk.umes.edu/sciences/index.html

Founded: NA

Description: Environmental Sciences (B.S.), Marine Sciences (B.S., M.S.), Marine, Estuarine, and Environmental Sciences (M.S., Ph.D), Environmental Chemistry (B.S., M.S.)

Contact(s):
Charles Hocutt, COASTAL ECOLOGY RESEARCH CENTER, CONTACT
Joseph Okoh, DEPT. CHAIR (ACTING)
Gian Gupta, ENVIRONMENTAL SCIENCE/MARINE SCIENCE, CONTACT
Phone: 410-651-6030
ggupta@umes_bird.umd.edu
Steve Rebach, MARINE, ESTUARINE, AND ENVIRONMENTAL SCIENCES (M.S., PH.D), CONTACT
Phone: 410-651-6013

UNIVERSITY OF MARYLAND BALTIMORE COUNTY

DEPARTMENT OF BIOLOGICAL SCIENCES
1000 Hilltop Cir.
Baltimore, MD 21250 USA
Phone: 410-455-2261   Fax: 410-455-3875
E-mail: ellis@umbc.edu
Website: www.umbc.edu/biosci

Founded: NA
Scope: Statewide

Description: Ecology and Environmental Biology focus with a strong emphasis on research, scientific approach, faculty contact and extensive lab offerings

Contact(s):
Lasse Lindahl, PROFESSOR AND CHAIR
lindahl@umbc.edu

UNIVERSITY OF MARYLAND CENTER FOR ENVIRONMENTAL SCIENCE

P.O. Box 775
Cambridge, MD 21613 USA
Phone: 410-228-9250   Fax: 410-228-3843
Website: www.umces.edu

Founded: 1925
Scope: International

Description: UMCES is an institution of the University System of Maryland, with a special mission in multidisciplinary environmental research on Chesapeake Bay, the mid-Atlantic region, and coastal systems around the world.

Contact(s):
Louis Pitelka, APPALACHIAN LABORATORY, DIRECTOR
301 Braddock Rd., Frostburg, MD 21532
Phone: 301-689-7101
Fax: 301-689-7200
pitelka@al.umces.edu
Kenneth Tenore, CHESAPEAKE BIOLOGICAL LABORATORY, DIRECTOR
P.O. Box 38, Solomons, MD 20688
Phone: 410-326-7241
Fax: 410-326-7263
tenore@cbl.umces.edu
Thomas Malone, HORN POINT LABORATORY, DIRECTOR AND PROFESSOR
Phone: 410-221-8406
Fax: 410-221-8490
malone@hpl.umces.edu
Donald Boesch, PRESIDENT

UNIVERSITY OF MARYLAND, COLLEGE PARK

GRADUATE SCHOOL
2123 Lee Bldg.
College Park, MD 20742-5121 USA
Phone: 301-405-4198   Fax: 301-314-9305
E-mail: gradmit@deans.umd.edu
Website: www.inform.umd.edu/grad/

Founded: NA
Membership: 75
Scope: Statewide
Publication(s): Graduate applications available on website

Contact(s):
Trudy Lindsey, DIRECTOR

UNIVERSITY OF MASSACHUSETTS

DEPT. OF NATURAL RESOURCES CONSERVATION
Holdsworth NRC
Amherst, MA 01003-9285 USA
Phone: 413-545-2665   Fax: 413-545-4358
Website: www.umass.edu/forwild/

Founded: NA

Membership: 25
Scope: Statewide

Contact(s):
William McComb, DEPARTMENT OF NATURAL
RESOURCES CONSERVATION, HEAD
Guy Lanza, ENVIRONMENTAL SCIENCES, PROGRAM
DIRECTOR
Phone: 413-545-3747
Richard Degraaf, U.S. FOREST SERVICE
Phone: 413-545-0357
Martha Mather, U.S. GEOLOGICAL SURVEY,
MASSACHUSETTS COOPERATIVE FISH AND WILDLIFE
RESEARCH UNIT, ASSISTANT LEADER
Phone: 413-545-4895
Kevin Friedland, U.S. NATIONAL OCEANIC AND
ATMOSPHERIC ADMINISTRATION COOPERATIVE
EDUCATION AND RESEARCH PROGRAM, DIRECTOR
Phone: 413-545-2842

UNIVERSITY OF MASSACHUSETTS
URBAN HARBORS INSTITUTE
100 Morrissey Blvd.
Boston, MA 02125-3393 USA
Phone: 617-287-5570   Fax: 617-287-5575
E-mail: urbanurban.harbors@usb..edu
Website: www.uhi.umb.edu

Founded: 1989
Membership: 7
Scope: International

Description: The Urban Harbors Institute was founded as a
center for the study of harbor, coastal and ocean issues. It
conducts multidisciplinary research on the policy and
management issues affecting the coastal area, with emphasis
on the urban waterfront. It also promotes linkages between
scientists, government, academic, and business communities
to improve decision-making. The institute publishes research,
sponsors seminars, conferences, and public forums to
disseminate and exchange information.

Publication(s): The Coastlines

Contact(s):
Richard Delaney, DIRECTOR
rich.delaney@umb.edu

UNIVERSITY OF MICHIGAN
SCHOOL OF NATURAL RESOURCES AND
ENVIRONMENT
Dana Bldg., 430 East University
Ann Arbor, MI 48109-1115 USA
Phone: 734-764-6453   Fax: 734-615-1277
Website: www.snre.umich.edu/

Founded: NA

Contact(s):
Daniel Mazmanian, DEAN
Donna Erickson, LANDSCAPE ARCHITECTURE (M.L.A.,
B.S., M.S), CONCENTRATION CHAIR
James Diana, RESOURCE ECOLOGY AND MANAGEMENT
(B.S., M.S., PH.D.), CONCENTRATION CHAIR
Bunyan Bryant, RESOURCE POLICY AND BEHAVIOR (B.S.,
M.S., PH.D.), CONCETRATION CHAIR

UNIVERSITY OF MINNESOTA AT CROOKSTON
NATURAL RESOURCES DEPARTMENT
2900 University Ave.
Crookston, MN 56716 USA
Phone: 218-281-8129   Fax: 2182-821-8603
Website: www.crk.umn.edu

Founded: NA
Membership: 6
Scope: Regional

Description: Offers a broadly-oriented natural resource major
which prepares students for entry-level resource management
positions. Practical and field instruction in integrated land
management is emphasized leading to a B.S. degree in
Natural Resource Management, Park Management, Soil and
Water Technology or Natural Resources Law Enforcement.

Contact(s):
Bobby Holder, DIRECTOR
Phone: 218-281-8135
bholder@mail.crk.umn.edu
Philip Baird, PARK & RECREATION PROFESSOR
Daniel Svedarsky, PROGRAM LEADER
dsvedars@mail.crk.umn.eu
John Loegerig, WILDLIFE MANAGEMENT PROFESSOR

UNIVERSITY OF MINNESOTA AT ST. PAUL
WISHERRIES WILDLIFE CONSERVATION BIOLOGY
200 Hodson Hall, 1980 Falwell Ave
St. Paul, MN 55108 USA
Phone: 612-624-1234   Fax: 612-625-5299
Website: www.cnr.umn.edu/

Founded: 1903
Scope: Statewide

Description: The mission of the College of Natural Resources is
to foster a quality environment by contributing to the
management, protection, and sustainable use of our natural
resources through teaching, research, and outreach.

Contact(s):
Dorothy Anderson, CENTER FOR ENVIRONMENTAL
LEARNING AND LEADERSHIP
Phone: 612-624-2721
Jim Perry, CENTER FOR NATURAL RESOURCES POLICY
AND MANAGEMENT, HEAD/WATER RESOURCES
SCIENCE GRADUATE PROGRAM
Phone: 612-624-9796
Francesca Cuthbert, CONSERVATION BIOLOGY
PROGRAM, DIRECTOR
Phone: 612-624-1756
David Smith, CONSERVATION BIOLOGY PROGRAM,
DIRECTOR
Phone: 612-624-5369
Dave Lime, COOPERATIVE PARK STUDIES UNIT, SENIOR
RESEARCH ASSOCIATE
Phone: 612-624-2250
Alfred Sullivan, DEAN OF COLLEGE OF NATURAL
RESOURCES
Jim Perry, DEPARTMENT HEAD
Robert Sterner, DEPARTMENT OF ECOLOGY, EVOLUTION
AND BEHAVIOR, HEAD
1987 Upper Buford Cir., St. Paul, MN 55108
Phone: 612-625-6790
Ira Adelman, DEPARTMENT OF FISHERIES AND WILDLIFE,
Phone: 612-624-3600

Alan Ek, DEPARTMENT OF FOREST RESOURCES, HEAD
Phone: 612-624-3400
Joseph Massey, DEPARTMENT OF WOOD AND PAPER
SCIENCE, HEAD
Phone: 612-624-5200
Jean Albrecht, FORESTRY LIBRARY, LIBRARIAN
B-50 Natural Resources Administration Bldg., 2003 Upper
Buford Cir., St. Paul, MN 55108
Phone: 612-624-3222
Anne Kapuscinski, INSTITUTE FOR SOCIAL ECONOMIC
AND ECOLOGICAL SUSTAINABILITY (ISEES)
Phone: 612-624-7719
Barbara Coffin, INSTITUTE FOR SUSTAINABLE NATURAL
RESOURCE MANAGEMENT
Phone: 612-624-4986
David Andersen, MINNESOTA COOPERATIVE FISH AND
WILDLIFE RESEARCH UNIT, LEADER
Phone: 612-624-3421
Steven Daley Laursen, OUTREACH AND EXTENSION,
ASSOCIATE DEAN
Phone: 612-624-9298
Bill Ganzlin, STUDENT SERVICES OFFICE, DIRECTOR
Phone: 612-624-6768
Patrick Brezonik, WATER RESOURCES CENTER,
DIRECTOR
Phone: 612-624-9282

UNIVERSITY OF MISSOURI

103 Anheuser-Busch Natural Resources Bldg.
Columbia, MO 65211-7220 USA
Phone: 573-882-6446   Fax: 573-884-2636
Website: www.snr.missouri.edu

Founded: NA
Membership: 50
Scope: Statewide

Contact(s):
Charles Rabeni, COOPERATIVE FISH AND WILDLIFE
RESEARCH UNIT, LEADER
302 Anheuser-Busch Natural Resources Bldg.
Phone: 573-882-3524
Albert Vogt, DIRECTOR
Jack Jones, FISHERIES AND WILDLIFE DEPT. (B.S., M.S.,
PH.D.), CHAIR
302 Anheuser-Busch Natural Resources Bldg.,
Phone: 573-882-3436
Carl Settergren, FORESTRY DEPT. (B.S., M.S., PH.D.),
CHAIR
302 Anheuser-Busch Natural Resources Bldg.
Phone: 573-882-2627
R. Hammer, SOIL AND ATMOSPHERIC SCIENCES
DEPARTMENT (B.S., M.S., PH.D.), CHAIR
302 Anheuser-Busch Natural Resources Bldg.
Phone: 573-882-6301

UNIVERSITY OF MONTANA SCHOOL OF FORESTRY

32 Campus Dr.
Missoula, MT 59812-0576 USA
Phone: 406-243-5521   Fax: 406-243-4845
E-mail: jilyon@forestry.umt.edu
Website: www.forestry.umt.edu

Founded: NA
Membership: 50
Scope: Statewide

Contact(s):
James Burchfield, BOLLE CENTER FOR PEOPLE AND
FORESTS
Phone: 406-243-6650
Jack Thomas, BOONE AND CROCKETT WILDLIFE
CONSERVATION PROGRAM
Phone: 406-243-5566
Kelsey Milner, INLAND NORTHWEST GROWTH AND YIELD
COOPERATIVE
Phone: 406-243-6653
Norma Nickerson, INSTITUTE FOR TOURISM AND
RECREATION RESEARCH
Phone: 406-243-5686
Robert Pfister, MISSION ORIENTED RESEARCH
PROGRAM
Phone: 406-243-6582
Perry Brown, MONTANA FOREST AND CONSERVATION
EXPERIMENT STATION, DIRECTOR
Phone: 406-243-5522
Steven Running, NUMERICAL TERRADYNAMIC
SIMULATION GROUP
Phone: 406-243-6311
Hans Zuuring, QUANTITATIVE SERVICES GROUP
Phone: 406-243-6465
Paul Hansen, RIPARIAN/WETLAND RESEARCH PROGRAM
Phone: 406-243-2050
Perry Brown, SCHOOL OF FORESTRY DEAN
Phone: 406-243-5522
Wayne Freimund, WILDERNESS INSTITUTE
Phone: 406-243-5184
Daniel Pletscher, WILDLIFE BIOLOGY PROGRAM,
DIRECTOR
Phone: 406-243-5272

UNIVERSITY OF MONTANA WILDLIFE

SCHOOL OF LAW
Building 32 Campus Drive
Missoula, MT 59812-6552 USA
Phone: 406-994-0211
Website: www.umt.edu/law/

Founded: NA
Scope: International

Contact(s):
Edwin Eck, DEAN

UNIVERSITY OF NEBRASKA

SCHOOL OF NATURAL RESOURCE SCIENCES
309 Biochemistry Hall, Box 830758
Lincoln, NE 68583-0759 USA
Phone: 402-472-9873   Fax: 402-472-3610
Website: www.snrs.unl.edu

Founded: NA
Scope: Statewide

Contact(s):
Marcy Tintera, GRADUATE PROGRAMS
Phone: 402-472-6622
fofw031@unlvm.unl.edu

UNIVERSITY OF NEVADA AT LAS VEGAS
ENVIRONMENTAL SCIENCE PROGRAM
4505 Maryland Parkway, Box 454030
Las Vegas, NV 89154-4030 USA
Phone: 702-895-3011   Fax: 702-895-3956
E-mail: biology@neveda.edu
Website:
www.unlv.edu/Other_Programs/Environmental_Studies/

Founded: NA
Membership: 40
Scope: Local

Contact(s):
Dawn Neuman, CHAIRMAN
newman@ccmail.nevada.edu
James Deacon, ENVIRONMENTAL STUDIES DEPARTMENT
(B.S., B.A., M.S., PH.D.)
jdeacon@ccmail.nevada.edu

UNIVERSITY OF NEVADA AT LAS VEGAS
WATER RESOURCES PROGRAM
4505 Maryland Pkwy.
Las Vegas, NV 89154-4029 USA
Phone: 702-895-4006   Fax: 702-895-4064
E-mail: wrmunlv@hotmail.com
Website: www.unlv.depts/wrm

Founded: NA
Scope: State

Description: The Water Resources Management Graduate
Program at the University of Nevada is an interdisciplinary
environmental program. The curriculum includes studies in
water quality and quantity; surface water and groundwater; and
water law, regulation, and management. Offers environmental
programs at graduate level.

Contact(s):
David Kreamer, DIRECTOR
kreamerd@nevada.edu

UNIVERSITY OF NEVADA AT RENO
DEPT. OF ENVIRONMENTAL & RESOURCES
SCIENCES
1000 Valley Rd.
Reno, NV 89512 USA
Phone: 775-784-6763   Fax: 702-784-4583
Website: www.unr.edu/

Founded: NA
Membership: 20
Scope: International
Publication(s): see publication web site

Contact(s):
James Sedniger, ASSOCIATE PROFESSOR
jsedniger@cabnr.unr.e
Watkins Miller, DEPARTMENT OF ENVIRONMENTAL AND
RESOURCE SCIENCES, CHAIRMAN
Glenn Miller, ENVIRONMENTAL SCIENCE (B.S., M.S.)
Phone: 775-784-4108
Fax: 775-784-1142
gcmiller@scs.unr.edu
John Guitjens, INTERDISCIPLINARY HYDROLOGY (B.S.,
M.S., PH.D.)
guitjens@unr.edu

Dale Johnson, NATURAL RESOURCE MANAGEMENT (B.S.,
M.S.)
Roger Walker, NATURAL RESOURCE MANAGEMENT
(B.S.,M.S.)
walker@unr.edu

UNIVERSITY OF NEW HAMPSHIRE
215 James Hall Natural Resources, 56 College Rd.
Durham, NH 03824 USA
Phone: 603-862-1234   Fax: 603-862-1234
Website: www.unh.edu

Founded: NA
Scope: Statewide
Publication(s): Campus Journal, New Hampshire, The

Contact(s):
Theodore Howard, DEPARTMENT OF NATURAL
RESOURCES, JAMES HALL, CHAIR
Robert Eckert, ENVIRONMENTAL CONSERVATION (B.S.),
PROGRAM COORDINATOR
Richard Weyrick, FORESTRY (B.S.) PROGRAM
COORDINATOR
Russell Congalton, NATURAL RESOURCES (M.S.),
PROGRAM COORDINATOR
John Aber, NATURAL RESOURCES (PH.D.), PROGRAM
COORDINATOR
Elizabeth Roschett, SOIL SCIENCE (B.S.), PROGRAM
COORDINATOR
Phone: 603-862-0713
roschett@cisunix.unh.edu
William McDowell, WATER RESOURCES MANAGEMENT
(B.S.), PROGRAM COORDINATOR
Peter Pekins, WILDLIFE ECOLOGY (B.S.), PROGRAM
COORDINATOR

UNIVERSITY OF NEW HAVEN DEPT. OF BIOLOGICAL & ENVIRONMENTAL SCIENCE
DEPT. OF BIOLOGICAL & ENVIRONMENTAL SCIENCE
300 Orange Ave.
West Haven, CT 06516 USA
Phone: 203-932-7101   Fax: 203-933-2036
E-mail: adminfo@charger.newhaven.edu
Website: www.newhaven.edu/

Founded: NA
Membership: 9
Scope: National

Contact(s):
Michael Rossi, CHAIR OF BIOLOGY DEPARTMENT
rossimj@charger.newhaven.edu
Charles Vigue, DEPARTMENT OF BIOLOGY AND
ENVIRONMENTAL SCIENCE (B.S., M.S.), CHAIRMAN
Phone: 203-932-7107
Fax: 203-931-6097
viguecl@charger.newhaven.edu

UNIVERSITY OF NORTH CAROLINA AT ASHEVILLE
ENVIRONMENTAL STUDIES DEPARTMENT
CPO 2330, One University Heights
Asheville, NC 28804-8511 USA
Phone: 828-251-6441   Fax: 828-251-6041
Website: www.unca.edu/envr_studies/

Founded: 1983

Membership: 7
Scope: Statewide

Description: The undergraduate environmental studies cirriculum offers degrees options in ecology, pollution control, earth science and natural resource management. There is a required internship program and numerous research opportunities in an environmentally diverse region in areas such as ecology, wetlands restoration, lead and pesticide contamination and waste analysis.

Contact(s):
Richard Maas, CONTACT

UNIVERSITY OF NORTH CAROLINA AT CHAPEL HILL
Campus Box 7400 Rosenau Hall
Chapel Hill, NC 27599-7431 USA
Phone: 919-966-1171   Fax: 919-966-7911
Website: www.unc.edu

Founded: NA
Membership: 45
Scope: Statewide

Contact(s):
Richard Kamens, AIR, RADIATION, AND INDUSTRIAL HYGIENE
Environmental Science and Engineering, 115 Rosenau Hall, Chapel Hill, NC 27599
Phone: 919-966-5452
kamens@unc.edu
Russell Christman, AQUATIC AND ATMOSPHERIC SCIENCES
Environmental Science and Engineering, 164 Rosenau Hall, Chapel Hill, NC 27599
Phone: 919-966-1683
russ_christman@unc.edu
Seth Reice, ECOLOGY (M.S., PH.D.), CHAIRMAN
244 Wilson Hall, Chapel Hill, NC 27599
Phone: 919-962-1375
sreice@biomass.bio.unc.edu
Louise Ball, ENVIRONMENTAL HEALTH SCIENCES
Environmental Science and Engineering, 4114 E McGavran-Greenberg Hall, Chapel Hill, NC 27599
Phone: 919-966-7306
lmball@sph.unc.edu
William Glaze, ENVIRONMENTAL SCIENCES AND ENGINEERING,
Environmental Science and Engineering, 105 Miller Hall, Chapel Hill, NC 27599
Phone: 919-966-9917
bill_glaze@unc.edu
Frederic Pfaender, INSTITUTE FOR ENVIRONMENTAL SCIENCE & ENGENEERING
Environmental Science and Engineering, 157 Rosenau Hall, Chapel Hill, NC 27599
Phone: 919-966-3842
fred_pfaender@unc.edu
Christopher Martens, MARINE SCIENCES
12-4A Venable Hall, Chapel Hill, NC 27599
Phone: 919-962-0152
martens@marine.unc.edu
Philip Singer, WATER RESOURCES ENGINEERING
Environmental Science and Engineering,110 Rosenau Hall, Chapel Hill, NC 27599
Phone: 919-962-3865
phil_singer@unc.edu

UNIVERSITY OF NORTH CAROLINA AT CHAPEL HILL
ENVIRONMENTAL RESOURCE PROGRAM
CB# 1105 Miller Hall
Chapel Hill, NC 27599 USA
Phone: 919-966-7754   Fax: 919-966-9920
E-mail: erp@sph.unc.edu
Website: www.sph.unc.edu/erp

Founded: 1985
Scope: Statewide

Description: The ERP was established to link the resources of the University with the citizens of North Carolina. Since its inception, the ERP has provided information, technical assistance, and training to citizen groups, local governments and school teachers, and has facilitated collaborative decision-making about environmental issues. The ERP has five main program areas: environmental education, community outreach and education, collaborative and policy research, Carolina Health and Environment Community Center (CHECC) website and sustainable development.

Publication(s): The Link, The Superfundscoop, Working Together, Community Based Approaches To Prevent Childhood Lead Poisioning, The Guide To North Carolina Environmental Groups

Contact(s):
Kathleen Gray, ASSOCIATE DIRECTOR
Frances Lynn, DIRECTOR

UNIVERSITY OF NORTH DAKOTA
BIOLOGY DEPARTMENT
Box 19019
Grand Forks, ND 58202-9019 USA
Phone: 701-777-2621
Website: www.und.nodak.edu/

Founded: NA
Membership: 16
Scope: International

Contact(s):
Steven Kelsch, FISHERY RESEARCH UNIT (B.S., M.S., PH.D.), LEADER
Richard Sweitzer, STAFF MEMBER

UNIVERSITY OF NORTH TEXAS
INSTITUTE OF APPLIED SCIENCES
NT Box 310559
Denton, TX 76203 USA
Phone: 940-565-2694   Fax: 940-565-4297
Website: www.ias.unt.edu/

Founded: 1976
Scope: Statewide

Description: An interdisciplinary unit whose primary research activities are oriented towards land and water resources. Research includes aquatic toxicology, surface and groundwater quality, archaelogy, remote sensing, geographic information systems and environmental modeling. In 1998, the institute took on an environmental outreach program which includes the Outdoor Environmental Learning Area (ODELA) and the Sky Theater. Over 200 scientists, support staff, volunteers, graduate and undergraduate students are involved in these activities.

Contact(s):

William Waller, AQUATIC TOXICOLOGY AND RESERVOIR LIMNOLOGY, DIRECTOR
Phone: 940-565-2694
Reid Ferring, CENTER FOR ENVIRONMENTAL ARCHAEOLOGY, DIRECTOR
Phone: 940-565-2694
Mike Nieswiadomy, CENTER FOR ENVIRONMENTAL ECONOMICS, DIRECTOR
Phone: 940-565-2573
Samuel Atkinson, CENTER FOR REMOTE SENSING, DIRECTOR
Phone: 940-565-2694
Andy Schoolmaster, CENTER FOR SPATIAL ANALYSIS, DIRECTOR
Phone: 940-565-2901
Thomas Lapoint, DIRECTOR
Phone: 940-369-7776
Rudi Thompson, ELM FORK EDUCATION CENTER, COORDINATOR
Phone: 940-565-2694
Steve Spurger, ELM FORK EDUCATION CENTER, COORDINATOR
Phone: 940-565-2694
Farida Saleh, ENVIRONMENTAL CHEMISTRY LABORATORY, DIRECTOR
Phone: 940-565-2694
Miguel Acevedo, ENVIRONMENTAL MODELING LABORATORY, DIRECTOR
Phone: 940-565-2091
Bruce Hunter, ENVIRONMENTAL VISUALIZATION LABORATORY, DIRECTOR
Phone: 940-565-2694
Tom Lapoint, EXPERIMENTAL STREAM DIRECTOR
Phone: 940-565-2694
Jan Dickson, FACULTY FOR ENVIRONMENTAL ETHICS, COORDINATOR
Phone: 940-565-2727
Eugene Hargrove, FACULTY FOR ENVIRONMENTAL PHILOSOPHY, EXECUTIVE DIRECTOR
Phone: 940-565-2266
Eugene Hargrove, GRADUATE PROGRAM IN ENVIRONMENTAL ETHICS (M.A.), COORDINATOR
Phone: 940-565-2266
Samuel Atkinson, GRADUATE PROGRAM IN ENVIRONMENTAL SCIENCE (M.S., PH.D.), COORDINATOR
Phone: 940-565-2694
Sandra Terrell, INTERDISCIPLINARY GRADUATE STUDIES, DEAN
Chris Littler, PLANETARIUM DIRECTOR
Phone: 940-565-2694
Kenneth Dickson, PROFESSOR
James Kennedy, WATER RESEARCH FIELD STATION, DIRECTOR
Phone: 940-565-2694
Robert Doyle, WETLANDS RESEARCH, DIRECTOR
Phone: 940-565-2694

UNIVERSITY OF NORTHERN BRITISH COLUMBIA

3333 University Way
Prince George, British Columbia V2N 4Z9 Canada
Phone: 250-960-5555   Fax: 250-960-5538
Website: www.unbc.ca

Founded: 1990

Scope: Statewide

Description: UNBC is a research intensive, small university founded in 1990 as "a university in the north, for the north." The faculty of Natural Resources and Environmental Studies develops managers and scientists to effectively meet the demands for natural resources products and services while maintaining a quality environment.

Natural Resource Management offers majors in Wildlife and Fisheries, Forestry (accredited by the Canadian Forestry Accreditation Board) and Resource Recreation.

Contact(s):

Winifred Kessler, NATURAL RESOURCES MANAGEMENT (B.S.)
Phone: 250-950-6664
winifred@unbc.ca
Jeff Zeiger, RESOURCE BASED TOURISM (B.A.), ENVIRONMENTAL STUDIES (B.A., B.S.), CONTACT
Phone: 250-960-5308
zeiger@unbc.ca

UNIVERSITY OF NORTHERN COLORADO

DEPARTMENT OF BIOLOGICAL SCIENCES
501 20th St.
Greeley, CO 80639 USA
Phone: 970-351-2921   Fax: 970-351-2335
Website: www.unco.edu/biology/

Founded: NA
Membership: 15
Scope: State

Description: The UNC Dept. of Biological Sciences offers undergraduate and masters degrees in Biology and a Ph.D. in Biology Education, which may emphasize environmental education research.

Contact(s):

Gerry Saunders, ASSOCIATE PROFESSOR, BIOLOGICAL SCIENCES
Curt Peterson, DEPT. CHAIR, BIOLOGICAL SCIENCES
Phone: 970-351-2923
cmpeter@bentley.unco.edu

UNIVERSITY OF OREGON

INSTITUTE FOR A SUSTAINABLE ENVIRONMENT
5247 University of Oregon
Eugene, OR 97403-5247 USA
Phone: 541-346-0675   Fax: 541-346-2040
E-mail: rribe@darkwing.uoregon.edu
Website: http://gladstone.uoregon.edu~evviro

Founded: 1994
Membership: 12
Scope: Regional

Description: Foster research and education at the University of Oregon with regard to environmental issues. The Institute's program encompass environmental themes in the natural sciences, the social sciences, policy studies, humanities and professional fields.

Publication(s): Organization Brochure

Contact(s):

Robert Ribe, DIRECTOR

UNIVERSITY OF OREGON
SCHOOL OF LAW
Eugene, OR 97403 USA
Phone: 541-346-3852   Fax: 541-346-1564
Website: www.law.uoregon.edu/home.html

Founded: NA
Scope: Statewide
Contact(s):
Rennard Strickland, DEAN
mholland@law.uoregon.edu
Michael Axline, WESTERN NATURAL RESOURCE LAW
CLINIC, DIRECTOR
maxline@law.uoregon.edu

UNIVERSITY OF PENNSYLVANIA
GRADUATE SCHOOL OF FINE ARTS
DEPT. OF LANDSCAPE ARCHITECTURE
119 Meyerson Hall, 210 S. 34th St
Philadelphia, PA 19104-6311 USA
Phone: 215-898-6591   Fax: 215-573-3770
E-mail: larp@pobox.upenn.edu
Website: www.upenn.edu/gsfa/

Founded: NA
Scope: National
Contact(s):
James Corner, DEPARTMENT OF LANDSCAPE
ARCHITECTURE, CHAIRMAN
corner@pobos.upenn.edu

UNIVERSITY OF PITTSBURGH
BIOLOGY DEPARTMENT
Langley Hall
Fifth & Luskin Aves.
Pittsburgh, PA 15260 USA
Phone: 412-624-4266   Fax: 412-624-4759
Website: www.pitt.edu/~biolohome/main.html

Founded: NA
Membership: 35
Scope: Statewide

Description: The Biology Department offers a major in Ecology
and Evolution, designed to provide the student with a selection
of courses covering various aspects of these two fields of
biology. The Department operates the Pymatuning Laboratory
of Ecology with laboratories and teaching facilities in northwestern Pennsylvania, offering year-round research opportu-
nites and summer courses.

Contact(s):
Stephen Tonsor, ASSOCIATE PROFESSOR
James Pipas, DEPARTMENT OF BIOLOGICAL SCIENCES,
CHAIRMAN
Phone: 412-624-4350
Fax: 412-624-9311
Gail Johnston, ECOLOGY AND PYMATUNING
LABORATORY OF ECOLOGY, DIRECTOR

UNIVERSITY OF PITTSBURGH
DEPARTMENT OF GEOLOGY AND PLANETARY
SCIENCE
321 Engineering Hall
Pittsburgh, PA 15260 USA
Phone: 412-624-8780   Fax: 412-624-3914
E-mail: geology@pitt.edu
Website: www.geology.pitt.edu

Founded: NA
Membership: 12
Scope: National

Description: B.A. in Environmental Science based on a strong
interdisciplinary framework of courses, co-requisites and a
variety of electives in natural and social sciences. B.S. in
Environmental Geology also offered.

Contact(s):
James Maher, ADVISORY BOARD CHAIR
Harold Rollins, CHAIRMAN
Phone: 412-624-8783
snail@pitt.edu
Mark Collins, UNDERGRADUATE COORDINATOR

UNIVERSITY OF PITTSBURGH
GRADUATE SCHOOL OF PUBLIC HEALTH
DEPT. OF ENVIRONMENTAL AND OCCUPATIONAL
HEALTH
260 Kappa Dr.
Pittsburgh, PA 15238 USA
Phone: 412-967-6500   Fax: 412-624-1020
Website: server1.ceoh.pitt.edu/

Founded: NA
Scope: International

Description: The mission of the Department of Environmental
and Occupational Health is to reduce the health risks
associated with exposure to chemical, physical, and biological
agents found in industry and nature. Three degrees and a
specialty certificate in risk assessment are offered.

Contact(s):
Herbert Rosenkranz
HERBERT S., CHAIRPERSON

UNIVERSITY OF RHODE ISLAND
DEPARTMENT OF NATURAL RESOURCES SCIENCE
105 Coastal Institute
Kingston, RI 02881 USA
Phone: 401-874-2495   Fax: 401-874-4561
E-mail: nrs@uri.edu
Website: www.edc.uri.edu/nrs/

Founded: NA
Scope: Statewide

Description: The teaching mission of the Department of Natural
Resources Science is to help students acquire the technical
knowledge and practical skills needed to understand and
wisely manage natural and disturbed ecosystems and their
basic components: soil, water, air and biota. The research
mission of the department is to use hypothesis-based methods
of scientific inquiry toward development of applicable solutions
to environmental problems.

Contact(s):
Dr. Thomas Husband, WILDLIFE BIOLOGY AND
MANAGEMENT/ENVIRONMENTAL SCIENCE AND
MANAGEMENT/WATER RESOURECES/DEPARTMENT
CHAIR
Thomas Husband, WILDLIFE BIOLOGY AND
MANAGEMENT/ENVIRONMENTAL SCIENCE AND
MANAGEMENT/ WATER RESOURCES AND SOIL
SCIENCE, CHAIRMAN

UNIVERSITY OF RHODE ISLAND
GRADUATE SCHOOL OF OCEANOGRAPHY AND
COASTAL RESOURCES CENTER
URI Bay Campus, South Ferry Rd.
Narragansett, RI 02882-1197 USA
Phone: 401-874-6246   Fax: 401-874-6889
E-mail: student_info@gso.uri.edu
Website: www.gso.uri.edu

Founded: 1972
Membership: 200
Scope: State

Description: Dedicated to advancing coastal ecosystem
management, nationally and internationally.

Publication(s): Publications on website

Contact(s):
Chip Young, CRC COMMUNICATIONS DIRECTOR
Phone: 401-874-6630
cyoung@gso.uri.edu
David Farmer, DIRECTOR
Scott Nixon, DIRECTOR, RI SEA GRANT COLLEGE
PROGRAM

UNIVERSITY OF SOUTH CAROLINA
P.O. Box 1630
Georgetown, SC 29442 USA
Phone: 843-546-3623   Fax: 843-546-1632
Website: www.baruch.sc.edu

Founded: NA
Membership: 25
Scope: Regional

Contact(s):
Dennis Allen, RESIDENT DIRECTOR
Phone: 843-546-3623
dallen@belle.baruch.sc.edu

UNIVERSITY OF SOUTH CAROLINA; MARINE SCIENCE PROGRAM
MARINE SCIENCE PROGRAM
Marine Science Program; University of South Carolina
Columbia, SC 29208 USA
Phone: 803-777-2692   Fax: 803-777-3935
E-mail: marisci@vm.sc.edu
Website: marine-science.sc.edu/

Founded: NA
Scope: Statewide

Description: The Marine Science Program, in the College of
Science and Mathematics at the University of South Carolina,
is an interdisciplinary educational program offering curricula
which lead to the bachelor of science, master of science, and
doctor of philosophy degrees.

Contact(s):
Bjorn Kjerfve, MARINE SCIENCE PROGRAM (B.S., M.S.,
PH.D.), DIRECTOR
Phone: 803-777-5288

UNIVERSITY OF SOUTHERN CALIFORNIA
DEPARTMENT OF CIVIL AND ENVIRONMENTAL
ENGINEERING
Los Angeles, CA 90089-2531 USA
Phone: 213-740-7832
E-mail: civileng@usc.edu
Website: www.usc.edu/dept/engineering/

Founded: NA

Publication(s): Publications

Contact(s):
L. Wellford, CHAIR, CIVIL AND ENVIRONMENTAL
ENGINEERING DEPARTMENT
Donald Lewis, ENVIRONMENTAL SOCIAL SCIENCES
PROGRAM, DEAN OF THE DIVISION

UNIVERSITY OF SOUTHERN CALIFORNIA ENVIRONMENTAL STUDIES PROGRAM
ENVIRONMENTAL STUDIES PROGRAM
Allan Hancock Foundation Bldg Rm M232
Los Angeles, CA 90089-0373 USA
Phone: 213-740-7770   Fax: 213-740-8566
E-mail: environ@usc.edu
Website: www.usc.edu/dept/LAS/envir

Founded: NA
Membership: 3
Scope: National

Description: B.A. combines basic science with the study of social
aspects of environmental issues. B.S. combines a concentration in Geology, Biology or Chemistry with social science. The
graduate program offers a choice of three concentrations:
Global Environmental Issues and Development; Law, Policy
and Management; and Environmental Planning and Analysis.

Contact(s):
Linda Duguay, DIRECTOR
duguay@usc.edu

UNIVERSITY OF SOUTHERN MISSISSIPPI
DEPARTMENT OF BIOLOGICAL SCIENCES
USM, P.O. Box 5018
Hattiesburg, MS 39406 USA
Phone: 601-266-4748   Fax: 601-266-5797
Website: www.biology.usm.edu

Founded: NA
Membership: 28
Scope: State
Publication(s): see publication web site

Contact(s):
David Beckett, AQUATIC BIOLOGY (M.S., PH.D.) AND
AQUATIC ECOLOGY
Stephen Ross, FISHERIES BIOLOGY (M.S., PH.D.) AND
AQUATIC ECOLOGY
Patricia Biesiot, MARINE BIOLOGY (M.S., PH.D.), CONTACT
Glenn Matlack, RESOURCE MANGEMENT AND
ENVIRONMENTAL PLANNING (B.S., M.S., PH.D.),
CONTACT

UNIVERSITY OF TENNESSEE AT KNOXVILLE DEPT. OF FORESTRY WILDLIFE & FISHERIES
DEPARTMENT OF FORESTRY, WILDLIFE AND FISHERIES
P.O. Box 1071
Knoxville, TN 37901-1071 USA
Phone: 865-974-7126   Fax: 865-974-4714
E-mail: fwf@utk.edu
Website: www.fwf.ag.utk.edu/

Founded: NA
Membership: 30
Scope: Statewide

Description: Offer B.S. degree in forestry, forest management concentration, wildland recreation concentration, M.S.in forestry, Thesis and Non-Thesis Option, and B.S. and M.S. degrees in wildlife and fisheries science.

Contact(s):
J. Wilson, ASSOCIATE DEPT. HEAD, FISHERIES
jlwilson@utk.edu\
George Hopper, PROFESSOR AND DEPARTMENT HEAD
ghopper@utk.edu
Richard Evans, SUPERINTENDENT, FORESTRY EXPERIMENT STATION
revans@utk.edu

UNIVERSITY OF TENNESSEE AT MARTIN
COLLEGE OF AGRICULTURE AND APPLIED SCIENCES
Martin, TN 38238 USA
Phone: 901-587-7250   Fax: 901-587-7968
Website: www.utm.edu/departments/agr/agr.html

Founded: NA
Scope: National

Description: Offers a B.S. degree in natural resources management with concentrations in wildlife biology, environmental management, soil and water conservation, and park and recreation administration.

Contact(s):
Jim Byford, DEAN

UNIVERSITY OF THE DISTRICT OF COLUMBIA
4200 Connecticut Avenue, NW
Washington, DC 20008 USA
Phone: 202-832-4888
Website: http://www.udc.edu/index-b.htm

Founded: NA

Contact(s):
Freddie Dixon, DEPARTMENT OF BIOLOGICAL AND ENVIRONMENTAL SCIENCES (A.A.S., B.S.), ACTING CHAIRPERSON
Bldg. 44
Phone: 202-274-7401

UNIVERSITY OF THE SOUTH (SEWANEE)
DEPARTMENT OF FORESTRY AND GEOLOGY
735 University Ave.
Sewanee, TN 37383 USA
Website:
www.sewanee.edu/Forestry_Geology/ForestryGeology.html

Founded: NA

Description: B.S. in Forestry, B.S. or B.A. in Geology or Natural Resources

Contact(s):
Stephen Shaver, CHAIRMAN
Phone: 931-598-1116
sshaver@seraph1.sewanee.edu

UNIVERSITY OF THE VIRGIN ISLANDS
DIVISION OF SCIENCE AND MATHEMATICS
CENTER FOR MARINE AND ENVIRONMENTAL STUDIES
No. 2 John Brewers Bay
Charlotte Amalie, St. Thomas, VI 00802-9990 USA
Phone: 340-693-1230   Fax: 340-693-1245
Website: www.uvi.edu

Founded: NA
Scope: State
Publication(s): Monitoring the Effects of Land Development on Near Shore Reef Environment of St. Thomas, US Virgin Islands written by R.S. Nemeth and J.S. Nowlis, The Effect of Natural Variation in Substrate Architecture on the Survival of Juvenile Bi-color Dansel Fish written by R.S. Nemeth

Contact(s):
Robert Stolz, DIVISION OF SCIENCE AND MATHEMATICS, CHAIR
Richard Nemeth, MACLEAN MARINE SCIENCE CENTER, CENTER FOR MARINE AND ENVIRONMENTAL STUDIES, DIRECTOR
Phone: 340-693-1381
Fax: 340-693-1385
rnemeth@uvi.edu

UNIVERSITY OF TORONTO
FORESTRY DEPARTMENT
33 Willcocks St.
Toronto, Ontario M5S 3B3 Canada
Phone: 416-978-6152   Fax: 416-978-3834
E-mail: gradprog@forestry.utoronto.ca
Website: www.forestry.utoronto.ca

Founded: NA
Scope: State

Description: Master of Forest Conservation Program (M.F.C.)

Contact(s):
Rorke Bryan, DEAN
r.bryan@utoronto.ca
D. Balsillie, M.F.C. COORDINATOR
Phone: 416-978-4638
david.balsillie@utoronto.ca

UNIVERSITY OF VERMONT, SCHOOL OF NATURAL RESOURCES
SCHOOL OF NATURAL RESOURCES
George D. Aiken Center, 81 Carrigan Dr.
Burlington, VT 05405 USA
Phone: 802-656-4280   Fax: 802-656-8683
Website: www.nature.snr.uvm.edu/

Founded: NA
Membership: 35
Scope: Statewide

Contact(s):
John Shane, CHAIR OF FORESTRY PROGRAM
jshane@nature.fnr.uvn.edu

Donald Dehayes, DEAN
Alan McIntosh, ENVIRONMENTAL SCIENCES PROGRAM (B.S.), DIR. OF WATER
Ian Worley, ENVIRONMENTAL STUDIES PROGRAM (B.S., B.A.), INTERIM DIRECTOR
Carlton Newton, FORESTRY (B.S., M.S.)
Patricia Stokowski, GRADUATE PROGRAM COORDINATOR
pstokows@nature.fnr.uvm.edu
Deane Wang, NATURAL RESOURCES PLANNING (M.S.)
Phone: 802-656-2694
dwang@snr.uvm.edu
Donald Dehayes, NATURAL RESOURCES (B.S., PH.D.)
Robert Manning, RECREATION MANAGEMENT (B.S.)
Alan McIntosh, WATER RESOURCES AND LAKE STUDIES CENTER (M.S.), DIRECTOR
David Hirth, WILDLIFE AND FISHERIES BIOLOGY PROGRAM (B.S., M.S.)

UNIVERSITY OF WEST FLORIDA

11000 University Parkway
Pensacola, FL 32514 USA
Phone: 850-474-2000
Website: uwf.edu

Founded: NA
Membership: 2000
Scope: International

Contact(s):
George Stewart, DEPARTMANT OF BIOLOGY, CHAIR
Phone: 850-474-2748
Joe Lepo, INSTITUTE FOR COASTAL & ESTUARINE RESEARCH, ACTING DIRECTOR
Phone: 850-857-6098
jlepo@uwf.edu

UNIVERSITY OF WISCONSIN AT EAU CLAIRE

Eau Claire, WI 54701 USA
Phone: 715-836-4166   Fax: 715-836-5089
Website: www.uwec.edu

Founded: NA
Membership: 26
Scope: Statewide

Contact(s):
Michael Weil, BIOLOGY (B.S., M.S.), CHAIRMAN
Paula Kleintjes, ENVIRONMENTAL SCIENCE; MINOR
Brady Foust, GEOGRAPHY (B.S.), CHAIRMAN

UNIVERSITY OF WISCONSIN AT GREEN BAY

2420 Nicolet Dr.
Green Bay, WI 54311-7001 USA
Phone: 920-465-2000   Fax: 920-465-2558
Website: www.uwgb.edu

Founded: NA
Membership: 5500
Scope: Statewide

Description: Natural and Applied Sciences Dept. offers a bachelor's degree in Environmental Studies. Areas of emphasis focus on Physical Systems and Ecology and Biological Resources. The Dept. of Public and Environmental Affairs offers a bachelor's degree in Environmental Studies and Planning with an emphasis in public policy or planning. The graduate school offers a M.S. in Environmental Science and Policy with an emphasis in Ecosystems Studies, Resource Management and Environmental Policy and Administration

Contact(s):
Robert Howe, COFRIN ARBORETUM CENTER FOR BIODIVERSITY, DIRECTOR
Phone: 920-465-2272
hower@uwgb.edu
Ronald Stieglitz, ENVIRONMENTAL SCIENCE AND POLICY PROGRAM, ASSOCIATE DEAN OF GRADUATE STUDIES
Phone: 920-465-2123
Fax: 920-465-2718
gradstu@uwgb.edu.
Hallett Harris, ENVIRONMENTAL SCIENCE, CHAIR
Phone: 920-465-2369
Fax: 920-465-2376
harrish@uwgb.edu
Denise Scheberle, INTERIM DEAN OF LIBERAL ARTS AND SCIENCE
Theatre Hall 335, 920-465-2595
scheberd@gbms01.uwgb.edu

UNIVERSITY OF WISCONSIN AT LA CROSSE

COLLEGE OF SCIENCE AND ALLIED HEALTH
1725 State St.
La Crosse, WI 54601 USA
Phone: 608-785-8218   Fax: 608-785-8221
Website: www.perth.uwlax.edu/sah/

Founded: NA
Membership: 1200
Scope: Local

Description: Biology, Chemistry and Geography offer B.S. degrees with a concentration in Environmental Science. Biology also offers a concentration in Aquatic Sciences for B.S. and M.S. degrees.

Contact(s):
Roger Haro, AQUATIC SCIENCE ADVISOR
Phone: 608-785-6970
haro_rj@mail.uwlax.edu
Robin Tyser, BIOLOGY DEPT./ENVIRONMENTAL SCIENCE, CONTACT
Phone: 608-785-8238
Fax: 608-785-6959
tyser@mail.uwlax.edu
Bruce Osterby, CHEMISTRY DEPT., CHAIR
Phone: 608-785-8266
Fax: 608-785-8281
oster_br@mail.uwlax.edu
George Huppert, GEOGRAPHY AND EARTH SCIENCE (B.S.), CHAIRMAN
Phone: 608-785-8333
Fax: 608-785-8332
huppert@mail.uwlax.edu
Mark Sandheinrich, RIVER STUDIES CENTER, DIRECTOR
Phone: 608-785-8261
sandhein@mail.uwlax.edu

UNIVERSITY OF WISCONSIN AT MADISON

INSTITUTE FOR ENVIRONMENTAL STUDIES (IES)
550 N. Park St., Science Hall
Madison, WI 53706 USA
Phone: 608-265-5296   Fax: 608-262-0014
Website: www.ies.wisc.edu/

Founded: NA
Membership: 150
Scope: International

Description: The IES is an intercollege unit of the University of Wisconsin-Madison that promotes, develops and administers interdisciplinary environmental instruction, research, and public service programs. The Institute offers graduate-level degrees in conservation biology and sustainable development, environmental monitoring, land resources and water resource management; optional graduate curricula in air resources management and energy analysis and policy; and an under-graduate certificate program in environmental studies.

Contact(s):
Thomas Yuill, DIRECTOR
tmyuill@facstaff.wisc.edu

UNIVERSITY OF WISCONSIN AT MADISON
SCHOOL OF NATURAL RESOURCES
1450 Linden Dr.
Madison, WI 53706-1562 USA
Phone: 608-262-4930   Fax: 608-262-4556
Website: www.cals.wisc.edu

Founded: NA
Scope: State

Description: The School of Natural Resources is within the college of Agricultural and Life Sciences. The school offers 13 undergraduate options in natural resources. Graduate instruction is available in many specialized and interdisciplinary areas.

Contact(s):
Elton Aberle, COLLEGE DEAN
Kevin McSweeney, DIRECTOR, SCHOOL OF NATURAL RESOURCES
1450 Linden Dr., Rm. 146, Madison, WI 53706
Phone: 608-262-6968

UNIVERSITY OF WISCONSIN AT STEVENS POINT
COLLEGE OF NATURAL RESOURCES
1900 Franklin Street
Stevens Point, WI 54481 USA
Phone: 715-346-2853   Fax: 715-346-3624
E-mail: uwsp.edu
Website: www.uwsp.edu/acad/cnr

Founded: NA
Scope: Statewide

Description: Located in Central Wisconsin, the College of Natural Resources began in 1946 with the nation's first conservation education major. The College now has over 60 faculty and staff, 1,750 undergraduates and 80 graduate students. The college offers 16 majors and 13 minors.

Publication(s): Becoming an Outdoors Woman
Contact(s):
Christine Thomas, ASSOCIATE DEAN
Phone: 715-346-4554
cthomas@uwsp.edu
Michael Bozek, COOPERATIVE FISHERY UNIT (M.S.), LEADER
Victor Phillips, DEAN
Randy Champeau, ENVIRONMENTAL EDUCATION (B.S., M.S.)
Robert Miller, FORESTRY (B.S., M.S.)
Michael Gross, RESOURCES MANAGEMENT (B.S., M.S.)
Ronald Hensler, SOIL SCIENCE (B.S., M.S.)
Stan Szczytko, WATER SCIENCE (B.S., M.S.)
Eric Anderson, WILDLIFE DEGREES (B.S., M.S.)
Michael Dombeck, PROFESSOR OF GLOBAL ENV. MGMT.
Phone: 715-346-3946

UNIVERSITY OF WYOMING
P.O. Box 3166
Laramie, WY 82071 USA
Phone: 307-766-5415   Fax: 307-766-5400
Website: www.uwyo.edu/

Founded: NA
Scope: Regional
Publication(s): Newsletter

Contact(s):
Nancy Stanton, DEPARTMENT OF ZOOLOGY AND PHYSIOLOGY, HEAD
Phone: 307-766-4207
Fax: 307-766-5625
Henry Harlow, NATIONAL PARK SERVICE RESEARCH CENTER, DIRECTOR
Phone: 307-766-4227
Fax: 307-766-5625
Joseph Meyer, RED BUTTES ENVIRONMENTAL BIOLOGY LABORATORY, DIRECTOR
Phone: 307-766-2017
Fax: 307-766-5625
Stanley Anderson, WYOMING COOPERATIVE FISH AND WILDLIFE RESEARCH UNIT, LEADER
Phone: 307-766-5415
Phone: 907-766-5400

UTAH STATE UNIVERSITY
BERRYMAN INSTITUTE FOR WILDLIFE DAMAGE MANAGEMENT
Logan, UT 84322-5270 USA
Phone: 435-797-2436   Fax: 435-797-1871
E-mail: conover@cc.usu.edu
Website: www.usu.edu/~cn r/fishwild/berry.htm

Founded: 1990
Scope: State

Description: The Jack H. Berryman Institute is a national non-profit organization which is centered at Utah State University. It engages in research, education, and extension activities aimed at resolving human and wildlife conflicts, enhancing the positive aspects of wildlife, and increasing human tolerance of wildlife problems.

Contact(s):
Michael Conover, DIRECTOR
conover@cc.uss.edu

UTAH STATE UNIVERSITY
COLLEGE OF NATURAL RESOURCES
5200 Old Main Hill
Logan, UT 84322 5200 USA
Phone: 435-797-2445   Fax: 435-797-2443
E-mail: www.cnr.us.edu
Website: www.cnr.usu.edu/

Founded: NA
Membership: 25
Scope: International

Description: Forest Resources offers programs in Forest Resources and Ecology, Environmental Studies (B.S. only) and Recreation Resource Management. The Geography Dept. offers concentrations in many fields including Environmental Education, River Processes and Rural Development/Land Use Planning.

Publication(s): see publication website

Contact(s):
Terry Sharik, DEPT. OF FOREST RESOURCES, HEAD
Phone: 435-797-3219
Fax: 435-797-4040
TLSharik@cc.usu.edu or forestry@cc.usu.edu
Ted Alsop, DEPT. OF GEOGRAPHY AND EARTH
RESOURCES, HEAD (ACTING)
Phone: 435-797-1790
Fax: 435-797-4048
tjalsop@cc.usu.edu
Chris Luecke, FISHERIES AND WILDLIFE DEPT., HEAD
Phone: 435-797-2463
Fax: 435-797-1871
luecke@cc.usu.edu
Derrick Thom, GEOGRAPHY DEPT. M.S. PROGRAM,
CONTACT
Phone: 435-797-1292
djthom@cc.usu.edu
Joanna Endter-Wada, NATURAL RESOURCE POLICY
INSTITUTE, DIRECTOR
Martyn Caldwell, USU ECOLOGY CENTER, DIRECTOR
(ACTING)
Phone: 435-797-2555
mmc@cc.usu.edu or ecol@cc.usu.edu

V

VANDERBILT UNIVERSITY
CIVIL & ENVIRONMENTAL ENGINEERING
Box 1831-Station B
Nashville, TN 37235 USA
Phone: 615-322-2697   Fax: 615-322-3365
Website: www.vanderbilt.edu/vuse/cee_dept

Founded: NA
Membership: 85
Scope: Local

Contact(s):
David Kossen, ENVIRONMETNAL ENGINEERING (B.E.,
B.S., M.E., M.S., PH.D.), CHAIRMAN

VERMONT LAW SCHOOL
ENVIRONMENTAL LAW CENTER
P.O. Box 96
South Royalton, VT 05068 USA
Phone: 888-277-5985
E-mail: elcinfo@vermontlaw.edu
Website: www.vermontlaw.edu/elc

Founded: 1973
Scope: International

Description: Administers three degrees of environmental law: the
Master of Studies in Environmental Law (M.S.E.L.), the
J.D./M.S.E.L. (joint degree), and the LL.M. in Environmental
Law. The curriculum at VLS includes approx. fifty courses in
environmental law, policy, science, and ethics. Its Summer
Session offers over thirty courses for law students, attorneys,
and non-lawyers interested in environmental policy and
management and public interest advocacy.

Contact(s):
Anne Mansfield, ASSISTANT DIRECTOR
Phone: 803-763-8303
Fax: 803-763-2940

VIRGINIA POLYTECHNIC INSTITUTE AND STATE UNIVERSITY
COLLEGE OF NATURAL RESOURCES
Attn: Peggy Quarterman, 324 Cheatham Hall
Blacksburg, VA 24061-0324 USA
Phone: 540-231-5481   Fax: 540-231-7664
E-mail: cfwr@vt.edu
Website: www.cnr.vt.edu

Founded: NA
Membership: 125
Scope: State

Contact(s):
John Cairns, CENTER FOR ENVIRONMENTAL AND
HAZARDOUS MATERIALS STUDIES, DIRECTOR
Phone: 540-231-7075
Richard Neves, COOPERATIVE FISH AND WILDLIFE UNIT,
LEADER
Phone: 540-231-5927
mussel@vt.edu
Gregory Brown, DEAN
browngn@vt.edu
Donald Orth, DEPARTMENT OF FISHERIES AND WILDLIFE
SCIENCES (B.S., M.S., PH.D.), HEAD
Phone: 540-231-5573
dorth@vt.edu
Harold Burkhart, DEPARTMENT OF FORESTRY (B.S., M.S.,
PH.D.), HEAD
Phone: 540-231-5483
burkhart@vt.edu
C. Dolloff, U.S. FOREST SERVICE COLDWATER AND
TROUT RESEARCH UNIT, LEADER
Phone: 540-231-4864
adoll@vt.edu
Paul Winistorfer, WOOD SCIENCE & FOREST PRODUCT

W

WASHINGTON STATE UNIVERSITY
Attn: William Budd, 305 Troy Hall
Pullman, WA 99164-4430 USA
Phone: 509-335-8536   Fax: 509-335-7636
Website: www.wsu.edu

Founded: NA
Scope: Statewide

Description: WSU has tripartite goals of providing higher education,
research, and outreach/service programs relevant to the needs of
Washington's citizens. The Department of Natural Resource
Sciences (NRS) (http://coopext.cahe.wsu.edu/~nrs/) and
Program in Environmental Science and Regional Planning
(ESRP) (http://www.sci.wsu.edu/envsci/) are the chief academic
units at WSU devoted to understanding, conserving, and
managing natural resources, the environments they create, and
the values such resources and environments provide to society.
Landscape Architecture (L.A.) (http://www.cahe.wsu.edu/~hortla)
offers related programs focused on planning and designing land
resources in an environmentally sound manner.

Contact(s):
William Budd, ENVIRONMENTAL SCIENCE AND REGIONAL
PLANNING (B.S., M.S., M.R.P., PH.D.), CHAIR
Washington State University, P.O. Box 644430, Pullman, WA
99164-4430
Phone: 509-335-8536
Fax: 509-335-7636

Charles Johnson, HORTICULTURE AND LANDSCAPE ARCHITECTURE (B.S., B.L.A.), CHAIR
Washington State University, P. O. Box 646414, 149 Johnson Hall, Pullman, WA 99164-6414
Phone: 509-335-9502
Fax: 509-335-8690
Edward Depuit, NATURAL RESOURCE SCIENCES (B.S., M.S., PH.D.), CHAIR
Washington State University, P.O. Box 646410, Pullman, WA 99164-6410
Phone: 509-335-6166
Fax: 509-335-7862

WASHINGTON UNIVERSITY, BIOLOGY DEPARTMENT

BIOLOGY DEPARTMENT
1 Brookings Dr., Campus Box 1137
St. Louis, MO 63110 USA
Phone: 314-935-6860   Fax: 314-935-4432
Website: www.biology.wustl.edu

Founded: NA
Scope: Statewide

Description: The laboratory is active in applying modern genetic techniques to problems in conservation biology such as conservation forensics (e.g., DNA fingerprinting of elephant tusks), systematics (identifying taxa that are significant evolutionary units), inter- and intraspecific hybridizataion, and genetic management of captive, translocated, and natural populations.

Contact(s):
Ralph Quatrano, DEPARTMENT OF BIOLOGY, HEAD
Phone: 314-935-6868
rsq@wustl.edu

WAYNE STATE UNIVERSITY DEPT. OF BIOLOGICAL SCIENCES

DEPARTMENT OF BIOLOGICAL SCIENCES
5047 Gullen Mall
Detroit, MI 48202-3917 USA
Phone: 313-577-2873   Fax: 313-577-6891
Website: biology.biosci.wayne.edu/biology/

Founded: NA
Scope: National

Description: Courses offered in such subjects as limnology, ornithology, mammalogy, biogeography, natural history of vertebrates, animal behavior, population genetics, population ecology, microbial ecology, aquatic botany, ecology, advanced ecology, and evolutionary ecology.

Contact(s):
Allen Nicholson, CHAIRMAN
Phone: 313-577-2783

WEST VIRGINIA UNIVERSITY

COLLEGE OF AGRICULTURE, FORESTRY AND CONSUMER SCIENCES
P.O. Box 6125
Morgantown, WV 26506 USA
Phone: 304-293-2941   Fax: 304-293-2441
Website: www.caf.wvu.edu/

Founded: NA
Scope: International

Contact(s):
Alan Collins, AGRICULTURAL AND RESOURCE ECONOMICS, UNDERGRADUATE COORDINATOR
P.O. Box 6108, Morgantown, WV 26506
Phone: 304-293-4832
acollins@wvu.edu
Joseph McNeel, DIVISION OF FORESTRY (B.S., M.S., PH.D.), DIRECTOR
P.O. Box 6125, Morgantown, WV 26506-6125
Phone: 304-293-2941
jmcneel@wvu.edu
Donald Armstrong, LANDSCAPE ARCHITECTURE (B.S.), CHAIR
Phone: 304-293-2142
Fax: 304-293-3752
darmstro@wvu.edu
Tim Phipps, NATURAL RESOURCE ECONOMICS (PH.D.), COORDINATOR
P.O. Box 6108, Morgantown, WV 26506
Phone: 304-293-4832
tphipps@wvu.edu
Steve Selin, RECREATION AND PARKS MANAGEMENT (B.S.R., M.S., PH.D.), PROGRAM COORDINATOR
Phone: 304-293-2941
sselin@wvu.edu
Robert Whitmore, WILDLIFE AND FISHERIES MANAGEMENT (B.S., M.S., PH.D.), PROGRAM COORDINATOR
Phone: 304-293-2941
u0eae@wvnvm.wvnet.edu
James Armstrong, WOOD INDUSTRIES (B.S.F., M.S., PH.D.), PROGRAM COORDINATOR
Phone: 304-293-2941
jarmstro@wvu.edu

WESTERN ILLINOIS UNIVERSITY

DEPARTMENT OF BIOLOGY SCIENCES
372 Waggoner Hall
Macomb, IL 61455 USA
Phone: 309-298-2408   Fax: 309-298-2270
E-mail: mibiol@wiu.edu
Website: www.wiu.edu/users/mibiol/

Founded: NA
Membership: 15
Scope: Regional

Description: Office of Aquatic Studies offers courses at the Shedd Aquarium in Chicago

Contact(s):
Sean Jenkins, DIRECTOR
Larry Jahn, FISHERIES (B.S., M.S.)
Phone: 309-298-1266
la-jahn@wiu.edu
Thomas Dunstan, WILDLIFE (B.S., M.S.)
Phone: 309-298-1752
thomas_dunstan@ccmail.wiu.edu

WESTERN MICHIGAN UNIVERSITY

ENVIRONMENTAL STUDIES PROGRAM
3930 Wood Hall
Kalamazoo, MI 49008-5419 USA
Phone: 616-387-2716   Fax: 616-387-2272
Website: www.wmich.edu/environmental-studies

Founded: NA
Membership: 200

Scope: Local,State

Description: This undergraduate interdisciplinary program provides intellectual and practical experience that provokes thought about the complex interrelationships between humans, the social and technological systems they develop, and the natural environment. The program encourages students to develop an appreciation for the many elements of planetary health and to devise creative solutions to environmental problems.

Contact(s):
Kathy Mitchell, PROGRAM COORDINATOR FOR ENVIRONMENTAL STUDIES

WESTERN WASHINGTON UNIVERSITY
HUXLEY COLLEGE OF THE ENVIRONMENT
516 High Mail Stop 9079
Bellingham, WA 98225 USA
Phone: 360-650-3520   Fax: 360-650-2842
E-mail: huxley@cc.wwu.edu
Website: www.ac.wwu.edu/~huxley/

Founded: 1968
Membership: 500
Scope: State Regional

Description: Principally a two-year, upper division and M.S. program; B.A., B.S. in environmental studies; M.S. in environmental science. Also cooperative programs, M.S. in marine and estuarine science and M.A. in political science and environmental studies.

Publication(s): available on web

Contact(s):
John Hardy, CENTER FOR ENVIRONMENTAL SCIENCE, DIRECTOR
Phone: 360-650-6108
Fax: 360-650-7284
jhardy@cc.wwu.edu
John Miles, CENTER FOR GEOGRAPHY AND ENVIRONMENTAL SOCIAL SCIENCE, DIRECTOR
Phone: 360-650-3284
Fax: 360-650-7702
jcmiles@cc.wwu.edu
Bradley Smith, DEAN
bfs@admsec.wwu.edu
Robin Matthews, INSTITUTE FOR WATERSHED STUDIES, DIRECTOR
Phone: 360-650-3510
rmatthews@wwu.edu
Wayne Landis, INSTITUTE OF ENVIRONMENTAL TOXICOLOGY AND CHEMISTRY, DIRECTOR
Phone: 360-650-6136
Fax: 360-650-6556
landis@cc.wwu.edu

WIDENER UNIVERSITY
DEPARTMENT OF CIVIL ENGINEERING
One University Place
Chester, PA 19013-5792 USA
Phone: 610-499-4042   Fax: 610-499-4059
E-mail: solid.waste@widener.edu
Website: www.widener.edu/solid.waste

Founded: NA
Scope: International

Description: The Department of Civil Engineering at Widener University offers undergraduate and graduate degrees which include courses in water resources, solid waste management, and environmental engineering. Continuing education seminars are taught in solid waste management and recycling. Research and development performed in solid waste and recycling. The department publishes The Journal of Solid Waste Technology and Management.

Publication(s): The Journal of Solid Waste Technology and Management, Proceedings of International Conference on Solid Waste Technology and Management

Contact(s):
Vicki Brown, CHAIR OF CIVIL ENGINEERING
Phone: 610-499-4607
vicki.l.brown@widener.edu
Ronald Mersky, WASTE MANAGEMENT PROGRAMS, COORDINATOR
Theresea Taborsky, WOLFGRAM MEMORIAL LIBRARY
One University Pl., Chester, PA 19013-5792
Phone: 610-499-4087

WILKES UNIVERSITY DEPT OF GEO-ENVIRONMENTAL SCIENCES & ENGINEERING
GEO-ENVIRONMENTAL SCIENCES/ENGINEERING DEPT.
Stark Learning Center 441
Wilkes Barre, PA 18766 USA
Phone: 570-408-4610   Fax: 570-408-7865
E-mail: gse@wilkes.edu
Website: www.wilkes.edu

Founded: NA
Membership: 7
Scope: Local

Description: The department offers two degree programs. The Environmental Engineering curriculum highlights a balance among the basic areas of water and waste-water engineering, water quality measurement, air pollution measurement and control technology, as well as the more recent demands in the areas of hazardous and solid waste management. The Earth and Environmental Science curriculum requires a concentration of departmental electives that can be used to create an area of specialization such as geology or environmental science.

Contact(s):
Dale Bruns, CO-CHAIRMAN
Phone: 717-408-4610
Fax: 570-408-7865
dbruns@wilkes.edu
Sid Halfor, CO-CHAIRMAN
Phone: 570-408-4611
Fax: 570-408-7865
shalfor@wilkes.edu

WILLIAMS COLLEGE
CENTER FOR ENVIRONMENTAL STUDIES PROGRAM
Box 632
Williamstown, MA 01267 USA
Phone: 413-597-2346   Fax: 413-597-3489
Website: www.williams.edu/

Founded: 1967
Scope: State

Description: The Center for Environmental Studies offers an integrated undergraduate program of studies to liberal arts students in combination with their major discipline. The Center also administers the 2,400 acre Hopkins Memorial Forest, a research and educational facility, as well as the Environmental Science Laboratory and the Matt Cole Memorial Library.

Publication(s): Field Notes-Newsletter

Contact(s):
Kai Lee, DIRECTOR
Andrew Jones, HOPKINS MEMORIAL FOREST MANAGER
Rachel Louis, PROGRAM ASSISTANT

WILLOW MIXED MEDIA INC.
P.O. Box 194
Glenford, NY 12433-0194 USA
Phone: 845-657-2914
E-mail: willowmx@ulster.net
Website: www.hudsonvalley.com/willow

Founded: 1979
Scope: Local

Description: Willow Mixed Media is a not for profit organization, dealing with documentary video and arts projects on issues of social concern—health, environment, the arts, criminal justice, ect.

Publication(s): The Hudson River PCB Story: A Toxic Heritage, Cancer: Just a word . not a sentence, Building the Ashokan Reservoir

Contact(s):
Tobe Carey, PRESIDENT
willowmx@ulster.net

Y

YALE LAW SCHOOL
CAREER DEVELOPMENT OFFICE
127 Wall Street
New Haven, CT 06511 USA
Phone: 203-432-1676   Fax: 203-432-8423
E-mail: cdolaw@yale.edu
Website: www.law.yale.edu/cdo
Scope: National

Contact(s):
Cathy Woods, CONTACT

YALE UNIVERSITY
SCHOOL OF FORESTRY AND ENVIRONMENTAL STUDIES
205 Prospect St.
New Haven, CT 06511 USA
Phone: 203-432-5100   Fax: 203-432-5942
Website: www.yale.edu/forestry/environmenatl

Founded: NA
Scope: Regional

Description: The mission of the school is to provide leadership in the science and management of natural resource and environmental systems. The school trains managers for governmental, non-governmental, and corporate institutions, and educates teachers and researchers

Publication(s): Yale School of Forestry and Environmental Studies Bulletin Series, Bulletin of Yale University, Bulletins

Contact(s):
Emily McDirmid, ADMISSIONS DIRECTOR
Phone: 202-432-5138
Gordon Geballe, ASSOCIATE DEAN

Phone: 202-432-5122

YORK UNIVERSITY
ENVIRONMENTAL STUDIES DEPARTMENT
355 Lumbers Bldg., 4700 Keele St.
Toronto, Ontario M3J 1P3 Canada
Phone: 416-736-5252   Fax: 416-736-5679
E-mail: fesinfo@yorku.ca
Website: www.yorku.ca/faculty/fes

Founded: 1968
Membership: 90
Scope: National

Description: The faculty of Environmental Studies offers interdisciplinary, flexible, individualized programs at both the undergraduate and graduate levels. FES is committed to a broad definition of environment, offering the opportunity to study natural, built, organizational, and social environments.

Contact(s):
Barbara Rahder, GRADUATE PLANNING PROGRAMS COORDINATOR
rahder@yorku.ca
Mora Campbell, GRADUATE PROGRAM DIRECTOR
Brent Rutherford, MES PROGRAM COORDINATOR
brentr@yorku.ca
Raymond Rogers, UNDERGRADUATE PROGRAM DIRECTOR
rrogers@yorku.ca

BUREAU OF LAND MANAGEMENT DISTRICTS

ALASKA
Anchorage District
THOMAS ALEXANDER
6881 Abbott Loop Rd., Anchorage AK 99507-2599
Phone: 907-267-1246   Fax: 907-267-1267
Glennallen District
CATHIE JENSEN
P.O. Box 147, Glennallen AK 99588
Phone: 907-822-3217   Fax: 907-822- 3120
Northern District
DEE RITCHIE
1150 University Ave., Fairbanks AK 99709
Phone: 907-474-2200

ARIZONA
Arizona State Office
222 N. Central Ave., Phoenix AZ 85004-2203
Phone: 602-417-9200   Fax: 602-417-9556
Kingman Field Office
JOHN CHRISTENSEN
2475 Beverly Ave., Kingman AZ 86401-3629
Phone: 520-692-4400   Fax: 520-692-4414
Lake Havasu Field Office
DON ELLSWORTH
2610 Sweetwater Ave., Lake Havasu City AZ 86406-9071
Phone: 520-505-1200   Fax: 520-505-1208
National Training Center
9828 N. 31st Ave., Phoenix AZ 85051-2517
Phone: 602-906-5500   Fax: 602-906-5555
Phoenix Field Office
MIKE TAYLOR
2015 W. Deer Valley Rd., Phoenix AZ 85027
Phone: 602-580-5600   Fax: 623-580-5580
Safford Field Office
WILLIAM CIVISH
711 S. 14th Ave., Safford AZ 85546-3321
Phone: 520-348-4400   Fax: 520-348-4450
San Pedro Project Office
1763 Paseo San Luis, Sierra Vista AZ 85635-2240
Phone: 520-458-3559
Tucson Field Office
JESSE JUEN
12661 E. Broadway, Tucson AZ 85748-7208
Phone: 520-722-4289   Fax: 520-751-0948
Yuma Field Office
GAIL ACHESON
2555 E. Gila Ridge Rd., Yuma AZ 85365-2240
Phone: 520-317-3200   Fax: 520-317-3250

CALIFORNIA
State Office
AL WRIGHT
2800 Cottage Way, RM W-1834, Sacramento CA 95825
Phone: 916-978-4400   Fax: 916-978-4305
Alturas Field Office
TIMOTHY BURKE
708 W. 12th St., Alturas CA 96101
Phone: 503-233-4666   Fax: 530-233-5696
Arcata Field Office
LYNDA ROUSH
1695 Heindon Rd., Arcata CA 95521
Phone: 707-825-2300   Fax: 707-825-2301

Bakersfield District
RON FELLOWS
3801 Pegasus Ave., Bakersfield CA 93308-6837
Phone: 661-391-6000   Fax: 661-391-6040
Barstow Field Office
TIMOTHY READ
2601 Barstow Rd., Barstow CA 92311
Phone: 760-252-6000   Fax: 760-252-6099
Bishop Field Office
STEVE ADDINGTON
785 N. Main St., Ste E, Bishop CA 93514
Phone: 760-872-4881   Fax: 760-872-2894
Eagle Lake Field Office
LINDA HANSEN
2950 Riverside Dr., Susanville CA 96130
Phone: 530-257-0456   Fax: 530-257-4831
El Centro Field Office
GREG THOMSEN
1661 S. 4th St., El Centro CA 92243
Phone: 760-337-4400   Fax: 760-337-4490
Folsom Field Office
DEANE SWICKARD
63 Natoma St., Folsom CA 95630
Phone: 916-985-4474   Fax: 916-985-3259
Hollister Field Office
ROBERT BEEHLER
20 Hamilton Ct., Hollister CA 95023
Phone: 831-630-5000   Fax: 831-630-5050
Needles Field Office
MOLLY BRADY
101 W. Spikes Rd., Needles CA 92363
Phone: 760-326-7000   Fax: 760-326-7099
Palm Springs / South Coast Field Office
JAMES KENNA
690 W. Garnet Ave., P.O. Box 1260,
North Palm Springs CA 92258-1260
Phone: 760-251-4800   Fax: 760-251-4899
Redding Field Office
CHARLES SHULTZ
355 Hemsted Dr., Redding CA 96002
Phone: 530-224-2100   Fax: 530-224-2172
Ridgecrest Field Office
HECTOR VILLALOBOS
300 S. Richmond Rd., Ridgecrest CA 93555
Phone: 760-384-5400   Fax: 760-384-5499
Surprise Field Office
SUSAN STOKKE
P.O. Box 460, 602 Cressler St., Cedarville CA 96104
Phone: 530-279-6101   Fax: 530-279-2171
Ukiah Field Office
RICH BURNS
2550 N. State St., Ukiah CA 95482
Phone: 707-468-4000   Fax: 707-468-4027

COLORADO
State Office
2850 Youngfield St., Lakewood CO 80215
Phone: 303-239-3600   Fax: 303-239-3933
Anasazi Heritage Center
27501 Hwy. 184, Dolores CO 81323
Phone: 970-882-4811   Fax: 970-882-7035

Glenwood Springs Field Office
 MIKE MOTTICE
 50629 Hwys. 6 & 24; P.O. Box 1009,
 Glenwood Springs CO 81602
 Phone: 970-947-2800   Fax: 970-947-2829
Grand Junction Field Office/Northwest Center
 CATHERINE ROBERTSON
 2815 H Rd., Grand Junction CO 81506
 Phone: 907-244-3000   Fax: 970-244-3083
Gunnison Field Office
 BARRY TOLLEFSON
 216 N. Colorado, Gunnison CO 81230
 Phone: 970-641-0471   Fax: 970-641-1928
Kremmling Field Office
 1116 Park Ave., P.O. Box, Kremmling CO 80459
 Phone: 970-724-3437   Fax: 970-724-9590
La Jara Field Office
 CARLOS PINTOS
 15571 County Rd. T5, La Jara CO 81140
 Phone: 719-274-8971   Fax: 719-274-6301
Little Snake Field Office
 JOHN HUSBAND
 455 Emerson St., Craig CO 81625
 Phone: 970-826-5000   Fax: 970-826-5002
Royal Gorge Field Office/Front Range Center
 3170 E. Main St., Canon City CO 81212
 Phone: 719-269-8500   Fax: 719-269-8599
Saguache Field Office
 TOM GOODWIN
 46525 Hwy. 114, P.O. Box 67, Saguache CO 81149
 Phone: 719-655-2547   Fax: 719-655-2502
San Juan Field Office
 CAL JOYNER
 15 Burnett Ct., Durango CO 81301
 Phone: 970-247-4874   Fax: 970-385-1375
Uncompahgre Field Office/Southwest Center
 ALLAN BELT
 2465 S. Townsend Ave., Montrose CO 81401
 Phone: 970-240-5300   Fax: 970-240-5367
White River Field Office
 JOHN MEHLHOFF
 73544 Hwy. 64, Meeker CO 81641
 Phone: 970-878-3601   Fax: 970-878-5717

IDAHO
State Office
 1387 S. Vinnell Way, Boise ID 83709-1657
 Phone: 208-373-4000   Fax: 208-373-3904
Bruneau Field Office
 JENNA WHITLOCK
 3948 Development Ave., Boise ID 83705-5389
 Phone: 208-384-3300   Fax: 208-384-3493
Burley Field Office
 THERESA HANLEY
 15 E. 200 South, Burley ID 83318
 Phone: 208-677-6641   Fax: 208-677-6699
Cascade Field Office
 JOHN FEND
 3948 Development Ave., Boise ID 83705-5389
 Phone: 208-384-3300   Fax: 208-384-3493
Challis Field Office
 RENEE SNYDER
 Hwy. 93, S., / Route 2, Box 610, Salmon ID 83467
 Phone: 208-756-5400   Fax: 208-756-5436

Coeur d'Alene Field Office
 ERIC THOMPSON
 1808 N. Third St., Coeur d'Alene ID 83814-3407
 Phone: 208-769-5030   Fax: 208-769-5050
Cottonwood Field Office
 GREG YUNCEVICH
 House 1, Butte Dr. / Route 3, Box 18, Cottonwood ID
 83522-9498
 Phone: 208-962-3245   Fax: 208-962-3275
Idaho Falls Field Office
 JOE KRAAYENBRINK
 1405 Hollypark Dr., Idaho Falls ID 83401
 Phone: 208-524-7500   Fax: 208-524-7505
Jarbridge Field Office
 EDDIE GUERRERO
 2620 Kimberly Rd., Twin Falls ID 83301
 Phone: 208-736-2350   Fax: 208-736-2375
Malad Field Office
 JEFF STEELE
 138 S. Main, Malad City ID 83252-1346
 Phone: 208-766-4766   Fax: 208-766-4087
Owyhee Field Office
 DARYL ALBISTON
 3948 Development Ave., Boise ID 83705-5389
 Phone: 208-384-3300   Fax: 208-384-3493
Pocatello Field Office
 JEFF STEELE
 1111 N. 8th Ave., Pocatello ID 83201
 Phone: 208-236-6860   Fax: 208-234-0246
Salmon Field Office
 DAVE KROSTING
 Hwy. 93, South / Route 2, Box 610, Salmon ID 83467
 Phone: 208-756-5400   Fax: 208-756-5436
Shoshone Field Office
 BILL BAKER
 400 W. F St., P.O. Box 2-B, Shoshone ID 83352-1522
 Phone: 208-886-2206   Fax: 208-886-7317

MISSISSIPPI
Jackson Field Office: Jurisdiciton for AL, AR, FL, GA, KY, LA, MI, NC, SC, TN and VA
 BRUCE DAWSON
 411 Briarwood Dr., Suite 404, Jackson MS 39206
 Phone: 601-977-5400

MONTANA
State Office for MT, ND and SD
 LARRY HAMILTON
 P.O. Box 36800, Billings MT 59107-6800
 Phone: 406-896-5012   Fax: 406-896-5299
Billings Field Office
 SANDRA BROOKS
 5001 Southgate Dr., Billings MT 59101
 Phone: 406-896-5013   Fax: 406-896-5301
Butte District
 MERLE GOOD
 106 N. Parkmont, Butte MT 59701
 Phone: 406-494-5059   Fax: 406-494-3474
Dillon Field Office
 SCOTT POWERS
 1005 Selway Dr., Dillon MT 59725-9431
 Phone: 406-683-2337   Fax: 406-683-2970
Lewistown Field Office
 DAVID MARI
 P.O. Box 1160, Lewistown MT 59457-1160
 Phone: 406-538-7461   Fax: 406-538-1904

Malta Field Office
  RICK HOTALING
  501 S. 2nd St. E., Malta MT 59538
  Phone: 406-654-1240   Fax: 406-654-2671
Miles City Field Office
  TIM MURPHY
  111 Garry Owen Rd., Miles City MT 59301-0940
  Phone: 406-232-4333   Fax: 406-233-2921
Missoula Field Office
  NANCY ANDERSON
  3255 Ft. Missoula Rd., Missoula MT 59804-7293
  Phone: 406-329-3914   Fax: 406-329-3712

NEVADA
State Office
  P.O. Box 12000, Reno NV 89520-0006
  Phone: 775-861-6586
Battle Mountain Field Office
  GERALD SMITH
  50 Bastian Rd., Battle Mountain NV 89820
  Phone: 702-635-4000   Fax: 702-635-4034
Carson City Field Office
  JOHN SINGLAUB
  5665 Morgan Mill Rd., Carson City NV 89701
  Phone: 702-885-6000   Fax: 702-885-6147
Elko Field Office
  HELEN HANKINS
  3900 E. Idaho St., Elko NV 89801
  Phone: 702-753-0200   Fax: 702-753-0255
Ely Field Office
  GENE KOLKMAN
  702 N. Industrial Way, HC 33, Box 33500, Ely NV 89301
  Phone: 702-289-1800   Fax: 702-289-1810
Las Vegas Field Office
  MIKE DWYER
  4765 W. Vegas Dr., Las Vegas NV 89108
  Phone: 702-647-5000   Fax: 702-647-5023
Winnemucca Field Office
  TERRY REED
  5100 E. Winnemucca Blvd., Winnemucca NV 89445
  Phone: 702-623-1500   Fax: 702-623-1503

NEW MEXICO
State Office for NM, TX, OK and KS
  MICHELLE CHAVEZ
  P.O. Box 27115, Santa Fe NM 87502-0115
  Phone: 505-438-7400   Fax: 505-438-7435
Albuquerque Field Office
  435 Montano Rd., NE, Alburquerque NM 87107
  Phone: 505-761-8700   Fax: 505-761-8911
Carlsbad Field Office
  620 E. Greene St., Carlsbad NM 88220-6292
  Phone: 505-887-6544   Fax: 505-885-9264
Farmington Field Office
  1235 La Plata Hwy., Suite A, Farmington NM 87401
  Phone: 505-599-8900   Fax: 505-599-8998
Las Cruces District
  1800 Marquess, Las Cruces NM 87005-3371
  Phone: 505-525-4300   Fax: 505-525-4412
Roswell District
  2909 W. Second St., Roswell NM 88201-2019
  Phone: 505-627-0272   Fax: 505-627-0276
Socorro Field Office
  198 Neel Ave., NW, Socorro NM 87801-4648
  Phone: 505-835-0412   Fax: 505-835-0223

Taos Field Office
  226 Cruz Alta Rd., Taos NM 87571-5983
  Phone: 505-758-8851   Fax: 505-758-1620

NORTH DAKOTA
North Dakota Field Office
  DOUGLAS BURGER
  2933 Third Ave., W., West Dickinson ND 58601-2619
  Phone: 701-225-9148   Fax: 701-227-8510

OKLAHOMA
Tulsa District
  7906 East 33 St., Tulsa OK 74145-1352
  Phone: 918-621-4100   Fax: 918-621-4130

OREGON
State Office for OR and WA
  1515 SW 5th Ave., Portland OR 92208-2965
  Phone: 503-952-6002
Burns District
  TOM DYER
  HC 74-12533, Hwy 20 West, Hines OR 97738
  Phone: 541-574-4400
Coos Bay Field Office
  SUE RICHARDSON
  1300 Airport Lane Rd., North Bend OR 97459
  Phone: 541-756-0100   Fax: 541-751-4303
Eugene Field Office
  2890 Chad Dr., Eugene OR 97408
  Phone: 541-683-6600   Fax: 541-683-6981
Lakeview Field Offices
  STEVE ELLIS
  HC10 Box 337, 1300 S. G St, Lakeview OR 97630
  Phone: 503-947-2177   Fax: 541-947-6399
Medford Field Office
  RON WENKER
  3040 Biddle Rd., Medford OR 97504
  Phone: 541-770-2200   Fax: 541-770-2400
Prineville District Field Office
  JAMES HANCOCK
  P.O. Box 550, Prineville OR 97754
  Phone: 541-416-6700   Fax: 541-416-6798
Roseburg District
  CARY OSTERHAUS
  777 NW Garden Valley Blvd., Roseburg OR 97470
  Phone: 541-440-4930   Fax: 541-440-4948
Salem District Field Office
  VAN MANNING
  1717 Fabry Rd., SE, Salem OR 97306
  Phone: 541-375-5646   Fax: 503-375-5622
  Phone: 503-952-6002
Vale District
  JUAN PALMA
  100 Oregon St., Vale OR 97918
  Phone: 541-473-3144   Fax: 541-473-6213

SOUTH DAKOTA
South Dakota Field Office
  PATRICK GUBBINS
  310 Roundup St., Belle Fourche SD 57717-1698
  Phone: 605-892-2526   Fax: 605-892-4742

TEXAS
Amarillo Field Office
> 801 S. Fillmore St., Ste. 500, Amarillo TX 79101-3545
> Phone: 806-324-2617   Fax: 806-324-2633

UTAH
State Office
> SALLY WISELY
> 324 S. State St., Salt Lake City UT 84145-0155
> Phone: 801-539-4001

Arizona Strip Field Office
> ROGER TAYLOR
> 345 E. Riverside Dr., St. George UT 84790-9000
> Phone: 801-688-3301   Fax: 435-688-3258

Cedar City District Field Office
> ART TATE
> 176 East D.L. Sargent Dr, Cedar City UT 84720
> Phone: 801-865-3053   Fax: 801-865-3058

Fillmore Field Office
> REX ROWLEY
> 35 E. 500 North, Fillmore UT 84631
> Phone: 435-743-6811

Kanab
> VERLIN SMITH
> 318 N. First East, Kanab UT 84741
> Phone: 435-644-2672

Moab District Field Office
> MAGGIE WHITE
> 82 E. Dogwood, Moab UT 84532
> Phone: 435-259-2100

Monticello Field Office
> KENT WALTER
> 435 N. Main, P.O. Box 7, Monticello UT 84535
> Phone: 435-587-1502

Price Field Office
> DICK MANUS
> 125 S. 600 West, Price UT 84501
> Phone: 435-636-3601

Richfield District Field Office
> JERRY GOODMAN
> 150 E. 900, N, Richfield UT 84701
> Phone: 435-896-1523

Salt Lake District
> GLENN CARPENTER
> 2370 S. 2300, W, Salt Lake City UT 84119
> Phone: 801-977-4300   Fax: 801-997-4397

St. George Field Office
> JIM CRISP
> 345 East Riverside Dr., St. George UT 84720
> Phone: 435-688-3200
> DAVE HOWELL
> 170 U. 500 Ut., Last, Vernal UT 04070
> Phone: 801-781-4400   Fax: 801-781-4410

VIRGINIA
Eastern States Office
> W. TIPTON
> 7450 Boston Blvd., Springfield VA 22153
> Phone: 703-440-1713

WASHINGTON
Spokane District
> 1103 N. Fancher, Spokane WA 99212
> Phone: 509-536-1200   Fax: 509-536-1275

WISCONSIN
Milwaukee Field Office: Jurisdiction for CT, DE, IL, IN, IA, ME, MD, MA, MI, MN, MO, NH, NJ, NY, OH, PA, RI, VT, WV and WI
> JIM DRYDEN
> 310 W. Wisconsin Ave., Suite 450, Milwaukee WI 53203
> Phone: 414-297-4400

WYOMING
State Office for WY and NE
> AL PIERSON
> 5353 Yellowstone; P.O. Box 1828, Cheyenne WY 82003
> Phone: 307-775-6256

Buffalo Field Office
> DENNIS STENGER
> 1425 Fort St., Buffalo WY 82834-2436
> Phone: 307-684-1100   Fax: 307-684-1122

Casper District
> JIM MURKIN
> 1701 East E St., Casper WY 82601
> Phone: 307-261-7600   Fax: 307-234-1525

Cody Field Office
> MIKE BLYMYER
> 1002 Blackburn; P.O. Box 518, Cody WY 82414-0518
> Phone: 307-587-2216   Fax: 307-527-7116

Kemmerer Field Office
> JEFFREY RAWSON
> 312 Highway 189 N., Kemmerer WY 83101-9710
> Phone: 307-828-4500   Fax: 307-828-4539

Lander Field Office
> JACK KELLY
> 1335 Main; P.O. Box 589, Lander WY 82520-0589
> Phone: 307-332-8400   Fax: 307-332-8447

Newcastle Field Office
> 1101 Washington Blvd., Newcastle WY 82701-2972
> Phone: 307-746-4453   Fax: 307-746-4840

Pinedale Field Office
> LESLIE THEISS
> 432 E. Mill Street; P.O. Box 768, Pinedale WY 82941-0768
> Phone: 307-367-5300   Fax: 307-367-5329

Rawlins Field Office
> KURT KOTTER
> 1300 Third St.; P.O. Box 2407, Rawlins WY 82301-2407
> Phone: 307-328-4200   Fax: 307-328-4224

Rock Springs Field Office
> STAN McKEE
> 280 Highway 191 N., Rock Springs WY 82901-3448
> Phone: 307-352-0256   Fax: 307-352-0329

Worland Field Office
> DARRELL BARNES
> S. 23rd St., PO Box 119, Worland WY 82401-0119
> Phone: 307-347-5100   Fax: 307-347-6195

NATIONAL ESTUARINE RESEARCH RESERVES

ALABAMA
Weeks Bay NERR
> L. ADAMS
> 11300 U.S. Highway 98, Fairhope AL 36532
> Phone: 334-928-9792   Fax: 334-928-1792

ALASKA
KACHEMAK BAY NERR
> 2181 Kachemak Dr, Homer AK 99603
> Phone: 907-235-6377   Fax: 907-267-4794

CALIFORNIA
Elkhorn Slough NERR
STEVE KIMPLE, BECKY CHRISTENSEN
1700 Elkhorn Rd., Watsonville CA 95076
Phone: 831-728-2822   Fax: 831-728-1056
Tijuana River NERR
JOANNE KERBAVAZ, PHIL JENKINS
301 Caspian Way, Imperial Beach CA 91932
Phone: 619-575-3613   Fax: 619-575-6913

DELAWARE
Delaware NERR
BETSY ARCHER
818 Kitts Hummock Rd, Dover DE 1901
Phone: 302-739-3436   Fax: 302-739-3446

FLORIDA
Apalachicola NERR
WOODWARD MILEY
Department of Environmental Protection, 350 Carroll St.,
Eastpoint FL 32328
Phone: 850-670-4783   Fax: 850-670-4324
Rookery Bay NERR
GARY LYTTON
Department of Environmental Protection, 300 Tower Rd.,
Naples FL 34113
Phone: 941-417-6310   Fax: 941-417-6315

GEORGIA
Sapelo Island NERR
BUDDY SULLIVAN
PO BOX 19, Sapelo Island GA 31327
Phone: 912-485-2251

MAINE
Wells NERR
JIM LIST
342 Laudholm Farm Rd., Wells ME 04090
Phone: 207-646-1555   Fax: 207-646-2930

MARYLAND
Chesapeake Bay NERR in Maryland
KATHY ELLETT
Department of Natural Resources, Tawes State Office Bldg.,
E-2, 580 Taylor Ave., Annapolis MD 21401
Phone: 410-260-8730   Fax: 410-260-8739

MASSACHUSETTS
Waquoit Bay NERR
CHRISTINE GAULT
Department of Environmental Management, P.O. Box 3092,
Waquoit MA 02536
Phone: 508-457-0495   Fax: 617-727-5537

NEW HAMPSHIRE
Great Bay NERR
NH DEPT. OF FISH AND GAME
PETER WELLENBERGER
Department of Fish and Game, 225 Main St, Durham
NH 03824
Phone: 603-868-1095   Fax: 603-868-3305

NEW JERSEY
Jacques Cousteau NERR Institute of Marine and Coastal
Sciences
71 Dudley Road, New Brunswick NJ 08901
Phone: 732-932-6555   Fax: 732-932-8578

NEW YORK
Hudson River NERR
ELIZABETH BLAIR
c/o Bard College Field Station, Annandale-on-Hudson,
Annandale NY 12504
Phone: 845-758-7010
North Carolina

NORTH CAROLINA NERR
JOHN TAGGART
1 Harvin Moss Ln., Willmington NC 28409
Phone: 910-962-2470   Fax: 910-962-2410

OHIO
Old Woman Creek NERR
EUGENE WRIGHT
2514 Cleveland Rd., East, Huron OH 44839
Phone: 419-433-4601   Fax: 419-433-2851

OREGON
South Slough NERR
MIKE GRAYBILL
P.O. Box 5471, Charleston OR 97420
Phone: 541-888-5559   Fax: 541-888-5559

PUERTO RICO
Jobos Bay NERR
CARMEN GONZALEZ
Department of Natural Resources, Call Box B, Aquirre
PR 00704
Phone: 787-853-4617   Fax: 787-853-4618

RHODE ISLAND
Narragansett Bay NERR
AL BECK
Department of Environmental Management, 55 South
Reserve Dr., Prudence Island RI 02872
Phone: 401-683-6780   Fax: 401-682-1936

SOUTH CAROLINA
ACE Basin NERR
MICHAEL McKENZIE
South Carolina Department of Natural Resources, P.O. Box
12559, Charleston SC 29412
Phone: 803-762-5412   Fax: 803-762-5412
North Inlet
WINYAH BAY RESERVE
DENNIS ALLEN
Winyah Bay NERR, Baruch Marine Field Lab, P.O. Box 1630,
Georgetown SC 29442
Phone: 843-546-3623

VIRGINIA
Chesapeake Bay NERR in Virginia
MAURICE LYNCH
Virginia Institute of Marine Science, PO BOX 1346,
Gloucester Point VA 23062
Phone: 804-684-7135   Fax: 804-684-7120

WASHINGTON
Padilla Bay NERR
TERRY STEVENS
10441 Bayview-Edison Rd., Mt. Vernon WA 98273
Phone: 360-428-1558   Fax: 360-428-1491

NATIONAL FORESTS

ALABAMA
National Forests in Alabama
JIM GOODER
2946 Chestnut St., Montgomery AL 36107
Phone: 334-832-4470

ALASKA
Chugach National Forest
DAVE GIBBONS
3301 C St., Ste 300, Anchorage AK 99503-3956
Phone: 907-271-2525
Tongass-Chatham Area National Forest
FRED SALINAS
204 Siginaka Way, Sitka AK 99835-7316
Phone: 907-747-4410   Fax: 907-747-4331
Tongass-Ketchikan Area National Forest
TOM PUCHLERZ
Federal Bldg., Ketchikan AK 99901-6591
Phone: 907-228-6281
Tongass-Stikine Area National Forest
CAROL JORGENSON
Box 309, Petersburg AK 99833-0309
Phone: 907-772-5800

ARIZONA
Apache-Sitgreaves National Forest
JOHN BEDELL
Federal Bldg., Box 640, Springville AZ 85938
Phone: 520-333-4301   Fax: 520-333-6357
Coconino National Forest
FRED TREVEY
2323 E. Greenlaw Ln., Flagstaff AZ 86004
Phone: 520-527-3600   Fax: 520-527-3620
Coronado National Forest
JAMES ABBOTT
300 W. Congress, Tucson AZ 85701
Phone: 520-670-4552   Fax: 520-670-4567
Kaibab National Forest
WILLIAM LANNAN
800 South 6th St., Williams AZ 86046
Phone: 520-635-2681   Fax: 928-635-8208
Prescott National Forest
COY JEMMETT
344 S. Cortez, Prescott AZ 86303
Phone: 520-771-4700   Fax: 928-771-4884
Tonto National Forest
CHARLES BAZAN
2324 E. McDowell Rd., P.O. Box 5348, Phoenix AZ 85006
Phone: 602-225-5200   Fax: 602-225-5295

ARKANSAS
Ouachita National Forest
ALAN NEWMAN
Box 1270, Federal Bldg., Hot Springs National Park
AR 71902
Phone: 501-321-5202   Fax: 501-321-5353
Ozark—St. Francis National Forest
CHARLES RICHMOND
605 West Main St., Russellville AR 72801
Phone: 501-968-2354   Fax: 501-964-7268

CALIFORNIA
Angeles National Forests
MIKE ROGERS
701 N. Santa Anita Ave., Arcadia CA 91006
Phone: 626-574-1613   Fax: 626-574-5233
Cleveland National Forest
ANNE FEGE
10845 Rancho Bernardo Rd., Ste 200, San Diego CA 92127
Phone: 619-673-6180
Eldorado National Forest
JOHN BERRY
100 Forni Rd., Placerville CA 95667
Phone: 530-622-5061   Fax: 530-621-5297
Inyo National Forest
JEFFERY BAILEY
873 N. Main St., Bishop CA 93514
Phone: 760-873-2400
Klamath National Forest
MICHAEL LEE
1312 Fairlane Rd., Eureka CA 96097
Phone: 530-842-6131   Fax: 530-841-4571
Lake Tahoe Basin Management Unit
ED GEE
870 Emerald Bay Rd., Ste. 1, South Lake Tahoe CA 96150
Phone: 530-573-2600   Fax: 530-573-2780
Lassen National Forest
H. SILVERMAN
55 South Sacramento St., Susanville CA 96130
Phone: 530-257-2151
Los Padres National Forest
JEANINE DERBY
6144 Calle Real, Goleta CA 93117
Phone: 805-683-6711
Mendocino National Forest
DANIEL CHISHOLM
825 N. Humboldt Ave., Willows CA 95988
Phone: 530-233-5811
Modoc National Forest
SCOTT CONROY
800 W. 12th St., Alturas CA 96101
Phone: 530-233-5811   Fax: 530-233-8709
Plumas National Forest
MARK MADRID
159 Lawrence St., Box 11500, Quincy CA 95971
Phone: 530-283-2050   Fax: 530-283-7746
San Bernardino National Forest
GENE ZIMMERMAN
1824 S. Commercenter Cir., San Bernardino CA 92408
Phone: 909-383-5588   Fax: 909-383-5770
Sequoia National Forest
ARTHUR GAFFREY
900 W. Grand Ave., Porterville CA 93257
Phone: 209-784-1500

Shasta-Trinity National Forest
SHARON HEYWOOD
2400 Washington Ave., Redding CA 96001
Phone: 530-244-2978   Fax: 530-242-2233
Sierra National Forest
JAMES BOYNTON
1600 Tollhouse Rd., Clovis CA 93611
Phone: 559-297-0706   Fax: 559-294-4809
Six Rivers National Forest
LOU WOLTERING
1330 Bayshore Way, Eureka CA 95501
Phone: 707-442-1721
Stanislaus National Forest
BENDEL VILLAR
19777 Greenley Rd., Sonora CA 95370
Phone: 209-532-3671   Fax: 209-533-1890
Tahoe National Forest
STEVE EUBANKS
631 Coyote St., Nevada City CA 95959-6003
Phone: 530-265-4531   Fax: 530-478-6109

COLORADO
Arapaho and Roosevelt National Forests
PETER CLARK
240 W. Prospect St., Fort Collins CO 80526
Phone: 970-498-1110
Grand Mesa, Uncompahgre and Gunnison National
Forests
ROBERT STORCH
2250 Highway 50, Delta CO 81416
Phone: 970-874-6600
Pike and San Isabel National Forests
1920 Valley Dr., Pueblo CO 81008
Phone: 719-545-8737
Rio Grande National Forest
1803 West Highway 160, Monte Vista CO 81144
Phone: 719-852-5941
Routt National Forest
JERRY SCHMIDT
19587 W. US40, Ste 20, Steamboat Springs CO 80487-9550
Phone: 970-879-1722
San Juan National Forest
CALVIN JOYNER
Federal Bldg., 15 Burhett Court, Durango CO 81301-3647
Phone: 970-247-4874
White River National Forest
MARTHA KETELLE
Old Federal Bldg., P.O. Box 948, Glenwood Springs
CO 81602
Phone: 970-945-2521

FLORIDA
National Forests in Forida
MARSHA KEARNEY
Woodcrest Office Park, 325 John Knox Rd., Ste F-100,
Tallahassee FL 32303
Phone: 850-942-9300

GEORGIA
Chattahoochee and Oconee National Forests
GEORGE MARTIN
1755 Cleveland Hwy., Gainesville GA 30501
Phone: 770-536-0541

IDAHO
Boise National Forest
DAVID RITTENHOUSE
1249 S. Vinnell Way, Ste. 200, Boise ID 83709
Phone: 208-373-4100
CARIBOU—TARGHEE NATIONAL FOREST
JERRY REESE
250 S. 4th Ave., Ste 172, Federal Bldg., Pocatello ID 93201
Phone: 208-236-7500
CLEARWATER NATIONAL FOREST
JAMES CASWELL
12730 Highway 12, Orofino ID 83544
Phone: 208-476-4541
Idaho Panhandle National Forests
DAVID WRIGHT
3815 Schreiber Way, Coeur d'Alene ID 83814-8863
Phone: 208-765-7223
Nez-Perce National Forest
BRUCE BERNHARDT
Rt. 2, Box 475, McCall ID 83530
Phone: 208-983-1950
Payette National Forest
DAVID ALEXANDER
Forest Service Bldg., Box 1026, McCall ID 83638
Phone: 208-634-0700
Salmon-Challis National Forest
GEORGE MATEJKO
Forest Service, RR 2, Box 600, Salmon ID 83467
Phone: 208-756-5100
Sawtooth National Forest
WILLIAM LEVERE
2647 Kimberly Rd., East, Twin Falls ID 83301-7976
Phone: 208-737-3200

ILLINOIS
Shawnee National Forest
901 S. Commercial St., Harrisburg IL 62946
Phone: 618-253-1000

INDIANA
Hoosier National Forest
KENNETH DAY
811 Constitution Ave., Bedford IN 47421
Phone: 812-275-5987

KENTUCKY
Daniel Boone National Forest
BRADLEY POWELL
1700 Bypass Rd., Winchester KY 40391
Phone: 606-745-3100

LOUISIANA
Kisatchie National Forest
LYNN NEFF
2500 Shreveport Hwy., Pineville LA 71360
Phone: 318-473-7160

MAINE
White Mountain National Forest
see New Hampshire

MICHIGAN
Hiawatha National Forest
CLYDE THOMPSON
2727 N. Lincoln Rd., Escanaba MI 49829
Phone: 906-786-4062
Huron-Manistee National Forest
1755 S. Mitchell St., Cadillac MI 49601
Phone: 616-775-2421
Ottawa National Forest
PHYLLIS GREEN
2100 E. Cloverland Dr., Ironwood MI 49938
Phone: 906-932-1330

MINNESOTA
Chippewa National Forest
LOGAN LEE
Rt. #3, Box 244, Cass Lake MN 56633
Phone: 218-720-5324
Superior National Forest
JIM SANDERS
8901 Grand Avenue Place, Duluth MN 55808-1102
Phone: 218-720-5324

MISSISSIPPI
Bienville, Delta, Desoto, Holly Springs, Homochitto, and
Tombigbee National Forests
KARL SIDERITS
National Forests in Mississippi, 100 W. Capital St., Ste. 1141,
Jackson MS 39269
Phone: 601-965-4391

MISSOURI
Mark Twain National Forest
RANDY MOORE
410 Fairgrounds Rd., Rolla MO 65401
Phone: 573-364-4624
Montana
Beaverhead—Deerlodge National Forest
420 Barrett St., Dillon MT 59725-3572
Phone: 406-683-3900

MONTANA
Bitterroot National Forest
RODD RICHARDSON
1801 N. 1st St., Hamilton MT 59840
Phone: 406-363-7121
Custer National Forest
NANCY CURRIDEN
P.O. Box 50760, Billings MT 59105
Phone: 406-657-6361
Flathead National Forest
CATHY BARBOULETOS
1935 3rd. Ave., E., Kalispell MT 59901
Phone: 406-758-5251
Gallatin National Forest
DAVE GARBER
10 E. Babcock Ave., Federal Bldg., Box 130, Bozeman
MT 59771
Phone: 409-58-76702
Helena National Forest
TOM CLIFFORD
2880 Skyway Dr., Helena MT 59601
Phone: 406-449-5201

Kootenai National Forest
BOB CASTENDA
506 US Highway 2 West, Libby MT 59923
Phone: 406-293-6211
Lewis and Clark National Forest
RICK PRAUSA
Box 869, 1101 15th St., N., Great Falls MT 59923
Phone: 406-791-7700
Lolo National Forest
DEBORAH AUSTIN
Bldg. 24, Ft. Missoula, Missoula MT 59801
Phone: 406-329-3797

NEBRASKA
Nebraska National Forest
MARY PETERSON
125 N. Main St., Chadron NE 69337
Phone: 308-432-0300

NEVADA
Humboldt—Toiyabe National Forest
GLORIA FLORA
2035 1200 Franklin Way, Sparks NV 89431
Phone: 775-331-6444

NEW HAMPSHIRE
White Mountain National Forest
RICK CABLES
Federal Bldg. 719 Main St., Box 638, Laconia NH 03247
Phone: 603-528-8721

NEW MEXICO
Carson National Forest
LEONARD LUCERO
Fed. Bldg., 208 Cruz Alta Rd., Box 558, Taos NM 87571
Phone: 505-758-6200
Cibola National Forest
JEANINE DERBY
2113 Osuna Rd. NE, Ste. A, Albuquerque NM 87111-1001
Phone: 505-761-4650
Gila National Forest
3005 E. Camino del Bosque, Silber City NM 88061
Phone: 505-388-8201
Lincoln National Forest
LEE POAGUE
Fed. Bldg., 1101 New York Ave., Alamogordo NM 88310-6992
Phone: 505-434-7200
Santa Fe National Forest
AL DEFLER
1220 St. Francis Dr., Santa Fe NM 87504
Phone: 505-988-6940

NORTH CAROLINA
Croatan, Nantahala, Pisgah and Uwharrie National
Forests
JOHN RAMEY
National Forests in North Carolina, P.O. Box 2750, Asheville
NC 28802
Phone: 828-257-4200

OHIO
Wayne National Forest
JOSE ZAMBRANA
219 Columbus Rd., Athens OH 45701
Phone: 740-592-6644

OREGON
Deschutes National Forest
SALLY COLLINS
1645 Highway 20 East, Bend OR 97701
Phone: 541-388-2715
Fremont National Forest
CHUCK GRAHAM
524 North G St., Lakeview OR 97630
Phone: 541-947-2151
Malheur National Forest
MARK BOCHE
139 NE Dayton St., John Day OR 97845
Phone: 541-575-1731
Mt. Hood National Forest
MIKE EDRINGTON
2955 Division St., Gresham OR 97030
Phone: 503-666-1700
Ochoco National Forest
TOM SCHMIDT
Box 490, Prineville OR 97754
Phone: 541-447-6247
Rogue River National Forest
JAMES GLADEN
Fed. Bldg., 333 W. 8th St., Box 520, Medford OR 97501
Phone: 541-776-3600
Siskiyou National Forest
MICHAEL LUNN
Box 440, Grants Pass OR 97526
Phone: 541-471-6500
Siuslaw National Forest
Box 1148, Corvallis OR 97339
Phone: 541-750-7000
Umatilla National Forest
JOHN KLINE
2517 SW Hailey Ave., Pendleton OR 97801
Phone: 541-278-3721
Umpqua National Forest
DON OSTBY
Box 1008, Roseburg OR 97470
Phone: 541-672-6601
Wallowa Whitman National Forests
ROBERT RICHMOND
Box 907, Baker City OR 97814
Phone: 541-523-6391
Willamette National Forest
DARREL KENOPS
Box 10607, Eugene OR 97440
Phone: 541-465-6521
Winema National Forest
BOB CASTANEDA
2819 Dahlia, Klamath Falls OR 97601
Phone: 541-883-6714

PENNSYLVANIA
Allegheny National Forest
JOHN PALMER
222 Liberty St., Box 847, Warren PA 16365
Phone: 814-723-5150

PUERTO RICO
Caribbean National Forest
PABLO CRUZ
Call Box 490, Palmer PR 00721
Phone: 787-888-1810

SOUTH CAROLINA
Francis Marion and Sumter National Forest
JEROME THOMAS
4923 Broad River Rd., Columbia SC 29212
Phone: 803-561-4000

SOUTH DAKOTA
Black Hills National Forest
JOHN TWISS
R.R. 2, Box 200, Custer SD 57730-9501
Phone: 605-673-2251

TENNESSEE
Cherokee National Forest
ANNE ZIMMERMAN
P.O. Box 2010, Cleveland TN 37320
Phone: 423-476-9700

TEXAS
Angelia, Davy Crockett, Sabine and Sam Houston
National Forest
RONNIE RAUM
National Forest in Texas, Homer Garrison Federal Bldg., 701
N. 1st St., Lufkin TX 75901
Phone: 409-639-8501

UTAH
Ashley National Forest
BERT KULESZA
355 N. Vernal Ave., Vernal UT 84078
Phone: 801-789-1181
Dixie National Forest
MARY WAGNER
82 N. 100 E St., Cedar City UT 84720-2686
Phone: 435-865-3701
Fishlake National Forest
ROBERT MROWKA
115 East 900 North, Richfield UT 84701
Phone: 435-896-9233
Manti-LaSal National Forest
JANETTE KAISER
599 West Price River Dr., Price UT 84501
Phone: 435-637-2817
Uinta National Forest
PETE KARP
88 West 100 North, Provo UT 84601
Phone: 801-342-5100
Wasatch-Cache National Forest
BERNIE WEINGARDT
8236 Federal Bldg., 125 S. State St., Salt Lake City UT
84138
Phone: 801-524-5030

VERMONT
Green Mountain and Finger Lakes National Forest
PAUL BREWSTER
Federal Bldg., 231 N. Main, Rutland VT 05701-0519
Phone: 802-747-6700

VIRGINIA
George Washington and Jefferson National Forests
> BILL DAMON
> 5162 Valleypointe Pkwy., Roanoke VA 24019
> Phone: 540-265-5100

WASHINGTON
Colville National Forest
> ROBERT VAUGHT
> 716 S. Main, Colville WA 99114
> Phone: 509-662-4335

Gifford Pinchot National Forest
> TED STUBBLEFIELD
> 6926 E. 4th Plain Blvd., Vancouver WA 98668-8944
> Phone: 360-891-5000

Mt. Baker-Snoqualmie National Forest
> 21905 64th Ave. West, Mountlake Terrace, Seattle WA 98043
> Phone: 425-775-4702

Okanogan National Forest
> SAMUEL GEHR
> 1240 S. Second, Okanogan WA 98840
> Phone: 509-826-3275

Olympic National Forest
> RONALD HUMPHREY
> 1835 Blacklake Blvd., SW, Olympia WA 98512
> Phone: 360-956-2300

Wenatchee National Forest
> SONNY O'NEAL
> Box 811, Wenatchee WA 98807
> Phone: 509-662-4335

WEST VIRGINIA
Monongahela National Forest
> CHUCK MEYERS
> USDA Bldg., 200 Sycamore St., Elkins WV 26241-3962
> Phone: 304-636-1800

WISCONSIN
Bridger-Teton National Forest
> TOM PUCHLERZ
> Forest Service Bldg., 340 N. Cache, Jackson WY 83001
> Phone: 307-739-5500

Chequamegon—Nicolet National Forest
> D. ROBERTS
> 1170 4th Ave., S., Park Falls WI 54552
> Phone: 715-762-2461

WYOMING
Bighorn National Forest
> ABIGAIL KIMBELL
> 1969 S. Sheridan Ave., Sheridan WY 82801
> Phone: 307-672-0751

Medicine Bow-Routt National Forest
> JERRY SCHMIDT
> 2468 Jackson St., Laramie WY 82070-6535
> Phone: 307-745-2300

Shoshone National Forest
> REBECCA AUS
> 808 Meadow Ln., Cody WY 82414-4516
> Phone: 307-527-6241

NATIONAL GRASSLANDS

CALIFORNIA
Butte Valley National Grassland
> Goosewest Ranger District, 37805 Hwy. 97, Macdoel
> CA 96058
> Phone: 530-398-4391

COLORADO
Comanche National Grassland
> 27162 Hwy. 287, P.O. Box 127, Springfield CO 81073
> Phone: 719-523-6591

Pawnee National Grassland
> STEVE CURREY
> 660 O St., Greenley CO 80631
> Phone: 970-353-5004

IDAHO
Curlew National Grassland
> P.O. Box 146, Malad ID 83252
> Phone: 208-766-64743

KANSAS
Cimarron National Grassland
> 242 Hwy. 56 E., P.O. Box 300, Elkhart KS 67950
> Phone: 316-697-4621

NEBRASKA
Oglala National Grassland
> 16524 Hwy. 385, Chadron NE 69337
> Phone: 308-432-4475

NEW MEXICO
Kiow / Rita Blanca National Grassland
> 714 Main St., Clayton NM 88415
> Phone: 505-374-9652

NORTH DAKOTA
Little Missouri National Forest, McKenzie Ranger District
> LESLEY THOMPSON
> HC 02, Box 8, Watford City ND 58854
> Phone: 701-842-2393

Little Missouri National Forest, Medora Ranger District
> LARRY DAWSON
> 161 21st St. W., Dickinson ND 58601
> Phone: 701-225-5151

Cheyenne National Grassland
> Box 946, Lisbon ND 58054
> Phone: 701-683-4342

OKLAHOMA
McClellan Creek/Black Kettle National Grassland
> Rt. 1, Box 55B, Cheyenne OK 73628
> Phone: 580-497-2143

OREGON
Crooked River National Forest
> BYRON CHENEY
> 813 SW Hwy 97, Madras OR 97741
> Phone: 541-475-9272

SOUTH DAKOTA
Buffalo Gap National Grassland, Fall River Ranger District
> 209 N. River, Hot Springs SD 57747
> Phone: 605-745-4107

Buffalo Gap National Grassland, Wall Ranger District / National Grasslands Visitor Center
> 708 Main St., P.O. Box 425, Wall SD 57790
> Phone: 605-279-2125

Cedar River / Grand River National Grassland
> FOREST MORIN
> 1005 5th Ave. W., P.O. Box 390, Lemmon SD 57638
> Phone: 605-374-3592

Fort Pierre National Grassland
> ANTHONY DeTOY
> 124 South Euclid Ave., P.O. Box 417, Pierre SD 57501
> Phone: 605-224-5517

TEXAS
Lyndon B. Johnson / Caddo National Forest
> 1400 N. US. 81/287 Hwy., P.O. Box 507, Decatur TX 76234
> Phone: 940-627-5475

WYOMING
Thunder Basin National Grasslands
> MALCOLM EDWARDS
> 2250 East Richards, Douglas WY 82633
> Phone: 307-358-4690

NATIONAL MARINE SANCTUARIES

CALIFORNIA
Channel Islands National Marine Sanctuary
> ED CASSANO
> 113 Harbor Way, Santa Barbara CA 93109
> Phone: 805-568-1582

Cordell Bank National Marine Sanctuary
> EDWARD UEBER
> Ft. Mason, Bldg. 201, San Francisco CA 94123
> Phone: 415-561-6622

Gulf of Farallones National Marine Sanctuary
> EDWARD UEBER
> Fort Mason Bldg. 201, San Francisco CA 94123
> Phone: 415-561-6622

Monterey Bay National Marine Sanctuary
> WILLIAM DOUROS
> 299 Foam St., Suite D, Monterey CA 93940
> Phone: 831-647-4201

FLORIDA
Florida Keys National Marine Sanctuary
> BILLY CAUSEY
> P.O. Box 500368, 5550 Overseas Hwy., Marathon FL 33050
> Phone: 305-743-2437

GEORGIA
Gray's Reef National Marine Sanctuary
> REED BOHNE
> 10 Ocean Science Cir., Savannah GA 31411
> Phone: 912-598-2345

HAWAII
Hawaiian Islands Humpback Whale National Sanctuary
> ALLEN TOM
> 726 South, Kihei HI 96753
> Phone: 808-879-2818

MASSACHUSETTS
Stellwagen Bank National Marine Sanctuary
> BRAD BARR
> 14 Union St., Plymouth MA 02360
> Phone: 508-747-1691

TEXAS
Flower Garden Banks National Marine Sanctuary
> SHELLEY DuPUY
> 216 W. 26th St., Suite 104, Bryan TX 77803
> Phone: 409-779

VIRGINIA
Monitor National Marine Sanctuary
> JOHN BROADWATER
> c/o the Mariners' Museum, 100 Museum Dr., Newport News VA 23606
> Phone: 757-599-3122

WASHINGTON
Olympic Coast National Marine Sanctuary
> GEORGE GALASSO
> 138 W. First St., Port Angeles WA 98362-2600
> Phone: 360-457-6622

NATIONAL PARKS

ALASKA
Denali National Park
> STEPHEN MARTIN
> P.O. Box 74680, Denali Park AK 99755
> Phone: 907-683-9581

Gates of the Arctic National Park
> DAVID MILLS
> 201 First Ave., Doyon Bldg., Fairbanks AK 99701
> Phone: 907-456-0281

Glacier Bay National Park
> TOMIE LEE
> One Park Rd., Gustavus AK 99826-0140
> Phone: 907-697-2230

Katmai National Park
> DEB LIGGETT
> One King Salmon Mall, King Salmon AK 99613
> Phone: 907-246-3305

Kenai Fjords National Park
> ANNE CASTELLINA
> 1212 4th Ave., Seward AK 99664
> Phone: 907-224-3175

Kobuk Valley National Park
> DAVID SPRITES
> P.O. Box 1029, Kotzebue AK 99752
> Phone: 907-442-3890

Lake Clark National Park
> DEB LIGGETT
> 4230 University Dr., Ste. 311, Anchorage AK 99508
> Phone: 907-271-3751

Wrangell-St. Elias National Park
JON JARVIS
P.O.Box 439, Copper Center AK 99573
Phone: 907-822-5234

AMERICAN SAMOA
National Park of America Samoa
CHARLES CRANFIELD
Pago Pago, American Samoa 96799
Phone: 684-633-7082

ARIZONA
Grand Canyon National Park
ROBERT ARNBERGER
P.O. Box 129, Grand Canyon AZ 86023
Phone: 520-638-7945
Petrified Forest National Park
MICHELLE HELLICKSON
One Park Rd., Petrified Forest AZ 86028
Phone: 520-524-6228

ARKANSAS
Hot Springs National Park
ROGER GIDDINGS
P.O. Box 1860, Hot Springs AR 71902
Phone: 501-624-3383

CALIFORNIA
Channel Islands National Park
TIM SETNIKA
1901 Spinnaker Dr., Ventura CA 93001
Phone: 805-658-5700
Death Valley National Park
RICHARD MARTIN
P.O. Box 579, Death Valley CA 92328
Phone: 760-786-2331   Fax: 760-786-3283
Joshua Tree National Park
ERNEST QUINTANA
74485 National Park Dr., Twenty-nine Palms CA 92277
Phone: 760-367-5500   Fax: 760-367-6392
Lassen Volcanic National Park
MARILYN PARRIS
P.O. Box 100, 38050 Hwy 36E, Mineral CA 96063
Phone: 530-595-4444
Redwood National Park
ANDREW RINGGOLD
1111 Second St., Crescent City CA 95531
Phone: 707-464-6101
Sequoia and Kings Canyon National Park
MICHAEL TOLLEFSON
47050 Generals Hwy., Three Rivers CA 93271-9651
Phone: 559-565-3341
Yosemite National Park
STANLEY ALBRIGHT
P.O. Box 577, Administration Bldg., Yosemite National Park CA 95389
Phone: 209-372-0200

COLORADO
Mesa Verde National Park
LARRY WIESE
P.O. Box 8, Mesa Verde National Park CO 81330
Phone: 970-529-4465

Rocky Mountain National Park
RANDY JONES
100 Hwy. 36, Estes Park CO 80517
Phone: 970-586-1200

FLORIDA
Biscayne National Park
DICK FROST
9700 SW 328th St., Homestead FL 33033
Phone: 305-230-1144
Dry Tortugas National Park
RICHARD RING
P.O. Box 6208, Key West FL 33041
Phone: 305-242-7700   Fax: 305-242-7711
Everglades National Park
RICHARD RING
40001 State Rd. 9336, Homestead FL 33034
Phone: 305-242-7700

HAWAII
Haleakala National Park
DONALD REESER
P.O. Box 369, Makawao, Maui HI 96768
Phone: 808-572-9306
Hawaii Volcanoes National Park
JIM MARTIN
P.O. Box 52, Hawaii Volcanoes HI 96718
Phone: 808-985-6025

KENTUCKY
Mammoth Cave National Park
RONALD SWITZER
P.O. Box 7, Mammoth Cave KY 42259
Phone: 502-758-2254

MAINE
Acadia National Park
PAUL HAERTEL
P.O. Box 177, Bar Harbor ME 04609
Phone: 207-288-0374

MICHIGAN
Isle Royale National Park
DOUGLAS BARNARD
800 E. Lakeshore Dr., Houghton MI 49931-1895
Phone: 906-482-0986

MINNESOTA
Voyageurs National Park
BARBARA WEST
3131 Hwy. 53, International Falls MN 56649
Phone: 218-283-9821

MONTANA
Glacier National Park
DAVID MIHALIC
West Glacier MT 59936
Phone: 406-888-7901

NEVADA
Great Basin National Park
REBECCA MILLS
Baker NV 89311
Phone: 702-234-7331

NEW MEXICO
Carlsbad Caverns National Park
FRANK DECKERT
3225 National Park Hwy., Carlsbad NM 88220
Phone: 505-785-2232

NORTH DAKOTA
Theodore Roosevelt National Park
NOEL POE
P.O. Box 7, 315 2nd Ave., Medora ND 58645-0007
Phone: 701-623-4466

OREGON
Crater Lake National Park
CHUCK LUNDY
P.O. Box 7, Crater Lake OR 97604
Phone: 541-594-2211

SOUTH DAKOTA
Badlands National Park
WILLIAM SUPERNAUGH
P.O. Box 6, Rt. 240, Interior SD 57750
Phone: 605-433-5280
Wind Cave National Park
JIM TAYLOR
RR 1, Box 190, Hot Springs SD 57747-9430
Phone: 605-7454600

TENNESSEE
Great Smoky Mountains National Park
KAREN WADE
107 Park Headquarters Rd., Gatlinburg TN 37738
Phone: 423-436-1200

TEXAS
Big Bend National Park
JOSE CISNEROS
P.O. Box 129, Big Bend National Park TX 79834
Phone: 915-477-1101
Guadalupe Mountains National Park
LARRY HENDERSON
HC 60, Box 400, Salt Flat TX 79847-9400
Phone: 915-828-3251

UTAH
Arches National Park
WALT DABNEY
P.O. Box 907, Moab UT 84532-0907
Phone: 435-259-8161
Bryce Canyon National Park
FRED FAGERGREN
P.O. Box 17001, Bryce Canyon UT 84717-0001
Phone: 435-834-5322
Canyonlands National Park
WALTER DABNEY
2282 SW Resource Blvd., Moab UT 84532-3298
Phone: 801-259-3911
Capitol Reef National Park
ALBERT HENDRICKS
HC 70, Box 15, Torrey UT 84775-9602
Phone: 435-425-3791

Zion National Park
DON FALVEY
Springdale UT 84767
Phone: 435-772-3256

VIRGIN ISLANDS
Virgin Islands National Park
RUSSELL BERRY
6310 Estate Nazareth #10, St. Thomas VI 00802-1102
Phone: 340-775-6238

VIRGINIA
Shenandoah National Park
DOUG MORRIS
3655 US Hwy 211-E, Luray VA 22835-9036
Phone: 540-999-3400

WASHINGTON
Mount Rainier National Park
WILLIAM BRIGGLE
Tahoma Woods, Star Route, Ashford WA 98304-9751
Phone: 360-569-2211
North Cascades National Park
BILL PALECK
2105 Hwy. 20, Sedro Woolley WA 98284-9314
Phone: 360-856-5700
Olympic National Park
DAVID MORRIS
600 E. Park Ave., Port Angeles WA 98362-6798
Phone: 360-452-4501

WYOMING
Grand Teton National Park
JACK NECKELS
P.O. Drawer 170, Moose WY 83012-0170
Phone: 307-739-3300
Yellowstone National Park
MICHAEL FINLEY
P.O. Box 168, Yellowstone WY 82190
Phone: 307-344-7381

NATIONAL SEASHORES

CALIFORNIA
Point Reyes National Seashore
DON NEUBACHER
Point Reyes CA 94956
Phone: 415-663-5100　　　Fax: 4156638132

FLORIDA
Canaveral National Seashore
ROBERT NEWKIRK
308 Julia St., Titusville FL 32796-3521
Phone: 321-2671-110　　　Fax: 3212642906
Gulf Islands National Seashore
JERRY EUBANKS
1801 Gulf Breeze Pkwy., Gulf Breeze FL 32561-5000
Phone: 850-934-2600　　　Fax: 8509329654

GEORGIA
Cumberland Island National Seashore
DENIS DAVIS
P.O. Box 806, Saint Marys GA 31558
Phone: 888-817-3421   Fax: 9126737747

MARYLAND
Assateague Island National Seashore
MARC KOENINGS
7206 National Seashore Ln., Berlin MD 21811
Phone: 410-641-1443

MASSACHUSETTS
Cape Cod National Seashore
MARIA BURKS
99 Marconi Site Rd., Wellfleet MA 02667
Phone: 508-349-3785   Fax: 5083499052

NEW YORK
Fire Island National Seashore
CONSTANTINE DILLON
120 Laurel St, Patchogue NY 11772
Phone: 631-289-4810   Fax: 6312894898

NORTH CAROLINA
Cape Hatteras National Seashore
ROBERT REYNOLDS
1401 National Park Dr., Manteo NC 27954
Phone: 252-473-2111   Fax: 2524732595
Cape Lookout National Seashore
KARREN BROWN
131 Charles St., Harkers Island NC 28531
Phone: 252-728-2250   Fax: 2527282160

TEXAS
Padre Island National Seashore
JACK WHITWORTH
P.O. Box 181300, Corpus Christi TX 78480-1300
Phone: 361-949-8173   Fax: 3619498023

NATIONAL WILDLIFE REFUGES

ALABAMA
Bon Secour
BILL GATES
12295 State Highway 180, Gulf Shores AL 36542
Phone: 334-540-7720   Fax: 334-540-7301
Choctaw
DOUGLAS BAUMGARTNER
P.O. Box 808, Jackson AL 36545
Phone: 334-246-3583   Fax: 334-246-5414
Eufaula
FRANK DUKES
509 Old Highway 165, Eufaula AL 36027
Phone: 334-687-4065   Fax: 334-687-5906
Wheeler (Blowing Wind Cave, Fern Cave,
Watercress Darter)
H. STONE
2700 Refuge Hq. Rd., Decatur AL 35603
Phone: 205-353-7243   Fax: 256-340-9729

ALASKA
Alaska Maritime (Alaska Peninsula Unit, Bering Sea Unit,
Chukchi Sea Unit, Gulf of Alaska Unit)
JOHN MARTIN
2355 Kachemak Bay Dr., Ste. 101, Homer AK 99603-8021
Phone: 907-235-6546   Fax: 907-235-7783
Alaska Peninsula (Becharof)
DARYLE LONS
P.O. Box 277, King Salmon AK 99613
Phone: 907-246-3339   Fax: 907-246-6696
Arctic
101 12th Ave., Box 20, Fairbanks AK 99701
Phone: 907-456-0250   Fax: 907-456-0428
EDWARD MERRITT
P.O. Box 69, McGrath AK 99627
Phone: 907-524-3251   Fax: 907-524-3141
Izembek
RICHARD POETTER
P.O. Box 127, #1 Izembek Dr., Cold Bay AK 99571
Phone: 907-532-2445   Fax: 907-532-2549
Kanuti
TOM EARLY
101 12th Ave., Box 11; RM. 262, Fairbanks AK 99701
Phone: 907-456-0329   Fax: 907-456-0506
Kenai
ROBIN WEST
P.O. Box 2139, Soldotna AK 99669-2139
Phone: 907-262-7021   Fax: 907-262-3599
Kodiak
JAY BELLINGER
1390 Buskin River Rd., Kodiak AK 99615
Phone: 907-487-2600   Fax: 907-487-2144
Nowitna (Koyukuk)
EUGENE WILLIAMS
P.O. Box 287, Galena AK 99741
Phone: 907-656-1231   Fax: 907-656-1708
Selawik
LESLIE KERR
P.O. Box 270, Kotzebue AK 99752-3799
Phone: 907-442-3799   Fax: 907-442-3124
Tetlin
RICHARD VOSS
P.O. Box 779, Tok AK 99780
Phone: 907-883-5312   Fax: 907-883-5747
Togiak
AARON ARCHIBERQUE
P.O. Box 270, Dillingham AK 99576
Phone: 907-842-1063   Fax: 907-842-5402
Yukon Delta
MICHAEL REARDEN
P.O. Box 346, Bethel AK 99559-0346
Phone: 907-543-3151   Fax: 907-543-4413
Yukon Flats
TED HEUER
101 12th Ave., Rm. 264, Fairbanks AK 99701
Phone: 907-456-0440   Fax: 907-456-0447

ARIZONA
Buenos Aires
WAYNE SHIFFLETT
P.O. Box 109, Sasabe AZ 85633
Phone: 502-823-4251   Fax: 520-823-4247
Cabeza Prieta
DONALD TILLER
1611 N. Second Ave., Ajo AZ 85321
Phone: 520-387-6483   Fax: 520-387-5359

Imperial
  MITCHELL ELLIS
  P.O. Box 72217, Yuma AZ 85365
  Phone: 520-783-3371   Fax: 520-783-0652
Kofa
  RAY VARNEY
  P.O. BOX 6290, Yuma AZ 85366-6290
  Phone: 520-783-7861   Fax: 520-783-8611
Lower Colorado River Complex
  WES MARTIN
  P.O. Box D, Yuma AZ 85364
  Phone: 520-343-8112   Fax: 520-343-8320
San Bernardino (Leslie Canyon)
  KEVIN COBBLE
  P.O. Box 3509, Douglas AZ 85607
  Phone: 520-364-2104   Fax: 520-364-2130

ARKANSAS
Bald Knob
  ROBERT ALEXANDER
  1439 Coal Chute Rd., Bald Knob AR 72010
  Phone: 870-724-2458   Fax: 870-724-2460
Big Lake
  CLARKE DIRKS
  P.O. Box 67, Manila AR 72442
  Phone: 870-564-2429   Fax: 870-564-2573
Cache River
  DENNIS WIDNER
  Rt. 2, Box 126-T, Augusta AR 72006
  Phone: 870-347-2614   Fax: 870-347-2908
Felsenthal (Overflow, Pond Creek)
  JIM JOHNSON
  P.O. Box 1157, Crossett AR 71635
  Phone: 870-364-3167   Fax: 870-364-3757
Holla Bend (Logan Cave)
  M. BLIHOVDE
  Rt. 1, Box 59, Dardanelle AR 72834-9704
  Phone: 870-229-4300   Fax: 870-229-4302
Overflow
  c/o Felsenthal NWR, PO Box 1157, Crossett AR 71635
  Phone: 870-364-3167   Fax: 870-364-3757
Pond Creek
  c/o Felsenthal NWR, PO Box 1157, Crossett AR 71635
  Phone: 870-364-3167   Fax: 870-364-3757
Wapanocca
  GLEN MILLER
  P.O. Box 279, Turrell AR 72384
  Phone: 870-343-2595   Fax: 870-343-2416
White River
  LARRY MALLARD
  P.O. Box 308, DeWitt AR 72042-0308
  Phone: 870-946-1468   Fax: 870-946-2591

CALIFORNIA
Antioch Dunes
  c/o San Pablo Bay NWR, PO Box 2012, Mare Island
  CA 94592-0012
Bitter Creek
  c/o Hopper Mountain Complex, PO Box 5839, Ventura
  CA 93005-0839
  Phone: 805-644-5185   Fax: 805-644-1732
Blue Ridge
  c/o Hopper Mountain Complex, PO Box 5839, Ventura
  CA 93005-0839
  Phone: 805-644-5185   Fax: 805-644-1732

Guadalupe- Nipomo Dunes
  c/o Hopper Mountain Complex, PO Box 5839, Ventura
  CA 93005-0839
  Phone: 805-644-5185   Fax: 805-644-1732
Hopper Mountain Complex (Hopper Mountain,
Battle Creek)
  MARC WEITZEL
  P.O. Box 5839, Ventura CA 93005
  Phone: 805-644-5158   Fax: 805-644-1732
Humboldt Bay
  KIM FORREST
  1020 Ranch Rd., Loleta CA 95551
  Phone: 707-733-5406   Fax: 707-733-1946
Kern (Blue Ridge, Pixley)
  DAVID HARDT
  P.O. Box 670, Delano CA 93216-0219
  Phone: 805-725-2767   Fax: 805-725-6041
Keterson
  PO Box 2176, Los Banos CA 93635-2176
  Phone: 209-826-3508   Fax: 209-826-1445
Klamath Basin Refuges (Bear Valley, OR; Bear Lake, OR;
Lower Klamath, OR & CA; Tule Lake; Upper Klamath,
OR; Klamath Forest, OR)
  THOMAS STEWART
  Rt. 1, Box 74, Tule Lake CA 96134-9715
  Phone: 530-667-2231   Fax: 530-667-2231
Modoc
  ANNE LAROSA
  P.O. Box 1610, Alturas CA 96101
  Phone: 530-233-3572   Fax: 530-233-4143
Sacramento (Butte Sink WMA, Colusa, Delevan, North
Central Valley, Sacramento River, Sutter, Willow Creek-
Lurline WMA)
  GARY KRAMER
  752 County Rd., 99W, Willows CA 95988
  Phone: 530-934-2801   Fax: 530-934-7814
San Diego Complex (Tijuana Slough, Sweetwater Marsh,
Seal Beach)
  W. RUNDLE
  2736 Loker Ave. W., Suite A, Carlsbad CA 92008-0524
  Phone: 760-930-0168   Fax: 760-930-0168
San Francisco Bay (Antioch Dunes, Don Edwards San
Francisco Bay, Ellicott Slough, Farallon, Humbolt Bay,
Marin Islands, Salinas River, San Pablo Bay)
  MARGARET KOLAR
  P.O. Box 524, Newark CA 94560
  Phone: 510-792-0222   Fax: 510-792-5828
San Luis (Grasslands WMA, Kesterson, Merced, San
Joaquin River)
  GARY ZAHM
  P.O. Box 2176, Los Banos CA 93635-2176
  Phone: 209-826-3508   Fax: 209-826-1445
Sonny Bono Salton Sea (Coachella Valley)
  E. BLOOM
  906 W. Sinclair, Calipatria CA 92233
  Phone: 760-348-5278   Fax: 760-348-7245
Stone Lakes
  THOMAS HARVEY
  1624 Hood-Franklin Rd., Elk Grove CA 95758-9774
  Phone: 916-775-4421   Fax: 916-775-4407

COLORADO
Alamosa NWR (Monte Vista NWR)
  MICHAEL BLENDEN
  9383 El Rancho Ln., Alamosa CO 81101-9003
  Phone: 719-589-4021   Fax: 719-587-6595

Arapaho (Hutton Lake, Bamforth, Mortenson Lake, Pathfinder)
　GREG LANGER
　P.O. Box 457, Walden CO 80480
　Phone: 970-723-8202　　Fax: 970-723-8528
Browns Park
　MICHAEL BRYANT
　1318 Highway 318, Maybell CO 81640
　Phone: 970-365-3613　　Fax: 970-365-3614
Rocky Mountain Arsenal
　RAY RAUCH
　USF&WS, Bld. 111, Commerce City CO 80022-1748
　Phone: 303-289-0232　　Fax: 303-289-0579
Two Ponds C/O ROCKY MOUNTAIN ARSENAL
　DANIEL JAMIEL
　USF&WS, Bld. 111, Commerce City CO 80022-1748
　Phone: 303-289-0232　　Fax: 303-289-0579

DELAWARE
Bombay Hook (Prime Hook NWR)
　PAUL DALY
　2591 Whitehall Neck Rd., Smyrna DE 19977
　Phone: 302-653-9345　　Fax: 302-653-0684
Prime Hook
　GEORGE O'SHEA
　R.D. 3, Box 195, Milton DE 19968
　Phone: 302-684-8419　　Fax: 302-684-8504

FLORIDA
Archie Carr
　1339 20th Street, Vero Beach FL 32960-3559
　Phone: 561-564-3909　　Fax: 561-564-7393
Arthur R. Marshall Loxahatchee NWR (Hobe Sound NWR)
　MARK MUSAUS
　10216 Lee Rd., Boynton Beach FL 33437-4796
　Phone: 561-732-3684　　Fax: 561-369-7190
Chassahowitzka (Crystal River, Egmont Key, Passage Key, Pinellas)
　ELIZABETH SOUHEAVER
　1502 S.E. Kings Bay Dr., Crystal River FL 34429
　Phone: 352-563-2088　　Fax: 352-795-7961
Crocodile Lake
　PO Box 370, Key Largo FL 33037
　Phone: 305-451-4223　　Fax: 305-451-1508
Florida Panther; Ten Thousand Island
　JIM KRAKOWSKI
　3860 Tollgate Blvd., Ste. 300, Naples FL 34114
　Phone: 941-353-8442　　Fax: 941-353-8640
Hobe Sound
　RYAN NOEL
　P.O. Box 645, Hobe Sound FL 33475-0645
　Phone: 561-546-6141
J.N. Ding Darling (Caloosahatchee, Island Bay, Matlacha Pass, Pine Island)
　LOUIS HINDS
　One Wildlife Dr., Sanibel FL 33957
　Phone: 941-472-1100　　Fax: 941-472-4061
Lake Woodruff
　HENRY SANSING
　P.O. Box 488, DeLeon Springs FL 32130-0488
　Phone: 904-985-4673　　Fax: 904-985-0926
Lower Suwannee (Cedar Keys)
　KENNETH LITZENBERGER
　16450 NW 31st Pl., Chiefland FL 32626
　Phone: 352-493-0238　　Fax: 352-493-1935

Merritt Island (Pelican Island, Archie Carr, Lake Wales Ridge, St. Johns)
　ALBRIGHT HIGHT
　P.O. Box 6504, Titusville FL 32782
　Phone: 407-861-0667　　Fax: 407-861-1276
National Key Deer Refuge (Key West, Great White Heron, Crocodile Lake)
　BARRY STIEGLITZ
　P.O. Box 43510, Big Pine Key Fl 33043-0510
　Phone: 305-872-2239　　Fax: 305-872-3675
St. Marks
　JAMES BURNETT
　P.O. Box 68, St. Marks FL 32355
　Phone: 850-925-6121　　Fax: 850-925-6930
St. Vincent
　TORRY PEACOCK
　P.O. Box 447, Apalachicola FL 32329-0447
　Phone: 850-653-8808　　Fax: 850-653-9893

GEORGIA
Okefenokee (Banks Lake)
　MALLORY REEVES
　Rt. 2, Box 3330, Folkston GA 31537
　Phone: 912-496-7366　　Fax: 912-496-3332
Piedmont (Bond Swamp)
　RONNIE SHELL
　718 Juliette Rd., Round Oak GA 31038
　Phone: 912-986-5441　　Fax: 912-986-9646
Savannah Costal Refuges
　SAM DRAKE
　1000 Business Center Dr., Ste. 10, Savannah GA 31405
　Phone: 912-652-4415　　Fax: 912-652-4385

GUAM
　ROGER DiROSA
　P.O. Box 8134, MOU-3, Dededo GU 96912
　Phone: 671-355-5096　　Fax: 671-355-5098

HAWAII
Hakalau Forest
　RICHARD WASS
　32 K Kincole St, Suite 101, Hilo HI 96720
　Phone: 808-933-6915　　Fax: 808-933-6917
Hawaiian and Pacific Islands NWR Complex
　JERRY LEINECKE
　P.O. Box 50167, Honolulu HI 96850
　Phone: 808-541-1201　　Fax: 808-541-1216
James Campbell NWR and Pearl Harbor NWR
　DONNA STOVALL
　66-590 Kamehameha Hwy., Rm. 2C, Haleiwa HI 96712
　Phone: 808-637-6330　　Fax: 808-637-3570
Johnston Island
　D. HAYES
　Box 396, APO, AP HI 96558-0396
　Phone: 808-421-0011　　Fax: 808-422-6905
Kealia Pond
　GLYNNIS NAKAI
　P.O. Box 1042, Kihei HI 96753-1042
　Phone: 808-875-1582　　Fax: 808-875-2945
Kilauea Point (Hanalei, Huleia)
　THOMAS ALEXANDER
　P.O. Box 1128, Kilauea, Kauai HI 96754-1128
　Phone: 808-828-1413　　Fax: 808-828-6634

Midway Atoll
ROBERT SHALLENBERGER
P.O. Box 29460, Midway Island Station #4, Honolulu
HI 96820-1860
Phone: 808-599-3914
Pacific/Remote Islands Complex (Hawaiian Islands,
Baker Island, Howland Island, Jarvis Island,
Rose Atoll)
DAVID JOHNSON
P.O. Box 50167, Honolulu HI 96850-5167
Phone: 808-541-1201   Fax: 808-541-1216

IDAHO
Bear Lake
RICHARD SJOSTROM
Box 9, Montpelier ID 83253
Phone: 208-847-1757   Fax: 208-847-1319
Camas
GERALD DEUTSCHER
2150 E. 2350 N., Hamer ID 83425
Phone: 208-662-5423   Fax: 208-662-5525
Deer Flat
ELAINE JOHNSON
13751 Upper Embankment Rd., Nampa ID 83686
Phone: 208-467-9278   Fax: 208-467-1019
Grays Lake
MICHAEL FISHER
74 Grays Lake Rd., Wayan ID 83285
Phone: 208-574-2755   Fax: 208-574-2756
Kootenai
DANIEL PENNINGTON
HCR 60, Box 283, Bonners Ferry ID 83805
Phone: 208-267-3888   Fax: 208-267-5570
Minidoka
MICHAEL JOHNSON
961 E. Minidoka Dam, Rupert ID 83350
Phone: 208-436-3589   Fax: 208-436-1570
Oxford Slough WPA
TERRELL GLADWIN
1246 Yellowstone Ave., Ste. A-4, Pocatello ID 83201
Phone: 208-237-6616   Fax: 208-237-6617
Southeast Idaho Complex
RICHARD MUNOZ
4425 Burley Dr., Ste. A, Chubbuck ID 83202
Phone: 208-237-6616   Fax: 208-237-8213

ILLINOIS
Crab Orchard
DANIEL DOSHIER
8588 Rt. 148, Marion IL 62959
Phone: 618-997-3344   Fax: 618-997-8961
Cypress Creek
MARGUERITE HILLS
137 Rustic Campus Dr., Ullin IL 62992
Phone: 618-634-2231   Fax: 618-634-9656
Illinois River National Wildlife and Fish Refuge
(Chautauqua, Emiquon, Meredosia)
ROSS ADAMS
19031 E. County Rd. 2105N, Havana IL 62644
Phone: 309-535-2290   Fax: 309-535-3023
Mark Twain
DICK STEINBACH
1704 N. 24th St., Quincy IL 62301
Phone: 217-224-8580   Fax: 217-224-8583

Mark Twain/Brussels District
HOWARD PHILLIPS
HRC, Box 107, Brussels IL 62013-9711
Phone: 618-883-2524   Fax: 618-883-2201

INDIANA
Muscatatuck
LELAND HERZBERGER
12985 E. U.S. Hwy 50, Seymour IN 47274
Phone: 812-522-4352   Fax: 812-522-6826
Patoka River National Wetlands Project
WILLIAM McCOY
510 1/2 W. Morton St., P.O. Box 217, Oakland City IN 47660
Phone: 812-749-3199   Fax: 812-749-3059

IOWA
De Soto (Boyer Chute NWR)
GEORGE GAGE
1434 316th Ln., Missouri Valley IA 51555
Phone: 712-642-4121   Fax: 712-642-2877
Mark Twain/Wapello District
KATHLEEN MAYCROFT
10728 County Rd. X-61, Wapello IA 52653-9477
Phone: 319-523-6982   Fax: 319-523-6960
Neal Smith
NANCY GILBERTSON
P.O. Box 399, Prairie City IA 50228
Phone: 515-994-3400   Fax: 515-944-3459
Union Slough (Iowa WMD)
BARRETT CHRISTENSEN
1710 360th St., Titonka IA 50480
Phone: 515-928-2523   Fax: 515-928-2230

KANSAS
Flint Hills (Marais des Cygnes)
JERRE GAMBLE
P.O. Box 128, 530 W. Maple, Hartford KS 66854
Phone: 316-392-5553   Fax: 316-392-5554
Kirwin
WILLIAM SCHAFF
R.R. 1, Box 103, Kirwin KS 67644
Phone: 913-543-6673   Fax: 913-543-5464
Quivira
DAVE HILLEY
R.R. 3, Box 48A, Stafford KS 67530
Phone: 316-486-2393   Fax: 316-486-2394

KENTUCKY
Clarks River
RICHARD HUFFINES
P.O. Box 89, Benton KY 42025
Phone: 502-527-5770

LOUISIANA
Bayou Cocodrie
JEROME FORD
P.O. Box 1772, Ferriday LA 71334
Phone: 318-336-7119   Fax: 318-336-5610
Cameron Prairie
PAUL YAKUPZACK
1428 Highway 27, Bell City LA 70630
Phone: 318-598-2216   Fax: 318-598-2492

Catahoula
ERIC SIPCO
P.O. Drawer Z, Rhinehart LA 71363-0201
Phone: 318-992-5261   Fax: 318-992-6023
Lacassine
VICKI GRAFE
209 Nature Rd., Lake Arthur LA 70549
Phone: 318-774-5923   Fax: 318-774-9913
Lake Ophelia (Grand Cole)
DENNIS SHARP
401 Island Rd., Marksville LA 71351
Phone: 318-253-4238   Fax: 318-253-7139
Louisiana WMD (Handy Brake)
JAMES OUCHLEY
1428 Hwy. 143, Farmerville LA 71241
Phone: 318-726-4400   Fax: 318-726-4667
North Louisiana Wildlife Refuge Complex (D'Arbonne, Upper Ouachita)
KENNETH BUTTS
11372 Hwy. 143, Farmerville LA 71241
Phone: 318-726-4222   Fax: 318-726-4667
Sabine
CHRIS PEASE
3000 Holly Beach Hwy, Hackberry LA 70645
Phone: 318-762-3816   Fax: 318-762-3780
Southeast Louisiana Refuges (Bayou Sauvage, Big Branch Marsh, Bogue Chitto, Breton, Delta, Atchafalaya, Shell Keys)
HOWARD POITEVINT
1010 Gause Blvd., Bldg. 936, Slidell LA 70458
Phone: 504-646-7555   Fax: 504-646-7588
Tensas River
GEORGE CHANDLER
Rt. 2, Box 295, Tallulah LA 71282
Phone: 318-574-2664   Fax: 318-574-1624

MAINE
Michigan WMD
JAMES HUDGINS
2651 Coolidge Rd., East Lansing MI 48823
Phone: 517-351-4230
Moosehorn
ROBERT PAYTON
R.R. 1, Box 202, Suite 1, Baring ME 04694
Phone: 207-454-7161   Fax: 207-454-2550
Petit Manan (Cross Island, Franklin Island, Seal Island, Pond Island MA)
STAN SKUTEK
P.O. Box 279, Millbridge ME 04658
Phone: 207-546-2124   Fax: 207-546-7805
Rachel Carson
WARD FEURT
Box 751, Wells ME 04090
Phone: 207-646-9226   Fax: 207-646-6554
Seney (Harbor Island, Huron, Kirtland's Warbler WMA)
MICHAEL TANSY
HCR #2, Box 1, Seney MI 49883
Phone: 906-586-9851   Fax: 906-586-3800
Shiawassee (Michigan Islands, Wyandotte)
DOUGLAS SPENCER
6975 Mower Rd., Saginaw MI 48601
Phone: 517-777-5930   Fax: 517-777-9200
Sunkhaze Meadows (Carlton Pond)
MARK SWEENY
1033 S. Main St., Old Town ME 04468
Phone: 207-827-6138   Fax: 207-827-6099

MARYLAND
Blackwater (Martin, Susquehanna)
GLENN CAROWAN
2145 Key Wallace Dr., Cambridge MD 21613
Phone: 410-228-2692   Fax: 410-228-3261
Eastern Neck
MARTIN KAEHNY
1730 Eastern Neck Rd., Rock Hall MD 21661
Phone: 410-639-7056   Fax: 410-639-2516
Patuxent Research Refuge
SUSAN MCMAHON
12100 Beech Forest Rd., Ste. 4036, Laurel MD 20708-4036
Phone: 301-497-5580   Fax: 301-497-5765

MASSACHUSETTS
Great Meadows (John Hay, NH, Massasoit, Nantucket, Norman's Long Island, Oxbow, Wapack, NH)
MANUEL OLIVEIRA
Weir Hill Rd., Sudbury MA 01776
Phone: 508-443-4661   Fax: 508-443-2898
Monomoy
SHARON WARE
Wikis Way, Morris Island, Chatham MA 02633
Phone: 508-945-0594   Fax: 508-945-9559
Parker River (Thatcher Island)
JOHN FILLIO
161 Northern Blvd., Plum Island, Newburyport MA 01950
Phone: 508-465-5753   Fax: 508-465-2807
Silvio O. Conte National Wildlife and Fish Refuge
LAWRENCE BANDOLIN
38 Ave. A, Turners Falls MA 01376
Phone: 413-863-0209   Fax: 413-863-3070

MINNESOTA
Agassiz
MARGARET ANDERSON
Rt. 1, Box 74, Middle River MN 56737
Phone: 218-449-4115   Fax: 218-449-3241
Big Stone
RONALD COLE
R.R. Box 25, Odessa MN 56276
Phone: 320-273-2191   Fax: 320-273-2231
Detroit Lakes WMD
RICK JULIAN
26624 N. Tower Rd., Detroit Lakes MN 56501-7959
Phone: 218-847-4431   Fax: 218-847-4156
Fergus Falls WMD
KEVIN BRENNAN
Rt. 1, Box 76, Fergus Falls MN 56537
Phone: 218-739-2291   Fax: 218-739-9534
Litchfield WMD
971 E. Frontage Rd., Litchfield MN 55355
Phone: 320-693-2849   Fax: 320-693-2326
Minnesota Valley
RICHARD SHULTZ
3815 E. 80th St., Bloomington MN 55425-1600
Phone: 612-854-5900   Fax: 612-854-3279
Morris WMD
STEVE DELEHANCY
Rt. 1, Box 877, Morris MN 56267
Phone: 612-589-1001   Fax: 612-589-2624
Rice Lake (Mille Lacs)
EUGENE PATTON
Rt. 2, Box 67, McGregor MN 55760
Phone: 218-768-2402   Fax: 218-768-3040

Sherburne (Crane Meadows)
CHARLES BLAIR
17076 293rd Ave., Zimmerman MN 55398
Phone: 612-389-3323   Fax: 612-389-3493
Tamarac
JAY JOHNSON
35704 County Hwy. 26, Rochert MN 56578
Phone: 218-847-2641   Fax: 218-847-9141
Upper Mississippi River W&FR/Winona District
JAMES FISHER
51 E. 4th St., Rm. 203, Winona MN 55987
Phone: 507-452-4232   Fax: 507-452-0851
Windom WMD
STEVEN KALLIN
Rt. 1, Box 273A, Windom MN 56101
Phone: 507-831-2220   Fax: 507-831-5524

MISSISSIPPI
Hillside
1562 Providence Rd., Cruger MS 35924
Phone: 662-235-4989   Fax: 662-235-5303
Mississippi Sandhill Crane (Grand Bay)
SABRINA KEEN
7200 Crane Ln., Gautier MS 39553
Phone: 601-497-6322   Fax: 601-497-5407
Mississippi WMD (Dahomey, Tallahatchie)
STEPHEN GARD
P.O. Box 1070, 16736 Hwy 8 West, Grenada MS 38902
Phone: 601-226-8286   Fax: 601-226-8488
Noxubee
JIMMY TISDALE
Rt. 1, Box 142, Brooksville MS 39739
Phone: 601-323-5548   Fax: 601-323-5806
Panther Swamp
W. STEVENS
13695 River Rd., Yazoo City MS 39194
Phone: 601-746-5060   Fax: 601-839-2619
St. Catherine Creek
JAMES HILL
P.O. Box 117, Sibley MS 39165
Phone: 601-442-6696   Fax: 601-442-8990
Yazoo (Hillside, Mathews Brake, Morgan Brake)
TIMOTHY WILKINS
Rt. 1, Box 286, 728 Yazoo Refuge Rd., Hollandale MS 38748
Phone: 601-839-2638   Fax: 601-839-2619

MISSOURI
Big Muddy National Wildlife and Fish Refuge
TOM BELL
4200 New Haven Rd., Columbia MO 65201
Phone: 573-876-1826   Fax: 573-876-1839
Mark Twain/Annada District
DAVID ELLIS
P.O. Box 88, Annada MO 63330
Phone: 314-847-2333   Fax: 314-847-2269
Mingo (Pilot Knob, Ozark Cavefish)
GERALD CLAWSON
R.R. 1, Box 103, Puxico MO 63960
Phone: 314-222-3589   Fax: 314-222-6343
Squaw Creek
RONALD BELL
P.O. Box 158, Mound City MO 64470
Phone: 816-442-3187   Fax: 816-442-5248

Swan Lake
JOHN GUTHRIE
Rt. 1, Box 29 A, Sumner MO 64681
Phone: 816-856-3323   Fax: 816-856-3687

MONTANA
Benton Lake
JAMES MCCOLLUM
922 Bootlegger Trail, Great Falls MT 59404
Phone: 406-727-7400   Fax: 406-727-7432
Bowdoin (Black Coulee, Creedman Coulee, Hewitt Lake, Lake Thibadeau)
DWAIN PRELLWITZ
HC 65, Box 5700, Malta MT 59538
Phone: 406-654-2863   Fax: 406-654-2866
Charles M. Russell (Hailstone, Halfbreed Lake, Lake Mason, UL Bend, War Horse)
MIKE HEDRICK
P.O. Box 110, Lewistown MT 59457
Phone: 406-538-8706   Fax: 406-538-7521
Lee Metcalf
PATRICK GONZALES
P.O. Box 257, Stevensville MT 59870
Phone: 406-777-5552   Fax: 406-777-5542
Medicine Lake (Lamesteer)
THEODORE GUTZKE
223 N. Shore Rd., Medicine Lake MT 59247-9600
Phone: 406-789-2305   Fax: 406-789-2350
National Bison Range (Nine-pipe, Pablo, Swan River)
DAVID WISEMAN
132 Bison Range Rd., Moiese MT 59824
Phone: 406-644-2211   Fax: 406-644-2661
Red Rock Lakes
DANIEL GOMEZ
Monida Star Rt., Box 15, Lima MT 59739
Phone: 406-276-3536   Fax: 406-276-3538

NEBRASKA
Crescent Lake
WILLIAM BEHRENDS
Phone: 308-762-4893
Crescent Lake/North Platte NWR Complex
STEVE KNODA
P.O. Box 1346, Scottsbluff NE 69363-1346
Phone: 308-635-7851   Fax: 308-635-7841
Fort Niobrara/Valentine
ROYCE HUBER
HC 14, Box 67, Valentine NE 69201
Phone: 402-376-3789   Fax: 402-376-3217
Rainwater Basin WMD
GENE MACK
P.O. Box 1686, Kearney NE 68848
Phone: 308-236-5015   Fax: 308-236-3899

NEVADA
Ash Meadows
ERIC HOPSON
P.O. Box 115, Amargosa Valley NV 89020
Phone: 702-372-5435   Fax: 702-372-5436
Desert Complex (Desert Moapa Valley)
RICHARD BIRGER
1500 N. Decatur Blvd., Las Vegas NV 89108
Phone: 702-646-3401   Fax: 702-646-3812

Pahranagat
    KEVIN SLOAN
    Box 510, Almo NV 89001
    Phone: 702-725-3417   Fax: 702-725-3389
Ruby Lake
    KIM HANSON
    HC 6, Box 860, Ruby Valley NV 89833
    Phone: 702-779-2237   Fax: 702-779-2370
Stillwater (Anaho Island, Fallon)
    P.O. Box 1236, Fallon NV 89407-1236
    Phone: 702-423-5128   Fax: 702-423-0146

NEW HAMPSHIRE
Lake Umbagog
    PAUL CASEY
    Box 240, Errol NH 03579
    Phone: 603-482-3415   Fax: 603-482-3308

NEW JERSEY
Cape May
    BRUCE LUEBKE
    24 Kimbles Beach Rd., Cape May Courthouse
    NJ 08210-4207
    Phone: 609-463-0994   Fax: 609-463-1667
Edwin B. Forsythe: Barnegat Division
    JEFFERY KING
    70 Collinstown Rd., P.O. Box 544, Barnegat NJ 08005
    Phone: 609-698-1378   Fax: 609-698-0109
Edwin B. Forsythe: Brigantine Division
    TRACY CASSELMAN
    P.O. Box 72, Great Creek Rd., Box 72, Oceanville NJ 08231
    Phone: 609-652-1665   Fax: 609-652-1474
Great Swamp
    WILLIAM KOCH
    152 Pleasant Plains Rd., Basking Ridge NJ 07920
    Phone: 201-425-1222   Fax: 201-425-7309
Supawna Meadows
    TOM WALKER
    197 Lighthouse Rd., Pennsville NJ 08070
    Phone: 609-935-1487   Fax: 609-935-1198
Wallkill River
    ELIZABETH HERLAND
    1547 County Rt. 565, Sussex NJ 07461
    Phone: 201-702-7266   Fax: 201-702-7286

NEW MEXICO
Bitter Lake
    WILLIAM RADKE
    P.O. Box 7, Rosewell NM 88202-0007
    Phone: 505-622-6755   Fax: 505-622-9039
Bosque do Apache
    PHILIP NORTON
    P.O. Box 1246, Socorro NM 87801
    Phone: 505-835-1828   Fax: 505-835-0314
Las Vegas
    JOE RODRIGUEZ
    Rt. 1 Box 399, Las Vegas NM 87701
    Phone: 505-425-3581   Fax: 505-454-8510
Maxwell
    JERRY FRENCH
    P.O. Box 276, Maxwell NM 87728
    Phone: 505-375-2331   Fax: 505-375-2332
San Andres
    P.O. Box 756, Las Cruces NM 88004
    Phone: 505-382-5047   Fax: 505-382-5454

Sevilleta
    TERRY TADANO
    P.O. Box 1248, Socorro NM 87801
    Phone: 505-864-4021   Fax: 505-864-7761

NEW YORK
Iroquois
    BOB LAMOY
    1101 Casey Rd., Alabama NY 14003
    Phone: 716-948-9154   Fax: 716-948-9538
Long Island Complex (Wetheim, Target Rock, Oyster Bay, Seatuck, Elizabeth A. Morton, Amagansett, Conscience Point)
    PATRICIA MARTINKOVIC
    P.O. Box 21, Shirley NY 11967
    Phone: 516-286-0485   Fax: 516-286-4003
Montezuma
    THOMAS JASIKOFF
    3395 Rt.5/20 East, Seneca Falls NY 13148
    Phone: 315-568-5987   Fax: 315-568-8835
St. Lawrence
    127 N. Water St., C/O US Customs House, Ogdensburg NY 13669
    Phone: 315-393-9002   Fax: 315-393-8570

NORTH CAROLINA
Alligator River (Pea Island)
    MICHAEL BRYANT
    P.O. Box 1969, Manteo NC 27954
    Phone: 919-473-1131   Fax: 919-473-1668
Mackay Island (Currituck)
    SUZANNE BAIRD
    P.O. Box 39, Knotts Island NC 27950
    Phone: 919-429-3100   Fax: 919-429-3185
Mattamuskeet (Cedar Island, Swan Quarter)
    DONALD TEMPLE
    Rt. 1, Box N-2, Swan Quarter NC 27885
    Phone: 919-926-4021   Fax: 919-926-1743
Pee Dee
    DAN FRISK
    Rt.1, Box 92, Wadesboro NC 28170
    Phone: 704-694-4424   Fax: 704-694-6570
Pocosin Lakes
    ELTON SAVERY
    3255 Shore Dr., Creswell NC 27928
    Phone: 919-797-4431   Fax: 919-797-7106
Roanoke River
    JERRY HOLLOMAN
    P.O. Box 430, Windsor NC 27983
    Phone: 919-794-5326   Fax: 919-794-5338

NORTH DAKOTA
Arrowwood Complex
    MARK VANIMAN
    7745 11th St. SE, Pingree ND 58476
    Phone: 701-285-3341   Fax: 701-285-3350
Audubon (Audubon WMD, Camp Lake, Hiddenwood, Lake Ilo, Lake Nettie, Lake Otis, Lost Lake, McLean, Pretty Rock)
    DAVID POTTER
    RR 1, P.O. Box 16, Coleharbor ND 59531
    Phone: 701-442-5474   Fax: 701-442-5546
Chase Lake
    MICK ERICKSON
    5924 19th St. SE, Woodworth ND 58496
    Phone: 701-752-4218   Fax: 701-752-4216

Crosby WMD, Lake Zahl
  TIM KESSLER
  P.O. Box 148, Crosby ND 58730-0148
  Phone: 701-965-6488   Fax: 701-965-6487
Des Lacs (Lostwood, Shell Lake, Lostwood WMD)
  FRED GIESE
  P.O. Box 578, Kenmare ND 58746-0578
  Phone: 701-385-4046   Fax: 701-385-3214
Devils Lake WMD (Brumba, Kellys Slough, Lake Alice, Lake Ardoch, Lambs Lake, Little Goose, Pleasant Lake, Rock Lake, Rose Lake, Silver Lake, Snyder Lake, Stump Lake, Sullys Hill NGP, Wood Lake)
  ROGER HOLLEVOCT
  P.O. Box 908, Devil's Lake ND 58301
  Phone: 701-662-8611   Fax: 701-662-8612
J. Clark Salyer (J. Clark Salyer WMD, Buffalo Lake, Cottonwood, Lords Lake, Rabb Lake, School Section Lake, Willow Lake, Wintering River)
  ROBERT HOWARD
  P.O. Box 66, Upham ND 58789
  Phone: 701-768-2548   Fax: 701-768-2834
Kulm WMD (Boone Hill Creek, Dakota Lake, Lake Patricia, Maple River)
  ROBERT VANDEN BERGE
  P.O. Box E, Kulm ND 58456-0170
  Phone: 701-647-2866   Fax: 701-647-2221
Long Lake (Long Lake WMD, Florence Lake, Slade, Appert Lake, Canfield Lake, Hutchinson Lake, Lake George, Springwater, Sunburst Lake)
  PAUL VAN NINGEN
  1200 353rd St. SE, Moffit ND 58560-9740
  Phone: 701-387-4397   Fax: 701-387-4767
Tewaukon (Lake Elsie, Storm Lake, Wild Rice Lake, Tewaukon WMD)
  SANDRA SIEKANICE
  9754 143 1/2 Ave. SE, Cayuga ND 58013
  Phone: 701-724-3598   Fax: 701-724-3683
Upper Souris
  DEAN KNAUER
  17702 212th Ave., NW, Berthold ND 58718-9666
  Phone: 701-468-5467   Fax: 701-468-5600
Valley City WMD
  HARRIS HOISTED
  11515 River Rd., Valley City ND 58072-9619
  Phone: 701-845-3466   Fax: 701-845-3482

OHIO
Ottawa (Cedar Point, West Sister Island)
  LARRY MARTIN
  14000 W. State, Rt. 2, Oak Harbor OH 43449
  Phone: 419-898-0014   Fax: 419-898-7895

OKLAHOMA
Deep Fork
  JON BROCK
  P.O. Box 816, Okmulgee OK 74447
  Phone: 918-756-0815   Fax: 918-756-0275
Little River (Little Sandy)
  BERLIN HECK
  P.O. Box 340, Broken Bow OR 74728
  Phone: 405-584-6211   Fax: 405-584-2034
Salt Plains
  RODNEY KREY
  Rt. 1, Box 76, Jet OK 73749
  Phone: 405-626-4794   Fax: 405-626-4793

Sequoyah (Ozark Plateau)
  STEPHEN BERENDZEN
  Rt. 1, Box 18A, Vian OK 74962
  Phone: 918-773-5251   Fax: 918-773-5598
Tishomingo
  JOHNNY BEALL
  Rt. 1, Box 151, Tishomingo OK 73460
  Phone: 405-371-2402   Fax: 405-371-9312
Washita (Optima)
  KENNETH SCHWENDT
  Rt. 1, Box 68, Butler OK 73625
  Phone: 405-664-2205   Fax: 405-664-2206
Wichita Mountains Wildlife Refuge
  SAM WALDSTEIN
  RR 1, Box 448, Indiahoma OK 73552
  Phone: 405-429-3221   Fax: 405-429-9323

OREGON
Ankeny
  RICHARD GUADAGNO
  10995 Hwy. 22, Dallas OR 97338-9343
  Phone: 503-623-2749   Fax: 503-623-7812
Hart Mountain National Antelope Refuge
  JENNY BARNETT
  P.O. Box 111, Lakeview OR 97630
  Phone: 541-947-3315
Malheur
  HC 72, Box 245, Princeton OR 97721-9505
  Phone: 541-493-2612   Fax: 541-493-2405
Mid-Columbia River Complex (Umatilla, Cold Springs, McKay Creek, McNary, Toppenish)
  GARY HAGEDORN
  P.O. Box 2527, Pasco OR 99301
  Phone: 541-545-8588   Fax: 541-545-8670
Oregon Coastal Refuges (Brandon Marsh, Cape Meares, Nestucca Bay, Oregon Islands, Siletz Bay, Three Arch Rocks)
  ROY LOVE
  2127 SE OSU Dr., Newport OR 97365-5258
  Phone: 541-867-4550   Fax: 541-867-4551
Sheldon
  MARK STRONG
  P.O. Box 111, Lakeveiw OR 97630-0107
  Phone: 541-947-3315
Sheldon/Hart Mountain Complex
  MICHAEL NUNN
  P.O. Box 111, Lakeview OR 97630
  Phone: 541-947-3315   Fax: 541-947-4414
Tualatin River
  RALPH WEBBER
  16340 SW Beef Bend Rd., Sherwood OR 97140-8306
  Phone: 503-590-5811   Fax: 503-590-6702
Western Oregon Complex (Ankeny, Naskett Slough, Tualatin River, William L. Finley, Brandon Marsh, Cape Meares, Nestucca Bay, Oregon Islands, Siletz Bay, Three Arch Rocks)
  JAMES HOUK
  26208 Finley Refuge Rd., Corvallis OR 97333-9533
  Phone: 541-757-7236   Fax: 541-757-4450

PENNSYLVANIA
Erie
  THOMAS MOUNTAIN
  11296 Wood Duck Ln., Guys Mills PA 16327
  Phone: 814-789-3585   Fax: 814-789-2909

John Heinz NWR at Tinicum (Supawna Meadows and Kilcohook Coordination Area)
RICHARD NUGENT
Ste. 104, Scott Plaza 2, Philadelphia PA 19113
Phone: 610-521-0662   Fax: 610-521-0611

PUERTO RICO
Caribbean Islands Refuges (Cabo Rojo, Desecheo, Laguna Cartagena, Sandy Point, Virgin Islands)
VAL URBAN
P.O. Box 510, Boqueron PR 00622
Phone: 809-851-7258   Fax: 809-851-7440
Culebra
TERESA TELLEVAST
P.O. Box 190, Culebra PR 00775
Phone: 787-742-0115

RHODE ISLAND
Ninigret Complex (Block Island, Pettaquamscutt Cove, Sachuest Point, Trustom Pond, Stewart B. McKinney)
CHARLES HERBERT
P.O. Box 307, Charlestown RI 02813
Phone: 401-364-9124   Fax: 401-364-0170

SOUTH CAROLINA
ACE Basin
JAMES BROWNING
P.O. Box 848, Hollywood SC 29449
Phone: 803-889-3084   Fax: 803-889-3282
Cape Romain (Santee)
GEORGE GARRIS
5801 Hwy. 17 N., Awendaw SC 29429
Phone: 843-928-3264   Fax: 843-928-3803
Carolina Sandhills
R. LANIER
Rt. 2, Box 100, McBee SC 29101
Phone: 803-335-8401   Fax: 803-335-8406

SOUTH DAKOTA
Huron WMD
MARK HEISINGER
200 4th St., SW, Rm. 317 Federal Bld., Huron SD 57350-2470
Phone: 605-352-5894   Fax: 605-352-6709
Lacreek (Bear Butte)
ROLF KRAFT
HC 5, Box 114, Martin SD 57551
Phone: 605-685-6508   Fax: 605-685-1173
Lake Andes (Karl E. Mundt)
SYLVIA PELIZZA
38027 291st St., Lake Andes SD 57356
Phone: 605-487-7603   Fax: 605-487-7604
Madison WMD
TOM TORNOW
P.O. Box 48, Madison SD 57042
Phone: 605-256-2974   Fax: 605-256-9432
Sand Lake (Pocasse)
JOHN KOERNER
39650 Sand Lake Dr., Columbia SD 57433
Phone: 605-885-6320   Fax: 605-885-6401
Waubay
DOUGLAS LESCHISIN
RR 1, Box 39, Waubay SD 57273
Phone: 605-947-4521   Fax: 605-947-4524

Waubay WMD
CONNIE MUELLER
RR 1, Box 39, Waubay SD 57273
Phone: 605-947-4521   Fax: 605-947-4524

TENNESSEE
Chickasaw
MICHAEL STROEH
1505 Sandy Bluff Rd., Ripley TN 38063
Phone: 901-635-7621   Fax: 901-635-0178
Cross Creeks NWR
643 Wildlife Rd., Dover TN 37058
Phone: 931-232-7477   Fax: 931-232-5958
Hatchie NWR
MARVIN NICHOLS
4172 Hwy 76 South, Brownsville TN 38012-8332
Phone: 901-772-0501   Fax: 901-772-7839
Lower Hatchie
EDWARD RODRIGUEZ
1505 Sandy Bluff Rd., Ripley TN 38063
Phone: 901-635-7621   Fax: 901-635-0178
Reelfoot
RANDY COOK
Fed. Bld. Rm. 129, 309 N. Church St., Dyersburg TN 38024
Phone: 901-287-0650   Fax: 901-286-0468
Tennessee
JOHN TAYLOR
P.O. Box 849, Paris TN 38242
Phone: 901-642-2091   Fax: 901-644-3351

TEXAS
Anahuac (Moody, McFaddin, Texas Point)
ANDY LORANGER
P.O. Box 278, Anahuac TX 77514
Phone: 409-267-3337   Fax: 409-267-4314
Aransas
J. GLEZENTANNER
P.O. Box 100, Austwell TX 77950
Phone: 512-286-3559   Fax: 512-286-3722
Attwater Prairie Chicken
TERRY ROSSIGNOL
P.O. Box 519, Eagle Lake TX 77434-0519
Phone: 409-234-3021   Fax: 409-234-3278
Balcones Canyonlands
DEBORAH HOLLE
10711 Burnet Rd., Ste. 201, Austin TX 78758
Phone: 512-339-9432   Fax: 512-339-9453
Brazoria (San Bernard, Big Boggy)
RONALD BISBEE
1212 N. Velasco, Ste. 200, Angleton TX 77515-1088
Phone: 409-849-7771   Fax: 409-849-5118
Buffalo Lake
LYNN NYMEYER
P.O. Box 179, Umbarger TX 79091
Phone: 806-499-3382
Hagerman
JAMES WILLIAMS
6465 Refuge Rd., Sherman TX 75092
Phone: 903-786-2826   Fax: 903-786-3327
Laguna Atascosa
STEPHEN LABUDA
P.O. Box 450, Rio Hondo TX 78583
Phone: 210-748-3607   Fax: 210-748-3609

Lower Rio Grande/Santa Anna Complex
  LARRY DITTO
  Rt. 2, Box 202A, Alamo TX 78516
  Phone: 210-787-3079   Fax: 210-787-8338
Muleshoe (Grulla, NM)
  DONALD CLAPP
  P.O. Box 549, Muleshoe TX 79347
  Phone: 806-946-3341   Fax: 806-946-3317
Trinity River
  STUART MARCUS
  P.O. Box 10015, Liberty TX 77575
  Phone: 409-336-9786   Fax: 409-336-9847

UTAH
Bear River Migratory Bird Refuge
  ALAN TROUT
  58 S. 950 West, Brigham City UT 84302
  Phone: 801-723-5887   Fax: 435-723-8873
Fish Springs
  JAY BANTA
  P.O. Box 568, Dugway UT 84022
  Phone: 801-831-5353   Fax: 801-831-5354
Ouray
  266 West 100 North, Ste. 2, Vernal UT 84078
  Phone: 801-789-0351   Fax: 801-789-4805

VERMONT
Missisquoi
  ROBERT ZELLEY
  P.O. Box 163, Swanton VT 05488
  Phone: 802-868-4781   Fax: 802-868-2379

VIRGIN ISLANDS
Sandy Point (Green Cay, Buck Island)
  MICHAEL EVANS
  3013 Estate Golden Rock, Ste. 167, Christiansted
  VI 00820-4355
  Phone: 809-773-4554

VIRGINIA
Back Bay (Plum Tree Island)
  JOHN STASKO
  4005 Sandpiper Rd., Virginia Beach VA 23456
  Phone: 757-721-2412   Fax: 757-721-6141
Chincoteague (Wallops Island)
  JOHN SCHROER
  P.O. Box 62, Chincoteague VA 23336
  Phone: 757-336-6122   Fax: 757-336-5273
Eastern Shore of VA (Fisherman Island)
  SUSAN RICE
  5003 Hallett Circle, Cape Charles VA 23310
  Phone: 757-331-2760   Fax: 757-331-3424
Great Dismal Swamp (Nansemond)
  LLOYD CULP
  P.O. Box 349, Suffolk VA 23434
  Phone: 757-986-3705   Fax: 757-986-2353
Potomac River Complex (Mason Neck, Featherstone)
  GREG WEILER
  14344 Jefferson Davis Hwy., Woodbridge VA 22191
  Phone: 703-490-4979   Fax: 703-490-5631
Rappahannock River Valley (James River, Presquile)
  HARRY BRADY
  P.O. Box 189, Prince George VA 23875
  Phone: 804-733-8042

WASHINGTON
Columbia
  ROBERT FLORES
  P.O. Drawer E, 735 E. Main St., Othello WA 99344
  Phone: 509-488-2668   Fax: 509-488-0705
Conboy Lake
  HAROLD COLE
  Box 5, Glenwood WA 98619-0005
  Phone: 509-364-3410   Fax: 509-364-3667
Hanford Complex (Saddle Mountain)
  DAVID GOEKE
  3520 Port of Benton Rd., Richland WA 99352
  Phone: 509-371-1801   Fax: 509-371-0196
Julia Butler Hansen Refuge for the Columbia White-tailed Deer
  JOEL DAVID
  P.O. Box 566, Cathlamet WA 98612-0566
  Phone: 509-795-3915   Fax: 360-795-0803
Little Pend Oreille
  LISA LANGELIER
  1310 Bear Creek Rd., Colville WA 99114-9713
  Phone: 509-684-8384   Fax: 509-684-8381
Nisqually (Grays Harbor)
  WILLARD HESSELBART
  100 Brown Farm Rd., Olympia WA 98516-2302
  Phone: 360-753-9467   Fax: 360-534-9302
Pierce (Franz Lake, Steigerwald Lake)
  JEFF HOLM
  Columbia River Gorge Refuges, 36062 SR 14, Stevenson WA 98648-9541
  Phone: 509-427-5208   Fax: 509-427-4707
Ridgefield
  THOMAS MELANSON
  P.O. Box 457, Ridgefield WA 98642-0457
  Phone: 509-887-4106   Fax: 360-887-4109
Turnbull
  NANCY CURRY
  26010 S. Smith Rd., Cheney WA 99004-9326
  Phone: 509-235-4723   Fax: 509-235-4703
Washington Maritime Complex (Copalis, Bungeness, Flattery Rocks, Protection Island, Quillayute Needles, San Juan Islands)
  KEVIN RYAN
  33 S. Barr Rd., Port Angeles WA 38362-9202
  Phone: 360-457-8451   Fax: 360-457-9778
Willapa (Lewis and Clark)
  CHARLES STENVALL
  3888 SR 101, Ilwaco WA 98624-9707
  Phone: 360-484-3482   Fax: 360-484-3109

WEST VIRGINIA
Canaan Valley
  KEVIN DESROBERTS
  P.O. Box 1278, Rt. 250 S., Elkins WV 26241
  Phone: 304-637-7312   Fax: 304-636-7824
Ohio River Islands
  JERRY WILSON
  P.O. Box 1811, Parkersburg WV 26102-1811
  Phone: 304-422-0752   Fax: 304-422-0754

WISCONSIN
Horicon Complex (Fox River, Gravel Island, Green Bay, Leopold WMD)
  PATTI MEYERS
  W 4279 Headquarters Rd., Mayville WI 53050
  Phone: 920-387-2658   Fax: 920-387-2973

Leopold
  STEVEN LENZ
  Phone: 920-387-0336   Fax: 920-387-2973
Necedah
  LARRY WARGOWSKY
  W. 7996 20th St., W., Necedah WI 54646-7531
  Phone: 608-565-2551   Fax: 608-565-3160
St. Croix WMD
  CHET MCCARTY
  1764 95th St., New Richmond WI 54017
  Phone: 715-246-7784   Fax: 715-246-4670
Trempealeau
  RICHARD FRIETSCHE
  W28438 Refuge Rd., Trempealeau WI 54661
  Phone: 608-539-2311   Fax: 608-539-2703

WYOMING

National Elk Refuge
  HARRY REISWIG
  675 E. Broadway, P.O. Box C, Jackson WY 83001
  Phone: 307-733-9212   Fax: 307-733-9729
Seedskadee (Cookeville Meadows)
  CAROL DAMBURG
  P.O. Box 700, Green River WY 82935
  Phone: 307-875-2187   Fax: 307-875-4425

NATIONAL WILDLIFE REFUGES REGIONAL DIRECTORS

REGION 1
CAROLYN BOHAN
911 NE 11th Ave., Eastside Federal Complex, Portland OR 97232-4181
Phone: 503-231-6214   Fax: 503-231-2364

REGION 2
DOM CICCONE
P.O. Box 1306, Albuquerque NM 87103
Phone: 505-248-7419   Fax: 505-248-6803

REGION 3
NITA FULLER
1 Federal Dr., Federal Bldg., Fort Snelling MN 55111-4056
Phone: 612-713-5401   Fax: 612-713-5288

REGION 4
DAVE HEFFERNAN
1875 Century Blvd., NE, Rm. 324, Atlanta GA 30345
Phone: 404-679-7166   Fax: 404-679-7081

REGION 5
TONY LEGER
300 Westgate Center Dr., Hadley MA 01035-9589
Phone: 413-253-8306   Fax: 413-253-8309

REGION 6
KEN MCDERMOND
134 Union Blvd., Lakewood CO 80228
Phone: 303-236-8145   Fax: 303-236-4792

REGION 7
TODD LOGAN
1011 E. Tudor Rd., Anchorage AK 99503
Phone: 907-786-3545   Fax: 907-786-3640

D

H

I

O

P

Q

R

Air Quality

Air Quality and Pollution

Air Resources

Alternative Agriculture

Ancient Forests

Arctic

Arid Lands

Biological Informatics

Biotechnology

Birds

Botanical Gardens

Botany

Brown Fields

Camp

Careers

Cave

Chemical Pollution Control

Chemistry

Chimpanzees

Cleanup

Coastal Construction And Erosion

Keyword Index

Conservation Biology

Conservation Districts

Conservation Easements

Dams

Deserts

Developing Countries

Development

Diseases

Dolphin

Drinking Water Protection

Ecoaction

Ecological Education

Ecology

Ecosystems

Ecotourism

Education

Endangered Resources

Endangered Species

Energy

Energy Conservation

Energy Efficiency

Enforcement

Engineering

Environment

Environmental And Conservation Education

Environmental And Humanitarian Education

Environmental Cleanup

Environmental Communication

Environmental Education Curriculum

Environmental Ethics

Environmental Health

Environmental Justice

Environmental Law

Environmental Legislation

Environmental Living

Environmental Planning

Environmental Protection

Factory Farms

Falconry

Family Planning

Family Recreation

Federalism

Feedlots And Pollution

Fieldwork

Fish

Fish Wildlife Management

Flood Control

Funding

Gap Analysis

Gardening And Horticulture

Genetics

Geographic Information Systems

Geography

Geology

Government Accountability

Grants

Grasslands

Great Plains

Green Building

Green Certification

Greenhouse Effect/Global Warming

Greenways

Ground Water Protection

Habitat Conservation

Harmful Algal Blooms

Hazardous Materials And Waste

Health And Nutrition

Healthy Home

Herpetoculture

Herpetology

Highly Migratory Species

Historic Preservation

Human Rights

Hunting

Hydropower Relicensing

Hyenas

Indigenous People

Inquiry Based Education

Insects And Butterflies

International Wildlife

Internships

Interpretation

Interpretive Center

International Environmental Law

International Trade And Environment

Islands

Justice

Lakes

Land Conservation

Land Management

Land Preservation

Land Protection

Land Purchase

Land Use Planning

Landscape Analysis

Landscape Architecture

Landscape Ecology

Law Enforcement

Legal Advocacy

Leisure

Librarians/Information Professionals

Libraries

Litter

Local Resource Conservation

Mammals

Management Plans

Manatees

Mangrove Habitats

Native Fish

Native Plants

Natural Areas

Natural Resource Conservation

Natural Science

Natural Systems

Nature Centers

Ocean Conservation

Oceanography

Oil And Gas

Oil Spill Response

Open Space

Otters

Outdoor Education

Outdoor Ethics

Outdoor Recreation

Overconsumption

Ozone Depletion

Pedestrian Environment

People Of Color In The Environment

Pest Management

Pesticides

Physiology

Planning Management

Plant Propagation

Plants

Politics And Government

Pollution Control

Pollution Prevention

Population Growth

Prairies

Precision Farming

Predators

Preservation And Protection

Preservation

Private Land Development

Professional Development

Professional Organization

Protected Areas

Protecting Special Places

Public Access

Public Farming

Public Health Protection

Public Information

Public Lands

Public Participation

Rainforests

Raptors

Recreational Boating

Recycling

Reforestation

Rehabilitation

Reintroduction

Remedial Action Plans

Renewable Resources

Reproductive Rights

Road Construction

Runoff

Rural Development

Salmon Recovery

Scholarships

Scholarships And Grants

Schoolyard Habitats

Sea Grass

Sea Turtles

Seabed Disturbance

Sustainable Energy

Sustainable Resources

Taiga

Terrestrial Habitats

Tortoises

Tourism

Toxic Reduction

Toxic Substances

Toxicology

Trade/Business

Trail

Training

Transportation

Trapping

Travel

Trees

Tropical Biodiversity And Conservation

Trout

Tundra

Turtles

Upstream Flood Prevention

Urban And Rural Development

Urban Environment

Urban Forestry

Utility Restructuring

Vision Quest

Volunteering

Waste Resources

Water And Air Quality

Water Conservation

Water Export And Diversion

Water Pollution Management

Water Quality

Water Quantity

Waterfowl

Watershed Protection

Watersheds

Wetland Habitat

Wetlands

Keyword Index

GEOGRAPHIC INDEX

Geographic Index

State Government Organizations

INDIANA

IOWA

UPDATE YOUR LISTING/CHANGE OF ADDRESS

Please help us keep the information in the directory up to date. Use this form to let us know of changes to your listing such as a new address or a new e-mail.

Please type or print clearly.

ORGANIZATION NAME ____

ADDRESS: STREET ____

CITY _______________________ STATE _______________ ZIP __________ - __________

COUNTRY _______________ E-MAIL ___________________________________

WEB SITE ____

PHONE NUMBER _________________ FAX NUMBER _______________________

PAGE NUMBER IN 2001 DIRECTORY ___________________________________

CHANGES TO YOUR DESCRIPTION _______________________________________

____

____

____

____

ADD CONTACT PERSON ____

REMOVE CONTACT PERSON ____

Please give us a contact name for the person we can obtain updates from.

NAME _________________________________ PHONE _________ - _________ - _________

Further updating materials will be sent to all organizations listed in the *2002 Conservation Directory* when updating begins for the *2003 Conservation Directory.*

Please mail form to:

**NATIONAL WILDLIFE FEDERATION
ATTN: CONSERVATION DIRECTORY
11100 WILDLIFE CENTER DRIVE
RESTON, VA 20190-5362
PHONE: 703-438-6000
FAX: 703-438-6061**

Information may be submitted on photocopies of this form.

Visit the *Conservation Directory* online at www.nwf.org to update your organization's information automatically at any time.

APPLICATION REQUEST

If you would like your organization to be listed in the *Conservation Directory* or you have a suggestion of an organization that should be listed in the directory, please let us know. An electronic version of this form is available at www.nwf.org/printandfilm/publications/consdir/infoform.html.

Please type or print clearly.

❏ Request for Listing   ❏ Suggested New Organization

ORGANIZATION NAME ____

ADDRESS: STREET ____

CITY _________________ STATE _____________ ZIP _________ - _________

COUNTRY _____________ E-MAIL _______________________________

WEB SITE ____

PHONE NUMBER _______________ FAX NUMBER _______________________

CONTACT PERSON ____

Please mail form to:

NATIONAL WILDLIFE FEDERATION
ATTN: CONSERVATION DIRECTORY
11100 WILDLIFE CENTER DRIVE
RESTON, VA 20190-5362
PHONE: 703-438-6000
FAX: 703-438-6061

Information may be submitted on photocopies of this form.

Visit the *Conservation Directory* online at www.nwf.org to add a listing automatically at any time.

Conservation Directory 2002

National Wildlife Federation's *Conservation Directory* is the only directory of its kind that is updated and verified every year. The *Conservation Directory* includes listings for over 4,000 dedicated individuals, museums, non-profit organizations, educational institutions, commercial businesses, habitat preserves, wildlife attractions, and government agencies worldwide. These groups range in size, scope, and budget from the smallest grassroots operations to extensive international organizations using their own methods of research, advocacy, preservation, and education to develop sound environmental solutions.

For the first time since it was launched in 1954, the National Wildlife Federation's *Conservation Directory* is available in both hard copy and online versions. The new and free online service will complement the printed directory and assist citizens to easily locate and work with conservation groups on issues affecting their local area. Conservation groups can learn more about their peers and gain new opportunities to generate collaborative efforts. Groups listed in the *Conservation Directory* can also update their organization's record automatically and immediately. New groups can apply online for inclusion in the *Directory*. To view the online version or to find out more about it, visit **www.nwf.org/conservationdirectory**.

The *Conservation Directory* continues to serve as an integral part of the National Wildlife Federation's mission to bring organizations and individuals together for the preservation of wildlife, wild places, and their shared environment. The progressive increase in conservation group listings shows the continual vigilance of communities and individuals to educate and promote sound environmental solutions to all generations. The National Wildlife Federation is proud that, in the 2002 edition, these outstanding efforts resulted in producing the most listings in the history of the publication.

The largest member-supported conservation education and advocacy group in the United States, the National Wildlife Federation unites people from all walks of life to protect nature, wildlife and the world we all share. The Federation has educated and inspired families to uphold America's conservation tradition since 1936.

ORDER FORM

Yes, I would like to order the National Wildlife Federation **Conservation Directory 2002**

_________ paperback copies @ $70.00 each
(ISBN: 1-55963-952-0)

_________ Total Book Price

_________ Sales Tax *(CA 7%; DC 5.75%)*

_________ Shipping & Handling
($5.75 for the first book, $2.00 for each additional)

_________ TOTAL

To place a standing order for future editions of the *Conservation Directory* at a 20% discount, please contact the Island Press customer service department at **1-800-828-1302** or by email at **service@islandpress.org**

For fastest service, order online at
www.islandpress.org/nwf1
or call **1-800-828-1302**
(Mon.–Fri., 8:00 A.M. –5:00 P.M., Pacific Coast Time)

Outside of the U.S., call **707-983-6432**
Fax orders to **707-983-6414**
Send inquiries to **service@islandpress.org**

Mail orders to:
ISLAND PRESS, PO Box 7, Covelo, CA 95428

Name/Address/City/State/Zip

____

____

____

____

❑ **Enclosed is my purchase order**
(universities, public libraries, and government agencies only)

Purchase Order #: _______________________________
The Island Press Federal ID Number is 94-2578166

❑ **Enclosed is my check.**

Please charge to my: ❑ **Visa**   ❑ **MasterCard**   ❑ **American Express**

Card #: _______________________________________

Expiration Date: _______________________________

Signature: _____________________________________

Phone #/E-mail: ________________________________
(in case we have a question about your order)

Join National Wildlife Federation Today!

Since 1936, National Wildlife Federation has been the nation's largest and most effective member supported conservation group. through our network of regional offices across this great land of ours we actively work to achieve our conservation goals in partnership with a wide range of regional, state, national and international groups. NWF has an effective common-sense conservation agenda and a membership including educators, scientists, gardeners, anglers, birders, hikers, campers, volunteer activists and wildlife enthusiasts who want to secure a lasting place for wildlife in the modern world.

NATIONAL WILDLIFE FEDERATION IS IN YOUR COMMUNITY

An NWF Natural Resource Center or regional office near you is always ready to address local problems that concern you and your neighbors…and to solve them with the help of NWF's tremendous nationwide resources. NWF educates children, families, community, political and business leaders. National Wildlife Federation equips grassroots activists with scientifically-based common sense solutions. And we reach out to people worldwide, to share information on how we can all work for a brighter tomorrow for wildlife, nature and ourselves.

YOUR MEMBERSHIP SUPPORTS NWF'S SIX-POINT ACTION AGENDA

Together, we can work to strengthen landmark legislation like the Endangered Species Act…safeguard fragile habitats…educate and inspire our children…save endangered species…preserve waters and wetlands…and help young and old alike enjoy the natural world through programs such as our *Backyard Wildlife Habitat*™ program.

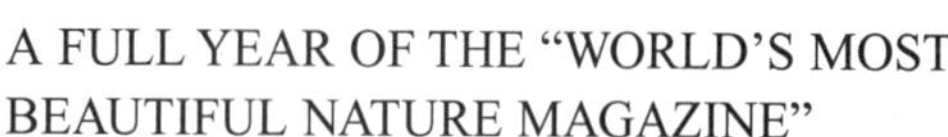

EXCLUSIVE MEMBER BENEFITS
OF NATIONAL WILDLIFE FEDERATION

A FULL YEAR OF THE "WORLD'S MOST BEAUTIFUL NATURE MAGAZINE"
Witness rare views of animals, visit wild places, discover nature's hidden secrets, and get the latest conservation updates.

A DISTINCTIVE NWF TOTE BAG
Holds your everyday items, conserves natural resources and expresses your concern for wildlife.

PLUS a membership card, decal, wildlife stamps, nature travel opportunities, wildlife alerts…and more!

CALL NOW! 1-800-822-9919

or

Visit our website at www.nwf.org for more information on how you can join the nation's largest conservation organization.

(89% of what you give to NWF goes directly to our vital conservation programs)

NATIONAL WILDLIFE FEDERATION®
11100 Wildlife Center Drive • Reston, Virginia 20190-5362 • www.nwf.org

YOU
CAN MAKE
A REAL
DIFFERENCE
FOR ONLY $15